Lecture Notes in Networks and Systems

Volume 411

AF892673

The series "Lecture Notes in Networks and Systems" publishes the latest developments in Networks and Systems—quickly, informally and with high quality. Original research reported in proceedings and post-proceedings represents the core of LNNS.

Volumes published in LNNS embrace all aspects and subfields of, as well as new challenges in, Networks and Systems.

The series contains proceedings and edited volumes in systems and networks, spanning the areas of Cyber-Physical Systems, Autonomous Systems, Sensor Networks, Control Systems, Energy Systems, Automotive Systems, Biological Systems, Vehicular Networking and Connected Vehicles, Aerospace Systems, Automation, Manufacturing, Smart Grids, Nonlinear Systems, Power Systems, Robotics, Social Systems, Economic Systems and other. Of particular value to both the contributors and the readership are the short publication timeframe and the world-wide distribution and exposure which enable both a wide and rapid dissemination of research output.

The series covers the theory, applications, and perspectives on the state of the art and future developments relevant to systems and networks, decision making, control, complex processes and related areas, as embedded in the fields of interdisciplinary and applied sciences, engineering, computer science, physics, economics, social, and life sciences, as well as the paradigms and methodologies behind them.

Indexed by SCOPUS, INSPEC, WTI Frankfurt eG, zbMATH, SCImago.

All books published in the series are submitted for consideration in Web of Science.

For proposals from Asia please contact Aninda Bose (aninda.bose@springer.com).

More information about this series at https://link.springer.com/bookseries/15179

Michael E. Auer · Thrasyvoulos Tsiatsos
Editors

New Realities, Mobile Systems and Applications

Proceedings of the 14th IMCL Conference

Springer

Editors
Michael E. Auer
CTI Global
Frankfurt, Germany

Thrasyvoulos Tsiatsos
Department of Informatics
Aristotle University of Thessaloniki
Thessaloniki, Greece

ISSN 2367-3370 ISSN 2367-3389 (electronic)
Lecture Notes in Networks and Systems
ISBN 978-3-030-96295-1 ISBN 978-3-030-96296-8 (eBook)
https://doi.org/10.1007/978-3-030-96296-8

This Springer imprint is published by the registered company Springer Nature Switzerland AG
The registered company address is: Gewerbestrasse 11, 6330 Cham, Switzerland

Preface

IMCL2021 was the 14th edition of the International Conference on Interactive Mobile Communication, Technologies and Learning.

This interdisciplinary conference is part of an international initiative to promote technology-enhanced learning and online engineering worldwide. The IMCL2021 covered all aspects of mobile learning as well as the emergence of mobile communication technologies, infrastructures and services and their implications for education, business, governments and society.

The IMCL conference series actually aims to promote the development of mobile learning, to provide a forum for education and knowledge transfer, to expose students to latest ICT technologies and encourage the study and implementation of mobile applications in teaching and learning. The conference was also platform for critical debates on theories, approaches, principles and applications of mobile learning among educators, developers, researchers, practitioners and policy makers.

IMCL2021 has been organized by Aristotle University of Thessaloniki, Greece, November 4–5, 2021.

This year's theme of the conference was "New Realities, Mobile Systems and Applications."

Again, outstanding scientists from around the world accepted the invitation for keynote speeches:

- Helen Crompton, Old Dominion University, Norfolk, VA, USA: **Mobile Learning: Integration Theories and a Social Ecological System Approach.**
- Marcus Specht, TU Delft, LDE-CEL, The Netherlands: **Mobile Learning 2.0, Toward Added Value and Cross-Media Collaboration.**

Furthermore, one very interesting workshop has been organized: "Navigating the Various Wired and Mobile VR Systems and VR Applications for a More Productive and Immersive Learning Experience" by Dimitrios Boglou, Cyprus University of Technology, Cyprus.

Since its beginning, this conference is devoted to new approaches in learning with a focus to mobile learning, mobile communication, mobile technologies and engineering education.

We are currently witnessing a significant transformation in the development of working and learning environments with a focus to mobile online communication.

Therefore, the following main topics have been discussed during the conference in detail:

- Mobile Learning Issues:
  - Dynamic learning experiences
  - Large-scale adoption of mobile learning
  - Ethical and legal issues
  - Research methods and evaluation in mobile learning
  - Mobile learning models, theory and pedagogy
  - Life-long and informal learning using mobile devices
  - Open and distance mobile learning
  - Social implications of mobile learning
  - Cost-effective management of mobile Learning processes
  - Quality in mobile learning
  - Case studies in mobile learning
- Interactive Communication Technologies and Infrastructures:
  - Wearables and Internet of things (IoT)
  - Tangible, embedded and embodied interaction
  - Location-based integration
  - Cloud computing
  - Emerging mobile technologies and standards
  - Interactive and collaborative mobile learning environments
  - Crowd sensing
  - 5G network infrastructure
  - Platforms to support students mobility
- Mobile Applications:
  - Augmented-, virtual-, mixed- and cross-reality apps
  - Smart cities
  - Remote and online laboratories
  - Serious games and gamification
  - Mobile health care, healthy lifestyle and training
  - Mobile apps for sports
  - Mobile credentials, badges and blockchain
  - Learning analytics
  - Mobile learning in cultural institutions and open spaces
  - Mobile systems and services for opening up education
  - Social networking applications
  - Mobile learning management systems (mLMS)

The following **Special Sessions** have been organized:

- **University – Industry – Developer "Creative Collaborations" In IoT, Mobile and Reconfigurable Technologies (IoT-MRT),** Chair: Doru Ursutiu, Transilvania University of Brasov – AOSR, Romania.
- **Let them Innovate: Developing Digital Competencies for Mobile STEAM Learning and Teaching by Utilizing Immersive and Adaptive Digital Technology (DigiCompImmersive),** Chairs: Petros Lameras, Coventry University, UK, and Nektarios Moumoutzis, Technical University of Crete, Greece.
- **User Experience with Remote Interaction and Fitting (uXRiF'2021),** Chair: Dionysios Politis, Aristotle University of Thessaloniki, Greece.
- **Skill-Labs for Skilled Learners (SL4SL),** Chairs: Jenny Pange, University of Ioannina, Greece and Eugenia Toki, University of Ioannina, Greece.
- **Digital Technology in Sports (DiTeS),** Chairs: Chair: Styliani Douka, Aristotle University of Thessaloniki, Greece, Lambros Lazuras, Sheffield Hallam University, UK.

Also, the "**4th IMCL International Student Competition for Mobile Apps**" has been organized in the context of IMCL2021. The winning teams were:

- **Winner**: "Smart Ladder" by Hippokratis Apostolidis, Crhristodoulos Tryphonidis, Nikolaos Politopoulos, George Psathas, Angeliki Mavropoulou, Aristotle University of Thessaloniki, Greece.
- **Runner-up**: "Robotics with Augmented Reality for Training and Rehabilitation" by José Carlos Rodrigues, LAETA-INEGI, Faculty of Engineering, University of Porto, Portugal, Paulo Menezes, Department of Electrical and Computer Engineering, Institute of Systems and Robotics, University of Coimbra, Portugal and Maria Teresa Restivo, LAETA-INEGI, University of Porto, A3ES, Portugal.

As submission types have been accepted:

- Full paper, short paper
- Work in progress, poster
- Special sessions
- Round table discussions, workshops, tutorials and students' competition

All contributions were subject to a double-blind review. The review process was very competitive. We had to review about **264** submissions. A team of about **180** reviewers did this terrific job. Our special thanks go to all of them.

Due to the time and conference schedule restrictions, we could finally accept only the best **104** submissions for presentation.

The **best papers** were the following:

- **Category "Full Paper"**: "A Board Game For Sustainable Development Education: Kindergarten Students As Game Designers" by Maria Tsapara, Tharrenos Bratitsis, University of Western Macedonia, Greece.

- **Category "Short Paper"**: "An Assessment of the Advantages using Smartphone – Based Tele-Audiology and its Effects on Hearing Care Professionals' Willingness for Integration into the Fitting Process" by Florian Ross, MATE Hungarian University of Agriculture and Life Sciences, Hungary.
- **Category "Work in Progress"**: "Hybrid Education: From Telecommuting with e-Learning to Broadcasting Ourselves" by Anastasios Nikiforos (1), Vasileios K. Vasileiou (2), Georgios H. Patronas (2), Stavros N. Dimitriadis (1). Organizations: (1) Dept. of Informatics, Aristotle University of Thessaloniki, Greece; (2) Dept. of Music Science and Art, University of Macedonia, Greece.

Our conference had more than 192 participants from 33 countries.
IMCL2023 will be held again at Aristotle University of Thessaloniki, Greece.

Michael E. Auer
IMCL Steering Committee Chair
Thrasyvoulos G. Tsiatsos
IMCL General Chair

Committees

Steering Committee Chair

Michael E. Auer	CTI, Frankfurt, Vienna, New York, Bengaluru, Hong Kong

General Conference Chair

Thrasyvoulos Tsiatsos	Aristotle University of Thessaloniki, Greece

International Chairs

Samir A. El-Seoud	The British University in Egypt (Africa)
Neelakshi C. Premawardhena	University of Kelaniya, Sri Lanka (Asia)
Alexander Kist	University of Southern Queensland, Australia (Astralia/Oceania)
Alaa Ashmawy	American University Dubai (Middle East)
David Guralnick	Kaleidoscope Learning New York, USA (North America)

Technical Program Chairs

Ioannis Stamelos	Aristotle University of Thessaloniki, Greece
Stavros Demetriadis	Aristotle University of Thessaloniki, Greece
Sebastian Schreiter	IAOE, France

IEEE Liaison

Russ Meier	IEEE Education Society Meetings Chair

Workshop, Tutorial and Special Sessions Chair

Andreas Pester	The British University in Egypt, Cairo, Egypt

Publication Chair

Sebastian Schreiter	IAOE, France

Local Organization Chair

Stella Douka	Aristotle University of Thessaloniki, Greece

Local Organization Committee Member

Christos Temertzoglou	Aristotle University of Thessaloniki, Greece

Program Committee Members

Abul Azad	Northern Illinois University, USA
Agisilaos Konidaris	Ionian University, Greece
Anastasios Economides	University of Macedonia, Greece
Anastasios Karakostas	Information Technologies Institute, Greece
Anastasios Mikropoulos	University of Ioannina, Greece
Apostolos Gkamas	University Ecclesiastical Academy of Vella of Ioannina, Greece
Barbara Kerr	Ottawa University, Canada
Carlos Travieso-Gonzalez	Universidad de Las Palmas de Gran. Canaria, Spain
Charalampos Karagiannidis	University of Thessaly, Greece
Christos Bouras	University of Patras, Greece
Christos Katsanos	Aristotle University of Thessaloniki, Greece
Christos Douligeris	University of Piraeus, Greece
Christos Panagiotakopoulos	University of Patras, Greece
Christos Pierrakeas	University of Patras, Greece
Daphne Economou	University of Westminster, UK
Demetrios Sampson	University of Pireaus, Greece
Despo Ktoridou	University of Nicosia, Cyprus
Dieter Wuttke	Technical University Ilmenau, Germany
Dimitrios Kalles	Hellenic Open University, Greece
Dionysios Politis	Aristotle University of Thessaloniki, Greece
Doru Ursutiu	University Transylvania Brasov, Romania
George Ioannidis	Patras University, Greece
George Magoulas	Birkbeck College, UK
George Palaigeorgiou	University of Western Macedonia, Greece
Giasemi Vavoula	University of Leicester, UK

Golberi S. Ferreira	CEFET/SC, Brazil
Helen Karatza	Aristotle University of Thessaloniki, Greece
Kostas Apostolou	McMaster University, Canada
Maiga Chang	Athabasca University, Canada
Manuel Castro	Universidad Nacional de Educación a Distancia, Spain
Maya Satratzemi	University of Macedonia, Greece
Michail Giannakos	Norwegian University of Science and Technology, Norway
Michalis Xenos	University of Patras, Greece
Monica Divitini	Norwegian University of Science and Technology, Norway
Nektarios Moumoutzis	Technical University of Crete, Greece
Nikolaos Avouris	University of Patras, Greece
Nikolaos Tselios	University of Patras, Greece
Olga Viberg	KTH Royal Institute of Technology, Sweden
Panagiotis Bamidis	Aristotle University of Thessaloniki, Greece
Panagiotis Petridis	Aston University, UK
Petros Lameras	The Serious Games Institute, UK
Petros Nicopolitidis	Aristotle University of Thessaloniki, Greece
Rhena Delport	University of Pretoria, South Africa
Santi Caballé	Open University of Catalonia, Spain
Stelios Xinogalos	University of Macedonia, Greece
Stavros Nikou	University of Strathclyde, UK
Stamatios Papadakis	The University of Crete, Greece
Tharenos Bratitsis	University of Western Macedonia, Greece
Ting-Ting Wu	National Yunlin University of Science and Technology, Taiwan
Vassilis Komis	University of Patras, Greece

4th IMCL Student International Competition for Mobile Apps Chairs

Andreas Pester	The British University in Egypt, Cairo, Egypt
Ioannis Stamelos Judges	Aristotle University of Thessaloniki, Greece
Antonios Bikas	Board of European Students of Technology (BEST), Greece
Bas Petrus Johannes Falkenburg	Huawei, The Netherlands
Christian Guetl	Graz University of Technology, Austria
Christos Katsanos	Aristotle University of Thessaloniki, Greece
Lambros Lambrou	Accenture, Greece
Petros Lameras	The Serious Games Institute, UK

Contents

Interactive and Collaborative Mobile Learning Environments

Serious Games and Gamification

Open and Distance Mobile Learning

Problems and Prospects of Using Remote Learning Technologies in Different Countries

Irina Makarova(✉), Larisa Fatikhova, Polina Buyvol, and Gleb Parsin

Kazan Federal University, Syuyumbike prosp., 10a, 423812 Naberezhnye Chelny, Russia
kamIVM@mail.ru

Abstract. Modern society is faced with such unprecedented challenges as COVID-19, which in the globalization context can lead to a collapse in all activity spheres, including education. During the lockdowns, which will be similar to those undertaken against the pandemic backdrop, there is a need for a forced transition from class teaching to online forms. At the pandemic beginning, enough time has passed to assess the accumulated experience, comprehend the problems, both those that were resolved and those that could not be resolved promptly. The study made it possible to compare the effectiveness of using different virtual environments for organizing training for engineers, as well as to analyze the training content and its differences from what is used in traditional class training. Different students' groups were selected for the case study in order to compare their participation in the process with traditional and online forms of education, as well as their performance. In addition, an analysis was carried out of how the education form affects the quality of their projects and recommendations were formulated for the further development of new education forms.

Keywords: Remote learning · Digital learning · Learning motivation

1 Introduction

Although the engineering education system has been improving in recent years through the use of new e-learning opportunities, the forced rapid transition of universities to online education in the pandemic context a required the active introduction of new digital resources, methods and technologies of distance learning into the educational process. In the event of a forced "switch", problems arise due to both the "internal" unpreparedness of the participants for changes, and the need for a quick transition to another way of implementing the process, which may not be prepared both from a technical and organizational point of view, and from regulatory legal. Thus, many university professors, in view of their unwillingness in a tight schedule to master and use the new forms of teaching, tried to transfer the existing experience of traditional teaching in the classroom to an online format, which caused a number of problems. To increase the educational system sustainability and its readiness to work under stress conditions, it is necessary to identify the risk situations causes and measures to manage the system in non-standard conditions. Therefore, this article goal is to study the experience of the online learning implementation and his perception by engineering

M. E. Auer and T. Tsiatsos (Eds.): IMCL 2021, LNNS 411, pp. 3–15, 2022.
https://doi.org/10.1007/978-3-030-96296-8_1

students in different countries to identify problems and factors affecting the educational motivation, as well as on successful acquisition of competencies necessary for professional activity in a digital society. Under these conditions, the relevance of analyzing the students position about new educational formats has increased. As a result of the forced transition to a remote learning format, students who previously studied in the traditional form gained experience of using these two education forms and could compare them. In turn, we, as researchers, used a unique chance to generalize students' perception of the learning process in two ways to explore the possibilities for the further development of these two formats in the higher education system, taking into account the positive and negative aspects. For the study, an online survey in the Google Forms format was used. The survey participants were undergraduate and graduate students of automobile department the Kazan Federal University and other universities. The study made it possible to assess the attitude of students to digital learning technologies. To find out the influence of online technologies on educational motivation and the formation of the required competencies for professional activity, the respondents were asked questions grouped according to features that allow making the following conclusions. First, to assess how prepared the students turned out to be in technical and technological terms for the new learning format. Secondly, what changes have occurred in the everyday teaching practices of students. Third, what are the advantages and disadvantages of the distance learning format adopted in the university? Fourth, what is the degree of satisfaction with the educational process when using distance technologies, depending on the hardware base chosen by the student and the educational platform, and how it affects the motivation for learning. Based on this analysis, recommendations were formulated for the further development of online technologies in the educational process.

2 Background: Online Education During COVID-19

During lockdowns, which will be similar to those announced a year ago against the pandemic backdrop, there is a need for a forced transition from traditional classroom learning to online forms. Now enough time has passed since the pandemic beginning, so many universities have accumulated experience, identified problems, as well as solutions that will help in the future.

2.1 Challenges of Transition to Online Education During COVID-19

The COVID-19 pandemic has triggered a global and sharp shift from regular face-to-face (F2F) classes to online in many educational institutions. According to the authors [1], measures are needed to mitigate the negative pandemic effects on engineering education, traditionally based on lectures, practical exercises, laboratory and projects. The authors surveyed 110 lecturers and 627 students from six engineering departments using forms that contained quantitative and qualitative questions to identify problems encountered during online learning in the spring of 2020. Negative issues included the following: logistical and technical, learning and teaching, privacy and security, and lack of sufficient hands-on training. Among the main problems, students noted lack of activity in the classroom, fatigue from online sessions and difficulty concentrating, as well as the difficulty of passing exams online. According to the authors, sharing the results of this study with other educators can help improve the effectiveness of online engineering education by choosing new teaching methods both during crises and after a pandemic. Article [2] describes a study to online teaching and learning (OTL) readiness of 739 higher education teachers in 58 countries during a pandemic by (a) defining teacher profiles based on a set of key readiness parameters; (b) explaining profile membership by individual teacher characteristics, contextual aspects of the transition to OTL, and country-level indicators representing educational innovation and cultural orientation. The authors found that teachers in higher educational institutions are not a homogeneous group: three profiles of teachers with consistently high or low training or an inconsistent profile of readiness were identified. Important aspects are key individual and contextual variables such as teacher gender and previous OTL experience, OTL shift context, educational innovation potential, and cultural orientation. Obviously, a deeper understanding of the profile of teachers' readiness is an important step towards understanding how best to support them during the transition to OTL.

Article [3] presents the research survey results of 61 students about their opinion on online support systems. This survey provides the first recommendations for developing an online support system for university students. Most of the students indicated that the system should be designed with user preferences in mind, interactive and personalized, with an emphasis on time and money management, relaxation exercises and social skills development. Research shows that university students will welcome online support in overcoming the challenges they face at university. Article [4] is devoted to the role of the engineer in the modern world and the he ethical duties. As the authors point out, engineering education in the United States has always valued technical competence over social or ethical competence. To assess ethical issues that are not being addressed in US engineering education, a survey was conducted among 165 graduates of engineering departments of the US State University. The form asked two open-ended questions: 1) How can engineers cope with the COVID-19 pandemic? 2) How important is it for engineering classes to focus on issues of modern society, such as the COVID-19 pandemic? The survey results showed that engineering students have a genuine interest in improving society and tackling the challenges posed by the pandemic. However, engineering education often focuses on technical knowledge rather than ethical development, which is reflected in the themes that emerged in the responses. The role of ethics in the engineering profession, whether in general or in specific circumstances, is often taught as a fluent lesson rather than woven into the curriculum.

The study [5] examined the problems of online engineering education in 4 higher education institutions (HEIs) in the Eastern Visayas, Philippines. Results of an online survey of 25 lecturers and 421 students using Google Forms show, that 98% of respondents were ready for online learning using gadgets. The majority of respondents (94%) believe that the education quality has suffered from the sudden shift to online learning, and 64% believe that it is not as effective as traditional face-to-face classroom interaction. After Covid-19, 60% of educators prefer blended education (BE); while students (65%) prefer traditional face-to-face communication in the classroom.

2.2 Capabilities of Remote Laboratories

The authors [6] believe that since as a result of the COVID-19 pandemic there have been radical changes in engineering education, and change in the classical learning process paradigm, the main task in these conditions is to teach future engineers to work with real equipment, which cannot be accessed online mode. The digital twin concept implementation for industrial equipment can partially solve this problem, which has arisen in the process of distance learning, as well as improve the professional training quality after the pandemic ends. The paper [7] analyzes the impact of remote classrooms and laboratories as a result of "social distancing" during the COVID-19 outbreak. In developing countries, there is a problem of the Internet services availability for the engineering courses online delivery and assessment methods used during this global pandemic. The authors note that the crisis has made it possible to use a variety of distance learning tools for teaching and assessment. It emphasizes that Online Certification decision matrix for Evaluation of Online Learning Readiness is critical to maintaining quality by certifying online courses at different levels and incorporating student feedback as a key performance indicator.

The report [8] states that the pandemic had to rethink and replace laboratory and practical training in engineering education, and in addition, practice and internships almost disappeared. Universities were forced to innovate, with the biggest problem being the loss of hands-on work and contact hours when students were suddenly unable to visit and use laboratories, equipment and manufacturing facilities. At the same time, some felt the courses became more effective and flexible, while many others struggled with difficulties, including low motivation, lack of in-depth training and problems with Internet connectivity. The overall article [9] goal is to propose a viable approach to quality education in the field of energy in the COVID-19 period, for which developments were used to introduce augmented reality (AR) technology to train future engineers. The authors analyzed the necessary paradigm shift to maintain the benefits of using AR, highlighting the usefulness of AR in energy education as part of online learning, and suggesting a possible way to expand the use of AR in education and learning.

Thus, developing virtual labs will allow students to study the design features of units and parts of systems, mechanisms' operation, technological processes of repair and transportation [10, 11], contemporary concepts like reverse logistics [12], modern technologies like Digital Twins [13], but collaboration more effectively is becoming another important issue for successful acquirement of skills.

2.3 Self-motivation During COVID-19

The study [14] goal is to find out how the global emergencies such as Covid-19 affects graduates studying supply chain management, as uncertainties due to such events affect the employment of graduates, causing anxiety and stress. The authors explore what employers look for in students at critical times. This study shows that one must actively seek opportunities for improvement, be flexible, ready to be creative and adapt to new situations in order to successfully enter the labor market at critical times.

This essay [15] explores the emotional experiences of students during online engineering education. Because change can lead to significant increases in student stress levels, the success of engineering education, both in the near and long term, depends on positive learning experiences. In addition, the online learning environment requires more self-motivation and self-regulation on the part of learners, as the computer-based learning environment causes difficulties. The emotional problems faced by students warrant further study, as new teaching methods must be taken into account as to how best to meet the academic intellectual and emotional needs of engineering students in order to ensure the quality education.

The article [16] focuses on a learning environment designed for e-learning for engineering students, studying student behavior, self-learning and self-assessment before and during COVID-19. While student engagement in online learning has increased, at the same time, teachers must interact with students, work together, and study student behavior across courses. The study [17] aims to examine the impact of the online learning climate on student engagement, as well as the relationship between online learning and student engagement according to basic psychological needs. The web survey involved 689 students attending online classes at ten (five public and five private) universities in Pakistan. The results show that the learning climate can stimulate more active student participation to get the most out of online learning, making online learning more resilient to similar challenges in the future. Such platforms provide more interactive and personalized experiences, and educators can regularly receive feedback from students to continually improve their online learning methods.

The study [18] goal is a statistical rating analysis of the students in the Department of Civil, Environmental and Civil Engineering (CCEE) of the San Diego State University (SDSU), obtained in the course of a self-standing online survey, in order to was to find out the scale of changes in the assessment of students of their own training in accordance with the requirements of ABET. The results showed that 81% of comparisons of student grades did not reveal statistically significant differences at the 5% level. According to student assessments, the core expectations for accreditation in the form of mandatory learning outcomes were satisfactorily met both before and under the COVID-19 restrictions. This transition led to a decrease in grades only in 21% of cases, and the transition had the most negative impact on the results of those courses in which the emphasis is on laboratory experience and teamwork. The most difficult and demanding adjustments were in courses with significant laboratory components.

The authors [19] argue that online courses, which are an integral part of basic higher education, often lack teacher feedback and all forms of communication. Student engagement is essential to effective learning and enhancing student satisfaction. There are online learning strategies that can improve students' perception of engagement. Educators should design courses using modern technology that will improve the online learning environment and increase student engagement, retention and satisfaction.

2.4 Problems of Monitoring and Evaluation of On-Line Learning Results

The article [20] discusses the relevance of evaluating student online learning processes for university degrees by assessing the students' perception of the online learning quality in five specific courses at the Faculty of Engineering at the University of Burgos, Spain. According to students' perceptions, teachers had the technical knowledge, social skills and personality to adapt their courses to online methodology during

COVID-19. The results showed that the quality scores of both F2F and online learning were very similar and highly dependent on the teacher's work, and in some cases the quality of learning improved according to the students. However, the general disadvantages of online learning are associated with the lack of direct contact between teacher and student, and between the students themselves, as well as with the personal qualities of students, their ability to self-regulate their learning. The use of web tools (Skype, Microsoft Teams) increased when it was needed to explain practical concepts.

Study [21] presents the impact of COVID-19 on the practice and implementation of Geotechnical and Geoecological Engineering (GGE) training modules, based on feedback from teachers from 14 countries. Key challenges for educators appear to be how to maximize learning flexibility and meet physical distance requirements without compromising learning outcomes, educational equity and interpersonal interactions in traditional F2F learning. Three future opportunities are identified, namely smart learning, flip learning and interdisciplinary education, which can provide learners with a more sustainable, engaging, interactive technology-driven learning environment, and equality of opportunity and interpersonal communication, can help GGE educators develop a more sustainable educational environment. Article [22] discusses a possible solution to help teachers track students' attention using a computer program for facial analysis. The article states that the program can automatically determine attendance by analyzing face detection data, and also determine if any student leaves the class early. The authors come to the conclusion that the program can provide attentiveness to all students in an online classroom. It is shown that a useful user interface will allow the program to be used more intuitively, and the development of a mobile application will allow teachers who use the phone for online classes to make their work easier.

3 Analysis of the Automotive Faculty Students Transition to Online Learning

The forced and rapid transition of universities to digital education in the context of the pandemic has actualized the issues of attitudes towards it among students who are focused on traditional forms of education at a university, and the possibilities for the further development of this scenario in the higher education system.

3.1 Problems of Transition: Student and Lecturer Assessment

About 15% of the entire faculty took part in the survey. The results of the study showed that teachers are organizationally ready for the transition to online learning formats, but psychologically they do not accept such a sharp break with traditional daytime education. A skeptical attitude towards what is happening is associated both with the peculiarities of the disciplines taught (for example, technical and experimental), and with conservative views on the educational process. Although, two positive social characteristics of the professional lecturer community were recorded: the adoption of a state policy to counter COVID-19 and the availability of skills and abilities to work in an online format. The main results of the survey are shown in Fig. 1. According to the majority of both teachers and students, a blended education form will be the most rational [23, 24], since not all types of distance learning are effective. The survey of students was carried out using online in Google Forms. In a pilot study of students' positions towards online learning, 344 full-time students (85.1% of bachelor's degrees and 14.9% of masters) took part. The age respondents' structure: 20 years old and younger - 25.6%, 20–25 years - 70.3%, 26–30 years - 1.7% and over 30 years - 2.3%. To determine the technical and technological preparedness of students for the transition to the online learning format, questions were asked about the problems and positive aspects. Among the most significant technical problems, students noted the communication instability during the classes (25%), the need for stable access to the Internet (24.9%), low Internet speed (19.8%), feedback lack, technical means lack and other reasons. The survey showed that only 65% of students were satisfied with the Internet quality when connecting to classes. At the same time, 74.4% of students were satisfied with the personal communication capabilities. The most frequently used for online training: laptop (desktop) - 49.4% of students, smartphone (tablet) - 49.1%. This shows that not all students can use computers for online learning.

Fig. 1. Results of the teacher survey

Although the online survey does not show the full picture of difference in access to online learning, the problem of "digital inequality" exists and its causes can be both material reasons and lack of access to broadband Internet in their living places.

3.2 Educational Process' Difficulties in the Online Format

The survey showed that the transition to online learning caused a number of problems for students: only 51% of the surveyed students adapted well to online learning. At the same time, in addition to technical difficulties, there were problems associated with the specifics of online learning itself, as well as the lack of the necessary skills. The main problems of online learning identified by students are shown in Fig. 2. Another problem, identified during the survey is low self-motivation and self-control, as well as inability to plan one's time and untimely feedback from the lecturer (Fig. 3). There were also problems with the lack of educational and methodological materials, especially for disciplines that require the acquisition of practical skills. The most frequently used methodological university guidelines (62.8%), methodological guidelines on the Internet (56.4%) and video materials provided by the university (40.4%) (Fig. 4).

At the same time, to search for educational resources and methods of their development, students turned to teachers - 33%, used the experience of other students - 32%, searched independently - 32.2% and "failed to find" 2.8% of students. To interact with teachers in online learning, students used: Microsoft Teams (56.7%), "Virtual classroom" of the university (26.2%), Zoom, Skype (13.4%) and other platforms (3.7%).

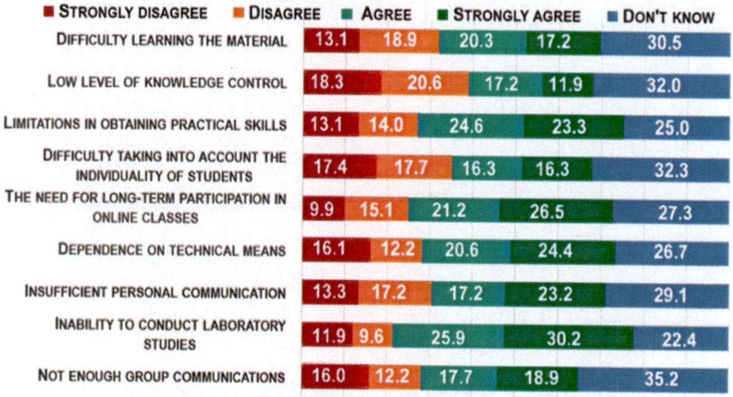

Fig. 2. The main problems of online learning

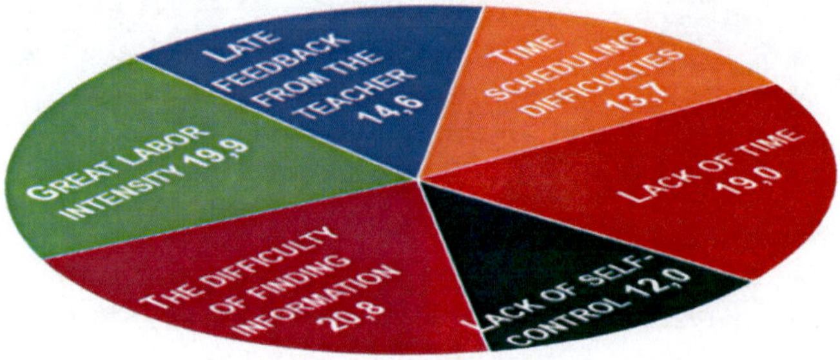

Fig. 3. The main problems at the self-dependent students work

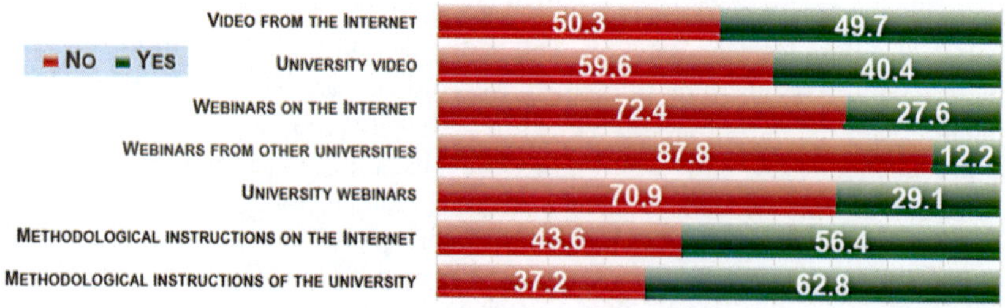

Fig. 4. Using additional teaching materials

Students see the such advantages of online learning as: the possibility of combining study with work; increasing the skills of independent work; the availability of materials and the ability to use them at a convenient time and in a familiar environment, and the such disadvantages as: the need to spend a lot of time at the computer during classes and independent work; lack of personal communication, both with teachers and with other students; inability to access the laboratory and gain practical skills; dependence on technical means and a decrease in the education quality.

3.3 Dependence of the Exam Session Results on Various Factors in Online Learning

To search for educational trends and assess the exams results after online training in the spring of 2020, were analyzed the summer session results of students-engineers of 1–4 courses of KFU (1191107-Materials Science and Technology of Materials; 1191109 - Mechanical Engineering; 1191111 - Design and technological support of engineering productions; 1191113 - Vehicle designer; 1191115 - Repair and maintenance of vehicles; 1191116 - Vehicle and automobile economy; 1191121 - Transport and logistic; 1191133 - Vehicle service). As a result, the average score in the theoretical (lecture) part of natural-sciences disciplines turned out to be higher than for the implementation of practical tasks. The resulting dependencies by disciplines: mathematics (M), structural materials technology (SMT), information technology (IT), descriptive geometry and computer graphics (DG & CG) are presented in scatter diagrams (Fig. 5, 6).

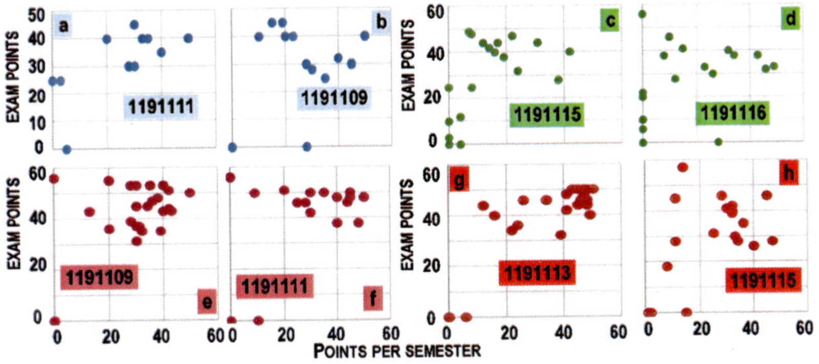

Fig. 5. Exam results in maths (a, b - teacher 1, c, d - teacher 2) and in SMT (e, f - teacher 3, g, h - teacher 4)

As can be seen from the diagrams, the teacher personality does not affect the grades, which depend only on the curriculum, therefore, the feasibility and effectiveness of online classes is higher for a lecture course, and personal communication between the student and the teacher is required to obtain practical skills. On the scatter diagram of results in physics (Fig. 7a, b), two clusters are clearly visible: students who received low scores both in the semester and on the exam, and students with scores above the average. Students of the first cluster note problems with the use of teaching materials, which forces professors them to solve the issue of using online technologies for courses related to obtaining practical skills.

Similar results for foreign languages (Fig. 6c–e) are explained by the complexity of organizing paired language practice when using some online platforms (MS Teams) or the lack of such an opportunity (Virtual classroom). Comparison of exam results in IT and other disciplines showed that IT proficiency directly affects the results of the session. For example, students of the fourth group (Fig. 7a) have a high average score in both IT and other disciplines (see also Fig. 5). Students from this group, when surveyed, indicated that they did not experience any difficulties in the transition to online learning.

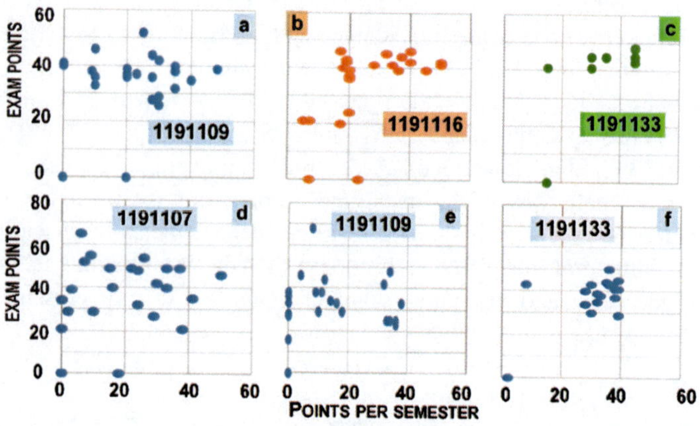

Fig. 6. Exam results in IT (a - teacher 1; b - teacher 2; c - teacher 3) and in DG & CG (d, e, f - teacher 4)

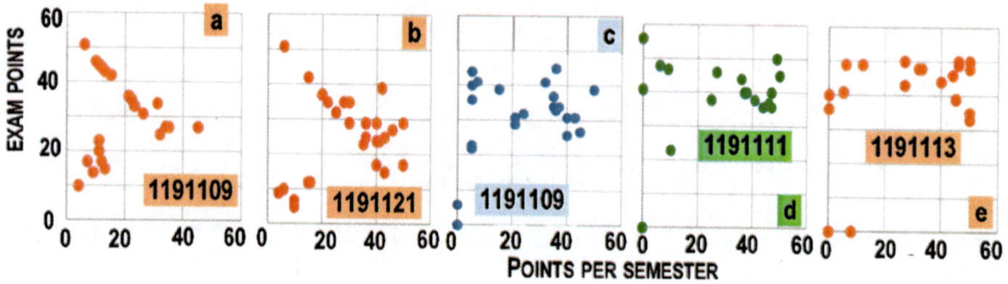

Fig. 7. Exam results in physics (a, b - teacher 1) and foreign languages (c - teacher 2; d - teacher 3; e - teacher 4)

In addition, the exams result in the same discipline of the same course to different teachers (Fig. 8b) demonstrate that students receive approximately the same scores in M and IT, with the exception of the last two groups in which M-classes were taught by another teacher, who, according to the "Teacher with the eyes of a student" survey results, has a low score. Thus, academic performance depends not only on the individual student qualities, but also on the teaching methodology.

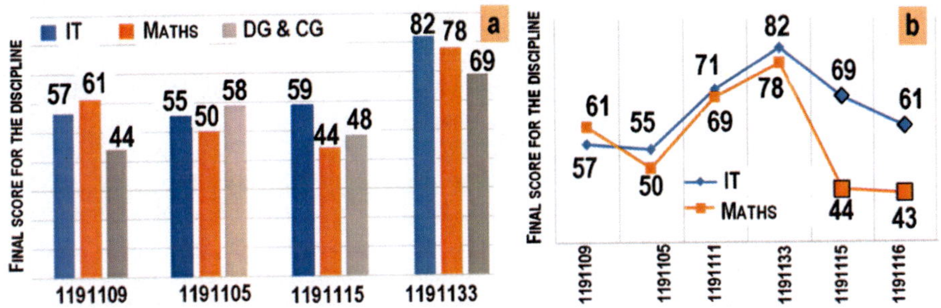

Fig. 8. (a) Average scores based on exams' results; (b) comparison of average scores in IT and Maths

4 Conclusion

Based on the online survey results, the advantages and disadvantages of online educational technologies were established, the main limitations and factors of reducing educational motivation in online learning were identified. Regardless of the university territorial location, the majority of students spoke in favor of the fact that the use of distance technologies is a necessity in modern society and will be actively used in the higher education system, due to a number of advantages. However, existing digital learning tools are insufficient to stimulate learning motivation, especially for students with a low initial training level. In addition, in engineering education, online learning cannot completely replace traditional laboratory and hands-on training in a traditional form. The educational content existing at the lockdown time could not ensure the proper education quality level and understanding of the real processes' essence, although at a certain elaboration level, practical exercises can be replaced by VR laboratories, therefore, it is necessary to develop specialized virtual laboratories. Based on the research results, we received answers to the following questions: (1) What difficulties arise in the implementation of online education forms, and what its advantages help to increase the motivation of engineering students. (2) How socio-demographic factors affect the mastering success and the distance learning perception. (3) What virtual environments are most effective in terms of process management, education quality, student engagement, and increasing their motivation. (4) For the implementation of what learning forms is best suited online form. (5) What resources and software should be developed to control the learning quality and process organization?

References

1. Asgari, S., et al.: An observational study of engineering online education during the COVID-19 pandemic. https://doi.org/10.1371/journal.pone.0250041. Accessed 29 June 2021
2. Scherer, R., et al.: Profiling teachers' readiness for online teaching and learning in higher education: who's ready? Comput. Hum. Behav. **118**, 106675 (2021)
3. Goozée, R., et al.: Survey to inform the development of an online support system for higher education students – higher education and online support. Health **10**, 351–364 (2018)
4. Roy, A., et al.: Ethics in engineering education during COVID-19 pandemic. Issues Educ. (2020). https://doi.org/10.1287/orms.2020.06.10
5. Perante, W., et al.: Challenges to online engineering education during the Covid-19 pandemic in Eastern Visayas, Philippines. Int. J. Learn. Teach. Educ. Res. **20**(3), 84–96 (2021)
6. Rassudov, L., Korunets, A.: COVID-19 pandemic challenges for engineering education. In: XI International Conference on Electrical Power Drive Systems (ICEPDS), pp. 1–3 (2020)
7. Khan, Z.H., et al.: Distance learning in engineering education: challenges and opportunities during COVID-19 pandemic crisis in Pakistan. Int. J. Electr. Eng. Educ., 1–20 (2021). https://doi.org/10.1177/0020720920988493
8. FEATURE: 5 ways the Covid-19 pandemic has changed engineering forever. https://www.imeche.org/news/news-article/feature-5-ways-the-covid-19-pandemic-has-changed-engineering-forever. Accessed 29 June 2021
9. Opriş, I., et al.: Challenges and opportunities to overcome. The impact Of COVID-19 pandemic on power engineering education. TEM J. **9**(4), 1687–1691 (2020)
10. Vakulenko, K., et al.: Designing optimal public bus route networks in a suburban area. Transp. Res. Procedia **39**, 554–564 (2019)
11. Galkin, A., et al.: Last-mile delivery for consumer driven logistics. Transp. Res. Procedia **39**, 74–83 (2019)
12. Makarova, I., et al.: The role of reverse logistics in the transition to a circular economy: case study of automotive spare parts logistics. FME Trans. **49**(1), 173–185 (2021)
13. Shubenkova, K., et al.: Possibility of digital twins technology for improving efficiency of the branded service system. In: Proceedings, GloSIC 2018 (2018)
14. Stratton, A., Curkovic, S.: Global emergencies: how do they affect supply chain management students? Creat. Educ. **12**, 231–264 (2021)
15. Park, M., et al.: Online engineering education under COVID-19 pandemic environment. Int. J. Multidiscip. Perspect. High. Educ. **5**(1), 160–166 (2021)
16. Jamalpur, B., et al.: A comprehensive overview of online education – impact on engineering students during COVID-19. Mater. Today Proc. (2021). https://doi.org/10.1016/j.matpr.2021.01.749
17. Shah, S.S., et al.: Online learning during the COVID-19 pandemic: applying the self-determination theory in the 'new normal.' Revista de Psicodidáctica **26**(2), 168–177 (2021)
18. Supernak, J., Ramirez, A., Supernak, E.: COVID-19: how do engineering students assess its impact on their learning? Adv. Appl. Sociol. **11**, 14–25 (2021)
19. Mitchell, A.: Online courses and online teaching strategies in higher education. Creat. Educ. **5**, 2017–2019 (2014)
20. Revilla-Cuesta, V., et al.: The outbreak of the COVID-19 pandemic and its social impact on education: were engineering teachers ready to teach online? Int. J. Environ. Res. Public Health **18**, 2127 (2021)
21. Ning-Jun, J., et al.: Geotechnical and geoenvironmental engineering education during the pandemic. Environ. Geotech. **8**(3), 233–243 (2021)

22. Ghani, T., et al.: Development and analysis of a machine learning based software for assisting online classes during COVID-19. J. Softw. Eng. Appl. **14**, 83–94 (2021)
23. Makarova, I., et al.: An integrated platform for blended learning in engineering education. In: Proceedings of the 9th International Conference on Computer Supported Education - Volume 2: CSEDU, pp. 171–176 (2017)
24. Makarova, I., Pashkevich A., Shubenkova, K.: Blended learning technologies in the automotive industry specialists' training. In: 32nd International Conference on Advanced Information Networking and Applications Workshops (WAINA), pp. 319–324 (2018)

Mobile Technology for Learning During Covid-19: Opportunities, Lessons, and Challenges

Oluwakemi Fasae[1(✉)], Femi Alufa[1], Victor Ayodele[1], Akachukwu Okoli[1], and Opeyemi Dele-Ajayi[2]

[1] Stemres Learning Initiative, Lagos, Nigeria
kemi.fasae@stemres.net
[2] Faculty of Engineering and Environment, Northumbria University, Newcastle upon Tyne, UK

Abstract. The Covid-19 pandemic has caused major disruptions to traditional teaching and learning activities globally. As governments continue to develop strategies and policies to tackle the impact of the pandemic on various sectors, teachers are at the forefront of several efforts to ensure that some form of learning and teaching activities are maintained throughout the period, essentially by leveraging on technology. In Nigeria, most of the teachers have not been trained in technology-enhanced teaching and are only able to deliver lessons within the traditional classroom. This paper presents the initial results of a pilot case study of a state government's response to training teachers in using EdTech to continue teaching during the Covid-19 pandemic. Using a set of criteria based on ease of use, usefulness, and availability, 10 mobile apps were selected and applied to support teaching activities like lesson planning and organization, creation and dissemination of lesson notes, lesson delivery, student assessment, and information management.

This paper highlights observations from the pilot training and provides valuable insights into how mobile devices and applications can be used in low-income and resource settings to support educational delivery during a crisis. The paper further provides valuable lessons from the training and challenges to delivery identified by the teachers. Finally, we provide some recommendations to governments and school administrators looking to support the professional development of their teachers with the use of mobile devices.

Keywords: Technology · Online teaching · Mobile apps

1 Introduction

It is common knowledge that the COVID-19 pandemic has caused a major disruption across various sectors in both developed and developing countries, not least in the educational sector. By mid-April 2020, 94% of learners worldwide were affected by the pandemic, representing 1.58 billion children and youths, from pre-primary to higher education, in 200 countries [1]. Many countries announced the temporary closure of schools to contain the spread of COVID-19 which invariably affected more than 91%

M. E. Auer and T. Tsiatsos (Eds.): IMCL 2021, LNNS 411, pp. 16–25, 2022.
https://doi.org/10.1007/978-3-030-96296-8_2

of students around the world leaving more than 1.7 billion children and young people out of school [2].

In response to the closure of schools, UNESCO and the United Nations recommended the use of distance learning programs to ensure the continuity of educational activities [1, 2]. According to the research report by [3], "consistent school attendance is a foundation of student learning. Whilst missing one or two days of school each year is not likely to have serious consequences, chronic absenteeism is related to significantly lower outcomes for students". Although missing school may be down to other factors and not primarily absenteeism, being away from the school or learning environment would have the same effects of being absent from school, consequently resulting in poor learning outcomes for the affected children. To mitigate this and ensure learning continuity, Government Agencies alongside international organisations and private organisations have sought ways to provide education remotely with the use of a combination of technologies [2].

Whilst some countries have facilities to easily migrate from classroom learning to online learning, the same cannot be said for developing countries such as Nigeria. As governments continue to develop strategies and policies to tackle the impact of the pandemic across various sectors, a key to successfully implement distance learning is: training and supporting teachers while encouraging collaboration amongst them as they are at the forefront of ensuring that learning and teaching activities are maintained during the period [2, 4].

The COVID-19 pandemic and the closure of schools presented a major challenge for teachers and educators to adapt their lessons and classes urgently and incorporate distance learning to maintain student learning with the same level of educational quality [5]. This transition was particularly challenging as teachers across the world were suddenly tasked with implementing online teaching without sufficient guidance, training, or resources.[4] For instance, according to a press release by UNESCO, in sub-Saharan Africa, only 64% of primary and 50% of secondary teachers have received minimum training, which doesn't include basic digital skills [1, 6]. Many educators, therefore, lack the most basic ICT skills needed to support the transition to online methods which means they would struggle with facilitating quality distance learning [1]. This shows that there is a need to train teachers adequately in new methods of education delivery, especially with the use of technology.

According to [7], it's not enough to train teachers on the use of technology in teaching, it is important for them to be trained on how to integrate its use and maximise the benefits from it. With the constraints imposed by the Covid-19 pandemic, it has become more pertinent for teachers in Nigeria to know how to use technology to perform tasks associated with lesson delivery and to teach content with ease. [7] indicates that, students are more skilled to take part in digital lessons as they only need to know how to access and understand the contents, but the creation and development of these lessons turns out to be challenging for the teachers. The transition to remote teaching for teachers would require a total migration from traditional methods to technologically enhanced methods. The general duties and activities of a teacher as highlighted by [8] are Classroom Management, Course Planning, Selection & Creation of materials, Instructional Teaching, Student Assessment/Test construction/Grading, Recording & Reporting student achievement and other class management and administrative tasks. To teach remotely successfully, a teacher requires knowledge of how to use technology/digital devices to carry out some or all of these activities which when utilised correctly, would create an effective online learning environment.

In the traditional method of teaching for instance, lesson planning and organisation require teachers to write out their timetables and ensure they recall lesson times, they also prepare and maintain hand-written lesson notes and curricula which they use in delivering lessons to students. When delivering lessons, teachers sometimes verbally dictate the notes and explain simultaneously, while some write the notes on a blackboard for the students to copy. Teachers often gather resources for their lesson contents from textbooks and hardcopy resources provided by their schools and write them out in lesson notes and also have hand drawn images in these lesson notes. In terms of assessment and grading, teachers often gather the notebooks of the students to assess and award marks to the notes or assignments submitted. They also physically monitor the assessments and award grades to them. All these activities are major duties of a teacher which must be carried out irrespective of if it's online or offline. However, to effectively perform these same responsibilities online, teachers need to acquire new skills and abilities.

For teachers to acquire the relevant skills needed to carry out their duties in an "online" environment, they need to be trained on how to use different technological tools. In Nigeria, most of the teachers have not been trained in technology-enhanced teaching and the current circumstances continue to expose the limitations of their current teaching methods. A few open-source e-learning materials and technologies such as Google classroom for classroom management are not practicable in the rural areas of Nigeria as internet coverage is still inadequate. Some urban areas where schools could have access to internet also suffer other related issues such as lack of power supply, lack of tech-inclined teachers and overcrowded classrooms. These challenges can however, be circumvented by making use of currently existing accessible low-cost technologies, which can be afforded by even middle-class families.

In response to these challenges, a pilot state government-led initiative in Ekiti state, Nigeria, was carried out to train teachers using a low-cost platform on how to deliver lessons and interface with students using already existing applications, technologies or channels. The study aimed to investigate:

1. What effect does teacher training with the use of online technology have on teachers' confidence and competence for successful lesson delivery?
2. What online tools and resources are suitable for conducting online lessons in low to middle-income settings?
3. What challenges plague the successful coordination of online lessons in remote areas in low- and middle-income countries.

This paper presents the initial results of the training. It discusses applications that can be used in low-income and low-resource settings to support teaching activities. It also provides recommendations to governments and school administrators concerning the adoption of mobile technologies to support distance learning and remote teaching.

2 Participants and Methods

2.1 Participants

A total of 20 teachers from all local governments in Ekiti state participated in this pilot study. The teachers were selected by the state government as representatives from all the local governments in the state. 40% of the teachers were primary school teachers who taught all the subjects and 60% were secondary school teachers, majorly science teachers.

2.2 Methods

An online survey method was used to collect data from the participants before and after the training. An initial evaluation of the technical devices and resources available to the teachers which could support online teaching was carried out through an initial online survey sent out to the teachers. From the evaluation and survey carried out as shown in Table 1, 70% had access to only smart phones, 20% had access to both smart phones and laptops and 10% had access to smart phones, laptops and tablets.

In deciding the mode of delivery and the stand-alone mobile and web applications utilized in the training and for the training, availability of technological resources to the teachers was one of the factors considered. Apart from the availability of the resources to the teachers, according to [9] where a modified Technology Acceptance Model for teachers was developed, a key factor to the adoption of technological tools by teachers is their preparedness and readiness to accept the technology. He further stated that teachers' readiness to accept and use a technological tool is based on how easy to use they perceive the tools to be. Considering this, our choices of mode of delivery and training tools were also determined by the ease of use and accessibility of these applications. We ensured they were low-cost resources which the teachers could easily afford to download or access online with their mobile data and on their phones since all the teachers had access to mobile phones.

Table 1. Teachers' access to digital devices

Technology/device	Percentage
Only smartphone	70
Smartphone and Laptop	20
Smartphone, Tablet and Laptop	10
3G network	65
4G network	25

2.3 Activities

The teachers were divided into 4 groups comprising 5 teachers per group with a training coach assigned to each group. The training was spread over a period of 5 days so to ensure the teachers had solid understanding and ample practice time with the applications in training. The training was conducted online using the Zoom platform and was focused on exposing the teachers to online teaching strategies and demonstrating to the teachers how to use existing mobile applications to support their teaching activities.

9 mobile applications were explored for the following teaching activities which the training focused on:

Lesson Planning and Organization
This included training the teachers on how to use electronic and mobile applications to plan their lessons, organise lesson materials and monitor the progress of each lesson. The selected applications used in this section were Planboard and Evernote. Planboard is a free online lesson planner that allows teachers to schedule their lessons, track their progress on their scheme of work and set up timetables with reminders. Evernote is a note-taking application that allows users to track their lessons plans for the day and to monitor progress in completing tasks scheduled for the day.

Lesson Illustration and Development
This session involved training on the use of electronic technologies to draw up lesson note content by making illustrations, attaching diagrams, attaching links to further resources, packaging lesson materials and providing the students with necessary audio-visual instructional materials needed for the lesson. It enables the teacher use real time resources or tools to describe what is being taught during lessons and assist the students in understanding lesson content. In this session, INKredible and XRecorder were used. The INKredible application was used to create illustrative and interactive designs. It's a freehand writing application which provides the teachers with the ability to create their own drawings and display to the students. It was discovered to be useful in subjects like Biology that require sketching of diagrams. It could also be used for drawing shapes, alphabets and figures for primary school pupils. The XRecorder is an audio and video recording mobile application which can be used by the teacher to make recordings of teaching illustrations which can be shared with the students. For example, while using the INKredible app to make illustrations and explanations, the XRecorder can be used to record the activities.

Resource Management
There was a need for teachers to record and keep track of materials developed and also find a means of sharing the lesson materials with the students or parents of the students without having to send information individually. To achieve this, the teachers were introduced to the use of mobile and web applications, specifically, Google Drive and Dropbox, to store and share lesson materials for immediate or future access. These applications give teachers the ability to organise their lessons or lesson materials using cloud storage which makes it accessible to students and parents from any location. These tools offer substantial amounts of free space for storage and collaborative purposes.

Lesson Delivery
As important as it is to create good lesson content for the students, it is equally important to have a good delivery mechanism. After preparing lesson content, teachers need to deliver and explain the material to the students. Therefore, the teachers were

trained on how to create and use Telegram channels and WhatsApp groups as a means to deliver class sessions to the students directly from their mobile devices. These applications have features that allow teachers to share texts, documents, audio files, video files, pictures etc. They also provide a platform for the teachers and students to interact.

Assessment
The teachers, on completing their online teaching sessions, need to assess and evaluate the students' learning by asking questions and getting feedback. To achieve this, the teachers were trained on how to use Google forms to create assessment questions and send out to the students. The students are able to access the forms, fill out their responses and send back to the teacher.

After every session, the teachers were tasked with responsibilities of creating sample content and instructional materials using the technologies they were introduced to, these were shared with their assigned instructors. This was to ensure that the teachers got the needed practical knowledge required to support the online teaching.

3 Results, Lessons and Challenges

At the end of the training, an online feedback survey was sent to the teachers to evaluate the impact of the pilot study.

3.1 Results

The teachers gave their feedback on the impact of the training on a scale of 1 to 10, with 1 being not useful and 10 being highly useful, 50% of the teachers rated 10 and said they were excited to apply the knowledge gained to their lessons, 30% teachers rated 9, 10% rated 8 and 10% rated 6 as shown in Fig. 1. None of the teachers rated between 1 and 5 or 7. The teachers were further asked about the level of knowledge they had in relation to the apps used and online teaching prior to the training and after the training on a scale of 1–10, the results from before and after the training are shown in Fig. 2. It shows a good improvement as over 50% of the teachers had moved from a level between 2 and 3 to a level above 6. This suggests that training often has a positive effect on the level and competence of teachers especially when it comes to using technology to teach. This could also be attributed to the fact that they went through training as students being taught using the technology. The first-hand experience from this helped them to have a better understanding of how to conduct online classes and manage the participation of students.

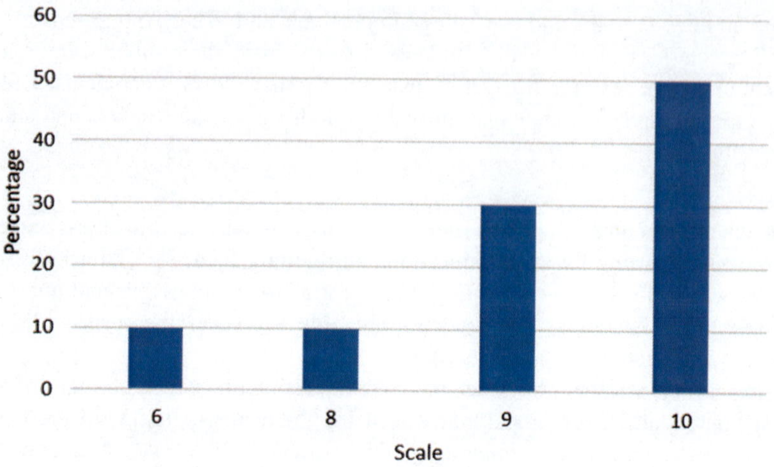

Fig. 1. Scale of impact of training on teachers

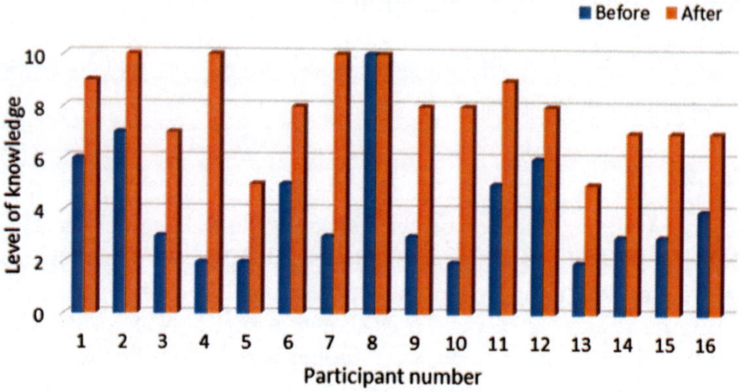

Fig. 2. Skill improvement before and after training

3.2 Lessons Learned

This section presents some of the findings from feedback collected from teachers and interactions during the training.

From the results of the survey and some of the comments made by the teachers, it was evident that the teachers needed some form of training on how to engage in online teaching especially since they had never been involved in any form of teaching online or the use of technology to teach. Some were scared of the possibility of online teaching as seen in this teacher's comment:

> "The training was really an eye opener; it was a wonderful session. I really enjoyed it. I thought teaching online was a big deal before, until I participated in the training, I now see online teaching as easy as ABC. Thanks to the organiser of this programme" – **Teacher 1**

This shows that teachers who didn't use technology to teach before the shutdown of schools didn't feel competent to successfully carry out online teaching. It is therefore essential that teachers receive training to fully engage the use of technology in their teaching practice. This finding is consistent with the works of [7, 10, 11] where the importance of teacher training and support for the use of ICT tools is identified.

Government educational institutions, school heads and general education policymakers need to design broad strategies that would prepare students and teachers to use e-learning tools and methods regularly in the school system. This could involve integrating the use of technology and providing regular training to teachers and students. This would enable them to cope with pandemic situations without so much hassle. The findings from [12] is in line with this finding that the extent and competence of technology use affects the perception of teachers and students to the success of online teaching.

Results also seem to suggest that the teachers were willing to improve themselves and also extend to other teachers the knowledge gained from learning to teach online using technological methods which is very commendable. And even though online teaching all seemed new to them, they were willing to learn new skills and tools while improving themselves to ensure that the students do not miss out on learning. [4] in his work supports this and recommends that teaching staff should be appreciated and compensated. This willingness and readiness are evident from the statement made by one of the teachers who is ready to introduce the concepts and experiences learned to his colleagues.

> "The training has been an eye opener I was thinking it's a big deal teaching pupil online, now I know better and now am a step above my colleagues and I can't wait to introduce as many teachers as possible so we help our pupils at a time like this". – **Teacher 3**

3.3 Challenges Faced by the Teachers

The teachers during the training, were however faced with some challenges. They were asked open-ended questions in the survey they completed about the challenges they faced. Their responses were grouped into 3 namely: Poor internet connectivity, cost of data and insufficient time to learn.

The study revealed that 95% of the teachers encountered network failure and unreliable connection as a major problem, as sometimes, it was difficult to follow through during the training sessions. One of the teachers expressed that "I have network challenges, sometimes the network is too bad that I have to get out of the house to look for network outside the house". Another also expressed the difficulty to go out to look for better network reception due to the lockdown- "Network problem is the major thing that affect us, because most of us didn't have the access to move out to where we can get network due to what is going on generally". Another expressed how internet connectivity accentuated the digital divide between people in different areas of the town -

> "The major problem is network failure that occurs during training because learners are not in the same area, some are favoured more than others".

The cost of data to engage in the training was another challenge experienced by some of the teachers, although this was duly compensated after the training. 25% of the teachers mentioned the problem of insufficient data; before the training, they only purchased a small amount of data needed for basic WhatsApp communication, but as the training was conducted via zoom, they had to top-up their data several times. One of the teachers expressed this- "Network problem and Insufficient money to buy data because I only bought small data before for WhatsApp, but it finished quickly because I was using zoom".

Another challenge faced by the teachers was the volume of things they had to learn in a day. The training spanned 5 days with an hour each day to learn about 4–5 applications. This put some sort of strain on the teachers. About 5% of the teachers express this and said that even though they learnt enough, there was information overload as they had to learn to use 10 new applications in just 5 days – "Time factor; the number of topics covered with in an hour are many". Another also said "Training duration is short". They further recommended that the training be regular and spread out over more hours and more days.

4 Limitation of the Study

The exercise involved a small group of teachers as it was a pilot programme, the first of its kind in Ekiti State, Nigeria. As the sample size of participants was small, we are unlikely to draw significant statistical relevance at this stage. It is anticipated that this initial phase will be expanded to accommodate a larger cohort of participants, which we envisage will give a clearer indication of teachers' perception towards online educational delivery and make a stronger case for policy-driven reforms in the educational sector.

The training was delivered via zoom, which is not optimised for online training and also made use of different stand-alone applications. Ideally the programme should have been delivered over a bespoke LMS platform with all the necessary resources embedded within the platform to give a more fluid training experience. The training was not subject-specific, but was a general orientation of how the different applications may be used in lesson delivery. Subsequently, training should be subject-specific to enable teachers understand how to transform or enhance lesson content.

5 Conclusions, Recommendations and Summary

Tasking teachers with implementing online teaching to cope with the problems of the pandemic without sufficient guidance, training, or resources was a problem which needed to be addressed. The pilot training study carried out revealed that teachers who didn't use technology in teaching before the shutdown of schools were worried and didn't feel competent. However, training and support helped them to gain more confidence and knowledge about how to proceed This illustrates the importance of training the teachers. Different mobile applications are also recommended to support the teachers in their different teaching activities.

We also intend to follow up on the initial cohort of participants to observe if there have been any attempts to implement the training received in the classrooms or if there have been any adjustments made to lesson delivery.

Acknowledgement. The authors would like to appreciate the Ekiti State Government Ministry of Education, Science and Technology for supporting the training. We also appreciate the input and contributions of the teachers involved, outreach staff and volunteers that worked on making the training a success.

References

1. United Nations: Policy Brief: Education during COVID-19 and beyond (2020)
2. UNESCO: Distance learning strategies in response to COVID-19 school closures (2020)
3. Ehrlich, S.B., et al.: Preschool Attendance in Chicago Public Schools Relationships with Learning Outcomes and Reasons for Absences (2014)
4. Noor, S.: Online teaching practices during the COVID-19 pandemic. Educ. Process Int. J. (2020). https://doi.org/10.22521/edupij.2020.93.4
5. Dietrich, N., et al.: Attempts, successes, and failures of distance learning in the time of COVID-19. J. Chem. Educ. (2020). https://doi.org/10.1021/acs.jchemed.0c00717
6. Startling digital divides in distance learning emerge (2020). https://en.unesco.org/news/startling-digital-divides-distance-learning-emerge
7. Zhu, C.: Pre-service teachers' perceptions of ICT integration in teacher education in Turkey. **14**(3), 97–110 (2015)
8. Scriven, M.: Duties of the teacher. J. Pers. Eval. Educ. **8**(2), 151–184 (1994)
9. Dele-Ajayi, O., Strachan, R., Sanderson, J., Pickard, A.: A modified TAM for predicting acceptance of digital educational games by teachers. In: IEEE Global Engineering Education Conference EDUCON, April 2017, pp. 961–968 (2017). https://doi.org/10.1109/EDUCON.2017.7942965
10. Adegbenro, J.B., Gumbo, P.M.T., Olakanmi, E.E.: In-service secondary school teachers' technology integration needs in an ICT- enhanced classroom. **16**(3), 79–87 (2017)
11. Gyimah, N.: Assessing technological innovation on education in the world of coronavirus (COVID-19). SSRN Electron. J. (2020). https://doi.org/10.2139/ssrn.3670389
12. Mailizar, Almanthari, A., Maulina, S., Bruce, S.: Secondary school mathematics teachers' views on e-learning implementation barriers during the COVID-19 pandemic: the case of Indonesia. Eurasia J. Math. Sci. Technol. Educ. **16**(7) (2020). https://doi.org/10.29333/EJMSTE/8240

The Abrupt Shift to Full Online and then Blended Learning at a French Engineering School: Difficulties and Practices, Reaction and Adaptation

Hanen Kooli-Chaabane[1,2], Antoine Lanthony[3(✉)], Aristide Boukaré[3], and Nicolas Peyret[3,4]

[1] University Paris Nanterre, Nanterre, France
[2] CEROS, 4429 Nanterre, EA, France
[3] ISAE-Supmeca, Saint-Ouen, France
antoine.lanthony@isae-supmeca.fr
[4] Laboratoire QUARTZ, 7393 Saint-Ouen, EA, France

Abstract. ISAE-Supmeca, a French engineering school located in the great Paris area, has been engaged in active learning, especially problem- and project-based learning (PBL), since decades. This experience, including some remote activities, gave background for the institution to get through lockdown and forced remote learning due to the covid-19 pandemic situation. Nevertheless, the shock of shifting in one week-end to full remote learning was abrupt. ISAE-Supmeca, and especially two of its units (the Learners' Affairs Department – LAD, and the Educational Innovation Unit – EIU), held discussion with learners and teachers in order to help them to cope with the situation. The two units also conducted a work in four phases (monitoring, analyzing, proposing and capitalizing) based on a series of surveys realized from June 2020 to June 2021. This paper presents the results of the monitoring, as well as a part of the analysis phase results. More precisely, in this work, the description of the actors' reactions and the adaptation of the learning process during the pandemic situation are detailed. Moreover, an attempt to compare the collected data of ISAE-Supmeca to a part of results from a study realized at California State University is made.

Keywords: Difficulties · Distance learning · Platforms

1 Introduction and Theoretical Framework

The whole world is likely to be more and more relaying on the mobility services and wireless communication [1]. The experts' forecasts show the acceleration of the growth of the number of connected devices. In 2025, the total number of connected devices in the world will be approximately 75.44 billion [1]. New technologies allow to overcome barriers of time and space [2]. The COVID-19 pandemic crisis has been a driving force for digitalization, especially in the education world. According to a recent study of McKinsey & Company [3], the pandemic has accelerated the adoption of digital

M. E. Auer and T. Tsiatsos (Eds.): IMCL 2021, LNNS 411, pp. 26–38, 2022.
https://doi.org/10.1007/978-3-030-96296-8_3

technologies by several years. The modern world has turned to the tools of digitalization to cope with lockdown to reinvent an environment of collaboration. Digital technology can support new pedagogies. They support learning and teaching communities with new tools such as remote and virtual labs, software for experimentation and simulation, social media and serious games [4].

The COVID-19 has already disrupted the lives of nearly 1.6 billion pupils and students in more than 190 countries around the world [5]. The measures adopted due to the pandemic situation have profoundly challenged the learning and teaching process [6]. Surely, such a deep and rapid evolution, affecting the processes and content of education was the trigger for abundant fears and anxiety for students, teachers, and educational stakeholders. Engineering school education has been challenged like no other time before, especially due to the practical and technical dimension of targeted skills and competencies fulfilled traditionally by face-to-face learning and learning by doing, generally done in classrooms. The COVID 19 crisis has raised fundamental questions about the online and blended learning in engineering schools. What have we learned from this experience? How active learning, especially problem- and project-based learning (PBL) where formal and informal exchanges between teacher, learners and other potential stakeholders is crucial, can evolve?

PBL are at the heart of the learners' experience at engineering schools. PBL approaches have been used since many decades [7] at engineering schools. They place learners at the centre of the learning process [8, 9]. In the framework of PBL approaches, learners are actively involved through prepared situations (problem-based learning) or real situations (project-based learning) [10]. PBL fosters self-regulated learning, develops effective problem-solving skills [9], helps learners to become effective collaborators and enhances their motivation [7, 11].

The first aim of this paper is to present the difficulties encountered and practices adopted by a French technical engineering school ISAE-Supmeca while coping with the COVID 19. An attempt to compare the collected data to a part of results from a study realized at California State University [12] is made. The second aim is to identify research directions to advance the re-imagination of the learning process. This article is organized as follows. In Sect. 2, we present the context of the study. In Sect. 3, an overview of the data collection and ISAE-Supmeca's answer to the pandemic crisis are provided. In Sect. 4, the results on learners' quality of life are exposed and compared to a California State University study. Section 5, focus on results, comparisons and actions on learning tools and formats. In Sect. 6, conclusions and future research directions are detailed.

2 ISAE-Supmeca Context and Specific Features

ISAE-Supmeca is a French public engineering school located in the north of Paris. It belongs to the great Paris area, more precisely to the rapid-changing location in which the 2024 summer Olympic and Paralympic games will take place. ISAE-Supmeca trains learners during a three-year curriculum through two ways: apprenticeship with part-time in a company (around 50 apprentices enrolled every year) and traditional with internships in companies and full-time at school (around 150 students enrolled every year). These two learning tracks are very common in the French context. The number of apprentices and students, i.e. the total number of learners, is around 600.

Since decades, ISAE-Supmeca has been engaged in active learning, especially PBL approaches. In the beginning of the 2010s, ISAE-Supmeca set up an Educational Innovation Unit (EIU), at the interface of teachers and learners. This unit works in close cooperation with the teaching community on the one hand and, for some activities, with the Learners' Affairs Department (LAD) on the other hand. Thanks to the engagement in educational research projects in which PBL has always been at the centre, ISAE-Supmeca has been developing an experience in multi-actor projects (with academic and industrial partners, for several semesters…). As a result of this type of complex projects, remote work, online meetings and asynchrony became common to many teachers and learners (mainly students) in a non-pandemic and mostly on-site context. Nevertheless, even in the context of a growing use of digital tools and asynchrony, only some PBL modules were concerned and not all teachers and learners were involved.

3 The ISAE-Supmeca Answer and Data Collection: From Reaction to Adaptation

ISAE-Supmeca, like all other French institutions, in mid-March 2020, was forced to switch from chosen, partial, well-known online activities to fully imposed remote activities, making this move very abrupt. In September 2020, everybody was back on-site with optimistic mood, but from October 2020 to June 2021, in a less abrupt move, a mix of full online and blended learning became the norm, with more on-site activities for some modules and limited gauges imposing different attendances depending on periods and locations of activities.

This new situation imposed by the pandemic led to a first phase of reaction. During this phase, teaching staff, administrative staff and learners tried to develop emergency solutions. Teachers, on their side, reacted with a huge diversity of initiatives. For instance, at ISAE-Supmeca, some created Discord servers [11] from scratch among other things to allow the continuity of the PBL and some other courses. Some filmed their whiteboards to explain to learners and keep some interactivity. Some others started to use multiple devices in order to have more screens at their disposal. At the institutional level, in parallel to the development of emergency solutions, a collective effort was deployed in order to start a formalization and a unification of digital tools. The implementation of the Microsoft Teams [12] collaborative solution and the proposal of digital tablets for technical illustration are examples of the deployed effort. This first phase that was prolonged until the summer holidays gave way to a process of feedback and analysis. To grasp the feedbacks and map issues encountered during this first phase, different surveys were conducted by both the LAD and the EIU. In this paper, we mainly present the results of four surveys, the aim of which was to capture feedback from learners (three surveys, targeting Bachelor 3, Master 1 and Master 2 students) and teachers (one survey).

Indeed, since the beginning of the pandemic, the main idea has been to follow the learners and learning situations as much as possible and as effectively as possible. To do so, large surveys were conducted in order to follow, analyze and draw conclusions.

The quantitative surveys also included open blanks in order to grasp qualitative feedback. Also, in addition to these surveys, a dozen of long semi-structured teachers' interviews were conducted from September 2020 to April 2021. Table 1 hereafter presents a summary of all surveys conducted.

Table 1. Summary of conducted surveys at ISAE-Supmeca on online education and practices from May 2020 to June 2021.

Surveys on online education and practices conducted at ISAE-Supmeca from May 2020 to June 2021				
# of survey	1	2	3	4
Date	May 13-28, 2020	Feb. 8 - March 1, 2021	Feb. 25 - March 5, 2021	May 26 - June 4, 2021
Department in charge	LAD	EIU	EIU	LAD & EIU
Objectives	Feedback on the remote learning experience	Feedback on teaching practices in the covid-19 period	Feedback on the hybridization of the learning experience	Feedback on learning experience in the covid-19 period
Target	Learners	Teachers	Learners	Learners
Permanent teachers	0	48	0	0
Students	291	0	474	474
Apprentices	128	0	136	136
Total of persons targeted	**419**	**48**	**610**	**610**
Number of answers	**315**	**26**	**148**	**212**
% of persons who answered	**75%**	**54%**	**24%**	**35%**

In May 2020, when learners were still fully online, a first survey was realized by the LAD. 75% of the 419 learners targeted answered it. Then, after an almost normal start of the school year in September 2020 due to the improvement of the sanitary situation during the summer, the latter quickly deteriorated and online education became again the norm from October 5, 2020. A mix of hybrid learning situations emerged from this date until the summer 2021. Indeed, during this second period (i.e. softer lockdown), some courses requiring the use of specific equipment or professional software were taught face to face in class with a reduced number of students. Also, a part of PBL modules and many exams took place on-site. This situation led to the creation of several surveys by the EIU. The idea was to have more formalized feedback from the two main actors who are learners and teachers, in order to adapt the answer to the situation and to better track the learners' situation. Two surveys were conducted in February and March 2021. The survey for teachers was designed to receive feedback on the educational practices, the learning processes and the use of digital tools. It targeted 48 permanent teachers (or assimilated) from ISAE-Supmeca. 54% answered it. The survey for learners was designed to receive feedback on the learning processes, the learning environment and the use of digital tools. The percentage of persons who answered was lower than the first survey, with 148 out of 610, i.e. 24% of respondents. Finally, the LAD and the EIU realized a common survey focusing on learners at the beginning of June 2021. It was designed to receive results allowing comparisons with both first and third survey. Items to be compared were, among others, students' conditions, use and satisfaction of online tools, as well as the quantity of work to produce.

Other new items focused on the preferences of students in terms of hybrid organization and formats, based on their experience with this new and long period. The percentage of persons who answered it was just better than the previous one, with only 212 out of 610, i.e. 35% of respondents.

The next section presents some of our main results. The choice was done not to be exhaustive. This section also presents comparisons on some data with the California State University study already quoted [12]. Even if the learners' population studied in the latter presents differences with ours (5 academic years instead of 3 notably), it presents strong common points: some hundred respondents, engineering education, supposed well-connected area, time of realization during the first wave of the pandemic (like our first survey). For these reasons, we chose to use it and make data comparisons.

4 Results and Comparisons on Learners' Quality of Life

The first result is related to the participation rate of the learners. The participation of the learners declined from 2020 to 2021 and was low beyond expectations in the third and fourth survey. It confirms the general impression of lack of implication and tiredness of many students, annoyed with this very long-standing situation impacting their studies. Even if the rate of respondents was not as high as expected, we chose to consider the surveys relevant and to use all the data, because the situation made it impossible to receive a better feedback and because, if we compare to the California State University response rate (12%) [12], it seems even more relevant.

The second result from our surveys is about the internet access of learners. It is surprisingly bad in a country like France. More surprisingly, our surveys show that there was no gain in quality of internet access for learners in one year. Indeed, 37% in May 2020 and 36% in May–June 2021 thought they had an "insufficient" internet access. Table 2 hereafter presents these results.

Table 2. Results on the question about the quality of the internet access of ISAE-Supmeca learners.

What is the quality of your internet connexion ? (to learners)	May 2020	May-June 2021
My connexion is shared and of good quality	41%	40%
My connexion is shared and insufficient for heavy videos/software	22%	16%
My connexion is shared and always insufficient		3%
My connexion is individual and of good quality	22%	24%
My connexion is individual and insufficient for heavy videos/software	10%	14%
My connexion is individual and always insufficient		3%
Other (not regular, depends on working place, depends on content...)	5%	0%
Sub-total bad quality of connection	**37%**	**36%**
Sub-total good quality of connection	**63%**	**64%**

It is notable that, if we compare these figures to figures coming from a study realized in another supposed well-connected area, they are almost the same: in Spring

2020, the study conducted on 627 engineering students at California State University highlighted that around 30% of them had no internet access or not reliable internet access [12].

A third result concerns the living place. Whereas during the first lockdown, most of learners, especially students, came back to their families, it was much less the case during the second lockdown (smoother) with few possibilities to be back on-site. We thought, that it would have an impact on the study conditions of learners, the reason why they were also asked about the space in which they work and study when online, no matter in which kind of housing they live (see Table 3). This fourth result shows that more than 30% do not have a dedicated room to work. This figure, which can seem relatively bad, seems in fact rather fair, or at least not so bad, if we compare it to the one from the California State University study, in which more than 50% of learners answered that they had no quiet or private space to study [12]. In additional comments, several students focused on the small-size apartments or on noise issue. Indeed, when it started to be permitted during the most restrictive periods, the school facilitated access to computer rooms for the students having the strongest difficulties and/or requested it.

Table 3. Results on the question about the working conditions of ISAE-Supmeca learners.

Working conditions of learners when online		
	May 2020	May-June 2021
In a dedicated room	63%	69%
In a shared space	30%	21%
Other (mix, including coming at school to find a working space...)	6%	10%

These first results form a first batch of results linked to material studying conditions. It shows that, even if they were rather good for a majority of learners, they were rather bad or difficult for many of them. It can result in a loss of time, more tiredness due to noise or stress, an impossibility to contribute in a team work and have negative consequences on the learning process of the learners concerned.

A second batch of results is made of data on physical and psychological conditions of learners. It shows globally a negative trend: in a period of one year, both the physical condition and the psychological condition of learners degraded, this context, with other elements, paving the way to a diminution of the amount of work by the learners.

The fifth finding presented in this paper is what we all heard about and experienced: fatigue. From 46% in May 2020, it jumped to 56% in May–June 2021. Some learners also highlighted in comments the difficulty to be concentrated and linked it to fatigue and to motivation, a word we will focus on later. Table 4 hereafter presents the figures about the physical condition of learners, with a stable low 32% of respondents saying that they feel good and most of them expressing one or more negative answer.

Table 4. Results on the question about the physical condition of ISAE-Supmeca learners.

How do you feel physically ? (to learners; multiple choices)		
	May 2020	May-June 2021
I feel good	32%	32%
I often have a headache	29%	26%
I often have pain in my eyes	37%	36%
My back hurts	26%	34%
I have a feeling of fatigue	**46%**	**56%**
Other	3%	2%

If we compare it to the 70% of respondents who mentioned fatigue or difficulty to maintain focus in the California State University study [12], this figure, which is high, may be considered as not so high but learning conditions and work demanded was certainly different. Nevertheless, the teachers who we interviewed confirmed the fatigue of students and highlighted their own tiredness.

A sixth result follows unfortunately the same trend: psychological condition of learners also worsened in one year. Less learners feel good or rather good and even 31% of respondents stated in a new item introduced in May-June 2021 that they "lost all motivation for classes". This item was several times also ticked by learners stating that they felt good or rather good. It means that the "I feel good" sub-total (43% in spring 2021) does not always mean "I feel good with my studies". Table 5 hereafter presents all the feedback on the psychological condition. It is interesting to note that, if we again compare results to the Californian study in which more than 50% of respondents highlighted a lack of motivation, the ISAE-Supmeca figures appear not so bad.

Table 5. Results on the question about the psychological condition of ISAE-Supmeca learners.

How do you feel psychologically ? (to learners; multiple choices in May-June 2021)		
	May 2020	May-June 2021
I feel good	22%	21%
I feel rather good	28%	22%
I am in a variable mood	**32%**	**44%**
I am not in the mood	**15%**	**17%**
I lost all motivation for classes		**31%**
Other negative standings	3%	2%

As for the physical condition, the teachers interviewed confirmed it and stated comments about their own rather variable mood and lack of motivation, one of the teachers resuming the feeling with these words: "I did not become a teacher to be in front of a computer talking to people behind turned-off cameras".

Also, probably in link or as an answer to these huge difficulties encountered, many learners seemed to work less or even much less in May–June 2021 than one year before. This seventh result is presented in Table 6, which presents impressive evolutions.

Table 6. Results about the feeling of the quantity of work done by ISAE-Supmeca learners.

When at-a-distance, do you have the feeling to work ? (to learners)	May 2020	May-June 2021
Much more than face-to-face	21%	6%
More than face-to-face	35%	16%
As well as face-to-face	14%	16%
Less than face-to-face	16%	22%
Much less than face-to-face	8%	36%
I did not mesure my working time	6%	4%
Sub-total more work than face-to-face	56%	22%
Sub-total less work than face-to-face	24%	58%

There can be several explanations to this finding, all related to the changes that happened in one year. In 2020, learners faced an abrupt shift and had to discover new distance learning tools, a new work rhythm and a total absence of face-to-face courses. In 2021, learners were able to come back to school for example for the courses involving tools difficult to learn alone or with a distance supervisor. Also, in 2021, the near standardization of the platforms at ISAE-Supmeca (later explained) and the acquired experience of both teachers and learners has made distance learning courses less time-consuming, with an adaptation of contents. Finally, the blended, hybrid, partly asynchronic formats adopted in 2021 allowed more possibilities for students to manage their time. The next and final section will propose a focus on results related to tools and formats, which have to be put in the context of the first changes and of ongoing and coming changes.

5 Results and Actions on Learning Tools and Formats

From the feedback concerning the digital collaborative used tools, we see a strong positive standing towards all of them, with a preference for one of them, as shown in Table 7 hereafter.

Table 7. Results on ISAE-Supmeca learners' satisfaction with the use of digital tools.

Are you satisfied with the following tools ? (to learners)	Moodle May 2020	Moodle May-June 2021	Teams May-June 2021	Discord May-June 2021
Very satisfied	15%	28%	33%	20%
Satisfied	73%	62%	56%	27%
Not very satisfied	7%	7%	8%	10%
Not satisfied	1%	2%	3%	5%
I did not use	4%	1%	0%	38%
Sub-total negative standing	8%	9%	11%	15%
Sub-total positive standing	88%	90%	89%	47%

The positive standing towards Microsoft Teams and Moodle is around 90% (which is very high for a tool) but the tool considered to be the most comfortable is Microsoft Teams (even if it is less the case now than one year ago; data not shown in tables). It has to be noted that these results come in the context of a use of Moodle [15] since many years, mainly for sharing and depositing materials between teachers and students, but also for some assessment and in rare cases for the full management of several modules. We can see that satisfaction with the use of this tool is high, regardless of the period of use, although its use has evolved somewhat and it is not the platform seen as the most comfortable to work with. Indeed the use as an exchange platform has lasted but many evaluations have been set up on this platform during the first lockdown and students have adapted well.

For live teaching, project sequences or any other synchronous pedagogical sequences, during the first lockdown, the Discord proprietary software was mainly used, Microsoft Teams and other tools, too. In September 2020, the choice was made to standardize the tools to avoid multiple installations and to respect data security. The choice was made to use Microsoft Teams for all remote exchanges. The interest of limiting the tools is that students no longer have to ask themselves the question "which platform is it for this course?", and they can use the tool for asynchronous group work.

Table 8. Results on ISAE-Supmeca learners' preferences on educational format.

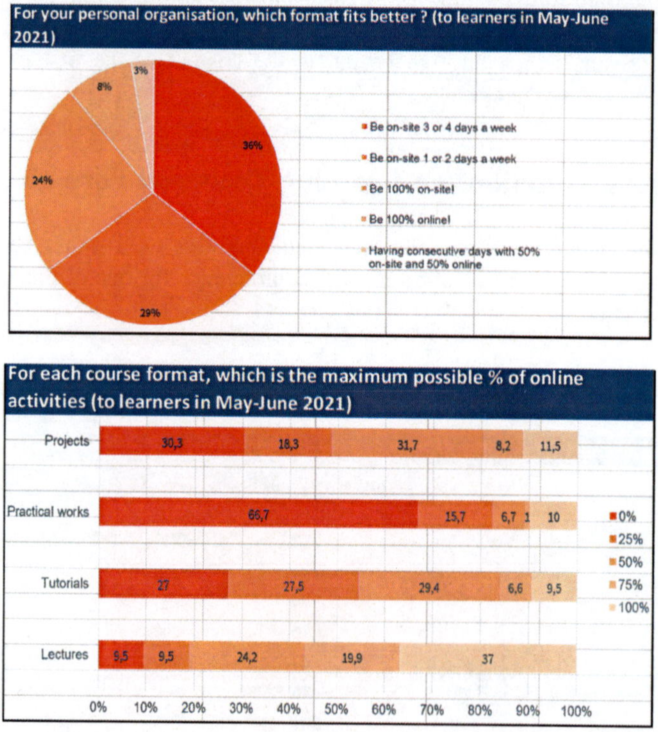

Microsoft Teams was used a lot by the students during the different project phases. Some rare pedagogical actions have persisted on Discord. Restricting the platforms to Moodle and Microsoft Teams is a strong evolution put in place thanks to the experience acquired during these two periods. The experience learned by both teachers and learners about the organisation of work helped dialogue. Capitalize on the experience is the first step to imagine the future of learning process. To do so, we questioned the learners on the rhythm of work that they would prefer. Table 8 presents these results, in general and depending on course format.

First, we asked learners about their general feedback about the use of their time. It turns out that if they mostly favour a return to face-to-face teaching, it seems that they have also appreciated being able to manage part of their own rhythm of work. One result also confirms the feeling of the teachers and the informal feedback from several learners, only 3% of learners favour "hybrid days" in which they are part-time online and part-time on-site. The other figures show that 24% of the learners would like to be 100% on-site, 60% of the learners would like to be 3 days a week or more on-site and 65% of the learners would like to be at least one day online. But, in fact, 3 or 4 days a week on-site, during the second and third year projects periods (several months), is already the situation for ISAE-Supmeca students, whereas, by definition, ISAE-Supmeca apprentices and part-time only at school. It means that the usual flexibility and habits at ISAE-Supmeca tend to show that this item gives an answer in favour of a comeback to the previous situation with improvement in order to better manage one's own time.

Then, in order to better analyse the formats desired by learners, we questioned them, for each type of format (lectures, tutorials, practical works and projects), about the proportion of distance learning that they would like to have. The results are:

- More than 80% of respondents would like to have at least 50% of lectures online,
- More than 66% do not want remote practical work at all, face-to-face is praised,
- For tutorials, the situation is more balanced: 45.5% of the learners would like to have 50% or more done at-a-distance, 27% of learners do not want remote activities and more than 54% of learners want a maximum of 25% of them done at-a-distance.
- For the projects, the situation is even more balanced than for tutorials: more than 30% of the learners would like to have 100% of their project periods on-site, while more than 50% of the learners would like to have 50% or more of the project activities (only students concerned) realized online. Our feeling is that the last projects experiments practiced in June 2021, with supervision in hybrid mode, were successful, as stated by some teachers, highlighting a greater interaction for students with both academic and industrial supervisors, a more insightful follow-up, and an increased possibility for students to come to the school to prototype solutions or meet teachers in-person.

We chose to compare these results to data in the study of Lappeenranta University [16] in which, among other, engineering students from master's thesis were asked about what they would want to add in the guided thesis process? They asked for content delivery, more specifically lecture video (29% of respondents), allowing you to approach the content at your own speed and when you want. But, much more, in the same idea as ISAE-Supmeca learners (and with a population having a profile, that we can consider close to the average ISAE-Supmeca learner), they asked for more concrete activities: 53% want more meetings with supervisors and 53% also want more workshops.

During and after these difficult periods, other practices, tools, remarks emerged. Some good practices are to be retained, and implemented in a return to a more classical training format. At ISAE-Supmeca, a redefinition of "what is a module?" is under study, with the general objective of freeing up time in small groups in order to give value to student/teacher exchange time. Work is also in progress to propose an online platform with spaces for share, remote meetings, collaborative work and access to the various tools used in our engineering training. This platform will be accessible from a distance or locally within ISAE-Supmeca. It should ease hybrid teaching in order to give all the possible educational freedom to the teachers to allow them to invent new forms of teaching and to maximize face-to-face with small groups of students. A first definition of the tool was done and it was presented in Spring 2021 [17]. Continuing our comparison with the California State University survey, we can see that one of their "proposed interventions" is the creation of a "virtual desktop environment", also "allowing faculty and students to access necessary software" [12].

6 Conclusion and Future Research Directions

The COVID 19 crisis has shaken up the entire French educational system. In emergency, students and professors had to find solutions to react to the shock. During this phase, teachers had to capitalize, train, and exchange in order to make their teaching evolve as well as possible. Students had to modify their aptitudes, their position in front of the work, and adapt to these new teaching modes. A lot of work and effort was required on both sides, not to mention all the administrative staff, IT department among all, which helped to get through these complicated periods. ISAE-Supmeca set up a phase of analysis from May 2020 in order to engage on long-term actions. We were, thus, able to identify material blocking points and the physical and psychological consequences of these difficult periods on the actors of education. These difficulties are not specific to France and to ISAE-Supmeca, we find them in other universities even if the comparisons made remain partial and can face limitations due to different factors: learning conditions, educational context, academic year of respondents, formats of curricula…Thus, new practices and new tools have been identified, and some good practices are to be retained and implemented in a return to a more classical training format. Also, remarks emerged not only on online tools and learning formats, but also on the classroom itself. At ISAE-Supmeca, as already stated, a redefinition of "what isa module" is under study, in the framework of a more general work about the evolution of the curriculum. Based on the results of the surveys, we were able to identify three areas for future work. The first is the integration of the informal dimension in the PBL in an increasingly digital context, notably the need to recreate the "being together" and "sharing together" situations to catalyze informal and friendly communication and exchanges. These exchanges allow students to intuitively benchmark the organization of their project thanks to the observation of other groups present in the space at the same time. In addition, they generate a competitive dynamics of value creation. The second is that today's "connected" learners and teachers are confronted with many challenging issues as information overload, violations of privacy, online bullying, media addiction… It sems important to have a deep analysis of how to achieve the balance and separation between the professional and personal sphere. The third is that it is a matter of fact that digital tools offer a great opportunity for learning. Digital technologies allow a more inclusive education. They make rapid feedback loops and real-time assessments possible, thus contributing to more personalized learning. Yet, the risk of dehumanizing the learning process and its consequences must be explored.

References

1. Alam, T.: A reliable communication framework and its use in internet of things (IoT). J. Sci. Res. Comput. Sci. Eng. Inf. Technol. **3**, 450–456 (2018)
2. McDonald, M.P., Rowsell-Jones, A.: The digital edge: exploiting information and technology for business advantage. Gartner, Inc., Stamford, CT (2012). https://www.myilibrary.com?id=835843
3. McKinsey & Company: How COVID-19 has pushed companies over the technology tipping point—and transformed business forever (2020)
4. OECD, Students, Computers and Learning: Making the Connection. OECD (2015). https://doi.org/10.1787/9789264239555-en
5. United Nations: Policy Brief: Education during COVID-19 and beyond (2020). https://www.un.org/development/desa/dspd/wp-content/uploads/sites/22/2020/08/sg_policy_brief_covid-19_and_education_august_2020.pdf. Accessed 30 June 2021
6. Almarzooq, Z.I., Lopes, M., Kochar, A.: Virtual learning during the COVID-19 pandemic. J. Am. Coll. Cardiol. **75**, 2635–2638 (2020). https://doi.org/10.1016/j.jacc.2020.04.015
7. Hmelo-Silver, C.E.: Problem-based learning: what and how do students learn? Educ. Psychol. Rev. **16**(3), 235–266 (2004). https://doi.org/10.1023/B:EDPR.0000034022.16470.f3
8. Sumarni, W.: The strengths and weaknesses of the implementation of project based learning: a review. Int. J. Sci. Res. **4**(3), 478–484 (2015)
9. Kokotsaki, D., Menzies, V., Wiggins, A.: Project-based learning: a review of the literature. Improv. Sch. **19**(3), 267–277 (2016). https://doi.org/10.1177/1365480216659733
10. Prince, M.J., Felder, R.M.: Inductive teaching and learning methods: definitions, comparisons, and research bases. J. Eng. Educ. **95**(2), 123–138 (2006). https://doi.org/10.1002/j.2168-9830.2006.tb00884.x(2006)
11. Peyret, N., Courtois, S., Chevallier, G.: Combined project-based learning and teacher-practical demonstrations to help acquire global engineering skills. In: Varietas Delectat... Complexity is the New Normality, SEFI Conference 2019 Proceedings, Budapest, pp. 871–878 (2019)
12. Asgari, S.: An observational study of engineering online education during the COVID-19 pandemic (2021). https://journals.plos.org/plosone/article?id=10.1371/journal.pone.0250041. Accessed 15 Aug 2021
13. https://discord.com
14. https://www.microsoft.com/en-us/microsoft-teams/group-chat-software
15. https://moodle.org/

16. Virkki-Hatakka, T., et al.: Working-life integrated engineering curriculum design and enhancing thesis process. In: SEFI Conference 2020 Proceedings, pp. 1146–1158 (2020)
17. Kooli-Chaabane, H., Peyret, N., Hammadi, M., Lanthony, A.: Virtual design office: proposition of problem- and project-based learning solution in the COVID-19 era and beyond. In: EDUCON2021 – IEEE Global Engineering Education Conference, Vienna, Austria (online) (2021)

Greek Parents' App Choices and Young Children's Smart Mobile Usage at Home

Stamatios Papadakis(✉), Foteini Alexandraki, and Nikolaos Zaranis

Department of Preschool Education, University of Crete, Crete, Rethymnon, Greece
{stpapadakis, alexandrak, nzaranis}@uoc.gr

Abstract. In the last decade, interactive touchscreen devices have become ubiquitous in young children, and toddlers first experience touchscreen technology before two. Although parents have a vital role in developing the home environment as a stimulus for development, they also have conflicting views on the appropriateness of using apps to deliver educational content for assorted reasons. The purpose of the study was to reveal various aspects of children's smart mobile use at home, such as the frequency of mobile device usage, preferred app types, and parent strategies on apps acquaintance. Three hundred twenty-five parents of kindergarten children took part in this study. The present study revealed that Greek homes are 'media rich' homes and that parents seek to support their children's learning at home via mobile devices. Furthermore, parents lack knowledge about app developmentally appropriateness and need further guidance. We expect the present study's findings to serve as a reference for researchers leading to better information for parents and creating apps with real educational value for children.

Keywords: Preschool children · Parents · Smart mobile devices · Apps

1 Introduction

In the last decade, interactive touchscreen devices are part of young children's lives [1]. Research links children's cognitive development with touchscreen devices and well-designed mobile applications (apps) [2]. As a result, self-proclaimed education apps are among the most accessed or purchased application categories in digital app stores [3]. Parents have a vital role in developing the home environment as a stimulus for development, and children's use of technology does not occur in a vacuum but within the context of family norms [4]. However, parents have conflicting views on the appropriateness of using apps to deliver educational content for assorted reasons, such as not having enough information about their children's development needs or controlling the content delivered via the apps [5]. Other factors include parenting style and socioeconomic status [6, 7].

Earlier research mainly focused on older children using interactive touchscreen technology, and thus there is not enough research focusing on young children [8]. This

M. E. Auer and T. Tsiatsos (Eds.): IMCL 2021, LNNS 411, pp. 39–50, 2022.
https://doi.org/10.1007/978-3-030-96296-8_4

research aims to fill this knowledge gap in Greece about mobile device apps usage among preschool-aged children at home.

2 Literature Review

Smart mobile device usage is increasing rapidly among young children due to the novel characteristics of these devices and the rapid development of mobile applications (apps) targeting these age groups [9]. Earlier research has pointed out mobile devices as the preferred technological tool for young children [10] due to the advantages of other older technologies. Researchers constantly worried that households might be positioned along a digital divide regarding the use of technology, a growing chasm 'between media-rich and media-poor homes' [11] (p. 926). The 'digital divide' refers to the gap between those who do and those who do not have access to technology, usually falling across low-income racial and ethnic minorities [12]. [13] notes that the previous years' gap discussed in various reports does not exist due to the proliferation of smartphones and tablets. Smart mobile device usage among young children is widening fast, even for disadvantaged backgrounds [14], providing significant opportunities for children from low-income families [15]. Regarding children's use of digital technology, earlier studies have revealed several reasons behind the contradictory parents' views, such as the lack of enough information on this subject or being unable to control the multitude of parameters of device usage.

It is well known that open-ended digital activities that support exploration and experimentation while offering cooperative and collaborative interaction opportunities contribute to children's learning [16]. Furthermore, a substantial body of research shows that children's participation in learning activities at home during their early years reflects an educational development in the later years [17]. Prior research on tablet apps in preschools has proved that severe educational apps can boost preschool children's various skills. These include literacy development, geography, art, science, technology, engineering, math, computational thinking, and cognitive and social control [18]. Although the 'app gap' discussed in the previous years had decreased [13], a quality app selection gap still exists. For instance, there is a strong link between low-income parents and their belief in marketing claims about the educational value of commercially available apps that could be downloaded from the Internet [14]. On the contrary, several studies have proven that most apps in the educational category for Android and iOS operating system devices (Apple App Store and Google Play) have no educational value based on rote learning and memorization. Furthermore, most self-proclaimed educational apps lack clear evidence of efficacy and are not scientifically established, having received no feedback from developmental specialists during their development.

Furthermore, studies revealed that young children mostly experience entertainment apps [19]. On the contrary to their addictive design features, these apps do not offer any learning benefits [19]. Further research has also shown that children play dozens of the most popular apps in games and do not use apps to get extra help on their reading, writing, and math skills [20]. Further concerns about parents' choice of apps involve the commercialization embedded in almost all of the freely available apps with many popups and inappropriate ads for children, disrupting their learning [19]. The companies make money on these free apps through advertisements, in-app purchases, or advertising paid apps on free ones.

Parent's role is essential in their children's technology-mediated activities [21, 22] as they control their children's interactive media experiences [23]. Parents must select appropriate apps to increase children's learning and enjoy reading, writing, and mathematics despite the abundance of low-quality educational apps. Taking this into account and considering the ubiquitous use of smart technology by children younger than six years, further research is needed to investigate whether parent app choices regarding children's learning at home.

3 Methods

3.1 Study Settings, Participants, Aims, and Research Questions

The study implemented the design of a method utilizing quantitative data. Parents with preschool-aged children, all enrolled in early childhood education classrooms, took part in the research. A stratified random sampling frame was implemented to ensure that the demographic composition is representative of national patterns. Kindergarten educators actively engaged in the process to increase participants' responses to the questionnaire. Parents who did not fully complete questionnaires were excluded from the sample. A total of 325 participating parents completed paper copies of the survey, a participation rate of 91%. The study was approved by the University of Research Ethics Committee to comply with ethical considerations. Furthermore, the participants were recruited ethically, without respect to their socioeconomic background. Confidentiality was maintained throughout the study.

The study aimed to examine parents' knowledge of apps, ownership of mobile devices; app purchasing habits; children's use of apps; and app usage contexts by parents and their children. The research questions that guided this study were as follows:

- What kind of access do Greek preschool children currently have at apps at home, and how are they used?
- What are the most popular app categories that Greek parents select for their preschool children?
- What factors influence parents' decisions when it comes to choosing which apps to use?
- What support do parents need to make these mobile tools more beneficial for child development?

3.2 Study Instrument

When designing this study questionnaire, we started with the research questions mentioned above that motivated the study and continued with existing reports and literature on children's media use. In addition, some of the questions were updated to reflect changes in technology and research on the content or context of early screen time. In the present study questionnaire development, we utilized an iterative process involving the research team for the item's creation, experts from the University of for the review of the items, and the research team for the revised item's creation. The questionnaire consisted of 28 questions, including dichotomous choice (yes/no), multiple-choice, and open-ended questions. The questionnaire included three parts. The first part focused on demographic information. The second part focused on the availability of technology to children at home and apps' frequency of use. The third part focused on parents' thoughts and concerns about their children's use of technology.

3.3 Threats to Validity

The approach in the present study was non-experimental. There was no interaction bias based on participant selection. Moreover, parents could fill out the questionnaire without spatial and time restrictions, so there were no cases of reactive arrangement. A small number of participants means that the external validity is limited, and thus the study results may not be generalizable to the overall population. To detect an effect of partial eta squared = .04 with 80% power in a one-way within-subjects ANOVA (three groups, alpha = .05, non-sphericity correction = 1), G*Power suggests we would need 119 participants [24]. Considering the final number of 325 participants, we can assume that the sample accurately represents a larger population [25].

4 Data Analysis

The data were analyzed using IBM SPSS statistical package version 26 (Chicago, Illinois, USA). There were no missing values. The results were determined to be statistically significant at the 5% level ($p < 0.05$). Parametric assumptions for all independent variables were examined, and they were not met. The majority of respondents (78.2%) consisted of mothers; almost all (95.1%) identified as Greeks, with a percentage of 4.0% identified as Albanians. Regarding parents' studies categorization, most parents reported being well educated, having at least a tertiary education diploma, and belonged to the 31–40 age group (67.1%) or the 41–50 age group (26.8%). However, only 24.3% had attended pedagogical studies (e.g., schoolteachers). Regarding the available family income, according to the most recent data (2019 Survey on Income and Living Conditions) [26], in this study, the risk of poverty or social exclusion was estimated at 23.1%. Children had access to several types of smart mobile devices at home. However, most important is that all children had access to at least an electronic device, while many children had access to two or more diverse types of devices.

Almost all children had access to a smart device daily or some days during the week. Three hundred nine parents (95.1%) declared that they use a restriction policy on their child's time with the mobile device. On the contrary, only 16 parents (4.9%) answered that they did not use any rule. In the question of whether the children play educational games (apps with a game-like format and an educational goal) on the smart mobile device, the parents in their majority (303, 93.2%) answered positively. The parents answered that their children play different mobile games during their engagement with smart mobile devices. As expected in their majority, children play with math apps (153, 47.1%) and read/writing apps (153, 47.1%). Surprisingly enough, the parents answered that children mostly play with apps that promote spatial reasoning skills (224, 68.9%) and coloring recognition apps (168, 51.7%). A small number of children play with coding and STEM apps (43, 13.2%). In an open question, on the other types of games that children play and not already mentioned in the questionnaire, the parents answered that their children play apps that help them learn the English language (23, 7.07%).

What is significant is that the parents in their majority answered that they download educational apps/games for their children (297, 91.4%). Only 28 parents (8.6%) answered that they do not download educational apps for their children. These numbers may follow the parents' answers; those 21 children play with smart mobile devices without an adult supervisor. Almost all parents (294, 90.5%) answered that they download and use only free apps. Only 31 parents (9.5%) said that they use both free and paid apps. None of the participants answered that they prefer to use only paid apps. Twenty-eight parents answered that they download apps for their children weekly, either two-three times per week (19, 5.8%) or one time per week (9, 2.8%). One hundred fourteen parents said they download monthly, either one app (50, 15.4%) or two-three apps (64, 19.7%). Thus, the parents in the majority (162) do not download apps for their children often. Similarly, 21 parents (6.5%) declared that they do not download apps. These numbers comply with the parents' previous answers that 21 children play with mobile devices without an adult supervisor. We can also suppose that in these families, children alone download apps.

When parents asked what motivates them to download apps for their children, they gave various answers. The most popular answers were 'as a reward for an achievement or a good behavior' (124, 38.2%), 'to support the child's learning (117, 36.0%), 'to satisfy the child's desire' (118, 36.3%), 'to encourage the child's play and creativity (82, 25.2%), 'as a gift' (36, 11.1%) and 'just because my child only needs a new app' (26, 8.0%). A question is how their child knows about a specific app. In the multiple questions the source's parents use to download an app, surprisingly enough, 139 parents (42.8%) answered that their child asks for the particular app. We suppose that they meant that they are looking together with their child for apps and might download apps that their child finds attractive due to colors, known heroes, etc. 136 parents (41.8%) answered that they download apps after a personal search in app stores. Sixty parents (18.5%) answered that they download apps recommended by the older member of the family or their husband or partner, their child educators (74, 22.8%), friends, colleagues (82, 25.2%), through social networks posts (49, 15.1%) and advertisements (22, 6.8%). These two last answers must problematize the researchers as other studies have found that user comments are primarily subjective and do not correlate with the actual educational value, while ads promote specific apps. We can also conclude that parents are informed mainly by their colleagues or friends and not by their children's teachers. However, in Greece, this can be explained by the low degree of penetration of mobile technology in schools and the lack of familiarity and knowledge of teachers about this new educational reality.

Parents seem to follow the same strategy regarding the criteria they use to download an app. Instead of following recommendations on specialized sites or blogs (89, 27.4%), they prefer to base their decision on friends or relatives' recommendations

(166, 51.1%), on comments and reviews on app stores (151, 46.5%), comments and reviews on social media (67, 20.6%) and apps downloads (92, 28.3%). Surprisingly, they do not base their strategies on app stars (18, 5.5%) and app prices (18, 5.5%). Their low-interest in-app price may be explained since they mostly download free apps (n = 294). There are no specific websites and blogs for app reviews in Greece, such as the Commonsense Media in the United States, so only 89 parents said they based their specialized apps criteria. Now another question emerges. Due to these apps' inexistence in Greece, what are these sites and blogs they refer to?

To further investigate parents' belief in digital technology, they answered questions on a 5 Likert scale. From the parents' answers, we can conclude that parents are positive about using mobile technology for their children's education in formal and informal settings (M = 3.55), although they feel unsure about their knowledge about this technology's utilization. They express their need for information from experts on how to find apps with real educational value to help their children learn (M = 3.86), on how to balance the time between apps usage and other activities for their children (M = 3.81), and the correct age that they will introduce mobile technology to their children (M = 3.85).

5 Study Limitations

On limitations, firstly, given that this study was a survey design, the causal direction between the dependent and the independent factors could not be ascertained. Furthermore, due to the correlational approach, extraneous variables such as individual differences (parents' personality, cultural values) might also influence the study outcomes. In addition, the data on this study were reported by parents. Thus, there might be a discrepancy between mobile media use reported by parents and the child's actual usage.

6 Discussion - Conclusion

The present study results suggest that all children at home have access to at least an electronic device, while many children have access to two or more diverse types of devices. Smartphones are the most popular device for children. Although still not widely available in kindergartens, smart mobile device adoption is increasing fast, and children at a younger age begin to use them early in Greek homes. Other studies have revealed that television had been the "go-to" device for parents of young children but noted that touchscreen and multi-use devices are gaining popularity [27, 28]. Compared to the results with another study in Greece, we can see the same trends [29]. These results are confirmed in other studies [30]. In [30] study, the researchers analyzed 0- to 3-year-olds children's use of touchscreen devices at home in three different countries, Japan, Norway, and Portugal. The study results revealed that touchscreen technologies dominate in these three different countries. Based on our study, we can conclude that most children in western countries, including Greece, live in homes that are "digitally fluent" environments [31]. Especially European children grow up in technology-rich homes [23], and printed books and educational television programs rapidly give way to digital content, shifting the learning environment at home for very young children [32]. In general, smartphones followed by tablets are popular in this age group due to their simplicity, portability, size of the screen, and ease of use [23], illustrating how deeply this technology has integrated into the daily parenting routines [6].

In Greece, young children use touchscreen devices (smartphones or tablets) daily or a few days per week, following [33] study. In her study, [33] investigated the digital technologies children under five use at home in four European countries: England, Greece, Malta, and Luxemburg. She found that many 3- to 5-year-olds use digital technologies (computer-based, internet-based) more than 30 min during the week and longer during the weekend. Based on the participants' responses, our results are in contradiction with previous studies. It seems that children in Greece use mobile devices less compared to other children around the globe. In Japan, almost half the two-year-old children watched videos or played games through their parents' smartphones for 60–80 min per day on average [30]. Similarly, in the United States, children as young as four years of age spend an average of an hour per day on an interactive screen device [13].

Regarding the impact of socioeconomic status, age, gender, and ethnicity on digital technology access and use, the previous year's studies found that socioeconomic background can influence how families incorporate digital media [11]. Socioeconomic status (SES) is a multidimensional vector that refers to a cluster of variables, such as lack of material resources, low parental education, or family financial pressures [34]. The present study found that even lower-income parents provide their children with versions of Apple and Android devices. These results followed other international study results. For instance, in a study in Canada with families from different socioeconomic backgrounds, social class did not seem to be a key indicator of technology practice or use [15]. In general, we can consider that in Greece, there is no digital divide in lower-income and ethnic minority children due to the overall increase in mobile device availability. While earlier studies have shown that children from low-income families have limited access to educational opportunities in digital content [35], in this study, children from both low and high SES have equal access to mobile content within the family environment. Several studies have also highlighted the parents' crucial role in guiding young children to use touchscreen mobile devices to gain educational benefits in informal educational settings [36]. This is considered important as mobile game-based learning in developmentally appropriate apps with educational value can help young children quickly develop their math, language, coding, and STEM skills [37, 38]. We also found that there is also no difference in apps usage and digital device access between boys and girls. This is important as the mobile ecosystem can create a supportive environment that can help children create positive attitudes and set clear learning goals [39]. Furthermore, gender equality in device and apps access can help females overcome technology phobias which can, in theory, impact the acquisition of CT skills and compromise their academic and professional future [40, 41]. Thus, the gap in which children of lower-income homeowners had substantially less access to smartphones, tablets, or educational apps described in older studies [42], does not exist.

Research suggests that children learn educational concepts from well-designed multimedia resources such as mobile applications (apps) [22]. An abundance of apps designated as educational (without evidence for this claim) are available in digital stores [7], and parents play an essential role in supporting their children in their learning with educational media [43] by choosing developmentally appropriate apps. However, reports indicate that children entertain themselves at home mostly with tablets [20]. In this study, we also found that children mostly use apps for entertainment purposes. Studies have also found that young children widely use three types of educational apps [19]: interactive gaming apps that have goals that progressively increase in difficulty, thereby piquing children's interest; applications that have tools for drawing or building to encourage children to join a constructive activity with may possible outputs; and electronic books with colorful, animated, and interactive features [18].

Although [13] discussed an "app gap" that mentioned the difference in the percent of parents who have downloaded apps for their children to use in this study, we can speak of an "app quality gap." In other words, the app's visual design, sound effects, and interface can distract children from the actual educational content. A balance between these characteristics is necessary to ease children's learning [7]. International statistics highlight ongoing concerns about the science education achievement gap that exists between kindergarten students. Girls worldwide are at higher risk for their STEM skill development than boys. Furthermore, studies highlight that early STEM engagement is crucial for young children of both genders to prepare for future STEM challenges [44, 45]. The present study found that parents do not download apps for their children's STEM skill development.

It is not expected that children can choose digital devices and apps with appropriate educational content. Research recognizes that parents play a critical role in children's technology introductory activities. Furthermore, the joint parent and child engagement with technology can improve children learning outcomes [3, 46]. Nevertheless, most parents worldwide do not know where to find appropriate tools and high-quality educational apps. For instance, it has been found that children play games instead of educational apps [3]. In the light of the research findings from the study in Turkey, the same researchers found that all the digital games were in the 'negative category,' characterized by inadequacy of their design and the education content [3]. In this study, similar to other studies [23], parents select and download apps and games that they consider educational and appropriate for their children.

Nevertheless, parents often perceive what is being taught as educational value, similar to early childhood curricula (quizzes, puzzles, and games) dismiss other aspects of context and skills [31]. Furthermore, in this study, the inadequate and often misleading information that parents receive from the media and other unofficial sources such as peers about digital content appears to be a significant factor in their decisions on children's touchscreen use [30]. In this study, parents mainly download and use free apps. Researchers have proven that the quality of an app may be reduced by features such as ads and popups that can divert children's attention away from the educational content [47].

Furthermore, these distracting features increase the cognitive burden, especially for younger children [48]. Additionally, researchers state that several free early literacy apps targeted at younger children imitate worksheets or flashcards promoting rote learning [19]. Fortunately, early childhood organizations (e.g., Resources for Early Learning or Zero to Three) provide parents with ideas and strategies for developmentally appropriate activities with smart screen technologies [7]. Unfortunately, Greece has no similar resources, although this lack has been already mentioned in the earlier studies [29].

Although digital learning can be used in diverse ways to enhance learning, parents still voiced uncertainty about adopting and implementing these digital resources [49]. In the present study, participants have mixed views about the potential of screen media for their children's development. Like other studies, most parents simultaneously recognize the potential benefits of using touchscreen technologies for their children, but they also expressed concern about potential risks, such as exposure to inappropriate content [13, 30]. Although the participants in this study, in their responses, did not mention the unfamiliarity and the high cost of monitoring software and parental control functions [50], almost all of them express their concerns regarding their lack of knowledge on selecting appropriate mobile educational content in the form of apps. However, regarding their concerns, in contrast to the study results of [31], parents do not get informed from scientific sources like doctors, educational reports, or experts. On the contrary, the participants answered that they mostly get recommendations from family and friends.

Like other studies in the present study, most parents, despite their willingness, mentioned a lack of scientific literacy to find educational apps in the digital stores [3, 51]. In addition, they are also being unaware of classifying young children learning needs and how they can use apps to scaffold children learning, such as encouraging engagement in meaningful activities [52] or higher-order skills such as problem-solving [31].

7 Conclusion

Parents are the primary mediators of children's home digital media, deciding children's digital presence, content, and activities [31, 46], especially in crises such as COVID-19 lockdown [53]. Providing parents with support services and strategies for coping with mobile media usage challenges can enhance learning and both learner and parent satisfaction with mobile technology. Educational organizations, stakeholders, and researchers should recommend or even better provide age-appropriate and developmentally appropriate content in apps. Unfortunately, there are no such digital resources in Greece and other countries in the parents' native language. Early childhood organizations must provide parents with advice for mobile developmentally appropriate activities to pursue their children's education.

Acknowledgments. The authors report financial support was provided by Greece and the European Union (European Social Fund- ESF) (MIS 5048168).

References

1. Burns, T., Gottschalk, F.: Educating 21st Century Children: Emotional Well-Being in the Digital Age. Educational Research and Innovation. OECD Publishing. 2, rue Andre Pascal, F-75775 Paris Cedex 16, France (2019)
2. Portugal, A.M., Bedford, R., Cheung, C.H., Mason, L., Smith, T.J.: Longitudinal touchscreen use across early development is associated with faster exogenous and reduced endogenous attention control. Sci. Rep. **11**(1), 1–12 (2021)
3. Gözüm, A.İC., Kandır, A.: Digital games pre-schoolers play: parental mediation and examination of educational content. Educ. Inf. Technol. **26**(3), 3293–3326 (2021). https://doi.org/10.1007/s10639-020-10382-2
4. Griffith, S.F., Arnold, D.H.: Home learning in the new mobile age: parent–child interactions during joint play with educational apps in the US. J. Child. Media **13**(1), 1–19 (2019)
5. Nikolopoulou, K.: Preschool children's use of tablet at home and parents' views. In: Mobile Learning Applications in Early Childhood Education, pp. 209–229. IGI Global (2020)
6. Lev, Y.B., Elias, N.: Digital parenting: media uses in parenting routines during the first two years of life. Stud. Media Commun. **8**(2), 41–48 (2020)
7. Radesky, J.S., Schumacher, J., Zuckerman, B.: Mobile and interactive media use by young children: the good, the bad, and the unknown. Pediatrics **135**(1), 1–3 (2015)
8. OECD: Impacts of technology use on children: Exploring literature on the brain, cognition and well-being. OECD Education Working Paper No. 195 (2019)
9. Hosokawa, R., Katsura, T.: Association between mobile technology use and child adjustment in early elementary school age. PloS One **13**(7), e0199959 (2018)
10. Brito, R., Dias, P.: The tablet is my BFF': practices and perceptions of Portuguese children under 8 years old and their families. In: The Digital Literacy and Multimodal Practices of Young Children: Engaging with Emergent Research. Proceedings of the First Training School of COST Action IS1410. University of Minho, Braga, Portugal, 6th–8th June 2016, Braga, Portugal (2016)
11. Livingstone, S.: Strategies of parental regulation in the media-rich home. Comput. Hum. Behav. **23**(2), 920–941 (2007)
12. Gutnick, A.L., Robb, M., Takeuchi, L., Kotler, J.: Always connected: the new digital media habits of young children. In: The Joan Ganz Cooney Center at Sesame Workshop, New York (2010)
13. Rideout, V.: The Common Sense census: Media use by kids age zero to eight. Common Sense Media, San Francisco, CA (2017)
14. Radesky, J.S., et al.: Overstimulated consumers or next-generation learners? Parent tensions about child mobile technology use. Ann. Family Med. **14**(6), 503–508 (2016)
15. Rizk, J., Hillier, C.: "Everything's technology now": the role of technology in home- and school-based summer learning activities in Canada. J. Children Media **15**(2), 272–290 (2021). https://doi.org/10.1080/17482798.2020.1778498
16. Eagle, S.: Learning in the early years: Social interactions around picture books, puzzles and digital technologies. Comput. Educ. **59**, 38–49 (2012)
17. Elliott, L., Bachman, H.J.: How do parents foster young children's math skills? Child Dev. Perspect. **12**(1), 16–21 (2018)
18. Liu, W., Tan, L., Huang, D., Chen, N., Liu, F.: When preschoolers use tablets: the effect of educational serious games on children's attention development. Int. J. Hum.-Comput. Interact. **37**(3), 234–248 (2021)
19. Neumann, M.M., Neumann, D.L.: The use of touchscreen tablets at home and preschool to foster emergent literacy. J. Early Child. Lit. **17**(2), 203–220 (2017)

20. Livingstone, S., Marsh, J., Plowman, L., et al.: Young Children (0–8) and Digital Technology: A Qualitative Exploratory Study National Report – UK. Joint Research Centre, European Commission, Luxembourg (2014)
21. Govind, M., Relkin, E., Bers, M.U.: Engaging children and parents to code together using the ScratchJr app. Visitor Stud. **23**(1), 46–65 (2020)
22. Sheehan, K.J., Pila, S., Lauricella, A.R., Wartella, E.A.: Parent-child interaction and children's learning from a coding application. Comput. Educ. **140**, 103601 (2019)
23. Chaudron, S., Di Gioia, R., Gemo, M.: Young children (0–8) and digital technology, a qualitative study across Europe. European Union: Joint Research Center (2018)
24. Faul, F., Erdfelder, E., Buchner, A., Lang, A.: Statistical power analyses using G*Power 3.1: tests for correlation and regression analyses. Behav. Res. Methods **41**(4), 1149–1160 (2009)
25. Bartlett, J.: Introduction to sample size calculation using G*Power. https://osf.io/zqphw/download. Accessed 21 June 2021
26. Hellenic Statistical Authority (ELSTAT): 2019 Survey on Income and Living Conditions. https://tinyurl.com/87x8wcn4. Accessed 22 June 2021
27. Kostyrka-Allchorne, K., Cooper, N.R., Simpson, A.: Touchscreen generation: children's current media use, parental supervision methods and attitudes towards contemporary media. Acta Paediatr. **106**(4), 654–662 (2017)
28. Parsons, D., Adhikar, J.: Bring your own device to secondary school: the perceptions of teachers, students and parents. Electron. J. E-learning **14**(1), 66–80 (2016)
29. Papadakis, S., Zaranis, N., Kalogiannakis, M.: Parental involvement and attitudes towards young Greek children's mobile usage. Int. J. Child-Comput. Interact. **22**, 100144 (2019)
30. Dardanou, M., et al.: Use of touchscreen technology by 0–3-year-old children: parents' practices and perspectives in Norway, Portugal and Japan. J. Early Child. Lit. **20**(3), 551–573 (2020)
31. Dias, P., Brito, R.: Criteria for selecting apps: debating the perceptions of young children, parents and industry stakeholders. Comput. Educ. **165**, 104134 (2021)
32. Furenes, M.I., Kucirkova, N., Bus, A.G.: A Comparison of Children's Reading on Paper Versus Screen: A Meta-Analysis. Rev. Educ. Res. **91**(4), 483–517 (2021). https://doi.org/10.3102/0034654321998074
33. Palaiologou, I.: Children under five and digital technologies: implications for early years pedagogy. Eur. Early Child. Educ. Res. J. **24**(1), 5–24 (2016)
34. Tonizzi, I., Traverso, L., Usai, M., Viterbori, P.: Fostering number sense in low SES children: a comparison between low- and high-intensity interventions. Math. Educ. Res. J. **33**(2), 345–363 (2020). https://doi.org/10.1007/s13394-019-00307-9
35. Ramani, G.B., Siegler, R.S.: Reducing the gap in numerical knowledge between low- and middle-income preschoolers. J. Appl. Dev. Psychol. **32**(3), 146–159 (2011)
36. Liu, C., Hwang, G.J.: Roles and research trends of touchscreen mobile devices in early childhood education: review of journal publications from 2010 to 2019 based on the technology-enhanced learning model. Interact. Learn. Environ. (2021). https://doi.org/10.1080/10494820.2020.1855210
37. Lehrl, S., Linberg, A., Niklas, F., Kuger, S.: The Home Learning Environment in the Digital Age—Associations Between Self-Reported "Analog" and "Digital" Home Learning Environment and Children's Socio-Emotional and Academic Outcomes. Front. Psychol. **12**, 592513 (2021). https://doi.org/10.3389/fpsyg.2021.592513
38. Vaiopoulou, J., Papadakis, S., Sifaki, E., Stamovlasis, D., Kalogiannakis, M.: Parents' perceptions of educational apps use for kindergarten children: development and validation of a new instrument (PEAU-p) and exploration of parents' profiles. Behav. Sci. **11**(6), 82 (2021)

39. Office of Educational Technology: Parent and Family Digital Learning Guide. https://tinyurl.com/39zhmstu. Accessed 21 June 2021
40. Relkin, E., Bers, M.: Factors influencing learning of computational thinking in young children. https://tinyurl.com/rzwrx48k. Accessed 21 June 2021
41. Papadakis, S., Vaiopoulou, J., Sifaki, E., Stamovlasis, D., Kalogiannakis, M.: Attitudes towards the use of educational robotics: exploring pre-service and in-service early childhood teacher profiles. Educ. Sci. **11**(5), 204 (2021)
42. Rideout, V.: Zero to eight: Children's media use in America 2013 (Rep.). Common Sense Media, San Francisco, CA (2013)
43. Yu, J., DeVore, A., Roque, R.: Parental mediation for young children's use of educational media: a case study with computational toys and kits. In: Proceedings of the 2021 CHI Conference on Human Factors in Computing Systems (CHI 2021), pp. 1–12. Association for Computing Machinery, New York (2021). Article no. 475
44. Stephenson, T., Fleer, M., Fragkiadaki, G.: Increasing girls' STEM engagement in early childhood: conditions created by the conceptual PlayWorld model. Res. Sci. Educ. 1–18 (2021). https://doi.org/10.1007/s11165-021-10003-z
45. Ampartzaki, M., Kalogiannakis, M., Papadakis, S.: Deepening our knowledge about sustainability education in the early years: lessons from a water project. Educ. Sci. **11**(6), 251 (2021)
46. Archer, K., Wood, E., De Pasquale, D.: Examining joint parent-child interactions involving infants and toddlers when introducing mobile technology. Infant Behav. Dev. **63**, 101568 (2021)
47. Northrop, L., Killeen, E.: A framework for using iPads to build early literacy skills. Read. Teach. **66**(7), 531–537 (2013)
48. Krcmar, M., Cingel, D.P.: Parent–child joint reading in traditional and electronic formats. Media Psychol. **17**(3), 262–281 (2014)
49. Livingstone, S., Mascheroni, G., Dreier, M., Chaudron, S., Lagae, K.: How parents of young children manage digital devices at home: The role of income, education and parental style. EU Kids Online, LSE, London (2015)
50. Marais, A.C.: Educational guidance for parents who are under pressure from their adolescent children. Doctoral dissertation, University of South Africa, Pretoria. http://hdl.handle.net/10500/8843. Accessed 21 June 2021
51. Lupton, D., Williamson, B.: The datafied child: the dataveillance of children and implications for their rights. New Media Soc. **19**(5), 780–794 (2017)
52. Hirsh-Pasek, K., Zosh, J.M., Golinkoff, R.M., Gray, J.H., Robb, M.B., Kaufman, J.: Putting education in "educational" apps: lessons from the science of learning. Psychol. Sci. Public Interest **16**(1), 3–34 (2015)
53. United Nations Committee on the Rights of the Child, Convention on the Rights of the Child. General comment No. 25 (2021) on children's rights in relation to the digital environment, https://tinyurl.com/aa4bypcs. Accessed 21 June 2021

Mobile Simulation Game for Learning Theory of Constraints Fundamentals

Igor Miladinovic(✉) and Sigrid Schefer-Wenzl

University of Applied Sciences Campus Vienna, Vienna, Austria
{igor.miladinovic,sigrid.schefer-wenzl}@fh-campuswien.ac.at

Abstract. Complex problem solving skills are increasingly important for engineering graduates, as they have to manage more and more complex problems in their work environments. To prepare students accordingly, we have integrated a complex problem solving course in our Master study program "Software Design and Engineering", which is based on the Theory of Constraints by Goldratt. One of the main challenges is motivating engineering students to delve into a non-engineering topic. Addressing this issue, we applied a game-based learning approach to teach the fundamentals of the Theory of Constraints with a dice game as a central element. In this paper, we present a digital mobile version of this game and discuss its potential benefits for students.

Keywords: Complex problem solving · Theory of constraints · Game-based learning · Serious game · Higher education

1 Introduction

Trends in the industry continue to show a strong shift from hardware towards software, resulting in an increased need for highly skilled software engineers with both technical and professional skills. Whereas technical skills refer to domain-specific knowledge, such as programming, mastering computer networks, and developing electronic components, professional skills (also referred to as soft skills, or 21st century skills) are competencies needed for any profession. They include complex problem solving, teamwork, and communication skills [10,17] and are relevant for different domains, jobs, and situations.

One of the central competencies in the 21st century is the ability for complex problem solving [3,6,9,15]. Complex problems are omnipresent in today's professional life. They have no clear problem or goal definition, so that progress towards the solution is not straight forward. The majority of software engineering and integration problems is of complex nature. They have multiple interacting systems, require many different competencies and involve several stakeholders, each of which with different goals that can even change over time. Therefore, developing strategies and techniques for solving complex problems among students should be one of the key objectives in an engineering degree program.

M. E. Auer and T. Tsiatsos (Eds.): IMCL 2021, LNNS 411, pp. 51–59, 2022.
https://doi.org/10.1007/978-3-030-96296-8_5

In our Master study program "Software Design and Engineering", we have integrated a dedicated course teaching methods and tools for complex problem solving. The course entitled "Complex Problem Solving" is based on a set of logical tools [5], which are designed to translate complex real-world situations into easy-to-read and absorb cause-and-effect diagrams [18,19].

One of the challenges was to find an optimal way to teach complex problem solving in engineering studies, since we were facing difficulties among our students to recognise the importance of professional skills. To address this issue, we considered the introduction of gamification or game-based learning elements as a promising approach to increase students' motivation. In this paper, we present a game-based mobile learning approach for our course "Complex Problem Solving". It illustrates the impact of constraints in a complex system based on the "Theory of Constraints" by Goldratt [7], which is a central part of our course.

The remainder of this paper is organised as follows. Section 2 introduces Theory of Constraints by Goldratt including the dice game, which is the central element of our game-based learning approach. It includes a non-digital and a digital version of this game we have applied so far. In Sect. 3 the idea of game-based learning is summarized, including examples of related work. Our mobile game-based learning approach is outlined in Sect. 4. Finally, Sect. 5 concludes the paper and provides an outlook to our future work.

2 Theory of Constraints

Our course "Complex Problem Solving" teaches one of the most desired professional skills [6,9]. We start this course with the introduction of Goldratt's "Theory of Constraints (ToC)" [7] to illustrate the impact of constraints in a complex system. According to this theory, a complex system can be considered a chain of subsystems which depend on each other (for example, the output of one subsystem is the input of another). In such a system there is always a constraint—a subsystem which limits the output of the overall system. The goal of ToC is to identify and improve this subsystem so that it does not constrain the system anymore. Following that, a new constraint within the system will need to be distinguished and improved, thereby creating an endless process of improvement.

A commonly used method to demonstrate the main idea behind the ToC is a dice game, first introduced in [8]. It simulates a chain of several production stations with constrained capacity. In this game, the capacity of each station is randomly determined by a die. The goal of this exercise is to show that the throughput of an unmanaged system is nearly always less than expected—which is the mean value of a die. For example, if the chain is composed of six stations, each of which can proceed between 1 and 6 units and the units are shifted from one station to another, the expected output after 10 rounds could be 35 units (the average value of a die times ten rounds), and after 20 rounds 70. However, after playing 20 rounds, the real output is usually around 50, which is far less than 70. This is caused by a high statistical fluctuation combined with the dependencies through the chain. In order to manage such a chaotic system, we first need to introduce a constraint and then apply the ToC. Consequently, this dice game is an example for game-based learning to provide the basic understanding of the ToC.

In on-site setup, we used to implement this game with real dice and tokens as production outcome. We divided students into teams each of which forming a production chain. Each student represented a production station, rolling a die on his or her turn to determine the current capacity. Tokens are then moved from one student to another according to the production capacity. At the end of the chain, we measured the produced items. After 20 rounds each team prepared a short presentation showing the achieved outcome and explaining the difference between the achieved and the expected outcome. In next steps, we introduced several variants of this game, for example with a single constraint in a chain (i.e., with one station that cannot proceed more than 4 tokens) or with the possibility to manage capacity of single stations by actively shifting production capacity from one station to another (so that one station can roll more than one die).

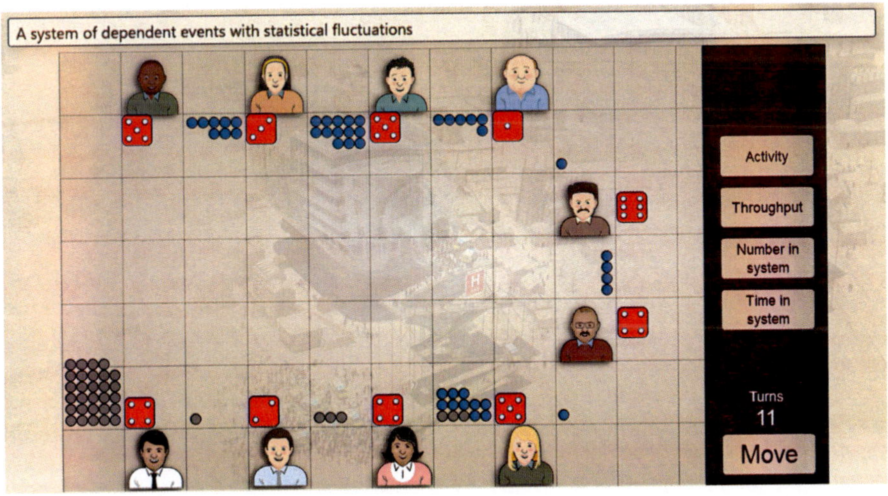

Fig. 1. Web-based version of the dice game [13]

Due to the Covid-19 pandemic, we changed the on-site setup to a remote setup and used a web application [13] to simulate this dice game. Figure 1 shows a screen-shot of the basic variant of this game with 10 stations and after 11 rounds. Whereas the demonstration of the impact of constraints in a system is possible with this remote setup, some limitations became obvious. For example, the web application supports only single-player mode for the complete chain, so that discussions and learning in the team were missing. Furthermore, only few variants of the game are available, limiting the possibilities to fully explore the importance of constraints. Each game variant consisted of exactly 10 stations and 20 rounds, preventing experiments with different lengths of the chain or longer system times.

3 Game-Based Learning

A promising approach to increase students' motivation is the introduction of gamification or game-based learning elements into courses in an engineering curriculum. Gamification is defined as the application of game design elements to non-game activities with the goal to increase user experience and engagement [4]. Using gamification in different educational contexts is known to be one way to enhance learner motivation and to improve learning outcomes by capturing the interest of learners and inspiring them to continue learning [2,14].

Game-based learning goes one step further than gamification. Instead of integrating game design elements into learning activities, game-based learning uses games to support the achievement of (some of the) learning objectives [1,16]. The central didactic tool of game-based learning is a custom-built game, which has a defined learning objective. The so-called serious games are usually digital, but can also be non-digital games [16].

The major potential benefit of serious games is an increased intrinsic motivation to make progress in the game and therefore to advance in achieving the learning objective. In an affective serious game, emotions are also triggered, which positively influences cognition, memory, attention and motivation of the students [20].

However, game-based learning also has potential disadvantages. An obvious one is the required effort to develop a serious game, since it usually has to be designed and developed towards a specific learning objective. The development also requires deep domain-specific knowledge in order to simulate a relevant real situation, which needs to be analyzed and improved during the game. In the lecture, organized support has to be provided for the game to run properly.

There are several examples of applying game-based learning in higher education area. Jesus et al. presented in [11] a game-based learning framework for the development of computational thinking skills. They evaluated the framework in a single group and in two complete classes. The presented results showed high and very high awareness of collaboration, learning and also fun for the majority of the students. Jing et al. implemented a game-based learning framework in a blended learning environment and evaluated its impact on students' test results in a course about Java programming [12]. The results revealed that serious games can enhance learning outcomes, as the students' interest to learn increased resulting in better test results than without the serious game.

4 Game-Based Mobile Learning Approach

The limitations of the web application described in Sect. 2 motivated us to develop a concept which would work equally efficiently in an on-site and a remote setup. In this mobile learning concept, where mobile devices of students are used as the central learning medium, we apply game-based learning to motivate students to deeply analyze the behavior of the system.

The central element of this concept is a novel version of the dice game described in Sect. 2. It is a mobile version of this simulation game, which runs distributed on several mobile devices. Each mobile device represents a production station in the chain which is controlled by a student. The students are able to communicate with each other so that the gained insights can be immediately discussed in the team. In the case that some production stations cannot be controlled by the students (e.g. too few students in a team), these stations can optionally be controlled by the AI of the system.

Figure 2 shows the architecture of the game with 6 stations, one of them controlled by the AI. For a game initialization, one of the stations sends a request to the server. In Fig. 2, it is Station 3 (1). This request contains at least an ID of the game, an ID of the group (for example, a study class), and the number of stations for this game. Other stations can join this game using the same set of IDs (2). The stations can be any type of mobile device, including smartphones, tablets and watches, or they can be simulated by the server (Station 5 in Fig. 2).

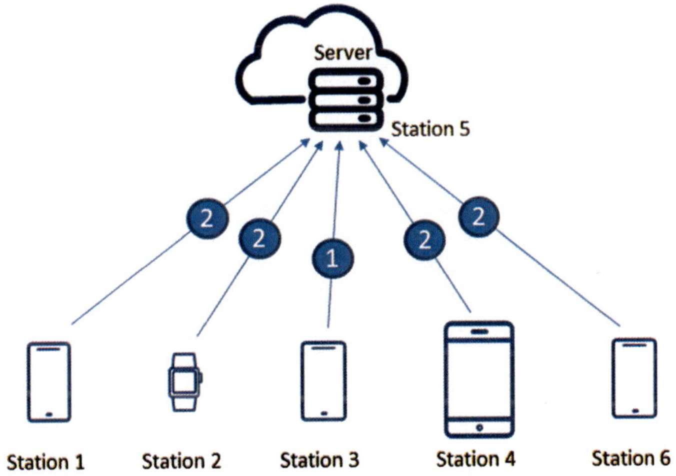

Fig. 2. Game architecture and initialisation

The message flow at the game start is depicted in Fig. 3. The server notifies the station whose turn it is (1), for example Station 1 at the beginning of the game. This station "rolls" the virtual die and communicates the result to the server (2). The server applies this result to determine how many resources are transferred to the next station, and notifies other stations about the new status (3). Afterwards, the server selects the next stations; these actions are repeated until the targeted number of turns is reached.

In the basic game variant, each station randomly produces between 1 and 6 items. Combining high statistical fluctuation with the dependencies through the chain, this system is categorized as a chaotic system with undefined outcome. In the constrained game variant, the students introduce a constraint in the system (for example, one station which cannot produce more than 4 items), and compare the outcomes with the first system. This is necessary for understanding the

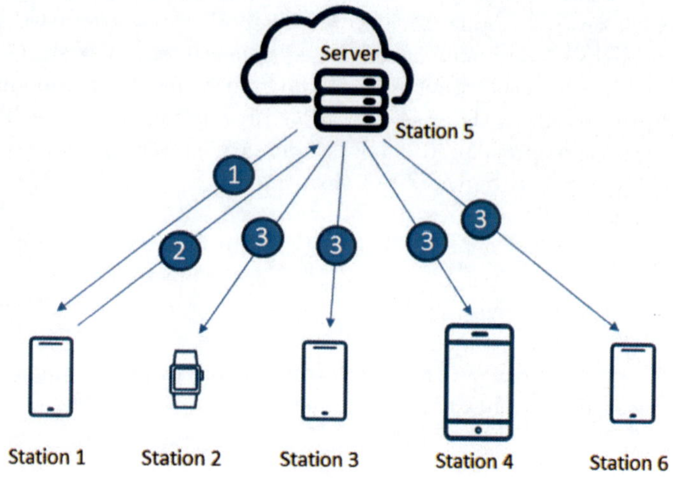

Fig. 3. Message flow at the start of the game

implications of constraints in a complex system. In the managed game variant, the students are able to manage the system by shifting the capacity from one station to another. This strongly influences the system outcome, and the students have the target to find out the optimal strategy for maximizing system outcome. In the lecture, each team presents the results from the three game variants, explaining the strategy and the achieved outcome.

Figure 4 shows a screenshot of Station 4 during the sixth turn of the basic game. The current active station is marked green, other stations are in orange. When it is the player's turn, he or she can roll the virtual die. This is indicated by the green color of the button "Roll" at the bottom. The virtual die is not limited to values between 1 and 6, and can be set arbitrarily at any range (for example, between 4 and 7 in Fig. 4). This provides additional flexibility in defining the constraints of the system and understanding their impact.

One of the main benefits of this interactive mobile learning approach is that it enables location-independent teams. The game can be played with the same user experience independently whether the team is completely on-site, completely remote or in a mixed setup. Different team sizes or even player mode with all other stations AI-controlled are possible as well. The interactions with the mobile phones and other students in the team ensure active participation in the game. Furthermore, all current and historical results are available on the server so that any kind of competition can be implemented.

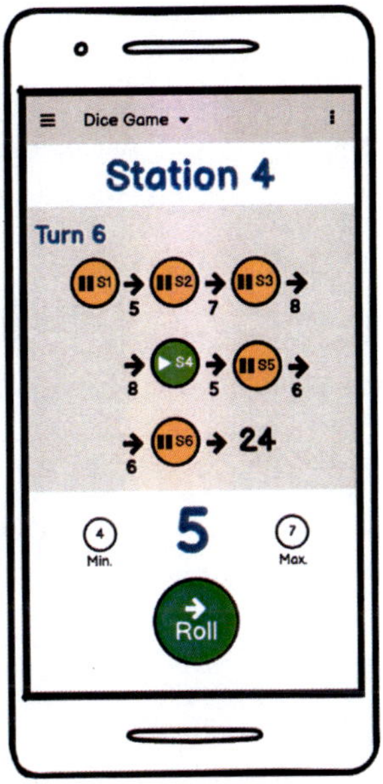

Fig. 4. Game screenshot at Station 4 in sixth turn

5 Conclusion

Complex problem solving is one of the most required skills in the 21st century. At the UAS FH Campus Wien, we have introduced a course, in which we teach methodology and tools for complex problem solving based on the Theory of Constraints by Goldratt in our Master study program "Software Design and Engineering". To demonstrate the impact of the Theory of Constraints in an easily understandable way, we applied game-based learning with a dice game as a serious game. However, the in-class dice game was not possible during the begin of the pandemic, which motivated us to design an digital version of this game.

The digital mobile version of the dice game presented in this paper is capable of supporting distributed team settings. Our goal was to keep all the benefits of the non-digital game, such as discussions between team members and learning from each other, and to enable the flexibility of a digital game, for example AI players, flexible constraints and production capacities, or teams partly on-site and remote.

In our future work, we plan to evaluate the effectiveness of this serious game by letting different groups of students work with different game types, including the real dice game, the web application, and the newly developed mobile version of the game. We will compare the learning outcomes of the groups and analyze benefits and challenges of our approach.

References

1. Anastasiadis, T., Lampropoulos, G., Siakas, K.: Digital game-based learning and serious games in education. Int. J. Adv. Sci. Res. Eng. **4**, 139–144 (2018)
2. Barata, G., Gama, S., Jorge, J., Goncalves, D.: Improving participation and learning with gamification. In: Proceedings of the 1st International Conference on Gameful Design, Research, and Applications, Toronto, Ontario, Canada (2013)
3. Demaria, M., Hodgsona, Y., Czecha, D.: Perceptions of transferable skills among biomedical science students in the final-year of their degree: what are the implications for graduate employability? Int. J. Innov. Sci. Math. Educ. **26**(7), 11–24 (2018)
4. Deterding, S., Dixon, D., Khaled, R., Nacke, L.: From game design elements to gamefulness: defining "gamification". In: Proceedings of the 15th International Academic MindTrek Conference: Envisioning Future Media Environments, Tampere, Finland, September 2011
5. Dettmer, H.W.: The Logical Thinking Process. A Systems Approach to Complex Problem Solving. American Society for Quality (2007)
6. Dörner, D., Funke, J.: Complex problem solving: what it is and what it is not. Front. Psychol. **8**, 1153 (2017). Article no. 115
7. Goldratt, E.M.: Theory of Constraints. North River Press (1990)
8. Goldratt, E.M., Cox, J.: The Goal: a Process of Ongoing Improvement. North River Press (2014). 30th anniversary edition
9. Griffin, P., Care, E., Wilson, M.: Assessment and Teaching of 21st Century Skills. EAIA, pp. 3–33. Springer, Cham (2015). https://doi.org/10.1007/978-3-319-65368-6
10. Jang, H.: Identifying 21st century stem competencies using workplace data. J. Sci. Educ. Technol. **25**(2), 284–301 (2016)
11. Jesus, Â.M.d., Silveira, I.F.: A collaborative game-based learning framework to improve computational thinking skills. In: 2019 International Conference on Virtual Reality and Visualization (ICVRV), pp. 161–166 (2019)
12. Jing, T.W., Yue, W.S., Murugesan, R.K.: Learning outcome enhancement via serious game: implementing game-based learning framework in blended learning environment. In: 2015 5th International Conference on IT Convergence and Security (ICITCS), pp. 1–3 (2015)
13. Knight, A.: The Dice Games as highlighted in Pride and Joy (2021). https://www.the-dice-game.com/game1.html
14. Kosa, M., Yilmaz, M., O'Connor, R., Clarke, P.: Software engineering education and games: a systematic literature review. J. Univ. Comput. Sci. **22**(12), 1558–1574 (2016)
15. National Research Council. Assessing 21st Century Skills: Summary of a Workshop. The National Academies Press (2011)
16. Sanchez, E.: Game-based learning. In: Kauffeld, S., Othmer, J. (eds.) Handbuch Innovative Lehre, pp. 249–253. Springer, Wiesbaden (2019). https://doi.org/10.1007/978-3-658-22797-5_18
17. Sánchez Carracedo, F., et al.: Competency maps: an effective model to integrate professional competencies across a stem curriculum. J. Sci. Educ. Technol. **27**(5), 448–468 (2018)
18. Schefer-Wenzl, S., Miladinovic, I.: Game changing mobile learning based method mix for teaching software development. In: mLearn 2017, 16th World Conference on Mobile and Contextual Learning (2017)

19. Schefer-Wenzl, S., Miladinovic, I.: Leveraging collaborative mobile learning for sustained software development skills. In: Auer, M.E., Tsiatsos, T. (eds.) ICL 2018. AISC, vol. 916, pp. 157–166. Springer, Cham (2020). https://doi.org/10.1007/978-3-030-11932-4_16
20. Wilkinson, P.: Affective educational games: utilizing emotions in game-based learning. In: 2013 5th International Conference on Games and Virtual Worlds for Serious Applications (VS-GAMES), pp. 1–8 (2013)

Exploring Pre-schoolers' Feelings During Online Learning with Elements of Educational Neuroscience

Sarah Vlachou, Spyridon Doukakis(✉), Elen Malliou, and Evangelia Filippakopoulou

Ionian University, Corfu, Greece
{cl9vlac,sdoukakis,cl9mall,cl9fli}@ionio.gr

Abstract. Online learning is a method of teaching and learning that takes place through the internet as a form of distance learning. Online learning is not a new phenomenon, however it became ubiquitous as a result of the COVID-19 pandemic during 2020. The closure of schools in most countries created new circumstances for almost the entire educational system. In the pre-school education domain, kindergarten teachers were summoned to manage new knowledge and overcome difficulties in order to continue the educational process by implementing distance online and offline teaching methods. The purpose of this paper is to explore the feelings and perceptions of preschool children who took part in these online learning practices. The literature review concluded that this subject has not been sufficiently researched. To fill in this gap this small-scale research was conducted on a semi-urban island region of Greece during the suspension of kindergartens in November and December 2020. The data collected derived from interviews and drawings, of 11 children-participants, which were subsequently analyzed in a qualitative study. The study demonstrates how the preschoolers experienced the online courses. According to the results, the children's perceptions of the online lessons were assessed as positive. Although their desire to return to school and face-to-face learning was evident, online learning offered them enjoyable learning experiences. The research, also, reveals the important factors according to educational neuroscience that need to be considered in order to promote effective online lessons for this particular age group.

Keywords: Online learning · Preschool education · Qualitative analysis · Educational neuroscience · Perceptions

1 Introduction

On 11 March 2020 the Director-General of WHO, Dr Tedros Adhanom Ghebreyesus, declared the coronavirus outbreak a pandemic. The daily lives of many people have been significantly affected since, considering that the new virus resulted in great changes in numerous fields and on numerous levels. Precautions needed to be taken in the field of education in order to avoid the spread of the virus. Many governments announced that educational institutions would have to apply emergency reforms

M. E. Auer and T. Tsiatsos (Eds.): IMCL 2021, LNNS 411, pp. 60–70, 2022.
https://doi.org/10.1007/978-3-030-96296-8_6

involving the implementation of distance learning solutions in order to keep the educational process running.

In November 2020 all educational institutions in Greece were required to suspend classroom lessons for the second time during that year. Online learning became the focus of educational activities. The development of technological infrastructures and the fact that most teachers and learners owned an electronic device made it possible to continue the educational process. Digital learning was supported by a blended learning model, which attempted to combine online and offline learning. This hybrid learning model aimed to effectively combine traditional teaching activities with distance learning methods, strategies and tools. In particular, in the field of preschool education, the Ministry of Education prepared an online distance learning program via a specially-designed digital platform. Teachers and pupils participated in the program in real time on a daily basis, for approximately two hours a day. At the same time, offline education tools were also developed, which offered pupils the opportunity to work at their own pace and in their own time, while also giving the teacher the opportunity to monitor the progress of pupils' assignments. The same preschool teachers who were in charge of their classroom learning undertook to teach the distance learning classes. The aim of the program was to maintain close contact between each pupil and their peers, their preschool teachers and the learning process.

Following the implementation of all these changes, there is now the question of how children in early childhood experienced learning from home. The shutting down of preschools and, consequently, the implementation of online learning deprived children of the opportunity to engage in free play, which is a key activity for children this age. Instead, this learning model provided structured activities, where the preschool teacher was required to structure children's play on an educational and social basis.

Based on research conducted in Britain and the US, structured activities have positive effects on children's development according to educational neuroscience findings [1, 2]. In recent years, the above findings are confirmed by research in the field of educational neuroscience [3]. These approaches can provide a caring, encouraging and pleasant environment. Furthermore, studies show that free play on preschool premises is not simply considered pedagogically and psychologically necessary for pre-schoolers, but it is a basic element for defining the pre-schooler as a pedagogical subject and also serves as a significant criterion for their assessment [4]. Children play at preschool and they must do so in order to grow, socialise, learn and build relationships [5]. This new situation, where free play is absent and structured activities prevail, has given rise to yet another concern regarding its effect on young pre-schoolers. This paper studies children's experiences during the online learning processes. In particular, it attempts to discover how children feel about the interaction between themselves and those involved, and also about the process of learning through an electronic device.

Qualitative researchers can reap benefits from involving children as informants, especially when examining childhood occupations. This involvement provides an opportunity for the researcher to learn about the children's worlds and perspectives, including the personal meanings children attribute to events and actions. The researcher can learn from the children's own experiences and from their knowledge. It is an important way to gain a better understanding of the phenomena being studied [6].

In order to identify the children's perceptions and feelings regarding the above approaches, their opinions were recorded through interviews and their thematic drawings titled "digital classroom" were analysed. The children's drawings were used as a tool for collecting additional information so as to have a more complete picture of their experiences during the online learning process. The analysis of these drawings focused on their visible elements and objects, since the process of analysing and interpreting the meaning of such drawings is a difficult and uncertain one, which depends on several factors. It would be very useful to draw additional information by observing the children while they drew, however this was impossible due to the restrictions on social contact in the period during which the survey was conducted.

2 Young Children's Online Learning During COVID-19 Pandemic

During the last decade, online learning has spread out rapidly due to the flexibility it provides in terms of time, place and pace of the study. Also, due to the easier and more affective access to a wider variety and greater quantity of information as well as the reduced financial costs [7]. Since online lessons can be taken from home more online programs have been developed and delivered to support young children with disabilities and allow children from remote locations or disadvantaged situations to take part in the lessons [8]. Online learning may have been existent but the spread of the Coronavirus disease in 2020 made it widespread and forced many educational institutes around the world to shift from traditional educational approaches to distance learning.

When performing a critical assessment of the relevant research [9–11], one reaches the conclusion that, when it comes to preschool-aged children, learning from home is a tough experience for all involved. The benefits of online learning are indeed recognised, however, there are many restrictions and conditions that must be met in order to make online teaching and learning effective. On the part of schools and teachers, what is needed is a good level of preparedness, proper lesson planning, development-appropriate approaches and the support of parents, whose roles change during home schooling as they are required to undertake different responsibilities. It is also necessary to take into account the different financial and social backgrounds of each family, in order to support the online learning process in terms of resources and equipment. The findings of the research also show that, at home, children easily loose interest in learning, they are easily distracted and they very often waste far too much time in front of screens. The above information has been drawn from studies conducted by scholars and from the views of teachers and parents. At the same time, however, the views of children themselves remain unexplored to a large degree. With the aim of filling this gap in the research conducted to date, this paper makes children the primary informants and attempts to understand how young children experience distance learning.

3 Method

3.1 Participants

The participating children were six boys and five girls aged 4–6 years old. All the children who participated in the research attended public kindergartens previously to the home quarantine, since Greece had imposed national lockdown for the second time in November 2020. Some of the demographic data are presented in Table 1.

During the suspension of face-to-face classes the children attended online approximately two-hour lessons by their two kindergarten teachers on a daily basis. These online lessons were held on from early November to late December 2020, through the 'Cisco Webex Meetings' platform. This was the first time these particular students had taken part in such a process. All the children involved were willing to answer the interview questions and draw what was asked of them. This sample was selected, mainly, because it was possible to communicate and cooperate with them and their parents as one of the authors of the current paper was their teacher at the time being. In this term, it was more possible the children would feel comfortable enough to express their feelings and provide accurate accounts of their experiences in regard to the online lessons.

The online lessons which the participants attended, included some daily routines (e.g. greetings, filling in a digital calendar), thematic learning, interactive games, story reading and sharing, drawing, workout, singing and dancing and two 20-min breaks. During the lessons the students were muted but kept their cameras on. Gradually, with the help of their caregivers, the children learned how to mute and unmute their microphones so that they could speak when wished. The presence of an adult was necessary throughout the lesson.

Table 1. Demographic variables of study population.

Participants	Gender	Age	Siblings	Family member	Device used
C1	Girl	5	No	3rd	P.C.
C2	Girl	5	No	3rd	Smart phone
C3	Girl	4	Yes	4th	Laptop
C4	Boy	5	Yes	3rd	Laptop
C5	Boy	5	Yes	4th	Laptop
C6	Boy	5	Yes	5th	Laptop
C7	Boy	5	Yes	3rd	Laptop
C8	Boy	4	Yes	3rd	P.C.
C9	Girl	4	No	3rd	Tablet
C10	Girl	4	Yes	3rd	Laptop
C11	Boy	5	No	3rd	Laptop

3.2 Procedure

This case study was conducted on a semi-urban island region of Greece in December 2021. The interest of our research focused on the analysis and understanding of the meaning of children's speech and drawings. The naturalistic paradigm was chosen as the research methodology in order to attempt a qualitative analysis of the children's view concerning online learning.

In order to identify the feelings and experiences of the children-participants, three tools were used: (a) a semi-structured interview, (b) a short story and (c) a drawing [12]. Specifically, the children were asked to answer five open-ended exploratory questions about their online lesson experiences, come up with a short story regarding the closures of schools and draw their digital classroom. The data was collected by one of the researchers via video calls, due to the COVID-19 disease restrictions. Concerning the drawings, instructions were given through the video calls and were subsequently sent by the parents [13, 14]. It was necessary to communicate with the parents in order to define the online meetings and to make sure they forward the children's drawings. After collecting the material, the textual and visual data were analyzed by adopting a qualitative approach. The children and their parents were informed that their participation in this research was voluntary, anonymous and that they could withdraw at any time if they wished. It was also made clear to the children that there were no right or wrong answers. The researcher kept a friendly and informal disposition creating a comfortable and secure communication environment in order to obtain the most honest answers and perspectives.

The semi-structured interview included the following 5 questions:

I. What is a digital class? What is an online course? What do you do there?
II. Is there anything that you enjoy during the lessons;/what do you like most about them?
III. Is there anything that you don't enjoy during the lessons;/what do you like least about them?
IV. If you were the teacher for one day, what would you change?/What else would you do? Would you do something differently?
V. Supposedly another child in another town started online lessons like you do. What would you like to tell him/her to prepare him/her? What must he/she know?

The storytelling was the following:

> "Once upon a time there was a little boy/girl named…, His/her school suddenly closed due to a pandemic after ... Continue the story, what happened next?"

Finally, the drawing was the following:

Please draw your digital classroom.

3.3 Data Analysis and Findings

The qualitative data collected from the open-ended questions, the storytelling and the drawing were analyzed using NVivo software. Useful information for our research were retrieved quickly and efficiently. The easy to handle functions of the software helped us to process the research material in a dynamic way, it raised new questions as we processed the material, and, consequently, new findings [15]. In addition to the new information that emerged, relationships and connections were found between the data which could be represented by graphs and diagrams providing a visual representation of the data and therefore a better understanding of the meaning.

To identify, analyze, and report patterns generated from the data thematic analysis was used [16]. The initial categories that had generated from the data was sorted into seven themes: a) online learning perceptions, b) positive comments about the online lessons, c) negative comments about the online lessons, d) requirements for participating in online lessons, e) suggestions for a better online lesson, f) stories and g) drawings.

a) *Online learning perceptions*: Participating for the first time in a distance education course, made it vital to consider the students' perceptions of the terms 'distance learning', 'digital class' and what these terms meant to them. Nine children presented online learning as the lesson through an electronic device. These children focused to describe the lessons as a type of school where they listen to their teachers through computers. Less often, in the same category, children talked about the curriculum, the learning materials, the activities or their classmates. Child 4 defined the online lessons as the lesson where you need to stay home due to the Coronavirus. Child 3 said she didn't know the meanings of these terms.
b) *Positive comments about the online lessons*: Particular importance was given to the analysis and study of the positive and the negative comments of the children. In order to elicit information about their positive feelings towards distance learning they were asked to describe what they liked. Most positive references were related to a particular lesson or activity that they enjoyed. Most of the positive comments in this sub-area were about specific experiences from the lessons they had done with their teachers. More specifically, they mentioned specific language and math interactive exercises, experiments, cutting something with scissors and other animal-related activities. The other positive comments were about the games they played during workout time, drawing time and the calendar they had to fill in every day. The fact that they could see and talk to their classmates was another comparatively positive comment. One child enjoyed the fact that that he could turn on and off his camera and microphone and press buttons. Child 6 commented "I like it with my camera turned off but with my microphone on because with a closed camera I listen better". Two students indicated the breaks as something they liked.
c) *Negative comments about the online lessons*: The negative references were more likely to be related to technical issues; five children disliked it when they either couldn't hear the teacher or the group could not hear them. For instance, child 2 commented "I have a good time all the time, only when they do not listen to me I do not have a good time. It bothers me when the teachers do not listen to me. They do not listen to me every time I speak and I say it very loudly". Other expressed displeasure when they had to wait due to poor internet connection. Child 6 found it frustrating when skipped to another page or area of the computer if a button was accidentally pressed. Some children also commented negatively on the fact that

they needed to sit in front of the screen for a long time. Finally, the homework given by the teachers was viewed as something negative and the fact that during class they could no talk whenever they pleased. The qualitative analysis found only one child (Child 1) who did not express any negative comments. When asked about what she disliked she commented "I enjoy the lessons every day and I would like to continue having lessons from the computer".

d) *Requirements for participating in online lessons*: In order to understand the children's views regarding the conditions required for attending the online courses, we asked them what was essential for participating in such a course or what we would say to another child to be prepared. Four children believe that the only essential thing is a computer, four children said the child must have coloring markers near him/her. Three children said he/she should sit quietly in a chair, and two children said he/she should know what time of day the lesson takes part.
e) *Suggestions for a better online lesson*: The qualitative data showed that the most mentioned suggestions were related to workout games, painting and completing the daily calendar. It is important to note that out of the 15 suggestions they made, 6 of them had never been implemented in the online lesson; they were new ideas. The new ideas involved different games, not using a device and being close to their friends. The other 9 suggestions were subjects that had often been applied during the existing online lessons.
f) *Stories*: When telling stories about school closures, it was noticed that most children's imagination did not 'travel' far from the real events they experienced during the quarantine period. The stories they came up with were mostly related to the distance lessons. They expressed negative feelings about staying home and the closure of schools and positive feelings about going back to school. Eight children ended their stories with the opening of the schools. They also made references about friends, events that took part in their homes and the coronavirus. It was only Child 1 and 2 whose events were located away from their home (e.g. the beach).
g) *Drawings:* The process of analyzing and evaluating the children's drawings focused on the description and distinction of the visible elements-objects of the images. The study focused on the qualitative analysis and interpretation of the elements contained in them in order to obtain additional information about their perceptions. No psychological interpretation was attempted. After studying the content and coding parts of the projects, as a way of examining where the children were focusing, seven thematic groups emerged. Specifically, the children drew 'people' and 'relationships between them', 'elements from the environment', 'objects', 'locations', 'symbols' and 'text'. High in the children's preferences was to draw faces. It was only child 8 who did not include any human elements in his drawings. In the drawings of eight children, children and adults (parents, teachers) were visibly drawn, while the rest two only drew themselves. Among them, only 2 children drew some kind of relationship or interaction between the representative faces. The rest illustrated the online interaction through lines connecting boxes (the screens). In some cases the children drew themselves outside a computer or a tablet, while others inside the screens as seen on the online lesson platforms. Such symbolic elements are found in almost all of the children's drawings. Two girls drew elements from the natural environment and gave some kind of information

about the location (e.g. garden). Three children added some kind of text or symbols into their drawings. In addition to the elements analyzed above, it was necessary to categorize the drawings into two areas, 'colorful' and 'not colorful'. As mentioned in relevant literature, color is an indicator for the analysis of children's paintings. Eight children used more than three different colors in their drawings, while three only one or two colors. The non-colorful designs were all from boys.

4 Discussion

The importance of this research lies in the fact that currently in Greece it is a unique study and one of a few internationally which looks into the emotional world of pre-school aged children during the period of distance education, by examining their views through their voices and artwork. It would be very useful to investigate the relation of the results obtained in this study with the international literature, in order to examine whether, and to what extent they are consistent alongside with the findings of educational neuroscience.

More specifically, the findings of educational neuroscience highlight the role that music can play in young children's self-regulation. Self-regulation enhances life-long learning, contributes to the obtainment of attention, to the resistance to distractions and helps to avoid conflicting behaviors. In this context, the use of music, rhythm and movement in online education could help strengthen the neurological development of self-regulation and enhance attention. More than that, in the context of online education, the required presence of parents or caregivers along with the quality interaction between them and the child can have an impact on the development of the brain and ultimately affect children's emotions and social development. Positive emotions can reinforce children and contribute in achieving in activities which are related to the next stage of literacy. Finally, the incorporation of playing within the context of online education can help create a conducive environment for children to develop self-regulation, to show good social behavior and to learn how to control their aggression. In this context and although playing was part of the online program, the lack of free play and physical activities had a negative impact on executive skills, such as attention and social skills which enhance the emotional and cognitive development [17].

The findings and results of the research can be used for future research and educational purposes, as they provide indications on how online teaching can be adapted to specific contexts, so that knowledge is approached in such a way that children are actively and happily involved in online learning processes. However, it would be useful for future research samples to be larger and randomly chosen. In the present study, the sample consists of a relatively small number of children. As a result there were significant limitations in generalizing the conclusions. Such a weakness was that it was not possible to draw conclusions that would make clear if some demographic variables of the study population influenced how the children experienced the lessons. For instance, it would be very helpful to identify if having siblings in the home during the lessons had an effect on how they experienced the process or whether their parents/caregivers were essentially by their side during the online lessons or were engaged by their own work at that same time. However, it became possible to understand individual situations and difficulties, as well as to thoroughly explore each student's perceptions about the lessons.

Furthermore, the data was collected online. There was no physical contact between the researcher and the children at any stage of the data collection. This alone, creates some limitations and difficulties. It would be very useful for the researcher to be present while the children drew the pictures to take notes of verbal statements whilst creating the pictures and thus gain a better understanding of the elements in the drawings. To elicit further information of depth and richness it would be beneficial to use questionnaires, completed by the children's caregivers [6]. These questionnaires would acquire information on verbal and non-verbal behaviors during the lessons. In this way a more spherical view of each student's experience would be obtained.

It is of crucial importance that similar investigations are conducted in the future so that further information about children's experiences regarding online lessons are obtained. It is necessary to maintain the learning process in the best possible way during periods that traditional face-to-face education is interrupted.

5 Conclusions

The qualitative methodology and more specifically the children's verbal reports, narratives and drawings contributed to the formation of a first image around the experience of distance education through the eyes of the students themselves.

By analyzing the data it was found that the children's perception regarding the lessons centered on the electronic device used by them and the teacher. The device was most often mentioned as a prerequisite for attending the courses in contrast to the lacking references about the pedagogical practices. It is possible that this approach is due to the fact that the lesson through a device was a new experience to them, but also the most noticeable difference between online and face-to-face learning.

This research found a greater variety of positive comments about the online lessons in comparison to the negative. The children explained in greater detail and were able to recall more pleasant moments and events rather than drawbacks. It also appeared that the five girls liked the online lesson more than the six boys, as they made more positive comments and less negative ones. The children mostly enjoyed the playful and interactive activities especially when motion was involved and to make use of the digital tools. They were, also, very pleased to see and talk to their classmates and teachers. We conclude, therefore, that teaching practices can be designed in such a way that compose interesting and enjoyable courses for preschool children. Such practices enhance their active participation and preserve their attention which makes the online program more effective. Unpleasant moments mainly included technical problems and struggling to sit still and not being able to play with their friends. Less often they reported the given homework as a negative element and the fact that during class they could not talk whenever they pleased. Therefore, it is important to design online courses that create plenty of opportunities for each student to speak and express his/her thoughts not only towards the educator but towards the rest of the group. For this reason, online teaching will work better with a small number of the children so that each child can share and interact as much as possible. The ratio between a teacher and the number of children is critical in distance education and must be taken into account.

By the analysis of the children's stories, it seems that the children connect and identify the pandemic with both their staying at home and learning online. However, it is important to note that the children's negative emotions were related to their staying at home, not the lessons. The research revealed that online learning was positively assessed, as it was a way to contact and communicate with their classmates and teachers. Beyond their feelings round the online lessons, it was shown that the children's desire was to return to their schools and face-to-face learning and playing.

Regarding the drawings, based on the superiority of the whole it was shown that the children were able to express their experiences and their perceptions very accurately. Through their designs, they managed to give us information and to see on paper their own mental images. The absence of interpersonal relationships was obvious, showing the isolation they experienced at the time. According to [18] the children who draw are happy children. These children drew and most of them drew in color and detail.

To sum up, this study revealed both positive and negative emotions about online learning. The children described enjoyable online learning experiences as well as difficulties. It is apparent that they wished to return to their school, classmates and teachers, however, the daily online contact and the pedagogical practices were beneficial to them. Under certain conditions and adaptations, distance education is a useful and necessary tool in times when students are forced to stay at home. Online learning can be applied successfully and can minimize the remoteness in circumstances such as the COVID-19 days.

Acknowledgments. This research is funded by the European Union and Greece (Partnership Agreement for the Development Framework 2014–2020) under the Regional Operational Programme Ionian Islands 2014–2020, project title: "Enhancing cognitive abilities of people with Mild Cognitive Impairment through measurable cognitive training—NEUROEDUCATION", project number: 5016113.

References

1. Miller, L.B., Dyer, J.L., Stevenson, H., White, S.H.: Four preschool programs: their dimensions and effects. Monographs Soc. Res. Child Dev. **40**, 1–170 (1975)
2. Osborn, A.F., Milbank, J.E.: The Effects of Early Education: A Report from the Child Health and Education Study. Clarendon Press, Oxford (1987)
3. Doukakis, S., Alexopoulos, E.: The role of educational neuroscience in distance learning. Knowledge transformation opportunities. In: Auer, M.E., Centea, D. (eds.) ICBL 2020. AISC, vol. 1314, pp. 159–168. Springer, Cham (2021). https://doi.org/10.1007/978-3-030-67209-6_18
4. Dafermou, C., Koulouri, P., Mpasagianni, E.: Guide for kindergarten teacher. Educational planning-creative learning environments. National Institution for Publishing Educational Books, Athens, Greece (2006)
5. Loizou, E., Avgitidou, S.: The Greek-Cypriot early childhood educational reform: introducing play as a participatory learning process and as children's right. Early Child Dev. Care **184**(12), 1884–1901 (2014)

6. Curtin, C.: Eliciting children's voices in qualitative research. Am. J. Occup. Ther. **55**(3), 295–302 (2001)
7. Khurana, C.: Exploring the role of multimedia in enhancing social presence in an asynchronous online course. Doctoral dissertation, The State University of New Jersey, Rutgers, U.S. (2016)
8. Smith, S.J., Burdette, P.J., Cheatham, G.A., Harvey, S.P.: Parental role and support for online learning of students with disabilities: a paradigm shift. J. Spec. Educ. Leadersh. **29**(2), 101–112 (2016)
9. Lau, E.Y.H., Lee, K.: Parents' views on young children's distance learning and screen time during COVID-19 class suspension in Hong Kong. Early Educ. Dev. **32**(6), 863–880 (2021)
10. Nores, M.: An exploratory study of the use of time diaries to understand young children's experiences during the pandemic. National Institute for Early Education Research (NIEER) (2020)
11. Solekhah, H.: Distance learning of Indonesian early childhood education (PAUD) during the Covid-19 pandemic. Int. J. Emerg. Issues Early Childhood Educ. **2**(2), 105–115 (2020)
12. Cohen, L., Manion, L., Morrison, K.: Research Methods in Education, 8th edn. Routledge, Milton Park (2017)
13. Kisovar-Ivanda, T.: Thematic analysis of the children's drawings on museum visit: adaptation of the Kuhn's method. World J. Educ. **4**(3), 60–67 (2014)
14. Poláková, P., Klímová, B.: The perception of Slovak students on distance online learning in the time of coronavirus. Preliminary study. Educ. Sci. **11**(2), 81 (2021)
15. Doukakis, S., Koilias, C., Chionidou-Moskofoglou, M.: An undergraduate primary education teaching practicum design and undergraduate primary teachers' satisfaction on developing technological, pedagogical and mathematical knowledge. Int. J. Teach. Case Stud. **3**(2–4), 180–195 (2011)
16. Braun, V., Clarke, V.: Using thematic analysis in psychology. Qual. Res. Psychol. **3**(2), 77–101 (2006)
17. Szécsi, T., Varga, L., Mak, V.: Current trends, dilemmas and future directions in neuropedagogy in the field of early childhood. Képzés És Gyakorlat Train. Pract. **16**(3), 51–58 (2018)
18. Farokhi, M., Hashemi, M.: The analysis of children's drawings: social, emotional, physical, and psychological aspects. Procedia Soc. Behav. Sci. **30**, 2219–2224 (2011)

Online Learning, Students' Assessment and Educational Neuroscience

Spyridon Doukakis[1(✉)], Maria Niari[2], Evita Alexopoulos[3], and Panagiotis Sfyris[3]

[1] Ionian University, Corfu, Greece
sdoukakis@ionio.gr
[2] Hellenic Open University, Patras, Greece
[3] Pierce-The American College of Greece, Aghia Paraskevi, Greece
{evitaalexopoulos,psfyris}@acg.edu

Abstract. Assessment is an important factor in any learning process or context. It is argued that educational assessment can serve as an indicator of the quality of students' performance and learning outcomes, as a facilitation means for students' learning, and as an essential parameter for the design and structure of a learning environment. Despite the debate on whether online learning assessment differs completely from the conventional assessment methods or not, it is accepted that it is a key aspect of learning in every context. Diagnostic, formative and summative assessment affect learning outcomes and have an impact on teacher and student roles. Educational neuroscience supports this argument, as well. The present paper investigates the ways and the extent to which educators in Greece used assessment techniques during the online teaching imposed by the pandemic crisis. A qualitative research methodology, based on semi-structured interviews was adopted. Research findings point to the lack of a structured assessment plan during this period and highlight the need for assessment techniques and tools to be part of teachers' educational design.

Keywords: Online learning · Assessment · Educational neuroscience

1 Introduction

Student assessment comprises a substantial component of the educational process. Furthermore, it can assist teaching as well as gauge its outcomes. Even though student assessment is considered to be a definitive element of online education, the educational community is trying to determine how it will be implemented so that, on the one hand, learning can be enhanced and, on the other, student learning can be assessed.

During the pandemic, learner assessment at all educational levels was brought to the foreground. In addition, due to the fact that the educational community had to employ emergency remote teaching, assessment was set aside so that educators and students could focus on technical issues, on discovering tools that would allow them to carry on their teaching and at the same time keep their students actively involved [1]. However, findings from educational neuroscience establish the importance/value of assessment in knowledge consolidation, accumulation, and retrieval. Therefore, as

M. E. Auer and T. Tsiatsos (Eds.): IMCL 2021, LNNS 411, pp. 71–82, 2022.
https://doi.org/10.1007/978-3-030-96296-8_7

online education was unfolding, so were conversations regarding student assessment and the ways in which this could be done since the lack of it led to the following question: teachers might achieve their teaching goals, but can they claim that the expected/anticipated learning outcomes they had set have been accomplished?

In the present study, assessment for learning and assessment of learning are firstly defined and these are followed by a discussion of online assessment. Subsequently, findings of neuroscience on student assessment, both in the physical learning environment and in the online learning environment, are presented. Finally, the qualitative research conducted with a sample of secondary education teachers is described and findings regarding assessment approaches employed and implemented by teachers are presented. These findings are discussed at the end of the article.

2 Assessment in the Educational Context

It is argued that assessment is an important aspect of any teaching and learning system [2]. It serves as an indicator of the quality of students' performance and learning outcomes [3], a facilitation means for students' learning [4], and an essential parameter for the design and structure of a learning environment [5]. There are certain types of assessment: diagnostic assessment, which includes all activities that teachers and students undertake to get information that can be used diagnostically to alter teaching and learning [4, 5]; formative assessment, which is used in order to give feedback to students, to support learning, and to adapt teaching to meet student needs [6]; and summative assessment, which is used to summarize students' achievements in order to award some kind of certification or course completion [7].

Elwood and Klenowski, as cited in [5], use the terms "assessment of learning" (assessment for the purposes of grading and reporting with its own established procedures) and "assessment for learning" (assessment whose purpose is to enable students, through effective feedback, to fully understand their own learning and the goals they are aiming for), respectively. The use of the concept "assessment for learning" places the student and learning in the center of assessment as an educational practice, while the term "assessment of learning" seems to be more teacher-centered. In the first case, both learners and educators share ownership and responsibility for assessing their own performance and learning outcomes, using peer and self-evaluation techniques, group assessment, and reflection [5, 8].

On the other hand, assessment is also linked to the educator's view of his/her role in educational practice [4] and it is argued that all forms of assessment can be regarded as "a reflection of the epistemological beliefs of each practitioner" [7]. In the same vein, educators decide on their assessment techniques and tasks according to the knowledge they consider important and their view of how this should be assessed in a particular context [8], focusing either on 'knowledge control', where the learning outcomes are seen as recollection of factual knowledge or skills [6, 9], or on 'assessment as learning', where the learning outcomes are seen as part of a competence-based curriculum which fosters the development of competence and capability [8, 9].

Meanwhile, the "hidden curriculum" notion introduced by Snyder in 1971, although considered dated and limited [10], still finds supporters in the 21st century

[11]. The term "hidden curriculum" describes how students infer what is important in a course based on the ways in which their learning is assessed [12]. According to [8], assessment as "hidden curriculum" sends a strong message to students about what counts as knowledge in a particular learning environment and subconsciously formulates their approach to learning based on their perceptions of assessment requirements: "Assessments which focus on recall of factual knowledge tend to steer students towards surface approaches to learning, whereas assessments which emphasize application and comprehension tend to encourage deep approaches to learning." (p. 4). In other words, assessment influences student learning in a number of ways, as it sends out messages about what counts as important knowledge, it has an impact on students' approach to learning, and, finally, it gives feedback to students about their learning [8].

A key factor in any form of assessment (especially in formative assessment) is for the educator to provide prompt feedback on student performance. Receiving prompt and detailed feedback helps students to improve learning as it enables them to understand - and therefore act on - the required level or performance standard [4, 13, 14]. In this way, it could be argued that assessment may lead to encourage students to adopt such deep learning strategies and skills, as acquiring and applying knowledge efficiently, thinking critically, analyzing, synthesizing and making inferences [15, 16].

3 Online Learning Assessment

It is argued that online learning "requires the reconstruction of student and instructor roles, relations and practices" [5] (p. 31) and therefore of online assessment, too. [5] underpin the need to identify effective assessment methods suitable to online learning and functional assessment techniques to make the feedback loop between instruction and assessment more meaningful. In the same vein, [17] suggests that assessment of online learning not be conducted the same way as in a traditional face-to-face classroom: "traditional assessment measures are unlikely to reveal the complexities of student-centered online learning environments that are radically different from the dominant teacher-centered instructional paradigm" (p. 109).

On the other hand, [2] supports the idea that "the principles of assessment do not change in an online environment" (p. 71). In the same vein, [5] state:

Assessment is expected to directly affect learning, whether online or traditional, by: communicating messages about how students should study and what things are most important to learn; providing opportunities for students about how to review, practice, and apply what they've learned; nurturing student ownership and promoting such skills as self-monitoring and self-evaluation (Brookhart, 1997, p. 164; Russell, Elton, Swinglehurst & Greenhalgh, 2006, p. 495)".

In order to develop self-monitoring and self-evaluation skills, the students need to take ownership of their learning, to engage in acts of metacognition, to reflect on and assess their own understanding, i.e., to take on an active role [18]. These processes require not only a learning environment that supports this active student role [5], but also an educator/teacher who is going to be a facilitator, a mentor, and a coach, by providing feedback and scaffolding student autonomy. Online discussion, written assignments, experimental assignments, problem assignments, portfolios and journals, projects, presentations, self-assessments, peer evaluations, timed tests and quizzes, groupwork, and exams are some of the online assessment techniques that are available to educators [19–21].

4 Educational Neuroscience and Student Assessment

Research in the areas of cognitive science, psychology, and education has contributed in the understanding of topics/issues related to perception, memory, language, and attention all of which, in turn, have significantly influenced the educational process. Over the past decades, research in these areas has advanced tremendously and has exploited neuroscientific findings concerning the function of the human brain [22]. Such findings indicate that student learning can be enhanced if a) learners become more actively involved in the educational process, b) the diagnostic, formative, and summative assessment that takes place aims at learning, c) obstacles students encounter during their learning are taken advantage of, d) multiple representations of knowledge are integrated to boost student involvement, and e) more creative activities and collaborative opportunities are offered to students in a suitable learning environment [23]. This section focuses on diagnostic, formative, and summative assessment for learning in an online environment while an attempt will be made to correlate it with the findings of this study.

In the context of scaffolding, consolidating, accumulating, and retrieving new knowledge, the teacher/educator can exploit the online tools on offer to carry out diagnostic, formative, and summative assessment. Using various assessment exercises/activities, the teacher can identify students' preconceptions, determine whether students have the required prior knowledge, or monitor their development. According to existing findings, assessment can play an instrumental role in the appraisal of a learner's knowledge as well as in its construction/forming [24]. Through the use of all available online tools and the incorporation of frequent, non-threatening assessment, the teacher can enhance student learning. A significant factor in this process is the appropriate feedback with regard to the understanding/comprehension of an idea or concept.

During synchronous online learning, the teacher has at his/her disposal a variety of tools to carry out all types of assessment. The use of online polling with True/False questions, multiple choice questions or matching exercises during the lesson or as homework assignment (asynchronous learning), offer learners opportunities to boost their memory, check their knowledge and understanding, as well as assess their personal learning needs. At the same time, assessment using gamification principles, in accordance with student abilities and abilities, leads to high participation and allows for differentiation in the instruction.

More specifically, an assessment in the form of a poll, where only the teacher can see the student answers encourages student participation. Furthermore, when the teacher asks questions that do not take a single answer or questions that are not commonplace, student curiosity is triggered and participation increased. As a follow-up to such activities and according to the answers students have given, the teacher can form new working groups as well as redefine his/her teaching practice so as to meet his/her students' needs. Moreover, grades from this type of assessment should either not be announced to the students or should contribute to their final grade so that it can form part of their formative assessment.

The development of the brain can also be promoted by exploiting learning obstacles students come across and errors that the teacher observes. In order to achieve this enhancement during online learning, learners should be actively involved in the learning process and should be asked to address issues defined/chosen by the teacher defines or derived from the group work. The aim is for students to review errors (whether their own or those of others) both individually and collaboratively, so as to overcome learning obstacles that can lead to misunderstandings/misconceptions [25]. Whether individually or in groups, students identify the errors, and they communicate them to their teacher via the private chat. This way the teacher gathers data and can evaluate students' progress. Likewise, the teacher can observe ways in which students respond to activities that require a multifaceted approach. The observations/comments above led to the present study which tried to answer the research questions presented in the next section.

5 Research Purpose and Research Questions

The purpose of this qualitative research study was to investigate the methods and implications of students' assessment techniques used during the emergency remote online teaching as a result of the pandemic crisis.

To this purpose, the following research questions were formulated:

1. Which type(s) of assessment do educators use in online teaching?
2. How and with which techniques do educators assess student learning outcomes in online education?
3. Which digital tools do they use for students' assessment?
4. How have these assessment types and techniques affected student learning?

6 Method

In order to explore the purpose and research questions, a qualitative research study was conducted, based on semi-structured interviews with open-ended questions. The main strands of the interviews were:

- assessment techniques and tools used by educators in online teaching during the pandemic.
- educators' perception of ease or difficulty of online teaching assessment.
- perceived effectiveness of teaching and assessment methods used during the pandemic.

The interviews were recorded and transcribed, with the consent of the participants, and the transcript data was analyzed based on a coding scheme. Anonymity was secured during the procedure and all the appropriate measures according to the GDPR code were taken.

Participants were selected based on convenience sampling; out of a pool of educators from a variety of subjects and, with varying qualifications (Bachelor's degree holders, Master's degree holders, PhD holders, etc.), eight (8) educators were selected and gave their consent to participate in the present empirical research. All educators work in private secondary schools, located in urban centers of the country (Greece) and fully equipped with the necessary tools to conduct online education. Their students used digital devices to participate in the online lessons, as well. In this context, both educators and students had solved technical problems of involvement in online education and had the opportunity to work in a rich online educational environment during the pandemic crisis. In the following table (Table 1) the educators' demographic characteristics are presented.

Table 1. Educators' characteristics.

No	Gender	Work experience (in years)	Education	Classes	Ease with tools (1–5)
E1	Male	6	BSc	7–12	4
E2	Female	10	PhD	8–12	4
E3	Male	13	MBA	8–12	4
E4	Female	22	PhD	7–12	4
E5	Female	22	BSc	7–12	5
E6	Female	20	MEd	7–9	4
E7	Male	8	BSc	7–11	3
E8	Male	4	MEd	12	4

7 Results

7.1 Type(s) of Assessment that Educators Use in Online Education

Even though the educators in this study do not define the type or types of assessment they used (diagnostic, formative, or summative), it seems that they were mostly concerned with summative assessment as in their interviews most of them do not talk about feedback, learning enhancement and adjustment of teaching to the needs of their students. On the contrary, five out of eight educators talk about testing students, requesting that students show their work, and homework without the provision of feedback, all of which point to an effort to assess learning but not assess for learning. It is also interesting to note that there is complete lack of diagnostic assessment, an element that suggests educators plan their lesson, carry it out, and, consequently, assess what they taught without assessing their students' needs.

7.2 Techniques Used by Educators to Assess Student Learning Outcomes in Online Education

Educators tried using various techniques in order to assess the expected learning outcomes in online education. As such techniques educators mentioned assigning homework with or without feedback given, oral testing during the online lesson, participation in the lesson, quizzes that could be conducted online, showing the work a student completed during the lesson, student participation through the private chat. Table 2 presents the overall results.

Table 2. Ways to assess student learning outcomes in online education.

No	Technique 1	Technique 2	Technique 3	Technique 4
E1	Assigning homework	Oral testing during online lesson		
E2	Lesson participation	Assigning homework	Oral testing during online lesson	
E3	Quiz during the lesson			
E4	Homework assignment – feedback given	Homework assignment – no feedback given; solutions presented in next lesson	Oral testing during online lesson	Demonstration of work completed during the lesson
E5	Demonstration of work completed during the lesson	Participation via private chat	Homework assignment – feedback given	
E6	Lesson participation	Creative projects		
E7	Oral testing during online lesson	Homework assignment – feedback given		
E8	Lesson participation			

7.3 Digital Tools Used by Educators to Assess Student Learning Outcomes in Online Education

Online education presupposes the use of a platform, a system that will support the online educational process. The educators of this study used either WebEx or Microsoft Teams for their online lessons. Furthermore, for the needs of asynchronous education, they used either Microsoft Teams or Moodle. However, it seems that not all eight educators in this research study used a platform for asynchronous education. Table 3 displays all platforms used by educators for the conducting of their lessons.

Despite the existence of the online learning management system and the synchronous communication platform, some of the digital tools exploited were the private chat, the homework assignment and the forms through which educators could create online quizzes. One educator mentioned an additional tool (Kahoot) which according to what s/he said "helps [me] understand how much kids in class have understood…".

Table 3. Learning management system and communication platform.

No	LMS	Communication platform
E1	Microsoft Teams (for assignments)	Microsoft Teams
E2	Moodle (for assignments)	WebEx
E3	Microsoft Teams (Microsoft Forms)	Microsoft Teams
E4	–	WebEx
E5	Microsoft Teams	Microsoft Teams
E6	–	WebEx
E7	–	WebEx
E8	Microsoft Teams	Microsoft Teams

Therefore, what is observed is that the tools that were employed were connected to assessment methods teachers used in the physical learning environment too. As one educator characteristically noted, "assessment is done in the same way here [online] too."

7.4 Assessment, Online Teaching and How it Affects Student Learning

There is no empirical data in the present research study to support the idea that a certain technique or tool of assessment may affect students' performance or learning outcomes. What educators in this study pointed out is that the digital means of communication and teaching has not only differentiated their teaching, but also this alternative student-teacher interaction may have different results for different types of students.

E8 states, "there are some students who were more hesitant in the classroom, and now, during the online teaching, they seem to feel a little more comfortable. This [one-on-one interaction in online teaching] seems to be helpful for them." E7 states that "very good students [the 'above average' ones] can follow the online learning process while the 'below average' students find it more difficult to keep up in online than they do in face-to-face teaching."

Not all educators agree on that. E6 supports that the learning outcomes are linked to the commitment and self-motivation of each student rather than the digital tool(s) used in learning. "Those who want to learn, learn in every way. I feel that… there are many categories. Students who lack motivation, do not learn in any way. There are students who are highly motivated and learn in whatever environment. And there are those students who are easily affected by the situation they are placed in - they cannot stay concentrated in the physical classroom while they tend to learn more easily on WebEx. On the other hand, there are students who perform better in the classroom and cannot concentrate on the screen for 7 h, so they become more passive."

Another aspect that has emerged from the interviews is the role of parents and of the environment/situation at home. During the pandemic crisis, parents also worked remotely, and, in many cases, the house was not a calm and supportive environment for students. "Learning also depends on the conditions in which the learner finds himself/herself, i.e. the conditions are not easy for all students at home," states E2. However, E3 finds a potential partner in parents. "The difficulty is that you do not know if the student is doing the homework/test alone or if he/she is getting help. This 'help' may come from a classmate or a private tutor; in both cases, assessment won't provide reliable results. This is something you cannot easily control. Perhaps, collaborating with parents - which is not always effective nor possible – could be a solution."

Educators also comment on the difficulties in assessing student learning in an online environment. E1 states that "there is great difficulty in choosing the most accurate assessment possible" while all educators point out that they cannot be sure if a student has completed the activity or/and test by himself/herself or with the help of others. In other words, the extent to which each assessment method is valid and reliable for measuring student learning outcomes remains a query. Consequently, they suggest close-ended or multiple-choice questions with a time limitation and activities which can be combined for an end-project.

8 Discussion

Educational assessment is an important factor for learning and teaching in any educational context. It can be used as an indicator of student achievement, of skills and knowledge development, and to assist in the learning process design. Assessment for learning is preferred to assessment of learning, yet most of the educators tend to use assessment methods in order to check student performance and report achievements. While in the literature review the collective use of the three types of assessment (diagnostic, formative and summative) is highly recommended, in the present study educators have employed the summative type in order to monitor and grade students, focusing either on 'knowledge control,' where the learning outcomes are seen as recollection of factual knowledge or skills [6, 9], or on 'assessment as learning,' where the learning outcomes are seen as a competence-based curriculum which fosters the development of competence and capability [8, 9].

The use of diagnostic assessment was not reported at all while attempts at formative assessment appeared weak in results and without proper design. Due to either lack of preparedness or lack of knowledge, educators did not use the effective results from a combination of several adequate assessment techniques but preferred to focus on the given curriculum so as to catch up with the Ministry goals. Subsequently, the types of assessment used were focused on controlling and grading the knowledge and skills acquired.

In addition, there is no evidence of the "hidden curriculum" [12] nor of students' active role or reflection upon the learning process since online assessment was not a key objective of educators' learning design. The fact that everyone expected this remote teaching period to be short in addition to the increasing need for student participation and engagement, aligns with educators' concern about the reliability and accuracy of online assessment. It should also be mentioned that many students faced several difficulties during the pandemic and it was the educators who had to deal with these problems, as well.

However, both neuroscience and online learning principles indicate the importance of prompt and explicit feedback as part of the assessment procedure [4, 13, 14]. Although it is highlighted as a key factor in deep learning and students' active role in the learning process [15, 16], none of the participants mentioned using or incorporating it in their online teaching design. Educators faced several difficulties during the abrupt transition to emergency remote teaching and, consequently, online assessment for learning was not a priority. Furthermore, it is a fact that educators kept their role as teachers and evaluators of knowledge and did not develop other roles needed in online learning, such as facilitator, mentor, coach [5].

All the above lead to the conclusion that, although students were asked to actively participate in synchronous online teaching, they were not, indeed, as active during the asynchronous and synchronous learning as expected. Educational neuroscience sheds light on the ways and techniques available for this purpose [23]. Students' active and deep learning, according to neuroscience, may be increased by the involvement of students in the educational and learning process, by the combined use of diagnostic, formative, and summative assessment, by overcoming the teaching obstacles encountered, by integrating approaches that enhance opportunities to engage students with multiple representations, and by integrating more creative and collaborative activities in an appropriately designed learning environment.

Another interesting finding is related to a debate that is mentioned in the literature: the differences –if any– between assessment in the physical and the virtual learning environment. As mentioned, there are theorists who underpin the need for differentiated assessment techniques in an online context due to 'new' roles, relations, and practices among educators and students [5, 17]. Others suggest that the basic principles of assessment do not change in an online context [2, 5] and any type of assessment used is expected to affect learning outcomes. Both sides were expressed in the present research study by the participants.

Another issue that was not pointed out in the interviews is the frequency of assessment. Frequent short activities, quizzes and tests presented as part of a gamification process could lead to increased student motivation and high engagement in the learning process. In the same vein, collaborative activities such as peer assessment of group projects could lead to deeper learning and high cognitive and soft skills development.

Finally, another interesting finding is that although educators and students had access to digital platforms and tools, still they did not take full advantage of them. Out of a quite large variety of assessment tools available, educators limited themselves to a very short number of tools and platforms, as reported earlier. One possible explanation relates the limited time they had at their disposal for educational design with the lack of knowledge of the available tools. As a result, they tried to incorporate conventional assessment methods and tools to online teaching.

9 Conclusions

Educators and teachers at all educational levels experienced a new reality when they were forced to abruptly transition to online education, which they did not know nor had been trained for. At that period when emergency remote teaching emerged due to the SARS-Cov-2 pandemic, conventional assessment methods (written short or hourly exams, oral assessment, homework, etc.) were difficult to implement in this new structure. However, the assessment techniques and methods used were not adapted to this new situation and learning context. Although relevant research findings demonstrate not only the value, but also forms, ways, and techniques adequate to online teaching and assessment, educators were either uninformed or reluctant to incorporate them in their educational plan.

The present qualitative research study focused on the ways and tools used during the pandemic crisis in Greece and attempted to connect learning outcomes with the learning assessment used online. What is important to mention is that, despite the fact that educational assessment–in the physical or virtual learning environment–is an important aspect of learning and teaching, during the emergency remote teaching period it was undervalued and neglected.

It seems that there is a need for effective and appropriate use of assessment techniques in order to improve students' learning process. The findings and principles from educational neuroscience support the idea of incorporating assessment in the learning and teaching design, both in the physical and the virtual learning environment.

Acknowledgments. This research is funded by the European Union and Greece (Partnership Agreement for the Development Framework 2014–2020) under the Regional Operational Programme Ionian Islands 2014–2020, project title: "Enhancing cognitive abilities of people with Mild Cognitive Impairment through measurable cognitive training—NEUROEDUCATION", project number: 5016113.

References

1. Hodges, C., Moore, S., Lockee B., Trust, T., Bond A.: The Difference Between Emergency Remote Teaching and Online Learning (2020). https://er.educause.edu/articles/2020/3/the-difference-between-emergency-remote-teaching-and-online-learning. Accessed 09 July 2021
2. Benson, A.D.: Assessing participant learning in online environments. New Dir. Adult Contin. Educ. **100**, 69–78 (2003)
3. Knight, P.: The local practices of assessment. Assess. Eval. High. Educ. **31**(4), 435–452 (2006)
4. Black, P., William, D.: Inside the black box: Raising standards through classroom assessment. Phi Delta Kappan **80**(2), 139–148 (1998)
5. Vonderwell, S., Liang, X., Alderman, K.: Asynchronous discussions and assessment in online learning. J. Res. Technol. Educ. **39**(3), 309–328 (2007)
6. Boston, C.: The concept of formative assessment. Pract. Assess. Res. Eval. **8**(9) (2002). https://scholarworks.umass.edu/pare/vol8/iss1/9/. Accessed 09 July 2021

7. Weurlander, M., Söderberg, M., Scheja, M., Hult, H., Wernerson, A.: Exploring formative assessment as a tool for learning: students' experiences of different methods of formative assessment. Assess. Eval. High. Educ. **37**(6), 747–760 (2012)
8. Samuelowicz, K., Bain, J.D.: Identifying academics' orientation to assessment practice. High. Educ. **43**, 173–201 (2002)
9. Carraccio, C., Wolfsthal, S.D., Englander, R., Ferentz, K., Martin, C.: Shifting paradigms: from Flexner to competencies. Acad. Med. **77**(5), 361–367 (2002)
10. Joughin, G.: The hidden curriculum revisited: A critical review of research into the influence of summative assessment on learning. Assess. Eval. High. Educ. **35**(3), 335–345 (2010)
11. Gibbs, G., Simpson, C.: Conditions under which assessment supports students' learning. Learn. Teach. High. Educ. **1**, 3–31 (2004–2005). http://www2.glos.ac.uk/offload/tli/lets/lathe/issue1/articles/simpson.pdf. Accessed 10 July 2021
12. Kearns, L.R.: Student assessment in online learning: challenges and effective practices. MERLOT J. Online Learn. Teach. **8**(3), 198–208 (2012)
13. Nicol, D.J., Macfarlane-Dick, D.: Formative assessment and self-regulated learning: a model and seven principles of good feedback practice. Stud. High. Educ. **31**(2), 199–218 (2006)
14. Sadler, D.R.: Formative assessment: revisiting the territory. Assess. Educ. **5**(1), 77–84 (1998)
15. Scouller, K.: The influence of assessment method on students' learning approaches: multiple choice question examination versus assignment essay. High. Educ. **35**(4), 453–472 (1998)
16. Gibbs, G.: Using assessment strategically to change the way students learn. In: Glasner, A., Brown, S. (ed.) Assessment Matters in Higher Education: Choosing and Using Diverse Approaches, pp. 41–53. SRHE & Open University Press, Buckingham (1999)
17. Reeves, T.C.: Alternative approaches for online learning environments in higher education. J. Educ. Comput. Res. **23**(1), 101–111 (2000)
18. Angelo, T.A.: Beginning the dialogue on promoting critical thinking: classroom assessment for critical thinking. Teach. Psychol. **22**(1), 6–7 (1995)
19. Arend, B.: Course assessment practices and student learning strategies in online courses. J. Asynchronous Learn. Netw. **11**(4), 3–13 (2007). http://www.sloanconsortium.org/sites/default/files/v11n4_arend_0.pdf. Accessed 10 July 2021
20. Gaytan, J., McEwen, B.C.: Effective online instructional and assessment strategies. Am. J. Distance Educ. **21**(3), 117–132 (2007)
21. Swan, K.: Virtual interaction: design factors affecting student satisfaction and perceived learning in asynchronous online courses. Distance Educ. **22**(2), 306–331 (2001)
22. Ansari, D., Coch, D., De Smedt, B.: Connecting education and cognitive neuroscience: where will the journey take us? Educ. Philos. Theory **43**(1), 37–42 (2011)
23. Doukakis, S., Alexopoulos, E.: The role of educational neuroscience in distance learning. Knowledge transformation opportunities. In: Auer, M.E., Centea, D. (eds.) ICBL 2020. AISC, vol. 1314, pp. 159–168. Springer, Cham (2021). https://doi.org/10.1007/978-3-030-67209-6_18
24. Hwang, G.-J., Chang, H.-F.: A formative assessment-based mobile learning approach to improving the learning attitudes and achievements of students. Comput. Educ. **56**, 1023–1031 (2011)
25. Moser, J.S., Schroder, H.S., Heeter, C., Moran, T.P., Lee, Y.H.: Mind your errors: evidence for a neural mechanism linking growth mind-set to adaptive posterror adjustments. Psychol. Sci. **22**(12), 1484–1489 (2011)

Cloud Technologies Application at English Language Studying and Self-control Realization for Maritime Branch Specialists

Vladlen Shapo[1(✉)], Oleksandr Shcheptsov[1], Serhii Kaznadiei[1], Yevhen Norokha[1], and Valeriy Volovshchykov[2]

[1] Institute of Naval Forces of the National University "Odessa Maritime Academy", Didrikhsona str. 8, Odessa 65029, Ukraine
vladlen.shapo@gmail.com
[2] National Technical University "Kharkiv Polytechnic Institute", Kyrpychova str. 2, Kharkiv 61002, Ukraine

Abstract. Last decade true jump in complex technical systems developing is occurred. In maritime field technologies based on Industry 4.0, IoT, IIoT concepts are implementing actively. Thanks to progress in wireless and satellite technologies it's possible to solve totally new tasks like fully unmanned ships creation. Majority of seafarers communicate with colleagues in the sea mostly in English language and have to improve its level nonstop. Without satisfactory level of English language knowledge it's impossible to get any job in maritime branch. Modern development directions also require improving of IT terms knowledge. By IMO statistics about 80% of accidents in the sea happen because of human errors. Some of them are caused by misunderstanding between people, including language problems. In this paper it's described LMS MOODLE implementation for automation of some formal aspects at English language studying without teacher's help, and realization of self-control. Proposed approach allows to use free base software environment and created additional software tool for flexible formation of studying material and self-control realization using possibility of tests creation.

Keywords: Industry 4.0 · IoT · IIoT · English language studying automation and self-control · Seafarers

1 Aim

It is absolutely evidently that English language dominates in the modern world as the mean of communications in business, culture, education, medicine, industry and so on or just in friendly relations. Such branch like transport in general and maritime transport in particular are also fully depends on knowledge of English because international ships' crews typically communicate exclusively in English.

That's why in National University "Odessa Maritime Academy" significant attention is devoted to English language learning. Cadets and students study English during all 4 years (8 semesters) at obtaining a bachelor degree or 11 semesters at

M. E. Auer and T. Tsiatsos (Eds.): IMCL 2021, LNNS 411, pp. 83–92, 2022.
https://doi.org/10.1007/978-3-030-96296-8_8

obtaining of the Master degree. Till summer 2019 there were 3 departments which worked with 3 technical specialties, "Navigation" specialty and "Management" and "Maritime law" specialties respectively. Only from September 2019 last 2 departments were integrated, but in the same time new project on all academic disciplines teaching in English for 3 academic groups (approximately 65–90 students, 55–75 of which are Ukrainian citizens (Ukrainian/Russian speaking, which got highest scores in the school) and 10–15 are citizens of Egypt, Ecuador, Turkey, India, etc., for which English is native or quite habitual). English is very necessary for navy cadets from the Institute of Naval Forces at communications with colleagues from NATO.

Also evidently that English language is the main in such fast developing field like information technology in general including the following.

1. Projects on programming, when programmers' teams have to communicate with customers around the world.
2. A lot of tasks connected with Internet and any other data transfer networks including theirs application in industrial environment.
3. Technologies connected with modern concepts like Life Long Learning (LLL), Industry 4.0, Internet of Things (IoT), Industrial Internet of Things (IIoT),
4. Cyber security aspects.

There are some online platforms which help to study English [1–4] but in the same time these platforms have another focus and have no flexibility.

But anybody who studies English (or any another) language, must work a lot of time himself, to try to automate some skills like conjugation of verbs and to widen own vocabulary. One of ways, which allows to realize this goal, is using of additional software. In any case trainee, domesticated some material, at some moment will need to change the content of mentioned software. The most flexible way is to use such own application software which will allow to change its content easily and quickly.

One of suitable program environments is famous learning management system (LMS) Modular Object Oriented Dynamic Learning Environment (MOODLE). Main positive sides of this application software are: gratuity for any goal for everybody;

a lot of different integrated possibilities; supporting of dozens of different interface and content languages; multiplicity of different additional program modules, created by thousands of independent developers; great experience in technical administrating and support; a lot of experienced professionals; popularity in academic environment (thousands of universities use MOODLE as LMS system); using of different platforms (operating systems) like Windows or Linux families; possibility of using locally (only on one personal or tablet computer, laptop, smartphone without Internet connection), in local area network (also without Internet connection) or using cloud approach with establishing of Internet connection when MOODLE and all necessary content are placed on remote server; regular new releases, etc. Some possibilities of MOODLE, deeply explored in practice, are analyzed in paper [5]. So the main goal of research work is the following: flexible and free of charge application of LMS MOODLE for English language studying and permanent self-control.

2 Findings

But one of the very useful MOODLE features is possibility to create different types of tests. Tests allow evaluating some knowledge of students or trainees, but also allow to automate the process of memorization (and sometimes cramming up) of some dates, facts, etc. It doesn't allow to check creativity of student or trainee but offers the opportunity of self-verification. In the case of foreign language studying such approach is very useful and convenient for cramming up of foreign words at any time, including forced waiting time in public transport, airports, custom service, etc.

Tests which were created once using any version of MOODLE and any operating system, may be used many times repeatedly after exporting on initial computer system and importing on any other computer system. MOODLE allows to export prepared test questions to the files (Fig. 1) using GIFT format, MOODLE XML format and XHTML format. These files will obtain txt, xml and html extensions respectively. GIFT format and XHTML format are supported by the variety of application software.

The GIFT (General Import Format Template) format is a "wiki-like" markup language for describing tests, created by Paul Shew in 2003. GIFT allows someone to use a text editor to create multiple-choice, true-false, short answer, matching, missing word and numerical questions. In this paper only "matching" questions creation is analyzed but "missing word" questions are also very actual because allow to check correctness of writing additionally to words' cramming up. This format is most convenient for automation of tests creating and language studying, because it's the simplest in comparison with other formats which are supported by MOODLE LMS. Quite complex structure of XML family formats is shown in Fig. 2.

Fig. 1. Appearance of test questions export window

After opening of the file with xml extension in any browser the window, presented in Fig. 2 will appear.

```
  <?xml version="1.0" encoding="UTF-8"?>
- <quiz>
        <!-- question: 0 -->
  - <question type="category">
    - <category>
          <text>$course$/top/За замовчанням для ИТ-ЕМ1_1</text>
      </category>
    - <info format="moodle_auto_format">
      - <text>
            <![CDATA[Категорія за замовчуванням для загальних питань у контексті "ИТ-ЕМ1_1".]]>
        </text>
      </info>
    </question>
        <!-- question: 1 -->
  - <question type="matching">
    - <name>
          <text>Q1</text>
      </name>
    - <questiontext format="html">
      - <text>
            <![CDATA[<p>Q1-text<br></p>]]>
        </text>
      </questiontext>
    - <generalfeedback format="html">
          <text/>
      </generalfeedback>
      <defaultgrade>1.0000000</defaultgrade>
      <penalty>0.3333333</penalty>
      <hidden>0</hidden>
      <idnumber/>
      <shuffleanswers>true</shuffleanswers>
    - <correctfeedback format="html">
          <text>Ваша відповідь правильна.</text>
      </correctfeedback>
    - <partiallycorrectfeedback format="html">
          <text>Ваша відповідь частково правильна.</text>
      </partiallycorrectfeedback>
    - <incorrectfeedback format="html">
          <text>Ваша відповідь хибна.</text>
      </incorrectfeedback>
      <shownumcorrect/>
    - <subquestion format="html">
      - <text>
            <![CDATA[<p>table<br></p>]]>
        </text>
      - <answer>
            <text>стол</text>
        </answer>
      </subquestion>
    - <subquestion format="html">
      - <text>
            <![CDATA[<p>room<br></p>]]>
```


Fig. 2. Structure of xml file containing exported from LMS MOODLE test questions

Example of two matching questions Q1 and Q2 in GIFT format is presented below.

```
// question: 0  name: Switch category to $course$/top/By default for IT-EM1_1
$CATEGORY: $course$/top/ By default for IT-EM1_1
// question: 1  name: Q1
::Q1::[html]<p>Q1-text<br></p>{
   =<p>table<br></p> -> стол
   =<p>room<br></p> -> комната
   =<p>car<br></p> -> автомобиль
   =<p>mouse<br></p> -> мышь
   =<p>list<br></p> -> список
   =<p>sun<br></p> -> солнце
}
// question: 2  name: Q2
::Q2::[html]<p>Q2-text<br></p>{
   =<p>vessel<br></p> -> судно
   =<p>engine<br></p> -> двигатель
   =<p>power<br></p> -> мощность
   =<p>spoon<br></p> -> ложка
   =<p>fork<br></p> -> вилка
   =<p>cup<br></p> -> чашка
}
```

After opening of the file with html extension (MOODLE Quiz XHTML Export) in any browser will appear the window, presented in Fig. 3. This format allows to check the appearance and correctness of the test very demonstrably.

After even brief analysis of three obtained files structures it's becoming apparent that GIFT format is the most clear and most convenient for automated generation by means of user written programming tool. Both XHTML and XML formats are very complex and redundant. Results of comparison are shown in Table 1. The smallest file (txt extension, GIFT format) is chosen as base measure. Ratio value is obtained by division of the corresponding format's file size (numerator) by GIFT format file size (denominator) multiplied by 100%. Redundancy value is obtained by subtraction of GIFT format ratio from the corresponding file format ratio.

So, the main conclusion of performed analysis is the following: GIFT format application at tests' exporting and further importing is the most suitable from all points of view. GIFT format is also available in the list of possible formats at importing test questions from the external file in LMS MOODLE (Fig. 4).

Possibilities of questions' export must be analyzed at the first stage, because after creation of trial test and its exporting it's become possible to analyze the structure of created by LMS MOODLE file and to automate formation of necessary structure by means of any high level programming language.

Next stage is the brief analysis of the formats supported by LMS MOODLE at test questions importing. Corresponding file formats presented on Fig. 4.

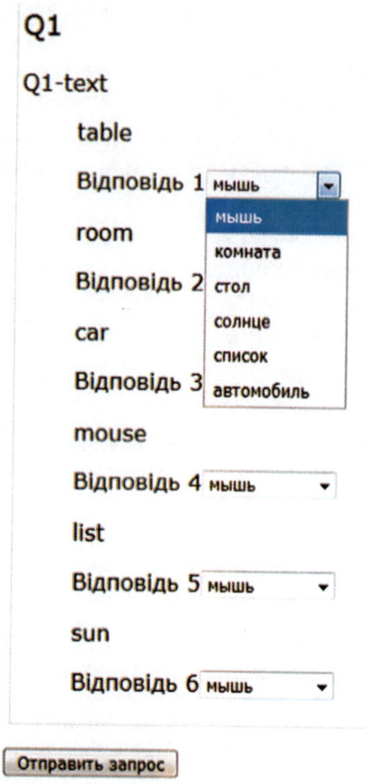

Fig. 3. Appearance of exported xml file, opened in the browser

Table 1. Comparison of the files format at test questions exporting in LMS MOODLE

File format/extension	File size, bytes	Ratio, %	Redundancy, %
XML/XML	2311	522	422
XHTML/HTML	4483	1012	912
GIFT/TXT	443	100	0

AIKEN (not analyzed above but it is presented on Fig. 4 and may be used) is a simple format which supports multiple choice questions with only one correct answer. Each question must be written in the separate line. Each answer must start with a single uppercase letter, followed by a period "." or a right parenthesis ")", then a space. The last line, containing correct answer, must be started with "ANSWER: " string followed by the correct answer's starting letter. The Aiken format lets to developer to create multiple choice or True/False questions in easy-to-read format that must be saved as a plain text file (in other words, ordinary text file) for further importing into the LMS MOODLE or another processing.

The word "ANSWER" and the answer letters (A, B, C and so on) must be capitalized as shown, otherwise the import will fail.

Fig. 4. The list of possible formats at importing test questions from the external file

Prepared file must be saved as plain text file with TXT extension in UTF-8 format (Unicode Transformation Format, which uses 8 bits, allows to save 256 symbols and is dominating in the Internet and at the any text saving in general). Examples of obtained text Aiken format file is shown below.

Which number completes the series 1, 2, 4, 8, 16…?

A. 132
B. 256
C. 32
D. 64
ANSWER: C

The Earth is bigger than the Moon?

A) True
B) False
ANSWER: A

But this format doesn't allow creation of matching or missing word tests and is inappropriate in at the solving of mentioned task.

At specific task solution Blackboard, Examview and WebCT formats (Fig. 4) are also not suitable because corresponding files must be firstly created in the corresponding application software, which must initially (or earlier) be bought, installed and used with some experience obtaining, question bank must be created and exported. These operations require additional time, forces and expenses.

MOODLE XML format is too complex and redundant for the current task which must be solved, as it was shown earlier.

Embedded answers (cloze) format is a multiline text, some phrases or sentences consisting of a portion of language with some items, words or signs removed beforehand where the trainee is asked to type the missing text. In general cloze test requires the ability to understand context and vocabulary in order to identify the correct language or part of speech that belongs in the deleted passages. This exercise is commonly administered for the assessment of native and second language learning and instruction. In other words, this is a test for diagnosing reading ability; words are deleted from a text or phrase, and the trainee has to fill in the blank places. This format is also very actual because allow to check correctness of writing additionally to words' cramming up but not analyzed in this paper.

MOODLE LMS may be installed on own computer and may be used without any internet connection and external administrating.

User must create ordinary text file (in style shown below) using text editors like Notepad or to save the file, originally created in doc or docx format in the style shown in Table 2, using plain text style. Procedure of saving doc/docx file in plain text format is shown in Fig. 5.

Table 2. Appearance of the file with English and translated words

Table	стол
Room	комната
Car	автомобиль
Mouse	мышь
List	список
Sun	солнце

Content of obtained file is shown below.

table
стол
room
комната
car
автомобиль
mouse
мышь
list
список
sun
солнце

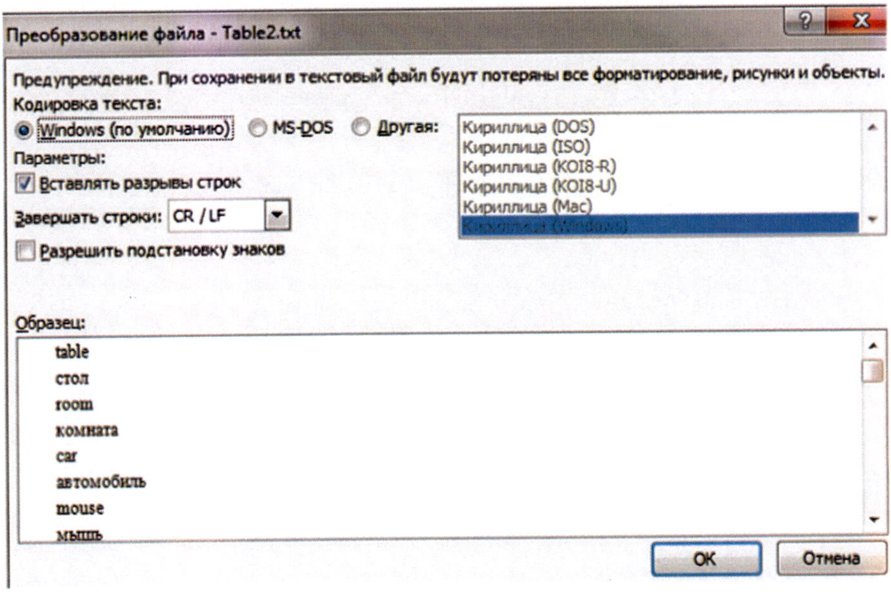

Fig. 5. Appearance of the window at file exporting from Microsoft word into plain text format

Such plain text file must be converted to the GIFT format using the proposed algorithm, which is described in written enumerated list below, and corresponding software tool created in any high level programming language (Visual Basic is implemented in described task).

In program interface which is created or is being created in the corresponding application software it's necessary to provide a certain field where user must define how many words will be included to the test which must be created (any integer value, but not bigger than number of lines in external file, divided by 2; this number must be even because 1 studied word occupies 2 lines in the file).

1. To open source plain text file containing the English words in odd lines, and translated English words in the even lines.

2. To check how many lines are present in opened initial file, to inform user about this value and to recommend to user how many words may be included in each question in created test and how many questions it's possible to create.

3. To open the final file and to write there the template similar to shown below.

// question: 0 name: Switch category to $course$/top/By default for IT-EM1_1.

$CATEGORY: $course$/top/ By default for IT-EM1_1.

4. In the loop with known number of repetitions (this value was calculated in item 2 depending on data or defined beforehand) to write the following template.

// question: N name: QN.

::QN::[html]<p>QN-text
</p>

where N is the number of the question in the test.

In the same loop to add to the formed final text file the following template:

```
{
  =<p>English_word_1<br></p> -> Translated_word_1
  =<p>English_word_2<br></p> -> Translated_word_2
…
  =<p> English_word_K<br></p> -> Translated_word_K
}
```

where K is the quantity of initial English and translated word.

The loop must be repeated necessary number of times.

5. To close initial and obtained files.
6. To import obtained file into the LMS MOODLE.

The algorithm described in paragraphs 1–6, automates the test creation procedure and adjusts the words which are being studied at foreign language studying very flexibly. Using of any computer platform is possible.

3 Conclusion

Approach devoted to automation of some aspects at English language studying and self-control realization is described in this paper. It allows to trainee to choose necessary material much more flexible in comparison with existing platforms and to adjust studying process for own need. Tests creation possibilities of popular free LMS MOODLE are used for reaching this goal. Algorithm and application software tool are created and described. Mentioned approach doesn't require any financing and may be realized in cloud version in the Internet, in local computer network or in any standalone computer system.

References

1. Marlins Test Platform [Electronic resource]: https://www.marlinstests.com/. Accessed 18 Feb 2021
2. SeaTALK Online Platform [Electronic resource]: http://www.seatalk.pro/. Accessed 18 Feb 2021
3. MarTEL Plus Maritime Tests of English Language [Electronic resource]: http://www.plus.martel.pro/. Accessed 18 Feb 2021
4. Maritime Training [Electronic resource]: http://www.maritimetraining.pro/index.php?option=com_content&view=article&id=114&Itemid=540. Accessed 18 Feb 2021
5. Shapo, V.: Distance learning system application for maritime specialists preparing and corresponding challenges analyzing. In: Auer, M.E., Zutin, D.G. (eds.) Online Engineering & Internet of Things. LNNS, vol. 22, pp. 1050–1057. Springer, Cham (2018). https://doi.org/10.1007/978-3-319-64352-6_97

Work-in-Progress: Development of VCDLN Model as Implementation of Distance Learning in the Era of the Covid-19 Pandemic in Indonesia

Deni Darmawan[1(✉)], Dinn Wahyudin[2], Dian Rahadian[3], and Andri Suryadi[4]

[1] Educational Technology Department, Universitas Pendidikan Indonesia, Bandung 40154, Indonesia
deni_darmawan@upi.edu
[2] Curriculum Development, Universitas Pendidikan Indonesia, Bandung 40154, Indonesia
dinn_wahyudin@upi.edu
[3] Informatics Science Department, Institut Pendidikan Indonesia, Indonesia
[4] Information System Department, Faculty Science and Technology, Universitas Terbuka, Jakarta, Indonesia
andri.suryadi@ecampus.ut.ac.id

Abstract. The development of the VCDLN (Virtual Community Digital Learning Nusantara) system in the era of the Covid-19 pandemic which is very much needed by educators in remote villages. The objectives of this research include: (1) Development of the VCDLN system mapped in the content developer network. Subjects to serve to learn at Elementary school, Junior High School, and Senior High School in remote areas: (2) Building communities for developing Distance Learning systems in the form of e-learning systems, mobile learning systems, and Blended learning systems; (3) Building Central Community Digital Learning by Community classification; (4) Evaluation Activity Learning from students Through VCDLN as PJJ in Pandemic Era. In this research, the research team will use a Mix Method (Qualitative and Quantitative). These three findings have been implemented through the support of television technology from TVUPI. This VCDLN Learning Model can be an alternative in Distance Learning Services in the Pandemic COVID-19 era throughout Indonesia.

Keywords: VCDLN · e-learning · m-learning · Distance learning · TVUPI

1 Introduction

Starting from a study of the pandemic phenomenon, difficulties in providing services and direct learning interactions. One day it may be recorded and it is certain that the development of the world of education has experienced a nearly 360-degree shift in the interaction system created between educators and students [1]. Of course, this is happening in worldwide, as reported by world research studies such as Finland,

M. E. Auer and T. Tsiatsos (Eds.): IMCL 2021, LNNS 411, pp. 93–101, 2022.
https://doi.org/10.1007/978-3-030-96296-8_9

England, Germany, New York, Japan [2] that the learning system will become a rapidly growing system so as to form a global digital learning community.

As one of the strategic thoughts on what is formulated through the policies of the Ministry of Education and Culture. The policy demands the participation and innovation of online digital learning [3], as a form of new strength in the learning revolution in the era of "New Normal Education Practices". From this condition, the research team felt compelled to conduct research on the development of "Virtual Community Digital Learning Nusantara (VCDLN) in the Covid-19 Pandemic era".

In particular, this study wants to answer research problems regarding efforts to: (1) Development of the VCDLN system mapped in the content developer network. Subjects to serve learning at Elementary school, Junior High School and Senior High School in remote areas: (2) Building communities for developing Distance Learning systems in the form of e-learning systems, mobile learning system, and Blended learning system; (3) Building Central Community Digital Learning by by Community classification (4) Evaluation Activity Learning from students Through VCDLN as PJJ in Pandemic Era. The objectives of this research are expected to: (a) be able to accommodate learning innovations and learning communication strategies within the VCDLN framework; (b) The acquisition of virtual, digital, online and mobile networks in realizing distance learning: (c) The establishment of learning centers in the form of Mobile Digital Television and e-learning developed and utilized by the VCDLN community; (d) Obtaining an overview of the level of Effectiveness and Efficiency of Distance Learning Services in the form of a VCDLN model in remote parts of the archipelago.

2 Literature Review

As discussed in the introduction section above, a number of innovations emerged and were unconsciously able to grow a "Culture Education Practices". For example, the analysis of Google Trends Indonesia (2020) to date has recorded 34% of the consultation process in education services carried out online, educational practices that have occurred have reached 33% [4].

2.1 VCDLN Quality Control and Regulation

Regulations for VCDLN implementation opportunities can be adopted from 2003 to 2012, for example [5–8]; and [7] that Distance Education, is a teaching and learning process carried out remotely through the use of various communication media. On April 3, 2020, in the conditions of the Covid-19 Pandemic, it seemed as if the New Ministry of Education and Culture had been entrusted with a strategic legacy of the ultimate weapon in keeping the education and learning process going. One of the 5 policies of the Ministry of Education and Culture, including urging all basic education institutions to universities to be able to form partnerships in obtaining a number of platform supports for providing online and digital learning systems. If you examine the New Normal conditions as stated by [9], where it is regulated in Phase III it is stated that "Educational Activities in Schools are carried out with a Shift system according to the number of classes", although this will risk the spread of Covid-19, but at least that VCDLN will still be able to be carried out by implementing regulations regarding "Blended Learning"[10].

2.2 Element of VCDLN

Some important elements in the implementation of VCDLN, can actually be analyzed in retrospect regarding a number of terms and objects or target subjects that we often refer to and use in educational practice [11]. For example, the terms software, hardware, brain ware and also environmental ware [12]. Likewise, in the analysis, it is likely that a new concept or model will soon be put into practice in the paradigm of implementing system services and educational communication strategies [13] and learning that the author named VCDLN. Where from the results of an analysis of these elements there are: (1) Software; (2) Hardware; (3) Brain ware; (4) Virtual Community; (5) Social Media; (6) Vicon-tech; (7) Environmental ware; (8) E-learning; (9) Learning Sustainable; (10) Learning-Practices.

2.3 Roadmap of VCDLN Based on Television Program

As discussed in the study of Regulations on VCDLN in the section above that the current regulations and demands return to normal in new conditions, the implementation of this VCDLN must be harmonized with the New Regulations [10]. Where as a form of VCDLN implementation in the context of the realization of this new normal condition, Blended Learning will be possible. The researcher recalled from one study regarding the level of digital literacy or skills possessed by millennials or generation-Z, where they were able to become designers of digital learning information system pathways in a number of universities [14].

If analyzed from the policy of the Ministry of Research, Technology and Higher Education regarding the targets for the implementation of full online distance learning at that time it was confirmed that it would produce 80%, educational success that touches all corners of the archipelago. Thus, if there is only 20% left, then through Blended Learning Television as one of the VCDLN models in the learning corridor in schools during the "New Normal". The atmosphere that will be built during their meeting will be a forum for exchanging experiences or working together to design the preparation and development of VCDLN learning content when they carry out inter-community learning services [15]. Gradually the production of online teaching materials to realize the Virtual Community Digital Learning Nusantara can be facilitated through satellite TVUPI access for dissemination to remote areas of the country.

3 Research Method

In this research, the research team will use the Mix Method [16], used for conducting research every year (2001, 2022 and 2023). The qualitative stage is needed when the team builds the VCDLN Model in the form of Mobile Digital Television Broadcast as an adaptive e-learning system with Elementary, Middle, and Higher Education. Then

Qualitative is also used in the process of developing a number of learning content at all levels of education, the analysis using the Design and Development method. For the application of a quantitative approach using Stratified Random Sampling which uses 42 teachers and 140 students from 14 districts/cities. Statistical Analysis using Path Analysis Diagram [17] used to test the level of (1) the speed of delivery of learning materials by the teacher; (2) the influence of the e-learning system in the form of television broadcasts as a VCDLN model on the level of skills development of teaching materials by teachers; and (3) the effect of both simultaneously on the acceleration of changes in student learning outcomes.

4 Result and Discussion

4.1 Development of the VCDLN System Mapped in the Content Developer

4.1.1 Network Infrastructure

The VCDLN system was developed using a server-client network system where the server is a data repository that serves requests from clients, [18]. The VCDLN system is prepared to be accessible anytime and anywhere in accordance with the concept of VCDLN which emphasizes the role of the community as well as research from [19]. Look at the picture below:

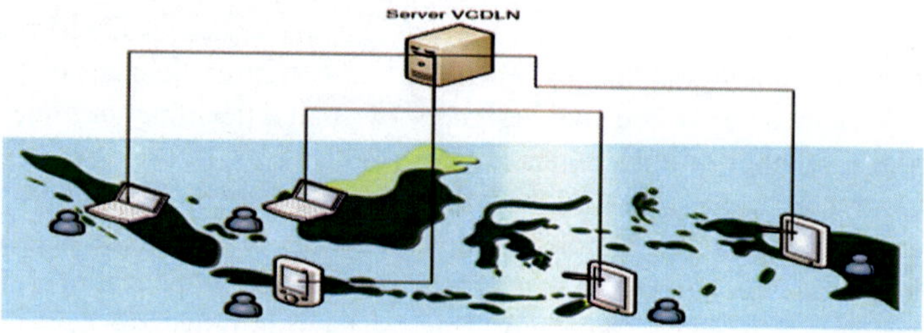

Fig. 1. VCDLN network mapping

In Fig. 1 the server is prepared to always be on standby to serve access from educators and students as users spread throughout Indonesia, this refers to the findings of [20]. The specifications in the development of network infrastructure are as follows (1) https://vcdlnlearning.com for VCDLN Domain; (2) Hosting Disk Usage 5120 MB for The disk capacity provided is 5120MB or about 5GB; (3) Unlimited Bandwidth for Bandwidth capacity provided is Unlimited; and (4) Access device for Access devices in the form of Laptops, Mobiles, Tablets, or others that can access web pages.

4.1.2 Application Infrastructure

In the application infrastructure, several things were developed to be able to support VCDLN. The recommended applications are general applications that are more user friendly for educators and students such as the Google browser that is accessed by mobile device. The following are the specifications of the application in the findings of this research, which are as follows:

Table 1. VCDLN application specifications

No	Specification	Information
1	Database Maria DB	Server version: 10.3.29-MariaDB-log-cll-lve – Maria DB Server
2	Browser	Chrome or similar
3	Moodle Learning System	Version 3

In Table 1 it can be seen that the database refers to research from [21] used is the type of Maria DB. Then a browser is also required to access the application. Browsers that can be used are chrome, mozilla, safari or the like, which support the application of innovations from MOOCs. The following is a display of the VCDLN application that has been successfully installed.

4.1.3 Human Infrastructure

Human infrastructure is a user who can perform activities on the VCDLN system. The activities contained in this infrastructure are described through a data flow diagram as follows:

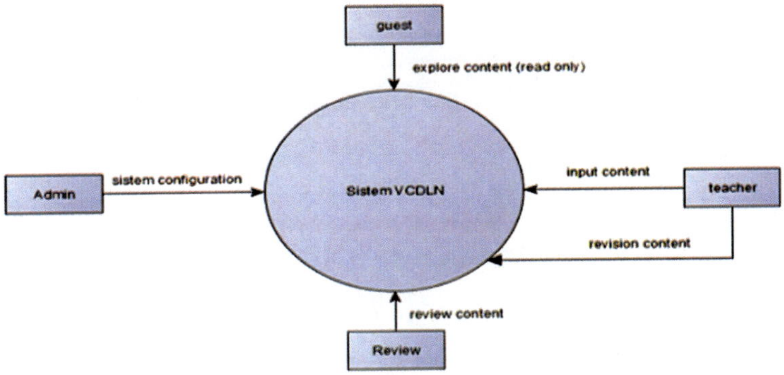

Fig. 2. User flow chart with system

In Fig. 2 there are four users involved in the system, namely admin, teacher, reviewer, and guest. Admin functions as the person who configures the system. The teacher functions as the person who enters the content and revises the content if the results of the review require the content to be revised. Review functions as the person in charge of reviewing content from the teacher and making comments on the content. The last one is guest, which is a user who is only allowed to browse published content from the VCDLN system.

4.2 Building Communities for Developing Distance Learning Systems in the Form of e-learning, Mobile Learning, and Blended Learning System

The formation of the VCDLN community consisting of teachers with their pedagogical literacy from the archipelago as the findings and products of this research as referring to research from [22]. As a result, it can be seen in the following chart, where a number of regions have been represented and have made virtual commitment.

In this development research, teachers who are members of the VCDLN Model community in the Implementation of PJJ in the Covid-19 Pandemic era as research conducted by [23]. These findings have produced interactive video-based learning materials. This video has a characteristic that is showing the teacher in the show on television.

4.3 Building a Central Community Digital Learning by Community

In addition to community members from the teacher element, in this study a learning organizer community was also formed through the television platform and other interactive platforms [11]. Community members consisting of the managers of the sub district office, Posyandu, Minimart, and Police Station offices, as well as parents of students, become the center of learning for students around them. This VCDLN community member may in the future become a model for future teachers. The representation of members of this community can be seen from the following graph (Fig. 3).

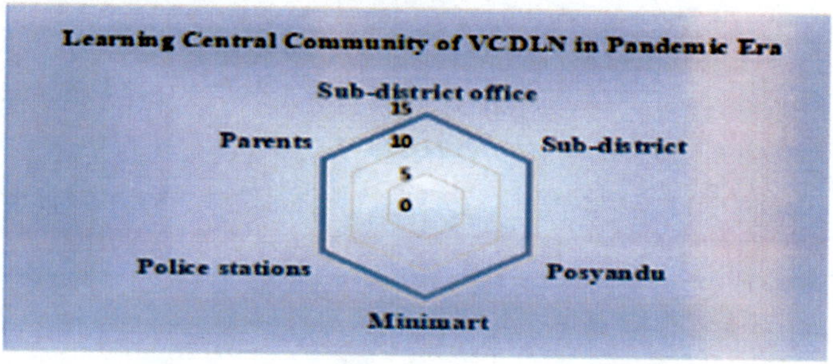

Fig. 3. Learning centers in the implementation of VCDLN in remote archipelago

Each of these learning centers is managed voluntarily to serve learning independently through VCDLN media in the form of learning tools [12]. The learning device is in the form of a tablet computer that is connected to a database of learning resources from the website http://vcdlnlearning.com. Students who are shopping, stop by the village hall or pass their Police station. During holidays there is usually a Posyandu event, so children aged early childhood education programs and Basic Education can receive learning services accompanied by their respective parents. These patterns will become a distance learning model in the pandemic era, if this pandemic condition continues, referring to the results of research from.

4.4 Evaluation Activity Learning from Students Through VCDLN

Measurement results using Path Analysis Diagram (1) The effect of the speed of delivery of distance learning materials by the teacher on the acceleration of changes in student learning outcomes; (2) the effect of the ability to develop teaching materials by teachers as e-learning in the form of VCDLN broadcasts through TVUPI on the speed of change in students' learning outcomes over long distances; and (3) the effect of both simultaneously on the acceleration of changes in student learning outcomes. These three findings refer to the results of the analysis of applied to the approaching case in Indonesia. The results can be illustrated in the following diagram (Fig. 4).

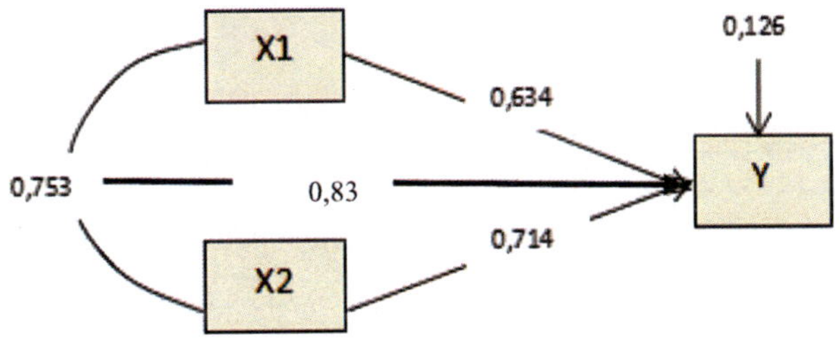

Fig. 4. Substructure of influence between research variables

From the calculation results show that the influence of the speed of delivery of learning materials by teachers on the acceleration of changes in student learning outcomes is 0.634 or 40.2% of its contribution, this finding contributes to research. As for the effect of the ability to develop teaching materials by teachers as e-learning in the form of VCDLN broadcasts through TVUPI on the speed of change in student learning outcomes by 0.714 or about 51% of their contribution. As for the two variables together, they have an effect of 0.832 or 69% on the speed of change in student learning outcomes.

5 Conclusion

From the results of research and discussion in this study, it can be concluded as follows; (a) the development of the VCDLN system is carried out in the form of 3 types of infrastructure as products which include network infrastructure, application infrastructure, and human digital competence, all three of which are the strengths of learning networking in the era of the covid-19 pandemic; (b) The formation of the VCDLN community from the element of educators has been mapped based on the category of district/city area with representation for the Western, Central and Eastern parts of Indonesia; and (c) Establishment of the VCDLN community classification as a Learning Center place during the Covid-19 pandemic. These three findings have been implemented through the support of television technology from TVUPI. This VCDLN Learning Model can be an alternative in Distance Learning Services in the Pandemic COVID-19 era throughout Indonesia.

Acknowledgment. This research was granted financial support from Ministry Education Culture and Research Technology of Republic Indonesia, Jakarta, 2021–2023.

References

1. Hallberg, S., Hirsto, L., Kaasinen, J.: Experiences and outcomes of craft skill learning with a 360° virtual learning environment and a head-mounted display. Heliyon **6**(8), e04705 (2020). https://doi.org/10.1016/j.heliyon.2020.e04705
2. Yuswohadi, D.: Online Home Schooling (2020). www.inventureknowledge.id
3. Scherer, R., Siddiq, F., Tondeur, J.: The technology acceptance model (TAM): a meta-analytic structural equation modeling approach to explaining teachers' adoption of digital technology in education. Comput. Educ. **128**, 13–35 (2019). https://doi.org/10.1016/j.compedu.2018.09.009
4. M.K. & Company. Online Education Course (2020). www.inventureknowledge.id
5. Ministry of Education and Culture: Ministry of Education and Culture, No. 24 of 2012 concerning Distance Education. Jakarta Ministry Education Culture (2012)
6. Ministry of Education and Culture: Minister of Education and Culture Regulation No. 109 of 2013 concerning PJJ. Jakarta Ministry Education Culture (2013)
7. Ministry of Education and Culture: Regulation of the Minister of Education and Culture Number 119 of 2014 concerning the Implementation of Distance Education at the Elementary and Secondary Education Levels. Jakarta Ministry Education Culture (2014)
8. Ministry of research and technology of higher education: Regulation of the Ministry of Research, Technology and Higher Education No. 50 of 2015 concerning Distance Learning and E-learning. Jakarta Ministry of Research Higher Education (2015)
9. President Jokowi: President's Statement May 15, 2020 regarding the New Normal. Jakarta Tempo.co world Health Organisation (2020)
10. Tempo.com. Preparing New Normal. Tempo.co World Health Organisation (2020)
11. Rodríguez-López, N.: Understanding value co-creation in virtual communities: the key role of complementarities and trade-offs. Inf. Manag. **58** (2021). https://doi.org/10.1016/j.im.2021.103487
12. Ahmady, S., Kohan, N., Bagherzadeh, R., Rakshhani, T., Shahabi, M.: Validity testing of classroom community scale in virtual environment learning: a cross sectional study. Ann. Med. Surg. **36**, 256–260 (2018). https://doi.org/10.1016/j.amsu.2018.08.021
13. Amponsah, E., Fusheini, A., Adam, A.: Influence of information, education and communication on prenatal and skilled delivery in the Tano North District, Ghana: a cross-sectional study. Heliyon **7**(6), e07245 (2021). https://doi.org/10.1016/j.heliyon.2021.e07245
14. Suryadi, E.: Post-election Digital Communication Studies Through the Official Website of Local Governments Based on the Perspective of the Millennial Generation. LPPM-UPI, Bandung (2019)
15. Strunga, A.: The integration of virtual learning communities into universities' knowledge management models. Procedia - Soc. Behav. Sci. **197**, 2430–2434 (2015). https://doi.org/10.1016/j.sbspro.2015.07.306
16. Creswell, J.W.: Research Design: Qualitative, Quantitative and Mixed Methods Approaches, 4th edn. CA Sage, Thousand Oaks (2014)
17. Stage, F.K., Carter, H.C., Nora, A.: Path analysis: an introduction and analysis of a decade of research. J. Educ. Res. **98**(1), 5–13 (2004). https://doi.org/10.3200/JOER.98.1.5-13

18. Models, C., Dubberly, H., Pangaro, P.: Introduction to Cybernetics and the Design of Systems Not for re-distribution, pp. 2001–2011, January 2010
19. Rimale, Z., El Habib, B., Tragha, A., El Guemmat, K.: Survey on the use of the mobile learning based on mobile cloud computing. Int. J. Interact. Mob. Technol. **10**(3), 35 (2016). https://doi.org/10.3991/ijim.v10i3.5672
20. Hussain, A., Mkpojiogu, E.O.C., Babalola, E.T.: Using mobile educational apps to foster work and play in learning: a systematic review. Int. J. Interact. Mob. Technol. **14**(18), 178–194 (2020). https://doi.org/10.3991/ijim.v14i18.16619
21. Prakancharoen, S.: Database secure manipulation based on paillier's homomorphic encryption (DSM-PHE). Int. J. Interact. Mob. Technol. **12**, 136–151 (2019). https://doi.org/10.3991/ijim.v13i12.11396
22. Strungă, A.: Using virtual learning communities in shaping the professional identity of primary and preschool pedagogy specialization students: a knowledge management approach. Procedia - Soc. Behav. Sci. **180**, 460–467 (2015). https://doi.org/10.1016/j.sbspro.2015.02.145
23. Tawafak, R.M., Malik, S.I., Alfarsi, G.: Impact of technologies during the COVID-19 pandemic for improving behavioral intention to use e-learning. Int. J. Inf. Commun. Technol. Educ. **17**(3), 137–150 (2021). https://doi.org/10.4018/IJICTE.20210701.oa9

Rethinking Audio-Haptic Perceptual Immersion from In-Person to Remote Testing During COVID-19

Guoxuan Ning[1,2], Quinn Daggett[1,2], Argyrios Perivolaris[3], Bill Kapralos[1,4,5](✉), Alvaro Quevedo[1,4], KC Collins[6], Kamen Kanev[7], and Adam Dubrowski[1,8]

[1] maxSIMhealth, Ontario Tech University, Oshawa, ON, Canada
bill.kapralos@ontariotechu.ca
[2] Faculty of Science, Ontario Tech University, Oshawa, ON, Canada
[3] Faculty of Neuroscience, University of Toronto, Toronto, ON, Canada
[4] Software and Informatics Research Centre, Ontario Tech University, Oshawa, ON, Canada
[5] Collaborative Human Immersive Interaction Lab (CHISIL), Sunnybrook Health Sciences Centre, Toronto, Canada
[6] School of Information Technology, Carleton University, Ottawa, ON, Canada
[7] Research Institute of Electronics, Shizuoka University, Hamamatsu, Japan
[8] Faculty of Health Sciences, Ontario Tech University, Oshawa, ON, Canada

Abstract. Auditory and haptic cues play an important role in medical simulation for developing cognitive and motor skills. For example, medical training whereby trainees practice drilling-related surgeries requires the use of force feedback haptic devices in conjunction with computer-based simulations that provide audiovisual cues. Traditionally, this practice requires trainees to be co-located and work in groups within a simulation laboratory. While virtual reality (VR) is providing opportunities for developing digital replicas that can be used without depending on access to the training site, VR equipment including haptic devices, and high fidelity head-mounted displays, are not widely available. Due to the need for specialized equipment and restrictions placed on in-person user testing due to the COVID-19 pandemic, here we describe an experiment that examined the simulation of a virtual drilling task conducted remotely. The experimental results are not discussed. Rather, we report our findings with respect to the various challenges faced when rethinking audio-haptic perceptual immersive user-based experiments during COVID-19. We also describe a future experiment that will be conducted to test the usability of a serious game and will also be conducted entirely online.

Keywords: Serious games · Virtual simulation · Assessment · Usability · COVID-19 · Human participants

1 Introduction

Examining multimodal interactions within virtual simulations and serious games for medical education with an emphasis on the effect of sound on visual and haptic fidelity perception remains an active area of research [1]. Simulation technology, and in

M. E. Auer and T. Tsiatsos (Eds.): IMCL 2021, LNNS 411, pp. 102–110, 2022.
https://doi.org/10.1007/978-3-030-96296-8_10

particular virtual reality (VR), is becoming widely embraced in education (and medical education in particular) for teaching or training purposes [2]. Prior work has studied how VR improves the efficiency of medical education and training [3]. Due to advances made in computer graphics, software, and hardware technology over the past decade, VR technologies/devices have become more affordable and available at the consumer level. This in turn has allowed them to be incorporated and accepted by the medical industry throughout a variety of areas, including medical education [4]. For instance, VR technology can be used to provide a dentist with a simulated practice scene as a low-cost and safe method; dentists can complete training in cutting and shaping teeth using a VR-based haptics simulation [5]. Another example can be seen in the work of Vapenstad et al. [6] where a VR environment with a surgical simulation and a cylinder-like device with force feedback, the Xitact ITP (Instrument Tracking Port), was used to simulate the laparoscopic surgery process.

Computer-based simulations rely on multiple sensory cues and the interaction of these cues and the resulting effects on fidelity perception has been previously studied [7]. However, these multimodal cues rely on specialized hardware such as force feedback haptic devices, computers capable of rendering and processing the simulation information, and in some applications VR-ready equipment. Since the time COVID-19 was declared a pandemic by the World Health Organization on March 11, 2020 [8], governments globally have issued lockdowns/stay-at-home orders that have led to the limited gatherings affecting educational institutions from K-12 to post-secondary institutions. Working and learning from home, and attending virtual meetings has become the norm throughout many forms of information distribution. With respect to education, this has resulted in an abrupt move from traditional face-to-face instruction to remote learning facilitated with a variety of internet-based tools. While this has allowed for the utilization of digital tools and virtual learning environments (VLEs) such as virtual simulations and serious games to provide the necessary instruction, it has also made evident the limitations with respect to immersion and presence associated with hands-on practices.

Simulations are able to provide a safe and risk-free environment to all users, whether they are refining already learned skills and education or they are acquiring the traits for the first time. Simulations (including virtual simulations and serious games) are effective at developing knowledge and skill training, especially in medical and public health fields where education and training continuance is critical for skills development and maintenance [9, 10]. In fact, the use of simulation given the COVID-19 pandemic has shifted from a "backburner training tool to a first choice strategy for ensuring individual, team, and system readiness" [11]. Virtual simulations and serious games provide the potential to revolutionize education and training, and this is evident particularly during the COVID-19 pandemic.

The adjustment to online learning has shown the potential and interest from educators and learners to adopt novel and more engaging tools for online learning. For example, when lectures focusing on the COVID-19 pandemic were delivered to medical students via an online lecture format or in the form of online virtual simulations or serious games, students who were exposed to these applications were able to retain the knowledge for a longer period of time [12]. In addition, when a virtual simulation/serious game was made to provide science-based information on preventing COVID-19, learners were engaged and successfully retained information that may that helped improved players' performance on the topic about the use of masks, which may reflect an increase in information about or adherence to mask use over time [13]. Furthermore, other studies have provided evidence that virtual simulation/serious gaming approaches are effective when compared to traditional online lectures when teaching medical students about COVID-19 for patient treatment [12]. Even after the COVID-19 pandemic recedes, simulations and virtual simulations–and serious games in particular– may become an optimal way of providing education and training to a wide demographic.

Aside from the disruption caused to the traditional in-person classroom experience, current COVID-19 preventive measures have affected some research laboratories by reducing their capacity or shutting them down. One such example is the Laboratory for Games and Media Entertainment Research (GaMER Lab) at Ontario Tech University in Oshawa, Canada. The GaMER Lab is a state-of-the-art experimental research space for games and immersive technologies (virtual and augmented reality) that includes researchers with interests in human-computer interaction, game science, gamification, interaction design, informatics, user experience (UX), serious games, virtual simulation, stereovision and robotics—all connected to each other in a common research space. A large part of the work taking place in the GaMER Lab focuses on the application of immersive technologies (including virtual simulation and serious gaming) for medical education, health, and well-being. This work has included conducting various experiments with human participants to examine, for example, the usability and effectiveness of developed immersive applications, including those that require VR-based hardware (e.g., head mounted displays, VR controllers, and haptic devices).

Conducting human-based experiments during the COVID-19 pandemic at Ontario Tech University has required us to modify and move the data collection process to a remote setting. Given the emphasis and need of hands-on and immersive applications to facilitate remote learning, such challenges must be overcome to ensure these immersive learning applications do in fact meet their intended objectives. It is possible to conduct a variety of experiments remotely where the participants and the experimenter are both present via a conferencing tool such as Google Meet or Zoom with minimal considerations. For example, usability experiments on non-immersive VR simulations without the need for any specialized hardware (e.g., head mounted displays (HMDs), haptic devices, and VR controllers) are possible by providing online access to the simulator followed by the completion of a questionnaire such as the System Usability Scale (SUS), a 10 item questionnaire with five response options for respondents; from Strongly agree to Strongly disagree, that is a reliable tool for measuring the usability of a system [14]. Additional information can be collected through open-ended questions about their experience. However, the situation is further complicated when specialized equipment such as higher fidelity HMDs and haptic devices are required to simulate virtual hands-on experiences, but inaccessible to the majority of participants as these are not readily available at the consumer level. In such a situation, although each participant can be sent their own equipment (e.g., via courier), such an approach can be cost-prohibitive, time consuming, and require sanitation protocols (e.g., time and cost to sanitize and ship the equipment to each participant, and to return it upon completion of the experiment).

In this paper we provide an overview of rethinking an experiment that examined the accuracy of a simulation of a virtual drilling task using standard computer equipment remotely that was redesigned from its in-person original form. We discuss the procedure taken to facilitate the experiment remotely, potential issues/problems we encountered, and lessons learned. We also discuss the procedure that will be taken to conduct a future usability experiment for a serious game being developed to educate youth about the dangers on vaping. The experimental results will not be discussed. Rather, a discussion will follow regarding the experimental process and issues we encountered while conducting this experiment.

2 Rethinking Virtual Drilling User Testing

Our ongoing work has examined the effects of audio on visual and haptic fidelity perception and task performance [7]. An in-person study examined the influence of sound on haptic performance through a virtual drilling task (using a Novint Falcon haptic device), revealing that sound can affect haptic fidelity perception [15]. Although we planned on expanding this experiment, due to the COVID-19 pandemic, access to the GaMER Lab where the haptic devices are housed, was not an option. We therefore shifted our focus on examining whether we can simulate a virtual drilling task without haptic devices, using standard computer hardware, such as a mouse, headphones, and a monitor/display commonly available to individuals within their household. See Fig. 1 for a comparison between real drilling and ideal in-person setup in contrast to the online setup. This process required remapping all haptic interactions to mouse movements. Haptic feedback relied on audio cues, and without VR, the simulation presented a monoscopic view from the back of the drill.

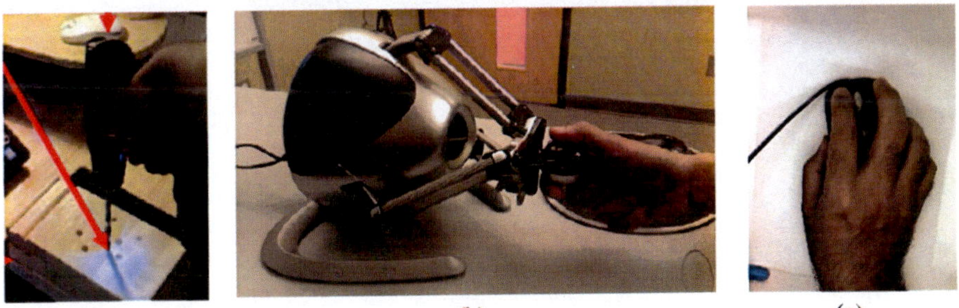

(a) (b) (c)

Fig. 1. Haptic interactive mode comparison. a) Real drilling, b) haptic drilling employing the Novint Falcon device, and c) online drilling using a mouse.

With respect to audio, the application presented pre-recorded drilling sounds captured while drilling through a block of wood. To ensure a higher-quality soundtrack, we would have recorded the drilling sounds using professional recording equipment in an audiometric room available within the GaMER Lab. Since the lab was not accessible, the recording was made inside the home of one of the experimenters using a Tascam field recorder. Effort was taken to limit any noise and reverberation (e.g., the recordings were made in a large room in the absence of others and without any additional sound sources turned on).

2.1 Scenario

The scenario presents a drilling task to be executed using visual (a drill with a drill bit and a block of wood – see Fig. 2), auditory (recorded sounds of a real drill drilling through a block of wood), and kinesthetic cues when operating the drill with mouse movements (to simulate virtual drilling through block of wood). The scenario was developed using the Unity Game engine and was built into a Windows-compatible application. The scenario was designed with limited fidelity and detail to ensure online participants could run it by meeting Unity's minimum system requirements.

To simulate the drilling and to convey that information in audiovisual format, a series of invisible layers were created inside the virtual wooden block. This allows triggering different audiovisual effects to convey resistance and penetration while drilling. These effects include slowing drill movement and dynamic changes in the sound recordings.

Fig. 2. Sample experiment environment (consisting of a drill and block of wood), presented to the participants.

2.2 Study Design

Our study focuses on examining how drilling simulation employing mouse movements, a regular computer monitor, and pre-recorded drilling sounds affect depth perception.

We also examine the interaction of these cues with respect to the accuracy of the simulated drilling task. The study presented four auditory conditions (i.e., no sound, drilling sound, dynamic sound, and continuous sound) to each participant five times for a total of 20 trials. The conditions were presented in random order to minimize carry-over effects. During each trial, participants were presented with one or a combination of these cues and tasked with stopping the drill when they believed it reached the target (and pre-fixed) depth of 12 cm.

Each of the participants completed the experiment in their own home and, informally, there appeared to be plenty of variation with respect to the actual room that each participant was in. More specifically, several participants appeared to be in smaller rooms (e.g., bedroom) while others in larger rooms (e.g., living room or open basement). We also did not have any control regarding any sources of sound (noise) in the environment, lighting, or potential distractions although such distractions did not appear to be a problem with any of the participants. That being said, during the experiment overview (prior to the start of the experiment), the experimenter did ask the participants to limit external sources of sound (those that they could control).

2.3 Limitations

Participants were required to have a mouse, headphones/earbuds, and were asked to set the resolution of their monitor/display at 1024 $\times$ 768. They were asked to sit on a chair in front of their display and to adjust the height according to their own habits but to keep their eyes at a similar height to the monitor. They were also asked to wear their headphones/earbuds and adjust the volume to a comfortable level. During the experiment, participants were required to be connected to the experimenter via Google Meet. The experimenter did not interfere with the participants and only answered any questions they may have had as they completed the experiment.

Given that the experiment was conducted in the home of each participant (i.e., not in the laboratory), there was no consistency with respect to the required hardware, and more specifically, the monitor, headphones, and mouse, each of which can vary widely with respect to quality, fit, etc. In an attempt to limit any effects from the hardware used by each participant, we did ask participants to set the resolution of their monitor to a specific setting (1024 $\times$ 768), and to use headphones (as opposed to loudspeakers), with the volume set to a comfortable level. Having volume levels set to a comfortable level is subjective, and thus there is no guarantee that the results are not influenced by any loudness effects. With respect to the mouse, various types were observed including a high-end gaming mouse which can result in a more comfortable grip and could potentially impact accuracy.

Since the experiment was conducted remotely and facilitated using Google Meet, it was difficult for the experimenter to observe the emotional state of the participants and more specifically, to gauge whether they were bored, interested, engaged, etc. It should be noted that the experimenter did not interact with the participants while they completed the experiment but was present (via Google Meet) in the event the participants had any questions, technical issues, etc. In the future, perhaps the experimenter could engage with the participants with, for example, simple questions such as "Is everything OK?".

3 Discussion

Conducting experiments that depend on laboratory equipment can be problematic even when such equipment is typically available for in-person testing. As a result of the COVID-19 restrictions, interest in online learning and teaching has been renewed sparking a number of innovations and workarounds to ensure access to teaching materials and hands-on experiences. As an alternative to specialized laboratory equipment, computer-based simulation has been facilitating the development of cognitive and motor skills in many fields. However, some simulations rely on computer hardware such as monitor/display, headphones, and a mouse (where the mouse is an integral part of the experiment).

It is likely that equipment will vary widely (e.g., a standard computer mouse vs. a gaming mouse). Although it is not clear whether the differences in hardware will impact the results, when the hardware is an integral part of the experiment it may have some impact (e.g., the use of a gaming mouse vs. a standard mouse to control the virtual drill in the experiment described here). The situation is further complicated when the required hardware is not available in the typical home (e.g., haptic devices and high fidelity HMDs). Although VR adoption has been increasing, HMDs are not readily available to every student or participant. A solution here includes shipping the equipment to the participants. This of course presents its own challenges, including those related primarily to cost and time. Purchasing the hardware for each participant is costly but will shorten the experiment duration, while purchasing one or a limited number of the required equipment may potentially reduce costs but will take more time (shipping to a participant, they ship it back upon completion of the experiment etc.). Additionally, due to the COVID-19 pandemic, sanitation protocols have to be implemented to ensure the wearable devices are safe to circulate. Sanitation adds another layer of complexity because if done incorrectly, the integrated circuits and sensors embedded in the headset can be damaged.

Aside from the technology-related challenges, task completion and support also suffered. For example, 10 participants had to be removed from the study since they did not complete the task. Unfortunately, we are not aware of the causes, but are aware that during COVID-19 there are situations outside of our control with respect to completion.

4 Conclusions

In this paper we provide an overview of an experiment that examined the accuracy of a virtual drilling task using standard computer equipment remotely. From this experience, we conclude the following: i) when developing immersive-based studies, when possible, consider integrating cross-platform compatibility since this ensures the application runs on multiple devices not constrained to a laboratory setting, ii) non-immersive VR can provide an alternative to immersive VR, however, consider adding motion parallax based on head tracking to add visual immersion that is lacking in regular monitors, iii) while VR-ready hardware is more affordable, it is not the norm, employing the WebXR device API (which provides the functionality needed to bring both augmented reality and virtual reality to the web [16]), can help deploy virtual experiences that run on a browser and can be experienced in immersive and non-immersive modes, iv) implement user customization options to compensate the lack of uniformity associated with hardware available at home, v) when focusing on auditory feedback, environmental calibration tools can help understand and manage levels of noise and non-contextual sounds that may affect the study, vi) due to the potential interruptions when conducting online studies, it is worth considering a pause and resume option since participants are likely to be working from home.

We are currently developing a serious game to educate children about the dangers of vaping and following an iterative development process whereby prototypes are developed, tested (for usability) and refined. With the completion of a prototype, we have recently submitted an application to the Research Ethics Board at Ontario Tech University to allow for the usability of the prototype to be explored with human participants once again following a fully online protocol. The usability study to be conducted is comprised of three parts. The first part will involve sending a form to those interested in participating that outlines the background and purpose of the study, how the study will be conducted, and the rights of participants with regards to consent and their data. Should the participant agree to the terms and procedures outlined in this document, they will be directed to a Google Form used to document their name and the date on which they provided their consent. Once the participant's consent has been obtained, they will be directed to a website that hosts the serious game in HTML5 format. They will then play the game (complete all five levels) and then complete the SUS questionnaire.

As with our previously described experiment, we will require participants to wear headphones and will ask them to adjust the volume to a "comfortable level". Based on our experience with our prior online experiment, during this experiment, the experimenter will take on a more pro-active role and will periodically engage with the participants by asking them whether there any issues/problems.

Acknowledgements. The financial support of the *Natural Sciences and Engineering Research Council of Canada* (NSERC), in the form of Discovery Grants to B. Kapralos and A. Quevedo, and the *Research Center for Biomedical Engineering and Research Institute of Electronics, Shizuoka University*, Japan is gratefully acknowledged.

References

1. Bourdot, P., Convard, T., Picon, F., Ammi, M., Touraine, D., Vézien, J-M.: VR–CAD integration: multimodal immersive interaction and advanced haptic paradigms for implicit edition of CAD models. Comput.-Aid. Des. **42**(5), 445–461, (2010)
2. Yu, P., Pan, J., Qin, H., Hao, A., Wang, H.: Real-time suturing simulation for virtual reality medical training. Comput. Anim. Virt. Worlds, **31**(4–5), e1940 (2020)
3. Rassie, K.: The apprenticeship model of clinical medical education: time for structural change. New Zealand Med. J. **130**(1461), 66 (2017)
4. Willaert, W.I.M., Aggarwal, R., Van Herzeele, I., Cheshire, N.J., Vermassen, F.E.: Recent advancements in medical simulation: patient-specific virtual reality simulation. World J. Surg. **36**(7), 1703–1712 (2012)
5. Xia, P., Lopes, A.M., Restivo, M.T.: Virtual reality and haptics for dental surgery: a personal review. Vis. Comput. **29**(5), 433–447 (2013)
6. Våpenstad, C., Hofstad, E.F., Langø, T., Mårvik, R., Chmarra, M.K.: Perceiving haptic feedback in virtual reality simulators. Surg. Endosc. **27**(7), 2391–2397 (2013)
7. Kapralos, B., Moussa, F., Collins, K., Dubrowski, A.: Levels of fidelity and multimodal interactions. In: Wouters, P., van Oostendorp, H., (Eds.) Techniques to Improve the Effectiveness of Serious Games, Springer Advances in Game-based Learning Book Series, Ch. 5, pp. 79–101 (2017)

8. Cucinotta, D., Vanelli, M.: WHO declares COVID-19 a pandemic. Acta Biomed. **91**(1), 157–160 (2020)
9. Lateef, F.: Simulation-based learning: Just like the real thing. J. Emerg. Trauma, Shock **3**(4), 348–352 (2010). https://doi.org/10.4103/0974-2700.70743
10. Esposito, C.P., Sullivan, K.: Maintaining clinical continuity through virtual simulation during the COVID-19 pandemic. J. Nurs. Educ. **59**(9), 522–525 (2020)
11. Brydges, R., et al.:. Lessons learned in preparing for and responding to the early stages of the COVID-19 pandemic: one simulation's program experience adapting to the new normal. BMC Adv. Simul. **5**(8), 1-10 (2020). https://doi.org/10.1186/s41077-020-00128-y
12. Hu, H., Xiao, Y., Li, H.: The effectiveness of a serious game versus online lectures for Improving medical students coronavirus disease 2019 Knowledge. Games Health J. **10**(2), 139–144 (2021). https://doi.org/10.1089/g4h.2020.0140
13. Gaspar, J.D., Lage, E.M., Silva, F.J., Mineiro, É, Oliveira, I.J., Oliveira, I., Reis, Z.S.: A mobile serious game about the pandemic (COVID-19 - did you know?): design and evaluation study. JMIR Ser. Games **8**(4), e25226 (2020). https://doi.org/10.2196/25226
14. Sauro, J.: A Practical Guide to the System Usability Scale: Background, Benchmarks & Best Practices. CreateSpace Independent Publishing Platform, California, USA (2011)
15. Melaisi, M., Rojas, D., Kapralos, B., Uribe-Quevedo, A., Collins, K.: Multimodal interaction of con-textual and non-contextual sound and haptics in virtual simulations. Informatics **5**(4), 43 (2018). https://doi.org/10.3390/informatics5040043
16. MacIntyre, B., Smith, T.F.: Thoughts on the future of WebXR and the immersive web. In: Proceedings of the 2018 IEEE International Symposium on Mixed and Augmented Reality Adjunct (ISMAR-Adjunct), pp. 16–20 (2018)

Work-in-Progress About Dynamicity as a Foundation for AMI, a Mobile Intelligent and Adaptive Learning System

Richard Hotte[1(✉)], Anis Masmoudi[1], Aymen Jaballah[2], Omar Masmoudi[2], and Alhoudourou Almaimoune Maïga[3]

[1] Institut de Recherche LICEF, Université TÉLUQ, Montréal, QC, Canada
{richard.hotte,anis.masmoudi}@teluq.ca

[2] ESPRIT, 1, 2 rue André Ampère - 2083 - Pôle Technologique - El Ghazala, Tunis, Tunisie
{aymen.jaballah,omar.masmoudi}@esprit.tn

[3] Université Alioune Diop de Bambey (UADB), Bambey, Sénégal

Abstract. If school, in its traditional form, cannot be accessible to all the children of the world, we believe that school can become accessible to them in the form of a smart learning system (smart system and smart learning), adaptative/personalized, and mobile. AMI, an intelligence-based learning system, could be a solution for children who are out of school.

AMI aims to enable learner self-learning. To do this, it must be dynamic. Its dynamicity stems from a close and sustained interaction between the learner and the system, which infers its adaptability. The system is then alive and, therefore, in constant reaction to the learner's activity. The continuous integration of new data from this learner/system interaction modifies the learner's profile and/or the learning path in progress. Therefore, how to provide the system with dynamic and sustainable self-learning capabilities, based on the learner's behaviors throughout his interaction with the system? More precisely, how to represent and interpret random events as messages to which the system can react to produce actions in continuous mode?

This paper presents a Work-in Progress on the implementation of two of the four intelligent components of the AMI system aiming at allowing a maximum adaptability of a personalized learning offer.

Keywords: Dynamicity · Adaptive learning system · Learner profile

1 Introduction

The problem of education access for millions of children around the world is far from new. Despite efforts to date, it is still challenging to catch children who start with the traditional classroom approach with a teacher present. The idea that school could be accessible to these children if it came to them involves notions of intelligence (smart system and smart learning), adaptability and personalization, and mobility.

In a context of mobility, the school is perceived not as a classroom, but as a mobile educational system, accessible everywhere and at all times: at home, in the family, on

M. E. Auer and T. Tsiatsos (Eds.): IMCL 2021, LNNS 411, pp. 111–119, 2022.
https://doi.org/10.1007/978-3-030-96296-8_11

the street, in the bush, and on the river, accessible to all, anytime and anywhere, adapted to local contexts and local traditional cultures, oriented towards the development of the child without any form of discrimination and finally safe, free from conflict. This type of learning system uses innovative mobile pedagogies.

1.1 Purpose or Goal

The Intelligence Mediated Learning (AMI in French) project asserts itself as a solution to the lack of education access. Modeling an intelligent adaptative/personalized mobile learning system as such is far from new (Brusilowsky et al. 2004; Chin 2001; Chen 2008; Limongelli et al. 2008; Kearney et al. 2012; Kim 2009; Yin 2010; Kim et al. 2013; Xie et al. 2019)., However, it is necessary to describe how the system can have dynamic and sustainable self-learning capabilities based on the learner's behaviors throughout the learner's interaction with the system. More precisely, we need to represent and interpret the messages of random events so that the system can react to produce actions in a synchronous and continuous mode. To be optimal, such adaptability must be dynamic, i.e., it must continuously take into account the progressive evolution of the learner's cognitive abilities and constantly update the proposed learning content according to this evolution.

This paper aims to present a Work-in-Progress on implementing two out of four intelligent micro-service-oriented components of the AMI system, aiming to allow maximum adaptability of a personalized learning offer. These components will allow, on the one hand, the intelligent selection of one or more types of learning paths and, on the other hand, the adaptation of these types of approaches to the evolution, interests, and needs of a learner throughout his interaction with the system and according to the evolution of the learner's profile. This paper is organized as follows. Section 2 gives some core concepts of intelligent, adaptative/personalized, and mobile learning systems. Section 3 introduces the notion of dynamicity related to the AMI system. Section 4 presents the practicality and relevance of a model-driven methodological approach for the development of the solution. Finally, Sect. 5 presents conclusions and further works.

2 Conceptual Background

Analysis of a set of journal articles related to trends and developments in personalized adaptative technologies published between 2007 and 2017 led Xie et al. (2019) to conclude that "adaptative personalized learning presents a significant number of opportunities of applications on smart devices such as wearable devices, smartphones, tablet computers, and computers with the rapid development of artificial intelligence, virtual reality, cloud computing, and wearable computing." Furthermore, "Adaptative/personalized learning has become possible by implementing intelligent learning systems, integrating learners' preferences, analyzing individual learning data, and so on" (Xie et al. 2019). For the latter, adaptative/personalized learning is informed by students' preferences and learning outcomes. According to Peng et al. (2019), "Personalized adaptative learning is formed by the combination of personalized learning and adaptative learning".

2.1 Smart Learning and Smart Learning Environment

When we talk about an intelligent learning environment, we are dealing with two realities simultaneously: smart learning and a smart learning environment.

Smart learning relies on smart devices: telephones, watches, tablets, etc., and smart technologies, i.e., technologies that use the Internet and advanced levels of automatization to operate effectively. In a literature review by (Zhu et al. 2016), it is indicated that Kim et al. (2013) considered that smart learning, which combines the advantages of social learning and ubiquitous learning, is a learner-centric and service-oriented educational paradigm, rather than one just focused on utilizing devices.

Moreover, the fact that smart learning is based on personal and smart technologies makes learners engage in their learning and increase their independence in more open, connected, and augmented ways through personally richer contexts. Thus, the challenge of intelligent learning is to feedback the control of learning to the learner, regardless of the context in which he or she is learning, with all other resources contributing to this autonomy, and to empower a system to respond dynamically, and thus vividly, to this challenge.

A smart learning environment aims to support effective, efficient, and meaningful learning for learners. "Smart learning environments supported by technologies should not only enable learners to digital resources and interact with the learning systems in any place and at any time, but also actively provides them with the necessary learning guidance, supportive tools. or learning suggestions in the right place, at the right time, and in the right form (Hwang 2014: 2)" (Zhu et al. 2016).

There are many different types of technologies used to support and enhance learning, including hardware and software. Hardware includes those tangible objects such as interactive whiteboard, smart table, e-bag, mobile phone, wearable device, smart device, sensors, which use ubiquitous computing, cloud computing, ambient intelligence, IoT technology, etc. Software includes learning systems, learning tools, online resources, educational games that use social networking, learning analytics, visualization, virtual reality, etc.

3 Dynamicity, the Core of the AMI Learning System

A review of the literature by Elghibari et al. (2015) leads to the observation that the limitation of the proposed systems lies in the fact that they present content adapted to the initial parameters of the learner's profile (preferences, learning style, score, etc…), without taking into account the evolution of his/her cognitive abilities. There is, therefore, the need for dynamic content that follows the transformations of the learner profile (cognitive abilities, preferences, learning style) during the use of the system by the user himself. According to the latter (2015), in order to understand the learners' behavior and current content tailored to their needs, adaptation must take into account the changes in the learner's profile during the learning process.

3.1 The Importance of Learner Profile

Learner profiles play an important role in the adaptation of the online learning environment. They are the key elements of system modeling and consist of a set of data and metadata based on one or more learners that influence the system's behavior (Herder 2016). By implementing intelligent systems, integrating learners' preferences, analysis of individual learning data, and adaptative/personalized learning have become possible (Xie et al. (2019). Everything hinges on the degree of dynamicity of such educational systems, i.e., their ability to act dynamically and synchronously with the learner's activity in progress. According to (Baldiris et al. 2011), "The concept of dynamic user characteristics has evolved to describe certain user characteristics that change continuously during the user's use of the system. The use of such characteristics in constructing user models defines the second type of user models, the dynamic user mode (DUM)." In the words of Elghibari et al. (2015), "There must be a continuous adjustment between the instructional resources and the learner's profile, producing a dynamic and flexible instructional process."

3.2 A Learner Master of His Choices

The primary function of intelligent selection is to make the learning system proactive by offering the learner-user a choice of courses, new courses or paths, aligned to his/her learning objective(s) as prescribed in the initial learner-user model. In this model, the learner profile is continuously modified throughout the completion of the selected typical course or par-course. The primary function of the course modification is to ensure the evolution of selected specific courses under the learner-user's action with the system in order to adapt to his/her real learning needs, thus to take the form of a course or a path adapted to both his/her learner profile (user model) and, also, to his/her cognitive style throughout his/her learning activity.

The modeling of the AMI solution in the form of a domain model has made it possible to position, among other things, the processes of 1) intelligent selection of typical courses or pathways, predefined according to typical learner user models and their evolution, and 2) modification of these typical pathways into a real pathway, i.e., a flexible and dynamic pathway (Elghibari et al. 2015), which follows the evolution of the user's learning, thus the learner throughout his/her interaction with the system.

4 AMI, A Model-Driven Software Solution

It seems very difficult to create a flexible learning environment without modeling learner profiles. Modeling typical learner profiles is the key to an adaptative learning system. As said by Peter Brusilovsky and Eva Millán (2007: 3): "*The user model is a representation of information about an individual user that is essential for an adaptive system to provide the adaptation effect, i.e., to behave differently for different users*".

A field study in a community in the Gao region of northern Mali provided the data needed to develop a user profile, the basis for modeling the AMI system as a solution to the problem of access to school for children in Gao. The personas method was used to collect information about a typical potential AMI user profile. According to Bornet and Brangier (2013), the persona is an instrument for stimulating prospective activity, making it possible to define a model adapted to a future user or learner because for them, Vygotski (1985)'s work emphasizes that effective learning should anticipate the development of the person and its advances. The results show that the profile of the randomly selected learner within a family that includes several characteristics related to the living environment and the local context is plural and is defined by taking into account his/her siblings, parents, and entourage.

4.1 The Domain Model

The domain model is derived from the organic architecture of the AMI system. Initially, this declarative model represents the initial state of the proposed solution to the problem of education access for GAO children. It makes that solution explicit. This model has three goals: 1) to create a learner-centered learning system that is accessible to anyone, anywhere, at any time, 2) to integrate this personalized system into the learner's environment, and 3) to adapt this learning system to the learner's family, social, economic, and cultural reality.

This solution is based on the following three assumptions: 1) the solution is built around a set of existing components or tools that are independent of each other and can be integrated into an ecosystem, 2) the collaboration, course management, instant communication, and learning content management components are proposed based on existing tools, and 3) the course management, learner profile management, learning path tracking and learning assistance components are redesigned around adaptive learning based on four types of intelligence that provide intelligent assistance to 1) the learner, his/her tutor or any other adult in charge of the child, 2) adapting the learning path to the learner's reality, 3) completing the path of the learner in difficulty and 4) injecting new data at the end of the learning path to update the learner's profile and, also, to enrich and diversify the path offer.

In addition, the solution must take into account two major constraints. The first is the enrichment of the learner profile adopted by the learner's family, cultural, economic, and social aspects. The second is the accessibility and logging of all the data on the learner's activities related to his or her learning path and interactions with the actors with whom he or she is in a support relationship in order to adapt the profile and path.

4.2 Towards the Dynamicity of AMI

AMI introduces dynamicity into both the learner's profile and the evolution of the learner's learning path by maintaining a sustained interrelationship between two components in a flexible and intelligent environment. This is in line with what is expressed by (Peng et al. 2019: 9) when they write, on the one hand, that "learner profile-based progression assesses learner progress by continuously measuring the individual performance of the learner's learning goals" and, on the other hand, that the flexible and intelligent learning environment can provide functional support for the adaptive adjustment of teaching strategies.

Figure 1 illustrates the dynamicity of the AMI system through a fictitious use case, that of the learner Sam Petros. It consists of nine steps. Step 0 is a preliminary step that reports the learner's profile which" aims to portray the individual characteristics of each learner's strengths, preferences, motivations, etc." (Peng et al. 2019).

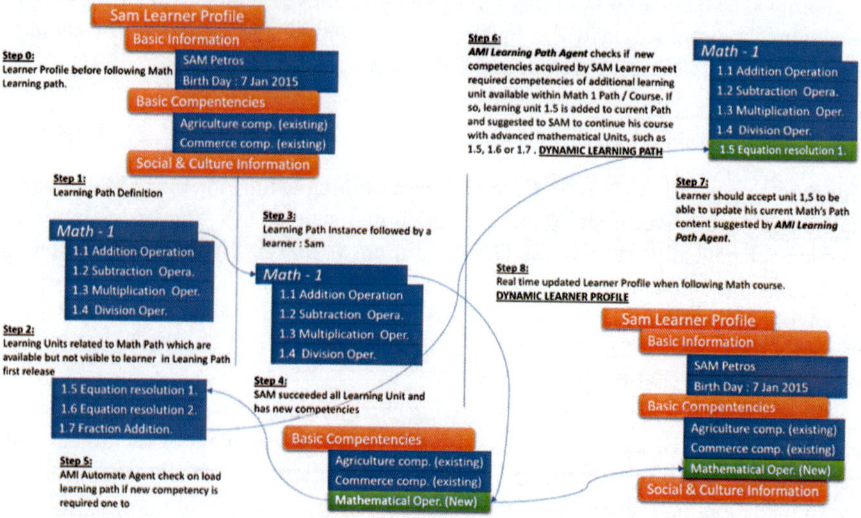

Fig. 1. Business use case

Then follow eight steps: 1) the knowledge model of the path, 2) its structure in learning units, 3) the instantiation of the learning path based on the enriched learner profile Sam, 4) Sam's success on the whole path, which allowed him to acquire new skills, 5) the AMI system's automaton agent checks if the learning path load requires new skills, 6) a DYNAMIC LEARNING PATH, where the path's automaton agent checks if the new skills acquired by the Sam learner match the skills required for the additional learning unit available in the Math 1 path/course. If so, learning unit 1.5 is added to the current pathway, and Sam is invited to continue with advanced math units, such as 1.5, 1.6, or 1.7, 7) the learner must accept unit 1.5 in order to update the content of his/her current math pathway suggested by the AMI Learning Pathway Agent, 8) a DYNAMIC LEARNER PROFILE, where the learner's profile is updated in real-time as his/she completes a math course.

The flow of information indicates that the dynamicity occurs at two points: when a learning unit is added to the initial course to match the learner's learning (step 6) and when the learner profile is updated in real-time as new skills are acquired (step 8). According to the SAM basic profile, steps 1 to 3 indicate a transition from a typical chosen path (step 1) to an instantiated path (step 3). Step 4 corresponds to an enrichment of SAM's skills, allowing him to enrich his learning path with units of his choice.

Figure 2 presents a diagram illustrating the implementation of use case as a schematic of the deployment of various AMI software components in different application servers.

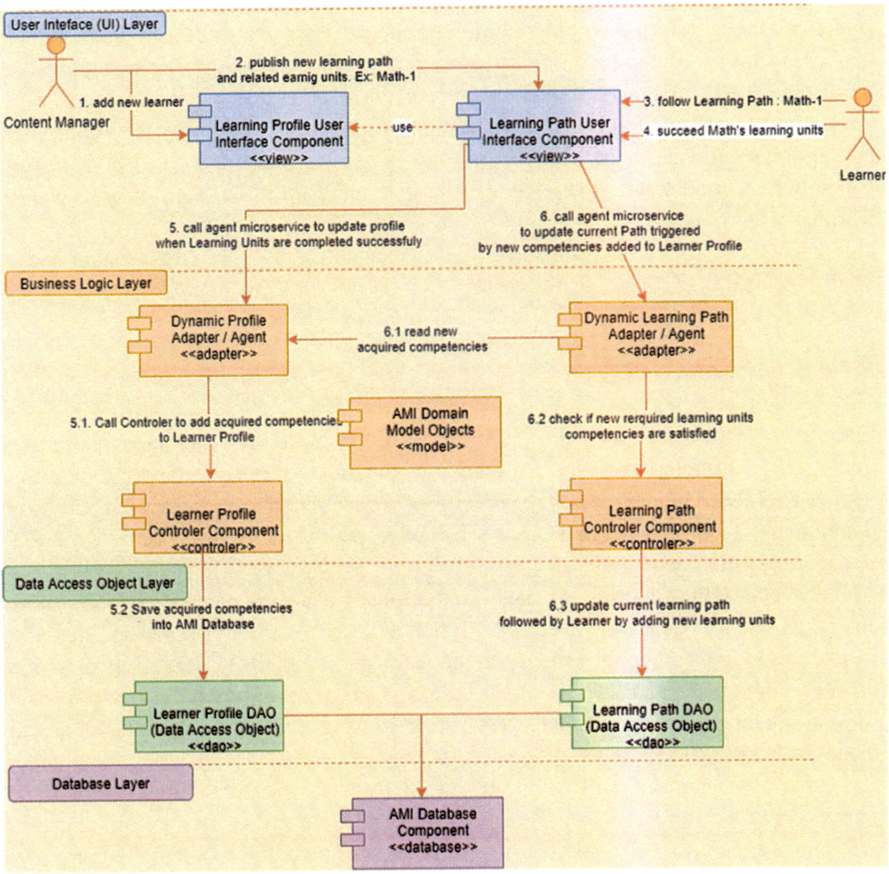

Fig. 2. Implementation of use case

Figure 2 shows how AMI uses the Model-View-Controller (MVC) software architecture through the various software components. AMI is deployed on three application servers: user interface (UI), learning path, and learner profile. The integration layer uses a microservices-oriented architecture. The UI layer shows that the UI component calls microservices provided by the learner profile and learning path adapters. The overall taxonomy of AMI is defined by four sub-domains of activity: Learning Path Selection, Learner Profile Management, Learning Path Management, and Learner Support. Each subdomain publishes a set of microservices that provide the necessary information to the other consumer components such as Learning Path, User Profile, and Learning Path Selection. Figure 2 shows all of the microservices of the above subdomains except those of the Learner Support subdomain. It is important to specify that the Learning Path Selection component uses two search sub-components corresponding to the two types of criteria: the existing path and the correlated learner profile. An existing path is a path completed by a learner, which enriches the bank of existing and available paths. A correlated learner profile results from an intelligent selection, i.e., a selection by keywords enriched by the correlation between the learner's existing profile and similar corresponding profiles. Each profile is weighted and validated. Some of these profiles will appear during the search for matching profiles. An aggregate component will integrate the search results of the two sub-components and assign weights to them. The user interface components (sequence calls 5, 6, and 8) call different components that manage the learner profile, the learning path, and the learning path's selection. Our goal is to reuse our adapters and controllers in order to be able to integrate different e-learning systems that will allow the reuse of their learning object metadata (LOM) or learning units.

5 Conclusions

Literature reviews and benchmarking analyses confirm that advances in artificial intelligence can help enrich mobile adaptive learning systems by adapting learning to learner profiles. The degree of dynamicity of these systems, i.e., their capacity to act dynamically according to the learner's behavior, will directly impact the evolution of the user's profile characteristics.

We believe that the system must continuously adapt to the learner's progress to ensure its vitality. Hence the need for a dynamic, learner-centered pedagogy that adapts to changes in the learner's profile when the learner is using the system. This dynamicity must be embodied in an intelligent selection of one or more typical learning paths and the modification of these paths according to the learner's needs and learning achievements.

The development of the two intelligent microservice-oriented components, i.e., the dynamic path selection and the dynamic profile adaptation, is nearing completion. We will prepare one or more test scenarios. In a first step, we will validate the components in the form of prototypes with the field researcher. In a second step, we will validate them with a sample of the targeted population in the GAO region in the north of Mali.

Acknowledgments. We would like to express our gratitude to the Canadian Commission for UNESCO and the Fonds de recherche du Québec for their support and funding to the UNESCO Chair in GSDL, to the MRIF - Québec for its financial support from the Québec-Sénégal bilateral cooperation program, and to the Fonds d'aide à la recherche de l'Université TÉLUQ. Finally, we would like to thank the team of students who contribute to the advancement of the research through their investment in internships and theses.

References

Baldiris, S., Graf, S., Fabregat, R.: Dynamic user modeling and adaptation based on learning styles for supporting semi-automatic generation of IMS learning design. In: 11th International Conference on Advanced Learning Technologies, pp. 218–220. IEEE, Athens, GA, USA (2011)

Bornet, C., Brangier, É.L.: méthode des personas : principes, intérêts et limites. Bull. de Psychol. **524**(2), 115–134 (2013)

Brusilovsky, P., Nejdl, W.: Adaptive Hypermedia and Adaptive Web. Chapman & Hall / CRC Press LLC, Practical Handbook of Internet Computing (2004)

Brusilovsky, P., Millán, E.: User models for adaptive hypermedia and adaptive educational systems. In: Brusilovsky, P., Kobsa, A., Nejdl, W. (eds.) The Adaptive Web. LNCS, vol. 4321, pp. 3–53. Springer, Heidelberg (2007). https://doi.org/10.1007/978-3-540-72079-9_1

Chen, C.-M.: Intelligent web-based learning system with personalized learning path guidance. Comput. Educ. **51**(2), 787–814 (2008)

Chin, D.N.: Empirical evaluation of user models and user-adapted systems. User Model. User-Adapt. Interact. **11**(1), 181–194 (2001). https://doi.org/10.1023/A:1011127315884

Elghibari, F., Elouahbi, R., Elkhoukhi, F., Chehbi, S., Kamsa, I.: Intelligent e-learning system model for maintenance of updates courses. In: International Conference on Information Technology Based Higher Education and Training (ITHET), pp. 1–3. IEEEXplore, Lisbon (2015)

Herder, E.: User modeling and personalization. 3: user modeling – techniques. In: Web Science – Investigating the future of Information and Communication. 55 slices (2016)

Hwang, G.-J.: Definition, framework and research issues of smart learning environments - a context-aware ubiquitous learning perspective Smart Learning environment – a context-aware ubiquitous learning perspective. Smart Learn. Environ. **1**(1), 1–14. SpringerOpen (2014)

Kearney, M., Schuck, S., Burden, K., Aubusson, P.: Viewing mobile learning from a pedagogical perspective. Res. Learn. Technol. **20**(1), 17 (2012)

Kim, P.H.: Action research approach on mobile learning design for the underserved. Educ. Technol. Res. Dev. **57**(3), 415–435 (2009)

Kim, T., Cho, J.Y., Lee, B.G.: Evolution to smart learning in public education: a case study of Korean public education. In: Ley, T., Ruohonen, M., Laanpere, M., Tatnall, A. (eds.) OST 2012. IAICT, vol. 395, pp. 170–178. Springer, Heidelberg (2013). https://doi.org/10.1007/978-3-642-37285-8_18

Limongelli, C., Sciarrone, F., Vaste, G.: LS-plan: an effective combination of dynamic courseware generation and learning styles in web-based education. In: Nejdl, W., Kay, J., Pu, P., Herder, E. (eds.) AH 2008. LNCS, vol. 5149, pp. 133–142. Springer, Heidelberg (2008). https://doi.org/10.1007/978-3-540-70987-9_16

Peng, H., Ma, S., Spector, J.M.: Personalized adaptative learning: an emerging pedagogical approach enabled by a smart learning environment. Smart Learn. Environ. **6**(9), 14. SpringerOpen (2019)

Xie, H., Chu, H.-C., Hwang, G.-J., Wang, C-C.: Trends and development in technology-enhanced adaptive, personalized learning: a systematic review of journal publications from 2007 to 2017. Comput. Educ. **140**, 16. SpringerOpen (2019)

Yin, C.: Samcco. Un système d'apprentissage mobile contextuel et collaboratif dans des situations professionnelles. (Thèse) Sciences de l'ingénieur, ÉcoleCentrale de Lyon, 225 (2010)

Zhu, Z.-T., Yu, M.-H., Riezebos, P.: A research framework of smart education. Smart Learn. Environ. **3**(1), 1-17. SpringerOpen (2016)

Augmented-, Virtual-, Mixed- and Cross- Reality Apps

Customer Journey: Applications of AI and Machine Learning in E-Commerce

Alexandros I. Metsai[1(✉)], Irene-Maria Tabakis[1], Konstantinos Karamitsios[1], Konstantinos Kotrotsios[1], Periklis Chatzimisios[2], George Stalidis[2], and Kostas Goulianas[2]

[1] My Company Projects O.E., Thessaloniki, Greece
{alexandros.metsai,tabaki,kk,kotrotsios}@mycompany.com.gr
[2] International Hellenic University, Thessaloniki, Greece
{pchatzimisios,stalidgi,gouliana}@ihu.gr

Abstract. In the past decade, ownership and usage of mobile devices has grown in a rapid manner, with users putting trust into these devices for online purchases. Corporations like Booking are reporting a higher number of interactions through mobile devices than on desktops, making these an important medium for online advertising and recommendations used by e-commerce applications. Coupled with the recent advances in artificial intelligence systems and machine learning algorithms, we aim to explore how these developments in the field affect the customer's journey, taking into account the aforementioned trends, as well as the personal user data that these may require to provide proper results. In this manner, we conduct a systematic literature review, using a transparent and thorough process for searching and analysing the recent bibliography, over the last couple of years, focused on intelligent applications in the customer journey.

Keywords: Artificial intelligence · E-Commerce · Mobile applications

1 Introduction

Artificial Intelligence (AI) and Machine Learning (ML) have become an integral part of everyday life, as they provide solutions in numerous different domains, with outstanding results being reported over the last few years. One of the areas where AI holds a prominent position is the broad field of e-commerce and online retailing. AI can be successfully applied to influence customers in the search of products and brands, in the evaluation of possible alternatives, in their decision making process and in their overall consumption, by providing an intelligent, adaptable and informed customer experience at any point of service.

For creating a highly valued business, a critical point to focus on is customer value. Genuinely understanding the customer's behaviour, needs, preferences and demands during their interaction with the platform is essential for managing a successful e-commerce platform. Customer experience encompasses every aspect

M. E. Auer and T. Tsiatsos (Eds.): IMCL 2021, LNNS 411, pp. 123–132, 2022.
https://doi.org/10.1007/978-3-030-96296-8_12

of a company's offering. The quality of customer care is an important aspect, but the product's advertisement, packaging, features, ease of use, and reliability have to also be taken into account [1]. Consequently, creating intensely positive experiences within the customer journey, in comparison with the competition, can improve customer loyalty and at the same time increase the possibility of customers recommending a platform to their closer circle (family, friends, co-workers, etc.) [2,3].

Driven by the above interest in the improvement of the customer journey, several different approaches have emerged that utilize techniques from the field of Artificial Intelligence and Machine Learning. Most of these are concerned with methods and applications where the service process is analysed, explained, modelled, managed and re-designed [4]. However, the continuous introduction of new and different approaches in improving the customer experience and the differences in the applications of various design, management and marketing methods may be overwhelming and confusing.

This paper aims to present the most recent applications of AI and ML in the improvement of the customer journey. Specifically, we contribute an overview of the existing intelligent approaches used in the relevant literature, which are focused on providing a personalised and optimal experience to customers. The structure of the report is the following: Sect. 2 presents this report's research question and goal. Sections 3 and 4 describe the research method and the key findings of this report, respectively. We finally conclude our study and discuss the opportunities and challenges in the field, with regard to the gathered insights, in Sect. 5.

2 Goal

The increased adoption of e-commerce platforms by retail stores, that usually provide mobile-friendly interfaces and apps, have lead to a rising interest in the research and development of methods for the analysis, optimisation, configuration and customisation of the customer's journey, that are based on AI algorithms. However, a recent literature survey of the different categories of these AI and ML approaches, that aim to optimise the customer journey, is lacking. This report aims to fill this gap by providing an overview of the newest intelligent approaches that belong to the aforementioned field, while identifying their use cases. The key research question and underlying goal that this paper tries to cover can be expressed as the following: "what are the intelligent applications used in the peer-reviewed literature during the past two years regarding the customer's journey?".

3 Approach

This survey was conducted as a systematic literature review, entailing a thorough, transparent and replicable process for the literature search and analysis. The research was conducted in April 2021 using the Google Scholar, Elsevier

Science Direct and IEEE Xplore literature search engines. The survey is targeted to articles published during the 1st of January 2018 to the 31st of March 2021, which make use of the keywords "Customer Journey" or "Buyer's Journey", and "Artificial Intelligence" or "Machine Learning" or "Deep Learning", and "E-Commerce" or "Retail". The main objective is the conduction a targeted survey of the most recent intelligent applications for improving the customer's journey.

The literature search led to a sum of approximately 70 peer-reviewed papers. After the removal of duplicates and the exclusion of studies where only the paper's abstract was available or accessible, as well as other works which were not in accordance with our research question, we finally conclude on ten (10) noteworthy works for the study. Table 1 presents the target industry, the scope and the underlying machine learning algorithms used in the selected scientific papers.

Table 1. The ML algorithms, the scope and the industry of the surveys selected papers.

Reference	Industry	Scope	Algorithm
Wu et al. [5]	e-commerce	Predict customer actions	LSTM, DNN
Karessli et al. [8]	Fashion	Visual sizing	CNNs
Goossens et al. [9]	e-commerce	Improving KPIs	Order-Aware RS
Damian et.al. [11]	e-commerce	Shopping history	RNN
Argal et al. [15]	Travel booking	Personalization	Collaborative RS
Gandi et al. [16]	e-commerce	Customer experience	DNN
Chan et al. [17]	Fashion	Customer analysis	Fuzzy logic
Jabbar et al. [18]	e-commerce	Customer analysis	SVM
Yiran et al. [19]	e-commerce	Predict customer actions	LDA models
Akour et al. [23]	Mobile learning	Customer analysis	Decision Tree

4 Results

The integration of AI and ML techniques can be useful in all of the three stages of the customer journey (pre-transaction, transaction, and post-transaction). The selection and use of the most prominent technology solutions depends on the purpose of these applications. The analysis of the possible paths for finding the most prolific ones, the identification of issues that occur during the customer's experience, as well as the redesign and the automated transform of the customer journey belong to the primary commercial objectives and goals of almost every online retail system. Furthermore, the measurement, analysis and further optimization of the customer journey contribute to the provision of a prominent and impactful experience.

The goals described above can be achieved with the analysis of each customer's value for a platform, using techniques like customer churn prediction, and with the adjustment and improvement of their experience, using AI and ML technologies. This personalisation and improvement of the user experience across the different paths can have a crucial role. Therefore, we highlight the importance that the integration of a continuous optimization of the customer journey has for a company active in e-commerce, which is a significant requirement for development and growth.

There are numerous of AI and ML technologies, regarding the customer journey optimization, that a company could include on its operational and marketing plan. The main categories that these technologies fall into, holding the majority of the provided solutions, are recommendation systems (RS), virtual reality systems (VR) and augmented reality systems (AR), and recommendation agents (RA). Furthermore, methods for customer analysis have been proposed to assist a company to better understand its clients, with the aim of facilitating the achievement of the above goals.

4.1 Recommendation Systems

Recommendation systems are essential tools that have been used in abundance by companies that provide online retail and streaming services. Two popular examples are the platforms provided by Amazon and Netflix [20,21], both of which are known for the personalized experience they offer to their customers. The methods deployed include the collection and the analysis of demographic data from their customers, and the combination of this data with information from previous purchases, product reviews and overall behaviour. By merging these features, it is possible to predict how customers will evaluate related products, or how likely they are to purchase a new product. At the same time, these contribute in leading the customer to the optimal journey, by guiding them towards the most prolific path. Several recommendation approaches have been proposed, with the most important ones described below [22]:

<u>Collaborative Filtering</u>: These are methods where automated predictions (filtering) associated with the user's interests are extracted by utilizing the preferences of multiple users (collaborating). The underlying premise of the collaborative filtering approach is that "if a person A has the same opinion as a person B on an item, A is more likely to have the same opinion with B on a different item than a randomly selected person". Recommendation systems that utilize collaborating filtering methods can yield exceptional results when there is a large record of previous user-item interactions. At the same time, they face a significant issue when the above requirement is lacking. This is expressed as the "cold start problem" in the relevant literature, and occurs when new users or new items are introduced. Another drawback of these approaches is that items that have a large track record are highly recommended, in contrast to those that do not have any, or have too few.

<u>Content-Based</u>: These methods utilize item features, such as the product description, and build a profile of the user, recommending similar items. In content-based recommendation systems, the algorithms attempt to recommend to the

user products similar to previous ones that they liked in the past. The key idea behind content-based filtering is that "if a user likes a product, they will also like a similar product", with examples being the recommendation of similar types of films or songs in corresponding applications. An essential problem in this approach is the diversity of the recommended items, which can fall into niche categories.

Utility-Based: This type of approach performs recommendations based on the utility of each object for the user. The main advantage of these systems is that they can make suggestions that are not related to product features, such as the availability and reliability of the supplier. However, the calculation of the utility that items have for individual users can be non-trivial and industry specific, a property that complicates the overall process.

Knowledge Based: A system of this type generates recommendations to users according to explicit information related to their preferences and needs about the item assortment. Using this knowledge, these methods discover correlations between products and customer needs, and avoid the cold-start problem. However, the processes of acquiring the desired knowledge can time-consuming and tedious.

Demographic-Based: This approach categorizes users based on a set of demographic information. Methods that follow this paradigm are trivial to implement if such data is available (which is the case for large companies such as social media platforms or large online retail stores). A big advantage is that no previous purchase information is needed. However, this information can include personal and sensitive data that falls into restrictions, and thus precautions regarding data leakage and ethical use should be taken.

Hybrid Approaches: These concern the combination of two, or more, of the above approaches. Hybrid systems can be used to improve the accuracy of the provided recommendations, and at the same time overcome the drawbacks that separate components may introduce. For example, by using a hybrid approach, challenges such as the "cold start problem" and the lack of diversity can be tackled with the use of a collaborative filtering approach coupled with a content-based one. The constitute methods can be combined in different manners, like the weighting or the switching of these. It is important to note that the most prominent industry solutions utilize hybrid approaches.

Concerning specific applications of recommender systems, in 2018 Goossens et al. presented OARA, an Order-Aware Recommendation Approach that focuses on predicting behaviours in the customer journey and recommend the ones that maximize an organization-specific key performance indicator (KPI). The approach allows for the prediction and recommendation on datasets which fit the concept of a customer journey. Moreover, it was shown how one could go beyond merely visualizing the journey, by utilizing the model for these tasks. Additionally, the order of events was taken into account in the previous parts of the journey, and leveraged the author's positive effects. Last but not least, it was shown that OARA could be further enriched by the effective use of additional sources of information. These predictions are then used in combination with the representative customer journeys during the recommendations, in order to find a recommendation that both increases the KPI and is well suited, based on the actions previously observed in the customer journey [9].

An approach for improving the recommendations of such systems was proposed in 2019 by Gandi et al. Since many e-commerce platforms provide products from third-party marketplace sellers, product listings can contain non-compliant images or offensive content. These include content that can damage the customer experience or lead to legal issues, which can both damage the company and the customer's trust in it. The authors propose a number of machine learning and computer vision strategies, based on deep neural networks (DNN), for tackling these problems, aiming to improve the customer journey [16].

4.2 Augmented and Virtual Reality

Current augmented reality (AR) and virtual reality (VR) technologies can have a thorough impact on a variety of customer experience centred applications, along with the customer journey, by enabling or enhancing consumer tasks and experiences. Simultaneously, machine learning techniques are continuously engaging with users and build smarter solutions, aiming to process data in real time and provide a customized experience based on user demand. There are three key areas in which machine learning is integrated with AR and VR applications. These concern the pre-purchase, purchase, and post-purchase stages in the areas of advertising, retailing and transactions.

Mobile AR advertisement enables the interactive exploration of products through the provision of information on content, design, reviews etc., and allows consumers to interact with printed and outdoor media by providing a QR code. Furthermore, product placement has been facilitated through AR to contextualize a product, such as showing it being used in a dreamy virtual context. Mobile AR and VR apps have been used to reduce friction and costs associated with returns in online retailing, by making it possible to try out products at home on avatars or on a visual representation of the customer. Last but not least, some AR applications allow customers to customize products to their specific needs.

As an example of the above applications, Nike created the mobile application Nike Fit [6], an AI solution which leveraged ML, data science and machine vision to find the appropriate shoe size for a unique foot shape. Another noteworthy example of a real-world mobile AR application in business is the IKEA Place app [7], which uses AI to offer users accurate and personalized home furnishing advice, reducing the customer's uncertainty before purchase and leading to an improved overall satisfaction after the purchase.

In 2019, Karessli et al. introduced a weakly supervised teacher-student approach, where an architecture called SizeNet, based on Convolutional Neural Networks (CNNs), acts as the student, in order to learn visual sizing cues from fashion images. A statistical model with access to articles sales is used, which has the role of the teacher. The authors demonstrated that fashion images contain valuable information regarding not only taste, but also size and fit issues that can be considered valuable assets in tackling the challenging "cold start problem" and consequently improve the customer journey [8].

Shorty after the above study, in 2020 Wedel et al. proposed an integrated framework for consumer marketing research based on VR that focuses on the improvement of customer experience. This is achieved by a) providing simulated multi-sensory product experiences, which are interactive processes that allow for dynamic adjustments in product viewing, trial and usage, b) specialization and contextualization of product experiences and c) social interactions around virtual products, services and experiences, by utilizing avatars or telepresence. These features provide the customers with a comprehensive experience of an augmented or virtual environment, and create the perception that they are present and active within these environments [10].

4.3 Recommendation Agents

Recommendation agents, although a relatively new technology that is still being evaluated by the market, can play a crucial role in e-commerce platforms. They can strongly affect the transaction and post-transaction stages, by providing recommendations, answering questions about products and their usages, and providing meaningful advices to customers in order to reduce uncertainty. Moreover, virtual assistants and chatbots can provide feedback and facilitate additional consumption [12]. Facebook, Google, Amazon and Apple are only a few examples of companies that have integrated recommendation agent technologies in their systems.

In 2018, Argal et al. proposed a system based in collaborative recommendations that can be applied to deploy services in the travel booking domain, so as to produce appropriately personalized and accurate search predictions for a user, using a stimulating and intelligent conversation in natural language. The lack of a visual interface was dealt via the use of information cards provided by the Alexa AI assistant and a web application. Authors concluded by showing that most of the basic features available on travel websites can be easily incorporated by a voice-enabled chatbot [15].

4.4 Customer Analysis

Plenty of customer analysis methods have been proposed in the literature, with most of them analysing information like the metadata of customers and items, as well as information regarding purchases, navigation history, etc. Focusing on IoT data applications, in 2018 Chan et al. proposed a platform for assisting fashion retail stores. By collecting data with sensor devices, customer behaviour can be captured and transmitted trough the wireless network towards the cloud. Using a fuzzy logic approach, insights can be gained regarding the customers, that can be used for future recommendations in the company's online platform, and future trends can be predicted [17].

In 2019, Wu et al. presented a two-step deep learning architecture. A long short-term memory recurrent neural network (LSTM-RNN) is used to handle time-series data, such as customer contact history, and a feed-forward deep neural network is used to handle static features. This framework aims to allow companies to predict future customer behaviour and understand the relationships between them. In addition to the framework, an abstract concept called "action" was introduced to describe a single-day customer behaviour. The framework enables the utilization of action embeddings to provide a better understanding of the underlying relationships among customer behaviour [5].

In the same year, Damian et al. proposed a hierarchical model in which two RNNs are combined, the parent RNN, which focuses on the customer's shopping history, and the child RNN, which targets transactions at the user session (basket) level. This approach allowed the child bi-directional RNN to obtain the actual embedding representation of each transaction content by combining the basket's items and by averaging the embeddings of these items. The model also utilized a decoder with the purpose of predicting complete future transactions for each customer [11]. In [18] a different, non neural, approach was proposed, where support vector machines (SVM) from classical machine learning were utilized for sentiment analysis, focused on e-commerce applications. The model made use of product reviews, and aimed to further improve customer experience with the gathered insights.

Authors of [19] argue that, while e-commerce platforms have become an easy and preferable way of shopping, most of the times information data is unstructured and difficult to arrange, in order to gather insights from it with the aim of understanding what potential consumers might desire as an improvement. The authors propose the utilization of LDA models for word clustering, and proceed to test their framework in a corpus of mobile phone reviews for the tasks of topic labelling and sentiment analysis. More recently, in 2021 machine learning algorithms were used for predicting the intention of using mobile learning platforms during the COVID-19 pandemic, with a decision tree method being reported as the best performing one [23].

5 Conclusions

Corporations and companies across different business domains and markets, of varying size and revenue, are using AI and ML technologies to improve their customer experience. This report provided an overview of the recent intelligent methods and the approaches taken. It is evident that recommendation systems, AR and VR technologies, recommendation agents and techniques for customer analysis have a significant impact on the personalization of the customer journey, with applications available in easy-to-use mediums like mobile phones. As a result, the utilization of these technologies constitutes important opportunities for such companies for increasing their profits and the overall customer satisfaction, but at the same time imposes multiple challenges. Future research initiatives could expand the capabilities of the existing intelligent applications, or lead to the emergence of new and innovative technologies.

Simply including an intelligent component in an online store is not enough to improve customer experience. It is necessary to constantly strive for optimization and personalization, along with the constant redefinition of objectives and goals, and to analyse the requirements that arise from the dynamic environments of both the consumer and the business world. Finally, important issues that emerge from the continuous development of the above, such as the security and integrity of personal and sensitive information, should be highlighted and addressed.

Acknowledgements. This work has been co-financed by the European Union and Greek national funds through the Operational Program Competitiveness, Entrepreneurship and Innovation, under the call RESEARCH - CREATE - INNOVATE (project code: T2EDK-03843).

References

1. Meyer, C., Schwager, A.: Understanding customer experience. Harvard Bus. Rev. **85**(2), 116 (2007)
2. Lemon, K.N., Verhoef, P.C.: Understanding customer experience throughout the customer journey. J. Mark. **80**(6), 69–96 (2016)
3. Homburg, C., Jozić, D., Kuehnl, C.: Customer experience management: toward implementing an evolving marketing concept. J. Acad. Mark. Sci. **45**(3), 377–401 (2015). https://doi.org/10.1007/s11747-015-0460-7
4. Følstad, A., Kvale, K.: Customer journeys: a systematic literature review. J. Serv. Theory Pract. **28**, 196–227 (2018)
5. Wu, Q., et al.: Speaking with actions - learning customer journey behavior. In: 2019 IEEE 13th International Conference on Semantic Computing (ICSC), pp. 279–286 (2019). https://doi.org/10.1109/ICOSC.2019.8665577
6. What is Nike Fit? Nike News 09 May 2019. https://news.nike.com/news/nike-fit-digital-foot-measurement-tool. Accessed 10 June 2021
7. IKEA Place app. IKEA mobile apps. https://www.ikea.com/gb/en/customer-service/mobile-apps/. Accessed 10 June 2021
8. Karessli, N., Guigoures, R., Shirvany, R.: SizeNet: weakly supervised learning of visual size and fit in fashion images. In: 2019 IEEE/CVF Conference on Computer Vision and Pattern Recognition Workshops (CVPRW), pp. 335–343 (2019). https://doi.org/10.1109/CVPRW.2019.00046
9. Goossens, J., Demewez, T., Hassani, M.: Effective steering of customer journey via order-aware recommendation. In: IEEE International Conference on Data Mining Workshops (ICDMW), pp. 828–837 (2018). https://doi.org/10.1109/ICDMW.2018.00123
10. Wedel, M., Bigné, E., Zhang, J.: Virtual and augmented reality: advancing research in consumer marketing. Int. J. Res. Mark. **37**(3), 443–465 (2020). https://doi.org/10.1016/j.ijresmar.2020.04.004
11. Damian, A., Piciu, L., Turlea, S., Tapus, N.: Advanced customer activity prediction based on deep hierarchic encoder-decoders. In: 22nd International Conference on Control Systems and Computer Science(CSCS), pp. 403–409 (2019)
12. Hoyer, W. D., Kroschke, M., Schmitt, B., Kraume, K., Shankar, V.: Transforming the customer experience through new technologies. J. Inter. Mark. **51**, 57–71 (2020)
13. Torres-Dávila, D., Porles-Arévalo, J., Mauricio, D.: The customer experience maturity model in the e-commerce processes. In: 2019 IEEE XXVI International Conference on Electronics, Electrical Engineering and Computing (INTERCON), pp. 1–4. IEEE (2019)

14. Terragni, A., Hassani, M.: Analyzing customer journey with process mining: from discovery to recommendations. In: 2018 IEEE 6th International Conference on Future Internet of Things and Cloud (FiCloud), pp. 224–229. IEEE (2018)
15. Argal, A., Gupta, S., Modi, A., Pandey, P., Shim, S., Choo, C.: Intelligent travel chatbot for predictive recommendation in echo platform. In: IEEE 8th Annual Computing and Communication Workshop and Conference (CCWC), pp. 176–183 (2018)
16. Gandhi, S., et al.: Scalable detection of offensive and non-compliant content/logo in product images. In: IEEE Winter Conference on Applications of Computer Vision (WACV), pp. 2236–2245. IEEE (2020). https://doi.org/10.1109/WACV45572.2020.9093454
17. Chan, C.O., Lau, H.C.W., Fan, Y.: IoT data acquisition in fashion retail application: fuzzy logic approach. In: International Conference on Artificial Intelligence and Big Data (ICAIBD), pp. 52–56 (2018). https://doi.org/10.1109/ICAIBD.2018.8396166
18. Jabbar, J., Urooj, I., JunSheng, W., Azeem N.: Real-time sentiment analysis on e-commerce application. In: IEEE 16th International Conference on Networking, Sensing and Control (ICNSC), pp. 391–396 (2019). https://doi.org/10.1109/ICNSC.2019.8743331
19. Yiran Y. and Srivastava S.: Aspect-based sentiment analysis on mobile phone reviews with LDA. In: 4th International Conference on Machine Learning Technologies (ICMLT), pp. 101–105. Association for Computing Machinery, New York, NY, USA (2019). https://doi.org/10.1145/3340997.3341012
20. The Amazon Recommendations Secret to Selling More Online. Rejoiner. https://rejoiner.com/resources/amazon-recommendations-secret-selling-online/. Accessed 2 June 2021
21. Netflix: Recommendations Worth a Million. KismetK. https://rpubs.com/kismetk/Netflix-recommendation. Accessed 2 June 2021
22. Ricci, F., Rokach, L., Shapira, B.: Introduction to recommender systems handbook. In: Recommender Systems Handbook, Boston, MA: Springer, US (2011). https://doi.org/10.1007/978-0-387-85820-3_1
23. Akour, I., Alshurideh, M., Barween, A.K., Amel, A. A., Salloum, S.: Using machine learning algorithms to predict people's intention to use mobile learning platforms during the COVID-19 pandemic: machine learning approach. JMIR Med. Educ. **7**(1), e24032 (2021). https://doi.org/10.2196/24032

LEXISGURU: Mobile Application for Learning Basic Lexis in English for Kids

M. Janeru Wageesha Jayasinghe,
W. H. M. Aruna Dilshan Hennayaka Hennayaka(✉),
M. Prakash Madhusanka Fernando,
K. Nethmini Umayangana Thilakarathne, Uthpala Samarakoon,
and Suriyaa Kumari

Faculty of Computing, Sri Lanka Institute of Information Technology,
Colombo, Sri Lanka
{it17165808,it17166270,it17166652,
it17169318}@my.sliit.lk,
{uthpala.s,suriyaa.k}@sliit.lk

Abstract. Lexis is an essential part of English vocabulary that puts a good foundation on a child's English knowledge. In this rapidly globalizing world, it is fundamentally essential to learn English from a young age. In recent years eLearning, mobile applications have been developed for teaching Lexis to children. The market of educational mobile apps, especially for English language learning, has been rapidly growing. Especially in a country like Sri Lanka, English is not the mother tongue, it is the second language. So, when that second language is not taught right the child will lose interest in learning that language. The problem is that the existing lexical learning mobile applications does not aim at keeping the child interested and interactive in the learning process and in Sri Lanka, children find it difficult to understand these lexical parts. As a result, teachers and parents had to spend a lot of time to teach them those lexical parts. We designed and developed a mobile application called "LexisGuru" that uses interactive and effective ways to teach three lexical parts that are homophones, synonyms, and antonyms to children aged between 8–10 in Sri Lanka. This mobile application uses Machine Learning (ML), Image Processing (IP), gamification that includes collaborative environments, and speech recognition techniques. The developed mobile application was introduced to primary level learners, and they were all very attracted and interested while using this application. The attractive user interfaces, the pretests, and posttests, notifying the child when he loses focus while learning, using interesting stories and activities to teach lexis, playing a game with multiple players, and asking questions from the lesson and taking the voice inputs gave a new experience and showed that making the mobile application interactive as possible is an effective way to teach lexis to children.

Keywords: Mobile application · Lexical learning · Machine Learning (ML) · Image Processing (IP) · Voice recognition

M. E. Auer and T. Tsiatsos (Eds.): IMCL 2021, LNNS 411, pp. 133–144, 2022.
https://doi.org/10.1007/978-3-030-96296-8_13

1 Introduction

Many Sri Lankan students find it difficult to cope with the English Language and it is evident by the statistics and school performance indices of the department of Examinations, Sri Lanka. In 2019, fail rates of English language are 37% and 46% at G.C.E. Ordinary Level examination and G.C.E. Advanced level examination, respectively [1]. Furthermore, in a research which involved a group of university students at Southeastern University of Sri Lanka shows that 40% of university students have passed the ordinary level English examination, while only 20% have passed the Advance level General English examination [2]. Lexis is one essential component of language and language development. Limited lexical knowledge can lead learners to frustration and demotivation. This has a negative impact on the student's future career [3].

According to the survey carried out with primary and secondary level English teachers in Government schools and Private schools, the most difficult and doubtful part of the English vocabulary is Lexis. Out of Lexical parts Homophones, Synonyms and Antonyms were identified as the most difficult parts. Lexical development has become an integral part of language acquisition. According to the "TKT, KAL Module of Cambridge university" Lexis refers to single words or sets of words that have a specific meaning [4]. Learners can build fluency in English by learning vocabulary systematically. Without grammar, people can talk little but without vocabulary, people can talk nothing. Hence lexis, help learners to develop their ability to use English in real contexts.

1.1 Mobile Based Learning

Mobile-based E-learning is the latest version of E-learning. Mobile learn encapsulates the ability to progress through course content on one's own personal devices such as smartphones and tablets. Mobile Learning offers ultimate accessibility & flexibility to learners especially for children as they can access learning courses anytime, anywhere via their parents' mobile devices.

Technology has a huge demand in the community and has a great relationship between education and technology. In the 21st century, the usage of technology is considered as the best solution for several common problems. Among those technologies that can be used, machine learning, speech recognition, collaborative learning, and face recognition have taken major positions.

1.2 Speech Recognition

Natural language communication with a computer adds to applications a new dimension. Speech recognition is a process that enables computers, mobile phones, and various devices to recognize and translate spoken language into text format. It improves the interaction between humans and machines by converting to speech to text. Applications which use speech dialogues are more user friendly than other applications. As per the research done by Heni and Hamam (2016), small children enjoy the applications embedded speech recognition [5].

1.3 Collaborative Learning

Collaborative learning technologies allow peer-to-peer virtual connection via virtual devices including real-time data synchronization for communication in a virtual environment. Machine learning is another technology that allows machines to learn from data. It brings together statistics and computer science to enable computers to learn how to do a given task without being programmed to do so.

1.4 Attention Detection

Attention is an important variable to be measured since it plays a fundamental role in the accumulation of information with the stimulus of the learner's memory during the integration of knowledge [6]. As per researchers, in e-learning, students often far away and out of teachers' control which may cause students do not have strong learning motivation and might feel fatigued and inattentive for learning. Hence, A real-time attention measuring approach can support better control the learning attention of students in unsupervised learning environments such as mobile applications [7].

2 Related Work

With the development of technology, many people tend to do research and develop applications based on mobile-based learning for the basics of English. Even though, most of the existing applications are focusing on gamification-based learning, a very smaller number of applications and research are focusing on collaborative learning and speech recognition systems to improve the user's interactivity and interest while learning. Also, it is important to check the knowledge, and to teach according to the level of knowledge is very important. Most of the research and applications are not focusing on knowledge level of users. Although the teaching methods are in a good standard, it is useless without a good concentration. Hence, student concentration on the lesson is another most important factor.

Karn et al. in (2019) has done a research to show the growth of knowledge level between the pre-test and the post-test among school students [8]. They identified a significant improvement in knowledge level in pre-test and post-test to school students regarding eye diseases. Overall knowledge before the test was 44% which was low and was increased to 71% after the test. Another research on automatic speech recognition (ASR) for second language learning examines how automatic speech recognition (ASR) can be used to improve pronunciation in a second language [9].

In a research on face detection and face recognition in mobile applications focuses to developing a mobile application for a smartphone to detect the human face [10]. As per the researcher, in the mid-1960, computers were used to detect the human face and lately robust facial recognition systems were developed. To have proper recognition, it is very important to understand the factors such as distance between eyes, the width of the nose, and the depth of eye sockets. There are myriad mobile applications related to Lexis available on the Google play store. Most of the existing applications mainly focused on teaching and learning one part of lexis, as only Homophones. Most of the applications provide some activities related to lexical words and there is no proper way of teaching theoretical parts of lexis. Most of the existing mobile applications focused on single-player game-based learning rather than using gamification-based collaborative learning. And, they have implemented level-based learning according to the knowledge level of most students. Further, checking the child's attention when using the application and child interaction with the application using speech recognition is very rarely used in existing applications.

"LexisGuru" application has more features over the other existing lexis learning applications. In this application, kids can learn all three lexical parts (Homophones, synonyms, and antonyms) as individual learning or collaborative learning. Individual learning contains gamification with speech recognition and teaching lexis using storytelling. Collaborative learning has two player-based collaborative gamification including interesting teaching methods for lexical learning. Before the learning and Gamification modules, this application would identify the child's knowledge level of lexis using pretest and post-test. Further, this application tracks the child's attention on the content while learning.

3 Methodology

As the first step, a questionnaire-based survey was conducted among primary English teachers to gather important facts to come up with a best solution for teaching English Lexis to children. According to the survey findings, building a mobile application was the best solution and as a result the "LexisGuru" mobile application was developed using Android Studio platform. The following Fig. 1 shows the system overview of the application.

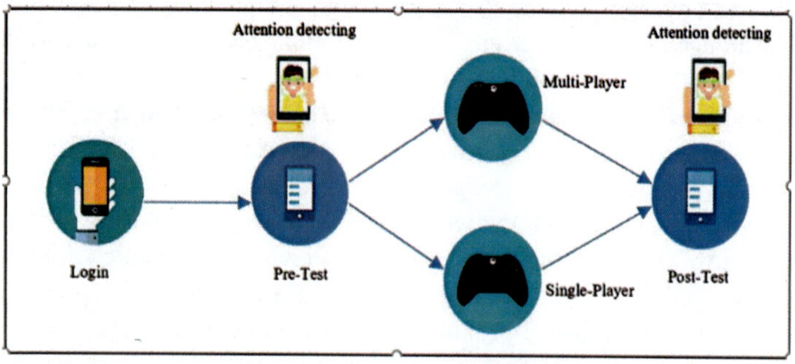

Fig. 1. System overview diagram

The application was designed to teach lexis in an interesting manner using gamification and speech recognition. Also, the application can detect user's attention while using the application. Hence, the "LexisGuru" application consists with four main functions; Identify and compare knowledge level of the user via pre and posttests,

Teaching lexis using storytelling embedded with speech recognition, Two-player collaborative Game-based lexis learning and Student attention monitoring.

3.1 User Knowledge Level Detection and Comparison via Pre and Post Tests

The knowledge level identification feature (Fig. 2) of the "LexisGuru" mobile application has been used to identify the appropriate knowledge level of the user and teach the lessons accordingly. Identifying the appropriate level for the child ensure the child start in the correct level of lexis according to his/her existing knowledge. Otherwise, child may feel boarded to learn already known concepts. Hence, machine learning techniques were used to identify the knowledge level. Due to the Covid-19 pandemic, data was gathered using online techniques such as Sharing google sheets. There are 10 questions for the pre-test covering all three lexical modules of Synonyms, Antonyms, and Homophones. Each user has 20 min for completion. The correct and incorrect answers were taken as 1s and 0s. Both accuracy of the answer and the time taken to complete the pre-test were included in the dataset. According to the pre-test child's knowledge will be categorize into three levels. Similarly, post-test was used to check the child's knowledge improvement after every lesson. Hence, three data sets were maintained for the three different sections of lexis.

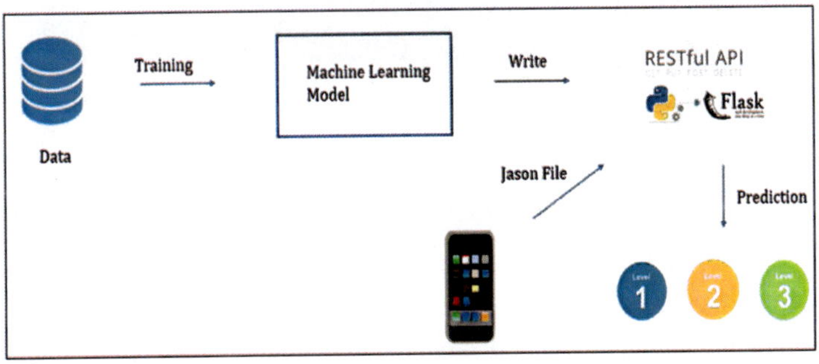

Fig. 2. Level identification high-level diagram

Analytically 70% of the data was used to train the Machine Learning (ML) model and 30% of the data used for testing the training data set. Initially, the model was created using both the Decision Tree (DT) algorithm and the Random Forest (RF) algorithm to select the best solution. When considering both models RF model was more successful for level prediction due to high accuracy level. Finally, successfully trained data set with approximately 2300 data was used for the pre-test and data set of approximately100 data was used for post-test.

When developing the "LexisGuru" application, both pre-test and post-tests were created using android studio IDE. PyCharm IDE was used to train the dataset. Data was passed through a JavaScript Object Notation (JSON) file with the time taken to complete both tests. Flask API was used to retrieve the data from the front-end and to receive the correct output from the server. If the child's response regarding the pre-test is like Level 1, the screen displays Level 1 as the suitable knowledge level for the child. This process demonstrates in Fig. 3 (Left). similarly, if it is Level 2 screen displays as Level 2 and if it is Level 3 screen displays as Level 3.

Upon the completion of the post-test, the "LexisGuru" will indicate whether the child has been passed or failed the corresponding level. The Fig. 3 (right) shows the interface of the post-test, and the result of post-test.

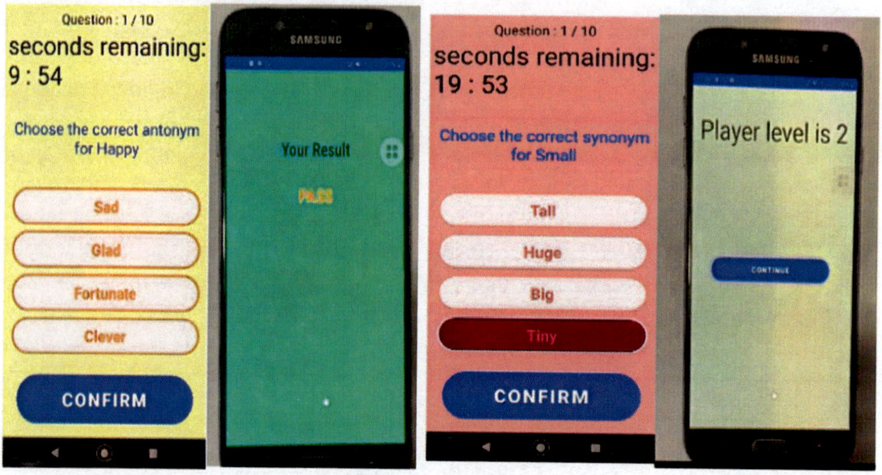

Fig. 3. Pre-test interface with output of pretest and post-test interface with post-test results

3.2 Teaching Lexis Using Storytelling Embedded with Speech Recognition

Maintaining curiosity and attention in teaching learning process is challenging. This component focuses on preserving user's interest and interactivity during the learning process using storytelling embedded with speech recognition. Android studio was used to develop this component.

This function is two folded, lexical teaching and single player game. An animated story with various characters used to teach lexis with many examples. The story is narrated as an audio output. The single player game uses speech recognition, audio outputs as well as attractive animations to keep the user interactive and fascinated while playing the game. Each level of the game has a pool of questions and random questions presented to the user when the user plays the game. The number of questions increased in each level to improve the complexity of the game. The questions presented as audio outputs using Android Text to Speech engine. Game instructions present as audio outputs prior to the game and the answer of the user has been taken as a speech input. The correctness of the given answer shown as the output with feedback for incorrect attempts (Fig. 4).

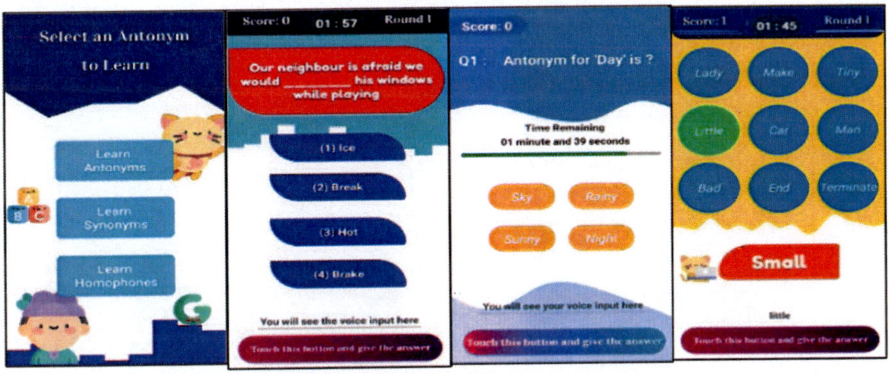

Fig. 4. Speech recognition game UI

Figure 5 shows the speech recognition process of Android and the speech recognition service in the android studio has been used. The speech recognition service grants the access to the speech recognizer and the API in the speech recognizer processes the speech input and convert to a text.

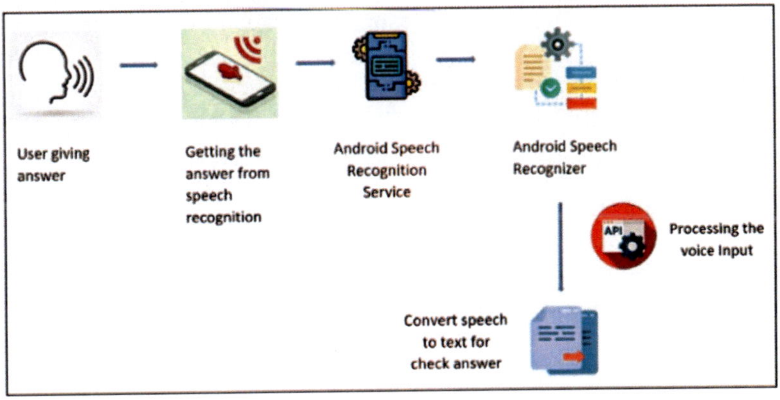

Fig. 5. Speech recognition process of android studio

3.3 Teaching Lexis via Two-Player Collaborative Game-Based Learning

This function used to increase the interactivity with the learners in the teaching learning process. This phase has used android studio IDE to reduce the complexity of the game while giving a smooth experience for those who use the application. The application has used real-time database functionality in Firebase to give real-time data synchronization between two devices. The application has 3 lexical modules as Homophones, antonyms, and synonyms. All these modules have a teaching part and a gaming part, and each module consists of different learning styles and different puzzle-based games. Matchmaking mechanism was used to select 2 players for the game with same knowledge level. Figure 6 (Left side) shows the mechanism of data synchronization in the firebase real-time database.

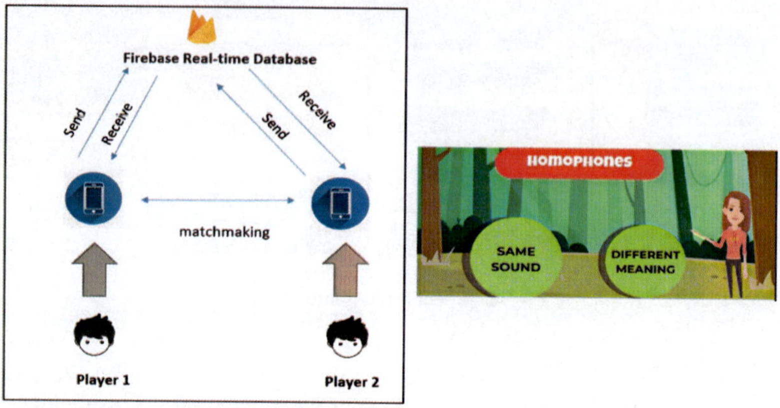

Fig. 6. Data synchronization in firebase (Left) and teaching session (Right)

Accordingly, event listeners catch the value change in one device and synchronize that value change in another device. In-room connection feature one device is creating a room and that is shown in the other devices as a room join invitation. When the second player accepts the invitation both players are connecting to start the game. Also, users could go through the teaching section before starting the game. The teaching section is consisting with a 3D animated video as shown in the of the Fig. 6 (Right side).

Gaming session begins with the voice-activated instructions, and it consists with puzzle-based mind games developed using Java language. Each game has real-time data synchronization modules developed using firebase event listeners and handlers. While playing the game, a tasked permed by one player has been caught by the data synchronization module and notify the other player and change the game state real time as shown in Fig. 7.

Fig. 7. Multiplayer game UI

Every level has 3 puzzle activities and at the end of each activity application prompts the marks and the winner. At the end of each level, the application prompts detailed progress of the activities and move to the next level in both devices real time.

3.4 Student Attention Detecting Using Google Mobile Vision API

Maintaining attention while teaching is challenging in teaching learning process. Hence, this component mainly focuses on detecting and rebuilding the user's attention using Google Mobile Vision API (Fig. 8). The technology used in this component tracks the orientation of facial landmarks and provide information related to facial features.

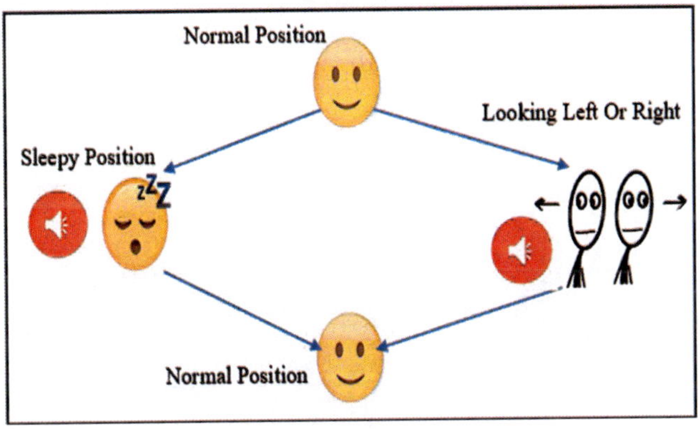

Fig. 8. Attention detecting overview diagram

In user attention detection process, focal point was used to identify whether the user is sleepy during the lesson. In here, eye opening probability was used as a measurement. Then left and right eye-opening probabilities were calculated separately. When human eyes are open, probability is closer to 1.0 and otherwise it is less than 0.2 (Fig. 9).

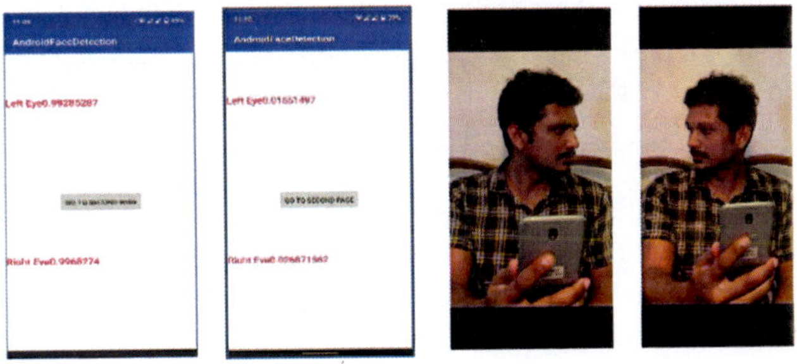

Fig.9. Eyes open and close position detection (Left) and looking left-side and right-side positions (Right)

Once the application identified a sleepy user, the system generates a real time voice output to draw the attention back to the lesson. Also, the application can identify distracting user as they are looking out of the screen (Fig. 9). This is done by recognizing the face orientation and head rotation of the user. The horizontal position of the face is measured by the command get position().x in the Google face API. When the value of the position X is between 90 and 140, the app identifies the user as concentrated to the lesson and otherwise as not. Here also the application gives a real time response using voice commands draw the attention back to the lesson.

4 Results and Discussion

Due to Covid19, the research team tested the application using few set of English teachers remotely. The APK file was sent to the teachers and feedback was taken using a google form. Accordingly, sixteen English teachers were involved in the testing phase.

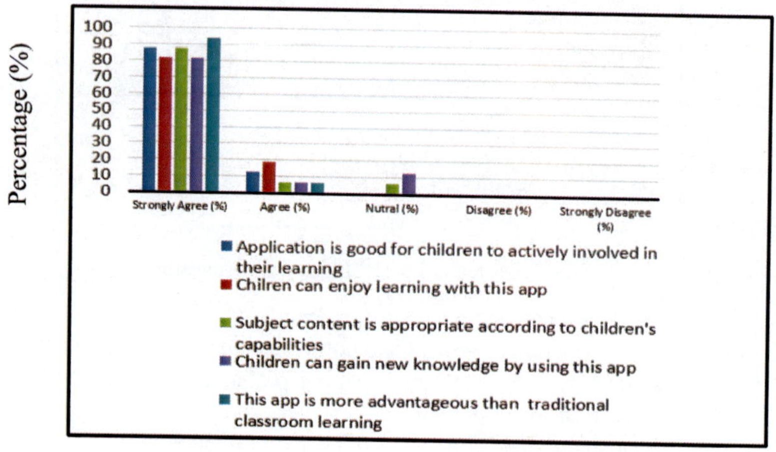

Fig. 10. English teachers feedback summary

The above Fig. 10 demonstrates the summary of statistics collected from the English teachers' feedback survey. Overall results indicate that most of the teachers are satisfied with the"LexisGuru" mobile application. Also, all the teachers agreed that application is good for active learning, children could enjoy the application and the application is better than traditional classroom-based learning. Only 3 teachers having unbiased impression on subject content appropriateness and ability of gaining new knowledge. None of the teachers involved in application testing phase were disagree or strongly disagree with any of the aspects mentioned above. The reason may be the low sample size due to Covid19 pandemic.

In addition to the above facts, the teachers were asked to comment on the application in an open-ended question in the questionnaire. One teacher mentioned that "This application seems very interesting". Another teacher's idea was that "This looks nice. I think student will like this". Another comment was that "This application is designed in a very interesting and attractive way. Also monitoring students' attention will be very useful too." A Few teachers have given some suggestions too, to improve the application such as adding more words to each lexical type and to include some other lexical sections too.

While considering all the facts it is evident that "LexisGuru" would be an application that can help children in gaining interest in English and obtaining higher grades for English in future.

5 Conclusion

Synchronization of latest mobile technologies and education contexts provide a fruitful learning experience for children since they be inclined to use mobile devices frequently. On the contrary to traditional classroom learning which gives board feeling mostly, mobile based learning implanted with latest technologies such as speech recognition, gamification and attention monitoring enables collaborative peer learning and fruitful and meaningful learning experiences. This research work studied the effectiveness and usage of mobile based learning over traditional classroom learning when teaching English lexis. Hence the mobile technologies and gamification with speech recognition techniques can moderate to give the best learning experience to children. This will lead to student centric active learning environments and children will get the chance to learn English lexis in a fascinating mode.

The testing results of qualitative data acquired through the survey shows higher teacher satisfaction in "Lexis Guru" application. As indicated in comments, "Learning lexis in this way will be very interesting to children", "This application will be very popular among children……" and "Attention monitoring is really good" teachers appreciated the proposed application. The overall system after development seemed promising at fulfilling its task but could be improved leaving room for future areas of research.

References

1. D.O. education, Statistics and School Performance Indices, Department of education, 2019. https://doenets.lk/statistics. Accessed 27 July 2020
2. Majeed, M.N.A.: ResearchGate, 2016. https://www.researchgate.net/publication/328627518_challenges_faced_by_students_in_english_medium_undergraduate_classes_an_experience_of_a_young_university_in_sri_lanka_introduction. Accessed 27 July 2020
3. Caro, K.: Lexis, lexical competence and lexical knowledge: a review. J. Lang. Teach. Res. **8** (2), 205 (2017)
4. C. University, Cambridge University Press 2020. https://www.cambridge.org/cg/cambridgeenglish/catalog/cambridge-english-exams-ielts/tkt-course-kal-module. Accessed 27 July 2020

5. Heni, N., Hamam, H.: Design of emotional educational system mobile games for autistic children. In: 2016 2nd International Conference on Advanced Technologies for Signal and Image Processing (ATSIP), pp. 631–637. IEEE, March 2016
6. Bos, A.S., et al.: Educational technology and its contributions in students' focus and attention regarding augmented reality environments and the use of sensors. J. Educ. Comput. Res. **57**(7), 1832–1848 (2019)
7. Liu, C.H., Chang, P.Y., Huang, C.Y.: Using eye-tracking and support vector machine to measure learning attention in elearning. In: Applied Mechanics and Materials, vol. 311, pp. 9–14. Trans Tech Publications Ltd. (2013)
8. Karn, R.R., Singh, S., Singh, S.K.: Awareness and knowledge level on eye health among students of government school of Udaypur district, Nepal. Int. J. Perceptions Pub. Health **3** (4), 82–86 (2019)
9. Ambra Neri, C.C.H.S.: www.researchgate.net, January 2003. https://www.researchgate.net/publication/228604457_Automatic_speech_recognition_for_second_language_learning_How_and_why_it_actually_works. Accessed 26 July 2020
10. Octavian DOSPINESCU, Iulian POPA, Face Detection and Face Recognition in Android Mobile Applications Informatica Economica, March 2016. https://www.researchgate.net/publication/303880721_Face_Detection_and_Face_Recognition_in_Android_Mobile_Applications. Accessed 26 July 2020
11. Hung, H.C., Young, S.S.C., Lin, C.P.: Constructing the face-to-face collaborative game-based interacted environment for portable devices in English vocabulary acquisition (2009)
12. Face Detection Concepts Overview | Mobile Vision | Google Developers, Google Developers, 2021. https://developers.google.com/vision/face-detection-concepts

Augmented Reality Smart Glasses in Education: Teachers' Perceptions Regarding the Factors that Influence Their Use in the Classroom

Georgia Kazakou(✉) and George Koutromanos

Department of Primary Education, National and Kapodistrian University of Athens, Athens, Greece
{gkazakou, koutro}@primedu.uoa.gr

Abstract. Augmented reality smart glasses are one of the emerging technologies in the field of education. As in the case of any type of emerging technology, so too in the case of smart glasses, acceptance by users is key. The purpose of this study was to explore in-service teachers' perceptions regarding the factors that influence their intention to use augmented reality smart glasses. The study was conducted in January 2021 and involved 91 primary and secondary education teachers. Initially, the technical characteristics and specific functions of smart glasses were presented online to teachers. They were then asked to fill out an online questionnaire with open-ended questions based on the theoretical framework of Technology Acceptance Models. The qualitative analysis of their answers revealed that the majority of teachers intend to use smart glasses in their teaching. Also, it was found that teachers perceive smart glasses as useful for teaching (e.g., by enriching it) and learning (e.g., by enhancing students' motivation). According to teachers' perceptions, other factors that may influence them to use smart glasses are their compatibility with them and facilitating conditions, such as infrastructure, technical and pedagogical training, provision of educational material, low cost, and support from educational leadership. Moreover, a small number of teachers referred to risks regarding privacy and health as factors that may inhibit them from using smart glasses. This study provides implications so that this new technology is effectively implemented in schools.

Keywords: Smart glasses · Teachers · Technology Acceptance Model

1 Introduction

Smart wearable devices or smart wearable technologies are considered to be the next generation of ubiquitous technology and are part of the IoT [1]. They are now widespread as they have already proven their potential and benefits in a variety of fields, such as the medical industry [2] and the work environment [3]. Smart wearable devices include smart glasses, which are considered to be among the most dynamic wearable devices [4]. Augmented reality smart glasses (ARSG) represent one category of smart

M. E. Auer and T. Tsiatsos (Eds.): IMCL 2021, LNNS 411, pp. 145–155, 2022.
https://doi.org/10.1007/978-3-030-96296-8_14

glasses. These are wearable devices that are worn like regular glasses and merge virtual and real information in the user's field of vision [4]. Education is among the areas where ARSGs have been used with positive results [5, 6].

One of the scientific community's research interests regarding smart glasses is the issue of their acceptance in different fields, such as the medical industry [7], tourism [8], and sports [9]. Well-known theories and acceptance models such as the Technology Acceptance Model (TAM), as well as its extensions [1, 10], have been used to investigate the acceptance factors of smart glasses. In the context of these studies, various factors that influence their acceptance, such as perceived usefulness, enjoyment, documentation, cost, and privacy, were identified (see related results in the next section).

Although there is abundant research on the acceptance of smart glasses in various fields [1, 10], a research gap exists in education, except for one study on their acceptance by university students [11]. However, as in the case of any type of emerging technology, so too in the case of augmented reality smart glasses, it is necessary to investigate teachers' perceptions on the factors that may influence their intention to use them. According to research literature in the field of social psychology, teachers' perceptions and views on educational innovations constitute an important factor for their successful introduction, implementation and continuation in education [12]. Several models and theories such as TAM [13], TAM2 [14], TAM3 [15], and Theory of Planned Behaviour (TPB) [16] have been developed to study the factors which influence teachers to use digital technologies in teaching. Compared to other technologies, augmented reality smart glasses have special characteristics such as portability, connectivity to other devices, and the merging of the real and the virtual world. What has yet to be studied is teachers' perceptions towards the implementation of this new technology in education. Such an investigation is likely to facilitate comprehension of the factors that can contribute to the incorporation of smart glasses into education. Furthermore, gaining knowledge of these factors can contribute to solving more practical issues related to the use of smart glasses in education. These issues include the design of educational applications for smart glasses, the development of didactic scenarios that utilize them, as well as the formation of a teacher training framework for the use of smart glasses in the educational process. All the above actions are crucial for the successful integration of this new technology in schools.

The purpose of this study is to fill the existing research gap regarding the acceptance of smart glasses in education by exploring primary and secondary education teachers' perceptions on factors that influence their intention to use augmented reality smart glasses in their teaching. The research questions answered here are the following: (1) What is the teachers' intention regarding the use of smart glasses in their teaching? (2) What are teachers' perceptions regarding the usefulness and compatibility of smart glasses in education? (3) Which are the factors that teachers believe will facilitate or inhibit the use of smart glasses in their teaching?

The article is organized as follows: Sect. 2 presents previous empirical research on the acceptance of augmented reality smart glasses, and Sect. 3 describes the research methodology. Section 4 presents the results, and Sect. 5 discusses the results and provides implications regarding the use of smart glasses in education. Section 6 provides the conclusions, and the last one refers to the study's limitations as well as to future work.

2 Previous Empirical Research on Smart Glasses Acceptance

Different theories and acceptance models have been used to study the acceptance of smart glasses, such as TAM [13], Unified Theory of Acceptance and Use of Technology (UTAUT) [16], TPB [17], Theory of Reasoned Action (TRA) [18], and Innovation Diffusion Theory (IDT) [19]. TAM stands out among them as it is an extremely popular theoretical model that can be applied to different types of technology across various fields [1], including education [20]. According to TAM, we can predict the use of a technology by an individual's intention to use the technology [13]. The core variables of TAM are attitudes towards technology, perceived usefulness and perceived ease of use. Perceived usefulness is defined as the degree to which one believes that the use of a particular technology will enhance his or her performance [13]. Perceived ease of use is defined as the degree to which one believes that the use of a particular technology can be effortless [13]. Perceived ease of use affects perceived usefulness and attitudes. The last two variables affect the intention, which, in turn, affects the actual use of the technology [13]. According to recent meta-analyses of researches that utilize TAM in education [20, 21], its model variables are still reliable in predicting the acceptance of a technology by teachers. However, in order to explain other factors that may affect intention to use a technology, many researchers have either extended or modified TAM by adding additional variables, such as facilitating conditions [22, 23] and compatibility [24]. Facilitating conditions are related to an individual's perceptions of the extent to which there are organizational and technological resources that will support the use of a technology [22]. According to [24], compatibility refers to whether a technology fits the way one works, one's existing work practices, one's previous experience and one's values.

In terms of research on the acceptance of smart glasses, the fields that have been studied using an acceptance model or an extension of it are: consumer behaviour [25–27], medical industry [7, 28, 29], tourism [9], sports [8], and education [11].

One of the first studies regarding consumer behaviour was conducted by [20], who, by extending TAM, found potential factors and barriers affecting intention to use ARSGs. Their findings were based on the answers to a questionnaire including 201 participants. According to the results, perceived usefulness, perceived ease of use, social norms and consumers' technology innovativeness determine their evaluation of Google Glass, as well as their intention to use it. Factors that influence the adoption of smart glasses by consumers were also investigated by [25]. Based on TAM and a sample of 122 participants, the study found that the factors that most influence consumers' perceived usefulness are perceived enjoyment, external influence and peer influence. Another indicative consumer-related study is the [26] study. According to the model they developed for the acceptance of ARSGs based on TAM, the decision to adopt smart glasses is guided by perceived usefulness, perceived ease of use, image and descriptive norms. Other findings of the study were that hedonic motivation, privacy risk and injunctive norms did not affect the intention to adopt smart glasses.

In the field of medical industry, all research is based on the theoretical framework of TAM. In particular, [7] found that the factors that positively affect the usefulness of smart glasses to doctors are compatibility, ease of reminding, speech recognition and ease of use. They also found that ease of learning, ease of medical education, external influence and privacy positively affect ease of use. Another study [28] investigated the use and acceptance of Google Glass by physicians. In this study, it was found that perceived usefulness was affected by ease of use, compatibility, ease of reminding, and speech recognition. It was also found that perceived ease of use was positively affected by ease of learning, ease of medical education and external influence, and negatively affected by privacy. Recently, [29] developed a model of acceptance of smart glasses by hospital physicians. The proposed model includes the following factors: integration of information systems, external effects, hands-free use, technological compatibility, and documentation. The latter factor was found to have the highest effect on attitude and intention to use smart glasses.

Furthermore, [8] developed a conceptual framework with variables from IDT, TRA/TPB, TAM and UTAUT. Based on interviews with 28 visitors to a UK art gallery, the researchers propose societal impact, perceived benefits, perceived attributes of innovation and visitor resistance as factors for the measurement of ARSG adoption in cultural tourism. Finally, in sports, [9] examined the factors that influence the acceptance of smart glasses by amateur cyclists. They relied on UTAUT2 and found that the low technological level and limited functionality of smart glasses are still barriers to their acceptance.

In the field of education, [11] developed a model for the adoption of Google Glass by university students in terms of their learning. This model was based on TAM and includes the following variables: motivation, functionality, and trust and privacy. A sample of 968 students from universities in Gulf countries was used to measure the model. According to the results, motivation, functionality, and trust and privacy have a positive effect on both perceived usefulness and perceived ease of use of Google Glass.

In summary, while reviewing the research activity for the acceptance of smart glasses, there seems to be a research gap regarding teachers' perceptions on factors that may influence their intention to use smart glasses.

3 Methodology

The study was conducted in January 2021 through the "Zoom" platform (due to the COVID-19 pandemic) and 91 primary and secondary in-service teachers participated. A similar methodology, i.e., remote presentation of the smart glasses to the participants, has been implemented by other researchers as well [29, 30], yet focusing on the use of smart glasses in other areas. As mentioned in previous section the study was based on the theoretical background of the Technology Acceptance Models. Also, it adopts a qualitative research design. A similar one, i.e., using qualitative data collection tools, was used by other studies [31, 32], and [33] that investigated pre-service and in-service teachers' beliefs.

3.1 Participants

To answer the research questions of the study, an invitation was e-mailed to a convenience sample of in-service teachers who worked in Attica, Greece. In total, 91 out of

126 primary and secondary education teachers responded to the invitation, from here onwards referred to as T1-T91. Fifty-nine of the teachers (64.8%) were female and 32 (35.2%) were male. Eighty (87.9%) were secondary education teachers and 11 (12.1%) were primary education teachers. Sixteen (17.58%) teachers possessed up to 10 years of teaching experience, 47 (51.65%) possessed 11 to 20 years of experience, and the remaining 28 (30.77%) possessed over 20 years of teaching experience. Thirty-four (37.36%) of them have been certified in basic ICT knowledge and skills for educational purposes, 39 (42.86%) have been certified in advanced ICT knowledge and skills for educational purposes, 10 teachers (10.99%) have received other ICT certifications, while the remaining 8 (8.79%) have not received any certification related to ICT. Epson Moverio AR BT-300 was the smart glasses device used.

3.2 Procedure

The research was conducted in three stages. In the first stage, the technical characteristics and basic functions of smart glasses were demonstrated to teachers in groups of 10. In the second stage, researchers wore the smart glasses (i.e., Epson Moverio AR BT-300) and projected the glasses' interface onto the computer screen so as to be viewed by the participants. Subsequently, the participants navigated through the applications of the smart glasses, such as camera, video recording and internet. Furthermore, the researchers provided Quick Response (QR) codes containing websites, videos, images and virtual tours. The educational content of the codes was related to History, Mathematics, Physics, Biology and Social Sciences. In the third stage, participants completed an online questionnaire and participated in unofficial discussions. The duration of the procedure was approximately an hour and a half.

3.3 Questionnaire

The study used an online questionnaire online in Google Forms which consisted of two parts. The first part contained questions on the demographic data of the sample (i.e., gender, age, level of education, and years of teaching service). The second part consisted of five open-ended questions. Specifically, the questions were: (1) Do you intend to use smart glasses in your teaching in the future? Why? (2) What do you think will be the usefulness of smart glasses in your teaching? (3) What do you think is the compatibility of smart glasses with your teaching? (4) What are the conditions or factors that will make it easier for you to use smart glasses in your teaching? (5) What are the conditions or factors that will inhibit you from using smart glasses in your teaching? Questions 1, 2, 4, and 5 were adapted from Technology Acceptance Models' variables (i.e., intention to use, perceived usefulness, and facilitating conditions respectively). Question 3 that refers to compatibility was adapted by IDT [19]. In order to examine the questions' validity of content, they were presented to three ICT experts and three teachers.

3.4 Data Analysis

The answers were coded in three phases: (1) open coding, (2) axial coding and (3) selective coding according to [34]. Coding was conducted by two ICT researchers in education.

4 Results

The findings of the study are presented in the following sections based on its research questions.

4.1 Intention to Use Augmented Reality Smart Glasses

Regarding teachers' intention to use smart glasses, 85 out of 91 (93.4%) stated that they intend to use them in the future in their teaching. Six teachers (6.6%) responded negatively regarding their intention to use smart glasses, referring to reasons such as the lack of internet connectivity in school, the lack of time for learning how to operate them, the high cost of obtaining them and their lack of usefulness.

4.2 Perceived Usefulness and Compatibility

The teachers of the sample believed that perceived usefulness is an important factor which will affect their intention. Almost all the teachers of the sample who stated that they intend to use them in their teaching (N = 85, 93.4%) believed that the use of smart glasses in the educational process is useful. This perceived usefulness is related to either the teaching or the learning process. In terms of teaching, teachers believe that the use of smart glasses will facilitate it, make it more efficient and enrich it. Regarding the learning process, teachers believed that the use of smart glasses will enhance students' motivation, increase their interest in learning and make their participation more active. Furthermore, teachers believed that the use of smart glasses in the educational process will offer a more interactive and experiential way of representing knowledge, thus contributing to its better comprehension.

According to the teachers, all of the above, in terms of smart glasses' usefulness, will be achieved thanks to smart glasses' technological capabilities. Teachers identified these as: direct access to information, access to various multimedia and applications, augmentation of the environment, and the hands-free feature, which is combined with either teacher's freedom of movement in the classroom or voice commands.

The following are indicative answers by the teachers regarding students' motives, the interaction offered through the use of smart glasses, the capability of augmentation, and the hands-free feature:

"I believe that they add value to teaching, as they can connect various learning objects with material that is not accessible to teaching as it has been done to date, and raise children's interest in both formal and informal learning environments" (T68).

"They will offer great interactivity and experiential involvement in teaching (formal or informal environment)" (T67).

"I think augmented reality is the most notable feature of the glasses and facilitates the process more than mobile devices" (T35).
"It would make it easier for me in situations where I would already be using my hands" (T41).

Another factor that 15 (16.48%) teachers believed will influence them to use smart glasses is their compatibility with this technology. More specifically, they believed that the use of smart glasses is compatible with the way they teach, their teaching experience, their teaching needs, but also with the values they serve as teachers.

"I am interested in exploring new technological achievements and incorporating them into my lesson" (T27).

"Smart glasses are perfectly compatible with the lesson I teach" (T72).

4.3 Factors Which Teachers Believe Will Facilitate or Inhibit the Use of Augmented Reality Smart Glasses in Their Teaching

The majority of teachers (N = 85, 93.4%) referred to a number of factors that may facilitate the use of augmented reality smart glasses in their teaching. A number of them included facilitating conditions.

The first facilitating condition mentioned by teachers (N = 40, 43.95%) was the infrastructure and equipment of the school. Teachers believed that having a fast internet connection in their school will make it easier for them to use smart glasses in their teaching. They will also be facilitated by the availability of smart glasses for themselves and their students, as well as other digital devices that could be connected to glasses (e.g., interactive whiteboards).

A second facilitating condition was the technical and pedagogical training of teachers. The teachers of the sample (N = 34, 37.36%) considered that their training in the technological characteristics of smart glasses and in their utilization through specific didactic practices that meet their teaching needs will facilitate their use.

A third facilitating condition was the provision of educational material. Teachers believed (N = 32, 35.16%) that providing them with appropriate educational material, such as educational applications and teaching scenarios that link each subject's teaching objectives with the use of smart glasses, will make it easier for them to use smart glasses.

The teachers of the sample (N = 21, 23.07%) referred to the affordability of purchasing smart glasses as a facilitating condition, stating that they need to acquire the device themselves in order to be properly prepared to use it in the classroom.

The support from educational leadership in using smart glasses was reported as the fifth most common facilitating condition. According to the teachers of the sample (N = 17, 18.68%), the support of institutions, such as the Ministry of Education and their school's head teacher in order to form a culture of adopting technological innovations in the classroom, is a facilitating condition for using smart glasses. They also believed that care should be taken so that students will develop digital skills in the use of smart glasses for educational purposes in order to avoid being addicted to the new technology.

Indicative answers that teachers gave regarding facilitating conditions are the following:

"There must be high-speed Wi-Fi in schools, accessible to all" (T37).

"I would be assisted by the appropriate training and technical support for their use in the educational process, and by the provision of properly designed educational applications per subject" (T85).
"All students and teachers should be able to get smart glasses. This would mean that their cost will be low" (T7).
"The school's head teacher should be open to new technologies" (T56).

In addition to the above factors that are likely to positively affect teachers regarding the use of smart glasses, two factors were found that may prevent them from using smart glasses. One of them is users' privacy risk. A small number of teachers of the sample (N = 7, 7.69%) considered that the use of smart glasses in the educational process may be linked to the collection of teachers and students' personal data, and express reservations about the security of this data.

"I would use them with caution against potential dangers that may arise from the easy recording of image and sound" (T78).

The second factor that a small number of teachers (N = 5, 5.49%) believe will prevent them from using smart glasses is potential health risks.

"I have to be convinced by research about their safety in relation to neurological and developmental disorders in young children" (T39).

5 Discussion

The present study aimed to explore the factors that primary and secondary school teachers believe will influence them to use smart glasses in their teaching. The answers showed that the majority of the teachers of the sample intend to use them. Also, various factors that teachers believe may influence them to use smart glasses were identified, with perceived usefulness being the most important. Other factors are compatibility, facilitating conditions, perceived privacy risk and perceived health risk. Facilitating conditions are related to appropriate infrastructure and equipment within the school, teachers' technical and pedagogical training, the provision of appropriate educational material, the affordability of acquiring smart glasses on their own, and the support from educational leadership.

All of the above factors are in accordance with previous research regarding the acceptance of smart glasses in various fields, as well as with research related to the use of ICT in education. More specifically, perceived usefulness has been found by [26, 28] and [29] to positively affect intention to use smart glasses. Similarly, compatibility was measured within the TAM by [7, 28, 29] and was found to positively affect the intention to use smart glasses through either perceived usefulness or perceived ease of use. Also, in the studies of [20] and [21] it was found that facilitating conditions influence teachers' intention regarding ICT use in education. In addition, privacy risks negatively affect the intention to use smart glasses through perceived ease of use according to [28], and perceived health risk was reported by [26] as a user intention characteristic of smart glasses.

This study enhances research activity regarding teachers' perceptions of the factors that may influence them to use augmented reality smart glasses and provides implications so that the integration of this technology in the educational system is more effective. The results show that teachers already recognize that the use of smart glasses will add value to their teaching and, therefore, intend to use them in the future. In order to further enhance teachers' positive perceptions of smart glasses, this technology needs to be included in educational policy decisions. This is proposed to be done initially through the formation of a framework of technological and pedagogical training for teachers in this new technology. The rapid pace of development of ICT in education implies a constant effort on teachers' part to keep up with it. Smart glasses are another technology that teachers will sooner or later be required to familiarise with and use in their classrooms. Educating teachers on how to use augmented reality smart glasses and how to create educational material which is compatible with smart glasses can greatly facilitate their work. Furthermore, the ability of teachers to recognize and understand the educational potential of smart glasses can lead to a positive attitude towards their use. In addition, it is very important to establish the conditions which will facilitate the use of smart glasses by teachers in their teaching.

6 Conclusion

The present study is one of the first to focus on the issue of teachers' perceptions about the acceptance of smart glasses. It is qualitative and it was based on the theoretical framework of Technology Acceptance Models' variables. It aimed to explore the factors that primary and secondary school teachers believe will influence them to use smart augmented reality glasses in their teaching. Its results reveal that teachers will use smart glasses if they are convinced that their integration is useful for their teaching and given that existing conditions will make it easier for them to use them. In addition, the factors identified by the present study (i.e., perceived usefulness, compatibility, facilitating conditions, perceived privacy risk and perceived health risk) are consistent with previous research activity on the acceptance of augmented reality smart glasses in different fields as well as research related to ICT in education. Therefore, the present study confirmed that the aforementioned factors could influence the intention to use augmented reality smart glasses in the field of education as well.

7 Limitations and Future Research

This study has two limitations. Firstly, the glasses presented to the participants were a specific model of smart glasses (Epson Moverio AR BT-300). This means that, perhaps, the use of another model (e.g., Microsoft HoloLens 2) would identify different perceptions of their use in teaching on the teachers' part. Secondly, the sample used in the study was convenient.

An extension of the present study is the measurement of the factors that emerged from its findings (i.e., perceived usefulness, facilitating factors, compatibility, perceived privacy risk and perceived health risk) in the context of acceptance models. This measurement will be performed using a larger sample with teachers having the opportunity to interact with the smart glasses. The ultimate goal of this extension will be to propose a model of acceptance of augmented reality smart glasses especially for the field of education.

References

1. Niknejad, N., Ismail, W.B., Mardani, A., Liao, H., Ghani, I.: A comprehensive overview of smart wearables: the state of the art literature, recent advances, and future challenges. Eng. Appl. Artif. Intell. 90, 103529 (2020)
2. Tully, J., Dameff, C., Longhurst, C.A.: Wave of wearables: clinical management of patients and the future of connected medicine. Clin. Lab Med. **40**(1), 69–82 (2020)
3. Khakurel, J., Melkas, H., Porras, J.: Tapping into the wearable device revolution in the work environment: a systematic review. Inf. Technol. People **31**(3), 791–818 (2018)
4. Ro, Y.K., Brem, A., Rauschnabel, P.A.: Augmented reality smart glasses: definition, concepts, and impact on firm values creation. In: Jung, T., tom Dieck, C. M. (eds.) Augmented Reality and Virtual Reality, pp. 169–181. Ed. Progress in IS. Springer, Cham (2017). https://doi.org/10.1007/978-3-319-64027-3_12
5. Havard, B., Podsiad, M.: A meta-analysis of wearables research in educational settings published 2016–2019. Education Tech. Res. Dev. **68**(4), 1829–1854 (2020). https://doi.org/10.1007/s11423-020-09789-y
6. Koutromanos, G., Kazakou, G.: The use of smart wearables in primary and secondary education: a systematic review. Themes eLearn. **33**, 33–53 (2020)
7. Göken, M., Başoğlu, A.N., Dabic, M.: Exploring adoption of smart glasses: applications in medical industry. In: Proceedings of the 2016 Portland International Conference on Management of Engineering and Technology (PICMET). Honolulu, HI, (2016)
8. Han, D.-I.D., tom Dieck, M.C., Jung, T.: Augmented reality smart glasses (ARSG) visitor adoption in cultural tourism. Leisure Stud. **38**(5), 618–633 (2019)
9. Berkemeier, L., Menzel, L., Remark F., Thomas, O.: Acceptance by design: towards an acceptance smart glasses-based information system based on the example of cycling training. In: Proceedings of Multikonferenz Wirtschaftsinformatik (MKWI). Lüneburg, Germany (2018)
10. Kalantari, M.: Consumers' adoption of wearable technologies: literature review, synthesis, and future research agenda. Int. J. Technol. Mark. **1**(3), 274–306 (2017)
11. Al-Maroof, R.A., Alfaisal, A.M., Salloum, S.A.: Google glass adoption in the educational environment: a case study in the Gulf area. Educ. Inf. Technol. **26**(3), 2477-2500 (2020)
12. Fullan, M.: The New Meaning of Educational Change. RoutledgeFalmer, London (2001)
13. Davis, F.D.: Perceived usefulness, perceived ease of use, and user acceptance of information technology. MIS Q. **13**(3), 319–340 (1989)
14. Venkatesh, V., Davis, F.D.: A theoretical extension of the technology acceptance model: four longitudinal field studies. Manage. Sci. **46**(2), 186–204 (2000)
15. Venkatesh, V., Bala, H.: Technology acceptance model 3 and a research agenda on interventions. Dec. Sci. **39**(2), 273–312 (2008)
16. Fishbein, M., Ajzen, I.: Belief, Attitude, Intention, and Behavior: An Introduction to Theory and Research. Addison-Wesley, Reading, Mass (1975)
17. Ajzen, I.: The theory of planned behavior. Organ. Behav. Hum. Decis. Process. **50**(2), 179–211 (1991)

18. Venkatesh, V., Morris, M.G., Davis, G.B., Davis, F.D.: User acceptance of information technology: Toward a unified view. MIS Q. **27**(3), 425–478 (2003)
19. Rogers, E.M.: Diffusion of Innovations, 4th edn. Free Press, New York (1996)
20. Scherer, R., Siddiq, F., Tondeur, J.: The technology acceptance (TAM): a meta-analytic structural equation modeling approach to explaining teachers' adoption of digital technology in education. Comput. Educ. **128**, 13–35 (2019)
21. Scherer, R., Teo, T.: Unpacking teachers' intentions to integrate technology: a meta-analysis. Educ. Res. Rev. **27**, 90–109 (2019)
22. Scherer, R., Siddiq, F., Tondeur, J.: All the same or different? revisiting measures of teachers' technology acceptance. Comput. Educ. **143**, 13–35 (2020)
23. Yousafzai, S., Foxall, G.R., Pallister, J.G.: Technology acceptance: a meta-analysis of the TAM: Part 1. J. Model. Manag. **2**(3), 251–280 (2007)
24. Karahanna, E., Agarwal, R., Angst, C.: Reconceptualizing compatibility beliefs in technology acceptance research. MIS Q. **30**(4), 781–804 (2006)
25. Rauschnabel, P.A., Ro, Y.K.: Augmented reality glasses: an investigation of technology acceptannce drivers. Int. J. Technol. Mark. **11**(2), 123–148 (2016)
26. Başoğlu, N.A., Ok, A.E., Daim, T.U.: What it will take to adopt smart glasses: a consumer choice based review? Technol. Soc. **50**, 50–56 (2017)
27. Kalantari, M., Rauschnabel, P.: Exploring the early adopters of augmented reality smart glasses: the case of microsoft hololens. In: Jung, T., tom Dieck, M. (eds.) Augmented Reality and Virtual Reality. Progress in IS. Springer, Cham. pp. 229–245 (2017). https://doi.org/10.1007/978-3-319-64027-3_16
28. Başoğlu, N.A., Göken, M., Dabić, M., Özdemir-Güngör, D., Daim, T.U.: Exploring adoption of augmented reality smart glasses: applications in the medical industry. Front. Eng. Manag. **5**(2), 167–181 (2018)
29. Özdemir-Güngör, D., Göken, M., Basoglu, N., Shaygan, A., Dabić, M., Daim, T.U.: An acceptance model for the adoption of smart glasses technology by healthcare professionals. In: Larimo, J., Marinov, M., Marinova, S., Leposky, T. (eds.) International Business and Emerging Economy Firms. Palgrave Studies of Internationalization in Emerging Markets, pp. 163–194. Palgrave Macmillan, Cham (2020). https://doi.org/10.1007/978-3-030-27285-2_6
30. Rauschnabel, P.A., Brem, A., Ivens, B.S.: Who will buy smart glasses? empirical results of two pre-market-entry studies on the role of personality in individual awareness and intended adoption of google glass wearables. Comput. Hum. Behav. **49**, 635–647 (2015)
31. Smarkola, C.: Efficacy of a planned behavior model: beliefs that contribute to computer usage intentions of student teachers and experienced teachers. Comput. Hum. Behav. **24**, 1196–1215 (2008)
32. Sadaf, A., Newby, T.J., Ertmer, P.A.: Exploring pre-service teachers' beliefs about using web 2.0 technologies in K-12 classroom. Comput. Educ. **59**, 937–945 (2012)
33. Koutromanos, G., Mavromatidou, E.: Augmented reality books: what student teachers believe about their use in teaching. In: Tsiatsos, T., Demetriadis, S., Mikropoulos, A., Dagdilelis, V. (eds.) Research on E-Learning and ICT in Education, pp. 75–91. Springer, Cham (2021). https://doi.org/10.1007/978-3-030-64363-8_5
34. Strauss, A., Corbin, J.: Basics of Qualitative Research: Techniques and Procedures for Developing Grounded Theory, 3rd edn. Sage Publications, Thousand Oaks, CA (2008)

"Aspects of Freedom" a Case of Design and Making of AR App for Interactive Communication in the Field of 3D Animation Production in Culture

Spyros Siakas(✉), Lamprini Trivella, Anastasia Lampropoulou, and Georgios Margaritis

University of West Attica, Agiou Spyridonos, 12243 Aigaleo, Egaleo, Greece
{sthsiakas,ltrivella,alampropoulou,gmargaritis}@uniwa.gr

Abstract. This research aims to provide the users with an experience of learning about tangible heritage items-as symbols of freedom through interactive mobile communication AR app. Some of the 3d models are the exact replicas of the genuine museum items, made with photogrammetry, from the National Historic Museum and the War Museum of Athens and its department in Tripoli. These three museums have collaborated with the University of West Attica for the digitalization of some interesting but unknown to public cultural heritage items. It is conducted in three parts, a preliminary stage of research and item collection, the stage of image target design and the scenario of the usage in many different printed materials as well as the making of AR app and the final stage of launching it. Also, specific material such as worksheets and table games were created as to supplement the learning experience.

So, in this paper, there is a synopsis of the concept, the major issues are addressed and the significance of the work is highlighted. Also, the theoretical and methodological approaches pursued are explained and finally data are presented and analyzed.

Keywords: Swot · Smart posters · AR app · Interactive learning · Museum items

1 Context

The anniversary of the 200 years from the Greek Revolution of 1821 together with an already existing collection of 3d models of museum items of that era, provided the basis upon which this project was constructed, in order to present the data along with the 3 D items in an innovative and more appealing way. The collection consisted of

Supplementary Information The online version contains supplementary material available at https://doi.org/10.1007/978-3-030-96296-8_15.

M. E. Auer and T. Tsiatsos (Eds.): IMCL 2021, LNNS 411, pp. 156–167, 2022.
https://doi.org/10.1007/978-3-030-96296-8_15

both 3D computer modelled and photogrammetred genuine cultural items, which were stored and not exhibited because of lack of space in the museums exhibiting areas. Having these models as the main context of the augmented reality app, smart posters were designed with specific image targets which revealed the 3D models through the use of mobile phones. In this way, an exhibition of these posters was designed to cater either for on-site or for on-line visitors. The aim for this exhibition was the interaction between the posters and the visitors through their mobiles so as to commute the tangible cultural heritage items, providing an innovative experience of learning about them. The idea behind it, is the concept of "Freedom" which bridges the past with the present. for this reason, these 3D replicas of genuine tangible heritage items, which were used in the era of Greek Revolution, Are Accessible to Visitors. Furthermore, specific worksheets were created to turn the learning experience into edutaining.

2 Purpose

The purpose of this production is the exhibition of 3D models of non-exhibited museum items in natural or virtual space because of the covid limitations. This can be done with the design of printed media with image targets, which are activated when the visitors download the AR app in their mobiles and aim at these targets. The 3D models and metadata emerging form the mobiles make learning fun both in-site and on- line.

3 The Importance of This Project

This project is significant as it makes a good use of mixed and augmented reality in cultural heritage, to offer engaging and immersive learning experience in informal education. Specifically, the project presents the non-exhibited museum items, to the public, to the researchers and scientists. It uses the mobile technology to present these items in an intriguing way to attract many different aged groups. In this frame, it gives an alternative to the museums to be sustainable in pandemic times, as the exhibition is transformed into an interactive, active informal learning experience and it provides prospects of interactive edutaining games.

4 Research Questions

According to the above purpose, some major issues appeared:

- Are the conditions proper to produce this project?
- How can this 3D animation production be managed strategically?
- What are the quality and functional characteristics, based on which, the target images will be designed to support exhibition in physical and digital way?
- What forms of printed media can be designed and combined with the image
- targets, and how playful activities can be included into them?
- Can an AR app be friendly, easily accessible and functional to users as to provide a qualitative interactive communication through mobiles?

5 Approach

Various methods and models are approached for having the best results with the minimum of time and effort.

First of all, the methodological analysis of SWOT is used to comprehend the conditions of its sustainability in relation to the environment, in which the AR app is about to be released.

Secondly, the organization and co-ordination of the project is approached by using a strategic synthesis of management methods taken from the creative industries of the audiovisual field. Thus, the tailored, 3-staged method with the distinct different stages of work is used. The batch work flow is used for allocating the task. Also, the Kaizen method is used, as it is based on improving the quality of the product in order to be appealing to the audience.

Finally, the development of the creative section of the project (second stage) is approached by using the waterfall model of multimedia app development. This waterfall or else the linearse quential model is widely used for the design and making of small or medium apps in short time. This model consists of the stages of analysis, design and making of an app with evaluation for its proper function and is aligned with the Kaizen method mentioned above.

6 Proper Conditions for Project Production

Research has been conducted about the circumstances in which the production would be sustainable. For this reason, the SWOT methodological analysis was used. According to this, the strengths of the project are the 3D models of genuine museum items, which are stored-away and are unknown to the visitors. In this way, these items can be accessible to the visitors and also, they can be a scientifically reliable source for researchers and students. Furthermore, the exhibition flexibility to be active on-line except for on-site is a strength due to the pandemic limitations.

The weak points of the project lie in the theme and in the new technologies' constant evolvement. The theme as a source is exhausted as far as it concerns tangible cultural heritage museum exhibits of two hundred years ago (1821). Also, the perpetual development of new technologies result in the launching of software new editions and updates, that change the basic data and cause technical issues.

One of the opportunities emerged for this project is that the current situation of the corona-virus has affected the sustainability of the museums, which turned into the new technologies to survive. The use of mobiles in this project is widely used. Thus, the use of an AR app for interactive mobile communication changes the learning experience and it offers new exciting experiences [3, 4].

The imminent threats in the environment are related to the bulk of offered on-line events, which are mostly appealing. These events are not related to museum cultural items exhibition and have a huge impact on the majority of mobile users.

Under these circumstances, the SWOT analysis indicated the proper conditions for the project to be developed.

7 Strategic 3D Animation Production Management

This project, as it evolves 3D animation models, metadata, the design of the image targets and finally the making of the AR app for the cultural items' exhibition, is a multi-level procedure that needs the strategic use of many management models. Thus, at first the tailored 3-staged type production from the creative industries of the audio-visual world is used [2–4]. This model is based on the 3 distinct stages of production according to the life circle of the product. The pre – production stage of research and material collection (3D models and metadata), the production stage which is the design of the targets and the making of the AR app and the post-production stage of launching the product.

As far as the work is concerned, the batch flow (cell production model) is followed as the work is segmented into different tasks, which are distributed to different kinds of experts [3, 4]. So, each expert has his own task to do and deliver the part of the project that he/she is committed to do and he/she has to co-operate with the others in order to make all the parts compatible with each other in the completed product.

As far as the quality is concerned, the model of Kaizen is used, as its main axes are the creativity and the quality of the function of the product [1]. This model is about the constant improvement of product quality of each part of the production. For this reason, this model is aligned with the waterfall model, which each sub-stage of the creative production stage uses to combine all the parts together successfully.

8 Design of AR Image Targets with the Quality and Functional Characteristics and the Creation of the Printed Material for Physical and Digital Exhibition

The use of augmented reality in museums is nowadays more and more frequent, because of the advantages it provides [5]. A cultural institution or artists, as well, can use this technology to highlight their exhibits with greater originality [6]. It is no coincidence that from research, on the question of which device they prefer to play electronic games, from students, parents and teachers, these portable devices were chosen with a great difference [7].

According to the proper use of Augmented Reality, this phase involves the design of targets and printed material that will support this kind of application. The purpose, as mentioned above, is the transformation of plain posters into smart posters. This can be done by adding augmentation, which it will enable to display 3D models of museum items in physical or virtual space due to the limitations of the pandemic.

Multipurpose Image Targets

In the Creation procedure of multipurpose image targets in the design part, the initial ideas for the targets, which would serve the Augmented Reality, were to be the posters themselves or symbolic targets. In the case of the physical exhibition, the use of the posters, was not the best choice as targets because it would demand longer distance between the poster and the visitor in order to have the best results in having the AR experience. In the case of creating printed and digital material such us booklets or card games, the use of the entire poster was again a problem as the texts did not appear in the small size of a booklet for example. The use of symbolic targets was preferred in this case as well.

Given the desired diversity the exhibition and the use of printed material, which may be available in digital form for physical and distance presentation, the study process for the design of the objectives began. Symbolic targets or even targets with letters would be effective. Furthermore, as the posters had rich visual material from the creation of the models in the blender, it was decided that it would be more useful to use images from pre-existing posters. This would serve the presentation in many ways. The target could be better conceptually integrated. There would be a clear visualization of the reference object in the printed material. And finally, the same target (Fig. 1, Fig. 2) could work from a functional and aesthetic point of view in a variety of printed materials such as: booklet, board games, card games, stickers, etc.

Following the above research, a first set of targets was created with images taken from the posters. After passing them into Vuforia, it was found that in the majority they were weak targets and only a few met the requirements set. The problem was that the images were screen shots from the blender program, which were not originally intended to work as targets in an Augmented Reality application. Even with the strengthening of the contrast they were still weak.

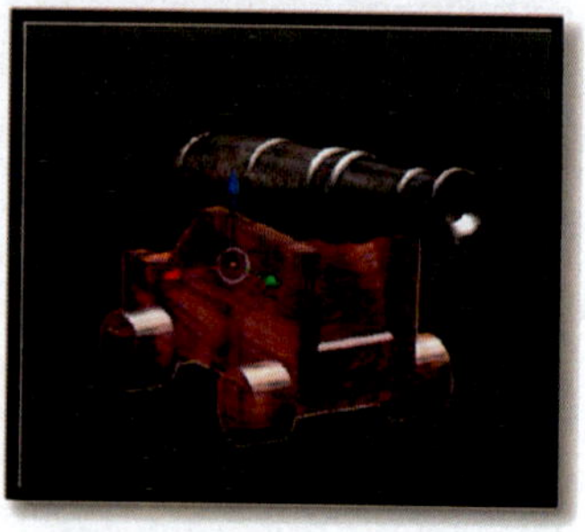

Fig. 1. Symbolic image target from 3D model

Fig. 2. Image target with caption

Smart Print Media with Vuforia Image Targets

A further investigation in the field of image targets indicated that there are many different types, such as simple targets with strong and bold frames, coded targets, images, text etc. The most basic target is a simple index with a wide-bold outline. The advantage of index targets is that they are easily recognized by the software with very little processing cost and they minimize the risk of running the application, for example, due to inconsistent ambient lighting or other environmental conditions. The ability to recognize and react showing images, is a huge boost to AR applications, as it avoids the need to create and distribute custom index in conjunction with specific applications. Image recognition falls into the category of Natural Feature Tracking (NFT) [8].

To evaluate the targets up to a score determines how well an image can be detected and tracked using the Vuforia Engine. This score is displayed in the Target Manager, the higher the score of an image target, the stronger its detection and tracking capability is. The image analyzer typically detects them as small yellow crosses, the higher this number the more successful the target is. Surfaces with angles give identification points while curved ones do not. Another feature that should be given special attention is the contrast that exists in the images that in many cases must be enriched [9].

Following all these data, it was necessary to reinforce the images with sharp letters that are very helpful in recognition, as all the images that were selected in the first attempts, were those that were accompanied by a text. It was from this perspective that the design of the targets proceeded, and they were reinforced with text that is semantically related to the three-dimensional model to which they were addressed and they were not simple combinations of letters.

Fig. 3. Final image target 1

Fig. 4. Final image target 2

Fonts with angles were used and in addition, the contrast of the images was increased, without further interference, with the aim not to lose its identity. These fonts with angles are surrounded with wide-bold border such as the ones of the simple index targets (Figs. 3, 4). For the upcoming integration in various types of forms, the functionality and aesthetics were reinforced in the next step with colors that differ in each target. The result of this gradual process was that the targets had the maximum score in the target manager. A similar procedure was followed in the targets of the funeral masks of the Heroes of 1821.

Summarizing the results of this process for the quality and functional characteristics of the image targets that can be used in many forms of printed media, that turn them into smart printed objects, are: the use of the image that exists on the poster that gives meaning even if the targets are independently shown, the use of fonts with angles to reinforce the recognition, the adding of wide-bold border round the fonts, the increasing of the contrast and last the use of colors that are combined with the posters.

Design of Printed Media that Are Combined with Image Targets with Playful Activities

After the design of the image targets was completed, there was a corresponding development in the process of creating the printed material. In the booklet (Fig. 5) there is text for each model that is accompanied by the image of the target. Whether the reader utilizes the AR function or not, he sees an image of the 3D model. Respectively, the further proceeding was the creation of a board game (Fig. 6) as well as a game with cards that is in progress (Fig. 7). Finally, the target itself can work individually in the form of a sticker which can be placed on any surface or object any time.

Fig. 5. Booklet with image targets and info

Fig. 6. Board game

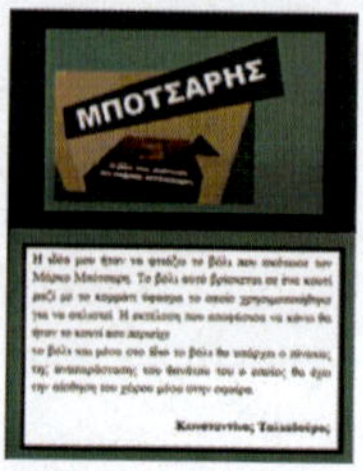

Fig.7. Card game

These possibilities give a variety of ways of presenting of the poster itself that can be more dynamic with the augmentation adapted on it through the target. This augmentation connects and feeds with material a variety of printed media that can work under different conditions in physical or digital way in pandemic times, in the field of formal or informal types of education.

All the above lead to the documented use of Augmented Reality for the development of the application, in order to be able to be used live or remotely, in formal or non-formal learning. To be able to function not only as a tool for highlighting the cultural heritage and acquiring new knowledge in a playful way but at the same time as a means of exhibiting works by artists. In order to achieve this, it is important to organize a scenario as the above that can be used in various forms of publications, starting with the pre-existing posters and the 3D content. Forms that adopt new approaches allow the interaction with their content and its dynamic renewal, enable the long-term survival and consequently their educational utilization [10]. AR gives a huge impetus to the creation of new publishing products with innovative and pioneering features 11].

9 AR Application as an Interactive Mobile Communicator/AR App as a User-Friendly Tool for Learning

Prologue
Due to the pandemic and the conditions following it, live events such as exhibitions or conventions are not feasible. Thus, the aim of our research team was to create an application able to function both live and from a distance.

Essential Factors for a Friendly, Easily Accessible and Functional AR App
The application requirements were set as follows:

1. Make use of the Image targets created in an earlier stage of production.
2. Make use the 3D models created with Blender and Photogrammetry software in an earlier stage of production.
3. Ability to project a model on screen, when an image target is tracked
4. Ability to project text information about the models in a proper way
5. Ability to play short explicatory videos, describing the development process and the ideas behind the models.
6. Provide basic model handling, such as translation, rotation and scale, in or- der to enhance viewing and interaction.

The following acceptances regarding application development derive from all the above:

The application must have a clear and user-friendly environment. The fact that the application may be used from a distance means that the potentials of this application ought to be self-explained. Therefore, no questions should arise regarding what actions the user should do. The user's interface ought to help towards this direction. In other words, user interface ought to be simple and clear while its purpose being to project objects without distracting the user's attention. The comprising elements of the interface have to be discreet and articulate about their function. Thus, wherever buttons are required, icons have been used instead for more clarity and less visual disturbance for the user.

A Game Engine as the Main Application Development Tool

Unity was chosen as the most suitable game engine for our purposes. The main reasons for that are its popularity and suitability for mobile applications development, the abundance of help material online, and also the fact that it works seamlessly with Vuforia. Furthermore, Unity is free for individual developers and educational purposes.

The Procedure

The 3D models that will be used, in some way form the raw materials of the application. Importing 3D models from Blender (or any other 3D modeling software) to Uni-ty is a crucial phase. It is the meeting point of different specialties (3D artist and programmer). In order to maintain a flowing production line, 3D models should meet a number of standards.

There are two ways to import models from Blender to Unity. The first way is to export the model from Blender in fbx format, and then insert it into Unity. The second is to directly import into Unity as Blender files. In this case Blender program must be installed, so that Unity will be able to communicate with Blender on the background, and fetch a copy of the model in a format that it can understand, which is fbx. But all that is being conducted fully automatically and the developer is not being bothered. In both cases, however, is important that the models meet some requirements. In particular, the following factors must be taken into consideration:

Textures: Models textures must exist as separate image files. Texture files should be kept low in size, so that they don't weigh down the build file produced.

Materials: It is good practice overall, that the object's materials are given unique names (and not random names such as Material001, Material002, etc.). Properly named materials ensure that there will be no mismatching between material and mesh. Baking the materials in Blender is also good practice, because it is easier for materials to be applied as textures onto the objects in Unity, and the result is more accurate. In this case however, the size of the external image files must be kept low.

Origin: Correct object origin is also important. While in Blender, the origin of the object must be set either to geometry, or to centre of mass. It also must be made sure that the object's position and rotation is zero in all axis, and the object's front and top faces are set in the program's top and front views, respectively.

Camera and Lights: There is no need for camera and lights to be included. A Blender camera is not going to work in Unity, so there is no reason it should be included in the. blender or.fbx file. Blender lights will be recognized and will work in Unity, but they usually create unwanted lighting conditions, and they also add up if more than one object with lights is inserted into Unity. Thus, lights management should be done in Unity.

Face Normals: Normals are directional vectors that are perpendicular to the face. Only the face of the normal side will be shown. Unity renders only one side of a plane, for example. In blender, it must be made sure that the normals are in the right direction, and if an object is to be seen from all directions, it must have a volume, e.g., sides which consist of two planes, with the plane's normals facing outwards.

Qualitative Interactive Communication Through Mobiles

The application works as follows: when a target image is recognized, scanned by the mobile's camera, the respective 3D model appears. The info icon-button is being activated, and its color changes to red, indicating to the user that it can be pressed. When the Info button is pressed, a semi-transparent 2D plane is activated, which reads text info about the 3D model. In the same way, if a 3D model has some escorting video that goes with it, a Play button appears next to the Info button. Pressing on the Play button activates another 2D plane, on which the respective video is played.

Other actions: Users can swipe the screen in order to rotate the 3D models, and also scale the 3D models (zoom in – zoom out) with a two-finger pinch gesture. There is also a button which reads "Be a creator" that leads to an external website, where users can upload their own content regarding the 200th anniversary of the Greek Revolution.

Another fact that has been taken into consideration is that AR is related to the Affordances theory [12]. The Affordances theory was established by J. Gibson in the late 1970s. According to this theory, environmental perception leads to certain actions in a number of possible manners, which are comprehended by the viewer without any specific visual analysis. For example, a user will easily figure out that the Affordance of a button in an application is that it can be pressed in order to trigger an action. According to this, there is no need for an instruction-filled interface, because users are capable to understand most of the possible options that are being given to them, without any further explanations.

The Outcome

The application contains both 2D and 3D elements. The user interface consists of 2D elements. 2D elements were used because they provide a cleaner, non-changing and more discrete environment.

So, the first step that has to be done is to import the Vuforia package to Unity. Vuforia creates a database with all the image targets and their characteristic features. This database can be exported as a Unity-ready package, and then imported into Uni-ty. The whole process is automated and takes a few clicks to be done.

Then, the second step is to set up the Unity scene. The scene in Unity includes all the objects that appear in the application. The scene is constructed in the following way:

Firstly, Image objects are created (Vuforia Engine Image). Each image target that has been imported from Vuforia is associated with an Image object. In order for Unity and Vuforia code to know which objects are going to be activated and shown during the augmentation, the objects are being nested into the Images. In this way, each Image is the parenting object to the object (or objects) that will appear when the Image is scanned with the mobile device's camera. The objects mentioned above will obviously ensue from the 3D models imported from Blender.

Secondly, the user interface is created. As mentioned before, the user interface consists of 2D objects, which are all nested under a Canvas parent object. The use of a Canvas is obligatory in Unity when 2D elements are used, and one is created automatically when a 2D user interface object (such as buttons, text, etc.) is being inserted into the scene.

For this application, the following 2D elements were used: buttons, images, text, video plane, logos. Lastly, the application code is being written. The code is written in the form of scripts, in C# language. The scripts are attached in objects that are included in the scene, and their role is to define the objects behavior and the way the application functions. In order for the scripts to be tested, Unity provides a "Play" capability, so the application can be run and tested within the engine.

Some Obstacles that had to be Overcome

Importing the 3D models from Blender to Unity. A number of difficulties occurred when importing 3D models from Blender to Unity. Those were mainly the following: Models being too large or too complex, and materials and textures that were not rendered properly or not at all. Solutions to these problems were granted by the following actions: downsizing of the models, dropping unnecessary features and geometry, and on the whole, adopting practices described earlier.

Creating a suitable User Interface. The main concern was to find a proper way to show and hide information. A 3D object was used in the beginning, so to project the text on its mesh, but the problem was that as the mobile was being moved and turned around, the object's angle of view and proportions changed, and the text could not be seen clearly. The solution was to use 2D objects for this purpose, as stated above. 2D objects can be anchored in certain areas of the screen, and they are overlaid on top of the other elements, so they are always visible. Furthermore, they are not stretched and their proportions remain stable as the mobile is being moved and turned. Apart from the text, the explicatory videos which escort the 3D models are also being projected on a 2D Video plane.

Three transforming options were proved to be too many. Providing all three transform options (rotation translation and scale) proved to be confusing and didn't work as expected. For the sake of simplicity, translation option was omitted.

10 Conclusion

The project was planned, organized and produced strategically according to the SWOT analysis of the proper conditions and the synthesis of multiple methods and models in order to save time, effort and money. The image targets were designed and functioned upon many different printed materials with playful elements, in an order that enhance active learning. They have the ability to work both ways physical and digital, in relation to specific quality and functional characteristics. Also, the AR app was used to connect the 3D models and metadata with the image targets and present the models through the mobile phones of the users/visitors, making the learning experience even more appealing and friendly to users. In this way, the new technology adds value to this project.

11 Further Prospects

This project provides a way of survive for the museums in the pandemic times or if there is a -long-time-closure for any reason. Also, it is a way of revealing the hidden-away cultural heritage treasures, which are important but stored due to lack of space, expanding the boundaries of the museums' exhibitions space.

Furthermore, it provides material to be used for museum education, such as board-card games and task-sheets a supplementary edutaining material, adapted to all ages of students.

References

1. Akdeniz, C.: Kaizen philosophy explained, Germany (2015). www.businesshacker.co. Accessed 28 June 2021
2. Farnes, N.: Modes of production, Fordism and distance learning, the journal of Open Distance and e-learning (10–20) (2006)
3. Rall, H.: Animation from Concept to Production. Taylor & Francis group, US (2018)
4. Štefanić, N., Križan, O., Cala, I.: Models and methods of production management. Strojarstvo **50**(3), (175–184) (2008). Accessed 15 Apr 2008
5. Billock J.: Five augmented reality experiences that bring museum exhibits to life (2017). https://www.smithsonianmag.com/travel/expanding-exhibits-augmented-reality180963810/AR features allow visitors to explore historical spaces and artifacts in new ways. Accessed 20 Apr 2021

6. Αναγνωστοπούλου, Α., Η χρήση της Επαυξημένης Πραγματικότητας (AR) σε εφαρμογές eLearning, Μελέτη περίπτωσης: Ίδρυμα «Κωνσταντίνος Γ. Καραμανλής» (Ι.Κ.Κ.), Τμήμα Γραφιστικής και Οπτικής Επικοινωνίας 3D Animation, Πανεπιστήμιο Δυτικής Αττικής, 1(1) (2020)
7. Λαμπροπούλου, Α. : Η αξιοποίηση της 3D σχεδίασης και της φωτογραμμετρίας στη δημιουργία παιγνιωδών εκπαιδευτικών δραστηριοτήτων. Open Journal of Animation, Film and Interactive Media in Education and Culture [AFIMinEC] 1(1) (2020)
8. https://library.vuforia.com/features/images/image-targets, last accessed 21/4/2021
9. https://subscription.packtpub.com/book/web_development/9781787286436/1/ch01lvl1sec11/types-of-ar-targets. Accessed 21 Apr 2021
10. Perey, C.: Print and publishing and the future of Augmented Reality. Inf. Serv. Use **31**, 31–38 (2011)
11. Hiwaizi, O.: (2016). https://www.campaignlive.com/article/why-augmented-reality-market-will-outpace-virtual-reality/1392252. Accessed 5 Apr 2021
12. Cheng, K.-H., Tsai, C.-C.: Affordances of Augmented Reality in Science Learning: Suggestions for Future Research (2012)

Building a General Purpose Educational Augmented Reality Application: The Case of ARTutor

George Terzopoulos, Ioannis Kazanidis(✉), and Avgoustos Tsinakos

International Hellenic University, Nea Moudania, Greece
kazanidis@teiemt.gr

Abstract. Augmented Reality (AR) technology is constantly evolving. Nowadays, there are AR platforms that allow educators to build AR experiences for their students. Most of these tools are not free and have specific domain orientation, allowing instructors to use them for specific domains and courses. The main aim of this study is to provide details on the approach, design and development frameworks that developers have to follow in order to build a general-purpose educational AR tool. The available technologies will be presented and an optimum approach and design will be proposed. In addition, this study presents the architecture and the adopted technologies of the general-purpose educational AR platform ARTutor (Version 3), a free tool for educators and their students. Compared to other AR platforms, ARTutor is a completely free and easy-to-use tool. With ARTutor, educators can create augmented books and add virtual content on top of printed material. The new version of ARTutor is based on state-of-the-art technologies such as Google's ARCore and Apple's ARKit. Evaluation results with pre-service teachers revealed that ARTutor is a valuable tool for education and it provides an easy way to implement AR experiences thus it can be easily adopted by various educational approaches.

Keywords: Augmented reality technologies · Educational augmented reality platforms · General purpose augmented reality platforms · Educational technology

1 Introduction

Augmented Reality (AR) technology brings the digital world into the real world. It enhances reality by adding virtual information and allowing users to interact with the real world and the AR experience simultaneously. An AR experience can be viewed on a mobile device (smartphone or tablet) or even on smart glasses. Smart glasses are considered the next big breakthrough for wearables and they will merge what we see in the real world with virtual information. AR technology can be used in various fields such as games, military, medical, industry and education.

Engaging with internationally focused literature, some of the most popular fields of education that use AR in teaching are Science [1], Ecology [2], Natural Sciences [3], Physics [4], Geometry [5], Chemistry [6], Mathematics [7], Writing Skills [8], and

M. E. Auer and T. Tsiatsos (Eds.): IMCL 2021, LNNS 411, pp. 168–179, 2022.
https://doi.org/10.1007/978-3-030-96296-8_16

Programming [9] are some of the most popular sectors of education where AR is employed in the classroom. AR technology can be provided either by augmenting current educational content such as books and notes [10, 11], creating in class AR activities such as AR escape rooms [12], mixed reality applications [13], mobile learning applications [9], and providing live lecture personalization and interaction [14]. Based on prior research [15, 16], the usage of augmented reality applications in education has the potential to be a valuable medium for teaching and learning. Recent research suggests that AR-based content can improve students' long-term memory, problem-solving skills, excitement, and collaboration capacities [17], as well as improve learning performance [18] and learning satisfaction. Using AR, an educator can develop new ways to teach educational concepts in the classroom, improve learning outcomes and excite students.

As it has been presented by various studies [15] in most of the cases AR is still applied using custom made solution oriented to the instruction of a specific field of study with a predefined way. These solutions are developed by programmers and require a considerable implementation time. However, teachers and educators need a tool that will allow them to instantly create educational AR experiences for their courses. In the past various tools were available either by payment or for free [19]. However, on one hand teachers cannot develop their own AR solutions since they are not programmers and they do not prefer to use paid solutions while on the other hand, previously free AR authoring tools have been either turned to paid (e.g. Aurasma, Blippar), withdraw (e.g. HPReveal) or discontinued (e.g. Metaverse). Therefore, there is a need for a free, general purpose and easy to use for non-programmers authoring tool for the development of AR educational experiences.

Building a general-purpose educational AR authoring tool is not a simple or straightforward process. There are various frameworks for developers to build AR tools that can be used for educational purposes. Therefore, it is difficult for developers to decide on the design, architecture and approaches they should follow in order to build general-purpose educational augmented reality applications. In addition, the provided functionality of an implemented AR authoring tool and the mobile application can be varied and follow difference procedures for the development of AR experience. This paper lists some of the most known tools for building AR applications that will help developers decide on the adoption of the most appropriate for their needs and knowledge, and also presents the case study of ARTutor (Version 3), a general-purpose AR application. The architecture of the system is analyzed, to provide valuable insights for AR developers and the functionality is discussed along with the results of its use by pre-service teachers.

The paper is structured as follows: Sect. 2 lists available tools and technologies for building AR applications, while Sect. 3 provides an overview a general-purpose AR platform, ARTutor. Section 4 presents the findings of the initial evaluation of ARTutor, while the conclusions of this study are listed in Sect. 5.

2 Tools for Building AR Applications

There are currently 2 ways to build an AR experience. The first one is to use a programming SDK. In a previous work [19], we identified 19 SDKs that provide AR features to developers. From these SDKs, only 4 were completely free, AR.js [20], ARCore [21], ARKit 5 [22] and XZIMG Augmented Vision [23]. AR.js runs through a browser and it supports Web AR (Augmented Reality on the Web). Web AR provide less smooth experience and higher lag compared to AR through mobile applications. XZIMG Augmented Vision supports less features than ARCore and ARKit and updates to the SDK are not frequent.

The other 15 offered prices up to 2000 euros for an application or required a monthly fee. In order to build a general-purpose educational AR application, it is essential to use an SDK that will have constant support by major tech companies.

Google's ARCore SDK provides APIs for all of the essential AR features like motion tracking, environmental understanding, and light estimation. With these capabilities developers can build entirely new AR experiences or enhance existing apps with AR features for Android devices. ARCore is completely free and is supported by many Android devices. However, in order for a device to be able to run ARCore applications, the device must be certified by Google. Certification is important for Google since the tech giant wants users to have a good experience with AR applications that are built with ARCore. To certify each device, Google checks the quality of the camera, motion sensors, and the design architecture to ensure it performs as expected. Also, the device needs to have a powerful enough CPU that integrates with the hardware design to ensure good performance. Google constantly updates the supported device models and brings new feature to the ARCore SDK.

ARKit is Apple's SDK for iOS devices and enables app developers to incorporate AR into their apps. ARKit handles many of the tough tasks associated with AR, including environment and movement detection, simplifying the process for developers to place virtual objects in the real world. ARKit requires iOS 11.0 or later and an iOS device with an A9 or later processor, while some ARKit features require later iOS versions or specific devices. For example, in order to display seamlessly AR content that realistically passes behind and in front of people in the real world, making AR experiences more immersive, requires better depth estimation provided on iPhone 12, iPhone 12 Pro, and iPad Pro devices. Apple also constantly improves the ARKit SDK bringing new capabilities for developers.

Another way to build an AR experience is to use an online platform with no coding skills. In a previous work [24] we identified 14 cloud-based platforms which enable rapid and simple creation and deployment of AR experiences using drag and drop items. These platforms don't require coding skills and are relatively easy to use. However, of the 14 platforms, only 3, ARTutor, Jig WorkShop [25] and Metaverse [26] were completely free. Jig WorkShop is only available for iOS devices only, while Metaverse does not provide many features regarding AR such as environmental understanding and motion tracking. Most platforms have also limitations in the content that can be uploaded and provide limited storage space. Furthermore, most online platforms are not targeted for educational use and provide solutions for commercial use.

3 The Case of ARTutor Version 3

ARTutor, is a free and simple to use tool that provides the ability to educators and students to bring books to life using AR. Educators and students can add digital content and augment specific parts of a book. Digital content can be images, videos, audio files and 3D objects.

The ARTutor platform consists of two modules. The first module is the web portal and the second is the mobile application. The web portal allows teachers to create educational AR experiences and mobile application enables end users (students) to scan trigger images of the educational content and watch the augmentations through the mobile screen.

ARTutor follows a simple architecture. A server is used to store the necessary files used as augmentations (images, videos, audio files and 3D objects) and books (in pdf format). A database is also used to store the appropriate information about the users, the books and its augmentations. The Web portal is hosted on the server, while the mobile application communicates with the server through JSON. A view of the ARTutor architecture is depicted in Fig. 1.

3.1 Web Portal

The web portal available at http://arturor.ihu.gr serves as the gateway to the available augmented books and the authoring tool. The web portal provides access to all the augmented books. Students and educators can visit the web portal and take a look at all augmented books that are available at the platform. Augmented books can be filtered by title, category or author. Users can view each book's details, including a list of the augmentations available in the book. They can also download each book in pdf format. Every book is associated with a QR code and users can scan the QR code with the mobile application in order to load the augmented book in their mobile device. All of the above functionality is provided to users without login credentials. Thus, non-registered users can view all available books and preview their augmentations.

Registered users have access to the authoring tool through the web portal. In order to register for free to the ARTutor platform, users can submit their information including their email and the desired password at http://artutor.ihu.gr/login/. After registration, access is provided to the authoring tool (Fig. 2). Users can create a book or edit an existing book created earlier by them. When creating a new book, users can provide basic information about the book (title, description, book category, language) and upload the book in PDF format. The uploaded book will be used as the source where the ARTutor mobile application will display the digital content (augmentations). After uploading the book, users can select parts of the book and assign augmentations to them. Augmentations can be images, videos, audio files and 3D objects.

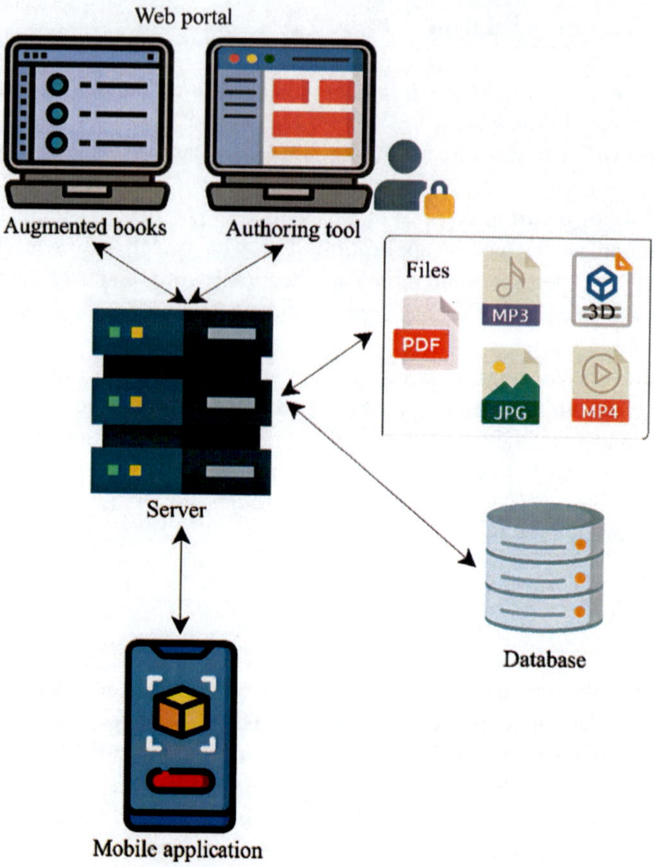

Fig. 1. ARTutor platform architecture

To sum up, the steps of creating an augmented book in the authoring tool are the following:

1. Log in to the web portal and access the authoring tool.
2. Upload a book in pdf format.
3. Select parts of the book and assign augmentations (images, videos, audio files or 3D models) to specific parts of the book
4. Save the book

Users can then create a new book and repeat the process from Step 2. After this process, the augmented book can be viewed in the ARTutor mobile application.

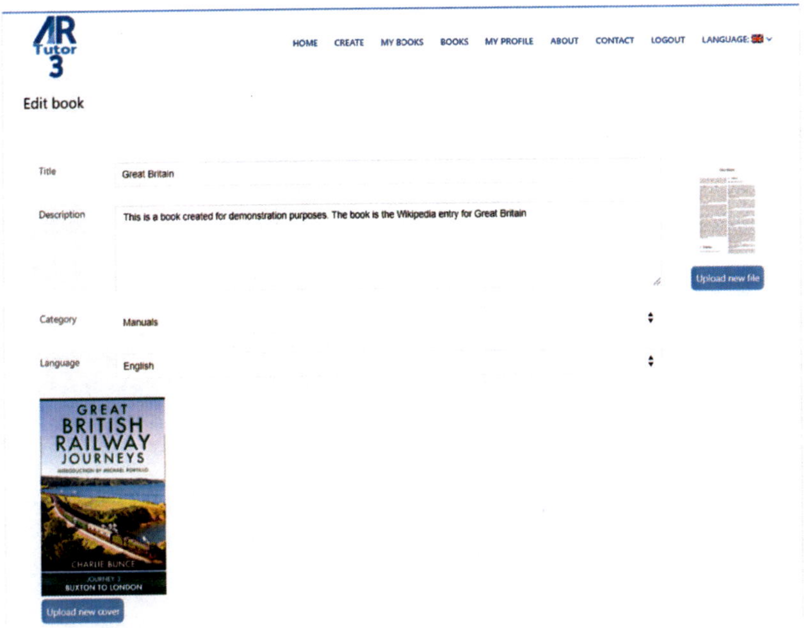

Fig. 2. Authoring tool

3.2 Mobile Application

The mobile application ARTutor is available at Google Play for Android devices and at App Store for iOS devices. When opening the application, the user will be navigated to app's home screen (Fig. 3). From here the user has two options: Find a book that is created/hosted on the ARTutor server or scan a QR code and load the augmented book instantly.

When a book is selected, all the augmentation objects that are assigned to the different areas of the book are retrieved from the server. After that, a new screen opens where the camera preview is displayed, and the device is ready to scan the various pages of the selected book and recognize the trigger images. As shown in Fig. 4 when an area of the book is recognized, the corresponding augmentation over the trigger image is displayed. When the augmentation is of type sound, the sound is played back. If the augmentation type is video, then the video is played on the top of the trigger image. The user is also able to interact with the augmentation objects by doing a pinching gesture on them. In this way, he can zoom in, zoom out and rotate them if it is supported by the augmentation itself.

Fig. 3. ARTutor's home screen

Fig. 4. Displaying augmentations

4 Evaluation of ARTutor Version 3

4.1 Methodology

Adopting the proposed approach and architecture, developers can build an educational AR platform with less effort and outstanding performance. However, there is need of evidence that confirm that on the one hand, the adopted development framework is considered as useful from the developers and on the other hand that the provided functionality and is easy to use for educators. For this reason, three experienced programmers commented the adopted approach for ARTutor development. Furthermore, pre-service teachers assessed the systems' provided functionality and its ability to create AR experiences for any domain.

Based on the previously presented tools for the development AR applications, the adopted development approach and the challenges we faced and the provided functionality, the research questions (RQ) are posited:

RQ1. Does experienced developers consider useful and meaningful the proposed development approach and tools?

RQ2. Do they consider ARTutor and AR as valuable technology for education?

RQ3. Can educators create with ease, AR experiences with ARTutor?

RQ4. What is their user experience in various issues regarding ARTutor provided functionality?

Research Design
Structured interviews were used to get information from experienced developers. The data from pre-service teachers was gathered using a quantitative approach; participants were asked to complete a survey instrument that included closed-ended and open-ended questions aimed at eliciting their opinions on ARTutor functionality in order to gain insight into their profound perceptions.

Participants
For the first part of the study three experienced programmers participated. Researchers presented them the architecture and implementation framework of ARTutor and discussed with them through a structured interview about the implementation process.

For the second part of the study, 95 pre-service student teachers were asked to use ARTutor in order to create educational experiences in various fields of study. All of them, were attending the course Information and Communication Technologies in Education, in the Computer Science department of International Hellenic University. The sample that used the ARTutor was at 70% students of the fourth semester of the department and the rest were students from other semesters. Fifty-three students (N = 53) were answered the questionnaire at the end of the semester which corresponds to the 56% of the students which utilized ARTutor during the semester. In addition, a focus group discussion took place to gather their experiences and discuss about their views regarding ARTutor. Thirty-seven of the participants were men (70%) and the rest women (30%). This is an indicative proportion between women and men in the current department. The majority of them (90.6%) have been user AR technology as end-users mainly in gaming (e.g. Pokemon Go, Mobile camera AR functionality). However, most of them had not implemented any AR experiences in the part nor had been used any AR authoring tools again (87%). The participants took part in the research voluntary, their answers were anonymous and would not affect by any mean their grades.

Instrument
The questionnaire used in this study consisted of three parts. Part A, consisted of 7 questions, asked about demographics and participants' socio-professional backgrounds, as well as their experiences with augmented reality. Part B consisted of 6 questions that probed participants' attitudes and opinions regarding the use the ARTutor as an authoring AR tool and the potential of the implemented AR experiences in the education. Part C, consisted of 17 agreement and open-ended questions assessed the ARTutor provided functionality and user experience in terms of usability and usefulness.

4.2 Results

Regarding RQ1, the programmers mentioned that the proposed approach gives a road path to programmers in order to develop educational AR applications. They stated that the proposed approach is valuable for them and insights on the path they have to follow in order to implement a general-purpose AR authoring tool for education. Additionally, using free technologies that are supported and promoted by major companies such as Google (ARCore) and Apple (ARKit) seems to be the only way to provide a free service to users.

On the other hand, pre-service students teachers initially answered the provided questionnaire. From the analysis of the data regarding RQ2, it appears that 84.9% of participants would like to use ARTutor for their futured classes and 90.6% considered that ARTutor could had a positive effect on students learning. In a fourth-grade scale (1 lowest grade and 4 highest grade) they considered as potential scenarios for ARTutor use for presentation of auxiliary educational content (2.36), development of interactive educational content (2.32), implementation of educational mixed reality games such as AR board/card games (2.19) and that they would engage their students on creating AR equational content (2.28). Overall, they considered ARTutor as a valuable tool for education (88.7%).

Regarding RQ3, the participants responded that ARTutor authoring tool uses a simple (73.2%) and fast (51.6%) process for AR experiences development. They stated that they do not encountered serious problems during the development phase (86.8) and they feel comfortable using ARTutor (79.2%) and they would suggest this tool for their colleagues (58.9%).

The analysis about RQ4, revealed various details about the use of ARTutor as an authoring tool. Participants agree that ARtutor messages are usefull (71.4%), they needed sometime to feel confident with the tool (57.1%), only 16.1% stated that they do not feel comfortable when using the tool and need some more time to learn it, 62.5% think that working with ARTutor is a satisfying process and 70.2% consider the way the system information is presented is clear and understandable.

At the focus group that took place, pre-service students teachers indicated that the result of the proposed approach, ARTutor, is an easy-to-use and useful platform independent of the domain. They commented that ARTutor enabled them to create meaningful AR augmented books and AR supported activities almost instantly with a very simple way. Also, they stated that educator imagination is the new limit on the development of AR experiences. On the other hand, they pointed out that many Android devices did not supported it (ARTutor is executed only in ARCore supported devices since AR technology requires multiple sensors available at the device) and sometimes ARTutor did not load correctly the augmentations (this happened occasionally when an augmentation is more than 30 MB and for this reason there is a strong suggestion not to upload large augmentation files). They proposed some useful functionality for the future that ARTutor could provide such as interactive questions, YouTube video presentation and webpage load. Finally, they commented that overall, this is a good tool that they would definitely are interested to use it in their potential classes.

5 Conclusions

AR technology has the potential of transforming the educational process. Nowadays there is a plethora of programming SDKs to develop AR applications, and programmers have to make hard decisions on the architecture that they will follow. One of the main problems for educator is choosing the educational AR application that suits their needs. Most of the educational AR applications available are domain specific and difficult to use in other domains and cases. Thus, it is beneficial to provide a framework for building general-purpose AR applications based on free SDKs and technologies. Adoption of an optimal technology and approach will allow the development of better educational AR applications with less effort and cost.

This research proposes an architecture for a general-purpose AR platform that supports most of the well-known functionalities, utilizing two of the most popular SDKs. ARCore and ARKit are Google and Apple's respective Augmented Reality frameworks for bringing more AR apps to their platforms (Android and iOS). Adopting the proposed architecture, functionality and database design a general-purpose AR application and authoring environment were implemented and evaluated. The evaluation outcomes show that the proposed technologies can be bound together in order to provide a successful solution for AR technology in education. The positive reactions of programmers and pre-service teachers show that the proposed approach can be useful for anyone who wants to develop an educational AR application and therefore the proposed implementation framework could be a valuable insight for potential AR developers and education stakeholders.

This study has some limitations. ARTutor was evaluated from a small number of pre-service teachers which are students of a national public university of a computer science department. Even the tool does not require programming knowledge, participants have good information technologies skills. Future experiments should be conducted with more participants and check whether ARTutor can be used by real teachers in various education levels. Furthermore, future works should study various scenarios that could applied through ARTutor and investigate further improvements on the provided functionality.

References

1. Atwood-Blaine, D., Huffman, D.: Mobile gaming and student interactions in a science center: the future of gaming in science education. Int. J. Sci. Math. Educ. **15**(1), 45–65 (2017)
2. Hwang, G.-J., Wu, P.-H., Chen, C.-C., Tu, N.-T.: Effects of an augmented reality-based educational game on students' learning achievements and attitudes in real-world observations. Interact. Learn. Environ. **24**(8), 1895–1906 (2016)
3. Chen, C.-H., Chou, Y.-Y., Huang, C.-Y.: An augmented-reality-based concept map to support mobile learning for science. Asia-Pac. Educ. Res. **25**(4), 567–578 (2016)
4. Chang, H.: How to augment the learning impact of computer simulations? the designs and effects of interactivity and scaffolding. Interact. Learn. Environ. **25**(8), 1083–1097 (2016)
5. Laine, T., Nygren, E., Dirin, A., Suk, H.-J.: Science spots AR: a platform for science learning games with augmented reality. Educ. Technol. Res. Dev. **64**(3), 507–531 (2016). https://doi.org/10.1007/s11423-015-9419-0
6. Cai, S., Chiang, F., Sun, Y., Lin, C., Lee, J.: Applications of augmented reality-based natural interactive learning in magnetic field instruction. Interact. Learn. Environ. **25**(6), 778–791 (2016)

7. Kazanidis, I., Pellas, N.: Developing and assessing augmented reality applications for Mathematics with trainee instructional media designers: an exploratory study on user experience. J. Univ. Comput. Sci. Special Issue. Immersive Learning Technologies: Research and Future Directions **25**(5), 489–514 (2019)
8. Wang, Y.-H.: Exploring the effectiveness of integrating augmented reality-based materials to support writing activities. Comput. Educ. **113**, 162–176 (2017)
9. Kazanidis, I., Tsinakos, A., Lytridis, C.: Teaching mobile programming using augmented reality and collaborative game based learning. In: Auer, M., Tsiatsos, T. (eds.) Interactive Mobile Communication Technologies and Learning, IMCL 2017. Advances in Intelligent Systems and Computing, Thessaloniki, vol. 725, pp. 850–859. Springer, Cham (2018)
10. Lytridis, C., Tsinakos, A., Kazanidis I.: Enhancing educational books with augmented reality using the ARTutor platform. In: The Annual Conference of International Council of Educational Media, 5–7 September 2018, Tallinn, Estonia (2018)
11. Kazanidis I., Sotiriadis G., Lytridis C., Tsinakos, A.: Using augmented reality application for childhood education in natural disasters. Novel approaches in Risk, Crisis and Disaster Management, vol. 10, pp. 337–354. Nova Science Publishers Inc.
12. Kazanidis, I., Fotaris, P., Gotzamanis, V., Tsinakos, A.: Educational escape room for disaster preparedness and response training. In: Proceedings of the 14th European Conference on Games Based Learning (ECGBL20), 24th – 25th September 2020, Brighton, pp. 832–839 (2020)
13. Kazanidis, I., Palaigeorgiou, G., Bazinas, C.: Dynamic interactive number lines for fraction learning in a mixed reality environment. In: 2018 South-Eastern European Design Automation, Computer Engineering, Computer Networks and Society Media Conference (SEEDA_CECNSM), 25–27 September, Kastoria, pp. 1–5. IEEE (2018)
14. Palaigeorgiou, G., Papadopoulou, A., Kazanidis, I.: Interactive video for learning: a review of interaction types, commercial platforms, and design guidelines. In: Tsitouridou, M., A. Diniz, J., Mikropoulos, T.A. (eds.) TECH-EDU 2018. CCIS, vol. 993, pp. 503–518. Springer, Cham (2019). https://doi.org/10.1007/978-3-030-20954-4_38
15. Pellas, N., Fotaris, P., Kazanidis, I., Wells, D.: Augmenting the learning experience in primary and secondary school education: a systematic review of recent trends in augmented reality game-based learning. Vir. Real. Spec. Issue: Virtual and Augmented Reality for Enhanced Experience in Education and Learning **23**, 329–346 (2019)
16. Pellas, N., Kazanidis, I., Palaigeorgiou, G.: A systematic literature review of mixed reality environments in K-12 education. Educ. Inf. Technol. **25**(4), 2481–2520 (2019). https://doi.org/10.1007/s10639-019-10076-4
17. Tobar-Muñoz, H., Baldiris, S., Fabregat, R.: Augmented Reality game-based learning: enriching students' experience during reading comprehension activities. J. Educ. Comput. Res. **55**(7), 901–936 (2017)
18. Wei, X., Weng, D., Liu, Y., Wang, Y.: Teaching based on augmented reality for a technical creative design course. Comput. Educ. **81**, 221–234 (2015)
19. Terzopoulos, G., Kazanidis, I., Satratzemi, M., Tsinakos, A.: A comparative study of augmented reality platforms for building educational mobile applications. In: Auer, M.E., Tsiatsos, T. (eds.) IMCL 2019. AISC, vol. 1192, pp. 307–316. Springer, Cham (2021). https://doi.org/10.1007/978-3-030-49932-7_30
20. GitHub, AR.js. https://ar-js-org.github.io/AR.js-Docs/. Accessed 26th June 2021
21. Google Developer, ARCore. https://developers.google.com/ar/. Accessed 6 July 2021
22. Apple Developer, Get Ready for ARKit3. https://developer.apple.com/augmented-real-ity/arkit/. Accessed 6 July 2021
23. XZIMG, Augmented Vision. https://www.xzimg.com/Products?nav=product-XAV. Accessed 6 July 2021

24. Terzopoulos, G., Tsinakos, A.: A comparative study of online augmented reality platforms for building educational experiences. In: 6th World Conference on Blended Learning (WCBL 2021), 6–7 May, pp. 80–87 (2021)
25. Jig WorkShop. AR presentations made easy. https://jig.space/workshop.html. Accessed 6 July 2021
26. Metaverse Studio. Metaverse Studio. https://studio.gometa.io/landing. Accessed 6 July 2021

IntraPlanet: An Embodied Approach of Teaching the Seasons Using Augmented Reality

Eirini Anastasiadou, Stefanos Xefteris, and George Palaigeorgiou(✉)

University of Western Macedonia, Florina, Greece
gpalegeo@gmail.com

Abstract. This paper evaluates an embodied learning intervention with augmented reality features, for teaching the phenomenon of the seasons. It was conducted at the Elementary Education Department of the University of Western Macedonia among 62 sixth grade students. The goal was to explore the learning effects of an easily reproducible environment created with accessible hardware.

Keywords: Embodied learning · Augmented reality · Mobile learning

1 Introduction

Elementary school students have difficulties in understanding the phenomenon of seasons as their preexisting representations are usually limited, incomplete or incorrect. Children interpret the natural phenomena in the most intuitive way [1]. Thus, they scarcely accept scientific explanations that contradict their naive beliefs [2]. Traditional educational interventions are often ineffective as students may temporarily embrace scientific explanations, but soon shoehorn them to their initial beliefs [3]. Understanding scientific issues effectively, requires learners to abandon past misunderstandings and replace existing cognitive structures with new ones [4]. This is usually achieved by involving students in situations where they can evaluate empirical data that contradicts their beliefs [2].

Studies have shown that interaction with computer simulations effectively facilitates learning and engagement in science education [5–7]. In recent years, however, there is an emerging literature that emphasizes the learning benefits of more expressive and embodied interaction [8, 9] as well as the value of hand-held interfaces and gesture-based interface [10, 11]. In a recent study [12] it is postulated that bodily interaction helps students create natural representations, which increases their conceptual understanding of abstract notions. Interaction with physical objects enables students to create conceptual anchors by becoming cogs in the systems they are exploring, their actions translating the system's inner mechanisms, the very things they are trying to understand.

This study endeavored to achieve conceptual change in the students' understanding of the phenomenon of seasons employing technologically enhanced embodied interaction affordances. The authors aimed at conceptual change through the contradiction of naïve beliefs with empirical data. Our main hypothesis was that students'

M. E. Auer and T. Tsiatsos (Eds.): IMCL 2021, LNNS 411, pp. 180–191, 2022.
https://doi.org/10.1007/978-3-030-96296-8_17

misconceptions would be shattered if they experienced in real time and from a different and wider point of view how the Sun illuminates the Earth. In a workflow that combines embodied learning with augmented reality, the pivotal point of this study is that when students actively experience the movement of the Earth's axis, by corresponding their body's stance to axial tilt and at the same time, observe the Sun's effects on Earth, they can more efficiently understand the mechanism behind seasonal change.

2 Literature Review

2.1 Misconceptions About How Seasons Occur and the Role of ICT

To explain the phenomenon of the seasons we need first to understand a series of other scientific principles and concepts (basic properties of light, the solar system, weather and climate) [13]. Research has highlighted a multitude of misconceptions about how seasons occur on Earth. The most prevalent misconception is that in winter the Earth is farther away from the Sun [3, 14–19]. There is a multitude of recorded misconceptions in literature. The list can be quite extensive, here we highlight a few prominent examples:

- Winter is caused by heavy clouds blocking the sun's rays [14, 16]
- We have winter because the sun moved to the other side of the earth, where it is now summer [14]
- The Earth's axis swings back and forth turning to the Sun in the summer and away from it in the winter [20]
- The Sun changes temperature naturally heating and freezing periodically [21].
- The Sun travels around the Earth at different speeds during the year [22]
- The Sun has different properties in summer and winter [20].
- The Earth rotates around its axis once a year [4]
- Aerial masses and sea currents cause temperature changes [3, 4]

The multitude of these misconceptions is indicative of the complicated challenge educators face when teaching the mechanism of seasons. Using traditional teaching approaches, literature shows that the cognitive results are not promising [23]. To improve efficacy of existing methods we need to devise new tools and environments. Two dimensional models have been commonly used to teach the phenomenon, providing useful visualizations, however two-dimensional models of three-dimensional actions and movements could actually prove to be hindrances rather than improvements.

In various studies, virtual reality (VR) and augmented reality (AR) applications were used in an attempt to teach complex scientific concepts. The authors of [3] proposed a three-dimensional virtual desktop environment that facilitated the students' personal engagement and close examination of the phenomena under study.. A virtual reality (VR) system was used [24] to visualize phenomena such as the four seasons. It concluded that learning complex phenomena relates to when and how students can manipulate virtual 3D objects. The Live Solar System (LSS) application [25] is an educational tool based on VR: In this application the human-computer interface was based on cubes and cards -objects the students were familiar with, rather than the usual mouse and keyboard. The use of 3D simulation software has shown significant results in teaching the mechanism for seasons, as visualization and physicalization of the phenomenon can lead to a deeper understanding.

2.2 Embodied Learning and ICT

Embodied interaction as a teaching framework for scientific concepts has shown promising results in recent research. New interaction technologies can prove an excellent basis for deploying teaching scenarios based on physical interaction which serves as conceptual leverage [9, 12, 26]. Under the umbrella of terms like embodied interaction, full-body interaction, motion-based interaction, gesture-based interaction, tangible interaction, bodily interaction, and kinesthetic interaction, several learning environments based on different interaction modalities have been developed. Following the embodied interaction precepts, these environments aim to:

1. Facilitate an embodied experience of a certain concept
2. Represent an abstract concept as a concrete instance
3. Operationalize actions as means to express specific content
4. Use space as a semiotic resource or even
5. Become embodied metaphors of the abstractions they represent.

In [12] the authors ascertained that the physicalization of concepts and whole body interaction with them leads to significant learning benefits, and a more positive attitude towards science. Compared to a group using the usual human-computer interfaces (mouse and keyboard), participants using a whole-body approach showed improved learning outcomes and a more positive attitude to the simulation experience and the learning environment. The authors in [11] created a computer simulation to teach the centripetal force. Participants who used their body by shaking a detectable object above their head showed higher long-term learning results in physics compared to students who used a mouse.

The new mediated environments seem to increase learner engagement since body-based experiences are more perceptually immersive and learners may feel that they are in a more authentic, meaningful educational space. However, designing embodied learning is an emerging and not yet systematized research area. Studies indicate that coupling knowledge with physical interactions has a strong effect on learning but not all interactions have learning impact. Congruency has been the focus on recent studies centered on embodied learning [12]: By congruency the authors define the condition under which, movements or body positioning are related to a particular conceptual domain. Further research is required for researchers to learn how to cue the body to enact certain actions and create physical representations that facilitate conceptual understanding.

3 The Learning Environment

In this study we deploy a solar system simulation using embodied learning affordances: In a dark room with a single light source (the Sun), we place two identical "Earth" spheres in opposite directions, depicting two different positions in the phases of the trajectory around the Sun, as in Fig. 1. The model Earths are sized so that one student fits inside them (from head to waist-level). When inside the Earth, the students observe the Sun through a mobile phone camera located at the latitude of our country (Greece). The students locate the position of the Sun on their virtual sky by looking at the screen of the mobile phone.

From the moment the students enter the sphere, the tilt of their body coincides with the Earth's axial tilt. We postulated that this environment, facilitating a real-time observation of the effect changes in axial tilt have on the Sun's position, would have a significant learning impact. Concurrently with the student inside the sphere, students were also able to study the phenomenon from the outside. The learning environment was designed to support up to four participants; two inside the Earth models and two outside. Those outside could observe differences in how light from the Sun illuminated the Earth models, depending on axial tilt (how the students inside angled their body).

Fig. 1. The sphere consisted of lightweight materials and full safety precautions were met

3.1 Learning Activities

The learning environment integrated activities and embodied interaction in a gamified framework. Students participated in a set of inquiry activities and interacted with the environment that prompted them to answer questions, make choices and reflect on observations.

Beginning the game, a character (Simon) appears on the ceiling guiding two students ("Earthlings") into each sphere, and urges the other two ("Astrals/Spacies") to observe how the Sun illuminates the Earth. "Earthlings" have the ability to change the tilt of the sphere into three possible positions by a) leaning forward b) sitting straight c) leaning backward. These positions were determined by three axles.

An example activity is this: Simon asks: *"It is known that the Earth constantly revolves around the Sun and rotates on its axis. And you* ***Spacies*** *are now in two opposite positions on its orbit. What is the true tilt of the Earth's axis?"* And then continues: "***Earthling*** *in position No1, grab the sphere from below and bend forward.* ***Earthling*** *in position No2 move the Earth exactly in the opposite direction, backwards.*". So, one sphere tilts forward and the other sphere, in the opposite position, tilts backward. After that, Simon asks: *"What is the season in Greece in position A and what in position B?"* Observing the amount of light falling in each position, students reflected on their observation and inferred the correct answer.

After completing all learning activities, students in groups used the Arlon Solar System tablet application and played a 5-min game to establish deeper understanding of the acquired knowledge.

4 Methodology

4.1 Participants

The participants were 62 sixth-grade students (11–12 years old). The research was conducted during an exhibition of educational technology applications for elementary school students, which took place at the University of Western Macedonia.

4.2 Procedure

Two pairs of students participated in each session. Before the intervention short instructions were given to the participants. The researcher explained the game and performed a brief demonstration of the game interactions. Each session lasted about 25–30 min, during which, help was provided when necessary. In the end, participants filled an online questionnaire and participated a semi-formal interview as focus groups.

4.3 Data Collection Instruments

Data collection was based on the completed questionnaires and oral semi-formal interviews.

The questionnaire consisted of 22 questions on the perceived value of the intervention in terms of its usability and attractiveness. The questions were obtained from the AttrakDiff questionnaire[27] and the Flow State Scale [28] and examined the following variables:

- Ease of Use (3 questions): How easy it is to learn and use the system;
- Autotelic experience (3 questions): The extent to which the system offers internal user satisfaction;
- User Focus (3 questions): Player concentration during the use of the system;
- Perceived learning (3 questions): Students' perceptions on the educational value of the system;
- Pragmatic Quality (4 questions): The extent to which the system allows a user to achieve their goals;

- Hedonic Quality-Stimulation (3 questions): The extent to which the system meets the user's need for innovation and whether it is of interest;
- Hedonic Quality-Identity (3 questions): The extent to which users can identify with the system.

All questions were adapted to a 5-point Likert scale.

Focus groups took place immediately after the end of the game, consisting of 10 questions. The focus groups intended to assess the students evaluation of the intervention giving them the opportunity to describe their experience in their own words.

Interview questions aimed to assess three basic aspects: a) Perceived learning processes and results (e.g. Did the use of your body as the Earth axis help you understand how Earth's axis contributes to the rotation of the seasons?) b) Environmental qualities (e.g., how experiential was the environment?) c) Comparison with typical learning apps (e.g., would the result be different if this activity was performed in a similar way on the computer?).

All audio-recorded focus groups were transcribed and then encoded and compared within and between cases using a QDA software and the constant comparative method. Afterwards the researchers collaborated to reach consensus for the commonly identified issues.

5 Results

5.1 Quantitative Data

Our findings indicate that variables had a satisfactory Cronbach's a (>0.65 with a maximum of 0.81) and can be considered as consistent.

As indicated in Table 1, students were easily accustomed to the new style of interaction. They claimed that the environment was easy to use and that they would like to interact with similar environments often (M = 4.18, SD = .84). Moreover, although the intervention took place in the midst of a noisy exhibition foyer, results indicate that students managed to maintain high levels of concentration showcasing increased engagement. We also note high scores on the autotelic experience variable, indicating the intrinsic pleasure students accrued from the activity.

Students were also positive in regards to the learning efficiency of the environment and the possibility of deployment inside a school classroom. They claimed that they would learn faster (M = 4.1, SD = .93) and would learn more (M = 4.00, SD = .90). However, there was a small group of 6 students (9,6%) which were negative towards the use of the learning environment in school conditions although similar responses or ideas did not surface during the interviews. It is important to note that such environments are deeply social which can be an asset regarding learning efficiency but also induce obstacles in the form of social anxiety and performance stress.

Table 1. Attitudes questionnaire results

	Minimum	Maximum	Mean	Std. Deviation	a
Easiness	2.67	5	4.28	0.61	0.66
Focus	1.33	5	4.44	0.77	0.81
Autotelic experience	2.33	5	4.47	0.69	0.79
Perceived learning	2	5	3.87	0.69	0.65
Pragmatic quality	1.75	5	4.02	0.78	0.72
Hedonic quality-identity	2.67	5	4.51	0.58	0.75
Hed. Qual.-Stimulation	3	5	4.66	0.54	0.81

Students' answers in the mini AttrakDiff questionnaire reveal that they considered the features of the learning environment as appropriate to achieve the goal of understanding the seasons' mechanism (pragmatic quality). Moreover, the hedonic quality variable, which is a measure of pleasure (fun, original, engaging) and avoidance of boredom and discomfort, had very high scores. Students' answers show that the identified with the environment (Hedonic Quality-Identity) and believed that it offered inspiring and novel features and interactions (Hedonic Quality-Stimulation).

5.2 Qualitative Data

The students focus groups answers corroborated the questionnaires findings. Analysis of the interviews revealed the students' perceptions about: 1. The learning experience 2. Their understanding of the phenomenon 3. Acquisition of new mental representations 4. Perceived learning effectiveness compared to computer simulation applications they were familiar with.

Experiential Learning

Participants insisted that the learning environment was strongly experiential. This attitude was shaped by the facilitation of active observation on a macroscopic level, with participants feeling in total control of the phenomenon from a point of view in space. According to the students these features formulated an immersive learning environment. Moreover, they were fascinated and found quite innovative, that they were operating the environment from inside the planet.

"I liked it because it's different to see the Earth rotate in the computer screen or through the projector than to act it yourself and see that if you lean back the season in the northern hemisphere is different than the season in the southern and the opposite."

"Yes, I felt I was in space and especially when I got into the Earth-sphere it was very nice."

Students enjoyed the activity. The novel features helped mark the experience as "nice" "stunning" and "real".

"It was nice because we could see the Sun through the Earth and we could answer the questions."

"I already knew the explanation, but it was a nice experience."

"Someone entered the sphere and through the cell phone could see the Sun as a real Sun."

Perceived Understanding

The majority of the students indicated that the arrangement helped them understand the underlying mechanism of seasonal change. At the same time, they claimed that this intervention helped them understand the phenomenon better than any other previous attempt in school. This reveals both the extant difficulties in teaching such complicated phenomena as well as the potential of the intervention.

"With it I understood how the seasons (winter and summer) work."

"Yes, we understood more than reading from our book because when someone moves the Earth they understand better when the Sun's rays fall and on which hemisphere."

"Yes, we did it in geography of the sixth grade this year. I learned more than what I learned at school."

"I feel I can explain it better."

"When we were tilting backward we did not see the Sun, but when we were moving forward we saw more of the Sun and so we understood what was happening with the axis."

"We learned the most when we were in the Earth-sphere."

The intervention consisted of three phases each with a considerable duration: 1. Observation outside the Earth 2. Observation from the interior of Earth and 3. Exploration of the phenomenon with the use of tablet apps. A majority of 90% of students answered that the feature that helped them improve their understanding of the phenomenon was the ability of observation from inside the planet and the embodied control of axial tilt.

This is an important finding as both the external observation of a three-dimensional illuminated model and the tablet simulation applications are used to teach the phenomenon of the seasons and are considered to be successful approaches. Students, on the other hand, seem to suggest that the representation from a point of view inside Earth is an more effective approach.

"[I understood] when we got to the ground and we got up and down."

"[I understood] with the help of the Earth's axis around the Sun when I entered the sphere."

"I understood a little better when I entered the sphere, because by watching the Sun I could explain better what is happening with the seasons.

At the same time, students noted that the phase of operation from inside the sphere was more enjoyable than the other phases. When asked their favorite part of the intervention, the vast majority replied that it was their time in the interior of the planet.

"I liked it when I was in the planet."

"I liked the point when we entered the spheres and made various moves."

Only two students claimed that the tablet application improved their understanding while one answered that both did. All others did not evaluate their interaction with the tablet as an important step in the activity.

Acquisition of New Representations

According to students the proposed embodied intervention had three main advantages:

A)Observing the Sun through a mobile phone.

They were able to observe the angle of incidence and evaluate the total brightness of the sun rays through the screen. The mobile phone camera induced a distortion effect to the light that resulted in a very convincing simulation of sunlight.

"Yes, because when we were in the sphere, we were watching the Sun from the camera and it was easier to understand than for a student who was out."

"The element I liked most was being in the planet, looking at the Sun and answering if we had winter or spring."

B)Embodied control of the axis.

In most proposed interventions, the Earth's axis is externally observed and controlled whereas in this case, participants controlled it with their own bodies, being a part of the mechanism itself. This embodied metaphor helped them understand how the Earth's axial tilt results in different seasons and what would happen if the axis had a different angle.

"My body's movement helped me understand the way the Earth's axis moves. When my body was going backward the Earth was going backward too, so its axis changed position."

"I understood that when I was going ahead it was summer. What helped me most was the movement of my body with the motion of the Earth because I felt like I was the Earth's axis and I was tilting the Earth. That gave me a better feeling."

C)Observation of the Sun from the perspective of a unique point on the surface of the Earth.

While students were inside the Earth-sphere, they were able to observe the way light from the Sun fell at a specific point of the globe in every axial position. Even slight changes in the incidence angle had dramatic and observable effects, and therefore provided strong empirical evidence resulting in conceptual changes.

"We could see from the interior of the planet we did not see the whole picture. We saw only one point. This helped us understand the phenomenon better because we did not have to see the whole planet."

"We saw firsthand how the Earth rotates, its axis and how the seasons occur…"

"When we were moving back, we couldn't see the Sun but when we were tilting forward we could see more of the Sun and so we understood what was happening."

Perceived Learning Effectiveness Compared to Simulation Applications

When students were asked if the learning experience would have been different on a computer screen, the majority claimed that the particular interface was more attractive and effective. According to the students the significant advantages of the proposed intervention were: (a) the plausibility and realism of the model; (b) its gamified character; (c) the hands-on approach; (d)the innovative features that made the intervention interesting and facilitated commitment e) embodied interaction facilitates seamless engagement with the intervention, in contrast to the conceptual obstacles that usual human-computer interfaces induce. Interestingly, students showed negativity towards computer-based educational applications and stated that authentic interaction models are preferable for them.

"There are many advantages with the model, it looks more attractive and it is like a game."

"We entered the Earth-sphere and we actually watched the rays fall while on the computer it would be just a video."

"On the computer, you only see a screen but when you are in the Earth-sphere it's like watching it live."

"It feels more real to go into the Earth-sphere. It isn't the same- seeing it on a screen and living it."

"It would not be enjoyable. Learn through such an activity is different than learning through a computer. In this way, you remember better what you've learned."

"It would be boring to watch it on a screen and just press a key."

"It feels more real and you are involved with your body."

6 Discussion

The majority of the students stated that they would enjoy having similar educational environments at school, showing appreciation for both the cognitive and entertainment aspects of the intervention.

The intervention was designed and deployed in such a way that a similar educational space can be easily reproduced and most certainly re-purposed for a broader spectrum of learning activities by researchers, teachers and students. Tangible and embodied learning interventions are now easily deployable through low-cost rapid prototyping hardware mixed with the uprising trend for arts and tinkering. This trend also helps in the design, development and effective exploitation of physical interfaces or in the identification of novel embodied metaphors and their implementation into concrete interaction instances.

The design approach moving the point of view in the interior of a planet provides an innovative stepping stone for the study of corresponding phenomena. Moreover, this design approach proved to be the most impressive and effective feature of the intervention, making students appreciate its ability to immerse them into the new point of view and evaluating it higher than the usual computer simulations they were accustomed to. Observing space phenomena from the point of view of Earth by filtering out our everyday perspective is a starting point for creating many different styles of interaction within the sphere. A simple extension of the proposed educational environment would be the addition of two more mobile devices in the interior so that students can simultaneously observe the Sun at different latitudes (e.g. one at the equator and one in Australia).

7 Conclusions

Overall, the study aimed to assess the efficacy and impact of the proposed setting, how an embodied interaction environment in a mixed reality context can foster new knowledge in an experiential learning framework. Offering a new perspective to students is a difficult task, especially when it clashes with preexisting misconceptions, a task which embodied interactions can positively affect. The effects of celestial body movement and spatial relations in the universe, how they are localized and affecting our immediate celestial neighborhood and our planet is a complicated task for younger students. Considering the study's outcomes, we assess those similar interventions, providing an embodied way of interacting with phenomena, while broadening the subject's perspective, can become a solid framework for acquisition of new knowledge and re-assessment of old. The authors acknowledge the small sample of participants and point the need for broader pilot studies with a wider user base. The use of multimodal technologies integrated on an embodied framework of reference using low-cost hardware, mixing reality with virtual representations paves the way to novel trends on educational technologies. The future focus will be effective implementation of relevant scenarios and deployment of embodied affordances as immersive and fascinating interaction modalities.

References

1. Driver, R.: Children's Ideas in Science. McGraw-Hill Education (UK) (1985)
2. Vosniadou, S.: Designing curricula for conceptual restructuring: lessons from the study of knowledge acquisition in astronomy. J. Curric. Stud. **23**(3), 219–237 (1991)
3. Bakas, C., Mikropoulos, T.: Design of virtual environments for the comprehension of planetary phenomena based on students' ideas. Int. J. Sci. Educ. **25**(8), 949–967 (2003)
4. Ojala, J.: Lost in space? the concepts of planetary phenomena held by trainee primary school teachers. Int. Res. Geogr. Environ. Educ. **6**(3), 183–203 (1997)
5. Rutten, N., Van Joolingen, W.R., der Veen, J.T.: The learning effects of computer simulations in science education. Comput. Educ. **58**(1), 136–153 (2012)
6. Smetana, L.K., Bell, R.L.: Computer simulations to support science instruction and learning: a critical review of the literature. Int. J. Sci. Educ. **34**(9), 1337–1370 (2012)
7. D'Angelo, C., Rutstein, D., Harris, C., Bernard, R., Borokhovski, E., Haertel G.: Simulations for STEM learning: systematic review and meta-analysis. Menlo Park SRI Int. (2014)
8. Richards, T.: Using kinesthetic activities to teach Ptolemaic and Copernican retrograde motion. Sci. Educ. **21**(6), 899–910 (2012)
9. Xefteris, S., Palaigeorgiou, G., Zoumpourtikoudi, H.: Educational robotics for creating tangible simulations: a mixed reality space for learning the day/night cycle. In: Auer, M.E., Tsiatsos, T. (eds.) IMCL 2019. AISC, vol. 1192, pp. 971–982. Springer, Cham (2021). https://doi.org/10.1007/978-3-030-49932-7_90
10. Chiu, J.L., DeJaegher, C.J., Chao, J.: The effects of augmented virtual science laboratories on middle school students' understanding of gas properties. Comput. Educ. **85**, 59–73 (2015)
11. Johnson-Glenberg, M.C., Birchfield, D.A., Tolentino, L., Koziupa, T.: Collaborative embodied learning in mixed reality motion-capture environments: two science studies. J. Educ. Psychol. **106**(1), 86 (2014)
12. Lindgren, R., Tscholl, M., Wang, S., Johnson, E.: Enhancing learning and engagement through embodied interaction within a mixed reality simulation. Comput. Educ. **95**, 174–187 (2016)
13. Sneider, C., Bar, V., Kavanagh, C.: Learning about seasons: a guide for teachers and curriculum developers. Astron. Educ. Rev. **10**(1) (2011)
14. Baxter, J.: Children's understanding of familiar astronomical events. Int. J. Sci. Educ. **11**(5), 502–513 (1989)
15. Danaia, L., McKinnon, D.H.: Common alternative astronomical conceptions encountered in junior secondary science classes: Why is this so?," Astron. Educ. Rev. **6**(2), 32–53 (2007)
16. Dunlop, J.: How children observe the universe. Publ. Astron. Soc. Aust. **17**(2), 194–206 (2000)
17. Parker, J., Heywood, D.: The earth and beyond: developing primary teachers' understanding of basic astronomical events. Int. J. Sci. Educ. **20**(5), 503–520 (1998)

18. Trumper, R.: Assessing students basic astronomy conceptions from junior high school through university. Aust. Sci. Teach. J. **41**(1), 21–43 (2001)
19. Tsai, C.-C., Chang, C.-Y.: Lasting effects of instruction guided by the conflict map: experimental study of learning about the causes of the seasons. J. Res. Sci. Teach. Off. J. Natl. Assoc. Res. Sci. Teach. **42**(10), 1089–1111 (2005)
20. Roald, I., Mikalsen, O.: Configuration and dynamics of the earth-sun-moon system: an investigation into conceptions of deaf and hearing pupils. Int. J. Sci. Educ. **23**(4), 423–440 (2001)
21. Sharp, J.G., Kuerbis, P.: Children's ideas about the solar system and the chaos in learning science. Sci. Educ. **90**(1), 124–147 (2006)
22. Kikas, E.: The impact of teaching on students' definitions and explanations of astronomical phenomena. Learn. Instr. **8**(5), 439–454 (1998)
23. Zhang, J., Sung, Y.T., Hou, H.T., Chang, K.E.: The development and evaluation of an augmented reality-based armillary sphere for astronomical observation instruction. Comput. Educ. **73**, 178–188 (2014)
24. Shelton, B.E., Hedley, N.R.: Using augmented reality for teaching earth-sun relationships to undergraduate geography students. In: The First IEEE International Workshop Agumented Reality Toolkit, p. 8 (2002)
25. Sin, A.K., Zaman, H.B.: Live Solar System (LSS): Evaluation of an Augmented Reality book-based educational tool. In: 2010 International Symposium on Information Technology, vol. 1, pp. 1–6 (2010)
26. Xefteris, S., Palaigeorgiou, G.: Mixing educational robotics, tangibles and mixed reality environments for the interdisciplinary learning of geography and history. Int. J. Eng. Pedagog. **9**(2), 82–98 (2019)
27. Hassenzahl, M., Monk, A.: The inference of perceived usability from beauty. Hum.-Comput. Interact. **25**(3), 235–260 (2010)
28. Jackson, S.A., Marsh, H.W.: Development and validation of a scale to measure optimal experience: the flow state scale. J. Sport Exerc. Psychol. **18**(1), 17–35 (1996)

A Hybrid Virtual-Physical Approach for Performing Control Theory Laboratories from Home

Mostafa Mohamed Soliman(✉)

W Booth School of Engineering Practice and Technology, McMaster University, Hamilton, Canada
Solimm12@mcmaster.ca

Abstract. A major challenge in the online delivery of engineering courses is the implementation of the lab component. With most campuses being closed due to COVID-19 pandemic, many of the engineering courses' labs are either cancelled, converted to a totally virtual, or remote lab setting. The paper proposes a novel hybrid virtual-physical laboratories for a control theory course that is delivered online. The virtual labs are implemented using a 3D animated digital twin of a motor. The physical labs are implemented using a low-cost take-home lab kit that was sent to students.

The use of the proposed hybrid approach achieves the benefits of both virtual labs and physical labs. Specifically, the virtual labs form a sandbox where the student can safely experiment and try new designs without worrying about damaging equipment. This will also form a gentle introduction to the utilization of the physical labs. The physical labs allow the student to see the actual control system components, and hardware troubleshooting.

Both the proposed virtual and physical labs can be performed by the student from home at any convenient time; and the lab is always available to the student through the whole semester.

The hybrid virtual-physical labs were utilized in the fall of 2020 within an introductory control theory course offered to a third-year undergraduate students at McMaster University. Based on the students' feedback, the designed labs were effective within the online course delivery and they provided deep insights and understanding of the concepts that were taught in the class.

Keywords: Virtual laboratories · Take-home laboratory kits · Digital twins · Remote learning · Control systems education

1 Introduction

Experiential and constructive learning are crucial for engineering education [1, 2]. By engaging students in interactive learning activities, students can have better knowledge retention, enhance their critical thinking, analysis, and creation. Furthermore, the different skills and competencies that are gained through hands-on learning offer an advantage for the student after graduation and working in the industry.

M. E. Auer and T. Tsiatsos (Eds.): IMCL 2021, LNNS 411, pp. 192–202, 2022.
https://doi.org/10.1007/978-3-030-96296-8_18

The usage of laboratories in automatic control education is a key element [3, 4]. Many of the control theory course topics are relatively hard to grasp and mathematically abstract. Providing hands-on experience is often a great aid for teaching these difficult concepts. Cognitive psychology studies show that people acquire better knowledge by doing things and thinking about the consequences of their action [3] rather than just visualizing a figure or reading a book.

In the wake of the COVID-19 pandemic, many universities requested the students to stay at home, and switched to online course delivery. This created a major challenge on how to deliver engineering laboratories online.

The recent advances in information and computational technologies has changed the educational laboratory landscape. In addition to traditional, face-to-face hands-on laboratories, three new formats have emerged: virtual/simulated laboratories, remote/tele-operated laboratories, take-home laboratories. A summary of these options is discussed below [5].

1.1 Review of Non-traditional Laboratory Delivery Options

Virtual Laboratories. Virtual laboratories mimic real experiments on computers. Using simulation models and 2D or 3D animations, students can conduct and visualize experiments. Simulations can reduce the time for teaching/learning and offer the students an active learning experience. However, data from virtual laboratories are not real, even with high performance simulators. Furthermore, there is no opportunity for the student to troubleshoot hardware problems.

Many papers in the literature suggest the use of virtual labs in teaching control theory. The experience varies from simple Excel simulations [6], 2D animations [7, 8], to immersive 3D animations such as in [9–13].

Remote Laboratories. Remote laboratories are characterized by mediated reality. Interaction between the student and the geographically remote lab equipment is provided through the Web, where students can visualize the system using a Web camera and observe data in real time from the remote system and download it [14, 15].

Many papers suggested the use of remote labs to teach control theory concepts [16].

Take-Home Hands-on Laboratories. With the continuous decrease in computing hardware price and size, the use of take-home lab kits has been recently introduced for teaching control theory [4, 17]. With this, it is now possible for students to have their own hardware and do the labs from home or any place. The use of the take-home laboratories reduces the burden on booking lab spaces and schedules. From the students view, performing the lab becomes extremely flexible in terms of timing and location.

Hybrid Laboratories

The use of "hybrid" or "blended" approach to laboratory learning is recently introduced. In this type of lab, both virtual/remote or non-traditional/traditional modalities are combined in an attempt to capitalize on the benefits of both.

Many recent papers discuss the use of hybrid virtual-remote labs [15, 18–20] in control systems education.

1.2 Paper Summary and Contribution

This paper proposed a novel hybrid virtual-physical laboratories to perform control theory labs from home. The labs were designed to be performed online as a result of COVID-19 pandemic. The first set of labs utilize a 3D virtual motor to teach students the fundamentals of Control Theory. This is a safe environment where the students can safely try and experiment ideas. As the students' competencies and knowledge build up during the semester, the labs are switched to physical labs where students can face real-life troubleshooting challenges. Both labs are carefully designed to complement each other and to offer a smooth transition for the students.

Another contribution of this paper is the design of a low-cost take-home motor control kit that is small in size and cheap in cost.

This paper is organized as follows. Section 2 describes the lab delivery prior to COVID-19 pandemic and the main lab learning objectives. The proposed hybrid virtual-physical labs are discussed in Sect. 3. The student learning experience is discussed in Sect. 4; and Sect. 5 concludes the paper.

2 Control Theory Laboratories Pre-COVID

The laboratories discussed in this paper are part of an introductory Control Theory course (PROCTECH 3CT3) that is offered to level 3a undergraduate students who are enrolled in the Automation Engineering Technology program at McMaster University. The course consists of 3-h/week of lecture/tutorials and a 3-h/week laboratory session.

The course is offered in the fall semester and it was fully delivered online for the first time in the fall of 2020 as a result of the COVID-19 pandemic. The course teaches standard classical control theory topics such as system modeling, transfer functions, system response, stability, control system design methods.

Laboratory work is a major component of the PROCTECH 3CT3 course, with a weight of 20% of the course grade.

Prior to the pandemic, the standard analog motor control kit in Fig. 1 was used to conduct the course laboratories. The kit consists of a brushed dc motor, sensors to measure position and speed, and an analog electronic board that is implementing a Proportional-Integral-Derivative (PID) controller. Students were capable to see the system response using oscilloscopes, and measure different performance indicators using cursor measurements.

The main learning objectives of these lab experiments are:

1. Tune PID controllers and understand the effect of the Proportional, integral, and derivative gains on the loop performance.
2. Design and implement a closed-loop position control system for a brushed DC motor using model-based design (MBD).
3. Model and simulate dynamical systems.
4. Utilize and connect different control components to construct a feedback control system

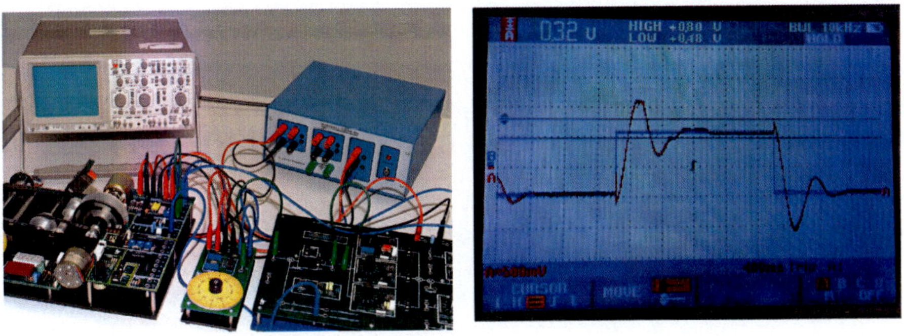

Fig. 1. Lab components used prior to the COVID-19 pandemic.

3 Online Control Theory Laboratories Using a Hybrid Virtual-Physical Approach

As a result of the COVID-19 pandemic, McMaster Campus was closed and the PROCTECH 3CT3 course was delivered fully online in the fall of 2020.

The course laboratories were fully redesigned to suite the online delivery. The new lab experiments used a virtual 3D digital twin of a motor and a take-home low-cost motor control kit that is sent to the students at home.

The first five labs utilized the 3D virtual motor that is animated using a 3D gaming engine. This platform offered the students an engaging and immersive control experience. After gaining sufficient knowledge, experience, and confidence from the first five labs, the students performed the last three labs on a low-cost motor control hardware that was sent to the students at home.

The different tools that are utilized in these labs are described in Sect. 3.1 and Sect. 3.2. In Sect. 3.3, the design and implementation of the hybrid virtual-physical lab is detailed.

3.1 Virtual Lab Component: Quanser QLabs

The virtual labs were implemented using Quanser Interactive Labs (QLabs) QUBE-Servo 2, shown in Fig. 2. The QLabs Virtual QUBE-Servo 2 is a dynamically accurate real-time digital twin of a physical hardware offered by Quanser to teach control theory. The digital twin can be interfaced and controlled using MATLAB®/Simulink® in the same way it is done with the actual physical system.

The QLabs Virtual QUBE-Servo 2 offers many advantages. First, the digital twin faithfully captures the physical system dynamics, non-linearities like dead zones, amplifier saturation, and many other real-world aspects. Furthermore, the digital twin was built using a powerful 3D gaming engine, the Unreal Engine 4, with an excellent rendering quality. This offers the students a very engaging and interactive experience.

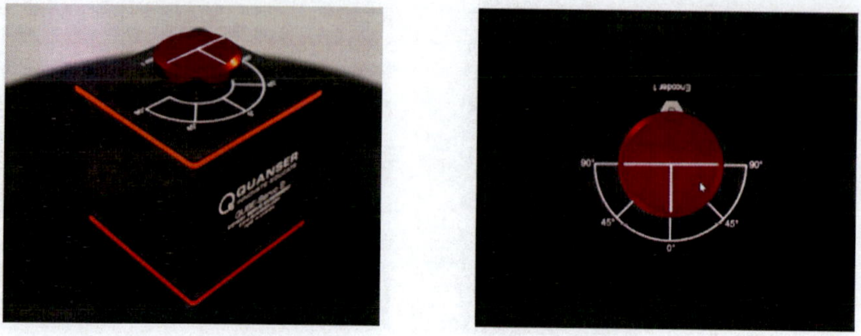

Fig. 2. A digital twin of the Quanser CUBE-Servo 2 Motor that is used to conduct the virtual laboratories.

3.2 Physical Lab Component: Take-Home Lab Kit

A personal low-cost take-home motor control kit, Fig. 3 and Fig. 4, is prepared and sent to the students to be perform the physical lab component. A listing of the components and their respective costs is given in Table 1. It should be noted that the total cost of the kit is around 110 $, which is comparable to the price of a textbook. Furthermore, all of the components were carefully selected to have overcurrent protection and to be extremely robust against wiring errors.

The main controller that was chosen is an Arduino Mega R3 board. The Arduino is widely used open-source prototyping platform with huge collection of compatible sensing and actuation devices. An Arduino-compatible motor driver shield was also used to allow for simple bidirectional motor control. Finally, a major advantage for the Arduino is that it can be programmed using MATLAB/Simulink, which is the same software used with the Virtual CUBE Servo 2 in Sect. 3.1. By using the same software in the Virtual and Physical laboratories, the student learning curve is gentle and the transition from the virtual to the physical labs is smooth.

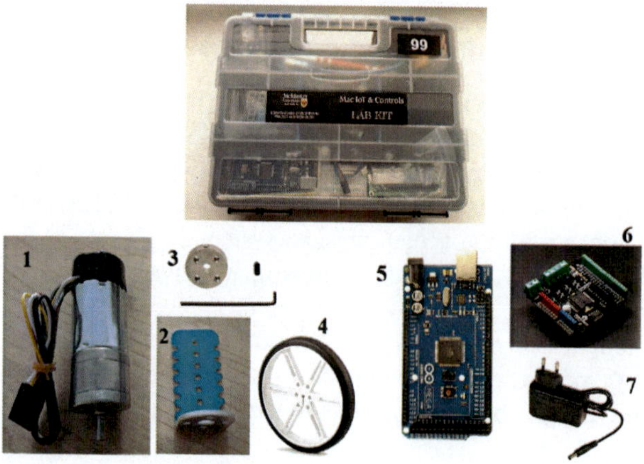

Fig. 3. A take-home motor control kit that is used to conduct the physical laboratories from home.

Fig. 4. Assembled take-home motor control kit that is used to conduct the physical laboratories from home.

Table 1. Take-home lab kit components

Item #	Description	Average cost
1	Motor + Encoder + Gearbox	~35 $
2	Motor L Bracket	~5 $
3	Mounting Hub	~10 $
4	Motor Wheel	~10 $
5	Arduino Mega 2560 R3 (Controller)	~20 $
6	Motor Driver Shield	~15 $
7	Power Adapter	~15 $
	Total Price	**~110 $**

3.3 Design and Implementation of a Hybrid Virtual-Physical Laboratories

The proposed hybrid laboratory curriculum consists of eight laboratories. The first five labs are virtual using the virtual QUBE Servo 2 and the last three labs uses the take-home motor control kit.

To allow student-student and student-instruction interactions, the labs were conducted using Zoom and every lab session contained a maximum of 20 students who are supervised by one lab instructor. The lab section is divided into breakout rooms containing two students in every room. The expectations are that students will work in groups of two, every student must do the work individually, collaborate and share the results together within the group, confirm that the results are matching, and to finally submit one a lab report per group.

For the laboratories to be effective, they must go hand-in-hand with the lecture. Table 2 shows the weekly labs, and the lecture topics discussed every week. In the first two weeks of the course, the students are requested to learn MATLAB and Simulink by completing the online MATLAB and Simulink onramp training. This allows time for the lecture to cover introductory control theory topics. The first virtual lab that uses the motor digital twin is performed in the third week. In Week 4, system modeling laboratory comes immediately after completing the modeling chapter in the lecture. Week 5 lab focuses on system first and second order response which is the same topic discussed in the same lecture week. Week 6 and 7 labs focus on P-only and PID tuning and block diagram reduction. After the students gained sufficient competencies in motor control using the virtual motor, they start using a real physical motor in week 8. Finally, in week 9 and 10, students systematically design feedback controllers using model-based design methodology.

Table 2. Control theory course weekly labs and lectures topics

W	Type	Lab	Lecture
W01		Matlab onramp	Introduction
W02		SIMULINK onramp	Modeling
W03	Virtual	Lab 1: DC motor and sensor interfacing	Modeling
W04	Virtual	Lab 2: Modeling using First-principles	1st order response
W05	Virtual	Lab 3: System Response	2nd order response
W06	Virtual	Lab 4: Proportional Control	Block diagrams
W07	Virtual	Lab 5: PID Control of a Servomechanism	Stability
W08	Physical	Lab 6: Arduino Motor Control	Steady State Errors
W09	Physical	Lab 7: Arduino Motor Speed Control	Root Locus
W10	Physical	Lab 8: Arduino Motor Position Control	Root Locus
W11			Frequency Response
W12			Review

To highlight the student's experience when performing a virtual and a physical lab, one example for each is discussed below.

Lab 5 (Virtual): PID Control of a Servomechanism

The main objectives of this experiment are

- Explore the effect of changing the Kp, Ki and Kd Terms on speed of response, overshoots & steady state error.
- Perform heuristic Tuning of a PID controller.
- To experience the effect when a control system goes unstable

In this exploration lab, the student is first requested to construct the PID controller in Fig. 5 in Simulink. After that, the student is requested to change the parameters of the PID controller and study the effect of changing these parameters on the performance of the system.

A major advantage for virtual laboratories is that it allows "smoke-free" testing. Students can explore new ideas and designs without damaging any equipment and in a safe virtual environment. As an example, within Lab 5, students will enter some of the controller values that will render the motor unstable. The students will be able to safely observe this phenomenon where the motor is going out of control, as shown in Fig. 6.

It should be noted that it is extremely undesirable for the student to observe the instability phenomenon in real-life since this can be associated with severe motor vibrations creating an unsafe incident.

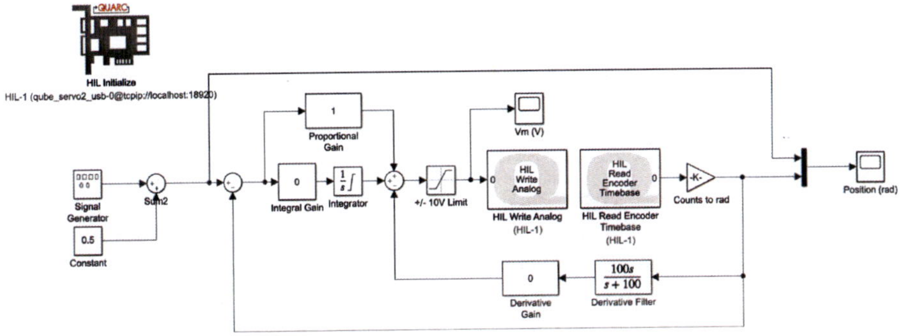

Fig. 5. Building a control system to control the Virtual CUBE Servo 2 Motor.

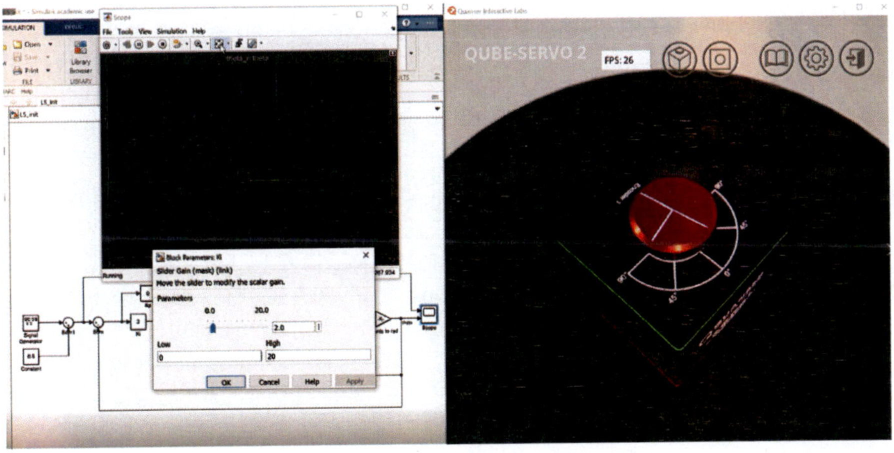

Fig. 6. Experimenting with the PID controller gains that will leads to instability in the motor response.

Lab 8 (Physical): Model-Based Design of a Motor Position Control System

The main objective of this lab is to systematically design a PD and a PID controllers using model-based design. In contrast with Lab 5, trial and error tuning is not used because it can be time consuming, unsafe, and result in equipment damage.

The student is first requested to identify a model for the take-home motor hardware. A controller is then designed based on that model to achieve certain design specifications. The controller is then simulated with the model to confirm its performance. The controller is implemented in Simulink, and deployed to the Arduino. Finally, the controller is tested with the physical motor to confirm its performance. A typical screenshot showing the student's results and the physical motor is shown in Fig. 7.

It is important to note that the controller implementation in both Lab 5 (virtual) and Lab (8) physical is very similar and it is done using the same software (MATLAB and Simulink). This is important so that the students will not feel extra software complexities when transitioning from virtual to physical laboratories.

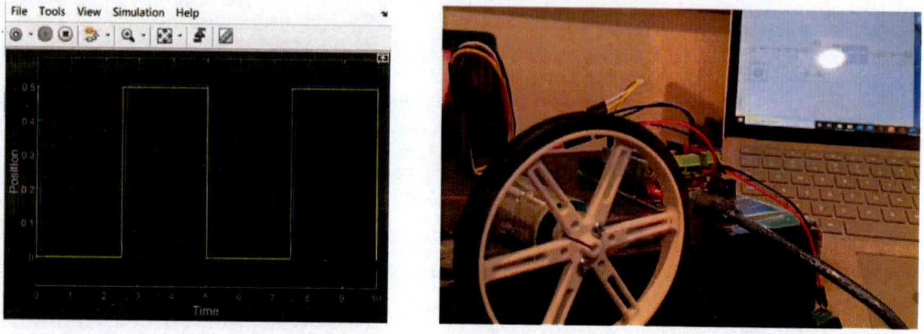

Fig. 7. Assessing the performance of a PID controller that is designed to meet certain performance specifications and observing the system behavior on a Scope and the physical motor.

4 Students' Feedback

In the fall of 2020, the hybrid virtual-physical labs in Table 2 were implemented. The labs were performed fully online with no face-to-face interactions.

The standard lab evaluation for all four lab sections includes the following two questions on student satisfaction: (Q1) how do you rate the value of this course compared with others you have taken at McMaster University and (Q2) independent critical judgment was encouraged. Figure 8 shows the students' response for the two questions. Both results indicate the student appreciation of the laboratories.

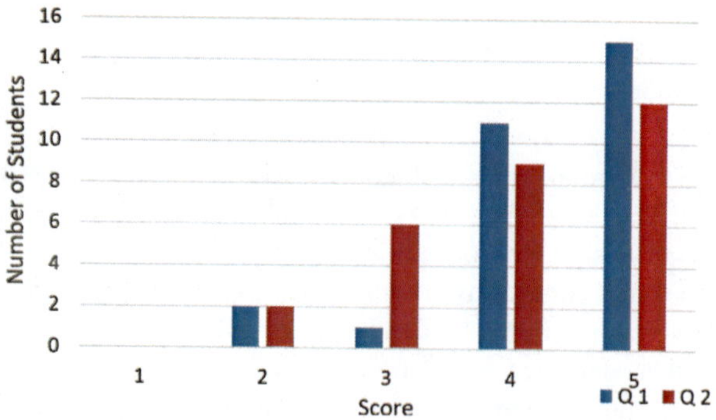

Fig. 8. Students score for Q1 and Q2 (score is from 1 to 5, 1: very poor and 5: excellent).

Some qualitative comments that were collected from the lab evaluation is shown below. Many of the comments stressed on the suitability of the labs for the online course format, and its effectiveness in supporting the theory taught in the lectures.

- *"the labs followed the content being taught in class and even aided with the comprehension of the class content"*
- *"lab format was very good for an online format"*
- *"the aspects are very valuable, like how to use matlab and simulink."*
- *"The labs are always connected to the concepts we learned from the lecture."*
- *"labs provided a lot of context to the theory we learned in class"*
- *"The labs themselves were exciting and engaging"*

5 Conclusions

The paper proposes a novel hybrid virtual-physical approach for performing control theory laboratories from home which is suitable for online course delivery.

The virtual labs used a digital twin that faithfully represented an actual motor and its associated dynamics and real-life phenomenon. Furthermore, the digital twin is animated using a gaming engine with excellent rendering and 3D visualization. The virtual labs offered an engaging and interactive experience for the students. The digital twin also formed a sandbox where students are capable to try and experiment safely without worrying about any equipment damage or hardware troubleshooting issues (e.g. loose wire, a failed component, …etc.).

Physical labs are utilized within the last three labs of the course, after the students gained enough competencies, knowledge, and experience with motor control. This is achieved by using low-cost take home lab kit.

The use of the proposed hybrid approach achieves the benefits of both virtual labs and physical labs. Furthermore, the proposed labs can be performed by the students at any time and any place, thus offering a huge flexibility for the students. Based on the students' feedback, the designed labs were effective within the online course delivery and they provided deep insights and understanding of the concepts that were taught in the class. Future research will be conducted to quantify the effectiveness of the hybrid virtual-physical versus virtual only and physical only laboratories that utilize the same tools.

References

1. Kolb, D.A.: Experiential Learning: Experience as the Source of Learning and Development. FT Press (2014)
2. Steffe, L.P., Gale, J.E.: Constructivism in Education. Psychology Press (1995)
3. Tejado, I., González, I., Pérez, E., Merchán, P.: Introducing systems theory with virtual laboratories at the University of Extremadura: how to improve learning in the lab in engineering degrees. Int. J. Electr. Eng. Educ. (2019). https://doi.org/10.1177/0020720919876815
4. Rossiter, J.A., Pope, S.A., Jones, B.L., Hedengren, J.D.: Evaluation and demonstration of take home laboratory kit. IFAC-PapersOnLine **52**, 56–61 (2019)

5. Brinson, J.R.: Learning outcome achievement in non-traditional (virtual and remote) versus traditional (hands-on) laboratories: a review of the empirical research. Comput. Educ. **87**, 218–237 (2015). https://doi.org/10.1016/j.compedu.2015.07.003
6. Yuen, T.: Low cost simulation lab for teaching control theory concepts. In: Auer, M.E., Centea, D. (eds.) ICBL 2020. AISC, vol. 1314, pp. 200–205. Springer, Cham (2021). https://doi.org/10.1007/978-3-030-67209-6_22
7. Rossiter, J.A.: Enthusing students to engage and learn control with flexible multimedia assignments. IFAC-PapersOnLine **48**, 211–216 (2015). https://doi.org/10.1016/j.ifacol.2015.11.239
8. Rossiter, J.A., Dormido, S., Vlacic, L., Murray, R.M.: Opportunities and good practice in control education: a survey. In: Abstract: IFAC World Congress, pp. 10568–10573 (2014)
9. Soliman, M.: Using virtual experiments in teaching control theory. In: Auer, M.E., Tsiatsos, T. (eds.) IMCL 2018. AISC, vol. 909, pp. 70–80. Springer, Cham (2019). https://doi.org/10.1007/978-3-030-11434-3_12
10. Li, C., Zhao, R., Wang, Y., Liu, Q., Dong, D.: Module-based virtual lab design model with matlab/simulink in automation control theory. In: ACM International Conference, Proceeding Series, pp. 6–10 (2020). https://doi.org/10.1145/3380625.3380638
11. Rossiter, J.A.: Education resources and teaching methodologies for the modern student and busy academic. IFAC-PapersOnLine **48**, 217–228 (2015). https://doi.org/10.1016/j.ifacol.2015.11.240
12. Rossiter, J.A., Shokouhi, Y.B.: Developing virtual laboratories for introductory control. In: Proceedings of 2012 UKACC International Conference on Control 2012, pp. 1025–1030 (2012). https://doi.org/10.1109/CONTROL.2012.6334773
13. Liang, Y., Liu, G.P.: Design of large scale virtual equipment for interactive HIL control system labs. IEEE Trans. Learn. Technol. **11**, 376–388 (2018). https://doi.org/10.1109/TLT.2017.2731772
14. Matijevic, M.S., Jovic, N.D., Nedeljkovic, M.S., Cantrak, D.S.: Remote labs and problem oriented engineering education. In: IEEE Global Engineering Education Conference EDUCON, pp. 1391–1396 (2017). https://doi.org/10.1109/EDUCON.2017.7943029
15. Morales-Menendez, R., Ramírez-Mendoza, R.A., Guevara, A.J.V.: Virtual/remote labs for automation teaching: a cost effective approach. IFAC-PapersOnLine **52**, 306–311 (2019). https://doi.org/10.1016/j.ifacol.2019.08.219
16. Cardoso, A., Gil, P., Urbano, D., Palma, L.B.: Benefits of using remote labs in intelligent control teaching. In: Auer, M.E., Guralnick, D., Simonics, I. (eds.) ICL 2017. AISC, vol. 716, pp. 79–85. Springer, Cham (2018). https://doi.org/10.1007/978-3-319-73204-6_11
17. Stark, B., Li, Z., Smith, B., Chen, Y.Q.: Take-home mechatronics control labs: a low-cost personal solution and educational assessment. In: Proceedings of the ASME Design Engineering Technical Conferences, vol. 4, pp. 1–9 (2013). https://doi.org/10.1115/DETC2013-12735
18. Sáenz, J., De La Torre, L., Chacón, J., Dormido, S.: A new architecture for the design of virtual/remote labs: the coupled drives system as a case of study. In: IEEE International Conference on Emerging Technologies and Factory Automation ETFA, September 2019, pp. 769–775 (2019). https://doi.org/10.1109/ETFA.2019.8869192
19. Wuttke, H.-D., Henke, K., Hutschenreuter, R.: Virtual control units in remote labs. In: Auer, M.E., May, D. (eds.) REV 2020. AISC, vol. 1231, pp. 223–231. Springer, Cham (2021). https://doi.org/10.1007/978-3-030-52575-0_18
20. Galan, D., Chaos, D., De La Torre, L., Aranda-Escolastico, E., Heradio, R.: PuzzlEx: an online experimentation environment for control engineering labs. In: Proceedings of the 2019 5th Experiment International Conference exp.at 2019, pp. 69–73 (2019). https://doi.org/10.1109/EXPAT.2019.8876580

Towards a Teachers' Augmented Reality Competencies (TARC) Framework

Stavros A. Nikou[1(✉)], Maria Perifanou[2], and Anastasios A. Economides[2]

[1] University of Strathclyde, Glasgow, UK
stavros.nikou@strath.ac.uk
[2] University of Macedonia, Thessaloniki, Greece

Abstract. Augmented reality, the technology that augments real environments with virtual components, is constantly evolving. Augmented reality (AR) has the potential to offer immersive, authentic, and meaningful learning experiences to students and therefore enhance learning. However, the effective integration of augmented reality into teaching requires from teachers to master a special set of digital competencies. The current study proposes a framework that defines the augmented reality competencies that teachers should have in order to effectively integrate augmented reality into their teaching. The framework comprises four dimensions: basic augmented reality literacies, create, use, and manage augmented reality learning resources. Based on the proposed framework, the study introduces also the Teachers' Augmented Reality Competencies (TARC) questionnaire that can help educators to self-assess and develop their AR competencies in order to integrate augmented reality in their practice.

Keywords: Augmented reality · Framework · TARC · Teachers' digital competencies · Teachers' digital skills

1 Introduction

Augmented Reality is an emerging educational technology that bridges the gap between the virtual and physical world by incorporating virtual components in real environments. The rapid growth of mobile and wireless technologies and other technologies such as motion tracking and sensors has facilitated the wider adoption of AR. AR (as being not fully immersive) is in the one end of the eXtended Reality (XR) or Mixed Reality (MX) spectrum, with the Virtual Reality (VR) to be in the other end (as being fully immersive). In AR, the physical world is linked with the virtual content through various triggers such as physical location triggers (e.g., latitude and longitude), application pre-defined triggers (e.g., an image or a QR code) or Artificial Intelligence services [1]. Therefore, users can navigate in the real world and through these triggers and can have augmented experiences with various overplayed digital assets. These assets can be text, 2D or 3D images, videos or other digital artefacts (e.g., holograms). The navigation is through either dedicated hardware or mobile devices loaded with appropriate AR apps.

M. E. Auer and T. Tsiatsos (Eds.): IMCL 2021, LNNS 411, pp. 203–212, 2022.
https://doi.org/10.1007/978-3-030-96296-8_19

2 Augmented Reality in Education

AR provides new possibilities for the educational domain and is a promising tool for teachers and learners. Studies have shown that there are many advantages associated with the use of AR in educational settings. AR enables visualization of invisible and abstract concepts, can provide interaction opportunities for students, increases student interest enhancing their satisfaction and engagement [2]. Moreover, AR enhances leaning achievement favouring long-term knowledge retention and increases learning motivation [3]. Despite the reported positive outcomes of AR in education, AR is still at an early stage and teachers are still hesitant to use AR. Asking educators to develop or use an augmented reality learning experience "might still illicit a questioning expression on their faces" [4]. One reason for this is the lack of AR competencies. Existing teachers' ICT competencies frameworks such as the UNESCO ICT Competency Framework for Teachers (ICT CFT) [5], the European Framework for the Digital Competence of Educators [6] and different technology integration frameworks provide only general guidelines. AR technology involves a wide range of elements that teachers need to master in order to make an effective use of it.

3 Teachers' AR Competencies Framework

The Association for Educational Communications and Technology defined Educational Technology as the "the study and ethical practice of facilitating learning and improving performance by creating, using, and managing appropriate technological processes and resources" [7]. Augmented reality, as an emerging educational technology can also be seen under the lens of this definition. "Creation" refers to the "research, theory, and practice involved in the generation of learning environments" [8]. The creation of augmented learning environments can involve the full instructional cycle of the analysis, design, development, implementation and evaluation. "Using" refers to "the theories and practices related to bringing learners into contact with learning conditions and resources" [8]. Using AR involves the deployment of augment learning experiences in the teaching practice. "Managing" AR involves both creation and using [8] as well other administrative tasks such as locating, classifying, evaluating or regulating. The current study proposes a framework that defines a set of competencies that teachers need to have in order to effectively integrate AR into their professional practice and enhance their teaching offering engaging and effective learning experiences. The proposed framework defines three main competency areas: *Create, Use* and *Manage* Augmented Reality Learning Experiences. We have also defined sub-dimensions for each of these areas. *Creation* involves the capacity to design and develop or modify augmented reality learning experiences. *Use* involves the capacity to *employ various pedagogies, teach (face-to-face or online), assess and provide feedback and communicate and collaborate* using augmented reality. *Management* involves the capacity to *find, classify and evaluate* augmented reality learning experiences as well as to consider *ethical implications (e.g., copyrights, privacy)* as well as *safety and security issues* with their use.

3.1 Basic Augmented Reality Literacies

Teachers should have the skills to use Augmented Reality and integrate it into teaching. There is no need for teachers to become experts in Augmented Reality. It is not required to have skills in tracking technologies, display technologies, image processing, computer vision, stereo rendering, etc. Without going too much into programming and software development details, teachers should have basic descriptive knowledge of the following areas: (i) basic understanding of AR definitions, terminology, functionalities as well as advantages and disadvantages, (ii) software needed to operate AR resources, (iii) available tools to create web-based and mobile-based AR off-the-self applications and the differences among these, (ii) required hardware and devices (e.g., headsets, smart-glasses, 360 cameras).

3.2 Creating AR

Design

Various issues need to be considered when designing AR learning experiences. The existence of many heterogeneous activities that can be integrated in AR design may result in an increased student cognitive load. Therefore, learning scenarios should allow a level of flexibility to accommodate unforeseen events and adapt to student needs [9]. Selecting the appropriate type of digital media is also important to avoid cognitive overload [10]. Most existing applications use only one type of digital element with the majority of them to use text and 2D images while animations, 3D objects and videos are used less frequently [11]. Educators should have the ability to design digital elements as overlay information to augment the digital world. Motivation design should be relevant, trigger students' attention, support their confidence and improve their satisfaction levels [12]. Educators should also consider accessibility features related to social inclusion. Following the Universal Design Principles [13], providing multiple means of representation, engagement, action and expression is recommended.

Development

Developing AR applications usually requires advanced digital skills and competencies (such as programming, tracking technologies, display technologies, image processing, computer vision, stereo rendering, etc.) that normally go beyond the basic AR literacy skills that have been previously discussed. *ARCore* for Android and *ARKit* for iOS are the most widely used frameworks in developing interactive AR. However, these frameworks require a rather specialised experience in software development, which most teachers do not currently have. This can keep teachers away from AR development resulting in AR experiences without clear learning objectives and curriculum integration [14]. However, there exists a range of closed end-user development tools (e.g., *BlippAR, Metaverse, ARCreator*) with easy-to-use templates and asset libraries, that teachers can use to create their own AR learning experiences. Usually, these tools do not require development of algorithms but only development of AR resources to be linked with objects or locations [15]. Teachers who want to start explore AR development should consider to use these tools to build AR experiences that are aligned with their own with instructional practices.

3.3 Use AR

Pedagogies

Deciding on the appropriate pedagogy that informs the instructional practices to be implemented when designing AR environments is important. AR learning experiences usually draw upon constructivism and situated learning theories and can support contextualized and student-centered learning [16]. Related literature suggests that the most common pedagogical approaches in AR interventions are collaborative learning, inquiry-based learning, situated learning, and project- based learning [17]. Studies have shown that while situated learning is the most common pedagogical approach, collaborative learning has the greatest impact on students' learning [18]. When designing three-dimensional (3-D) virtual learning environments, there are several contextual variables that need to be considered such as locus of control, class dynamics, level of interactivity, source of information [19]. Providing learning experiences that can be teacher or student controlled, have an appropriate level of interaction and scaffolding while being focused on the learning objectives is a matter of adopting each time the appropriate pedagogy. However, researchers agree that there is a lack of clear guidelines describing pedagogical considerations regarding the development and use of augmented reality environments [20]. Educators, who develop, modify or simply use AR applications should be aware of this and try not to simply use the technology but better integrate pedagogical strategies in their AR interventions.

Teaching

Teachers can always take advantage of the superiority of AR learning environments not only by developing new AR resources but also by using existing AR resources. There are many AR applications readily available that teachers can use to harness AR content. These applications can facilitate the representation of abstract information in an interactive way by allowing teachers/students to create AR content in different subject areas. AR applications such as *Google Expeditions* or *Thinglink* empower teachers and students to create their own AR experiences and can offer authentic and contextualized learning directly connected to the real-world contexts. Despite the extra workload required, the educational benefit of these AR apps can be rewarding [21].

Assessment

Learning assessment is an important part of the educational process. Learning outcomes can be evaluated by assessing cognitive (acquisition of knowledge and information and intellectual skills), behavioural (engaging with the learning activities) or affective outcomes (learners' perceptions of their learning) [22]. The ongoing evolution of educational technologies allow the use of new assessment types. The new modes of information representation made available through educational technologies allow innovative digital classroom assessments types. There is a variety of assessment elements that can be incorporated in AR such as game elements or multimodal assessment types. Multimodal assessment can go beyond the conventional paper-based/digital text assessments and require students to combine two or more representational modes using digital technologies [23]. AR technologies can successfully enrich low-stake assessment practices such as formative assessment [24], peer-assessment [25] and self-assessment [26] with impact on learning performance and learning motivation.

Simulation-based assessment can be effectively used in assessing problem-solving skills [27]. Medical students assess their clinical skills using AR simulated environments [28].

Feedback
It is very well known that timely informative feedback enhances learning outcomes [29]. However, providing instant individual feedback to students especially in complex learning tasks such as project-based learning activities can be a challenge for teachers. AR learning environments can provide automated meaningful real-time feedback making a significant difference in terms of learning benefits. This is due to its timeliness and support for visualization [30]. AR interfaces can provide multimodal feedback ranging from visual and auditory to haptic feedback through sensors and control devices. Since feedback can be timely and interactive, learners can be immersed in the learning experience. A properly configured AR-based feedback strategy can help students to better understand the learning tasks. For example, studies have shown that in virtual simulations the provision of explanation type feedback can be more appropriate for declarative tasks whereas for procedural tasks knowledge of correct response feedback is suitable [14]. AR-facilitated feedback can be helpful for teachers as well to identify students' misunderstandings and weaknesses. For example, through an AR interface, teachers receive immediate, private and individualized feedback for each student as well as aggregated feedback for the whole class [31].

Communication
Communication in virtual environments can involve a variety of activities: creation of digital and virtual artefacts in various forms to convey ideas, verbal or text communication with peers and teachers, interactions through avatars, role playing, visiting each other's' virtual spaces while working in the process of solving a problem or navigating through augmented and virtual spaces. Users can have multiple choices to make connections using interactive interfaces [32] and communicate in group tasks [33]. Emerging assistive technologies, such as smart glasses, facilitate social communications among learners with autism spectrum disorder addressing such the problem of communication deficiencies [34]. Studies have shown that AR facilitates the development of communication competencies [35] and therefore teachers should be able to take advantage of the enhanced interactions offered in virtual environments to build connections and to develop students' communication skills.

Collaboration
Augmented and virtual reality environments provide a wide range of opportunities for cooperative and collaborative work. Collaboration can include group work, peer review and social negotiation during the development of AR resources or other community-based learning and teaching activities; this works well especially in location-based AR environments [36]. Moreover, in cloud collaboration platforms, learners can also share their local environment remotely in order to collaborate on spatial tasks in shared virtual spaces [37]. AR technologies can enhance non-virtual or virtual collaborative tasks and teachers face the challenge to be able to support and take advantage of these kind of interactions.

3.4 Manage AR

Find

The capacity to find the appropriate AR educational resources that can be used in specific learning scenarios aiming to address specific learning objectives is fundamental. Teachers are not always in position to develop their own applications and should be able to find ready applications that are stored in various AR repositories (e.g., *GitHub*). Usually, AR software keeps on a library of apps that have been developed with the specific software and teachers can re-use these resources by modifying them according to their needs (e.g., *Google expeditions*).

Classification

It would be useful for educators to be able to classify different AR resources based on different criteria on order to be able to effectively use them in class. We would argue that a meaningful classification should be based on the curriculum content knowledge that AR resources can offer and support: declarative, procedural or conceptual. According to [38], educational resources can be: information display and presentation (declarative knowledge), practice resources (procedural knowledge) as well as concept representation and data display resources (conceptual knowledge). Similarly, AR resources can be classified with respect to the above criteria.

Evaluation

Teachers to know how to evaluate AR educational resources with respect to usability, usefulness, credibility, appropriateness/suitability, enjoyment, safety, mobility, educational outcomes/students' performance. Most importantly, teachers should be able to evaluate the feasibility of integrating AR resources in their teaching (e.g. usability) and their educational impact (e.g. knowledge retention, cognitive skills, motivation) as well [39].

Ethics

There may be several ethical challenges arising from the use of AR in Education. For example, in in pervasive mobile augmented reality the users' private space can be exposed to the outer world [40]. Moreover, facial recognition or geolocation feature can put learners' privacy at risk. Sometimes also, AR and MR application can manipulate emotions and create unrealistic expectations [41] that can have negative mental and social negative effects. Ethical challenges associated with the use of AR should always be taken into consideration especially for minor learners who are more vulnerable.

Security and Safety

Educators should be aware that augmented reality application could put learners at physical or mental risk [41]. The use of AR devices such as headsets should be avoided to be used outdoors (e.g. in the street) because of their immersive nature. Educators should always promote safe and healthy behaviours.

4 Conclusions and Future Work

The current study proposes a framework that defines a set of competencies that teachers need to have in order to effectively integrate AR into their teaching. Based on the aforementioned dimensions (Create, Use, Manage) of the proposed framework, the Teachers' AR Competencies (TARC) Questionnaire has been developed (Appendix). In line with our previous efforts to define specific actions someone needs to do in order to improve their digital competencies [42, 43] and introducing a progression model with different levels of expertise [44], our future research aims to further improve the granularity of the proposed framework and validate it through a survey to be organized among European teachers.

Appendix

Teachers' AR Competencies (TARC) Questionnaire

CREATE:

DESIGN: I can design AR educational experiences using AR applications and tools to meet specific educational objectives.
DEVELOPMENT: I can develop AR educational resources using easy-to-use AR templates and asset libraries. I can adapt AR educational resources to my teaching goals.

USE:

PEDAGOGIES: I can use/adapt AR educational resources employing various pedagogies and teaching methods.
TEACHING: I can use AR educational resources to teach (e.g., present, demonstrate, explain) my students.
ASSESSMENT: I can use AR educational resources (e.g., AR and multimodal game-based and simulation-based assessments) to assess the students' progress.
FEEDBACK: I can use AR educational resources (e.g., avatars, multimodal interfaces) to guide, feedback, advise, support, and inspire students.
COMMUNICATION: I can use AR educational resources (e.g., avatars, AR spaces) to interact and communicate with students and enable students' interactions and communication.
COLLABORATION: I can use AR educational resources (e.g., avatars, AR spaces) to collaborate with students and enable students' collaboration.

MANAGE:

FIND: I can use search engines, digital repositories, and databases to find existing AR educational resources using appropriate criteria, metadata filters, and recommender systems.

EVALUATION: I can evaluate AR educational resources using appropriate criteria.
CLASSIFICATION/ORGANIZATION/SCHEDULING: I can classify the AR educational resources to information display and presentation, practice resources as well as concept representation and data display resources. I can organize and schedule the most appropriate AR educational resources for achieving specific educational objectives.
ETHICS: I can control the ethical and responsible use of AR resources by all participating in the educational activities (e.g., respecting participants' personality, privacy, rights).
SECURITY & SAFETY: I can secure the safe use of AR resources by all participating in the educational activities (e.g., securing participants' resources, safety, health).

References

1. MacCallum, K., Parsons, D.: Teacher perspectives on mobile augmented reality: the potential of Metaverse for learning. In: Proceedings of World Conference on Mobile and Contextual Learning 2019, pp. 21–28. IAmLearn (2019)
2. Akçayır, M., Akçayır, G.: Advantages and challenges associated with augmented reality for education: a systematic review of the literature. Educ. Res. Rev. **20**, 1–11 (2017)
3. Radu, I.: Augmented reality in education: a meta-review and cross-media analysis. Pers. Ubiquit. Comput. **18**, 1533–1543 (2014)
4. Tinti-Kane, H., Vahey, P.: xR in EDU survey 2018: benchmarking adoption trends in K12 and higher education. Report series, EdTech Times (2018)
5. UNESCO: UNESCO ICT Competency Framework for Teachers, version 3. United Nations Educational, Scientific and Cultural Organization (2018). https://unesdoc.unesco.org/ark:/48223/pf0000265721. Accessed 02 July 2021
6. European Commission: JRC Science for Policy Report. European Framework for the Digital Competence of Educators (DigCompEdu) (2017). https://ec.europa.eu/jrc/en/publication/eur-scientific-and-technical-research-reports/european-framework-digital-competence-educators-digcompedu. Accessed 02 July 2021
7. Januszewski, A., Molenda, M.: Definition. In: Januszewski, A., Molenda, M. (eds.) Educational Technology: A Definition with Commentary, pp. 1–14. Routledge, New York (2008)
8. Florence, M., Betrus, A.K.: Digital Media for Learning, Theories, Processes and Solutions. Springer, Cham (2020). https://doi.org/10.1007/978-3-030-33120-7
9. Cuendet, S., Bonnard, Q., Do-Lenh, S., Dillenbourg, P.: Designing augmented reality for the classroom. Comput. Educ. **68**, 557–569 (2013)
10. Thees, M., Kapp, S., Strzys, M.P., Beil, F., Lukowicz, P., Kuhn, J.: Effects of augmented reality on learning and cognitive load in university physics laboratory courses. Comput. Hum. Behav. **108**, 106316 (2020)
11. Garzón, J., Acevedo, J.: Meta-analysis of the impact of augmented reality on students' learning effectiveness. Educ. Res. Rev. **27**, 244–260 (2019)
12. Keller, J.M.: ARCS model of motivation. In: Seel, N.M. (eds.) Encyclopedia of the Sciences of Learning. Springer, Boston (2012). https://doi.org/10.1007/978-1-4419-1428-6_217
13. Meyer, A., Rose, D.H., Gordon, D.: Universal Design for Learning: Theory and Practice. CAST, Wakefield (2014)

14. Merchant, Z., Goetz, E.T., Cifuentes, L., Keeney-Kennicutt, W., Davis, T.J.: Effectiveness of virtual reality-based instruction on students' learning outcomes in PreK-12 and higher education: a meta-analysis. Comput. Educ. **70**, 29–40 (2014)
15. Herpich, F., Guarese, R.L.M., Tarouco, L.M.R.: A comparative analysis of augmented reality frameworks aimed at the development of educational applications. Creat. Educ. **8**(09), 1433 (2017)
16. Leighton, L.J., Crompton, H.: Augmented reality in K-12 education. In: Kurubacak, G., Altinpulluk, H. (eds.) Mobile Technologies and Augmented Reality in Open Education, pp. 281–290. IGI Global (2017)
17. Saltan, F., Arslan, Ö.: The use of augmented reality in formal education: a scoping review. Eurasia J. Math. Sci. Technol. Educ. **13**(2), 503–520 (2017)
18. Garzón, J., Kinshuk, Baldiris, S., Gutiérrez, J., Pavón, J.: How do pedagogical approaches affect the impact of augmented reality on education? A meta-analysis and research synthesis. Educ. Res. Rev. **31**, 100334 (2020)
19. Fowler, C.: Learning activities in 3-D virtual worlds. Br. J. Educ. Technol. **46**, 412–422 (2015)
20. Garzón, J., Pavón, J., Baldiris, S.: Systematic review and meta-analysis of augmented reality in educational settings. Virtual Reality **23**(4), 447–459 (2019). https://doi.org/10.1007/s10055-019-00379-9
21. Romano, M., Díaz, P., Aedo, I.: Empowering teachers to create augmented reality experiences: the effects on the educational experience. Interact. Learn. Environ., 1–18 (2020, ahead-of-print). https://doi.org/10.1080/10494820.2020.1851727
22. Wei, X., Saab, N., Admiraal, W.: Assessment of cognitive, behavioral, and affective learning outcomes in massive open online courses: a systematic literature review. Comput. Educ. **163**, 104097 (2021)
23. Fjørtoft, H.: Multimodal digital classroom assessments. Comput. Educ. **152**, 103892 (2020)
24. Bhagat, K.K., Liou, W.-K., Spector, J.M., Chang, C.-Y.: To use augmented reality or not in formative assessment: a comparative study. Interact. Learn. Environ. **27**(5–6), 830–840 (2019)
25. Lan, C., et al.: Mobile augmented reality in supporting peer assessment: an implementation in a fundamental design course. In: IADIS International Conference on Cognition and Exploratory Learning in Digital Age (CELDA 2013), CELDA (2013)
26. Lai, A.S.Y., Wong, C.Y.K., Lo, O.C.H. : Applying augmented reality technology to book publication business. In: 2015 IEEE 12th International Conference on e-Business Engineering, pp. 281–286 (2015)
27. Bergeron, B.: Augmented assessment as a means to augmented reality. Stud. Health Technol. Inform. **119**, 49–51 (2006)
28. Campisi, C., Li, E., Jimenez, D., Milanaik, R.: Augmented reality in medical education and training: from physicians to patients. In: Geroimenko, V. (ed.) Augmented Reality in Education. SSCC, pp. 111–138. Springer, Cham (2020). https://doi.org/10.1007/978-3-030-42156-4_7
29. Spector, J.M.: Foundations of Educational Technology: Integrative Approaches and Interdisciplinary Perspectives, 2nd edn. Routledge, New York (2015)
30. McNamara, D.S., Jackson, G.T., Graesser, A.: Intelligent tutoring and games (ITAG). In: AIED 2009 14th, pp. 44–65 (2009)
31. Zarraonandia, T., Aedo, I., Díaz, P., Montero, A.: An augmented lecture feedback system. Br. J. Educ. Technol. **44**, 616–628 (2013)
32. Zhao, Q.: The application of augmented reality visual communication in network teaching. Int. J. Emerg. Technol. Learn. **13**(7), 57–70 (2018). https://doi.org/10.3991/ijet.v13i07.8780

33. Chen, C.-H., Yang, C.-K., Huang, K., Yao, K.-C.: Augmented reality and competition in robotics education: effects on 21st century competencies, group collaboration and learning motivation. J. Comput. Assist. Learn. **36**, 1052–1062 (2020)
34. Sahin, N.T., Abdus-Sabur, R., Keshav, N.U., Runpeng, L., Salisbury, J.P., Arshya, V.: Case study of a digital augmented reality intervention for autism in school classrooms: associated with improved social communication, cognition, and motivation via educator and parent assessment. Front. Educ. **3**, 57 (2018)
35. Fernández-García, C.: Effect of augmented reality on school journalism: a tool for developing communication competencies in virtual environments. Electron. J. Inf. Syst. Dev. Countries **87**, e12169 (2021)
36. Czerkawski, B., Berti, M.: Pedagogical considerations for mobile-based augmented reality learning environments. In: Auer, M.E., Tsiatsos, T. (eds.) IMCL 2019. AISC, vol. 1192, pp. 215–222. Springer, Cham (2021). https://doi.org/10.1007/978-3-030-49932-7_21
37. Piumsomboon, T., Day, A., Ens, B., Lee, Y., Lee, G., Billinghurst, M.: Exploring enhancements for remote mixed reality collaboration. In: SIGGRAPH Asia 2017 Mobile Graphics & Interactive Applications (SA 2017), pp. 1–5. Association for Computing Machinery, New York (2017). Article no. 16
38. Churchill, D.: Digital resources for learning. STE. Springer, Singapore (2017). https://doi.org/10.1007/978-981-10-3776-4
39. da Silva, M.M.O., Teixeira, J.M.X.N., Cavalcante, P.S., Teichrieb, V.: Perspectives on how to evaluate augmented reality technology tools for education: a systematic review. J. Braz. Comput. Soc. **25**(1), 1–18 (2019). https://doi.org/10.1186/s13173-019-0084-8
40. Brinkman, B.: Ethics and pervasive augmented reality: some challenges and approaches. In: Pimple, K. (eds.) Emerging Pervasive Information and Communication Technologies (PICT). Law, Governance and Technology Series, vol. 11, pp. 149–175. Springer, Dordrecht (2014). https://doi.org/10.1007/978-94-007-6833-8_8
41. Pase, S.: Ethical considerations in augmented reality applications. In: Proceedings of the 2012 EEE International Conference on e-Learning, e-Business, Enterprise Information Systems, and e-Government (2012)
42. Perifanou, M., Economides, A.A.: The digital competence actions framework. In: Proceedings of ICERI2019, the 12th Annual International Conference of Education, Research and Innovation, Seville (Spain), 11–13 November. IATED (2019)
43. Parmaxi, A., Demetriou, A.A.: Augmented reality in language learning: a state-of-the-art review of 2014–2019. J. Comput. Assist. Learn. **36**(6), 861–875 (2020)
44. Nikou, S.A.: Mobile learning teachers' competencies framework. In: Proceedings of the 12th International Conference on Education and New Learning Technologies, 7 July, pp. 2726–2731. IATED (2020)

Employing Mozilla Hubs as an Alternative Tool for Student Outreach: A Design Challenge Use Case

Ryan Brown[1], Samin Habibi-Luevano[1], Gil Robern[2], Kody Wood[3], Sharman Perera[1], Alvaro Uribe-Quevedo[2](✉), Callan Brown[1], Khalid Rizk[1], Filippo Genco[1], Jennifer McKellar[1], Kirk Atkinson[1], and Akira Tokuhiro[1]

[1] Faculty of Energy Systems and Nuclear Sciences, Ontario Tech University, Oshawa, ON L1G0C5, Canada
{ryan.brown,samin.habibiluevanol,sharman.perera,callan.brown,khalid.rizk, filippo.genco,jennifer.mckellar,kirk.atkinson, akira.tokuhiro}@ontariotechu.ca

[2] Faculty of Business and IT, Ontario Tech University, Oshawa, ON L1G0C5, Canada
{gil.robern,alvaro.quevedo}@ontariotechu.ca

[3] Faculty of Science, Ontario Tech University, Oshawa, ON L1G0C5, Canada
kody.wood@ontariotechu.net

Abstract. Due to the 2020 COVID-19 pandemic, online virtual gathering platforms have risen at all levels as alternatives to traditional videoconferencing. The implementation of physical distancing and limited capacity implemented to reduce the spread of COVID-19 changed how academic activities are conducted. For example, student outreach presents students with experiential learning opportunities and teamwork on campus. While video conferencing tools have thrived over the pandemic, these lack immersion and presence, lead to fatigue, and lack engagement. In this paper, we present the development and hosting of a Design Challenge employing the open-source virtual reality (VR) platform, Hubs by Mozilla. Usability perceptions from five out of ten participants were gathered and analyzed employing a simplified version of the System Usability Scale questionnaire. The process of developing the Mozilla Hubs environment allowed us to identify technical issues associated with performance and audio quality.

Keywords: Collaborative design · Experiential learning · Virtual reality

1 Introduction

In recent years, virtual reality (VR) online gathering tools have become an alternative to traditional videoconferencing [1]. Using internet-connected immersive technologies allows for more engaging experiences than conventional video conferencing by providing spatial awareness through visual, auditory, and haptic

M. E. Auer and T. Tsiatsos (Eds.): IMCL 2021, LNNS 411, pp. 213–222, 2022.
https://doi.org/10.1007/978-3-030-96296-8_20

immersion [2]. Due to the COVID-19 pandemic, in-person meetings have moved to video conferencing formats as social distancing and limited indoor capacity prevent people from being physically co-located in the same room. VR safely allows gatherings, thus respecting physical distancing procedures put in place by governmental authorities due to the pandemic. The use of interactive VR technology provides experiential learning that would not otherwise have been possible through the exclusive use of currently available videoconferencing services such as Google Hangouts, Zoom, Jabber Video, Adobe Connect, or similar [3].

The use of VR in education has been associated with high levels of immersion, presence, engagement, knowledge retention, and skills transfer [4]. Currently, consumer-level VR adoption has seen a spike in applications other than entertainment including health care, training, education, and tourism among others [5]. For example, [6] investigated how using VR may produced greater student motivation and improved learning outcomes, by comparing the instructional effectiveness of immersive virtual reality (VR) versus a desktop slideshow as media for teaching scientific knowledge. Early VR installments presented learners with isolating experiences due to technical limitations that sparked the interest in developing multiuser solutions for collaborative environments [7]. The mass adoption of VR is enabling research on a larger scale where understanding its impacts on learning is now possible [8]. Because of the perceptual cues and levels of realism provided by VR environments, ongoing research is focusing on understanding the risks of extraneous cognitive load that can cause distractions [6,8].

Due to COVID-19 procedures reducing in-person interactions and capacity for social gatherings, events and competitions have focused on virtual and online events employing video conferencing and VR. While videoconferencing tools remain the most common choice, VR has gained momentum as an engaging alternative for meetings that produce high interactivity and participation in events [1]. Educational and academic institutions rely on numerous outreach activities to increase awareness regarding programs, courses, and community collaborations that used to take place in person. Shifting from traditional videoconferencing or virtual tours based on 360 still photographs, and 360 videos requires creating virtual experiences that enable attendees to interact with each other and the environment itself while ensuring the achievement of the expected outcome in a usable manner [9]. This process includes designing a VR environment, placing relevant resources and materials for the attendees to use, coordinating participants during the event, and ensuring technical issues are rapidly solved to improve participant engagement during the event. Currently, there are a number of solutions that can are being used in academic settings for teaching, open houses, and conferences among others, including open source and commercial tools such as AltSpace VR, VirBela [10], and VR Chat [11], among others.

This paper presents the design and implementation of a virtual space for conducting an online Design Challenge employing Mozilla Hubs as an alternative to videoconferencing tools like Zoom or Google Meet. Our goal was to understand the technical challenges and usability perceptions of participants using the Mozilla Hubs platform for the first time. The participants are required to conceptualize and design an energy system to supply electricity to a future Mars colony in VR breakout rooms in addition to access to a four-part lecture series covering the fundamentals of energy production, nuclear power, space missions, and VR.

2 Methods

Due to the COVID-19 outbreak, an online format was considered to substitute for the traditional in-person event. VR was chosen as it allows to retain a similar structure where speakers and attendees can interact and share content, thus promoting active student participation, but virtually [2].

2.1 About Hubs by Mozilla

Hubs by Mozilla is a virtual reality collaboration open-source platform that runs in a browser. It allows creating 3D spaces or rooms for synchronous online gatherings. Rooms can be shared by generating a web link, or by generating a meeting link with a passcode. Accessing a Mozilla Hubs room requires configuring the microphone and web camera on desktop and mobile devices. Once inside the virtual room, attendees can edit their avatar's name and appearance. Mozilla Hubs is compatible with stand-alone and Desktop VR headsets. Editing rooms for Mozilla Hubs can be done through Spoke, the built-in scene editor for creating environments that can be used in Hubs rooms. Similar to Hubs, the editor runs entirely in the browser and allows uploading 3D models, images, and files among others. However, it is worth noting that Hubs makes the following recommendations for the room: i) a maximum of 50,000 triangles, ii) 25 unique materials, iii) 256 MB of video RAM (Random Access Memory) textures, iv) a maximum of three lights, and v) a maximum file size of 16 MB.

While in Hubs, attendees can share their screens, web cameras, microphone, upload files such as PDFs, images, and share web links. Hubs allow the attendees to navigate the environment by employing the arrow keys and mouse to move and look around. A chat box facilitates messaging in addition to a pointer and marker tool that can be used on surfaces or in the air.

2.2 VR Room Design

The Design Challenge requires the participants to conceptualize and design an energy system to supply electricity to a future Mars colony. For this purpose, the VR environment required a central area for hosting the opening and closing ceremonies, and breakout rooms for each group to work on the challenge. A hierarchy room layout was chosen to facilitate the navigation between the central and group rooms.

2.3 Iterative Design

The design process used to create and test the Design Challenge in Mozilla Hubs is depicted in the flowchart in Fig. 1a. This approach allowed us to ensure meeting the recommended room configuration to ensure all participants have access to the virtual room without issues.

2.4 First Iteration

Initially, the Mars One scene available in the Mozilla Hubs gallery was chosen because of its alignment with the Design Challenge theme (Fig. 1b). The scene was customized in Mozilla Hubs by adding breakout room hyperlinks for the participants to gather. The access was placed manually on the Martian surface. Objects placed manually were located directly in front of the user's avatar, oriented normally to the avatar's line of sight. This posed design problems, as objects would begin to drift out of alignment unless the user avatar remains in alignment while placing the objects.

The physical size of the central VR room environment proved to be too large for the event, and the boundary of the environment floor was not properly defined, which caused some participants to get trapped underneath the ground level. Additionally, the scene had 271,362 polygons, using 524 MB of video RAM, and a file size of 64 MB. A preliminary test was performed having 20 simultaneous users. While individuals accessing the room with mobile devices including iOS and Android, Mac computers, and Windows computers did not experience issues, when having all participants joining at the same time, the attendees expressed having a problem with audio, staying connected, and drops in frames per second.

2.5 Second Iteration

To address the challenges faced during the first iteration, the VR environment was optimized using the Mozilla Spoke web editor. The iteration design led us to create a second VR environment (Fig. 1c), where a well-defined border was included to contain attendees in the form of mountains to evoke a Martian landscape. When tested, the second iteration presented issues associated with slow navigation due to the drop in frames per second even with only one person present. The VR model would not load when multiple participants were present in the central VR room due to a high polygon count. The second VR room had 54 objects made out of 8,351 polygons. This number is more than twice the number of objects recommended by Mozilla Hubs.

To reduce the data size of the second design, the pre-made 3D model of the environment by Mozilla was exported from Spoke as a glTF file (Graphics Library Transmission Format). This file format was imported into Blender, an open-source software package for creating 3D models, and the polygon count was reduced via the decimate function. Upon exporting the model as a glTF from Blender, it lost its textural data. It was then opened into Microsoft Paint 3D to apply a stamp of the texture tinted red to reflect the Mars environment.

This was then saved and re-imported into Spoke. The edited model had reduced texture resolution as it also became smaller in scale and had to be scaled up to use.

The second central VR room design without additional objects had a significantly reduced file size from 21 MB to 17 MB, thereby allowing the VR environment to be loaded for more than one user. However, when attempting to add video links and the banners from event sponsors, the VR environment once again had issues loading. Attempting to open the environment with more than one user present resulted in the software crashing. While the polygon count in the environment contributes to the file size, the number of objects placed in the VR environment and the pixel count of included images also affect the file size.

2.6 Final Iteration

The third and final iteration of the central VR room was built to address the recurring issues with the VR environment file size (Fig. 1d). This third design was significantly smaller than the first template, with the intent of further reducing polygon count and file size. The pre-made design chosen had a smaller enclosure resembling a geodesic dome with no other discerning features. As space was limited in the dome design selected, two central VR rooms were created using pre-made smaller enclosures. One version served as the main lobby during the opening ceremony and contained the promotional material and lecture series links, and the second version featured the gallery of participant designs. Both versions included the altered landscape from the previous iteration so the environment exterior to the dome resembled the Martian landscape. The web browser that best performed with Mozilla Hubs was Firefox. From the previous iterations, we found that Firefox presented fewer technical issues than other web browsers leading to significant improvements in audio and quality for the majority of participants joining the test sessions.

In addition to Mozilla Hubs features, the following online tools were provided to facilitate the ideation process: i) an online whiteboard, and ii) a Google Meet link to troubleshoot any issues. Additionally, a facilitator was present in each room to help with any inquiries.

2.7 The Design Challenge Event

Information about the event was made available online and promoted by the Ontario Tech University in Oshawa, Ontario, Canada. The number of participants was 20, including the organizers and staff (12) and the students (8).

Before the participants joined the Design Challenge event, they were provided with lectures and tutorials to familiarize themselves with engineering and virtual reality concepts. Additionally, a Mozilla Hubs tutorial was provided to reduce any possible entry barrier associated with the novelty of using Mozilla Hubs to collaborate with others. For example, the tutorial teaches about navigation, screen sharing, file sharing, and general troubleshooting.

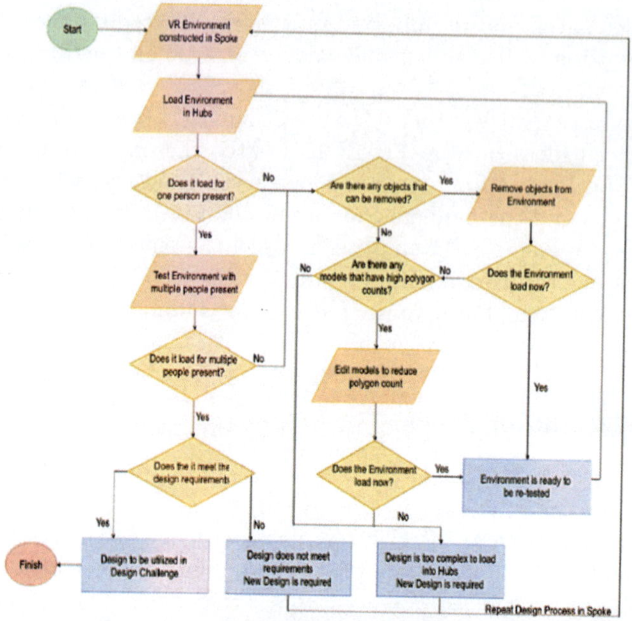

(a) Iterative design flowchart

(b) Mars room available in Hubs

(c) Room surrounded with mountains

(d) Final Design Challenge design

Fig. 1. Iterative design process and outcomes

On the Design Challenge day, the participants were welcome to an opening ceremony of the event held in the central VR room (https://hub.link/rjDTpe2). After the ceremony was finished, the students were divided into groups to work on the design challenge inside group breakout rooms. Once finalized, the designs were collected and judged by a panel of experts leading to a closing ceremony held in the central VR room during which the winning design was presented.

For assessing the experience, two surveys were presented at the end of the Design Challenge to the participants. The first survey was comprised of questions pertaining the most common technical issues found during the iteration design process. The technical questions included the following: i) no issues, ii) the audio would frequently cut out, iii) the audio was difficult to hear, iv) the audio was distorted, and v) there were issues with audio from other participants. The second survey presented six questions extracted from the System Usability

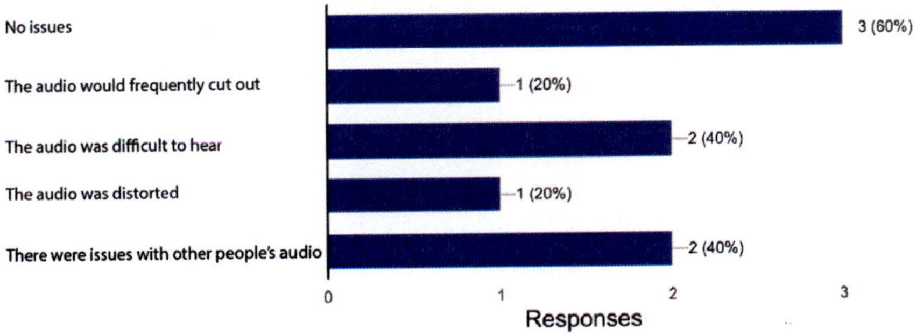

Fig. 2. Results of the post-event technical difficulties survey.

Scale questionnaire (SUS) [12]. With respect to the modified SUS questionnaire, we purposely left out three questions associated with needing technical support, inconsistency of the system, and the need to learn new things since only five of the organizers were familiar with Mozilla Hubs and everyone else was required to follow the tutorials and stay connected to the facilitators. Here our goal was not to establish a usability score, but rather to understand specific areas of improvement. The following questions were used: i) How was the overall experience? ii) I felt confident using the system, iii) I felt awkward using the system, iv) I felt it was easy to learn, v) I felt the system was easy to use, vi) I felt the system was Unnecessarily complex, and vii) I would frequently use the system.

3 Preliminary Results

Only five participants completed the surveys at the end of the Design Challenge. While this number of participants allows identifying some usability and technical issues, a larger number is preferred. As a result, the collected data does not allow us to draw any significant conclusions.

3.1 Technical Issues

Audio quality was identified as a significant issue (Fig. 2), while no other technical problems were reported. For example, one attendee expressed that "there were no audio issues for me, but it was hard to hear some other people due to their mic issues," and another expressed that "audio would sometimes become distorted when one person would talk continuously for a long period of time." Figure 2 presents a summary of responses associated with the technical issues reported by five participants.

3.2 Usability

Usability perceptions were captured on a 5-point Likert scale employing six questions from the System Usability Scale [12]. The usability questionnaire showed

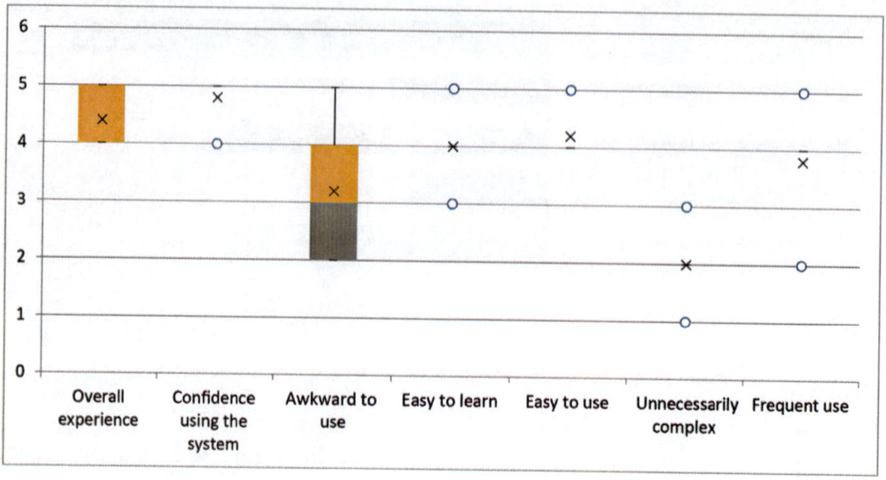

Fig. 3. Results of the post-event survey: VR environment usability.

mixed results regarding the use of the Mozilla Hubs environment, as shown in Fig. 3. Results indicate that the participants found the overall experience well integrated (Median = 4), felt confident interacting inside the Mozilla Hubs room (Median = 5), found the navigation awkward to use (Median = 3), indicated that using Mozilla Hubs is easy to learn (Median = 4), perceived interacting with others easy to use (Median = 4), did not believe the room was unnecessarily complex (Median = 2), and only one person did not think of Mozilla Hubs as something to be used frequently (Median = 4).

3.3 Open Feedback

The participants were asked to provide any additional feedback, where they expressed finding the experience provided a sense of a near-normal in a very different modern-day educational environment as they were able to virtually gather with others and looking at their interactions through the avatars.

4 Discussion

The participants' feedback from the post-event survey aligns with the technical issues found during the iterative development. Despite our efforts to mitigate the technical difficulties, challenges associated with the variability of computer hardware and bandwidth even when meeting the Mozilla Hubs minimum requirements are difficult to predict and manage. For example, during the Design Challenge, we had a couple of participants following the activities through Google Meet as they were unable to join the Mozilla Hubs rooms. Interestingly, unlike our various iterations, audio issues were minimum during the Design Challenge, and those who experienced them were able to solve them by refreshing their web browser tab. However, the participants were required to customize the individual audio level for each speaker due to the spatial audio effects embedded in Mozilla Hubs that made some people being heard louder or quieter than others.

With respect to the usability perceptions, the navigation mechanics requiring mouse and keyboard resulted in participants spending more time getting familiar with virtually walking, teleporting, and interacting with media embedded in the room. Although a tutorial was provided in advance, the participants required additional assistance when joining the central room. This may require a redesign of the on-boarding process to ensure those new to Mozilla Hubs have a basic understanding of how the environment works. Other than the navigation, the participants reported finding the Mozilla Hubs room easy to use, learn, thus feeling confident in using it frequently.

5 Conclusions

Here we have presented our experience hosting an Engineering design challenge using Mozilla Hubs. While our results are preliminary due to the small number of participants, our main contribution lies in the iterative process presented to troubleshoot and design a virtual environment with Mozilla Hubs. This is relevant as VR becomes available to all, thus facilitating connecting with others in immersive and engaging environments. Taking into consideration the physical distancing protocols resulting from COVID-19, students and teachers have access to virtual platforms that allow them to collaborate and challenge themselves in a new environment. Virtual reality is enabling the creation of online events, competitions, and challenges without physical limitations. Mozilla Hubs allow creating a customized environment that allows the inclusion of promotional material, event resources, and collaborative space that students can use in a collaborative remote manner.

The limitations found in the central VR room design and the breakout VR conference rooms for the Energy Solutions Challenge were consistent throughout the design process. These limitations must be considered and re-evaluated when designing future virtual events. Technical issues such as bandwidth limitations and audio connectivity can create a substantial disturbance. While Mozilla provides documentation on addressing possible issues, we found ourselves conducting several iterations to fine-tune the experience to run the event efficiently.

Future work will focus on conducting a more extensive study to better understand the effects of Mozilla Hubs on task completion and employing standardized usability, engagement, cognitive load, motion sickness, and knowledge retention tests for those joining the experience on their mobile, desktop, or virtual reality head-mounted display. Additionally, we will create interactive guides to improve on-boarding to have participants with no experience in VR be able to engage with others in the platform. Finally, we will compare Mozilla Hubs to other open source and commercial platforms such as AltSpace VR, Oculus Workrooms, and Virbela among others to understand their effects on attendees' performance.

Acknowledgments. The support of the Natural Sciences and Engineering Research Council of Canada (NSERC), [funding reference number RGPIN-2018-05917 to Dr. Uribe-Quevedo supporting Mr. Wood and Mr. Robern]. The support of the Sigma Lab from the Faculty of Energy Systems and Nuclear Sciences, participating high schools, and students is acknowledged.

References

1. Gunkel, S.N., Stokking, H.M. , Prins, M.J., van der Stap, N., Haar, F.B.t., Niamut, O.A.: Virtual reality conferencing: multi-user immersive VR experiences on the web. In: Proceedings of the 9th ACM Multimedia Systems Conference, pp. 498–501 (2018)
2. Le, D.A., MacIntyre, B., Outlaw, J.: Enhancing the experience of virtual conferences in social virtual environments. In: 2020 IEEE Conference on Virtual Reality and 3D User Interfaces Abstracts and Workshops (VRW), pp. 485–494. IEEE (2020)
3. Triyason, T., Tassanaviboon, A., Kanthamanon, P.: Hybrid classroom: designing for the new normal after COVID-19 pandemic. In: Proceedings of the 11th International Conference on Advances in Information Technology, pp. 1–8 (2020)
4. Di Natale, A.F., Repetto, C., Riva, G., Villani, D.: Immersive virtual reality in k-12 and higher education: a 10-year systematic review of empirical research. Br. J. Edu. Technol. **51**(6), 2006–2033 (2020)
5. Lee, J., Kim, J., Choi, J.Y.: The adoption of virtual reality devices: the technology acceptance model integrating enjoyment, social interaction, and strength of the social ties. Telematics Inform. **39**, 37–48 (2019)
6. Parong, J., Mayer, R.E.: Learning science in immersive virtual reality. J. Educ. Psychol. **110**(6), 785 (2018)
7. Greenwald, S.W., Corning, W., Maes, P.: Multi-user framework for collaboration and co-creation in virtual reality. In: 12th International Conference on Computer Supported Collaborative Learning ... (2017)
8. Madden, J., Pandita, S., Schuldt, J., Kim, B., Won, A.S., Holmes, N.: Ready student one: exploring the predictors of student learning in virtual reality. PLoS ONE **15**(3), e0229788 (2020)
9. Won, A.S., Bailey, J.O., Yi, S.: Work-in-progress—learning about virtual worlds in virtual worlds: how remote learning in a pandemic can inform future teaching. In: 2020 6th International Conference of the Immersive Learning Research Network (iLRN), pp. 377–380. IEEE (2020)
10. Franks, P.C.: Work-in-progress—developing criteria for virtual reality courses based on virtual world experiences. In: 2020 6th International Conference of the Immersive Learning Research Network (iLRN), pp. 369–372. IEEE (2020)
11. Jouet, P., Alleaume, V., Laurent, A., Fradet, M., Luo, T., Baillard, C.: AR-Chat: an AR-based instant messaging system. In: 2020 IEEE International Symposium on Mixed and Augmented Reality Adjunct (ISMAR-Adjunct), pp. 153–157. IEEE (2020)
12. Lewis, J.R.: The system usability scale: past, present, and future. Int. J. Hum.-Comput. Interact. **34**(7), 577–590 (2018)

Mobile Learning Models, Theory and Pedagogy

Exploring the Utilization of Online Open-Source Environments for Mobile Applications Development in the Vocational Education and Training (VET) Curriculum

Dimitrios Magetos, Dimitrios Kotsifakos(✉), and Christos Douligeris

Department of Informatics, University of Piraeus, 80 Karaoli & Dimitriou Street, 18534 Piraeus, Greece
{dmagetos,kotsifakos,cdoulig}@unipi.gr

Abstract. In this paper, we focus on the pedagogical use of the *thinkable* online environment in mobile applications development courses for Vocational Educational and Training (VET) specialties. The *thunkable* environment provides an easy-to-use interface for students and users with minimal knowledge of mobile applications programming and it can be also used in distance online education. This teaching approach proves to be particularly effective for teaching programming in individuals with minimum programming skills. We evaluate this approach in a distance learning environment in combination with a flipped class and asynchronous and online synchronous approaches. Moreover, we provide ideas about planning introductory courses, programming mobile applications, and designing practical activities for teaching elementary computer programming lessons for all VET specialties.

Keywords: Thunkable · Mobile applications · Flipped classroom · Vocational Education and Training (VET)

1 Introduction

The Internet has a variety of free coding platforms that are available for students of all ages to begin learning and practicing coding. However, students are most motivated when they can immediately observe the results of their efforts with hands-on experiences [27]. In recent years, young people have been living in a world flooded with smart mobile devices and have become fanatical users of both these devices and their applications, with many of them showing considerable interest in developing their mobile applications [29]. Since the middle of the last decade (2010–2020), research findings show that students are more motivated to use new technologies and that carefully designed pedagogical activities using smart mobile devices [18] motivate students to fully engage with them [19].

Essentially, the reported environments take advantage of the rapid impact that mobile devices have on the young population to stimulate students' interest in programming. Research also shows that school-age students show interest in visual block

M. E. Auer and T. Tsiatsos (Eds.): IMCL 2021, LNNS 411, pp. 225–236, 2022.
https://doi.org/10.1007/978-3-030-96296-8_21

programming languages because they allow them to make their programs by dragging and dropping graphic blocks rather than by writing the code of a program [3].

To increase the students' interest in computer science, considerable effort has been put into developing tools and activities for novice programmers in all levels of education [28]. The Internet has a variety of free coding platforms that are available for students of all ages to begin learning and practicing coding. However, students are most motivated when they can immediately observe the results of their efforts. In our teaching intervention during the covid-19 pandemic, we used the inverted classroom model [5] in which we combined modern asynchronous education tools via the internet. However, many teachers find it extremely difficult to design practical activities for teaching computer programming, especially in Vocational Education and Training (VET) [23].

Creating a mobile application that runs on at least two major operating systems requires the creation of two different native apps. The Android project should be developed in Java while the iOS in Swift (Objective C). The developer must acquire the appropriate knowledge on both platforms to create quality applications. The growing popularity of Android and the need for developers to implement their applications on two different platforms created, from a very early stage, the need for development tools regardless of the destination platform (cross-platform tools).

From the beginning, web technologies (HTML 5, CSS, JavaScript) have been an ideal candidate for applications development, targeting more than one platform with the ability to reuse part of the code [11]. These technologies were also open and part of the existing knowledge of many developers, who could use them directly without the risk of being trapped in closed application development ecosystems. This was happening because cross-platform development approaches allow the development of applications for different platforms in a single step, avoiding duplication and increasing productivity. There exist many platforms for developers that make them able to create their cross-platform applications. Some of them are: "Twixl" [37], "Linx" [9], "Claris FileMaker" [8], "Salesforce Platform" [10], and "Caspio" [7]. A teacher or a student can make changes online and see them on their device instantly using a mobile app [32] in Scratch, Alice, PencilCode, or the App Inventor [21] for Android. Nevertheless, with the proliferation of smart devices, new tools have been proposed and made available for educational use [14]. In block-based programming, which is a form of visual programming, students can select the appropriate blocks of code [30]. With drag and drop functionalities a student can create a full program [15].

Visual programming and especially block-based programming environments [2] have been cited in many research studies as important tools for learning to program and organize problem-solving projects in education [6]. These efforts include performing kinesthetic activities as well as developing visual programming environments for Beginners (Initial Learning Environments - ILEs) such as Scratch, Alice, and most recently the App Inventor for Android (AIA) [23], to successfully reduce the barrier of the first entry into programming. These environments facilitate software development within a context that is fun and does not deter the novice developer. The most popular of these environments used in early programming training is Scratch [31], but there are many other environments available, such as App Inventor [20], Alice [1], and PencilCode [26].

In the next section, we discuss the specific benefits of the *thunkable* environment [32]. *Thunkable* is a cross-platform technology that offers application development for Android and iOS devices [28]. Apps developed with *thunkable* can run on Android and iOS mobile devices, while similar development environments like App Inventor only allow it on Android devices. Developing applications for mobile devices as well as developing applications for computers means developing code in a specific programming language for a specific system [25]. From our research, we found that *thunkable* is a cross-platform technology and is suitable for teaching mobile application development courses for VET's students and novice developers. These programming environments provide the necessary support in introducing students to programming and relieve students from the difficulties of traditional complex text-based languages [1]. In this type of introductory programming tool, coding blocks are organized into colored categories that can help students figure out how to choose the right block, thus reducing some of the obstacles to programming [33]. The right choice in the categories of sizes and colors standardizes the thinking and consciousness of the student because the reduction to common similarities is considered as a first conception of the organization of the local and global variables of programming [12].

Covid -19 has set an unprecedented stage for all the organized education systems. The educational community was invited in a short time to take advantage of online environments and to use advanced tools, like flipped classes [24], and other distance learning mechanisms for supporting teaching and learning. The need for a complete rethinking and redesign of teaching and learning scenarios in the covid – 19 period, based on the use of online environments, raised as a primary concern the issue of modeling the new teaching practice. The added value of didactic modeling is significant and concerns real scientific research questions rather than just technical choices. Our exploration approaches a paradigm shift in the teaching of programming in VET and concerns all existing specialties. In addition, in this article, we formulate effective instructions for teaching programming and explore aspects of scientific perspectives on how professional knowledge is created and developed and what consequences this has for a comprehensive review of the curriculum of all VET specialties. The article summarizes individual achievements in the fields of Human-Computer Interaction, Internet Technologies, Software Technology, and finally, a new approach to Education Science with new prototype lesson plans for VET.

The technical training programs have mainly included experiential parts in the learning of skills as learning methodologies. VET teachers believe that the students should acquire professional skills that are valued by employers as a goal of VET. VET programs for each specialty have integrated theory and laboratory experience with the prospect of providing students with an important opportunity to learn. The primary purpose of the article is to connect integrated theory and programming with laboratory experience.

The structure of the paper is as follows: Sect. 2 poses the scientific questions and the purpose of the research, while Sect. 3 introduces issues about programming, specialties, and VET. Section 4 presents the model of the inverted class as the first step of the lesson, while Sect. 5 presents the didactic strategy by presenting the conceptions part as a module of the whole project. Section 6 discusses the educational scenario and, finally, Sect. 7 presents the conclusions and future work.

2 Purpose of Research

Thunkable is the no-code platform to build powerful native mobile apps. App Inventor and YCombinator's W2016 batch, as a part of *thunkable*, allows users to build powerful native apps [38]. Users do not need to download large development suites, command-line kits, or simulators. They can use the many built-in app templates as they are, or by making some minor changes. One of the secondary purposes of this study is to determine the educational suitability of the *thinkable* platform as well as the learning benefits that result from its use. Because VET school curricula must cover different specialties, it is difficult to apply in-depth programming learning for all. The core questions to uncover the pedagogical content knowledge of teaching programming for a mobile device as a part of Informatics education are:

- Does the *thinkable* platform have the functions that VET students and inexperienced users can use to design the user interfaces of mobile applications?
- Does the *thinkable* platform have features suitable for mobile application programming by VET's students and inexperienced programmers?
- Is it possible to run applications immediately without the need for a mobile phone in a VET classroom?
- What did a teacher and his students have to do to be initiated in the context of building mobile application environments?

These or similar questions regarding *thunkable* have been posed by other researchers [35], but as far as we know, never for VET. Because the answers to the four questions are not somehow related to each other, we believe that this field remains an unexplored field in Informatics and therefore we need research efforts to study this field. The originality of our research lies in the fact that we focus on teaching the use of *thinkable* for developing mobile applications, both to VET students and trainees in the context of distance education in the period of the pandemic of covid-19 [32].

3 Programming, Specialties, and VET

The exploration for the Utilization of Online Open-Source Environments for Mobile Application Development in the VET curriculum extends the well-established laboratory teaching methodologies. The first purpose of our research is to present new knowledge about mobile learning as well as the emergence of technologies, infrastructures, and mobile communication services and their impact on education. On the other hand, the article is about a new message of technology-driven change and development examples within the educational community. Through the modeling of an intervention for the design of educational activities and the utilization of the *thunkable* environment, a relatively simple application was built. Starting from the use of the flipped classroom [34], we moved on to the "development of mobile applications" as required by many specialties in the context of upper secondary and post-secondary VET. It should be emphasized that the use of mobile technology goes beyond the traditional teaching of computers which is no longer sufficient to manage the issues that arise from mobile applications. This is also a key point for the total change of example in the teaching of technological applications and innovations.

In addition, we argue that the second major problem - an obstacle to the development of learning patterns in VET is related to a one-sided rational epistemology in which the basic idea is that professional knowledge consists of theoretical principles and processes that must be applied elsewhere, in some "external" practical situations. This idea influences educational traditions and structures and leads to a separation between theoretical and practical topics and areas of learning. Programming in these environments takes the form of transfer and storage blocks, while their assembly results in the creation of programs. A helpful feature of these environments is the feature that if two blocks cannot be joined together to form a valid syntax, the interface prevents them from joining. The most common syntax errors found in text-based languages are avoided or may not occur. In this context, the combined use of graphics and visualization is recognized as an effective teaching tool in computer science to help students effectively connect algorithms with programming commands [16].

Whether our focus is on "mobile applications development", on "teaching and learning mobile programming", or on "programming as teaching as a whole" our intention is one: the total restoration of hands-on teaching and learning about with the student's professional identity. The central goal of this paper is to provide some effective solutions to the focus issues related to the teaching of programming in VET for mobile. The fundamental direction of the article is to equip teachers so that in this teaching they can transform their knowledge about a specific subject of the specialty, in a learning context accessible to their students. The analysis and the proposal aimed at enhancing the students' problem-solving skills, the acquisition of programming knowledge and the acquisition of programming strategies with advanced technologies, and the general orientation of learning mobile device programming.

4 The Model of the "Inverted" Class ("Flipped Class") and Its Combination with *Thunkable*

In the mobile applications development course, we used the "inverted" class as an introductory part of the course and the WebEx platform (https://webex.sch.gr/) for synchronous teaching. The "inverted" class (flipped classroom), already known since 2012 [36] is a modern technique teaching subjects using Information Communication Technology (ICT). This technique, which is widely used for teaching a wide range of subjects in many countries around the world, is the subject of scientific interest worldwide. Bergmann and Sams [4] explaining the "traditional" model of the inverted order emphasized that "what should be done at school based on a teacher-centered model takes place at home and work to be done at home, completed in class." In an "inverted classroom" lessons are always given in the form of video lectures and the teacher does not teach the lesson directly in the classroom, but students have the opportunity to learn by talking. In this model, the students come to class having watched the video related to the lesson of the day [24].

In our case, the students watch videos and appropriate multimedia material of the next teaching unit through the asynchronous class platform, while during our synchronous communication we use the Cisco WebEx platform. The synchronous lesson begins with short questions and answers. If there are points in the video that the students have watched that are not understood, then these are explained in detail by the teacher in the classroom. The rest of the time, the teacher suggests question-based activities to the students and provides them with personalized support.

By connecting to a platform, students can videotape lesson “lectures”, watch modern lesson sessions, discuss lesson content with their teacher and classmates, and get feedback. All student learning activities are recorded in the platform system diary, while the lessons that students must attend as well as the materials that must be processed and the activities that they must complete, before physical education in the classroom, are uploaded in a cloud computing environment so that students can access them from any mobile or non-mobile device. The students can also attend online classes through learning platforms to carry out activities in the virtual classroom under the guidance of the teacher. For the implementation of our didactic proposal, the inverted class is necessary because even at the primary level of the teaching of programming the student will have to accept and adopt some introductory steps, regardless of the subsequent application.

5 Description of the Instructional Strategy

Inverted classroom models [5] combine our scenarios with asynchronous education tools. The implementation of our proposal will be through an asynchronous e-class platform in conjunction with the platform Cisco WebEx for the needs of synchronous distance education. Through synchronous lessons, we use the *thunkable* environment, which can be run through mobile devices in both Android and iOS environments. To be as precise as possible, about using the *thunkable* environment for teaching purposes we present three types of conceptual modeling [17] in modular ways: a use case diagram type (Fig. 1), an interaction diagram type (Fig. 2), and an activity diagram type (Fig. 3). To understand the meaning of the diagrams, the teaching scenario should be considered as a system. From this point of view, we distinguish the information of the individual diagrams. These diagrams allow for the change of the flow and the integration of different thematic areas depending on the teaching interests and the corresponding subjects of the VET Specialties. Finally, conceptual diagrams help for the collection of tasks (teaching scenarios) that a teacher wants to achieve for his classroom while managing a course.

5.1 Use Case Diagram Type as a Scenario Hypothesis

In Fig. 1 we describe three basic roles (actors) - Research/Administrator – Teacher, and Student. This use case diagram (Fig. 1) presents the interactions in the system (the teaching scenario) and analyzes how the specific users respond to various events. A Research/Administrator actor sets up the online environments. A teacher can analyze the construction and organize the presentation of the whole project, implement and support the students during the project, and, finally, evaluate the students’ conclusions and presentations.

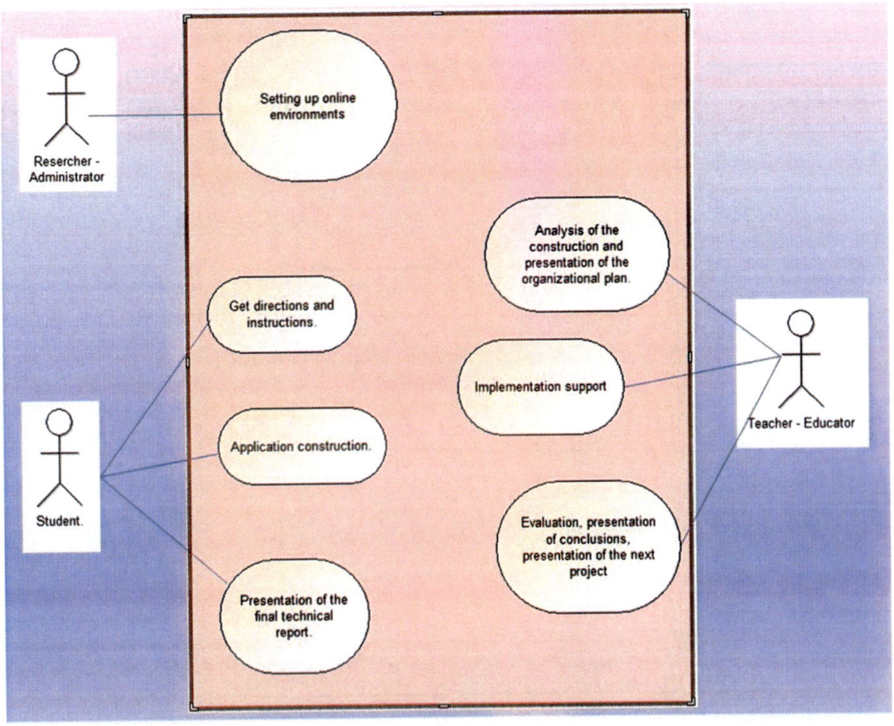

Fig. 1. Use case diagram type.

A Student as the third actor gets directions and instructions (flipped class), constructs the application following the teacher's directions (synchronous distance teaching), and, finally, organizes the presentation of the final technical report.

5.2 An Interaction Diagram Type for Learning Analytics

From the Teaching and Learning Analytics perspective, we present an interaction diagram for the synchronization of the basic phases of the scenario. Figure 2 shows exactly where a teacher must intervene to implement what to pursue. Teaching and Learning Analytics is a new theoretical approach, which combines teaching expertise, visual analytics, and design-based research to support teacher's diagnostic pedagogical ability to use data and evidence to improve the quality of teaching [22]. This diagram presents the idea of using patterns in the classroom to model the actual processes (practices and techniques) of teaching analysis and design a lesson. This diagram helps teachers to accomplish the various systems analysis and design practices and techniques creatively [20].

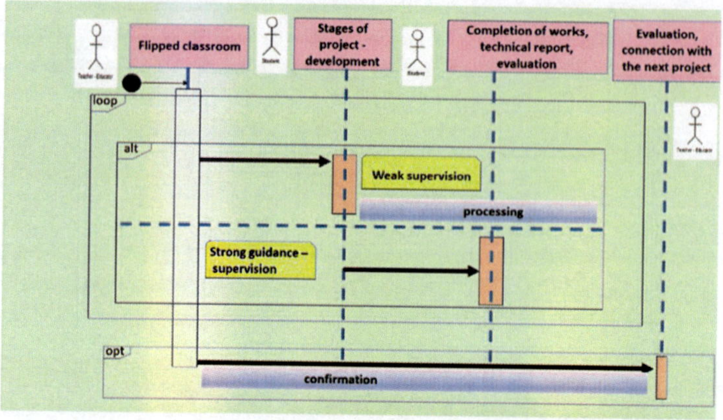

Fig. 2. Interaction diagram type.

5.3 An Activity Diagram Type for Teaching Analytics

The activity diagram in Fig. 3 is about teacher's and student's preparations for the lesson.

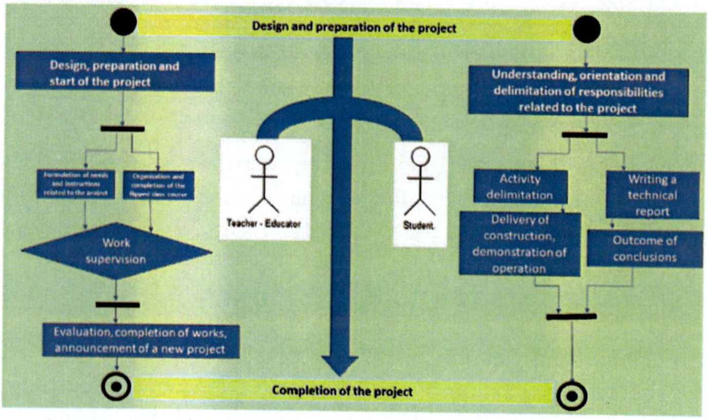

Fig. 3. Activity diagram type.

These three behavioral diagrams [13] are intentionally incomplete and provide important lesson guidance for the whole working group, admin, teachers, and students. These diagrams could be discussed as part of the teaching planning in a technical school before the implementation of the teaching, and, then, they could be used for the evaluation of the overall project. The keyword for the effective use of the above diagrams is selectivity. With the diagrams as a guide, a group of teachers from a technical high school who want to teach programming, discuss and face specific problems in how to organize their teaching script.

The discussion of the diagrams and their interpretation may involve a group of qualified teachers. As diagrams focus on communication, often the burden of discussion on them falls on communication rather than on the technical implementation specifications. In addition, the goal of a team of teachers may be to use the diagrams to reinforce alternatives to what is to be done. The discussion can be extended to colleagues from specialties who have not delved into programming. From the visualization of the teaching plans, a new scenario can start for all the students of the technical high school. The meetings of the teachers involved can be very short (a ten-minute session at a break), to discussions of a few hours (in the teachers' associations for the planning of the subject). Conversely, diagrams can work the other way around: teachers can use them to explain or delve into a functional part of the diagram. According to the above, diagrams can function dynamically and collaboratively, as a means of communication like a table.

6 The Experimental Part of the Lesson

Students build mobile apps that run on either Android or IOS devices by simply dragging and dropping components of code and connecting them in blocks using the website's platform. The first step in creating apps with *thinkable* is to register for a free *thinkable* account [34]. The next step is to ask for students to download the *thunkable* Live app from the Play (Android) or App (Apple) store. Once the *thunkable* app is installed and linked, students create the app using blocks of code on the *thunkable* web platform on a computer. Then they can test the code through an app on their mobile device. At any time during the creative process of blocking code, they can execute the code on the mobile device linked with the account.

The free account does have some restrictions. Projects that are created with free accounts are public. Other *thunkable* members can view and copy the app code associated with a free account. However, one advantage of this is that students can view projects created by others, analyze how others organized their components, and modify others' code to fit their own needs.

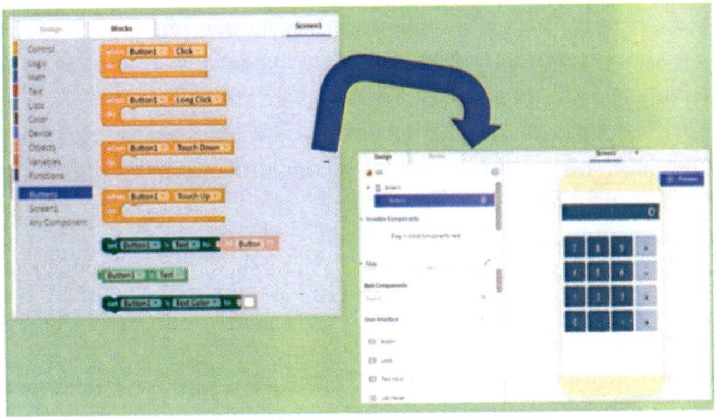

Fig. 4. Coding block options, and design screen sections.

To set up the capacity to test the code students create on the web-based platform, they link their mobile devices to their accounts. The *thunkable* website has simple instructions for linking. As changes are made online using the *thunkable* platform, the changes are reflected in the linked app running on the mobile device. When we select Button1, a list of possible coding commands (arranged by categories) for Button1 appears (see Fig. 4). We will want to select a Control, Logic, and Color command (which we refer to as blocks) from the list. The term "blocks" is used to signify that these times are combined to build the app's functions. The different types of command blocks have different colors. The Control blocks are orange, the Logic blocks are green, and the Color blocks are gray.

7 Conclusions – Evaluation – Future Work

This study is empirical research and can serve as a useful guide for programming lessons for VET specialties. In this paper, we explored the pedagogical suitability of the *thunkable* environment as a tool for teaching mobile applications programming. We found that thunkable provides an easy-to-use interface for students and users with minimal knowledge of mobile applications programming. Using Thuncable, the application under development is previewed without the need to use a mobile device in the classroom. We argued that using the *thunkable* online environment in mobile applications can serve as a useful tool for teaching programming in various VET specialties and also, be used to be integrated as an educational tool for teaching mobile development courses to students we show in the experimental part of the project. From the empirical experience point of view, we found that the application of the inverted classroom during the period of covid-19 was an innovative teaching tool. As part of this, we utilized the asynchronous class platform for posting educational materials as well as the Cisco Webex platform for synchronous distance learning.

Teaching scenarios with thinkable provides a suitable design interface for the user of mobile applications simply and pleasantly. From the practical side of teaching, we found out that the *thunkable* platform has the functions that students and inexperienced users can use to design the user interface of mobile applications. Also, we observed that the *thunkable* platform has features suitable for planning and organizing mobile applications, especially for VET's students and inexperienced programmers.

One of the most important conclusions of the project is that online learning laboratories and web environments should be organized with clear, comprehensive, and effective guidelines for a programming lesson and lead to the implementation of new attractive applications for VET students. In our didactic proposal, it must be systematically checked whether authentic didactic practice and experiential learning are implemented and to what extent. The added value of our scenario is that it delivers an engaging approach to teaching the topics and skills students need to be digitally literate because as mentioned in [31] "using practical content, hands-on projects, and interactive simulation activities students are engaged in learning".

As future extensions that may have some educational value, we mention that *thunkable* is capable of creating an app to translate words from one language to another, use text-to-speech to talk or use a mobile device's camera to recognize objects.

Students can create apps that use any of their mobile device's special features, such as the camera, microphone, gyroscope, accelerometer, and timer. Finally, comparing the *thunkable* environment with other similar environments like App Inventor or Kodular could be the subject of future research.

Acknowledgment. This work has been partially supported by UPRC (University of Piraeus Research Center).

References

1. Alice: Tell Stories. Build Games. Learn to Program. https://www.alice.org/. Accessed 14 Aug 2021
2. Bak, N., Chang, B.M., Choi, K.: Smart block: a visual programming environment for smart things. In: 2018 IEEE 42nd Annual Computer Software and Applications Conference (COMPSAC), vol. 2, pp. 32–37. IEEE, July 2018
3. Bak, N., Chang, B.M., Choi, K.: Smart block: a visual block language and its programming environment for IoT. J. Comput. Lang. **60**, 100999 (2020)
4. Bergmann, J., Sams, A.: Before you flip, consider this. Phi Delta Kappan **94**(2), 25 (2012)
5. Bergmann, J., Sams, A.: Flip your classroom: Reach every student in every class every day. International Society for technology in education (2012)
6. Broll, B., et al.: A visual programming environment for introducing distributed computing to secondary education. J. Parallel Distrib. Comput. **118**, 189–200 (2018)
7. Caspio: The No-Code Platform for Custom Business Applications. https://www.caspio.com/. Accessed 14 Aug 2021
8. Capterra: Claris FileMaker. https://www.capterra.com/p/162784/FileMaker/. Accessed 14 Aug 2021
9. Capterra: Linx Low Code Development. https://www.capterra.com/p/160547/Linx-Low-Code-Development/. Accessed 14 Aug 2021
10. Capterra: Salesforce Platform. https://www.capterra.com/p/164356/App-Cloud/. Accessed 14 Aug 2021
11. Douligeris, C., Mavropodi, R., Kopanaki, E., Karalis, A.: Technologies and Programming on the World Web. New Technologies Publications, Athens (2017). ISBN 9789605780319
12. Evans, A., Martin, K., Poatsy, M.A.: Technology in Action. Pearson Prentice Hall, Hudson Street (2008). ISBN 13: 978-0-13-543519-9. https://tinyurl.com/5xr8fh74
13. Fowler, M.: UML Distilled: A Brief Guide to the Standard Object Modeling Language. Addison-Wesley Professional, Boston (2004)
14. Grewal, D., Motyka, S., Levy, M.: The evolution and future of retailing and retailing education. J. Market. Educ. **40**(1), 85–93 (2018)
15. Hu, Y., Chen, C.H., Su, C.Y.: Exploring the effectiveness and moderators of block-based visual programming on student learning: a meta-analysis. J. Educ. Comput. Res. **58**(8), 1467–1493 (2021)
16. Joy, J.: Review on different types of drag and drop mobile app development platforms. Int. J. Comput. Sci. Eng. **11**(6), 864–866 (2018)
17. Karagiannis, D., Buchmann, R., Burzynski, P., Reimer, U., Walch, M.: Fundamental conceptual modeling languages in OMiLAB. In: Karagiannis, D., Mayr, H., Mylopoulos, J. (eds.) Domain-Specific Conceptual Modeling, pp. 3–30. Springer, Cham (2016). https://doi.org/10.1007/978-3-319-39417-6_1

18. Kearney, M., Burden, K., Schuck, S.: Disrupting education using smart mobile pedagogies. In: Daniela, L. (ed.) Didactics of Smart Pedagogy, pp. 139–157. Springer, Cham (2019). https://doi.org/10.1007/978-3-030-01551-0_7
19. Ladias, A., Ladias, D., Karvounidis, T.: Categorization of requests detecting in scratch using the SOLO taxonomy. In: 4th South-East Europe Design Automation, Computer Engineering, Computer Networks and Social Media Conference (SEEDA-CECNSM), pp. 1–7. IEEE, September 2019
20. Lending, D., May, J.: Teaching tip: using activity diagrams to model systems analysis techniques: teaching what we preach. J. Inf. Syst. Educ. **24**(2), 91 (2013)
21. MIT App Inventor: With MIT App Inventor, anyone can build apps with global impact. https://appinventor.mit.edu/. Accessed 14 Aug 2021
22. Ndukwe, I.G., Daniel, B.K.: Teaching analytics, value and tools for teacher data literacy: a systematic and tripartite approach. Int. J. Educ. Technol. High. Educ. **17**(1), 1–31 (2020)
23. Ouhbi S., Pombo, N.: Software engineering education: challenges and perspectives. In: IEEE Global Engineering Education Conference, pp. 202–209 (2020). https://doi.org/10.1109/EDUCON45650.2020.9125353
24. Ozdamli, F., Asiksoy, G.: Flipped classroom approach. World J. Educ. Technol. Curr. Issues **8**(2), 98–105 (2016)
25. Papadakis, S.: Evaluating the efficiency of two programming environments in shaping novices' attitudes, perceptions, beliefs, and knowledge in programming: a comparison between Scratch and App Inventor. Int. J. Teach. Case Stud. **10**(1), 31–52 (2019)
26. PENCIL code: Dream it. Code it. https://pencilcode.net/. Accessed 14 Aug 2021
27. Pérez-Álvarez, J.M., Mos, A.: From abstract specifications to application generation. In: 2020 IEEE/ACM 42nd International Conference on Software Engineering: Software Engineering in Society (ICSE-SEIS), pp. 11–20. IEEE, October 2020
28. Price, T.W., Barnes, T.: Comparing textual and block interfaces in a novice programming environment. In: Proceedings of the Eleventh Annual International Conference on International Computing Education Research, pp. 91–99, August 2015
29. Saraubon, K., Anurugsa, K., Kongsakpaibul, A.: A smart system for elderly care using IoT and mobile technologies. In: Proceedings of the 2018 2nd International Conference on Software and e-Business, pp. 59–63, December 2018
30. Seralidou, E., Douligeris, C.: Learning with the AppInventor programming software through the use of structured educational scenarios in secondary education in Greece. Educ. Inf. Technol. **24**(4), 2243–2281 (2019)
31. Siegle, D.: There's an app for that, and i made it. Gifted Child Today **43**(1), 64–71 (2020)
32. Syauqi, K., Munadi, S., Triyono, M.B.: Students' perceptions toward vocational education on online learning during the COVID-19 pandemic. Int. J. Eval. Res. Educ. **9**(4), 881–886 (2020)
33. Tamilias, A.G., Themelis, T.J., Karvounidis, T., Garofalaki, Z., Kallergis, D.: B@ SE: blocks for@rduino in the students' educational process. In: 2017 IEEE Global Engineering Education Conference (EDUCON), pp. 910–915, IEEE, April 2017
34. Thunkable Platform: Create Your Native Apps with No-Code. https://thunkable.com/. Accessed 14 Aug 2021
35. Tlili, F., Haddad, R., Bouallegue, R., Shubair, R.: Shoulder and trunk posture monitoring system over time for seating persons. In: Barolli, L., Woungang, I., Enokido, T. (eds.) AINA 2021. LNNS, vol. 226, pp. 250–260. Springer, Cham (2021). https://doi.org/10.1007/978-3-030-75075-6_20
36. Tucker, B.: The flipped classroom. Educ. Next **12**(1), 82–83 (2012)
37. Twixl: Don't just make an app. https://twixlmedia.com/. Accessed 14 Aug 2021
38. YCombinator: https://www.ycombinator.com/companies/thunkable. Accessed 14 Aug 2021

Learning Analytics - Survey and Practical Considerations for Intelligent Education

Malinka Ivanova[1(✉)], Valentina Terzieva[2], Tatyana Ivanova[1], and Katia Todorova[2]

[1] Technical University of Sofia, 8 Kl. Ohridski Blvd, 1000 Sofia, Bulgaria
{m_ivanova, t.ivanova}@tu-sofia.bg
[2] Institute of Information and Communication Technologies, Bulgarian Academy of Sciences, Acad. G. Bonchev Street, Bl. 2, 1113 Sofia, Bulgaria
{valentina.terzieva, katia.todorova}@iict.bas.bg

Abstract. Todays, learning analytics is a vital component in the concept of intelligent education. Analytical tools give many benefits for educators and enable a smart educational process. Our research hypothesis is that learning analytics will provide essential information for achieving adaptable learning paths, better learning performance, and as a result – a more efficient educational process. The paper aims to present a survey of the used learning analytics tools in contemporary education and their applicability for personalization and learning in groups. It also shows how clustering methods can be applied for grouping students with similar learning performance or interests to support learning in groups. Another issue that is considered is how learning analytics can work together with knowledge-based intelligent approaches to achieve personalized and adaptive tutoring.

Keywords: Learning analytics · Students learning performance · Intelligent learning environment

1 Introduction

The recent expansion of distance and mobile education and the need for personalized education have prompted educational institutions to adopt new technologies to provide greater adaptability and flexibility. Usually, universities and even a rising number of schools have learning management systems (LMS) and systems containing student profile information, including personal socio-demographic data, learning portfolio, and enrollment data. These technologies augmented with learning analytics (LA) and semantics can be considered as a core of an intelligent learning environment (ILE) [1]. Learning analytics is becoming a phenomenon with a wide application in different educational stages from preschool to postgraduate level [2, 3].

Nowadays, LA is crucial for modern education systems and could be in support of mobile learning and teaching [4]. In the context of this paper, we accept a broad definition of mobile learning – it is considered as a part of distance education that can take place at any time, in any place, and on any mobile device (smartphone, tablet, laptop). Learning-related data collected within an educational environment can be used

M. E. Auer and T. Tsiatsos (Eds.): IMCL 2021, LNNS 411, pp. 237–249, 2022.
https://doi.org/10.1007/978-3-030-96296-8_22

to provide many different analyzes. Usually, comparative analyses, correlations between many parameters of the learning process, correlations between students' profiles and their learning performances are made. Also, different metrics concerning learning interactions (type, number, and usage duration of learning resources; time for complexion learning tasks, tests or assignments; parameters of collaboration and discussions, etc.), an educational process such as pedagogical approaches or scenarios used, learning paths, indicators for comparing traditional and smart education are explored. LA can be combined with some additional advanced tools such as recognizing facial expressions or monitoring small movements to determine the involvement of students in the educational process. The LA data are usually anonymized and can be further statistically aggregated by many relevant criteria. The extensive analyses in the ILE environment facilitate a deep understanding of factors influencing the teaching-learning process, such as academic, pedagogical, motivational, etc. [1, 3, 5].

The paper aims to explore LA approaches used in intelligent education and their role to provide more efficient and personalized learning paths by applying a combination of LA and Semantic web technologies in the teaching-learning process. The paper continues with a presentation of the survey methodology and a state-of-the-art review of LA and its application in intelligent learning environment. The fourth section shows the results of the thematic literature survey carried out. Next, some practical considerations concerning the application of LA to propose personalized learning services for students from College of Energy and Electronics at Technical University are provided. The paper concludes with a discussion of findings and outlines directions for future works.

2 Methodology of Survey

The used methodology in current research consists of three main parts:

First, exploration of the recently published indexed in world databases scientific papers by formulating relevant keywords within the topic of interest. Evaluation of findings through bibliographic and other methods to outline the conceptual framework regarding research in the fields of learning analytics and learning paths, mobile learning, and intelligent technologies. The bibliometric data explicitly show the relations among the most utilized keywords in these research areas. Thus, the essentials in the topic of interest of the current research and directions for further examination are revealed.

Second, a survey of the student performance within the context of an online learning platform by gathering data about their assessment and usage of provided learning resources is performed. The tools used are those provided in the non-commercial version of the learning analytics system, which is expected to give meaningful information for the main factors for learning performance improvement. The findings are processed, summarized, and prepared for in-deep analysis to solve possible issues.

Third, for analysis and interpretations of the collected data, semantics-based, statistical, and machine-learning-driven approaches are combined to model students' learning performance. The standard methods for creating the analytical models in the context of tutoring and assessment related to specific concepts or achieving specific tutoring goals are applied. Gathering, pre-processing, and semantics-based modeling of data are made regarding the current learner's knowledge, thus demonstrating the usefulness of combined usage of these approaches for learning results improvement.

3 Current State of Learning Analytics Application

This section presents the examined recent research studies concerning the state of the art of learning analytics. It provides an overview of used methods and techniques and outlines the most used ones. The potential challenges for learning analytics implementation in mobile and distant intelligent educational environments are discussed.

Considering the given in [2] definition, learning analytics embraces the phases of measurement, collection, analysis, and reporting of data related to learners in the learning contexts where they occur. The goal of LA is to use a data-driven approach to understand the process essence, and based on this, to enhance and optimize the management and control of the teaching-learning process. LA itself comprises the following facets: technical, educational, and privacy [2]. The technical aspect is related to data processing – tracking, gathering, and analyzing. The significant issue of privacy is to determine who is responsible for access and management of sensitive personal data and information, how much data to collect, and how long to store them. The educational aspect covers the scope and amount of data collection, considering pedagogical approaches, matching tutor purposes, and closely related to the requirements of stakeholders. However, this paper does not discuss technical and privacy issues. Here the main focus is on the use of LA methods in the context of mobile education in an ILE and to show the gained benefits for the performance of students.

LA is one of the key tools to support the design and adaptation of learning paths in intelligent educational systems [1, 4]. Such systems can be both fully electronic and blended, and they also usually provide distant and mobile opportunities for learning. However, the manner of how LA supports adaptation of learning paths and learning design in general, as well as teacher practice to do this, vary significantly.

3.1 Learning Analytics Methods

LA concerns the analysis of educational data to draw conclusions and therefore present paths to make decisions [6]. The continuous collection and analysis of large data sets derived from learning interactions and actions in an e-learning environment help identifying students' patterns of behaviors and thus improve the level of learning and retention of students. LA methods embrace visual data analysis techniques, social network analysis, semantic, and educational data mining. The last is applied for various purposes in an educational context, such as clustering, relationship mining, model discovery, prediction, etc., and results in extracting hidden knowledge about pattern discovery [7]. Another technique LA uses is predictive modeling for providing actionable information. The aim is to adapt the educational goals to the needs and abilities of each learner by providing feedback, instructional content, and automatic responses to students [8]. Some universities have already applied LA in higher education by predictive modeling to enhance learning outcomes. They use approaches such as: identifying students at risk by analyzing data gathered from LMS; improving student retention by examining the learners' demographics data set; predicting which students have to be provided with particular learning resources, based on correlations between resources used, student achievement, and level of risk [3].

3.2 Learning Analytics in Intelligent Learning Environments

As a natural consequence of the continuous technology development and changing the way people acquire knowledge, ILEs have gained more popularity. ILE provides an intelligent education based on the concept of a smart classroom [1], where educational services are more engaging, adaptive, personalized, and effective.

ILE offers innovative opportunities for the overall improvement of education through the increasingly widely used analysis of educational data sets to better understand how to make the learning process more efficient. For that purpose, numerous techniques are used – artificial intelligence, LA, semantic technologies, etc. LA is applied to students learning interaction data sets, gathered at different levels – resources, courses, curricula, LMS, institution, etc. [6]. The extraction of this educational data allows performing various analyses that enable intelligent decision-making and management, personalized knowledge delivery, assessment, creating and updating learner profiles, etc. All these services are based on continuous student interaction tracking, with a view to access student performance, predict student completion, provide proactive support and prompt feedback, and increase retention. The role of LA in ILE includes support of curriculum design, implementation of tutoring strategies for improving knowledge acquisition, facilitation of student progress, etc. [9].

4 Literature Survey

The bibliometric analysis for reviewing the published scientific papers is the preferred method for outlining the key points from a thematic framework [10]. It gives a summarized view of the research interests and connections between related topics. The bibliometric approach is used in this work to describe the scientific production in the area of LA and ILE. The bibliographic data exploration is conducted on 16 June 2021 in the scientific database Scopus according to the keywords *learning analytics*, *mobile learning*, *learning paths,* and *intelligent*. The extracted data for the keywords *learning analytics* and *learning paths* are used for the bibliometric map construction (Fig. 1). The search performed in the articles title, abstract, and keywords returns 52 documents. The type of analysis is the co-occurrence of author keywords (items) in papers.

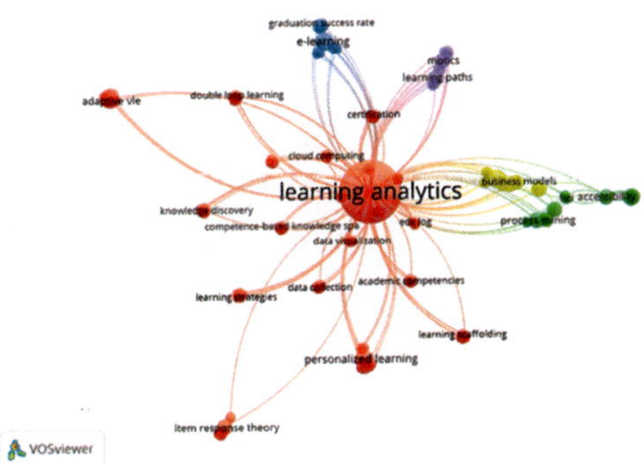

Fig. 1. The bibliometric map concerning keywords *learning analytics* and *learning path.*

The similar items with a strong connection between them are grouped in clusters, which are shown in different colors. Five clusters are formed as the bigger one is around the keyword learning analytics. Table 1 presents the distribution of the keywords in the clusters. It appears that the domain of learning analytics is connected to various research topics like personalized learning, (dynamic) learning paths, learning strategies, big data, artificial intelligence, adaptive VLE, learning performance.

Table 1. Keywords distribution in clusters.

Cluster 1 (78 items)	Cluster 2 (17 items)	Cluster 3 (12 items)	Cluster 4 (11 items)	Cluster 5 (11 items)
Learning analytics	Process mining	Big data	Artificial intelligence	Adaptive MOOCs
Personalized learning	Accessibility	Blended learning	Dynamic learning path	Learning behavior
Adaptive VLE	Behavioral analytics	e-learning	Neural network	Learning paths
Learning strategies	Learning path discovery	Information visualization	Predictive modeling	Learning personalization
Academic competences	Learning analytics metrics	Self-study	Educational technology	Learning styles
Learning scaffolding	Technology-enhanced learning	Learning management system	Algorithms-based learning	Learning performance

The map constructed through the method co-authorship can present the strength of the co-authorship between countries and the numbers of citations. The results show that authors from Malaysia have 80 citations, Indonesia - 67, Germany - 60, United Kingdom - 57, Canada - 53, Spain - 51, USA - 35, Netherlands -30, Italy – 28.

The second query uses the keywords *learning analytics* and *mobile learning* and returned 99 documents. On the bibliometric map 6 clusters are formed (Fig. 2).

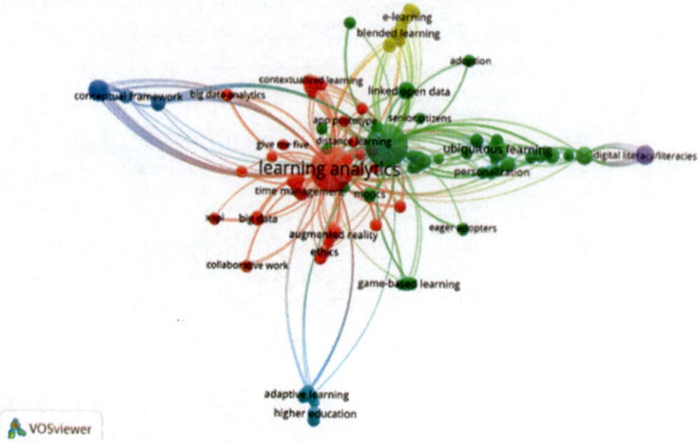

Fig. 2. The bibliometric map according to keywords *learning analytics* and *mobile learning*.

It is seen the direct and strong connection between the keywords *learning analytics* and *mobile learning* as well as with the close relationship to the keywords: *ubiquitous learning* (with total link strength 40), *personalized system of instruction* (19), *mobile personal learning environment* (16), *semantic web* (12), *blended learning* (12), *self-regulated learning* (10), *adaptive learning* (8), *intelligent learning systems* (8), etc. The main contributions to the topic, according to the number of papers, make authors from the countries: Germany (14 documents), USA (14), Australia (12), Spain (7), etc.

Another query with the keywords *intelligent* and *learning analytics* returns 294 documents. Figure 3 shows the constructed bibliometric map. The observed relationships are among the keywords *learning analytics*, *intelligent learning management*, *artificial intelligence in education*, *intelligent tutoring*, *intelligent support*. The connection with the keywords *personalized learning*, *adaptive learning*, *academic performance* is also revealed. Six clusters are formed, where the main keyword and occurrences in each cluster are as follows: *learning analytics* (163), *intelligent tutoring system* (34), *adaptive learning* (10), *predictive analytics* (3), *e-learning* (9), and *artificial intelligence* (7).

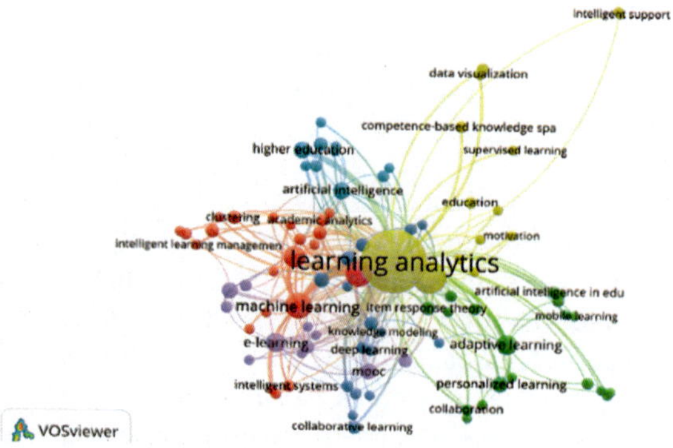

Fig. 3. The bibliometric map according to keywords *learning analytics* and *intelligent*

5 Practical Considerations for Intelligent Education

5.1 Students' Learning Performance Clustering

The aim of this experiment is to group students with similar learning performances. The data are gathered from their assessment tasks in three courses in the College of Energy and Electronics at TU - Sofia. For this purpose, three unsupervised machine learning algorithms are evaluated: k-Means, k-Medoids, and X-Means to find the best students segmentation. The data similarity is measured with the distance between every two records. The explored number of clusters is chosen to be from k = 2 to k = 7, because one students' group is relatively small (consists of 29 students). More clusters will hamper the educator to make the correct decision when some learning paths have to be planned. Table 2 summarizes the results. It can be observed that increasing the number of clusters leads to more precise segmentation. In addition, the comparison shows that k-Means and X-Means algorithms are characterized with better N values than k-Medoids.

Table 2. Evaluation of clustering algorithms.

Number of clusters	Algorithm		
	Avg. within centroid distance		
	k-Means	k-Medoids	X-Means
k = 2	266.933	348.172	273.060
k = 3	194.119	251.621	193.882
k = 4	134.185	169.000	130.738
k = 5	90.009	138.862	90.009
k = 6	68.647	117.793	68.647
k = 7	**57.750**	102.690	58.700

For further exploration is chosen k-Means algorithm with k = 7 (Avg. within centroid distance – 57.750). The distances form the root set with clusters are shown in Table 3.

Table 3. Formed clusters after applying k-Means algorithm.

Cluster	0	1	2	3	4	5	6
Avg. within centroid distance	51.580	45.250	92.500	59.673	36.000	89.438	0.000
Number of students	10	2	2	7	3	4	1
Ratio of total, %	34.48	6.89	6.89	24.14	10.34	13.79	3.45

Figure 4 shows the clusters with the students and their learning performance in courses. Cluster 0 includes 10 students and their learning performance during the courses is kept almost the same with the higher value in comparison to students from the other clusters. The student from cluster 6 has almost the same learning performance during three courses, but its value is much smaller than the students from cluster 0. The students from other clusters possess different learning performance in the three courses.

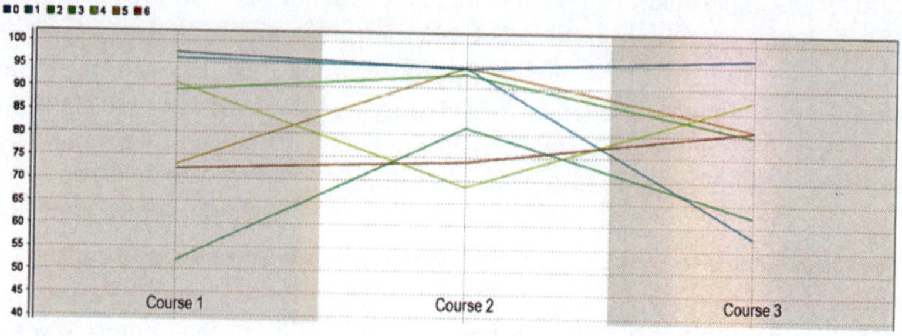

Fig. 4. Clusters with students and their learning performance in the courses

Clustering methods are applied for grouping students with similar learning interests or performance. However, they are not appropriate for the diagnosis of specific learning problems of individual students. Our idea for the solution of the problem "how to help the students, having insufficient learning performance" is to use semantic technologies for proposing needed personalization and adaptation of the learning process.

5.2 Combining Learning Analytics and Semantic Technologies

The aim is to explore possibilities for combining learning analytics and Semantic web technologies to propose personalized help to the students with difficulties or low levels of satisfaction during learning. First, we have searched related research in scientific libraries, and we have found only a few papers discussing both ontologies and LA. An Integrated Ontology of Learning Analytics (IOLA) that represents the knowledge within the LA area is proposed in [11]. This ontology is a comprehensive description of the LA domain. The mobile learning environment MeLOD applies LA methods that support Linked Open Data (LOD) [12]. The ontological specification in MeLOD represents interactions in the learning system. This system also uses statistical methods and techniques as Classification Clustering and Correlation Analysis. In the latest systematic literature review of the LA technologies [13], the coordinated use of LA and Computational Ontologies to support educators is discussed, though only two relevant papers are cited: [12] and [14]. An ontological network that models students and their state of knowledge, assessments, and learning units is used in the ON-SMMILE model [14]. All these researches don't report results about coordinated use of LA and Computational Ontologies for monitoring and controlling student's academic performance.

To examine the potential for better personalization of combined use of learning analytics and semantic web technologies, first, particular problems of individual learners have to be diagnosed, and then solutions to be proposed. Initially, we discuss how to realize this concept, then the results from putting into practice our ideas are shown.

The development of computer programing skills is a complex task. The knowledge needed cannot be learned in only one university course. It is result of hard-working through years on many related subjects. Good tests should not only perform average estimation of knowledge and skills but answer the questions such as: "What exactly every student should learn to increase his knowledge and skills." In many cases, students need to refer to external knowledge (e.g., in previously attended courses) to understand some issues of course learning content. Learning data and analytics should give sufficient information about what exactly (and from what course) should study the students to increase their knowledge and skills. Precise semantics-based models of all these related courses and relations between concepts in different learning courses help generate personalized recommendations about filling learners' knowledge gaps.

5.3 Development of Semantic Models of the Tutoring Course

To improve learning and teaching support, we develop and use semantic models of the tutoring course and its prerequisites, and show how these models can be used by discussing examples. The methodology is as follows:

1. Preparation of domain ontology network:
   a. Development of domain ontologies, classifying learning domain terminology.
   b. Defining of alignments between domain ontologies.
2. Preparation and annotation of learning objects, using domain ontology network:
3. Analysis of learning data:
   a. Selection and preliminary analysis of learning analytics data, generated and stored in the e-learning system NEO.
   b. Classification of the possible sources of the learning problems.
   c. Finding of causes for errors of every student in the tests.
4. Proposing recommendations to every student for further learning.

The experiment covers three closely related programming courses taught at the College of Energy and Electronics at TU – Sofia: Programming 1, Programming 2, and Programming languages. We have developed three ontologies that represent the learning content of each of these courses. Two questionnaires – Test 1 with 30 and Test 2 with 40 questions on the Programming languages subject are created. All the questions have only yes/no answers and are tagged by concepts and course names, where they are defined. Correct answers of 10 questions in every test require knowledge, studied in the "Programming 1" or "Programming 2" courses. In our experiment, all tests are done by 32 students, working in two groups (16 students in every group).

Preparation of Domain Ontology Network. We have developed two types of ontologies: simple hierarchical ontologies, classifying course terminology, and small ontologies, representing semantic models of every concept described in the tutoring course. Also, alignments between small ontologies and corresponding hierarchies are defined. Figure 5 presents the concept hierarchy of the courses "Programming 1" and "Programming 2." the concept hierarchy of the course "Programming languages" is shown in Fig. 6a). The small ontology describing the loop "For" is shown in Fig. 6b).

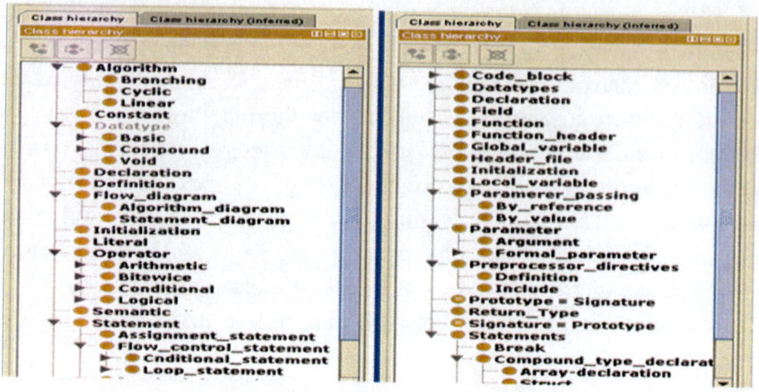

Fig. 5. Concept hierarchy of a) "Programming 1" and b) "Programming 2"

Preparation and Annotation of Learning Objects in the Domain Ontology Network. All the learners use protégé and can view the developed ontologies, including concept hierarchies of all the subjects and small ontologies, defining concepts. Ontologies also are saved in JPG files for mobile access by devices that cannot run Protégé. The best learning sources for specific concepts are listed in the ontology annotation properties.

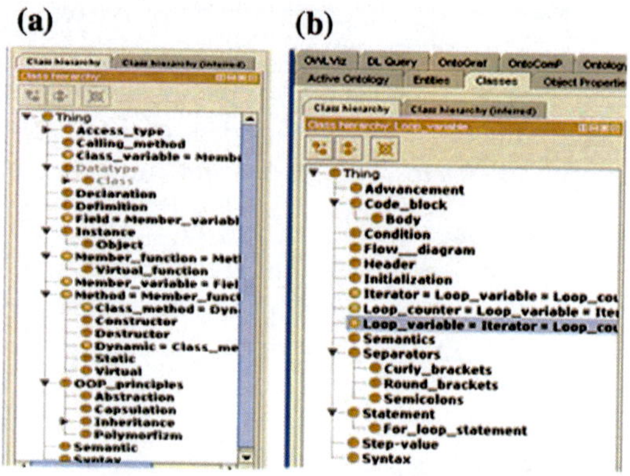

Fig. 6. a) PL concept hierarchy and b) The ontology for "Loop For".

Analysis of Learning Data. We have used built-in learning analytics tools in the e-learning environment NEO to extract data and statistics about learners and the learning process. Students' data extracted from tests are analyzed and classified. Then they are used for personalized diagnosis of learning problems of every learner and the group of learners as a whole and further control of the learning process. We categorized test questions into three categories according to the type of knowledge needed for answering. 1) Q3 – require some prior knowledge from the two previous courses; 2) Q2 – require some knowledge from one of the previous courses; and 3) Q1 – not related to previous courses. The results shown in Table 4 clearly explain the main idea of the analysis. It presents the main concept, which understanding is significant for answering the questions and the percentage of the correct answers to the questions.

Table 4. Examples of question types and percentage of correct answers.

Questions	Programming 1		Programming 2		Pr. Languages	
	Concept	% Correct	Concept	% Correct	Concept	% Correct
Q1					Class	75%
Q2	Loop For	68%			Structure	68%
Q3	Loop While	44%	Function	44%	Inheritance	44%
Average		56%		44%		66%

This type of classification and statistics gives the tutor information about the complete knowledge of the learners related to both the tutoring subject and prerequisites, which is essential for successful learning. Such an analysis of learning data about the student groups can give the tutor important information concerning the overall knowledge related to the learning course and the necessary prior knowledge. LA can give an evaluation of the overall results of a group of students. For this purpose, we calculated the average results of the question type Q1. If this result for the group is less than the previously accepted gap (e.g., 40%), the group should attend the course again. Further analysis of the answers of every student, based on semantic links between concepts (using ontologies), can give essential information for personal learning guidance.

Experiment and Results. We subdivided the learning content of the Programming languages subject into two modules. All students are divided into two groups. After each tutoring module, students in the first group did the test, then are motivated by the teacher to prepare once again and redo the test for better results. Students in the second group, besides motivation, receive personalized help in the way specified above. Table 5 shows results from the Tests. Considering the experiment results, we can conclude that additional work on the learning content can increase test success, though it decreases student's satisfaction (maybe because reading the same content again is perceived as boring). When learners receive personalized help and additional resources as immediate feedback, the learning process is more interesting, and results are significantly higher.

Table 5. Tests results.

Test	Group	Programming 1 questions	Programming 2 questions	Pr. languages questions	Average	Satisfaction (questionnaires)
1	1	54%	58%	48%	52%	88%
1	2	48%	54%	56%	54%	75%
Retake						
1	1	57%	64%	54%	56%	38%
1	2	62%	66%	68%	65%	100%
2	1	58%	66%	44%	51%	75%
2	2	56%	64%	46%	53%	62%
Retake						
2	1	61%	64%	51%	56%	54%
2	2	68%	72%	62%	66%	88%

6 Discussion and Conclusions

The performed bibliometric analysis outlines a direct connection between the keywords *intelligent*, *learning analytics*, *mobile learning*, and *learning path* – hot research areas. This paper is in line with these topics and shows how LA could facilitate educators in providing a teaching process that enhances students' performance. The revealed correlations and tendencies are analyzed in detail to make suggestions for appropriate interventions. Their focus is an enhancement of students' learning performance, identification of students at risk, etc. The proposed methodology for exploring students' learning performance using LA could assist educators in creating more engaging and efficient learning paths. The analyzed use case reveals some correlations and dependencies between students' performance and learning resources usage. The findings can be a strong base for optimizing the way students learn by suggesting better learning paths in compliance with personal achievements. The derived correlations help the improvement of learning and teaching practice in the considered local case at the College of Energy and Electronics at TU - Sofia. However, some of them could be applied broadly in a similar context. The findings show that in the context of an ILE, meaningful recommendations for improvement and adaptation of the learning process can be made. LA offers new insights into education and can give some directions for allocating resources, developing competitive advantages, thus improving the quality of learning. The experimental combining LA with semantic technologies, shows that better personalization and adaptation can be achieved, thus learning to be more enjoyable and results to be better.

Acknowledgments. This research is supported by the Bulgarian FNI fund through the project "Modeling and Research of Intelligent Educational Systems and Sensor Networks (ISOSeM)", contract KП-06-H47/4 from 26.11.2020.

References

1. Aguilar, J., Sánchez, M., Cordero, J., Valdiviezo-Díaz, P., Barba-Guamán, L., Chamba-Eras, L.: Learning analytics tasks as services in smart classrooms Univ. Access Inf. Soc. **17**(4), 693–709 (2017). https://doi.org/10.1007/s10209-017-0525-0
2. Siemens, G. (ed.): Learning analytics: envisioning a research discipline and a domain of practice. In: Proceedings of the 2nd International Conference on Learning Analytics and Knowledge. ACM (2012)
3. Avella, J., Kebritchi, M., Nunn, S., Kanai, T.: Learning analytics methods, benefits and challenges in higher education: a systematic literature review. Online Learn. **20**(2), 13–29 (2016). https://doi.org/10.24059/olj.v20i2.790
4. Kaliisa, R., Kluge, A., Mørch, A.I.: Combining checkpoint and process learning analytics to support learning design decisions in blended learning environments. J. Learn. Anal. **7**(3), 33–47 (2020)
5. Mangaroska, K., Sharma, K., Gasevic, D., Giannakos, M.: Multimodal learning analytics to inform learning design: lessons learned from computing education. J. Learn. Anal. **7**(3), 79–97 (2020)
6. Picciano, A.G.: The evolution of big data and learning analytics in American higher education. J. Asynchronous Learn. Netw. **16**(3), 9–20 (2012)
7. Hung, J.L., Hsu, Y.C., Rice, K.: Integrating data mining in program evaluation of K-12 online education. Educ. Technol. Soc. **15**(3), 27–41 (2012)
8. Scheffel, M., Drachsler, H., Stoyanov, S., Specht, M.: Quality indicators for learning analytics. Educ. Technol. Soc. **17**(4), 117–132 (2014)
9. Prinsloo, P., Slade, S., Khalil, M.: Stuck in the middle? Making sense of the impact of micro, meso and macro institutional, structural and organisational factors on implementing learning analytics. In: Volungeviciene et al. (eds.) EDEN 2018, p. 41 (2018)
10. Ellegaard, O.: The application of bibliometric analysis: disciplinary and user aspects. Scientometrics **116**(1), 181–202 (2018). https://doi.org/10.1007/s11192-018-2765-z
11. Nguyen, A., Gardner, L.A., Sheridan, D.: Building an ontology of learning analytics. In: PACIS, p. 158 (2018)
12. Fulantelli, G., Taibi, D., Arrigo, M.: A framework to support educational decision making in mobile learning. Comput. Hum. Behav. **47**, 50–59 (2015)
13. Costa, L.A., Pereira Sanches, L.M., Rocha Amorim, R.J., Nascimento Salvador, L.D., Santos Souza, M.V.D.: Monitoring academic performance based on learning analytics and ontology: a systematic review. Inform. Educ. **19**(3), 361–397 (2020)
14. Yago, H., Clemente, J., Rodriguez, D., Fernandez-de-Cordoba, P.: On-SMMILE: ontology network-based student model for multiple learning environments. Data Knowl. Eng. **115**, 48–67 (2018)

Fearful to Fearless: Design of ICT Based Learning Tools to Combat Extremism, Terrorism and Violence

Rani Parvathy Venkitakrishnan(✉)

TCS Lab, Spoken Tutorial Project, Department of Chemical Engineering, Indian Institute of Technology Bombay, Powai, Mumbai 400076, India
rani.rmc.bl@gmail.com

Abstract. Violence comes in many forms. Governments across the globe spend a lot, on protecting their citizens from violence and studying violence. Modern forms of violence extensively borrow from culture, religion, history and mythology: which is violence towards women, children, and the vulnerable. The population most affected by violence are also the poor, who are without resources or organized knowledge to resist violence. They also lack resources to practice skills of non-violence. Though education is hailed as a unanimous choice to combat extremism, terrorism and violence, current education-job system is not sufficient to combat or reduce violence. Currently, there is no clear practical adaptable methods in the public domain, for non-violence education. Short audio-video tutorials have been successfully used to improve health and nutrition of newborns in India. This article will explore the extension of this method to adapt to the creation of skills needed to practice non-violence in adverse conditions by self-learning. Considering the large scale of violence across the globe, coming up with an adaptable mobile technology based resource is of importance. Online mobile learning can reach the remote and dangerous places of the planet, where one-to-one interaction is not possible.

Keywords: Large scale adoption of mobile learning · Life-long and informal learning using mobile devices

1 Introduction

Even though only a fraction of the population tends to turn to extremism, violence or terrorism, the result has devastating consequences for the rest of the population. It is known that enacting laws or presence of police is not sufficient to decrease violence [1]. Some of the most devastating violence are the results of (i) War or civil war, (ii) Religious unrest and related violence (iii) Culture, beliefs of a society, prevalent social norms, (iv) Laws and regulations and (v) Biology and Evolution. Skills required to conquering fear, can be helpful for people facing organized violence. Even though ICT blended learning tools are suited to promote skills and practices of non-violence, the many websites of WHO or UN [2], do not provide a self-learning step-by-step guide to develop the skills needed to practice non-violent behaviour. Currently, there are no clear, scaled up adaptable methods in the public domain, for non-violence

M. E. Auer and T. Tsiatsos (Eds.): IMCL 2021, LNNS 411, pp. 250–258, 2022.
https://doi.org/10.1007/978-3-030-96296-8_23

education. In high crime regions of the world, it is difficult and even dangerous for volunteer programs or support group to have one-on-one access or meetings. When people join groups of organized violence, they are fearless of the consequence. A detailed analysis By Gandhi Foundation [3] indicates that, those who resist violence by methods of non-violence are also fearless. Thus, when the state of being fearless is associated with non-violence practice, instead of violence we progress to less violence.

A common factor in all the above said forms, is the violence towards women and children [4–6]. The UN rightly points out that [1], violence towards women is not only about women, it is about the perpetrator too. Women comprise of about 50% of the population. The violence towards women is prevalent in households across all continents, cultures, religion and even mythology [7–9]. Women are the majority of primary caregivers of children even in western societies. Terrorists and conquerors routinely turn to violence towards women to cultivate fear, to reach their goals even in 21st century. This indicates, religion, culture and citing of social practices is often used to curtail rights and freedom of 50% of population. Even though most countries have some rules to protect their women and children from violence, domestic violence still is a leading cause of 'day-to-day violence'. Multiple studies have shown that, innate aggression is different from violence, which is a learnt skill [6]. We know that, life forms have evolved with ability for fight or flight ability when facing danger, identify danger and aggression if different from this. For example, violence becomes a skill learnt from watching domestic abuse which also doubles as a psychological abuse of children paragraphs, however, are indented.

Education is hailed as the unanimous choice to combat extremism, terrorism and violence. However, the current education-to-job system is not sufficient to combat or reduce violence. Present educational policies are not geared to teach skills to reduce violence, even in violence prone societies. Students study subject matters in Science, arts, language, economics and so on which are geared towards finding a job. The present education system is not geared to reduce gender-based violence either. There are many examples for this point. Census of the Bay area region, Los Angeles region and/or the East coast around New Jersey in USA, show that immigrants from across the globe stay here for job purposes [10]. These regions have large number of people from India, East Asia and other regions. This can be attributed to the mass global migration of millions for job opportunities in the last 3 decades. Most of the immigrants in the Bay area (Silicon Valley for example), have high educational qualification, have high income due to the nature of the jobs in cutting edge technology and a work visa in skilled categories. They move to take up lucrative job opportunities involving several modern technologies. Still there is an acute shortage to help the women from Indian or South East Asian communities in these regions. The disconnection and complete non-relation between higher education and domestic violence is reflected well with the presence of multiple NGOs, battered Asian women help and women assistance organizations in these regions [11–13]. The help sites such as, Maitri, Sakhi, Sahara, Shuddi, Saheli and so on translates woman friend, pure (purity) and dear friend in Indian languages.

There is no consistent violence discouragement program in these societies. Further, violence and organized violence in societies is a sensitive topic. It is also a cultural, economic and national policy issue [14, 15]. Further violence towards women are justified in myriad of ways. Many a times, the focus is on free speech or equality, and initiatives deviates from purpose, lose momentum, lack sustenance or turn violent. This gets more complex if/when countries practice gender neutral laws, which do not reflect the social norms and life (here for the 3 new reference).

Some of this can be attributed to lack of knowledge to sustain and/or maintain the momentum on non-violence movement to bring about changes or to practice non-violence. When people from societies in Asia or South America move to North America or Europe, for job opportunities, their cultural values remain the same even though they live in western societies. Though UK is a smaller country compared to USA, it has a large number of Indian/Asian immigrants. They also have several NGOs, changing laws and regulations to help women caught up in such domestic abuse situations far from their support system. Such scenario, leaves the most vulnerable, in the most difficult situations and exposes the loop holes in the law system even in the absence of external violence [16–22].

ICT tools can provide insightful details for people in regions of mass migration, when people leave their countries and families in caravans or in smaller groups to escape violence. For example, larger numbers cross the border to USA, leaving everything they have behind, to escape drug related violence or national policies. However, this adversely affects their families and ways of life in homeland and many head to a lone unknown future, in lands with different language and culture. Many in Asian, African or South American countries flee homes to religion, sectarian or terrorism related violence or political instability. Women are seen less in the forefront of indulging in war. Even in the armed forces, the majority of them are in nursing, in support, or administration. Only a smaller percent of women are seen to be prone to committing violent crimes compared to men. Further, they face more violence even in the safety of their homes, irrespective of region, color, religion or culture. Hence an approach to combat violence starting with addressing the prevalent forms of abuse towards women and children, will reduce the overall violence.

Most of the people involved in traumatic violence related events, do not have access to information to resist violence, on how to adopt, adapt and practice skills for non-violence movements, or on going about to rebuild their lives. It is the ability of resilience and coping methods of oneself and of people around [23, 24], which act as a guide. Many eventually become a refugee, an illegal migrant or have worse outcome.

We will discuss about extending the proved, practical, time tested and working model methodology frameworks of skill training to mitigate violence. The last decade, has seen a sharp rise in people who use smart phones. Long Covid lockdowns have pushed even the least skilled of the population, to train themselves to use mobile phone or smart phone features. Majority of people who use the mobile technology, are on board with skills to view video, listen to audio and social media. This along with internet and easy translation availability can be coupled to deliver the knowledge worldwide in multiple languages. We ask, if the skills of mobile use and prevalence of mobile use can be combined with known and proven skill training programs, to create a working model to self-train large populations at the bottom of skill sets towards non-violence [25, 26]. We will address a methodology, examples available for creation, topics on violence which are adaptable, concept of fear, and adaptability of the methods to various languages and address the varied facets of the issue. The chosen methods can also be tailored by regional authorities to suit local needs specifically.

2 Methodology

Series of short audio-visual tutorials have been successfully used for skill training by the Spoken Tutorial project of IIT-Bombay [27], for more than a decade. In this methodology, learner uses short audio-video tutorials, and practices it side-by-side to learn a skill. Each video is not more than 10–15 min long. The methodology considers a lower ability of any given learner and the absence of 1-on-1 training for a given session. A novice reviewer first practices the skill. If the novice is unable to complete successfully, the contents are iteratively improved to reach the level of understanding for the novice. Detailed methodology of this process is available in the spoken tutorial website. It is known that, learners grasp concepts and understand it better with visual learning. Step-by-step demonstrations in the video, allows the learner to repeat and practice the skill till they are comfortable with the process. With more practice the skill gets better. So far, more than 3 million students have been trained for various types of IT skills, using this method in a decade. Many of the video series is hosted by multiple sites, such as skill training and governmental government web sites, added to YouTube channels or in social media for wider outreach.

Use of open source software (or free apps for smart phone) is of importance for creation or use of these methods allowing to bring down production costs. We will discuss the several softwares that are suitable for this purpose. Choice of software such as audacity for audio, Blender to create graphics or animation, inskscape or synfig for media, Freeplane to visually show concept maps, Kdenlive for video editing, Kazam for screencasting, ELAN for annotation, translation or language analysis makes this methodology adaptable across all languages and continents.

The same methodology is utilized, in the field of health and nutrition to address the malnutrition during infancy and early childhood [1]. In this framework, open source software animation/graphics software are utilized to create a series of 10–15-min tutorials, to train mothers in self-care and to nurse newborn infants. The website also hosts skill tutorials on not to practice corporal punishment, it's downside and alternate methods to handle student issues in schools. Since the method addresses the large number of poorer and less skilled population, in the life protecting skills and methods, they are basic in nature, to build a strong foundation.

The ability of the methodology adapt to a wide variety of need brings in wider acceptance. If the learner and trainer is required to learn new methodology for different topics the interest and success may not be high. Here, learner has the skills to view a video or listen to an audio on the phone. With the onset of Covid-19 induced lock-downs ICT tools have received wide acceptance in educational methods and world have adopted and adapted to online modes of education. The videos can be played repeatedly to instill the importance of non-violence, to keep persistence and keep motivation. The videos can be translated and dubbed into the various languages for the various populations. These videos can be viewed in privately, without an individual to be physically present in a classroom. Individuals can explore, understand and imbibe the world of non-violence, without having to announce their desire for non-violence which may put them in danger in regions of high-risk violence.

The above-mentioned methodology has high chances of success when extended to create self-learning educational resources to address the topic on violence. Here, the perpetrator and the victim can be trained towards non-violence [26, 28, 29]. Both sides of the coin are on the same platform, like the teacher and the student learning by online modes. Some of the possible topics for videos that can address non-violencc skills are given in Table 1. The target audience who would benefit from non-violence skills training are from the poorer section of the society. Since the methodology, learning skills and approach and target audience are similar, the method will likely lead to similar efficacy in terms of ability to learn the skill. Differences in timescale to practice the change of mindset to non-violence and presence or absence of government policies may change the duration or efficacy to see the results on the field. A practical implementation will need setting up an active website, advertisement, and monitoring it to get quantitative efficacy information.

Table 1. Various issues involving violence, which are suitable for audio-video tutorial creation for large population. The first two skills involve the victims. Third skill is aimed towards the perpetrator of violence (ref 31, 33). Next two are the skills that history shows that are needed and utilized. The last, is on one's rights and can be tailored to suit the situation.

Issue	Skill to develop
Fearful of consequence when facing violence or organized crime	Loss of Fear, going from fearful to fearless
Lack of support in high risk violent areas	Building support group online, using online forums
De-escalation of violence (the opj.gov reference goes here from Sect. 3.1)	Motivational interviewing [30]
Suffrage Movement	Unification, rights, moving away from marginalized section of society
Bonus Army	Skills of Persistence, motivation
Reaching out to the mass population with information for non-violence	Example is India's independence movement. Gandhi's non-violence movement and mass mobilization
Civil rights education	Knowing one's rights

3 Discussion

The ability to train a person remotely, for skills, who is in a violent region gives an edge to such ICT based methods. If a person has difficulty or questions, a trainer can remotely address the questions and not risk being on the field in high risk violent areas. Some of the key points, brought forth is (i) Peace is different from practice of non-violence (ii) Aggression and violence is a learned skill (Butler reference) (An example is a child who watches domestic abuse and learns the skill) and (iii) Extended violence can wipe out societies and civilizations. Freedom struggles (an example is India's independence from British rule) or right to vote (suffrage) for women have utilized several skills needed to steer away from violence. Violence resulting from (i) organized crime (ii) lack of women's rights or (iii) cultural or religious context (either here for the 3 new reference) are a bit different from the non-violence movements which utilized the methods earlier. Over the last few decades, there is a tremendous increase in human population, migration pattern, floating populations, methods of communication and the methods utilized in crime have also changed. However, laws to prosecute and laws determining rights have changed very little. Methods to counter violence needs utmost support from the governing agencies, who are in charge, where the method is to be applied.

3.1 Peace is Different from Practice of Non-violence

An important concept towards peace is the de-escalation of violence. There are journals dedicated to the topics and methods utilized. One of the methods utilized to change the mindset a violent person is termed motivational interviewing. The article by Clark et. al. [31] and the references therein discuss ways to address the violent perpetrators.

Several types of aggressive violence is a learned skill: The bernardvanleer website, their documents and references therein stresses on 'responsive parenting' and gives several examples for it. It is an excellent reference for families, who want to protect their children from falling to ways of violence, to better communicate with them and to improve the quality of life for their children. This is especially useful for children, who are not exposed to violence within home. Ways of better communication for parents with children. Some of the testimonials show, parents are able to instill hope, courage, bonding and felling of security in the child, instead of distress and violence. This in turn will produce generations of people, who are less prone to violence.

3.2 Violence and Fear are not Sustainable

Another concept to convey to the perpetrator is that, continuous violence leads to collapse of societies. When facing many crimes, the accountability does not fall on the crime or the perpetrator. Studies focus on the resilience of the victim. The first targets are mostly women and children. When a society turns violent, it self-destructs with lowering the living conditions. A violence ridden society does not have sufficient resources to provide for self-betterment, a natural calamity or face an epidemic, since their expertise is in violence. Savagery and barbarism become prominent and fear is on the forefront.

Their ability to build health or education facilities, enact sustenance policies or environment protection programs does not exist even when a disaster strikes. Over the years, the able part of the population leaves the violence ridden regions and the society is further depleted of expertise to set up infrastructure to handle adverse situations. Such societies live under the fear, consequences of the actions. Many countries trying to cope with the current Covid pandemic lacking even medical resources or expertise, indicate this key aspect. Several violence prone countries, from Afghanistan to the countries which are across the globe in the west are examples. They have to rely on help from outside for expertise, infrastructure, finances and medical commodities due to lack of infrastructure. When 50% of the population live in fear any region, the development is the first casualty.

Based on the contents, videos can be categorized into three broad categories. They are, (i) Common ideas related to developing non-violence practice (ii) Region and problem specific and (iii) Improving conditions for women and children. Examples for common topics for creation of videos are shown in Table 2. It is beyond the scope of this article, to create videos with specific examples and for regions. Due to the sensitivity of the topic, support of a government body or an NGO or funding agency is needed on this matter. An example of such sensitivity is where women face violence or lack of rights due to culture, religion or regional circumstance. This also brings to fore that, most governments are on board and support working methods, when it becomes popular and it shown to give results. In this case, to show the feasibility and get the desired results, one needs government support in terms of funding to push it forward. Endorsement and support from government is essential for promotion and success of non-violence programs even in western societies.

Table 2. Common topics which can be made to short videos

Methods to practice non-violence	Methods used for resistance to bring about change	General skills
Make sure not to turn violent in front of children	Be courageous	Responsive parenting
Make sure children do not watch violence	Protection of children	Expanding education - Productive and critical Thinking
Character building	Learn one skill at least per person	Communication skills Computer skills Health and Education
Non-violent protest methods seen in several situations	Picketing mass parade and mass boycott	Stress Management Non-violence Persistence Decision Making

3.3 Region and Problem Specific

Often, we hear about, caravan migration from South/Central America, where large number of people attempt to leave the organized violence. They find safety in numbers and travel together northwards to USA thinking it is for better future. Travelling to unknown future, country, language, destination comes with lot of perils. Most are not skilled for sought after jobs in the foreign countries, pay the ultimate price during the journey, face extortion, face more violence, and so on. For those who make it across, a journey back is equally or more perilous. They are separated from their culture, family, language, familiarity and support system. They are removed from things that are familiar and the ways of life. Hence an attempt to inform them of the equally or more perilous journeys, and training them in the options to de-escalate the violence may likely bring positive outcome in the regions. The videos are suitable for areas of escalated violence, where volunteer programs or assembly of people could be risky.

Acknowledgment. We thank Prof. K Moudgalya for discussions. Spoken tutorial project work is supported by grants to Prof. K, Moudgalya, by Government of India.

References

1. Tonry, M.: Building safer societies. Eur. J. Crim. Policy Res. **5**, 49–60 (1997). https://doi.org/10.1007/BF02677650
2. https://www.who.int/violence_injury_prevention/violence/norms.pdf
3. https://www.mkgandhi.org/ebks/GandhiAutobio_morallessons.pdf
4. https://eca.unwomen.org/en/news/stories/2019/01/10-myths-about-violence-against-women-and-girls
5. Kalras, G., Bhugra, D.: Sexual violence against women: understanding cross-cultural intersections. Indian J. Psychiatry **55**(3), 244–249 (2013). https://doi.org/10.1007/BF02677650
6. Bowden, B.: Civilization and its consequences political science, international relations, political theory. Oxford (2016). https://doi.org/10.1093/oxfordhb/9780199935307.013.30
7. https://statusofwomendata.org/explore-the-data/violence-safety/
8. https://now.org/resource/violence-against-women-in-the-united-states-statistic/
9. Walby, S.: Violence and society: Introduction to an emerging field of sociology. Curr. Sociol. **61**(2), 95–111 (2013)
10. http://www.bayareacensus.ca.gov/bayarea.htm
11. Fogarty, A., Woolhouse, H., Giallo., R, Wood., C, Kaufman., J, Brown. S.: Violence, mothers' experiences of parenting within the context of intimate partner violence: unique challenges and resilience. J. Interpersonal Violence (2019). https://doi.org/10.1177/0886260519883863
12. Kalokhe, A., et al.: Domestic violence against women in India: a systematic review of a decade of quantitative studies. Glob. Public Health **12**(4), 498–513 (2017). https://doi.org/10.1080/17441692.2015.1119293
13. Liao, M.S.: Domestic violence among asian indian immigrant women: risk factors. Acculturation Intervention, Women Therapy **29**(1–2), 23–39 (2006). https://doi.org/10.1300/J015v29n01_02
14. Adam, N.M., Schewe, P.A.: A multilevel framework exploring domestic violence against immigrant Indian and Pakistani women in the United States. Journal of Muslim Mental Health **2**(1), 5–20 (2007). https://doi.org/10.1080/15564900701238468
15. Edna Erez, E.: Migration/immigration, domestic violence and the justice system. Int. J. Comp. Appl. Crim. Just. **26**(2), 277–329 (2002). https://doi.org/10.1080/01924036.2002.9678692
16. Montgomery, B.E.E., et al.: Violence against women in selected areas of the United States. Am. J. Public Health **105**, 2156–2166 (2015). https://doi.org/10.2105/AJPH.2014.302430
17. Hester, M., Westmarland, N.: Domestic violence perpetrators. Criminal Justice Matters **66**(1), 34–35 (2006). https://doi.org/10.1080/09627250608553400
18. Violence, D.: Criminal Justice Matters (1994). https://doi.org/10.1080/09627259408552666
19. Mahapatra, N., Rai, A.: Every cloud has a silver lining but… "pathways to seeking formal-help and South-Asian immigrant women survivors of intimate partner violence." Health Care Women Int. **40**(11), 1170–1196 (2019). https://doi.org/10.1080/07399332.2019.1641502

20. Mann, S.K., Roberts, L.R., Montgomery, S.: Conflicting cultural values, gender role attitudes, and acculturation: exploring the context of reproductive and mental health of Asian-Indian immigrant women in the US. Issues Ment. Health Nurs. **38**(4), 301–309 (2017). https://doi.org/10.1080/01612840.2017.1283376
21. Peja, T.: Domestic Violence Among Asian Indian Immigrant Women in the United States Master's Thesis (2017). https://ecommons.luc.edu/luc_theses/3699
22. https://www.aph.gov.au/Parliamentary_Business/Committees/Senate/Legal_and_Constitutional_Affairs/DowryAbuse/Report/c03
23. Koci, A.F., McFarlane, J., Nava, A., Gilroy, H., Maddoux, J.: Informing practice regarding marginalization: the application of the koci marginality index. Issues Ment. Health Nurs. **33** (12), 858–863 (2012). https://doi.org/10.3109/01612840.2012.713081
24. Inman, A.G., Rao, K.: Asian Indian women: domestic violence, mental health, and sites of resilience. Women Therapy **41**, 1–2, 83–96 (2018). https://doi.org/10.1080/02703149.2017.1324189
25. Stephens, W., Sieckelinck, S.: Working across boundaries in preventing violent extremism: towards a typology for collaborative arrangements in PVE policy. J. Deradicalization **20**, 272–313 (2019)
26. Summy, R.: Understanding Nonviolence in Theory and Practice, UNESCO-EOLSS: Peace, Literature, and Art, vol. II
27. The Spoken-tutorial Project, IIT-Bombay. https:www.spoken-tutorial.org
28. Brian, M.: The dynamics of nonviolence knowledge" Faculty of Law, Humanities and the Arts - Papers (Archive). 2468 (2015)
29. Baldoli, R.: Reconstructing Nonviolence: A New Theory and Action for a Post-Secular Society, Routledge (2018). ISBN: 9781138553897
30. Clark, M.D.: Motivational interviewing for deradicalization: increasing the readiness to change. J. Deradicalization **20**, 47–74 (2019)
31. https://www.ojp.gov/pdffiles1/nij/grants/208551.pdf

The Learners' Perceptions of Learning Design for Mobile MOOCs

Anna Mavroudi[1(✉)] and Angelika Kokkinaki[2]

[1] Norwegian University of Science and Technology, P.O. Box 8900, 7491 Trondheim, Torgarden, Norway
anna.mavroudi@ntnu.no

[2] University of Nicosia, P.O. Box 24005, 1700 Nicosia, Cyprus

Abstract. Massive Open Online Courses (MOOCs) and the use of mobile learning are two learning/teaching forms that have received much attention in recent years. Despite the popularity of both approaches and the efforts to integrate or combination them, there is still a lack of papers that are focusing on the aspect of learning design for mobile MOOCs. To compensate for this lack, this paper is examining the preferences of MOOC learners between mobile and non-mobile MOOCs with respect to a number of crucial learning design parameters. An online anonymous survey was answered by 68 MOOCs participants, The results indicate that on the one hand, they think that MOOCs' accessibility and flexibility are better supported using a mobile MOOC format. On the other hand, they think that the ability to concentrate, motivation, self-study/self-learning, and self-regulated learning are better supported in a non-mobile, conventional MOOC. The main conclusion drawn is that when it comes to learning design parameters that can promote deep learning, the participants perceive the non-mobile format as being superior. Yet, flexibility and accessibility are also important when it comes to creating inclusive learning environments. Thus, it seems that the creation of an inclusive mobile learning environment that would also support deep learning using MOOCs is a challenging problem for the learning designers.

Keywords: MOOC · Learning design · Mobile learning · Learners' perceptions

1 Introduction

Massive Open Online Courses (MOOCs) and mobile learning are two emerging learning approaches that are constantly evolving [9]. The numbers of MOOC enrollments during the COVID19 pandemic were staggering [9] and each year there are thousands of MOOCs enrolling tens of millions of potential learners [18]. Also, as of late January 2019, there were more than 5 billion unique mobile users out of 7.7 billion people on the planet [9]. Thus, it is not surprising to read in the recent literature of a trend towards combining MOOCs with mobile learning or MOOCs intended to be accessible through mobile devices [12]; and that major MOOC platforms, such as Coursera and edX developed mobile applications to provide learners more access to their courses via tablet and smartphone [16].

M. E. Auer and T. Tsiatsos (Eds.): IMCL 2021, LNNS 411, pp. 259–268, 2022.
https://doi.org/10.1007/978-3-030-96296-8_24

Some researchers claimed that this combination can be advantageous for several reasons: for example, it has been suggested that since mLearning is more present than computer-based learning in many developing regions, it would be worthwhile to explore the MOOC format in combination with mLearning in developing regions to overcome the lack of trainers in these regions [2]. In relation to that, it has been claimed that advances of technologies like ubiquitous computing have had a positive impact on the scope and the accessibility of mLearning [11]. Another reason for motivating research on the intersection of MOOCs and mLearning is that the design philosophy of both these learning environments is similar in the sense that it shares some common characteristics (something that it is discussed in the next section). Also, MOOCs allow learning to happen across space and time due to the fact that course participants can learn asynchronous and online. Similarly, mLearning has been characterized by its ability to facilitate learning that takes place "anytime, anywhere" [2, 16]. Furthermore, both these learning forms have a great potential for informal and lifelong learning allowing knowledge construction to occur over time without being tied to a particular space and contexts [2]. In addition, it has been suggested that MOOCs combined with mobile learning have the potential to enhance dialogue, collaboration and interaction among learners [5]. Finally, it has been argued that mobile learning acts as a promoter of incorporating MOOCs into the learning experience [5] and that MOOCs, due to the emergence of cloud computing, are the next evolution of both online and mobile learning [8].

In contrast to the advantages mentioned above, the combination of MOOCs and mobile learning pertains to several challenges. For instance, it has been suggested that MOOCs suffer from a careful consideration of the relation between technology and pedagogy [16] and that many MOOCs simply wrap up old pedagogies in new technologies [12]. Also, there is a lack of data that can provide insight on what extent MOOC learners actually use mobile devices in their learning [16]. Furthermore, it has been claimed that the mobile learning experience might be hindered by learners' distraction and mind wandering [10].

In addition to these challenges, designing a MOOC for mobile learning deserves theoretical research and practical exploration [4]. There is a need for adapted learning designs that create additional and improved learning opportunities by utilizing the special characteristics of mobile devices [16]. To address that need, this case study is focusing on learning design for mobile MOOC. In particular, the research questions examined herein are: what are the perceptions of experienced MOOC users on crucial aspects of learning design of a mobile MOOC? Would these users prefer a mobile MOOC over a conventional (i.e. non-mobile, stationary) MOOC in conjunction with a number of crucial learning design aspects? What is their preference between mobile MOOC and a conventional non-mobile MOOC format, in general?

2 Previous Projects that Combined mLearning with MOOCs

Several attempts on integrating MOOCs and mobile learning have been observed during the last decade [5]. My LearningMentor was a mobile application developed in 2015 to support learners participating in MOOCs [1]. The application was built upon research in two fields, mobile context-aware recommender systems and adaptive learning planners. The aim was to enhance the learning experience by providing personalization in terms of planning that facilitates that the learners are following up with the MOOCs in which they are enrolled in. This planning was adapted to learners' profiles, preferences, priorities, and previous performance, measured in time devoted to each task. The application scaffolded the learning experience by providing tips and hints aimed at helping learners develop efficient study habits and skills, so that they will eventually become self-learners. The paper is describing the architecture and technical implementation of the application. Although the effort described is promising, the authors herein could not identify follow-up research work of empirical research with MOOC participants.

The paper of [14] examined how the usage of mobile devices influences the learning behavior and outcome in MOOCs. The authors presented a proof-of-concept implementation of advanced mobile Learning Analytics (LA) which tracked and mirrored learners' behavior in the MOOC platform and additionally tracked contextual properties of the used devices and applications. The mobile LA metrics that were exploited included tracked interaction events like: visited items, average session duration, quiz performance, video plays, video downloads, slide downloads and forum activity. Four courses from two real-world MOOC platforms were used to test the effectiveness of mobile LA. The findings revealed that learners who additionally learnt with the mobile LA visited more items, performed better in quizzes, and watched and downloaded more videos, which resulted in a relevant increase on average course completions.

A comparison between mobile and stationary MOOCs was attempted in a recent empirical study [17] that investigated the impact of the learning condition (i.e. mobile vs stationary) to learners who participated in two mini MOOCs. The researchers were motivated by the fact that empirical studies focusing on comparisons between mobile and stationary MOOCs and conducted in realistic settings are scarce. The findings of the study revealed that participants' learning gains were slightly lower in the case of mobile learning.

3 Theoretical Framework

Due to their popularity, definitions for MOOCs and mobile learning abide in the recent literature. In the context of this study, a MOOC is defined as an online course with the option of free and open registration, an open curriculum, and open-ended outcomes [18]. When a learner is enrolling in a MOOC, (s)he obtains access to the resources, and (s)he can interact, share knowledge and reflect with peers [18].

Mobile learning is defined as learning that happens when the learner is not at a fixed, predetermined location, or when (s)he exploits the affordances of mobile technologies to learn [6]. The concept of mobile learning has become more sophisticated throughout the last years something that pertains to a shift from content-oriented to interaction-oriented definitions with an emphasis in active pedagogies and especially, collaborative learning [13]. More recent conceptualizations focus on the mobility of learners and thus define mobile learning as learning that happens across contexts while learners move along contexts and situations. Finally, some researchers stress the nature of "mobile" as a larger concept that alters the nature of learning and presents a new learning paradigm [16].

In the intersection of mobile learning and MOOCs some researchers have defined mobile MOOC (or mMOOC) as a MOOC intended to be accessible through mobile devices [12]. Although this definition seems not relevant to deep learning, yet it is meant to emphasize that advancements in digital technologies have enhanced the scope of mLearning and its accessibility, as well as, the interplay between these two [11]. Other researchers emphasize the flexibility of mLearning as a crucial parameter of a mobile MOOC. For instance, the authors of [5] argue that mLearning with its flexible nature facilitates active learning and consequently, MOOCs may be considered as a strong learning milieu that allows students to be part of a mobile system; while others link it to the concept of today's connected society focusing on individualization, instant and universal access, and flexibility. Some researchers focus on learners' autonomy (e.g. self-study, self-learning) and factors that can motivate or demotivate them, claiming that such parameters can also impact the learning process and outcomes [11]. In relation to that, recent reviews on MOOCs have explored self-regulation with encouraging results (Lee et al. 2019), while a review on the relationship between mobile learning and self-regulated learning conducted recently [11] revealed that mLearning enhanced SRL but also that the interplay between them is dynamic and complex. Finally, some student-focused studies on MOOCs focus on self-regulated learning, motivation, engagement, and interaction, but they also suggest that more research is still needed with respect to these parameters [18].

The theoretical framework used herein guided the development of the research instrument, as described in the next section. The theoretical constructs of the framework are learning design parameters that are important in the case of mobile MOOCs. These parameters include: flexibility, accessibility, ability to concentrate, interaction with other learners, motivation, self-study/self-learning, and better direct/regulate learning. The rationale for selecting these specific parameters to focus upon pertains to the fact that they have been strongly associated with MOOCs and/or mobile learning in the recent relevant literature. For instance, researchers emphasize the flexibility of mLearning as a crucial parameter of a mobile MOOC. Consequently, it was included in the theoretical framework,

4 Materials and Methods

An online survey was distributed to users who had participated in a MOOC. The survey was targeting 234 professionals from more than 58 countries worldwide who successfully completed the MOOC and received their certificates in accordance with the European and national qualification frameworks and the European credit system for vocational education. In terms of gender, 32% of the professionals are females, 63% are males (and 5% other). In terms of age, 29% are between 25 and 34 years old, 29% between 35 and 44 years old, 16% between 45 and 54 years old, and the remaining percentage is outside the range between 25 and 54 years old. The estimated effort for the course participants was 5–10 h per week and approximately 80 h in total.

The MOOC was hosted by the University of Nicosia and it was aiming to cultivate competences relevant to DevOps for Smart Cities among the participants. It consisted of 11 modules and it was created within the Smart DevOps project (https://smartdevops.eu/dev/) which was funded with the support of the European Commission. The focus of the project was the professional development, through appropriate training, of the personnel who works or will work in municipalities that are gradually transformed into Smart Cities. Each module consisted of learning materials in the forms of videos and presentations for individual learning, a social learning component realized through online forums, and an assessment component realized through online quizzes. Two of the 11 modules were especially designed to promote reflections on behalf of the course participants.

The survey was open for 3 weeks on June 2021, it was online, anonymous, and optional; thus, the sample consists of self-selected course participants. The data collection process was in line with the new EU GDPR. The link to the survey was disseminated via email. The survey was asking the perceptions of the course participants regarding a mobile MOOC as opposed to a conventional MOOC in conjunction with several important learning design parameters. These parameters are in line with the theoretical framework discussed in Sect. 3 and the challenges associated to MOOCs and/or mobile learning mentioned in Sect. 1. Furthermore, it was asking whether participants would prefer a mobile MOOC over a non-mobile one. It also included some basic relevant demographics, such as gender, age, ethnicity, education level; and, finally it was asking on the use of mobile device while participating in the MOOC. The results were analyzed statistically to understand participants' relative perceptions on the comparison between a mobile MOOC and a conventional MOOC.

5 Results

5.1 Participants' Profile

The participants' group was quite diverse in terms of age and gender, as can be seen in Figs. 1 and 2 respectively. Approximately 2 out of 10 participants are between 25 and 34 years old, 4 out of 10 are between 35 and 44 years old and 2 out of 10 are between 45 and 54 years old.

In terms of gender, approximately 2 out of 3 participants are men and 1 out of 3 are women. In addition, the participants' group is quite diverse in terms of residence and highest education level achieved as depicted in Figs. 3 and 4 respectively. In particular, 6 out of 10 participants reside in Europe, 1 out of 8 reside in Asia and 1 out of 8 reside in America. In terms of highest education level achieved, more than 8 out of 10 have a university degree. All of them own and use a mobile device (i.e. smartphone, tablet) with internet connection.

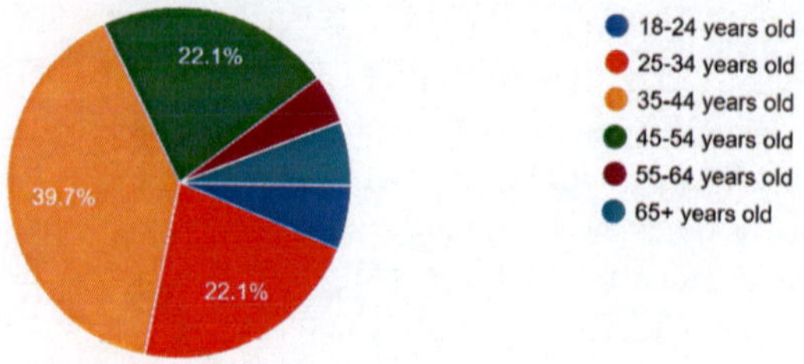

Fig. 1. Research participants by age group

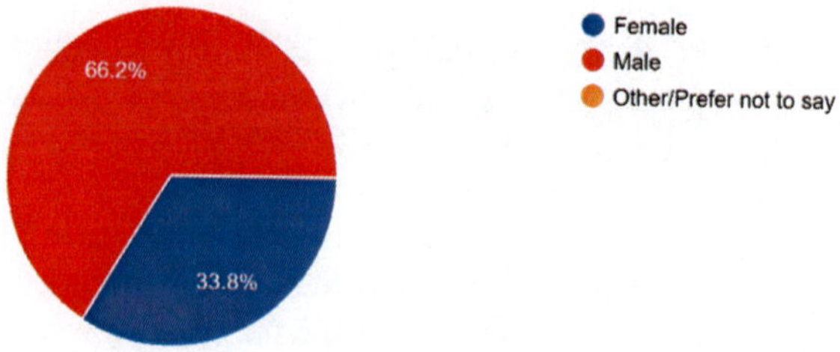

Fig. 2. Research participants by gender

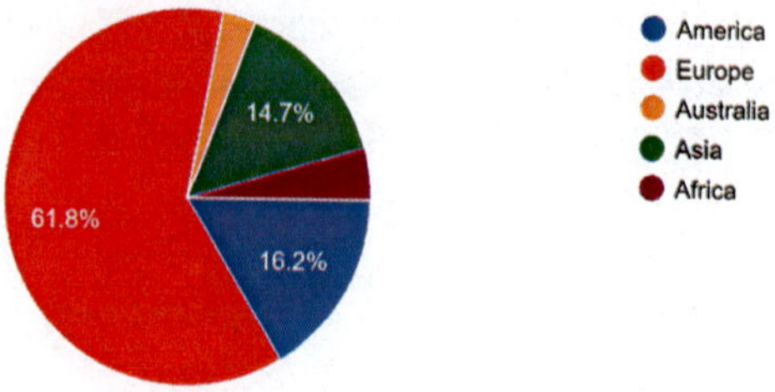

Fig. 3. Research participants by residency

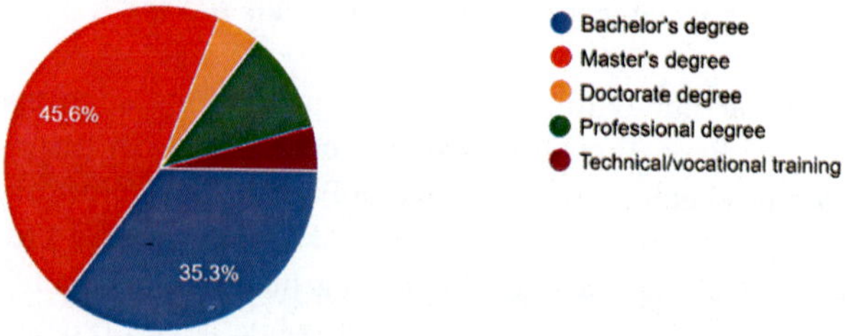

Fig. 4. Research participants by highest education level

Finally, they are also diverse with respect to how often they use a mobile device (i.e. smartphone, tablet) to complete any activities of the MOOC that they follow. As can be seen in Fig. 5, one out of three participants claimed that used their mobile devices often or always for that purpose, while two out of three use it seldom or never.

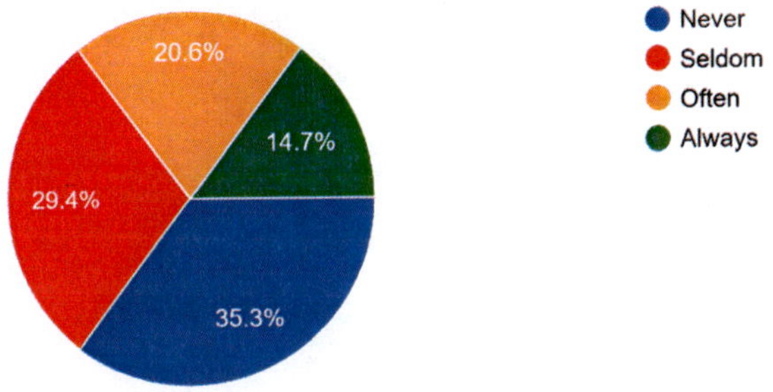

Fig. 5. Research participants by the use of a mobile device to complete MOOC activities

5.2 Preferences with Respect to a Mobile MOOC Format

As can be seen in Fig. 6, almost 4 out of 10 participants would prefer a mobile MOOC format over a non-mobile, traditional format for the MOOC that they have completed. It is also worth noticing that only almost 2 out 10 participants would prefer the traditional MOOC format.

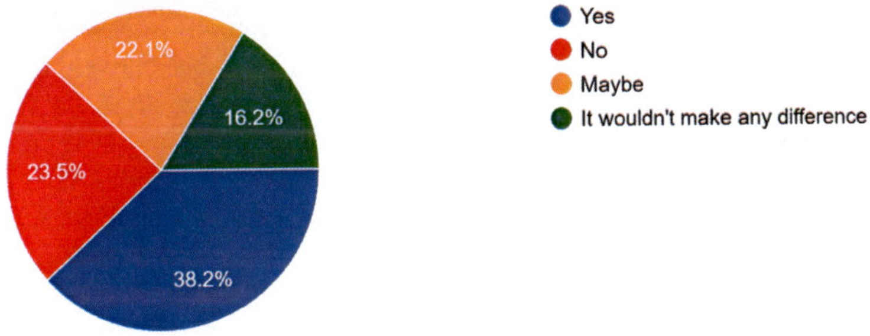

Fig. 6. Research participants by preference of a mobile MOOC format

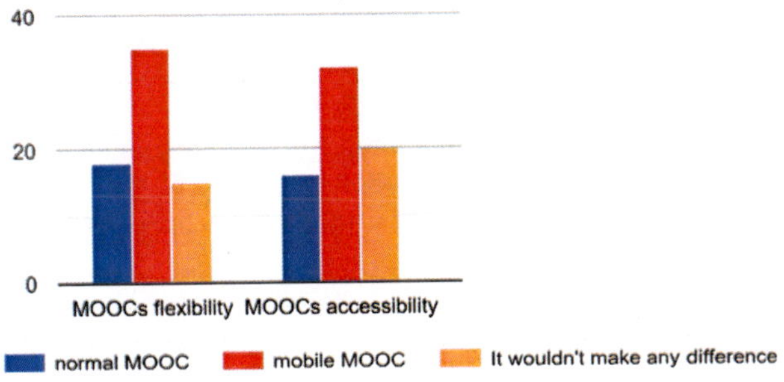

Fig. 7. Answers on preference with respect to flexibility and accessibility

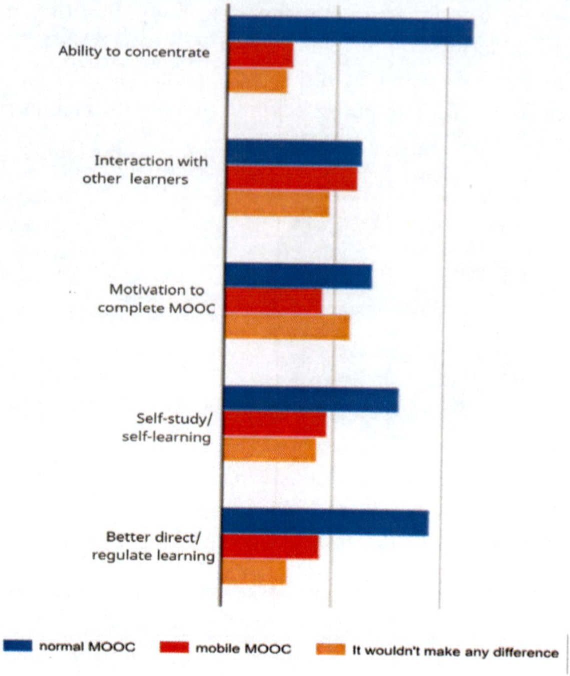

Fig. 8. Answers on preference with respect to several learning design aspects

Mobile MOOC is the preferred format with respect to both flexibility and accessibility (Fig. 7), whereas non-mobile MOOC is clearly the preferred format with respect to the learners' ability to concentrate and to be able to self-study as well as to self-regulate their learning more effectively (Fig. 8). There is no preferred format when it comes to the participants' interaction with other learners; and finally, there is a small tendency towards mobile MOOC when it comes to learners' motivation to complete the course.

6 Conclusions and Discussion

The aim of this study was to reveal the preferences of the learners between mobile and non-mobile MOOC with respect to influential parameters of the learning experience. In particular, with respect to flexibility, accessibility, ability to concentrate, motivation to complete the MOOC, self-study/self-learning, and the ability to better direct or regulate one's learning. The results can be categorized into three groups. Firstly, the results are rather inconclusive with respect to the learners' interaction, their motivation to complete the MOOC, and to self-study or to learn independently. Secondly, the results indicate that mobile MOOC is the preferred format in conjunction with the flexibility and the accessibility aspects. In this category the non-mobile format is slightly ahead in terms of preference, but there are numerous learners for which the MOOC format wouldn't make any difference. Thirdly, the non-mobile, traditional MOOC format is clearly the preferable one when it comes to the ability to concentrate and to better regulate one's learning. In addition, the study provides some insight on what extent MOOC learners actually use mobile devices for their learning, something that is lacking from the current literature.

On the one hand, it is not surprising that the participants of this study think that accessibility and flexibility are better supported via a mobile MOOC format. These perceptions are in line with the literature that suggests that advances of technologies like ubiquitous computing have had a positive impact on the scope and the accessibility of mLearning. On the other hand, it seems that when it comes to learning design parameters that can deeply influence learning, the participants perceive the non-mobile format as being superior. This finding is partially in line with the work of the authors in [17] that conducted a comparison between mobile and non-mobile MOOC format and found that the latter was associated with slightly higher learning gains.

Yet, flexibility and accessibility are also important when it comes to creating inclusive learning environments. Thus, it seems that the creation of an inclusive mobile learning environment that would also support deep learning using MOOCs is a difficult problem for the learning designers. More research on that particular problem could lead to a better understanding of how learning designers can create more effective MOOCs.

A limitation of this study pertains to the fact that the participants were self-selected, and the sample is small. Yet, the sample of participants in terms of basic demographic characteristics resembles the pool of the MOOC participants, thus it can be considered as representative. More research is needed to understand whether these results can be generalised. If so, then we need to understand the reasons behind these answers.

An ensuing recommendation would be to create mobile MOOCs by thoughtfully integrating the already existing learning design principles for mobile learning with those that touch upon MOOC design. Two specific recommendations that emerge are: firstly to cater especially for the fact that it seems that learners think that it is more difficult to concentrate in a mobile MOOC. This is in line with the existing literature which posits that learners might get easily distracted in mobile learning. The literature suggests specific instructional techniques to alleviate this problem; these techniques could also be embedded in the learning design of the mobile MOOC. Secondly, the creation of specially designed scaffolds that aim to help learners better regulate their learning. These scaffolds could be embedded in the mobile MOOC learning environment and help learners enhance their metacognitive skills. There are solid theoretical works in the educational research field that cater upon metacognitive scaffolds that can be relevant in this problem. Existing applications like The MyLearning Mentor application that is mentioned in Sect. 2 could be also be useful.

The findings of these results can be relevant to several stakeholders, but mostly to the learning designers of MOOCs and to researchers that are interested in the intersection of MOOCs and mobile learning.

Acknowledgements. The authors would like to acknowledge that the study was carried out in the context of the "DevOps Competences for Smart Cities" project (Project No.: 601015-EPP-1–2018-1-EL-EPPKA2-SSA Erasmus + Program,

KA2: Cooperation for innovation and the exchange of good practices-Sector Skills Alliances).

References

1. Alario-Hoyos, C., Estévez-Ayres, I., Pérez-Sanagustín, M., Leony, D., Kloos, C.D.: MyLearningMentor: a mobile app to support learners participating in MOOCs. J. Univers. Comput. Sci. **21**(5), 735–753 (2015)
2. De Waard, I., et al.: Exploring the MOOC format as a pedagogical approach for mLearning. In: Proceedings of 10th World Conference on Mobile and Contextual Learning, pp. 138–145, October 2011
3. Goksu, I.: Bibliometric mapping of mobile learning. Telemat. Inform. **56**, 101491 (2021)
4. Jia, J., Zhang, B.: Design guidelines for mobile MOOC learning—an empirical study. In: Cheung, S.K.S., Kwok, L.-F., Kubota, K., Lee, L.-K., Tokito, J. (eds.) ICBL 2018. LNCS, vol. 10949, pp. 347–356. Springer, Cham (2018). https://doi.org/10.1007/978-3-319-94505-7_28
5. Kaypak, P.E., Canbek, P.N.G., Chih-Hsiung, T.U.: Mobile learning and MOOCs. Int. J. **8** (3), 01 (2017)
6. Krull, G., Duart, J.M.: Research trends in mobile learning in higher education: A systematic review of articles (2011–2015). Int. Rev. Res. Open Distrib. Learn. **18**(7) (2017)
7. Lee, D., Watson, S.L., Watson, W.R.: Systematic literature review on self-regulated learning in massive open online courses. Australasian J. Educ. Technol. **35**(1) (2019)
8. Machun, P.A., Trau, C., Zaid, N., Wang, M., Ng, J.: MOOCs: is there an app for that? expanding mobilegogy through an analysis of MOOCS and iTunes university. In: Proceedings of the 2012 IEEE/WIC/ACM International Conferences on Web Intelligence and Intelligent Agent Technology, vol. 3, pp. 321–325. IEEE, December 2012
9. Martin, F., Dennen, V.P., Bonk, C.J.: A synthesis of systematic review research on emerging learning environments and technologies. Educ. Tech. Res. Dev. **68**(4), 1613–1633 (2020). https://doi.org/10.1007/s11423-020-09812-2
10. Palalas, A.: Mindfulness in mobile and ubiquitous learning: harnessing the power of attention. In: Mobile and Ubiquitous Learning, pp. 19–44. Springer, Singapore (2018)
11. Palalas, A., Wark, N.: The relationship between mobile learning and self-regulated learning: a systematic review. Australas. J. Educ. Technol. **36**(4), 151–172 (2020)
12. Pegrum, M. : Future directions in mobile learning. In Mobile Learning Design (pp. 413–431). Springer, Singapore (2016)
13. Rodger, H., Glover, I.: The death of ‘mobile learning.’ In: Crompton, H., Traxler, J. (eds.) Mobile learning and higher education, pp. 94–103. Routledge, New York (2018)
14. Rohloff, T., Bothe, M., Renz, J., Meinel, C.: Towards a better understanding of mobile learning in MOOCs. In: 2018 Learning With MOOCS (LWMOOCS), pp. 1–4. IEEE, September 2018
15. Sanchez-Gordon, S., Luján-Mora, S.: Research challenges in accessible MOOCs: a systematic literature review 2008–2016. Univ. Access Inf. Soc. **17**(4), 775–789 (2018)
16. Stöhr, C.: Anywhere and anytime? An analysis of the use of mobile devices in MOOCs. In: Proceedings of INTED2017 Conference, pp. 8933–8943, March 2017
17. Zhao, Y., Robal, T., Lofi, C., Hauff, C.: Stationary vs. non-stationary mobile learning in moocs. In: Adjunct Publication of the 26th Conference on User Modeling, Adaptation and Personalization, pp. 299–303, July 2018
18. Zhu, M., Sari, A.R., Lee, M.M.: A comprehensive systematic review of MOOC research: research techniques, topics, and trends from 2009 to 2019. Educ. Tech. Res. Dev. **68**(4), 1685–1710 (2020). https://doi.org/10.1007/s11423-020-09798-x

An Extended Technology Acceptance Model in the Context of Mobile Learning for Primary School Students

Sadjad Eskandari and Juan Pedro Valente(✉)

Department of Languages. Information Systems and Software Engineering, School of Computer Engineering, Universidad Politécnica de Madrid, Campus de Montegancedo, s/n, Boadilla del Monte, 28660 Madrid, Spain
Sadjad.eskandari@alumnos.upm.es, jpvalente@fi.upm.es

Abstract. In recent years, the use of mobile learning (M-learning) has significantly increased in educational and academic settings with the growth of mobile technology (MT) and many studies have been realized in this field. But the literature reveals that most of the previous researches presented in the field of MT acceptance in high school and university. On the other hand, lack of sufficient research is clearly seen on the acceptance of MT and the use of the M-learning in primary school due to the corona pandemic and its negative effects e.g. the absence of learners in educational settings. Hence, this lack has motivated us to present an extended technology acceptance model (TAM) in the context of M-learning for primary school students. This quantitative non-experimental research has investigated the possibility of using TAM for identifying effective factors on M-learning acceptance by primary school students as a user acceptance model for academic intentions and determined these factors through testing TAM in the school setting. Experimental findings prove that perceived ease of use (PEOU), perceived enjoyment (PE), perceived convenience (PC), and perceived usefulness (PU) had effect on attitude toward using MT by primary school students. Also, continuance intention to use (CI) was directly influenced by perceived usefulness and attitude toward using (AT) constructs.

Keywords: M-learning · TAM · D-learning · Mobile technologies · Primary education

1 Introduction

Today, we live in a world where the acceptance of two important facts is inevitable: the rapid growth of technology, followed by the advancement, development, and diversity of learning tools. Second, the prevalence of coronary pandemic disease in the world and the occurrence of disruption in educational processes based on learners' presence. According to the two facts mentioned above, human beings need to apply novel approaches to learning with the benefit of new technology such as mobile learning technology (MLT).

M. E. Auer and T. Tsiatsos (Eds.): IMCL 2021, LNNS 411, pp. 269–280, 2022.
https://doi.org/10.1007/978-3-030-96296-8_25

Recently, the use of MLT in various fields of application, especially educational and research spaces, has attracted the attention of many mobile users due to its quick and easy access to modern science [1–3]. M-learning is now a well-known positive approach, enabling learners and educators to learn lessons and other educational goals at any time and place in educational settings over the past 20 years [4]. The main and key reason to choose virtual settings by learners in different applied areas lies in the fact that learners and educators often perform more motivating and interesting activities in the virtual environment compared to the traditional education system. Also, the previous studies show that there are three main advantages for applying the M-learning approach in educational settings including the facilitation of the learning process, the sharing of thoughts and ideas, supporting the possibility of learners to collaborate between each other, etc. [5]. Therefore, the application of MLT in educational environments has been especially welcomed among learners that are mobile user.

It can be said that the acceptance or rejection of technologies such as MLT by different users depends on different factors that reflect the failure or success of this technology. Thus, investigating the factors affecting the acceptance of the use of a new technology or its non-acceptance by different users in different fields of application, can attract the attention of researchers as a challenging issue. For this purpose, various models of technology acceptance (TAM) are employed by researchers during last years [5–8]. The focus of TAM is on the end-users of the technology. The most important goal of TAM is considered to employ the theory of reasoned action model (TRA) to show how the relationship between attitude and intention affects the end user's adoption of technology. In fact, the aim of TAM is considered explaining and predicting the TA behavior of the users [9, 10]. Regarding TAMs, literature review shows that most of the TAM studies involving M-learning focused on extending the TAM with external variables. Also, it is inferred that the current students are digital natives who got exposed to technology at earlier ages in their life [11].

So far, many studies have been conducted on the adoption of MLT in different learning environments such as universities, high schools, and research institutes [4, 7, 9, 11]. But, unfortunately, the volume of these studies is few on MLT and technology acceptance models usage in primary education for students. On the other hand, education in today's society has been affected by the negative effects of the corona epidemic. In other words, in today's world, educational institutions prevent the collective presence of learners in educational centers. We think that this educational deficit reveals the need to expand studies to provide efficient models for adopting MLT in primary schools, as well as other presence-based learning environments.

Hence, this paper proposed an extended TAM of quantitative non-experimental for MLT in primary education. The proposed model combines two intrinsic variables, namely perceived enjoyment (PE), perceived convenience (PC). Accordingly, we examined the factors influencing the acceptance of MLT in the school educational environment. Also, we tried to increase the acceptance percentage of this technology by identifying effective factors in the context of acceptance of MLT in primary education that can be lead to develop good teaching ways in M-learning.

Other sections of the paper are followed as follows: hypotheses and research model are presented in Sect. 2, then our extended TAM model is presented for M- learning in primary education in Sect. 3. In Sect. 4, results and finding of this research are reported. We attempted to provide a discussion of the obtained results in Sect. 5. In the final section, some key points are remarked, which concluded from the performed research.

2 Hypotheses and Research Model

So far, many studies have been conducted on the factors affecting the adoption of mobile technology among different personals. By reviewing these researches, it can be concluded that various factors can effect on percentage of the use of MT in the learning process. Some previous studies have stated that the predictive power of TAM is restricted to the productivity-oriented utilitarian systems due to its primary focus on extrinsic stimulus [12]. But other works emphasize the importance of intrinsic motivations that are often ignored by the TAM. In fact, they believed that intrinsic motivation is the motivation type leading to exhibiting a behavior that is optimal or desired by nature [13].

Hence, this paper integrated two separate intrinsic variables known as 1) PC and 2) PE into the proposed TAM model in order to better understand the M-Learning perception on the part of the students. On this basis, ten hypotheses are established, which presented as follows. Also, the extended TAM of quantitative non-experimental is depicted in Fig. 1.

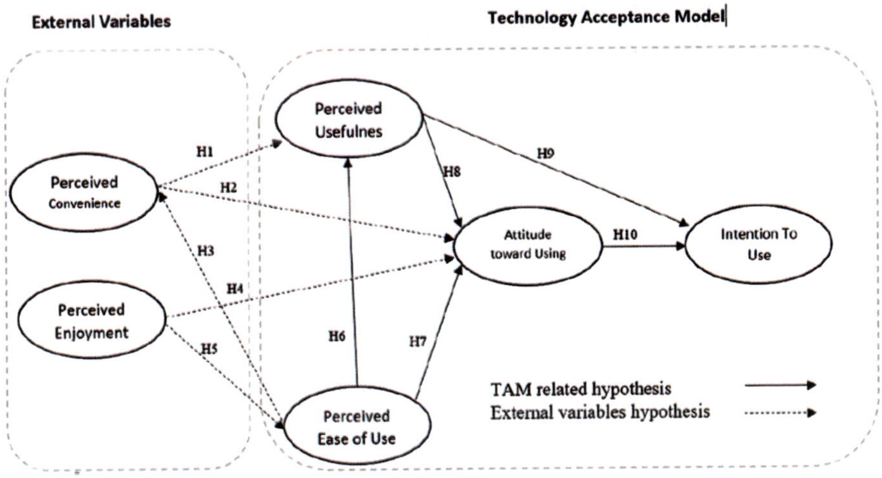

Fig. 1. The extended TAM of quantitative non-experimental.

- **H1**: PC has a significant positive affect on students' AT of mobile technologies in primary school curriculums.
- **H2:** PC has a significant positive affect on students' PU of mobile technologies in primary school curriculums.

- **H3**: PE has a significant positive affect on students' AT of mobile technologies in primary school curriculums.
- **H4**: PEOU has a significant positive affect on students' PC of mobile technologies in primary school curriculums.
- **H5**: PE has a significant positive effect on students' PEOU of mobile technologies in primary school curriculums.
- **H6**: PEOU has a significant positive effect on students' PU of mobile technologies in primary school curriculums.
- **H7**: PEOU has a significant positive effect on students' AT of mobile technologies in primary school curriculums.
- **H8**: PU has a significant positive effect on students' AT of mobile technologies in primary school curriculums.
- **H9**: PU has a significant positive effect on students' CI of mobile technologies in primary school curriculums.
- **H10**: AT has a significant positive affect on students' CI of mobile technologies in primary school curriculums.

3 Method

The main goal of this section is to present the methodology of our proposed model, which organized as follows:

3.1 Research Participants and Procedure

In this paper, students entered the study from only two schools in in Madrid, Spain. The reason for the selection of these schools is considered similarities in their size and regional characteristics that serve a similar student population. These schools have been selected for convenience, and given the similarities, it can be concluded that the findings can be generalized to other schools in the region.

Also, data was collected using Google Forms, an online review tool. The survey included criteria for measuring students' demographic information: final grade, age and gender. Participants participated voluntarily with no outcomes for not participating. The participants' parents firstly filled the informed consent form and expressed agreement for participation and then participants could see the items of the survey. The Google database was used for storing the gathered data. Then, the data was downloaded to the researcher's computer and loaded in IBM SPSS AMOS 24.

After designing the English version of the test, several experts translated it to Spanish version. Then 10 students were asked to complete a questionnaire. Students' opinions about the questionnaire were collected and subsequent corrections were made. An email was sent to all students in the participating schools. Online lists of email addresses provided by the school. The research email will be provided to the participants with a link at the end of the consent agreement form and display of items on the front pages. Following the expression of the agreement for participation and responding to the screening questions, the survey was presented to the participants. The online survey was constructed using Google Forms. Data was gathered from the students that completed the survey and it was employed in the analysis. The Google database was used for online data storage, and data was downloaded on the researcher's computer. Then, the IBM SPSS AMOS 24, SEM, including CFA are used to run.

3.2 Measure and Research Design

A five-point Likert scale was used for measuring all variables, which range from (1) totally disagree, (2) disagree, (3) no idea (neither agree or disagree), (4) agree, (5) totally agree. As Helen Doron (HD) software is used by the students for learning English through M-learning, the items were adopted according to it. In this paper, confirmatory factor analysis was performed for confirming that the items measure the constructs of the model. The number of the survey items related to each construct is given with the definitions below. We should mention that Appendix A provides the items pertaining to these constructs.

i. *PC*: The survey contained 4 items as indicators for measuring this construct.
ii. *PE*: The survey contained 3 items as indicators for measuring this construct.
iii. *PEOU*: The survey contained 4 items as indicators for measuring this construct.
iv. *PU*: The survey contained 4 items as indicators for measuring this construct.
v. *AT*: The survey contained 3 items as indicators for measuring this construct.
vi. *CI*: The survey contained 3 items as indicators for measuring this construct.

We followed research design used in other studies [14, 15] that have used the TAM presented in [16]. Also, a non-experimental quantitative analysis is conducted on the data collected from a cross-sectional survey of the chosen population in the selected settings.

3.3 Data Analysis

In this paper, a SEM path analysis is used to investigate the effects of hypothesized models in phylogenetic studies. This method is defined as a statistical technique that allows users to examine effect patterns in a system of variables [20]. Path analysis is usually employed along with TAM since correlation between the independent variables is crucial in specifying what predicts the purpose to adopt a system. In fact, to better understand the correlation between each independent variable and the dependent variable, one should be understood the effect of the independent variables on other items [21].

4 Results

In this section, we have provided the results of our the extended TAM evaluation. An online survey performed on Google Form was sent to 230 respondents between October, November and December 2020. 183 responses were received. Only 162 questionnaires were completed correctly. Incomplete questionnaires were removed and not included in the analysis. Evaluation result of hypotheses are shown in Table 1.

Table 1. Evaluation result of hypotheses

Hypotheses	Result
H1	PC has a positive and significant effect on students' AT of mobile technologies ($p < 0.001$, 35 = 0.358)
H2	PC has not a positive and significant effect on students' PU of mobile technologies ($p = 0.541$)
H3	PE has a significant positive affect on students' AT of mobile technologies ($p < 0.001$, ß = 0.472)
H4	PE has a significant positive effect on students' PEOU of mobile technologies (ß = 0.312, $p < 0.05$)
H5	PE has a significant positive effect on students' PEOU of mobile technologies (ß = 0.312, $p < 0.05$)
H6	PEU has a significant positive affect on students' PU of mobile technologies ($p < 0.05$, ß = 0.358)
H7	PEU have a positive effect on students' AT of mobile technologies (ß = 0.572, $p < 0.001$)
H8	PU has a significant positive affect on students' AT of mobile technologies (ß = 0.553, $p < 0.001$)
H9	PU has a significant positive effect on students' CI of mobile technologies ($p < 0.001$, ß = 0.378)
H10	AT has a significant positive effect on students' CI of mobile technologies ($p < 0.05$, ß = 0.418)

The SEM performed by IBM SPSS AMOS Graphics 24 to check the hypotheses then results are shown in Fig. 2.

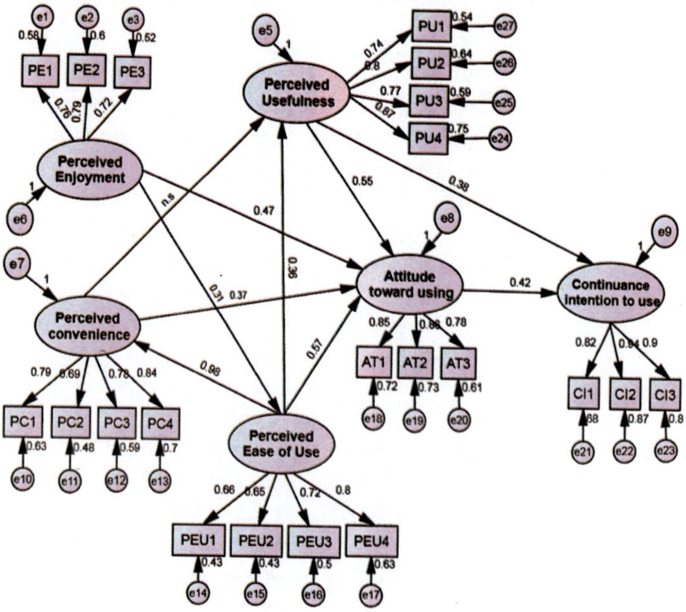

Fig. 2. SEM Model (IBM SPSS AMOS 24)

According to the final values evaluated using the CFA, the model was fit. A set of such indicators is used to evaluate the model fit. According to the objectives of the current research, we have used four incremental fit indices (TLI, CFI, NFI, IFI), 4 absolute fit indices (SRMR, Relative Chi-Square, Chi-Square, RSMEA), and one parsimonious fit index (AIC), which are the common fit indicators [22]. The values used to determine the fit are given in Table 2.

Table 2. Model fit indices with recommended level of fit

Recommended level of fit	
Absolute fit indices	
SRMR	<= 0.05 [22]
RSMEA	<= 0.06 [22]
Relative Chi-Square	<5 [23]
Chi-Square	Not significant at $p < 0.05$ [22]
Parsimonious fit index	
AIC	Smaller the value the better [22]
Incremental fit indices	
TLI	>= 0.95 [22]; >= 0.90 [23]
NFI	>= 0.95 good and .90 and 0.95 acceptable [23]
IFI	>0.90 [23]
CFI	>= 0.95 good [22] >= 0.90 acceptable [23]

The reliability values are estimated using Cronbach's alpha test for evaluation of internal reliability as shown in Table 3. Also, convergent validity is obtained using the average variance extracted in Table 4 for investigating effect of multiple items in measuring a construct.

Table 3. Reliability Statistics - CR and Cronbach's alpha measurements of all items

Construct	Items	CR	Cronbach's Alpha
PC	4	0.811	0.853
PE	3	0.915	0.796
PE	4	0.867	0.799
PU	4	0.870	0.865
AT	3	0.797	0.865
CI	3	0.798	0.861

As is seen in Table 3, the CR values obtained were higher than 6, indicating confirmation of the composite reliability.

Table 4. Discriminate validity and convergent

	AVE	AT	PU	PEU	PE	PC	CI
AT	0.591	**0.441**					
PU	0.782	0.388	**0.401**				
PEU	0.686	0.423	0.370	**0.633**			
PE	0.627	0.320	0.272	0.387	**0.382**		
PC	0.527	0.366	0.299	0.506	0.269	**0.517**	
CI	0.569	0.407	0.373	0.316	0.257	0.262	**0.594**

As is seen in Table 4, the results are greater than 0.5 (.591−.569), which confirms the convergent validity.

5 Discussion

In this research, the constructs included perceived usefulness, perceived ease of use, perceived convenience, attitude toward using, continuance intention to use, and perceived enjoyment. In the beginning, an informed consent form was presented to participants for reading and agreeing with participation prior to rating other items. Data was analyzed using confirmatory factor analysis, PLS-SEM. The data was gathered pertaining to 6 constructs employed in the model. The research aimed at determining whether the TAM/TAM2 can be applied for prediction of usage of mobile devices for academic intentions in the primary school setting.

In general, the fit index provided the appropriate model fit except for $\chi 2$ ($\chi 2$ = 442.572, df = 174, $p < 0.001$, CMIN/DF = 2.544, RSMEA = 0.000, SRMR = 0.045, AIC = 504.000, CFI = 0.999.123, IFI = 0.999, NFI = 0.998, TLI = 0.926). Due to the small size of the studied sample and the sensitivity of χ 2 the volume of the sample, the fit of one of the four absolute fit indices is absolutely suitable for continuation with the model.

The analysis of the results obtained from performing SEM techniques shows that exist six out of ten hypotheses with $p < 0.001$ and three with $p < 0.05$. Also, investigating these results indicates that only one hypothesis was not supported by the SEM method. H1, H3, H4, H7, H8, and H9 were confirmed with $p < .001$. H5, H6, and H10 were confirmed with $p < .05$. H2 was rejected. According to the SEM findings presented in this study, PC did not have a significant effect on PU. The PEU, PE, PC, and PU had effect on AT. These constructs did account for 58% of the variance of AT (R2 = .584).

As a general conclusion from the PERFORMED evaluation, it can be said that the proposed model in the field of MLT has provided acceptable performance to identify effective factors in an acceptable way.

6 Conclusions and Implications

Technology has grown significantly in various applied fields, especially in the context of mobile devices, and human beings are forced to adapt their living conditions to the different conditions of today's world. On the other hand, the emergence of new technologies in educational fields, leads human to use the novel learning approaches. For example, MLT has attracted the attention of many mobile device users as an efficient learning solution.

But what is important about the effectiveness of this technology among learners, identify effective and determinative factors in the percentage of acceptance and usage of MT and M-learning for academic aims in primary schools using effective TAMs.

In this paper, we have tried to provide an extended TAM related M-learning for primary school students. The main purpose of this model is to identify the factors influencing the acceptance of MT among primary school students in the corona period. We used SEM techniques such as CFA to analyze the results of this study. The findings indicate the acceptable performance of the extended TAM model in this field.

According to the performed research, there are six out of ten hypotheses with $p < 0.001$ and three with $p < 0.05$. Also, our research can be used as one of the pioneers of research on the adoption of technology in primary school.

Despite the many advantages in our developed model, some limitations can be seen in this model. Students attended from only two primary schools. This is a limitation on the sample population that threatens external validity. In the performed research, generalizability is limited because the respondents were randomly selected. We think that investigating each item can help improve the quality of our model in future.

Appendix A – Survey Instrument

Items to measure model constructs-Responses on 5-pint Likert scale from 1-strongly disagree, 2-disagree, 3-neutral (neither agree or disagree), 4-agree, 5-strongly agree.

PC:

- I can learn English at any time via the HD application
- I can learn English at any place via the HD application
- The HD application is convenient for me to engage in English learning.
- I feel that the HD application is convenient for me to learn English.

PE:

- I enjoy learning English via the HD application
- Helen Doron application is enjoyable and fascinating
- I have fun using the HD application

PEOU:

- My interaction with the HD application does not require a lot of my mental effort
- I would find it easy to get the HD application to do what I want it to do.
- My interaction with the HD application would be clear and understandable.
- I find the HD application to be easy to use.

PU:

- Using the HD application would enable me to accomplish my learning English more quickly.

- Using the HD application would improve my performance in learning English.
- Using the HD application would enhance my effectiveness in learning English.
- I find the HD application to be useful in my learning English.

AT:

- Learning English via the HD application is a good idea
- I would like to use the HD application
- In my opinion, using the HD application is very favorable

CI:

- In the future, I would like to learn English via the HD application
- In the future, I predict that I will learn English via the HD application
- In the future, I plan to learn English via the HD application

Table A.1 Research variables

	In/Dependent	Adapted from		
PC	Dependent/ Independent	[17]	PC1	I can learn English at any time via the HD application
			PC2	I can learn English at any place via the HD application
			PC3	The HD application is convenient for me to engage in English learning
			PC4	I feel that the HD application is convenient for me to learn English
PE	Independent	[19]	PE1	I enjoy learning English via the HD application
			PE2	HD application is enjoyable and fascinating
			PE3	I have fun using the HD application
PEOU	Dependent/ Independent	[16]	PE1	My interaction with the HD application does not require a lot of my mental effort
			PE2	I would find it easy to get the HD application to do what I want it to do
			PE3	My interaction with the HD application would be clear and understandable
			PE4	I find the HD application to be easy to use
PU	Dependent/ Independent	[16]	PU1	Using the HD application would enable me to accomplish my learning English more quickly
			PU2	Using the HD application would improve my performance in learning English
			PU3	Using the HD application would enhance my effectiveness in learning English
			PU4	I find the HD application to be useful in my learning English

(*continued*)

Table A.1 (*continued*)

	In/Dependent	Adapted from		
AT	Dependent/ Independent	[16, 17]	AT1	Learning English via the HD application is a good idea
			AT2	I would like to use the HD application
			AT3	In my opinion, using the HD application is very favorable
CI	Dependent	[17, 18]	CI1	In the future, I would like to learn English via the HD application
			CI2	In the future, I predict that I will learn English via the HD application
			CI3	In the future, I plan to learn English via the HD application

References

1. Pirhonen, A., Rousi, R.: Educational technology goes mobile: Why? a case study of Finland. In: Mobile Devices in Education: Breakthroughs in Research and Practice, pp. 171–179. IGI Global (2020)
2. Al-Dmour, R., Dawood, E.A.H., Al-Dmour, H., Masa'deh, R., et al.: The effect of customer lifestyle patterns on the use of mobile banking applications in Jordan. Int. J. Electron. Marketing Retailing **11**(3), 239–258 (2020)
3. Bernacki, M. L., Greene, J. A., Crompton, H.: Mobile technology, learning, and achievement: advances in understanding and measuring the role of mobile technology in education. Contemporary Educ. Psychol. **60**, 101827 (2020)
4. Klimova, B.: Impact of mobile learning on students' achievement results. Educ. Sci. **9**(2), 90–112 (2019)
5. Lai, J.W., Bower, M.: How is the use of technology in education evaluated? a systematic review. Comput. Educ. **133**, 27–42 (2019)
6. Mugo, D. G., Njagi, K., Chemwei, B., Motanya, J. O.: The technology acceptance model (TAM) and its application to the utilization of mobile learning technologies. J. Adv. Math. Comput. Sci., 1–8 (2017)
7. Scherer, R., Siddiq, F., Tondeur, J.: The technology acceptance model (TAM): a meta-analytic structural equation modeling approach to explaining teachers' adoption of digital technology in education. Comput. Educ. **128**, 13–35 (2019)
8. Al-Emran, M., Mezhuyev, V., Kamaludin, A.: Technology acceptance model in m-learning context: a systematic review. Comput. Educ. **125**, 389–412 (2018)
9. Marangunić, N., Granić, A.: Technology acceptance model: a literature review from 1986 to 2013. Univ. Access Inf. Soc. **14**(1), 81–95 (2014). https://doi.org/10.1007/s10209-014-0348-1
10. Granić, A., Marangunić, N.: Technology acceptance model in educational context: a systematic literature review. Br. J. Edu. Technol. **50**(5), 2572–2593 (2019)
11. Al-Emran, M., Elsherif, H.M., Shaalan, K.: Investigating attitudes towards the use of mobile learning in higher education. Comput. Hum. Behav. **56**, 93–102 (2016)

12. Roca, J.C., Gagné, M.: Understanding E-lEarning continuance intention in the workplace: a self-determination theory perspective. Comput. Hum. Behav. **24**(4), 1585–1604 (2008). https://doi.org/10.1016/j.chb.2007.06.001
13. Deci, E.L., Ryan, R.M.: Conceptualizations of intrinsic motivation and self-determination. Intrinsic Motivation and Self-Determination in Human Behavior, pp. 11–40 (1985). https://doi.org/10.1007/978-1-4899-2271-7_2
14. Almaiah, M.A.: Acceptance and usage of a mobile information system services in University of Jordan. Educ. Inf. Technol. **23**(5), 1873–1895 (2018). https://doi.org/10.1007/s10639-018-9694-6
15. Raza, S.A., Umer, A., Qazi, W., Makhdoom, M.: The effects of attitudinal, normative, and control beliefs on m-learning adoption among the students of higher education in Pakistan. J. Educ. Comput. Res. **56**(4), 563–588 (2017)
16. Davis, F.D.: Perceived usefulness, perceived ease of use, and user acceptance of information technology. MIS Q. **13**(3), 319–340 (1989)
17. Lai, H.C., Chang, C.Y., Wen-Shiane, L., Fan, Y.L., Wu, Y.T.: The implementation of mobile learning in outdoor education: application of QR codes. Br. J. Edu. Technol. **44**(2), E57–E62 (2013)
18. Huang, L.: Acceptance of mobile learning in classroom instruction among college English teachers in China using an extended TAM. In: 2017 International conference of educational Innovation through technology (EITT), pp. 283–287. IEEE (2017)
19. Leong, L.W., Ibrahim, O., Dalvi-Esfahani, M., Shahbazi, H., Nilashi, M.: The moderating effect of experience on the intention to adopt mobile social network sites for pedagogical purposes: an extension of the technology acceptance model. Educ. Inf. Technol. **23**(6), 2477–2498 (2018). https://doi.org/10.1007/s10639-018-9726-2
20. Garson, G.D.: Path Analysis. Statistical Associates Publishing, Asheboro (2013)
21. Hair Jr, J. F., Sarstedt, M., Ringle, C. M., Gudergan, S. P.: Advanced issues in partial least squares structural equation modeling: SAGE publications (2017), ISBN: 9781483377391
22. Bryne, B.M.: Structural Equation Modeling with AMOS: Basic Concepts, Applications, and Programming. Routledge, New York (2016)

Student Acceptance and Perceptions of Mobile Learning: An Introspection to the Pedagogical Exigencies and Psycho-Physical Hazards of Student Community

Sherine Akkara, Jiby Jose E(✉), and Ebin V. Francis

Hindustan Institute of Technology and Science, Chennai, India
rp.20603017@student.hindustanuniv.ac.in

Abstract. Digital learning has been ubiquitous for almost all students in the new normal following the pandemic that has padlocked all schools and universities. With the second wave that claimed more than three lakh lives in India, online classrooms have replaced the conventional classrooms that once used to be the abode of knowledge consumption and generation. This transitional shift has affected the perception of both teachers and students towards the teaching-learning process in digital learning mode. M-learning has influenced the entire globe as it eases mobility, affordability, and accessibility. The current study has explored the students' perceptions of mobile learning in various remote parts of southern India. The researchers used descriptive research to understand the nuances of the mobile learning process. The respondents of the research were chosen through the purposive sampling technique, and the data was collected from 489 college students in the outskirts of the southern states of India. A questionnaire that is based on student happiness and frustration with mobile learning was circulated online, and the student responses were analyzed using statistical tools. Students acknowledged comfortable learning conditions, seamless communications, and effective time management, while network volatility, unilateral interactions, and decreased focus were found to be the sources of student grievances. The findings of this report divulge the quality of mobile learning, the acquiescence of this remote learning method, sequels of m-learning in the scholastic life, and the innumerable psycho-physical repercussions of this pedagogical practice which proposed restricted usage of mobile phones.

Keywords: COVID 19 · Mobile learning · Student perceptions · Impact of M-Learning · Digital transformation

1 Introduction

The recent educational environment unraveled different versions of fringe benefits and challenges. The pandemic and perilous season generated multifaceted impacts concerning the visualization and execution of mobile learning and teaching. The forced shutdown of the educational system brought the culture of e-learning, and educators were forced to deviate from traditional methods to an online mode of teaching culture

M. E. Auer and T. Tsiatsos (Eds.): IMCL 2021, LNNS 411, pp. 281–292, 2022.
https://doi.org/10.1007/978-3-030-96296-8_26

(Aleksandar 2020). The practice and procedures of online teaching and learning conceded unprecedented effects in the academic scenario. The precarious situation invoked ingenious contributions in teaching-learning practices. The digital transformation in imparting knowledge witnessed a magnificent exposition of creativity and inventiveness. However, this buoyant picture did not make an exemption to the defeatist nature. The system was exposed to harsh effects of monetary and technical issues, and raised challenges like interaction with the instructor, time to respond, and absence of traditional classroom and on-campus socialization [1]. The pandemic has turned old learning techniques and traditional education systems by exposing varied dimensions of students' learning. Smart devices and mobile phones govern human lives, from entertainment and communication to learning methods. A study by Ofcom states that handheld devices, including smartphones and tablets, are now the most preferred gadgets for accessing the internet, and with the easy availability of internet connection and incremental improvements in both the design and affordability of mobile devices, it is highly preferred for learning processes. Mobile learning postulates the possibility of learning whenever and wherever you want, as long as you have a modern mobile device connected to the Internet [31].

The paper disentangles the challenges involved in mobile learning precisely among the students of rural areas who are more likely sequestered and without access to technological advancements. The study also peruses the responses and perspectives of rural students towards the new-fangled m-learning practices. This academic exercise becomes an instrument in understanding student perspectives on mobile teaching-learning experience and to bring about transformation in devising better online platforms by experimenting on the gaps that exist in the availability of platforms for teaching learning and the ease of using the platforms. The paper unravels students' perspectives regarding the user-friendliness of the system, how it replaces traditional teaching-learning practices, and sets standards for excellence. The study results unveil the failure of the existing remote learning system to promote whole-person development and suggest an urgent need to transform the prevalent learning method into user-friendly, interactive, and student-centered teaching-learning mechanisms. Also, it focuses on identifying the gaps between the traditional method and the remote learning experience of the students. It throws light into the urgency of framing techniques that cater to the needs of both teachers and students for a better experience of teaching and learning. It guides the educationalists in devising suitable systems of remote teaching-learning methods based on participatory, collaborative, user-friendly approaches. It also advocates policy level changes required from governments on education considering the emergencies that may rupture the envisioned values that education will bring to the life of the student community.

1.1 M-learning and Higher Education During COVID-19

The pandemic had a significant impact on higher education, students' academic work, and lives. The number of students required to stay home during the pandemic was 1.59 billion (UNESCO 2020). All the higher educational institutions faced the challenge of continuity of the teaching-learning process without physically attending teachers and students. The only solution for the problem was to depend on online platforms and support the students through an online mode (Pravat 2020). Social networking sites like Twitter, Instagram, Facebook, WhatsApp, and online platforms such as Google meet, Zoom, WebEx, etc., were beneficial to the teacher and students to have supportive, collaborative learning with knowledge sharing (Elumalai 2020). This transition was sudden, and the effectiveness and satisfaction differed from person to person [33]. The online platforms gained popularity and acceptance due to the ease of use, the flexibility of learning, and the controllable environment. It helped the educators to use these platforms for teaching, assessing, and evaluating the students (Mohammad 2020). M-learning can be perceived as the natural evolution of e-learning with effective communication and personalized mechanisms or a powerful platform for distant learning (M. Al-Emran and K. Shaalan 2015). Mobile devices can be used as an effective learning tool because of their mobility and approachability.

1.2 Effectiveness and Benefits of M-learning During COVID 19

M-learning enhanced the educational process during an emergency time like the COVID 19 pandemic situation. It provided flexibility in delivering education and accessing the content and resources though it lacked face-to-face learning (Bakia 2012). [10] says that the use of m-learning in the teaching and learning process would be a more natural and effective way of learning for this generation. A study conducted by [20] and his teams says that students use mobile devices mainly for learning, social interaction, entertainment, and work. Mobile usage in the education scenario presents many opportunities as well as many challenges.

M-learning allows learners to adapt existing mobile features to meet their needs, develop their interests, and construct their own learning (Crompton 2016). M-learning helps in increasing the active learning process. The use of mobile devices permits the students to move and access content and information from anywhere. They are also considered a tool for accessing content by storing it locally on the device or in cloud services (Hamidi & Nitschi, 2018, 2013). Having the flexibility to learn anywhere at a feasible time is considered the greatest advantage of m-learning. It enables knowledge building by learners. The context of mobile learning surpasses time and space constraints. M-learning can take learning outside the classroom, far from the reach and supervision of the teacher, which generates resistance to the use of m-learning by educational institutions and teachers. (Barren 2014). The digital transformation has enabled the students to acquire the study materials on their smartphones, get information online to meet their requirements with the help of learning management systems, access academic databases, etc. The works of Masiu & Chukwuere [23] have emphatically stated that the smartphone has made students' academic lives easier, as the candidates can access the information on the gadgets through electronic learning [e-learning] and mobile learning [m-learning].

1.3 Problems and Challenges of M-learning During COVID 19

An article published in India Today (2020) showed that young people, school children, used their mobile phone six h/day [hours per day] on average before lockdown, increasing to over eight h/day on average during the lockdown. A study conducted by

Mehdipour and Zerehkafi [2013] says about the different technical, social, and educational challenges in m- learning. Technical challenges for m-learning include connectivity and battery life, screen size and key size, having required bandwidth for nonstop/fast streaming, multiple standards, multiple screen sizes, multiple operating systems, reworking existing e-learning materials for mobile platforms, and limited memory. Social and educational challenges for m-learning include accessibility and cost barriers for end-users, digital divide, security or pirating issues, no demographic boundary, disruption of students' personal and academic lives, and risk of distraction.

An online article by Hindustan Times (2020) reported that the longer hours children spent online classes using mobile phones affect their health in different ways, including headaches, eye problems, and stress. As children depend more on mobile phones, they get addicted to this device, which is worse than cannabis addiction. The study also reported that increased mood swings could be observed in the children due to the increase in screen time. An article written by Balram [2020] about the usage of small-screen devices in e-learning states that the usage of mobile phones affects the sleep cycle and also the physical and mental health of children. The increase in screen time usage for other activities along with academic activities tends to increase anxiety and depression. Nandy [27] says that the mental health of the children is highly affected due to the prolonged online classes, and they succumb to health issues like eye strain, headache, and fatigue due to lengthy screen time interaction. Physical health is also deteriorating in many aspects (Narayana, 2020). The troubles of eye strain, obesity, mood swings, depression, etc., are treated as part of excessive screen time (Bruce 2020).

2 Methods

The study was focused and conducted among the students of rural areas who were challenged with inaccessibility to technological headway. The study's major objectives were to understand the problems of the mobile learning experience of the rural students, know their level of satisfaction in m-learning, and analyze the physical and psychological problems among the students due to the increase of screen time. Through the quantitative method of research using descriptive design, researchers tried to understand the problems and needs of online learning. Using the purposive sampling technique, a Student Satisfaction Index/scale based on various criteria was sent to 489 [N = 489] students, and responses were collected through a google form. Statistical tools of Percentile and Carl Pearson's Correlation were used for data analysis and interpretation. Inventory on physical problems and inventory on psychological problems were developed. The gravity of the physical and psychological problems was assessed on a five-point scale of "Nil-Mild-Moderate-Severe and Acute/ Profound." Researchers consulted practicing psychiatrists, psychologists, and other professionals who are working with people having psychological problems during COVID and prepared the inventory on physical and psychological problems. Ethical considerations of the research were keenly taken into account during all the phases of research, especially during data collection by obtaining the consent of the respondents for the data collection, and confidentiality of the data was maintained to the maximum. The dignity of the respondents was given priority by avoiding questions that might hurt the sentiments of the respondents.

3 Results and Discussion

3.1 Level of Satisfaction of M-learning Among the Students

This statistical reflection became an eye-opener for all of us as it depicted students' level of satisfaction regarding the m- learning process. 27.20% of the total respondents belonged to the average level, 22.10% low, and 8.20% very low level of satisfaction, which pointed towards the need to rethink and restructure the existing mobile teaching-learning process. Only 26.40% of the respondents treated m-learning seriously, and 16.1% of the respondents are highly satisfied with the existing process. It required the necessity to transform the existing process into a participative and collaborative system of learning.

3.2 Academic Support

5.3% of respondents strongly agreed, and 43.8% of students agreed that proper academic support was given to the students during m-learning. 40.9% of students were neutral on the academic support given to them, 5.1% of the respondents strongly disagreed, and 4.9% of students disagreed that they were not given proper academic support. The varied reasons for the poor academic support included ignorance of the technical aspects, poor quality of data network, and unplanned lectures.

3.3 Meeting Academic Expectations

Only 8.8% of the respondents strongly supported this concept. 29.7% of the total respondents were of the opinion that the academic expectation was met properly in the mobile learning process. 45.6% of the respondents were neutral on their views on meeting the academic expectation, 11.2% of the respondents said online teaching did not meet their expectations, and 4.7% strongly disagreed with it. Poor quality of data network, lack of involvement, quick completion of syllabus, and lack of face-to-face communication were the major reasons for the poor materialization of academic expectation.

3.4 The Technological Expertise of Students

5.9% of the total respondents were excellent in the field of technological expertise, 32.7% had high expertise, 51.7% proved to be average in the field, 6.5% had low expertise, and 3.2% had no technical know-how regarding the digital device and its usage. As the majority of the respondents belonged to average or poor expertise in technological know-how, m-learning had severe implications on academic growth, holistic development, and self-confidence.

3.5 The Inability of M-learning to Promote Cohesiveness

Only 6% of the respondents agreed that they had very high comprehension and cohesiveness, 37.2% had high comprehension, 44.8% were neutral on the concept, 3% said low cohesion, and 9% said there was no proper cohesion in mobile learning. The majority of the students felt that due to the lack of face-to-face interaction, inability to listen to the teachers for a longer time using gadgets, distractions from the family, and lack of proper academic environment, mobile teaching and learning failed in meeting the purpose.

3.6 Lack of Interest in M-learning

10.8% of the respondents had a very high interest in m-earning, 16.8% had high interest, 44.6% had average interest, 13.9% had low interest, and 13.9% had a meager interest in m-learning. Most of the students had no interest or little interest in m-learning as it did not become attractive, and there were more attractions on other technical usages than learning.

3.7 Engagement and Involvement

Only 9% of the respondents were actively involved in m-learning. 16.8% of the respondents somehow managed their involvement, but 53.4% of the respondents had minimum or no involvement in m-learning. 12.8% of respondents had fair involvement, and 8% of the respondence had poor involvement in m- learning. The major reasons were; lack of interest in learning, different online games as the substitutes for online classes, and the implicit assignments, tests, and seminars.

3.8 The Inability of Promoting Intellectual and Holistic Growth

M-learning, to a great extent, failed to promote the intellectual growth and holistic development of the students. 13.5% of the respondents had very high growth and development, 19.8% had high growth and development, 46.6% were neutral in their opinion, 4.8% had low growth, and 15.3% had no growth and development. Scholastic empowerment of the students did not occur as m-learning could not impact much on students' interest and involvement in the online class.

3.9 Screen Time of the Respondents

5% of the respondents had 0–3 h of screen time, 28% of the respondents used their mobile for 3–6 h, 42% of the respondents used their mobile 6–9 h per day, and 25% of them used the device 9–12 h. Screen time of the respondents varied from the range of 3–12 h per day. The majority of the respondents' screen time was between 6–9 h as they spent four hours of online classes, two hours of homework and self-study, and another two hours for mobile games and accessing social media. 42% of the total respondents spent a minimum of eight hours per day on mobile, which might cause severe implications on their physical and psychological growth and development. 25% of the respondents [male] spent more than 10 h a day on online gaming, social media, homework given by the teachers, self-study, and online classes.

3.10 Physical Problems

The researchers prepared an inventory for understanding the gravity of the physical problems of the students during the pandemic, which was set on a 5-point scale measuring acute, severe, moderate, mild, and no physical problems.

Table 1. Showing Health issues due to m-learning and allied activities

Sl. No	Physical Problems faced	Nil [%]	Mild [%]	Moderate [%]	Severe [%]	Acute [%]
1	Eye problems	10	13	28	35	14
2	Neck Problems	13	17	29	29	12
3	Backache	16	14	24	36	10
4	Headache	14	09	28	35	14
5	Nervousness	08	14	21	35	22
6	Sleep disorders	15	05	29	32	19
7	Muscular pain	14	10	18	38	20
8	Fatigue	07	08	26	29	30
9	Obesity	18	13	32	24	13

The table shows the urgency of immediate response towards the physical problems faced by the respondents. Lack of physical exercise, longer hours of mobile use, changes in the sleeping pattern, radiation from the device, inappropriate learning practices, eating patterns, etc., are the major reasons for the physical problems faced by the respondents. Increasing screen time, blue light emission, concentrating on the screen for a long time has caused problems.

3.11 Psychological Issues

Significant symptoms of mental ill-health were taken for the data collection, and a mental health inventory on a five-point scale of 0–4 was prepared by the researchers and circulated among the students.

Table 2. Showing psychological problems due to m-learning and allied activities

Sl. No	Mental issues faced	Nil [%]	Mild [%]	Moderate [%]	Severe [%]	Profound [%]
1	Anger	15	18	24	26	17
2	Mood swings	12	20	23	29	16
3	Stress	18	20	24	24	14
4	Anxiety	12	19	25	27	17
5	Irritability	10	16	26	30	18
6	Hatred	20	24	22	22	12
7	Intolerance	10	15	27	30	18
8	Addictions	28	20	18	25	9
9	Lack of interaction	10	17	23	37	13
10	Social isolation	11	15	23	35	16

Behavioural changes were noticed among the students due to m-learning and the possible impacts on their psychological development. The majority of the students undergo mental trauma as m-learning creates psychological imbalances due to the long hours of mobile usages. Many times, respondents have shown symptoms of acute stress, anxiety, and depression which require immediate attention.

3.12 Correlations Between Cohesiveness and Satisfaction of Student-Friendly M-learning

The variables were analyzed with the help of the Carl Pearson correlation test. The cohesiveness of the students and satisfaction of the student-friendly m-learning were closely related [P = .493] at the significance level [p] of 0.000. The level of online teaching enhanced the cohesiveness of the students [7%], and the level of satisfaction in making the m-learning process student-friendly [7%] was related to each other. Since the m-learning was not student-friendly, the cohesiveness of the students was badly affected. As students found this mobile learning process very difficult to manage, they were dissatisfied and could not comprehend what was taught to them.

3.13 Correlation of Interest in Online Classes and Involvement and Engagement of the Student in Online Classes

The interest in mobile learning and involvement and engagement in the m-learning classes were related [P = .747] at the significance level [p] of 0.000. Interest in mobile learning would affect the participation of students in online classes. The results explained that the students were not interested in the m-learning process, and thereby, they had a low level of involvement and engagement in the same.

3.14 Correlation of Academic Expectation with Intellectual Growth

The academic expectation of the students and intellectual growth through mobile learning were correlated [P = .341] at the significance level [p] of 0.000. The students who achieved their academic expectations [4%] and the ability to experience intellectual growth by m-learning [15%] were related to each other. Therefore, from the derived result, it was clear that m- learning did not meet the students' expectations, and thus they lacked intellectual growth through the mobile-assisted learning process.

3.15 Correlation of Screen Time and Physical Problems

The screen time and physical problems encountered by the respondents were correlated [P = .584] at the significance level of 0.000. As the screen time increased, physical problems like eye problems, obesity, headache, back pain, neck pain, nervousness, sleep disorders, fatigue, muscular pain, etc., were high. The increase in screen time aggravated the physical problems of the respondents drastically.

4 Recommendations

4.1 Engagement and Interest

The students' ennui to attend the mobile-assisted learning process deteriorated the ambiance of classroom learning. Most of the students were not interested in m-learning due to other attractions of the internet, mobile games, social media, etc., and hence teachers needed to create interest among the students by making their classes interactive, attractive, and fun-filled. Activities could be incorporated to make the class interactive and interesting.

4.2 Holistic Growth

As the m-learning process was unilateral and mainly focused mostly on the subject coverage, it might not fulfill the holistic growth of the students. It was recommended to integrate more exercises to ensure the holistic growth of each student by the participatory teaching-learning process, counselling, mentoring, and involvement in different online social sensitization programs.

4.3 Technical Know-How

The majority of the students were not technically efficient in using online platforms; mainly to submit assignments, write exams, upload exam sheets, and so on. Proper training should be given to the students to ensure their maximum participation.

4.4 Screen Time

Students use mobiles on an average time of 8 h per day. It causes severe health issues and psychological problems. Thus, measures are to be taken to limit the screen time of the students by restricting the syllabus, controlling social media usage and gaming, etc.

4.5 Health Issues

M-learning and allied use of mobile for social media usage and online gaming cause serious health hazards due to overuse of the mobile for longer hours. Students, parents, and teachers are to be aware of the problems that may occur due to the excessive use of mobile for learning and restrict the use of the usage of mobile. Alternative measures like yoga, meditation, physical exercises, proper diet control practices, etc., are to be resorted for solving health issues.

4.6 Psychological Issues

Psychological issues are to be taken seriously, and proper care should be taken for behavioural changes and emotional imbalances. Counselling services should be made available to the students to solve their psychological problems.

5 Conclusion

The current pedagogical practices and learning strategies have created increased ingression of mobile learning, which evolve unparalleled and revolutionary effects in the academic exercises. Also, the challenging time and unpropitious effects ingrain an inclination to the overdependence on mobile phones and other digital devices for educational purposes. The m- learning strategies can, to a certain extent, reconcile the rescindments, aloofness, snapping off the on-campus socialization, etc., in the educational environment. However, the promising picture unravells hostile and imperilling aspects too. The m- learning process marks diminishing satisfaction among the students as it cannot substitute the traditional classroom environment and corroborate the holistic growth and development it promoted down the centuries. The practice also delineates the backlashes of physical and psychological predicaments. Since the current educational system demands an intense proclivity to cybernated and technological media, the mobile learning exercises are to be imbued more with quality techniques and approaches to promote academic curiosity, holistic growth, and intellectual erudition.

References

1. Anwar, K., Adnan, M.: Online learning amid the COVID-19 Pandemic: Students perspectives. J. Pedag. **1**(6), 45– 51 (2020)
2. Umek, L.: Impacts of the COVID-19 Pandemic on Life of Higher Education Students: A Global Perspective. Sustainability **12**(4) (2020). https://doi.org/10.3390/su12208438

3. The Health impacts of online learning (2020). https://nashuproar.org/36711/features/the-health-impacts-of-online-learning
4. Schmitz, L., Stanciu, C.: Online teaching and learning in higher education during the coronavirus pandemic: students' perspective. Sustainability **12**, 10367 (2020)
5. Nycz, M.: Learning Objects and E-Learning: An Informing Science Perspective, Interdiscip J.ESki." Lifelong Learn. 2 (2006)
6. Cheung, C., Cable, J.: Eight principles of elective online teaching: a decade-long lessons learned in project management education. Project Manage. World J. **6**, 1–16 (2017)
7. Get the facts about coronavirus (2020). https://www.cdc.gov/coronavirus/2019-ncov/index.html
8. C. " The Use of Mobile Learning in Science: A Systematic Review, J." Sci. Educ. Technol (2016): 160
9. Dhawan, S. " Online Learning: A Panacea in the Time of COVID-19 Crisis, Journal of educational technology systems, 49, 5–22." (2020):
10. Dingli, A., Seychell, D.: The New Digital Natives, Stuttgart, Germany (2015)
11. Engelbrecht, E.: Adapting to changing expectations: Post-graduate students experience of an e-learning tax program. J. Comput. Educ. **45**, 217–229 (2005)
12. Retrieved from article: Studies show students' smartphone usage became excessive during COVID-19 lockdown (2020)
13. Iqbal, S., Hatti, Z.A.: An investigation of university student readiness towards m-learning using technology acceptance model. Int. Rev. Res. 33 Open Distr. Learn. **16**(4) (2015). https://files.eric.ed.gov/fulltext/EJ1082185.pdf
14. Ifeanyil, P., Chukwuere, J.E.: The impact of using smartphones on the academic performance of undergraduate students. Knowl. Manage. E-Learn. **10**(3), 290–308 (2018)
15. Impact of Covid-19 on Higher Education in India, International journal of advanced education and research (2020). https://https://doi.org/10.31235/osf.io/jg8fr
16. "Impact of pandemic COVID-19 on education in India, Int J Curr Res, 12[7] (2020)
17. Students' Perception towards E-Learning duringCOVID-19 Pandemic in India: An Empirical Study. Sustainability **13**, 57 (2021). /https://doi.org/10.3390/su13010057
18. Electronic Learning Management System: Relevance, Challenges and Preparedness. J. Emerging Technol. Reservations **7**, 471–480 (2020)
19. Effect of mobile phones on Children's health, Innohealth magazine (2019)
20. Kukulska - Hulme, A., Viberg, O.: Mobile collaborative language learning: State of the art. Br. J. Educ. Technol., 218 (2018)
21. Psychological State of College Students During COVID-19, Epidemic, Lancet (2020)
22. Miller, C.: Trapped in a Double Bind: Chinese Overseas Student Anxiety during the COVID-19 Pandemic (2020)
23. Masiu, M., Chukwuere, J.E.: The effect of smartphones on students' academic life: a perceptive from a South African University. In: International Conference on Business and Management Dynamics ICBM-2018 (2018)
24. Mobile Learning for Education: Benefits and Challenge. Int. J. Comput. Eng. Res. **3**(6) (2013)
25. Muhammad, S., Salman, M.: COVID-19 Pandemic and environmental pollution: ablessing in disguise. Sci. Total Environ. **728**, 138820 (2020)
26. The impact of online learning during COVID-19: students' and teachers' perspective. International Journal of Indian Psychology **8**(2), 783–793 (2020). DIP:18.01.094/20200802
27. Nandy, Sukanya, 23 December 2020. https://www.educationworld.in/impact-of-online-classes-on-childrens-mental-health

28. Nitsche, K.: Acceptance of mobile, classroom-bound e-learning systems from the perspective of students and lectures. In: Proceedings of the 13th International Conference on Advanced Learning Technologies, Beijing, China (2013)
29. Online learning during lockdown period for covid-19 in India. International Journal of Educational Research 9 (2020)
30. Factors Affecting the Quality of E-Learning During the COVID-19 Pandemic from the Perspective of Higher Education Students, Journal of Information Technology Education: Research, 731–753 (2020)
31. Priscila, 17 July 2020. https://www.easy-lms.com/knowledge-center/lms-knowledge-center/mobile-learning/item10388
32. Report by UN. (4 August 2020). UN Chief Outlines' Bold Steps' for Education in the Face of COVID-19 Disruption. https://news.un.org/en/story/2020/08/1069442./08/1069442
33. Sahu, P.: Closure of universities due to Coronavirus Disease 2019 [COVID-19]: Impact on education and mental health of students and academic sta. IJRAR, **12**, e7541 (2020)
34. Stone, K.: Zoom for educators: How to set up virtual classrooms for distance learning (2020). https://getvoip.com/blog/2020/04/08/zoom-for-educators

Digital Didactics in Professional Education: Limitations, Risks and Prognosis

Aida Nurutdinova(✉)

Kazan (Volga Region) Federal University, Kazan, Russian Federation
AiRNurutdinova@kpfu.ru

Abstract. The relevance of the digitalisation in professional education is caused by the global transition to a digital economy and society. Building a digital economy and digital education are significant priorities of state policy in the Russian Federation. It is necessary to take into account that the education digitalization has two sides: firstly, the formation of a digital educational environment as a set of digital learning tools; second, the profound modernization of the educational process. The digitalization of educational process is a reciprocal transformation, on the one hand, and the digital technologies and tools used in the educational process, on the other. The aim of the educational process transformation is to make the fullest possible use of the potential didactic capabilities of digital technologies. The aim of digital technology transformation is to maximize its adaptation to effectively meet the pedagogical objectives. The expected educational outcomes of the digitalisation in professional education are related to identifying and maximising the potential of digital technology.

Keywords: Transition process · Electronic educational resources · Digital education · Reciprocal transformation · Educational transformation · Digital technologies

1 Introduction

Digital technologies make it possible to create an environment with a variety of educational resources, almost unlimited in content. Under these conditions, it is up to the individual (perhaps with some help from teachers, tutors and/or the adaptive learning systems) to solve a number of educationally important tasks, the first of which is to comprehend and formulate his/her own educational needs and on this basis to form an individual educational track. The first task is to think about and formulate one's own educational request and, on that basis, to create an individual learning path. In the conditions of distance learning online courses, the learner is required to be able to independently organise his or her learning activities at all stages of the educational process. Consequently, the digital learning environment is a complex of conditions and opportunities for human learning, development, socialisation and education.

Digitalisation, as a priority project of higher education, aims to improve the quality of professional education through digital technologies [1]. The regulatory process calls for higher education institutions:

M. E. Auer and T. Tsiatsos (Eds.): IMCL 2021, LNNS 411, pp. 293–305, 2022.
https://doi.org/10.1007/978-3-030-96296-8_27

- to update the education and training system;
- to form a digital learning environment;
- to introduce digital learning tools;
- to prepare a specialist in accordance with the requirements of the digital economy.

In this research paper, we view digital didactics as a process of constructing a set of digital educational technologies and teaching methods, electronic resources, which allow the budget and rapid implementation of integrative and competence-based approach to learning and the professional competencies formation and professional activities readiness. The digital education development is one of the top priorities worldwide [2–4]. Higher education is being integrated into "going digital" and transferring educational activities online, while analogue, traditional education is becoming the part of the past. Higher education institutions are at the stage of rethinking and upgrading the educational process and didactics, searching for effective digital educational technologies that will prepare graduates for the modern labour market requirements [5]. With digitalisation, the spread of telecommunication and networked technologies and learning tools, the subject matter of digital didactics is expanding considerably. This expansion takes place in the following directions:

- From teaching and learning limited to the classroom to learning in different environments and spaces including the networked and virtual world;
- From the learning process in an educational organisation to learning in an educational network and self-learning in an educational environment;
- From the teaching and learning activities to the processes of designing, shaping and mastering educational tracks;
- From teaching as the leading activity to the variety of pedagogical functions in the digital learning process.

Having analysed the online education resources, we found out that Russian universities are:

- actively modelling and creating a digital educational environment, the learning process digitalization (digital libraries, campuses, teaching laboratories, etc.);
- information and communication resources are enriched and expanded;
- new educational technologies, methods and techniques of work, interaction formats are developed and implemented;
- training and retraining of "digital" teachers, providing throughout the learning process.

2 Methods and Results

The empirical basis for the study was set of regulatory documents:

- the programme "Digital Economy of the Russian Federation";
- Presidential Decree No. 203 of 09.05.2017 "On the Strategy for the Development of Information Society in the Russian Federation for 2017–2030";

- the federal project “Modern Digital Educational Environment in the Russian Federation”;
- the materials of a number of international and national conferences and forums, analysis of foreign and Russian experience in online education;
- analysis of publications (articles, theses, interviews, research results), online platforms and distance learning systems.

We have studied the topical issues of digital didactics in professional education based on:

- the review of digital educational resources;
- analysis of professional community problems in the era of digitalization;
- the study of digital educational environment;
- the main trends of digital education in higher education.

The work was carried out based on the integrative-competence approach, which considers the multifaceted pedagogical process as a single whole, giving a new qualitative result, as an integrated characteristic, determining the readiness of a future specialist for professional activity and his/her professional competence [7].

The research methods are:

- bibliographical review and analysis;
- retrospective and terminological analysis;
- questionnaire and personal survey;
- monitoring, observation of students’ activity products.

The digitalization prerequisites and its evolutionary trajectory are considered in a number of sources:

- The process of informatisation and computerization and large-scale Internet connectivity took place in 1980–1990. Computers became a didactic component of the training process.
- From 2000 to 2006, information resources were increased, and distance education was introduced.
- Between 2016 and 2020, global telecommunications developed and digital educational services emerged.
- In the future, digital campuses in higher education institutions and a new generation are expected to be completed by 2035.

3 Discussion and Results

3.1 Professional Education Digitalisation: Limitations

1. Social inertia is a limitation related to the unpreparedness of society (public opinion) and its institutions (legislative system, public administration, training system) for the changes that the digitalization process entails. A particular case of this limitation is the unwillingness of pedagogical staff to quickly abandon traditional

pedagogical approaches and teaching methods and to "digitally transform" their professional activity. The overall effect of social inertia on the digitalization process is that change does not happen as quickly as its initiators would like it to. At the same time, the social and psychological mechanisms of social inertia are aggravated by a lack of resources for change, including lagging regulatory and methodological updating [8].

2. The importance of the human factor in the educational process is a limitation due to the impossibility of permanently excluding live interpersonal communication from the educational process and its complete automatization or transfer to a network format. Human beings are social beings, and they need a process of live communication to develop fully. Contrary to a recent popular belief, human beings, and not digital media, are the most powerful source of motivation in learning. The inefficacy of "machine learning" was exposed as early as the turn of the century 60s and 70s, by experiments in various countries that introduced the programmed-learning model [9, 10]. Under a programmed learning environment, the individual loses subjectivity and essentially becomes a machine-like element of an automated educational system, strictly acting according to a pre-determined algorithm. The use of modern digital technology, in itself, is only able to create an ad hoc, short-term motivation to learn. On the other hand, after exposure to online courses, a number of students have or become reluctant to attend classroom (traditional) classes [11].

The "human factor limitation" is least important for short programmes of vocational training and further vocational education, and most important for long programmes of secondary vocational education, which inevitably include educational (value-oriented) and personality-developmental components [13]. In addition, it should be taken into account that many universal competences can be fully formed only on the basis of personally significant, meaningful activity experience, obtained by the student in the real environment of human communication, full of emotions, struggle of interests, conflicts, requiring empathy, included reflection, immediate and accurate human reaction, etc. Similarly, developing professional competences in most cases requires immersing the learner in a real professional context, with its inherent interpersonal relationships, role positions, communications and interactions.

3. Practice-oriented - the requirement to organise the practical part of the educational process on a full-time basis in some professions and specific professional education, due to the necessity of personal contact between a teacher and a student to form complex professional skills and competences. In Russia, this restriction is enshrined in legislation. Being on the one hand a didactic principle and on the other hand a limitation, practice-orientation is the central factor which determines the specificity of the digitalisation process in professional education and training and the specific parameters of this process (orientation, dynamics, breadth).
4. The quality of technical resources which support the digital education process may be a significant limitation which limits its pedagogical effectiveness. For example, according to one survey conducted with students in the system of distance education, the biggest obstacle to their learning is internet interference (53.2%) and the quality of sound (20.6%) [14].

5. A set of health and hygiene restrictions requires that the educational process take into account the negative effects of digital technologies on the health, functional and emotional-psychological state. Negative results of work with computers include: decreased vision, various signs of fast fatigue, the occurrence of neurological symptoms, etc. [15]. Hygiene and sanitary restrictions are of particular importance when working with young children, including adolescents studying in vocational programmes.

3.2 Professional Education Digitalisation: Risks

There is a danger of the social institutions dehumanization, which in professional education is manifested as a loss of educational and personality-development goals and a focus on the narrow functional training of the future employee [16].

Computerization inevitably affects the social fabric, with great risks associated with:

- the phenomenon of direct competition between the human being and the computer in human-machine systems, which forces the human being either to be excluded from such systems or to become "computer-like";
- with the development of technocratic thinking, characterised by the domination of means over ends, and of technology over man;
- the spread of irrationalism, the loss of capacity to think critically and to perceive reality adequately against a background of information noise, flaming and massive misinformation dumping.

1. The risk of distorting thinking, attitudes and value systems.
2. Risk of excessive "digital optimism" - exaggerated possibilities assessment of digital learning environment, digital resources and learning tools, combined with underestimation of human factor significance in educational process. The flip side of "technocratic optimism" is always "humanitarian pessimism", where humans are seen as the least effective component of human-machine systems. As a result, at the design stage of such systems there is a tendency to displace humans (in the case of educational systems - the teacher, as well as opportunities for live communication between learners).

The transition to digital education is expected to lead to a significant reduction in the role of the teacher in the educational process with a dramatic increase in the importance of self-learning through digital technologies. At the same time, the digital educational environment, the system of online courses and other educational resources are considered as self-sufficient means to ensure high efficiency of the educational process. However, in the countries that are ahead of Russia in their development, the hopes for the transition to mass distance education have not been fulfilled.

According to a study by the University of Pennsylvania (2013), of all registered users on Coursera, from 27 to 68% have watched at least one lecture, and only from 2 to 14% have completed more than half of the course or the entire course [14]. Characteristic of this is the recognition by an international group of futurists that "human-to-human learning is likely to continue to be a key development process, the most effective form of learning for both guided and 'self-directed' learning".

3. The risk of replacing education with digitisation. Pedagogically ineffective 'digitised' didactic practice is characterised by a combination of the following:

- Firstly, the use in digitised form of traditional didactic elements of the educational process (class and lesson system, content, forms and methods of teaching, the former system of evaluation and control of knowledge) without any fundamental transformation;
- Second, the use of universal information and communication technologies that are not focused on solving specific pedagogical tasks;
- Thirdly, there is a lack of scientific understanding of the first two points.

Thus, the "digitized" didactic practice is based on the empirical mutual selection of available didactic support (content, forms and methods) and the most accessible information and communication technologies. A digitised textbook, for example, is a traditional, digitised text, with hyperlinks and links to external resources, but perhaps also "live images" - animations and video clips. Its use may be able to create a slightly higher learning motivation in today's children than in a conventional educational process, but this motivation is external and short-lived. Using an electronic textbook instead of a printed book considerably increases the strain on the eyesight. But the main problem with "digitised" didactic practice in this case is that the learning strategy of the textbook does not change, or changes for the worse, losing its humanistic component. The latter happens because the teacher, trusting in the power of the digitized textbook, increasingly withdraws from the educational process, making room for dialogue "learner-computer" [15].

Even the individualised learning opportunities that the digitisation of traditional educational media brings with it often have a negative impact on development: the learner is locked into a personal computerised learning environment, to the detriment of group forms of work. In contrast to "digitized" traditional didactics, digital didactics involve rethinking and essential transformation of the existing educational process and its elements (see Table 1).

Table 1. Contrasting the features of "digitised" didactic practice and digital didactics

Approach	Educational process objectives	Learning content	Dominant sub-process of learning	Learning forms	Learning tools
Traditional didactics	Traditional: learning certain social experiences represented in a didactically adapted form of "knowledge", "skills", "attitudes" (worldview)	Products of social experience represented in the iconic form of learning information ("knowledge")	Teaching	A dominance of frontal forms of classroom work and individual forms of independent work	Textbook, printed visual aids, occasional use of real objects
"Digitised" didactics (transitional phase)		The different ways of doing things to be learned		Frontal, with attempts at individualisation and a pronounced lack of group forms	Digital, occasional use
Digital didactics	Preparing to live effectively in a digital economy	Different ways of doing things to be learned	Learning	Dominance of group and individual forms of learning, dynamic forms	Digital (information and communication technology, pedagogical technology)

There is a risk of "digitization" learning in professional education, when using a set of simulators, simulators and other meta-digital technologies (software and hardware complexes), the process of professional competence formation is transferred from a real professional context to a virtual one.

Achieving the professional education goals - mastering professional competences, supporting the professional processes and personal self-determination, professional identification, socio-professional adaptation of a young person - requires a flexible combination of digital, material and pedagogical technologies [17].

4. The pressure of digital developers is due to a lack of activity in the education sector in the role of digital educational products. The digital products' developers, being unfamiliar with the educational process and having little understanding of pedagogical goals and didactic principles, view education in the superficial context of a "service". As a result, many promoted digital products that are positioned as educational do not aim to achieve meaningful pedagogical goals, but instead provide secondary objectives, sometimes not directly related to learning objectives.

Overcoming this risk requires a systematic analysis of the digital learning products' development for professional education:

– educational needs and goals;
– the digital generation features, the learners' and teachers' capabilities;
– actual and potential didactic properties of the different digital technologies;

- didactic principles and characteristics of the educational process in professional education and training.

The solution to this problem requires the introduction of a new professional position, that of methodologist-architect of digital learning tools, who acts as a qualified intermediary between the pedagogical community, familiar with didactics, and the developers of digital products. The main task of the Digital Learning Resource Architect is to identify the current gaps in the practice of teaching and to develop terms of reference, in a language that the developers can understand, for the development of digital learning tools that are actually needed to solve the pedagogical problems at hand. They should have an in-depth understanding of didactic theory and educational practice, a good understanding of digital technologies, including the most modern, systematic analysis and constructive communication skills [18].

5. The ethical digitalization risks in the educational process are primarily due to the accumulation of large amounts of personal information about learners (including that related to their health, individual psychological characteristics, value preferences, social contacts, success rate in various activities). In essence, all relevant aspects of a learner's life are tracked in the digital learning process. There are inevitably risks related to the transparency of this information for the different actors involved in the educational process (teachers, parents, administration, digital footprint analysts, other support staff) as well as its possible leakage. Thus, when developing digital learning platforms and systems, special attention should be paid to information security, both in technical and pedagogical terms (identifying who has access to what information, establishing appropriate contractual arrangements, etc.).
6. Management risks related to the process of digital education:

- Digitalization for the education purpose, training and development, formation of socially and professionally important competences demanded by the digital society, digitalization for utilitarian purposes of making the educational process cheaper, simpler and more manageable;
- haphazardness and haste in innovation (resulting in the risk of psychological and functional unpreparedness of teachers to work in the digital education process), voluntarism, lack of scientific validity in the proposed approaches and solutions;
- focus exclusively on formal indicators of "administrative quality" of education (availability of high-speed Internet, provision of digital technology and the ability of teachers to use it, the number of developed online courses, the place of Russia and its educational organizations in international ratings, etc.), ignoring or underestimating the social and semantic indicators of social and didactic quality of the digital educational process [16].

To minimise management risks, it is necessary to.

- organisation of scientific research into the process of digitalisation and the digital educational process of professional education and training, including by organising

a network of experimental sites on the basis of educational organisations, educational networks, professional and educational clusters;
- organisation of comprehensive substantive monitoring of the process of digitalisation of vocational education and training;
- organisation of systematic professional development pedagogical and managerial staff of professional education for developing new competencies, providing readiness to work in the conditions of digital education process;
- development of a set of methodological recommendations for the heads of professional educational organisations, teachers, industrial training supervisors, teachers of additional education, curators of groups (class teachers), educational psychologists to work in the conditions of the digital educational process.

3.3 Digital Didactics Development: Prognosis

Building digital didactics for professional education involves solving a set of new tasks which require full-fledged scientific and experimental research. The following are some of the research tracks.

1. Development of dynamic and open model of expected educational outcomes of professional education ("floating goals" of educational process) as well as model of personalized educational process flexibly adjusting for continuously changing goals.
2. Construction of the compensatory model of the educational process, providing levelling the negative features of the digital generation.
3. Advance in learning autonomy as a readiness to independently, actively and effectively use the competences of the digital learning environment.
4. Determining the pedagogical balance between the didactic principles of personalization (freedom of choice) and flexibility (adaptability), i.e. between electivity and selectivity, in constructing individual educational tracks and in other aspects of individualization of the digital educational process.
5. To reflect on the didactic potential of new and constantly improving digital technologies, as well as ways for using them to achieve pedagogical goals and solve current problems in the educational process. Development of new digital-based pedagogical technologies, testing and improvement.
6. Identifying deficiencies in the educational process of professional education, creating new and upgrading existing digital learning tools and electronic educational resources. This includes the development:

   - Adaptive learning models that provide automated personalized adjustment of the digital learning process to individual characteristics and take psychophysiological conditions into account.
   - Approaches to the learning experience design in a digital environment.
   - Virtual models of substitution communication, ensuring the achievement of the educational objectives.
   - Digital tools which enable the automation of routine elements and, at the same time, prevent "monotony effects" in the consolidation process.

- Approaches, methods and tools for managing learning motivation at different stages of the digital learning process.
- Digital tools for inclusive assessment for different types of learning activities.

7. Determining the ratio and alternation of virtual and real professional components in practice-oriented professional education. Conditions' identification for pedagogical effectiveness of existing meta-digital learning complexes, the pedagogical demand for their modernization, ensuring the professional skills creation, abilities and competencies for the digital economy.
8. Pedagogically appropriate infographics use in the educational process, as well as methods and tools for the combined figurative-logical thinking.
9. Teacher's role in the digital education process; formation, description and continuous updating of a dynamic set of his/her competences; identification and description of his/her new job functions, including in the form of new pedagogical professions for digital education.

4 Conclusions

The transition to the digital learning process significantly transforms the professional activities of pedagogical staff in professional education. Three groups of roles are actualized, which provide different levels of interaction in the digital education process:

1. Educator (specialist) ↔ learner (group of learners): organizer and motivator of learning, trainer, game technician, project activity specialist, developer of environments for group project work, developer of educational trajectories, manager of individual educational tracks (interdisciplinary tutor (the student's personal trajectory is supervised for a long period of time and is involved in solving individual educational problems, coordinating the activities of teachers working within the individual curriculum; on the basis of longitudinal observation of the student and analysis of his/her learning success he/she gives recommendations for further development of the educational trajectory.), etc.;
2. Educator (specialist) ↔ digital technology and tools learner (group of learners): integrator-mediator between virtual world and real world, network educator-curator (online tutor (Compiles online courses for distance learning, adapting the requirements of specific disciplines to the online environment, and administers the operation of the online education platform.)), internet navigator, digital footprint analyst-corrector, web psychologist, etc.;
3. Specialist ↔ digital technologies and tools: methodologist-architect of digital learning tools, developer of educational digital environments, electronic educational resources expert, etc.

At the same time, in the digital education process, many of the traditional roles of the educator – "knowledge-bearer", informer, explainer, checker, reprimanded and "punisher" for non-compliance, etc. - are losing their importance. In general, multidisciplinary, "convergent" professionals are increasingly in demand in digital education as in many other sectors of the digital economy. Practitioners with experience in various social, industrial, business projects will be in more demand in the digital educational process of professional education and training than traditional "mono-professional" educators.

Terminological analysis has identified a bank of new terms for the problem: "digital pedagogy", "digital learning environment", "digital educational resources", "digital learning", "digital culture", "digital marketing in education", "blended education", "flipped learning", "digital educator", "online professor", "digital student", "interactive education", "digitalisation", "smart university", "artificial intelligence", "digital cluster", "digital communication", "E-didactics".

Having analysed the main tools of digital didactics in professional education are:

1. *Personalized education process* - as the transformation of the educational process into a set of individual educational trajectories that take into account, on the one hand, the personal educational needs and demands of the students and, on the other hand, their individual psychological, pedagogical and medical (for students with disabilities) characteristics. The personalisation of learning is achieved by:
   - Individual educational trajectories;
   - The use of distributed forms of the educational process;
   - The use of adaptive technologies;
   - Creating an educational environment for independent work, self-education and self-development.
2. *Digital pedagogical technologies* - has the potential to provide an almost infinite number of pathways for individualisation of learning (including: content, learning pace, complexity, delivery mode, activity form, group size, repetition rate, external assistance degree, openness and transparency for other participants, etc.). It is important that all these areas can be implemented simultaneously, which allows adjusting the educational process to each specific student (adaptability principle), ensuring a high level of learning motivation and full assimilation of the given educational outcomes. Individualization of vocational education and training on the basis of digital technologies allows for an organic transition to multiprofessionalism - a post-industrial model of professionalism, when a profession ceases to be a standardized set of labour functions and actions, in-demand knowledge, skills and abilities - and becomes a dynamic personalized set of competencies.
3. *Meta-digital educational (hardware and software) complexes*, both training (simulators, augmented reality tools, sensors capturing the quality of a single work action, etc.) and used directly in the production process of enterprises, are of particular importance in the digital educational process of vocational education and training. The use of such complexes is a prerequisite for shaping a learner's professional skills and competencies needed to work in the chosen occupation (speciality) or within the framework of the work function being mastered. Under the conditions of digitalization, the partnership between a professional educational organization and employers acquires the form of a unified production and training digital environment. For example, students can be trained in a situation centre where they can remotely observe real production processes, participate in

discussion and analysis of emerging production situations (including problematic ones), and develop solutions.

References

1. Russia's Digital Economy Program. [Electronic resource]. http://www.static.government.ru/media/files/. Accessed 10 Mar 2020
2. Decree of the President of the Russian Federation of 09.05.2017 N 203 "On the Strategy for the development of information society in the Russian Federation for 2017–2030". [Electronic resource]. http://www.kremlin.ru/acts/bank/41919. Date of reference: 10.03.2020
3. Federal project in the field of education "Modern digital educational environment in the Russian Federation". [Electronic resource]. http://neorusedu.ru/about. Date of accession: 09. 03.2020
4. Towards change: seven tasks of digitalization of Russian education. 2019. [Electronic resource]. https://www.rbc.ru/trends/education/5d9ccba49a7947d5591e93ee. Accessed 19 Mar 2020
5. Can online education replace traditional education? Opinions of scholars on distance learning, online formats of university courses and knowledge control. [Electronic resource]. https://postnauka.ru/point-of-view/155079. Accessed 05 July 2020
6. The main trend of Russian education - digitalization. [Electronic resource]. http://www.ug.ru/article/1029. Accessed 20 Feb 2020
7. Andriukhina, L.M., Sadovnikova, N.O., Utkina, S.N., Mirzaahmedov, A.M.: Digitalization of professional education: prospects and invisible barriers. Educ. Sci. **22**(3), 116–147 (2020)
8. Astratova, G.V.: Key development trends of modern market of higher education online services. World of Science. Pedagogy and Psychology. No. 3. [Electronic resource] (2020). https://mir-nauki.com/PDF/33PDMN320.pdf. Date of accession: 23.06.2020
9. Korshunova, O.V.: Main trends in the development of modern didactics. Vestnik of Vyatka State University, pp. 70–80 (2019)
10. Nurutdinova, A.R., Dmitrieva, E.V.: Using social networking sites in higher education context: differences in learning outcomes between Moodle and Facebook. Modern Problems Sci. Educ. **1**, 19 (2018)
11. Semenova, L.M.: Digitalization in modern higher education: realities and prospects of development. In: NIR. Modern communicativism. M.: Limited Liability Company Scientific-Publishing Centre INFRA-M. - No4 (41), pp. 9–14 (2019)
12. Popova, O.I.: Digitalization of education and brand of higher education institution: students' attitude to the processes. Manage. Issues **3**(39), 245–251 (2019)
13. Semenova, L.M.: The dynamics of digital didactics in the context of the transformation of higher education. Part I. World of Science. Pedagogy and psychology, [online] 3(8) (2020). https://mir-nauki.com/PDF/87PDMN320.pdf. (in Russian)
14. Blended learning and motivation crises. https://newtonew.com/higher/motivation-in-blended-learning. Accessed 02 Feb 2019
15. Luksha, P., Rabinovich, P., Asmolov, A. (eds.): Education for a complex society: Global Education Futures Report, p. 77 (2018)
16. Nurutdinova A., Fazlyeva Z., Dmitrieva E., Panfilova E., Vasallo Baez I.: The distinctive features of online learning and learning using distance learning technologies in extreme pandemic conditions. Case study: approaches to assessing the distance learning effectiveness. In: EDULEARN21 Proceedings: 13th International Conference on Education and New Learning Technologies, 5–6 July 2021, pp. 6703–6712 (2021)

17. Sherbina, E., Shmurygina, O.V., Utkina, S.N.: Digital didactics of vocational pedagogical education: basic components. Bus. Educ. Law **2**, 411–418 (2020). https://doi.org/10.25683/VOLBI.2020.51.268
18. Ainoutdinova, I., Blagoveshchenskaya, A., Nurutdinova, A., Dmitrieva, E.: A paradigm shift in distance education in Russia towards open, massive and experiential modes of training. In: INTED2019 Proceedings: The 13th Annual International Technology, Education and Development Conference, Valencia, Spain, 11–13 March 2019, pp. 5519–5525. ISSN: 2340-1079; ISBN: 978-84-09-08619-1. https://doi.org/10.21125/inted.2019

Hybrid Tools and Blended Learning for the Pedagogy of Clinical Courses in Special Education

Dimos Charidimou[1(✉)], Dionysios Politis[1], Georgios Chamouroudis[1], and Georgios Kyriafinis[2]

[1] Aristotle University of Thessaloniki, Thessaloniki, Greece
dcharidim@csd.auth.gr
[2] AHEPA University Hospital, Cochlear Implantation Centre, Thessaloniki, Greece

Abstract. Educational activities and other on-line interactions have often aimed at teaching peripheral subjects related to the science of medicine or to the treatment of illnesses and injuries. Through the experience of the Aristotle University of Thessaloniki and other Institutions, this research points out key figures on how "environmental" issues of clinical ecology have set up the previous two academic years the instructional profile during the Covid-19 lockdowns. It investigates the behaviorism associated with the technical substrate of its deployment and the possibility of applying this type of education in special institutions such as technical schools, conservatoirs or schools with mandatory laboratories. The monitored interactive educational tools, alongside a list of pedagogical and technological practices, intended to keep up with the quality, performance and reliability needed for clinical courses in general. Qualitative factors give indices about "Quality of Life" measurements, used in psycho-kinetic evaluations, while quantitative indices present metrics for the actual working conditions.

Keywords: On-line counseling · Synchronous and asynchronous tools · Quality of Life indices

1 Introduction

During the COVID-19 pandemic, schools and universities were under closure, leading to an unprecedented shift to Distance Learning. In Greece, as in most countries of the world, immediate actions have been taken to adapt to the new reality, so that the educational systems respond immediately to the requirements of Emergency Distance Learning and on-line learning.

One generic issue is whether favorable learning conditions have been deployed, facilitating learners to attend and focus on essential material in order to increase their productivity and efficiency.

Another one, however, is if the suspension of physical education may be a potential source of danger for the coverage of entire curricula based on laboratory or clinical studies rather than theoretical ones.

M. E. Auer and T. Tsiatsos (Eds.): IMCL 2021, LNNS 411, pp. 306–321, 2022.
https://doi.org/10.1007/978-3-030-96296-8_28

Distance Learning, the protagonist in tertiary education for this period, is defined as an organized form of education where there is separation between teacher and learners and where electronic communication means are used to bridge the "natural" gap that exists. The methods and techniques of its implementation considerably vary depending on the period and the level of technology in perspective [1].

The "Internet-based Distance Learning" model is part of the new and modern way of life that emphasizes the extensive use of the Internet not only for education, but also for a variable width of social and consumer practices. Through the use of Information and Communication Technologies (ITC) and the Internet, the conceptual deconstruction of space and time is achieved, resulting in the immediate transfer of information and educational material, in a multitude of languages simultaneously [2]. In particular, Distance Learning has two sides, the physical and the temporal, which relate to the segregation between:

I. Teacher – trainee
II. Training with each other
III. Teaching and learning processes

As Lionarakis has pointed out since 2001, Distance Education is based on three important axes, the trainer, the trainee and the educational material in contrast to the binary relationship between trainer and trainee that characterizes conventional education. The pedagogical characteristics of Distance Learning do not simply instruct students how to learn on their own but also how to function autonomously in an exploratory course for knowledge acquisition [3].

Based on the above principles, the trainee takes responsibility for his studies and acquires a more accountable role, while at the same time the teacher is called to take on the role of counselor and mediator, with the responsibility to motivate the trainees and organize the training programs ensuring maximum efficiency in the learning process.

Indeed, during this survey it was noticed that classes with large audiences recorded more consistent and frequent attendance, as limiting conditions connected with student mobility, availability and amphitheater capacity were annihilated.

Technological tools are en masse used in education because they help and support learning, while providing students with opportunities to gain faster new knowledge and learning experiences. In order to be able to properly support educational activities, the deployed technological tools should incorporate various teaching strategies, allowing the student to explore and interact, and be as interdisciplinary as possible [4].

ICTs are defined as the forms of technology used for the purpose of transporting, processing, storing, creating, presenting, sharing or exchanging information through electronic media. More specifically, the ICTs agenda refers to the use of computers for delivering individual functions through software, usually with the extensive use of Internet or similar resources. Also, the conceptual merit of this agenda is directly related with how successfully machinery, software and equipment are incorporated into the educational process.

ICTs essentially correspond to a subset of media such as New Media, Telecommunications and mainly Multimedia. They may include media with data in the form of text, image and sound combining in parallel synchronous and asynchronous communication and information technologies such as video conferencing, forums, e-mail, blogs, computer software, radio, television, video, telephone satellite systems, portable and mobile gear, and many other devices that allow the interconnection and transmission of any amount of data in a very short time.

2 The Impact of Covid-19 on Professional Counseling

During The provision of assistance and guidance in resolving physical or psychological problems has been kept high in the agenda of public health and counseling. The difficulty to approach *in situ* the clinics, the laboratories (for the cases where direct physio-kinetic measurements are needed) and the professionals themselves have been issues not thoroughly resolved.

The COVID-19 pandemic has forced institutions to find new ways to stay safe. Many people turned to technology to be able to simulate normal life with virtual life. This was achieved to some extent through video communication and teleconferencing, which allowed most parties involved to continue working and training in digital on-line environments [5].

Nowadays the Internet of Things (IoT), Artificial Intelligence (AI), cloud services and automation are some of the concepts that express the fourth industrial revolution that we are going through (as stated in many European Union publications in 2020). Technology and digitization are transforming the way people work and educate, and this has resulted in the proliferation of both products and services. In education, technology influences teaching in many ways. Initially, it gives a more modern way to the teaching methodology and introduces new practices in the teaching models [6].

Health literacy or related courses are considered difficult by nature, as they contain modules, such as biology, natural chemistry, biochemistry laboratories and their branches, physiology, anatomy, surgeries and many more. Health literacy is known as a complex concept. It has multiple definitions, underlying meanings, and conceptual models, which learners will have to understand from the beginning, to be able to build knowledge on them [7].

The rise of the COVID-19 pandemic in conjunction with advances in technology has brought the Hybrid content delivery standard to the forefront in almost all education systems. The hybrid model of educational content delivery has created new possibilities in education, with the appearance of on-line, blended, and "flipped' learning models, that supplement or replace face-to-face instruction with distance learning.

More and more researches are monitoring and highlighting ways in which on-line, portable and hybrid learning enrich active learning outcomes. Research has shown that medical students prefer hybrid learning over Distance Learning or traditional face-to-face teaching [7, 8].

In face-to-face learning the training takes place at a specific time and place while in on-line learning the whole lecture takes place entirely remotely either synchronously or asynchronously, using ICT tools. In hybrid education, an effort is made to combine and utilize both live and on-line education. In blended learning, asynchronous communication is essentially used to enhance face-to face teaching and learning using a variety of educational strategies. One of the most commonly used learning strategies by teachers is the Flipped Classroom in which students receive content, usually through multimedia lectures to convey useful information and save valuable teaching time [8].

On-line multimedia websites have become useful resources for transferring information; however, the challenge is to create professional videos that are unambiguous, clear and apply pedagogical principles of multimedia development so that targeted and meaningful learning can be ensured [9]. According to the research of Grønlien et al. [7], blended learning is more beneficial for learners with good digital skills who have self-knowledge, self-confidence and can self-regulate and take responsibility for their own learning. In any case, these cognitive parameters also have their importance, as learning is a multidimensional and complex phenomenon which can be influenced by several factors such as learning motivation, learning environment, instructor aptness, educational activities, teaching techniques, the effort, perseverance, and emotional reactions of students, etc.

3 The Challenges for the Educational Environment: Training Programs, Supportive Policies, Technological Solutions

The double-standards commitment for most therapeutical staff members to function simultaneously as clinicians and as educators as well has resulted in insufficient time for the face-to-face lessons and counseling. To run smoothly such high level sessions ample resources should be allocated, alas not adequately appropriate in some cases during the pandemic.

As a result, experiential teaching methods were used to increase the receptivity of the involved institutions. In many cases, however, a plethora of field exercises, clinical projects and simulations were sufficiently transferred to the sphere of on-line pedagogy [10].

The methodologies set up for coping with this difficult situation, included the following expedients, with obvious plus or minus signs:

I. Since students, patients on recovery route and their families were restricted from attending face-to-face sessions, they were relieved from all transportation and relocation costs. Unless direct medical care was necessary or attendance to clinical courses, involved beneficiaries were transferred to the digital zone of communications.

II. Observation and distributed remedial treatment of actual patients and learners was as well transferred to the virtual reality channels for interaction. Physical, mechanical or therapeutic treatment of actual patients, along with instruction on laboratory matters, rather than theoretical deployments, was also allocated to far-flung practices like telematics engineering, telemedicine or Remote Fitting, to name a few approaches.

III. Training and restoration for students in special education is not unemotional; on the contrary, clinical sessions rely heavily on creating affective learning conditions, sometimes involving family members in the rehabilitation learning processes. It is not clear how in Remote Fitting, for instance, this may emerge.

IV. Fees, administrative costs, detached clinical functionality charges are encountered for the first time in the sphere of hybrid communiqué. As it happens with this conference as well, it is not clear how the newly-rendered services should be marketed as public admissions.

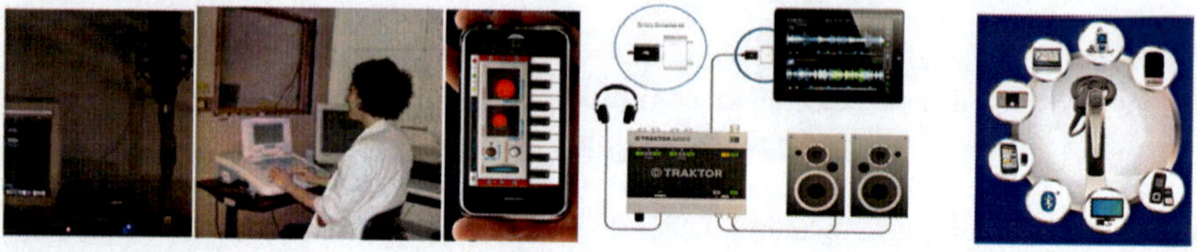

Fig. 1. Synopsizing visually remote home-studio recording, ENT Audiometry, and other distributed audiovisual applications (like the Tracktor HiFi system) over WiFi or other 2.4 GHz interconnections, for rehabilitation and learning.

This research seeks to decipher under which circumstances remote interaction may be a viable alternative to local face-to-face fitting and other clinical sessions either for rehabilitation or learning. It is not as easy as it sounds, compared for instance to remote lecturing, for objective MAP parameters in audiological fitting may mingle with subjective feedback data cropped by instructors, monitoring clinicians, beneficiaries of rehabilitation and special education counselors.

In any case, the hybrid environment for offering such services is reliant on the manufacturers' recommendations, when expensive, specialized equipment is used and the recipients' ability to simulate within the VR environment the department where outpatients are given medical treatment or advice.

Suggested procedures have to prove safe in terms of trustworthiness for the medical treatments, the patient ethics - as far as his medical record's personal data protection is concerned, the reliable and reputable recovery to a normal state of health, attainable to restituting the social and professional status for learners, and the credibility of the academic marks, to name a few.

Since the 1950s, it has been acknowledged that there is a connection between instruction and personality. As Hartley has put it, distinctive qualities that form an individual's idiosyncratic character may be shaped by teaching and education in general [11].

Alternative norms of education emerging through the Covid-19 confining become prevalent features envisionaged to remain even after restitution is encountered. How these techniques may be adapted in a timely manner to the clinical routine is an important topic for academic discussion.

4 Deploying Experiential Learning Methods as Structural Arrangements for Instruction

Management, coaching, team collaboration are essential parameters for the normal development and advancement of physical, mental, or societal growth conditions. It's not only a matter of fulfillment or satisfaction, but, predominantly of state conditions guaranteeing that subjects remain free from pain, illness or injury.

During the pandemic, these values acquired a Distance Learning approach dependent, however, on the creativity of the platforms used for remote interventions alongside the content of the programs used to improve knowledge retention [12].

If the involved patient, learner and his family are not adequately motivated to participate, under whatsoever circumstances, then the program is bound to fail [13].

The COVID-19 pandemic made distance work and training a necessity, and as it turns out this trend has come to stay. Most educational institutions, such as the Aristotle University of Thessaloniki, have already realized that and for this reason they upgrade their wireless network and install video conferencing systems in the rooms with large screens allowing a more natural cooperation and a considerable improvement of its asynchronous communications.

The main disadvantage for remote methods of instruction as well as rehabilitation is that learners do not mix socially with their peers. Therefore, they animate the predicament of isolation, i.e. they lack entirely the experience of cropping up interaction stances acceptable by our society.

In other words, it is not certain anymore that students may socially conduct in accordance to their BS's standards and therefore the transition of academic ranks, conferred by a college or university after the completion of a specific course of study, are not guaranteed to persistently pledge as professional activity, e.g. that of an engineer or a doctor.

Even further, things may become worse, as university studies stretch overall a young professional how to expand himself over the narrow family horizon and become an accomplished expert; lockdowns, on the other hand, recede him or her to the intimacy of the household unit, thus downgrading his social & professional outwardness, the true substance for conducting studies in an organized campus.

Therefore, the aim of this research is to diagnose unsuitable manners of direction in remote instruction having the possibility of promoting misguided assessments (like those connected with not adequate inspections of the examination procedures) and inaccurate judgments about the abilities of course infrastructures to transfer expert skills and knowledge.

It is essential, as universities distinguish themselves in their educational pursuit by vigorous effective properties on how estimations of quality learning may proceed, in the examination period for instance, to have rapid estimations of the learning status of their studentship on the on-line world.

Thus far, a multitude of quantitative factors have been put into the discretion of the academic community and its units, with a resolution expected for the academic year 2021–22 after consideration on the actual parameters of massive on-line instruction.

For the Aristotle University of Thessaloniki (AUTh) the shape of things is presented in Fig. 2 for the spring semester of 2020–21.

Type of activity	Frequency of appearance for a certain activity
Κατάθεση εργασιών (Assignments)	36.57
Κουίζ (Quiz)	31.31
Ομάδες Συζητήσεων (Forum)	29.12
Επιλογή (Choice)	2.07
Λεξικό (Glossary)	0.36
Wiki	0.57

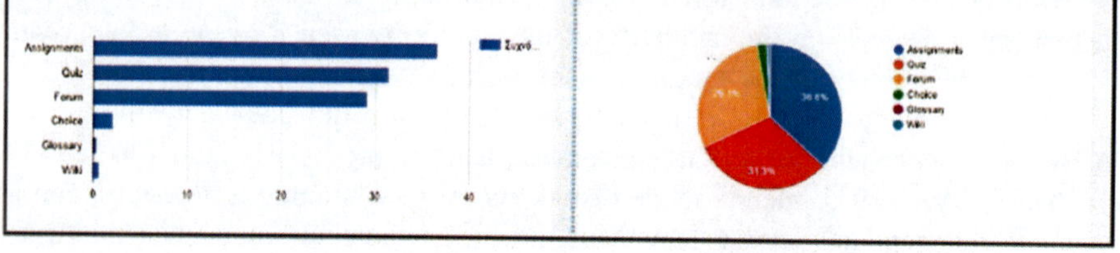

Fig. 2. Quantitative indices displaying how AUTh characteristically performed student evaluations the 2nd semester of the academic year 2020–21.

In terms of Interactivity, university IT centers track down the components users exchange with its e-Learning systems, seeking to morph student activity and decipher the conditions under which (social, educative or other) things are happening.

E.g., a new categorization of activities inferring information about the scale of involvement in digital recreation, as far as the learners are responding to on-line instruction may be seen in Fig. 3.

Type of content	No. of data sources	No. of data sources created/modified spring semester 2021	% of data sources created/modified spring semester 2021
Αρχεία (Files)	10971974	1329761	12.12
Φάκελοι (Folders)	17877	4010	22.43
Ετικέτες (Labels)	19608	1854	9.46
Σελίδες (Pages)	34017	2146	6.31
Βιβλίο (Book)	302	26	8.61

Fig. 3. User participation in conferring digital content within AUTh's digital learning IT systems during the 2nd semester of the academic year 2020–21.

However, as it has been found out by clinical researches, the formal introduction of computer activity or automated software generated statistics does not have thus far the vigilance to produce personalized clinical-like data that have pieces of information concerning user attitudes, affection stances and motivation.

To examine and record such features, a rather extensive survey investigating opinions and experience was set-up. Thus far, a multitude of quantitative factors have been put into the discretion of the academic community and its units, with a resolution expected for the academic year 2021–22 after consideration on the actual parameters of massive on-line instruction.

5 Results

As special education had been transferred abruptly to the on-line sphere of communications, only a few cases were probed, resulting in findings characterized in their phenomenology having rather qualitative metrics than quantitative ones.

Its croppings, especially those focused on family support and involvement, where parallelized with a survey conducted on healthy subjects who were forcibly, under the extensive lockdowns, put into similar sociotechnical arrangements.

100 participants took part in this survey, conducted between April and June 2021. 17 were instructors and tutors, 10 were parents or guardians of children actively participating into their sibling's on-line unusual experience, and 73 were pupils or students.

Their age distribution and their attitudes towards this unforeseeable adventure, as far as fulfillment is related to user experience and ease of use concerning the technologies involved may be seen in Fig. 4.

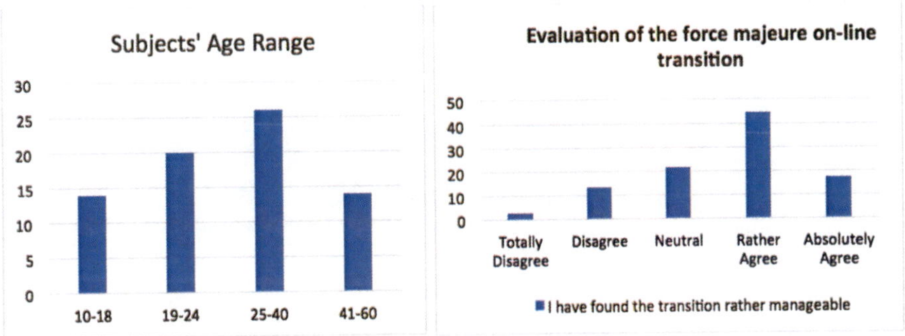

Fig. 4. Left, age variation, and right, an evaluation by all involved subjects on the rigidity/discomfort the prolonged lockdown has provoked as far as the technological manageability of on-line education is concerned.

The equipment used and its resulting ease of use are presented in Fig. 5. The special situation where both set of devices (mobile-tablet, laptop-pc) were simultaneously used was not encountered in this survey, perceived as a hybrid situation usually met in student sessions with tension due to adverse or demanding situations. Qualitatively, it has been estimated that these cases count for some 1–2% of the overall cases examined. For the sake of simplicity, they were not put in the charts as a recognizable, self-attained paradigm. However, experts in the field estimate that if a new lockdown is enforced, such cases may be normality, at least for very demanding interactions in user-experience terms.

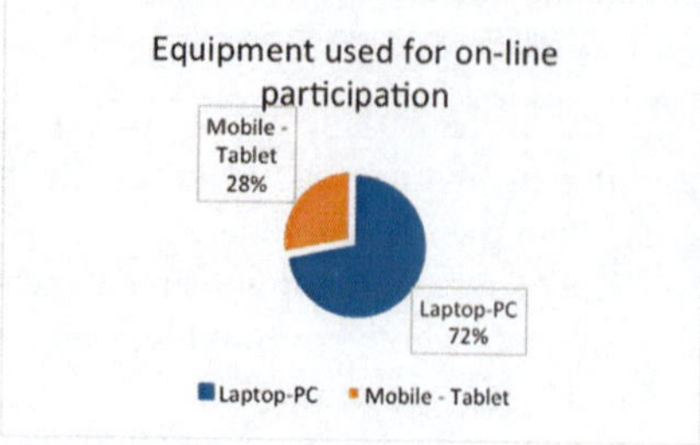

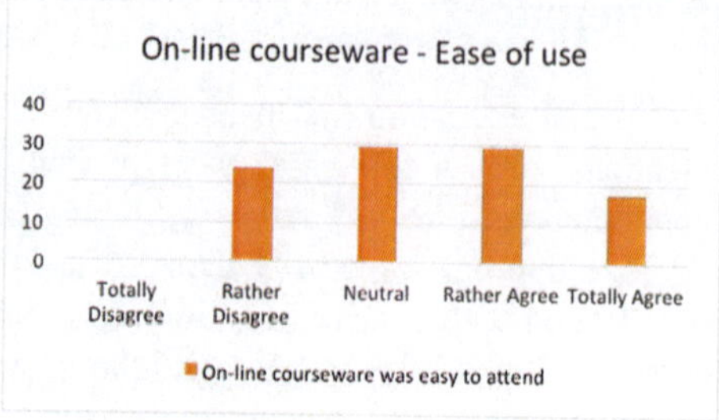

Fig. 5. User Left, the kind of equipment used for particular fitting to on-line learning interactivity. Right, statistical data for inferring the ease of adaptation to massive and total e-Learning practices.

Apart from computer hardware, a very pivotal device for conferring visual images or video signals is the camera deployed for teleconferencing sessions (Fig. 6).

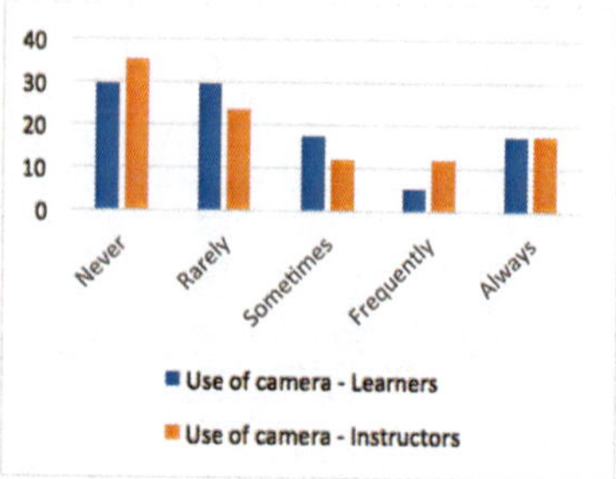

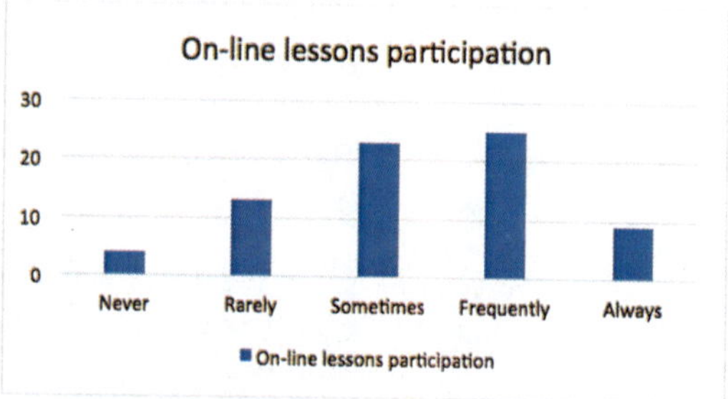

Fig. 6. Left, the use of camera during lecture broadcasts. Right, how students frequented to lessons during lockdowns.

This use of cameras within a household requires care and skill in its handling, since it seems that vital personal data may be endangered from becoming widespread without the prior consent of the copyright holders.

This is another topic of concern for the successful delivery of compulsory on-line education. Being a multidimensional issue, with significant leverage for medical practices some years ago [14], it has raised recently polemic against its hidden hazards [15], concerning not only the university's IT Center but the telecommunication companies involved as well; for this reason, it was left aside as it is a significant matter of research by itself.

Surpassing these safety warnings, the next step in evaluating the services offered, is whether the Quality of Educational Services retained its status during lockdowns (Fig. 7). At this point the survey visualizes the opinions or reactions of the users involved; it does not proceed further in analyzing what the expression of considerably varying views may mean in practical terms - apart from those that express explicit postures, attitudes or facts, like noisy conditions due to housing activities.

The only notion that the authors of this research transfer to the IMCL conference participants is that certain aspects of on-line sessions during closures will be certainly favorable to stay when normal education resumes, while some others will be formally put to an end as institutional practices [16].

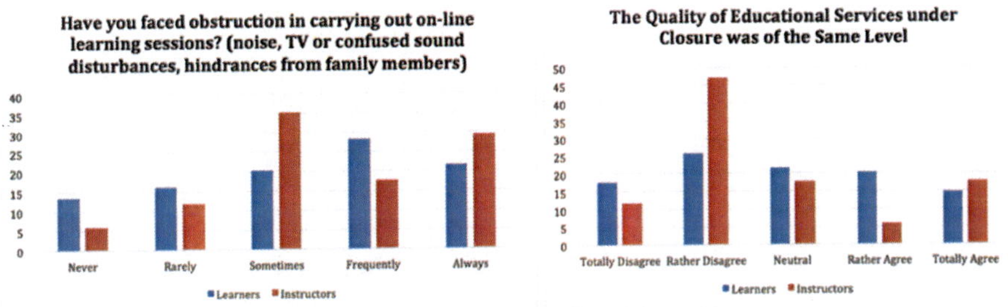

Fig. 7. Evaluation of the Quality of the Educational Services offered before and after the closures.

Apart from the issue of Quality for the offered instructional services, the regularity, longevity and resilience of the e-Learning sessions committed during the strenuous lockdowns, brought to surface another concern that should be addressed with practical care: it is not the first time that employees and the general public use teleworking, but thus far such workers had the privilege being warned; consequently, they could organize their home environment accordingly, bearing in mind that it would serve both for residence and working place.

The abruptness of turning the rooms in which domiciled students, instructors and members of the family advocated to working places offering schooling did not provide much opportunity to vest in proper restorations of the living habitat.

As seen in Fig. 8, students and instructors to a certain degree adopted a relaxing stance towards organizing their learning commitments.

It seems that instructors were more adept in conforming to generally accepted standards on how an appropriate working corner should function, already a prerequisite, more or less, for exercising the particular profession, while students were not that much confined. It is true that students having relocated from home are the most possible ones to go around with libraries, along with tutors or learners that for various reasons could not afford a decent place for telecommuting - not that much in terms of room space as for lack of disturbance, noise or the usual home activity that a family exerts.

In any case, the potential to massively attend or provide learning from his bedroom or couch provides new features in the learning process, whose linear extension to the future of education is more crucial than what the statistical data in Fig. 8 denote.

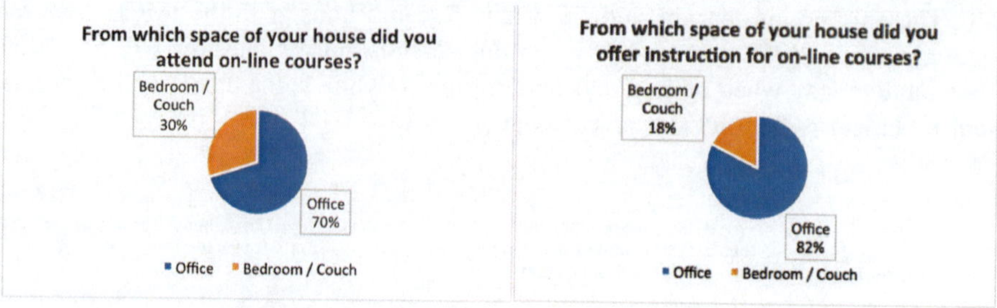

Fig. 8. Spatial data affecting user behaviorism during attendance of on-line instruction from home, left for students, right, for instructors.

At first, the schoolwork that has to be committed is obviously facilitated by economizing funds and other resources thus far indispensable for relocation or commuting from the residence to the schooling unit.

However, a relaxed stance adopted necessarily sometimes in the domiciles of e-Learning users, brings into surface other factors, like listening to radio or navigating through the Internet while attending educative discourse, that would normally be not the case in attending lectures within the campus.

As seen in Fig. 9, left, there is a potential threat for the commitment of learners to their moral obligations, or, even further to the schoolwork that they are required to do at home. This threat, at first consideration, was not highly or widely regarded [17].

In moral obligations many also consider the fact that assessments and examinations are virtually unsupervised; however, this serious phenomenon, undermining the nature of academic performance - with serious implications for what could be the safe practice of a profession like engineering, medical work, administration, etc., is not perceived but merely as an instance of a particular extravagant situation. However, if lockdowns prove to be something more permanent than a sporadic case, this issue would be of prime seriousness.

In any case, studying from home is not an easy-going task. As residencies have poor orientation for providing isolation, a range of behavioral disorders may crop up, plunging primarily children, with main symptoms the following: poor concentration, tediousness, uncontrolled hyperactivity, impulsivity, frailty in attending for long hours and some others, more specific to the medical record of each attendee.

As expected, a very important factor for attending perpetually university courses via e-Learning and massive ZOOM-like sessions is Internet stability and speed. As, all of a sudden, relocated students retreated back to their hometown residencies, it became obvious that telecommunication providers were not all in a suitable state to provide fast Internet, especially for residencies dispersed in remote rural areas (Fig. 9, right).

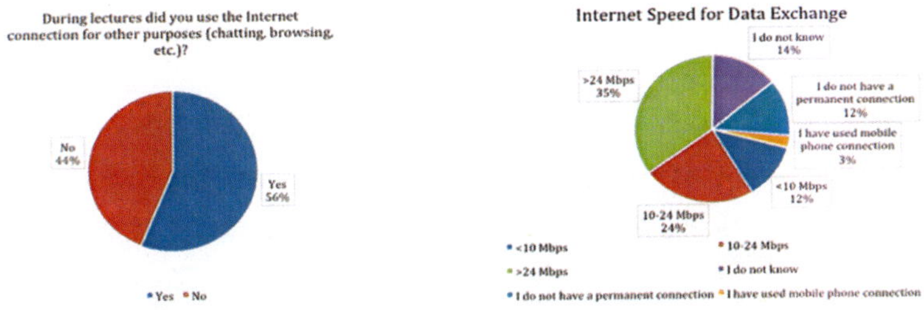

Fig. 9. Left, spatial data affecting user behaviorism during attendance of on-line instruction from home. Right, Internet speed reported by the users participating in this survey.

Related to mobile communication and learning is the fact that cellular data seem to be an indispensable ally for guaranteed functionality. Furthermore, it seems that expansion to 4G+, 5G technologies may be a viable solution to the cases where the expansion of fiber technology may take some time to reach remote settlements or problematic neighborhoods, restrictive for digging into them networks (Fig. 9, right).

As expected, a very important factor for attending long sessions of on-line instruction is the physical ability to withstand continuous irritations of the globular areas, pains in the head, neuralgias, irritations and so on. No serious conditions were reported as side effects, but, as seen in Fig. 10, left, for many subjects the situation was at least annoying in terms of physical suffering, not lest discomfort.

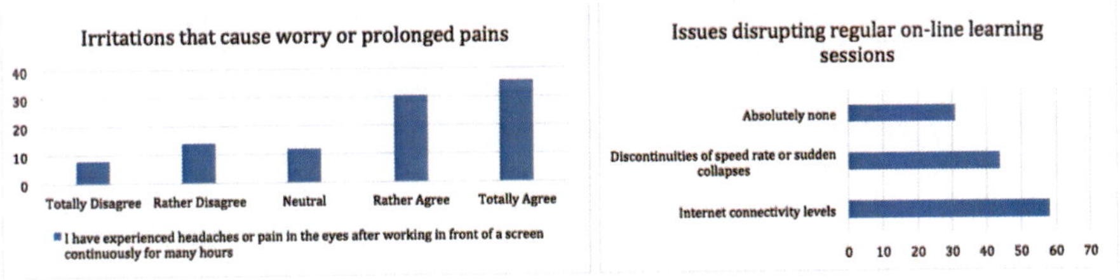

Fig. 10. Left, intense physical suffering or discomfort reported by the users of on-line education. Right, disturbances drastically reducing Quality of Services for on-line education.

Last, but not least, in recent years working on-line had been extended to the medical field with practices and results expressing approval for a bright future, at least some ten years ago [18].

However, critical medical practices rely heavily on active processes with limited discharges of their functions.

As a precursor stage, massive on-line education may provide facts about the probability of undesirable concurrencies reducing the Quality of offered Services (Fig. 10, right).

When extending the deportments previously analyzed to pure audiometric rehabilitations, restoration and training as therapeutic processes follow the scheme seen in Fig. 11.

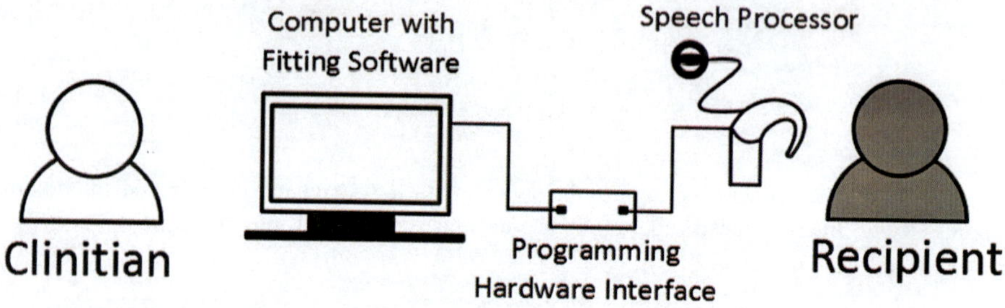

Fig. 11. A mild medical remote interaction, cochlear implant programming setup.

Another responsibility is to support the recipient in any case that he or her may require assistance during the fitting session [19]. While the Remote Fitting setup in Fig. 11 takes advantage of advances in telecommunication technology, it is prone to errors that may be caused by disturbances like those reported in Fig. 10.

Although Remote Fitting is a mild session in medical care terms, as far as its consequences are concerned, it has some aspects that require subtle probing, as the actual speech processor is not in front of the monitoring clinician but at the home of the patient. The connection pod, allowing Internet programming, is also required in the remote location so to allow the speech processor to be connected with the computer and the fitting session to take place. A stable Internet connection is also required in order for the two remote systems to communicate one both sides [19].

The use of audiologic hardware is appended with a video camera in both locations so that all parties can visually communicate with each other. The resolution of the video camera is also an important factor during the session. Although modern cameras may offer high-definition video (HD: 1080 × 720, FullHD:1920 × 1080) transmitting a live video feed at these resolutions is not viable especially in the case of slow or unsteady connections. As such the resolution must be adjusted in the video conferencing software to the value that offers acceptable image quality, while not causing delays in the connection between the two computers. Visual communication is required so that the clinician is able to see the reactions of the recipient to the new mapping of his implant and also communicate visually with the monitoring clinician in the remote location [19].

Furthermore, a set of good quality speakers must be available in both locations so that all parties can communicate verbally. Seen as a pure diagnostic treatment telemedicine it is perceived to be rather safe when conducted on the grounds of stable telecommunication technology [20]. However, it is a highly engaging activity, as both the fitting clinician or the patient with his supporting family members should be adequately trained to achieve the desired attains. For instance, the clinician as in a local fitting would ask for feedback from the recipient about the T and C Levels as well as if he has any problems with the new mapping [19, 21].

Apart from the hardware a set of adequate apps is also required to successfully complete a fitting session. In the clinician's fitting room a remote control software is required. This software will enable the clinician to remotely control the computer in the remote location on which the recipients' speech processor is connected. In the location of the remote computer an installation of the cochlear implant programming software must be available. Apart from the above software, both computers must have video conferencing software that will enable the clinician and recipient to communicate through the Internet. There is a number of video conferencing applications that are commercially available.

ZOOM or Skype seem to be the mostly preferred for special education. Whilst selecting one the clinician must take into account the cost of such software, although the majority is available free of charge, the compatibility with the operating system on each computer, the ease of use and learn ability of the conferencing software as well as the robustness of the software [16]. This means that the software should not lag often and will also have a small chance of crashing and terminating the connection prematurely. Other features may also be helpful such as the ability to communicate with text messages, the ability to share a screen and also the option of filesharing. Although these features are not required as part of the medical session, they may appear helpful as secondary communication methods. While setting such a meeting it is also necessary to make sure that the proper drivers are installed that will support all the hardware connected to the computers, including the speaker system and microphone that are usually connected through the computers sound card.

It sounds unnecessary for most students to take such precautions. However, for the general public involved in tele-therapy, it is advised that the connection between the clinician's fitting room and the remote location is initiated before the recipient arrives for the fitting. This will save time and frustration for the recipient and any problems that may occur can be addressed before the beginning of the session.

Even the ability to exchange documents on-line is crucial for progressively limiting the administrative load.

6 Conclusion: Attainability, Technological Maturity and Behavioral Patterns

This research does not extend to medical treatment practices given on-line by a health center or a doctor's office for instance, although medical personnel has been involved in gathering data for the statistics performed.

Usually, medical data *per se* need significant processing time due to double-checks for determining their accuracy, quality or conditions under which they were cropped-up. Therefore, this study evaluates how close to telemedicine practices on-line tutoring has been deployed close by effort or skill, mentally and physically, as far as both ends are concerned.

Management, coaching and team collaboration are essential parameters for the normal development and advancement of physical, mental, or societal growth conditions. It's not only a matter of fulfillment or satisfaction, but predominantly of state conditions guaranteeing that subjects remain free from pain, illness or injury.

During the pandemic, these values acquired a Distance Learning approach dependent, however, on the creativity of the platforms used for remote interventions alongside the content of the programs used to improve knowledge retention [11].

If the involved patient, learner and his family are not adequately motivated to participate, under whatsoever circumstances, then the program is bound to fail [12].

The COVID-19 pandemic made distance work and training a necessity, and as it turns out this trend has come to stay. Most educational institutions, such as the Aristotle University of Thessaloniki, have already realized it and for this reason they upgrade their wireless network and install video conferencing systems in rooms equipped with large screens that allow more natural cooperation and the improvement of modern synchronous and asynchronous communication.

The application of new technologies in education and especially in Distance Learning offers new possibilities, as these innovative tools provide rich communication and interaction. The AUTh IT center offered the educational community different technological options so that on a case-by-case basis, favorable learning conditions were offered.

However, this case study has limitations, which require further research so to reach a safe conclusion about which technologies can be effective in times of crises. The pandemic brought to the foreground the least guided educational methods of teaching, which, although very popular, are not considered by themselves effective approaches. In other words, to be constructive they should get related to some learning theory and good pedagogical practices, yielding thus substantial learning outcomes.

Therefore, it would be interesting through further research to explore and take into account not only the collaborative platforms, but also the teaching methods assorted with these technologies. Thereby, instructors may understand how to facilitate learning and how to receive successful outcomes via accreditation and certified experiental learning practices.

The findings of this research correspond to two distinct ways for approaching data-driven results:

Qualitative factors give indices about "Quality of Life" measurements, while quantitative indices present metrics for the actual working conditions.

Both are used to characterize the appropriate technological infrastructure as well as the appropriate human capital involved.

References

1. Politis, D., Stagiopoulos, P., Aleksić, V. (eds.): Advanced Technologies and Standards for Interactive Educational Television: Emerging Research and Opportunities. IGI Global, Hershey (2020)
2. Charidimou, D., Politis, D., Aleksić, V.: Multilingual conversational communication tools for distance learning - synchronous & asynchronous teaching during the Covid-19 Pandemic. In: 8th International Scientific Conference Technics and Informatics in Education, Čačak, Serbia, 18–20 September 2020

3. Lionarakis, A.: Opinions and concerns for Open and Distance Education, Athens (2001)
4. Jimoyiannis, A.: E-Learning: theoretical approaches and educational design, Athens (2017)
5. Zoom.: A Story of Agility and Innovation: Findings from the Impact of Video Communications During COVID-19 Report (2021). https://blog.zoom.us/findings-from-the-impact-of-video-communications-during-covid-19-report/. Accessed 07 July 2021
6. Zygouris, F., Mavroidis, I.: Communication between teachers and distance learning trainees. Case study in the Instructor Training Program of K.E.N.E.AP, Open Education: the magazine for Open and Distance Education and Educational Technology, 7(1), pp. 69–86 (2011)
7. Grønlien, H., Christoffersen, T., Ringstad, Ø., Andreassen, M., Lugo, R.: A blended learning teaching strategy strengthens the nursing students' performance and self-reported learning outcome achievement in an anatomy, physiology and biochemistry course – A quasi-experimental study. Nurse Education in Practice, 52 (2021)
8. Gagnon, K., Young, B., Bachman, T., Longbottom, T., Severin, R., Walker, M.: Doctor of physical therapy education in a hybrid learning environment: reimagining the possibilities and navigating a "new normal." Phys. Ther. (2021). https://doi.org/10.1093/ptj/pzaa096
9. Sofos, A., Costas, A., Paraschou, V.: Online distance education. Athens: Association of Greek Academic Libraries (2015). http://hdl.handle.net/11419/182
10. Zimmerman, B.J.: Self-efficacy: an essential motive to learn. Contemp. Educ. Psychol. **25**, 82–91 (2000)
11. Hartley, E.L., Hartley, R.E.: Fundamentals of Social Psychology. Knopf, New York (1959)
12. Schwartz, M.: Flexible Learning - Instructional Design and Research Strategist, for the Learning & Teaching. Ryerson University, Toronto (2016)
13. Stars, I.: Health literacy as a challenge for health education. SHS Web Conf. **40**, 02004 (2018). https://doi.org/10.1051/shsconf/20184002004
14. Armstrong, A.W., Kim, R.H., Idriss, N.Z., Larsen, L.N., Lio, P.A.: Online video improves clinical outcomes in adults with atopic dermatitis: a randomized controlled trial. J. Am. Acad. Dermatol. **64**(3), 502–507 (2011). https://doi.org/10.1016/j.jaad.2010.01.051. [PubMed: 21236514]
15. Newlands, G., Lutz, C., Tamo-Larrieux, A., Fosch-Villaronga, E., Harasgama, R., Scheitlin, G.: Innovation under pressure: Imlications for data privacy during the Covid-19 pandemic. Big Data Society, SAGE (2020)
16. Margounakis, D., Pachidis, T., Politis, D.: A rubric-based evaluation of video conferencing services for educational use. In: 8th International Scientific Conference Technics and Informatics in Education, Čačak, Serbia, 18–20 September 2020
17. Reimers, F., Schleicher, A., Saavedra, J., Tuominen, S.: Supporting the continuation of teaching and learning during the COVID-19 Pandemic. OECD, Annotate resources for online learning (2020)
18. Keulers, B.J., Welters, C.F., Spauwen, P.H., Houpt, P.: Can face-to-face patient education be replaced by computer-based patient education? A randomised trial. Patient Educ Couns. **67**(1–2), 76 82 (2007). https://doi.org/10.1016/j.pec.2007.03.012. [PubMed: 17448621]
19. Chriskos, P., Kyriafinis, G.: Cochlear implant programming through the Internet. In: Politis, D., Tsalighopoulos, M., Iglezakis, I. (eds.) Digital Tools for Computer Music Production & Distribution. IGI Global, N. York (2016)
20. Ramos, A., Rodríguez, C., Martinez-Beneyto, P., et al.: Use of telemedicine in the remote programming of cochlear implants. Acta Otolaryngol. **129**(5), 533–540 (2009)
21. Wesarg, T., et al.: Remote fitting in nucleus cochlear implant recipients. Acta Otolaryngol. **130**(12), 1379–1388 (2010)

Integration of Software and Hardware AI Learning Models in the SEPT Learning Factory

Dan Centea[1(✉)], Ishwar Singh[1], Anoop Gadhrri[1], Sean Hodgins[2], and Reiner Schmidt[3]

[1] McMaster University, Hamilton, ON, Canada
centeadn@mcmaster.ca
[2] Idle Hands Development, Ottawa, ON, Canada
[3] Roboteurs Inc., Hamilton, ON, Canada

Abstract. The learning factory developed in the School of Engineering Practice and Technology at McMaster University is an academic entity that focuses on education, applied research and training. One of the current focuses of this learning factory is the integration of Artificial Intelligence (AI) learning models that support the delivery of AI-related curricula. The development and use of applications related to prediction systems are presented, and an approach to develop an AI prediction model for machine health monitoring used for education and training is described. Four small cyber-physical systems that have been designed, built, and put in operation for demonstration and applied research on AI technology topics are described. AI based vision systems for quality monitoring, object detection, gesture recognition, and facial recognition are also described. The readers can contact the authors to get detailed information about the hardware, software modules, libraries and procedures used to teach the neural networks and develop the prediction models.

Keywords: Artificial intelligence · AI · Machine health · SEPT learning factory

1 Introduction

AI systems receive large amounts of labelled training data. They analyse the data, use algorithms to discover patterns, and use them to make predictions about the future states of the analysed system. Machine Learning (ML) is a subcategory of AI, and supervised machine learning is a subcategory of ML that uses labelled datasets to train the algorithms used to predict the outcomes of the system.

The number of AI and machine learning implementations in different industries have been expanding at a fast pace. Machine learning and AI algorithms are used in many fields such as manufacturing [1–3], food industry [4], marketing [5] – to name just a few. In the case on the manufacturing industry, AI is expected to augment human intelligence through different implementation such as machine vision for quality checks and control; prediction of failure modes and predictive maintenance; generative design;

M. E. Auer and T. Tsiatsos (Eds.): IMCL 2021, LNNS 411, pp. 322–330, 2022.
https://doi.org/10.1007/978-3-030-96296-8_29

use of digital twins; reduction of environmental impacts; improvement of business operations including price forecasts; and hyper-automation.

There is a latency between the pace of introducing AI application in industrial environments and the pace in which academic curricula can be changed or adapted to provide qualified graduates. The knowledge of the graduates should be ahead of the knowledge used in industry – and not the other way around. The graduates with interest in the AI fields need to be able to understand and maintain industrial existing AI application, to improve them, to implement new applications, and ultimately to develop in-house applications specific to their workplace.

Although AI learning modules have been applied to all level of education to encourage deep learning, from elementary school [6] to university level [7, 8] and for lifelong learning [9], there is a shortage of talent in this field as evidenced by AI based company acquisitions, large employment opportunities, and hundreds of new AI start up companies looking for talent. The School of Engineering Practice and Technology (SEPT) at McMaster University developed an academic entity called Learning Factory (LF) that focuses on education, applied research and training. The SEPT LF has adopted a multi- faceted approach in addressing the challenge created by the shortage of experts in AI by putting an emphasis on AI-related elements such as the use of AI tools in its operation; implementation of AI in training, education, and applied research [10, 11]; and creating an Information Technology (IT) infrastructure for developing AI models.

The purpose of this paper is to present a change in academic practice aimed at introducing to engineering education practical knowledge related to developing and using basic AI systems and to provide experiential learning skills to students with interest in self-developing AI applications. The use of a LF for education and for a safe development and use of AI experiments is described in Sect. 2. Section 3 presents the AI-related hardware and software modules that can be used to develop and use four machine-health monitoring and prediction systems. Three AI experiments that use vision systems for quality monitoring, object detection, gesture recognition, and facial recognition are presented in Sect. 4, while Sect. 5 includes a summary of the paper.

The AI based systems and analysis stations presented in this paper include the hardware and software learning modules that are expected to provide the intended AI knowledge and experiential learning skills to groups of learners such as undergraduate students performing laboratory experiment, graduate students engaged in research, and industrial employees interested in training related to AI applications. Although the systems are described with a reduced level of detail, the readers interested in developing similar applications for their students or trainees are encouraged to contacts the authors to request detailed specifications of the hardware modules, suggested software applications, and step-by-step procedures for all the AI systems described in this paper.

2 SEPT Learning Factory

The SEPT LF is an educational unit in the Faculty of Engineering at McMaster University that delivers programs with a strong emphasis on student-centered learning. SEPT provides undergraduate programs with emphasis on engineering technologies and specialized graduate programs with focus on engineering practice.

The success of the undergraduate programs offered within SEPT relies on the School's ability to respond to industry needs and to the challenges of delivering graduates with the knowledge and skills required by the constantly-evolving modern technologies. Targeting niche educational markets and industry sectors, these programs have the flexibility to modify, enhance, and add new curricula to address the changing needs of industry. The approaches proposed by SEPT to address these challenges is to use a LF as an innovation hub supported by a series of specialized labs used for education, training, and research [12–14]. These labs are designed to address the shortage of skilled personnel with knowledge in designing and using modern technologies and applications, one of them related to AI, through a life-long learning approach implemented in an academic environment. This life-long learning starts at undergraduate level and continues through graduate studies or through trainings offered to industry employees.

Beyond acting as an academic leaning and training centre, the LF is a safe environment for students engaged in industrial projects and employees from industry to have hands-on experience with modules related to digital technologies. Instead of developing or modifying AI-related applications in industrial environments, the AI systems can be developed and tested in an environment that does not affect industrial production.

The SEPT LF is an ideal space for students who can learn modern systems and methods and develop experiential learning skills. Evidence of the learning that occur in the LF is the fact that most of the work described in this paper is developed by the students who are the co-authors of this paper. Moreover, these projects were accomplished as a part of setting self-learning goal set up in the LF and described in [15]. The following sections describe AI-related systems developed and tested in the SEPT LF that are in use for education and research, and are expected to be used for training.

3 Machine-Health Monitoring and Prediction Systems

Many industrial processes need systems able to detect mechanical problems and predict possible failures. Before being applied to real production machines, it is useful to develop systems that mimic the possible errors at a smaller scale. This section of the paper describes four machine-health systems that can be developed and tested in an academic environment, can be programmed by students to predicts failures, can be used for training employees from industry, and can be presented in academic industry-focused forums. If needed, similar systems built at a larger scale can be implemented in industry for similar machines health applications.

3.1 Excel-Based Neural Network Model Development

Consider a system that includes a small massager for sore and aching muscles known as jitter critter massager - ladybug (LBM). This inexpensive mechanical palm-size device that looks like a ladybug has four feet which, when moving, work human's muscles and relieve stress. Consider a hardware module mechanically attached to the LBM that is connected to an Arduino board and communicates with it using the Inter-Integrated Circuit communication protocol I^2C. If the hardware module is an inexpensive micro electro-mechanical system, such as GY-521 MPU 6050 sensor module, it sends to the Arduino board sensor data from three sensors: a 3-axis gyroscope, a 3-axis accelerometer, and temperature sensor. These values are sent to Microsoft Excel with Excel Data Streamer add-in enabled and are processed with Excel's Fast Fourier Transform (FFT). The output to the FFT analysis can be processed to obtain an ordered list of frequency vs FFT magnitude. This data set is available for neural networks (NN) training for a given condition. Three sets of data can be collected: LBM off (to acquire base vibration data); LBM turned on and placed on a hard floor; and LBM turned on and placed on a soft surface. The results obtained from the NN model can predict the condition of the LBM running on a hard or soft surface, or mixed mode when two legs of the LBM are on the hard and two on the soft surface. An advantage of this experiment is that it uses tools that are free and open source and can be used for small to medium sized data sets.

The experiment provides easy-to-use tools for beginners in the areas of neural network, prediction, and classification problems. The students can put their own data sets for different acceleration axis in the tools, play with various parameters, and study the effect of result. Furthermore, since the indicated sensor modules also provides gyroscopic data and temperature, sensor data can be included in model building and tested for prediction use. The experiments are easy to setup for classroom teaching and can be used to build small prototypes.

The programming code, written in Visual Basic for Applications (VBA - Microsoft's programming language for Excel) is available for students to view and edit. Students are encouraged to make modifications for gaining experience in building a NN model and testing NN applications.

Microsoft Excel is used as it represents the quickest way to introduce the supervised leaning model development and prediction without writing a single line of code. In the model presented in this section the code saved in the microcontroller is used for data collection and streaming to Excel, while Excel is used for calculating FFT and providing data for developing a supervised learning model. The model is used to predict the state of the machine that generates the vibration data. All other programming methods to accomplish this task require a lot of coding (Python and various AI/ML libraries) for software and hardware integration.

Due to COVID 19 restrictions for in-person labs, the experiment provides an alternative to in-class labs. Students can purchase these parts or the School can ship them to their home. An approach for students who do not have the hardware at home is to send them collected vibration data for preprocessing, model building, and prediction tasks.

Another version of this experiment involves the use of a sensor that measures 3-axis angle, angular velocity, acceleration and magnetic field, and transmits data either by Bluetooth or USB (for instance the accelerometer and inclinometer BWT901CL that uses an altitude and heading reference system AHRS and inertial measurement units IMU). Its strength lies in the algorithm which can calculate the three-axis angle accurately. The robust housing and the small outline make it perfectly suitable for industrial retrofit application such as motion capturing, engineering machine monitoring, etc. This sensor is mounted on the top of the LBM and turned to continuously stream the data to a PC with Windows 10 data capture app supplied by the vendor. The collected data is saved as a CSV file and then opened in Excel to develop the NN model for prediction.

3.2 IoT Vibration Analysis Station Using MQTT

A vibration analysis station has been designed, built, and is used in the SEPT LF to predict mechanical failure using AI. A mechatronic system that includes a small electric motor controlled by a Programmable Logic Controller (PLC), a capacitive touch trigger, and sensors for acceleration, DC current, and temperature is connected to an Arduino board. The purpose of the vibration analysis station is to use NN to build a failure prediction model. To simulate a failure event, a counterweight is attached to the motor shaft to create a counterbalance needed to generate data for building a predictive model.

The experiment includes the following steps: (1) vibration data collection (load and no-load); (2) perform FFT to convert vibration time-waveform to frequency domain; (3) generate spectrum data from FFT calculations; (4) select the data for training the NN; (5) transform the data so that its distribution has a mean value 0 and a standard deviation of 1; (6) create the NN model for prediction; (7) plot the model during the training progress; (8) save the model for prediction. The model is then loaded into a program running on the Arduino to determine the existence of an imbalance. The program continuously collects data every two seconds, converts into FFT and frequency values, performs a failure prediction, and displays it on a dashboard developed using a Node-Red platform. This platform connects with a broker of the MQTT standard messaging protocol for IoT to publish the prediction information data.

3.3 Fan Fault Detection and Diagnosis Using Machine Learning

The purpose of the described fan fault and diagnosis system is to determine faults in a spinning fan and to detect the part of the fan that is faulty. The system utilizes vibration and current measurements in combination with machine learning. The circuit board that contains a microcontroller (SAMD21G18A), an I^2C accelerometer (ADXL345), an op-amp circuit for measuring current, and LEDs (Fig. 1) is connected to a Raspberry Pi (RPi) through the Serial Port via USB. Upon request from the RPi, the microcontroller records vibration (in FFT) and current data from the fan. A program collects acceleration and current sensor data and transforms the vibration data using FFT library function. The RPi uses the data to train a NN to make assumptions about the state of the fan. A program runs several times to detect the faults using the trained NN model.

When training is finished, the program waits for the user to start running the system so that it can collect data continuously in real-time and perform the predictions. The system can detect the several fan conditions that include "healthy", "air blockage", and "balance issue". It is interesting to note that, although the NN was not trained for a combination of the air blockage and balance issue, it can predict this condition correctly.

3.4 Wireless Machine Health Monitoring and Prediction System

The advanced predictive learning station shown in Fig. 2 uses an embedded vibration sensor in combination with other sensors to analyze a mechanical system for faults and transmit the information wirelessly. The system can be easily extended to further application demonstrations such as bearing faults.

The system can be used at different academic levels - from hands-on undergraduate labs to applied research. The undergraduate students are be able to identify errors such as general imbalances, mechanical failure, resonance, electrical faults, and critical speeds. The graduate students can develop AI models using multiple sensor inputs and perform various analyses.

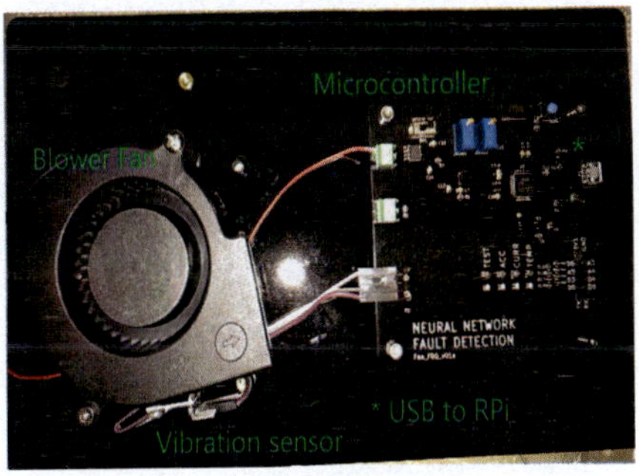

Fig. 1. Fan fault detection and diagnosis system

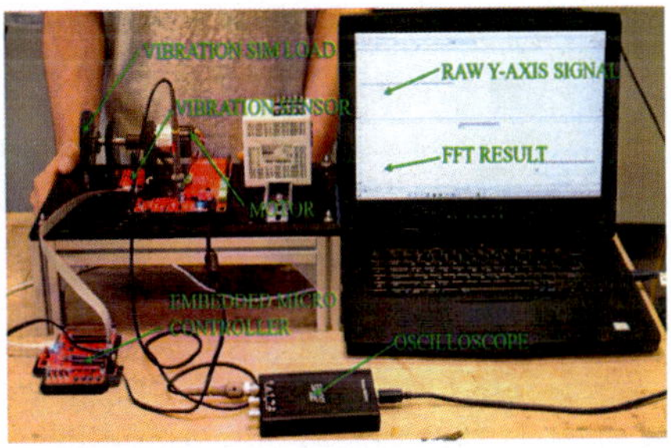

Fig. 2. Machine health monitoring and prediction system

4 AI Based Vision Systems

4.1 RFID and Facial Recognition-Based Security and Monitoring System

Many buildings have rooms that require strict access. The traditional approach of using radio frequency identification (RFID) systems has flaws as it provides access to anyone having the card. A stolen RFID tag can be used to get unauthorized access. This issue can be avoided by combining a RFID system with a NN based facial recognition.

A deep neural network is used on a 2-D image to extract facial features and represent them as numerical values. A program compresses pixel data to data that represents features such as skin tone, eye spacing, and so on. The numerical numbers can be used to detect faces. When the same face is run through the model twice the two outputs have similar numerical outputs, and a comparison algorithm can be used to identify the person. This system runs on a Linux platform and communicates with a server when a face detection is triggered. The server checks the database to identify if the person has access or not and sends a true or false condition to the Linux system.

4.2 Vision-Based Cobot for Parts Assembly System

One of the challenges in the SEPT LF is to provide to a collaborative robot (cobot) the ability to pick up a Kanban container from the rack and deliver it to an assembly station. The cobot, equipped with a pickup fork and mounted on mobile-intelligent-robot (MIR), contains a camera to guide its movement when the MIR brings it near the Kanban container. A machine vision system guides the robot to pick up the container. Machine vision algorithms, developed in Python using OpenCV software library, includes functions to detect corners and use them for pose estimation and camera calibration. The students using this system spent most of their time using these modules for camera calibration and detection tasks. The types of cameras tested so far with the systems are a web camera, a standard machine vision camera, and an intel real sense camera. The last camera gave the best results.

4.3 Gesture Recognition Systems

AI approaches have been used in the SEPT LF to develop two human-machine interfaces based on gesture recognition systems. Each system includes a digital ambient light/proximity/gesture sensor (APDS 9960), a muscle sensor, a gesture control arm-band, and a vision system. These gesture recognition systems are used for controlling the operation of a cobot at an assembly station and to control the open/close status of the door of a CNC machine tool. For controlling the cobot operations, a python program acquires gesture data from the gesture recognition system and sends commands to the cobot to perform the motion related to the arm movement. In a similar fashion, the CNC door is opened or closed based on the gesture data received by the python program. The gesture recognition software component is based on building NN models from scratch. Students are initially asked to develop different models and test the output of the NN models. The students who manage to develop a working NN model, which is not a trivial task, are given more complex models and are asked to edit them to get outcomes with a higher accuracy and precision.

5 Summary and Conclusions

Four different types of physical models (Excel-based NN model, IoT vibration analysis station, a fan fault detection and diagnosis system and a machine health Learning Station) developed and in use in the SEPT Learning Factory have been presented. A method of accelerating NN model development for machine health prediction using Excel along with a realistic vibration data collection example has been demonstrated. It has been observed that the Excel model development requires no programming, while other programming tools require deeper understanding of the software. Model development and deployment require the careful consideration of many factors and have limitations. AI based vision system applications been explored with good results.

Acknowledgement. The projects presented in this paper are supported by Future Skills Centre, Canada.

References

1. Ghahramani, M., Qiao, I., Zhou, M.C., O'Hagan, A., Sweeney, J.: AI-based modeling and data-driven evaluation for smart manufacturing processes. IEEE/CAA J. Automatica Sinica **7**(4), 1026–1037 (2020). https://doi.org/10.1109/JAS.2020.1003114
2. Li, Bh., Hou, Bc., Yu, Wt., Yang, C.W.: Applications of artificial intelligence in intelligent manufacturing: a review. Front. Inf. Technol. Electron. Eng. **18**, 86–96 (2017). https://doi.org/10.1631/FITEE.1601885
3. Yao, K., Zhou, J., Zhang, J., Boër, C.R.: From intelligent manufacturing to smart manufacturing for Industry 4.0 driven by next generation artificial intelligence and further on. In: 5th Int. Conference on Enterprise Systems (ES), pp. 311–318 (2017). https://doi.org/10.1109/ES.2017.58
4. Kakani, V., Nguyen, V.H., Kumar, B.P., Kim, H., Pasupuleti, V.R.: A critical review on computer vision and artificial intelligence in food industry. J. Agric. Food Res. **2**, 100033 (2020). Doi:https://doi.org/10.1016/j.jafr.2020.100033
5. Ma, L., Sun, B.: Machine learning and AI in marketing – connecting computing power to human insights. Int. J. Res. Mark. **37**(3), 481–504 (2020). https://doi.org/10.1016/j.ijresmar.2020.04.005
6. Ryu, M., Han, S.: AI education programs for deep-learning concepts. J. Korean Assoc. Inform. Educ. **23**(6), 583–590 (2019). https://doi.org/10.14352/jkaie.2019.23.6.583
7. Cui, W., Xue, Z., Shen, J., Sun G., Li, J.: The item response theory model for an AI-based adaptive learning system. In: 18th International Conference on Information Technology Based Higher Education and Training (ITHET), pp. 1–6 (2019). https://doi.org/10.1016/j.ijresmar.2020.04.005
8. Perrotta, C., Selwyn, N.: Deep learning goes to school: toward a relational understanding of AI in education. Learn. Media Technol. **45**(3), 251–269 (2020). https://doi.org/10.1080/17439884.2020.1686017
9. Poquet, O., De Laat, M.: Developing capabilities: Lifelong learning in the age of AI. British J. Educ. Technol. **52**(4), 1695–1708 (2021). https://doi.org/10.1111/bjet.13123
10. Gao, Z., Wanyama, T., Singh, I.: Project and practice centered learning: a systematic methodology and strategy to cultivate future full stack artificial intelligence engineers. Int. J. Eng. Educ. **36**(6), 1760–1772 (2020)
11. Jiang, W., Singh, I., Gao, Z.: Comparative studies of supporting vector machines and artificial neural networks for scheduling optimization. Int. J. Mech. Electr. Comput. Technol. (IJMEC) **10**(38), 4749–4751 (2020)
12. Elbestawi M, Centea D, Singh I, Wanyama T. SEPT learning factory for Industry 4.0 education and applied research. Procedia Manufacturing 23:249–254 (2018), https://doi.org/10.1016/j.promfg.2018.04.025
13. Singh, I., Centea, D., Elbestawi, M.: IoT, IIoT and cyber-physical systems integration in the SEPT learning factory. Procedia Manuf. **31**, 116–122 (2019). https://doi.org/10.1016/j.promfg.2019.03.019

14. Centea, D., Singh, I., Elbestawi, M.: SEPT approaches for education and training using a learning factory. Procedia Manuf. **31**, 109–115 (2019). https://doi.org/10.1016/j.promfg.2019.03.018
15. Hsiao, Y.-C., Al-emara, S., Gadhrri, A.S., Singh, I., Gao, Z.: Self-directed learning compared to traditional engineering approach: case studies in developing machine learning capabilities to solve practical problems. In: Auer, M.E., Centea, D. (eds.) ICBL 2020. AISC, vol. 1314, pp. 132–144. Springer, Cham (2021). https://doi.org/10.1007/978-3-030-67209-6_15

Emerging Mobile Technologies and Standards

Mobile Apps in Retail: Usage Frequency Before, During, and After the SARS-CoV-2 Pandemic – Insights from the German Market

Atilla Wohllebe(✉)

MATE Hungarian University of Agriculture and Life Sciences – Kaposvár Campus, Kaposvár, Hungary
atilla.wohllebe@gmail.com

Abstract. In the context of the SARS-CoV-2 pandemic and the increasing digitization of stationary retail, mobile apps in retail (MAR) have gained attention among many consumers. This study examines how the frequency of MAR usage changed before and during the pandemic and how consumers estimate how MAR usage will evolve after the pandemic. Additionally, the role of affinity to mobile devices usage (AMU) in this context is examined. Furthermore, the relevance of pandemic-specific MAR features is elaborated. Based on an online questionnaire with 167 responses, it can be shown that MAR have gained significant relevance in the context of the pandemic and will continue to be an important tool for customers when shopping stationary after the pandemic. A moderating role of AMU cannot be confirmed. Click & collect and online shopping are identified as the most important MAR functions in the context of the pandemic. Information about rules for pandemic compliant shopping doesn't play much of a role for consumers. The results are an important signal for retailers to continue their efforts in digitization and MAR development.

Keywords: Smartphone Apps · Retail · Corona · Pandemic · Covid · Germany

1 Introduction

The advancing digitalization has significantly increased the relevance of mobile apps in retail (hereinafter: "MAR") in recent years [1, 2]. Particularly against the backdrop of the high prevalence of smartphones, consumers are increasingly using MAR in stationary retail, for example as a digital shopping assistant [3, 4].

In parallel, the SARS-CoV-2 pandemic (hereafter: "the pandemic") has massively changed society and the economy [5]. Thus, as the pandemic has progressed, not only has e-commerce, which was growing anyway, grown even stronger [6–8]. As a catalyst, the pandemic has made mobile apps even more popular in several areas, including directly for pandemic response (contact tracing, symptom reporting, etc.), but also for communication via video telephony or mobile payment [9–12].

Stationary retail has also had to respond to the pandemic in many places. On the one hand, the stationary shopping experience has been adapted in light of the risk of infection, e.g., by introducing distance rules (1.5–2 m), one-way aisles and "safe shopping" times for vulnerable groups [13]. Shopping for groceries and other everyday goods is

M. E. Auer and T. Tsiatsos (Eds.): IMCL 2021, LNNS 411, pp. 333–341, 2022.
https://doi.org/10.1007/978-3-030-96296-8_30

increasingly also taking place online [8, 11, 14]. Many retailers now also offer Click & Collect, whereby the customer orders the desired products in advance, often digitally, the order is assembled in the store and the customer can pick it up there [15, 16].

Against the backdrop of the digitalization of the retail industry, which has been intensified by the pandemic, and the steadily increasing relevance of mobile apps, especially in the retail sector, this paper seeks to understand how the pandemic has affected the frequency of use of MAR. Specifically, it will examine how the pandemic has affected MAR usage frequency and what consumers expect to see happen to their usage frequency post-pandemic. Additionally, the role of consumer's affinity for mobile device usage (AMU) in this context will be explored. Furthermore, we will find out (explorative) which MAR functions are particularly relevant for consumers in the context of the pandemic.

In terms of frequency of usage, the existing literature suggests that MARs are used more during the pandemic than previously [8, 11, 17].

H1: MAR are used more during the pandemic than before the pandemic.

Furthermore, it can be assumed that consumers will continue to use MAR even after the pandemic has passed [8].

H2: MAR are used more after the pandemic than before the pandemic.

Once the pandemic is over, mobile apps will probably initially lose popularity again, as some use cases are no longer or less relevant. In this respect, the particularly high app usage at present could also in part merely be a temporary phenomenon [7, 17, 18].

H3: MAR are used more during the pandemic than after the pandemic.

We hypothesize that consumers who used MAR frequently before the pandemic will use MAR more frequently during the pandemic because they already know and appreciate the benefits of MAR when they shop [19–21].

H4: The usage of MAR before the pandemic increases the usage of MAR during the pandemic.

Research has shown that an affinity for mobile shopping has a positive effect on consumers installing a retailer's app [22]. From this, it is hypothesized that consumer's affinity to mobile device usage acts as a moderator on the ratio of usage before and during the pandemic.

H5: The effect of usage before on usage during the pandemic is moderated by AMU.

Assuming that many consumers use MAR more intensively for the first time during the pandemic and recognize its benefits in this context, it is assumed that these perceived benefits have a positive effect on the expected continued use even after the pandemic [19, 23, 24].

H6: The usage of MAR during the pandemic increases the usage of MAR after the pandemic.

Analogous to H5, it is assumed that.

H7: The effect of usage during and usage after the pandemic is moderated by AMU.

Because consumers may also increasingly avoid stationary retailers during the pandemic due in part to the risk of infection [7, 25], but already know the benefits of MAR, pre-pandemic usage frequency could also have a positive impact on expected post-pandemic usage frequency [19, 23, 24].

H8: The usage of MAR before the pandemic increases the usage of MAR after the pandemic.

Also, as in H5 and H7, AMU is assumed to be moderated.

H9: The effect of usage before on usage after the pandemic is moderated by AMU.

Regarding the usefulness of pandemic-specific MAR functions, a total of seven functions are identified that are likely to be of particular relevance in the context of the pandemic.

1. Ordering via **click & collect**, because customers want to avoid long stays in stores [11, 16].
2. Ordering from **online shop** (by package delivery), because customers prefer online shopping and stay at home [8, 17].
3. Making an **appointment**, e.g. for product advice in the store, because customers want to make more targeted purchases and spend less time in the store [11].
4. Using an **online chat**, e.g. for virtual product advice, because customers (increasingly) want to shop online or avoid store visits completely [7, 8, 11].
5. Viewing **contact** options (e.g. via e-mail and telephone), because e.g. currently applicable rules, opening hours or advice are requested [7, 8, 25]
6. Seeing the **opening hours**, because there are e.g. special "safe shopping' times for vulnerable groups [25].
7. Getting information on the pandemic **rules** in the store, because the rules on shopping in stationary retail are dynamic and interpreted differently [25].

2 Material and Methods

A three-part questionnaire is developed to test the hypotheses stated and to find out which of the functions identified as relevant to the pandemic are perceived by consumers as useful. In the first part, consumers are asked about their age, gender and affinity for using mobile devices. In the second part, consumers are asked to indicate how often they used MAR before, use MAR during and think they'll use MAR after the pandemic. The last part lists the features that have been outlined previously; respondents are asked to rate how useful they find them in the context of the pandemic. All questions (except age and gender) are answered on a Likert scale from 1 (not at all affine/very rarely/not at all useful) to 5 (very affine/very often/very useful) [26].

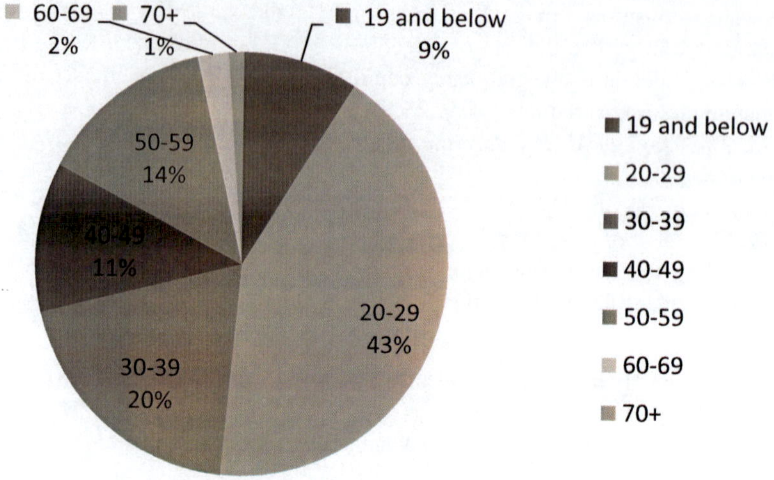

Fig. 1. Distribution of age

The questionnaire is sent to the users of a MAR of a German retailer. A total of 167 users answer the questionnaire. 109 respondents are male, 58 female. Regarding age, the majority of the respondents are rather young. Figure 1 shows the distribution of age.

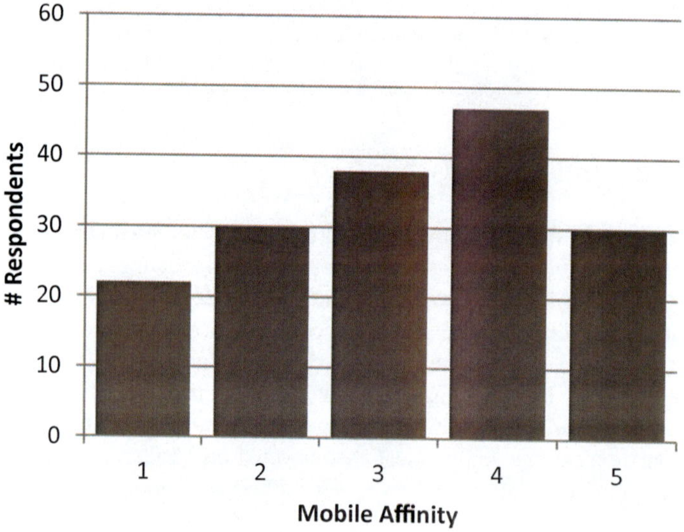

Fig. 2. Affinity of respondents to use mobile devices (AMU)

Figure 2 shows how respondents rate their affinity for using mobile devices. Most respondents rate their affinity as rather high (4).

Based on the survey of users of a MAR, the respondents probably have a higher AMU than the overall population (M = 3.19, SD = 1.29).

Table 1. MAR usage - descriptive statistics.

Usage	Mean	Standard deviation
Pre-pandemic	2.92	1.27
During pandemic	3.40	1.29
Post-pandemic	3.17	1.15

Table 1 shows the descriptive statistics, how frequent the respondents used MAR before the pandemic, use it during the pandemic and think they will use MAR after the pandemic. Respondents were asked to indicate this on a scale of 1 to 5.

3 Results

In the context of the presentation of the results, the hypotheses are processed first, and then the respondents' assessment of the usefulness of MAR functions is presented.

Variance homogeneity is present for MAR usage frequency before and during the pandemic (Variance Ration Test: $p = .79$). Comparison of means using a two-sample t-test shows that frequency of use is significantly greater during the pandemic ($t(332) = 3.46, p < .001$). This confirms H1.

There is also variance homogeneity for frequency of usage before and after the pandemic ($p = .19$). Respondents expect MAR to be used significantly more often after the pandemic than before the pandemic ($t(332) = 1.95, p < .05$). The results confirm H2.

Variance homogeneity can also be demonstrated for frequency of use during and after the pandemic ($p = .12$). Comparison of means provides evidence for acceptance of H3 ($t(332) = 1.70, p < .05$).

A simple linear regression is calculated for H4. The regression model, which attempts to explain frequency of usage during the pandemic by pre-pandemic usage, explains usage during the pandemic to a small to moderate extent, $R^2 = .09, F(1, 165) = 15.84, p < .001$ [27]. The influence of frequency of usage before the pandemic is significant ($\beta = .30, t(165) = 3.98, p < .001$). Thus, H4 is confirmed.

Inclusion of AMU as a moderator improves the previously established model in terms of model fit ($R^2 = .13, F(3, 163) = 7.91, p < .001$). However, the individual regression coefficients and especially AMU as moderator are not significant ($\beta = .02, t(163) = 0.33, p = .75$). Accordingly, H5 is rejected.

Inspection of H6 shows that users who use MAR more during the pandemic expect to do so in the post-pandemic period ($R^2 = .25, F(1, 165) = 54.12, p < .001$); use during the pandemic has a significant impact ($\beta = .44, t(165) = 7.36, p < .001$). H6 is accepted.

Introducing AMU as a moderator yields an even more significant regression model ($R^2 = .30, F(3, 163) = 23.75, p < .001$), with only frequency of use during the pandemic remaining significant (*β* = *.47, t(163)* = *2.87, p* < *.05*). AMU as a moderator is not significant (*β* = *−.03, t(163)* = *−0.64, p* = *.52*). Therefore H7 is rejected.

Analogous to H2, we test whether pre-pandemic usage also influences expected post-pandemic usage. A significant influence can be demonstrated (R^2 = *.23, F(1, 165)* = *49.04, p* < *.001; β* = *.43, t(165)* = *7.00, p* < *.001*). Thus, H8 is confirmed.

In addition to H8, the influence of AMU as a moderator is also tested. The regression model is significant (R^2 = *.26, F(3, 163)* = *19.21, p* < *.001*). No significance can be demonstrated for AMU as a moderator (*β* = *-.02, t(163)* = *−.033, p* = *.74*). Accordingly, H9 is rejected.

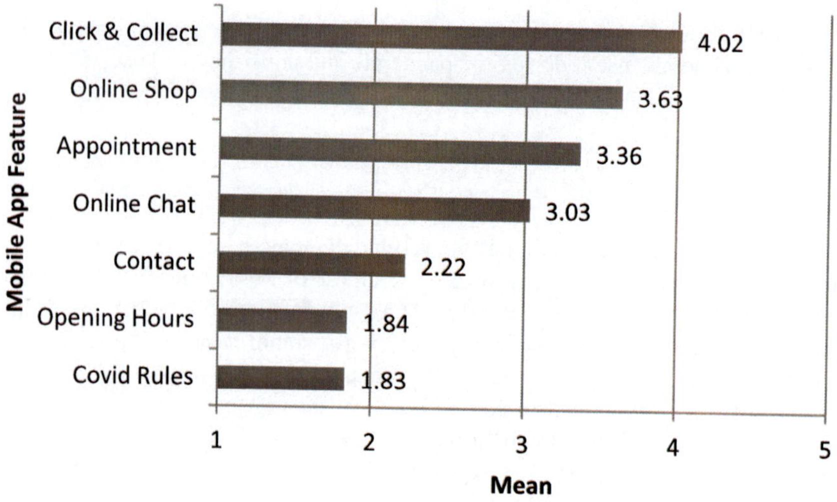

Fig. 3. Usefulness of pandemic-related MAR features

With regard to the usefulness of the pandemic-related MAR functions, the respondents find transaction-related functions particularly useful (cf. Figure 3). Click & Collect as a contact-avoiding option for shopping directly in stationary retail stores is of greatest relevance to respondents (M = 4.02), followed by the option to shop online (M = 3.63). Respondents find functions that inform them about opening hours (M = 1.84) or the pandemic rules that apply in the store (M = 1.83) least useful.

4 Summary and Discussion

In summary, six of the nine hypotheses regarding the frequency of MAR usage can be confirmed. The three hypotheses that assume a moderating role of AMU in the frequency of MAR usage are not confirmed. Thus, the results of hypotheses H1-H4, H6, and H8 are in line with the literature. Table 2 presents the results of the hypotheses.

Table 2. Hypotheses and results.

Hypothesis	Results
H1: More during than before	Confirmed
H2: More after than before	Confirmed
H3: More during than after	Confirmed
H4: Usage before increases usage during	Confirmed
H5: Effect of usage before on usage during moderated by AMU	Rejected
H6: Usage during increases usage after	Confirmed
H7: Effect of usage during on usage after moderated by AMU	Rejected
H8: Usage before increases usage after	Confirmed
H9: Effect of usage before on usage after moderated by AMU	Rejected

The rejection of H5 could be explained by the fact that respondents with a high AMU may have switched directly to online shopping during the pandemic and therefore did not use MAR more than people with lower AMU [7]. Analogously, the rejection of H7 could, for example, be attributed to the fact that people with high AMU think that they will continue to shop online more strongly after the pandemic and therefore MAR usage after the pandemic is not moderated by AMU as expected. This applies analogously to H9, so that overall it could happen that people with high AMU do not become digital-savvy customers of stationary retail, but directly become online shopping customers [5, 8].

This assumption would also be in line with the results of the question about the relevance of MAR features in the context of the pandemic: Online shopping plays the second most important role after Click & Collect. Against the background of avoiding contacts and striving for the shortest possible times in the store, also unsurprising is the relevance of the functions for making appointments and online chat [7, 8, 11]. In the context of MAR features, what is surprising - especially in light of the findings from the literature - is the low relevance of features that provide information on opening hours and pandemic-compliant behavioral rules for shopping [11, 25]. One possible explanation for this could be that many people have already become accustomed to the rules since the beginning of the pandemic or obtain information on current rules from sources other than MAR.

Like any research, this one comes with some limitations. At the same time, these also offer the opportunity for further research projects. In particular, the limited transferability of the results due to the sample composition needs to be mentioned. The sample is geographically limited to Germany, and all respondents use the MAR of a specific retailer. Also, the age group 50 + is only weakly represented (17%), the age group 60 + practically not at all (3%) and the sample is male dominated (65%). To better relate the results to the overall population, similar research could be conducted with larger or different sample compositions.

In addition, the models and procedures used here are relatively simple and in some cases have little explanatory power (e.g., cf. H4; R^2 = .09). More comprehensive models that also take into account, for example, perceived usefulness, friends of use, or use within the customer journey could provide even better explanatory power [13].

5 Conclusion

In this study, the effect of the SARS-CoV-2 pandemic on retail mobile app usage was examined. Using an online questionnaire, 167 consumers were asked to rate their frequency of usage of retail mobile apps before, during, and after the pandemic. The data is consistent with the hypotheses derived from the literature in six out of nine cases. Frequency of usage increased as a result of the pandemic and is likely to be higher after the pandemic than before the pandemic. This confirms the significant increase in relevance of digitally assisted retail and mobile apps in this context. Additionally, consumers were asked how useful they found certain pandemic-related app features. In particular, click & collect and an online store, but also options for making appointments or online advice are valued by consumers. The findings make an important contribution to the understanding of mobile apps in retail and their relevance especially in the wake of the pandemic. In particular, the findings are an important signal for retailers to continue developing MAR to reach customers digitally in the long term. Future research should verify and complement the results found here in more comprehensive models and with larger samples.

References

1. van Heerde, H.J., Dinner, I.M., Neslin, S.A.: Engaging the unengaged customer: the value of a retailer mobile app. Int. J. Res. Mark. **36**, 420–438 (2019). https://doi.org/10.1016/j.ijresmar.2019.03.003
2. Shukla, P.S., Nigam, P.V.: E-shopping using mobile apps and the emerging consumer in the digital age of retail hyper personalization: an insight. PBR **10**, 131–139 (2018)
3. Jäger, R.: Multi-Channel im stationären Einzelhandel: ein Überblick. Springer Gabler, Wiesbaden (2016)
4. Wohllebe, A., Dirrler, P., Podruzsik, S.: Mobile apps in retail: determinants of consumer acceptance – a systematic review. Int. J. Interact. Mob. Technol. **14**, 153–164 (2020). https://doi.org/10.3991/ijim.v14i20.18273
5. Brandtner, P., Darbanian, F., Falatouri, T., Udokwu, C.: Impact of COVID-19 on the customer end of retail supply chains: a big data analysis of consumer satisfaction. Sustainability. **13**, 1464 (2021). https://doi.org/10.3390/su13031464
6. Adler, M., Wohllebe, A.: Consumers choosing retailers on online marketplaces: how can retailers differentiate apart from the price? – an exploratory investigation. IJARBM **1**, 27–36 (2020). https://doi.org/10.51137/ijarbm.2020.1.1.3
7. Dannenberg, P., Fuchs, M., Riedler, T., Wiedemann, C.: Digital transition by COVID-19 pandemic? the german food online retail. Tijdschr. Econ. Soc. Geogr. **111**, 543–560 (2020). https://doi.org/10.1111/tesg.12453
8. Goddard, E.: The impact of COVID-19 on food retail and food service in Canada: preliminary assessment. Canadian J. Agric. Econ./Revue canadienne d'agroeconomie. (2020). https://doi.org/10.1111/cjag.12243
9. Almeida, F.: COVID-19 and the Digitalization Pace. Academia Letters (2021). https://doi.org/10.20935/AL644
10. Kondylakis, H., et al.: COVID-19 Mobile apps: a systematic review of the literature. J. Med. Internet Res. **22** (2020). https://doi.org/10.2196/23170

11. Martin-Neuninger, R., Ruby, M.B.: What does food retail research tell us about the implications of coronavirus (COVID-19) for grocery purchasing habits? Front Psychol. **11** (2020). https://doi.org/10.3389/fpsyg.2020.01448
12. Okpara, K., Hecht, J., Wohlfeiler, D., Prior, M., Klausner, J.: Evaluation of TellYourContacts.org, a patient-initiated digital COVID-19 contact notification tool (Preprint). JMIR Formative Res. **5** (2020). https://doi.org/10.2196/23843
13. Roggeveen, A.L., Sethuraman, R.: Customer-interfacing retail technologies in 2020 & beyond: an integrative framework and research directions. J. Retail. **96**, 299–309 (2020). https://doi.org/10.1016/j.jretai.2020.08.001
14. Zhao, Y., Bacao, F.: What factors determining customer continuingly using food delivery apps during 2019 novel coronavirus pandemic period? Int. J. Hospitality Manage. **91**, 102683 (2020). Doi:https://doi.org/10.1016/j.ijhm.2020.102683
15. Bug, P., Gordon, N., Staudenmaier, A.-S.: Fashion business case study on the German click & collect situation (2018)
16. Davies, A., Dolega, L., Arribas-Bel, D.: Buy online collect in-store: exploring grocery click&collect using a national case study. Int. J. Retail Distribution Manage. **47** (2019). https://doi.org/10.1108/IJRDM-01-2018-0025
17. Zhuang, M., Fang, E., Wang, R., Han, Y.: The effects of COVID-19 on mobile app usage. SSRN Electron. J. (2020). https://doi.org/10.2139/ssrn.3740433
18. Shankar, V., et al.: How technology is changing retail. J. Retail. **97**, 13–27 (2021). https://doi.org/10.1016/j.jretai.2020.10.006
19. Kim, S., Baek, T.H.: Examining the antecedents and consequences of mobile app engagement. Telematics Inform. **35**, 148–158 (2018). https://doi.org/10.1016/j.tele.2017.10.008
20. Pantano, E., Priporas, C.-V.: The effect of mobile retailing on consumers' purchasing experiences: a dynamic perspective. Comput. Hum. Behav. **61**, 548–555 (2016). https://doi.org/10.1016/j.chb.2016.03.071
21. Parker, C.J., Wang, H.: Examining hedonic and utilitarian motivations for m-commerce fashion retail app engagement. J. Fashion Mark. Manage. Int. J. **20**, 487–506 (2016). https://doi.org/10.1108/JFMM-02-2016-0015
22. Wohllebe, A., Ross, F., Podruzsik, S.: Influence of the net promoter score of retailers on the willingness of consumers to install their mobile app. Int. J. Interact. Mob. Technol. **14** (2020). https://doi.org/10.3991/ijim.v14i19.17027
23. Li, C.-Y., Fang, Y.-H.: Predicting continuance intention toward mobile branded apps through satisfaction and attachment. Telematics Inform. **43**, 101248 (2019). https://doi.org/10.1016/j.tele.2019.101248
24. Wu, L.: Factors of continually using branded mobile apps: the central role of app engagement. IJIMA. **9**, 303 (2015). https://doi.org/10.1504/IJIMA.2015.072884
25. Shumsky, R.A., Debo, L., Lebeaux, R.M., Nguyen, Q.P., Hoen, A.G.: Retail store customer flow and COVID-19 transmission. PNAS **118** (2021). https://doi.org/10.1073/pnas.2019225118
26. Likert, R.: A technique for the measurement of attitudes. Arch. Psychol. **22**, 1–55 (1932)
27. Cohen, J.: Statistical power analysis for the behavioral sciences. Elsevier (1977). https://doi.org/10.1016/C2013-0-10517-X

Proposal for a Deployment of a Non-standalone 5G Mobile Network Architecture for Developing Countries: Case of Senegal

Latyr Ndiaye(✉), Samba Diouf, Kéba Gueye, and Samuel Ouya

Laboratory LIRT, Higher Polytechnic School, University Cheikh Anta Diop of Dakar, Dakar, Senegal
{latyrndiaye, keba.gueye}@esp.sn

Abstract. Mobile networks have evolved a lot in recent years, moving from 2G to 5G. This transition is marked in Africa by a delay compared to the West, especially for 4G. Today, developed countries have deployed 5G either in standalone mode or in non-standalone mode. This deployment allows huge technological advances even if the networks are more and more massive. In this context, SDR (Software-Defined Radio) coupled with SDN (Software-Defined Networking), for the control of deployed equipment, are a major asset to ensure performance and quality of service. Our paper is positioned to help African countries avoid a delay with 5G. To achieve this, we propose an architecture and deployment of 5G in non-standard mode for developing African countries where 4G deployment is in its final stages. This transition from 4G to 5G will be based on SDR technology for a better management of the radio part with a lower cost deployment.

Keywords: Open5gs · gNB · UE · Non-standalone 5G · 4G

1 Introduction

The availability of new mobile services as well as the very large number of mobile objects has led to a very high data usage. In addition, the new types of services are very demanding in terms of QoS, bandwidth, speed and latency which are fundamental elements for the fifth generation of 5G mobile network.

5G is deployed in standalone or non-standalone mode and is an environment that combines many networks with different protocols and requirements thanks in part to software engineering. Virtualization has changed the economics of mobile networks and given the chance for network service market players to become independent of proprietary solutions and apply an open and flexible network architecture. Therefore, it becomes very important to create such a network that includes low capital and operational expenses and grows revenues through the introduction of new services and the increase of network infrastructure. In this context, three technology trends are emerging in telecom networks. These are Software Defined Radio (SDR), Software-defined Networking (SDN), and Network Function Virtualization (NFV), which are expected

M. E. Auer and T. Tsiatsos (Eds.): IMCL 2021, LNNS 411, pp. 342–360, 2022.
https://doi.org/10.1007/978-3-030-96296-8_31

to converge under the umbrella of the 5G Software-Defined Wireless Network (5G-SDWN) concept to meet the growing demands of various services [1].

The transition of the RAN from analog to digital has also enabled the use of components with software-defined functionality rather than hardware components. This is the case with software defined radio. SDR allows a radio access network to be deployed using only an RF-baseband transceiver and a digital processor or FPGA. This approach allows for simplified cross-platform communication and flexible configuration of the radio network architecture. It has also enabled the deployment of 5G in a non-standalone mode and saves time from the technology gap. The fifth generation mobile network also relies on the considerable contribution of software defined networking (SDN) and function virtualization (NFV).

Therefore, it becomes very important to create such a network that includes low capital and operating expenses and at the same time increases revenues through the introduction of new services and the increase of the network infrastructure. Among all the new contexts, three emerging technology trends in telecommunication networks Software Defined Radio (SDR), Software-defined Networking (SDN) and Network Function Virtualization (NFV) are expected to converge under the umbrella of 5G Software-Defined Wireless Network (5G-SDWN) concept to meet the growing demands of different services.

The software defined virtual radio application can be provided as a service and managed by a virtual controller [2]. In addition, interaction, flexibility and agility are managed.

Although the 5G network is a radio access environment, it is also a terrestrial network that allows for managing the interaction of modules in a virtualized environment. That's why the software-defined network (SDN) approach is a perfect choice to manage these interactions. A network based on this approach is more flexible, easily saleable and has the ability to manage the services provided through a public virtual controller [3].

The combined use of SDN SDR in 5G network deployments enables more efficient service delivery while ensuring QoS. In this paper, we propose a non-standard 5G deployment for developing African countries for which 4G deployment is in the finalization phase. In order to avoid a heavy investment at a loss, we propose this solution which would allow them to catch up with the technology despite the problems of means and infrastructure.

2 Related Works

In this section we present the studies related to the deployment of 5G in autonomous and non-autonomous mode taking into account the concepts of SDR and SDN. The concept of programmable networks has been studied in detail in [4]. The idea behind the concept of SDN was discussed as well as how SDN enables innovation through network flexibility and adaptability. The authors presented the SDN architecture and the OpenFlow standard. SDN-based protocols, services and schemes were also provided and potential uses and directions of SDN were highlighted. The development of the SDN concept in wireless networks was presented in [5]. Four classes of popular wireless networks were discussed, namely cellular, sensor, mesh and home networks. The work in [6] introduced several SDN-based extensions to control platforms, switches and base stations in cellular networks. The authors study how subscriber mobility, fine-grained measurement and control, and real-time adaptation are handled using SDN functions. In [7], network virtualisation was discussed from the perspective of cloud computing. In [8], the integration of heterogeneous wired and wireless networks using SDN, NFV and SDR technologies is focused [9] experimentally evaluates two distinct hybrid architectures applied to FiWi 5G New Radio (NR) systems and based on different optical fronthaul approaches. The first architecture operates in non-autonomous mode (NSA), as defined by the 3rd Generation Partnership Project (3GPP), to transmit 4G and 5G technologies simultaneously through a single FiWi system. The second architecture uses Radio over Fibre (RoF), Free Space Optical (FSO) and wireless technologies converging to a heterogeneous network (HetNet). We [10] propose a development and evaluation platform capable of integrating QoS and investigating ways to enable 5G performance at a much earlier stage of the generation transition. In the experimental setup, an LTE air interface is complemented by a software defined radio (SDR) [11] analyses and compares the SA NR and NSA NR deployment modes in terms of coverage, network capacity, interworking between 4G and 5G, network deployment complexity and cost, and the latest industry developments. NSA NR performs better in terms of interworking in the initial phase, while SA NR performs better in terms of network capacity, device performance, simplicity of network deployment and cost effectiveness [12] presents a non-autonomous 5G ETSI MEC architecture for mission-critical push-to-talk (MCPTT) services. Its proposal is based on a hierarchical distributed MCPTT architecture that allocates the user plane to the edge, keeping the control plane (CP) centralised for synchronisation and support purposes [13] focuses on the early deployment of the new 5G Non-Standalone (NSA) radio (NR) over eMBB to achieve the required throughput. The [14] attempts to identify all the important obstacles and to review the state of the art of research regarding the place of SDN in 5G. His study focuses on the most important issues and examines the solutions proposed in and outside the SDN literature. Particular work is done on fronthaul, backward compatibility, the supposedly disruptive nature of SDN deployment, business cases and monetisation of SDN-related upgrades, general purpose processor (GPP) latency and the additional security vulnerabilities that softwarisation brings to the mobile network. It also provides a summary of architectural developments in the SDN-based mobile network landscape, including options for deploying SDN in the NFV framework [15] addresses access to radio resources across heterogeneous networks for roaming mobile users. It proposes a unified service architecture that enables seamless handover between a 5G (New Generation Core) and a 4G (Evolved Packet Core) service via the network slicing paradigm [16] presents the work done in the Open Air Interface project which is an open source initiative providing today a 3GPP compliant reference implementation of LTE, eNB, UE and EPC. LTE, eNB, UE and EPC, which runs on general purpose (×86) computing platforms as well as on COTS SDR cards such as the ETTUS USRP. Open Air Interface has been presented in the Standard and Non-Standard scenarios mainly for the description of its evolution regarding 5G technologies.

Our paper, although close to some, brings a new element. It proposes a deployment of 5G in non-standard mode for developing African countries where the deployment of

4G is in the final stages. In order to avoid a heavy investment at a loss, we propose this solution, which would allow them to catch up with the technology despite the problems of means and infrastructure.

3 Our Approach

Today we note a very important evolution in the telecommunications sector and mobile network. With the advent of the fifth generation (5G) which constitutes an important turning point for the access to Internet, the African states must be able to appropriate this technology. The deployment of a mobile network should no longer be a problem for African states to offer innovative services to the population. Some operators have even started to deploy the 4G network in some countries such as Senegal where the telephone operator Sonatel has already deployed 4G.

African developing countries must understand the need to have a functional operator network to meet the demands of the population. In the context of covid 19 that we live, many state services do not function properly because the staff is put on technical unemployment or forced to telework. However, the African states do not have a high speed access to the Internet which would allow them to carry out this telework. In addition to this, students are still claiming difficulties to follow their courses online because of covid.

It is time that African states take their destiny in hand to have their own cell phone network. This will allow them to deploy dedicated services to address the problems of the populations.

In this article, we propose to the states the fluid mechanisms to have a cell phone network by relying on the existing infrastructure, namely the mesh, for some countries, in optical fiber. This network is based on purely open source software for the radio and core network parts. This allows a significant reduction of the deployment costs.

In this paper, we propose a 5G network deployment based on free and open source software, so the radio part will be maintained as 4G entities and the core network will be a 5G implementation.

This paper will first show how a state can deploy its own 4G network, then we will show how the same state that has a 4G network can migrate to 5G without changing its existing infrastructure what we call here a non-autonomous 5G network.

We will present different solutions obtained in relation to the different cases mentioned.

4 Technologies Used

In this section, we will present the different tools and technologies that we have used to implement first a 4G network and then to migrate to a non-autonomous 5G network.

4.1 srsLTE

srsLTE is a free and open source 4G LTE software suite. With srsLTE, we can create an end-to-end software-defined mobile radio network.

The srsLTE suite includes [17]:

- srsUE - a complete SDR LTE UE (user equipment) application;
- srsENB - a complete SDR LTE eNodeB (base station) application;
- srsEPC - a lightweight LTE EPC (Core Network) implementation with MME, HSS and S / P- GW (Fig. 1).

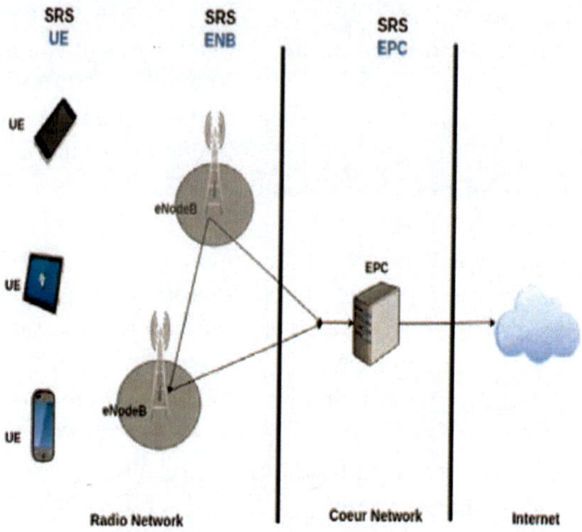

Fig. 1. srsLTE architecture

4.2 srsUE

srsUE is a fully software-implemented LTE UE modem. Running as an application on a standard Linux-based operating system, srsUE connects to any LTE network and provides a standard network interface with high-speed mobile connectivity. To transmit and receive radio signals over the air, srsUE requires SDR hardware such as Ettus Research's USRP [17].

LTE srsUE includes the following features [1]:

- LTE Release 10 aligned with features up to release 15;
- TDD and FDD configurations;
- Tested bandwidths: 1.4, 3, 5, 10, 15 and 20 MHz;
- Transmission modes 1 (single antenna), 2 (transmit diversity), 3 (CCD) and 4 (closed loop spatial multiplexing);
- Manually configurable DL/UL carrier frequencies;
- Soft USIM supporting XOR/Milenage authentication;
- Hard USIM support via PC/SC;
- Snow3G and AES integrity/encryption support;
- TUN virtual network kernel interface integration for Linux OS;
- Detailed logging system with layer-based logging levels and hexadecimal dumps;

- MAC and NAS layer wireshark packet captures;
- Command line trace measurements;
- Detailed input configuration files;
- Advanced Multimedia Broadcast and Multicast Service (eMBMS);
- Frequency-based ZF and MMSE equalizers;
- Highly optimized turbo decoder available in Intel SSE4.1/AVX2 (+100 Mbps) and Standard C (+25 Mbps);
- Supports Ettus USRP B2×0/X3×0, BladeRF, LimeSDR families (Fig. 2).

• **UE architecture**

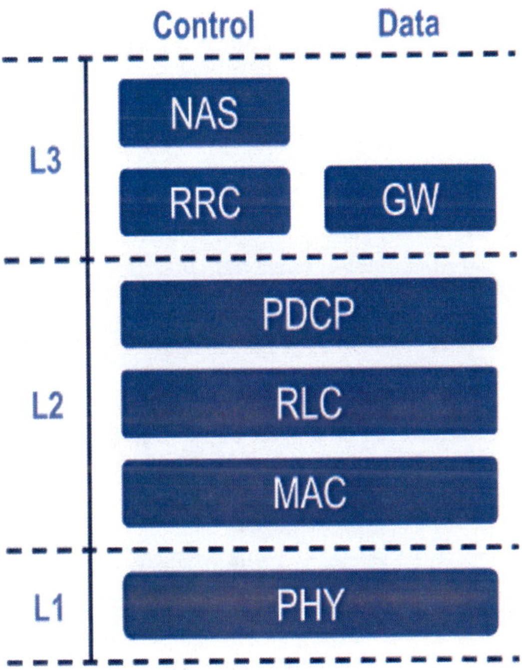

Fig. 2. UE architecture

4.3 srsENB

srsENB is a fully software-based LTE eNodeB base station [18]. Running as an application on a standard Linux operating system, the srsENB connects to any LTE core network (EPC) and creates a local LTE cell. It can transmit and receive radio signals over the air and requires SDR hardware such as Ettus Research's USRP.

The eNodeB LTE srsENB includes the following features:

- LTE Release 10 aligned;
- FDD configuration;

- Tested bandwidths: 1.4, 3, 5, 10, 15 and 20 MHz;
- Transmission mode 1 (single antenna), 2 (transmit diversity), 3 (CCD) and 4 (closed-loop spatial multiplexing);
- Frequency-based ZF and MMSE equalizer;
- Advanced multimedia broadcast and multicast service (eMBMS);
- Highly optimized turbo decoder available in Intel SSE4.1/AVX2 (+100 Mbps) and Standard C (+25 Mbps);
- MAC, RLC, PDCP, RRC, NAS, S1AP and GW layers;
- Detailed logging system with layer-based logging levels and hexadecimal dumps;
- MAC layer wireshark packet capture;
- Command line trace metrics;
- Detailed input configuration files;
- Channel simulator for EPA, EVA and ETU 3GPP channels;
- ZeroMQ-based fake RF driver for I/Q on IPC/network;
- Round Robin MAC scheduler with FAPI-like C++ API;
- SR Support;
- Support for periodic and aperiodic I/Q feedback;
- Standard S1AP and GTP-U interfaces to the core network;
- 150 Mbps DL in 20 MHz MIMO TM3/TM4 with commercial UEs;
- 75 Mbps DL in SISO configuration with commercial UEs;
- 50 Mbps UL in 20 MHz with commercial UEs;
- User plane encryption (Fig. 3).

  - eNodeB architecture

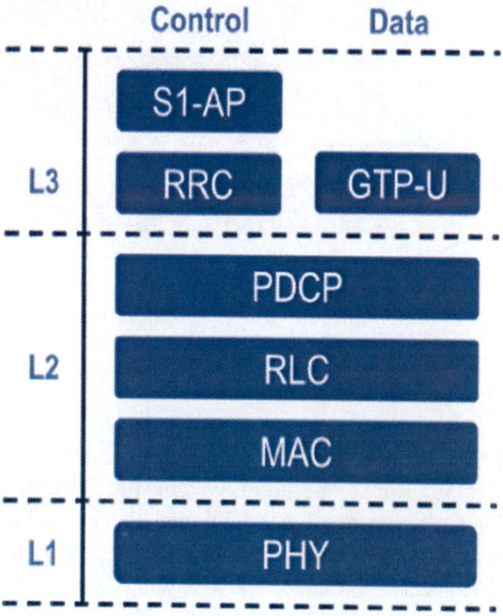

Fig. 3. eNodeB architecture

The srsENB application includes layers 1, 2 and 3 as shown in the figure above.

At the bottom of the protocol stack, the Physical (PHY) layer carries all information from the MAC over the air interface. It is responsible for link adaptation and power control.

The Medium Access Control (MAC) layer multiplexes data between one or more logical channels into Transport Blocks (TBs) which are passed to/from the PHY layer. The MAC is responsible for scheduling uplink and downlink transmissions for connected UEs via control signalling, retransmission and error correction (HARQ) and priority handling between logical channels.

The Radio Link Control (RLC) layer can operate in one of three modes: Transparent Mode (TM), Unacknowledged Mode (UM) and Acknowledged Mode (AM). The RLC manages multiple logical channels or bearers for each connected UE. Each bearer operates in one of these three modes. Transparent Mode bearers simply pass data through the RLC. Unacknowledged Mode bearers perform concatenation, segmentation and reassembly of data units, reordering and duplication detection. Acknowledged Mode bearers additionally perform retransmission of missing data units and resegmentation.

The Packet Data Convergence Protocol (PDCP) layer is responsible for ciphering of control and data plane traffic, integrity protection of control plane traffic, duplicate discarding and in-sequence delivery of control and data plane traffic to/from the RRC and GTP-U layers respectively. The PDCP layer also performs header compression (ROHC) of IP data if supported.

The Radio Resource Control (RRC) layer manages control plane exchanges between the eNodeB and connected UEs. It generates the System Information Blocks (SIBs) broadcast by the eNodeB and handles the establishment, maintenance and release of RRC connections with the UEs. The RRC also manages security functions for ciphering and integrity protection between the eNodeB and UEs.

Above the RRC, the S1 Application Protocol (S1-AP) layer provides the control plane connection between the eNodeB and the core network (EPC). The S1-AP connects to the Mobility Management Entity (MME) in the core network. Messages from the MME to UEs are forwarded by S1-AP to the RRC layer, where they are encapsulated in RRC messages and sent down the stack for transmission. Messages from UEs to the MME are similarly encapsulated by the UE RRC and extracted at the eNodeB RRC before being passed to the S1-AP and on to the MME.

The GPRS Tunnelling Protocol User Plane (GTP-U) layer within srsENB provides the data plane connection between the eNodeB and the core network (EPC). The GTP-U layer connects to the Serving Gateway (S-GW) in the core network. Data plane IP traffic is encapsulated in GTP packets at the GTP-U layer and these GTP packets are tunneled through the EPC. That IP traffic is extracted from the tunnel at the Packet Data Network Gateway (P-GW) and passed out into the internet.

4.4 srsEPC

srsEPC is a lightweight implementation of a complete LTE Core Network (EPC). The srsEPC application runs as a single binary but provides the key components of the EPC, namely the Home Subscriber Service (HSS), the Mobility Management Entity (MME), the Service Gateway (S- GW) and the Packet Data Network Gateway (P-GW) [19].

- **EPC overall architecture**

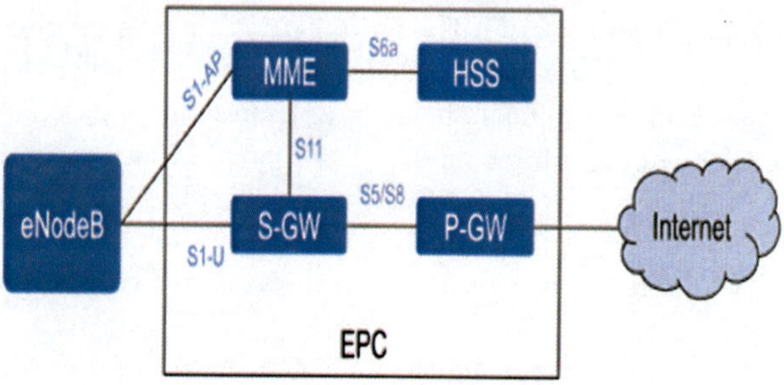

Fig. 4. EPC overall architecture

The figure above illustrates the main components of the EPC, along with the main interfaces between them (Fig. 4).

HSS: The Home Subscriber Service (HSS) is the user database. It stores information such as the user's id, key, usage limits, etc. It is responsible for authenticating an authorizing the user's access to the network.

MME: Mobility Managment Entity (MME) is the main control element in the network. It handles mobility and attach control messages. It is also responsible for paging UEs in idle mode.

S-GW: The S-GW is the main dataplane gateway for the users, as it provides the mobility anchor for the UEs. It works as an IP router and helps setting up GTP sessions between the eNB and the P-GW.

P-GW: The Packet Gateway (P-GW) is the point of contact with external networks. It enforces the QoS parameters for subscriber sessions.

4.5 Open5gs

The 5G System consists of a Radio Access (NG-RAN: Next Generation RAN) and a Network Core (5G Core) [20] (Fig. 5).

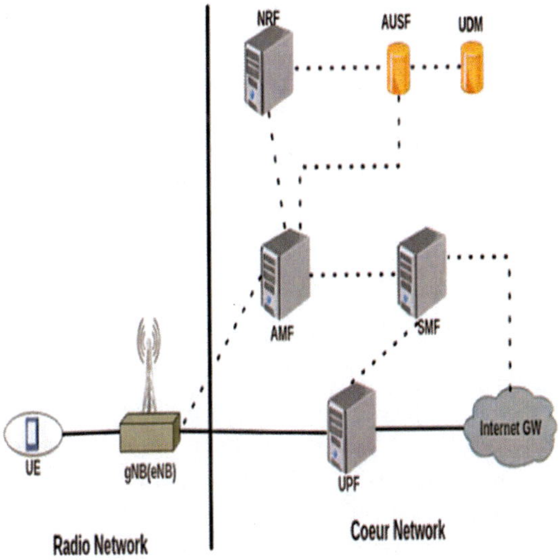

Fig. 5. 5G architecture with open5gs

– **5G radio access**

The 5G radio access consists of next generation base stations that form the connection node of the mobiles with the 5G core network (5GC) [20].

The UE mobiles communicate with the base stations either through a 5G radio link or through a 4G radio link. If the communication is in 5G, the base station is called gNB (next Generation Node Base Station), if the communication is in 4G, the base station is a 4G eNB base station evolved to interconnect with the 5G core network. The base station is called ng-eNb (Next Generation eNb).

The functions of the gNb base station are quite similar with the eNB entity. However, the differences concern the management of the quality of service by flow and not by bearer and the management of the network slices (Slices) on the radio interface.

As a reminder, a slice is composed of logical instances of the mobile network allowing to provide a network service in an efficient way in order to meet a QoS specific to this service (refer to the Network Slicing article).

– **The 5G network core**

The 5G core network is suitable for network virtualization and is based on the Control Plane and User Plane partitioning defined in the CUPS architecture.

In comparison to 4G CUPS, we could say that:

The Access and Mobility Managmenent Function (AMF) entity takes over the role of the MME entity. The AMF entity establishes a NAS connection with the UE mobile and has the role of registering (attaching) the UE mobiles and managing the location of the mobiles on the 3GPP and/or non-3GPP network.

The SMF entity (Session Management Function) takes over the role of the SGW-C and PGW-C entities. The SMF entity allows to control the PDN sessions. The SMF entity is chosen by the AMF entity since the AMF entity manages the NAS signaling with the mobile. The SMF entity is responsible for managing the control plane.

The SMF entity has an interface with the entity that manages the flow policy (PCF: Policy Charging Function).

The transport plane is composed of data gateways that perform measurements on the transported data and interconnect with the data networks (PDN). In the CUPS architecture, the transport plane functions are managed by the SGW-U and PGW-U entities. For the 5G core network, the transport plane functions are handled by the User Plane Function (UPF) entity. The UPF entity communicates with the SMF entity through the Sx interface and using the PFCP protocol. See the article presenting the CUPS architecture.

The PCRF entity of the 4G architecture allows to define control rules and flow policies with the SGW/PGW entity. In 5G, the PCRF entity is renamed PCF and allows to control the flows at the level of the SMF entity but also at the level of the AMF entity in order to bring a better granularity on the authorized flows by taking into account the location of the UE mobile.

The user profile are saved in a UDR database accessible via the UDM entity (Unified Data Management). The UDM entity keeps the data session profiles (PDU sessions) and the AMF entity to which the UE mobile is attached (possibly the AMF entities for a 3GPP and non 3GPP access on another operator).

The registration of the mobile requires a double authentication performed at the level of the AMF entity and the UE mobile from authentication vectors provided by the AUSF entity (AUthentication Server Function).

Finally, the NSSF (Network Slice Selection Function) is an entity that assists the AMF entity in the selection of logical network instances for a defined network slice.

Figure 6 shows the 5G architecture and the interfaces between each entity.

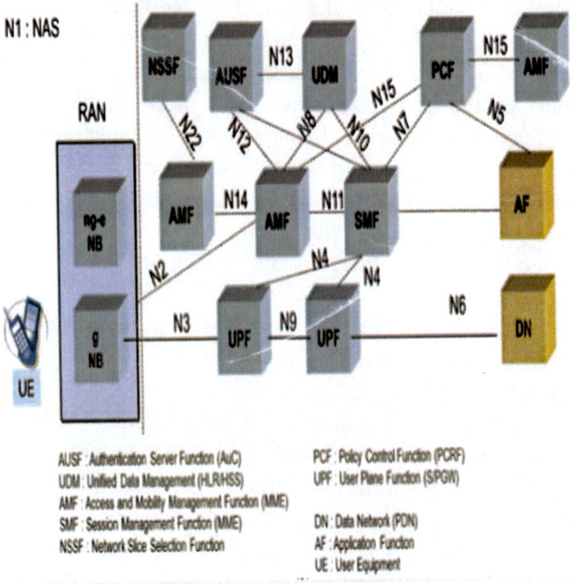

Fig. 6. The 5G architecture and the interfaces between each entity.

5 Proposed Architecture

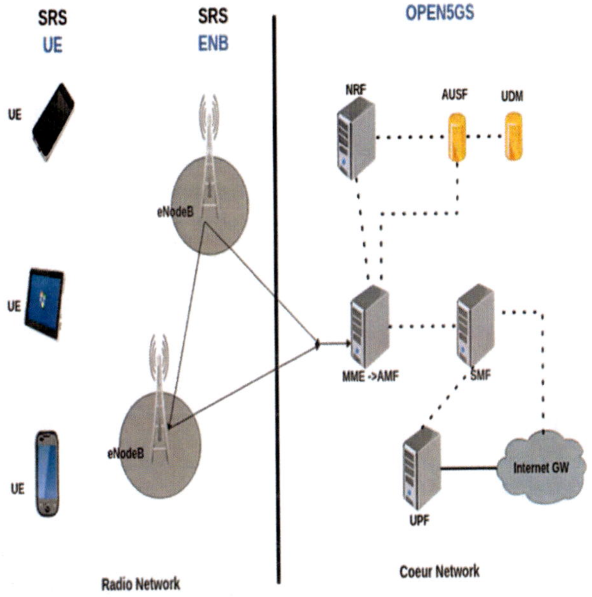

Fig. 7. Architecture of the non-autonomous 5G solution

In this architecture, we have the radio part which consists of the srsUE and srsENB entities (Fig. 7). They will ensure the access of the mobile to the core network which is made up of the Open5gs technology.

A UE makes a connection to its associated eNodeB. The latter is interconnected to the Open5gs MFA, which is responsible for establishing the mobile's attachment to the network.

The information of a mobile is stored in the UDM database. The SMF entity will then manage the NAS signalling with the mobile. It is this entity that will allow the mobile to have Internet access.

The transport plane is made up of data gateways which carry out measurements on the transported data and interconnect with the data networks (PDN). In the CUPS architecture, the transport plane functions are managed by the SGW-U and PGW-U entities. For the 5G core network, the transport plane functions are handled by the User Plane Function (UPF) entity. The UPF communicates with the SMF via the Sx interface and the PFCP protocol [20].

The PCRF entity in the 4G architecture is used to define control rules and flow policies with the SGW/PGW entity. In 5G, the PCRF entity is renamed PCF and allows to control the flows both at the SGW entity level and at the MFA entity level in order to provide a better granularity on the authorised flows by taking into account the location of the UE mobile.

The user profile is saved in a UDR database accessible via the UDM (Unified Data Management) entity. The UDM entity keeps the profiles of the data sessions and of the MFA entity to which the UE mobile is attached.

The registration of the mobile requires a double authentication performed at the level of the AMF entity and the UE mobile from authentication vectors provided by the AUthentication Server Function (AUSF) entity [20].

Finally, the NSSF (Network Slice Selection Function) is an entity that assists the AMF entity in selecting the logical network instances for a defined network slice.

6 Implementation of the Solution

For the implementation of the solution, we will approach it in two phases:

6.1 Radio Access

In this part, we implement the radio access part defined in the proposed architecture. We will proceed to a configuration of the entities of the radio part which are the srsENB and the srsUE.

For the configuration of the srsENB, the following figure shows the different configuration parameters grouped into two major sections.

In the enb section, we have among others the parameters like the mcc (mobil country code) set here to 901 and the mnc (mobil network code) to 70.

We also have the mme_addr which will establish the connection between the radio part and the network part.

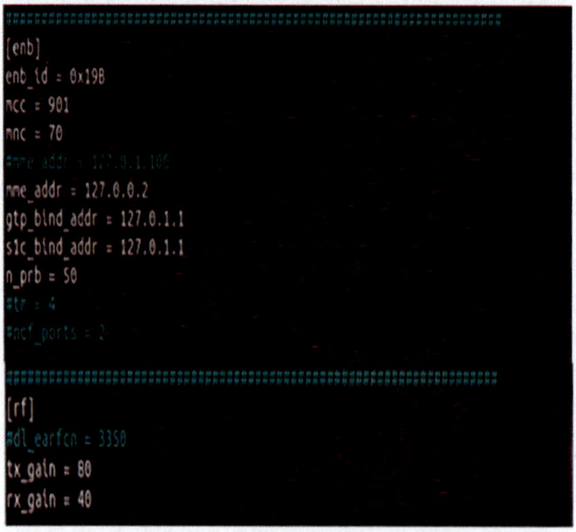

Fig. 8. Configuration of the ENB

For a mobile (UE) configuration, we have the following figure which shows how we have done this. We have the different parameters of the UE allowing it to authenticate itself to the network in order to register.

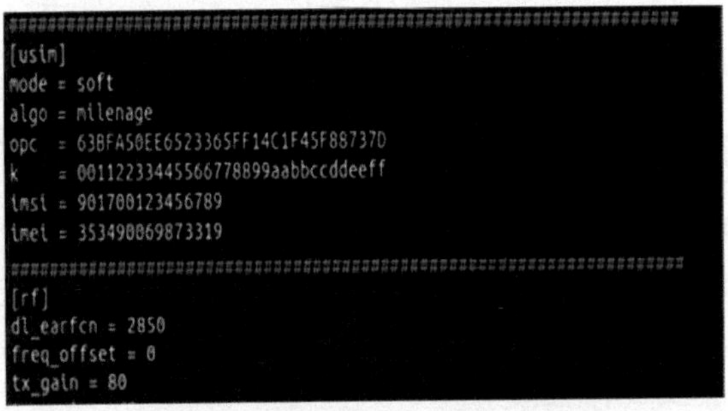

```
##############################################################################
[usim]
mode = soft
algo = milenage
opc  = 63BFA50EE6523365FF14C1F45F88737D
k    = 00112233445566778899aabbccddeeff
imsi = 901700123456789
imei = 353490069873319
##############################################################################
[rf]
dl_earfcn = 2850
freq_offset = 0
tx_gain = 80
```

Fig. 9. Configuration of the UE

6.2 Core Network

After having configured the radio access part, we will now proceed to a configuration of the network part, which is 5G with open5gs. This configuration will be done in the following way:

We will first configure the MME entity which allows to link the two parts. The following figure shows a configuration of this entity.

```
mme:
    freeDiameter: /etc/freeDiameter/mme.conf
    s1ap:
      - addr: 127.0.0.2
    gtpc:
      - addr: 127.0.0.2
    gummei:
      plmn_id:
        mcc: 901
        mnc: 70
      mme_gid: 2
      mme_code: 1
    tai:
      plmn_id:
        mcc: 901
        mnc: 70
      tac: 7
    security:
        integrity_order : [ EIA2, EIA1, EIA0 ]
        ciphering_order : [ EEA0, EEA1, EEA2 ]
    network_name:
        full: Open5GS
    mme_name: open5gs-mme0
#
```

Fig. 10. The configuration file mme.yaml

Now we have to configure the parameters of the mobile on the web interface of open5gs. After connection, we obtain this interface where we will fill in all the parameters of the UE.

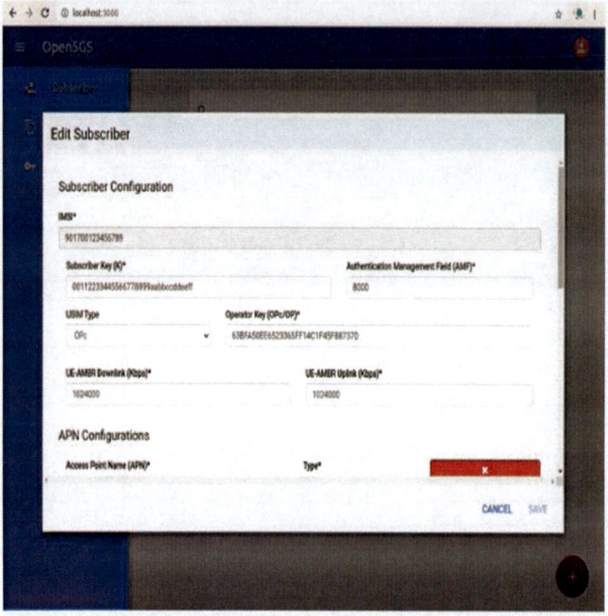

Fig. 11. Subscriber configuration

7 Results Obtained

To proceed to the functional tests, we will proceed as follows:

First, proceed to the start-up of the entities of the radio part. This is done as shown in Fig. 8 for the ENB entity and Fig. 9 for the UE mobile (Fig. 12).

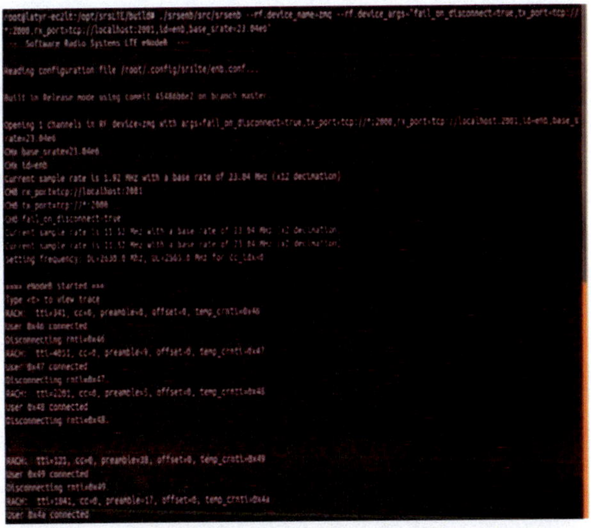

Fig. 12. Start-up of the ENB entity

We can see that the mobile has successfully connected to the network and obtained an IP address of 10.45.0.2 (Fig. 13). We will now show whether the mobile can contact the gateway to allow it to exit for internet access. Figure 10 shows that the ping to the gateway was successful (Fig. 14).

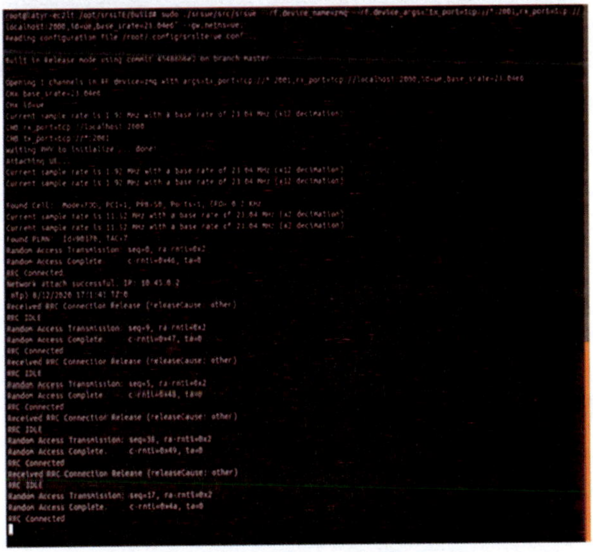

Fig. 13. Start-up of the UE entity

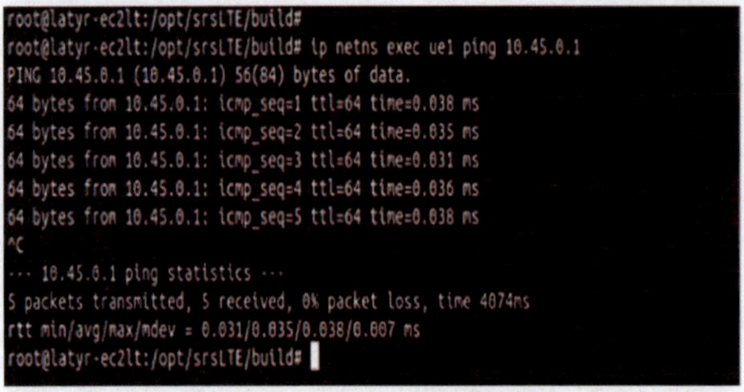

Fig. 14. Ping the network gateway

Now that the mobile is able to reach the gateway, we will see if it will be able to contact the google DNS server with a ping to 8.8.8.8. Figure 11 shows that the mobile has a very high speed internet access (Fig. 15).

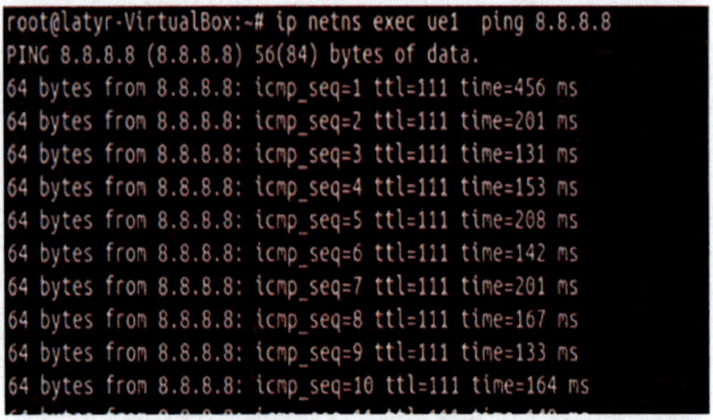

Fig. 15. Ping to a google DNS server

Figure 12 shows the messages exchanged between entities on wireshark. The generated traffic is also shown (Fig. 16).

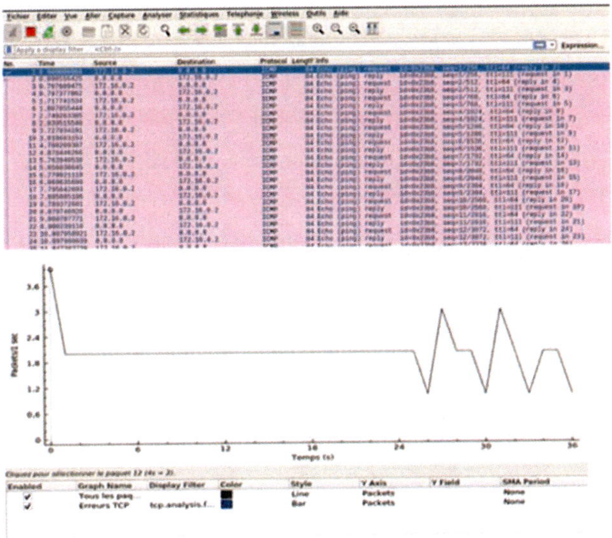

Fig. 16. Trafic wareshark

8 Conclusion

The benefits offered by 5G will allow developing African countries to have the opportunity to build their mobile network architecture to ensure a reliable and more efficient service offering. Some African countries do not have the means to move directly to a full 5G rollout when they have just finalised their 4G rollout. Today, this article shows how these countries can combine the two into a non-standalone 5G architecture. This article will allow them not to lose all the investment made for the 4G deployment. It allows them to keep the radio access part all being 4G entities and to proceed to a 5G deployment for only the network part.

This article contributes to a reduction of the costs of deployment of a last generation network because all the implementation is done only by free software. Today it should be noted that the entire infrastructure of a next-generation network can be emulated from within IT.

References

1. Klymash, M., Beshley, H., Masiuk A., Strykhalyuk, I.: Concept for ensuring effective functioning of mobile communication system in heterogenous 5G infrastructure. In: 2017 14th International Conference The Experience of Designing and Application of CAD Systems in Microelectronics (CADSM), pp. 272–274 (2017). https://doi.org/10.1109/CADSM.2017.7916132
2. Barakabitze, A.A., Ahmad, A., Mijumbi, R., Hines, A.: 5G network slicing using SDN and NFV: a survey of taxonomy, architectures and future challenges. Comput. Netw. **167**, 106984 (2020)

3. Khakimov, A., Ateya, A.A., Muthanna, A., Gudkova, I., Markova, E., Koucheryavy, A.: IoT-Fog based system structure with SDN enabled. In: Proceedings of the 2nd International Conference on Future Networks and Distributed Systems, ser. ICFNDS 18. Association for Computing Machinery, New York, NY, USA (2018)
4. Nunes, B.A.A., Mendonca, M., Nguyen, X.N., et al.: A survey of softwaredefined networking: past, present, and future of programmable networks. IEEE Commun. Surv. Tutorials **16**(3), 1617–1634 (2014)
5. Haque, I.T., Abu-Ghazaleh, N.: Wireless software defined networking: a survey and taxonomy. IEEE Commun. Surv. Tutorials **18**(4), 2713–2737 (2016)
6. Li, L.E., Mao, Z.M., Rexford, J.: Toward software-defined cellular networks. In: 2012 European Workshop on Software Defined Networking, pp. 7–12. Darmstadt (2012). https://doi.org/10.1109/EWSDN.2012.28
7. Jain, R., Paul, S.: Network virtualization and software defined networking for cloud computing: a survey. IEEE Commun. Mag. **51**(11), 24–31 (2013)
8. Sarigiannidis, P., Lagkas, T., Bibi, S., Ampatzoglou, A., Bellavista, P.: Hybrid 5G optical-wireless SDN-based networks, challenges and open issues. Inst. Eng. Technol. **6**(6), 141–148 (2017)
9. Celso Henrique, de S.L., et al.: Non-standalone 5G NR fiber-wireless system using FSO and fiber-optics Fronthauls. J. Lightwave Technol. **39**(2), 406–417 (2021)
10. Heimann, K., Gorczak, P., Bektas, C., Girke, F., Wietfeld, C.: Software–defined end–to–end evaluation platform for quality of service in non–standalone 5G Systems. In: IEEE International Systems Conference (SYSCON2019), Orlando, Florida, USA (April 2019). https://doi.org/10.1109/SYSCON.2019.8836743
11. Guangyi, L., Yuhong, H., Zhuo, C., Liang, L., Qixing, W., Na, L.: 5G deployment: standalone vs nonstandalone from the operator perspective. In: IEEE communications Magazine, November 2020 0163-6804/20/$25.00©. IEEE (2020)
12. Solozabal, R., Sanchoyerto, A., Atxutegi, E., Blanco, B., Fajardo, J.O., Liberal, F.: Exploitation of mobile edge computing in 5G distributed mission-critical push-to-talk service deployment. IEEE Access **6**, 37665–37675 (2018). https://doi.org/10.1109/ACCESS.2018.2849200
13. Udoh, S.J., Srivastava, V.M.: Analytical modeling of radio network performance for 5G (non-standalone) and it's network connectivity. J. Commun. **15**(12), 886–895 (2020)
14. Zainab, Z., Vasilis, F., Zarrar, Y., Simon, F., Mischa, D., Hamid, A.: Will SDN be part of 5G?. 1708.05096v2 (7 Feb 2018)
15. Balasubramanian, V., Zaman, F., Aloqaily, M., Ridhawi, I.A., Jararweh, Y., Salameh, H.B.: A mobility management architecture for seamless delivery of 5G-IoT services. In: ICC 2019 - 2019 IEEE International Conference on Communications (ICC), Shanghai, China, pp. 1–7 (2019). https://doi.org/10.1109/ICC.2019.8761658
16. Kaltenberger, F., Silva, A.P., Gosain, A., Wang, L., Nguyen, T.-T.: OpenAirInterface: democratizing innovation in the 5G era. Comput. Netw. **176**, 107284 (2020)
17. SrsUE [Online] available on https://docs.srslte.com/en/latest/srsue/source/1_ue_intro.html#ue-intro
18. srsENB [Online] available on https://docs.srslte.com/en/latest/srsenb/source/1_enb_intro.html#enb-intro
19. srsEPC [Online] available on https://docs.srslte.come/n/latest/srsepc/source/1_epc_intro.html#epc-intro
20. Open5gs [Online] available on http://blogs.univ-poitiers.fr/f-launay/2018/08/24/le-reseau-5g-5gs/

Recommending a Retailer's Mobile App – Influence of the Retailer and the Mediating Role of Push Notifications

Atilla Wohllebe[1(✉)], Dirk-Siegfried Hübner[1], Uwe Radtke[1], and Szilárd Podruzsik[2]

[1] MATE Hungarian University of Agriculture and Life Sciences – Kaposvár Campus, Kaposvár, Hungary
atilla.wohllebe@gmail.com
[2] Centre for Economic and Regional Studies, Institute of Economics, Budapest, Hungary

Abstract. Against the backdrop of increasing digitization, many retailers are trying to better reach their existing customers with the help of mobile apps. For the growth of an app and the satisfaction with it, the willingness to recommend this app by existing users is of great importance. This paper develops a structural equation model that relates the willingness to recommend a retailer's mobile app to consumer perceptions of that retailer, including the role of push notifications as a mediator. The collected data show a significant relevance of the perception of the retailer on the willingness to recommend the app to others. It can be shown that this relationship is partially mediated by the perception of the retailer's app push notifications. In this respect, all hypotheses stated can be confirmed. Further research questions are proposed and include in particular a validation of the model in other socio-demographic compositions as well as in the context of a field study and the expansion of the model.

Keywords: Retail · Mobile apps · Consumer behavior · Mobile technologies

1 Introduction

With the digitalization in retail, not only e-commerce is growing but also consumers expect retailers to offer both online and offline sales channels and to network them with each other [1, 2]. Mobile devices are fundamentally changing consumers' shopping behavior, making mobile shopping the new standard of shopping [3]. Accordingly, the relevance of mobile apps is growing not only in retail but also in many other sectors [4–6]. Studies show a positive impact on customer sales [7]. Push notifications to reach app users with advertising draw customers' attention to exclusive promotions or discounted products, for example [8]. At the same time, however, retailers risk disturbing their users with the notifications and thus harming their business [9].

In order to meet high expectations of retailers for mobile applications, user acquisition is a key challenge. As customers being satisfied with a company are more likely to recommend it to others, it is conceivable that satisfaction with a retailer may

M. E. Auer and T. Tsiatsos (Eds.): IMCL 2021, LNNS 411, pp. 361–371, 2022.
https://doi.org/10.1007/978-3-030-96296-8_32

lead existing users to recommend that retailer's app to others. Thus, satisfied customers of a retailer who use the retailer's app can contribute to its growth.

This study therefore investigates how the perception of a retailer and the perception of its push notifications affect consumer's willingness to recommend that retailer's app.

2 Literature Review

2.1 Recommendation of Mobile Apps

The role of recommendation of products, brands and companies is interesting in several ways. In a direct way, recommendation is a form of word-of-mouth and leads to more people being aware of what is being recommended. In addition, there are - quite controversial - approaches that proclaim the management of an entire company based on the willingness to recommend [10]. Furthermore, recommendation is an indirect expression of customer satisfaction; recommendation and satisfaction are closely related [11, 12].

Recommendation as a dependent variable is repeatedly investigated in the context of mobile apps. Factors such as respect for privacy and the added value can be named as indirect influencing factors on recommendation [13]. In particular, however, customers who are particularly satisfied with an app are more likely to recommend it to others [13, 14]. Iyer et al. [15] demonstrate congruence between a retailer image and a retailer mobile app.

H1: The app user's perception of a retailer positively influences the probability of the app to be recommended.

2.2 Perception of Retailers

The retailer's brand is of central importance in determining how consumers perceive a retailer. A strong brand or a positive attitude of the consumer towards the retailer's brand promotes the purchase intentions of the consumer. [16]. Customer satisfaction is of great importance in this context [17]. Also there is a congruence between retailer image and mobile app of a retailer [15]. Customer satisfaction manifests itself in the question of whether a consumer likes to buy from a retailer.

Another key factor in the perception of a retailer is the extent to which customers have had good experiences with it. Good experiences are not only important for customer satisfaction [18]. Furthermore, good experiences lead to brand attachment and can thus increase the intention to continuously use a retailer's mobile app [19]. Customer satisfaction is a mediator of the influence of the corporate image on the behavioral intention to repurchase. [20]. In the long term, good experiences and thus trust built up over time lead to loyalty [21].

Particularly in connection with customer satisfaction, the willingness to recommend is repeatedly emphasized. Recommendation is a consequence of customer satisfaction [22]. Perceived retail service quality impacts the intention to recommend a company to others [20]. In addition, the recommendation of a retailer is considered an important driver of purchase intention [23]. Customer satisfaction and loyalty have a positive long-term effect on consumer acceptance of apps [14, 24].

Derived from the literature, the items enjoyment, recommendation and experiences are identified as dimensions for determining the perception of a retailer (cf. tbl. 1, 1.1 - 1.3).

2.3 Perception of Push Notifications

In the context of push notifications, the importance of the relevance of the content for the perception of push notifications is emphasized. For example, customized content of push notifications tailored to the app user increases the relevance and subsequently the probability of interaction [25]. Relevance is important in connection with research on the effects of frequency in that the frequencies tolerated by users increase with the relevance of content [26]. The use of location data in the sense of location-based marketing increases the relevance of the notifications because a reference to the context of the respective user is established [27].

The relevance of notifications can be achieved both by using personal data of app users and by incorporating the individual context. Users are more likely to perceive content as relevant or helpful accordingly [28]. Notifications are perceived as helpful if they refer to current events. For news and social media in particular, the topicality of the content is especially important so that users find it helpful [29].

In the context of shopping apps and mobile apps in retail, monetary incentives play an important role in the positive perception of push notifications. Incentives, such as coupons and discounts, have a positive influence on the basic attitude toward push notifications [30]. In online auctions, push notifications lead to an increase in the bidders' chances of winning the auction [31].

The frequency with which app users receive push notifications appears to play an important role. On the one hand, push notifications with higher frequency can generate higher user engagement in the short term [32]. On the other hand, depending on the use case, target group and content, different frequencies are perceived as too high [33]. In general, especially with regard to the long-term perception of push notifications, the frequency should not be too high [9], as they can always be perceived as a disturbance [26, 34].

After reviewing the relevant literature, usefulness, offers, relevance and frequency are defined as items for the perception of previous push notifications (cf. tbl. 1, 2.1 - 2.4).

Push notifications are an essential functionality of mobile apps. The content communicated via push notifications therefore has a high influence on how the push notifications themselves and the app that sends them are perceived [30, 35]. Consequently, push notifications have an effect on the satisfaction with an app and correspondingly on the likelihood of recommendations [13, 14, 24].

H2: The app user's perception of push notifications from a retailer's app positively influences the probability of the app to be recommended.

Furthermore, it can be assumed that push notifications are perceived more positively if the sender is perceived positively. This assumption corresponds with the assumption that the sender (here: a retailer) has a positive effect on the perception of the app itself [15, 36, 37].

H3: The app user's perception of the retailer positively influences the app user's perception of push notifications from the retailer's app.

Accordingly, it can be assumed that:

H4: The effect of the perception of the retailer on the probability of the retailer's app to be recommended is partly mediated by the perception of push notifications.

3 Material and Methods

3.1 Sampling and Data Collection

A quantitative research approach is used for this study. Consumers are interviewed via an anonymous online survey. Consumers are first informed that a prerequisite for participating in the survey is that they have installed the mobile app of at least one retailer. The respondents had to indicate which mobile application of which retailer they will refer to. Figure 1 shows how often each retailer was mentioned. Most consumers referred to their experiences with Amazon and the corresponding app.

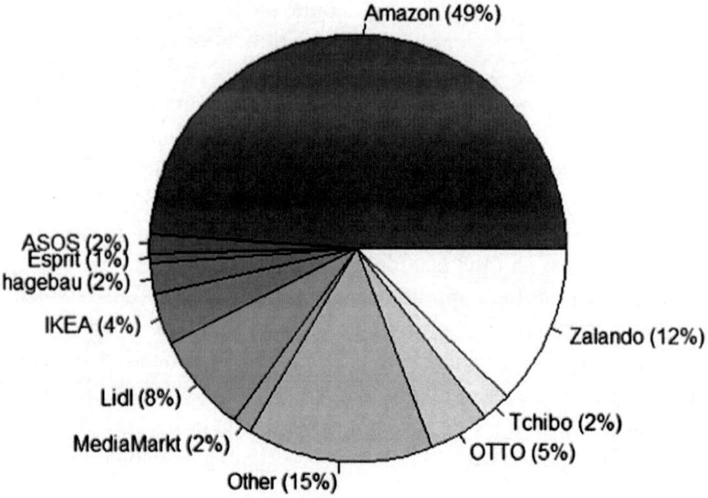

Fig. 1. Retailer's referred to in questionnaire

The survey is distributed in German-speaking countries and is conducted in German. A total of 131 people take part in the survey.

3.2 Measures

The questionnaire consists of three sections. First, consumers are informed about the survey and asked to define which retailer they will be referring to for their answers. In the second section, consumers are asked to provide information about their perception of the retailer and the push notifications they have received from the app so far. In the third part, consumers are asked to indicate whether they would recommend the app of the retailer to friends or colleagues. All items are measured from 1 (completely disagree) to 5 (completely agree) using a Likert scale.

Table 1 lists the questionnaire items corresponding to the perception of the retailer and the perception of push notifications. Furthermore, it shows the rotated factor loadings and uniqueness of each item as well as Cronbach's Alpha for each factor.

Table 1. Measures.

	Factor loading	Uniqueness	Cronbach's Alpha
Perception of retailer			
1.1 Enjoyment	.7838	.3531	.88
1.2 Recommendation	.7951	.3182	
1.3 Experience	.7994	.3519	
Perception of push notifications			
2.1 Usefulness	.8171	.3218	.73
2.2 Offers	.6919	.4728	
2.3 Relevance	.8141	.2925	
2.4 Frequency	.1962	.9610	

Based on Cronbach's Alpha, the reliability testing of the measured variables shows good and excellent levels of internal consistency [18, 38].

The factor loadings indicate a high relevance of all items for the respective factors. Only frequency plays a separate role; the factor loading is relatively low in comparison. However, the item remains as part of the factor, as the relevance in the context of push notifications has been emphasized as critical by researchers in the past [9, 33].

In addition, also on a Likert scale of 1 to 5, respondents were asked how likely they would be to recommend the retailer's app to others (cf. Fig. 2).

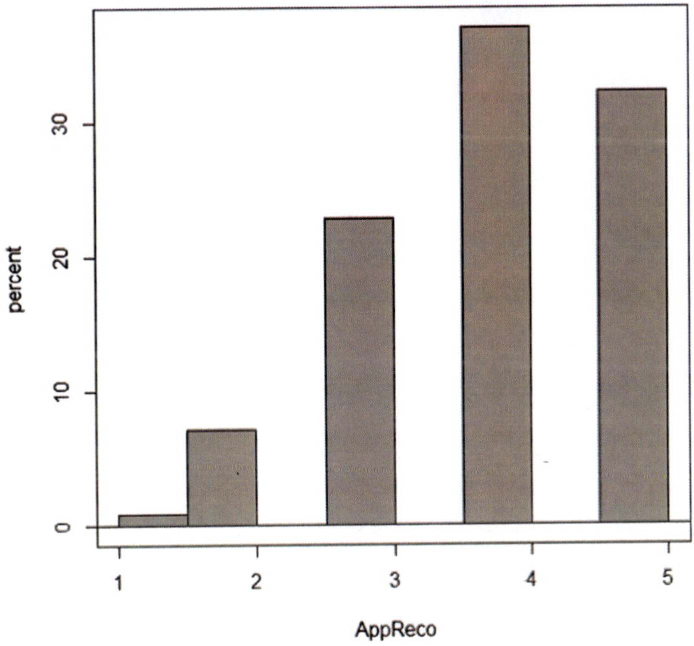

Fig. 2. Distribution of likelihood to recommend the retailer's app

When looking at the distribution, it should be noted that consumers who would not recommend a retailer's app to others would probably be dissatisfied with it. In this respect, they would probably uninstall it and would therefore not be able to refer to it in the survey.

4 Results

The hypotheses are tested using a structural equation model. For this purpose, the "lavaan" package in R is used. In addition to testing the individual hypotheses, the direct and indirect effects are calculated. Thus, the role of the perception of push notifications as a mediator can be discussed. The model is estimated using a robust maximum likelihood with Yuan-Bentler correction of the Chi Square statistics and Huber-White standard error estimation [39]. Figure 3 shows the final estimated model.

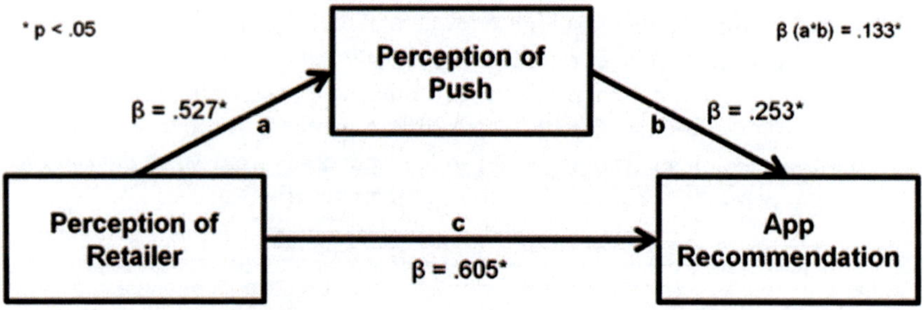

Fig. 3. Final estimated model, standardized parameters.

The model fit is determined on the basis of the common goodness of fit indices, such as those proposed by Kline [40] and Hu & Bentler [41] (cf. Table 2). The Chi Square test shows that the proposed model is better than the baseline model. The values for CFI, TLI, RMSEA and SRMR are all in the very good range [41].

Table 2. Model fit statistics.

Index	Value
Chi square	21.515
p (Chi square)	.254
CFI	.990
TLI	.985
RMSEA	.038
P (RMSEA)	.569
SRMR	.044

All previously assumed effects can be confirmed as significant on the basis of the data (cf. Table 3). The positive perception of push notifications has a significant effect on the willingness to recommend the app to others (cf. effect *b*), however significantly weaker than the positive perception of the retailer (cf. effect *c*). The effect of the retailer's perception on the app recommend (cf. effect *c*) is similar in size to the effect on positive perception of push notifications (cf. effect *a*).

Table 3. Regression results.

Effect	Estimate	Std. Err	z-Value	p
a	.527	.107	4.937	.000
b	.253	.106	2.397	.017
c	.605	.121	4.985	.000

Since the indirect effect (*ab*) and the direct effect (*c*) are significant, partial mediation is present. Mediation analysis results are display in Table 4.

Table 4. Direct and indirect results.

Effect	Estimate	Std. Err	z-Value	p
Indirect (ab)	.133	.058	2.305	.021
Direct (c)	.605	.121	4.985	.000
Total	.738	.103	7.195	.000

According to the results of the structural equation model analysis, the results of the hypotheses to be tested are summarized in Table 5. All four hypotheses can be confirmed.

Table 5. Summary of hypotheses.

Hypothesis	Effect	Interpretation
H1: The app user's perception of a retailer positively influences the probability of the app to be recommended	c	Confirmed
H2: The app user's perception of push notifications from a retailer's app positively influences the probability of the app to be recommended	b	Confirmed
H3: The app user's perception of the retailer positively influences the app user's perception of push notifications from the retailer's app	a	Confirmed
H4: The effect of the perception of the retailer on the probability of the retailer's app to be recommended is partly mediated by the perception of push notifications	ab, c	Confirmed

5 Discussion and Implications

The model proposed in this paper puts the retailer and its mobile app into one context and additionally considers the role of push notifications for the first time. The study provides a supplement to the existing findings on the antecedents of the willingness to recommend. With regard to the perception of retailers, previous studies have considered the sub-aspects used here separately from one another or as a consequence of one another. Although the retailer perception factor is much broader than retailer image, the confirmation of H1 is unsurprising in light of prior work finding congruence between retailer image and mobile app of a retailer [15]. The mediating role of push notifications on app recommendation confirms previous research results connecting push notification and mobile app user behavior [31–33]. The model shows that consumers seem to be generally more positive about push notifications when they are associated with a retailer that consumers perceive positively.

From a managerial perspective, mobile app marketing in retail should not just focus on the app as a technology. Obviously, the retailer itself and how it is perceived plays a major role in marketing: If a customer has a positive perception of a retailer, the probability to recommend the retailer's app to others increases. In addition, the positive perception affects how customers perceive the push notifications. These are additionally relevant for the app recommendation probability. The study therefore contributes to a better understanding of mobile apps in retail.

At the same time, it is not without limitations, which can be seen as possible further research questions. Firstly, the model of the study is based on data from a survey. There may be differences in the survey of willingness to recommend between behavior that respondents report and behavior that respondents would actually exhibit in a field study. This has already been criticized, particularly in the context of recommendations [42]. The survey data comes from a clearly defined geographical region (Germany). For other regions, the statistical analyses could turn out differently in the result [42]. Shifts in the gender composition of respondents can lead to changes in [43]. In this respect, it is suggested that the model be validated by further surveys. In addition, the model presented here is reduced to two latent variables and the target variable. Among other things, more complex relationships are conceivable, especially of the variables that are considered exogenous and equivalent within the framework of the model proposed here.

6 Conclusion

The goal of this paper is to develop a model that relates the willingness to recommend a retailer's mobile app to consumer perceptions of that retailer, including the role of push notifications as a mediator. The literature review shows that previous research rarely combines the retailer and mobile app perspectives. In particular, mobile applications have so far been considered primarily in an encapsulated way as technological platforms. The context to the company publishing the app is rarely established. In this respect, the proposed model represents a simple but much more context-rich approach compared to existing literature. The collected data show a significant relevance of the perception of the retailer on the willingness to recommend the app to others. Also, it can be shown that this relationship is partially mediated by the perception of the retailer's app push notifications. In this respect, all hypotheses stated can be confirmed. Further research questions include in particular a validation of the model in other socio-demographic compositions as well as in the context of a field study and the expansion of the model.

References

1. Ewerhard, A.-C., Sisovsky, K., Johansson, U.: Consumer decision-making of slow moving consumer goods in the age of multi-channels. Int. Rev. Retail Distrib. Consum. Res. **29**, 1–22 (2019). https://doi.org/10.1080/09593969.2018.1537191
2. Adler, M., Wohllebe, A.: Consumers choosing retailers on online marketplaces: how can retailers differentiate apart from the price? – An exploratory investigation. IJARBM **1**, 27–36 (2020). https://doi.org/10.51137/ijarbm.2020.1.1.3
3. Pantano, E., Gandini, A.: Exploring the forms of sociality mediated by innovative technologies in retail settings. Comput. Hum. Behav. **77**, 367–373 (2017). https://doi.org/10.1016/j.chb.2017.02.036
4. Ross, F.: Hearing aid accompanying smartphone apps in hearing healthcare. A systematic review. Appl. Med. Inf. **42**, 189–199 (2020)
5. Wohllebe, A., Dirrler, P., Podruzsik, S.: Mobile apps in retail: determinants of consumer acceptance – a systematic review. Int. J. Interact. Mob. Technol. **14**, 153–164 (2020). https://doi.org/10.3991/ijim.v14i20.18273
6. Papadakis, S., Kalogiannakis, M., Zaranis, N.: Educational apps from the Android Google Play for Greek preschoolers: a systematic review. Comput. Educ. **116**, 139–160 (2018). https://doi.org/10.1016/j.compedu.2017.09.007
7. van Heerde, H.J., Dinner, I.M., Neslin, S.A.: Engaging the unengaged customer: the value of a retailer mobile app. Int. J. Res. Mark. **36**, 420–438 (2019). https://doi.org/10.1016/j.ijresmar.2019.03.003
8. Ahrholdt, D., Greve, G., Hopf, G.: Online-Marketing-Intelligence: Kennzahlen, Erfolgsfaktoren und Steuerungskonzepte im Online-Marketing. Springer Fachmedien Wiesbaden, Wiesbaden (2019). https://doi.org/10.1007/978-3-658-26562-5
9. Wohllebe, A.: Consumer acceptance of app push notifications: systematic review on the influence of frequency. Int. J. Interact. Mob. Technol. **14**, 36–47 (2020). https://doi.org/10.3991/ijim.v14i13.14563
10. Reichheld, F.F.: The One Number You Need to Grow (2003). https://hbr.org/2003/12/the-one-number-you-need-to-grow
11. Luo, A., Baker, A., Donthu, N.: Capturing dynamics in the value for brand recommendations from word-of-mouth conversations. J. Bus. Res. **104**, 247–260 (2019). https://doi.org/10.1016/j.jbusres.2019.07.015
12. Siqueira, J.R., Peña, N.G., ter Horst, E., Molina, G.: Spreading the word: how customer experience in a traditional retail setting influences consumer traditional and electronic word-of-mouth intention. Electron. Commer. Res. Appl. **37**, 100870 (2019). https://doi.org/10.1016/j.elerap.2019.100870
13. Xu, C., Peak, D., Prybutok, V.: A customer value, satisfaction, and loyalty perspective of mobile application recommendations. Decis. Support Syst. **79**, 171–183 (2015). https://doi.org/10.1016/j.dss.2015.08.008

14. Peng, K.-F., Chen, Y., Wen, K.-W.: Brand relationship, consumption values and branded app adoption. Ind. Manage. Data Syst. **114**, 1131–1143 (2014). https://doi.org/10.1108/IMDS-05-2014-0132
15. Iyer, P., Davari, A., Mukherjee, A.: Investigating the effectiveness of retailers' mobile applications in determining customer satisfaction and repatronage intentions? A congruency perspective. J. Retail. Consum. Serv. **44**, 235–243 (2018). https://doi.org/10.1016/j.jretconser.2018.07.017
16. Park, H.H., Jeon, J.O., Sullivan, P.: How does visual merchandising in fashion retail stores affect consumers' brand attitude and purchase intention? Int. Rev. Retail Distrib. Consum. Res. **25**, 87–104 (2015). https://doi.org/10.1080/09593969.2014.918048
17. Pappu, R., Quester, P.: Does customer satisfaction lead to improved brand equity? An empirical examination of two categories of retail brands. J. Prod. Brand Manage. **15**, 4–14 (2006). https://doi.org/10.1108/10610420610650837
18. Vakulenko, Y., Shams, P., Hellström, D., Hjort, K.: Online retail experience and customer satisfaction: the mediating role of last mile delivery. Int. Rev. Retail Distrib. Consum. Res. **29**, 306–320 (2019). https://doi.org/10.1080/09593969.2019.1598466
19. Li, C.-Y., Fang, Y.-H.: Predicting continuance intention toward mobile branded apps through satisfaction and attachment. Telematics Inform. **43**, 101248 (2019). https://doi.org/10.1016/j.tele.2019.101248
20. Yu, W., Ramanathan, R.: Retail service quality, corporate image and behavioural intentions: the mediating effects of customer satisfaction. Int. Rev. Retail Distrib. Consum. Res. **22**, 485–505 (2012). https://doi.org/10.1080/09593969.2012.711250
21. Mainardes, E.W., Cardoso, M.V.: Effect of the use of social media in trust, loyalty and purchase intention in physical stores. Int. Rev. Retail Distrib. Consum. Res. **29**, 456–477 (2019)
22. Eger, L., Mičík, M.: Customer-oriented communication in retail and Net Promoter Score. J. Retail. Consum. Serv. **35**, 142–149 (2017). https://doi.org/10.1016/j.jretconser.2016.12.009
23. Beneke, J., de Sousa, S., Mbuyu, M., Wickham, B.: The effect of negative online customer reviews on brand equity and purchase intention of consumer electronics in South Africa. Int. Rev. Retail Distrib. Consum. Res. **26**, 171–201 (2016). https://doi.org/10.1080/09593969.2015.1068828
24. Bellman, S., Treleaven-Hassard, S., Robinson, J., Varan, D., Potter, R.: Brand communication with branded smartphone apps: First insights on possibilities and limits. GFK Market. Intell. Rev. **5**, 24–27 (2013)
25. Bidargaddi, N., Pituch, T., Maaieh, H., Short, C., Strecher, V.: Predicting which type of push notification content motivates users to engage in a self-monitoring app. Prev. Med. Rep. **11**, 267–273 (2018). https://doi.org/10.1016/j.pmedr.2018.07.004
26. McGookin, D., Tahiroğlu, K., Vaittinen, T., Kytö, M., Monastero, B., Vasquez, J.C.: Investigating tangential access for location-based digital cultural heritage applications. Int. J. Hum Comput Stud. **122**, 196–210 (2019). https://doi.org/10.1016/j.ijhcs.2018.09.009
27. Banerjee, S., Xu, S., Johnson, S.D.: How does location based marketing affect mobile retail revenues? The complex interplay of delivery tactic, interface mobility and user privacy. J. Bus. Res. (2020). https://doi.org/10.1016/j.jbusres.2020.02.042
28. Kim, E., Lin, J.-S., Sung, Y.: To app or not to app: engaging consumers via branded mobile apps. J. Interact. Advert. **13**, 53–65 (2013). https://doi.org/10.1080/15252019.2013.782780
29. Sahami Shirazi, A., Henze, N., Dingler, T., Pielot, M., Weber, D., Schmidt, A.: Large-scale assessment of mobile notifications. In: Proceedings of the 32nd annual ACM conference on Human factors in computing systems - CHI 14, pp. 3055–3064. ACM Press, Toronto, Ontario, Canada (2014). https://doi.org/10.1145/2556288.2557189

30. Shankar, V., Venkatesh, A., Hofacker, C., Naik, P.: Mobile marketing in the retailing environment: current insights and future research avenues. J. Interact. Mark. **24**, 111–120 (2010). https://doi.org/10.1016/j.intmar.2010.02.006
31. März, A., Lachner, M., Heumann, C.G., Schumann, J.H., von Wangenheim, F.: How you remind me! the influence of mobile push notifications on success rates in last-minute bidding. J. Interact. Mark. **54**, 11–24 (2021). https://doi.org/10.1016/j.intmar.2020.08.002
32. Pham, X.-L., Nguyen, T.-H., Hwang, W.-Y., Chen, G.-D.: Effects of push notifications on learner engagement in a mobile learning app. In: 2016 IEEE 16th International Conference on Advanced Learning Technologies (ICALT), pp. 90–94. IEEE, Austin, TX, USA (2016). https://doi.org/10.1109/ICALT.2016.50
33. Freyne, J., Yin, J., Brindal, E., Hendrie, G.A., Berkovsky, S., Noakes, M.: Push notifications in diet apps: influencing engagement times and tasks. Int. J. Hum. Comput. Interact. **33**, 833–845 (2017). https://doi.org/10.1080/10447318.2017.1289725
34. Fischer, J.E., Yee, N., Bellotti, V., Good, N., Benford, S., Greenhalgh, C.: Effects of content and time of delivery on receptivity to mobile interruptions. In: Proceedings of the 12th international conference on Human computer interaction with mobile devices and services - MobileHCI 10, pp. 103. ACM Press, Lisbon, Portugal (2010). https://doi.org/10.1145/1851600.1851620
35. Stocchi, L., Michaelidou, N., Pourazad, N., Micevski, M.: The rules of engagement: how to motivate consumers to engage with branded mobile apps. J. Mark. Manage. **34**, 1196–1226 (2018). https://doi.org/10.1080/0267257X.2018.1544167
36. Newman, C.L., Wachter, K., White, A.: Bricks or clicks? Understanding consumer usage of retail mobile apps. J. Serv. Mark. **32**, 211–222 (2018). https://doi.org/10.1108/JSM-08-2016-0289
37. Wohllebe, A., Ross, F., Podruzsik, S.: Influence of the net promoter score of retailers on the willingness of consumers to install their mobile app. Int. J. Interact. Mob. Technol. **14**, 124 (2020). https://doi.org/10.3991/ijim.v14i19.17027
38. Schmitt, N.: Uses and abuses of coefficient alpha. Psychol. Assess. **8**, 350–353 (1996). https://doi.org/10.1037/1040-3590.8.4.350
39. Steinmetz, H., Matiaske, W., Berlemann, M., Fantapié Altobelli, C., Seidel, W.: Lineare Strukturgleichungsmodelle: eine Einführung mit R. Hampp, München Mehring (2015)
40. Kline, R.B.: Principles and Practice of Structural Equation Modeling, 4th edn. Taylor & Francis Ltd., New York (2015)
41. Hu, L., Bentler, P.M.: Cutoff criteria for fit indexes in covariance structure analysis: conventional criteria versus new alternatives. Struct. Equ. Modeling **6**, 1–55 (1999). https://doi.org/10.1080/10705519909540118
42. Fisher, N.I., Kordupleski, R.E.: Good and bad market research: a critical review of net promoter score. Appl. Stochast. Models Bus. Ind. **35**, 138–151 (2019). https://doi.org/10.1002/asmb.2417
43. Eskildsen, J.K., Kristensen, K.: The gender bias of the net promoter score. In: 2011 IEEE International Conference on Quality and Reliability, pp. 254–258 (2011). https://doi.org/10.1109/ICQR.2011.6031720

Recommendation Engine of Learning Contents and Activities Based on Learning Analytics

Adelina Aleksieva-Petrova[1](✉) and Milen Petrov[2]

[1] Technical University of Sofia, Sofia, Bulgaria
aaleksieva@tu-sofia.bg
[2] SU "St. Kliment Ohridski", Sofia, Bulgaria
milenp@fmi.uni-sofia.bg

Abstract. Recommendation engines are being increasingly deployed into the e-learning systems. This paper proposes a software architecture for recommending learning content and learning activities, which has been validated by means of a case study. The main goal of that architecture is to achieve better recommendations of learning content and learning activities not only in systems, but also in similar e-learning environments.

Keywords: Learning analytics · Learning management system · Recommendation engines

1 Introduction

Nowadays, students' motivation and high student results are one of the biggest challenges in education. Different interactive learning approaches are implemented in systems, tools or services. The use of recommendation engines of learning content or learning activities are methods which could be applied in the learning environment. Recommendation engines usually provide suggestions to users for different items that they may be interested in, based on the analysis of the behaviors of that user or other users.

Learning data can be regarded as big data with unique characteristics which have potential in the design of education systems [1]. Learning analytics (LA) allows students' behaviors to be analysed by using tracking or content analysis, which should be used in recommendation engines.

Within the Aptitude project the recommendation engine, called Recommender, is one of the main components in the system architecture [2]. The main project goal is to design and implement the platform, which is based on learning analytics from learning management systems (LMS) and educational games, to recommend and adapt the learning content and activities.

Thus, the main goal of the study is defined, which is to propose software architecture for a recommendation engine based on learning analytics, which has an impact on two main parameters: learning content and learning activities. This approach should be validated using a case study of an e-learning system.

M. E. Auer and T. Tsiatsos (Eds.): IMCL 2021, LNNS 411, pp. 372–378, 2022.
https://doi.org/10.1007/978-3-030-96296-8_33

The methodology of our research can be broken down into the following main phases, which follow the paper structure. The first phase involves a review of related works on this problem. The next phase is to propose the software architecture of the recommendation engine at micro and macro levels based on LA. The next phase is a proof-of-concept implementation of the proposed architecture, which allows executions to be compared based on an abstract performance indicator.

2 Related Works

Recommendation systems are usually implemented by using two types of algorithms: Content-Based Filtering (CBF) and Collaborative Filtering (CF) [3]. They differ according to the type of data used. The CF calculates predictions using a set of user feedback data on items (usually a rating) and CBF uses only the main features of the items.

A new recommendation approach is needed to develop future recommendation systems, e.g. semantic technology such as an ontology and resource description framework [4]. One example of CF is the Case-based Learning Assistant System [5], whose adaptation mechanism has a combination of value comparison based on requested word association profiles and manual adaptation based on user collaborative recommendation to find references and learning materials. On the other hand, CBF is used for the self-organisation of the learning objects based on an e-learning recommendation approach, where learning objects are modeled as intelligent entities and related metadata are extended to describe their state [6].

Most systems actually use hybrid recommendation algorithms which combine those two algorithms and try to improve them. For example, in order to improve efficiency and accuracy, a heuristic hybrid recommender method is proposed, which uses memory-based and model-based approaches, such as improving the ontology in the CF part and the ontology structure by eliminating the uniformity of the edges of the hierarchical relations between the concepts in the item ontology in the CBF part, and creates a new method for measuring semantic similarity [7].

Also, the quality of recommendation systems, particularly the accuracy, can be improved significantly by incorporating additional sources of information about learners [4]. The attributes of individual students for learning path assignments, according to their track record in past learning experiences and performance, are used for the classification of learners and recommendation of appropriate course materials. Clustering techniques and a long short-term memory neural networks are applied to build a personalized learning full-path recommendation model [1].

In study [8] a learning path recommendation model based on a multidimensional knowledge graph framework is proposed, which creates separately stored learning objects organized in several classes and six main semantic relationships between learning objects in the knowledge graph. A learning path recommendation model is designed for satisfying different learning needs based on the multidimensional knowledge graph framework, which can generate and recommend customized learning paths according to the e-learner's target learning object.

The characteristics of the individual learner and their mapping to the learning objectives and facilitating the achievement of these objectives are two elements intended to support the design of curricula which are used to provide learners with recommended learning-resource options [9]. Also, at the level of seeking appropriate courses in a real-world setting, the virtualized agent-based recommendation system is built from semantic recommendations with the aid of virtual agents based on user requirements and preferences. That system improves user learning skills and makes course selection easier, depending on users' interests and preferences [10].

Most of the recommendation methods are implemented as a plug-in in different LMS. A similar system is MoodleRec, which can sort through a set of supported standard compliant Learning Object Repositories and suggest a ranked list of Learning Objects following a simple keyword-based query [11].

3 Aptitude Recommendation Engine

In an Aptitude project we define the recommendation as a process which affects two variables: learning content and learning activity. It is also divided into two levels: micro and macro. The micro level is related to the learning content and activities within the boundary of the system itself, while the macro level is related to the content and activities that are outside the system in similar learning environments (see Fig. 1).

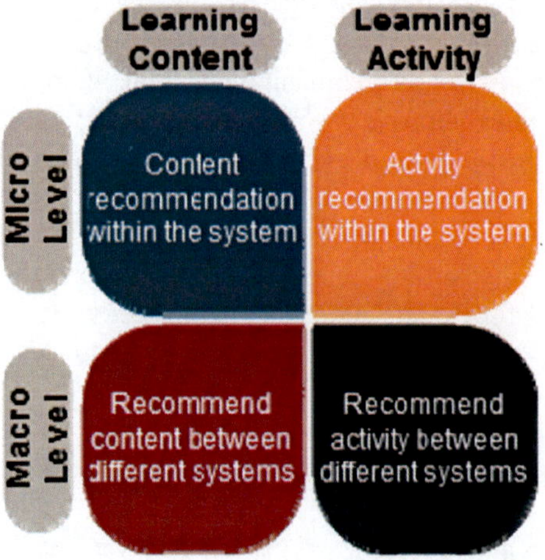

Fig. 1. Aptitude recommendation process

An Aptitude Recommendation Engine, called Aptitude Recommender, is designed to suit the process described. As a result, a software architecture is proposed. In order to standardize the architecture, the UML component diagram will be used as a notion. The main system components are defined and explained, specifically their interfaces to provide or request services (see Fig. 2).

The Text Extractor component refers to the processing of various learning resources and the initial initialization of the main keywords in the learning content. Its main functionality is to extract the content from the file by reading, filtering and loading the plain text. If this is not possible, for example in the case of video and audio files, then the metadata for them is extracted. This text is provided to the Domain Knowledge Controller component, which has to extract the ontology from the subject area (if any) via the Domain Ontology Controller component.

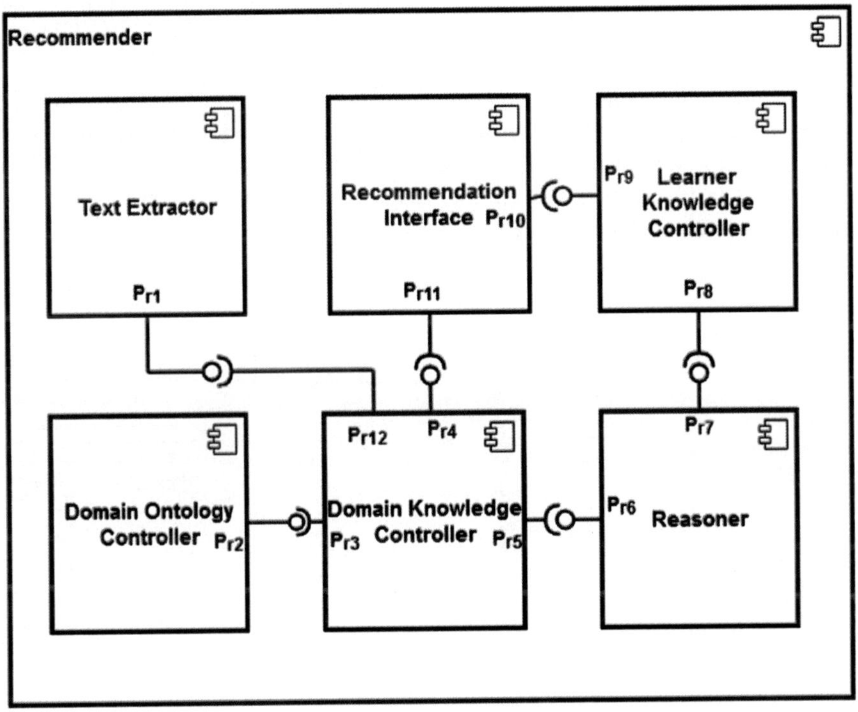

Fig. 2. Aptitude recommendation engine architecture

The Reasoner Component contains a collection of rules (Additional rules) that give additional expressiveness to the ontology and serve to issue some recommendations.

The Recommendation Interface is a component which provides the recommendations using CBF for learning content and CF for learning activities.

The Learner Knowledge Controller component is basic for performing LA for a specific learner. By using the provided logs from different LMSs or tools with student activities and the student results from courses, the so-called "student log" is compiled with learner preferences for learning content and activities.

4 Case-Study

In order to validate the proposed Recommendation Engine architecture, a prototype of the system was designed and implemented. It was designed as an external web-based platform with responsive design in order to be used by both mobile devices and computer stations.

A remote REST service was implemented in the platform in order to read and insert learning content. The source of the learning resources was the Moodle system, which is one of the most widely used LMS. In order to display the recommendation results, Moodle becomes the client of the web service and the Aptitude Recommender runs as a server.

The learning resources are two types: learning content and learning activity. The learning content is every file or web page which contains reading material in a specific domain. For instance, such files are PowerPoint presentations, Word files or Acrobat Reader files. The learning activities are usually some resources, such as assignments, quizzes, wikis, forums, etc.

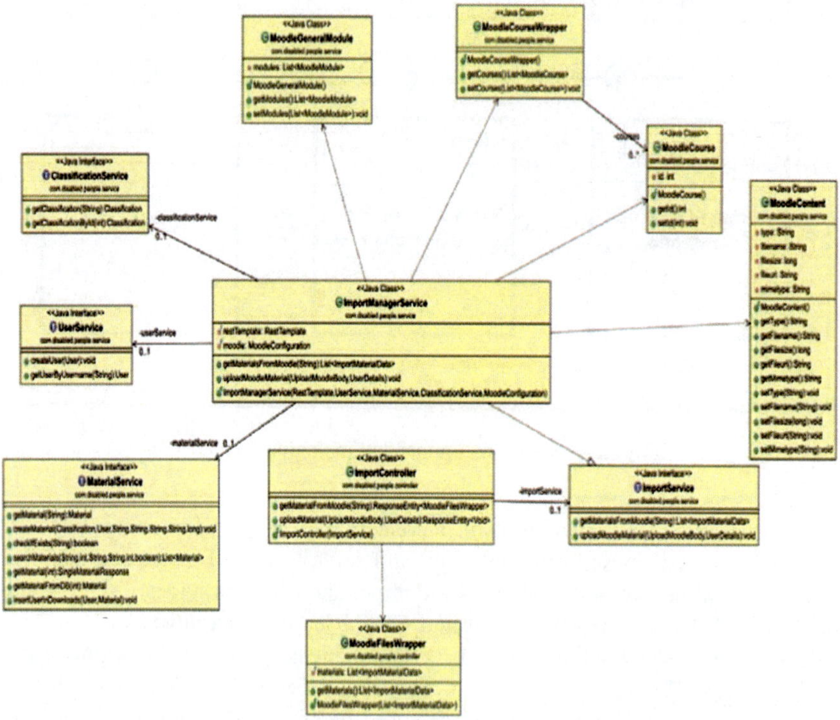

Fig. 3. Class diagram of Moodle resources extract service

The class diagram of Moodle resources extract service is displayed in Fig. 3. ImportManagerService class takes care of all Moodle-related operations. It supports two main methods - retrieving resources from Moodle by a short course name, and inserting a specified Moodle resource into the application. It interacts with almost all other services to accomplish its purpose. Moodle data is retrieved using the Moodle API and is sent in JSON format. All requests coming to this service are handled by the ImportController.

RecommendationManagerService manages all functions related to learning resources recommendations. RecommendationController in turn accepts the requests from the user interface, transforms them as operations and delegates them to RecommendationManagerService.

Using some Moodle databases and system logs for students the following metrics are selected for measuring the behaviour of learners using tracking and content analysis:

- Getting information about courses
- Getting summary information from the log table about course visits
- Getting information about tests
- Receiving information about test grades
- Getting information about learning resources
- Getting detailed information from the log table on course attendance
- Getting information from the log table on resource views
- Loading resource texts

Each record is processed and recorded as fact in the data repository in order to be used in the recommendations. There are three main services:

- Semantic recommender - provides relevant learning activities calculated by the Apriori method;
- Semantic Search - enables learning content searching by terms in the subject area ontology;
- Semantically similar resources - provides a list of similar learning resources according to their cosine similarity.

These services make recommendations on learning resources for the relevant LMS. Using this approach, it is possible to include other systems, tools or services as resources and make recommendations both inside (micro level) and outside of the system (macro level).

5 Conclusion

Using learning analytics in the recommendation process of learning content and activities will contribute to achieving high student results through the delivery of appropriate learning resources. The proposed software architecture tries to be independent from other systems, tools or services which are sources of learning content and activities.

The main advantage of this architecture is that it allows the recommendation of both learning content and learning activities which are internal or external to the system. This defines both the micro and macro level of the recommendations. To validate this architecture, a web system has been designed and developed in order to be used by different mobile and other devices. There are three main recommendation services provided by this system: semantic recommender, semantic search and semantically similar resources.

Further research is needed to carry out more experiments with other systems or tools and different learning content and activities in different application domains.

Acknowledgment. The research reported here was funded under a project entitled "An innovative software platform for big data learning and gaming analytics for a user-centric adaptation of technology enhanced learning (APTITUDE)" - research projects on societal challenges – 2018 by the Bulgarian National Science Fund with contract №: KP-06OPR03/1 from 13.12.2018.

References

1. Zhou, Y., Huang, C., Hu, Q., Zhu, J., Tang, Y.: Personalized learning full-path recommendation model based on LSTM neural networks. Inf. Sci. **444**, 135–152 (2018)
2. Aleksieva-Petrova, A., Petrov, M.: Formal Specification of Aptitude Architecture for Recommendation and Adaptation of Learning Contents and Activities Based on Learning Analytics. Research Book Series: Transactions on Computational Science & Computational Intelligence (2021) in print
3. Gunawardana, A., Shani, G.: A survey of accuracy evaluation metrics of recommendation tasks. J. Mach. Learn. Res. **10**(12), 2935–2962 (2009)
4. Rawat, B., Samriya, J.K., Pandey, N., Wariyal, S.C.: A comprehensive study on recommendation systems their issues and future research direction in e-learning domain. In: Materials Today: Proceedings (2020)
5. Nasiri, S., Zenkert, J., Fathi, M.: Improving CBR adaptation for recommendation of associated references in a knowledge-based learning assistant system. Neurocomputing **250**, 5–17 (2017)
6. Wan, S., Niu, Z.: An e-learning recommendation approach based on the self-organization of learning resource. Knowl. Based Syst. **160**, 71–87 (2018)
7. Bagherifard, K., Rahmani, M., Nilashi, M., Rafe, V.: Performance improvement for recommender systems using ontology. Telematics Inform. **34**(8), 1772–1792 (2017)
8. Shi, D., Wang, T., Xing, H., Xu, H.: A learning path recommendation model based on a multidimensional knowledge graph framework for e-learning. Knowl. Based Syst. **195**, 105618 (2020)
9. Neville, K.J., Folsom-Kovarik, J.T.: Recommendation across many learning systems to optimize teaching and training. In: International Conference on Applied Human Factors and Ergonomics, pp. 212–221. Springer, Cham (2018)
10. Ali, S., Hafeez, Y., Humayun, M., Jamail, N.S.M., Aqib, M., Nawaz, A.: Enabling recommendation system architecture in virtualized environment for e-learning. Egypt. Inf. J. **23**, 33–45 (2021)
11. De Medio, C., Limongelli, C., Sciarrone, F., Temperini, M.: MoodleREC: a recommendation system for creating courses using the moodle e-learning platform. Comput. Hum. Behav. **104**, 106168 (2020)

SELFIE Helper, an Automated Support Chatbot for the SELFIE Platform

Ioannis Kazanidis[1(✉)], George Terzopoulos[1], Panagiotis Kanakakis[2], Avgoustos Tsinakos[1], and Vasilis Tsoukalas[1]

[1] International Hellenic University, Kavala, Greece
kazanidis@cs.ihu.gr
[2] GFOSS Open Technologies Alliance, Athens, Greece

Abstract. SELFIE (Self-reflection on Effective Learning by Fostering the use of Innovative Educational technologies) is a free tool designed to assist schools in embedding digital technologies into teaching, learning and assessment. However, in many cases, school representatives have questions regarding the registration process. They have to ask SELFIE representatives in each involved country of the project and wait for their response. This approach does not allow them to finish the registration process before they get the answer and causes more work for SELFIE experts per country. SHERPA (SELFIE HElpeR & Pedagogical innovation Assistant) is a two-year Erasmus + KA3 project with a mission to enhance innovation in schools by supporting self-assessment processes for making better use of digital technology in teaching and learning. One of the key outputs of the SHERPA project is the SELFIE Helper, a component based on chatbot technology that is responsible for providing automated step-by-step assistance to schools when registering to the SELFIE platform. The goal of SELFIE Helper is to enable schools to further exploit SELFIE to its full potential as a tool for Digital Competence development through self-elevation. This paper presents the architecture of the SELFIE Helper tool and the technologies used for building it. Initial internal evaluation results show that the proposed approach can lead to the development of chatbots that can help users and provide automated support when this is necessary, thus improving processes, such as SELFIE registration.

Keywords: Chatbot · SELFIE · SHERPA · Education · Databases

1 Introduction

Chatbots were initially introduced to improve customer service operations and are part of many systems nowadays. Chatbots are computer programs that act as conversational agents and can conduct conversations with the users using natural language speech [1]. Chatbots rely on Artificial intelligence (AI), natural language processing (NLP), and machine learning technologies. Nowadays, chatbots are incorporated into many systems. By 2024, Insider Intelligence predicts that consumer retail spend via chatbots worldwide will reach $142 billion, while in 2019 it was only $2.8 billion [2].

M. E. Auer and T. Tsiatsos (Eds.): IMCL 2021, LNNS 411, pp. 379–389, 2022.
https://doi.org/10.1007/978-3-030-96296-8_34

Chatbots have many benefits compared to traditional operators. They provide 24-h availability and can answer to many questions simultaneously, while an operator has limited capabilities and can assist only a few customers. Furthermore, chatbots provide consistent answers while in traditional systems, customers are tempted to try again in order to get different assistance by another operator. Moreover, chatbots have infinite patience and can provide instant answers. Finally, chatbots cost significantly less than traditional operators and can be leveraged to increase customer engagement with timely tips and offers, resulting in increased sales.

SELFIE (Self-reflection on Effective Learning by Fostering the use of Innovative Educational technologies) is the European Commission's free, customizable self-evaluation tool helping schools to better understand their progress on digitally-enabled teaching and learning [3]. SHERPA is a two-year Erasmus + project with a mission to strengthen the adoption and systematic use of SELFIE across Europe [4]. SHERPA will provide schools with vitally-needed scaffolding for on-boarding SELFIE and will help them gain full benefits from their SELFIE report in terms of both strategy and practices.

When schools are attempting to register to SELFIE, in many cases, school representatives have questions regarding the registration process. They have to ask SELFIE experts (SELFIE representatives in each involved country of the project) and wait for their response. This is a slow process and causes significant additional workload to the SELFIE experts per country.

One of the key outputs of the SHERPA project is the SELFIE Helper, a chatbot system that assists schools to use the SELFIE platform, providing them with real-time help in resolving their specific user issues during registration to the SELFIE platform.

This paper presents the technologies and algorithms used in the "SELFIE" Helper tool. It provides the proposed architecture for educational chatbots, explains the key technologies and methodologies that were applied, and presents the case of SELFIE Helper architecture and its consisting modules. For every module, it describes and discusses the problems and challenges that were addressed, the adopted methodologies for their design and development, and provides operation details. Special emphasis is given to the machine learning algorithms that were adopted from the Case Based Reasoning (CBR) Inference Engine, the Knowledge Base structure and contained data and the Chatbot interface functionality and logic. Finally, it presents the results of an informative internal evaluation of the prototype along with limitations and future work. This information will hopefully assist designers and developers to select the appropriate tools in building a chatbot designed for educational purposes.

The remainder of this paper is organized as follows: An overview of the SELFIE Helper is given in the next section. Then, all the SELFIE Helper modules are presented in the following sections. Section 3 depicts the Chatbot Interface, where questions by the users are asked. Section 4 describes the CBR Inference Engine, where questions submitted to the Chatbot Interface are forwarded to, while Sect. 5 presents the Knowledge Base, where queries/answers are stored and serves as the CBR Inference Engine training data set. Section 6 lists information regarding the Backend Management system where all administrative procedures take place. Finally, Sect. 7 presents some early findings from the initial internal evaluation of the SELFIE Helper Helper followed by the conclusions of this study.

2 An Overview of SELFIE Helper

The SELFIE Helper system consists of 4 modules:

- The Chatbot Interface
- The CBR Inference Engine
- The Knowledge Base
- The Backend Management System

Queries posted by SELFIE school representatives via the Chatbot Interface are forwarded to the CBR Inference Engine which activates a diagnosis process capable of identifying any existing, appropriate similar case/answer stored in the Knowledge Base. The SELFIE Helper architecture is depicted in Fig. 1.

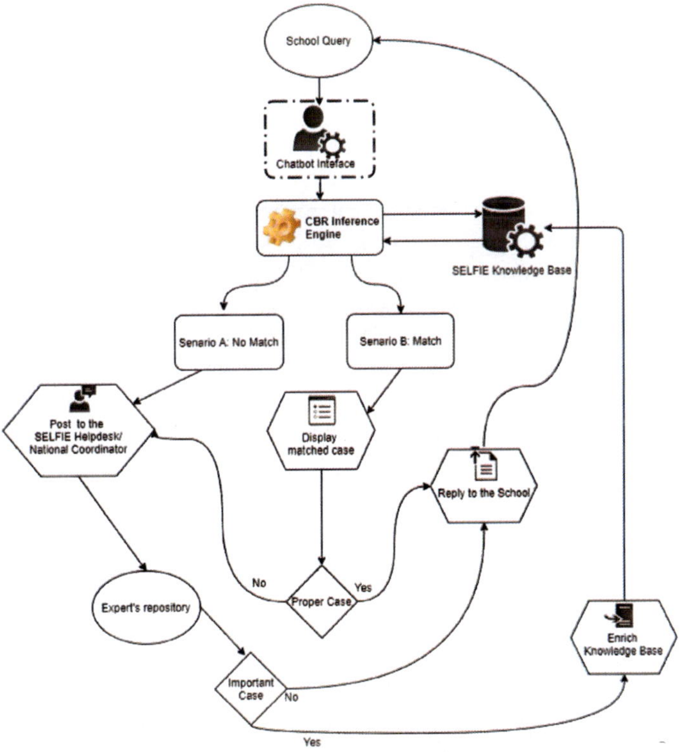

Fig. 1. SELFIE helper architecture

The system can be proven useful and time saving for the SELFIE helpdesk service and the schools, considering that there is a strong likelihood of repeated problems emerging from initial use of SELFIE by schools.

Two scenarios are possible when a school representative submits a question to the system:

- **Scenario A:** where "No relevant case is found" (e.g. a novel issue/case has been posted), the question is redirected to the SELFIE helpdesk team and/or to the SELFIE national coordinator. If the SELFIE helpdesk team deems the query to be an important new case, the system allows them to add the data to the Knowledge Base, thereby enriching it with new content. On the contrary, where such query is deemed to be a minor/routine matter, the SELFIE helpdesk reply is simply communicated to the school without being registered.
- **Scenario B:** where "One or more relevant cases are found" a ranked set of relevant cases is automatically retrieved and the case with the best score is provided to the school (note: all query/response cases are stored in a totally anonymous manner). The school's SELFIE representative (i.e., the user) can report if she/he is satisfied with the provided answer. If the user cannot find a suitable response among those proposed, then "Scenario A" is triggered.

3 Chatbot Interface

Chatbot Interface is a web application that serves as a publicly available front-facing interface to interact with CBR Inference Engine application programming interface (API) and Knowledge Base service. CBR Inference Engine is used for submitting questions and getting a response, provided that an algorithm is able to find a suitable answer from available data stores. The Chatbot Interface will suggest asking the question again, when no suitable answer can be found. The next step is to offer the school representative the possibility to submit the question as a suggestion that will be handled by Selfie Experts. Suggested questions would then be submitted to the Knowledge Base as a Pending Question type.

A screenshot of the Chatbot Interface in English is depicted in Fig. 2. It consists of two main parts: a few example questions on the left of the screen and the conversation on the right. The left side content consists of some indicative popular that the user can asked to the system. Clicking on one of the questions will quickly ask the system the specific question. If the conversation has several questions and answers with scrolling enabled, then the interface will make sure that the conversation area is scrolled to the bottom with both answer and input becoming visible. The school representative can also change the language of the User Interface (UI), though that will not only change the interface, but would also make sure that the system is being asked questions while providing the language context according to the currently selected language. Users can type their question into an input below the conversation and hit Enter or press the button to the right. When the answer is provided, users can state their satisfaction through three well known faces (happy, neutral, sad) providing feedback to the system about the corresponding answer.

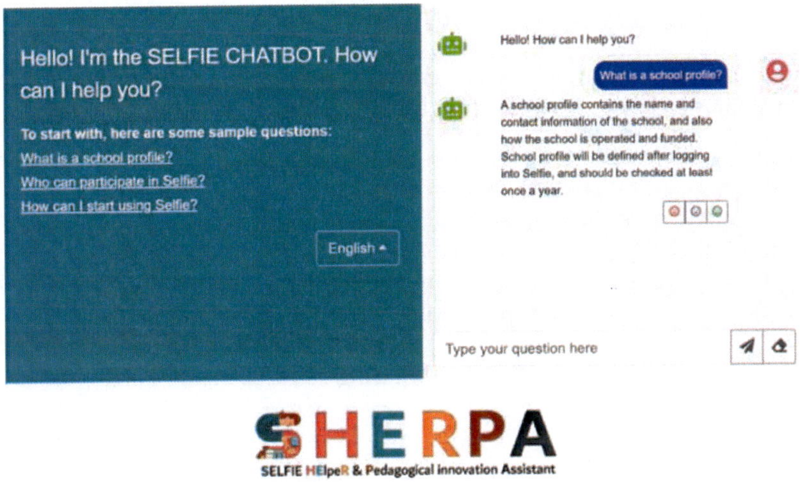

Fig. 2. Chatbot interface

Chatbot Interface was generated with the use of a tool called Vue CLI [5] and uses TypeScript [6]. Bootstrap User Interface framework [7] provides some of the reusable components and styling. BootstrapVue [8] is used to get Vue.js specific flavor of those components. The Chatbot Interface web application is a static asset package that is being built from the source code and includes configuration for services (CBR Inference Engine and Knowledge Base).

4 CBR Inference Engine

The CBR Inference Engine, is the module which is capable of answer selection from a predefined closed set of Questions and Answers (Q&A) pairs, based on user's questions. Its main core is based on the ChatterBot platform [9] which is a machine-learning based conversational dialog engine generates responses based on collections of known conversations. It uses a selection of machine learning algorithms to produce different types of responses. The more input that ChatterBot receives increase the accuracy of its responses.

NLP is a field of artificial intelligence in which computers analyze, understand, and derive meaning from human language in a smart and useful way [10]. This human-computer interaction enables real-world applications like automatic text summarization, sentiment analysis, topic extraction, named entity recognition, parts-of-speech tagging, relationship extraction, stemming, and more. NLP uses all kinds of algorithms to identify linguistic rules, extract meaning, and uncover the structure of a text. Depending on the task the algorithms can be categorized to:

- Text classification algorithms, which are used for organizing unstructured text into predefined categories.

- Text extraction algorithms, which are used for extracting specific pieces of data from a text.
- Topic modelling algorithms where a method is used for clustering groups of words and similar expressions within a set of data.

The CBR Inference Engine works in a multilingual environment and due to the use of machine learning algorithms achieves high accurate results on user's questions. The implementation covers two main scenarios based on the existence of the appropriate response. Given a question, the CBR Inference Engine determines the similarity with the predefined closed set of questions from the Knowledge Base. In case the similarity or confidence value is over a predefined threshold (0.7) the answer of this question is given as a response to the user. In any other case, the engine returns an empty result and the Chatbot Interface handles the following steps, notifying a SELFIE expert to handle the unanswered question.

The issue of understanding users' queries and intent and providing appropriate responses was faced as a text and semantic similarity problem. Thus, the CBR Inference Engine takes into account not only the surface closeness of two pieces of text but also the meaning of them. In order to create a more concrete solution and achieve accurate results, the implementation combines results from two different algorithms. Thus, the user response comes from the algorithm who achieved the highest score. The algorithms used are the following:

- Levenshtein distance [11] which is a string metric for measuring the difference between two sequences. Informally, the Levenshtein distance between two words is the minimum number of single-character edits (insertions, deletions, or substitutions) required to change one word into the other.
- FastText algorithm [12], which is a library for learning of word embeddings and text classification created by Facebook's AI Research (FAIR) lab. The model represents questions as word vectors leveraging the fact that neighboring words in a sentence affect the semantic meaning of that word.

Each algorithm is trained based on the closed set of questions located on the Knowledge Base. As new questions are populated on the Knowledge Base the models are retrained once a day via a cron job. Both algorithms gather the Q&A pairs via REST calls, transform and apply them as an input. The FastText algorithm creates as many models as the languages supported by the system. Each model is located on the local file system and is used when needed. The CBR Inference Engine exposes a REST API to communicate with the Chatbot Interface and based on the input parameters, language code and question, respond to the user.

5 Knowledge Base

SELFIE Helper, includes a database that contains a set of questions and answers, described as Knowledge Base. The Knowledge Base consists of a set of questions and answers. Questions and answers are organized into categories or topics. Each category includes several questions and answers, while each question-answer corresponds to exactly one category or topic. The concept category or topic is considered equivalent.

There are four categories of users. The Master Selfie Expert, the Country Selfie Experts and the End Users (school representatives registering to SELFIE) who suggest - in their language - a question to be registered "as suggested" in the database. Finally, there is the Chatbot Service User who reads data from the database to regularly update its training.

Master Selfie Experts have the ability to dynamically create as many categories of questions-answers as they want and can add to each category as many questions-answers as needed. Country Selfie Experts do not have the ability to add new questions or answers, but they can insert translations on their corresponding language. They can also suggest new questions (pending questions). Each pending question may have a set of rephrased questions. In this case, a set of rephrased question belong to a group (etc. groupNo). Some questions may not have answers (at least for a while) - until they are answered (if answered). It is also possible that some questions do not belong (at least for a while) to a topic.

Each category has a unique identifier, a description and creation date. Each question has a unique identifier, creation date, status and should be translated in all the languages that exist in the database. The status can be inTraslation, Translated and Published.

A suggested question (pending question) has the same attributes that questions have and also may have translated in some languages. The status of a suggested question may have the values Propagated, Completed or Canceled. A suggested question -if approved- can become a normal question. In this case its ID is stored in the corresponding normal question (Pend) for later reference.

Each answer has a unique identifier, a response date, status - should be translated in all the languages that exist in the database- and its existence depends on the question it answers. However, there may be similar questions -formulated differently by different end users- that lead to the same answer. The status of an answer can be inTraslation, Translated or Published.

The SELFIE Helper KB provides multilanguage support for End Users. To address the questions/answers in the supported languages, with the aid of a back-end management system that is presented at the next section, the following scenario is suggested:

- When an End User does not get a satisfactory answer from the chatbot then he will be able to enter the question as a "pending question" in his/her language (eg. Greek).
- Then, the Country Selfie Expert of the respective language (eg. Greek) - who has access to the pending questions - will translate in English and insert a row in the PendingQuestionLanguage TABLE with the description in English language. The related pending question will be marked as propagated. He/She may also rephrase the pending question to a set of similar questions belonging to the same group.
- Then the Master Selfie Expert will create -if the question is approved (along with the rephrased set of pending questions)- a unique question ID (and a unique

question ID for each rephrased question) and will add the question -as a normal question. He/She will mark the new question as in Translation.

- The Master Selfie Expert will also create a unique answer ID and add the answer in the database. The related answer will be marked as in Translation.
- Then the Country Selfie Experts of each corresponding language (eg. Greek) will translate the questions and answers in their language.
- Finally, the Master Selfie Expert, verifies that the new question and answer have been translated in –all- languages, and changes the state of the new normal question (and answer) as Published (or just Translated if he/she doesn't want to publish the question yet).
- The Master Selfie Expert will also mark the status of the corresponding pending question as Completed. If the pending question is not approved, it will be marked as Canceled.
- After that the published question-answer is available in the Chatbot Service User, in order to update its training.

SHERPA, by its nature, is a project that provides services that do not involve the processing of any personal data. As such, SHERPA does not expect to collect any information that is related to the identity of a specific user, i.e. Personally Identifiable Information (PII) and, therefore, it is not asking for any such information. Users are specifically advised not to use any personal data or information that might be related to their identity to the SELFIE Helper.

Even in this case though, users might ignore or, mistakenly, provide PII to SHERPA, through a question submitted to the system. The approach that should be taken in this case, is to anonymize the supplied data before any processing takes place on the question. This is accomplished through an initial screening process where system's representatives, namely Country Selfie Experts, need to check for any PII and properly anonymize it, prior to proceeding with any translation.

A second layer of screening against any personal data will be accomplished by the Master Selfie Expert to make sure that no personal data will be stored on the SELFIE's database or accidentally be included in the answer and provided to unauthorized third parties, as a result of subsequent questions placed through the system.

Data Anonymization is conducted manually by deleting any personal information, such as Name, contact details (telephone, email, address), id numbers, prior to being processed to provide an appropriate answer. If such an action might semantically alter the question, and make it unprocessable, alternative options might be used, such the replacement of personal data by dummy data or keywords that cannot disclose information about a specific subject. Examples are the replacement of data subject's name with the keyword "NAME", contact details by the keyword "CONTACT DETAILS", email by "EMAIL" and tel.no by "PHONE NUMBER".

6 Backend Management System

The web-based Backend Management System (BMS) supports the Master Selfie Expert and the Country Selfie Experts categories of users. Overall, the BMS supports several roles, with some of those being explicit and some implicit. The roles are as follows:

- Anonymous - anonymous user that will only be able to see the landing page with a link to login page, not an explicit role or a role as such.
- Administrator - a user role that should mostly deal with user account management tasks, though there also is a full access for the Knowledge Base specific content.
- Authenticated - a user with an account that will be able to authenticate and access the User Home with no content, not an explicit role
- Country Selfie Expert - a user that has ability to review Pending Questions, add new Questions and Answers, translate existing Questions and Answers, mark Questions and Answers as translated and send those for review. This role requires the language field to have a value, making a user an expert for that specific language.
- Master Selfie Expert - a user that has the ability to perform Country Selfie Expert actions for any of the languages, reviews Questions and Answers before those could be published.

Users with credentials can login in order to gain access to the BMS features (Fig. 3).

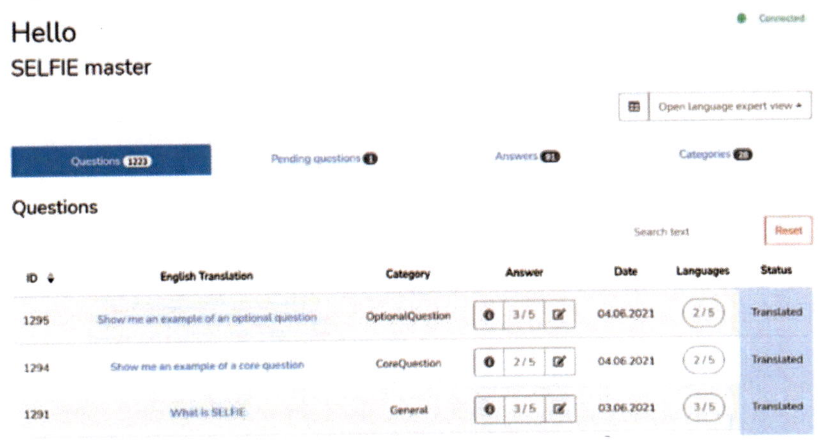

Fig. 3. Backend management system interface

User Management can be accessed through the dropdown menu at the top right of the page where the name of the current user is shown. Corresponding menu entry will only be available to users that have sufficient rights for user management. As the system does not allow self-registration, the only way to create new accounts would be for one of the administrators to do that through User Management. User roles can be assigned at creation time or later on through the user account edit dialog.

The Administrator can perform tasks regarding user management, such as adding a new user and assign roles, viewing all existing users and editing existing users' settings. The Country Selfie Expert can add new Questions and Answers or edit existing Questions and Answers. Country Selfie Expert views for Questions and Answers also include a creation functionality. The system will also allow for any of the content types to be searched/filtered by both text in English and one for the currently selected language or language of the expert. The Country Selfie Expert can also view Pending Questions and edit pending questions in order to provide translations. Pending Question status could be changed to Propagated and thus send it to Master Selfie Expert for review. The Master Selfie Expert can perform Country Selfie Expert actions for any of the languages. Specifically, the Master Selfie Expert can view the same content as a Country Selfie Expert, see a list of all questions and answers for a specific language and see a list of pending questions for a specific language. The Master Selfie Expert can also review Questions and Answers, before those could be published. Finally, he/she can view and review all pending questions and access statistics data about the Knowledge Base.

7 Initial Internal Evaluation Results

By adopting the proposed approach and architecture, developers can build an educational chatbot that can manage a variety of user queries about a specific domain. However, one of the project objectives was to check whether the SELFIE Helper is efficient both for the end users and for the SELFIE experts. In order to get an initial feedback about the implemented prototype, internal evaluation took place. The goal of the internal evaluation was to test SELFIE Helper tool and collect participants' feedback and comments on functionality, usability and quality of the provided answers. All SHERPA team members along with external experts participated in this internal trial, by the following three actions: a) register to SELFIE tool for a virtual school, b) use the SELFIE Helper whenever any concept or step in the SELFIE registration process was not clear for them and they do not know how to proceed and c) give feedback on the provided answers and SELFIE Helper functionality, efficiency and usability.

The initial results were promising and proposals for new implementations were made. In particular, evaluation results and the provided feedback suggested that SELFIE Helper is easy and straightforward to use tool, although the SELFIE Helper did not always provide the required support. However, after the analysis of these cases, it was concluded that this was happening because the questions were general, or when there were not any related answers into the database, thus there was a generic recommendation to further enrich the Knowledge Base with more questions/answers. The following improvement proposals were provided by trial users for SELFIE Helper language processing: support for synonyms, support for plural mode, support for lower/upper case letters. Improvement proposals for the basic SELFIE Helper user interface suggested that an answer window could be scrollable, buttons need to have linked info text and to provide feedback buttons after each answer. These proposals were taken into account and added in the basic functionality of the chatbot.

Overall, the internal evaluation results revealed that SELFIE Helper succeeds in guiding the SELFIE users through the registration process. Thus, chatbots in general can be a useful tool for providing guidance in various similar processes.

8 Conclusions

While chatbot technologies have been widely used in business and commercial sectors, they are a relatively novel approach for improving educational services. Chatbots can help their users with a variety of processes both simple or even more complicated, by automatically responding to their questions. A good case study, however, necessitates the resolution of numerous challenges. As a consequence, providing a technique and a case study for developing a chatbot that assists school representatives in the registration process (i.e. SELFIE), is useful since the provided results may be applied to a variety of different educational procedures. A new prototype was created for the Erasmus + project SHERPA, and the technologies and techniques that were used were presented. The overall positive results of the case study internal evaluation show that the proposed approach can lead to the development of chatbots that can help users and provide automated support when this is necessary, thus improving processes such as SELFIE registration.

Funding. The present work was supported by grant No. 612867-EPP-1-2019-1-EL-EPPKA3-PI-FORWARD from the EC EACEA Agency under the Erasmus + project, SHERPA – SELFIE HElpeR & Pedagogical innovation Assistant.

References

1. Abdul-Kader, S.A., Woods, J.: Survey on Chatbot design techniques in speech conversation systems. Int. J. Adv. Comput. Sci. Appl. **6**, 72–80 (2015)
2. Business Insider. https://www.businessinsider.com/chatbot-market-stats-trends. Accessed 6 July 2021
3. SELFIE. https://ec.europa.eu/education/schools-go-digital/about-selfie_en. Accessed 6 July 2021
4. SHERPA project homepage. https://sherpa4selfie.eu/. Accessed 6 July 2021
5. Vue Cli homepage. https://cli.vuejs.org/. Accessed 6 July 2021
6. TypeScript language homepage. https://www.typescriptlang.org/. Accessed 6 July 2021
7. Bootstrap library homepage. https://getbootstrap.com/. Accessed 6 July 2021
8. BootstrapVue framework homepage. https://bootstrap-vue.org/. Accessed 6 July 2021
9. ChatterBot: Machine learning, conversational dialog engine documentation. https://chatterbot.readthedocs.io/en/stable/. Accessed 6 July 2021
10. Algorithmia, What is natural language processing? Introduction to NLP. https://algorithmia.com/blog/introduction-natural-language-processing-nlp. Accessed 6 July 2021
11. Wikipedia, Levenshtein distance. https://en.wikipedia.org/wiki/Levenshtein_distance. Accessed 6 July 2021
12. Wikipedia, FastText. https://en.wikipedia.org/wiki/FastText. Accessed 6 July 2021

NB-IoT Technology Benefits in Educational Institutes

Apostolos Gkamas(✉)

University Ecclesiastical Academy of Vella, P.O. Box 1144,
45001 Ioannina, Greece
gkamas@aeavellas.gr

Abstract. The Internet of Things (IoT) is the extending of Internet connectivity to physical devices and common objects. Narrow Band-Internet of Things (NB-IoT) is a component of the 5G network infrastructure that focuses on providing connectivity to IoT devices. NB-IoT is a standards-based Low Power Wide Area Network (LPWAN) technology that allows a wide range of new IoT devices and services to be developed. This paper presents the benefits of using NB-IoT technology in Educational Institutes. The aim of this paper is to answer the following question: "How has the emergence of NB-IoT technology affected the Education Process and an Educational Institute Operation?". The paper begins with a brief overview of IoT networks and NB-IoT technology, followed by a discussion of the potential benefits of NB-IoT technology in Educational Institutions. In addition, we examine the impact of NB-IoT technology on the educational process and the functioning of an Educational Institute. Moreover, we compare NB-IoT against other similar technologies like LoRa in the context of Educational Institute Operation.

Keywords: 5G · NB-IoT · IoT · LoRa · Impact on education

1 Introduction

Due to the significant effect of technology, education has evolved from a knowledge-transfer approach to a more collaborative self-directed model. Many Educational Institutions have been pushed to reconsider how they teach and learn as a result of this. Technology has an impact on many aspects of education, including student involvement in learning, content development, and the deployment of tailored content, among others.

The Internet of Things (IoT) is the expansion of Internet connection to physical devices and commonplace items. These gadgets, which are equipped with electronics, Internet connectivity, and other hardware (such as sensors), can communicate and interact with others through the Internet, as well as be monitored and controlled remotely.

IoT began as a tiny market for amateurs and has now grown into a significant business. Smart lighting, online cameras, access-control devices, biometric readers, thermostats, voice-activated assistants like Amazon Alexa, and other gadgets that can connect to the Internet and communicate are examples of IoT.

M. E. Auer and T. Tsiatsos (Eds.): IMCL 2021, LNNS 411, pp. 390–399, 2022.
https://doi.org/10.1007/978-3-030-96296-8_35

The difference between an IoT device and a typical device such as a computer or a smartphone is that an IoT device has a very restricted user interface and is regarded a device that measures and transmits (usually modest amounts) of data. The overall number of IoT devices is enormous, and it is growing at a breakneck pace, with some forecasts predicting significant future increase in the short to medium term. According to IoT analytics report [1] in 2018 there was 7 billion IoT devices with a trend to become 21.5 billion devices by 2025. This significant increase in IoT devices necessitates the present Internet infrastructure's support, and it also creates certain difficult challenges that we must address, particularly in the area of Internet security.

IoT brings new capabilities for both industry and end users. Adoption of IoT technology has resulted in significant benefits for industry. For example, costly PLC (Programmable Logic Controller) modules that utilize outdated technology and require proprietary software to read can be replaced with less priced IoT devices that perform identical duties while providing considerably more connectivity choices. Consumer IoT devices may also be used to build a smart house that monitors things like lighting, temperature, security systems, and power consumption, among other things.

5G is the fifth-generation technological standard for broadband cellular networks, which cellular phone operators began installing globally in 2019. It is the expected successor to 4G networks, which offer connection to the majority of today's cell phones. According to the GSM Association, 5G networks would have more than 1.7 billion customers worldwide by 2025. Narrow Band-Internet of Things (NB-IoT) is a subset of 5G mobile networks that focuses on providing connection to IoT devices. NB-IoT is a standards-based Low Power Wide Area Network (LPWAN) technology that allows a wide range of new IoT devices and services to be enabled. NB-IoT improves device power consumption, system capacity, and spectrum efficiency, particularly in deep coverage. An NB-IoT device's typical battery life is more than 10 years, making it suitable for a wide range of applications. NB-IoT provides extensive coverage in both rural and urban areas. The initial cost of NB-IoT modules is equivalent to that of GSM/GPRS. The underlying technology, on the other hand, is far less complex than GSM/GPRS, and its cost is projected to fall fast as demand grows. All major mobile equipment, chipset, and module manufacturers support NB-IoT, and it can coexist with 2G, 3G, and 4G networks. It also takes use of all of mobile networks' security and privacy capabilities, such as support for user identity confidentiality, entity authentication, confidentiality, data integrity, and mobile equipment identification.

The NB-IoT is a new technology that aims to improve communication between computer systems and other devices, and it's quickly finding its way into classrooms and education in general. NB-IoT improves education by adding a lot of value to the physical teaching environment and organized learning. Smart Educational Institutes feature well-functioning facilities and highly individualized learning techniques.

The advantages of adopting NB-IoT technology in Educational Institutions are presented in this article. The purpose of this article is to respond to the following question: "How has the emergence of NB-IoT technology influenced the educational process and the operations of an Educational Institute?" Furthermore, in the context of Educational Institute Operations, we compare NB-IoT to other comparable technologies such as LoRa.

Before we provide information regarding NB-IoT technology benefits in Educational Institutes, we will answer the question why NB-IoT technology is so important: The two main approaches to provide data access in IoT have been based on either multi hop mesh networks using short-range communication technologies in the unlicensed spectrum, or long-range legacy cellular technologies operating in the corresponding licensed frequency bands. These reference models have recently been challenged by a new form of wireless connection, typified by low-rate, long-range transmission technologies in the unlicensed sub-gigahertz frequency bands, utilized to create low-power WANs (wireless access networks with a star topology) (LPWANs). NB-IoT technology is one of the most promising LPWANs technologies, and it also transmits in licensed frequency bands, which is essential since an Educational Institute does not need to operate its own infrastructure and can instead utilize a cellular operator's infrastructure. The cellular operators providing the necessary tools for the management of the NB-IoT infrastructure used by an organization like an Educational Institute (e.g. Vodafone Tailored Narrowband-IoT [2]). Furthermore, mission-critical IoT communications necessitate safe and dependable connections between utilities and devices, as well as a high level of service (QoS). Due to the congested license free band, this is difficult to implement for unlicensed LPWAN technologies. NB-IoT is a licensed LPWAN technology that provides cellular-level QoS and is built on current long-term evolution standards and infrastructure.

The NB-IoT technology has impact in the various aspects of an Educational Institute operation including:

- Organization operation benefits
- Student's health monitoring.
- NB-IoT technology provides many education benefits on the operation of an Educational Institute.
- E-Learning benefits.
- Research benefits.

This paper has the following structure: In Sect. 2, we present the related work available in the literature and Sect. 3 provides information about the NB-IoT technology. After that we present the benefits of NB-IoT technology in Educational Institutes in Sect. 4. Next section, Sect. 5 presents a comparison between NB-IoT and LoRa. Finally, we conclude our paper and we present our future work in Sect. 6.

2 Related Work

As we discussed in the previous section, the IoT is a true progression of our lives, and as a result, it has an impact on the educational process, particularly the functioning of Educational Institutions. In [15] the authors describe contemporary IoT initiatives in education and examine and analyze how IoT has altered the Education Business Model and brought new value propositions to such organizations.

IoT has a significant influence on higher education, according to the authors of [3], who provide a study on the impact of IoT on higher education, particularly universities. IoT has the potential to transform the way universities operate and improve student learning in a variety of fields and at all levels. If adequately prepared to assure widespread and successful implementation by leadership, staff, and students, it has enormous potential for universities and other Educational Institutions. The IoT is in need of development, and colleges can help. Academics, researchers, and students are in an ideal position to drive the development of IoT systems, devices, applications, and services.

Paper [4] focuses on a research associated with the predictable impact of IoT in the superior education. This work examines a theoretical analysis as well as a statistical investigation. The findings of the theoretical and statistical analyses revealed that the IoT has a significant impact on the educational ecosystem in terms of learning and management elements. The effect is significant in some situations, such as hyper-connectivity, cooperation, and research prospects. Emergent technologies, such as IoT, are currently transforming traditional education systems into scalable, adaptable with rapid dynamic changes, flexible, and more efficient e-learning with a topology that involves a large number of physical and virtual interacting learning objects.

Furthermore, the authors of [5] emphasize the necessity of incorporating IoT into higher education. Several practical approaches for incorporating IoT characteristics in academia are identified in the study, particularly in the areas of teaching and learning enhancements. Furthermore, article [6] investigates how higher education and the IoT should work together. The IoT is supporting changes in higher education, such as changes in education and teaching, learning changes, management changes, experiment and training changes, school changes, and so on.

Because the IoT has an impact on not only higher education but also undergraduate education, the authors of [7] propose that science, technology, engineering, and mathematics core courses be transformed by incorporating an IoT-based learning framework into their corresponding lab projects. The article summarizes the design difficulties of the new learning framework. The article also includes a case study that incorporates an IoT-based learning framework into a Software Engineering (SWE) embedded system analysis and design course.

The impact of IoT is not only significant for Educational Institutions, but it also has an impact on educational platforms. The authors of [8] present an integrated framework for improving educational quality by leveraging sophisticated IoT devices. The capabilities and characteristics of many IoT controller boards are explained and compared for implementing an IoT solution in an educational platform. The proposed platform's goal is to link the teaching process to a real-life environmental situation. A novel paradigm for incorporating items into Virtual Academic Communities (VAC) is also presented in [9]. The proposed model was put to the test through the use of a case study, and the results show that using IoT provides a more engaging learning environment for students as well as more data about the learning process to help teachers improve their knowledge of their students' learning pace and learning difficulties.

NB-IoT has been considered as candidate technology for the implementation of various smart campuses. Paper [10] explores the representative applications in the smart grid and analyze the corresponding feasibility of NB-IoT. Moreover, the performance of NB-IoT in typical scenarios of the smart grid communication environments, such as urban and rural areas, is carefully evaluated via Monte Carlo simulations. Moreover, authors in [11] discuss the deployment and report the practical performance of a single-cell NB-IoT deployed as a part of the 5G Test Network (5GTN) and controlled by a smart-campus micro-operator. Paper results demonstrate that the NB-IoT technology is a viable connectivity solution for various non-critical machine-based applications deployed indoors, highlight the practical performance of this technology, and reveal some practical specifics and challenges of acting as an IoT micro-operator. Finally, authors in [12] present ongoing research project's assumptions, led within the frames of business-university cooperation, focused on Narrowband IoT technology.

In the next section we provide a summary of the NB-IoT technology.

3 NB-IoT Technology

NB-IoT [13] is a standards-based low power wide area (LPWA) technology that has been created to allow a variety of new IoT devices and applications. NB-IoT improves user device power consumption, system capacity, and spectrum efficiency, particularly in deep coverage. The NB-IoT modules are projected to be equivalent in price to GSM/GPRS modules at first. However, the underlying technology is considerably simpler than GSM/GPRS today, and the cost is projected to drop significantly as demand grows.

Indoor coverage, low cost, extended battery life, and high connection density are key priorities for NB-IoT. The NB-IoT standard is based on a subset of the LTE standard, however the bandwidth is limited to a single narrow band of 200 kHz. The 3GPP has standardized NB-IoT as a radio access method for cellular wireless connectivity. It's ideal for a large number of low-complexity devices in difficult radio settings like deep building or subterranean radio signal penetration.

NB-IoT allows for the deployment of low-cost, low-complexity devices with a long battery life (>10 years), in large numbers (50.000 connections cell), and with a wide geographic cell range (up to 40 km or even up to 100 km which is a special Ericsson standard compliant solution). Fleet management, logistics, asset management, and smart metering are examples of use cases that need very little data use in networks.

A wide range of chipsets and modules are available for NB-IoT. (pre-integrated chip-sets for easy integration into whatever device). Water, power, parking, lighting, and agriculture are examples of devices aimed for specific use cases. A processing chipset, memory, extra radio, and other functions, such as a location module like GPS, are all incorporated in modules.

4 Benefits in Educational Institutes

NB-IoT technology, like IoT technology in general, has influenced or will influence every aspect of human life and business operations. The advantages of NB-IoT technology in the administration of Educational Institutes are presented in the following paragraphs. The advantages are organized into five categories:

- Organization operation benefits
- Students and Personnel health monitoring
- Education benefits
- E-Learning benefits
- Research benefits

4.1 Organization Operation Benefits

NB-IoT technology provides many benefits on the operation of an Educational Institute including the following:

- Energy management and eco-system monitoring on campus: NB-IoT technology may be utilized in energy management and eco-system monitoring to deliver energy efficiency and a more sustainable future. By decreasing CO_2 footprint, monitoring and managing energy and water, and so on, an Educational Institution may establish a green campus environment. By adding intelligence to the current infrastructure, the Educational Institute can efficiently balance power generation and energy use to enable more efficient operations.
- Secure campus and classroom access control: NB-IoT technology may be used to regulate access to classrooms, labs, and other areas in universities for students and professors. As a consequence, NB-IoT technology may be utilized to increase university security and simplify access control. For example, NB-IoT technology may be used to build a classroom access control mechanism.
- Smart Fire Evacuation System: In a fire scenario, emergency data transfer is critical. NB-IoT Technology develops a cutting-edge IoT application use case that benefits cities and communities. This technology will help to make public spaces and huge buildings safer and more pleasant. This system employs dynamic exit signs that are connected with NB-IoT technology, allowing users to navigate complicated hazardous zones in real-time by indicating the shortest and safest escape route.
- Smart Buildings: NB-IoT Technology enables IoT connectivity in big, dense settings with hundreds of linked devices that must be controlled autonomously. Administrators are able to make better decisions as a result, resulting in more efficient energy usage and lower energy expenses. The scalability and capacity to penetrate thick building materials of NB-IoT Technology make it an excellent platform for IoT connected smart building applications of any size.

4.2 Students and Personnel Health Monitoring

The NB-IoT technology can play an important role in a wide range of healthcare applications, from monitoring people health to preventing disease. Furthermore, such technology has the potential to cut healthcare costs. The usage of wearable devices in conjunction with NB-IoT technology can be utilized to augment healthcare monitoring solutions. A wearable gadget can help discover prospective medical situations by monitoring physiological signals over lengthy periods of time in a non-invasive and unobtrusive manner. This wearable gadget, in addition to displaying the user's vital signs and condition, may also be used to track down the person who is wearing it. NB-

IoT is a popular technology for these applications since it has a large range, low energy consumption, and a long battery life. This service is especially useful for those who have a compromised mental state or are unable to react when they are away from their normal surroundings. Indicatively: (a) People in the autism spectrum disorders for whom, it has been reported that, at a percentage of 49%, they have disappeared or have been at risk due to a tendency to flee, at least once since the age of 4. (b) People suffering from some form of dementia who statistically have at least 60% chance of being lost outdoors.

4.3 Education Benefits

NB-IoT Technology may help Educational Institutions improve the quality of teaching and learning by offering a richer learning experience. It also has a lot of potential to improve the teaching-learning process, making the notion of "anytime, anywhere" a reality. NB-IoT Technology, in particular, can help to improve teaching and learning in the following ways:

- Interactive learning: Most textbooks are now available online, and professors may engage students in the classroom by using IoT software to provide a variety of supplementary resources, exams, videos, and interactive learning experiences.
- Different kinds of learning sources: Students can study in a variety of ways with the help of sophisticated technologies. Teachers can utilize a variety of management technologies to develop additional sources for student learning.
- Learners with disabilities: The NB-IoT technology can be very useful in training and helping individuals with disabilities with special requirements to work and be self-sufficient. Regardless of their impairments, NB-IoT technology may provide these people with a dignified way of life in which they can gain skills and subsequently find professions to support themselves.
- Personalized learning: A great deal of study has gone into laying the groundwork for customized learning, which allows students to learn at their own speed, in their own time, and in their own place while still achieving the greatest outcomes. Educators may now place customized education at the core of a student-centered teaching strategy because to technology developments and their integration into teaching. Through the integration of technology after thorough planning and study, NB-IoT technology gives an efficient solution to such difficulties.
- Attendance monitoring system: Educational Institutes benefit from an Attendance Monitoring System in a variety of ways. It allows academics, for example, to rapidly enter important information.

4.4 E-Learning Benefits

The major benefit of NB-IoT technology for e-learning services is that it expands the learning ecosystem by integrating physicality and virtuality, bringing the learning process closer to susceptibility [4]. Intelligent Agents, which may be used to communicate with a computer program, can be implemented using NB-IoT technology. By utilizing Intelligent Learning, the constraints of E-learning in terms of communication, collaboration, and "face-to-face" interactions between participants may be minimized. In the learning environment, agents based on NB-IoT technology sensors allow interaction between the virtual e-learning space and the users' various physical locations.

4.5 Research Benefits

NB-IoT networks, like other IoT networks, link billions of everyday objects to the Internet and gather massive amounts of data that may be used in almost every area of modern life. For industry, education, production, and business, the above presents several new opportunities and difficulties. The expansion of IoT applications and services involves the most significant issues of research and development science groups across the world, including nearly every scientific area that is being pursued by a large number of corporations and universities around the world. IoT, as mentioned, is expected to promote significant improvements in the production, education, manufacturing, healthcare, energy, transportation, security, communication, government, and economic growth which means generating innovative modern challenges finding solutions.

5 Comparison Between NB-IoT and LoRa

The LoRa and NB-IoT are compared in this section based on their technological differences [14]. LoRa is an asynchronous protocol that utilises unlicensed spectrum. Although LoRa based on CSS modulation can tolerate interference, multipath, and fading, it cannot deliver the same level of quality of service as NB-IoT. This is due to the fact that NB-IoT operates on a licensed spectrum, and its time slotted synchronous protocol provides the best QoS. This benefit of QoS, however, comes at the penalty of cost. Because of the trade-off between QoS and high spectrum costs, applications that require QoS should choose for NB-IoT, while those that do not require high QoS should choose for LoRa. Because LoRa is an asynchronous, ALOHA-based protocol, devices can sleep for as little or as long as the application wishes. In NB-IoT, because of infrequent but regular synchronization, the device consumes additional battery energy. Because of these higher energy needs, the battery life of NB-IoT devices is less than that of LoRa devices. These requirements, on the other hand, provide NB-IoT the advantage of low latency and high data rate. As a result, LoRa is the ideal solution for applications that don't care about latency and don't have a lot of data to transmit. NB-IoT is the superior option for applications that demand low latency and larger data rates. The main benefit of LoRa is that it can cover an entire city with just one gateway or base station. NB-IoT deployment is restricted to 4G/LTE base stations. As a result, it is not suited for rural or suburban areas without 4G service. The flexibility of the LoRa ecosystem is a big benefit. The network coverage of LoRa may be greater than that of the NB-IoT network. NB-IoT may be implemented by repurposing and updating existing cellular networks, however deployments are limited to the cellular network's coverage region. On the other hand, the LoRa components and ecosystem are now mature and ready for production. Various cost factors must be considered, including spectrum costs, network costs, device costs, and deployment costs. It can be observed that LoRa has a significant cost advantage.

6 Conclusion – Future Work

The IoT refers to the expansion of Internet connectivity to physical devices and ordinary things. The 3GPP (3rd Generation Partnership Project) established the NB-IoT radio technology standard to allow a wide range of cellular devices and services. The advantages of adopting NB-IoT technology in Educational Institutes are presented in this article. The potential for NB-IoT technology to add considerable value to Educational Institutions by engaging and inspiring students and staff, as well as increasing learning speed, is enormous. The NB-IoT technology serves Educational Institutes in a variety of ways, including improving organizational operations, improving the learning process, improving E-Learning services, and conducting research.

Our future work includes the study of the impact on Educational Institute operation of other LPWANs technologies like Sigfox or the impact on Educational Institute operation by the combination of LPWANs technologies. In addition, we will be researching Green IoT in the future (G-IoT). The goal of a totally smart classroom has never been closer to reality thanks to the evolution of the Internet of Things. Despite the numerous advantages of IoT development, IoT infrastructure is energy and resource expensive, as well as contributing to waste and hazardous contamination. As a result, Green IoT (G-IoT) looks to be the best approach for maximizing benefits while minimizing harm to people and the environment. The G-IoT process entails designing, fabricating, and deploying IoT devices and infrastructures in an economically and ecologically responsible manner. The G-IoT paradigm, which must be implemented in educational institutions to address energy and resource consumption, e-waste, hazardous emissions, and toxic contamination, has the potential to develop cost-effective, adaptable, and engaging teaching practices.

References

1. State of the IoT 2018: Number of IoT devices now at 7B – Market accelerating. https://iot-analytics.com/state-of-the-iot-update-q1-q2-2018-number-of-iot-devices-now-7b/. Accessed 22 June 2021
2. Specialised Narrowband-IoT services. https://www.vodafone.com/business/iot/managed-iot-connectivity/nb-iot. Accessed 22 June 2021
3. Aldowah, H., et al.: Internet of Things in higher education: a study on future learning. In: Journal of Physics: Conference Series, vol. 892, no. 1. IOP Publishing (2017)
4. Abbasy, M.B., Enrique, V.Q.: Predictable influence of IoT (Internet of Things) in the higher education. Int. J. Inf. Educ. Technol. **7**(12), 914–920 (2017)
5. Banica, L., Burtescu, E., Enescu, F.: The impact of internet-of-things in higher education. Sci. Bull. Econ. Sci. **16**(1), 53–59 (2017)
6. Tianbo, Z. The internet of things promoting higher education revolution. In: 2012 Fourth International Conference on Multimedia Information Networking and Security. IEEE (2012)

7. He, J., et al.: Integrating Internet of Things (IoT) into STEM undergraduate education: case study of a modern technology infused courseware for embedded system course. In: 2016 IEEE Frontiers in Education Conference (FIE). IEEE (2016)
8. Tew, Y., Tiong, Y.T., Yoon, K.L.: A study on enhanced educational platform with adaptive sensing devices using IoT features. In: 2017 Asia-Pacific Signal and Information Processing Association Annual Summit and Conference (APSIPA ASC). IEEE (2017)
9. Marquez, J., et al.: IoT in education: Integration of objects with virtual academic communities. In: New Advances in Information Systems and Technologies, pp. 201–212. Springer, Cham (2016)
10. Li, Y., Cheng, X., Cao, Y., Wang, D., Yang, L.: Smart choice for the smart grid: narrowband Internet of Things (NB-IoT). IEEE Internet Things J. **5**(3), 1505–1515 (2018). https://doi.org/10.1109/JIOT.2017.2781251
11. Yasmin, R., Pouttu, A., Mikhaylov, K., Niemelä, V., Arif, M., Liinamaa, O.: NB-IoT micro-operator for smart campus: performance and lessons learned in 5GTN. IEEE Wireless Commun. Netw. Conf. (WCNC) **2020**, 1–6 (2020). https://doi.org/10.1109/WCNC45663.2020.9120621
12. Brdulak, A.: Characteristics of Narrowband IoT (NB-IoT) technology that supports smart city management, based on the chosen use cases from the environment area. J. Decis. Syst. **29**, 489–496 (2020). https://doi.org/10.1080/12460125.2020.1791481
13. Ratasuk, R., Vejlgaard, B., Mangalvedhe, N., Ghosh, A.: NB-IoT system for M2M communication. IEEE Wireless Commun. Netw. Conf. **2016**, 1–5 (2016). https://doi.org/10.1109/WCNC.2016.7564708
14. Sinha, R.S., Yiqiao, W., Hwang, S.-H.: A survey on LPWA technology: LoRa and NB-IoT. ICT Express **3**(1), 14–21 (2017)
15. Bagheri, M., Movahed, S.H.: The Effect of the Internet of Things (IoT) on Educational Business Model. In: 12th International Conference on Signal-Image Technology & Internet-Based Systems (SITIS), pp. 435–441 (December 2016)

On Digitizing the Greek Music Tradition: Designing the Cretan Lute for Mobile Devices

Georgios Tsotakos[1], Dimitrios Margounakis[2(✉)], Theodore Pachidis[2], and Dionysios Politis[3]

[1] Hellenic Open University, Patras, Greece
[2] International Hellenic University, Kavala, Greece
dmargoun@emt.ihu.gr, pated@cs.ihu.gr
[3] Aristotle University, Thessaloniki, Greece
dpolitis@csd.auth.gr

Abstract. In this paper, an application, which simulates the playing of Cretan lute, is presented. This study is the continuation of a greater project, which started with the simulation of the musical instruments "Cretan lyre" and "mandolin" for mobile devices [1].The aim of the project is to create the appropriate mobile applications for Greek traditional musical instruments, both from an educational and cultural point of view, as well as for entertainment. The purpose of the application is not only the representation of the "Cretan lute" and the user's familiarity with it, but also its use in a real play environment in combination with a (digital or real) Cretan lyre. Through the application the user becomes familiar with the form and sound of the Cretan lute, learns the two basic metrics used by musicians in Crete to perform the most well-known traditional rhythms *("syrtos", "siganos", "pentozali", "maleviziotis", "sousta", "anogianos pidihtos")* and has the ability to record and reproduce his own compositions. Also, the connectivity of the device to an audio console enables the use of the instrument in combination with other instruments, thus creating an environment equivalent to that of a real concert or a traditional festival.

Keywords: Computer Music · Mobile musical instruments · Android apps · Cretan lute · Musical instrument simulation

1 Introduction

In recent years the widespread use of mobile devices to meet the needs of our daily lives at educational, professional and recreational level has led to the creation of many corresponding applications. A significant part of those, for the educational and entertainment sectors, belongs to applications that simulate musical instruments. Users' interest in music and musical instruments is generally strong, resulting to the popularity of the respective applications. Typical examples are applications that simulate the musical instruments piano, guitar and drums [2–4]. Also, the increased speed and connectivity capabilities of mobile devices, as well as the interactivity allowed by their touch screens, enable the programmers to create environments to simulate the musical

M. E. Auer and T. Tsiatsos (Eds.): IMCL 2021, LNNS 411, pp. 400–409, 2022.
https://doi.org/10.1007/978-3-030-96296-8_36

instruments, almost equivalent to the way the musician is playing the actual instrument. Another interesting, and challenging task is the use of a mobile app in real-time performing conditions, for example in a concert [5]. Finally, the evolution of the Internet of Musical Things has also recent applications in Augmented Reality research [6], as well as Virtual Reality Musical Instruments (VRMIs) [7], which is the next big thing in the field of Computer Music.

The aim of this work is to create a fully functional application for mobile devices that will enable the user to approach the musical instrument "Cretan lute", familiarize himself with its playing technique and its sounds, compose songs, store them and also use it in real-time in combination with other musical instruments digitally or not.

The paper has the following structure: In Sect. 2, the project of digitizing the Greek traditional musical instruments and related work is presented, with special focus to the instrument of interest, the Cretan Lute. Section 3, which is the core of the study, presents crucial parts of the design and development of the application processes, and multi-device support is particularly emphasized. Section 4 presents the use of the applications in real-time playing conditions, while limitations and future work are discussed in Sect. 5. Finally, Sect. 6 contains the conclusions of the study.

2 The Cretan Lute

2.1 Digitization of Greek Traditional Musical Instruments

The project of digitizing the instruments of the Greek musical tradition began with the simulation of musical instruments used in the Cretan tradition and in particular with the creation of the prototype application cretanlyra.apk [1], where the digitization of the dominant musical instrument of the chosen region was the Cretan lyre. The purpose of this application was both educational and recreational. Specifically, in the first edition of the application, the user can be familiarized with the musical instrument "lyra" and its forms that exist in the region of Crete, understand its sound, learn its playing technique and become familiar with some well-known melodies from the tradition of the island of Crete. Later, in its second edition (Fig. 1), the application was enriched by the introduction of a digitized mandolin. The functions implemented for the mandolin are analogous to those of the lyre. Thus, except for learning the elements of the mandolin and how to play it, the user also acquires a more complete picture of the tradition and musical instruments used in Crete.

2.2 The Cretan Lute as a Digital Musical Instrument

The digitization of the Cretan Lute was the natural continuation of the attempt to digitize the tradition of Crete. Through the application cretanlyra.apk, users of mobile devices had the ability to hear the Cretan lute, as well as to play well-known Cretan melodies in lyre or mandolin accompanied by it. However, it was necessary to create an integrated environment in which the user would have direct contact with the sound and the use of the instrument. Also, in the context of diversity and innovation, it was necessary to test new techniques and ways of representing the digital instrument. The development of a new application was chosen geared more to its practical use, which would not lag in the fixation and performance of the style of the musical instrument.

In order to achieve full functionality of the application in most mobile devices, its implementation was done with the Unity programming platform. In addition, the application was chosen to be functional on devices running Android version 4.4 (Kit Kat) and above, thus covering 99% of the devices currently in use.

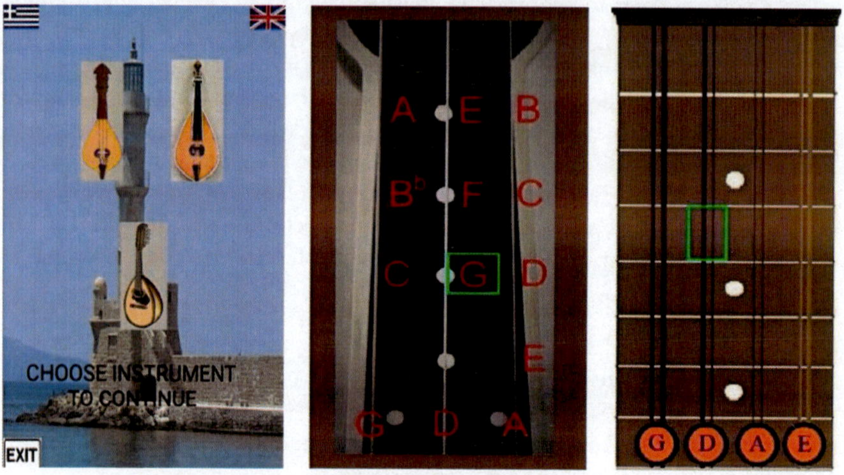

Fig. 1. Screens from cretanlyra.apk

The Cretan Lute. The lute is an instrument of the Cretan musical tradition as important as the lyre. In addition, it is necessary for the accompaniment of Cretan lyre. Famous is the Cretan "zigia", which is the musical group consisting of a musician who plays the Cretan lyre and a musician who plays the Cretan lute. The purpose of the *cretanlute.apk* is the accurate duplication of the Cretan "zigia" and the accompaniment of the lyre from the lute using the "uncharacteristic" chords.

The Cretan lute (Fig. 2) is a stringed instrument which is structurally larger than the land and island lutes. It has 8 strings (4 pairs) and its bottom-up tuning per pair is E, A, D and G. It is played with a plastic or handmade pen. Concerning its material, there are some rules regarding the type of wood used by manufacturers. In the *cretanlute.apk*, the lute whose sound is simulated, is made with a soundboard from fir, a fingerboard of ebony and a body of wood from mulberry tree.

The technique and way of playing the instrument resembles that of the guitar or even more so with that of the oud instrument, which belongs to the same string family. In the *cretanlute.apk* a specific way of playing the instrument, that is called "vourtsa", is the rhythmic repetition of chords (simultaneous impact of multiple strings together) which can be reproduced in different musical patterns. Besides, this is the most common accompanying way of playing on the Cretan "zigia" while the lute plays simultaneously with the lyre.

Fig. 2. Design of the Cretan lute

2.3 Related Work

During the search for applications implementing musical instruments, mainly in the official Android distribution area (Google Play Store), but also in other academic sources, it was found that there was no previous attempt to simulate the lute in the form displayed in Greece. However, there are applications that simulate its related instrument (in terms of shape and sound) "oud". Our application *cretanlute.apk* can also be used in real time alongside with other digital or real instruments.

With the aim of achieving exactly this capability, *cretanlute.apk* was chosen to focus on this exact direction. For that reason, in its first version, the user has the option to play "uncharacteristic" chords and select and change their playing tempo. In addition, the user can compose songs, save them and perform them in real time, while at the same time specifying its duration according to the needs of the live performance.

3 Modeling and Programming

Figure 3 shows graphically the basic functions of *cretanlute.apk*, according to the architectural structure of applications implemented in Unity environment. Scene 2, which allows the user to create his/her own melodies and compose songs, also supports *save*, *load* and *play* file functions.

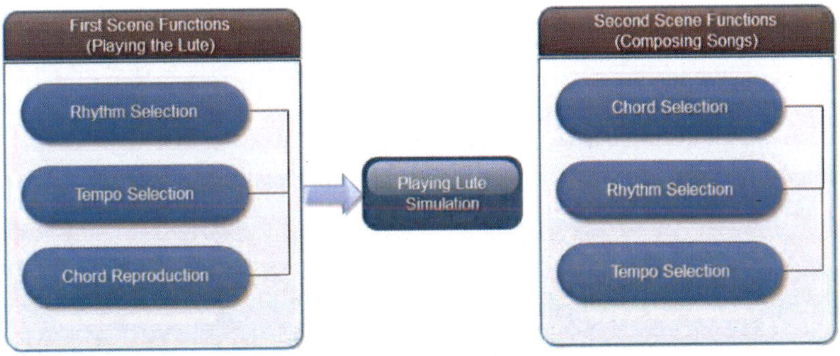

Fig. 3. Basic functions of the application *cretanlute.apk*

3.1 Multi-device Support

The design of the application was done with the aim of running on most mobile devices, having a user-friendly interface and being functional, providing additionally the possibility of using it in real time. For that reason, it was chosen to be implemented in the Unity programming environment and run on Android mobile devices. Specifically, the creation of two scenes (screens) was chosen. In the first one the user can play the chords in the rhythm and tempo he/she desires, while in the second one he/she can compose his/her own songs by selecting their name, chords, tempo and rhythm. In addition, in the second scene the user can play the songs and store them on his/her device. When playing them, he/she can alternate the chords back and forth in any way he/she wants, change the elements of the songs, store them again and of course pause and start playing again. It is also worth noting at this point the significant advantage of Unity environment to adapt the application interface to each kind of density and size of the screen of each mobile device.

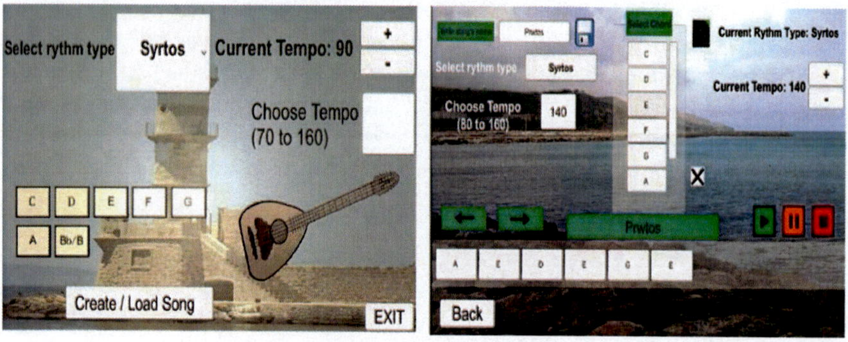

Fig. 4. Scenes 1 and 2.

3.2 Design

Based on the scenario decided, two scenes (screens) were designed (Fig. 4).

First Scene. It is the first impression to the user. For this reason, a beautiful image was chosen (the lighthouse of Chania), while during the user's stay in the menu the sound of a nice improvisation in lute is heard. Particular attention was paid to the correct placement of objects on the screen so that the user can easily perceive and use its elements.

Second Scene. On this screen the user can create his own composition. The image of the background also refers to a beautiful landscape of the island of Crete. The chords are selected from the center of the screen, while at the bottom of the screen and horizontally the selected chords are displayed. At the top of the screen there are the corresponding buttons to render the name of the song, its rhythm and tempo and of course to save it. There is also a loading button for the already saved tracks. While the user is pressing the load button, a table with 22 storage locations pops up. From this table the user can load the file he/she wants. In addition, there are buttons on the screen to delete selected chords, toggle chords, play, pause, and stop them.

3.3 Development

Each scene of the application consists of a background, lighting and a "Canvas" on the foreground on which the objects (menu and buttons) are placed. Each object is managed through its "Inspector", which defines its basic features such as its position on the screen, its size, the ability to produce audio, and the ability to interact with other objects on the screen. In addition, "Inspector" also connects objects to JavaScript files, in which the desired functions of the objects are developed for the program flow.

Moreover, audio files, materials and fonts, which are collected and stored in the corresponding folders, as well as scripts, play an important role in the implementation of the scenario per screen.

Sampling. The process of collecting and editing audio files has been one of the most important steps in implementing the application. As the aim was to recreate the most accurate sound output of the instrument, the recordings were made with a Cretan lute, in a studio environment, using a condensing microphone and sound card (via PC). Each chord was recorded separately for each rhythm ("syrtos" and "grigoros") at intermediate speed (90) for greater efficiency and sharpness. The files were then further processed through sound editing software.

Implemented Functions. The necessary functions for the user that had to be implemented on the first screen of the application are:

1. The choice of rhythm.
2. Selecting, reproducing and switching chords.
3. The tempo switch.
4. The visual playing performance from the screen lute.

The selection of a rhythm was made through an object dropdown. In addition to its basic settings through "Inspector", a java file was also connected to this object. Through this file, a selection variable is created and updated.

For the user to choose the chord he/she wishes to reproduce, seven buttons with appropriate names have been created. The first six are for the natural, uncharacteristic chords C, D, E, F, G, A, while the seventh is for B recession when performed at a "syrtos" rhythm and B natural when performed at a "grigoros" rhythm. In the "Inspector" of each button, the corresponding audio file that has come from sampling has been added and it can be played through an AudioMixer that exists in Unity. The selection of the correct rhythm for each button, as well as the playback speed of the corresponding audio file, are controlled by the appropriate java file that has been added.

One of the most interesting programming functions is the switching of tempo playback. As the user selects the tempo in two ways (spin buttons or number fill field), the appropriate "Update" function (Fig. 5) is called in the script of the corresponding button which alters the tempo but maintains the correct tonality of the sound by configuring the Pitch appropriately.

Update 's "change tempo" algorithm

```
selectedType = rythmTypeScript.x;
newTempo = InputFieldScript.temp;
newSpeed=newTempo/90;
audioSourceC1.pitch = newSpeed;
audioSourceC1.outputAudioMixerGroup.audioMixer.SetFloat("Pitch", 1f / newSpeed);
audioSourceC2.pitch = newSpeed;
audioSourceC2.outputAudioMixerGroup.audioMixer.SetFloat("Pitch", 1f / newSpeed);
tempo = tempo * newSpeed;
```

Fig. 5. Tempo change algorithm

It is worth noting that the numeric range that the user can choose is from 70 to 160. This was so limited for two reasons. The first is because outside these limits there may be exclusions in the tonality of the sound and the second is because within these limits most Cretan songs are performed. Specifically, songs that are in the rhythm of "syrtos" are usually performed in tempo ranging from 130 to 150, songs in a "siganos" rhythm ("kontylies") in tempo from 85 to 105, while the "grigoroi" ones ("sousta", "pentozali", "maleviziotis") in tempo from 140 to 155.

The visual rendering of the lute is done through the appropriate JavaScript that has been connected to the object of the lute. During the execution of the audio files from the chord buttons, the appropriate color switching on the screen is made, depending on the tempo, which gives the feeling that the strings are pulsating.

The necessary functions for the user that had to be implemented on the second screen of the application are:

1. The choice of rhythm.
2. Select, add, delete and place chords in a virtual table based on a selection order.
3. The tempo switch.
4. Start, pause and stop playing the song.
5. The name rendering in the song and saving the screen data.
6. Loading saved files.

The selection of rhythm is done as on the first screen. However, there is a difference in the selection and reproduction of chords. The user now selects chords one after the other through a "ScrollRect" and creates "Prefabs" (copies of the original objects) which are placed in a horizontal line at the bottom of the screen, thus creating a virtual table. Additionally, via an appropriate button, the user can delete the last chord each time and then add it again. The virtual table has 9 locations. The tempo change is done in a similar way as in the first scene. The playback of the song is fully controlled by a

JavaScript file ("PlayModeScript"). The corresponding code allows the user to play (play button) the song by performing the chords one after the other, to alternate (back and forth) the chord that is performed (via arrow buttons on the screen, or through selected keyboard buttons on the mobile device), pause (pause button) and stop (stop button).

The name rendering of the song and the storage of all screen data is done through the storage system with json files. Thus, by running the file DataManager.cs (activated by pressing the save or load buttons) the currently valid data from the second screen of the application is collected, stored in appropriate variables and json files are created. These files are placed appropriately in 22 locations (table 11 $\times$ 2). To retrieve the data (song) the user selects through the table the song he/she wants, and the data is retrieved from the corresponding json file.

4 Using the App in Real-Time Playing Conditions

The target that was set before the creation of the application was not only to construct a functional application that would simulate the Cretan lute in the context of the acquaintance and learning of the instrument, but also, and most importantly, to be used in real playing conditions (Fig. 6). The existence of speakers on mobile devices that deliver good sound quality and the ability to connect to consoles to mix the sound from their output with other instruments, enable the application to be exploited and used in real playing conditions. Thus, the user can use his/her mobile device by accompanying his/her playing on an instrument (usually lyre, lute, mandolin, thiaboli) with Cretan lute. This would be very useful for an apprentice musician when practicing because the arpeggio of the lute is proportional to the use of a metronome, while the existence of an accompaniment from a tuned instrument helps for proper tonality to be achieved. Of course, the application is also very useful for the professional musician as he/she could have the desired accompaniment of lute, avoiding the use of recorded files (playback). In addition, the connection of the application to a central console and its handling by a user in real time of a concert, offers the possibility to the band that performs in a concert to integrate into its sound, the sound of the Cretan lute, thus accompanying either exclusively or ancillary the existing musicians.

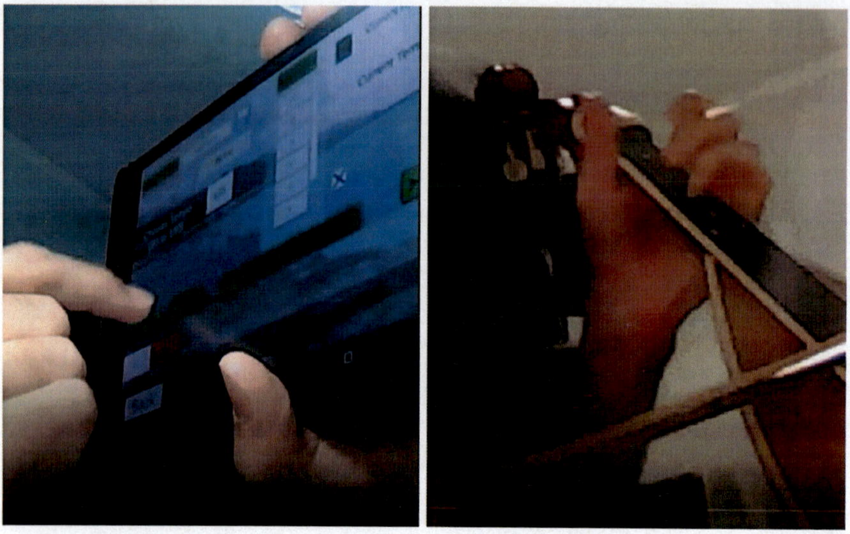

Fig. 6. Demonstration of playing the Cretan Lute with *cretanlute.apk*.

5 Limitations and Future Work

While the musician is playing the lute, he often differentiates his dynamic playing during the song, alternating at the same time the volume of sound that the instrument brings out. The reason is either for facilitating the playing of the lyre which accompanies, or because he/she wants to create a special timbre in his playing. This is not easy to implement in the application functions as the digital files are recorded with specific dynamics.

This limitation could be overcome in a subsequent version of the application, by differentiating the implementation of the chord's reproduction. For example, a file for each note of any chord could be recorded separately and then the user could take over its playing by specifying the tempo and the way of arpeggio. In this way he/she could determine the intensity according to the pressure on the device's screen. Ideally, this could be implemented using a virtual reality environment for more accurate performance.

Also, although that we have collected some qualitative and constructive comments about the app from actual Cretan lute players that have tested the app, a specially designed user study has not been conducted by the time of this manuscript writing. It is an immediate priority for us to quantitively evaluate the Cretan Lute app by designing and conducting a suitable user study in the near future.

There are many ideas for a next version of *cretanlute.apk* but also for the continuation of the project. About the application for lute, it would be useful for the user to have more options and controllers especially for real-time performances (e.g. more than 9 chords, timer, tempo changers, songs switching). Also, more traditional Cretan rhythms and chords with more features could also be added to enrich the Cretan music experience.

Musical tradition of Greece is very rich and extends from end to end of the country. Also, mobile interfaces can recreate musicalities from different worlds, unusual music systems and dissimilar semantics to the Common Music Notation of staff music [8]. As the richness of timbre, chromaticism and playing and designing techniques for the traditional musical instruments is tremendous [9], it would be very interesting and intriguing to simulate traditional instruments from other parts of the country in subsequent versions as well.

6 Conclusions

In this paper the *cretanlute.apk* was presented. The application was implemented for presenting the Cretan lute as well as its use in real time during the practice of the musician or during a concert. The application runs in Android environment, which is very popular and installed in a variety of mobile devices. In addition, *cretanlute.apk* constitutes a sequence for the project of digitization of the musical instruments of Crete and generally the traditional musical instruments of Greece.

References

1. Margounakis, D., Tsotakos, G., Floros, A.: On digitising the Greek music tradition: simulation of the Cretan lyre for mobile devices. Int. J. Arts Technol. **12**(2), 103–117 (2020)
2. Hwang, B.K.: An implementation of smartphone-based multiple musical instruments application supporting social playing. J. Digit. Contents Soc. **12**(4), 575–583 (2011)
3. Reis, T., Carriço, L., Duarte, C.: Interaction design: the mobile percussionist. In: Altinsoy, M. E., Jekosch, U., Brewster, S. (eds.) HAID 2009. LNCS, vol. 5763, pp. 109–118. Springer, Heidelberg (2009). https://doi.org/10.1007/978-3-642-04076-4_12
4. Ng, S.C., Lui, A.K.F., Kwok, A.C.H.: Easy-to-learn piano: a mobile application for learning basic music theory and piano skill. In: Lam, J., Ng, K.K., Cheung, S.K.S., Wong, T.L., Li, K. C., Wang, F.L. (eds.) ICTE 2015. CCIS, vol. 559, pp. 103–112. Springer, Heidelberg (2015). https://doi.org/10.1007/978-3-662-48978-9_10
5. Wang, G., Essl, G., Penttinen, H.: The mobile phone orchestra. Oxford Handb. Mob. Music Stud. **2**, 453–469 (2014)
6. Tan, K.L., Lim, C.K.: Development of traditional musical instruments using augmented reality (AR) through mobile learning. In: AIP Conference Proceedings, vol. 2016, no. 1, p. 020140. AIP Publishing LLC, September 2018
7. Serafin, S., Erkut, C., Kojs, J., Nilsson, N.C., Nordahl, R.: Virtual reality musical instruments: state of the art, design principles, and future directions. Comput. Music J. **40**(3), 22–40 (2016)
8. Politis, D.: Reconstructing digitally instruments and scales in the synchrony and diachrony of music. In: Interdisciplinary Digital Preservation Tools and Technologies, pp.103–118. IGI Global, Hershey (2017)
9. Politis, D., Margounakis, D.: Modeling musical Chromaticism: the algebra of cross-cultural music perception. Int. J. Acad. Res. **2**(6), 20–29 (2010)

User Experience and Music Perception in Broadcasts: Sensory Input Classification

Dionysios Politis[1](✉), Rafail Tzimas[1], Dimitrios Margounakis[2], Veljko Aleksić[3], and Nektarios-Kyriakos Paris[4]

[1] Aristotle University of Thessaloniki, Thessaloniki, Greece
dpolitis@csd.auth.gr
[2] International Hellenic University, Kavala, Greece
[3] Faculty of Technical Sciences-Čačak, University of Kragujevac, Kragujevac, Serbia
[4] University of Macedonia, Thessaloniki, Greece

Abstract. Mobile devices of any sort, size and construction have increased drastically the sensory input of the Internet in unprecedented levels. Massive recordings of every day artistic activities, in the form of sound archives, photographs, movie films footages or video signals have proved to be particularly successful and influential within the social services' immense cloud. Mobile devices prove to be the most convenient means to input in digital format what millions of correspondents notice with their physical senses happening in their everyday cognitive walk within their natural habitat. As a result, the amount of sensory input, within each user's walkthrough in his digital habitat, has increased drastically for hearings that have been not that much promoted by the established media hubs - which are more or less using digital footages highly influenced by the US movie industry and the lifestyles of people within each country that are behaving accordingly to the prevalent model.

Keywords: Learning with audiovisual tools · Music perception · Broadcasted learning

1 Introduction

Multimedia Learning as an enabler for facilitating music understanding depends heavily on the operational environments that overwhelmingly supply endless streams of melodious hearings [1].

The heaps of tunes that overpower the music web within the 21st century intrude the listener's personal sphere of comprehension posing two serious questions:

- are users capable of understanding music different from their artistically established styles and techniques?
- would they pose themselves in a state of excited curiosity capable of producing interest and feedback in such an extend so to make music networks commercially or artistically sustainable?

M. E. Auer and T. Tsiatsos (Eds.): IMCL 2021, LNNS 411, pp. 410–419, 2022.
https://doi.org/10.1007/978-3-030-96296-8_37

Since the times of Pythagoras music has been interrelated with Mathematics, and thus, reconstruction from sources dating back to ancient past for great civilizations having flourished in Southeastern Europe, the Eastern Mediterranean basin and the Middle East is more or less feasible.

Nowadays the American model of music prevails, as USA is the hub for delivering relatively inexpensive cloud computing infrastructure. However, also the rest of the world has the ability for producing actual performances with high quality aesthetic content within cultural boundaries and national enclaves that keep the long-term memories of artistic excellence [2, 3].

Therefore, in computer-science terms, the decision-making and "translation" from one language or culture to another, as far as music intended meaning is concerned passes the following stages [4]:

- Attention and Perception - are the listeners able to have an understanding of the sensation, notion or intuition expressed in a non - English speaking musical environment?
- Encoding by Intrinsic or Extrinsic Motivation - i.e., furthering intuitive understanding of the previous stage, may listeners sufficiently decipher the acoustic stimuli in terms of American or Western European Music semiology?
- Cognition - may audiophiles recognize the timbral elements, the building blocks (the motifs), and the intended character (the scales)?
- Behavioral reaction - could the listeners turn themselves from passive receivers to energetic broadcasters within a music network?

2 Some Historical Facts: An Overture to Contemporary Music

2.1 Historical Evolution in the Western World

The 20th century has been characterized by significant changes for nearly all elements of music production and distribution. Harmony, rhythm, melody, sound color are some of the predominant elements heavily influenced by ardent technological changes [5, 6].

When the famous Russian composer Igor Stravinsky (1882–1971), resident of the U.S. since 1939, shocked with his ballets *The Firebird* (1910) and *The Rite of Spring* (1913) his Paris audiences - with their irregular rhythms and frequent dissonances, he was simply signaling the postmodern era that would conquer the world having its epicenter within the American audiovisual industry.

What started in New Orleans in early 1900s, where a type of music of black American origin characterized by improvisation, syncopation, and usually a regular or forceful rhythm was coming to the frontend, proved to be the emerging trend after World War II when USA, less hurt by warfare and financial disaster, became the hub of innovation. As a result, brass, woodwind instruments and piano particularly associated with jazz, along the (electric) guitar and rock-n-roll and occasionally violin-like string tools were the protagonists of an emerging musical paradigm - not forgetting of course wind instruments like the clarinet and following the intended route, the saxophone (Fig. 1).

Fig. 1. Left, Buddy Bolten's jazz band, depicted in 1905, New Orleans. Right, Juke Joint in Clarksdale, Mississippi, November 1939.

It is of great interest to historians to decipher what emigrants were actually listening to in USA during the 1920s. For instance, how the jazz revolution is linked to the mandolinata preferred by Italian and Greek orchestras moving henceforth between Asia (Constantinople, Smyrna), Europe (Rome, Milan, Athens) and the Americas for live performances.

Of course, Europe had a say in these groundbreaking changes. For more than 3 centuries, up to 1950, serious polyphonic music was following the long-established principles set in Germany, Austria and France, and to a lesser degree in UK or Italy. Italy, on the other hand, was dominant in developing the opera rather than a folk or popular tradition. Dramatic works in one or more acts were also promoted as operatic works in Germany and Russia, with the latter excelling in the establishment of ballet as a kinetic alternative to operatic performance [5].

Along with Stravisky, another 20th century US theorist, Arnold Schoenberg (1874–1951), born in Austria, excels in smoothing the transition to the new world (of music). He introduces atonality into his 2nd string quartet (1907–08), while *Serenade* (1923) becomes the first example for the technique of serialism. Therefore, he gives room for jazz and similar new style forms to breathe deeply in the general principles of independent theoretic setups for the beauty of form, harmony, and expression of emotion.

Last, but not least in this very concise review, Iannis Xenakis (1922–2001), French composer and architect, of Greek descent, gives theoretical insight with Mathematics and Computer Music, during his travels to Indiana University, for promoting both in Europe and the New World electronic and aleatory techniques [6].

2.2 Contemporary Music in a Worldwide Perspective

As music marches triumphantly from 1950s and onwards to the 21st century, changes in style become more often: melodic arrangement are linked to technological revolutions.

Rock and Roll in the USA combines Gospel, Blues, Jazz, Country, and new dancing paradigms flourish, where participants move rhythmically to music, typically following a set sequence of steps.

While the piano is used pretty much in the early beginnings, the electric guitar makes its triumphant entry and establishes itself as the leading instrument. Alongside synthesized music and electric drum sets, since the 1970s, new forms appear and catch-up user preferences: heavy metal, punk and funk, hip-hop music.

Computerized music commences to evolve at the dawn of the new millennium: techno, house, rap and similar are continually catching up with their audiences and progressing accordingly [2]. Early synthesizers, like the Minimoog and Yamaha's DX7 have been exhibited as the instigation force for composing jazz, funk, pop, hip-hop and generally speaking styles of popular dance music typically using synthesized drum and bass lines, sparse repetitive vocals, and a fast beat (Fig. 2, left).

In the Ottoman or Persian-Arabic traditions these dynamic variations are known as maqams, and characterize as a whole both the ethnic traditions of the peoples of Asia Minor, Middle East, Central Asia alongside Northern Africa and simultaneously, the spiritual music of Islam. However, prior to this enlargement of state territory by encroaching cultural uniformity on several nations, pursued as a political or religious strategy after the 7th century AD, traces of such different cultures in music were traced in the Roman world, the Byzantine world or even the Indian-Pakistani sphere of interest, in Afghanistan, Tajikistan and other Turkic regions [9].

Therefore, it seems that nearly all the early Indo-European continuum hosted ancestral tunings that were somewhat different to those that have characterized the prevalent Western-European musical paradigm of the last centuries.

As these traditions are boosted by the means of modern technology, they come to a realization of where their tradition differs from the prevalent music theory, for starters regaining consciousness on how their scripts figuratively differ to the CMN scheme, the very same way their language uses different notations than those of the dominant Latin alphabet.

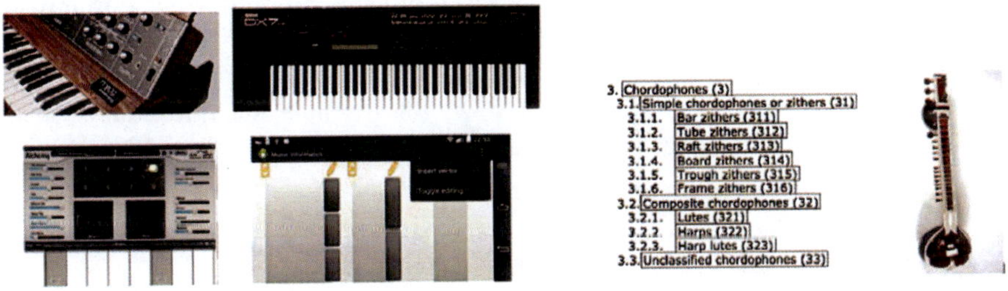

Fig. 2. Left: Top row, legendary synthesizers that have reshaped the music industry. Bottom row, mobile device interfaces alternatives that have given the possibility to perform in alternate modes. Right: A classic ethnic instrument classified according to the Hornbostel-Sachs scheme.

3 Problem Formulation and Experimental Results: From Hearing to Sensing, from Sensing to Perceiving

Using AI techniques within the Web, like the Google™ tools for navigating the ocean of Internet music or the YouTube™ streams, users become directed to a plethora of hearings that they would never have dreamt of purchasing at their closest music store.

One thing in detecting and consequently understanding what specific meaning is relayed via the broadcasts heard. It is something like detecting in a conversation heard what language the speakers are using.

Even further, in music, for proper sensing it is essential to come to some recognition for the instruments used in a particular melodic rendition. The most widely model used for such a purpose is the Hornbostel-Sachs methodology, similar in structure to Computer Science ontologies and object-oriented analyses with classes. It uses numbers to denote the hierarchical structures used in classifying the world-wide spectrum of musical instrumentation.

When a new instrument is bound to be identified, a recursive process is initiated aiming to spot the closest similar appliances already in use (Fig. 2, right).

In the classic model, the systematics for instrument taxonomy are based in 4 main classes: Idiophones (1), membranophones (2), chordophones (3) and aerophones (4).

A 5th taxonomy has been added recently, that of electrophones (5) to stock the electrical and electronic equipment. In practical terms, however, in the 21st century this last addition has superseded most instruments found in contemporary synthesizers, by for far more numerous than the classic sounds simulated by generating and combining signals of different frequencies.

Sensing a music instrument is an important step for the mental faculty by which the body perceives an external stimulus.

Sensing is also connected with feeling. Although from the 5 basic senses, that is the faculties of sight, smell, hearing, taste, and touch, only the third is directly employed in music aesthetics, melodic awareness is also concerned with specific states of human perception.

When for instance one hears a music piece in an auditorium simulating how one feels when observing a river flow along the prairies, of course he does not see any water flowing through. Even further he has no sense of any humidity. However, he has some kind of perceptive simulation of how, for example the direction of motion simulated by the mathematical arrangement of music objects may sound like the effect of the river downfall.

3.1 Spotting Performance Deficiencies in Quantitative Terms

As music composers and performers constantly see to expand their repertoire by engulfing new hearings and prime melodies, they incorporate sounds or melodies from traditions outside their dominant musicality [10].

For this research, Aristotle University students, randomly formed groups of 2 or 3 members, and participated in a survey focused on the musical piece analysis of the song *Soleil*, as performed by Dalida in 1984.

Dalida (1933–1987) was a distinguished artist born to Italian parents in Egypt. At the age of 21 she moved to Paris and followed a career in modeling and singing. Her international repertoire, in French predominantly, depicts her Egyptian background and is expressive of the difficulties of her childhood during the WWII military campaigns in the region she was living in - troubles that forged her distinctive performance style.

Therefore, this specific song, albeit its fast-dancing style, has some melancholic nuances expressed accordingly by the predominant C minor harmonic scale used for its deployment.

18 student groups tried, in their first approach with the song, attempting with their perceptive awareness for 21st century music to classify what genre and style was the song Soleil (Fig. 3, left).

To a smaller extent, i.e., expressed by 1 or 2 groups at most, it was perceived as vocal, new wave, electronic, dance, pop-rock.

As in the video footage used for the taxonomy of this song only Dalida appears and no performing instruments are visible on screen, the students participating in this survey were asked to identify by the sound heard what the performing appliances were.

However, distinguishing the instruments or the genre of a song it is not a very difficult in perceptive terms - it is related to the users' capacity to have awareness of what sensory-neural content was meditated.

To add some mental processing to this, the 40 + users participating in this survey, all of them born around the turn of the millennium, were asked to find similar songs in style and performing characteristics.

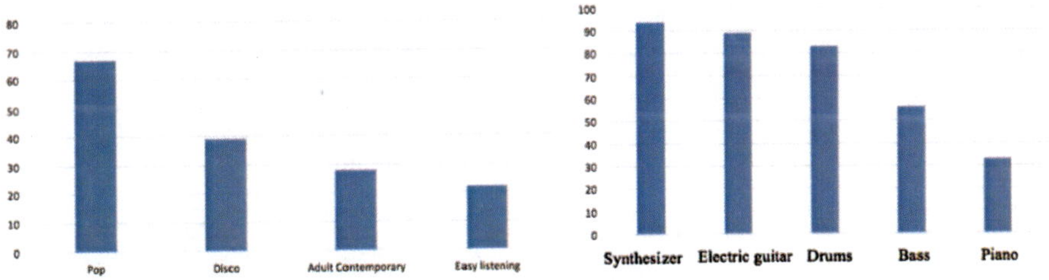

Fig. 3. Left, how students reviewed the song Soleil using 21st century genre classification schemes (%) and right, what instruments they perceived performing with Dalida.

Their answers included:

- Dancing Queen - ABBA
- Gimme Gimme Gimme - ABBA
- Se Mi Lasci Non Vale - Julio Iglesias
- Voyage Voyage - Desireless
- Diva - Dana International (Eurovision): - pointed out by 3 groups
- Shadilay - Pepe
- Besame Mucho - Dalida
- If I Can't Have You - Yvonne Elliman
- Cambodia - Kim Wilde

- Sage Comme Une Image - Lio
- Mamma Mia - Ricchi e Poveri
- Sara Perche Ti Amo - Ricchi e Poveri
- You Can Win If you Want - Modern Talking
- Moonlight Densetsu - Sailor Moon
- You Spin Me Round - Dead or Alive
- 男と女 - Junko Ohashi
- Girls Just Wanna Have Fun - Cyndi Lauper
- I Need A Hero - Bonnie Tyler
- Et Si Tu N'Existais Pas - Joe Dassin

In statistical terms this uncontrolled variation in giving answers in diverse directions indicates that some things may not be readily estimated by sensory input evaluation metrics. It seems that there is no considerable number of quantitative measurements leading to obvious taxonomies.

3.2 Shifting Taxonomies to Qualitative Characteristics

As quantitative indices demonstrate a considerable dispersion in evaluation criteria, it becomes obvious that genre taxonomies precipitate sensory inputs in premature stages of music perception.

They were performed by musicians recently awarded a BS in Music Science and Art, with specialization in Byzantine Music. All of them were mature students, who have received another degree prior to their commitment with the BS in Music.

They performed three compositions. The first two were kalophonic chants, i.e., melodies, according to Oxford Music Online, that have long melodic motifs in order to demonstrate the beautiful musical sounds the singing voice can perform.

The third was a well-known Greek pop-song, of the "rebetiko" style (Fig. 4).

Fig. 4. Three master class performances of Greek "ethnic" melodies, recorded with mobile paraphernalia.

A well-known Computer Music expert and certified pianist undertook the task to listen carefully to the melodies and reproduce them with his piano.

The pianist had not any view of semiological writing of the melodies and performed by hearing the melodies simultaneously playing what he perceived of them with his piano (Fig. 5).

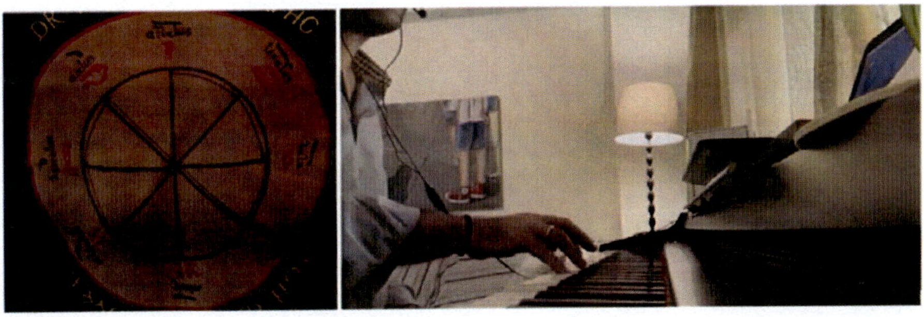

Fig. 5. Left, the "apichimata" used in the original melodies. Right, the expert rendering the melodies in his piano while hearing them.

The results of this simulation may be clearly seen in the report filed by the expert, following next.

`MELODY 1 TRANSCRIPTION (LEVEL OF DIFFICULTY: 9/10)`

The first melody proved to be extremely difficult to convey in an equal tempered musical instrument, like the piano. Although the apichima at the beginning of the performance reveals the basic components of the melody (base, intervals, dominant sounds etc.), the performance itself is particularly demanding and sometimes "awkward" in the ears of a western-trained musician [9]. There are many reasons for that.

The main difficulty of the chanting performance lies in its melismatic nature with the continuous phrases, vocal vibratos, and the same syllable flowing through many notes [10]. Although the piece is considered slow, performers' master-class expertise results in outstanding continuous chromatic effects and microtonalities, which are difficult or even impossible to apply to the classical piano.

Many pitches from characteristic intervals of Byzantine Music, although perceived by the ear, can neither be supported by the western-oriented Common Music Notation, nor, of course, by an instrument such as the piano. Finally, the presence of just a few lyrics, as well as the complex harmony that is created in many cases by the polyphony from the professional musicians, make it even more difficult to accurately comprehend a par excellence "exotic" mode for a western-trained musician.

`MELODY 2 TRANSCRIPTION (LEVEL OF DIFFICULTY: 8/10)`

The second melody has all the elements of church vocal music performance in common with the first melody. The musicians perform the art of chanting exceptionally, making the devout nature of the melody even more intense.

What was written for the first tune also applies here. Perhaps this melody is a little slower than the first one, which makes some tones easier to perceive and record. However, there is great difficulty in accurately performing the melody by playing the piano. Melismatic elements and the strong chromaticism applied by the singers cannot be attributed to piano performance. Trills, auxiliary notes and other CMN techniques and their corresponding symbols prove to be inadequate for precise performance of melodies of music genres, like Byzantine Music.

On western notation level, microtones can be recorded to some degree with the help of extra symbols (half-flat and half-sharp signs), but not with the desirable absolute pitch accuracy [9, 11]. Moreover, even if special notation is recruited for the vocal abilities that emerge in the master-class performances, it is impossible to be applied in a well-tempered musical instrument, like the piano. It should be noted, finally, that there are several more problems in such kinds of conversions, even by specialized computer music software for audio-to-CMN transcription, concerning other musical parameters too (rhythm, music bars creation, scale finding, expression features, etc.).

MELODY 3 TRANSCRIPTION (LEVEL OF DIFFICULTY: 3/10)
Melody 3 is the case of the well-known Greek folk tune «Το κόκκινο φουστάνι» ("The red dress" - composer: Stavros Kougioumtzis, lyrics: Kostas Kindinis), which was first performed by Eleftheria Arvanitaki in 1985. The song is performed by the same musicians here and the transcription is easier than the previous two melodies.

The main reason is the existing familiarity with the tune. Also, the song has been performed many times with equal tempered western musical instruments and the piano transcription (and thus the use of CMN) is considered satisfactory [12].

However, as in the previous cases, the vocal abilities of the expert singers during the performance prevail over those of the classical instrument and create "decorative" phenomena (chromaticism, vibrato, etc.) that cannot be attributed to the piano.

Yet, in the case of this melody, there are specific lyrics throughout the song, which are sung per syllable. This minimizes the extensive melismatic performance and reduces chromaticism at the syllable level, thus facilitating the comprehension of the musical phrases and the transcription to Common Music Notation.

4 Conclusion

The enthusiastic use of mobile device paraphernalia for use with musical recording and audiovisual productions has resulted in very detailed renditions of popular melodies or songs. Not counting amateur performances, the "master class" level renditions comprise already a big number of broadcasts.

Learning out of the Internet proves to be a rather vulnerable to side-effects process if a tutor is not available to become aware of the student's shortcomings.

For the wide public, the more melodies the Web can host, the more complicate the learning process becomes, since there is hysteresis in perceiving the depth of melismatic motifs, thus having difficulty in either memorizing such sequences of notes collectively or being able to reproduce with any kind of music machinery.

There seems to be a gap in the perceptive mechanism for sequences not thoroughly deciphered as familiar melodies - as far as their building blocks are concerned.

References

1. Markaki, E., Kokkalidis, I.: Interactive technologies and audiovisual programming for the peforming arts: the brave new world of computing reshapes the face of musical entertainment. In: Politis, D., Tsalighopoulos, M., Iglezakis, I. (eds.) Digital Tools for Computer Music Production and Distribution, pp. 137–157. IGI Global, Hershey (2016)
2. Cox, C., Warner, D.: Audio Cultures, Readings in Modern Music. Continuum, London (2007)
3. Riedl, M., Young, R.: Narrative planning: balancing plot and character. J. Artif. Intell. Res. **39**, 217–268 (2010)
4. Shneiderman, B., Plaisant, C.: Designing the User Interface, Strategies for Effective Human-Computer Interaction. Pearson International Edition (2005)
5. Landormy, P.: A History of Music. C. Scribner's Sons, New York (1923)
6. Hatten, R.: Musical Meaning in Beethoven: Markedness, Correlation, and Interpretation (Advances in Semiotics). Indiana University Press, Bloomington (2004)
7. Cook, P.R.: Music, Cognition and Computerized Sound, An Introduction to Psychoacoustics. MIT Press, Boston (1999)
8. Cumming, N.: The Sonic Self: Musical Subjectivity and Signification (Advances in Semiotics). Indiana University Press, Bloomington (2001)
9. Kouroupetroglou, G., Papadakos, C., Kamaris, G., Chryssochoidis, G., Mourjopoulos, J.: Optimal acoustic reverberation evaluation of byzantine chanting in churches. In: Karagounis, K., Kouroupetroglou, G. (eds.) The Psaltic Art as an Autonomous Science: Scientific Branches – Related Scientific Fields – Interdisciplinary Collaborations and Interaction. Academy for Theological Studies of Volos, Department of Psaltic Art and Musicology, Greece (2015)
10. Barsky, V.: Chromaticism. Harwood Academic Publishers, Amsterdam (1996)
11. The New Penguin Dictionary of Music. Penguin, New York (1980)
12. The SAGE International Encyclopedia of Music and Culture. SAGE Publications, Thousand Oaks (2019)

Interactive and Collaborative Mobile Learning Environments

Blockchain as an IoT Intermediary

Matija Šipek[1,2], Martin Žagar[2(✉)], Nikola Drašković[2], and Branko Mihaljević[2]

[1] CS Computer Systems, Zagreb, Croatia
[2] RIT Croatia, Zagreb, Croatia
martin.zagar@rit.edu

Abstract. Blockchain technology provides a private, secure, transparent decentralized exchange of data. Also, blockchain is not limited to a particular area, but it has a wide range of applications and can be integrated into a variety of Internet interactive systems. For example, the Internet of Things (IoT), supply chain tracking, Electronic Health Records (EHR), digital forensics, identity management, trustless payments, and other key business elements will all benefit from its implementation. Next layer solutions such as Ethereum 2.0, Polkadot, Cardano, and other Web 3.0 technologies provide developers versatility. Moreover, these platforms utilize smart contracts which are similar to standard, traditionalized software during development but offer key utilities to end-users such as online wallets, secure data with transparent rules.

Blockchain is receiving a lot of attention in educational technology (EduTech) as it aims to achieve a more transparent and multipurpose educational system. In addition to smart contract technology which defines how data should be registered, gathered and processed, blockchain can be used as an IoT intermediary for mobile usage.

Therefore, we implemented an educational learning platform powered by blockchain technology to examine feasibility in industry and academic environment. In essence, this is a web application which is adapted to mobile platform and connected to blockchain for crucial data exchanges. In this paper we want to emphasize the potential of blockchain technology in multiple sectors as well as the need to really understand the underlying principles which are allowing disruptability of traditional centralized software solutions.

Keywords: Blockchain · Smart contracts · Data security · Web 3.0 · IoT

1 Introduction

In this paper, we are presenting a decentralized system powered by blockchain technology also used as an IoT system. While there are research articles that use blockchain as an IoT intermediary [1] most of them do not go into enough detail regarding needed technical specification as there is no one-size-fits-all. Every blockchain network has the *blockchain trilemma*: decentralization, security, and scalability. These factors are interconnected, meaning if you try to improve one facet, another one will decrease as a trade-off.

M. E. Auer and T. Tsiatsos (Eds.): IMCL 2021, LNNS 411, pp. 423–430, 2022.
https://doi.org/10.1007/978-3-030-96296-8_38

Due to a lack of transaction speed some IoT blockchain systems [2] tend to eliminate particular parts, for example removing consensus mechanisms which just makes the need of a blockchain completely redundant and without the needed level of security.

On the other hand, in order to achieve a similar level to legacy systems of transaction processing, blockchain networks must be capable of handling an enlarging number of users, transactions and critical data. For instance, a common comparison is Bitcoin which can handle around 4.6 transactions per second while Visa handles around 1700. The problem is obvious, if blockchain networks were to compete with legacy systems they need to at least match the current numbers.

1.1 Centralized System

Most of today's internet systems and applications are centralized, meaning that a central authority is in control of data and functions an individual is using. An example would be an email service; the provider knows when you send the message and to whom, the data is allegedly privately stored, still, the email service has a copy of your data to which it is not impossible to access.

It can be claimed [3] that a centralized system offers better security, as there is less surface of attack compared to decentralized systems; however, there is always a possibility of a *coup d'état* whereas central entity goes rogue and maliciously uses data for its own benefit, also not an uncommon event.

Traditional database systems can have a reasonable amount of privacy and security with additional measures like firewalls, but the centralized nature and lack of user control raise concerns about the true level of privacy.

Still, it can be said that centralized systems are more efficient since there is a clear command structure; it is simpler to plan since network and infrastructure requirements can be predetermined, and quick changes and decisions can be made.

1.2 Decentralized Systems

The widespread adoption of blockchain and distributed ledger technology has resulted in a workable solution that presents a consensus of synchronized digital assets that are not owned by a single entity and are instead stored on a public digital ledger.

The shift from a centralized to a decentralized approach to application deployment opens up new possibilities for several industrial disciplines where access management, integrity, immutability, and non-discriminatory governance of data is imperative.

The key safety mechanism is a consensus protocol and the most common type of this verification is Proof-of-Work (PoW), used by the majority of modern cryptocurrencies. The process is known as mining, and the construction of a new block is completed by solving a complex mathematical puzzle that yields a 256-bit number hash, which is a unique identifier of data within the block.

Furthermore, the distribution of computational power and storage is an important security feature of a blockchain. As a result, because each node preserves a complete replicated transaction history, so if an attacker tries to profit through a double-spend attack, a 51% attack, or any other malevolent incentive, he will need an enormous amount of resources in relation to the respected consensus algorithm.

Even so, the high level of security PoW provides is diminished by the high resource consumption and slow transaction speed, making it unfeasible for large-scale adoption.

1.3 Layer-1

Layer 1 blockchain is a set of solutions that are incorporated in the blockchain's core protocol to improve the blockchain's functionality and scalability. Consensus protocol modifications and sharding are the two most popular Layer-1 scalability solutions.

The new consensus protocol Proof-of-Stake (PoS) tries to remove costly mining operations with the introduction of *Staking*. To be able to participate in the verification and creation of blocks on a blockchain, you must stake tokens. In a Proof-of-Stake blockchain, an individual or group is picked at random to verify transactions. The algorithm weighs the number of tokens they have staked on the network as collateral. Those chosen to confirm a block are routinely rewarded with the transaction fees associated with that block. The stake serves as a deterrent to malicious behavior.

Sharding is a way of splitting the database horizontally into smaller distinct datasets called "shards". This way, not all nodes need to maintain a copy of the entire network, thus allowing parallel processing so sequential work can be done on numerous transactions.

1.4 Layer-2

Layer-2 solutions refer to the collective pool of technologies designed to run on top of an underlying blockchain network in order to enhance its efficiency. The base layer blockchain becomes less congested and ultimately more scalable by abstracting the majority of data processing to auxiliary networks. A great example of this is *Rollups,* the heavy-computational transaction execution is performed outside the main chain, but data is posted on Layer-1, meaning its secured by Layer-1 and also achieves much better performance. There are several similar concepts in Layer-2 sphere: Nested Blockchains, State Channels, and Sidechains, and not to go into too much detail, these are all trying to fix the limitations of layer-1 technologies [4].

2 Internet Learning Platform

Smart contracts are a key component of programmable distributed ledgers like Ethereum, Cardano, and Polkadot. Smart contracts are programs that run on the global blockchain; the code, as well as all of the data controlled within the transactions, is public, resulting in a system that is trustworthy and cannot be cheated if properly designed. In our system implementation, we took into consideration the lack of scalability and designed our system to handle transactions in a similar way to Layer-2 Rollups; namely, separation of concerns so that only light-weight end result data is put on-chain. Main communication with the user is done on a simple website and all the heavy computation is done on a separate.NET API platform as can be seen from Fig. 1.

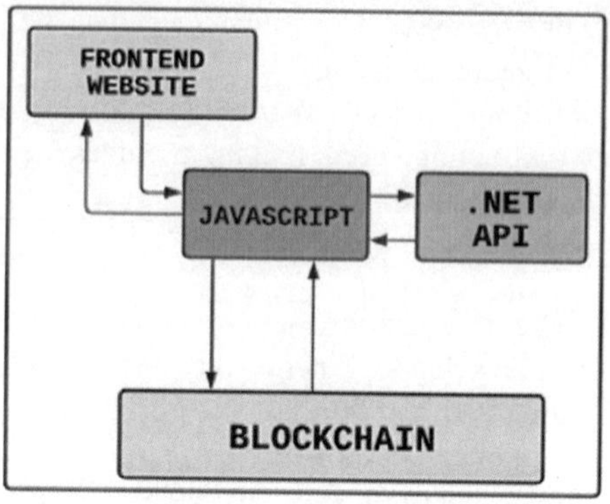

Fig. 1. High-level system overview

2.1 Ethereum Network-Based Solution

In our initial system design, we selected Ethereum Virtual Machine (EVM)-based system replica to emulate the real network. Our architecture connects three main sectors with clear duties that are linked by an intermediary level that manipulates and shares data. In order to replicate EVM, we had to employ multiple technologies: Truffle suite, Ganache, Web3.js, MetaMask, Solidity and other technologies.

Truffle suite is a development environment that includes smart contract compilation, linking, and deployment to different networks, as well as an interactive debugger, numerous libraries, and automated testing, as well as scriptable deployment and migration frameworks.

Ganache is a local in-memory blockchain environment that mimics the behavior of distributed ledgers in the reality. Solidity-related files, artifacts, and migrations are the three fundamental pieces of the execution model that split smart contract functionalities. The system's starting point is composed of solidity contracts, solidity-related libraries, and dependencies.

Web3.js is a suite of libraries that allow communication with Ethereum nodes, both local and remote, using various network protocols like HTTP, IPC, and WebSocket.

MetaMask is a browser plugin that acts as a crypto wallet, allowing websites to request Ethereum accounts and therefore run Ethereum decentralized applications (dApps).

Solidity is an object-oriented high-level language for smart contract development. A smart contract is a collection of code, functions, variables, and state that is recorded and exists on the blockchain at a specific address.

2.2 WIP Cardano Network-Based Solution

The current state of smart contract implementation in the Cardano network can be defined as development in progress [5]. The team behind Cardano is implementing the Plutus Platform for running smart contracts both on-chain and off-chain code, and the update is named the 'Alonzo' hard fork. Recently, an alpha test net, Alonzo Blue, has been launched to a testing community with exercises in order to assess solutions with incorporate next areas: Hard Fork, Addresses and Transactions, Managing Native Tokens, Interaction with the Wallet etc.

While all of this sounds great, the key factor of compiling and submitting Plutus scripts is not implemented yet, meaning our tests on the Cardano network will have been marked for future work.

2.3 WIP Polkadot Network-Based Solution

Polkadot is a next-generation blockchain technology that can connect numerous specialized chains in a universal network, with a focus on building Web 3.0 infrastructure. When comparing Polkadot to other blockchains, several new concepts are introduced, most importantly *Parachains* and *Parathreads* and it changes the way smart contracts are interrelated with these [6]. The main Polkadot Relay Chain does not natively support smart contracts, however, the concept of Parachains will allow for more precise and fast execution of extensive logic than a smart contract platform could provide. For example, Custom fee and Monetary Policy structure, Decentralized Autonomous Organization (DAO) Governance, Shared Randomness etc. Arguably, with a more complex runtime logic, an abstraction layer is added on top of it as developers will be able to declare a whole environment of their chain, also allowing smart contracts to be written on top of them [7]. Both, Polkadot and Cardano present potential future benefits based on the concept of om- and off-chains codes in universal networks, implementation is still not possible due to the lack of network details. That's why we proceeded with the implementation described in Sect. 2.1.

3 Results

Our educational learning platform comprises several interconnected technologies which together create a useable, precise, and responsive system with agility for further operations. One of the main goals was to focus on privacy guarantees for both students and professors such as decentralization and differential privacy.

We have built our educational learning platform based on smart contracts to examine all potential benefits and flaws in the technology. In order to be able to use smart contracts, we compiled and deployed them using contract creation transactions. In this process, each contract is associated with an address derived from the creating transaction, more specifically from the invoking account and the nonce, which is a timestamp. After deployment, contracts are ready to be called and run by externally owned accounts (EOAs), in our case EOA's calls will be users, that will call smart contracts after each simulation iteration circle finishes. Our system uses solidity v0.5.0 with which we are deploying contracts to the Ethereum blockchain. The key communication line for blockchain reporting is the web3.js connected to a Solidity object.

Our learning platform simulates a simple business environment of IoT on Blockchain where students (representing nodes in IoT) are supposed to simulate digital marketing efforts for a hypothetical smartphone brand with a limited range of three devices with different technical specifications and a targeted market. The simulation consists of several rounds in each of which students are planning the budgets per digital platform and decisions about other parameters (e.g., keywords strategy). After each round, activity reports (i.e., digital platform insights/analytics) as a result of execution of smart contracts are generated, providing students with valuable feedback on their actions. During the start of the simulation, pre-determined data entered by the admin is loaded on the blockchain. This data is the starting benchmark of globally defined values which define successfulness of each user's overall situation e.g., Number of likes, Post engagement, Page views, etc. and they are equally set for all users.

Since the platform is based on the concept of a smart contract, all the activity reports are immutable, showing data integrity and enabling everyone in the simulation is dealing with the same datasets. The simulation consists of 16 iterations where the initial iteration is for instructional setup and creating the environment. Reporting at the end of each round is based on smart contracting and for executing the contract, it is required to pay a gas fee. Our basic solution is built on the previously mentioned EVM-based environment with Truffle suite, Ganache, Web3.js, and MetaMask, and smart contracts developed in Solidity. In order to compare our proposed solution on Ethereum, with possible Cardano and Polkadot solutions, we adjusted the predicted amount of activities per iterations of each possible solution.

As a major benefit of our solution, we have proved that our solution is applicable for use in any kind of smart contract. The time to the finality of executing the smart contracts on our platform is presented in Fig. 2. It could be seen that the average time to finality is denoted in a couple of seconds which is acceptable time for proving and executing business contracts, comparing to the real world where such operations could take even a few days.

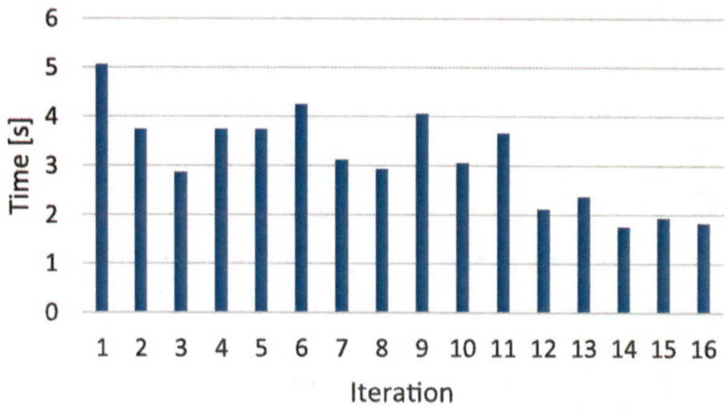

Fig. 2. Time to finality per simulation iteration on our platform

As an additional outcome, as presented in Fig. 3 on average transaction costs, while our solution based on Ethereum is constantly showing costs around 0.7 gas (which is currently 15.80 Gwei), predicted gas price based on [8] and [6], for reporting based on Polkadot and Cardano technologies will be some 3 times lower while enabling even broader and cheaper usage of our platform.

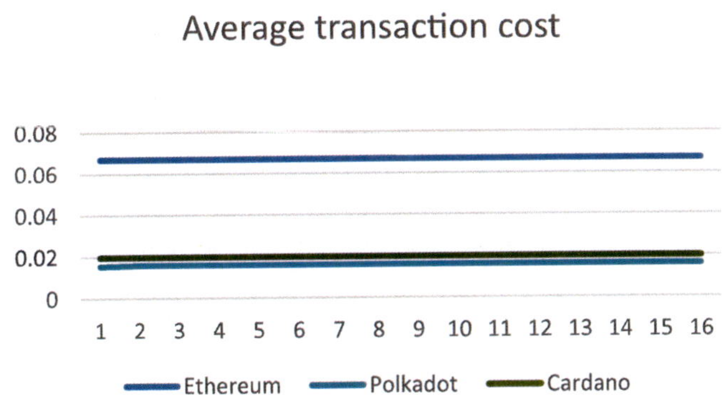

Fig. 3. Comparison of average transaction costs in gas

4 Conclusion

According to revised Bloom's taxonomy, the creation of this new and original educational learning platform, for the educational process and proving Blockchain technologies as IoT intermediaries (in our case, students represented nodes in IoT) is able to cover all six cognitive processes (which is in our case valid for both Business and IT students). Business students get an insight into strategic knowledge and conceptual knowledge by using the platform, and knowledge of Internet marketing specific techniques and methods, knowledge about the terminology, together with knowledge about the criteria for determining when to use appropriate procedures when designing business logic in the background of this application. IT students acquire knowledge of distributed application development skills and Blockchain and smart contract technologies, knowledge about the classification when acquiring the initial input data from Business students, and knowledge about the terminology in building the distributed applications from their instructor.

We also proved that our approach based on EVM is applicable to different business domains that is based on the immutability and data integrity provided by the smart contracts. We anticipate gas fees as the only downside, although the relevant numbers are counted in cents, which would not be a problem in real business application for replacing and approving the business contracts with the smart contracts.

Our result datasets show that this new approach of gamification in learning platforms is inclusive towards different perspectives of knowledge sharing and knowledge gaining, and in the near future when other frameworks like Cardano and Polkadot will be fully available, with additional options and lower gas prices and off-chain code management in universal networks, our proposal could be defined as a benchmark for educational learning platform for business and IT curricula, and also a showcase for proving the applicability of smart contracts in business applications.

References

1. Pranto, T.H., Noman, A.A., Mahmud, A., Haque, A.B.: Blockchain and smart contract for IoT enabled smart agriculture. PeerJ Comput. Sci. 1–29 (2021)
2. Christidis, K., Devetsikiotis, M.: Blockchains and smart contracts for the internet of things. IEEE Access **4**, 2292–2303 (2016) . IEEE
3. Patsonakis, C., Samari, K., Roussopoulos, M., Kiayias, A.: Towards a smart contract-based, decentralized, public-key infrastructure. In: Capkun, S., Chow, S.S.M. (eds.) CANS 2017. LNCS, vol. 11261, pp. 299–321. Springer, Cham (2018). https://doi.org/10.1007/978-3-030-02641-7_14
4. Saraph, V., Herlihy, M.: An empirical study of speculative concurrency in ethereum smart contracts. ArXiv, pp. 1–13. Cornell University (2019)
5. Aydinli, K.: Performance Assessment of Cardano. Independent Study – Communication Systems Group. pp. 1–39 (2019)
6. Parity Technologies: A Brief Summary of Everything Substrate and Polkadot. https://www.parity.io/blog/a-brief-summary-of-everything-substrate-polkadot/. Accessed 07 Jan 2021
7. Polkadot Builders Starter's Guide. https://wiki.polkadot.network/docs/build-build-with-polkadot/. Accessed 07 Jan 2021
8. Burdges, J., et al.: Overview of Polkadot and its Design Considerations. ArXiv, abs/2005.13456 (2020)

A Mobile Educational Application for Enhancing Cognitive and Language Skills of Children with Disabilities

Matthaios Gerakis[1] and Christina Volioti[1,2](✉)

[1] Department of Information and Electronic Engineering, International Hellenic University, Thessaloniki, Greece
chvolioti@gmail.com
[2] Department of Applied Informatics, University of Macedonia, Thessaloniki, Greece

Abstract. The current paper aims at the development of a mobile educational application for enhancing cognitive and language skills of children with disabilities. During their development, they receive a variety of stimuli that influence and shape these skills. Therefore, they should be helped and supported as much as possible with all available means, such as ICT tools. Additionally, the identification of language and cognitive disorders is a major challenge to special education teachers. It is thus important to develop a specially designed application aiming at the enhancement of skills of children with disabilities and will be available to all stakeholders (special education teachers, children, parents). For the evaluation, an intervention was designed in collaboration with two special education teachers and took place in a training center for children with disabilities in Thessaloniki (Greece), called Redditus. The results were very promising, since children with disabilities greatly improved their skills during the intervention. Moreover, a semi-struc,*tured interview was carried out with the two teachers. They supported that it is useful, simple to use, enjoyable, fun and innovative. Summarizing, the proposed application could be an auxiliary support tool to special education teachers in order children with disabilities to improve their cognitive and language skills.

Keywords: Children with disabilities · Skills · Mobile educational application

1 Introduction

It is well known that the development of cognitive and language skills is essential to children. During their development, they receive a variety of stimuli that influence their skills. Therefore, typically developing children, even more children with disabilities, should be helped and supported as much as possible with all available means, such as ICT tools. ICT offer children new learning opportunities and experiences [1], helping them acquire and develop these skills [2]. Educational applications and games should not only provide an attractive design, but also should be based on the principles of cognitive neuroscience and research [3]. So, a well-designed application with

M. E. Auer and T. Tsiatsos (Eds.): IMCL 2021, LNNS 411, pp. 431–442, 2022.
https://doi.org/10.1007/978-3-030-96296-8_39

educational activities could become very useful and effective for children's education and proper mental development [2].

The identification of language and cognitive disorders is a major challenge to professionals (such as special education teachers). So, in the case where children face learning disabilities, the need to develop these skills becomes even greater. It is thus important to develop specially designed applications by using more efficient educational approaches which are personalized according to children's needs and will be available to all stakeholders (special education teachers, children, parents).

This paper is structured as follows, reflecting its purpose to both develop and evaluate the mobile educational application for improving cognitive and language skills of children with disabilities. The next section introduces the learning disabilities that the application aims to improve. Section 3 outlines related research on the use of applications for both typically developing children and children with disabilities. Section 4 describes the mobile educational application. The method of the intervention is presented in Sect. 5. Section 6 analyses the findings as well as the discussion of the research work and finally in Sect. 7, the conclusions along with the future work are referred.

2 Learning Disabilities

This section presents the different learning disabilities that the proposed mobile educational application takes into consideration. The two main categories that the application aims to improve through its specially designed activities are a) specific language impairments and b) cognitive disorders. More specifically, Specific Language Impairment (SLI) is one of the most common learning difficulties encountered in children [4]. SLI is a form of developmental disorder in which children face difficulties in acquiring language skills [5] and in mastering the medium of communication [6]. SLI may affect speaking, listening, writing and reading of a child [7]. The second category that the present work also incorporates is cognitive disorders. Cognitive disorders are defined as deficits or disruptions that concern problem solving, attention, perception and language [8]. In other words, children with cognitive disorders show difficulties in understanding and performing verbal and nonverbal instructions and tasks [9].

3 Literature Review

Several research works have been conducted aiming at improving skills of typically developing children or children with disabilities with the use of technology. Papadakis et al. [10] investigated whether tablet and computer affect early childhood children in learning mathematics by focusing on the following skills, number word sequence, counting, enumeration, quantities, basic addition and subtraction skills. The use of digital technologies showed significantly positive results, meaning students who used tablet-based applications had even higher scores that those who used computer-based ones. This is due to the fact that touch screens let students to interact in a more intuitive way and therefore have a more real-life experience. Also, concluded that stimuli-rich multimedia environment enhances students' mathematical thinking and sensory-motor skills. Watkins et al. [1] implemented a software, called ELAN, which supports literacy acquisition through phonics. They concluded that a software which can adapt to child's needs and focuses on the repetition of decoding skills, it could have beneficial results in the learning process by improving syllable and word reading skills as well as semantic skills. Additionally, Schaefer et al. [11] conducted a pilot study to explore if receptive vocabulary and semantic skills in monolingual and multilingual children can be screened in a tablet application with the use of four-choice picture paradigms. The findings were satisfactory, since the application was easy to use increasing children's motivation and therefore their performance. Volk et al. [12] examined the use of tablet in 3rd grade students for teaching maths. The results indicated that students had better learning outcomes, and specifically in a) cognitive domain (spatial orientation, instructions, sequences, comparison), b) psychomotor domain (fine motor skills) and c) affective-social domain (perseverance, engagement, cooperation).

All the aforementioned studies referred exclusively to typically developing children. Regarding children with disabilities, Chmiliar [13] implemented iPad into interventions with preschool children with disabilities. Different applications were used to improve different skills. The use of iPad seemed to be a useful tool to children with disabilities, helping them to improve and extend their skills, such as language skills, play skills, trace shapes and letters, fine motor skills, etc. Additionally, Skiada et al. [3] developed a mobile application, called EasyLexia, for children with dyslexia to improve their skills with respect to comprehension (categorization), orthographic coding, short-term memory and mathematical problem solving (shapes, sequences). The application helped children to concentrate and be more focused on the touch screen. Also, children with disabilities showed progress and some of them improved their skills. Chai et al. [14] developed an iPad application to learn and receptively identify the phonemes to young children with disabilities. Children demonstrated an increase in learning phonemes and managed to generalize their literacy skills after the intervention ended.

Apart from the above research works, a variety of applications which are available to Google Play Store are also aim to practice different skills of children. For example, ABC Kids [15], ABC Kids Games [16], Educational Game 4 Kids [17], Intellijoy Early Learning Academy [18], Kids Brain Trainer [19], and many more, aim to improve different skills, such as cognitive skills, literacy skills, memory skills, motor skills, etc. mainly to typically developing children of different ages.

Summarising, several research studies aim to enhance skills of both typically developing children and children with disabilities. However, these studies are mainly research works and the applications are not publicly available to children with disabilities, or other stakeholders (such as special education teachers or even their parents). On the other hand, there is a majority of applications, available on Google Play, mainly of typically developing children and not exclusively of children with disabilities. Therefore, there is a need for an application that will link the aforementioned, aiming at the development and enhancement of skills entirely of children with disabilities and will be publicly available to all stakeholders. In addition to this, the current work aims to answer the following two research questions:

RQ1. *Do children with disabilities manage to improve their skills by using the mobile educational application?*
RQ2. *What is the view of special education teachers towards the application?*

4 Description of the Mobile Educational Application

The mobile educational application was developed to contribute to the education of children with disabilities in an easier and a more entertaining way. The purpose of the application is both educational and entertaining. Therefore, it is designed to be friendly and easy to use with simple graphics in order the child not only to learn by improving his/her skills, but also to play and interact with it through using it as a pleasant and creative game. The mobile application aims to improve language and cognitive skills of children with disabilities. The selection criteria of the desired skills were based on the literature review and the guidance of two special education teachers who are working in a training center for children with disabilities in Thessaloniki (Greece), called Redditus. Because the needs of the children with disabilities are well-known to them and in combination with the lack of a specially designed application aimed at improving these skills of children with disabilities, the specific application was proposed to be used as an auxiliary support tool in the interventions.

Emphasis was placed on monitoring the performance of each child. In order to succeed this, it is necessary for the user, who could be either the special education teacher or the child's parent or the child with disabilities himself/herself, depending on his/her ability, to create an account. When the application is launched and the user is logged in, s/he is directed to the main menu where s/he has the option to go to either a category of activities or to the settings menu which includes functionalities such as log out, delete account, change password, view the scores of each activity, delete all the scores and change details (Email and username). Each category of activities consists of three sub-categories which contain two activities each one. More specifically, the main two categories are the a) "Language Activities" and b) "Cognitive Activities". Each category includes three subcategories and each subcategory contains three activities (see Table 1).

Table 1. Mapping between the activities and the corresponding skills.

Category	Subcategory	Name of activity	Skill(s)
Language Activities	*Description*	Select those who match	Semantics [1, 11]
		Select the correct description	Categorization/Classification [3]
		Select the right one	Receptive vocabulary [11, 14]
	Instructions	Find the path	Shapes [3, 13]
		Select the arrows	Spatial orientation [12]
		Drag and drop the objects	Instructions [12]
	Numbers	Drag and drop the numbers	Quantities [10]
		Find the path with numbers	Arithmetic sequences [10]
		Drag and drop the images	Quantities [10]
Cognitive Activities	*Comparison*	Select the bigger one	Comparison [12]
		Write the correct height	Counting [10]
		Find the bigger number	Comparison [12]
	Arithmetic	Add the animals	Arithmetic operations [10]
		Select the following number	Arithmetic sequences [10]
		How many animals are there	Quantities [10]
	Patterns	Complete the pattern	Sequences [3, 12]
		Complete the shape	Mirroring [3, 13]
		Select the one which differs	Categorization [3]

It is worth mentioning that the third activity in each subcategory is appeared to the user only after the successful completion of the first two activities. The third activity is more difficult and depends on the user's choice whether s/he will select it or will move on to the next subcategory. The categorization to language and cognitive activities has been made, as it is already mentioned, to improve language and cognitive skills respectively. As a result, every and each activity is mapped to a skill which aims to contribute to its improvement (see Table 1).

Regarding the user interface design, many challenges emerged. However, taking into consideration the design guidelines arising from the literature review in combination with the guidance from the two special education teachers, a specially designed application was developed for children with disabilities. More analytically, the principles that were implemented in the proposed mobile application summarized below:

a) simple and clear interface should be necessary to increase children's attention [20]
b) consistency should be existed throughout the interface by placing all the elements (e.g., titles, headings) on the same part of the screen [21]
c) clear and simple font styles should be used to help children easily recognize numbers and text [21]
d) colors must be carefully selected [22]
e) feedback with instructional cues should be used [23]
f) audio could be used to hold children's attention [22]

From a technical point of view, Android Studio software and Java programming language were used for the development of the application. For online data storage, a Google Firebase service was used and specifically Firestore. Cloud Firestore is a service provided by Google and is an easy-to-use non-relational database. It consists of a hierarchy of collections, documents and fields that interact with each other based on the rationality that collections contain documents, which in turn contain fields. The Google Authentication service was used to identify users. To each user a unique identifier, called User UID, was given. All operations that require communication with the database are performed after user authentication, which is implemented using the unique identifier (User UID). Finally, in order for the application to communicate with this Firestore project, a token was used.

The proposed application was firstly designed for smartphones. However, there is a wide range of mobile devices (both smartphones and tablets) with a large variety of available options in both screen resolutions and size. To deal with this, it was necessary to adapt the mobile educational application to the needs of appearance in different devices (for example to display the application in a screen larger than 7 in.).

Last but not least, regarding the proper design of the database, a very important issue is that the user should be able to use all the functionalities without any restriction. That means that the user can move on to any activity, check which ones are completed and receive the necessary feedback. All the data, including scores for each activity, is stored in the database, which is available through the personal account. This is the added value of the application, which provides the ability to the special education teacher and/or parent to monitor the performance of the child through the settings menu. Therefore, some of the functionalities provided in the settings are more directed to the special education teacher and/or the child's parent, such as deletion of account or scores. However, the application can be used by children without the need of adult supervision, because in order to perform any significant action (such as logging out), the confirmation from the user is mandatory.

4.1 Indicative Example

In this section, an indicative description of an activity is demonstrated. The name of the activity is "Select those who match" and the skill it improves is "Semantics".

Fig. 1. An activity from the mobile educational application.

In the toolbar (see Fig. 1), the instructions of the activity are written and rolling over. On the right, the previous score (if the activity has been completed before) along with the evaluate button (which is pressed by the user when s/he has completed the activity in order to perform the evaluation) are displayed. On the left, there are 2 buttons to return to the main menu or to the previous menu (the buttons with the home and arrow icons). The activity contains two images and four descriptions for each image. Each description is accompanied by a speaker shaped button which plays the corresponding audio message, when is pressed by the user. The user has to select the description or descriptions (meaning sometimes the correct answer is more than one) that fit(s) on each image and when s/he is ready, s/he has to press the evaluate button. After the evaluate button is pressed, a feedback message will appear stating the score achieved as well as a reward message depending on the rating the user achieved.

Depending on the child's performance, the next time the button that leads to this activity will have one of the following colors (see Fig. 2):

- Red color: When his/her score ranges from 0% to 33%
- Yellow color: When his/her score ranges from 34% to 66%
- Green color: When his/her score ranges from 67% to 100%

In case the activity has not been completed by the user yet, the color of the button will be blue.

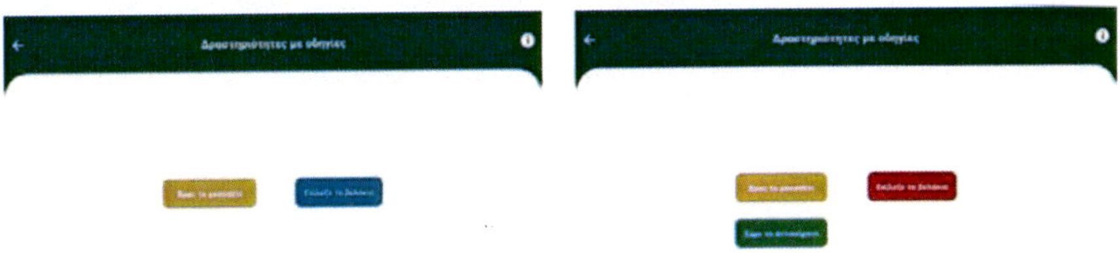

Fig. 2. The colored buttons of the activities.

5 Method

5.1 Design of Intervention

The educational intervention was designed in collaboration with two special education teachers. An initial training phase was included, where the special education teachers learned how to use the application in order to become familiar with the system. For the purposes of the intervention, the special education teachers had to arrange two face to face sessions with each child with disabilities in the training center Redditus. In the first basic session, each child had to run the application, playing all the activities with the presence of his/her teacher. Since each activity is mapped to a skill, the teacher had the advantage of monitoring each child's progress and assessing in which areas s/he needs improvement by checking his/her score in each activity. Then, a number of intermediate sessions were arranged between the special education teacher and the child in order to improve the specific skills. After these intermediate sessions, the second basic session took place, where the child had to play the entire application again. Finally, each child's scores between the two basic sessions were noted for the teacher to easily monitor his/her progress. It is worth mentioning, that a) the time duration for each child to complete all the activities in each session, b) the time between the two sessions as well as c) the number of intermediate sessions that was required, was completely different for each child, because it depends on what kind of impairments and disorders s/he was faced and how many areas s/he needs to improve according to his/her scores.

5.2 Instrument

A combination of qualitative and quantitative data was collected to evaluate the efficiency of the mobile educational application. Specifically, regarding the two special education teachers, a semi-structured interview was conducted. The interview included seven (n = 7) questions in which each respondent was asked to give his/her opinion on whether the application is easy to use, whether it helped the educational process and whether the application could be integrated into the educational process. Regarding the children with disabilities, a descriptive statistical analysis was applied to the scores that were collected in all the activities during the intervention.

5.3 Participants

In total, two (n = 2) special education teachers took part in the research, one of whom was man and one woman. Both of them have at least ten years of experience with children with disabilities. Additionally, fifteen (n = 15) children with disabilities participated in the intervention, whose age ranges from 10 to 14 years old. The main disabilities that children had in this specific intervention, were mild autism, intellectual disability, language impairments and cognitive disorders.

6 Results and Discussion

The purpose of this study was to examine whether the proposed mobile educational application could improve specific skills as well as whether it has the potential to be a useful and supportive educational tool to special education teachers for interventions to children with disabilities. To accomplish this, a statistical analysis was conducted in the final scores of the children who participated in the intervention while using the application and a semi-structured interview was conducted to the two special education teachers. Below are the research questions which answer each aforementioned point:

RQ1. *Do children with disabilities manage to improve their skills by using the mobile educational application?*

In general, the use of the proposed application during the intervention seems to significantly enhance the skills of children with disabilities. The children showed improvement in their scores in all activities comparing the 1st with the 2nd session. Although many of the activities aimed to improve more than one skill; each activity was mapped to only one skill, the one that is cultivated most intensively, according to teachers' suggestions.

The following figures illustrate the performance of children by category, meaning the Fig. 3 presents the results of the "Language Activities" while the Fig. 4 of the "Cognitive Activities". In each figure, the part A. shows the average score of all children per activity and per session. The numbers 1–9 represent the 9 activities of each category, while the blue bar corresponds to the mean of all the first attempts in each activity (1st session) and the red to the mean of all the second attempts (2nd session). Moreover, the part B. depicts the mean of all the first attempts from all activities compared to the mean of all the second attempts.

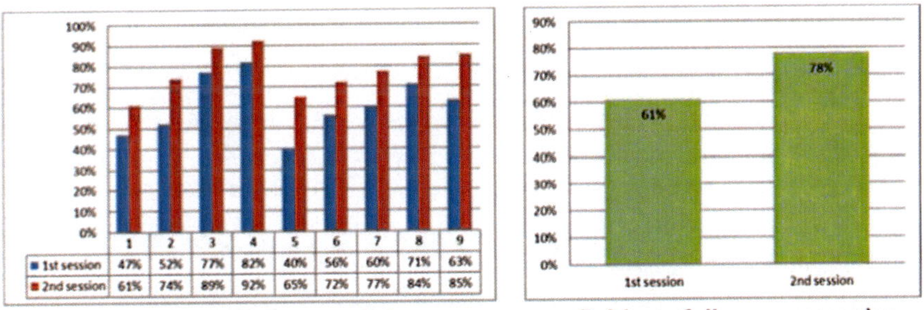

A. Mean of scores per activity B. Mean of all scores per session

Fig. 3. Mean of scores of "Language Activities".

In Fig. 3, the average values of scores in all the activities of the category "Language Activities" of the 1st session are above 40%, while in the 2nd session, are over 60%. In this category, there seems to have been skills' improvement of 17% to all children with disabilities (see Fig. 3B.).

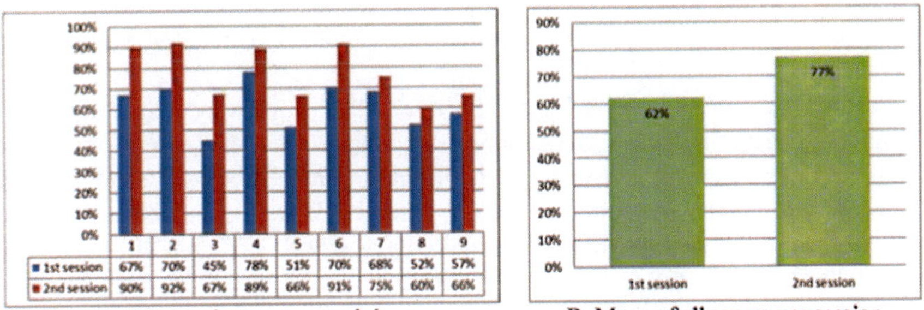

A. Mean of scores per activity B. Mean of all scores per session

Fig. 4. Mean of scores of "Cognitive Activities".

Similarly, in Fig. 4, the mean values of scores in all the activities of the category "Cognitive Activities" of the first attempts are above 50%, except one (scores in activity 3). While of the second attempts, are over 60%. It is distinguishable that there is an improvement of about 15% (see Fig. 4B.).

After conducting the above descriptive statistical analysis, the special education teachers were asked to interpret the aforementioned results. Regarding the diversity of the scores, they claimed that is due to the fact that each child has different disorders as well as different cognitive and language level, and therefore some activities seemed to be easier for some children, while others were more difficult. Another reason could be the different age of children, since each child, according to his/her age and his/her ability, acquires and practices different skills.

Regarding the cause of the improvement in their second attempts, teachers considered that is due to the application's user interface which is friendly and easy to use as well as to the visual representations that helped them to understand more easily each activity. This finding is consistent with previous works [11, 20, 21], which indicated that an easy to use, simple and consistent application can increase children's motivation and therefore their learning outcomes. Finally, they supported that the interaction with a specially designed application for the children with disabilities, helped them to improve their skills by playing with great interest the educational application. This is reinforced by the majority of the relevant research [1–3, 11–14], which revealed that a well-designed application can positively affect children's performance as well as help them to learn and extend different skills.

RQ2. *What is the view of special education teachers towards the application?*

The interviews showed that the special education teachers were also helped by the application since its main advantage was the ability that provided to them to automatically monitor and assess specific skills of children with disabilities. They stated that the application is easy to use and user friendly. Specifically, based on their answers, they claimed that such an application could make the educational process more efficient and entertaining and it could be integrated into the intervention since most of the children with disabilities are already familiar with such devices. Additionally, it could be beneficial if it is used on a daily basis, either during the intervention or at home with the supervision of their parents. However, they emphasized that such applications should not replace the current educational process but it could be an auxiliary support tool.

Regarding the attitude of children with disabilities towards the application, special education teachers mentioned that children tended to concentrate more when they used the application in comparison with the traditional teaching process, a statement that is also confirmed by the literature review [3]. Characteristically, the children showed a huge interest in the application's use and the teachers said that it "attracts their attention". Therefore, the use of such applications could provide encouraging results for improving children's skills as well as their psychology.

7 Conclusions and Future Work

Summarizing, the proposed mobile educational application aimed at enhancing cognitive and language skills for children with disabilities. In general, the results from both qualitative and quantitative analysis of the application are very positive. Special education teachers supported that it is very beneficial for both children and themselves. They stated that it is simple to use, enjoyable, fun, innovative and useful since it provides an easy way of monitoring through the mapping between each activity and the corresponding skill that is developed. It became evident that involving special education teachers in the design of the educational application had served well the purpose of improving the design features, before its pilot application with real users. Last but not least, the proposed application could be an auxiliary support tool to special education teachers of children with disabilities in order to improve their cognitive and language skills.

Future work will include the expansion of the application by adding more activities that mapped to different skills. Features such as a) multiple profiles in one account (meaning that the special education teacher will be able to add different profiles in his/her account according to the number of children that is participated into the intervention) and b) displaying the scores of the activities into the form of chart for better and easier understanding from teachers, could be also added. Moreover, it could be tested to a larger sample and to children with different disabilities in order to generate more valuable conclusions for the proposed mobile educational application.

Acknowledgements. Authors wish to thank all children as well as special education teachers who took part in this study. This study would not have been made possible without their valuable participation and contribution.

References

1. Watkins, C.P., Caporal, J., Merville, C., Kouider, S., Dehaene, S.: Accelerating reading acquisition and boosting comprehension with a cognitive science-based tablet training. J. Comput. Educ. **7**(2), 183–212 (2020). https://doi.org/10.1007/s40692-019-00152-6
2. Zaranis, N., Kalogiannakis, M., Papadakis, S.: Using mobile devices for teaching realistic mathematics in kindergarten education. Creat. Educ. **4**, 1–10 (2013)
3. Skiada, R., Soroniati, E., Gardeli, A., Zissis, D.: EasyLexia: a mobile application for children with learning difficulties. Procedia Comput. Sci. **27**, 218–228 (2014)
4. Prelock, P.A., Hutchins, T., Glascoe, F.P.: Speech-language impairment: how to identify the most common and least diagnosed disability of childhood. Medscape J. Med. **10**(6), 136 (2008)
5. Tomblin, J.B., Records, N.L., Buckwalter, P., Zhang, X., Smith, E., O'Brien, M.: Prevalence of specific language impairment in kindergarten children. J Speech Lang. Hear. Res. **40**(6), 1245–1260 (1997)
6. Bishop, D.V.M., Leonard, L.B. (eds.): Speech and Language Impairments in Children: Causes, Characteristics, Intervention and Outcome. Psychology Press, Hove (2000)
7. Specific Language Impairment. https://www.nidcd.nih.gov/health/specific-language-impairment. Accessed 28 June 2021
8. Berryhill, M.E., Peterson, D., Jones, K., Tanoue, R.: Cognitive disorders. In: Ramchandran, V.S. (ed.) Encyclopedia of Human Behavior. Elsevier, Amsterdam (2012)
9. Ohta, M.: Cognitive disorders of infantile autism: a study employing the WISC, spatial relationship conceptualization, and gesture imitations. J. Autism Dev. Disord. **17**(1), 45–62 (1987)
10. Papadakis, S., Kalogiannakis, M., Zaranis, N.: The effectiveness of computer and tablet assisted intervention in early childhood students' understanding of numbers. An empirical study conducted in Greece. Educ. Inf. Technol. **23**(5), 1849–1871 (2018). https://doi.org/10.1007/s10639-018-9693-7
11. Schaefer, B., Bowyer-Crane, C., Herrmann, F., Fricke, S.: Development of a tablet application for the screening of receptive vocabulary skills in multilingual children: a pilot study. Child Lang. Teach. Therapy **32**(2), 179–191 (2016)
12. Volk, M., Cotič, M., Zajc, M., Starčič, A.I.: Tablet-based cross-curricular maths vs. traditional maths classroom practice for higher-order learning outcomes. Comput. Educ. **114**, 1–23 (2017)
13. Chmiliar, L.: Improving learning outcomes: the iPad and preschool children with disabilities. Front. Psychol. **8** (2017)

14. Chai, Z., Vail, C.O., Ayres, K.M.: Using an iPad application to promote early literacy development in young children with disabilities. J. Spec. Educ. **48**(4), 268–278 (2015)
15. ABC Kids. https://play.google.com/store/apps/details?id=abc_kids.alphabet.com. Accessed 28 June 2021
16. ABC Kids Games – Phonics to Learn alphabet Letters. https://play.google.com/store/apps/details?id=com.kids.learn.reading.app. Accessed 28 June 2021
17. Educational Game 4 Kids. https://play.google.com/store/apps/details?id=com.pescapps.game4kids. Accessed 28 June 2021
18. Intellijoy Early Learning Academy. https://play.google.com/store/apps/details?id=com.intellijoy.kids.academy. Accessed 28 June 2021
19. Kids Brain Trainer. https://play.google.com/store/apps/details?id=brilliant.sari. Accessed 28 June 2021
20. Jerrett, D.: The Inclusive Classroom: Math and Science Instruction for Students with Learning Disabilities. Northwest Regional Educational Laboratory, Portland (1999)
21. Bley, N.S., Thornton, C.A.: Teaching Mathematics to Students with Learning Disabilities, 4th edn. Pro-Ed, Austin (2001)
22. Higgins, K., Boone, R., Pierce, T.B.: Evaluating software for use by students with disabilities to foster inclusion in general education. In: Proceedings of the International Special Education Conference (ISEC), Glasgow, Scotland (2005)
23. Miller, S.P., Hudson, P.: Using evidence-based practices to build mathematics competence related to conceptual, procedural, and declarative knowledge. Learn. Disabil. Res. Pract. **22**, 47–57 (2007)

Automated Essay Feedback Generation in the Learning of Writing: A Review of the Field

Paraskevas Lagakis(✉) and Stavros Demetriadis

Software and Interactive Technologies (SWITCH) Lab, Computer Science Department, Aristotle University of Thessaloniki, Thessaloniki, Greece
{plagakis, sdemetri}@csd.auth.gr

Abstract. Despite being investigated for over 50 years, the task of automated essay scoring continues to draw a lot of attention in the natural language processing community, in part because of its commercial and educational values, as well as due to the associated research challenges. Although the importance of automating the holistic scoring of an essay is indisputable, such systems could potentially have a wider impact if they could help students improve their essay writing skills while providing them with some sort of feedback. A mere low score could not possibly help a student understand the reason for scoring weakly neither would it show the areas where the student could improve and how. In light of this deficiency, researchers have begun anew to work on scoring specific aspects of text quality such as consistency, technical mistakes, and relevance to prompt. Automated essay scoring systems that offer didactic feedback along multiple dimensions of essay quality have also begun to arise. This paper critically reviews the recently published scientific literature on AES systems that provide automated student feedback and their impact on student learning and the teachers/students attitude towards using such systems in the learning procedure.

Keywords: Automated essay feedback generation · Automated essay scoring · Interactive learning environments · Learning of writing

1 Introduction

This paper's goal is to document an overview of the Automated Essay Scoring (AES) systems that also provide automated feedback, outlining firstly the approaches used, current trends, practices and developments in such systems, as well as how these systems are evaluated. From our research, only one recent article [1] summarizes the state of the art in providing automated feedback, focusing on the scientific methodology of the architecture of those systems, whereas our survey focuses mostly on the learning impact of case studies that utilize such systems.

M. E. Auer and T. Tsiatsos (Eds.): IMCL 2021, LNNS 411, pp. 443–453, 2022.
https://doi.org/10.1007/978-3-030-96296-8_40

The approach is threefold:

1. Provide a short overview of the field of AES and automated essay feedback,
2. analyze the case studies that utilize AES systems with automated feedback, with respect to certain characteristics in their design and organize them under a unifying classification scheme,
3. present research evidence on the impact of these systems on student learning, and, identify current trends and open research questions in the field.

To research the literature, we used Google Scholar, ResearchGate and ERIC with the search terms "Automated Essay Scoring", "Automated Essay Feedback" and "Automated Essay Scoring with Feedback Generation". For the purposes of this paper, any systems that implemented automated evaluation and feedback generation for items other than written essays, e.g. coding exercises or selected-response questions [2], were not included in the review.

After systematically searching online bibliographic databases, more than 35 articles were included in the review, reporting both novel approaches on developing AES systems with automated feedback as well as case studies with concrete evaluation data on the learning impact of AES systems with feedback generation.

2 Automated Essay Scoring Overview

AES, which is the usage of artificial intelligence in order to score any written document in an automated manner, was initially introduced as a concept in 1966 by E. B. Page with his work [3] on computer aided grading systems. Currently, AES is considered one of the most prominent academic applications of Natural Language Processing (NLP), with most AES researchers up until recently being attracted to perfecting the system to rate the quality of any essay in a holistic manner. Holistic scoring is popular as a research field for two reasons, the main one being the abundancy of datasets using manual holistic scores that are available online, which created a fertile ecosystem for holistic scoring approaches and systems to "learn" from. Secondly, holistic scoring mechanisms are in high commercial demand.

AES systems that offer didactic feedback along multiple dimensions of essay quality have also been developed in the last 20 years, for example Criterion [4] was one of the first ones. Other dimension-specific approaches using hand-crafted features to evaluate essays are also still very popular among researchers, for example approaches scoring coherence [5], technical mistakes, relevance to prompt [6], organization [7], thesis clarity [8] and argument persuasiveness [9].

There are many dimensions of an essay that a machine can accurately evaluate, but most of them refer to its technical aspects, like the use of correct spelling and grammar, punctuation and word usage. Aspects that refer to its content quality and are an important part of a human grader's evaluation, are still very difficult to be effectively estimated by an AES system.

Features and Approaches
Early AES systems where developed with handcrafted features in mind. For example the pioneering Project Essay Grader (PEG) from Page, used an approach of supervised learning with linear regression, and made use of features based on length, such as the average length of words in the text and the length of the text, as well as lexical features, like features that encode the number of occurrences of a particular punctuation in a text. Some of the other well-known systems that were developed early on, with handcrafted features approaches, include IEA, E-rater, IntelliMetric and BETSY.

Lexical features have also been used together with syntactic features like part-of-speech (POS) tag sequences, parse trees and grammatical error rates. For example, Chen and He [10] in their system, which was based on a ranking approach with the use of LambdaMART, used parse tree depth as a representation of the syntactic complexity of the sentences in a text.

The other category of AES systems consists of models that erase the need for feature engineering, by following neural approaches, for example [11–13]. Finally, one of the current trends in NLP research in general, is to use transformer-based models with fine-tuning in tasks having to do with supervised learning, with more traditional linear models losing popularity.

More specifically, the use of deep neural networks has been widely popularized in the NLP community in recent years, and in particular, the use of the transformer architecture that was introduced in 2017 [14], and established by BERT (Bidirectional Encoder Representations from Transformers) [15] has been the current trend, replacing older recurrent neural network (RNN) models such as the long short-term memory (LSTM).

3 Automated Feedback Generation

In this chapter, we will describe various recently published AES systems that focus on providing automated feedback, different models and approaches, as well as the issues that currently hinter their evolution. Subsequently, we will examine a number of case studies of such systems, and we will identify their common characteristics and taxonomize them under a common classification schema.

As mentioned before, AES's popularity has risen over the last decade, mostly due to the need to scale scoring tasks on written essays, but in the context of this increased research interest, researchers also explored the possibilities of, rather than just scoring an essay, providing formative feedback to the student that can improve their writing quality. Systems that combine AES together with automated feedback can be found in literature as automated essay evaluation systems or AEE [16].

Balfour [17] suggests that MOOCs can use AES for providing students with automated feedback for specific dimensions of the essay in the form of several rounds of drafts that get feedback, while using human-graded (calibrated peer review) evaluation for the final results. In that way, issues with the essay, specifically ones that refer to the mechanics of it, can be corrected earlier on in the writing process.

There is research like [18, 19] that indicates that, when compared with human feedback, automated essay feedback can have mixed results, especially regarding the usefulness of the feedback, whether or not it is incomprehensible or too general to be utilized, or whether it is accurate in identifying the parts of the essay that have errors or need correcting.

Recent efforts in the field, have tried to tackle such issues, some with promising results. For example, Zupanc and Bosnić [20] developed a novel method of providing specific, sentence-level feedback. Using semantic analysis, they extract coherence attributes of the essay by transforming sequential parts of it into the semantic space. The changes between them are then used to estimate the dimension of coherence for the essay. These attributes are used to detect and report semantic errors in the form of feedback, for example a student writing in an essay about a girl named Lisa and then writing in another sentence that Lisa is a boy, will result in an error because the classes "boy" and "girl" are subclasses of "female" and "male" classes, respectively, thus a relation in the ontology contradicts the extraction from the written sentence.

Other research efforts focus on providing feedback for a specific dimension of an essay, trying to do so even for dimensions of an essay that may require an understanding of the essay's content. Carlile et al. [21] attempt to provide feedback for specific attributes of the document that refer to the essay's argument persuasiveness. Another attempt to provide dimension-specific feedback [22] revolves around the essay's thesis strength. It tries to identify the attributes that could impact thesis strength and annotate the essays accordingly, which could provide feedback to the student about their score. Fiacco et al. [23] developed a model for evaluating the rhetorical structure in academic writing, and they approached the issue as a structured prediction task, using different algorithms that make use of local and global cues of the essay. In their work, they present a hierarchical neural architecture that performs this task, with very promising results that are close to those of a human evaluator.

The common difficulty that hinters such approaches, besides the inherent fact that modeling such aspects of a document is in itself largely still a very challenging task even for the most advanced NLP systems, is the limited provision of appropriately annotated datasets leading some researchers to provide their own annotated corpora [22].

As we discussed in the previous chapter, novel approaches to AES involve deep learning models and transformer-based models. These approaches and their usages in feedback generation have also been researched as of late, for example [24] explores how deep learning in AES, which is usually considered a black box and provides a holistic score of an essay with largely unexplained internal workings, can be used to provide personalized feedback to a student. Altoe and Joyner [25] proposed an artificial intelligence-based approach, aiming to mitigate two problems in machine learning-based approaches to auto-graders: the requirement of a large number of training essays to generate high-quality feedback and the need for annotation by a human grader. The approach involved the utilization of 10 pre-graded exemplary examination essays to be used as input for a custom algorithm for automatic concept map generation to create a rubric. The algorithm was custom made because even the most widely utilized Automatic Concept Map Generation (ACMG) from text tools (e.g. Stanford Open Information Extraction Tool) did not show satisfactory results for multi-concept sentences composing short texts. Generation feedback was then presented to students in the form of two concept maps; one showing the concepts that were included in both the rubric and the student's essay and one showing all missing concepts from the student's essay.

In the following chapter we will examine some recent case studies of AEE systems that were tested in actual student samples. Some of them use common AEE software, like PEG Writing from Measurement Incorporated and Revision Assistant from Turnitin, while others test their own novel solutions.

The systems classification will refer to five basic characteristics that were identified as most appropriate to distinguish them. These characteristics are the target audience, the system approach, the scoring method, the feedback generation method, and finally their evaluation. The evaluation metrics of such case studies might differ greatly from study to study, and thus it is difficult to provide a unified evaluation schema, however the most prominent metrics in the scope of evaluating such studies have to do with the student improvement, which is the student's growth and improvement in writing after revisions of their essay drafts, taking the automated feedback into account, as well as student acceptance, which is the evaluation of the engagement of the students with the system and how well they receive it. Thus, in Table 1, these two indicators will be used for the systems, using the abbreviations S.A. for student acceptance and S.I. for student improvement. The results are labeled POS for positive results, NEUT for neutral, or NEG for negative results, according to the authors conclusions. Finally, when applicable, the teachers' perception and acceptance of such a system is also documented using the abbreviation T.A., for teacher acceptance.

Case Studies

Wilson et al. [26] studied the effects of automated feedback in scaffolding growth in students' writing quality. They studied the improvement of students' essay quality across multiple revisions of their drafts, using the feedback they got, and whether or not this improvement is transferable to improved independent performance on a new essay (first draft). The students' sample was from primary and secondary education level. For evaluating the improvement of essay quality, a sample of 955 students that completed at least two revisions on their original draft, was tracked, whereas for researching transfer skills, 739 students, that completed at least one draft to a new prompt, were retained. The findings were that, whereas the students showed some consistent improvement in their overall quality score after multiple drafts, the improvements were not transferable in new subsequent essays.

Woods et al. [27] used Revision Assistant, an educational software product, that provides immediate, rubric-specific, sentence-level feedback to students to supplement teacher guidance. In their implementation they suggested the use of ordinal logistic regression (OLR), so as to avoid the issues of linear regression approaches, which they thought suffered from violations of modeling assumptions, or classification approaches, that lose ordering information. Their scoring performance, using the ASAP dataset, is on par with similar state-of-the-art solutions. Finally, their approach, when deployed and tested by approximately 79,000 students through the web interface of the software, showed improvements of an average of 2.6 points after 7 drafts (which was the average number of drafts). Finally, regarding the students' acceptance, out of 8% of the students that did evaluate the feedback comments, 88% evaluated positive comments as helpful and 72% evaluated negative comments as helpful.

Liu et al. [28] used a machine learning approach to develop an AES system that provides automated feedback. They tested their system using a student sample of 110 students, 56 of which used the system for feedback and 54 used direct teacher-generated feedback. The results showed that for specific dimensions of the essay like structure, organization, supporting ideas, and coherence, the automated feedback can prove useful, approximating the direct comments of the teachers.

A very interesting case study was performed by Roscoe et al. [29], where the correlation between the perception of students about an automated feedback generation system - claims about the accuracy and quality of the automated scoring and feedback – and the acceptance of the students for the system. The study resulted that negative perception of the system had a negative impact on students' reusing the system in the future and students' experiences with the software were strongly associated with their per- caption for it, even though they revised and improved their essays regardless.

Wilson [30] tested the PEG Writing system in a student sample of 1196 students, where 598 of them were students with disabilities and the rest 598 were typically-developing students. Results document a positive association between the use of PEG Writing and growth in writing quality for students with disabilities. More specifically, whereas students with disabilities tended to perform on average lower than typically-developing students on their initial drafts, their overall writing quality improved at a statistically significantly faster rate, and they were able to close the performance gap after five revision drafts.

Palermo et al. [31] evaluated the effects of automated feedback on the essay quality of secondary education students, using the system NC Write, which was also part of the family of products from Measurement Incorporated that includes PEG Writing, for three different conditions: an automated feedback and teacher combination condition (NC + TRAD), an automated feedback together with Self-Regulated Strategy Development [32] instruction condition (NC + SRSD), or a traditional teacher instruction condition for comparison (COMP). The results suggested that the latter condition was the one where students composed longer essays with better overall quality, with more key elements of argumentation included than the other two conditions. Also, regarding the students acceptance of the systems, their opinions were mostly favorable.

Zhang et al. [33] created eRevise; a system that uses natural language processing to provide formative feedback to the student on how to better utilize evidence in Response to Text Assessments (RTA). The system was applied in two public elementary schools in Louisiana. 143 students from 7 classrooms, from 5th and 6th-grade, wrote essays, received feedback and revised them. They explored the impact of the system in the improvement of the students' writing quality of their drafts in regards to human graders, as well as the improvement in the usage of the source text, using NLP features. Experimental results showed that eRevise helped students improve their text evidence usage after receiving formative feedback and engaging in essay revision.

O'Neill and Russell [34] presented the results of a study where students' responses on Grammarly usage and traditional non-Grammarly usage were compared. Grammarly is an automated proofreading system which can identify errors related to 250 grammar rules. Most frequent errors are dealt in descending order and in general Grammarly encourages students to think about their choices and decide whether to accept or reject a change to their draft. In the study, 96 university students were involved, where 54 were given access to Grammarly and 42 received traditional, not-automated feedback. Results showed that the students receiving Grammarly feedback responded more positively to 9 out of 15 survey evaluation areas compared to non-Grammarly feedback students and they were significantly more satisfied with the grammar advice that they received compared with non-Grammarly students. However, there were some low-percentage negative feedback showing that not all Grammarly's errors were valid (especially when complex sentences and vocabulary were chosen or when pre-settings were set to American English). Students also complained that Grammarly suggestions were at times difficult to understand and that there were compatibility issues when uploading documents to the system. Only for 2 students the program's faults outweighed the positives.

Ng et al. [35] created the Automated Essay Scorer with Feedback (AESF) mechanism; an automated essay scoring system to help students improve their writing skills and teachers to mark essays in an easier and more efficient way. The mechanism was based on the Malaysian University English Test essay marking criteria and used a Design-based research (DBR) approach. A series of prototypes were implemented to improve the system because at its preliminary phase it could only grade essays with an accuracy of 74.7% without providing any kind of feedback. However, due to time constraint and limited resources to provide guidance on how to use the prototype successfully, it was difficult to establish specific results. It was also mentioned that it was limited to only two topics of feedback which made it insufficient to be used extensively. All in all, the mechanism at its current state leaves great room for evolution and improvement.

Wilson and Roscoe [36] used PEG Writing and compared it with a word processing condition that used Google Docs on a total of 114 middle school students (56 and 58 respectively). The effectiveness of the tools was evaluated with multiple metrics such as writing self-efficacy, writing quality etc. while they also evaluated three teachers' perceptions on the acceptance of the tools. All participating teachers used both tools and were trained to use PEG since they had no prior experience with it.

Ariyanto et al. [37] used a free version of ProWritingAid (PWA), a web based AWE program that has features such as style, structure, sentence correction etc., and evaluated the perspectives of six ESP (English for Specific Purposes) university student's about it. The students were chosen from a range of writing achievement based on the scoring their teachers gave them according to 5 writing criteria. By analyzing the students' responses on specific topics about the usefulness and effectiveness of PWA, it became clear that although students with different writing achievements utilized PWA differently to improve their drafts, they perceived it positively in general, as it gives them useful feedback and it gives them confidence in their drafts and also, through the usage of the system, teachers can also spend their time in the class in a more productive manner.

Table 1. AES case studies with automated feedback

System	Students sample	Students education	Feedback dimensions	Evaluation
Wilson et al. (2014) – PEG Writing	955 students	Primary & Secondary education	Ideas and development, organization, support, sentence structure, word choice, mechanics	S.I.: POS
Woods et al. (2017) – Revision Assistant	79,000 students	Secondary education	Support, language, claim, organization	S.A.: POS S.I.: POS
Ming Liu et al. (2017)	56 students with automated feedback; 54 students with human feedback	University education - second language learners	Sentence diversity, structure, organization, supporting ideas, coherence, conclusion	S.A.: POS S.I.: NEUT
Wilson (2017) – PEG Writing	1196 students (598 students with disabilities, 598 typically developing students)	Primary & Secondary education	Ideas and development, organization, support, sentence structure, word choice, mechanics	S.A.: POS S.I.: POS
Palermo et al. (2018) – NC Write (PEG)	829 students (COMP: 270 students, NC + TRAD: 272 students, NC + SRSD: 287)	Secondary education	Ideas and development, organization, style, language, sentence structure, conventions	S.A.: POS S.I.: POS
Zhang et al. (2019) – eRevise	143 students	Upper primary education	Elaboration of evidence, explanation of evidence, connection with main idea	S.I.: POS
O'Neil and Russell (2019) – Grammarly	96 students (54 with Grammarly feedback and 42 non-automated feedback)	University education	Contextual spelling, grammar, punctuation, sentence structure, style, vocabulary enhancement	S.A.: POS S.I.: POS
Wilson (2020) – PEG Writing & Google Docs	114 students (PEG: 56 students, Google Docs: 58 students)	Secondary education	Ideas and development, organization, support, sentence structure, word choice, mechanics	S.A.: POS S.I.: POS T.A.: POS
Ariyanto et al. (2021) – ProWritingAid (PWA)	6 students (3 writing assessments each)	University education	Grammar, vocabulary choice, sticky sentence, punctuation, capitalization, spelling	S.A.: POS S.I.: POS

4 Summary

In summary, we examined the field of AES with automated feedback generation. The field of AES has, since its inception more than 50 years ago, made a lot of progress, following the developments in the more general research field of NLP. Be that as it may, as we mentioned in previous chapters, the core fundamental approaches of AES, like hand-crafted features used in AES systems, are still valid today, and should not be ignored in future work. Furthermore, these dimension specific scoring and hand-crafted features are very important in automated feedback generation, since by using them, feedback can be more specific and understandable by humans.

We examined the more educational applications of AES, that aim to provide meaningful feedback to the students, in order for them to improve their writing ability. Developing and utilizing such systems, has become increasingly popular and important, especially as technological developments, remote working and latest changes in our living conditions like the COVID pandemic call for remote learning platforms like MOOCs that can support massive amounts of students. In that context, the idea that scalable feedback-providing automated solutions can be utilized to help improve such platforms is of great interest, and aspect-specific AES systems can provide a more explainable and transparent way of evaluating written text and providing feedback to the author.

In recent years, such systems have continued to develop rapidly, trying to incorporate novel techniques and methodologies for improved results, like more specific feedback, both in the aspect of *which* parts of an essay need corrections, as well as *how* to correct or improve them. Also, there are continuous attempts to provide targeted feedback to aspects of an essay that were previously largely unexplored, due to their complexity and need of understanding the content of the essay. These efforts are still limited and have inherent issues that hinder their development, like lack of appropriately annotated datasets, but still recent results are hopeful for future developments.

We finally examined case studies of such systems actually being used in classroom scenarios, with mostly positive results. While there are still a lot of issues to be further researched, mostly how to improve students' and teachers acceptance of such systems, the context on how these systems are used from the students, the incentives involved in using them, and whether improvements correlated with the usage of such systems are actually transferable skills in other essays or are specific to the essay they are applied to and its revisions, still their usage has started to consolidate as a viable solution for immediate feedback and a useful tool for the self-improvement of a student in the learning of writing.

References

1. Larrondo, P., Frank, B., Ortiz, J.: The state of the art in providing automated feedback to open-ended student work. In: Proceedings of the Canadian Engineering Education Association (CEEA) (2021)
2. Isaacs, T., Zara, C., Herbert, G., Coombs, S.J., Smith, C.: Key Concepts in Educational Assessment. SAGE, Thousand Oaks (2013)

3. Page, E.B.: The imminence of... grading essays by computer. The Phi Delta Kappan **47**, 238–243 (1966)
4. Burstein, J., Chodorow, M., Leacock, C.: Automated essay evaluation: the criterion online writing service. AI Mag. **25**(3), 27 (2004)
5. Somasundaran, S., Burstein, J., Chodorow, M.: Lexical chaining for measuring discourse coherence quality in test-taker essays. In: Proceedings of COLING 2014, The 25th International Conference on Computational Linguistics: Technical Papers, pp. 950–961. Dublin City University and Association for Computational Linguistics, Dublin, Ireland (2014)
6. Persing, I., Ng, V.: Modeling prompt adherence in student essays. In: Proceedings of the 52nd Annual Meeting of the Association for Computational Linguistics (Volume 1: Long Papers), pp. 1534–1543. Association for Computational Linguistics, Baltimore (2014)
7. Persing, I., Davis, A., Ng, V.: Modeling organization in student essays. In: Proceedings of the 2010 Conference on Empirical Methods in Natural Language Processing, pp. 229–239. Association for Computational Linguistics, Cambridge (2010)
8. Persing, I., Ng, V.: Modeling thesis clarity in student essays. In: Proceedings of the 51st Annual Meeting of the Association for Computational Linguistics (Volume 1: Long Papers), pp. 260–269. Association for Computational Linguistics, Sofia (2013)
9. Ke, Z., Carlile, W., Gurrapadi, N., Ng, V.: Learning to give feedback: modeling attributes affecting argument persuasiveness in student essays. In: IJCAI International Joint Conference on Artificial Intelligence, pp. 4130–4136 (2018)
10. Chen, H., He, B.: Automated essay scoring by maximizing human-machine agreement. In: Proceedings of the 2013 Conference on Empirical Methods in Natural Language Processing, pp. 1741–1752. Association for Computational Linguistics, Seattle, Washington (2013)
11. Alikaniotis, D., Yannakoudakis, H., Rei, M.: Automatic text scoring using neural networks. In: 54th Annual Meeting of the Association for Computational Linguistics, ACL 2016 - Long Papers, vol. 2, pp. 715–725 (2016)
12. Farag, Y., Yannakoudakis, H., Briscoe, T.: Neural automated essay scoring and coherence modeling for adversarially crafted input. In: NAACL HLT 2018 - 2018 Conference of the North American Chapter of the Association for Computational Linguistics: Human Language Technologies - Proceedings of the Conference, vol. 1, pp. 263–271 (2018)
13. Dong, F., Zhang, Y., Yang, J.: Attention-based recurrent convolutional neural network for automatic essay scoring. In: CoNLL 2017 - 21st Conference on Computational Natural Language Learning, Proceedings, pp. 153–162 (2017)
14. Vaswani, A., et al.: Attention is all you need. In: Advances in Neural Information Processing Systems, pp. 5999–6009 (2017)
15. Devlin, J., Chang, M.-W., Lee, K., Toutanova, K.: BERT: pre-training of deep bidirectional transformers for language understanding. arXiv:181004805 (2019)
16. Shermis, M. D., Burstein, J.: Handbook of Automated Essay Evaluation: Current Applications and New Directions. Routledge/Taylor & Francis Group, New York (2013)
17. Balfour, S.: Assessing writing in MOOCs: automated essay scoring and calibrated peer ReviewTM. J. Res. Pract. Assess. **8**, 40–48 (2013)
18. Dikli, S.: The nature of automated essay scoring feedback. CALICO J. **28**, 99–134 (2010)
19. Dikli, S., Bleyle, S.: Automated essay scoring feedback for second language writers: how does it compare to instructor feedback? Assess. Writ. **22**, 1–17 (2014)
20. Zupanc, K., Bosnić, Z.: Automated essay evaluation with semantic analysis. Knowl.-Based Syst. **120**, 118–132 (2017)

21. Carlile, W., Gurrapadi, N., Ke, Z., Ng, V.: Give me more feedback: annotating argument persuasiveness and related attributes in student essays. In: ACL 2018 - 56th Annual Meeting of the Association for Computational Linguistics, Proceedings of the Conference (Long Papers), vol. 1 pp. 621–631 (2018)
22. Ke, Z., Inamdar, H., Lin, H., Ng, V.: Give me more feedback II: annotating thesis strength and related attributes in student essays. In: ACL 2019 - 57th Annual Meeting of the Association for Computational Linguistics, Proceedings of the Conference, pp. 3994–4004 (2020)
23. Fiacco, J., Cotos, E., Rosé, C.: Towards enabling feedback on rhetorical structure with neural sequence models. In: ACM International Conference Proceeding Series, pp. 310–319 (2019)
24. Kumar, V., Boulanger, D.: Explainable automated essay scoring: deep learning really has pedagogical value. Front. Educ. **5**, 186 (2020)
25. Altoe, F., Joyner, D.: Annotation-free automatic examination essay feedback generation. In: Proceedings of 2019 IEEE Learning With MOOCS, LWMOOCS 2019, pp. 110–115 (2019)
26. Wilson, J., Olinghouse, N.G., Andrada, G.N.: Does automated feedback improve writing quality? Learn. Disabil. Contemp. J. **12**, 93–118 (2014)
27. Woods, B., Adamson, D., Miel, S., Mayfield, E.: Formative essay feedback using predictive scoring models. In: Proceedings of the ACM SIGKDD International Conference on Knowledge Discovery and Data Mining Part F, vol. 1296, pp. 2071–2080 (2017)
28. Liu, M., Li, Y., Xu, W., Liu, L.: Automated essay feedback generation and its impact on revision. IEEE Trans. Learn. Technol. **10**(4), 502–513 (2017)
29. Roscoe, R., Wilson, J., Johnson, A., Mayra, C.: Presentation, expectations, and experience: sources of student perceptions of automated writing evaluation. Comput. Hum. Behav. **70**, 207–221 (2017)
30. Wilson, J.: Associated effects of automated essay evaluation software on growth in writing quality for students with and without disabilities. Read. Writ. **30**(4), 691–718 (2016). https://doi.org/10.1007/s11145-016-9695-z
31. Palermo, C., Thomson, M.M.: Teacher implementation of self-regulated strategy development with an automated writing evaluation system: Effects on the argumentative writing performance of middle school students. Contemp. Educ. Psychol. **54**, 255–270 (2018)
32. Graham, S., Perin, D.: Writing Next: Effective Strategies to Improve Writing of Adolescents in Middle and High Schools. A Report to Carnegie Corporation of New York. Alliance for Excellent Education (2007)
33. Zhang, H., et al.: eRevise: using natural language processing to provide formative feedback on text evidence usage in student writing. AAAI **33**, 9619–9625 (2019)
34. O'Neill, R., Russell, A.M.T.: Stop! Grammar time: university students' perceptions of the automated feedback program grammarly. Australas. J. Educ. Technol. **35**, 42–56 (2019)
35. Ng, S.Y., Bong, C.H., Hong, K.S., Lee, N.K.: Developing an automated essay scorer with feedback (AESF) for Malaysian University English Test (MUET): a design-based research approach. Pertanika J. Soc. Sci. Humanit. **27**, 1451–1468 (2019)
36. Wilson, J., Roscoe, R.D.: Automated writing evaluation and feedback: multiple metrics of efficacy. J. Educ. Comput. Res. **58**, 87–125 (2020)
37. Ariyanto, M.S.A., Mukminatien, N., Tresnadewi, S.: College students' perceptions of an automated writing evaluation as a supplementary feedback tool in a writing class. Jurnal Ilmu Pendidikan **27**, 41–51 (2021)

Multidisciplinary Problem-Based Learning (MPBL) Approach in Undergraduate Programs

Amin Reza Rajabzadeh(✉), Moein Mehrtash, and Seshasai Srinivasan

W Booth School of Engineering Practice and Technology, McMaster University, Hamilton, Canada
{rajaba,mehrtam,ssriniv}@mcmaster.ca

Abstract. In the 21st century, education is about training graduates with a variety of competencies and reducing the gap between the classroom and the real-world environment via professional practice and simulating a work environment in the curriculum. In this paper, the authors explore the pedagogical benefits of implementing multidisciplinary open-ended research-based projects in an undergraduate curriculum to improve students' understanding of the concepts and develop meta-skills. Specifically, this paper discusses how such projects could facilitate experiential learning by providing students with an opportunity to actively participate in real-world research projects that are multidisciplinary in nature. The proposed project is a typical real-world problem that draws on competencies from various disciplines. The authors' goal is to develop deep content knowledge, foster critical thinking, engage in collaboration, and promote creativity and communication skills. Such meta-skills are a crucial component to succeed in today's workplace. The other significant objective of this new teaching and learning methodology is to support collaborative and concurrent competency by involving students with various backgrounds working on multidisciplinary projects.

Keywords: Experiential learning · Engineering education · Problem-based learning · Multidisciplinary projects

1 Introduction

1.1 Experiential Learning

The experiential and collaborative learning paradigm has been known as a global best practice in engineering education across different disciplines as it facilitates active learning and provides an opportunity for students to engage in discussion and reflect on their experience and performance within the team [1]. According to Kolb, "*Experiential learning is a powerful and proven approach to teaching and learning that is based on one incontrovertible reality: people learn best through experience*" [2]. However, the experience could not automatically lead to learning [3] or in another words "*the richness of Dewey's concept of experience is lost if it is reduced to simply learning by*

M. E. Auer and T. Tsiatsos (Eds.): IMCL 2021, LNNS 411, pp. 454–463, 2022.
https://doi.org/10.1007/978-3-030-96296-8_41

doing" [4]. Therefore, a structured reflection and practices of former understanding are required to promote the continuity of experiences which could eventually lead to learning [5–8]. Kolb's experiential learning theory, which is concerned with the learner's internal cognitive processes, suggests that learners transform the experience to knowledge by completing a four-stage learning cycle of concrete experience, reflective observation, abstract conceptualization, and active experimentation [9–11]. Although experience is central to Kolb's theory, the learning model requires the experience to be supported by reflection and analysis, and ultimately something to be generated from experience, as shown in Fig. 1.

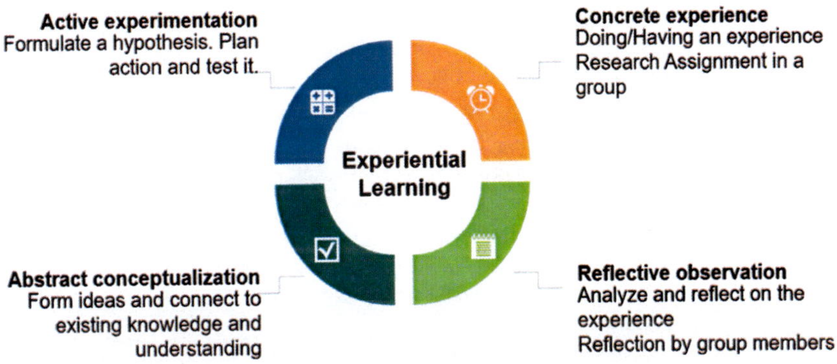

Fig. 1. The four-stage learning cycle of Kolb's experiential learning theory.

Experiential learning has long been employed as part of engineering education in various forms, such as applied research projects [12–16], capstone projects [17, 18], interactive simulation, and explicit use of technology [19–22], case studies [23–25], labs [26, 27], and co-op and internships, which are also referred to as *work-integrated learning* according to the experiential learning guideline issued by the Ministry of Training, Colleges and Universities in September 2017 [28–30].

1.2 Collaborative and Multidisciplinary-Based Learning

Engineering education scholars have identified that collaborative education in various forms of problem-based learning [31–34], project-based learning [35], or research-based learning could promote critical thinking, especially in a small team environment [1, 36–39], and produces higher achievement and greater productivity [40]. In collaborative education, it is aimed that learners often work in groups to develop a solution for authentic or ill-structured problems. It is important to note that problem-based and project-based learning are two categories of experiential learning with slightly different definitions. In problem-based learning, the problems lack a well-defined answer, and learners typically work in groups and apply critical thinking to examine and solve the problems, fostering learners' metacognitive skills, while there is no one correct answer. In project-based learning, the goals are typically set, and through structured instructions, students learn by investigating complex and often authentic problems. To better differentiate various types of learning, which are used in this paper, Table 1 has been prepared.

Table 1. Definition of various learning pedagogical approaches.

Term	Definition	Reference
Experiential learning	*"the process whereby knowledge is created through the transformation of experience. Knowledge results from the combination of grasping and transforming experience."*	[9]
Experiential education	*"a teaching philosophy that informs many methodologies in which educators purposely engage with learners in direct experience and focused reflection in order to increase knowledge, develop skills, clarify values, and develop people's capacity to contribute to their communities."*	[41]
Problem-based learning	*"The learning that results from the process of working toward the understanding or resolution of a problem."*	[42]
Project-based learning	*"a systematic teaching method that engages students in learning essential knowledge and life-enhancing skills through an extended, student-influenced inquiry process structured around complex, authentic questions and carefully designed products and tasks."* *"an instructional technique in which meaningful tasks, often in the form of problems, serve as the context and stimulus for knowledge-building and critical thinking."*	[43, 44]
Cased-based learning	*"a set of learning and teaching models that uses real or realistic events holding multifaceted issues and complexity as part of learning resources, which engages students in individual and/or group inquiry on the given events with other relevant information, and which promotes students' reflections on their own learning and problem solving"*	[45]
Collaborative learning	*"giving students an opportunity to engage in discussion, take responsibility for their own learning, and thus become critical thinkers"*	[46]

Brame (2019) has also identified collaborative learning, when it goes well, as one of the most effective teaching approaches in the classroom regardless of whether the instructors seek to enhance deep learning or meta-skills [47]. Researchers have also recognized that collaborative learning could go poorly due to the lack of student contribution, engagement, and motivation and several other factors often leading to arguments and failing to achieve the desired outcome [47]. As a result, several researchers have argued for the importance of explicitly teaching students how to work collaboratively at the undergraduate level and provide students with an opportunity to practice teamwork [48, 49].

Several industrial surveys and reports have argued for the need for graduates with more experience in a multidisciplinary team environment. At the same time, engineering education literature has recognized the need for engineering curriculums with multidisciplinary capabilities to address industry requirements and increase graduates' employability. In the end, "*it is sometimes forgotten that industry is an important customer of engineering education*" [50]. Brassler and Dettmers have also indicated that problem- and project-based learning are two of the most suitable pedagogical approaches that could enhance learners' interdisciplinary competence [37]. Multidisciplinary research projects allow learners to understand better the relationships between various disciplines, which encourages a higher level of thinking and innovation, leading to solving complex real-life problems [37].

This paper adds to experiential learning literature and provides an overview of problem-based learning in second-year and third-year Automation Engineering Technology and Biotechnology programs. The overall goal is to transform a traditional deterministic lab-based setting into an open-ended active learning environment. This paper builds on a pilot project started in Fall 2019 involving second-year students designing their biosensing platform to detect an analyte/biomarker of their choice [15, 16].

2 Framework

Students from two programs (*Biotechnology* program, thorough the 3rd year Bioprocess Control and Dynamics course, and *Automotive and Vehicle Engineering Technology* program, through the 2nd-year Fluid Mechanics course) collaborate to design robotic endoscopic capsules with closed-loop drug delivery systems, including the design and fabrication of capsule robots and the development of the closed-loop system (sensors, actuators, and microcontrollers) for drug delivery systems. The students from the Automotive stream focused on the design and dynamic modeling of capsule robots, including practicing fluid mechanics principles such as external fluid analysis, computational fluid dynamics (CFD) study, and fluid similitude for experimental tests. The biotechnology stream students collaborate closely with automotive stream students and emphasize designing and modeling a closed-loop drug delivery system integrated with capsule robots. This includes practicing sensors and actuators selection and dynamic modeling, closed-loop response analysis, and control logic analysis.

Students' activities comprise both self-study and in-class activities. Due to the open-end nature of the project, students are motivated to complete self-study activities. These activities are aligned with weekly in-class laboratory session activities. Students are required to perform a minimum of 3-h lab work every other week. In the first few weeks, the students undertake a literature review to design their protocol. Students present their progress every four weeks in addition to a detailed midterm and final report. Students also reflect on their learning experience in the final report. The students' activities were designed based on the four-stage learning cycle of Kolb's experiential learning theory. The implementation details of the MPBL in Automotive and Biotechnology stream courses are outlined in the ensuing paragraphs.

Implementation of Kolb's Experiential Learning

In this subsection, we present the details on the planning of laboratory sessions employing the four stages of Kolb's Experiential Learning Theory in the Fluid Mechanics course. From the course design perspective with the MPBL approach, the learning activities need to be aligned with a course learning outcome. The learning outcomes (LO) for the fluid mechanic course are:

- LO1: Relate real-world components in fluid systems with theoretical and numerical models
- LO2: Evaluate and analyze fluid systems performance
- LO3: Optimize fluid system performance with numerical and experimental methodologies
- LO4: Design and build fluid systems to achieve the required system performance

The first activity is Kolb's concrete experience. The fluid mechanic course has six classic laboratory experiments that include fluid statics, fluid dynamics, internal and external flow, and turbomachines. These six laboratory experiments, two hours laboratory sessions for each experiment, provide concrete experiences on the application of the theoretical concepts of fluid mechanics. As an example, Fig. 2 demonstrates the experimental measurement of fluid impact force on a flat surface. The second activity is Kolb's reflective observation; students in groups of two discuss their observations from the experiment and compare the experimental measurement with the theoretical calculations using the principles of fluid mechanics' (Fig. 2B shows one such sample analysis done by students).

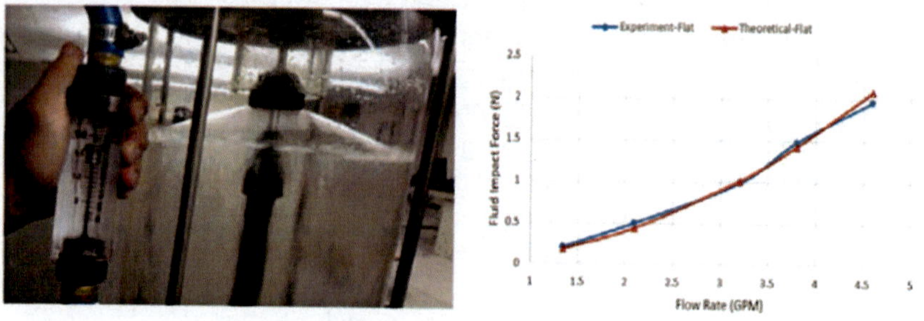

(A) (B)

Fig. 2. A) Fluid impact force measurement experiment. B) Comparison of experimental data with theoretical calculations, as a reflective observation.

The third activity is Kolb's abstract conceptualization; students are involved in thinking and using fluid mechanics principles in a real-world and multidisciplinary project. The final project is defined as the design of robotic endoscopic capsules with controlled drug delivery. The Automotive stream students need to determine the specifications of a robotic endoscopic capsule, including the dimension, speed, propulsion system, and the drug dispensing mechanism. Automotive students use the principle of fluid mechanics to approximate the required force for the motion of the endoscopic capsule robot (Fig. 3).

The Biotechnology students are responsible for researching a closed-loop drug delivery system that could be integrated with the capsule robot. This requires the students from both programs (biotechnology and automation) to work in close collaboration to understand the design limitations. While automotive students need to perform an extensive literature review to determine the required performance for a robotic capsule robot, the biotechnology students need to focus on the disease or the disorder of interests that could be regulated using a closed-loop drug delivery system. For this, students identify the associated biomarker, select a biosensor to detect the biomarker, select a drug, determine the controlled and manipulated variables and potential disturbances, and eventually propose a mechanism of the closed-loop delivery system (self-regulated administration), including the control diagram.

Some of the projects that biotechnology students selected for this open-end research project included a *milrinone drug delivery system* to treat congestive heart failure, *regulation of heparin and thrombin* to control blood coagulation, and a *closed-loop drug delivery system regulating the cortisol levels* associated with depressive disorder.

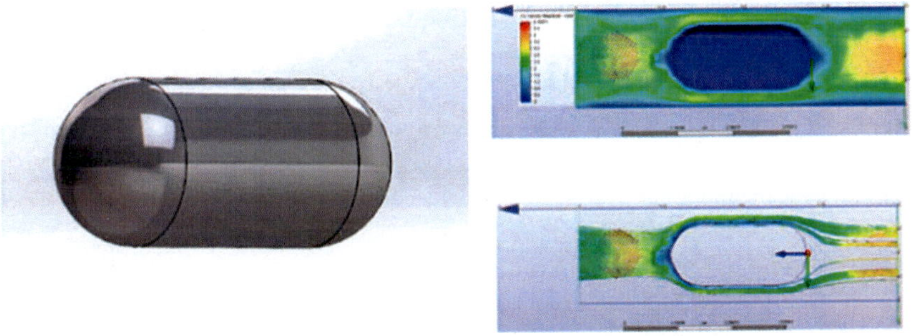

(A) (B)

Fig. 3. A) endoscopic capsule robot designed by one group of students, B) Using CFD to estimate the fluid velocity, pressure, and forces acting on capsule robot.

The biotechnology students are responsible for developing a dynamic model (by writing transient mass balance equations based on the drug release, and drug uptake and metabolism knowing the drug's pharmacokinetics) to model and predict the concentration of the drug in the body. It is important to note that the biotechnology students learn how to formulate a steady-state and dynamic process in two different courses of Chemical Engineering Concepts (BIOTECH 2EC3) and Bioprocess Control and Dynamics (BIOTECH 3BC3). Using a first-order dynamic model for the actuator and the sensor, students develop a block diagram based on their proposed closed-loop delivery system and obtain the transfer functions of each block using Laplace Transform. Biotechnology students are then asked to solve the block diagram and analyze the stability of their closed-loop system using different types of controllers. Students are asked not only to perform hand calculations but also to set up a Simulink model in MATLAB, and compare their results. Using the developed and verified Simulink model, students will now be able to 1) shift their focus on analyzing and comprehending the transient response when, for example, there is a change in the setpoint value or if there is any disturbance, and 2) better reflect on their understanding of the theoretical concepts.

The fourth activity is Kolb's active experimentation: In this, students actively deal with a real-world problem. Since testing of the endoscopic capsule in an actual situation is beyond the scope of an undergraduate-level course, in vitro experiments can be performed. The students can use 3D printer technology, print the scaled model, use wind tunnel measurement to validate their theoretical calculations and numerical simulation, build the control loop hardware (mechanical and electrical), and obtain drug release profiles using spectrophotometry methods. Table 2 has been prepared as a summary that demonstrates the mapping of the students' activities and alignment with Kolb's experiential learning cycle and course learning outcomes.

As mentioned earlier, to support student learning based on Kolb's theory, each group (from both programs) is asked to present their progress every four weeks in addition to a detailed midterm and final report. The latter includes a section in which students reflect on their learning experience.

Table 2. Implementation of Kolb's experiential learning in the MPBL.

Learning outcome	Mapping to Kolb's cycle	Activities
LO2, LO3	Concrete experimentation, Reflective observation	Six classic laboratory sessions, result analysis with theoretical concepts
LO1, LO4	Abstract conceptualization	Apply knowledge to a real-world problem, collaborate with biotechnology students, literature review, preliminary design, and analysis
LO3, LO4	Active experimentation	Build and test, model the experimentation, optimize the design

3 Limitations

While this pilot study explores the importance of multidisciplinary problem-based learning (MPBL) in undergraute education, future studies are required to evaluate the success of such a learning strategy. In the future studies, questionnaires will be provided to students, and their feedback will be collected and analyzed to better understand student's perceptions around MPBL. To validate the success of MPBL, learning objective criteria will be defined in the future studies and a comparison with a control group using a different learning method will be made.

4 Conclusions

This paper sheds light on best practices around simulating the real-world environment in the undergraduate curriculum using multidisciplinary open-end problems, based on a four-stage Kolb's experiential learning theory. Specifically, we present the details on a multidisciplinary research-based curriculum across two programs that enables students from two very different backgrounds to collaborate on a research project of real relevance. Using this pedagogical approach, the authors seek to foster long-term retention of content and improve students' interest, engagement, and motivation in learning technical concepts and soft skills.

References

1. Göl, Ö., Nafalski, A.: Collaborative learning in engineering education. Glob. J. Eng. Educ. **11** (2007)
2. Kolb, D.A.: Experiential Learning- Experience as the Source of Learning and Development, 2nd edn (2015)
3. Verdonschot, S.G.M.: Methods to enhance reflective behaviour in innovation processes. J. Eur. Ind. Train. **30** (2006). https://doi.org/10.1108/03090590610715004
4. Smith, T.E., Knapp, C.E., Seaman, J., et al.: Experiential education and learning by experience. In: Sourcebook of Experiential Education, pp. 15–26. Routledge, London (2011)
5. Rodgers, C.: Defining reflection: another look at John Dewey and reflective thinking. Teach. Coll. Rec. **104** (2002). https://doi.org/10.1111/1467-9620.00181
6. Ambrose, S.A.: Undergraduate engineering curriculum: the ultimate design challenge. Bridge Linking Eng. Soc (2013)
7. Turns, J.A., Sattler, B., Yasuhara, K., et al.: Integrating reflection into engineering education. In: ASEE Annual Conference and Exposition, Conference Proceedings (2014). https://doi.org/10.18260/1-2-20668
8. Dewey, J.: Experience and education. Educ. Forum **50** (1986). https://doi.org/10.1080/00131728609335764
9. Kolb, D.A.: Experiential Learning: Experience as The Source of Learning and Development. Prentice Hall, Inc., Hoboken (1984). https://doi.org/10.1016/B978-0-7506-7223-8.50017-4
10. Kayes, A.B., Kayes, D.C., Kolb, D.A.: Experiential learning in teams. Simul. Gaming **36** (2005). https://doi.org/10.1177/1046878105279012
11. Healey, M., Jenkins, A.: Kolb's experiential learning theory and its application in geography in higher education. J. Geogr. **99** (2000). https://doi.org/10.1080/00221340008978967
12. Noguez, J., Neri, L.: Research-based learning: a case study for engineering students. Int. J. Interact. Des. Manuf. (IJIDeM) **13**(4), 1283–1295 (2019). https://doi.org/10.1007/s12008-019-00570-x
13. Almanza-Arjona, Y.C., Vergara-Porras, B., García-Rivera, B.E., et al.: Research-based approach to undergraduate chemical engineering education. In: IEEE Global Engineering Education Conference, EDUCON (2019). https://doi.org/10.1109/EDUCON.2019.8725195
14. Sharma, K., Sharma, Y., Mantri, A., et al.: Inculcating the spirit and passion for research among engineering students at Undergraduate level. In: Procedia Computer Science (2020). https://doi.org/10.1016/j.procs.2020.05.162
15. Bogoslowski, S., Geng, F., Gao, Z., Rajabzadeh, A.R., Srinivasan, S.: Integrated thinking - a cross-disciplinary project-based engineering education. In: Auer, M.E., Centea, D. (eds.) ICBL 2020. AISC, vol. 1314, pp. 260–267. Springer, Cham (2021). https://doi.org/10.1007/978-3-030-67209-6_28
16. Srinivasan, S., Rajabzadeh, A.R., Centea, D.: A project-centric learning strategy in biotechnology. In: Auer, M.E., Hortsch, H., Sethakul, P. (eds.) ICL 2019. AISC, vol. 1134, pp. 830–838. Springer, Cham (2020). https://doi.org/10.1007/978-3-030-40274-7_80

17. Khan, M.M., Ahmed, M.E.: Capstone project based experiential learning – case of institute of business management (IoBM). J. e-Learn. High. Educ. (2018). https://doi.org/10.5171/2018.412705
18. Ritenour, A.P., Ferguson, C.W., Gardner, P., et al.: Collaborative project-based learning capstone for engineering and engineering technology students. In: ASEE Annual Conference and Exposition, Conference Proceedings (2020). https://doi.org/10.18260/1-2-34300
19. Chernikova, O., Heitzmann, N., Stadler, M., et al.: Simulation-based learning in higher education: a meta-analysis. Rev. Educ. Res. **90** (2020). https://doi.org/10.3102/0034654320933544
20. Mehrtash, M., Centea, D.: Experiential learning approaches in automotive engineering: implementing real world experiences. In: Auer, M.E., Tsiatsos, T. (eds.) ICL 2018. AISC, vol. 916, pp. 532–541. Springer, Cham (2020). https://doi.org/10.1007/978-3-030-11932-4_51
21. Mehrtash, M., Yuen, T., Balan, L.: Implementation of experiential learning for vehicle dynamic in automotive engineering: roll-over and fishhook test. In: Procedia Manufacturing (2019). https://doi.org/10.1016/j.promfg.2019.02.284
22. Rajabzadeh, A.R., Long, J., Couper, R.G., et al.: Using engineering design software to motivate student learning for math-based material in biotechnology courses. Int. J. Eng. Educ. **36** (2020)
23. Alani, F.: Development of case-based learning (CBL) in engineering technology education. Int. J. Eng. Educ. **36** (2020)
24. Koehler, A.A., Fiock, H., Janakiraman, S., et al.: Asynchronous online discussions during case-based learning: a problem-solving process. Online Learn. J. **24** (2020). https://doi.org/10.24059/olj.v24i4.2332
25. Vivas, J.F., Allada, V.: Enhancing engineering education using thematic case-based learning. Int. J. Eng. Educ. **22** (2006)
26. Lam, G.: A design-centered laboratory curriculum for first-year biomedical engineering students. In: Proceedings of the Canadian Engineering Education Association (CEEA) (2019). https://doi.org/10.24908/pceea.vi0.13753
27. Shahriar, H., Whitman, M., Lo, D.C.T., et al.: Experiential learning: case study-based portable hands-on regression labware for cyber fraud prediction. In: Proceedings - 2019 IEEE International Conference on Big Data, Big Data 2019 (2019). https://doi.org/10.1109/BigData47090.2019.9005713
28. Rampersad, G., Zivotic-Kukolj, V.: Work-integrated learning in science, technology, engineering and mathematics: drivers of innovation for students. Int. J. Work-Integr. Learn. **19** (2018)
29. Karim, A., Campbell, M., Hasan, M.: A new method of integrating project-based and work-integrated learning in postgraduate engineering study. Curriculum J. **31** (2020). https://doi.org/10.1080/09585176.2019.1659839
30. Paull, M., Lloyd, N., Male, S.A., et al.: Engineering work integrated learning placements: the influence of capitals on students' access. J. High. Educ. Policy Manag. **41** (2019). https://doi.org/10.1080/1360080X.2019.1646382
31. Centea, D., Srinivasan, S.: Enhancing student learning through problem based learning. In: PBL, Social Progress and Sustainability (International Research Symposium on PBL), pp. 376–385. Aalborg Universitetsforlag, Aalborg (2017)
32. Sidhu, G., Srinivasan, S., Centea, D.: Implementation of a problem based learning environment for first year engineering mathematics. In: PBL, Social Progress and Sustainability (International Research Symposium on PBL), pp. 201–208. Aalborg Universitetsforlag, Aalborg (2017)
33. Centea, D., Srinivasan, S.: A comprehensive assessment strategy for a PBL environment. Int. J. Innov. Res. Educ. Sci. **3**, 364–372 (2016)

34. Srinivasan, S., Centea, D.: Problem based learning in finite element analysis. In: Auer, M.E., Centea, D. (eds.) ICBL 2020. AISC, vol. 1314, pp. 240–246. Springer, Cham (2021). https://doi.org/10.1007/978-3-030-67209-6_26
35. Muhammad, N., Srinivasan, S.: A problem solving based approach to learn engineering mathematics. In: Auer, M.E., Hortsch, H., Sethakul, P. (eds.) ICL 2019. AISC, vol. 1134, pp. 839–848. Springer, Cham (2020). https://doi.org/10.1007/978-3-030-40274-7_81
36. Zhu, X., Xiong, Z., Zheng, T., et al.: Case-based learning combined with science, technology, engineering and math (STEM) education concept to improve clinical thinking of undergraduate nursing students: a randomized experiment. Nursing Open **8** (2021). https://doi.org/10.1002/nop2.642
37. Brassler, M., Dettmers, J.: How to enhance interdisciplinary competence—interdisciplinary problem-based learning versus interdisciplinary project-based learning. Interdisc. J. Prob.-Based Learn. **11** (2017). https://doi.org/10.7771/1541-5015.1686
38. DebBurman, S.K.: Learning how scientists work: experiential research projects to promotecell biology learning and scientific process skills. Cell Biol. Educ. **1** (2002). https://doi.org/10.1187/cbe.02-07-0024
39. Balan, L., Yuen, T., Mehrtash, M.: Problem-based learning strategy for CAD software using free-choice and open-ended group projects. In: Procedia Manufacturing (2019). https://doi.org/10.1016/j.promfg.2019.02.223
40. Laal, M., Ghodsi, S.M.: Benefits of collaborative learning. In: Procedia - Social and Behavioral Sciences (2012). https://doi.org/10.1016/j.sbspro.2011.12.091
41. AEE: What is Experiential Education? Association for Experiential Education
42. Barrows, H.S.: A taxonomy of problem-based learning methods. Med. Educ. **20** (1986). https://doi.org/10.1111/j.1365-2923.1986.tb01386.x
43. Buck Institute for Education: Project based learning. http://www.bie.org
44. Howard, J.: Technology-enhanced project-based learning in teacher education: addressing the goals of transfer the problem of transfer. J. Technol. Teach. Educ. **10** (2002)
45. Luckin, R., Puntambekar, S., Goodyear, P., et al.: Handbook of design in educational technology (2013). https://doi.org/10.4324/9780203075227
46. Gokhale, A.A.: Collaborative learning enhances critical thinking. J. Technol. Educ. **7** (1995). https://doi.org/10.21061/jte.v7i1.a.2
47. Brame, C.J., Brame, C.J.: Chapter 5 - Group work: using cooperative learning groups effectively. In: Science Teaching Essentials (2015)
48. Crebert, G., Bates, M., Bell, B., et al.: Developing generic skills at university, during work placement and in employment: graduates' perceptions. High. Educ. Res. Dev. **23** (2004). https://doi.org/10.1080/0729436042000206636
49. Long, J., Rajabzadeh, A.R., MacKenzie, A.: Teaching teamwork to engineering technology students: the importance of self-reflection and acknowledging diversity in teams. In: Proceedings of the Canadian Engineering Education Association (CEEA) (2017)
50. Todd, R.H., Sorensen, C.D., Magleby, S.P.: Designing a senior capstone course to satisfy industrial customers. J. Eng. Educ. **82** (1993). https://doi.org/10.1002/j.2168-9830.1993.tb00082.x

Automatic Code-Switched Lecture Annotation

Amjad Mohamed[1], Nada Nasser[1], and Nada Sharaf[2](✉)

[1] The German University in Cairo, New Cairo, Egypt
amjad.ahmed@student.guc.edu.eg, nada.nasser@guc.edu.eg
[2] The German International University, New Cairo, Egypt
nada.hamed@giu-uni.de

Abstract. Speech Recognition is one of the advanced technologies that has various usages in our everyday lives. By taking advantage of this technology, there is a chance to help make learning for deaf and hard-of-hearing students easier. This paper presents a lecture annotation tool that can facilitate the learning process to these students by proving post-lecture annotated slides. The system does this by utilizing the lecture slides, audio recording of the lecture and a Speech Recognition API. The system goes through two phases, the transcription phase and the slide matching phase. After recording the speech of the lecturer, the audio file is taken as an input to the transcription module. Afterwards, the transcription module transcribes it to text. The main challenge was that lecturers use code-mixed language (English/Arabic). Thus, the transcription module transcribes bilingual audio simultaneously by chunking up the audio file. Afterwards, the text and the lecture slides are given to the slide-matching module. The slide-matching module outputs a PDF file with the text aligned on the PDF slides accordingly. This PDF lecture file can later be viewed by students to help them study and understand the lecture.

1 Introduction

Modern Technology has made a serious impact on our day-to-day lives and society has dramatically changed with the introduction and evolution of technology. With the increasing of accessibility to audio-visual recording equipment and the simplicity of recording, it has become a common practice for organisations to record presentations such as lectures or seminars and upload them online for the students to look at and study later in their private time [1]. These lectures can also be utilised by the learning students who are physically remote to their institution. A portion of these students are the deaf and hard-of-hearing students that are around us in campus. By taking advantage of advanced technology like Automatic Speech Recognition (ASR) and Natural Language Processing (NLP), we can transcribe lecture audio recording and contextually match the transcript to the lecture slides in order to aid deaf and hard-of-hearing students in their educational journey.

M. E. Auer and T. Tsiatsos (Eds.): IMCL 2021, LNNS 411, pp. 464–477, 2022.
https://doi.org/10.1007/978-3-030-96296-8_42

Previous work has been done to synchronise text with lecture slides aimed to help hearing-impaired students. AutoNotes: Automatic Lecture Annotation Application [2] focuses mainly on synchronising the speech of the instructors as text with the lecture slides in real-time. One of the main problems that arose was code-switching during lectures that failed the systems' accuracy and provided weak results. Our paper focuses mainly on the code-switching challenge and can be considered a continuation of AutoNotes. It was a major problem in AutoNotes and our paper explores this problem and introduces new ways to solve it. In AutoNotes, the synchronisation of text was done in real-time while our work uses a post-lecture annotation approach.

2 Background

A student who is deaf or hard-of-hearing has a hearing loss aided or unaided, that impacts the processing of linguistic information which adversely affects their performance. Hearing impairment is a vast term that refers to individuals with hearing losses of varying degrees to total deafness.

The extent to which hearing loss occurs in deaf and hard-of-hearing students ranges from mild to moderate to profound. These labels are a reflection of real life qualitative distinctions between students across a wide bunch of educational and personal experiences. Among deaf and hard-of-hearing students identified for special education, 3 of every 18 students have a "mild" hearing loss (17%), 7 of every 18 have a "moderate" hearing loss (39%), and 8 of every 18 have sever hearing loss (44%), according to Blackorby and Knokey(2006) [3].

Automatic Speech Recognition (ASR) has impacted different aspects of life. With the availability of more powerful machines making use of multi-core processors, general purpose graphical processing units (GPGPUs), and CPU/GPU clusters, complex computational models were produced. Such models were able to reduce the error rate of ASR systems especially with the availability of more data. Moreover, ASR systems were used in a lot of wearable devices, intelligent home systems. In such systems, ASR is much more convenient and efficient than using a mouse or a keyboard [4].

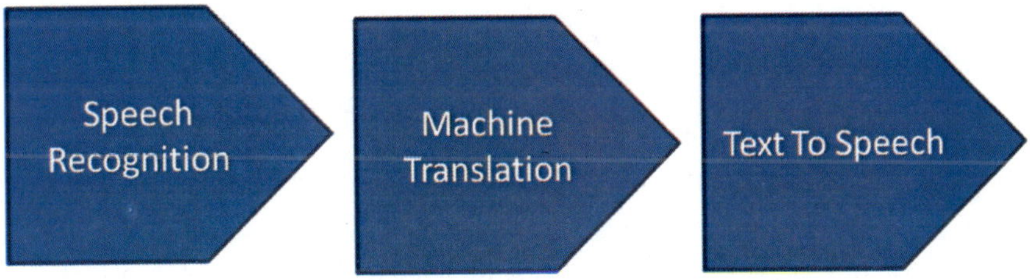

Fig. 1. Components in a typical speech to speech translation system [4]

Natural Language Processing (NLP) is a branch of artificial intelligence. It aims at enabling machines to understand and process human languages.

Computers try to simulate the tasks the humans do to understand and process sentences. The common and basic tasks done include tokenization and stemming and language identification among other tasks. Such techniques split sentences into smaller units that make it easier for machines to then process and draw relationships [5].

Code-switching is a practice where the speaker changes between different languages or dialects [6]. Code switching complicates the tasks that machines have to do to understand the sentences.

3 Methodology

This section introduces the architecture of the Post Lecture Annotation System. The implementation of the system and the description of the modules used.

There are two types of lecture annotations: Real-Time Captioning and Post Lecture Annotation. In our system we implement a Post-Lecture Annotation method. This means that we first record the professors' lecture, transcribe it later and then perform the slide matching algorithm. Finally, the system outputs the lecture with annotations matched on the slides.

3.1 System Architecture

Our system is divided into two phases: the transcription phase, in which the system transcribes an audio file and processes and analyzes the text. Then, the slide-matching phase, after transcribing the audio file, the module matches the text onto slides according to the term weighting algorithm.

Our post lecture annotation system is designed to aid Deaf and Hard-of-Hearing students in their education and make their learning process easier and smoother. We do that by matching what the lecturer said with the closest slide relating to the context of the speech.

The flow of the system is as follows: First, we record the speech of the instructor during the lecture and input this audio file to our transcription module with WAV format file extension. The transcription module transcribes the audio file and passes it on to the next module, along with the PDF slides of the lecture. After performing several information retrieval techniques and several algorithms, the slide-matching module matches the correct piece of text onto the corresponding lecture slide. Finally, after matching the whole transcript, the system outputs a new lecture PDF ready for students to use.

Phase 1 - Transcription Module. The transcription module takes a recorded audio file as an input (WAV format) and outputs text. The biggest challenge faced was with the code-switching during the speech, that is because there is no Automatic Speech Recognition (ASR) system that can transcribe two different languages simultaneously. To tackle this challenge, the audio file is split into small chunks that are sent to the ASR API to process. This method aims to try to transcribe one language at a time as much as possible. For example,

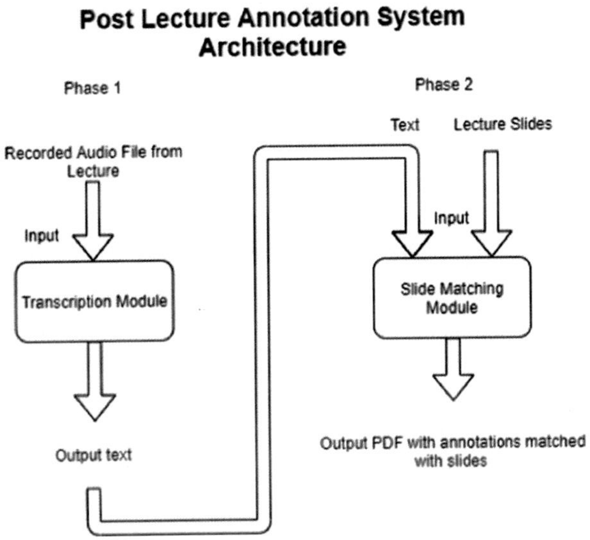

Fig. 2. Post lecture annotation system architecture flowchart

if a speaker records an English sentence followed by an Arabic sentence, the perfect scenario would be to send each sentence individually for transcription. This scenario however is not very applicable in real applications since it is hard to predict beforehand if the presenter will speak in different languages or not. Several attempts were made to find the most suitable duration to split the audio file to be able to separate the different languages spoken. The best suitable duration was found to be 5 s. Because 5 s was the closest duration to allow only one language input to be transcribed at a time and concurrently allowed a reasonable amount of words to be transcribed.

Another minor challenge that we faced was that when we sliced up the audio every 5 s, some words were cut off in the middle. This caused either losing the word, transcribing half of the word or transcribing a different word. To overcome this obstacle, we overlapped 0.5 s of the previous audio chunk into the new chunk. This way, the system transcribes the upcoming 5 s chunk with 0.5 s overlapped from the previous chunk.

Phase 2 - Slide-Matching Module. In phase 2 of our system, we introduce the slide matching algorithm and the Natural Language Processing (NLP) principles that we used to perform slide-matching. The slide matching module takes as an input the lecture transcription, with timestamps, and the lecture slides. The slide-matching module outputs the same lecture slides but with the text annotated on the corresponding slides.

The implementation that we adopted in the slide matching algorithm is majorly based on the approach that Jones, Gareth and Edens, Richard took in their paper, Automated Alignment and Annotation of Audio-Visual

Presentations (2002) [7]. We implemented some of the key concepts as they are and altered some of the other concepts.

After obtaining the transcribed lecture, we now have textual data that we can process and analyze. We apply core NLP principles to this textual data so that we can match the text to the slides. The first step after obtaining the code-mixed transcribed audio file is to translate all the Arabic sentences to English sentences, so that we have a monolingual body of text. After translating the Arabic sentences in the transcript, we split the transcript into documents of length 25 words per document. Now is the time to perform information retrieval techniques. Stop word removal, suffix stripping and term weighting IR techniques were implemented in our system. These are the typical techniques used in order to enhance the IR system.

Stop Word Removal. One of the major types of pre-processing texts is filtering out useless data. In Natural Language Processing, useless words (data) are referred to as stop words. Stop words include words such as “the”, “a”, “an” and “in”. We would not want these words to take up space in our database or hinder the results.

Suffix Stripping. The goal of suffix stripping, or stemming algorithm, is to reduce the many different forms of a word to a common stem form. Removing suffixes from a word is an automatically effective procedure in the field of Information Retrieval. In a typical IR environment, one has a number of documents, each of these documents is labelled and described by the words in the title, abstract and the bodies of the text in those documents. Ignoring where the words originate from, we can say that a document is represented by a vector of words, or terms, and terms with a common stem will usually have meanings close to each other [8]. For example:

DEPART
DEPARTED
DEPARTING
DEPARTURE
DEPARTURES

The efficacy of an IR system will be enhanced if term groups such as this are converged into a single term. This may be done by the removal of the various suffixes, such as -ED, -ING, -URE, -URES, to leave this single stem DEPART. Moreover, the suffix stripping process will cut down the total number of terms in the IR system, and thus, reducing the size and complexity of the data in the system, which is always advantageous and desirable [8].

Term Weighting. Term weighting is a very crucial part of the IR system and the overall slide-matching module. Term weighting is an operation that occurs during the text indexing process in order to weigh the value of each term to the document. Term weighting is the distribution of numerical values to terms that mirror their influence in a document in order to enhance retrieval effectiveness [9].

The objective in our system is to give the high values to the terms that are able to select the relevant documents from the many, general, non-relevant documents. Our system uses the standard Okapi BM25 combined weight (cw) probabilistic term weighting term model which is widely used in current IR systems [7]. BM25 is a ranking function that lists a number of documents based on the query terms appearing in each document, disregarding the inter-relationship between the query terms within a document. BM25 is not a single function, but instead it's a whole family of scoring functions, with somewhat different components and parameters. It is run by different search engines to rank matching documents according to their relevance to a given search query it is often referred to as "Okapi Bm25" [9].

The bm25 cw weight for a term is calculated as follows,

$$cw(i,j) = \frac{cfw(i) \times tf(i,j) \times (K1+1)}{K1 \times ((1-b) + (b \times ndl(j))) + tf(i,j)}$$

where *cw(i, j)* represents the weight of term i in document *j*, *cfw(i)* = log N/n(i) is the standard collection frequency (inverse document frequency) weight, where N is the total number of documents in the collection, *n(i)* is the total number of documents containing term *i*, *tf (i, j)* is the document term frequency, and *ndl(j)* = *dj(j)/Av. dl* is the normalised document length, where *dl(j)* is the length of *j*. *K1* and *b* are empirically selected tuning constants for a particular collection. Investigation of values for these constants in our algorithms showed that best results were found with $K = 2.0$ and $b = 0.75$. A matching score is computed between the query and each document by summing the weights of terms common to both the query and the document [7].

Time Factor. In our system, we also took into consideration the time factor to increase the accuracy of the slide-matching algorithm. We assume that the lecturer moves through the lecture in a sequential manner. Based on this assumption, we divide the total time of the lecture by the number of slides, giving an average of how many seconds are spent per lecture slide. Now that we have our estimated average, we check to see the document being queried timestamp and the lecture slide it is supposed to be on. The weight of the other lecture slides decrease the further away they are from the current lecture slide. For example, if the lectures' duration is 40 min and there are 10 lecture slides. Then, given our assumption, the lecturer speaks about each slide for 240 s. If the current timestamp of the current document is 800 s, it means that the lecturer is speaking about the third slide (or somewhere close to it). It makes more sense that the lecturer is speaking around lecture slides 2–5 rather than slides from 15–18. That's why the slides from 15–18 are given less weight than slides from 2–5.

The time factor is done by calculating which slide is the current timestamp supposed to be on and then we calculate how far away each of the other slides are far from the current slide. Then, we multiply its' weight by a factor. This factor decreases the further away it is from where the document is supposed to actually be. We do not want to massively alter the BM25 scores so that the data is corrupt, so the factors that we multiply by the scores are small. That's why before multiplying the weights by the factor, we divide the factor by 1/8 first and then multiply it by the weight.

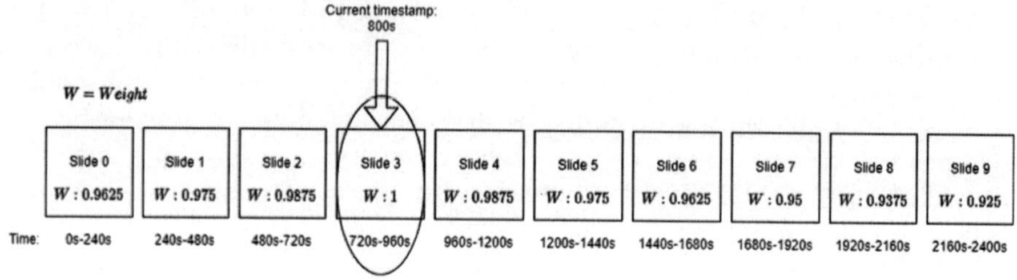

Fig. 3. The further away the slide is from the timestamp, the less weight it has

$$\boldsymbol{Slide1} : (1 - (0.3 - 1/10) \div 8) = 0.975$$

$$\boldsymbol{Slide2} : (1 \quad (0.3 \quad 2/10) \div 8) = 0.9875$$

$$\boldsymbol{Slide3} : (1 - (0.3 - 3/10) \div 8) = 1$$

$$\boldsymbol{Slide9} : (1 - (9/10 - 0.3) \div 8) = 0.925$$

We multiply the weights by a small factor because we do not want to alter the data too much. While our assumption is logical, it is not always the case. We want to make it that the document is more likely to match in a logical time position than an illogical one. This process is repeated each time a document is being queried.

Slide-Matching. Now we have our processed PDF text and they're ready for slide-text alignment. Each transcript document goes through processing. They get processed by lower-casing all words, removing special characters and multiple spaces. We remove all the stop-words from the documents, tokenize the document, and porter stem all the words. Now the document is ready to be queried. Queries here are very rich containing a number of very sufficient search terms linked with the topic of the slide. The Slides are then ranked in decreasing query-document matching score. The highest slide is picked as the matched slide.

4 Results and Discussion

In this section, we present the results of the testing phase, how we measured these results and the techniques used. The aim of this accuracy testing is to evaluate the accuracy of the system. We tested the accuracy of the transcription module and the slide-matching module separately.

Originally, we were going to test our system in a real-life lecture setting. We were going to record lecturers and test the transcription and slide-matching modules with those audio files. But Unfortunately due to the COVID-19 pandemic, we were not able to go ahead with our original testing plan because all universities and educational institutes were halted. Instead, we tested on mock lectures and online video lectures. We would have liked to test our system in a more real life setting but this was not possible due to the pandemic.

Word Error Rate. Word Error Rate is a common metric used to measure the accuracy of the transcripts produced by speech recognition APIs.

A 25% word error rate is about average for regular speech recognition, for speech recognition APIs like IBM Watson and Google Speech. This average is likely to drop down when the speech data gets more technical, is accented, more industry-specific, or noisier.

The method to calculate WER is a pretty simple method. Simply put, WER is the number of errors divided by the total words. To get WER, we calculate the number substitution and deletion of recognized words, and insertion of new words, and add all of them up. And then, we divide that number by the total number of words transcribed.

To put it in a simple formula, Word Error Rate = (Substitutions + Insertions + Deletions)/Numbers of Words Spoken.

4.1 Transcription Module Accuracy Results

We tested from 6–12 audio files. Their duration ranged from 7 min to 20 min. The audio files were mock lectures and included technical terms and terms relating to the course slides. We classified the audio files into 4 difficulties, easy, medium, hard, and extremely hard. The easy audio files meant very minimal code-switching, almost one language being spoken only, speaking pace is slow, word mouthing is clear, and sentences directly related to the lecture slides. The medium difficulty audio files meant a little more code-switching, inter-sentential code-switching, speaking pace is fair, and sentences were a little vague to the lecture slides. The hard difficulty audio files meant a lot of code-switching, intra-senential code-switching, speaking pace is faster, and sentences may not relate to lecture slides directly. The extra-hard difficulty were audio files where the code-switching was extra-sentential, the speaking pace was very fast and the code-switching was dominant during the speakers' lecture. We also should not forget the pronunciation of the words, if the speaker is native or not to the language being spoken, and how good the English or Arabic of the speaker is.

Easy Difficulty. The first audio file we tested was 7 min long. It was ranked as easy level difficulty because it contained very minimal code-switching and the technical course-related terms were directly related to the course slides. It was a mock lecture of an Operating Systems course lecture. For the 5 s chunks, the test resulted in an WER of 0.0972, which meant 90.27% transcription accuracy. 85 words were either substituted, deleted or inserted, over the total number of words, 874. Meanwhile the 10 s audio chunks resulted in 0.1055 WER, which meant 89.7% transcription accuracy. 90 words were either substituted, deleted, or inserted, over the total number of words, 874.

The second audio file was also 7 min long, and it was another mock lecture. It was ranked as easy level difficulty and contained course-related technical terms that related directly to the lectures' slides. The 5 s chunks resulted in WER of 0.095, which meant transcription accuracy of 90.5%. Meanwhile, the 10 s chunks resulted in WER of 0.098, which meant transcription accuracy of 90.2%.

Medium Difficulty. These audio files we tested were ranked as medium because of the inter-sentential code-switching and how the content of the audio file may not directly relate to the lecture slides.

The third audio file was an actual lecture. It was taken from a video recorded lecture and uploaded on the German University in Cairo's Video on Demand service. We played the lecture on the laptop and used a microphone to record and transcribe it. It was 20 min audio lecture and ranked medium level difficulty, because although the speaker did not code-switch, the spoken words did not relate directly to the slides and there were slight vagueness in the sentences. This also ranks up the difficulty for later on in the slide matching process. The 5 s chunks resulted in WER of 0.095, which meant transcription accuracy of 90.4%. On the other hand, the 10 s chunks resulted in WER of 0.093, which meant transcription accuracy of 90.67%.

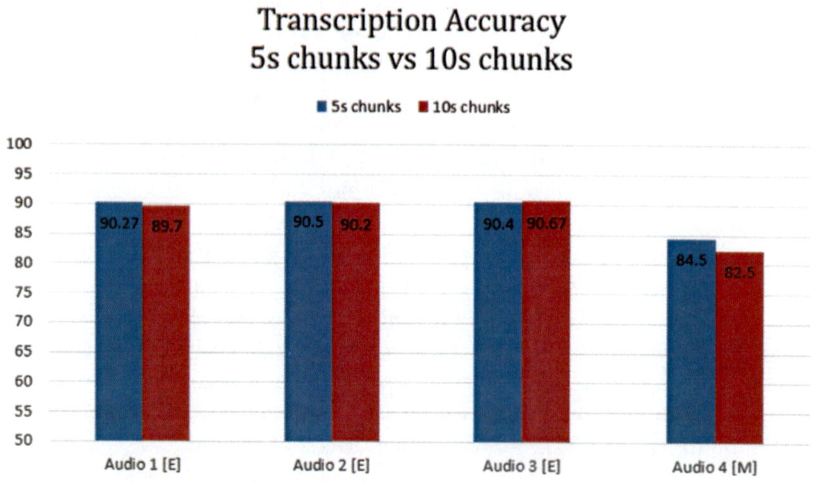

Fig. 4. Transcription Accuracy Results Chart of 5 s vs. 10 s. E - easy. M - medium.

A fourth audio file with medium difficulty generated accuracy results as follows: 5 s chunks resulted in WER of 0.153, which meant transcription accuracy of 84.5% and the 10 s chunks resulted in WER of 0.174 which meant transcription accuracy of 82.5%.

Seeing as the 5 s chunks produced better results. We decided to move forward with testing using the 5 s chunks method.

A fifth audio file was tested with it's 5 s chunks WER resulting in 0.23, which meant transcription accuracy of 77%.

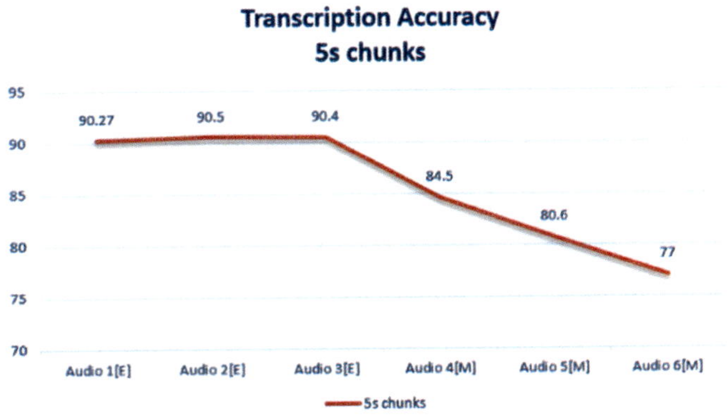

Fig. 5. 5 s chunks Transcription Accuracy Results Chart. The difficulty of the audio files increases gradually. E - Easy. M - Medium.

Hard and Extra-Hard Difficulty. These audio files were ranked hard and extra-hard because of the intra-sentential and extra-sentential code-switching that made it hard for the transcription service to produce results as good as the medium and easy audio level files.

A harder difficulty audio file was tested, which was also a mock lecture, of a Digital Signal Processing course, that included speaking English and Arabic and describing equations, which made it a little harder for the transcription service. This audio file also included inter-sentential code-switching and it resulted in WER of 0.185, which meant transcription accuracy of 80.6%.

Another hard-ranked difficulty audio file resulted in WER of 0.32, which meant transcription accuracy of 68%

Two more extra-hard difficulty ranked audio files were tested for transcription and their WER resulted in 0.42 and 0.44, which meant transcription accuracy of 58% and 56%, respectively.

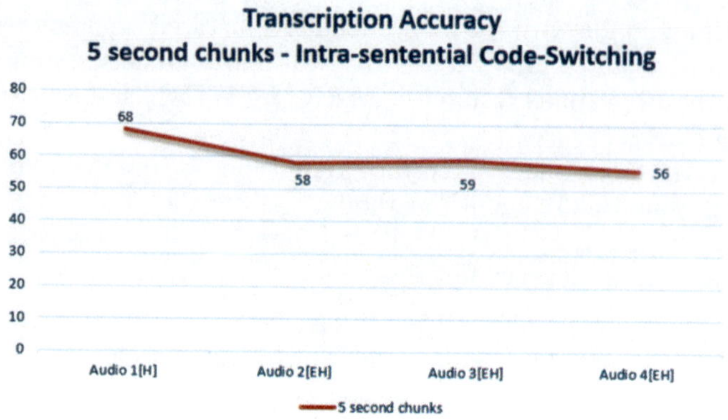

Fig. 6. 5 s chunks Transcription Accuracy Results Chart for hard and extra-hard difficulties.

4.2 Slide-Matching Module Accuracy Results

In this section, we present and discuss the results obtained from testing our slide-matching algorithms after obtaining the transcription from the transcription module.

We calculated the accuracy of the slide-matching module by checking if the document matched the correct slide or not. We added up the number of the correctly matched documents over the total number of documents, giving us the accuracy percentage. A document is considered matched correctly if the words said in the document were spoken over that slide. If a document is linked between two slides and it matched either of them, its' considered a correct match. For example, the first half of the document was spoken over slide 8 and the other half was over slide 9 and it matched either of them, its' considered a correct match.

We adopted the approach that Jones, Gareth and Edens, Richard took in their paper, Automated Alignment and Annotation of Audio-Visual Presentations (2002), regarding the first slide-matching algorithm. The papers' approach was that the contents of the PDF were the query and it was queried against the documents to find out the best match.

Slide-Matching Algorithm. We first tested the slide-matching algorithm on 5 s chunks and 10 s chunks. The 5 s chunks matched with accuracies of 95%, 88%, 61% and 69%, respectively. The 10 s chunks matched with accuracies of 91%, 86.6%, 55%, and 62%, respectively.

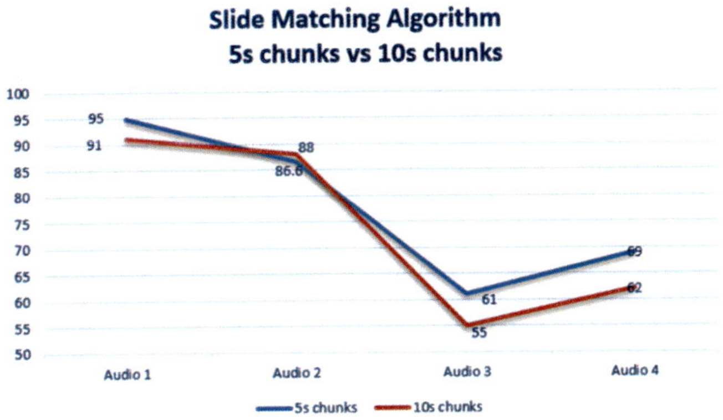

Fig. 7. The results of the second slide-matching algorithm.

Moving forward, we decided to stick to the 5 s chunk approach and because it matched with better accuracy results. We ran tests on higher difficulty and more challenging audio files. This graph shows how the 5 s chunk audio files matched with the PDF slides. As we can see, the accuracy decreases along the x-axis because the audio file difficulty increases along the x-axis.

4.3 Discussion

In this research, we were able to attain good transcription results on English-only audio, Arabic-only audio and English and Arabic code-mixed audio. The results were excellent for monolingual audio files and were good for code-mixed audio. We were able to go as further as inter-sentential code-mixed audio files and generate acceptable transcription results. However, for the harder audio files,

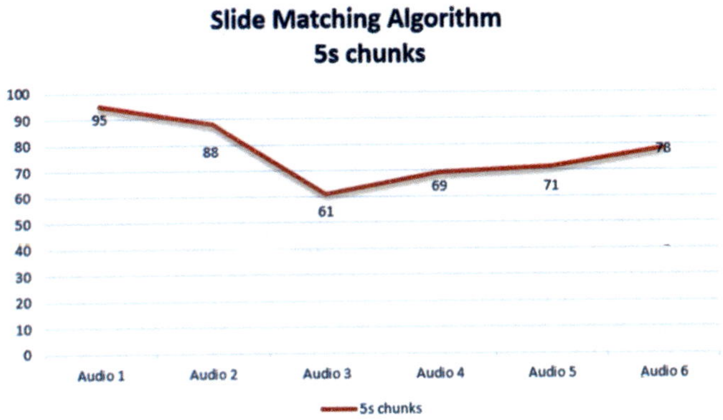

Fig. 8. The results of the second slide-matching algorithm testing 5 s chunk audio files only. The audio file difficulty increases along the x-axis.

such as intra-sentential and extra sentential code-mixed audio files the results were not as expected and needs further enhancement.

We were able to reach the most optimal results using the 5 s chunk approach and the slide-matching algorithm where the documents were queried against the PDF files. The results show encouraging signs that, with little improvement, this system could be implemented and used practically. With 77% average slide matching accuracy and 75% average transcription accuracy, including the Hard and Extra Hard audio files. Excluding the hard and extra hard audio files, the average transcription accuracy of the system is 86%. These results show that this system can be helpful to students who need assistance in note taking.

5 Conclusion

Our system utilizes lecture PDF slides and audio recordings of the lecture to help deaf and hard-of-hearing students. By transcribing the audio recording and then performing NLP techniques on the transcribed text for it to be ready to be matched on lecture slides. This system runs on two modules, the transcription module and the slide-matching module. The transcription module takes as an input an audio file and returns the transcription of the audio file. The slide-matching module takes as an input the transcription of the audio file and the lecture PDF slides. After processing the text using various NLP techniques aligning the text on the slides, the slide-matching module outputs a PDF file with the text aligned on the slides.

The final results show that the average transcription accuracy, excluding intra-sentential and extra-sentential audio files, is 86%. Meanwhile if we include intra-sentential and extra-sentential audio files, the average accuracy becomes 75%. The average slide-matching accuracy of the system was 75%. These are promising results that show that the system can be enhanced for the future for practical use.

6 Future Work

With enhancement and further research, this system can become a very handy tool for students in taking notes, but especially to those who rely on it in their education, deaf and hard-of-hearing students.

To help enhance this system, further work and research needs to be done on transcribing intra-sentential and extra-sentential code-mixed audio files. Furthermore, if the Arabic parts of the transcript were translated in a way that retains the context of the text in Arabic, it would greatly improve the slide-matching accuracy. Another area that needs enhancing is dealing with images in lecture slides. When the instructor is discussing an image, a figure or a diagram, this part of the audio does not get aligned with any of the slides. This area was not handled in this research but could be an interesting area to explore in future research as it will vastly enhance this system and make it viable. Finally, handling mathematical equations when transcribing the audio was also a problem that we encountered and is an area that needs to be an improved in the transcription phase. Another feature that could be added to the lecture annotation system is to summarize these notes and then align them with the lecture slides.

References

1. Jones, G.J.F., Edens, R.J.: Automated alignment and annotation of audio-visual presentations. In: Agosti, M., Thanos, C. (eds.) ECDL 2002. LNCS, vol. 2458, pp. 276–291. Springer, Heidelberg (2002). https://doi.org/10.1007/3-540-45747-X_21
2. Nasser, N., Salah, J., Sharaf, N., Abdennadher, S.: Automatic lecture annotation. In: IEEE Frontiers in Education Conference, FIE 2020, Uppsala, Sweden, 21–24 October 2020, pp. 1–9. IEEE (2020)
3. Karchmer, M.A., Mitchell, R.E.: Demographic and achievement characteristics of deaf and hard-of-hearing students. Oxford Handb. Deaf Stud. Lang. Educ. **38**, 21–37 (2003)
4. Yu, D., Deng, L.: Automatic Speech Recognition. Springer, London (2016). https://doi.org/10.1007/978-1-4471-5779-3
5. Chowdhury, G.G.: Natural language processing. Annu. Rev. Inf. Sci. Technol. **37**(1), 51–89 (2003)
6. Gardner-Chloros, P., et al.: Code-Switching. Cambridge University Press, Cambridge (2009)
7. Porter, M.F.: An algorithm for suffix stripping. Program **14**(3), 130–137 (1980)
8. El-Khair, I.A.: Term weighting. In: Liu, L., Özsu, M.T. (eds.) Encyclopedia of Database Systems, pp. 3037–3040. Springer, Boston (2009). https://doi.org/10.1007/978-0-387-39940-9_943
9. Amati, G.: BM25. In: Liu, L., Özsu, M.T. (eds.) Encyclopedia of Database Systems, pp. 257–260. Springer, Boston (2009). https://doi.org/10.1007/978-0-387-39940-9_921

The Integrating Face-to-Face Learning, Distance Learning Technologies and M-Learning Technologies: Effectiveness

Aida Nurutdinova[1(✉)], Dilyara Shakirova[1], Guluysa Ismagilova[1], Zulfiia Fazlyeva[1], Evgeniya Panfilova[2], Dina Sheinina[1], Gulnara Galeeva[1], and Sabina Ilminbetova[1]

[1] Kazan (Volga Region) Federal University, Kazan, Russia
AiRNurutdinova@kpfu.ru
[2] IACLD CPE ANCO, Kazan, Russia

Abstract. The process of higher education digitalization is driven by global transition to a digital economy and society. Building a digital economy and digital education are significant priorities of state policy in the Russian Federation. Mobile learning is a modern educational trend that allows knowledge to be acquired anywhere, anytime using portable devices. It has attracted the attention of many researchers from different disciplines who are aware of the high potential of mobile technology to enhance learning. In this research paper we will focus on the issue of improving learning outcomes using mobile technologies. An example will be the work in English as a second language classes in short-term language courses. Information technologies in teaching foreign languages have shown their effectiveness. The technical and psychological readiness of learners to use mobile technologies in learning is analyzed as well. In this research paper, the effectiveness criteria were defined as: providing learners with additional conditions for self-realization; mastering new ways of comprehending life, culture, history in the university and urban space around them, diversity of communication between teacher and students, individualization of the educational process; expanding the arsenal of learning tools, preparedness of pedagogical developments with the use of mobile learning. Teachers and students should no longer be limited to being able to teach and learn at a particular place and time. Mobile Learning and wireless technology will become an everyday part of learning, both inside and outside the auditorium. With the education standardization, mobile technology may be a chance to maintain the personalized approach to learning and bring to life the adage that the whole world is an auditorium. Conclusions and a discussion of these outcomes are offered as well as some inferences and speculation regarding the future of M-Learning in the classroom and beyond.

Keywords: Mobile learning · Teaching methods and tools · Mobile technologies · Mobile learning systems · Life-long education

M. E. Auer and T. Tsiatsos (Eds.): IMCL 2021, LNNS 411, pp. 478–486, 2022.
https://doi.org/10.1007/978-3-030-96296-8_43

1 Introduction

At the moment, higher education institutions in the Russian Federation are implementing such initiatives as the Strategy for the Development the Information Society of the Russian Federation for 2017–2030. Such initiatives as the Strategy for Development the Information Society in the Russian Federation for 2017–2030 and the Program for Modern Digital Educational Environment in the Russian Federation are being implemented to create necessary conditions for the development, implementation and use of digital technologies in education, which should improve the country's competitiveness and the quality of educational services. In order to fulfil a number of objectives, it is necessary to modernize the education system, align educational programs with the needs of the digital economy, widely implement digital educational tools in educational activities and integrate them into a unified information environment, and ensure such an individual learning path that a learner can learn throughout life, at any time and in any place.

Mobile learning (m-learning) broadly refers to the use of mobile phones, smartphones, tablets, laptops and other portable devices and data storage and transfer technologies for the direct delivery and organisation of learning. The possibility of knowledge accumulation and transfer via mobile devices already exists today poses an urgent and pressing challenge to didactics to develop teaching methods and learning technologies using such devices. It has to be acknowledged that mobile learning technologies are currently limited either by the didactic vision of their creators or by the private practice of individual teachers, usually young and following technological innovations. Such rapid development of the mobile industry cannot but affect various spheres of human activity, including education. In the field of education, mobile learning is becoming quite popular today, which is developing at a great pace due, first, to the constant growth in the number of mobile devices in use and, second, to the growth in the number of applications for organizing educational activities and learning itself.

In Soviet pedagogy, the term "mobile learning" is widely understood and rarely used. As far as the practice of teaching foreign languages is concerned, this area has hardly been explored. Today there are a large number of educational platforms, electronic textbooks and dictionaries, virtual excursions and online services. All these tools are designed to intensify the learning process, further motivate students, facilitate the effectiveness of independent extracurricular activities, aimed at systematizing and consolidating the acquired knowledge. Taking this into account, we can talk about the relevance of the problem of using mobile technologies in English as a foreign language classes within the framework of short-term language courses. The task before us is to prove the effectiveness of mobile learning technologies in relation to the discipline of English as a foreign language, supporting it with the analysis of the results of the experiment.

2 Literature Review

Mobile learning theory is well established in international pedagogy. The UNESCO Institute for Information Technologies in Education has developed the UNESCO Policy Recommendations on Mobile Learning [1]. The document provides a succinct definition of mobile learning: "Mobile learning refers to the use of mobile technology, either alone or in combination with other information and communication technologies (ICTs), to deliver educational experiences wherever and whenever it is needed. Learning can take many forms, with mobile devices enabling learners to access educational resources, connect with other users, and create content in and out of the classroom. Mobile learning encompasses the activities necessary to achieve learning goals, e.g. effective management, improved interaction between educational institutions and students' families".

Most researchers [M. Bransford, J. Douglas, D. Kelly, T. Reckedal, S. Gedess] [2–4] distinguish between E-Learning and M-Learning. In their view, the main difference between E-Learning and M-Learning is that the latter is not tied to a specific time and place. This in turn individualizes the learning process, making it informal.

Scientists [S. Hargadon, S. Geddes, J. Douglas, D. Kelly, M. Bransford, etc.] [2–4] consider M-Learning as one of the most promising areas of modern pedagogy. C. Robinson [5], the British international adviser on development of creative thinking, education systems and innovations in public and social organizations, is convinced that development of M-Learning technology aimed at optimizing the learning process to meet the needs of students who daily deal with modern technical means, will certainly lead to a revolution in education.

M. Kumari and S. Vikram [6], speaking about advantages of M-Learning, notes that such learning is truly individualized, taking into account interests, needs and capabilities of a particular student.

M.A. Goryunova and M.B. Lebedeva, analyzing M-Learning note the following features:

- "clear focus on meta-disciplinary learning outcomes: the use of mobile devices helps to form all kinds of universal learning activities (cognitive, regulatory, communicative);
- opportunity to implement new approaches to evaluation - involvement of students in evaluation process, increasing role of reflective evaluation tools, use of computer based evaluation tools;
- orientation towards increasing students' independent work;
- ensuring a wider range of information resources used in teaching (electronic textbooks, electronic educational resources, cloud tools and services)". [7].

Researchers note a certain connection and complementarity of traditional and M-Learning. For example, M.B. Fine is convinced that "the most promising way of introducing mobile devices in education is in a competent combination of new forms of learning (interactive lectures, webinars, simulations, trainings, discussions), new types of learning tasks (slide presentations, web projects, educational podcasts) and traditional ones" [8].

Of particular interest is a study by I.N. Golitsyna and N.L. Polovnikova [9] who highlight the strengths and weaknesses of M-Learning technology.

The strengths include:

- students' interaction with each other and the teacher;
- the simplicity of providing the classroom with technical devices (e.g. much more space is needed for a computer than for a mobile phone or tablet);
- space saving nature of mobile learning technologies;
- ability to exchange information between students and teacher;
- the ability to use mobile devices at any place and time;
- greater motivation to learn by appealing to the interests and needs of those actively using mobile devices.

But weaknesses are also noted:

- mobile screens are capable of presenting a limited type of information correctly;
- limited storage capacity for mobile phones and PDAs;
- special attention should be paid to the operation of the device itself, as its incorrect operation may lead to loss of information;
- less reliability (in comparison with desktop computers);
- there may be problems with the use of graphics;
- quickly obsolescence of the device;
- reduced bandwidth when actively using a wireless network.

The authors also note that many of these weaknesses, which for the most part relate to the technical side, are solvable due to the active development of technology in this area.

I.N. Golitsyna and N.L. Polovnikova [9] cite the following forms of mobile technology (mobile phone) implementation in the learning process:

1. Internet access to sites with learning information;
2. a means of playing back files of different formats (audio, video, audio, text);
3. opportunity to use adapted electronic textbooks and training courses.

V. A. Kuklev singles out the following means of mobile learning:

1. "mobile means for studying mobile content (mobile textbook, electronic book, mobile dictionary, interactive translator, mobile TV facilities, mobile excursion, online presentation, set of bookmarks for resources, mobile guide, podcast, network storage of multimedia objects);
2. tools for mobile communication with learners (mobile chat, mobile email, mobile videoconferencing, mobile forum, mobile blog);
3. tools for mobile control of knowledge (SMS-testing tools; SMS polling tools, voting tools; mobile forum and chat tools; mobile testing tools on PDAs, smartphones and communicators; knowledge testing tools for mobile Internet devices);
4. mobile skills-building tools (mobile games and simulations; mobile training, mobile group projects, mobile research);
5. tools to support mobile learning (mobile information and management system; tools for mobile access to information in computer networks)" [10].

Teaching English as a second language using M-Learning technology is discussed in an article by I.V. Nefedov and K.A. Popova [11]. The authors are convinced that in English as a foreign language classes it is possible to use specialized educational programmes (today Google Play can offer about 80), create virtual excursions, have students create video clips and work with them in groups, use QR codes in the learning process. So, despite the minor disadvantages of M-Learning, mostly related to the technical capabilities of modern mobile devices, researchers are convinced of the necessity and feasibility of its use in the learning process.

3 Materials and Method

The following research methods were used in this work:

1. analysis of previous pedagogical experience on the problem under study in order to assess the relevance and feasibility of applying mobile technologies in English as a foreign language classes in short-term language courses;
2. making changes in the model of Russian as a foreign language lessons;
3. observing the changes made and assessing their appropriateness and effectiveness;
4. analysis and selection of didactic material forming the content component of the experiment;
5. creating learning activities using mobile technologies.

Taking into account the strengths and weaknesses of mobile learning, noted by both foreign and Russian researchers, we came to the following conclusion:

- mobile learning meets all the principles and features of short-term learning (intensive nature of classes);
- additionality in comparison with the main course and practical orientation of the educational process;
- the corrective character as compared to a fixed-term course;
- the heterogeneity of the learners with respect to their age, psychological and social characteristics;
- narrow learning objectives and narrow communicative orientation; priority of aspect-complex approach.

Short-term ESL courses are characterised by a communicative approach and an individualised learning process. The possibility of using mobile learning for self-study is also of great importance, as the number of class hours is very low due to the specific nature of the courses. The new forms of lessons, designed with mobile devices, help to keep the learning motivated, diversify the learning process and remove psychological barriers. Tasks for experimenters are designed to be used on mobile devices, which each student uses on a daily basis. In this way, a comfortable learning environment is created that takes into account the individual capabilities and needs of the student, his or her temperament and cognitive abilities (comfortable pace of the task, speed of reaction and memorisation, etc.).

4 Results

The introduction of Mobile-Learning technologies in "English as a second language" classes was conducted at Kazan Federal University, as well as IACLD CPE ANCO, from 2018 to 2020.

We would like to point out some special features of the summer language courses:

1. an average course which lasts four weeks and consists of 80–100 academic hours;
2. the main aim of course participants is mastering of communication skills in Russian;
3. training groups are formed on the basis of the results of an entrance test designed to determine the level of language proficiency, as well as the professional interests of the participants (students of philological specialities and trainees constitute separate groups);
4. the level of English language proficiency of the trainees varies: from elementary A1 to B;
5. the composition of groups is usually multinational;
6. taking into account the main goal, the main method is communicative, which implies mastering a large amount of lexical material in a conversational context. Classes are focused mainly on oral communication;
7. the task of language immersion is solved by additional courses (elective courses): "English and American Literature", "Country Studies", etc.;
8. the philological profile of education has a number of specific features, in particular, it has a set of special courses designed to improve the professional skills of students: English Phraseology, English Lexicology, Syntax, etc.

Elements of mobile technologies were applied to the disciplines "English as a second language" and "English phraseology". The elective course "English Phraseology" includes an introduction to terminology, a study of Russian phraseology, and assignments. The language proficiency is B2 level.

There were 114 participants in the experiment, including 61 in the control group and 53 in the experimental group. Extracurricular work for the participants of the experiment was organised with the help of an interactive trainer "English phraseology", which could be accessed from a mobile phone or tablet connected to the Internet.

The simulator is thematically divided into 6 parts: four groups of phraseological units (relations between people, with verbs of motion, character of a person, about time) and repetition exercises. The students received lexical comments on the idioms in the classroom and the simulator is designed for extracurricular self-study. After repeating their previous knowledge of English phraseological expressions, the students start performing tasks, which are educational mini-games, quizzes, puzzles, etc.

At the end of the interactive trainer-book there is a vocabulary of the studied idioms. Thanks to the system of hyperlinks, the student can at any time refer to the dictionary, where he or she will find the definition of a phraseological unit, its translation into English, an example of its use in a sentence and its pronunciation.

The arithmetic mean test results of the control and experimental groups are shown in the diagram (see Fig. 1). The control group performed traditional tasks: insert phraseological units omitted in the text, connect phraseological units in pairs, correct mistakes in the use of phraseological units, invent situations in which certain phraseological units should be used, etc.

Throughout the course and at the end of it, tests were conducted. The maximum score for each interim test was 50, the maximum score for the final test was 100.

The average score of the experimental group was higher in both the interim and final tests. The experimental group also showed a higher percentage of assimilation of the material (the difference was 10.9%). We can therefore conclude the effectiveness of mobile technologies in teaching ESL (elective course "English Phraseology") within the framework of short term language courses.

The second part of the experiment was conducted on the basis of an ESL course. The level of participants was B1. The average age of students in the group is 21 years old. The ESL course is designed to form students' speech competence. The communicative learning principle implies a strong focus on oral expression. Taking into consideration the individual approach to learning, the choice of topics and communicative situations is in line with the interests and wishes of the students themselves. Accordingly, the topic "Social networks, computers and modern technology" is proposed.

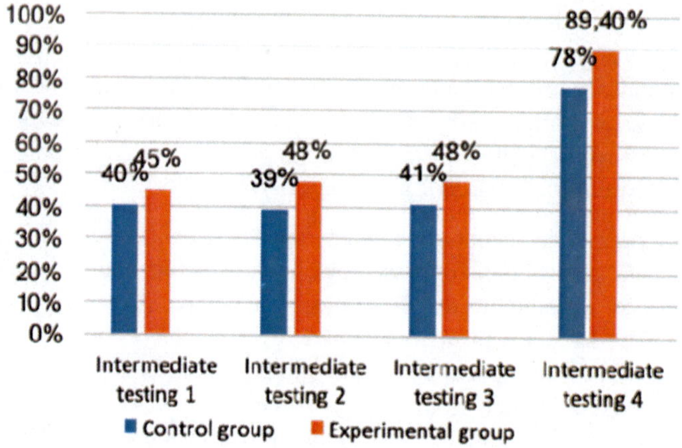

Fig. 1. Distribution Diagram of the test results in the control and experimental groups

A total of 125 students took part in the experiment: 68 in the control group and 57 in the experimental group. Self-work on the topic for the experimental group was based on a video clip and the tasks for it. At the beginning of the work the students were introduced to the vocabulary relevant to the topic being studied. In the picture the students see objects or actions on objects. In order to find out the lexical meaning of a word they need to click on the marker with the mouse.

The vocabulary check is done using a "puzzle": the student selects a new vocabulary item at the top of the screen and clicks on the picture corresponding to that word. If the task is completed successfully, the thematic picture appears. After viewing the video on the mobile device (available on Youtube: 10 comic strips on how social media has changed our lives for ever) is followed by a text comprehension check. The students work in a dialogue simulator. A character asks them questions about the content of the video. Each correct answer is followed by a follow-up question. If they answer incorrectly, the character advises them to review the episode of the video. All of the character's remarks are spoken with corresponding facial expressions and gestures.

The control group works with the traditional tasks for learning new vocabulary and understanding the video.

At the end of the independent work on the topic "Social networks in our life" the students in the control and experimental groups were given a final test on understanding the video and assimilating the new vocabulary (see Fig. 2).

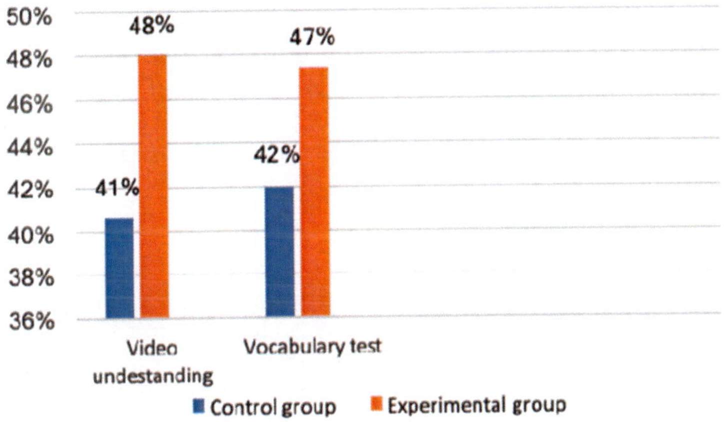

Fig. 2. Distribution Diagram of the test results in the control and experimental groups on topic "Social networks in our life"

It is revealed that the percentage of comprehension of the material of the experimental group is higher by 13.8%. This gives us the right to speak about the appropriateness of using mobile learning technologies in ESL independent work in short-term language courses.

5 Conclusion

Thus, mobile learning has a high didactic potential and its technologies are being integrated into education, creating a new model of training. The implementation is possible with the effective use of interactive, innovative learning methods, methods based on learner autonomy. It should be noted that the most promising way of introducing mobile devices in education is to combine new forms of learning (interactive lectures, webinars, simulations, trainings, discussions), new types of learning tasks (slide presentations, web projects, educational podcasts) and traditional ones. Thus, mobile learning should be based on the principle of interactive guided self-learning, which will reduce the destructive impact of information and communication technologies on social and cognitive activity of the learner.

The realities of the twenty-first century are forcing pedagogy to rethink all educational paradigms. New technical possibilities inevitably lead to radical changes in technology and teaching methods. The modern life is inconceivable without mobile devices. This is particularly true for the young generation at schools and universities. Taking into account individual peculiarities, appealing to their interests and desires can increase learning motivation and lead to better assimilation of acquired knowledge.

The main strength of mobile learning as a modern educational technology are its universality, an individual approach to every student, and the possibility to receive education at any time and in any place. Mobile learning has proven its potential, but there are a number of serious problems that hinder its implementation in the educational process: the material and technical provision, the ability to supply mobile devices to students, low computer and technical literacy of teaching staff, lack of basic research on the problem in relation to specific branches and levels of knowledge, insufficient software for the educational process built on the principles of mobile.

In our opinion, mobile learning has great potential in working with students at different stages of education and with different levels of language proficiency.

References

1. UNESCO Policy Recommendations on Mobile Education. https://iite.unesco.org/pics/publications/ru/files/3214738.pdf. Accessed 7 Dec 2018
2. Bransford, M., Douglas, J.: How People Learn: Brain, Mind, Experience, and School. Washington (D.C.) (2000)
3. Rekkedal, T., Dye, A.: Mobile distance learning with PDAs [Electronic resource]. In: Ally, M. (ed.) Mobile Learning: Transforming the Delivery of Education and Training (2009). http://www.aupress.ca/index.php/books/120155. Accessed 07 Mar 2017
4. Geddes, S.: Mobile Learning in the 21st Century: Benefit for Learners (2004). http://knowledgetree.flexiblelearning.net.au/edition06/download/geddes.pdf. Accessed 12 Oct 2015
5. Bawden, D., Robinson, L.: The dark side of information: overload, anxiety and other paradoxes and pathologies. J. Inf. Sci. 1–12 (2008)
6. Kumari, M., Vikram, S.: Mobile learning: an emerging learning trend. In: HiTech Whitepaper, vol. 11, pp. 18–29 (2009)
7. Goryunova, M.A., Lebedeva, M.B.: Mobile learning in the context of FGOS implementation. In: ChiO, vol. 4, no. 49, pp. 91–95 (2016)
8. Fain, M.B.: Mobile learning in the educational process: foreign experience. Mod. Sci. Res. Innov. **1**(Ч), 3 (2015). http://web.snauka.ru/issues/2015/01/43006. Accessed 23 Sept 2018
9. Golitsyna, I.N., Polovnikova, N.L.: Mobile learning as a new technology in education [Electronic resource]. OTO **1** (2011). https://cyberleninka.ru/article/n/mobilnoe-obuchenie-kak-novaya-tehnologiya-v-obrazovanii. Accessed 29 Nov 2018
10. Kuklev, V.A.: Formation of the system of mobile learning in open distance education: author's abstract. D. In: Pedagogical Sciences: 13.00.01: Defended 26.10.10. Ulyanovsk, 46 p. (2010)
11. Nefyodov, I.V., Popova, K.A.: M-learning as an innovative tool in teaching RKI. In: Proceedings of Southern Federal University, vol. 3, pp. 170–178 (2016)

Virtual Reality Against Doping: The Case of Project VIRAL

Vassilis Barkoukis[1(✉)], Anne-Marie Elbe[2], Lambros Lazuras[3], Louis Moustakas[4,5], Nikos Ntoumanis[6], George Palamas[7], and Monica Stanescu[8]

[1] Aristotle University of Thessaloniki, Thessaloniki, Greece
bark@phed.auth.gr
[2] Leipzig University, Leipzig, Germany
[3] Sheffield Hallam University, Sheffield, UK
[4] European Network of Sport Education, Vienna, Austria
[5] German Sport University, Cologne, Germany
[6] University of Southern Denmark, Odense, Denmark
[7] Aalborg University, Aalborg, Denmark
[8] National University of Physical Education and Sport, Bucharest, Romania

Abstract. Doping is considered a major threat of modern sport. Since the establishment of the anti-doping system emphasis was placed on the detection and punishment approach in the fight against doping. Recently a shift to education as a preventive tool against doping has been made. In this effort several anti-doping educational interventions have been developed. These interventions have been found modestly effective in educating athletes against doping. To move education, forward a virtual reality game is proposed. Virtual reality has been found effective in changing attitudes, intention and behavior. Therefore, virtual reality can be suitable in addressing the appearance and performance related reasoning underlying the decision making towards the use of performance and appearance enhancing substances. The present paper describes the conceptual basis of VIRAL project that aims to develop the first virtual reality game for doping prevention in competitive and recreational sports. The VIRAL project will a) utilize cutting-edge behavioural science research about the risk and protective factors against doping use to inform the development of an anti-doping virtual reality program, b) use an "open innovation" framework to co-design the anti-doping virtual reality program and c) apply and evaluate the effectiveness of the doping prevention VR program in changing young people's learning, motivation, beliefs and behaviour towards doping. Overall, the VIRAL project is expected to develop anti-doping educational material that will address the needs of young athletes and will be able to educate against doping through innovative learning pedagogies.

Keywords: Doping · Performance enhancement · Virtual reality

M. E. Auer and T. Tsiatsos (Eds.): IMCL 2021, LNNS 411, pp. 487–496, 2022.
https://doi.org/10.1007/978-3-030-96296-8_44

1 Introduction

1.1 Doping Use in Sports: Evidence from Elite and Amateur Athletes

Doping use is defined as the use of performance and appearance enhancement substances (PAES) that are prohibited by the World Anti-Doping Agency (WADA), such as synthetic forms of human growth hormone, testosterone and related derivatives, masking agents, stimulants and other drugs that were originally designed to treat diseases in humans and/or animals, as well as designer synthetic drugs that have been developed to improve athletic performance [5, 16]. In elite sports, doping is an ongoing issue, and the scandals involving elite athletes (e.g., use of meldonium by Sharapova) and the Russian anti-doping agency in the Rio Olympics cast a pall on sports and shake athletes' and the public's confidence about the rules of fair play and clean sports [3]. According to studies that used indirect measures and anonymous and confidential self-reported surveys, doping use in competitive and elite sports ranges from 15% to 39% [10, 15]. Recent studies also show that doping use is increasingly becoming a crisis in amateur and grassroots sports as well. In particular, a study showed that doping use can found among amateur athletes as young as ten years old [21]. Accordingly, data from recreational athletes showed that, on average, 1 out of 5 young amateur athletes and exercisers aged between 16–25 years have used doping substances at least once in their lifetime, with higher prevalence rates being reported in South-East European countries like Greece (27.6%) and Cyprus (28.9%), and lower prevalence rates in Germany (17%) and the UK (14.6%) [17].

1.2 Risk and Protective Factors for Doping Use

Following a systematic review of 51 studies, Nicholls et al. [21] identified nine key risk factors for doping use among young athletes aged between 10–21 years: age, gender, participation in sports, sport type, beliefs/behaviours of coaches and athlete's entourage, as well as psychological variables, and use of nutritional supplements. Clearly, not all of these variables are amenable to interventions against doping use, but there are numerous psychological and social aspects associated with doping use that can be directly targeted by tailor-made educational interventions. In this respect, another meta-analysis of 63 independent studies on doping behaviour in adolescent and adult athletes showed that doping behaviour is better understood as a goal-directed, intentional process and that variables such as attitudes, self-efficacy, and perceived social norms (e.g., social approval from referent others such as fellow athletes and coaches; the perceived prevalence of doping among referent others) directly predicted athletes' intentions to use doping substances in the near future [22]. Previous evidence also showed that an urgency to seek immediate performance and appearance benefits and to recover quickly from heavy training or injuries during training were among the top five reasons for doping use in young amateur athletes and exercisers [17]. Another line of research has highlighted the psychological and social factors that act protectively *against* doping use, that is, the factors that can be targeted by educational interventions in order to strengthen attitudes against doping use and empower athletes to "stay clean" even in the face of internal (e.g., performance anxiety and stress) or external situational pressures and temptations (e.g., peer pressure, coach pressure). These protective factors include health beliefs and awareness of the adverse health consequences of doping use; factual knowledge about the actual and alleged effects of doping use on athletic performance (and on physical appearance where exercisers are concerned); self-regulation and resilience against social pressures; and a "self-determined" approach to exercise and sport participation, whereby athletes are motivated to participate in sports for the sake of participation and intrinsic motivation and not for external rewards and the need to outperform others [17, 9, 8, 13, 20]. Taken together, these findings indicate the reasons and motivations that would "push" athletes into the dark side of performance enhancement (doping use), as well as the factors that would act protectively to prevent doping use. This evidence can be utilized to inform, design, and evaluate tailored anti-doping educational interventions.

1.3 Education Against Doping in Sports: Where we are and Where we Need to Go

The need to incorporate empirical evidence from the social and behavioural sciences in concerted and systematic efforts to educate young athletes against doping use has been emphasized by scholars and researchers in anti-doping [4, 29], as well as global leaders and stakeholders of sport, such as WADA, the International Olympics Committee (IOC). Researchers have also noticed that many existing efforts in anti-doping education focus on the negative aspects of doping use (e.g., increasing health concerns) while neglecting a "positive" approach to anti-doping education, such as promoting a "clean sport" identity and culture and empowering athletes to make informed decisions against doping use in amateur, grassroots and elite sports [11]. So far, however, there have been limited efforts to develop and evaluate the effectiveness of evidence-based anti-doping education efforts. A recent study showed that a theory-driven and evidenced-based intervention strengthened anti-doping attitudes in adolescents with varying experiences and engagement in sports and promoted the spirit and values of sport [2]. Another study showed that a school-based media literacy intervention against doping use bolstered more negative attitudes towards doping and reduced use of legal performance enhancement substances (i.e., nutritional supplements) among Italian adolescents [18]. However, although they are based on state-of-art research designs, methods and evidence, those interventions targeted young people who do not necessarily engage in sports. More concerted educational interventions that target athletes are needed. To this end, WADA introduced the Anti-Doping Education and Learning Platform (ADEL), an online learning platform where elite athletes can access educational resources about the health effects of doping use and information about banned substances. ADEL does not directly address risk and protective factors for doping use but largely focuses on doping control procedures, and it is also unclear if ADEL successfully strengthens the psychological skills and capacities that can protect athletes from (realistic) doping pressures, such as resilience to peer (and coach) pressure, and self-regulation.

Anti-doping education is currently at an early stage, and there are several needs that must be addressed with respect to the design, implementation, and evaluation of anti-doping education programmes. Firstly, anti-doping education should be based on state-of-art learning pedagogies that will enable effective engagement, learning and retention of the learned material. Existing anti-doping interventions [2] are allegedly educational but have not been developed on the basis of learning pedagogies - rather, they have adopted a traditional, lecture-like, one-way communication approach, where the "learner" has little to do with the learning process, and is simply expected to independently interact with the learning material and accordingly change his/her mindset and behaviour towards doping use. Learning pedagogies, such as problem-based learning (PBL), enable the learners to actively engage with the learning content (e.g., anti-doping education), set their own learning objectives, and accordingly engage in self-directed and independent learning [31]. This approach enables learners to more effectively acquire and reflect on their knowledge and use this knowledge to resolve problems in real-life situations [26]. Secondly, anti-doping education is concerned with changing mindsets, intentions and behaviours towards doping use by reducing the risk factors and promoting the protective factors against doping. Therefore, a system needs to be in place to assess how effectively anti-doping educational interventions fulfil their goals with respect to the different aspects of behaviour change. In the behavioural and health sciences, effective behaviour change is indicated by changes in three main aspects: beliefs about the behaviour (e.g., perceived health risks of doping use), intentions to change the behaviour (e.g., intentions to avoid doping use; intentions to become or remain "clean" from PAES use), and actual behaviour change (e.g., changes in doping related-behaviours, such as abuse/misuse of nutritional supplements or other ergogenic drugs that are not banned from WADA; long-term avoidance of doping; and reductions/abstinence from doping use among former users) - the effectiveness of interventions can, therefore, be evaluated against these three criteria [12]. Nevertheless, currently, there is a lack of such a systematic approach for evaluating the behaviour change outcomes of anti-doping educational interventions. Thirdly, with the exception of SAFEYOU Tool (www.safeyou.eu; which is the only known contemporary anti-doping educational resource that explicitly promotes clean sports mentality), there are no other known anti-doping educational interventions that promote a positive approach to doping prevention, such as developing a drug-free performance enhancement culture, and a "clean sport" identity [11]. In addition, other educational interventions, such as ADEL, target primarily a narrow group of elite athletes who will independently seek to find information about anti-doping on WADA's website - this does not represent the bigger population of athletes in grassroots and amateur sports who are at risk for doping use but currently lack educational provisions and resources to avoid doping. Finally, anti-doping educational interventions should incorporate the learning process in the context of new learning technologies that can also facilitate behaviour change outcomes, such as virtual reality.

1.4 Virtual Reality as an Innovative Way to Promote Clean Sport Education

Virtual Reality (VR) represents an innovative technological approach to education and training strongly influencing both teachers and students. The core advantage of using VR is that the user can be "another person". Two broad factors are essential for an exciting and vivid experience: the sense of experiencing a virtual body as the own body (embodiment) and the feeling of presence in a virtual space (immersion). A typical VR system consists of a wearable display that allows the user to see the world and hear the sounds. Along with the senses of seeing and hearing, tactile sensors will be used to generate the illusion of touching virtual objects. The tactile sensors can extend the information coming from the environment (exteroception) but also from someone's own body (interoception), thus enhancing the embodied experience. For example, a tactile sensor can be used to feel the "touch" of virtual objects but can also be used to provide a feeling of someone's own "heartbeat". Immersive technologies are powerful learning tools because they provide a safe way to explore a variety of stressful situations, and learners can experience failure without real-world consequences. VR and tactile sensors can stimulate a much wider spectrum of senses that other screen-based media cannot reach. Because of this vital role of VR in the modern training landscape, great opportunities exist to accurately deploy, measure and validate anti-doping educational programs. In the context of a game based on social interaction, a VR simulator appears to be a promising and motivating platform to overcome limitations related to the exposure of real-life situations.

Secondly, VR can instigate behaviour change. Early studies showed that virtual reality and immersive technologies were successfully used to treat clinical conditions, such as phobias and other anxiety disorders, by showing sufferers how to effectively regulate their emotional responses [24] and by improving their self-efficacy to cope with phobic symptoms [6]. Contemporary psychotherapy and psychiatry advocate the use of VR applications for the treatment of anxiety disorders and related behaviour change [7, 19]. Outside clinical settings, VR applications have been recently used to change attitudes, intentions and behaviour in diverse applied domains, such as tourism and leisure [30], pro-environmental behavior [14], and social interactions [25]. Another study recently showed that VR can be effectively used to change people's attitudes, intentions and behaviours and also boosted implicit learning processes by developing a sense of a "virtual body ownership" that moves and feels like the body of the VR user - much like an avatar [28]. These findings suggest that VR applications can provide a very useful, innovative, and relevant alternative to existing anti-doping and clean sport education initiatives. First of all, if VR experiences can change attitudes, intention and behaviour (core elements of the behaviour change process) [1], then it is possible that learning about clean sports through VR applications can lead to significant changes in young exercisers' and athletes' attitudes, intentions and behaviour towards the use of performance-enhancing drugs, such as steroids. Secondly, the "virtual body ownership" can be especially relevant to the motivations underlying performance-enhancing drug use in young athletes in grassroots, fitness and amateur sports. Specifically, physique improvement, in the form of drive for muscularity and muscle dysmorphia (i.e., pathological preoccupation with muscularity and weight gain), serves as the prime motivation for using doping substances in recreational sport and exercise settings [23, 32]. A common characteristic in muscle dysmorphia and drive for muscularity is extreme anxiety and maladaptive emotional responses to personal and social/environmental triggers (e.g., social media exposure to unrealistic muscular ideals) [27]. As such, and given VR's success in reducing anxiety symptoms and regulating emotional responses [7], it is possible that VR applications can be effectively used to teach young exercisers and athletes how to cope adaptively with muscularity issues and reduce their anxiety symptoms related to muscularity, thereby decreasing the risk for doping use in the future. In other words, VR applications can provide a psychological "safehouse" where young people can learn the skills needed to effectively cope with demands and situations that elicit anxiety over muscularity and performance enhancement.

At the moment, there is only one VR program related to doping worldwide. The Australian Anti-Doping Agency (ASADA; https://www.sportintegrity.gov.au/what-we-do/anti-doping) has developed such a program to educate athletes in Australia about doping control procedures. Similarly to ADEL, the ASADA VR game is addressed to competitive level athletes that participate in doping controls, thus neglecting grassroots level athletes. Furthermore, the ASADA program aims to inform about doping control procedures rather than address the risk and protective factors against doping and therefore contribute to the protection of athletes from doping. Against this background, there is an important gap in the application and evaluation of the effectiveness of VR technology for "learning" how to do sports without drugs and how to cope with the psychological (e.g., anxiety over muscularity) and social triggers (social comparison with unrealistic muscularity ideals) of performance-enhancing drug use.

1.5 The VIRAL Project

Based on the abovementioned literature review, there is a need to advance anti-doping education intervention targeting both competitive and recreational athletes. Furthermore, there is a need to move forwards and transform the way anti-doping education is designed, delivered and evaluated through the development of a VR program. Importantly, VR programsare able to provide an effective mechanism for educating the public about the health, psychological and social risks involved with doping. Moreover, the high degree of immersion allows the simulation of situations that can have a strong impact on the learning of the taught material (e.g., a doping offer by a teammate). The benefits of virtual reality education are generated from two pedagogical features. The first involves the participation of the senses of seeing, hearing, touching and sense on the skin, while the second targets participant's reaction to such stimuli. Therefore, it is expected that virtual reality education would enhance learning, as compared to typical educational approaches, through the use of experiential learning. Based on these ideas, the project (VIRAL; www.viral-project.eu) lays out a clear roadmap for how to improve previous anti-doping educational efforts. In particular, the project will deliver a much stronger educational experience than that offered via desktop or mobile application, along with a wider set of tools and methods to assess trainee performance and provide valuable feedback for future improvement.

The VIRAL project aims to utilize: a) updated research from the social and behavioural sciences on doping use; b) state-of-art learning pedagogies; and c) cutting-edge virtual reality gaming design and technology in order to deliver an innovative and impactful anti-doping educational intervention. Updated evidence from the social and the behavioural sciences will help in designing problem-based learning scenarios that can realistically depict risk-conducive situations and contingencies in a virtual reality environment. This can maximize the relevance, engagement and impact of the envisaged anti-doping virtual reality program. Furthermore, blending learning pedagogies with virtual reality gaming technology will result in an effective learning approach that facilitates and stimulates independent and active learning, better retention of information, and greater likelihood for persuasion, attitude and behaviour change. Finally, applying behaviour change indicators (e.g., changes in beliefs, intentions, and actual behaviour) will allow us to validate the effectiveness and impact of the envisaged anti-doping virtual reality program and also to recommend specific policy actions that are needed to optimize the effectiveness of anti-doping education. More specifically, behaviour change can be seen as the dynamic interaction between Capabilities (e.g., role modelling, environmental barriers or facilitators of the target behaviour), Opportunities for behaviour change, Motivation, and actual Behaviour (Michie et al. 2014). This COM-B approach helps us identify the key intervention areas as well as different policy-making categories that are relevant to optimizing anti-doping education and intervention outcomes. The main goal of the VIRAL project is to utilize the above-mentioned information in order to develop an effective virtual reality program. By effective, it is meant a program that has the potential to strengthen the anti-doping mindset of clean athletes, help athletes at risk resolve the dilemma of using prohibited substances in a positive way (i.e., abstain from use) and provide athletes who use or intent to use doping with information that may help them avoid doping. On the whole, it is the first time an anti-doping virtual reality program with these features will be developed, and overall the objectives of the VIRAL project are to:

1. Utilize cutting-edge behavioural science research about the risk and protective factors against doping use in amateur and grassroots sports to inform the development of an anti-doping virtual reality program.
2. Use an "open innovation" framework to co-design the anti-doping virtual reality program through the active collaboration of VR designers and young people engaged in amateur and grassroots sports.
3. Apply and evaluate the effectiveness of the doping prevention VR program in changing young people's learning, motivation, beliefs and behaviour towards the use of PAES, and in promoting a more positive mentality about drug-free and health-enhancing physical activity and sports.
4. Train the trainers on how to promote the virtual reality program into several of the project's target groups, namely adolescents and young competitive and recreational athletes.

1.6 Evaluation of the Virtual Reality Program

Following its development, the virtual reality program will be implemented and evaluated. The goal of the evaluation is twofold:

a) Understand and evaluate the user experience: How players reflect on and assess their experience with playing the virtual reality program.
b) Evaluate the effects of playing the virtual reality program on psychological dimensions and factors related to behaviour change - such as, willingness towards doping, temptation, attitudes and moral justifications.

For these purposes, we will employ both quantitative (structured surveys) and qualitative (semi-structured interviews) methods, and utilize a universal approach across the countries involved in our project. The evaluation approach will include two phases: the pre-intervention evaluation (i.e., evaluating basic background characteristics, and beliefs towards doping before playing the virtual reality program), and the post-intervention evaluation (i.e., evaluating user experience and beliefs towards doping after playing the virtual reality program). Based on previous research with virtual reality programs it is expected that the VIRAL program will have a positive influence on athletes beliefs towards doping and will undermine any thoughts for using prohibited substances in the near future.

Conclusions

Virtual reality can be a very powerful tool in the education of athletes, coaches and ASP against doping. Towards this end, project VIRAL envisages to develop a virtual reality program against doping in sport. This program will be evaluated for its usability, acceptability and effectiveness in influencing athletes' beliefs about doping. Overall, the VIRAL project is expected to develop anti-doping educational material that will address the needs of young athletes and will be able to educate against doping through innovative learning pedagogies.

Acknowledgement. This research has been funded with support from the European Commission. This publication reflects the views only of the author, and the Commission cannot be held responsible for any use which may be made of the information contained therein. The VIRAL project (Project No: 603479-EPP-1-2018-1-EL-SPO-SCP) is coordinated by Aristotle University of Thessaloniki and in the consortium participate Aalborg University, European Network of Sport Education, Leipzig University, National University of Physical Education and Sport, Sheffield Hallam University, and University of Southern Denmark.

References

1. Ajzen, I.: The theory of planned behaviour is alive and well, and not ready to retire: a commentary on Sniehotta, Presseau, and Araújo-Soares. Health Psychol. Rev. **9**(2), 131–137 (2015)
2. Barkoukis, V., Kartali, K., Lazuras, L., Tsorbatzoudis, H.: Evaluation of an anti-doping intervention for adolescents: findings from a school-based study. Sport Manag. Rev. **19**(1), 23–34 (2016)
3. Barkoukis, V., et al.: A serious game approach in mitigating performance enhancement culture in youth (GAME project). In: Auer, M.E., Tsiatsos, T. (eds.) ICL 2018. AISC, vol. 916, pp. 733–742. Springer, Cham (2020). https://doi.org/10.1007/978-3-030-11932-4_68
4. Barkoukis, V.: Moving away from penalization: the role of education-based campaigns. In: Barkoukis, V., Lazuras, L., Tsormpatzoudis, H. (eds.) The Psychology of Doping in Sport, pp. 241–255. Routledge, London (2015)
5. Baron, D.A., Martin, D.M., Magd, S.A.: Doping in sports and its spread to at-risk populations: an international review. World Psychiatry **6**(2), 118 (2007)
6. Botella, C., Baños, R.M., Villa, H., Perpiñá, C., García-Palacios, A.: Virtual reality in the treatment of claustrophobic fear: a controlled, multiple-baseline design. Behav. Ther. **31**(3), 583–595 (2000)

7. Botella, C., Fernández-Álvarez, J., Guillén, V., García-Palacios, A., Baños, R.: Recent progress in virtual reality exposure therapy for phobias: a systematic review. Curr. Psychiatry Rep. **19**(7), 1–13 (2017)
8. Chan, D.K.C., Dimmock, J.A., Donovan, R.J., Hardcastle, S.A.R.A.H., Lentillon-Kaestner, V., Hagger, M.S.: Self-determined motivation in sport predicts anti-doping motivation and intention: a perspective from the trans-contextual model. J. Sci. Med. Sport **18**(3), 315–322 (2015)
9. Chan, D.K.C., et al.: Modal salient belief and social cognitive variables of anti-doping behaviors in sport: examining an extended model of the theory of planned behavior. Psychol. Sport Exerc. **16**, 164–174 (2015)
10. De Hon, O., Kuipers, H., van Bottenburg, M.: Prevalence of doping use in elite sports: a review of numbers and methods. Sports Med. **45**(1), 57–69 (2015)
11. Englar-Carlson, M., Gleaves, J., Macedo, E., Lee, H.: What about the clean athletes? The need for positive psychology in anti-doping research. Perform. Enhancement Health **4**, 116–122 (2016)
12. Epton, T., Harris, P.R., Kane, R., van Koningsbruggen, G.M., Sheeran, P.: The impact of self-affirmation on health-behavior change: a meta-analysis. Health Psychol. **34**(3), 187 (2015)
13. Erickson, K., McKenna, J., Backhouse, S.H.: A qualitative analysis of the factors that protect athletes against doping in sport. Psychol. Sport Exerc. **16**, 149–155 (2015)
14. Fonseca, D., Kraus, M.: A comparison of head-mounted and hand-held displays for 360 videos with focus on attitude and behavior change. In: Proceedings of the 20th International Academic Mindtrek Conference, pp. 287–296 (2016)
15. Laure, P.: Epidemiologic approach of doping in sport. A review. J. Sports Med. Phys. Fitness **37**(3), 218–224 (1997)
16. Lazuras, L., Barkoukis, V.: Performance enhancing drugs In: Ecklund, R., Tenenbaum, G. (eds.) Encyclopedia of Sport Psychology. SAGE Publication, London (2015)
17. Lazuras, L., et al.: "I want it all, and I want it now": lifetime prevalence and reasons for using and abstaining from controlled Performance and Appearance Enhancing Substances (PAES) among young exercisers and amateur athletes in five European countries. Front. Psychol. **8**, 717 (2017)
18. Lucidi, F., et al.: The effectiveness of a new school-based media literacy intervention on adolescents' doping attitudes and supplements use. Front. Psychol. **8**, 749 (2017)
19. Maples-Keller, J.L., Bunnell, B.E., Kim, S.J., Rothbaum, B.O.: The use of virtual reality technology in the treatment of anxiety and other psychiatric disorders. Harv. Rev. Psychiatry **25**(3), 103 (2017)
20. Mohamed, S.A., Bilard, J., Hauw, D.: Qualitative and hierarchical analysis of protective factors against illicit use of doping substances in athletes calling a national anti-doping phone-help service. Montenegrin J. Sports Sci. Med. **2**(2), 21–25 (2013)
21. Nicholls, A.R., et al.: Children's first experience of taking anabolic-androgenic steroids can occur before their 10th birthday: a systematic review identifying 9 factors that predicted doping among young people. Front. Psychol. **8**, 1015 (2017)
22. Ntoumanis, N., Ng, J.Y., Barkoukis, V., Backhouse, S.: Personal and psychosocial predictors of doping use in physical activity settings: a meta-analysis. Sports Med. **44**(11), 1603–1624 (2014)
23. Parent, M.C., Moradi, B.: His biceps become him: a test of objectification theory's application to drive for muscularity and propensity for steroid use in college men. J. Couns. Psychol. **58**(2), 246 (2011)

24. Parsons, T.D., Rizzo, A.A.: Affective outcomes of virtual reality exposure therapy for anxiety and specific phobias: a meta-analysis. J. Behav. Ther. Exp. Psychiatry **39**(3), 250–261 (2008)
25. Qu, C., Ling, Y., Heynderickx, I., Brinkman, W.P.: Virtual bystanders in a language lesson: examining the effect of social evaluation, vicarious experience, cognitive consistency and praising on students' beliefs, self-efficacy and anxiety in a virtual reality environment. PloS ONE **10**(4), e0125279 (2015)
26. Savery, J.R.: Overview of problem-based learning: definitions and distinctions. Essential readings in problem-based learning: exploring and extending the legacy of Howard S. Barrows **9**(2), 5–15 (2015)
27. Schneider, F.M., Zwillich, B., Bindl, M.J., Hopp, F.R., Reich, S., Vorderer, P.: Social media ostracism: THE effects of being excluded online. Comput. Hum. Behav. **73**, 385–393 (2017)
28. Slater, M.: Implicit learning through embodiment in immersive virtual reality. In: Liu, D., Dede, C., Huang, R., Richards, J. (eds.) Virtual, Augmented, and Mixed Realities in Education. SCI, pp. 19–33. Springer, Singapore (2017). https://doi.org/10.1007/978-981-10-5490-7_2
29. Tsorbatzoudis, H., Lazuras, L., Barkoukis, V.: 16 next steps in doping research and prevention. The Psychol. Doping Sport **230** (2015)
30. Tussyadiah, I.P., Pesonen, J.: Drivers and barriers of peer-to-peer accommodation stay–an exploratory study with American and Finnish travellers. Curr. Issue Tour. **21**(6), 703–720 (2018)
31. Wood, D.F.: Problem based learning. BMJ **326**(7384), 328–330 (2003)
32. Zelli, A., Lucidi, F., Mallia, L.: The relationships among adolescents' drive for muscularity, drive for thinness, doping attitudes, and doping intentions. J. Clin. Sport Psychol. **4**(1), 39–52 (2010)

Towards a Smart Classroom Enabled Sustainability Education: A Conceptual Model

Maria Eftychia Angelaki(✉), Theodoros Karvounidis, and Christos Douligeris

Department of Informatics, University of Piraeus, Piraeus, Greece
{epiang, tkarv, cdoulig}@unipi.gr

Abstract. This work proposes a conceptual model for integrating education for sustainable development in a smart classroom learning environment. This environment may support the education that enables a more sustainable society for all the students and actively promotes the cultivation of the students' environmental awareness. The model, which has in its center the students and the teachers, consists of four core components, namely the infrastructure, the sustainability content, the evaluation – feedback and the motivation. The smart classroom is viewed as a kind of system implemented in a school that could realize high learning experiences, high-quality content, and high teaching efficiency, focusing on sustainability. Furthermore, the proposed model implies that the integration of the sustainability educational part needs a holistic approach related to the classroom (educational programme, teaching methods and the infrastructure) instead of adding supplementary topics or themes on sustainability into an existing curriculum or educational projects that a smart classroom may offer.

Keywords: Sustainability education · Smart classroom · Smart learning environment · Conceptual model · Infrastructure

1 Introduction

In the information age, smart cities all over the world use Information and Communication Technologies (ICT) to resolve their urban sustainability issues such as the need for responsible resource management and energy efficiency and various other crucial problems in traffic, environmental pollution, city crowding, and poverty [1]. These technologies may gradually contribute to a more sustainable and green future; thus, it is becoming clear that creative skills to work with ICT must be taught to upcoming generations in order to empower citizens and facilitate a more active role in smart cities [2]. The level of education and learning opportunities are among the factors which correlated with urban smartness and are drivers for a city's sustainability and competitiveness [3]. A proper, sustainability-focused and innovative learning environment that offers creativity, education, knowledge and learning, should become a vital part of the smart cities future. Therefore, smart cities initiatives should include more investments in training and continuing education to foster the city's learning and innovation capacity [4]. In addition, smart cities should develop the tools and frameworks for

M. E. Auer and T. Tsiatsos (Eds.): IMCL 2021, LNNS 411, pp. 497–509, 2022.
https://doi.org/10.1007/978-3-030-96296-8_45

thinking in a way that will help people unravel and resolve the complexities of sustainability issues that encounter their communities.

Nowadays, with the sufficient development of sensors, networks, media and artificial intelligence technologies, the classroom environment has the right tools to optimize teaching content presentation, facilitate the learning process and obtain and promote classroom interaction development. This environment may support the integration of an education that encourages changes in learning, abilities, values and attitudes in order to enable a more sustainable society for all the students. Furthermore, it may promote the cultivation of the students' environmental awareness and contribute to their education in sustainable development (ESD). The concept of ESD defined by the United Nations [5] supports the development of the knowledge, skills, understanding, and actions required to create a sustainable world, which ensures environmental protection and conservation, promotes social equity and encourages economic sustainability.

The availability, accessibility and use of open-source software and technologies such as 3D printing give rise to the students' participation in activities where they are the center of the learning process. Moreover, they can be taught STEM (Science, Technology, Engineering and Mathematics) curriculum content using project-based learning technologies in ways that comply with the real-world sustainable practices [6]. The high growth and ubiquity of the Internet and the rapid 5G technology deployment provide the necessary bandwidth capacity to deploy high-tech educational content anywhere and anytime. The teaching methods have already begun to change from the traditional one-way teaching into a bidirectional interaction. In addition, the existing learning environment should be adapted to provide smart learning opportunities for all students to meet their diverse learning needs [7]. Consequently, the learning environment is rapidly transformed from face-to-face teaching into virtual classroom settings, and the delivery of the educational content is revolutionized through the use of multiple mobile applications.

As the smart classroom is a relatively new and emerging research area, it is challenging to offer concrete specifications, implementation techniques, and evidence of its practicality in these early development phases [8]. In addition, it is essential to integrate innovative methodologies to implement sustainability education in the learning environment [9]. For this reason, we propose a conceptual model for integrating ESD in the smart classroom learning environment, which takes the students and teachers in its center and has at its peripheral four core components: the infrastructure, the sustainability content, the evaluation – feedback and the motivation. This model considers the smart classroom as a living mechanism implemented in a school environment aiming to provide high learning experiences, content suitability, and teaching efficiency with a focus on sustainability by taking advantage of modern science and technologies.

The rest of this paper is organized as follows: in Sect. 2, the background information provides the smart classroom concept and its components and the connection with ESD. In Sect. 3, the description of the proposed conceptual model for integrating the ESD in the smart classroom learning environment is provided. The paper closes with the conclusions, limitations and future steps of the presented work.

2 Background Information

The smart or intelligent classroom as a concept was introduced several years ago. In the existing literature, we may identify several definitions and good practices of implementing the smart classroom concept, which utilizes various software and hardware tools to achieve its educational goals. A smart classroom may be seen as a technology-enhanced classroom that fosters teaching and learning opportunities by integrating learning with technological tools [7]. It should not be connected with the conventional classroom, equipped with certain education technologies, such as projectors and overheads, as it was in the early days. It may also be viewed as a transformative strategy to transition from traditional ways of working to a meaningful, engaging and connected digital way of working [8]. The smart classroom concept has evolved from a broader concept: the distance education paradigm that utilized the Internet as a medium to transform a conventional classroom into an intelligent space equipped with several hardware and software components [10]. It could also be seen as an intelligent classroom that enables teachers to use a real classroom-type teaching approach to teach distant students [11].

The first generation of smart classrooms, during the years 2001–2007, was primarily focused on the synchronous delivery of educational content to the students in the actual physical classroom using the face-to-face teaching mode or in remote locations using the online teaching mode [12]. The second generation of smart classrooms, from 2008 until now, is mainly based on the active use of ICT by the students and the automatic communications between them and the smart classroom environment. Within a smart classroom, the students do not simply browse information passively. Instead, they can create, attach, connect, and distribute educational content from one location to another and from one group to the next [13].

In this way, a smart classroom is an educative space equipped with technology in different senses, from the incorporation of digital devices and learning software to the inclusion of sensor networks. This technology may help to provide more convenient teaching and learning conditions for educators and students through the processes of tracking the classroom, gathering data and offering insights to aid the decision making for better and faster learning [14]. Many smart classroom researches realize that with the help of ICT, inquiry, collaborative, group, mobile and ubiquitous learning are established [15, 16]. Moreover, several pedagogical cases stress the adaptive abilities of the smart classroom for the support of the individual and interactive learning simultaneously [17].

Table 1 presents the components that constitute a smart classroom in terms of hardware (devices and technological equipment), software (applications and emerging technologies), environmental conditions of the classroom, implementation of activities during the teaching process and, finally, the types of learning or pedagogy that may be used [14, 18–23]. A smart classroom is mainly associated with the teaching of learning content related to science. Moreover, Table 1 shows that smart classroom employs action-oriented, innovative pedagogy such as cooperative inquiry and collaboration amongst students, as well as experimentation, project-based learning, and learning by doing. These methods allow such flexibility and adaptability of the learning space;

thus, methodologies that address sustainability challenges can easily be adopted. Some established pedagogies related to sustainability and transformative education include community service-learning, critical emancipatory pedagogy, environmental education, participatory action research, pedagogy for eco-justice and community, and traditional ecological knowledge [24].

Table 1. Components of a smart classroom.

Hardware - Equipment	Software – Application	Environmental conditions	Activities	Types of learning or pedagogy
Audio-video elements – 3D projectors - 3D printers	Learning management systems	Architecture (promotion of green materials)	Communication between the local and remote students	Collaborative learning
Internet and Bluetooth enabled devices	Security system for log-in and log-out of a registered student	Functional design (reduce energy consumption)	Collection data from sensors and export reports related to the students' performance, participation and interest	Participatory learning
PCs with touch screens	Software systems to support students with special needs	Control lighting (natural or artificial) - Acoustic (outside and inside)	Voting regarding an issue in a student group with local and remote students synchronously	Learning by doing
Interactive whiteboards, digital pens	Applications for drawing on smart whiteboards and navigating	Classroom temperature - Humidity	Adjusting by voice commands or automatically the classroom environment, e.g., lights, temperature	Flipped classroom pedagogy
Wearable devices (smartwatches)	Advanced software for multimedia, control and processing	Air quality (oxygen levels, carbon dioxide concentration, smell)	Learning and discussing educational content using multimedia and communications tools and devices	Storytelling-based learning - Games-based learning
Wireless sensor and recognition systems	Recognition software and motion or gesture stabilizing software	Recycled material use - Spatial arrangement	Learning and discussing educational content using multimedia and communications tools and devices	Systemic thinking and analysis with the use of real-world case studies

Moreover, a smart classroom could enable students to develop knowledge and awareness and take action to transform society into a more sustainable one. These expected results could be accomplished by including sustainable development issues, such as climate change and biodiversity, into teaching and learning. It will be easier to stimulate students' environmental knowledge motivation in a smart classroom environment, promote the students' active sustainable behavior, and achieve good learning performance [25]. Students would, thus, be encouraged to be responsible persons who resolve challenges, respect cultural diversity and contribute to creating a more sustainable world. Moreover, a smart classroom may promote learning through experience, communication and sharing amongst the students and help them to adapt the environmental conditions (e.g., furniture made from recycled materials, high performance of natural lighting, reuse items and the use of the resources efficiently) in order to contribute to their local community sustainability goals [26].

Therefore, in terms of integrating the principles and practices of sustainable development into all aspects of education and learning, a smart classroom may lead to the reorientation of the curriculum. In this way, educational communities need to identify and select the appropriate knowledge, issues, perspectives, skills, and values addressed to sustainable development in each of the three components of sustainability, namely the environment, society, and economy and integrate them into the curriculum [27]. In addition, it is generally accepted [5, 14] that specific characteristics related to educational content and pedagogy are essential for the successful implementation of ESD. This implementation reflects the equal importance of both the learning process and the outcomes of the education process. The digital facilities of a smart classroom may support this fundamental behavioral shift to sustainable development and stimulate the students' learning motivation. Furthermore, they could provide opportunities for the students to engage in individualized and social learning activities and practices such as:

- The promotion of critical thinking, problem-solving and action, which develop self-confidence in addressing the challenges to sustainable development.
- The utilization of literature, art, and drama to illustrate the projects and allow students to participate in the design process of sustainable educational programs.
- The focus on educational and learning dimensions of sustainable development.
- The contribution in the innovative development of new and creative solutions to common local as well as global planet issues.
- The benefit of the potential for replication of environmental friendly educational content.
- The support of an evaluation in terms of innovation, success and sustainability.

3 A Conceptual Model for the Integration of the Sustainability Education in a Smart Classroom Environment

In order to support sustainability education with the means that a smart classroom could offer, we should reconsider the physical and virtual learning space in line with sustainable development. This development implies the transformation of the classroom and the curriculum toward inquiry-based learning, a student-centred process, and an active and experiential learning approach that provides opportunities to learn from and in real sustainable practices [28].

In the literature, one may find theoretical frameworks as a base for smart classroom environment development and frameworks for ESD implementation. An open-source one develops a smart classroom in several layers, using a range of devices and approaches [29]. The first layer includes an online database of student-generated, tagged, and socially connected learning objects. The second layer integrates the learning devices with the content. The third supports the visualization, sorting, and sharing of collaboratively generated artefacts.

A complete conceptualization of smart classrooms identifies three dimensions that must coexist in such spaces: technology, environment, and the processes carried out [30]. Technology includes hardware and physical technology, and software. Environment consists of architecture and environmental factors. The processes include the learning content, the processes performed by the actors and the processes and features that helped from the system.

Another conceptual model is based on three main phases: interaction learning phase, collaborative learning phase, and smart analytics phase [31]. The interaction learning phase includes the interaction between a lecturer and students and between students and devices. The collaborative learning phase focuses on the interaction between students for advancing to a higher level of understanding and knowledge. The smart analytics phase collects data to monitor and analyze student behaviors in the classroom and develop and choose the appropriate learning method.

The "SMART" conceptual model implies that an innovative classroom is a typical materialization of an intelligent learning environment and is the high-end form of network classroom, where the "intelligent" involves five dimensions: Showing, Manageable, Accessible, Real-time interactive and Testing [32]. Therefore, a smart classroom involves and relates to the optimization of teaching content presentation, the convenient access of learning resources, the deep interactivity of teaching and learning, the contextual awareness and detection and the classroom layout and management.

UNESCO presents a new global framework on ESD called ESD for 2030 to address the new opportunities and risks on sustainable development posed by emerging technologies [5]. The first main feature of ESD for 2030 is the emphasis given to education's role that needs to transform itself. ESD aims to raise knowledge, awareness and action from three perspectives: the cognitive learning dimension in order to understand sustainability challenges and their complex interlinkages, explore disruptive ideas and alternative solutions; the social and emotional learning dimension in order to build core values and attitudes for sustainability, cultivate empathy and compassion for other people and the planet, and motivate to lead the change; the behavioral learning dimension in order to take practical action for sustainable transformations in the personal, societal and political spheres.

In terms of learning outcomes, the transformative sustainability learning (TSL) framework offers an organizing model to explicitly unite and embody the cognitive engagement, practical application and emotional connection practices of sustainability within academic and applied fields. After the development and conduction of three case studies on courses related to sustainability and citizenship, it concludes that systems that were engaging students in a cognitive, psychomotor and efficient sphere enhanced TSL [24].

In accordance with the frameworks for smart classroom development mentioned above and the sustainable development goals, we propose a conceptual model to enhance the different types of classroom environments and increase the contribution of education to build a more fair and sustainable world. This model places the students and the teachers in the center and four core components at its peripheral (Fig. 1).

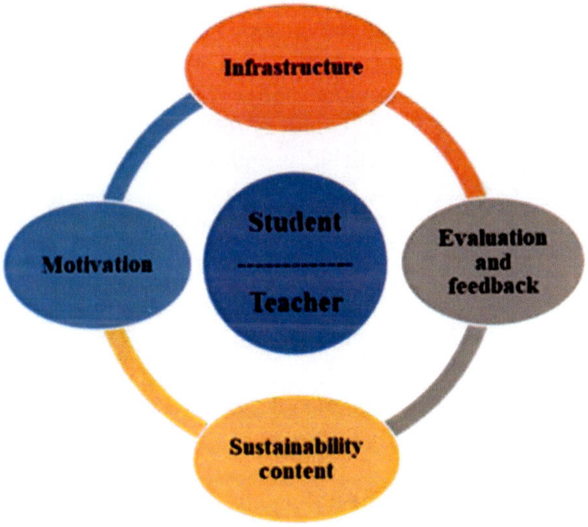

Fig. 1. A conceptual model for the integration of the sustainability education in a smart classroom environment.

The first component, the infrastructure, refers to the hardware components, devices or equipment. This kind of equipment may allow the real-time students' collaboration, research, and effective integration of information technology and digital resources into classroom teaching [33]. The multimedia classroom environment should be upgraded and equipped with smart devices such as multi-screen touch displays and interactive whiteboards. In addition, wireless sensors could measure and monitor the students' and teachers' cognitive and behavioral processes or the environmental factors inside the room to provide input on the air quality, temperature, lighting and acoustics.

Moreover, a smart classroom should have recognition systems for face, eye and motion tracking and the appropriate virtual simulation equipment [34]. This equipment may help to teach abstract and complex concepts associated with sustainability since simulations reduce complexity and highlight salient aspects. At the same time, simulations reflect real-world situations that face communities, give a sense of reality, motivate students, and promote higher-order thinking skills [35].

In addition, the "seeding" and "U-layout" classroom spatial arrangements should be used to support instructive classroom teaching [36]. Moreover, the architecture and functional design must respect the basic principles of sustainability-related to all the environmental factors and promote ecological methods and materials for the reduction of energy consumption, the maximization of sound absorption and sustainable temperature control.

The second component, the evaluation - feedback, allows the tracking of the various activities' evolution in the classroom and the customization of the students' learning path related to their own learning profile in order to satisfy the different learning needs in the form of social equity, which is a core concept of sustainability. It involves students in the optimization of the educational process according to their answers in order to record the students' knowledge and experience, make the content relevant and use different teaching and learning processes. In some cases, the automatic assessment, generated by the appropriate software, is more objective as the teacher is not involved in the grading process, is faster because a repeating task is more adapted to computers and finally is more reliable as anti-plagiarism features are included [37]. The provision of the evaluation - feedback includes - but is not limited - the following actions which may take part from the remote students in real-time with the local students:

- Complete in-classroom assignments and submit corresponding files.
- Discuss and evaluate the presented learning content in each student's preferred language.
- Participate in a poll regarding a particular environmental issue.
- Collect immediate feedback from students in terms of interest and the likeability of an activity.
- Gather data from sensors automatically and record the students learning behavior, performance, interest, participation, etc.

The third component, the sustainability content, includes the applications and emerging technologies used for the creation of the appropriate content that should be included in a curriculum addressing sustainable development. Sustainability education should support eco-friendly daily actions in school life, along with the educational contents of various school subjects, in order to enable the students to develop knowledge and awareness and take the appropriate actions to transform the whole society into a more sustainable one. Students need basic knowledge from the natural, social, and humanities sciences to understand the principles of sustainable development, the way they can be implemented, the values involved, and the consequences of their implementation. Students also must be provided with practical skills that will enable them to keep learning after they are through with school, find out a sustainable living, and live in an eco-friendly environment.

For this reason, we point out indicative curricular themes that may be used and adapted as "trigger points" to develop sustainability education further. These themes are: environmental, economic and social sustainability; natural resources management; travel, transport and urban mobility; climate change; health and wellbeing; sustainable and ethical tourism [38]. Furthermore, sustainability in the curriculum requires a shift towards active, participative, and experiential learning methods that engage the students and make a real difference in their understanding, thinking, and ability to act.

The utilization of this content includes pedagogical approaches or methods such as critical reflection (learning journals, discussion groups), systemic thinking and analysis (real-world critical incidents, project-based learning, stimulus activities), participatory learning (group or peer learning developing dialogue, experiential learning), thinking creatively for future scenarios (role play, real-world case studies) and collaborative learning (contributions from invited speakers, work-based learning, interdisciplinary projects).

The fourth component addresses the motivation of the students. It targets their potential to foster healthy environmental behavior in terms of easiness, engagement, and effectiveness of the learning environment since the learning outcome in the smart classroom almost meets or even is higher than the expectation generated for this learning environment [39]. Therefore, this environment should provide a fast and convenient learning atmosphere and make students' process of completing learning targets easy.

In addition, the learning environment should stimulate and facilitate the students' learning interest and keep this interest and the corresponding level of participation at a relatively high state [40]. Extra-curricular activities and special events should enhance and practice classroom learning about sustainability. The design and the development of the appropriate sustainability content should ensure an immediate and positive interplay and collaboration both between students and teachers, as well as between students, and motivate their enthusiasm. Furthermore, it should support personalized learning and present different learning resources tailored to the specific characteristics of each student. This desirable result may occur by tracking the student's learning progress and learning behavior.

The implementation of this conceptual model may reform the school education process by applying emerging information technologies to facilitate the efficacy and efficiency of the education system. This cultivation and education of students meet the needs for the transformation towards a sustainable society. Furthermore, the smart classroom use aims to inspire students to develop lifelong learning habits of continuous acquisition of knowledge and eco-friendly attitudes to adapt themselves to society's evolution green paces.

4 Conclusions, Limitations, Future Steps

This paper proposes a conceptual model for the support of sustainability education with the means that a smart classroom could offer. By taking advantage of modern science and technologies in different settings or approaches, it can provide and support sustainability learning services and experiences for students and teachers. These settings involve the incorporation of digital devices and learning software related to one or more sustainability elements, the inclusion of sensor networks that help with tracking classroom processes, and simulation equipment that reflect real-life problems.

A smart classroom can promote education equity by meeting all the students' learning needs and increasing learning performance by gathering and analyzing data generated from the students' learning process and learning activities. Using a variety of pedagogies associated with ESD also stimulates students to ask questions, explore, think critically, and offer insights to help decision-making for better and faster learning on sustainability. In addition, an intelligent classroom can provide more convenient teaching conditions for students and teachers and support active participation in modelling and transforming future earth scenarios towards sustainable practices.

Therefore, the infrastructure and resources must be adapted to the underlying pedagogical and sustainable aspects, giving a response to the educational needs for sustainability rather than being merely innovative but unconnected solutions. Furthermore, this situation implies that the integration of the sustainability educational part needs a holistic point of view related to the whole organization of the classroom (educational programme, teaching methods and the infrastructure) instead of the adding of supplementary topics or themes on sustainability into the existing curriculum and educational projects.

One limitation to the implementation of a smart classroom with sustainable orientation and practices may be the investment and technology development costs, which are always high at the beginning, and lower over time. In addition, technology constraints such as potential problems with technology updates, upgrades and system maintenance, too many interlinked devices, data exchange protocols used, etc., may lead to unexpected significant complexity and consequent failures of the entire system. At the same time, new educational content and teaching and learning resources that integrate sustainability into the smart learning environment should be developed and tested.

For this reason, teacher training programs should focus on the development of the proper teaching and digital competencies in order to support the smart classroom sustainable nature. Another limitation is the assumption that a smart classroom is a space that copes equally with any student's education level. However, it should be noted that some differences might exist, and research is needed to report on the actual effects of smart classrooms on students with different cognitive styles and levels.

It is essential in the future to create smart classroom spaces according to the proposed conceptual model in order to explore their impact and influence on the building of core values and attitudes for sustainability and the cultivation of environmental awareness to students and teachers. Furthermore, the students should perform summative and formative evaluations to collect sufficient data about the quality of smart classroom main components - hardware, software, technologies, services – and contact the necessary changes.

Acknowledgment. This work has been partially supported by UPRC (University of Piraeus Research Center).

References

1. Dameri, R.P.: Searching for smart city definition: a comprehensive proposal. Int. J. Comput. Technol. **11**(5), 2544–2551 (2013)
2. Wolff, S., Schulp, C.J.E., Verburg, P.H.: Mapping ecosystem services demand: a review of current research and future perspectives. Ecol. Ind. **55**, 159–171 (2015)
3. Hudson, L., et al.: Supporting urban change: using a MOOC to facilitate attitudinal learning and participation in smart cities. Comput. Educ. **129**, 37–47 (2019)
4. Neirotti, P., De Marco, A., Cagliano, A.C., Mangano, G., Scorrano, F.: Current trends in smart city initiatives: some stylized facts. Cities **38**, 25–36 (2014)
5. Education for Sustainable Development: A roadmap. United Nations Educational, Scientific and Cultural Organization (2020)
6. Marouli, C.: Sustainability education for the future? Challenges and implications for education and pedagogy in the 21st Century. Sustainability **13**(5), 2901 (2021)
7. Kwet, M., Prinsloo, P.: The 'smart' classroom: a new frontier in the age of the smart university. Teach. High. Educ. **25**(4), 510–526 (2020)
8. Al-Sharhan, S.: Smart classrooms in the context of technology-enhanced learning (TEL) environment. In: Alshahrani, K., Ally, M. (eds.) Transforming Education in the Gulf Region: Emerging Learning Technologies and Innovative Pedagogy for the 21st Century, Routledge, (2016)
9. Tang, K.H.D.: Correlation between sustainability education and engineering students' attitudes towards sustainability. Int. J. Sustain. High. Educ. **19**(3), 459–472 (2018)
10. Shi, Y., et al.: The smart classroom: merging technologies for seamless tele-education. In: 4th International Conference Proceedings on Multimodal Interfaces (ICMI 2002), pp. 429–434. IEEE CS Press (2002)
11. Pishva, D., Nishantha, G.G.D.: Smart classrooms for distance education and their adoption to multiple classroom architecture. J. Networks **3**(5), 54–64 (2008)
12. Uskov, V.L., Howlett, R.J., Jain, L.C. (eds.): Smart Education and Smart e-Learning. SIST, vol. 41. Springer, Cham (2015). https://doi.org/10.1007/978-3-319-19875-0
13. Simon, B., Haghirian, P., Schlegelmilch, B.B.: Enriching global marketing education with virtual classrooms an effectiveness study. Mark. Educ. Rev. **13**(3), 27–39 (2003)
14. Cebrián, G., Palau, R., Mogas, J.: The smart classroom as a means to the development of ESD methodologies. Sustainability **12**(7), 3010 (2020)
15. Gligorić, N., Uzelac, A., Krco, S.: Smart classroom: real-time feedback on lecture quality. In: 2012 IEEE International Conference Proceedings on Pervasive Computing and Communications Workshops (PERCOM Workshops), pp. 391–394. Lugano, Switzerland (2012)
16. Lin, Y.T., Huang, Y.M., Cheng, S.C.: An automatic group composition system for composing collaborative learning groups using enhanced particle swarm optimization. Comput. Educ. **55**, 1483–1493 (2010)
17. Li, B., Kong, S.C., Chen, G.: Development and validation of the smart classroom inventory. Smart Learn. Environ. **2**(1), 1–18 (2015). https://doi.org/10.1186/s40561-015-0012-0
18. Kim, Y., Soyata, T., Behnagh, R.: Towards emotionally aware AI smart classroom: current issues and directions for engineering and education. IEEE Access **6**, 5308–5331 (2018)
19. MacLeod, J., Hao-Yang, H., Zhu, S., Lee, Y.H.: Understanding students' preferences towards the smart classroom learning environment: development and validation of an instrument. Comput. Educ. **122**, 80–91 (2018)

20. Isaksson, E., Naeve, A., Lefrère, P., Wild, F.: Towards a reference architecture for smart and personal learning environments. In: Popescu, E., Khribi, M.K., Huang, R., Jemni, M., Chen, N.S., Sampson, D.G. (eds) Innovations in Smart Learning. Springer, Berlin, Germany (2017). https://doi.org/10.1007/978-981-10-2419-1_13
21. Liu, D., Huang, R., Wosinski, M.: Smart learning in classroom environment. In: Smart Learning in Smart Cities. Lecture Notes in Educational Technology. Springer, Singapore (2017). https://doi.org/10.1007/978-981-10-4343-7_5
22. Saini, M.K., Goel, N.: How smart are smart classrooms? A review of smart classroom technologies. ACM Comput. Surv. **52**(6), 1–20 (2019)
23. Al-Hunaiyyan, A., Al-Sharhan, S., Alhajri, R.: A new mobile learning model in the context of the smart classrooms environment: a holistic approach. Int. J. Interact. Mob. Technol. **11** (3), 39–56 (2017)
24. Sipos, Y., Battisti, B., Grimm, K.: Achieving transformative sustainability learning: engaging head, hands and heart. Int. J. Sustain. High. Educ. **9**, 68–86 (2008)
25. Liu, M., Horton, L., Olmanson, J., Toprac, P.: A study of learning and motivation in a new media enriched environment for middle school science. Educ. Tech. Res. Dev. **59**, 249–265 (2011)
26. Brundiers, K., Wiek, A., Redman, C.L.: Real-world learning opportunities in sustainability: from classroom into the real world. Int. J. Sustain. High. Educ. **11**, 308–324 (2010)
27. Jormanainen, I., Toivonen, T., Nivalainen, V.: A smart learning environment for environmental education. In: Chang, M. (ed.) Challenges and Solutions in Smart Learning. Springer, Berlin, Germany (2018). https://doi.org/10.1007/978-981-10-8743-1_2
28. Zhu, Z.-T., Yu, M.-H., Riezebos, P.: A research framework of smart education. Smart Learn. Environ. **3**(1), 1–17 (2016). https://doi.org/10.1186/s40561-016-0026-2
29. Tissenbaum, M., Slotta, J.D.: A new framework for smart classroom research: co-designing curriculum, research and technology. In: 8th International Conference Proceedings on Computer Supported Collaborative Learning, CSCL 2009, vol. 2. Rhodes, Greece (2009)
30. Palau, R., Mogas, J.: Systematic literature review for a characterization of the smart learning environments. In: Cruz, A.M., Aguilar, A.I. (eds.) Propuestas Multidisciplinares de Innovación e Intervención Educativa, pp. 55–71. Universidad Internacional de Valencia, Valencia, Spain (2019)
31. Alhaboobi, Z.A.F., Yousif, S.T., Shawkat, A.R.: Intelligent classroom a conceptual model for the effective use of Internet of Things technique. In: 2nd Scientific Conference Proceedings of Computer Sciences (SCCS), pp. 116–120. University of Technology – Iraq, Baghdad, Iraq (2019)
32. Huang, R.H., Hu, Y.B., Yang, J.F., Xiao, G.D.: The concept and characteristics of smart classroom. Open Educ. Res. **2**, 22–27 (2012)
33. Mrabet, H.L., Moussa, A.A.: Smart classroom environment via IoT in basic and secondary education. Trans. Mach. Learn. Artif. Intell. **5**(4) (2017)
34. Carson, P.P., Harder, N.: Simulation use within the classroom: recommendations from the literature. Clin. Simul. Nurs. **12**(10), 429–437 (2016)
35. Tissenbaum, M., Slotta, J.D.: Developing a smart classroom infrastructure to support real-time student collaboration and inquiry: a 4-year design study. Instr. Sci. **47**(4), 423–462 (2019). https://doi.org/10.1007/s11251-019-09486-1
36. Liu, D., Huang, R., Wosinski, M.: Characteristics and framework of smart learning. In: Smart Learning in Smart Cities. Lecture Notes in Educational Technology. Springer, Singapore (2017). https://doi.org/10.1007/978-981-10-4343-7_3
37. Nowotny, J., et al.: Towards global sustainability: education on environmentally clean energy technologies. Renew. Sustain. Energy Rev. **81**(2), 2541–2551 (2018)

38. Tilbury, D., Wortman, D.: Engaging people in sustainability, commission on education and communication. IUCN, Gland, Switzerland and Cambridge, UK (2004)
39. Sauvé, L.: Education as life. In: Jickling, B., Sterling, S. (eds.) Post-sustainability and environmental education: Remaking education for the future, pp. 111–124. Palgrave Macmillan, Cham (2017)
40. Cebrián, G., Junyent, M.: Competencies in education for sustainable development: exploring the student teachers' views. Sustainability **7**(3), 2768–2786 (2015). https://doi.org/10.3390/su7032768

Serious Games and Gamification

Designing a Serious Game to Teach Pre-analytical Phase for Medical Technologist Students

Carla Toro Opazo[1](✉), María Natalia Véliz Olivos[1], Pablo Ignacio Rojas Valdés[2], and Karina Vergara Reyes[3]

[1] Tecnología Médica, Facultad de Salud, Universidad de Talca, Talca, Chile
ctoro@utalca.cl

[2] Ingeniería en Desarrollo de Videojuegos y Realidad Virtual, Facultad de Ingeniería, Universidad de Talca, Talca, Chile

[3] Doctorado en Sistemas de Ingeniería, Facultad de Ingeniería, Universidad de Talca, Talca, Chile

Abstract. Teaching theoretical knowledge of the processes related to phlebotomy and the processing of medical exam samples is a delicate task, since many of the errors that are made when processing medical exams are due to the preanalytical phase. The students of the Medical Technology career at the University of Talca have indicated that it is necessary to modernize the teaching processes related to technical knowledge and in this work the creation of a video game is proposed to reinforce the technical knowledge involved in the pre-analytical phase of the laboratory testing process. This video game is subsequently evaluated by a group of students who, at the end of the semester of studies, carry out a satisfaction survey on the use of the video game as an educational tool.

Keywords: Serious game · Pre-analytical phase · Medical Thechnologist · Case studies · Gamification · Mobile learning

1 Introduction

Regarding the educational model of the University of Talca, it has declared that the institutional educational model will be the training model by competencies, where it aims to train a competent professional, that is, capable of knowing how to act in a particular context, putting into play the personal and contextual resources for the solution of a specific problem, with a process of reflection on what is being done.

In evaluating the model, the university signed a Curricular Harmonization Performance Agreement, where it declared its second specific objective, to update the curricular plans in line with the results of the evaluation of the first cohorts of graduates formed by competencies, enhancing achievement of learning and favoring the development of innovation and entrepreneurship skills.

Within the declared graduation competencies for the Medical Technology career at the University of Talca, the following are related to the disciplinary field:

M. E. Auer and T. Tsiatsos (Eds.): IMCL 2021, LNNS 411, pp. 513–524, 2022.
https://doi.org/10.1007/978-3-030-96296-8_46

- Perform rigorous laboratory tests from samples of blood, urine, stool and other biological fluids, to contribute to the diagnosis and monitoring of pathologies.
- Implement quality assurance programs, based on current national and international standards, to evaluate transversal processes of laboratories and Transfusion Medicine units.
- Apply management tools, such as planning, organization, control and direction, to perform in administrative positions in health institutions.
- Manage a Clinical Laboratory and Transfusion Medicine Units, in accordance with current regulations, to fulfill their professional care role.

The learnings that account for these competencies are developed in the Pre-Clinical modules, fundamental bases of the clinical laboratory, Clinical Biochemistry I and II, Hematology I and II, which form the curricular training plan of this career since 2015.

From the second year of studies, the students of the Medical Technology career begin their clinical practices, in which they attend health centers where they apply and reinforce the knowledge learned during their learning process.

Since October 1, 2012, Law 20,584 of the Republic of Chile came into force, which regulates the rights and duties that people have in relation to actions related to their health care. This law brought with it that patients could discern in relation to the care provided by students in practice, thus reducing the amount of care provided by students and, therefore, the possibilities of applying their learning.

From the surveys evaluating the teaching of the courses that perform the training plan of the Medical Technologist career, applied to students, in the self-evaluation process of the career developed during the first semester of 2018, it was detected that one of the criteria of the survey was manifested with a poor rating (less than 5, using a scale of 1 to 7) with respect to the other criteria that made up the survey, this criterion is related to the way in which the modules of the curricular plan of the career integrates the contents, methodologies and forms of evaluation to promote the development of skills in students.

Having a digital tool as a methodology to support traditional teaching activities, easily accessible to students, which includes the learning of the disciplinary modules in a playful way, with a design adapted for different electronic devices, allowing them access in any place and at any time, it would allow to achieve the integration of the contents of these modules, which according to the perception of the students at present are manifested as a weakness in the training plan. It would strengthen the knowledge of the students declared in the syllabus of the Pre-Clinical, Clinical Biochemistry II, Hematology II and Professional Seminar modules. It would create a learning environment that would allow students to experience real clinical laboratory problems through a video game. It would also develop in the student the autonomy to make decisions about the management of the laboratory, improving the management of the student's non-contact time with digital resources.

The learning requirements of the new generations will continue to drive the demand for more immersive and attractive interactive tools and environments, as well as the demands of patients who are cared for in the world's health systems, which is why it is necessary to recognize the limitations of the technology available to health educators, not only the perceived benefits, in any pedagogical framework constructed, making the success of the experiences of the practical and simulation activities depend on the participation and commitment of the entire educational institution involved, the literature describes that this should be from the planning phase [1, 2].

Currently, games are also presented in their technological or digital version, which also contributes to the learning process of other skills and abilities related to the field of ICT, such as working in multiple contexts or becoming familiar with different languages simultaneously [2]. In this sense, gamification, as an innovation that brings play and learning closer together, also favors the development of various capacities in students such as autonomous and cooperative learning, creativity, and taking turns, among others [3].

Software has no value by itself without the figure of the teacher, who selects and integrates them to turn the activity in the classroom or outside it, into a motivating experience, presenting itself as a learning support tool.

The generations of students change every year and with the passage of time, the way of educating must also adapt. Several investigations suggest that as a result of the interaction that children establish with technology from birth –in that they know or learn reality from the digital world–, they would have developed the ability to carry out activities simultaneously, which is known as "multitasking" [24,25]. According to the researchers, the new generations would have the ability to interrupt their activities and return to focus on their work immediately, without negatively affecting performance in the tasks they perform [4].

One of the essential tasks performed by clinical laboratories is to analyze the various parameters that are measured in examinations to control diseases within the population. The disease process corresponds to the alteration or deviation of the physiological state in one or more parts of the body, for generally known causes, manifested by characteristic symptoms and signs, whose evolution is more or less predictable [5], according to what the World Health Organization (WHO).

Laboratories are very helpful diagnostic support units and have an important role in people's health, mainly because of the responsibility they fulfill when conducting tests on the population that are guaranteed under a quality system and that provide reliability to users.

In addition, the laboratory is present in providing data for the prevention of disease, diagnosis or control of possible health problems, treatment of diseases, physiological states or parentage conditions [6] that occur in patients.

Laboratory work is a complex process, since the quality of clinical examinations is mainly focused on the analysis of a sample, leaving aside other stages of the process that takes place in the laboratory [7]. For the complete analysis of an exam, the laboratory work is divided into three main phases: preanalytical, analytical and post-analytical phase. Each of these phases are fundamental to deliver a correct result, in which is the phase that involves the procedures before the laboratory analysis, starting with the choice and request of the tests, going through the registration of patients and their exams and collecting and transporting samples [6]. Within the analytical phase, is the processing and analysis of biological fluids. Finally, the post-analytical phase consists of the delivery of results, which must be reliable, so it is necessary to rigorously comply with the aforementioned phases.

Although there are 3 phases, in recent years, various authors have defined two additional stages, a pre-preanalytical stage that corresponds to the entire process contained from the examination order to receipt of the sample in the clinical laboratory, and the post-post-analytical stage related to the interpretation of results [8], given the importance of identifying each of the stages to maintain the quality of the exams delivered to the patient.

The analytical phase has decreased its percentage of error as time passes, since over the years improvements have been established in the quality control of the techniques used and in addition a greater automation of the equipment has been achieved, achieving with this that the source of error in that stage has been reduced.

Technology has helped health personnel a lot to carry out examinations in a timely manner, but these have the limitation that they only process samples and deliver results, so the previous stage "taking samples" is a fundamental step that, if done correctly, it will enforce the effectiveness and efficiency of these teams.

On the other hand, the preanalytical phase contains a higher percentage of error in relation to the procedures associated with the laboratory, since it is the one that contains the greatest human intervention, either by professionals or users.

Based on what was presented above, this work proposes the design, development and evaluation of a video game that allows the use and evidence of the technical learning involved in the preclinical phase to be integrated into a Moodle platform.

2 Background

The preanalytical phase corresponds to the entire process involved before the analysis of a sample, which includes the request for examinations, requirements that the patient must meet to undergo said examination and its correct identification, in addition to including the taking of the sample as such., conservation of the samples and the transport of these, which must be carried out under certain rules and instructions according to the examination carried out, contemplates until before the examination is executed, that is, before the beginning of the analytical phase [9–11].

By means of a sample of biological fluids, it is possible to obtain results of different biological parameters of the patient (biochemical, hematological, immunological, microbiological, among others) and thus obtain a diagnosis or monitoring of an adequate pathology. For this, the preanalytical phase is essential and therefore all the processes that are related to taking the sample, the requirements according to the examination, the correct procedure and transport in optimal conditions, thus allowing to obtain true results on the health of the patients. This is of utmost importance, since it must be taken into account that the health status of people is evaluated through examinations, and any error can lead to an equivocal diagnosis and consequently damage the health of the patient.

Within the laboratory procedure, the phase that contains the highest percentage of error is the preanalytical phase, followed by the post-analytical phase and ultimately the analytical phase, because the latter presents greater automation, improvement in quality control and less human intervention in its process [10, 11], which is contradicted by the preanalytical phase, whose processes always require human intervention.

Some studies mention that of the total errors that can be made in the laboratory, it is identified that between 40% and 70% are attributed to the preanalytical phase [12–14]. However, everyone agrees that it is this phase where the highest percentage of errors are made, most of which can be preventable to provide a better quality of examination results and have a direct impact on the patient's health [10]. Table 1 shows different types of errors that can be made in the preanalytical phase.

Table 1. Main reasons for errors in the preanalytical phase according to their authors [11, 15, 16].

Quiroz C. 2018	Gil P. et al. 2016	Carraro P. & Plebani M. 2006
• Coagulated samples • Hemolyzed samples • Inadequate volume	• Error in exam request • Inadequate volume • Hemolyzed samples • Coagulated samples	• Inadequate volume • Error in patient identification • Wrong container

In relation to the post-analytical phase, the errors associated with it vary between 18.5 and 47% of the total of laboratory errors, some authors agreeing with a percentage of 20% [8, 10].

Although the complete process within the laboratory is generally divided into three main phases (pre, intra and post analytical), Plebani, M. in 2006 [7] emphasized the importance of the different stages of the laboratory process, since that today despite the fact that laboratory services are safe, they are not as safe as they could or should be, mentioning that the pre-analytical stage is the one in which more errors are committed within the laboratory process and that if it were in a better controlled way, the risks that patients may suffer when attending health services could be reduced. They point out that the analysis of the beginning and the end of the cycle reveals that currently, the pre- and post-analytical steps are more prone to errors than the intra-analytical processes.

These errors bring some consequences such as a sample that cannot be analyzed due to factors that mostly depend on the official in charge or the patient, which will imply requesting the patient again to attend the health facility and perform a venipuncture, or in other cases, events of greater impact on the patient may occur, such as administering equivocal treatments due to poor performance of the procedure, thus affecting the safety of the person. Therefore, it is important that, in each facility, whether in public or private health, they comply with standardized procedures with scientific support, which reduce the possibility of errors, since these can be preventable by the staff of work and health professionals to prevent possible damage and improve the quality of this phase.

That is why it is in this phase where different processes that are required to be carried out correctly in order to provide optimal blood samples to the clinical laboratory professional for subsequent analysis, and consequently obtain accurate results on the health status of patients.

The correct performance of the preanalytical phase is becoming more and more relevant, since this phase is part of the accreditation process of clinical laboratories, therefore, complying with the quality of this phase as one of the criteria, can allow them to access to said accreditation, which will give them certain guarantees such as demonstrating that they have a quality management system and therefore the results of the analytical tests will be reliable, and demonstrating that they comply with good practices within the laboratory [17] to promote safe and quality care. In addition, an accredited laboratory, in the event that it performs the entire procedure related to medical examinations, is competent to carry out the activities it performs, including the reduction of errors in the pre-analytical, analytical and post-analytical processes, facilitating diagnoses. accurate, fast and efficient treatment for the user [18]. The Manual of the General Standard of Accreditation for Clinical Laboratories published by the Superintendency of Health of Chile since 2010 establishes a series of indicators, with which the campus must comply mostly to be able to acquire it, within these indicators is the scope process management, where the institutional provider must supervise compliance with standard precautions and the correct use of antiseptics, and must also comply with other requirements such as "regulate, apply and periodically evaluate the processes of the pre-analytical stage" in order to provide quality care and safe health benefits.

It is vital that great importance is attached to this stage by both the physician and the laboratory staff, since it is the one that will largely determine the quality of the results to be obtained [19].

The pre-analytical phase is the stage in which the greatest number of errors occur, it explains that the main source of error during the collection of samples is mainly human, with most of these errors being preventable in a simple way, such as offering continuous training to the staff through courses that update on how to correctly perform the sampling technique, reinforce knowledge about the fundamentals and importance of each stage, always ensure the improvement of patient safety in healthcare [20].

The video game is today a tool that has an incipient development in education, but it is considered a promising resource due to its widespread use and acceptance among today's student and university generations. For Kim S. et al. [21] a game is an action or a set of actions, which includes one or more people, objects or animals, general.

3 Methods and Materials

To complement the learning process of the knowledge involved in the preanalytical phase, a video game was designed and developed, in which students use their knowledge and can independently validate the various decisions they make.

A. Technologies

This video game was integrated into the LMS Moodle platform. Due to the development architecture of the Moodle plugins and the information display restrictions that this platform provides, HTML 5 was used to develop the game and it was published on an independent Moodle web page. To add this game to Moodle, you must add a new resource, specifically a URL and configure a set of parameters that are available on the platform to be added to said URL.

To validate access to the video game, the parameters provided in the URL are verified, which correspond to the URL of the Moodle platform, the course identifier and the user identifier.

As the game was not developed as a Moodle plugin, a Middleware was developed for the integration between Moodle and the video game. For this, a service-oriented software architecture was used, which delivers the necessary endpoints so that Moodle can display the processed information to the teacher.

Additionally, a Moodle plugin was developed to review and control the progress of the students by the teacher. The integration scheme is presented in Fig. 1.

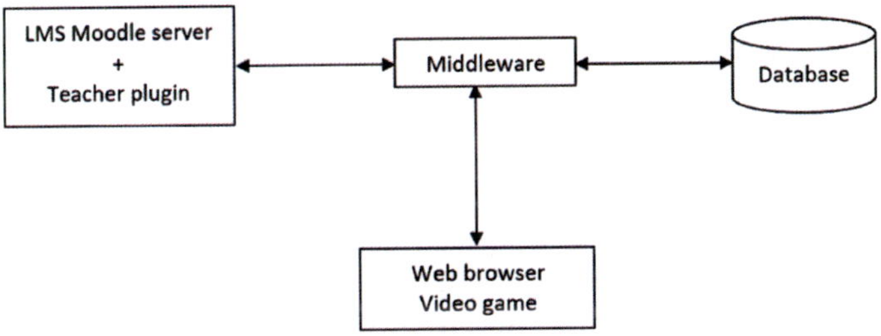

Fig. 1. Services architecture diagram

Finally, to develop the video game, the Game Engine Unity 3D was used, although only the characteristics for the development of 2D video games were used. It was then exported to HTML5 so that students could play using a web browser, either from a computer, smartphone or tablet.

B. Application Design

The video game developed simulates a clinical laboratory, in which a simplified process of the preanalytical phase was modeled, adjusting as much as possible to reality. Every day that the laboratory works, patients arrive with a medical order to perform one or more clinical laboratory tests. The video game has a database of exams and patient information and randomly generates an order for medical exams. Due to the complexity of some types of exams, the first days of the game are generated those exams and exam combinations that have a low difficulty and as the student progresses within the game, new types of exams and new combinations are enabled that require the student handle more complex concepts.

At the reception, the person in charge checks that the order data corresponds to the patient's identification and that, according to the examinations, it complies with the prerequisites to be able to perform said procedures. Here the student must put into practice the knowledge of the prerequisites of the exams and the procedures to apply the contents committed in their course. Figure 2 shows how the reception process is represented.

Once the patient has checked his data at the reception and fulfilled the prerequisites, the patient is referred to the process of taking exams, where a Medical Technology professional performs each of the actions required in each exam. Depending on the test,

Fig. 2. Reception process design

a blood, urine or stool sample will be taken. Only in the case of blood samples does the Medical Technology professional carry out their intervention.

For the phlebotomy process, the student must once again apply the acquired knowledge related to the sampling processes. Figure 3 shows how the phlebotomy process is represented. Once the student has completed this procedure, the samples are sent for processing. To do this, in the video game this validation was implemented in a machine that processes the exams and later returns the result. If all the processes where the student had to apply the knowledge acquired in the course were correct, the machine will give a positive result, otherwise, it will throw an error message indicating the incorrect procedures.

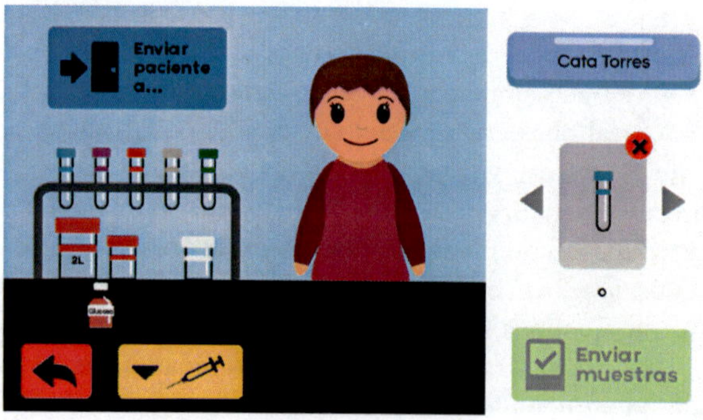

Fig. 3. Phlebotomy process design

Additionally, with the student's identification data, each of the interactions that the student performs with the game is recorded, in this way it is possible to track the playing times, the most used functionalities within the game, the most common errors made. by the players and thus the teacher can determine the most complex exams and other critical learning/teaching processes.

4 Results

The result of this work is a video game aimed at strengthening the learning of students in the preanalytical phase. Access to this video game is done through a Moodle platform which provides user information. From Moodle you can access a URL that shows the game in a web browser and can be played on a computer, smartphone or tablet. The teacher, through a plugin developed for Moodle, can check the progress of the students, the time they have played and can check the access times to this video game.

A pilot test was carried out in the course "Preclinical: Bases for the clinical laboratory" which is in the fourth semester of studies of the Medical Technology career at the University of Talca. 29 students participated in this course, who used the game and later completed a satisfaction survey on the use of the video game. The objective of this survey is to validate the use of the video game as a complementary tool for student learning. Next, Table 2 presents the results obtained when applying the survey. On the scale used, 4 corresponds to the best evaluation and 1 corresponds to the lowest.

Table 2. Results of the satisfaction survey, where 4 corresponds to the best evaluation and 1 to the lowest.

Question	1 (%)	2 (%)	3 (%)	4 (%)
1. Do you believe that the use of the video game is presented as a facilitating resource for the learning of the module	3.4	0	13.8	82.8
2. The usage of this methodology allows the understanding of concepts at an abstract level (theoretical), which makes them better retained by the student	0	0	31	69
3. Considers that the use of the video game is an innovative methodology	0	0	20.7	79.3
4. At the end of the last week of reporting (week number 8), you noticed an advance in your knowledge with respect to the initial week (week 1)	0	0	13.8	86.2
5. Do you believe that the use of the video game helps to supplement the clinical practice that cannot be carried out due to contingency reasons (COVID-19 pandemic)	13.8	34.5	37.9	13.8
6. The use of the video game is a methodology that you would recommend continuing using in the module for future generations	3.4	3.4	17.3	75.9
7. The use of the video game made distance education more welcoming and didactic for you	3.4	3.4	24.2	69

Additionally, in the survey, students were asked about the advantages and disadvantages that they appreciated using the video game. Among the main advantages, the students pointed out that the video game is easy to use, that they could easily apply the knowledge learned without having the complication that by making mistakes they damage a medical sample and that it served to learn some concepts that they had not been able to learn with the classes traditional. As disadvantages, the students pointed out that this video game does not replace clinical practices, that an internet connection was needed each time it was used and that it only reinforced theoretical knowledge.

5 Discussion and Conclusion

During the second semester of 2021, the video game was applied to 29 students of the course "Preclinical: Bases for the clinical laboratory" belonging to the fourth semester of the Medical Technology career at the University of Talca. Results of the satisfaction survey on the use of the video game, the students expressed their comments related to the advantages and disadvantages of the video game together with their appreciations of whether the video game helps the learning process of the preclinical phase.

As can be seen in the results, the criterion evaluated with the lowest evaluation of the survey applied to the students coincides with one of the disadvantages mentioned by the majority of the students who participated in this experience, and this is related to the statement "Do you believe that the use of the video game helps to supplement the clinical practice". This is explained given the importance of clinical practice activities for the training of health students.

Clinical practice is the process declared in the training curricula of health professionals that historically constitutes one of the main methodologies for the acquisition and application of knowledge, development of skills, making autonomous and collaborative decisions that are necessary for their professional training.

Today practice in training is conceptualized as a reflective action mediated by practice and not as the mere application of previously acquired knowledge [22, 23]. In medical education the concept of learning in context is used, which considers that students learn more if they do it in the same context in which they will use the knowledge.

In addition to this, in clinical practice, students face situations with dynamic, random and complex characteristics which cannot be reproduced by the video game, as it was developed, however this disadvantage opens a window to make improvements in this area in the tool developed.

Finally, it should be noted that a large percentage of the students indicated that the video game is presented as a resource that facilitates learning and that allows the understanding of theoretical learning and recommends its future use.

Due to the limitations of the Moodle plugins development, it was needed an API of this game to correctly integrate data to Moodle, this will be helpful to use another LMS tool.

References

1. Pulman, A., Scammell, J., Martin, M.: Enabling interprofessional education: the role of technology to enhance learning. Nurse Educ. Today **29**(2), 232–239 (2009). https://doi.org/10.1016/j.nedt.2008.08.012

2. Guerra-Antequera, J., Revuelta-Domínguez, F.I.: Videojuegos precursores de emociones positivas: propuesta metodológica con Minecraft en el aula hospitalaria. Int. J. Educ. Res. Innov. **3**, 105–120 (2015)
3. Lopez, C.: El videojuego como herramienta educativa. Posibilidades y problemáticas acerca de los serious games. Apertura. Rev. innovación Educ. **8**(1), 1–15 (2016). http://www.scielo.org.mx/pdf/apertura/v8n1/2007-1094-apertura-8-01-00010.pdf
4. Sweller, J., Kirschner, P.A., Clark, R.E.: Why minimally guided teaching techniques do not work: a reply to commentaries. Educ. Psychol. **42**(2), 115–121 (2007). https://doi.org/10.1080/00461520701263426
5. Herrero Jaén, S.: Formalización del concepto de salud a través de la lógica: impacto del lenguaje formal en las ciencias de la salud. Ene **10**(2) (2016). Accessed 12 Jul (2021). https://scielo.isciii.es/scielo.php?script=sci_arttext&pid=S1988-348X2016000200006&lng=es&nrm=iso&tlng=es
6. Martins, J.M., Rateke, E.C.M., Martinello, F.: Assessment of the pre-analytical phase of a clinical analyses laboratory. J. Bras. Patol. e Med. Lab. **54**(4), 232–240 (2018). https://doi.org/10.5935/1676-2444.20180040
7. Plebani, M.: Errors in clinical laboratories or errors in laboratory medicine? Clin. Chem. Lab. Med. **44**(6), 750–759 (2006). https://doi.org/10.1515/CCLM.2006.123
8. Toledo, F., Rassetto, M.: Introducción a la etapa pre-analítica en el análisis bioquímico. vol. 2014, no. 2, pp. 1–8 (2017)
9. Esteban, M., Ruiz-Moraga, M., Pérez-Gómez, B., Castaño, A.: Aspectos prácticos de la fase preanalítica del estudio de biovigilancia BIOAMBIENT.ES. Gac. Sanit. **27**(1), 77–80 (2013). https://doi.org/10.1016/j.gaceta.2012.07.004
10. San Miguel Hernández, A., de la Fuente Alonso, P., Garrote Adrados, J.A., Lobo Valentin, R., Lurueña, M.L., Eiros Bouza, J.M.: Minimización de errores preanalíticos y su repercusión en el control del laboratorio clínico. Rev. del Lab. Clin. **11**(1), 51–58 (2018). https://doi.org/10.1016/j.labcli.2017.02.001
11. C1, Q.-A., Florez, O., Salcedo-Cifuentes, M.: Errores pre analíticos en el laboratorio clínico de un hospital de tercer nivel: prueba piloto
12. Marzana Sanz, I., et al.: Recomendaciones para el diseño e implementación de un programa de aseguramiento de la calidad de la fase preanalítica. Rev. del Lab. Clínico **12**(4), e54–e65 (2019). https://doi.org/10.1016/J.LABCLI.2019.01.003
13. Rodríguez Fraga, O., et al.: Recomendaciones preanalíticas para la medición del equilibrio ácido-base y los gases en sangre. Recomendación (2018). Rev. del Lab. Clínico **12**(4), e66–e74 (2019). https://doi.org/10.1016/J.LABCLI.2018.12.001
14. Ibarz, M., Gómez-Rioja, R.: Recomendaciones conjuntas EFLM-COLABIOCLI para la extracción de muestras de sangre venosa; armonizar desde la base. Rev. del Lab. Clínico **12**(2), 61–63 (2019). https://doi.org/10.1016/J.LABCLI.2019.02.001
15. Gil, P., Franco, M., Galbán, G.: Evaluación de errores preanalíticos en el laboratorio de planta del HIGA O. Alende de Mar del Plata. Acta Bioquímica Clínica Latinoam. **50**(3), 463–468 (2016). http://www.redalyc.org/articulo.oa?id=53549173015
16. Carraro, P., Plebani, M.: Errors in a stat laboratory: types and frequencies 10 years later. Clin. Chem. **53**(7), 1338–1342 (2007). https://doi.org/10.1373/CLINCHEM.2007.088344
17. Am, S., et al.: Survey of national guidelines, education and training on phlebotomy in 28 European countries: an original report by the European Federation of Clinical Chemistry and Laboratory Medicine (EFLM) working group for the preanalytical phase (WG-PA). Clin. Chem. Lab. Med. **51**(8), 1585–1593 (2013). https://doi.org/10.1515/CCLM-2013-0283
18. Zima, T.: Accreditation of medical laboratories – system, process, benefits for labs. J. Med. Biochem. **36**(3), 231 (2017). https://doi.org/10.1515/JOMB-2017-0025

19. Herrera, Y.C., Rivero, M.C., Correa, M.A., Savón, K.G., Martínez, O.F., Herrera, A.G.: Importancia de la fase preanalítica en el laboratorio clínico de la Atención Primaria de Salud. Rev. Med. Isla la Juv. **15**(1), 3–21 (2014). http://www.remij.sld.cu/index.php/remij/article/view/89/188. Accessed 12 July 2021
20. Cuadrado Cenzul, M.A., Collado Yurrita, L., Gonzalz Estecha, M., Pedro Moro, J.A., Arroyo Fernandez, M.: Impacto de los errores del laboratorio clínico en la asistencia sanitaria y seguridad del paciente. Roche Diagnostics Inf., pp. 2–8, October 2015
21. Kim, S., Song, K., Lockee, B., Burton, J.: What is a game? In: Gamification in Learning and Education. AGL, pp. 15–23. Springer, Cham (2018). https://doi.org/10.1007/978-3-319-47283-6_3
22. Cochran-Smith, M., Lytle, S.L.: Relationships of knowledge and practice: teacher learning in communities. Rev. Res. Educ. **24**, 249–305 (1999). https://doi.org/10.3102/0091732X024001249
23. Pérez, Á.I., Resumen, G., Ángel, C., Pérez Gómez, I.: Aprender a educar. Nuevos desafíos para la formación de docentes. Rev. Interuniv. Form. del Profr. **68**(2), 37–60 (2010)

Natural Language Processing Environment to Support Greek Language Educational Games

Aristides Vagelatos[1(✉)], John Stamatopoulos[2], Maria Fountana[1], Monica Gavrielidou[1], and Christos Tsalidis[2]

[1] Computer Technology Institute & Press, 10563 Athens, Greece
vagelat@cti.gr
[2] Neurocom S.A., 15124 Maroussi, Greece
info@neurocom.eu

Abstract. In this paper, we present the so far implemented infrastructure of the "Lexipaignio" project, a research project co-financed by EU and Greek national funds, where an innovative and state-of-the-art NLP (Natural Language Processing) environment is being developed for the creation of digital educational games for Greek students. An initial, brief presentation of the the position of digital games in the today's educational system is followed by a more detailed presentation of the implemented NLP infrastructure for the Greek language. Examples of the games that have already been implemented are also provided. The paper concludes with the current stage of the project and the pending steps towards its completion.

Keywords: Digital educational games · Natural Language Processing · Game-based learning · Mobile learning · Open and distance learning

1 Introduction

Lately, with the integration of new digital accomplishments into both educational and everyday life of students, important changes are under progress in the educational and learning processes, where Information and Communication Technology (ICT) plays a critical role. The situation was boosted even more, during the COVID-19 pandemic, where more and more pupils started to utilize digital technology in order to participate in distance learning classes. From this perspective, the use of digital games to support (game-based) learning through an alternative, more attractive way is rapidly developing in both European and worldwide level.

"Digital games" is a rapidly developing field, as they are amongst the most popular technologies young people use to amuse themselves. The educational potential of digital games is correlated to the properties of motivation, amusement and the trigger of interest, which are considered consistent with positive learning results. According to related research [12], computer games provide a quick and interesting learning pace in contrast to the conventional teaching methods and in this perspective, they introduce modern alternatives in digital learning. Furthermore, the embedding of Artificial

M. E. Auer and T. Tsiatsos (Eds.): IMCL 2021, LNNS 411, pp. 525–536, 2022.
https://doi.org/10.1007/978-3-030-96296-8_47

Intelligence (AI) solutions (in the form of machine learning, big data analysis, as well as "modern" NLP techniques) within the school setting is becoming more intense, providing a wide variety of tools and opportunities. In this context, the use of Natural Language Processing (NLP) can be very useful for the creation of educational language games.

Within a project, called "Lexipaignio", an innovative and state-of-the-art computational environment is under development, targeting the creation of digital educational games for Greek students (primary and secondary level) in order to: a) improve language competence and overall level of students' knowledge and b) develop various vocabulary and linguistic skills, while understanding the context of specific school subjects (biology, geography etc.).

The new environment supports the automated production of questions (e.g., in the form of quizzes) related to the various levels of the Greek language structure and use as for example spelling, morphology, vocabulary, as well as terminology found in school textbooks which is integrated into the overall environment and narration of digital educational games.

Additionally, it enables teachers to introduce new content and define additional areas of interest regarding the Greek language structure. This will result to an automatic creation of an extended volume of questions/tasks, supervised by educators, through their intervention in educational games (crosswords, match games, multiple choice, games where students have to find the correct order of the mixed letters). At the same time, teachers are able to adjust specific game parameters taking under consideration aspects such as: a) educational level, b) school subject (biology, geography, literature etc.), c) school grade and teaching module, d) the general class of grammatical phenomena (conjugation, spelling, syntax, vocabulary).

In this paper, we present the so far implemented infrastructure of the "Lexipaignio" project. Firstly we discuss briefly the position of digital games in today's educational system, then we present the implemented NLP infrastructure for the Greek language, following by a few examples of the games that have already been implemented and finally we give some conclusions.

2 Games in Education

Digital games are becoming a useful tool in the teaching process. The multiplicity of their qualities results in considering digital games as a pedagogical support to teach, train, solve, provide examples and raise awareness among students [3].

Mechanics, one of the main constituents of digital games [15] determine the rules and procedures of the gameplay. Thus, player interaction, immersion and sense of engagement are made possible. In the context of Artificial Intelligence (AI), Natural Language Processing (NLP) is being increasingly used in digital game design. Although current research on educational digital games does not treat NLP techniques as distinct game features but as a facilitator of game objectives [11], nowadays NLP claim a pivotal role in the creation of educational games. In "Lexipaignio" project NLP is connected to the dynamic creation of a series of language games integrated in the Greek school curricula.

Effectively, the current curricula for the teaching of Modern Greek in primary and secondary education promote the study of grammatical phenomena through the text-centered approach [7]. Thus, grammatical phenomena are studied in textual environments where students are trained in recognizing the function of grammatical structures and language mechanisms. Similarly, in the context of vocabulary teaching, words are not treated individually, but are included in the text and are studied as structural elements of speech. With focus to the computer-human interactions through human language, NLP technologies can prove to be particularly useful in the development of educational language games where language challenges will be produced dynamically, utilizing a specific body of text and other resources.

3 NLP for the Greek Language

At the point of convergence of computer science and linguistics, Natural Language Processing (NLP) constitutes a challenging area for numerous applications in our everyday life. Focusing on education, the "Lexipaignio" project aims at the use and further development of a series of Natural Language Processing infrastructure tools (Morphological Lexicon, Lemmatizer, "Mnemosyne" language processing system, corpus of Greek school subjects, etc.), for the implementation of dynamically created gamified educational material.

In this section, the NLP infrastructure that has been developed is presented, alongside some basic Greek language characteristics, in order to exemplify the peculiarities that had to be addressed.

3.1 Some Facts About the Greek Language

Greek is the official language of Greece, as well as one of the 24 official languages of the European Union (EU). In the 23rd edition of Ethnologue - a language reference database published by SIL International [1], it is described as having almost 13 million native language speakers[1], whereas it has been ranked at the 89th position among the 200 most common languages worldwide. The Greek alphabet consists of 24 letters and two diacritics, the stress mark (e.g., «ί») and the diaeresis, consisting of two dots above a vowel letter and indicating a separate syllable (e.g., «ϊ»).

Certain characteristics of the Greek language have challenged its computational processing, with the most significant being the following:

(a) Historical orthography: Certain vowel phonemes are represented with multiple orthographic graphemes: /o/ can be spelled with either ο or ω, /i/ with *η*, *ι*, *υ*, *ει*, οι, or *υι*; and /e/ can be spelled with either ε or *αι*. This has an effect not only on lexemes (e.g., '*φύλλο*' (leaf), '*φίλο*' (friend), '*φύλο*' (gender) are all pronounced /filo/), but also on inflectional affixes (e.g., '*πολλοί*' (many) (masculine plural), '*πολλή*' (much) (feminine singular), '*πολύ*' (neuter singular) are all pronounced /poli/)).

[1] Besides, Greek is spoken by around five (5) million people, members of the Greek communities (the Diaspora) who live outside Greece.

(b) Morphology and inflection: Greek language is characterized by an extensive morphological structure, especially in the noun and verb inflectional systems (i.e., three genders, two numbers and four cases for nouns and adjectives, and a sheer number of discrete stems and endings for verbs which indicate mood, aspect, voice, tense, and person).
(c) Morphosyntactic ambiguities: These are due to homographs, as in the case of most adverbs (e.g., '*καλά*' (well)) vs adjectives in plural ('*καλά*' (good)).
(d) Word order and syntax: Greek deploys a subject-verb-object pattern (SVO) in syntactic structures. However, unlike languages like English, the Greek word order is not fixed and the constituents of a clause or of a nominal phrase are characterized by a great flexibility.

In the next subsections, we provide an overview of the standard handling methods developed for the Greek language at different levels of text processing: (a) segmentation and tokenization, (b) lemmatization, stemming, spell checking, and morphosyntactic tagging, (c) morphosyntactic disambiguation, syntactical analysis and/or parsing, d) named entity recognition (NER), and term extraction.

3.2 Segmentation and Tokenization

Text analysis follows a two-phase process. In the first phase, the text is separated to paragraphs, sentences, and words or other tokens, such as the punctuation symbols. In the second phase, the synthesis, the tokens are compiled to syntactic structures that facilitate the recognition of named entities and meanings. Tokenization is usually the first component in every NLP pipeline. The main problem in sentence splitting is the recognition of dot ('.') as a sentence delimiter in distinction with other uses (e.g. in abbreviations). As far as Greek language is concerned, there is no secure way to deal with the task of abbreviations' recognition. The most effective approach has proved to be the use of techniques that combine dictionaries of abbreviations with heuristic patterns.

3.3 Word Processing

One important feature in processing textual data, especially from social media platforms and chatbots, is the ability to recognize misspelled words. Spelling errors are divided in two main categories, the typographic and the orthographic ones [6]. During a previous project, a spelling correction system for the Greek language has been implemented some years ago [18]. Based on this implementation, and for the needs of the current project, correction algorithms are implemented for the opposite task: to produce misspelled words from a correct one. Thus, typographic errors which are common to all languages and are usually created by mistake, are now introduced in correct word forms in order to produce misspelled ones e.g., wrong letter, missing letter, extra letter, wrong order on a pair of letters, etc. Notably, the orthographic errors are usually due to the lack of knowledge of the correct word. The misspelled word sounds or looks very similar to the correct one and the correction process must consider the similarity factor and order of the proposed words accordingly.

3.4 Sentence Based Techniques

Ambiguity is an inherent property of natural languages, as it expresses the level of uncertainty about the meaning of a word, phrase, or sentence, and is present in all phases of content analysis. Word ambiguity can be lexical, syntactic, and semantic. Lexical ambiguity is due to polysemy (i.e., the fact that there is more than one meaning for a word in a natural language). The process to resolve the correct meaning is called "word sense disambiguation" and is presented in advanced NLP pipelines. A subcase is considered the Part of Speech (POS) disambiguation, where the process is referred as tagging and the program that accomplishes it is called tagger. Usually, a Greek tagger uses a mixed scheme with three phases [10]. In the first phase, the word is looked up in the morphological dictionary and, if it is found and it has a single entry, all morphological attributes are returned. In case no single entry is found, a set of rules is applied on the context of the word, trying to distinguish the correct one. If a word is unknown, a set of rules examine the suffix and other characteristics of the word, trying to guess its POS. To support this processing, a morphological dictionary has been developed [16] with $\sim$ 100.000 lemmas and 1.200.000 words, containing rich morphological and semantic information for lemmas and inflectional or derivational types.

3.5 Semantic Analysis

Named Entity Recognition (NER) as well as term recognition, are widely used in many NLP applications, such as classification, semantic indexing and searching, virtual assistants, etc. The task of recognition of the syntactic structure of a phrase or a sentence is complex and nondeterministic. Although for multi-word terms the task is simpler because they follow specific morphosyntactic patterns, nevertheless a grammar formalism is necessary to describe these patterns, as well as a resolution algorithm that will apply the grammar to a text. A context sensitive grammar formalism has been developed, called Kanon [17], along with an efficient resolution algorithm that applies the grammar to the text and recognizes syntactic structures. The resolution algorithm applies a surface parsing technique that permits the recognition of parts of a sentence without requiring the syntax recognition of the whole sentence.

3.6 NLP Applications

Modern NLP emerges from the opportunity to better utilize the high volume of open data created in the Web [2]. Using ML and deep learning technologies, this knowledge is encoded to multidimensional numeric vectors that are used in a plethora of classification applications. For the Greek language, the volume of data that are open and in appropriate format is restricted and cannot lead to efficient models. A combination of rule-based technologies utilizing the existing solid knowledge stored on prefabricated language resources, and ML technologies with auto-adaptation characteristics utilizing the knowledge residing inside texts, has proved to deliver the best results [4].

Using the technologies mentioned above, several useful components have been developed which can be used either as stand-alone applications, or in suites that implement complex NLP tasks and support the "Lexipaignio" infrastructure. Some of these components are presented below:

Stemming & Lemmatization. Stemming is an important function in NLP applications, when about the main interest is the meaning/concept and less the nuances of morphology. Stemming is achieved by using a process called "Suffix Stripping", which returns the word stem by removing the end part of the word that bears morphological information (i.e., inflection or derivation). Stemming is not considered an efficient technique to deal with Greek words, since in many cases it produces non-existent words. Therefore, it is used only in cases of unknown words. Lemmatization is a similar function with the exception that the produced word, called lemma, is a correct word. Lemma is usually the headword in electronic or printed dictionaries and is used for citation. Accurate lemmatization for Greek language requires the existence of a morphological lexicon. For the medical sector, two additional problems arise: (a) unknown vocabulary that is not negligible, (b) the mixed morphology (Katharevousa vs Demotic forms) that virtually doubles the size of the words.

Keyword Extraction. This component applies tokenization, lemmatization and stemming to the words of the documents and the result becomes the input to statistical algorithms as TF-IDF [8] and BM25 [13]. The output is the tagging of the documents in a collection with the most "important" words, i.e., keywords that identify the document in the collection. Keywords are useful in indexing, searching, and classifying documents as presented in components below.

Related Documents. This component utilizes the "Keyword extraction" and NER components [9] to find the attributes that characterize the document and through them to find related documents in a collection. The component can utilize dictionaries or vocabularies, named entities, terms and events with a customized weighting scheme to compute the level of relation between documents.

Clustering. For profiling documents and mining the knowledge therein contained, the clustering component implements known algorithms as K-Means [8] and Hierarchical [14]. These algorithms are unsupervised learning techniques which can be successfully used for clustering collections with medical diagnoses or medicines without needing other resources.

Word Embeddings. Word representation with a vector of numbers opens a window to the utilization of modern technologies of machine and deep learning emerging from the application of mathematical sectors as linear algebra, statistics, probability theory, etc. These vectors are created from the processing of large corpora and encoding the semantic characteristics of the words examining the different context environments they appear. According to Distributional Hypothesis [5], two words have similar meaning if they can appear in the same context. The word vectors map the words pointing in a high dimensional space; having these properties means that the Euclidean distance of the points corresponding to two words with similar contexts is small. In our systems, word embeddings are also used inside grammar rules to codify the meaning of a word.

4 Example Games

4.1 Configuration

The above described NLP infrastructure was extended and configured appropriately in order to serve as the basis for dynamic production of linguistic games. More specifically the following steps were made:

- A corpus was compiled in order to serve as the input for the games. The collection of documents used includes more than 180 chapters from the educational books used in upper primary and lower secondary education in Greece, cleared from images, tricky typographical and layout related elements.
- The corpus was processed and annotated, using some of the NLP tools described earlier.
- Regarding each individual game preferences, data input was created dynamically. For example, in the case of "fill in the blanks" type of games, certain sentences were selected (see Fig. 1) from the corpus, based on a number of requirements (e.g. specific language textbook, school level, grade, etc.). For each of these sentences, various information has been retrieved and is ready to be presented whenever needed (e.g. input document, link to the digital book, etc.).
- Having all the morphosyntactic information for these sentences, and according to the game characteristics, it is easy to find out certain part of speech (e.g. verbs), hide them, and ask the user to put these words in the correct boxes (see Fig. 3). In a similar yet different game type, the educator might select to ask the user to find certain word types (e.g. verbs) in a sentence (see Fig. 4).
- Changing to a different game type like "find the correct word form", through which the teacher seeks a way to motivate students to improve their language competency, for a certain grammatical phenomenon e.g. adjective inflection system and more specifically adjectives ending in «–ης» (επίθετα σε –ης), the infrastructure selects an adjective ending in "-ης", removes it from the sentence, and produces three incorrect – although valid- word forms and present them to the user together with the correct form. In this manner, games of the type "who wants to be a millionaire" can be produced as can be seen in Figs. 5 and 6.

It has to be stressed once again, that the input data for the games are produced dynamically, based on a) the corpus collected, b) the type/needs of the selected game and c) on the educator's preferences/needs.

1. The educator has to choose the grammatical/language phenomenon that he wants to teach with the utilization of the game (e.g. use of adjectives, ending in "–υς").
2. The appropriate processing of the corpus (e.g. language school textbooks) has to be realized. The output of this processing becomes the input of the selected game.
3. The teacher gets the games that are produced and integrates them in the way that he/she chooses in his/her class lesson.

In the following section, we present examples of the abovementioned procedure.

4.2 Input Data Samples

Following the above described procedure, the elementary school's 6th grade language text book, has been processed in order to produce the appropriate output that can become the input data for the first experimental games.

The above described NLP infrastructure was extended and configured appropriately in order to serve as the basis for dynamic production of linguistic games.

A sample output is presented in Fig. 1, where the document title can be found in the first column, the selected sentence in the second one and a link to the actual textbook in the third column.

Document Title	Selected sentence(s)	Link to the textbook
Η χελώνα και ο Ρεβιθάκης	Άμα την είδε αυτός, τρελάθηκε από την ομορφιά της. Βλέπει και το πέπλο που είχε κεντήσει κι ήταν το καλύτερο απ' όλα, και είπε να την πάρει και αυτή αποκρίνεται πως είναι παντρεμένη μ' έναν ψαρά.	https://tinyurl.com/bdsj4c6y
Αιολική γη	Αργά τα δάχτυλα του γέροντα ανοίγουν το μαντίλι όπου είναι φυλαγμένο το χώμα. Ψάχνουν κει μέσα, ψάχνουν και τα δάχτυλα της γιαγιάς, σαν να το χαϊδεύουν. Τα μάτια τους, δακρυσμένα, στέκουν εκεί.	https://tinyurl.com/8kmax5p2
Σπίτι μας είναι η γη	Ένα πράγμα που ξέρουμε εμείς –και που ο λευκός άνθρωπος θα ανακαλύψει ίσως κάποτε– είναι ότι ο Θεός μας είναι ο ίδιος Θεός. Ίσως να σκέφτεστε να τον αποκτήσετε τώρα, όπως θέλετε να αποκτήσετε τη γη μας, αλλά δεν μπορείτε. Είναι ο Θεός του ανθρώπου, και το έλεός του είναι το ίδιο και για το λευκό και για τον ερυθρόδερμο άνθρωπο. Αυτή η γη είναι πολύτιμη για τον άνθρωπο, και όταν τη βλάπτει είναι σαν να δείχνει περιφρόνηση στον Δημιουργό.	https://tinyurl.com/8kmax5p2
Τα Χριστούγεννα του υπολογιστή	Επειδή ο Ντίνος ήταν πάντα πρόσχαρος και γελαστός, οι συνάδελφοί του νόμισαν ότι αστειευόταν, όπως συνήθως. Χαμογέλασαν μ' αυτά που έλεγε και συνέχισαν τη δουλειά τους.	https://tinyurl.com/2ndsnvx6

Fig. 1. Sample output of the preprocessing stage.

In the next table (Fig. 2) a sample input for the "millionaire" game is presented: following the selected lemma (first column), the main sentence is presented followed by the correct answer as well as three incorrect answers that have been produced by the system. The incorrect answers in this game, have been produced by utilizing the thesaurus of the infrastructure where for each entry, synonyms, as well as antonyms are included. The algorithm selects only lemmas that have adequate synonyms, as well as antonyms in order to be able to produce incorrect answers.

Selected Lemma	Selected sentence(s)	Correct answer	Incorrect answers
άβολος	«Αυτή η καρέκλα είναι μικρή και ... για το θείο Αλέκο» είπε ο Κώστας. Ο θείος έχει μεγάλα πόδια και δε βολεύεται.	άβολη	εξυπηρετική, άνετη, βολική
ανόητος	«Τι ... που είναι ο Νίκος» σκέφτηκε η Αθηνά. «Κυνηγάει τη Ροζαλία που δεν τον πείραξε ποτέ σε τίποτα».	ανόητος	εύστροφος, πονηρός, έξυπνος
βγάζω	Ο θείος Τάκης είχε ταξίδι. ... τα ρούχα του από την ντουλάπα και τα έβαλε στη βαλίτσα του.	έβγαλε	ενέταξε, σημείωσε, έβαλε
δειλός	Ο Κώστας είναι ... και φοβάται να πάει στον οδοντίατρο.	δειλός	θαρραλέος, γενναίος, ανδρείος
καθαρίζω	Ο κύριος Μιχάλης ... τα γυαλιά του κάθε μέρα για να βλέπει καλά.	καθαρίζει	ρυπαίνει, βρομίζει, λερώνει
μυστικός	Στον πύργο της κακιάς βασίλισσας υπήρχε ένα ... δωμάτιο που δεν το ήξερε κανείς.	μυστικό	φανερό, σαφές, εμφανές
συχνός	Οι ... επισκέψεις του θείου Τάκη έδωσαν μεγάλη χαρά στον κύριο Γιάννη: ήρθε τρεις φορές μέσα σ' ένα μήνα.	συχνές	σπάνιες, αραιές, νερουλές

Fig. 2. Sample output of the preprocessing output.

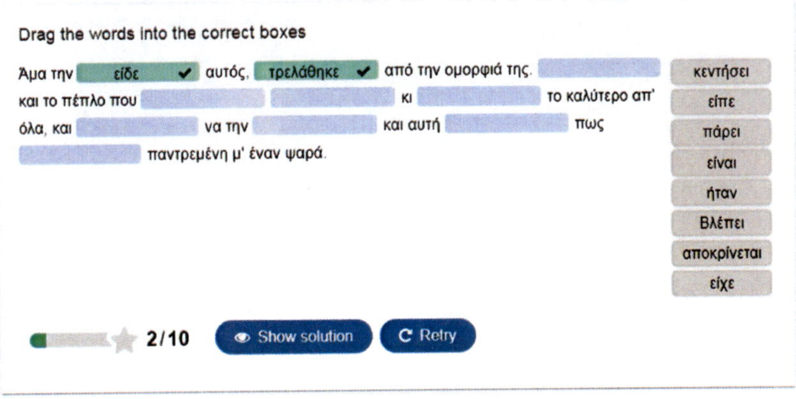

Fig. 3. An example of drag and drop tasks, showing the validation process of a correct selection.

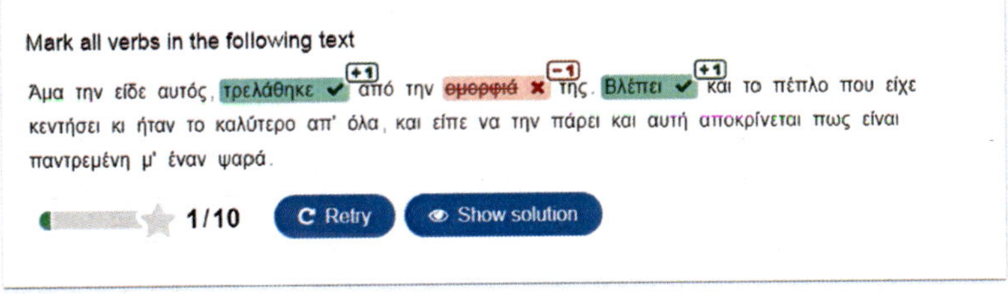

Fig. 4. An example of drag and drop tasks, showing the validation process of both correct and incorrect selections along with the scoring mechanism.

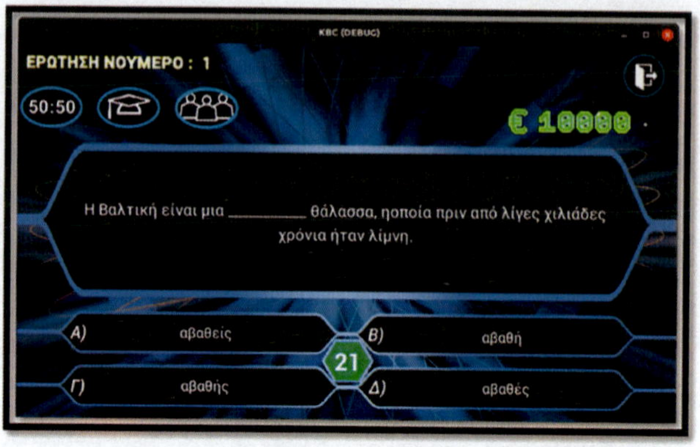

Fig. 5. A sample input for the “millionaire” game is presented followed by the correct answer as well as three incorrect answers that have been produced by the system.

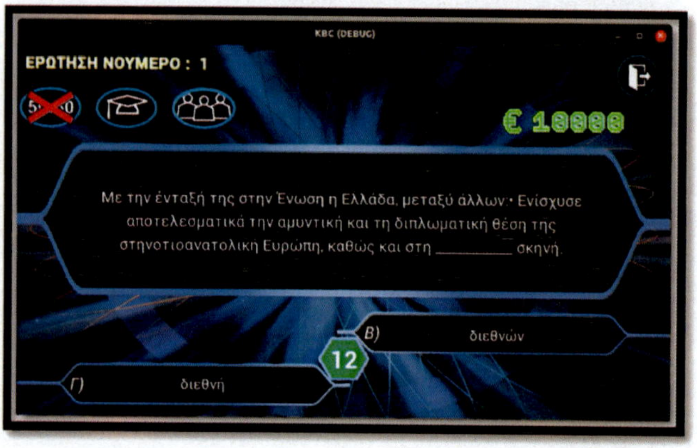

Fig. 6. A sample input for the “millionaire” game is presented followed by the correct answer and one incorrect answer after having applied the option of eliminating half of the possible answers.

5 Conclusions

The NLP environment that is being implemented within the “Lexipaignio” project can become a successful tool to support and enrich the educational process in an appealing and attractive way. As far as language discipline is concerned, the traditional approach, solely giving emphasis to the teaching of rules and the monotone problem-solving on the premise that language is a one-dimensional teaching object, seems not to convey the expected results and should be redefined on the basis of modern functional and communicative teaching approaches.

It should be pointed out that the school textbooks, constitute the testing field for the previously described technology. Consequently, the produced results will almost immediately be useable by everybody involved in the learning process (teachers-parents-students), in various platforms (mobile devices, desktop/laptop computers, etc.). Last but not least, the outcomes of the project will be freely available under open source license.

The project is currently passing its final stages with really promising, initial results. The thus far implemented infrastructure has already enabled the development of pilot digital games for the Greek language, offering users the option to customize (by selecting / setting various parameters) or use them directly for beta-testing purposes. The challenge now is to finalize and introduce the outcomes to educators in order to receive feedback from the actual application both inside and outside the classroom.

Acknowledgment. This research has been co-financed by the European Regional Development Fund of the European Union and Greek national funds through the Operational Program Competitiveness, Entrepreneurship and Innovation, under the call RESEARCH – CREATE – INNOVATE (project code: T1EDK-05094).

References

1. Eberhard, D.M., Simons, G.F., Fennig, C.D., (eds.) Ethnologue: Languages of the world, 23rd edn. SIL International, Dallas (2020). https://www.ethnologue.com/
2. Eisenstein, J.: Introduction to Natural Language Processing. The MIT Press, Cambridge (2019)
3. European Games Developer Federation (2020). Using Digital Games in Education. Retrieved from www.egdf.eu Using digital games in education (egdf.eu)
4. González-Carvajal, S., Garrido-Merchán, E.C.: Comparing BERT against traditional machine learning text classification (2020). https://arxiv.org/abs/2005.13012
5. Harris, Z.S.: Distributional structure. Word **10**(2–3), 146–162 (1954). https://doi.org/10.1080/00437956.1954.11659520
6. Hládek, D., Staš, J., Pleva, M.: Survey of automatic spelling correction. Electronics **9**(10), 1670 (2020). https://doi.org/10.3390/electronics9101670
7. Iordanidou, A.: Κειμενοκεντρικές Προσεγγίσεις του Σχολικού Εγγραμματισμού: Κείμενο, Συμφραζόμενα, Γραμματική. In: Ματσαγγούρας, Η. (ed.) Σχολικός Εγγραμματισμός. Λειτουργικός, Κριτικός, Επιστημονικός. Athens, Grigori (2007). (in Greek)
8. Manning, C.D., Raghavan, P., Schutze, H.: Scoring, term weighting, and the vector space model. In: Introduction to Information Retrieval, pp. 100–123. Cambridge University Press, Cambridge (2008). https://doi.org/10.1017/CBO9780511809071.007
9. Nadeau, D., Satoshi, S.: A survey of named entity recognition and classification. Lingvisticae Investigationes **30**, 3–26 (2007)
10. Orphanos, G., Christodoulakis, D.: Part-of-speech disambiguation and unknown word guessing with decision trees. In: Proceedings of the 9th EACL Conference, Bergen, Norway, 8–12 June, 1999, pp. 134–141 (1999)
11. Picca, D., Jaccard, D., Eberle, G.: Natural language processing in serious games: a state of the art. Int. J. Serious Games **2**(3), 77–97 (2015)
12. Reinders, H.: Digital Games in Language Learning and Teaching. Palgrave Macmillan Publishing (2012)

13. Robertson, S.E., Walker, S., Jones, S., Hancock-Beaulieu, M., Gatford, M.: Okapi at TREC-3. In: Proceedings of the Third Text Retrieval Conference (TREC 1994), Gaithersburg, USA, November 2–4, 1994. NIST Special Publication 500–225, 109–26. Gaithersburg: National Institute of Standards and Technology (NIST) (1994)
14. Camastra, F., Vinciarelli, A.: Clustering methods. In: Machine Learning for Audio, Image and Video Analysis. AIKP, pp. 131–167. Springer, London (2015). https://doi.org/10.1007/978-1-4471-6735-8_6
15. Schell, J.: The Art of Game Design – A Book of Lenses. Carnegie Mellon University, Elsevier (2008)
16. Tsalidis, C., Vagelatos, A., Orphanos, G.: An electronic dictionary as a basis for NLP tools: the Greek case. In: Proceedings of 11th Conference on Natural Language Processing, Fez, Morocco (2004)
17. Vagelatos, A., Mantzari, E., Pantazara, M., Tsalidis, C., Kalamara, C.: Developing tools and resources for the biomedical domain of the Greek language. Health Inform. J. **17**(2), 127–139 (2011)
18. Vagelatos, A., Triantopoulou, T., Tsalidis, C., Christodoulakis, D.: A Spelling Correction System for Modern Greek. Int. J. Artif. Intell. Tools **3**(4), 429–450 (1994)

Studying the Ancient Civilizations on the Balkan Peninsula Through Serious Game and Storytelling

Desislava Paneva-Marinova[1(✉)], Maxim Goynov[1], Lilia Pavlova[2], Lubomir Zlatkov[1], and Detelin Luchev[1]

[1] Institute of Mathematics and Informatics at the Bulgarian Academy of Sciences, Sofia, Bulgaria
dessi@cc.bas.bg, {goynov, lyubcho}@gmail.com, dml@math.bas.bg

[2] Laboratory of Telematics at the Bulgarian Academy of Sciences, Sofia, Bulgaria
pavlova.lilia@gmail.com

Abstract. This paper presents a new serious game, exploring the multilayered archeological excavations in the town of Aquae Calidae (from Latin, 'warm waters'), situated in the Burgas mineral baths region. The history of Aquae Calidae (Therma, Thermopolis) is related to the famous hot mineral springs at the site and covers a long period of time from the 1st millennium BCE to the XVIth century CE. It is linked with events of the history of Ancient Greece, Roman and Byzantine times, Medieval Bulgarian period, The Crusades and the Ottoman period, and was visited by famous historical figures. The rich history of the place makes it possible to develop educational games, related particularly to the ancient civilizations. Through immersing in the virtual three-dimensional reality of the ancient complex, and playing intuitive educational mini-games, students from elementary schools improve their knowledge and understanding of the ancient civilizations on the Balkan peninsula.

Keywords: Serious games · Storytelling · Cultural heritage · eLearning

1 Introduction

In the technology-enhanced education, new approaches, methods, and tools for innovative learning process, are introduced. They assist the teacher in educating more efficiently by stimulating the desire to learn, and the creative and the logical thinking of the learners, thus improving different abilities and competencies.

This paper presents the design process of a new serious game, exploring the multilayered archeological excavations in the town of Aquae Calidae (from Latin, 'warm waters'), situated in the Burgas mineral baths region. The history of Aquae Calidae (Therma, Thermopolis) is related to the famous hot mineral springs at the site and covers a long period of time from the 1st millennium BCE to the XVIth century CE. It is linked with events of the history of Ancient Greece, Roman and Byzantine times, the Medieval Bulgarian period, The Crusades and the Ottoman period, and was visited

M. E. Auer and T. Tsiatsos (Eds.): IMCL 2021, LNNS 411, pp. 537–546, 2022.
https://doi.org/10.1007/978-3-030-96296-8_48

by famous historical figures. The rich history of the place makes it possible to develop related educational games.

Through immersing in the virtual three-dimensional reality of the ancient complex, and playing intuitive educational mini-games, students from elementary schools improve their knowledge and understanding of the ancient civilizations on the Balkan peninsula. The playing learner initially encounters the Thracian ruler Rhoemetalces II, king of the Sapaeans and the Odrysian Kingdom from 18 to 38 CE, who tells the story of the creation of the temple of the goddess Demeter near the magical springs of Aqua Calidae. The player visits a hall with artefacts of Thracian treasures from the period, and, in gaming mode, acquires crucial aspects of the Thracian culture and civilization. The game development is related to the heyday of Aqua Calidae under the rule of the Roman emperors. In the beginning of 1st century CE, Emperor Nero issues a decree for the initiation of building of the Roman thermae at Aqua Calidae, in the Thracian province, where the water was believed to be magical.

During the development, the storytelling and serious games methods are used in a suitable learning scenario, for better understanding of the historical facts and recreation of the era. The combination of these methods contributes to a more efficient education, stimulating the desire for learning, curiosity, creative and logical thinking, and to the development of diverse abilities and competencies of the learners. To address these issues, the mini-games quizzes and puzzles inspire the Aqua Calidae game players to carefully gather the pieces of educational content providing the necessary information.

2 New Pedagogical Approaches for History Study. Serious Games and Storytelling

In most of the reviews of the history and studies of the "serious games" [1–3], the term is accepted as introduced by C. Abt in his book "Serious games" [4], were it is presented as a "game having explicit and carefully crystallized educational purpose, as the main goal is not entertainment". From another point of view, by using the entertainment, serious games can elicit significant engagement from learners and further the effectiveness of the learning process [5]. In the digital era, edutainment [6] became accepted as a useful combination of traditional content and teaching methods in the context of new technologies [7, 8].

Activities in grades 1–4 for studying ancient history and civilization are mainly based on methods such as storytelling, demonstrations, guided-discovery, simulations, games, etc., which are typical for the study of Humanities in this stage of education. The educational environment has rapidly changed. Modern technology allows new possibilities for the development of innovative methods, scenarios, and tools for deeper understanding, for attracting attention, for applying learning-by-doing, and learning-by-authoring, as well as creative thinking, which are little represented in traditional education [9, 10]. A survey of the publications about computer games and serious games, shows a game-based approach to learning, as used across many different curricular areas, to be enjoyable and motivational for the players [11], although the question "does gamification work?" still needs more empirical answers from the educational area [12]. Following the idea of "making learning fun", seven factors presented in serious games: challenge, curiosity, control, fantasy, competition, cooperation and recognition, endorse the intrinsic motivations of students for learning [5]. Serious games can assist, facilitate and support very well the affective goal of the learning process for acquiring new knowledge, skills, and/or attitudes by the students after a learning episode [13].

The increased interest in the positive impacts and outcomes of the serious games and the use of term interchangeably with "games for learning", is illustrated by a literature overview of impacts and outcomes of serious games [14]. At its most basic, historical games studies can be defined as 'the study of those games that in some way represent the past or relate to discourses about it' [15]. The education is presented as personal, fun, collaborative, relevant, multimodal, technical and open-minded in the contemporary trends, where "gamification should be treated like a tool: very efficient and precise, but above all, comprehensive" [16]. All serious games in cultural heritage and history learning can be used at school (except the ones located in exhibitions or used for augmented reality visits), but only a few of them are adapted for a specific students' level or curriculum [13], although the game-based learning approach might be effective in facilitating students' 21st century skill development [17]. The use of computers and video games for teaching history at school resulted in a shift from a traditional teacher-centred learning environment to a student-centred environment with much more activity and engagement by them [18]. The digital game-based learning seeks to maintain a balance between learning and games elements [19, 20]. Like small group work, primary source analysis and historical role-plays and simulations, the serious educational games can be included among the other student-centred methods of history studying [21]. The "serious games" method is accepted as a research, pedagogic, and evaluative tool in the technology-enhanced education and generates good levels of comprehension and unconscious processing of content of relatively great difficulty [22].

Digital storytelling as a practice of using digital technologies to tell a short story [23] is used in a number of contexts and for different purposes [24–26]. The genre of digital history is often associated with the telling of personal stories, often of cultural or historical significance to the author [27]. Such stories often focus on interesting experiences or memories of some past event, or present overcoming personal and collective challenges or achieving goals [24]. With regard to historical education, two important types of digital stories can be identified - those that inform or instruct and those that explore historical events [23]. Classroom practice, which combines the use of digital media with the art of history – using both the skills and preferences of digital students and the inherent human interest in history – is a potentially powerful pedagogical environment [28]. Digital storytelling can be used to engage, inform, explore and transform and thus lend itself to educational contexts [29]. The nature of the engagement goes beyond simple entertainment, although the value of entertainment in an educational context is not to be underestimated, and the use of digital storytelling in the curriculum can provide real educational benefits [26]. Firstly, the multimedia nature makes the content of the digital story more accessible to technology-oriented students, many of whom are alienated from traditional textual forms [30], and, secondly, the combination of text integrated with visual images improves students' understanding [31]. The visual component helps to place the story in a recognizable context.

According to Bruner's theory of situational knowledge, this increases the time during which students are able to retain and understand information [28], as well as allows students to better organize information into manageable pieces.

The new serious game, exploring the multilayered archeological excavations in the town of Aquae Calidae, follows the ideas of the game "The Thracians", previously developed by the team [32, 33], in the frameworks and models for interactive learning content, based on the serious games and storytelling [34–36]. The main goal of interactive game-based educational practices is for students to learn the concepts, facts and specifics of history, civilization, and traditions of the people living on the Balkan Peninsula. The best practices, assessing the effectiveness of digital game-based learning are widely investigated in [37]. A version of the game "The Thracians" was tested in classes with 9–11-year-old students (third and fourth grade). At the end of the game, students answered 22 questions in an anonymous survey. It was found that students show great interest in telling stories, and thus, in games that introduce them to culture and history in this novel way. After the game and the analysis of the questionnaire, a special session was organized with the history teachers to clarify the results achieved by the students in their class. The discussion was based on the ten main obstacles to the application of games in education [38]. As a conclusion, we could point out the need to include such resources in the learning process, and to implement "learning +" through such games, in order to provide opportunities to expand and deepen the knowledge of students, especially in the humanities [39].

3 Aquae Calidae Game Presentation

3.1 Game Interior

The hot mineral springs, known in Roman times as Aqua Calidae, were a revered location for the ancient Thracians. With its temple and the sanctuary of the Three Nymphs, it was an attractive place for worship. Since the middle of the first millennium BCE, the area was first controlled by the Odrysian Kingdom, and then (since the middle of the first century BCE) by the Sapaean kingdom – a Roman vassal state [40]. After the annexation by the Roman Empire and the creation of the new Roman province Thracia in the middle of the first century CE, the Romans further improved the site with new baths and temples.

The game is designed to assist the students in their understanding of the ancient Thracian culture and civilization, by letting them explore the Aqua Calidae mineral baths. The player moves through halls, mirroring the architecture of the ancient Thracian burial mounds. By observing the digitally recreated artefacts from the period and by solving the necessary mini-game puzzles, the player advances further, and ultimately reaches the main hall – a rectangular room with a mineral water pool in the middle (Fig. 1).

Fig. 1. A screenshot from the Aqua Calidae game – The main hall with the mineral water pool.

3.2 The Game Content, Scenario and Gameplay

The game avatar is seen from a third-person perspective and explores the map with the help of the assigned keyboard keys for movement and the mouse cursor for adjusting the camera view. The different objects and artefact found on throughout the halls are activated with a cursor click. These objects could be of two types - visual (*i.e.*, text, maps, pictures) educational content and mini-games, which must be solved to allow further progression. The activation of the educational content provides the information, needed later for solving the mini-games. The mini-games are designed to challenge the students' knowledge award it (*i.e.*, with unlocking new areas of the map, or giving additional information, required for the successful solution of other mini-games). Thus, along with the short-term goals of progressing through and completing the game, the interdependence between the informational and playable components also strives to achieve the long-term goal of improving the students' knowledge on the subject. Already visited locations and puzzles, and the information they have provided, remain available for consultation.

Fig. 2. A screenshot from the Aqua Calidae game –A mini-game requiring solution for further advancement.

Figure 2 provides a glimpse into the gameplay. To open the iron grate door, the player roll (by clicking) the wooden cylinders, so that they spell correctly the name of the Thracian ruler pictured on the left ("CEBT" – Bulgarian for "Seuthes"), a piece of information, already gathered from previous exploration. The additional content, that could be seen in the next hall includes a map of the Thracian tribes on the Balkans on the left, an Orpheus mosaic on the right, and the puzzle required to open the door to the next level at the end of the hall.

3.3 Game Core

The game is web based, so no installation is needed and users can play it using any device having a modern web browser and mid- to high-range graphic card (PCs, Smart TVs, smartphones, tablets, VR devices, *etc.*)

The 3D engine is based on one of the most popular, stable and well documented libraries for 3D in web – THREE.JS. Using the WebGL technology and HTML Canvas API, the engine is able to utilize the graphic card of the device with great efficiency.

The programming language used for the web development is JavaScript (ES2020 standard). In order to serve the game optimized and compatible for most browsers, the code is bundled, minified, and polyfilled using WebPack 5.

One of the main technological objectives is to serve the game as fast as possible, to consume less resources (hardware – CPU, GPU, memory; network load, especially for devices with limited connection). Thus, every game asset – textures, 3D objects and other media are compressed and optimized, properly scaled or sampled. Lazy-loading technologies are also applied.

Depending on the device (mobile phone, desktop, tablet, *etc.*), users can also choose between the following graphic modes.

Standard Mode. Suitable for all kind of targeted devices. The player can use any convenient input – mouse, gamepad, keyboard. For touchscreen devices, on-screen controls are implemented (Fig. 3).

Fig. 3. Standard mode.

Anaglyph Mode. Suitable for all screens (especially large ones) using anaglyph glasses (passive glasses with cyan/red filter, filtering different image for the left and right eye). With this stereoscopic technology the user can feel the depth of the objects and can acquire more realistic view of the scene and the artefacts. The drawback of anaglyph technology is that the cyan and red filters can distort the color palette of the view (Fig. 4).

Fig. 4. Anaglyph mode.

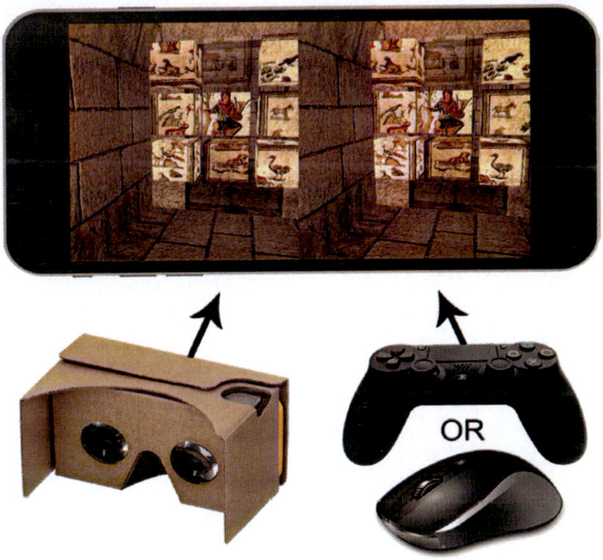

Fig. 5. Virtual reality (VR) mode.

Virtual Reality (VR) Mode. Another stereoscopic technology, providing two different images for the left and the right eye, without color or any other distortions (Fig. 5). The technology can be consumed using mobile phone and VR glasses, or using another VR device. Our VR implementation is not based on the Google (WebVR, WebXR) or

Apple (ARKit) implementations. It is based on standard web browser APIs and technologies like Gamepad API, DeviceOrientation API (using the device sensors – gyroscope and accelerometer). The stereoscopic effect is achieved using two scene cameras recording from two points – for the left and the right eye. The player can use bluetooth mouse, gamepad, or any other pointing device to control the game.

4 Conclusion

The integration of ICT in the primary school (grades 1–4) ancient history curriculum allows – by means of game playing, interactive interface, visualization, video, and animation – presentation of the material in a fun and accessible way. It makes it easier to explain connections, relationships, and influences among ancient civilizations and will improve the students' understanding of the evolution of a civilization. The first-person point-of-view interface and the high-quality textures and graphic images, used for the creation of the game, offer an authentic and exciting exploration experience. The basic mapping of the site is augmented with realistic interactive 3D objects, avatars and light shade effects. The mini-games' graphic outlook is designed to be smoothly incorporated in the environment. The user experience is enhanced by adequate in-game media effects. The game engine is implemented in a way that it allows other scenarios, environments, mini-games, and other assets to be easily added in a new or existing serious game.

Acknowledgments. This research is partially supported by the Bulgarian Ministry of Education and Science under the National Research Programme "Cultural heritage, national memory and development of society" approved by DCM №. 577/17.08.2018.

References

1. Djaouti, D., Alvarez, J., Jessel, J.-P., Rampnoux, O.: Origins of serious games. In: Ma, M., Oikonomou, A., Jain, L.C. (eds.) Serious Games and Edutainment Applications, pp. 25–43. Springer, London (2011). https://doi.org/10.1007/978-1-4471-2161-9_3
2. Susi, T., Johannesson, M., Backlund, P.: Serious Games: An Overview [Technical Report HS-IKI-TR-07-001] University of Skövde, Sweden (2007)
3. Wilkinson, P.: A brief history of serious games. In: Dörner, R., Göbel, S., Kickmeier-Rust, M., Masuch, M., Zweig, K. (eds.) Entertainment Computing and Serious Games. LNCS, vol. 9970. Springer, Cham (2016). https://doi.org/10.1007/978-3-319-46152-6_2
4. Abt, C.: Serious Games. University Press of America (1987). https://doi.org/10.1007/978-3-319-19126-3
5. Malone, T., Lepper, M.: Making learning fun: a taxonomy of intrinsic motivations for learning. In: Snow, R.E., Farr, M.J. (eds.) Aptitude, Learning and Instruction: III. Conative and Affective Process Analyses, pp. 223–253. Routledge, Milton Park (1987)
6. Okan, Z.: Edutainment and learning. In: Seel, N.M. (eds.) Encyclopedia of the Sciences of Learning. Springer, Boston, MA (2012). https://doi.org/10.1007/978-1-4419-1428-6_1938
7. Patrício, R., Moreira, A.C., Zurlo, F.: Gamification approaches to the early stage of innovation. Create. Innov. Manage. **27**(4), 499–511 (2018)

8. Nikolova, A., Georgiev, V., Paneva-Marinova, D.: Using applications with digital cultural-historic content for children development. In: Proceedings of the International Conference on Education and New Learning Technologies, pp. 3954–3957 (2018)
9. Paneva-Marinova, D., Rousseva, M., Pavlov, R., Luchev, D.: The ancient Thracian civilization and modern youth convergence by serious games. In: Proceedings of the 10th annual International Conference on Education and New Learning Technologies, pp. 3940–3945 (2018)
10. Paneva-Marinova, D., Stanchev, P., Pavlov, R., Goynov, M.: Towards efficient and continuing usage of the digital culture ecosystem for learning purposes. In: Proceedings of 13th annual International Technology, Education and Development Conference (IATED 2019), pp. 362–366 (2019)
11. Paneva-Marinova, D., Pavlov, R., Kotuzov, N.: Approach for analysis and improved usage of digital cultural assets for learning purposes. Int. J. Cybern. Inf. Technol. **17**(3), 140–151 (2017)
12. Connolly, T.M., Boyle, E.A., MacArthur, E., Hainey, T., Boyle, J.M.: A systematic literature review of the empirical evidence on computer games and serious games. Comput. Educ. **59** (2), 661–686 (2012)
13. Hamari, J., Koivisto, J., Sarsa, H.: Does gamification work? – A Literature Review of Empirical Studies on Gamification. In: HICSS '14 Proceedings of the 47th Hawaii International Conference on System Sciences, pp. 3025–3034, HI, USA (2014)
14. Patrício, R., Moreira, A.C., Zurlo, F.: Enhancing design thinking approaches to innovation through gamification. Eur. J. Innov. Manag. (2020). https://doi.org/10.1108/EJIM-06-2020-0239
15. Boyle, E.A., et al.: An update to the systematic literature review of empirical evidence of the impacts and outcomes of computer games and serious games. Comput. Educ. **94**, 178–192 (2016)
16. Chapman, A.: Digital Games as History: How Videogames Represent the past and Offer Access to Historical Practice. Routledge, New York (2016)
17. YDP, Young Digital Planet, The Book of Trends in Education 2.0 (2016). http://www.ydp.eu/assets/pdf/The-Book-of-Trends-in-Education-2.0-YDP.pdf
18. Qian, M., Clark, K.R.: Game-based Learning and 21st century skills: a review of recent research. Comput. Hum. Behav. **63**, 50–58 (2016)
19. Watson, W.R., Mong, C.J., Harris, C.A.: A case study of the in-class use of a video game for teaching high school history. Comput. Educ. **56**(2), 466–474 (2011)
20. Nussbaum, M., Beserra, V.D.S.: Educational videogame design. In: 2014 IEEE 14th International Conference on Advanced Learning Technologies, ICALT 2014, pp. 2–3. IEEE Computer Society (2014)
21. Chapman, A., Foka, A., Westin, J.: Introduction: what is historical game studies? Rethinking Hist. **21**(3), 358–371 (2017)
22. Maloy, R.W., LaRoche, I.: Student-centered teaching methods in the history classroom: ideas, issues, and insights for new teachers. Soc. Stud. Res. Pract. **5**(2), 46–61 (2010)
23. Huotari, K., Hamari, J.: Defining gamification: a service marketing perspective. In: Proceedings of the 16th International Academic MindTrek Conference, Tampere, Finland, pp. 17–22 (2012)
24. Georgieva-Tsaneva, G.: Serious games and innovative technologies in medical education in Bulgaria. TEM J. **8**(4), 1398–1403 (2019)
25. Georgieva-Tsaneva, G., Noev, N., Bogdanova, G.: Serious educational games and the study of the military historical heritage. Digital Present. Preserv. Cult. Sci. Heritage **8**, 133–140 (2018)

26. Noev, N., Valev, I., Kancheva, S., Sapundjiev, V.: A model of content structure for a serious educational game related to the military and historical heritage presented through the "Mission Opalchenets." Digital Present. Preserv. Cult. Sci. Heritage **8**, 151–158 (2018)
27. Roby, T.: Opus in the classroom: striking CoRDS with content-related digital storytelling. Contemp. Issues Technol. Teach. Educ. **10**(1), 133–144 (2010)
28. Lambert, J.: Digital Storytelling Cookbook. Digital Diner Press (2010)
29. Caballero-Hernández, J.A., Palomo-Duarte, M., Dodero, J.M.: Skill assessment in learning experiences based on serious games: a systematic mapping study. Comput. Educ. **113**, 42–60 (2017)
30. Yuksel, P., Robin, B., McNeil, S.: Educational uses of digital storytelling all around the world. In: Society for Information Technology & Teacher Education International Conference, pp. 1264–1271. Association for the Advancement of Computing in Education (2011)
31. Gunter, G., Kenny, R.: Digital booktalk: Digital media for reluctant readers. Contemp. Issues Technol. Teach. Educ. **8**(1), 84–99 (2008)
32. Burmark, L.: Visual presentations that prompt, flash & transform. Media Methods **40**(6), 4–5 (2004)
33. Paneva-Marinova, D., Rousseva, M., Dimova, M., Pavlova, L.: Tell the story of ancient Thracians through serious game, digital heritage. In: Progress in Cultural Heritage: Documentation, Preservation, and Protection. LNCS, vol. 11196, pp. 509–517 (2018)
34. Dimova, M., Paneva-Marinova, D., Pavlova, L.: Towards better understanding of ancient civilizations by storytelling and gaming. TEM J. **7**(3), 658–661 (2018)
35. Luchev, D., Paneva-Marinova, D., Dimova, M.: Digital storytelling and digital book trailer applications for educational purposes in Bulgaria. In: Proceedings of 13th annual International Technology, Education and Development Conference (IATED 2019), pp. 529–534 (2019)
36. Van der Vegt, W., Westera, W., Nyamsuren, E., Georgiev, A., Ortiz, I.M.: RAGE architecture for reusable serious gaming technology components. Int. J. Comput. Games Technol. (2016). https://doi.org/10.1155/2016/5680526
37. Márkus, Z.L., et al.: Interactive game development to assist cultural heritage. Digital Present. Preserv. Cult. Sci. Heritage **8**, 71–82 (2018)
38. All, A., Castellar, E.P.N., Looy, J.V.: Assessing the effectiveness of digital game-based learning: best practices. Comput. Educ. **92–93**, 90–103 (2016)
39. Shapiro, J.: Games in the Classroom: Overcoming the Obstacles (2014). https://ww2.kqed.org/mindshift/2014/09/12/games-in-the-classroom-overcoming-the-obstacles/
40. Luchev, D., Paneva-Marinova, D., Senka, G., Pavlova, L., Pavlov, R.: Conceptual models for the development of online learning games in cultural heritage field. Pedagogy, **92**(5), 620–633 (2020). (in Bulgarian)
41. Kiachkina, P., Karayotov, I.: Aquae Calidae. Centre de la culture thrace. In: Thracia Pontica, VI.2, Series 1. La Thrace et les sociétés maritimes anciennes, pp. 125–130 (1994)

Designing and Developing a Learning Analytics Platform for the Coding Learning Game sCool

Alexander Steinmaurer(✉), Anil Kumar Tilanthe, and Christian Gütl

Graz University of Technology, Graz, Austria
alexander.steinmaurer@tugraz.at
https://codislabgraz.org/

Abstract. Learning Analytics are a valuable way for teachers and educators to support their student's learning progress. Data can be used to identify issues, provide personalized learning, and improve the overall quality of learning. There are diverse types of educational data which are available for instance log data, given answers, or source code. Just as diverse as the data are, so are the data sources such as learning management systems (LMS) or serious games. However, analyzing and visualizing a huge amount of data in a valuable educational way can be challenging. A learning analytics tool should support learners and educators as good as possible and find a balance between an optimal overview and complex evaluations and interpretations. In this paper, we describe our approach to create a learning analytics software for the serious coding learning game sCool. We outline the concept, architecture, and development of the application. The tool was evaluated in two separate evaluations with i) 31 technical experts and ii) 8 domain experts (teachers). The evaluations showed, that the users rated the usability of the system high above average (M = 79.02; SD = 10.69) in terms of System Usability Scale. Nevertheless, the evaluation with the domain experts revealed, that there is room for improvements regarding educational usage. Finally, we summarize our experience and results of the evaluation to outline relevant requirements and implementation details for a learning analytics platform. In this way, we want to help educators, researchers, and developers when designing and implementing learning analytics tools.

Keywords: Learning analytics · Game-based learning · Serious game · Computer science education

1 Introduction

The COVID-19 pandemic has highlighted the importance of effective distance learning with widespread school closures across the world [1]. Serious games

M. E. Auer and T. Tsiatsos (Eds.): IMCL 2021, LNNS 411, pp. 547–558, 2022.
https://doi.org/10.1007/978-3-030-96296-8_49

provide an effective, engaging, and immersive learning environment where students can participate in the learning process remotely as well. The term serious games refer to *"games that do not have entertainment, enjoyment, or fun as their primary purpose"* [2]. Most academics refer to the use of digital games for educational purposes by the term serious games [3,4].

The need for modern learning technologies emerges among many educational fields. Computational thinking and coding are essential skills in today's digital world [5,6]. Many novice programming learners face various difficulties while learning coding skills [7]. These difficulties have to be acknowledged and recognized to support students as well as possible.

There are several ways to help students when learning to program such as personalized code review by educators or pair programming [8]. Getting feedback on a learner's source code is necessary to encounter possible improvements and to increase coding skills. Therefore, educators need to identify students' progress and analyze submitted programming solutions to understand how students learn, plan improvement of the course, and also support and guide the learning experience. Educators and developers could also discover various strategies undertaken by students for course completion. Our study tries to identify how a learning analytics application could assist educators in improving course content based on student's performance [9–11]. The study also attempts to analyze how a learning analytics application assists educators to gain an overview of students' performance. Additionally, the user interface of the learning analytics application is evaluated using System Usability Scale and NASA-TLX questionnaires for its usability and acceptability.

In this work, we present a learning analytics tool for a serious game to learn to code. Besides students, such learning systems should also support educators to get a quick overview of a class's learning progress. For this reason, we conducted two studies with different objectives. The first study was conducted with 31 participants with computer science background. The second study was an expert's evaluation with 8 teachers investigating the educational benefit from our system. The following research objectives were defined to evaluate the developed platform.

- Does the platform support all requirements teachers have in a learning analytics tool?
- Are teachers capable to analyze students' learning progress?
- Are there any major usability issues in the current prototype?

2 Related Work

Digital games can collect vast amounts of detailed information about students learning process compared to traditional methods. The gathered information can be used for behaviour and learning analysis and assessments. Digital games can capture student's inputs, the number of attempts, strategies used for course progress, time allocation to various course stages, problem-solving sequences, or programming solutions submitted [12]. The data can be analyzed to provide feedback to various stakeholders involved, improve course content and instructions, gain an overview of student's performance, or for discovering how students learn to program [9]. Digital games are also able to capture student's actions as they work on the course tasks or programming assignments, such as individual keystrokes, programming errors, code edits, and compilations [13].

The insights from the data gathered by educational games can be displayed to different groups of people such as learners, educators, parents, or developers via learning dashboards. Learning dashboards integrate information from learning tools to provide a comprehensive visual representation of student's progress [14]. The insights and feedback ensure that relevant data inform decisions about learning and course content.

There are various serious games for teaching programming skills focusing on learning to code, algorithmic thinking, or creating games for learners of various age groups and previous experiences [15]. *CodeMonkey*[1] is an app and web-based educational game where kids can learn to code with CoffeeScript and Python. Players can also use a block-based course, where they can drag-and-drop blocks of code to control an avatar. The players can learn many programming concepts such as variables, objects, conditionals, function calls, or loops. CodeMonkey also provides a dashboard where educators can track the learning progress of students and see code submitted. The teachers can see the progress of each of the participating students, and export results and progress, or receive a more detailed analysis. They can also see all programming concepts such as loops, functions, and objects and student's proficiency in each of them respectively.

Ozaria[2] is another web-based serious game for learning to code. It is an immersive story-based fantasy learning environment where programming courses are taught in JavaScript and Python programming languages. It is designed for both in-person and remote learning settings. The players control an avatar using code to fulfill tasks. The players are also provided with audio and textual hints to help them complete programming tasks. Ozaria also provides a dashboard for students where they can see their courses and progress. The dashboard is also available for teachers showing a class view of students and their progress. Teachers can see all levels completed by students and their in-game progress and assigned levels as well as the code submissions of students.

The mobile learning game *sCool*[3] is an educational game for learning programming skills [16,17]. sCool provides an immersive and engaging experience to students following a narrative of the escape of a space shuttle and its crew members from an alien planet. The players learn various programming concepts such as sequencing, loops, or data types in a concept-learning part. It is followed by a practical programming part where the players apply the previously learned programming concepts using the Python programming language to control a robot avatar. The programming section consists of draggable code blocks which get converted to editable Python commands. In additional, the code can be adapted

[1] https://app.codemonkey.com/.
[2] https://www.ozaria.com/.
[3] https://scool.codislabgraz.org/.

using the device keyboard. The game is highly flexible since game-related data is sent and received using a web application for educators. In this way, teachers can create courses and get an overview of the class' learning progress.

3 Design and Implementation

A web-based learning analytics application was developed to provide insights into student's activities and course progress to stakeholders, comprising educators, teaching assistants, or game developers, of the mobile learning game sCool. The application focuses on empowering insights into student's activities and analysis of courses, rather than making a decision tool. The empowered educators could utilize their knowledge to improve course content or gaining an overview of their class students. The students can see the structure of the entire course and their course progress within the mobile game itself.

3.1 Data Collected

The sCool mobile game captures player's game-related data at a granular level such as dragged code blocks, compiled code, deleted code, interpreter errors, submitted answers, or results code execution. This data is collected and stored in the sCool database for further analysis [13]. The time duration of every game session and information such as points are transmitted to the web server using a REST API. All the data is stored in a relational database. Some of the log data is also stored as JSON.

3.2 Requirements

The players' game-related data are collected in the sCool database should be processed to extract meaningful information which must be provided in simplistic ways [18,19]. The information should empower educators with insights into student's course progress and performance. Interactive data visualization tools could be used to make information easily available. The focus of the application should be on lower-dimensional plots to make the visualization easy to understand. For simple statistics, bar charts, or scatter plots could be used to display game metrics of student's game-related data. Graphs such as time series, or Gantt charts could be used to visualize student's game timelines and strategies undertaken for course completion.

Users should be able to just access data that is relevant for his or her learning activity. Within the learning analytics tool, educators create so-called *learning activities* which are in-class activities with a certain group of students. The educators should be able to access data of only their learning activities. Only administrators should have access to the data of all students on the application. Hence, the users were divided into two main groups of educators and sCool game administrators representing the stakeholders. An administrator can gain additional insights by comparing various learning activities, whereas an educator can only compare within their learning activities. Both serious game and web application do not store sensitive data of students, such as names, email addresses, or phone numbers.

3.3 Development Details

Plotly's Dash[4] is used as the application framework as it is an open-source full-stack framework for building web analytic applications with interactive visualization. The application connects with the sCool database to access game-related data. The queried data is used to create a Pandas[5] DataFrame, which is a 2-dimensional data structure containing labeled axes (rows and columns), for storing and processing data [20].

The code solutions submitted by students are parsed to create an Abstract Syntax Tree (AST) to identify programming concepts learned and used by the students for each of the practical tasks. Python's *ast* module is used to generate the AST. The log data stored as JSON string in the database is used to extract information such as UI interactions in the game. Besides that, each code execution within the game is transmitted to the server giving detailed information on the code producing progress.

When a user selects a learning activity, the related data is displayed using various visualization techniques. The important numbers and information are highlighted and presented. Interactive horizontal bar charts are used to display sorted information as a simple visualization technique for easy interpretation.

The user interface is designed to be responsive and adaptive to various screen sizes. It is divided into three main sections. The *details* section comprises an overview of a single learning activity. The users can gain information such as the number of student participants, points collected, task-wise completion rates, student-wise task completion, and programming concepts learned and used by students. The *student's* section provides a detailed overview of a specific student. This compromises detailed information such as points collected, programming concepts learned, course progress, game interactions, and a timeline. In the *custom* section, users can create custom graphs by selecting features, graph types, and other information. Information such as task-wise errors by the students, or course-wise errors faced by students can be easily discovered using the custom section. The application components follow Model View Controller (MVC) design pattern [21].

4 Evaluation

Within this research project we have conducted two evaluations with a different focus. The first evaluation (study 1) mainly involved technical experts such as programmers, researchers, or graduated computer science students to get perspectives and opinions on usability, system design, and interaction. The second

[4] https://dash.plotly.com/introduction.
[5] https://pandas.pydata.org/docs/index.html.

evaluation (study 2) focused on domain experts and so teacher's conducted the evaluation to get information on the system's characteristics in an educational setting. Both groups were asked to work on several tasks on the system and additionally fill-out a questionnaire regarding system-related properties such as usability.

4.1 Participants

Study 1. The first evaluation (system's usability evaluation) was conducted with 31 participants in total. A vast majority of the participants (25 people) have a background in computer science with different professional experiences. Table 1 shows that most of the participants (20 people) obtain a Master's degree in computer science. Seven people hold a Bachelor's degree in computer science and two people have a PhD and a high school graduation. In terms of the gender distribution 23 males and 8 females participated the evaluation. A predominant part of the participants (14 people) is working in computer science-related areas.

Table 1. Overview of the participants of study 1.

Gender		Highest degree		Profession	
23	Male	20	Master's degree	14	Expert (CS-related)
8	Female	7	Bachelor's degree	8	Other employment
		2	PhD	5	Full-time student
		2	High school	4	Teacher

Study 2. The domain expert's evaluation covers eight teachers, 5 female and 3 male. The evaluation was taken in the scope of a teaching training for junior teachers. The mean professional experience is 1.06 years (SD = 0.63). All of them hold a Bachelor's degree from a university (n = 2) or a college of education (n = 6). Seven are teacher's at a vocational schools and one as a teacher at the college of education. All of them have their degree in computer science education (with a strong focus on multimedia and design).

4.2 Instruments

For the evaluation in study 1 the participants were asked to complete 11 tasks within the learning analytics environment. The tasks were designed as multiple-choice questions covering different aspects of the system such as analysis on course, user, or concept level. Each question was worth 1 point so participants could receive 11 points in total when answering all questions correct. The tasks were assigned to one of four categories covering: i) tasks solved in the game (3 tasks), ii) comparing students (2 tasks), iii) used coding concepts (2 tasks), iv) student information (3 tasks), and v) custom plots (1 task).

After each group of tasks (five times in total) the participants received a modified version of the NASA-TLX [22] to investigate i) mental demand, ii) temporal demand, iii) overall performance, iv) effort, and v) frustration level. This assessment instrument's goal was to evaluate the workload of certain tasks to get a better understanding on the system's complexity. The questions of the NASA-TLX were answered on a five-items Likert scale ranging from very low to very high.

The System Usability Scale (SUS) [23] was used to easily receive an overview of the user's perceived usability. Within this 10 items questionnaire the participants rated the items on a 5 point Likert scale from strongly disagree to strongly agree. Using the SUS questionnaire we wanted to get a quick evaluation on the overall usability and some possible issues to tackle them in future improvements. The value of the system usability scale can be between 0 and 100, and it is common [23] that values above 68 are considered as above average usability.

The focus of the evaluation in study 2 was different, since it relates to the system's educational usage. The users were asked to solve five different tasks within the learning analytics platform. Table 2 contains all given tasks. Additionally the teacher's were asked *"I found task x ..."* after each question rating from very easy to very hard. Finally, they received questions whether they would use the system and where they see room for improvements.

Table 2. The participants of the expert's evaluation were given five tasks to different in-game analysis. The tasks covered different analytics where various parts of the web application should be tested. A task was classified as solved, when a participant answered it fully correct, if this is not the case it is counted as wrong. Participants that were not able to solve a task (e.g. provided no answer) are classified as N/A. The difficulty is calculated using the mean of a 5-items grading scheme from 1 (very easy) to 5 (very hard).

#	Question	Correct	Wrong	N/A	Difficulty
T1	Find out which practical task the fewest players were successful at. What is the number of this task?	2	3	3	3.5
T2a	In which practical task were variables used most frequently? What is the number of the task?	5	1	2	3.63
T2b	How many players used variables in this task?	3	3	2	2.63
T3	Which players were able to solve the level *"Calculate the Fibonacci Sequence and check it with the storage"* (Task 64)?	6	–	2	2.13
T4	How many attempts did the player *'dabod'* need to complete the level *"Decide what you have to print to the console"* (Task 63)?	5	1	2	3
T5	Which players had the most syntax errors when creating their codes?	1	3	4	3.63

4.3 Procedure

The system evaluation was sent out to 60–80 people in computer science-related fields. The questionnaire was provided using Google Forms and shared via communication channels of various research groups, schools, and tech companies. 31 people participated and completed the evaluation. At the beginning of the survey, the participants were introduced to the key concepts of the video game. Afterwards, the users were asked to watch a brief video (5:30 min) that introduces the participants to the web application. The next step covered 11 different data analysis tasks which are assigned to a certain category where participants should use the tool to get some information on the student's or the group's progress. Each category was completed by answering five NASA-TLX questions regarding workload. After all tasks in the learning platform, the participants additionally completed the System Usability Scale. The questionnaire did not query any personal information and thus is fully anonymous.

The SUS data showed that there might be some potential outliers. They were detected using standard deviation method. Due to the number of participants we decided to choose a less conservative approach where two standard deviations [24] from the mean were used as cut-off. Using this approach three outliers could be detected which were removed from the observations of the system usability (see Fig. 1).

The participants from the expert's evaluation are mainly vocational school teacher's. The evaluation was conducted unsupervised and online. In total 14 teachers were contacted, and 8 people completed the questionnaire. The questionnaire consisted of four parts: i) demographic and professional-related questions, ii) general questions about learning analytics tools, iii) tasks on the learning analytics platform, and iv) post questions about the system. All teachers were familiar with the serious game sCool, but an additional video was provided within the survey. They did not get any introduction to the learning analytics platform to avoid any advantages regarding the system's usage.

All data from both questionnaires was analyzed using the Python programming language and the Data Analysis library Pandas. All categorical data was coded and analyzed using basic statistic techniques such as calculating mean and standard deviation. Numerical or textual data from the tasks was manually evaluated and assigned to Boolean values (true/false) depending on the provided answer. All qualitative data such as open-ended questions were evaluated manually and grouped into different categories.

5 Results and Discussion

Usability. The usability evaluation conducted with the SUS showed a well-rated usability. The SUS value is 79.02 (SD = 10.69) which is rated as A- [25] and thus, means well satisfying usability. Figure 1 depicts the SUS results, including even three outliers (P25, P28, and P29) which are not considered in the calculation of the overall SUS value. The mean Raw NASA-TLX (RTLX) [26,27] score is

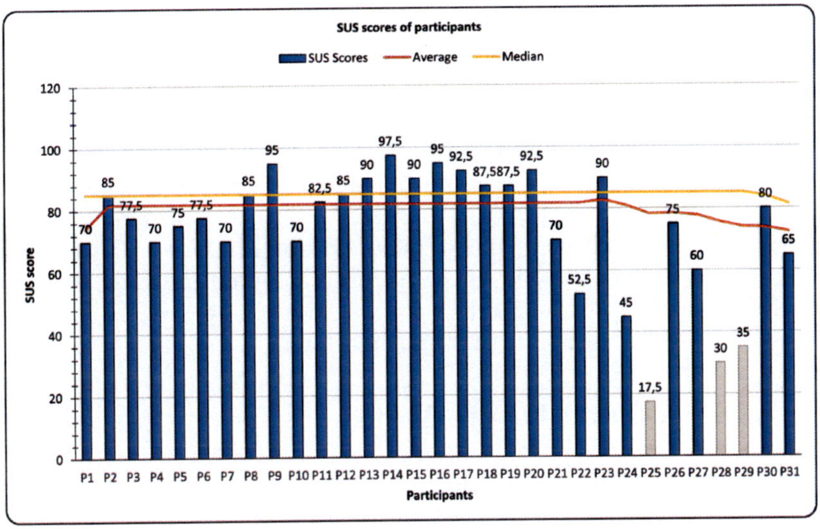

Fig. 1. In total 31 participants (P) respond to the System Usability Scale. The mean system usability scale value is 79.02 (SD = 10.69), which is considered to be a good grade in terms of usability. It is in the percentile 85–89 which is graded as A- within the scope of SUS [25]. The participants P25, P28, and P29 were identified as outliers and therefore not included in the calculation.

2.38 (SD = 0.46) on a scale of 1 to 5 (very low to very high), implying low to moderate perceived workload.

When reviewing the open-ended questions the participants mentioned that the system gives comprehensive functionalities, but they had issues dealing with this complexity. They desire a simpler and more intuitive user interface that is self-explaining. This evaluation is similar to the results of the teacher's evaluation. However, study 2 did not focus on usability aspects in particular, but using the open-ended questions the majority of the participants had some issues solving tasks due to usability reasons.

Further, the evaluations showed that the user experience and usability highly depend on the considered group and the tasks [28,29]. Study 1 was conducted with participants that are highly familiar with interactive systems and data science. They seemed to be more familiar with the systems architecture. On the other hand teachers (study 2) commented that the system is rather complex and confusing at first glance. Another reason for the different evaluation results could have to do with the design of the tasks. While the first study's aim was to evaluate the system's usability, we focused on learning analytics aspects on the expert's evaluation. However, since many teachers had issues with the tasks (see Table 2) the user interface and system design should be reconsidered in future versions.

Analysis and Visualization. Each teacher responds that he or she wants to use a learning analytics tool with a learner-oriented approach. They want to keep track of the students learning progress and possible issues. Table 2 gives an overview of the results of the evaluation. The number of correct and wrong tasks revealed that some sections of the application are understandable and can be used for a meaningful analysis. This mainly concerns the tasks related to the student's sections (Tasks T2a/b, T3, and T4). Due to some ambiguous wrong answers the system might provide too many possibilities for analytics. The fact that just one person was able to complete Task 5 (and the high rating of the difficulty) showed, that the custom data visualization needs to be optimized in terms of the system's usability and user interface.

6 Conclusion and Future Work

In this paper, we presented an approach for a learning analytics environment for a serious game in computer science education. The goal was to design and evaluate the prototype. The results showed that there is a high demand for learning analytics platforms to support both students and teachers [9,10,30]. The design of such a platform has to fit the educators' needs. We encountered several aspects that are important in the design of a platform:

Lightweight Design. One key element of a learning analytics platform is a lightweight design that is well structured and easy to use. The tool should have a flat hierarchy that can be easily navigated. The provided information should just compromise the most important data.

Easy Usage. Strongly linked to a lightweight design is a simple usage of the system. Teachers want to have easy navigation and easily receive all data. The system should have a good balance between comprehensive data analysis and intuitive usage. The teachers responded, that they enjoyed using the system after they understood its depth. An on-boarding via an interactive tutorial or expressive labels/tooltips can improve usability and can have considerable benefits over explanation videos or manuals [31–33].

Learning Progress. Obviously, the tool should provide meaningful data over the students learning experience and progress. The system should be open for both students and educators in order to make the results as transparent as possible. Custom data analysis and visualization are perceived as a good possibility for individual support measures. In addition, the system should also help teachers to grade. Therefore, a flexible assessment and evaluation should be provided that can be exported easily.

Privacy. Teachers are hardly concerned about their student's privacy. They like to use systems that are already used in other educational institutions and are GDPR conform. Since data about children are classified as particularly worthy of protection [34], the transmitted data should contain as little personal data as possible. A good approach would be the usage of pseudonyms where just teachers can identify certain students (especially for individual guidance) [18].

Future Work. For further improvements we want to provide player's data on their learning progress. Learners should get access to a dashboard where they get comprehensive information about their performance within the learning environment. On the one hand, this should cover game-related information such as passed levels or points and on the other hand, they should see which concepts and skills they already acquired and which are left. Predicting student's knowledge gain, classifying students based on similar behavior and requirements, and the ability to analyze a large number of students with easy-to-understand reporting could also be considered [10,11].

The results showed, that teacher's demand for usability improvements. After the improvement of the mentioned usability issues, further studies should be conducted on a larger population. The focus of this further observation will be the interaction with the system to identify workflows and features. Therefore, web analytics tools such as Matomo will be used to track user's behaviors.

References

1. UNESCO: COVID-19 impact on education (2021)
2. Michael, D.R., Chen, S.L.: Serious games: Games that Educate, Train, and Inform. (2006)
3. Djaouti, D., Alvarez, J., Jessel, J.P., Rampnoux, O.: Origins of serious games. Serious Games Edutainment Appl. 25–43 (2011)
4. Abt, C.C.: Serious Games. International Series of Monographs on Physics. The Viking Press, New York (1981)
5. Barr, V., Stephenson, C.: Bringing computational thinking to k-12: what is involved and what is the role of the computer science education community? ACM Inroads **2**, 03 (2011)
6. Kanaki, K., Kalogiannakis, M.: Introducing fundamental object-oriented programming concepts in preschool education within the context of physical science courses. Educ. Inf. Technol. **23**(6), 2673–2698 (2018). https://doi.org/10.1007/s10639-018-9736-0
7. Lahtinen, E., Ala-Mutka, K., Järvinen, H.: A study of the difficulties of novice programmers, vol. 37, pp. 14–18 (2005)
8. Albluwi, I., Salter, J.: Using static analysis tools for analyzing student behavior in an introductory programming course. Jordanian J. Comput. Inf. Technol. (JJCIT) **6**, 215–233 (2020)
9. Chaudy, Y., Connolly, T., Hainey, T.: Learning analytics in serious games: a review of the literature (2014)
10. Hauge, J.B., et al.: Implications of learning analytics for serious game design. In: 2014 IEEE 14th International Conference on Advanced Learning Technologies, pp. 230–232. IEEE (2014)
11. Verbert, K., Duval, E., Klerkx, J., Govaerts, S., Santos, J.L.: Learning analytics dashboard applications. Am. Behav. Sci. **57**(10), 1500–1509 (2013)
12. United States Department of Education: National education technology plan. 2010a
13. Ihantola P., et al.: Educational data mining and learning analytics in programming: Literature review and case studies. In: ITiCSE-WGP 2015 - Proceedings of the 2015 ITiCSE Conference on Working Group Reports, pp. 41–63 (2015)

14. United States Department of Education. Reimagining the role of technology in education: National education technology plan (2017)
15. Combéfis, S., Beresnevičius, G., Dagiene, V.: Learning programming through games and contests: overview, characterisation and discussion **10**, 39–60 (2016)
16. Kojic, A., Kojic, M., Pirker, J., Gütl, C., Mentzelopoulos, M., Economo, D.: sCool - a mobile flexible learning environment. In: iLRN 2018 Montana, pp. 72–84. Verlag der Technischen Universität Graz (2018)
17. Steinmaurer, A., Pirker, J., Gütl, C.: sCool - game-based learning in computer science class a case study in secondary education. Int. J. Eng. Pedagogy **9**(2), 26–50 (2019)
18. Freire, M., Serrano-Laguna, Á., Manero, B., Martinez-Ortiz, I., Moreno, G.P., Fernández-Manjón, B.: Game Learning Analytics: Learning Analytics for Serious Games, pp. 1–29 (2016)
19. Owen, E., Baker, R.: Learning Analytics for Games (2019)
20. Pandas DataFrame: Pandas dataframe (2021)
21. Leff, A., Rayfield, J.T.: Web-application development using the model/view/controller design pattern. In: Proceedings fifth IEEE International Enterprise Distributed Object Computing Conference, pp. 118–127. IEEE (2001)
22. NASA NASA: Task load index (tlx) v. 1.0 manual. NASA, NASA-Ames Research Center Moffett Field (1986)
23. Brooke, J.: SUS: a quick and dirty usability scale. Usability Eval. Ind. **189**, 11 (1995)
24. Leys, C., Ley, C., Klein, O., Bernard, P., Licata, L.: Detecting outliers: do not use standard deviation around the mean, use absolute deviation around the median. J. Exp. Soc. Psychol. **49**(4), 764–766 (2013)
25. Lewis, J., Sauro, J.: Item benchmarks for the system usability scale **13**, 158–167 (2018)
26. Moroney, F.W., Biers, D.W., Eggemeier, F.T., Mitchell, J.A.: A comparison of two scoring procedures with the NASA task load index in a simulated flight task. In: Proceedings of the IEEE 1992 National Aerospace and Electronics Conference (at NAECON) 1992, pp. 734–740, vol. 2 (1992)
27. Miller, S.: Workload measures. National Advanced Driving Simulator. Iowa City, United States (2001)
28. Renz, J., Staubitz, T., Pollack, J., Meinel, C.: Improving the onboarding user experience in MOOCS. In: Proceedings EduLearn, pp. 3931–3941 (2014)
29. Morris, M.G., Dillon, A.: How user perceptions influence software use. IEEE Softw. **14**(4), 58–65 (1997)
30. Romero, C., Ventura, S.: Educational data mining and learning analytics: an updated survey. Wiley Interdisc. Rev. Data Mining Knowl. Disc. **10**(3), e1355 (2020)
31. Yoganathan, S., Sherburn, N., Halupka, V., Li, J.: Pedagogy, practice, participation: reimagining Edtech onboarding. In: 29th Australasian Association for Engineering Education Conference 2018 (AAEE 2018), p. 705. Engineers Australia (2018)
32. Schneider, M.: Interface usability testing to maximize user onboarding experience. PlotProjects-Geofencing Notifications for apps (2016)
33. Davis, F.D.: Perceived usefulness, perceived ease of use, and user acceptance of information technology. MIS Q. **13**(3), 319–340 (1989)
34. Slade, S., Prinsloo, P.: Learning analytics ethical issues and dilemmas. Am. Behav. Sci. **57**, 1510–1529 (2013)

A Game-Based Smart System Identifying Developmental Speech and Language Disorders in Child Communication: A Protocol Towards Digital Clinical Diagnostic Procedures

Eugenia I. Toki[1(✉)], Victoria Zakopoulou[1], Giorgos Tatsis[1], Konstantinos Plachouras[1], Vassiliki Siafaka[1], Evangelia I. Kosma[1], Spyridon K. Chronopoulos[1], Despina Elisabeth Filippidis[2], Georgios Nikopoulos[3], Jenny Pange[4], and Anastasios Manos[3]

[1] Laboratory of New Approaches in Communication Disorders, Department of Speech and Language Therapy, School of Health Sciences, University of Ioannina, Ioannina, Greece
{toki,vzakop,gtatsis,kplachouras,siafaka,e.kosma, spychro}@uoi.gr

[2] Teleglobal, Ioannina, Greece
elizabeth@teleglobal.gr

[3] DotSoft, Ioannina Branch, Ioannina, Greece
nikopoulos@bauboproductions.com, tmanos@dotsoft.gr

[4] Laboratory of New Technologies and Distance Learning, School of Education, University of Ioannina (UoI), Ioannina, Greece
jpagge@uoi.gr

Abstract. Although a high prevalence of developmental speech/language disorders (3–17%) has been reported in the current literature still many children are underdiagnosed resulting to miss out on effective interventions that could be of more impact if administered early. The utilization of digital and mobile technologies in health and learning has evolved, presenting new opportunities for monitoring, decision making, classification and assessment procedures. This study focuses on reporting and justifying a protocol for the design and development of a digital approach intended to support and enhance screening and early detection procedures of developmental speech/language difficulties in child communication using smart computing models, sensors, and early diagnostic speech and language deficiencies indicators. The proposed solution will be designed and developed in phases. The design consists of (i) an interactive game-based digital approach for the child, (ii) an online environment to collect necessary data from parents, and clinicians (iii) the full functional specification of the game-based activities together with the overall architecture of the proposed innovative system. The proposed smart innovative system has the potential to support digital health care on children's communication skills, suggesting a positive economic impact according to current digital trends.

M. E. Auer and T. Tsiatsos (Eds.): IMCL 2021, LNNS 411, pp. 559–568, 2022.
https://doi.org/10.1007/978-3-030-96296-8_50

Keywords: Developmental speech and language disorders · Child · Game-based system · Evaluation · Screening

1 Introduction

1.1 Clinical Background: Speech and Language Development

According to the DSM-5 [1], Neurodevelopmental Disorders (ND), include intellectual disability (ID) (also known as Intellectual Developmental Disorder), communication disorders (speech, language, social/pragmatic, etc.), autism spectrum disorder (ASD), attention deficit hyperactivity disorder (ADHD), specific learning disorder (SLD), and motor disorders [2]. They usually occur in the developmental period and include deficits that cause reductions in the personal, social, academic/professional life of the individual. The extent and severity of these deficits varies from focused difficulties in executive functions (or learning) to generalized social skills disorders or intellectual impairment. In addition, ND may co-occur (e.g., many children with ASD have ID, and often children with ADHD have SLD) and/or present high rates of comorbidity with other disorders, possibly resulting in cognitive impairment, reduced quality of life for both individuals and their families and poor long-term prognosis [3, 4].

Speech and language development, as a useful early indicator of a child's overall development and cognitive ability [5], with evidence of a developmental continuity from early childhood to adulthood can reveal the close link between language and cognitive development [6]. When a child struggles to understand, speak and master the language milestones at the time that other peers have already acquired these skills, it may be a sign of a language or speech delay or disorder. Children may deal with [3, 4, 7, 8]: (i) speech disorders (childhood apraxia of speech, dysarthria, orofacial myofunctional disorders, speech sound disorders (articulation and phonological processes, stuttering, voice), (ii) language disorders (preschool language disorders, learning disabilities, selective mutism), (iii) medical and developmental conditions (attention deficit/hyperactivity disorder, autism, cleft lip and palate, right hemisphere or/and traumatic brain injury).

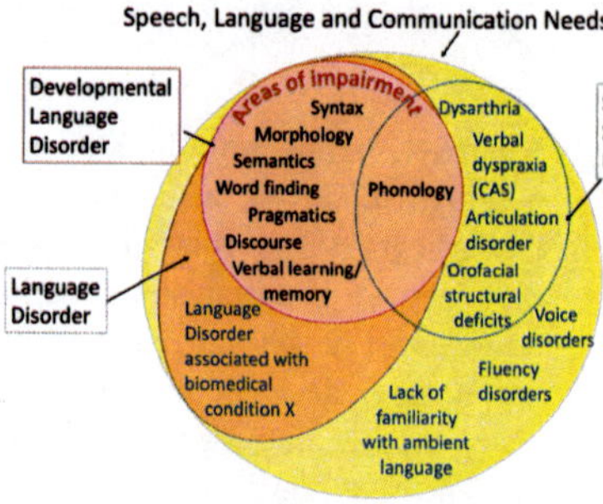

Fig. 1. Diagnostic terms [8, 12]

Clinical screening and assessment procedures can identify a language or speech delay or disorder. Screening, "*the process of identifying healthy people who may have an increased chance of a disease or condition*" [5], enables early detection and provides explanations in functioning and/or behavior, allowing the individual to be involved in care/intervention planning when rather little damage has been done and thus (i) improve quality of life and (ii) reduce the chance of developing a serious condition or its complications [5, 9, 10]. Assessment, "*the ongoing procedures used by qualified personnel to identify the child's unique strengths and needs and ...the early intervention services...and includes the assessment of the child...and...the child's family..."* [11], must include information such as parents interview/questionnaires, clinician's observations, and standardized age-normed tests or criterion-based assessments, speech and language sample as well as expressive and receptive language [4]. Finally, Fig. 1. illustrates diagnostic terms [8, 12].

1.2 Technology, Sensors and Wearables as Assets in Diagnostic Procedures

In recent decades, technology has dominated the daily lives of both children and adults [13]. Young children as they grow older are regularly in contact with a variety of modern technological tools, receiving a lot of stimuli and technology experiences [14]. With the integration of technological means in the speech therapy process, the possibility of strengthening the diagnostic and intervention procedures is acquired, especially when it is addressed to children [13, 15].

Various types of technologies and devices have been used such as, laptops, touchscreens, smartphones, digital games, videos, robots, avatars, cameras, virtual and augmented reality and sensors. Mobile devices with touch screens are used especially in cases of autistic spectrum disorder, as communication is the key element and presenting images, sounds and even text-to-speech services and vice versa is easy [16].

Moreover, computer games were exclusively designed for fun. Serious games are games prioritizing other goals in various different fields (i.e. military, government, corporate, health-care, and education) [17]. Developing a serious game can be challenging as it entails to combine the serious contents and at the same time to keep the fun characteristics [17, 18]. Serious games based on their ability to offer chances for (i) students' active involvement, (ii) various learning styles facilitation and (iii) fostering decision making, can be a vital complement to traditional learning and a primary driver in the current digital shift in education [19–21]. Serious games for Health, turned out to be also a promising tool that contributes towards health issues regarding awareness, as well as physical, and mental issues [22, 23]. The design and development of health games usually requires experts for the game production due to the need to perform very specialized tasks, and to target specific and possibly vulnerable populations [22].

Serious games in speech and language therapy are one of the most favorable children's practices, used for screening and diagnostic purposes children (i.e. autism, dyslexia, apraxia, stuttering) and to plan intervention for strengthening skills [13, 15, 24].

As such, there are games integrated with automatic speech recognition (ASR) offering increased motivation with speech productions for gameplay control [25]. Games, that can be classified as pediatric Computer-Aided Speech Therapy (CAST) tools focus on screening, assessment, intervention and even on individualized intervention [24]. Furthermore, games implementing automated analysis of speech disorders in children investigate various challenges in processing child disordered speech [26] and employ several artificial intelligence (AI) approaches including expert systems, neural networks, deep learning, machine learning [27–31].

Alongside, eye tracking can offer clinicians and educators insights of what attracts attention, what is overlooked, in which order elements are spotted, and relations between them at a clinical assessment of at risk of autism, in children under 3 years of age [32]. Eye monitoring technology was used in the detection of social communication disorders (and in children with ASD), as a means of early detection [33].

A variety of wearable devices are available on the market today, aiming to monitor both the psychological state and the emotions, such as wristbands and watches, chest belts, vests, blouses, pads and sleeves [33, 34]. The use of sensors implemented in mobile phones is a very convenient way of data collection, used by interdisciplinary studies combining both quantitative and qualitative derived values [35]. A smart phone can alert in real-time about the person's health status, such as stress levels [36]. Heart rate, temperature, respiration rate, electrodermal activity, movement, body posture are some of the biomarkers that can be recorded with such wearable devices [27].

Today, eye tracking technology, wearable and microsensor solutions for health monitoring, offer opportunities for innovation in developing solutions to measure complex health outcomes in non-specialist and remote settings. According to systematic literature review there is adequate knowledge for the implementation of these approaches to provide additional valuable insights to justify their inclusion in clinical study protocols [27, 37, 38]. Also, eye tracking has the potential to offer clinicians and educators insights of what attracts attention, what is overlooked, in which order elements are spotted, and relations between them [39].

This study aims to contribute to the increasing research area of digital health care and particularly in early detection of developmental speech and language disorders in child communication with the utilization of serious games, to enhance children's and families' quality of life. Specifically, being focused on reporting and justifying a protocol for the design and development of a digital approach, this study intends to support and enhance screening and early detection procedures using smart computing models, sensors, and early diagnostic speech and language deficiencies indicators.

2 The Process of Smart System Design

This study reports on the design of a smart system, that is part of work under a research project with the acronym SmartSpeech, funded from the Region of Epirus and supported from European Regional Development Fund (ERDF). SmartSpeech, in a systemic approach, accommodates speech and language pathology screening/assessment procedures for monitoring, classification, and decision making towards developmental speech and language disorders.

The review of the current literature on clinical approaches regarding screening and diagnostic issues on speech and language developmental disorders determines the clinical requirements of the proposed system. The interaction with an interdisciplinary group of clinicians/specialists, children and their parents in speech and language therapy, offers insights into the range of speech and language difficulties faced. The target group aims for preschoolers/early elementary school children. Digital solutions reported in the latest literature used by researchers and clinicians provides an overall understanding of the market. Ethical, privacy, and data security issues are considered.

The proposed solution has been designed and developed in the following phases. The design consists of (i) an interactive video game app for the children, (ii) an online environment to collect, and display data from parents, and clinicians and (iii) the functional specification of the game-based activities together with the overall architecture of the proposed innovative system.

3 Defining Systems Requirements

The system requirements used to identify the functionality needed to satisfy the user's requirements, guide the serious game design, and processes are listed in Table 1.

Table 1. SmartSpeech system requirements

	System requirements' description
	the video game should…
1	… target the child player
2	… adopt positive characteristics of the widely-used screening tasks
3	… run on mobile devises using Android, IoS or Windows
4	… take in consideration levels of difficulty in tasks
5	… be an entertaining game that can stimulate speech and language skills screening procedures
6	be easy to understand Graphical User Interface (GUI) elements
7	3D animation with detailed clear and expressive faces for the characters in the game
8	clear and simple audio instructions for the game tasks
9	the system should record users' responses and tasks' performance in the main database
10	Tasks should address: Verbal & NonVerbal Communication, Psychomotor ability, Audiometry, Executive Functions, Psychoemotional State

4 System Design

4.1 System's Specifications

The detailed design spec of the game activities, the methodology to be followed to interpret the results, and flow diagrams illustrating how the proposed system operates is presented next in Fig. 2.

This innovative system consists of sub-systems:

o A subsystem supports and serves the users' interface. The users of the system are: - the child, the clinician (speech therapist, psychologist, special educator), and the caregiver/parent. It can be used functionally in order to collect and combine data in relation to the child's abilities from multiple sources, i.e. the child's responses thought the game, the bio-sensors data, the clinician's observations and responses, and the caregiver/parent reports.

p Throughout the child interaction processes a subsystem supports the collection of data from biosensors during specific processes.

q A subsystem supports the synchronization of the above measurements in combination with the activities, real time and their registration in a properly configured database using timestamps.

The above subsystems support the data collection throughout the research with the support of users' interfaces. The operation of the interface is supported by a mobile application that is to be used by children and are in the form of an interactive game accommodating specially designed tasks for the purpose of speech and language evaluation collecting biometric data via sensors. The design and implementation of digital game is based on the objectives of the speech and language evaluation.

Next, follows a call to various populations (parent and child) to participate in data collection. Once the parents are informed and have given a written consent, they can register and answer online questions regarding their child's development and skills. Then the child can play the game under the clinician's supervision, while the system collects responses and biometric data (biodata). These data are analyzed and statistically processed and thus while using AI, conclusions are drawn regarding game task responses and biometric measurements, classifying the child's communication profile and guiding the clinician towards the detection/ diagnosis of communication disorders. The goal is to extract important values that identify the various communication deficits.

All the above activities of this innovative system are targeted to acquire knowledge that is based both on the evaluation of the communication skills of the child and on accompanying biometrics. Then the AI strategy is to be established so the system can make automated decisions. In more detail, this strategy can include new knowledge in combination with existing knowledge of experts, literature, and diagnostic criteria, constituting the "knowledge" of the smart system with automatically issuing decision on child's profile/skills on communication deficits. Testing validation and verification of the innovative game-based system, is be conducted on various populations through a pilot study, following by deployment and maintenance of the final system.

4.2 System's Administration and Architecture

The main game data management panel is used mainly for user registration actions, the recording of data on child's health and developmental history, as well as the visualization of the results from the children's – players' game responses. In particular, the features supported by the game are: User registration (parents, clinicians, administrator), Entry child's data (ID, nickname, physical data -height, weight, etc., date of birth, school, family status, spoken languages), Child medical history, Demographic Elements, Parents perceptions on child's verbal & non-verbal communication aspects, orofacial issues, voice, articulation/phonology, expression of Language, psychomotor development, auditory comprehension and memory, Storing data in main database, Collection of data from the database on children's responses to tests, Data retrieval & Visualization of players' responses per game, episode and task, and Clinician's feedback. Also, the fully functional specification of the game-based smart innovative system together with the overall architecture are visualized in Fig. 2.

5 Conclusions

The current study describes process for the design of a smart system contributing to the implementation of new innovative tools to assist clinicians, and/or other professionals, towards complicated screening and diagnostic procedures of communication deficiencies in children. The main objectives of this study and the methodology behind the proposed design is to:- (i) provide an accurate estimation of the child's communication profile at an early developmental age, (ii) include additional physiological data that may provide further insights on the child's development and skills and difficulties, (iii) assist the clinician in clinical decisions and treatment plan, (iv) reduce clinical cost, (v) add with a new tool contributing to current research in and (vi) create new educational and training opportunities for students and lifelong learners to gain skills in clinical labs in the field of digital health in communication disorders.

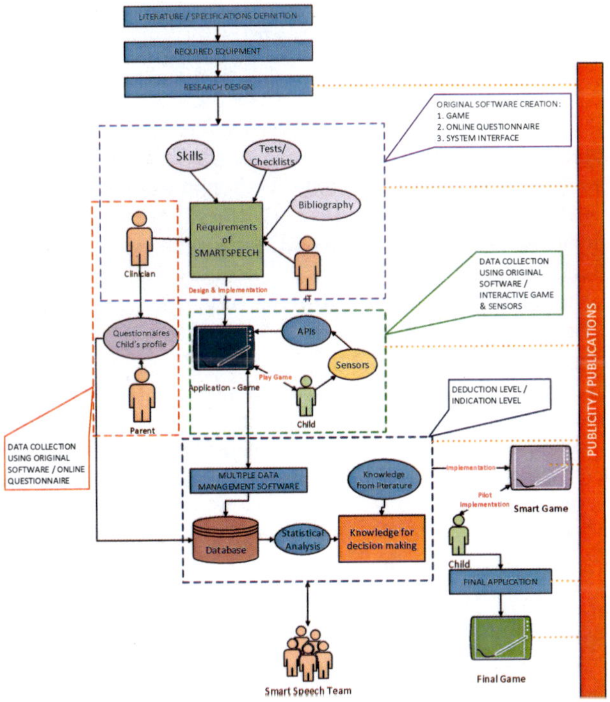

Fig. 2. SmartSpeech: system's architecture

Acknowledgements. We wish to thank the Region of Epirus for funding this project titled "Smart Computing Models, Sensors, and Early diagnostic speech and language deficiencies indicators in Child Communication", acronym "SmartSpeech" with code HΠ1AB-28185, supported from European Regional Development Fund (ERDF).

References

1. American Psychiatric Association, American Psychiatric Association eds: Diagnostic and statistical manual of mental disorders: DSM-5. American Psychiatric Association, Washington, D.C (2013)
2. Harris, J.C.: New classification for neurodevelopmental disorders in DSM-5. Curr. Opin. Psychiatry **27**, 95–97 (2014). https://doi.org/10.1097/YCO.0000000000000042
3. Oyono, L.T., Pascoe, M., Singh, S.: The Prevalence of Speech and Language Disorders in French-Speaking Preschool Children From Yaoundé (Cameroon). https://pubs.asha.org/doi/10.1044/2018_JSLHR-L-16-0400
4. Bishop, D.V.M., Snowling, M.J., Thompson, P.A., Greenhalgh, T., CATALISE consortium: CATALISE: a multinational and multidisciplinary Delphi consensus study. identifying language impairments in children. PLOS ONE. 11, e0158753 (2016). https://doi.org/10.1371/journal.pone.0158753
5. Public Health England: NHS population screening explained. https://www.gov.uk/guidance/nhs-population-screening-explained
6. Flensborg-Madsen, T., Mortensen, E.L.: associations of early developmental milestones with adult intelligence. Child Dev. **89**, 638–648 (2018). https://doi.org/10.1111/cdev.12760
7. American Speech-Language-Hearing Association: Child Speech and Language. https://www.asha.org/public/speech/disorders/childsandl/
8. Bishop, D.V.M., Snowling, M.J., Thompson, P.A., Greenhalgh, T.: CATALISE-2 consortium: Phase 2 of CATALISE: a multinational and multidisciplinary Delphi consensus study of problems with language development: Terminology. J. Child Psychol. Psychiatry. **58**, 1068–1080 (2017). https://doi.org/10.1111/jcpp.12721
9. Asvestopoulou, T., et al.: Towards a robust and accurate screening tool for dyslexia with data augmentation using GANs. In: 2019 IEEE 19th International Conference on Bioinformatics and Bioengineering (BIBE), pp. 775–782. IEEE, Athens, Greece (2019)
10. Ooi, C.Y., Ng, C.J., Sales, A.E., Lim, H.M.: Implementation strategies for web-based apps for screening: scoping review. J. Med. Internet Res. **22**, e15591 (2020). https://doi.org/10.2196/15591
11. Assessment and Evaluation of Speech-Language Disorders in Schools. https://www.asha.org/slp/assessment-and-evaluation-of-speech-language-disorders-in-schools/
12. Archibald, L.: Developmental Language Disorder (DLD): A persistent language disorder of unknown aetiology (2018)
13. Toki, E.I., Zakopoulou, V., Pange, J.: Preschoolers' learning disabilities assessment: new perspectives in computerized clinical tools. Sino-US Engl. Teach. **11**, 401–410 (2014)
14. Liu, X., Toki, E.I., Pange, J.: The use of ICT in preschool education in Greece and China: A comparative study. Procedia - Soc. Behav. Sci. **112**, 1167–1176 (2014). https://doi.org/10.1016/j.sbspro.2014.01.1281
15. Anzulewicz, A., Sobota, K., Delafield-Butt, J.T.: Toward the autism motor signature: gesture patterns during smart tablet gameplay identify children with autism. Sci. Rep. **6**, 1–13 (2016)
16. Koumpouros, Y., Kafazis, T.: Wearables and mobile technologies in Autism spectrum disorder interventions: a systematic literature review. Res. Autism Spectr. Disord. **66**, 101405 (2019)
17. Guðmundsdóttir, K., Wille, A.L., Savu, A.: Designing games for children with cerebral palsy. In: Schouten, B., Fedtke, S., Bekker, T., Schijven, M., Gekker, A. (eds.) Games for Health, pp. 99–116. Springer Fachmedien Wiesbaden, Wiesbaden (2013).https://doi.org/10.1007/978-3-658-02897-8_8

18. Guadalupe, J.G.S., Alberto, O.Z.C.: Serious games: fundamental concepts and their importance in health care 13 (2011)
19. Lekka, A., Toki, E., Tsolakidis, C., Pange, J.: Literature review on educational games for learning statistics. In: 2017 IEEE Global Engineering Education Conference (EDUCON), pp. 844–847. IEEE, Athens, Greece (2017)
20. Mavridis, A., Tsiatsos, T., Terzidou, T.: Designing and deploying 3D collaborative games in education: Int. J. Game-Based Learn. **6**, 43–57 (2016). https://doi.org/10.4018/IJGBL.2016010104
21. Mavridis, A., Tsiatsos, T.: Game-based assessment: investigating the impact on test anxiety and exam performance: game-based assessment. J. Comput. Assist. Learn. **33**, 137–150 (2017). https://doi.org/10.1111/jcal.12170
22. Boletsis, C., McCallum, S.: The smartkuber case study: lessons learned from the development of an augmented reality serious game for cognitive screening. In: De Paolis, L.T., Bourdot, P., and Mongelli, A. (eds.) Augmented Reality, Virtual Reality, and Computer Graphics, pp. 457–472. Springer International Publishing, Cham (2017). https://doi.org/10.1007/978-3-319-60922-5_36
23. Chaldogeridis, A., Politopoulos, N., Tsiatsos, T.: Designing exergames for working memory training using MaKey MaKey|SpringerLink. In: Interactive Mobile Communication Technologies and Learning. IMCL 2017. Springer, Cham (2018). https://doi.org/10.1007/978-3-319-75175-7_62
24. Zakopoulou, V., et al.: Evaluating new approaches of intervention in reading difficulties in students with dyslexia: The ilearnRW software application. J. Educ. Pract. **8**(27), 36–52 (2017)
25. Ahmed, B., Monroe, P., Hair, A., Tan, C.T., Gutierrez-Osuna, R., Ballard, K.J.: Speech-driven mobile games for speech therapy: user experiences and feasibility. Int. J. Speech Lang. Pathol. **20**, 644–658 (2018). https://doi.org/10.1080/17549507.2018.1513562
26. Shahin, M., Zafar, U., Ahmed, B.: The automatic detection of speech disorders in children: challenges, opportunities, and preliminary results. IEEE J. Sel. Top. Signal Process. **14**, 400–412 (2020). https://doi.org/10.1109/JSTSP.2019.2959393
27. Alcañiz Raya, M., et al.: Application of supervised machine learning for behavioral biomarkers of autism spectrum disorder based on electrodermal activity and virtual reality. Front. Hum. Neurosci. **14** (2020). https://doi.org/10.3389/fnhum.2020.00090
28. Georgoulas, G., Georgopoulos, V.C., Stylios, C.D.: Speech sound classification and detection of articulation disorders with support vector machines and wavelets. In: 2006 International Conference of the IEEE Engineering in Medicine and Biology Society, pp. 2199–2202 (2006)
29. Xu, D., Gilkerson, J., Richards, J., Yapanel, U., Gray, S.: Child vocalization composition as discriminant information for automatic autism detection. In: 2009 Annual International Conference of the IEEE Engineering in Medicine and Biology Society, pp. 2518–2522 (2009)
30. Georgopoulos, V.C.: Advanced time-frequency analysis and machine learning for pathological voice detection. In: 2020 12th International Symposium on Communication Systems, Networks and Digital Signal Processing (CSNDSP), pp. 1–5 (2020)
31. Pahwa, A., Aggarwal, G., Sharma, A.: A machine learning approach for identification amp; diagnosing features of Neurodevelopmental disorders using speech and spoken sentences. In: 2016 International Conference on Computing, Communication and Automation (ICCCA), pp. 377–382 (2016)
32. Özdemir, S., Kök, İ., Bülbül, I.A., Özdemir, S., Töret, Z.: An Eye-Tracking Data Based Decision Support System for Autism Risk Assessment (2018)
33. Yamamoto, H., Sato, A., Itakura, S.: Eye tracking in an everyday environment reveals the interpersonal distance that affords infant-parent gaze communication. Sci. Rep. **9**, 1–9 (2019)

34. Bacon, E.C., Moore, A., Lee, Q., Barnes, C.C., Courchesne, E., Pierce, K.: Identifying prognostic markers in autism spectrum disorder using eye tracking: Autism (2019).https://doi.org/10.1177/1362361319878578
35. Ueberham, M., Schmidt, F., Schlink, U.: Advanced smartphone-based sensing with open-source task automation. Sensors **18**, 2456 (2018)
36. Gradl, S., Wirth, M., Richer, R., Rohleder, N., Eskofier, B.M.: An overview of the feasibility of permanent, real-time, unobtrusive stress measurement with current wearables. In: Proceedings of the 13th EAI International Conference on Pervasive Computing Technologies for Healthcare, pp. 360–365 (2019)
37. Byrom, B., McCarthy, M., Schueler, P., Muehlhausen, W.: Brain monitoring devices in neuroscience clinical research: the potential of remote monitoring using sensors, wearables, and mobile devices. Clin. Pharmacol. Ther. **104**, 59–71 (2018). https://doi.org/10.1002/cpt.1077
38. Ness, S.L., et al.: An observational study with the Janssen autism knowledge engine (JAKE®) in individuals with autism spectrum disorder. Front. Neurosci. **13** (2019). https://doi.org/10.3389/fnins.2019.00111
39. Shaked, K.B.-Z., Shamir, A., Vakil, E.: An eye tracking study of digital text reading: a comparison between poor and typical readers. Read. Writ. **33**, 1925–1944 (2020). https://doi.org/10.1007/s11145-020-10021-9

SkyWords: A Serious Game To Enhance Typing and Spelling Skills

Lampros Karavidas(✉), George Topalidis, and Grigorios Zilidis

Aristotle University of Thessaloniki, Thessaloniki, Greece
karavidas@csd.auth.gr, george.top@windowslive.com

Abstract. The last few decades serious games, a genre that does not have entertainment, enjoyment, or fun as its primary purpose, have emerged to be a great way to educate, especially since more and more people are familiar with playing games and are fond of them. The purpose of our study was to create a serious game designed and developed to help young students enhance both their typing and spelling skills, based on the expert's recommendations and guidelines. As this game is destined to be played by kids in primary school, apart from being educational it should also be entertaining enough for them and not frustrate them. To achieve this, after developing the game it was evaluated by 30 children and an experts' evaluation was held as well. All things considered there were no major problems that emerged from the evaluation of the game. However, there are some future work that should be done to make the game meet the user's needs. At first, a beginners' vocabulary could be useful for children that do not have basic knowledge of the English language and maybe a progress bar of the levels remaining to be played so as to help the children not get discouraged while playing it. Moreover, several alterations to the difficulty level of the game will be made incorporating a dynamic difficulty adjustment solution.

Keywords: Serious games · Typing · Evaluation

1 Introduction

Nowadays, digital competence which consist of technical skills to use technology for working, studying and in everyday life among others, is of great importance [7] both for adults and children. This was made clear, throughout the whole pandemic, as most of the educational activities were done virtually.

The last few decades serious games, a genre that does not have entertainment, enjoyment, or fun as its primary purpose [8], have emerged to be a great way to educate. Serious games are designed in a way to help their users gain knowledge or enhance a skill of theirs passively by playing the game. Serious games have been applied to a great variety of areas such military, education, corporate and healthcare [9].

Serious games have been used widely as they can provide users with simulated environments and systems and enable them gain valuable experience that would be hard to get in the real life for reasons of safety, cost, time etc. Schlickum (2009) proved that by playing computer games and mastering them, a certain number of students have managed to transfer their skills and have better performance in real life challenges

M. E. Auer and T. Tsiatsos (Eds.): IMCL 2021, LNNS 411, pp. 569–579, 2022.
https://doi.org/10.1007/978-3-030-96296-8_51

related to their job, such as surgical performance in advanced virtual reality endoscopic simulators in this case.

More and more people are familiar with playing games and are fond of them as stated by the Entertainment Software Association (2020). Especially, the same report states that, games have been a great success in kids under 18, as 51.1 million kids only in the US have been playing a game or more in 2020. Thus, it is safe to say that serious games will help young students be more engaged in learning as games as a medium will be familiar to them.

Engagement in video games is notion linked to many others as immersion [3] and the flow state which is a state where the player is neither bored nor frustrated while playing a game [4]. Calleja (2010) states that games are engaging by their own nature as they are a great way for users to escape from the reality. Video games tend to keep players immersed and lose track of time as they provide them with pattern recognition, problem solving and decision making challenges and help them practice their cognitive abilities. Therefore, serious games as a video game genre seem to be a great choice to educate young students.

Finally, technology seems to have a great positive impact on learning a foreign language and sometimes technology-based language instruction may have the same results as teacher-delivered instruction [6].

The purpose of our study was to create and evaluate a serious game designed to help young students enhance their typing and spelling skills, which are connected to digital competence and foreign language learning, respectively. The rest of the paper is organized as follows: In the next section we present the design decisions that were made to create an engaging serious game and the gameplay of the final product. We then describe the evaluation process and present and discuss the results deriving from it. Finally, we describe the future changes that will be made according to the evaluation results analysis and conclude with the effectiveness of the serious game that was created.

2 More About SkyWords

2.1 Game Design

The game is designed to be a 2D falling words one with rogue lite elements, as no progress is saved and some of the game elements are randomly generated. The game is designed to be played by children in primary school, 7–12 years old as they should already have some basic knowledge of the English alphabet.

In order to create such a game, we contacted a primary education informatics teacher to help us realise the level of typing skills of her students, throughout the last years. According to her experience, children know how to read in the second grade of the primary school and are familiar with the layout of a keyboard until then. However, most of them are usually confused when they have to press a character on the keyboard which is similar both on the English (foreign language) and on the Greek (mother language) alphabet. Finally, she emphasized that younger students usually get easily frustrated when they can not find a letter on the keyboard and usually ask for help from their teacher.

To keep the users' interest, the game will gradually show words that are more difficult than the previous ones. The words that are shown in the game are randomly picked from a pool of random English words. However, in order to the gradual difficulty increase all of the words should be accompanied by a metric indicating their complexity. To generate such a number, we created the following formula (1) where x is the amount of times each character can be found in a certain word.

$$\text{Difficulty} = \sum\nolimits_{\forall Character} \left(1 + \frac{1}{x^{(x-1)}}\right) \quad (1)$$

Our formula does not take into consideration the place of the character on the keyboard layout as the game's purpose is to help the users enhance their typing skills, so we take as a fact that they are not familiar with the keyboard layout or how to place their fingers on it. Moreover, words that are common in the English language don't have a lower metric by default, because kids that are part of our target group will not necessarily have knowledge of the basic English words, so all of the words are considered to be new to them. The following table (Table 1) is a demonstration of our formula on some random English words.

Table 1. Difficulty formula application on English words

Word	Difficulty Metric
and	3
three	4.5
majestic	8
representative	10,62

As the game is designed to be played by young children a simple, yet familiar style will be adopted. The font that will be used will have rounded edges, the colours will be bright and the game's graphics will be influenced by some common themes that many children are fond of, such as pirates.

2.2 Gameplay

The game was created using the Unity 3D game engine.

When the user opens the game, he can see the main menu. The main menu consists of the "PLAY" and the "QUIT" button, that ends the app. If the user presses the "PLAY" button then he/she has to choose the difficulty of the game, which is related to the falling speed of the words in the game. Each of these difficulty options are made to make the game fun and interesting to children with a great variety of typing skills and knowledge of English words. The easier one is suitable for children with basic knowledge of the English alphabet and the keyboard layout. The difficulty can be chosen either by the child on its own or by his/her teacher's recommendations.

Then he/she starts playing the game. The goal of the game is to type as many words as possible in a certain amount of time and finally gather a great amount of points. The words spawn at random spots on the top of the screen and then start falling. They are available to be typed while they can be seen and once the user presses one of the letters of the word then it is labeled as active and turns green, in order to help the user focus on it (Fig. 1). However, once the word is active the player must complete typing it before start typing another one. Each time the user types correctly a letter of the word then the letter disappears. The words disappear when they reach the bottom of the screen. If an active word reaches the bottom of the screen then the active word slot is freed and the user can start typing a new word.

Fig. 1. Screenshot of the game with active word

SkyWords consists of ten distinct levels and each of them displays gradually more difficult words. When a level is completed, the user gets to decide if he/she wants to proceed playing the following more difficult level or go to the main menu of the game. Once the ten levels of the game are completed the user can see the number of points, he/she managed to gather and goes back to the main menu.

The points are gathered by typing correct letters. Each letter translates into a point. If the user types something wrong or if he/she does not complete typing a word either active or not before disappearing there are no penalties. That was a design decision that was made as we do not want the players to get discouraged and get frustrated, based on the expert's recommendations (Fig. 2).

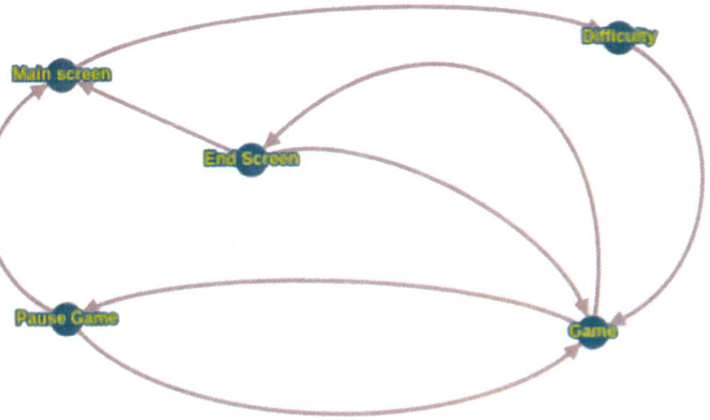

Fig. 2. Flow of the game's screens

3 Evaluation

Two different evaluations were held to highlight possible problems of the game and provide useful feedback to the game's creators. The first one was a user evaluation with the game's target group and the other one was an expert evaluation to spot even more improvements that could be made to the game.

3.1 User Evaluation Methodology

Participants. The participants were children (n = 30) in the primary school. Half of them were girls. One third of them play digital games at least once per week, as it can be seen in the following figure and 47% of them play board games at the same frequency (Fig. 3).

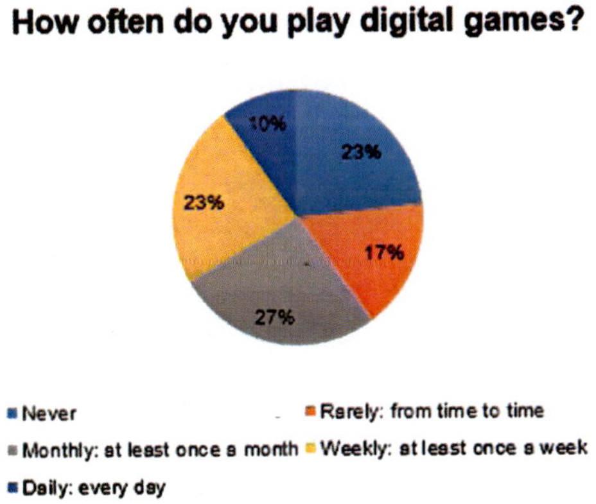

Fig. 3. Amount of time playing digital games

Questionnaires. A valid with high internal consistency questionnaire, named MEEGA +kids [10] was used to evaluate the overall quality of our serious game and therefore the enjoyment as perceived by our users. The 26 questions in the questionnaire attempted to address different aspects of the game. The dimensions the are evaluated through the questionnaire are the following 12: 1) Aesthetics, 2) Learnability, 3) Operability, 4) Accessibility, 5) Confidence, 6) Challenge, 7) Satisfaction, 8) Social Interaction, 9) Fun, 10) Focused Attention, 11) Relevance and 12) Learning. The statement in each question was rated on a 5-point Likert scale, where −2 was Strongly Disagree, −1 was Disagree, 0 was Indifferent, 1 was Agree and 2 was Strongly Agree.

Procedure. The participants were all part of a class in the primary school. The evaluation took place in the school's computer lab. The creators of the game and a primary education teacher were present to provide further explanation and instructions on how to play the game to the users. The questionnaire was answered after completing ten levels of the serious game. Moreover, some qualitive data were also gathered from the school teacher.

Data Analysis. The data analysis was done using datasheets and the results of the analysis were presented through descriptive statistics.

3.2 User Evaluation Results

Questions Related to the Usability. All these questions were intended to investigate, from different perspectives, how usable the game was according to its users. In this context, usability is defined as the level of effectiveness and efficiency that the users managed to achieve their goals using the game [11].

Aesthetics. The questions related to aesthetics showed us that there were no issues related to the font or the design of the game such as the interface and the graphics. The vast majority replied positively as more than half of the users strongly agreed with the questions they were asked (Table 2).

Table 2. Responses related to aesthetics

Question	−2 (Strongly Disagree)	−1 (Disagree)	0 (Indifferent)	1 (Agree)	2 (Strongly Agree)
The game design is attractive	0%	3%	17%	23%	57%
The font and colors of the game match	7%	10%	20%	10%	53%

Learnability. The answers for the learnability of the game confirmed us that it was easy for the kids to play the game as more that 50% said that they agreed or strongly agreed, and we it perceive as a positive feedback. Some of them, that stated the opposite might be affected by the difficulty of the actual game (Table 3).

Table 3. Responses related to learnability

Question	−2 (Strongly Disagree)	−1 (Disagree)	0 (Indifferent)	1 (Agree)	2 (Strongly Agree)
Learning to play this game was easy for me	23%	20%	3%	17%	37%

Operability. The questions that were about the operability of the game made it clear that the rules of the game are quite simple for children of young age to understand, which was an important factor that was taken into consideration while designing the game. However, even though half of the students replied positively that the game is easy to play it is obvious that the difficulty level of the game might be a little high for some of the students (Table 4).

Table 4. Responses related to operability

Question	−2 (Strongly Disagree)	−1 (Disagree)	0 (Indifferent)	1 (Agree)	2 (Strongly Agree)
I think that the game is easy to play	17%	23%	10%	27%	23%
The game rules are clear and easy to understand	7%	3%	10%	7%	73%

Accessibility. Almost all the participants, stated that the colors and the fonts of the game were readable and meaningful, which is important as the game is a word game and the users must read to be successful in it (Table 5).

Table 5. Responses related to accessibility

Question	−2 (Strongly Disagree)	−1 (Disagree)	0 (Indifferent)	1 (Agree)	2 (Strongly Agree)
The fonts (size and style) used in the game are easy to read	7%	0%	3%	17%	73%
The colors used in the game are meaningful	10%	0%	7%	10%	73%

Questions related to the Player Experience. These questions were asked to have a better picture of the user experience while playing the game. In other words, player experience is described by deep involvement of the user in the game that results in him/her gaining knowledge, being gratified for his try, and interacting with the game or even with other players [11] (Table 6).

Table 6. Descriptive statistics related to the dimensions evaluated

Dimensions	−2 (Strongly Disagree)	−1 (Disagree)	0 (Indifferent)	1 (Agree)	2 (Strongly Agree)
Confidence	10%	7%	13%	17%	53%
Challenge	24%	12%	18%	17%	29%
Satisfaction	5%	4%	13%	17%	60%
Social Interaction	12%	4%	14%	11%	57%
Fun	17%	5%	12%	13%	53%
Focused Attention	30%	7%	17%	3%	43%
Relevance	10%	2%	10%	18%	60%
Learning	10%	0%	20%	13%	57%

It can be safely assumed that the users felt confident, satisfied, interacted with their classmates, and had fun while playing SkyWords. Moreover, the serious game was highly relevant with what the children were taught till then in their English class and it contributed to their learning in this particular course, that was learning English as a foreign language.

However, even though half of the participants stated that they had no problem paying attention while playing the game. Most of them forgot about their surroundings, which is a sign of being immersed, as 60% of the participants strongly agreed with this statement. However, the focused attention ratings were low as the users wanted something to grab their attention at the start of the game.

Table 7. Responses related to challenge

Question	−2 (Strongly Disagree)	−1 (Disagree)	0 (Indifferent)	1 (Agree)	2 (Strongly Agree)
The game is appropriately challenging for me	40%	17%	10%	10%	23%
The game provides new challenges at an appropriate pace	10%	13%	23%	20%	33%
The game does not become monotonous	23%	7%	20%	20%	30%

Finally, the lowest rating can be seen in the challenge factor. After looking thoroughly, the questions (Table 7) that are part of the challenge dimension, it is prominent that most of the users felt that the game was not appropriately challenging for them. More than half of them provided positive feedback to the pace of the new challenges of the game and stated that it did not become monotonous while playing it.

Educator's Comments. The primary school educator also provided us with feedback based on her observations of the children while playing the game and a conversation she had with them right after the evaluation. She stated most of the users asked her if they were able to play again the game in the computer lab or play it in their free time, indicating that it was a pretty enjoyable experience for them. Moreover, she mentioned that the participants seemed to be more confident at typing after the session and that the serious game would be a great activity to be integrated in the class activities.

However, she was concerned about the children that had almost no English knowledge that the game seemed to be more difficult for them and seemed to be dissatisfied at first, although soon enough they started playing the game and were immediately immersed.

3.3 Expert Evaluation

To find more aspects of the game to improve an expert evaluation was held. The focus point of this evaluation was the usability of the serious game SkyWords and for this purpose the heuristic evaluation method was used. This kind of evaluation is based on empirical rules and findings, so in order to ensure impartial judgement five evaluators were used, as this has been found to be an optimal number [12]. The following table (Table 8) presents the evaluators' demographics.

Table 8. Demographics of the evaluators

Gender		Size	Percentage (%)
	Female	2	40.0
	Male	3	60.0
Age		Size	Percentage (%)
	25–34	4	80.0
	35–44	0	0
	>45	1	20.0
Educational Level		Size	Percentage (%)
	MSc Graduate	5	100.0
	PhD Graduate	0	0

Table 9. Heuristic evaluation results

No.	Heuristic rule	Ev. 1		Ev. 2		Ev. 3		Ev. 4		Ev. 3	
		Sev	Fq	Sev	Fq	Sev	Fq	Sev	Fq	Sev	Fq
1	Aesthetic and minimalist design	1	1	1	1	2	1	1	1	1	1
2	Match between system and the real world	1	1	1	2	5	1	1	1	1	1
3	Recognition rather than recall	1	1	3	1	4	2	1	1	1	1
4	Consistency and standards	1	1	1	1	5	5	1	1	1	1
5	Visibility of system status	3	4	4	1	4	3	4	4	3	3
6	User control and freedom	1	1	1	1	4	4	1	1	1	1
7	Flexibility and efficiency of use	1	1	1	5	5	4	1	1	1	1
8	Help users recognize, diagnose, and recover from errors	3	5	4	5	4	3	1	1	1	1
9	Error prevention	4	5	1	1	5	4	1	1	1	1
10	Help and documentation	1	1	5	4	4	1	4	4	4	2

For the evaluation purposes, the experts played the game for the proper amount of time to have a better understanding of it and afterwards completed the questionnaire based on then ten heuristic rules of Table 9. Each rule had two indicators, the severity of a problem ("S") and the frequency ("F") it emerged while playing the game. Moreover, to justify their ratings most of the evaluators made comments to help the development realise what they should fix in the future versions of the game.

Most of the problems that emerged from the expert evaluation were related to error prevention and recognition and documentation. Specifically, all the evaluators commented that there was a lack of feedback when the player typed a wrong word. Some of them proposed an error sound to be heard or to change the colour of the screen instantly, providing a visual feedback in that way. In addition, all of them stated that a guideline or introductory level would be beneficial to help the user realise what he/she should right after opening the game. Finally, an expert proposed to add more metrics to the game in order to make it more interesting and competitive for the users. The rest of the aspects that were evaluated were found to be adequate and well designed and executed.

4 Conclusions

All things considered there were no major problems that emerged from the evaluation of the game. The user evaluation provided the development team with great results as most of the aspects of the game were evaluated high by the majority of the participants. However, there are some future work that should be done to make the game meet the user's needs.

At first, the difficulty level of the game should be tailored to meet the user's need. The current menu that is developed to choose the difficulty level that seems fit for each user, is found to be inadequate. To overcome this issue a dynamic difficulty adjustment solution is proposed and will be incorporated in the following version of the game. The falling speed of each level will be adjusted to the typing rhythm of the player.

A further design option to consider is the visual feedback in case that the player presses a wrong key. Staying loyal to the recommendations of the primary education educator and taking as a fact that the users enjoyed playing the game, small changes will be made however not radical ones. Several metrics will be added and probably a visual feedback in case that a wrong key is pressed, without making the user lose any points. A progress bar of the levels remaining to be played, will be incorporated as well to help the children not get discouraged while playing it.

Finally, to be more friendly to users with little to no knowledge of the English alphabet a beginners' vocabulary could be useful.

All in all, taking everything into consideration the SkyWords serious game is found to be a solid solution to help young students practice their typing and spelling skills and its development will continue to have a product ready to be a valuable tool for any child.

References

1. Schlickum, M.K., Hedman, L., Enochsson, L., Kjellin, A., Felländer-Tsai, L.: Systematic video game training in surgical novices improves performance in virtual reality endoscopic surgical simulators: a prospective randomized study. World J. Surgery **33**(11), 2360–2367 (2009)
2. 2020 Essential Facts About the Video Game Industry: Entertainment Software Association, 15 July 2020. https://www.theesa.com/resource/2020-essential-facts/
3. Brown, E., Cairns, P.: A grounded investigation of game immersion. In: Conference on Human Factors in Computing Systems - Proceedings, pp. 1297–1300 (2004). https://doi.org/10.1145/985921.986048
4. Csikszentmihalyi, M., Nakamura, J.: Flow theory and research. In: The Oxford Handbook of Positive Psychology, pp. 195–206. Oxford University Press, Oxford (2009)
5. Calleja, G.: Digital games and escapism. Games Culture **5**(4), 335–353 (2010)
6. Zhao, Y.: Recent developments in technology and language learning: a literature review and meta-analysis. CALICO J., pp. 7–27 (2003)
7. Ilomäki, L., Kantosalo, A., Lakkala, M.: What is digital competence?. Linked portal (2011)
8. Chen, S., Michael, D.: Serious Games: Games that Educate, Train and Inform. USA, Thomson Course Technology (2005)
9. Susi, T., Johannesson, M., Backlund, P.: Serious games: an overview (2007)
10. von Wangenheim, C.G., Petri, G., Borgatto, A.F.: Meega+ kids: a model for the evalua-tion of educational games for computing education in secondary school. In: INCoD-Brazilian Institute for Digital Convergence (2018)
11. von Wangenheim, C.G., Petri, G., Borgatto, A.F.: MEEGA+ KIDS: a model for the evaluation of games for computing education in secondary school. RENOTE **18**(1) (2020)
12. Nielsen, J.: 10 usability heuristics for user interface design. Nielsen Norman Group, **1**(1). (1995)

Dynamic Serious Game for Developing Programming Skills

Georgina Skraparli(✉), Lampros Karavidas, and Thrasyvoulos Tsiatsos

Aristotle University of Thessaloniki, Thessaloniki, Greece
{skraparl, karavidas, tsiatsos}@csd.auth.gr

Abstract. Learning programming is a difficult process and usually many people give up thereon. In order to motivate people who desire to acquire programming skills, a serious game named Friend Me was developed. It offers the opportunity to obtain programming skills via an interactive and fun environment. The main aim of the game is the adaptation in every programming language and the regular addendum of questions relevant to a programming language in an effortless way. During gameplay, players could learn a programming language from the beginning as the theoretical part is included and continue by practicing it. Otherwise, there is the option to play it only to improve their present knowledge. Friend me provides constantly new challenges to the learners as they have to answer programming questions to accomplish some missions.

The evaluation results show that the game has the potential to be used as an additional useful tool in learning programming.

Keywords: Serious game · Dynamic game · Programming skills

1 Introduction

Serious games are games that offer a solution to a real-world problem leveraging the benefits of a pleasurable environment. Their goal is the person either to acquire a new knowledge/skill or to improve their already obtained. It is proven that they could be used as effective educational tools (Serrano-Laguna et al. 2017). A serious game is a motivating factor to engage the user in something thanks to the interaction and challenges that exist during gameplay. The most common feelings due to games are being absorbed, committed and the feeling of achievement (Subhash and Cudney 2018).

In our days, many people try obtaining programming skills, but it is a difficult process. Learning programming demands to be understood abstract concepts. The traditional way of teaching is not very helpful because of the lack of learner's active participation and lack of immediate feedback. Moreover, it is a long-term process to get familiar with a programming language at a primary level. So, a significant number of people give up on learning a programming language (Bhateja et al. 2020).

The combination of a serious game and learning programming is a promising solution to this problem. In this paper, a serious game was developed to be a tool about

M. E. Auer and T. Tsiatsos (Eds.): IMCL 2021, LNNS 411, pp. 580–592, 2022.
https://doi.org/10.1007/978-3-030-96296-8_52

programming learning in which constantly will be added questions and it will be easily adaptable to every programming language.

1.1 Serious Games – Design Principles

The first step for creating a serious game is to analyze the scientific frame so it is properly designed and provides the desired results. However, the primary goal of a digital game is to cause engagement and immersion to the player. Those are necessary for educational purposes too. The general characteristics to achieve these are the existence of a goal, interaction, rewards and feedback to the player (Clark et al. 2016).

According to Tekinbas and Zimmerman (2003) to create a successful game is necessary to exist a connection between the player's actions and the system's correspondence. In other words, there is the need to be understood that the player affects the game and his/her progress thereon (Burke et al. 2009).

In a game, the player should experience both successes and failures (Perron and Wolf 2008). Furthermore, it is necessary to be evoked emotions in him/her during the gameplay. Responsible for this action is the implementation of the game with a story, graphics, interesting missions, and sounds.

Another significant characteristic of a digital game is the competition that it could offer. This is a way for the player to stay active (Vorderer et al. 2003). It is known that the competition works encouragingly by forcing the individual to try harder to achieve his/her goal. Leaderboards are the most used element of competition. When the person copes with the competition, he/she has a feeling of pleasure.

2 Related Work

The existing serious games developed by others focus mainly on learning at first stage concepts of a specific programming language or learning the main concepts of programming in a logic-related way. The evaluation results of these games show that learners have a positive position on this way of learning considering other ways.

As an example, ProBot (Cadavid 2012) is a digital competition web game to improve learners' skills related to sequencing, defined iteration and nesting. The player as a boxer has to beat his/her opponents, who are non-player characters, by putting in the right order the available instructions (high punch, low block, etc.). In other words, an algorithm is defined. This is a parallelism to a source code in a programming language. Before the match begins, the player can see the opponent's movements to be prepared for his/her move. The match ends when someone's resistance level drops to zero.

This game detects and informs the player about errors and misconceptions. More specifically, it checks if the order of iteration instructors is wrong or if they do not match the closures. Additionally, there are seven difficulty levels and the player has a score that is affected by the efficiency of the algorithm. There is, also, a web application where students can see each other's scores to compare themselves.

The evaluation results of 123 students aged 16–18 years old show that the game is attractive and it is a tool to comprehend better programming concepts for them. Nevertheless, there is no statistical clue if the game impacts their retention.

On the other hand, the serious game Py-rate Adventures (Sideris and Xinogalos 2019) aims on learning basic programming concepts using Python language. There is no need for previous knowledge to play the game. The game has six difficulty levels where every level represents a different programming concept in Python. This is a platformer action game in which the player is a pirate who collects treasures and tries to avoid or kill the enemies to earn points. Each treasure contains theory and examples of the level's learning goal.

During gameplay, some obstacles appear which are theoretical or code-based quizzes based on treasure's theory. Those need to be correctly answered by the player to move on. By completing a level, the player selects if he/she would continue to the next level or will play again the same level.

The player has five lives in total. The way to lose one is by falling from a platform, falling to the spikes, being touched by an enemy or answering a quiz wrong. From the beginning, there is the option to start the game from the level the player wants to.

This game was evaluated by 31 graduated and postgraduate students. There were positive results about the player experience and short-term learning. Furthermore, most of them stated they prefer this way of learning compared to other approaches. However, the game needs to be improved on focused attention.

In this paper, the serious game's focus is to offer the ability to be an easily adaptable game to any programming language and update immediately new learning challenges that are provided by certified professors.

3 Game Design

The serious game Friend Me was designed in order to motivate people who desire to learn a programming language without giving up on it due to the difficulties that seem to appear. During gameplay, players could learn a programming language from the beginning as the theoretical part is included and continue by practicing it. Otherwise, there is the option to play it only to improve their present knowledge.

The main aim of the game is the adaptation in every programming language and the regular addendum of relevant questions in an effortless way.

3.1 Basic Characteristics

Friend Me is a 2D game for mobile devices in which the player as a new student to a school tries to make new friends. The way to make this happen is by helping them accomplish a mission assigned to the player.

Before game development, some specific elements have been defined to exist:

- a realistic scenario to give meaning to the game and evoke player's interest
- interaction with the game's environment to explore it and stay active
- immediate feedback to have knowledge of his/her actions

- game elements such as rewards, competition and challenges to be motivated
- personal account to store data and information on player's performance so he/she has awareness of personal progress
- theory and questions derived from a database
- difficulty about learning concepts increases gradually
- not an actual ending to it

3.2 Scenario

As mentioned before, the game idea is that player's character is a new student to a school and wants to fit in this new place by meeting his/her classmates. To make new friends needs to be very helpful to them. Above classmates' (non-player characters) heads, there are face icons that represent the classmate's relationship with the player. At first, all the classmates have a neutral face. The goal is for each classmate to have three happy faces which can be earned by accomplishing the missions they assign to the player.

During the execution of a mission, the player is asked to answer one or more questions, either theoretical or code-based. In addition, he/she could attend courses to learn programming concepts of the programming language.

3.3 Target Group

The target group of this serious game is people aged 12 years old or above and there is no need for previous knowledge in programming. Although it is necessary to exist some basic knowledge of using technology and playing a game.

3.4 Graphics and Sound

As the serious game targets a wide range of ages, for game's graphics has been chosen the pixel art style. The reason why this is appropriate is it can approach from the younger to the older generation. Nowadays, pixel art remains a popular choice. It is appealing and aesthetically pleasing. Frequently, it is used for mobile devices because they do not require so much space and memory. Moreover, the older generation feels nostalgic because it was used in classic video games. This is a retro style. Plus, the game's sound follows the same retro style.

3.5 Structure

In Fig. 1, it is demonstrated the generic interactions between the player and the game. The diagram describes the basic game's operations with the database communication needed to extract and store data.

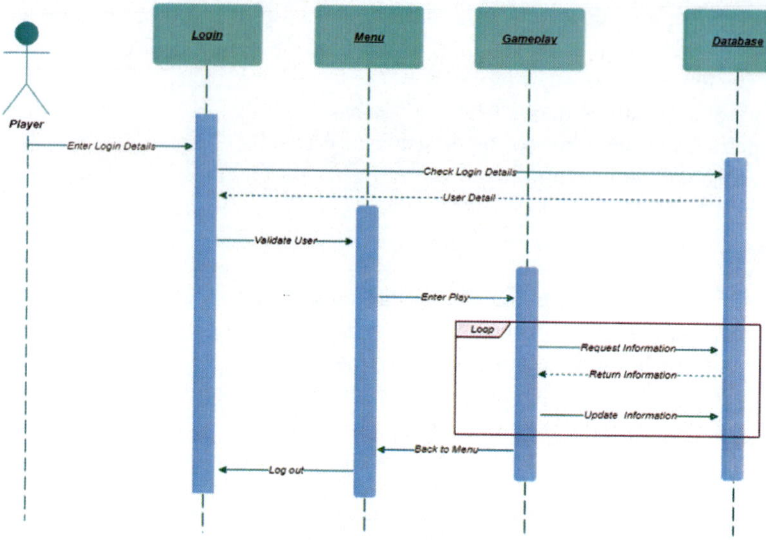

Fig. 1. Game's structure

Firstly, the user needs to login to his/her personal account. After validation from the database, he/she transfers to the menu screen where he/she can click the play button to start the actual game. During gameplay, there is constant data exchange between the player's actions and the database. In particular, the game requires information about the player's progress and sends information to update it. Lastly, the player has the option to return to the menu while he/she is playing and to log out if he/she wants.

4 Implementation

Friend Me was developed with Unity game engine for mobile devices and it is connected with a database to store data. As there is constant communication with a database, the user needs to be connected to the internet while playing. Also, it is only implemented in the English language.

The first time the user launches the game's app has to create an account and the next times has to login with his/her login details.

4.1 Gameplay

Friend Me starts by clicking the play button from the menu. If the player has not finished the theory of the programming language he/she is learning, he/she transfers to the classroom. There, programming concepts are taught through brief theory and examples (see Fig. 2). Otherwise, the player is inside the school where he/she can move around to interact with the environment and talk to his/her classmates. Moreover, there is the school's yard to explore and the principal's office, a mostly locked area.

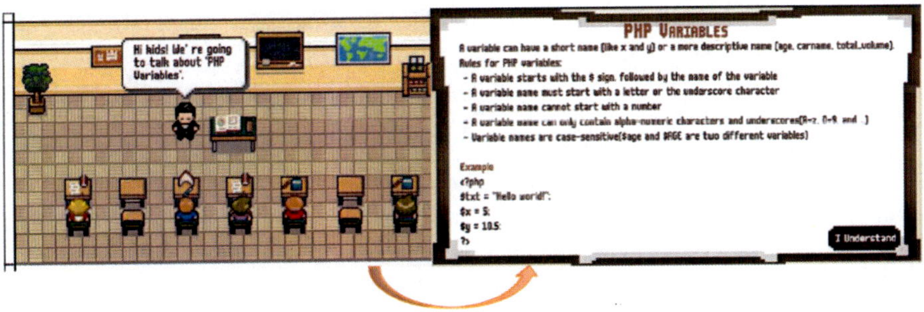

Fig. 2. Theory screens

From the third-person perspective, the character can walk and interact with the game's environment to find objects and communicate with his classmates by pressing the buttons that are at the bottom of the screen. Whenever the player gets closer to another student, a dialog box appears in which the student asks for help and the player decides if he/she wants to help (see Fig. 3). But when the player is already helping someone, messages from other students do not appear unless it is a part of the mission. Additionally, the whole time of gameplay, messages are displayed to keep the player informed. For instance, there are some points of the environment where he/she can look for objects, which will help him/her to complete some missions. Therefore, a question mark appears above the player's head at these points and by interacting he/she will get informed if there is something there.

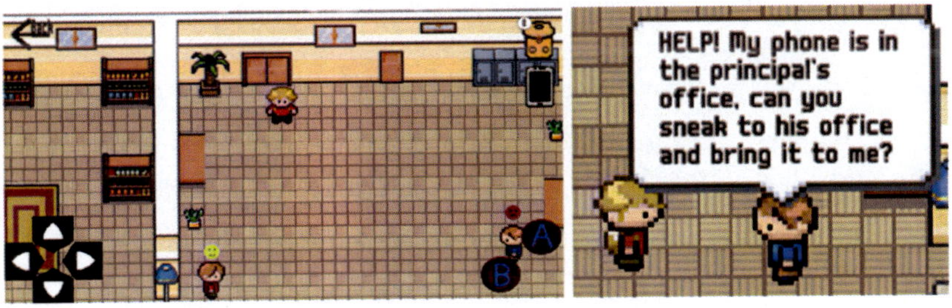

Fig. 3. Player's starting position and communication with a classmate

The found objects are placed in the player's backpack. He/she can have access to them by clicking the backpack icon that is at the up and right part of the screen. Inside the backpack, there are all the collected items where the player can see them and use them when there is the need by clicking on them.

Above students' heads, there are face icons that show if they like the player. At first, all the classmates have a neutral face. Every student can have up to three happy faces or one neutral face or up to three sad faces. The goal is for every student to have three happy faces and to succeed that, the missions that are assigned to the player have to be accomplished. If the player fails to help a classmate, the classmate's face icons fall on a level. For example, from a neutral face to a sad face. Otherwise, classmate's face icons go to the next level.

At this moment, it has been implemented five missions (see Table 1) that are repeated during gameplay and they appear in arbitrary order. During a mission, the player needs answering some questions related to the programming concepts that have been learned. There are three types of questions (see Fig. 4): a) multiple choices, b) putting code blocks in the right order, c) writing code's result. The type of question that shows is randomly selected except the questions that appear when the player needs to unlock something where writing code's result questions display.

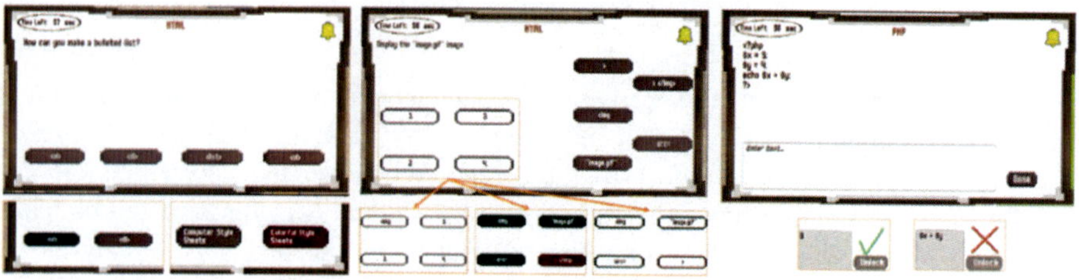

Fig. 4. Questions

The player has a specific time to answer a question and there is immediate feedback to his/her response. If the answer is incorrect or the time ends, the mission has failed. In case of putting code blocks in the right order questions, the correct answer displays. Moreover, a bell icon exists to be clicked whenever the player needs help with a specific question. By clicking it, he/she transfers to the associated theoretical part.

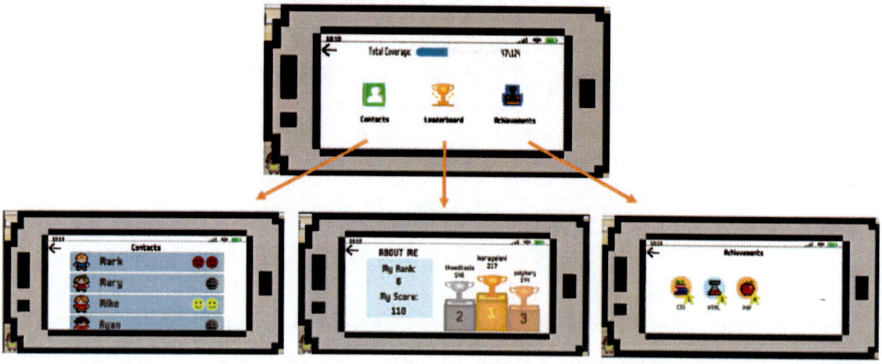

Fig. 5. Mobile phone

In the game, the player has a mobile phone where he/she can have awareness about his progress through it and sometimes it is useful for missions. At first, the phone needs to be unlocked to use it by answering correct a writing code's result question. Then, the player can be informed about the total coverage of the questions answered correctly by him/her to all the questions that exist so far (see Fig. 5). On the same screen, there are three apps: a) contacts, b) leaderboard and c) achievements.

Contacts contain all of the students' pictures, names and moods towards the player. To be added a student to the contacts, the player should at least once gain a happy face icon during the game. Additionally, the contacts app is useful for some missions.

Leaderboard app is where the user can see his/her rank and score (the number of questions answered correctly) compared to other users. Moreover, the top three users with the highest scores are shown at the winners' podium. The existence of this element is to provoke competition between the players to try harder.

Achievements app reveals which learning goals have been achieved and at which level. The learning goals are the chapters that have been set for the theoretical part by the certificated professors. The learner's achievement is measured by the questions answered correctly in the relevant chapters. The player earns badges and at the right and bottom corner of a badge, there is a star icon in which a number is included. This number represents the player's level at the specific learning goal.

Table 1. Missions

No	Places	Need to be used	Classmate's message
1	Inside school or yard	–	"Hey, can you help me with my homework?"
2	Inside school or yard, principal's office	Backpack (key, phone)	"HELP! My phone is in the principal's office, can you sneak to his office and bring it to me?"
3	Inside school or yard, principal's office	Backpack (key, grades)	"HELP! My grades are in the principal's office, can you bring them to me before my parents come?"
4.1	Inside school or yard	Mobile phone (contacts)	"I need [classmate name] 's phone number. Can you give it to me? If you don't have it please get it first."
4.2	Inside school or yard	Mobile phone (contacts)	"Hey, it wasn't cool to give my number. Please delete my number from your phone!"
5.1	Inside school or yard, PCs area	Backpack (usb)	"Hey, can you move a file from usb stick to computer's folder: Files? Find the usb stick first, of course."
5.2	Inside school or yard, PCs area	-	"Well… I changed my mind, this file is pretty important for me. Would you mind deleting it from the computer?"

It is important to be mentioned that the theory, the questions, the ranking of the users and the earned badges with their level are retrieved entirely through the database. The reason why this happens is to easily adapt the game to any programming language and to constantly adding questions. Moreover, Friend Me could be extended by adding new areas, classmates or missions.

5 Evaluation

In order to measure the effectiveness of the developed serious game, a pilot evaluation was made with questionnaire method. Friend Me was evaluated in terms of the user experience and the perceived learning. The evaluation results also have the aim to be found errors and considering users' suggestions to improve the game in the near future.

5.1 Methodology

For the evaluation has been selected a questionnaire based on the MEEGA+ model (Petri et al. 2016). This questionnaire aims to evaluate educational games that are used in computing learning. Its evaluation is about the player experience and perceived learning. There were 42 questions and the participants could submit their own comments about the game.

Questions about the player experience (33 questions) are divided into the following categories: usability, confidence, challenge, satisfaction, social interaction, fun, focused attention and relevance. Questions about perceived learning are divided into these of short-term learning (2 questions) and these of specific learning objectives (3 questions) that have been set. There are, also, 4 questions about demographics.

The closed-ended questions are rated with a 5-point Likert scale from strongly disagree to strongly agree (-2: strongly disagree, −1: disagree, 0: indifferent, 1: agree, 2: strongly agree).

5.2 Procedure

Friend Me was evaluated voluntarily by 33 people that were undergraduate students at the Aristotle University of Thessaloniki and they attended the course "Internet Educational Environments". In this course, students were taught HTML, CSS and PHP. That is the reason why HTML, CSS and PHP questions were added to the game and the theoretical part was abstracted. For evaluation purposes, Friend Me was exported also for the web to be more accessible to everyone. The participants played the game and afterward, they answered the questionnaire. This process was done from their home.

5.3 Results

From demographic data, it is observed that most of the participants are males (24 males, 9 females). Only 9% of them never play digital games.

The results about player experience are presented in Fig. 6. Firstly, in the usability category, participants think it is easy to learn how to play the game (51% agree or strongly agree and 27% have a neutral opinion). Specifically, users mentioned it was easy for them to understand how to play it, but some of them require supplemental instructions on how to use the items added to the backpack. The 54% of them are agreed that there is a need for previous knowledge of programming which is reasonable as the theoretical part has been removed for this pilot evaluation. In addition, from users' comments seem to be appreciated the design of the game like the pixel art graphics and the music. Although some of them complained about the font. To be mentioned, the game is not designed to allow customization in its appearance that is the reason this question has a negative result.

Nevertheless, there is the need for consideration the prevention from making mistakes as 81% of participants does not feel to be any prevention from them. Maybe if the bell button were enabled on the questions the results would be better. Some of them suggest their answer not to be immediately locked in these types of questions: multiple choices and putting code blocks in the right order.

Fig. 6. Player experience results

Regarding the confidence category, the percentages are high. Only a small percentage of people disagree.

The 48% of participants find that the game is appropriately challenging for them and 15% have a neutral opinion. However, a lot believe that the game becomes monotonous (66%) because it ends up being repeated. From the comments, it seems that the players want more questions and missions to be added to the game. This is not a negative result as the main purpose of this serious game is to constantly be added questions and extra missions.

In the satisfaction category, the 57% of players felt satisfied by completing game tasks and the 69% of them think that they advance the game because of their effort.

The results in the social interaction category are negative as the game was not designed to offer this kind of interaction. There is only the element of competition between players as a ranking exists.

The largest percentage of the participants (39%) is observed to be neutral about the fun of the game and the 30% of them considering it fun.

In the focused attention category, the 48% of the players agree or strongly agree something was interesting at the beginning of the game. Nevertheless, the 51% of them did not lack the sense of time during gameplay.

Positive are the results of the relevance between the game and their course. The 66% of the users think the contents of the game are relevant to their interests and the 67% of them find this teaching method as adequate for this course. It is significant to be noticed that the 85% of them think the relevance is clear between the game and the course. However, only 27% of the learners think this method is preferable to other ways of learning.

In Fig. 7, the results about perceived learning are presented. One of the most important results is that the 54% of the participants believe the game contributes to their learning in the course. Although the 42% of them do not think that is more efficient than other course's activities.

Regarding the specific learning objectives that were set, many of the users think the game contributed to learning better PHP, HTML and CSS. A higher percentage is observed about learning PHP (42% agree and 12% strongly agree) which is normal because of the fact the questions about PHP were better distributed in the types of questions that exist. In addition, there were many comments which stated that due to this game they gained new knowledge and tested their already acquired knowledge related to PHP, HTML and CSS.

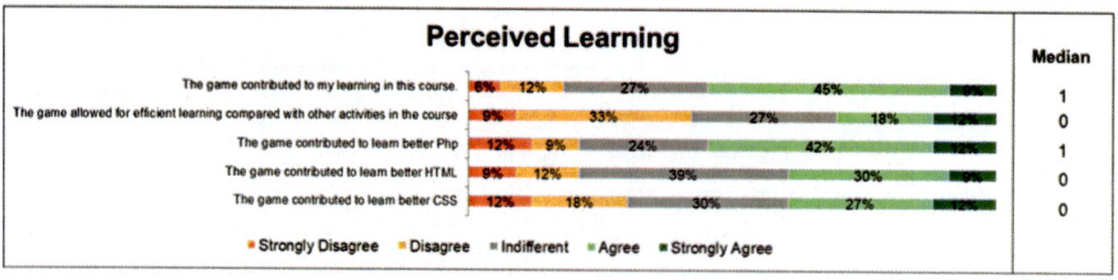

Fig. 7. Perceived learning results

In general, the evaluation results were optimistic about programming learning. Considering participants' comments there could be a better future version of the game to be more effective and fun.

6 Conclusions and Future Work

In conclusion, the developed serious game seems to have a positive impact on learners and it could be an effective tool in obtaining programming skills. Evaluation results, also, show that it could be used as an additional tool to programming learning and for practicing a programming language. As the theoretical part was not included, it is unclear if it could stand alone in programming learning from the beginning. So, in future work, it should be assessed with users with no knowledge of a specific programming language. However, many comments from users mentioned that the game has the potential for improvement to achieve its purpose.

As future work, it is going to be developed an improved version of the game based on players' comments. A valuable addition could be the ability to select which programming language want to be taught. Furthermore, an online form will be created for easily adding theory and questions by verified professors.

References

Bhateja, V., Satapathy, S.C., Satori, H. (eds.): Embedded Systems and Artificial Intelligence: Proceedings of ESAI 2019, Fez, Morocco. Springer Singapore (2020). https://doi.org/10.1007/978-981-15-0947-6

Burke, J.W., McNeill, M.D.J., Charles, D.K., Morrow, P.J., Crosbie, J.H., McDonough, S.M.: Optimising engagement for stroke rehabilitation using serious games. Vis. Comput. **25**(12), 1085 (2009). https://doi.org/10.1007/s00371-009-0387-4

Cadavid, J.: Digital competition game to improve programming skills. Educ. Technol. Soc. **15** (2012)

Clark, D.B., Tanner-Smith, E.E., Killingsworth, S.S.: Digital games, design, and learning: a systematic review and meta-analysis. Rev. Educ. Res. **86**(1), 79–122 (2016). https://doi.org/10.3102/0034654315582065

Perron, B., Wolf, M.J.P. (eds.): The Video Game Theory Reader 2, 1st edn. Routledge, Reading (2008)

Petri, G., Gresse von Wangenheim, C., Borgatto, A.: MEEGA+: An Evolution of a Model for the Evaluation of Educational Games (2016)

Serrano-Laguna, Á., Martínez-Ortiz, I., Haag, J., Regan, D., Johnson, A., Fernández-Manjón, B.: Applying standards to systematize learning analytics in serious games. Comput. Stand. Interf. **50**, 116–123 (2017). https://doi.org/10.1016/j.csi.2016.09.014

Sideris, G., Xinogalos, S.: PY-RATE ADVENTURES: a 2D platform serious game for learning the basic concepts of programming with python. Simul. Gaming **50**(6), 754–770 (2019). https://doi.org/10.1177/1046878119872797

Subhash, S., Cudney, E.A.: Gamified learning in higher education: a systematic review of the literature. Comput. Hum. Behav. **87**, 192–206 (2018). https://doi.org/10.1016/j.chb.2018.05.028

Tekinbas, K.S., Zimmerman, E.: Rules of Play: Game Design Fundamentals. The MIT Press, Cambridge (2003)
Vorderer, P., Hartmann, T., Klimmt, C.: Explaining the enjoyment of playing video games: The role of competition. In: Proceedings of the Second International Conference on Entertainment Computing, 1–9 (2003)

Work-in-Progress: Escape the Experiment – A Serious Game for Teaching Youth About the Dangers of Vaping

Quinn Daggett[1(✉)], Bill Kapralos[1], Cindy Baker-Barill[2], Tracey Burnet-Greene[2], and Melissa van Zandvoort[2]

[1] Ontario Tech University, Oshawa, ON L1G 0C5, Canada
{quinn.daggett,bill.kapralos}@ontariotechu.net
[2] Simcoe Muskoka District Health Unit, Barrie, ON, Canada
{Cindy.Baker-Barill,Tracey.Burnet-Greene,
Melissa.vanZandvoort}@smdhu.org

Abstract. With vaping becoming a bigger issue amongst youth aged 12–17 in Ontario, there is a need for more innovative teaching methods when it comes to educating youth regarding the physical and psychological effects of vaping, and on how peer pressure and marketing techniques employed by the industry can affect their choices. In collaboration with the Simcoe Muskoka District Health Unit, we have developed a point-and-click escape room style serious game for the purpose of teaching youth aged 12–17 about the risks associated with vaping and the pressure-based marketing techniques used to promote vaping. The serious game features a series of puzzles to be solved by the player in preparation for a quiz that takes place during the final level. Topics covered within the game include the physical effects of vaping on a developing body and brain; marketing tactics employed by the industry; and how media and peer influences can affect youth decision-making. It is anticipated that through its interactivity and high levels of engagement, the serious game will better educate youth about the dangers and consequences of vaping. We are currently in the process of preparing to conduct usability tests with the serious game and refine it based on the usability testing outcomes. Upon completion of the usability testing and any refinements that may follow, the aim is to deploy the serious game in school boards throughout Ontario.

Keywords: Serious game · Vaping risks · Escape room

1 Introduction

According to the Not an Experiment website [1] to vape, a person inhales and exhales an aerosol produced by the vaping device/e-cigarette (battery operated devices that heat a liquid (e-liquid) until it vaporizes into an aerosol). Aerosol is created when e-liquid is heated by the device. The heating element within the vape heats the e-liquid until an aerosol is produced. There is a misconception that this aerosol is just a harmless water vapour, but in fact it is a mixture containing many fine particles. These particles are inhaled through the mouth, down deep into the lungs and then are absorbed into the

M. E. Auer and T. Tsiatsos (Eds.): IMCL 2021, LNNS 411, pp. 593–600, 2022.
https://doi.org/10.1007/978-3-030-96296-8_53

bloodstream where they get circulated to all parts of the body, including the brain. The remaining aerosol is then exhaled back into the air and appears as a white smoke or cloud. In recent months in the United States, there have been hundreds of reported cases and even several deaths from severe lung disease associated with the use of vape products among youth. The first Canadian case of severe lung illness linked to vaping was reported in Ontario in September 2019. New information and research about vaping and related health effects is constantly emerging, in part from watching people who are vaping now. Considering the emerging cases of lung disease linked to vaping, it is safe to say that for people who have never smoked, vaping is not a safe or healthy addiction to start.

Also from the Not an Experiment website, the biggest concern with youth vaping is the potential for nicotine addiction. Nicotine addiction is extremely powerful, can develop quickly, and is very difficult to quit. Research is finding that youth who vape are more likely to switch to cigarettes once addicted to nicotine. This is concerning since young people don't always under-stand the danger or potential consequences to their future health. Furthermore, the developing teenage brain is more susceptible to nicotine addiction compared to an adult brain and this addiction can develop at lower exposure levels then what would be required for an adult. Aside from addiction, nicotine can change how the teenage brain develops affecting cognitive functions like memory, concentration as well as reduce impulse control and cause behavioural problems. Finally, an addiction to nicotine at an early age may also predispose youth to developing addictions to other drugs.

As the number of adolescents are using vapes and other electronic cigarettes [3] increases, there is an increased need for modern education resources to teach youth about the various risks associated with vaping. These risks include physical and psychological health effects, as well as marketing tactics employed by the industry to appeal to younger consumers. This paper introduces Escape the Experiment, a serious game, adapted from the in-person Not an Experiment game (see Sect. 2.1), designed to teach youth aged 12–17 years about the dangers of vaping while keeping them engaged as they solve a series of interactive puzzles. The paper will describe the design of the Escape the Experiment serious game and outline an evaluation plan that includes usability testing along with anticipated outcomes.

2 Background

2.1 Simcoe Muskoka District Health Unit (SMDHU)

The Simcoe Muskoka District Health Unit is a provincially funded public health unit in Ontario whose mission is to prevent disease and injury, and promote and protect health for all in Simcoe Muskoka region in the province of Ontario [1]. One of the public health topics that the SMDHU focuses on is Smoking and Vaping and educating the youth about the dangers of vaping. To this end, they have created the "Not an Experiment" website which provides information and resources regarding the dangers of vaping. The website also includes the Not an Experiment activity, an escape room style in-person game that allows the youth (ages 12–17) to learn about vaping in a fun and interactive manner. Not an Experiment was originally designed to be played in small groups, although due to physical distancing requirements during COVID-19 it has been adapted for individual play. A full discussion of the game in addition to all the resources required to facilitate and play the game are available via the Not an Experiment website [2]. Being an in-person game, it was decided to create a video game version (e.g., a serious game) of Not an Experiment and leverage the benefits of serious games while ensuring it is available and playable online.

3 Escape the Experiment Overview

The method that was used to determine the best interactive medium to deliver this information in was a combination of consultation with the Simcoe Muskoka District Health Unit and consideration of usability and accessibility for its target demographic. As the information in Not an Experiment is currently distributed in the form of a PDF, it must be able to be split up into different segments for the players. Unity3D was chosen as the engine to develop the serious game in, due to its flexibility and the team's prior experience in using it.

4 Application Design

The serious game is designed as a point-and-click puzzle game, meaning that all interactions within the game can be performed solely using a mouse or other pointing device. The current prototype was developed using the Unity game engine and the final version will be available in a web browser using HTML5. The game is divided into different segments (referred to as "levels") which the player progresses through while playing.

4.1 Main Menu

Upon starting the game, the player is presented with the main menu which allows them to start the game, set/change options (e.g., volume levels for any music and sound effects), and quit the game. Players are also sent back to this screen after having completed the game.

4.2 Level Selection

Upon choosing the "Start" option from the main menu, the player is taken to the level selection ("central hub") area, where they can choose which level they wish to play. Initially, only the first level (Level 1) is accessible. Attempting to access levels which have not been completed ("unlocked") results in the output of a "locked door" sound letting the player know they cannot access the level. Each time the player successfully completes a level, they will be taken back to this scene with the next level unlocked (except for Level 6, which, once completed, takes they player back to the main menu).

4.3 Level 1: Chemistry Classroom

This level takes place in a chemistry classroom (see Fig. 1(a)), as its purpose is to teach players about the various chemicals that are present in e-liquid such as vegetable glycerin and formaldehyde. Before solving any puzzles, the player will receive a message from Dr. Ada Mizer, a professor that has tasked the player with recovering her research on vaping, and who will communicate with the player throughout the game through a series of messages. The main activity of this level focuses on a whiteboard, which the player must populate with letters to spell out two key words, by operating Bunsen burners and observing the effects they have on the beakers of liquid. Once they have collected enough letters, a hint on the whiteboard (Fig. 1(b)), in addition to messages given to the player by Dr. Ada Mizer will lead them to spell out "fantasy" and "reality" on the whiteboard. Once they do this, they will be given a lab book containing the key message they learned in the level (i.e., "The industry targets youth with candy and fruit flavoured e-liquid. Don't be tricked."), and returned to the level select screen.

(a) (b)

Fig. 1. (a) The chemistry classroom, and (b) whiteboard that the player uses to spell out keywords.

4.4 Level 2: Biology Classroom

This level takes place in a biology classroom as its purpose is to teach players about the health effects of vaping on a developing body and brain (see Fig. 2(a)). The player begins by collecting puzzle pieces cut out from a beaker label and matching the correct piece to the label on a jar with lungs submerged in fluid. Once the player solves the puzzle by forming the sentence "less harmful does not mean harmless", the main activity will unlock, where players must form a different sentence on a whiteboard using words highlighted in a series of medical reports (See Fig. 2(b)). The player must pick up a magnifying glass to see the reports clearly, and once they have read through each one and clicked on its appropriate key word, they return to the whiteboard. Using clues given throughout the level, the player will arrange the words into the phrase "vaping is linked to severe lung disease". Once the player has done this, they will receive the next lab book and be taken back to the level select scene shown in Fig. 2.

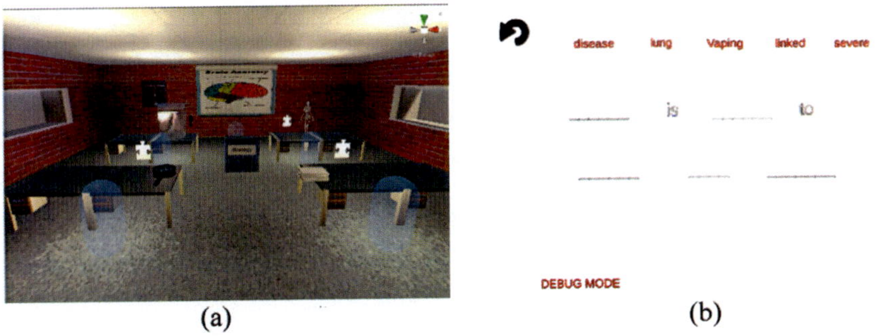

Fig. 2. (a) The biology classroom, and (b) whiteboard puzzle.

4.5 Level 3: Gymnasium (Gym)

This level takes place in a school gym (See Fig. 3(a)), as its purpose is to teach players about the risks and long-term effects of addiction. The player must solve a series of puzzles to obtain the combination to a locked box, with the primary one being a crossword puzzle that uses words associated with nicotine addiction. After solving the crossword puzzle, the player must look at a banner on the wall to find the numbers they need to open the locked box, with the order of the numbers being related to the hints in the crossword. After opening the locked box, the player will receive a blacklight, which they must turn off the lights using a light switch to use, revealing the key message written in invisible ink on the gym wall (see Fig. 3(b)). After clicking on the message ("Most e-liquids contain nicotine which is very addictive and can change how the brain develops"), the player will receive their next lab book.

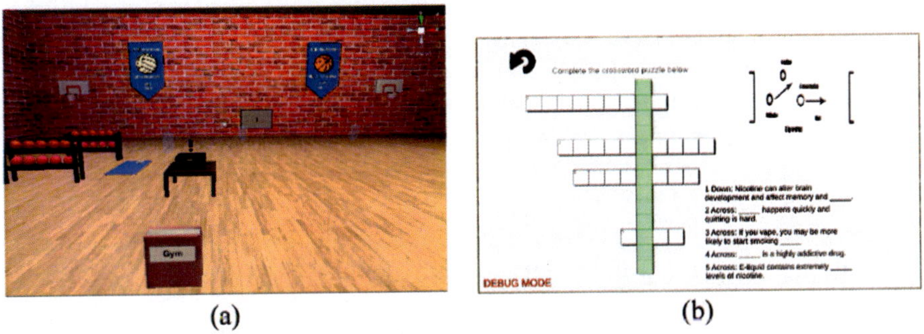

Fig. 3. (a) The gymnasium, and (b) crossword puzzle.

4.6 Level 4: Business and Marketing Classroom

This level takes place in a school computer lab (see Fig. 4(a)), as its purpose is to teach the player about the marketing tactics used by the industry to sell their products. The player must first figure out the password to a computer using hints given to them via posters on the wall, and the computer's screen saver (see Fig. 4(b)). Once the player correctly enters "target marketing" as the password into the computer, the text on a bunch of sticky notes around the room will become visible to the player, who must collect them to proceed to the next puzzle. Once the player has collected all the sticky notes, the projector at the front of the room will display a word search for the player to complete, featuring the words from the sticky notes they collected which all have to do with advertising. When the player has found all the required words in the word search, they will be given their next lab book containing the key message "The industry makes billions of dollars each year by addicting people to nicotine. They care about profits, not people".

(a) (b)

Fig. 4. (a) The business and marketing classroom (lab), (b) and the computer screen saver that provides the player with hints.

4.7 Level 5: Bathroom

This level takes place in a school bathroom (see Fig. 5(a)), as its purpose is to teach the player about how peer pressure can influence their choices when it comes to vaping, and its consequences. The level begins covered in a thick haze of smoke, which the player must disperse by solving a jigsaw puzzle (see Fig. 5(b)), using pieces shaped from the smoke. After the smoke is gone, the player will see a series of posters in the bathroom, including ones describing the different fines associated with supplying and/or using vapes on school property. Using hints found in messages written in fog on the bathroom mirrors, the player must select the correct poster associated with each activity. Each time the player chooses the correct poster, the three ominous figures that are watching the player will stumble, and after the player has selected all the correct posters, they will circle around the bathroom ceiling and explode. The player will be given their final lab book containing the room's final message: "When you share or sell vapes you are helping the industry addict people to nicotine. Don't do their dirty work".

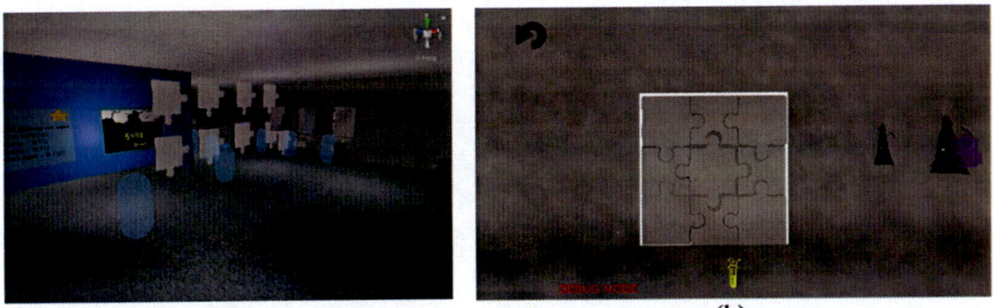

(a) (b)

Fig. 5. (a) The Bathroom, and (b) jigsaw puzzle that the player must solve.

4.8 Level 6: Principal's Office

This final level takes place in the principal's office, where the player is introduced to Principal Bunsen, who the player has been tasked with seeking out and submitting their findings on vaping to. It is where players will be tested on what they learned in the previous levels. The main task of this level is a quiz that covers all the previous levels the player completed, but before having them complete this, the player is instructed on how their answer data will be used and asked for their consent. If they do not provide their consent, they will still complete the quiz, however their answers will not be saved by the game. After successfully completing the quiz, the player is given a letter of congratulation from Dr. Ada Mizer and a gold microscope. They are then taken back to the main menu.

5 Evaluation Plan: Usability Study

5.1 Study Design

With the completion of the Escape the Experiment serious game prototype, the next step is to conduct a usability study to examine the ease of use of the game and its user interface. Results of the usability study will be used to refine the serious game before it is deployed in classrooms and made available to the public at which point, effectiveness testing will be conducted using a pre- and post-study. The usability study will be conducted remotely (facilitated by Google Meet) and will include 10 Ontario Tech University student participants from the Game Development and Computer Science programs ensuring that the participants have the appropriate human-computer interaction and gaming background and expertise. The reason for choosing a sample size of 10 participants is to maintain a logistically feasible number of participants while still having a large enough sample size to gather data from. Although the educational content of the game is targeted at a younger demographic (youth aged 12–17 years), the usability study will be conducted with university students who are older than 18 enrolled in Game Development or Computer Science programs since they will be able to provide quality feedback on both the gameplay and usability of the game.

As part of the usability study, participants will play the game. After having completed the first five levels of the game, the participants will complete the Level 6 (the quiz level), and their correct/incorrect answers are stored in a.txt file for analysis. After completing the game, participants will also complete the 10 question System Usability Scale (SUS) questionnaire (a 10-item questionnaire with five response options for respondents; from Strongly agree to Strongly disagree [4]) and a post-game questionnaire, answering how much they strongly agree or strongly disagree with a series of statements regarding the game.

6 Conclusion and Future Work

As vaping amongst youth aged 12–17 is increasing, public health units are also increasing their efforts to educate youths about the risks and dangers of vaping. Smoking and Vaping is one of the topics that the Simcoe Muskoka District Health Unit (SMDHU), a province of Ontario health unit, focuses on. Part of their education includes educating youths about the dangers of vaping. In this paper, we have described the development of the Escape the Experiment serious game prototype that was designed in conjunction with the SMDHU to educate youth about vaping and its risks and dangers. The Escape the Experiment serious game is based on the Not an Experiment escape room activity, an in-person game that allows the youth (ages 12–17) to learn about vaping in a fun and interactive manner.

With the completion of the prototype, we will be conducting a usability study to examine the ease of use of the game and its user interface. The results of the usability study may lead to refinements of the serious game. Following the usability study and any refinements, the game will be made freely available and be deployed within Ontario classrooms.

References

1. "Home Page." Not An Experiment. https://www.notanexperiment.ca/. Accessed 11 July 2021
2. "Downloads & Educator Resources." Not An Experiment. https://www.notanexperiment.ca/downloads/. Accessed 11 July 2021
3. Hammond, D., et al.: Prevalence of vaping and smoking among adolescents in Canada, England, and the United States: Repeat national cross sectional surveys. BMJ **365**, May 2018, 2019. doi: https://doi.org/10.1136/bmj.l2219
4. Sauro, J.: Practical Guide to the System Usability Scale: Background, Benchmarks & Best Practices. CreateSpace Independent Publ. Platform, North Charleston (2011)

Serious Game Concept to Promote Citizen Engagement for the Energy Transition Using Virtual Reality and Web Platforms

Felix Longge Michels, Laura Müller, Victor Häfner, and Polina Häfner(✉)

Karlsruhe Institute of Technology, Karlsruhe, Germany
{felix.michels,victor.haefner,polina.haefner}@kit.edu,
laura.mueller@student.kit.edu
www.imi.kit.edu

Abstract. The energy transition in Germany is planned to take place by 2045. It includes the complete abandonment of fossil and nuclear energy sources, a nationwide expansion of renewable energy sources such as solar energy, wind and hydropower as well as the reduction of energy consumption. Those measurements will contribute significantly to the reduction of greenhouse gas emissions. Our work aims to give citizens an intuitive insight into the energy transition and provide a better understanding through visualization and playful, realistic interaction.

In this paper, the concept of an educational game for citizens is explained, in which the topic of renewable energies is to be conveyed to them playfully. Particular attention is paid to the calculations necessary for the game logic and the three-dimensional interactive visualization. The 3D environment and the application logic is developed with the virtual reality engine PolyVR and ported for its experience within a web browser. The implementation of this work focuses on wind power and the German province Baden-Württemberg but can be extended analogously for other forms of renewable energies to fully depict the energy transition in Germany.

Keywords: Vitual reality · Serious game · 3D web game · 3D simulation · Green games · Environmental games · Public engagement · Energy transition · Climate change · Climate goals · Renewable energy sources

1 Introduction

In order to reduce greenhouse gas emissions and achieve the climate neutrality planned by Germany by 2045, the expansion of renewable energies is an important aspect. Currently, 28% of the electricity generated in the German federal state of Baden-Württemberg is generated from renewable energies [23]. To achieve a climate-neutral energy supply, a massive expansion of renewable

M. E. Auer and T. Tsiatsos (Eds.): IMCL 2021, LNNS 411, pp. 601–612, 2022.
https://doi.org/10.1007/978-3-030-96296-8_54

energies is required in Baden-Württemberg. On 24 March 2021, the Federal Constitutional Court ruled that parts of the Federal Climate Protection Act are incompatible with fundamental rights. As a result, the reduction targets had to be adjusted, whereby the climate crisis is also assigned increasing relevance from the legal side. Following the legal response, the importance has been raised to address the issue of the climate crisis even further, especially in the field of public education. Researchers found evidence that serious games in the area of energy consumption, conservation and efficiency are motivating, as they can improve the domain knowledge and even positively influence the users towards a pro-environmental behaviour [7,16].

This paper, therefore, describes the conception and implementation of a serious game, through which the topic of renewable energies and their relevance for climate neutrality is to be communicated. Furthermore, the citizens should be engaged in the energy transition process through playful planning and decision-making on renewable energy sources. The technical basis for this work is the realistic visualization of the earth's surface with topography, satellite view, national borders and cities. Since the concept of the serious game is based on all renewable energy sources, the technical solution is exemplary for wind power. Thus, energy data, from wind turbines, as well as wind potentials and the representation of day and night, shadows and weather conditions are visualized. The interactive three-dimensional game is developed using an open-source virtual reality (VR) engine and can thus be deployed on visualization clusters for CAVE environments as well as on desktop and VR headsets. Furthermore, the game was ported for the web and could be played online. The 'serious' character of the game simulation comes not only from the importance of the topic but also the usage of real data, resources and assets.

2 Related Works

A serious game is a game designed for a primary purpose other than pure entertainment [3]. The adjective 'serious' is generally prefixed to refer to video games used by industries such as education, research, defence, or politics, and explicitly emphasizes the added educational value of fun and competition. Serious games are a sub-genre of Serious Storytelling, where storytelling is applied outside the context of entertainment, with the narrative progressing as a sequence of patterns which are part of a thoughtful progression [14].

In the literature review by Stanitsas et al. the authors analyse 77 serious games that aim to facilitate the transition to sustainability. The games differ greatly in their sustainability subthemes, target groups and goals, as well as in their type and the technology used [20]. The majority of serious games in the field of the energy transition are in the domain of domestic energy consumption and the pro-environmental behaviour for energy efficiency [7,16]. Few publications deal simultaneously with the planning and decision-making on renewable energy sources (e.g. adoption of solar energy [19]) and even less can be played online and with virtual reality hardware [20].

Furthermore, the combination of serious games and data visualization of geographical data has been discussed and researched for various purposes. Typical use cases name the training aspect for emergency management during natural catastrophes. Other use cases study the educational aim for building spatial awareness and knowledge through connecting educational information and visual data [1,2,25].

The uniqueness of the concept and implementation of the serious game, described in this paper, consists of the combination of a virtual reality simulation with a 3D geographical data visualization and the aim to engage citizens through the planning of renewable energy sources based on a real problem, real data and real resources.

3 Concept

3.1 Theoretical Basics

The serious game is being developed to familiarize adults with the topic and relevance of renewable energies and to allow them to install them in Baden-Württemberg. Since the game conceived in the context of this work is a serious game, the main aim is to impart knowledge through play, in addition to entertainment.

In order to illustrate the impact and potential of renewable energies, especially wind power, it is important to know that the share of wind power in gross electricity generation in Baden-Württemberg was 6.8% in 2019 and the share of renewable energies 31%. These values represent the initial situation of the game. Currently, the percentage shares of the energy forms in Baden-Württemberg are as follows: Wind 33.6%, Solar 39.6%, Geothermal 1.6% and Biomass 25.6%. Due to the characteristics of renewable energies, e.g. their volatility, it is advisable to focus on the development of diverse technologies, as this is the only way to provide a reliable energy supply.

3.2 Gameplay

The game starts with a description of the initial situation and a short explanation of renewable energies and their relevance in simple language. At the beginning of the game, the player can choose whether to play until the year 2030, 2040 or until climate neutrality in 2045 (see Fig. 2). In addition, a budget is available, which is displayed in an overview bar. This can be used to build individual power plants or parks for the generation of renewable energies. After each construction, the remaining budget and the distance to the target are displayed. It also shows how many tonnes of carbon dioxide (CO_2) have been saved by the plants already built and how many households have been supplied with electricity, as a result. The game aims to reduce the CO_2 emissions, which are emitted through energy production. The construction of plants for the generation of renewable energy thus contributes to the climate neutrality of the region.

3.3 Time Course

The starting point of the game is the year 2020. Due to the time horizons 2030, 2040 and 2045 defined in the Climate Protection Act with corresponding emission reduction targets, one can choose between these three possibilities as the time frame for the game. The game runs in a 2-year rhythm, so each level corresponds to two years. A corresponding budget is available for each level. A level ends as soon as the budget is exhausted, or the decision is made to continue with the next level. The planning time for a wind power plant is 1 to 2 years, the construction approx. 6 months [9]. Multiple wind power plants or parks can be planned and completed within one level.

3.4 Roles

The player has the role of an employee of a planning and implementation office for renewable energies and is therefore responsible for the entire process from planning and construction to operation and maintenance of the plants. All areas of responsibility, such as the choice of location, the selection of wind turbine models and the handling of the budget, therefore lie with the players.

3.5 Decisions

The first decision to be made is which form of renewable energy should be used. The subsequent decisions are explained below using wind energy as an example. After setting the time frame, a location for a single wind turbine or a wind farm is selected by clicking on the map. Based on the wind potential, the distance to buildings and other legal restrictions, feedback is given whether the selected site is suitable for the planned project or whether a new one must be chosen. In addition, it must be determined whether the project involves a single wind turbine or a wind farm. In the case of the latter, the number of wind turbines included must be specified so that the calculations (see Sect. 3.7) can be carried out accordingly. Subsequently, a choice can be made between five different wind turbine models, each with different outputs of 1.5 to 4.2 megawatts (MW). For the choice made, the tonnes of CO_2 saved, the contribution to achieving the target and the average number of households supplied with electricity as a result, are displayed in an overview. In addition, the wind turbine or wind farm is visually displayed on the map. Finally, the decision can be made to employ a consultant. By using part of the budget, this person is responsible for ensuring that no negative scenario occurs for the selected period.

3.6 Scenario

In the course of the game, unforeseen events may occur. In these cases, the players are notified and have the opportunity to decide how to proceed. The following scenarios can occur in the course of the game, even several times:

- **Change of the minimum distance:** The regulations for the minimum distance of a wind turbine are changed. For instance, the minimum distance to settlements is reduced from 700 m to 650 m. This makes it possible to build new wind turbines in additional areas.
- **Citizen participation:** The citizens of the surrounding communities have decided to operate the wind turbine or wind farm in the form of a citizens' energy cooperative. Therefore, there are no costs for the players, because the citizens pay for the construction and operation. In return, they can purchase the electricity from the wind turbine or wind farm that has been built. This scenario can occur as soon as a suitable area has been selected.
- **Citizens' protest movement:** Citizens from a neighbouring municipality have started a protest against the planning of a specific wind turbine or wind farm. Due to a lawsuit filed, the construction is stopped for the time being and there is a time delay. The construction stop and the lawsuit result in additional costs.
- **Ecological reasons for construction stop:** Environmentalists have discovered that strictly protected animals live on the land to be built on [27]. For this reason, the Higher Administrative Court in Mannheim has halted further construction for the time being in summary proceedings and the project is delayed. Either the specific project can continue or it will be abandoned due to the animal population.
- **Subvention of private investments:** In addition to the construction of large wind turbines and wind farms, it is also possible for private individuals to set up small wind turbines for self-sufficiency in their gardens [8]. It is possible to subsidize the construction of private wind turbines and thus come a little closer to the goal of CO_2 reduction. It is up to the players to decide how much of their remaining budget they want to use for the subsidy.

3.7 Calculation Methodology

The following list provides an overview of the data and values relevant to the serious game and their basis for calculation. The calculations are explained and the sources of the data needed to carry out the calculations are given in the following:

- **Budget:** In 2020, €1.97 billion was invested in onshore wind power plants throughout Germany. Since 2.5% of the wind turbines installed in Germany are located in Baden-Württemberg, a proportionate investment amount of €49.25 million can be estimated in 2020 [24]. Based on this budget and taking into account inflation of 1.49% [22], the total budget is €0.584 billion by 2030, €1.204 billion by 2040 and €1.55 billion by 2045. The costs of investment per wind turbine as well as the expected annual maintenance and operating costs were calculated. The investments here amount to €1200 per kilowatt (kW) of generator output. The maintenance and operating costs are estimated at 3% of the investments for an average service life of 20 years [18]. For wind farm with an amount greater than tree turbines, the investments corresponds to the sum of the individual wind turbines. A factor of 0.95 takes into account the reduced planning effort due to the joint planning of the turbines.

- **Yield of a wind turbine:** The annual yield of a single power plant is calculated from the product of the mean capped wind power density, the rotor area of the plant and the running time of the plant in hours per year [18]. The mean capped wind power density can be taken from the Energy Atlas Baden-Württemberg [11]. Due to the location-dependency of the average capped wind density, this is dependent on the choice of location of the players. The rotor area of the turbine depends on the model. The annual yield of a wind farm is calculated from the sum of the annual yields of all the wind turbines it contains. The field efficiency is also taken into account by the factor of 0.95.
- **Profit of a wind turbine:** The annual profit of the individual wind turbine is calculated from the revenue minus all fixed costs. Fixed costs include both the maintenance and operating costs and the imputed depreciation, whereby the investment costs are divided by the average useful life of the plant of 20 years. The revenue is the product of the energy generated per year and the cost of electricity. The energy generated per year is the product of the generator output and the annual full load hours. The electricity production costs vary between 3.99–8.23 ct/kWh, therefore the average of 6.11 ct/kWh is assumed for the calculations [10].
- **Carbon dioxide savings:** The CO_2 saved by the use of a wind turbine or wind farm is calculated in tons per year. They are the product of the emission factor of the current electricity mix[1], the generator output of the plant and the annual hours of full load. However, the CO_2 emissions released in the upstream chain, i.e. the production and construction of the wind turbine, must be deducted from this. They are calculated by multiplying the generator output of the plant and the annual hours of full load by the emission factor of wind power[2].
- **Number of households supplied with electricity:** The number of households that can be supplied with electricity through the construction of a wind turbine or wind farm serves as a reference value for the game players to understand the effects of their actions in the game. The calculation is made as the product of the generator output of the plant, the annual hours of full load and the average consumption of a German two-person household[3].

4 Implementation

4.1 Data Resources

The data for the visualization of the relevant information was collected from publicly available sources in addition to data from the Energy Atlas Baden-

[1] The emission factor of the German electricity mix was 408 kg CO_2/MWh in 2019 [15].

[2] The emission factor of wind power was 11 kg CO_2/MWh, as 2019 [15].

[3] The average annual consumption of a German two-person household was 3.4 MWh/a [21]. A two-person household was assumed as it corresponds to the average size of a German private household.

Württemberg [11]. This data was obtained in heterogeneous formats and further analysed and prepared for processing and integration. The currently available information for the visualization of the serious game includes elevation data, satellite photos, city borders and municipal borders, as well as data for the energy-relevant topics in the region. Actually installed wind turbines with their meta-information (e.g. location, municipality, hub height, rotor diameter, maximum power output, responsible operators) were also integrated (see Fig. 1).

For the elevation model, the publicly available dataset of the 'Shuttle Radio Topography Mission' (SRTM) in GeoTIFF format was utilized. On top of the elevation model, satellite photos taken by 'Landsat 8' were superimposed. From the platitude of photos, the ones with the least amount of cloud formations were chosen. These photos were acquired originally as multiple photos divided by their frequency bands in Tiff format, recombined into true colour pictures and converted into PNG format. In the area of Karlsruhe, both elevation data and overflight imagery in higher resolution compared to the aforementioned sources were provided by the LUBW within the timeframe of the ViewBW project [12,13].

Semantic information for the visualization of city and municipal borders was extracted from the publicly available geoinformation system 'Open-Street-Map' (OSM) and parsed in XML format [17]. The data for Baden-Württemberg was filtered for relevant information, providing city locations as well as their borders.

For the planning and placement of new wind turbines within the gameplay of the serious game, the wind potential in Baden-Württemberg was combined with legal requirements (e.g. distance to the nearest settlements, water protection areas, nature reserves) for placement and integrated into the visualization as polygon areas and stored in GeoJSON format (see Fig. 2).

Fig. 1. 3D visualization of an existing wind turbine with displayed user interface and meta information on the left

4.2 3D Environment Development

The data sources as described in Sect. 4.1 were utilized to implement a 3D environment providing a topographical map superimposed with satellite imagery and semantic information as well as the energy-related points of interest for the serious game. The majority of geographical relevant data in the formats GeoTIFF, TIFF and GeoJSON as well as their underlying spatial reference systems and therefore needed conversions were handled by utilizing the library GDAL [4].

For the base layer of the map, the elevation data was calculated and transformed alongside the earth's curvature. The aforementioned satellite photos were superimposed on this elevation data. Additionally, the wind potential areas combined with legal requirements of wind turbine placement was implemented as a visual option, which could be displayed by choice. This was realized by deploying a stencil buffer to combine the calculated geometry mesh of the potential areas with the topographical map. The resulting overlay currently shows two distinctions. A yellow hue on top of the map as limited available area and a green hue as fully available area (see Fig. 2).

On top of the resulting map, the positions of currently existing wind turbines were located. On these spots models of wind turbines were generated dynamically, which were constrained to real-life values. As a result, the virtual turbine was scaled according to its real twin and therefore the hub height and rotor diameter were identical in real and virtual terms.

Fig. 2. 3D visualization of wind potentials on top of the map with the overlayed user interface (Color figure online)

4.3 PolyVR Engine and Deployment

The serious game was developed in the open-source VR engine and authoring tool PolyVR [5]. PolyVR is a VR authoring system that allows the user to dynamically create immersive and interactive virtual worlds. PolyVR is based on open-source libraries like OpenSG and OpenGL. OpenSG is the backbone of the solution and features scene graph management with a focus on clustering and threading. The clustering capabilities of OpenSG allow the flexible software abstracted from the hardware configuration for distributed visualization on computer clusters. PolyVR can thus be deployed on visualization clusters for CAVE environments as well as on desktop, VR headsets or any 3D display. Applications are realized through Python scripts for dynamic and interactive content. Python bindings provide access to all integrated C++ functions of PolyVR. This allows fast and flexible integration of external libraries or other resources and is a very intuitive language with a low learning curve.

PolyVR supports a multitude of different hardware setups. These range from the standard Desktop PC with the operating systems Microsoft Windows and Linux-Ubuntu, over high-end fidelity virtual reality setups such as a CAVE system with distributed rendering setup, to personal virtual reality setups such as VR headsets. In addition, within the scope of the research project ViewBW, an advance into a device-independent solution was developed to run PolyVR within a web browser environment.

WebAssembly is a new technology that opens up many possibilities for running complex applications in web browsers. WebAssembly enables native applications (i.e. written with C/C++) to run in the user's browser. This does not require any software to be installed on the user's side. The WASM file is simply obtained from the server like any other resource, and native functions can be executed from JavaScript. The PolyVR engine was compiled as a WebAssembly within the framework of the project. For this, all software components, so-called dependencies, were compiled and packaged for WebAssembly. Further steps included the port of the GDAL library for the processing of geodata, as well as the adaptation of the shaders for the visualization of the topography and other elements of the virtual environment.

4.4 User Interaction and User Interfaces

The navigation on the 3D map is a typical orbit around a fixed point paradigm on the middle mouse button. The right mouse button displaces the fixed point. The left mouse button is reserved for interaction with the user interface (UI) and virtual objects. When deploying in VR the navigation changes to a fly-through paradigm and interaction with UI elements and virtual objects through a hand-held controller to point, trigger and drag and drop.

The UI elements are sprites attached to the virtual camera. Each sprite is a website rendered to texture using the Chromium Embedded Framework (CEF). This feature is approximated in the web port of the PolyVR engine by overlaying the UI websites as IFrames over the WebGL canvas.

The main UI consists of widgets showing the game progress, the available budget for the level as well as the main menu. With the help of the main widget, the user can select energy source, control the timeline, configure the parameters (e.g. the parameters of a wind park) and can get feedback during the game (as depicted in Fig. 1 and Fig. 2).

The interaction with the virtual objects and terrain uses the height maps and geo-localization system to correctly place wind turbines, as well as validating potential placements in real-time. Renewable energy sources can be selected, displaying meta-information, and removed from the map.

4.5 Performance

To keep the performance of the virtual map sufficiently high and handle the amount of data integrated into the visualization process, a level-of-detail system for textures and images (elevation data, satellite photos) was designed and implemented. This downgrades the resolution at higher distances from the view to the map. The results of varying data were integrated into a procedural asynchronous loading process, which would enable a smoother operation of the application on the varying hardware and software configurations as listed in Sect. 4.3.

5 Conclusion and Outlook

Through the proposed concept and developments of the serious game, the authors believe that the topic of renewable energies can be brought closer to the players. By explaining the situation through scenarios, the individual effects and the respective consequences of one's actions in simple language will familiarize the users with the topic in a playful way. The calculation of the budget, the CO_2 savings and the households supplied with electricity should also help the players to get a better feeling for the magnitudes and effects of the individual actions and to be able to classify them better. The VR version of the game could not be evaluated under laboratory conditions because of the pandemic situation, but a study with the web version is planned.

Further development, improvements and an outlook regarding the works on the here depicted serious game can be divided into the development aspect, the content aspect and lastly use cases. From the implementation and data integration viewpoint, further improvements can be considered regarding the currently deployed resolution of the elevation data and satellite imagery. Cooperation with local authorities or local cartography providers could be aimed to gain access to data in higher fidelity.

Currently, it is not reasonable and realistic to achieve the defined climate goal of reducing emissions with only one form of renewable energy generation. Future content expansions should therefore include other energy sources such as hydro, solar, biomass, geothermal and nuclear power. In the context of further scientific work, the necessary data for the other forms of renewable energy as listed could be determined, analogous to the data designed and calculated in this work for wind power. The serious game can find its application for educational purposes, both for the general public as well as for schools. Through a varying level of detail and set complexity, it can be adapted to the users' prior knowledge and familiarity with the topic of renewable energy.

Acknowledgements. This publication describes the authors' work during the ViewBW research project conducted by various university institutes in Karlsruhe and funded by the Ministry for the Environment, Climate and Energy Management of the German province Baden-Württemberg. The concept and implementation were done by the authors, supported by colleagues Melanie Groß and Michail Gebhardt.

References

1. Bureau, S.: Arcwatch Serious Game Play (2020). https://www.esri.com/about/newsroom/arcwatch/serious-game-play/. Accessed 22 Aug 2021
2. Cheng, Z., Hao, F., JianYou, Z., Yun, S.: Research on design of serious game based on GIS. In: 2010 IEEE 11th International Conference on Computer-Aided Industrial Design & Conceptual Design vol. 1, pp. 231–233 (2010). https://doi.org/10.1109/CAIDCD.2010.5681365
3. Djaouti, D., Alvarez, J., Jessel, J.: Classifying serious games: the G/P/S model. In: Handbook of Research on Improving Learning and Motivation through Educational Games: Multidisciplinary Approaches. IGI Global (2011)
4. GDAL/OGR contributors: GDAL/OGR Geospatial Data Abstraction software Library. Open Source Geospatial Foundation (2021). https://gdal.org
5. Häfner, V.: PolyVR - a virtual reality authoring framework for engineering applications. Dissertation, Karlsruher Institut für Technologie (KIT) (2019). https://dx.doi.org/10.5445/IR/1000098349
6. Hoeft, C., Messinger-Zimmer, S., Zilles, J.: Bürgerproteste in Zeiten der Energiewende: Lokale Konflikte um Windkraft, Stromtrassen und Fracking. transcript Verlag, Bielefeld (2017). https://doi.org/10.14361/978-3-839-43815-2
7. Johnson, D., Horton, E., Mulcahy, R., Foth, M.: Gamification and serious games within the domain of domestic energy consumption: a systematic review. Renew. Sustain. Energy Rev. **73**, 249–264 (2017)
8. Köck, W., Bovet, J.: Perspektiven der Windenergieplanung. First UFZ Energy Days 2012. Leipzig (2012)
9. Kohlmann, J.: Projektierungs- und Genehmigungsprozess bei Windenergie Onshore. EnBW (2014)
10. Kost, C., Shammugam, S., Jülich, V., Nguyen, H., Schlegl, T.: Stromgestehungskosten: Erneuerbare Energien. Fraunhofer-Institut für Solare Energiesysteme ISE, Freiburg (2018)
11. Landesanstalt für Umwelt Baden-Württemberg, Energieatlas Baden-Württemberg. https://www.energieatlas-bw.de/. Accessed 03 July 2021
12. Landesanstalt für Umwelt Baden-Württemberg: Grundlage: Daten aus dem Räumlichen Informations- und Planungssystem (RIPS) der Landesanstalt für Umwelt Baden-Württemberg; 28 Jan 2021 (2019). http://www.lubw.baden-wurttemberg.de
13. Landesanstalt für Umwelt Baden-Württemberg. (2019). Geobasisdaten copyright Landesamt für Geoinformation und Landentwicklung Baden-Württemberg, Az.:2851.9-1/19 www.lgl-bw.de

14. Lugmayr, A., Sutinen, E., Suhonen, J., Sedano, C.I., Hlavacs, H., Montero, C.S.: Serious storytelling – a first definition and review. Multimedia Tools Appl. **76**(14), 15707–15733 (2016). https://doi.org/10.1007/s11042-016-3865-5
15. Ministerium für Umwelt, K., Baden-Württemberg, E.: Erneuerbare Energien in Baden-Württemberg 2020 - Erste Abschätzungen, Stand April 2021. Stuttgart (2021)
16. Morganti, L., Pallavicini, F., Cadel, E., Candelieri, A., Archetti, F., Mantovani, F.: Gaming for earth: serious games and gamification to engage consumers in pro-environmental behaviours for energy efficiency. Energy Res. Soc. Sci. **29**, 95–102 (2017)
17. OpenStreetMap contributors (2019). Planet dump https://planet.osm.org
18. Quaschning, V.: Regenerative Energiesysteme: Technologie - Berechnung - Klimaschutz (10. Hanser, München, Auflage) (2019). https://doi.org/10.3139/978-3-446-46114-7
19. Rai, V., Beck, A.L.: Play and learn: serious games in breaking informational barriers in residential solar energy adoption in the United States. Energy Res. Soc. Sci. **27**, 70–77 (2017)
20. Stanitsas, M., Kirytopoulos, K., Vareilles, E.: Facilitating sustainability transition through serious games: a systematic literature review. J. Clean. Prod. **208**, 924–936 (2019)
21. Statistisches Bundesamt: Stromverbrauch der privaten Haushalte nach Haushaltsgrößenklassen (2020). https://www.destatis.de/DE/Themen/Gesellschaft-Umwelt/Umwelt/UGR/private-haushalte/Tabellen/stromverbrauch-haushalte.html. Accessed 03 July 2021
22. Statistisches Bundesamt (2021). Verbraucherpreisindizes. https://www.destatis.de/DE/Themen/Wirtschaft/Preise/Verbraucherpreisindex/Tabellen/Verbraucherpreise-12Kategorien.html;jsessionid=D0B7CD9723BE6C8E23884A1D5FC92B48.live722?nn=214056. Accessed 03 July 2021
23. Statistisches Landesamt Baden-Württemberg: 28% des erzeugten Stroms aus erneuerbaren Energien. Pressemitteilung 314/2019. Stuttgart (2019). https://www.statistik-bw.de/Presse/Pressemitteilungen/2019314. Accessed 05 July 2021
24. Strom-Report: Windenergie in Deutschland: Daten & Fakten zur Wind-energie, Stand 2021 (2021). https://strom-report.de/windenergie/. Accessed 04 July 2021
25. Tomaszewski, B., et al.: GIS and Serious Games (2017). https://doi.org/10.1016/B978-0-12-409548-9.09623-8
26. Untersteller, F.: Pressemitteilung des Landes Baden-Württemberg. 1.000-Meter-Mindestabstand ist vom Tisch. Stuttgart (2021). https://www.baden-wuerttemberg.de/de/service/presse/pressemitteilung/pid/1000-meter-mindestabstand-ist-vom-tisch/. Accessed 03 July 2021
27. Voigt, C.C. (ed.): Evidenzbasierter Fledermausschutz in Windkraftvorhaben. Springer, Heidelberg (2020). https://doi.org/10.1007/978-3-662-61454-9
28. Wood, G., et al.: Serious games for energy social science research. Technol. Anal. Strategic Manage. **26**(10), 1212–1227 (2014)

Dynamic Learning Experiences

Web 2.0 Digital Marketing Tools in the Ecuadorian Tourism Sector Against of the COVID-19 Pandemic

Leonardo Ballesteros-López[1(✉)], Santiago Peñaherrera-Zambrano[1], Sonia Armas-Arias[2], and Sonia López-Pérez[3]

[1] Facultad de Ciencias Administrativas, Grupo de investigación Marketing C.S, Universidad Técnica de Ambato, Ambato, Ecuador
{lg.ballesteros, spenaherrera}@uta.edu.ec

[2] Facultad de Ciencias Humanas y de la Educación, Grupo de investigación Marketing C.S, Universidad Técnica de Ambato, Ambato, Ecuador
sp.armas@uta.edu.ec

[3] Centro de Idiomas, Universidad Técnica de Ambato, Ambato, Ecuador
sda.lopez@uta.edu.ec

Abstract. The objective of this research was based on analyzing the behavior of the Ecuadorian tourist specifically from the province of Tungurahua, which was focused on the use of digital marketing tools during the COVID-19 pandemic. The problem was the low influx of tourists in the tourist destinations of the country's province, as well as the low and zero income from economic activities related to tourism. It is a descriptive-prospective study with an exploratory approach and quantitative data. Two samples were considered per canton: tourists (potential of the economically active population) and companies related to this type of activity to which two surveys validated by Cronbach's alpha statistic were applied. In this process, the Cronbach's Q test, the KMO test and a factor analysis were used. The results of the study were obtained, in this way tourism companies use digital media as marketing tools to promote their services, their attractions and show the biosecurity measures that are implemented in each of the cantons. Finally, during the pandemic, tourists have a young profile with an affinity for social networks, who mostly prefer destinations considered safe and prefer to be in contact with nature.

Keywords: ICT · Digital Marketing · Tourism · Tourism industry · COVID-19

1 Introduction

The global pandemic had negative consequences not only for people due to the different mobility restrictions and the high probability of contagion, but also for companies because the low number of customers and sales decreased and consequently their profitability was affected. Tourism is a sector that depends entirely on the mobility and continuous travel of people, as a result of the pandemic, government institutions

M. E. Auer and T. Tsiatsos (Eds.): IMCL 2021, LNNS 411, pp. 615–626, 2022.
https://doi.org/10.1007/978-3-030-96296-8_55

closed international and national access until the outbreak of contagion ceases, so the different businesses related to the Tourism were seriously affected in the income of these activities.

In this way, digital marketing tools become a fundamental axis to reactivate the tourist economy because they support the strengthening of the image of each destination. It focuses on the biosecurity measures used to ensure optimal virus management. In addition to contributing to the attraction of national tourists in times of pandemic, also people from the same localities or close to them have the option of visiting them or doing natural tourism.

The problem of Ecuadorians in the tourism sector during the pandemic is the low number of people in different tourist destinations. As well as the low and zero income in activities related to this sector that generates an impoverishment of the country's economy. Therefore, this work aims to analyze the behavior of Ecuadorian tourists, specifically from the province of Tungurahua that focuses on the use of digital marketing tools during the COVID-19 pandemic.

The results of this work highlight the profile of the tourist in times of COVID-19, a young female audience who works on their own but with salaries relatively lower than those of the pandemic. Its behavior reflects a considerable decrease in visits within the province due to vehicle restrictions and mobility measures, however the most visited destinations are Pelileo and Baños, thanks to their commercial characteristics and natural attractions, the good quality of service and attention in these places. On the other hand, the actions implemented must be oriented to digital media, in order to take advantage of these young segments through attractive and interesting content from each of the tourist sites.

Regarding the management of biosafety protocols against COVID-19 in the province of Tungurahua, it has been good because companies and businesses are aware of the importance of protecting the well-being of their customers. Thus, the use of a mask, social distancing and the frequent use of alcohol are applied mainly in the cantons and the contents related to its promotion on social networks.

Digital marketing tools in tourism companies allow sharing positive information about promotions, discounts, biosecurity measures for destinations and attracting tourists from the localities themselves. However, despite these efforts, trained personnel are needed in these areas to define optimal strategies according to commercial and business activities. Consequently, a brief theoretical approach is made about the objective of the study, the methodology used, the findings and its conclusions.

2 State of the Art

In 2020 began with a great pandemic that revealed the lack of preparedness to face this type of emergencies throughout the world. Thus, the pandemic paralyzed most of the economic and social activities. This generated the collapse of the health systems of many countries [1]. Thus, tourist activity is no stranger to this crisis and was seriously affected by its almost generalized paralysis worldwide. It is starting with air transport, tourist services such as accommodation, restaurants, leisure and entertainment [2].

Ecuador was no exception, and it was that as of March 2020, the Ecuadorian State declared a health emergency due to the arrival of the COVID-19 pandemic through decree No. 1017 proposed by [3] as president of Ecuador. This caused the paralysis of almost the entire productive sector, then with a very slow and gradual return of activities, which is why it is entering what has been called “New Normality”.

In that sense, the tourism sector of Ecuador before the pandemic in 2019, contributed directly to GDP with 2.2%. The arrival of foreigners without counting the migratory movements of Venezuelans was 1,471,968 of foreigners, with an estimated income from foreign exchange from tourism of $ 2,287,500,000. The accommodation and food services activities had 477,382 employees and the tourist establishments registered in the 2019 Catastro Nacional were 24,257 [4].

However, during the 2020 pandemic there is an average decrease of 42%, in the overall number of arrivals. Even during the months of total confinement, there was no arrival of tourists from the country or income related to tourism [4]. In this context, the pandemic has had negative effects on the tourism sector because in the country. Currently, despite no longer being in a state of exception, biosecurity measures and mobility restrictions are in force that directly affect local and national tourism.

It should be noted that the pre-COVID-19 pandemic scenario in the province of Tungurahua was characterized by the fact that it benefited directly and partially from tourist activities economically. Thus, according to [5], this allowed a more equitable distribution of income. The Gini coefficient for the tourism sector positioned Tungurahua at 0.403, that is, 0.63 points below the coefficient for Ecuador. In this way, there was an income distribution that generated economic well-being in the population. A better satisfaction of needs is allowed by equating opportunities with a notable improvement in their standard of living.

The tourism industry, and then according to estimates by the Ecuadorian government, registers a loss of 540 million dollars for the three months of standstill due to the COVID-19 pandemic. This situation has forced the national government to look for a series of alternatives and mechanisms that help to reactivate companies dedicated to tourist activities, such as refinancing loans or credits in financial institutions, advertising campaigns to consolidate the confidence of tourists as destinations, insurance and training on biosafety measures or digital tools for companies [6].

In addition, as part of the recovery, reactivation and continuity needs of the 24,257 registered tourist establishments and of which 98% are MIPYMES and to counteract the effects caused by the total closure of their activities. The National Government has considered that it requires access to financing lines, the implementation of security measures for the protection of tourists and the adequate economic response for the gradual recovery of activities in the tourism sector [7] and [8].

The authors [9]; [10] state that currently tourism focused on natural areas and outdoor activities provides an outlet or escape for national tourists who live in stressful places such as cities. It is being an advantage to support the economies that have been affected by the COVID-19 pandemic. Thus, for [9]; [11] globalization has allowed companies and businesses through digital media to share different activities, attract local tourists and promote their tourism services that provide the opportunity to generate economic income.

While it is difficult to imagine a world without tourists, but despite the research interest in the negative effects that are already perceived. There are divergences with the approach that is analyzed both negatively for requiring changes in sustainability actions [12] and positive as new technologies and digital media. [13] to attract tourists and promote Ecuadorian culture.

On the other hand, the digital transformation due to the current health situation and mobility restrictions has been a key aspect for companies to remain in the market and survive a difficult period of recession [14]. As a result, social networks such as WhatsApp, Facebook and Instagram directly support the acquisition of customers and the acquisition of products or services that were previously carried out mostly physically [15].

Thus, among the digital marketing strategies for tourism according to [16] and [17] are search or content marketing, media marketing (actions with influencers and on social networks), mailing and viral marketing (promotions, launches, sending videos, animations and music), online advertising (banners, podcasts)), videocasy., online games); online research (searches, market and media monitoring). Therefore, it is emphasized that the content must be attractive, fun and must inform or raise awareness about the problem of the pandemic, so a viable option is the use of influencers on social networks in order to naturally transmit a message clear to tourists regarding travel safety and the application of biosecurity standards.

3 Methodology

This research was descriptive-prospective and was presented under an exploratory-quantitative approach. Two data collection instruments were also used. Then, the behavior of tourists from the province of Tungurahua in the face of the new normality by COVID-19 was analyzed. In this sense, two surveys were conducted aimed at tourists and tourism companies in each canton as shown in Table 1. Potential tourists were the economically active population (EAP) aged 18 and over and the tourism companies that are registered in 2018 business directory. The sample calculation used the finite formula with a confidence level of 95% and a margin of error of 5%.

Table 1. Population and Sample in Tungurahua

Canton	Population		Sample	
	Potential tourists EAP	Business	Potential tourists EAP	Business
Ambato and Tisaleo	228959	711	385	256
Baños	13257	501	374	219
Cevallos	5609	126	361	96
Mocha	5203	115	359	90
Patate	8847	145	369	106
Pelileo	38233	93	381	76
Quero	13779	78	375	66
Total	322250	1867	2972	988

Source: Own elaboration based on data from Ministerio de Turismo (2016); Gobierno Provincial de Tungurahua (2021); INEC (2018)

It should be clarified that an intentional non-probabilistic sampling was applied, excluding the Píllaro canton and unifying the Ambato and Tisaleo cantons for having variable populations and a very small sample.

Within the reliability analysis, Cronbach's alpha statistic was used, which uses a scale of 0.70 to 0.79 as acceptable, 0.8 to 0.89 as good, and 0.9 to 1 as excellent. In the two instruments, its 21 elements or questions were evaluated and present an internal consistency close to 1. It is being shown that the closer it is to this value, the greater its feasibility. Thus, the results were 0.91 for the survey aimed at tourists and 0.82 for the business survey, which shows that the questionnaires had excellent and good reliability, respectively, making them suitable for their application.

In addition, Cronbach's Q test was used to determine if the differences between the frequencies of the variables are statistically significant. In addition, a cluster analysis to obtain homogeneous groups of tourists with similar characteristics during the pandemic, graphed in a dendodrama by canton. In turn, the KMO test was used to compare the observed correlation coefficients with the partial results within a range of 0 to 1 and a factorial analysis by regression to determine the behavior of the tourist through a scatter diagram.

4 Results

Six sections were analyzed on: A) Profile of the tourist in times of COVID; B) Behavior of the tourist; C) Management of biosafety protocols against COVID in the tourism industry of Tungurahua; D) Management of digital marketing in tourism companies; E) Statistical analysis to analyze the behavior of the tourist before the new normality.

4.1 Profile of the Tourist in Times of COVID

In the province of Tungurahua during the pandemic, the majority are Ecuadorian tourists (90.85%) who visit their own cantons where they are eradicated. Furthermore, Fig. 1 shows that the majority are women, with Ambato, Tisaleo, Cevallos and Pelileo standing out with percentages above 60%. Most have tertiary education (51.85%) and secondary education (39.23%). In addition, they are married in five cantons with percentages higher than 50% and single in two of them. It was evidenced that women with tertiary education frequently travel between cantons as they take their families to spend time in recreation.

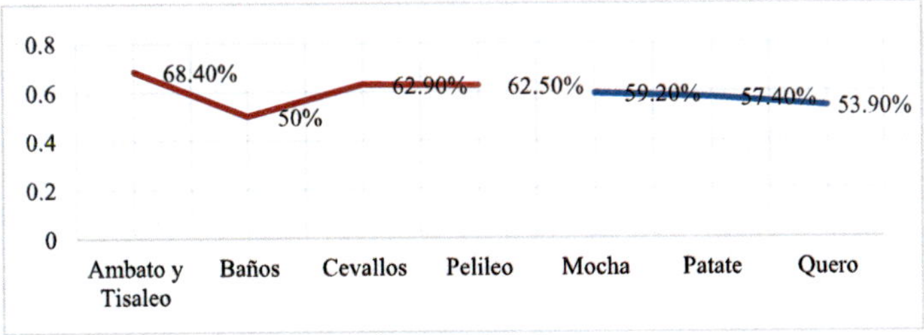

Fig. 1. Gender of the tourist

Ambato, Tisaleo and Patate are between 26 and 34 years old (56.7%), while in Baños, Cevallos, Pelileo and Quero the average age of tourists is between 18 and 25 years old (52.07%). This shows that the majority of the surveyed tourists are self-employed (58.13%), but they are also students (49.85%). Thus, the tourist in the province of Tungurahua is in a young profile that requires actions in different media, especially digital.

COVID has had negative effects on the generation of economic income for tourists in the province of Tungurahua since only two cantons register incomes above $ 801 and two cantons have incomes below $ 400 where Baños does not even register income during the pandemic (Fig. 2). For this reason, tourists have been forced not to travel as often as they used to do it.

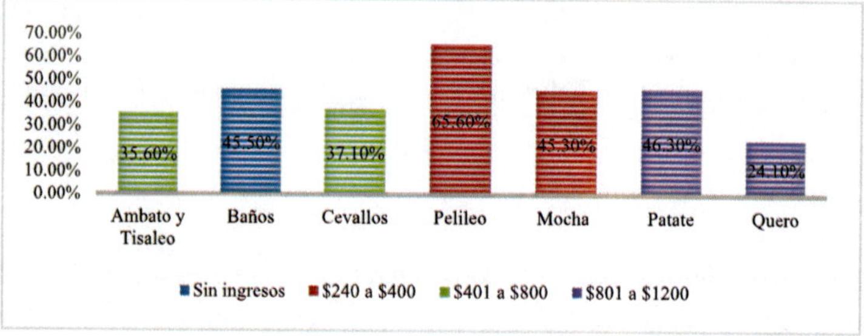

Fig. 2. Tourist income

4.2 Tourist Behavior

Before the pandemic, the Tungurahuense tourist traveled mainly to Pelileo with 93.80% of tourists, followed by Baños with 96% and Cevallos with 85.70%. However, between March and December 2020 there has been an average 7% decrease in trips in the cantons, highlighting a negative variation of 19% in Baños and a positive variation of 10.50% in Ambato and Tisaleo (Table 2). It is evident that the behavior of tourists has been affected by COVID-19, the use of biosafety standards is becoming necessary to protect their health and safety as well as the support of the government with strategies that allow this sector to be promoted.

Table 2. Tourist visits by canton

City	Before COVID	During COVID
Ambato and Tisaleo	77.00%	87.50%
Baños	96.00%	77.10%
Cevallos	85.70%	84.40%
Pelileo	93.80%	84.40%
Mocha	70.00%	65.40%
Patate	61.00%	56.90%
Quero	73.80%	69.30%

In this sense, during the pandemic the duration of the visit in the cantons Pelileo (75%) and Cevallos (57.1%) stands out, where it has been less than 24 h while in Baños (29.50%) it was more two days. In this way, tourists generally travel a single day to their destination, it is essential to define promotions and discounts that motivate people to visit the canton and focus on activities that take place in a few hours.

Figure 3 shows that the tourist associates several characteristics according to each canton, among which the following stand out: buying or buying with Pelileo (90.60%), because in this place jeans are sold in its traditional fairs (81.30%); the tranquility with Cevallos (65.70%), thanks to its mountains and natural spaces (62.90%); as well as fun with Mocha (47.30%), for its rich and varied cuisine (33%). Consequently, these cantons are more representative during the pandemic, however, due to biosafety standards and mobility restrictions, Baños has had a relatively lower rating than the others (10.50%). Although its attractions such as waterfalls, thermal pools and the possibility of practicing extreme sports (24.30%) are valued.

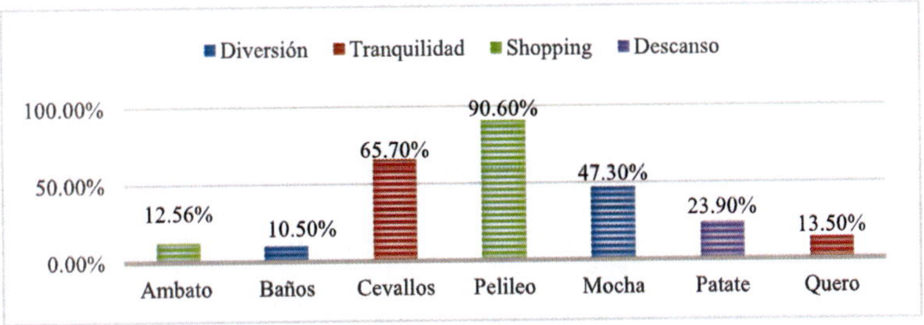

Fig. 3. Characteristics that the tourist associates with the canton

Several aspects are fundamental for the decision to visit the tourist, among which are: good service and quality of care (45.86%), friendliness (38.26%) and care for the environment (51.73%). This shows that companies and businesses seek quality care and are concerned about the cleanliness of their environments, it is being essential during the time of a pandemic. For this reason, 51.81% of the tourists surveyed would recommend the cantons of the province as an optimal travel destination since the tourist's perception of their different products and services is positive (Fig. 4).

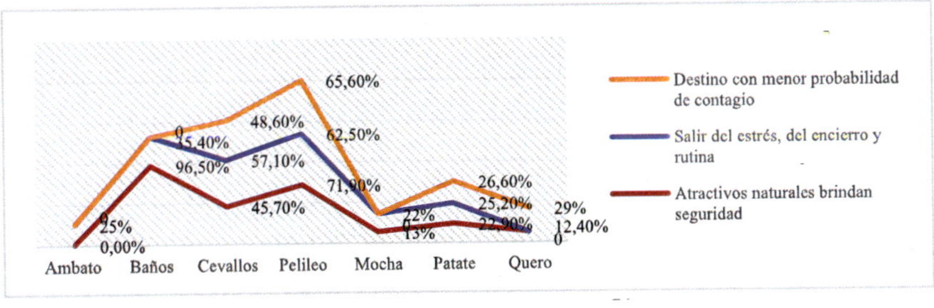

Fig. 4. Biosafety measures implemented in companies.

4.3 Management of Digital Marketing in Tourism Companies

80.5% of Tungurahua companies consider it important and very important to know the origins of COVID-19. The symptoms, how to act if a person is infected, the use of a mask and safety protocols for the well-being of tourists through digital media. In addition, 90.3% of the companies agree or totally agree that it is very important to apply contagion prevention measures, so that tourists visit the cantons safely.

Thus, 80.5% of those surveyed consider that the digital marketing they use for their businesses in the different cantons was what they expected and even much better than they expected. It is leading the WhatsApp Business, Facebook and Instagram applications while 12.6% were not satisfied with this type of advertising it did not favor them. Among the reasons are that only 69% of companies have personnel trained in technology and marketing areas, 34% mishandle promotional content and 35% do not have the necessary technological tools. Content planning, the definition of business objectives and the most appropriate strategies for their business models and activities are necessary.

The social networks that have been used the most to share positive information about promotions, discounts and biosafety measures are: *Facebook* (37.14%), *Instagram* (14.23%), *YouTube* (8.19%) and *WhatsApp* (7.29%)). Therefore, Fig. 5 highlights that the cantons of Quero (72.40%) and Patate (50.50%) mostly use the social network Facebook. Although networks such as *Twitter, TikTok* and *Snapchat* are not channels used by the surveyed companies since their market segments are found in the aforementioned social networks that are usually used by Ecuadorian citizens.

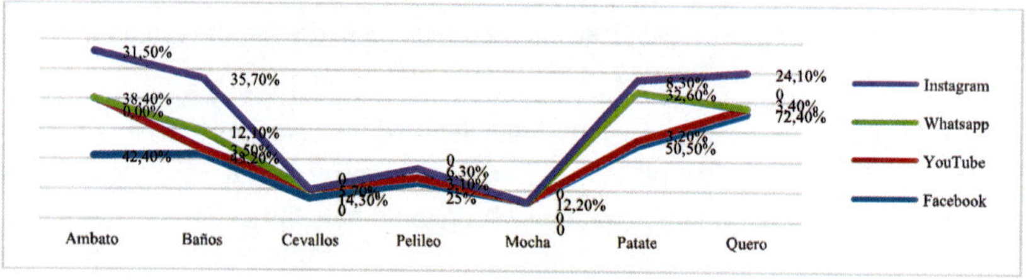

Fig. 5. Use of social networks by canton

Among the viable options as strategies to implement them in the tourism sector of Tungurahua are email marketing (37.9%), so that tourists know information about attractive sites in the cantons. As well as online advertising through multichannel (51, 7%) and promotions, discounts or raffles carried out by companies to attract new customers (51.7%) through digital channels. Companies in the tourism sector in the province of Tungurahua are in a migration from traditional to digital advertising where one of the reasons is the current pandemic that has affected the Ecuadorian economy. For this reason, digital marketing tools help attract new tourists or clients who want to leave their places of confinement to be distracted and relax.

This trend of looking for natural places to visit is an opportunity that companies should take advantage of. Nowadays and with a very busy daily life, people require rest times away from the cities and the natural landscapes of the cantons are ideal for this type of tourism. Therefore, through digital media such as social networks, companies can promote their tourist attractions to provide all the necessary security to people and support the economic growth of the areas in question.

4.4 Statistical Analysis

Innovations in digital marketing in the tourism sector of Tungurahua and the management of biosafety protocols against COVID are statistically similar for tourists (194,742 Q/0,000 Cochran's p-value), (Table 3) and for companies (196,020 Cronbach's Q/0.000 p-value), (Table 4). Consequently, both parties involved point out that digital marketing tools support the proper management of biosafety standards to preserve people's health and therefore provide a good service.

Table 3. ANOVA with the Tourist Cronbach test

		Sum of squares	gl	Mean square	Cronbach'Q	Sig.
Inter-people		17644,348	277	63,698		
Intra-people	Inter-elements	1098,424	47	23,371	194,742	,000
	Residual	72599,242	13019	5,576		
	Total	73697,667	13066	5,640		
Total		91342,014	13343	6,846		
Global mean = 4.532						

Source: data obtained with the SPSS

Table 4. ANOVA with the Cronbach company's test

		Sum of squares	gl	Mean square	Cronbach'Q	Sig.
Inter-people		5266,149	209	25,197		
Intra-people	Inter-elements	947,511	20	47,376	196,020	,000
	Residual	19354,204	4180	4,630		
	Total	20301,714	4200	4,834		
Total		25567,863	4409	5,799		
Global mean = 4.468						

Source: data obtained with the SPSS

A cluster analysis was carried out for each canton to analyze the behavior of tourists in the face of the new normal. It is showing that most of them are obtained between 2 to 5 groups that have similar characteristics. The dispersion of tourists is minimal in almost all the cantons, except in Cevallos and Pelileo, where there is a varied profile of the tourist. Below is a detail by canton with its main analyzes.

In Ambato and Tisaleo, eight groups were obtained from which three of them were selected at a distance of 10 points that have similar characteristics between them. Conglomerate 1 stands out with less dispersion in its behavior, which has adequately implemented security measures and prefers advertising on Facebook and Instagram because they have an average age between 18 to 34 years and is being a young audience (Fig. 6). Furthermore, in the KMO test a value of 0.931 was obtained close to 1 with a significance level of 0.000, so is showing that factor analyzes are valid.

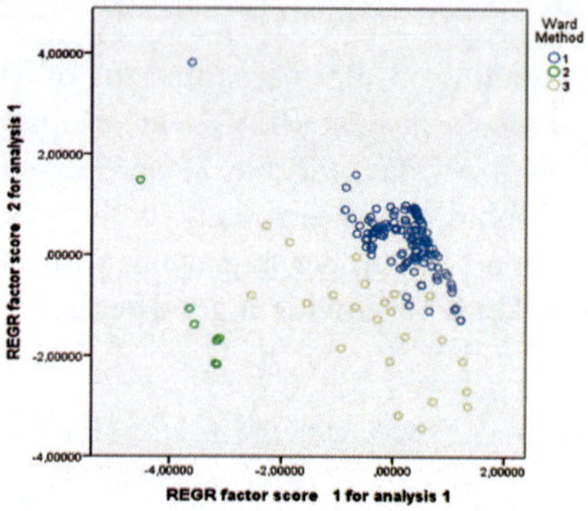

Fig. 6. Tourist conglomerates in the Ambato and Tisaleo pandemic

In Baños de Agua Santa canton, four clusters were obtained at a distance of 10 points, it is observed that groups 1 and 4 are the ones with the greatest similarity and the lowest level of dispersion with respect to the other groups (Fig. 7). In the case of the content on the social networks of the Baños companies, these should focus on activities in nature and extreme sports that are preferred by young audiences between 18 and 25 years old. In the KMO test, a value of 0.931 was obtained, which is close to 1 at a significance level of 0.000, which is why it is showing that factor analyzes are valid.

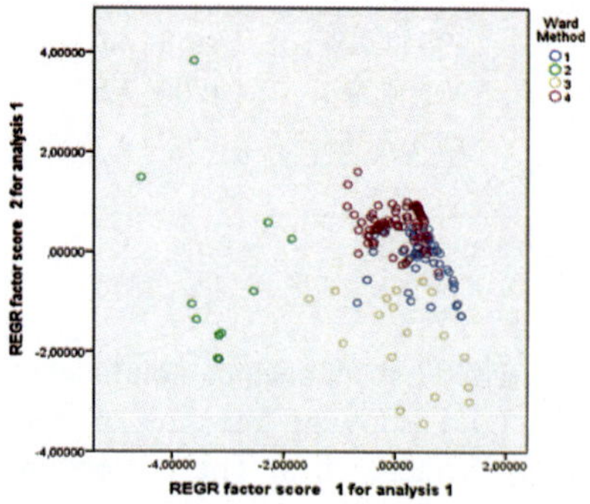

Fig. 7. Tourist conglomerates during COVID Baños

5 Conclusion

The COVID-19 pandemic has directly impacted the tourism sector worldwide, but specifically in Ecuador, being a country where this sector contributes to the productive matrix, its economy has been affected. It follows that the actions that companies implement to get afloat are of vital importance, along with government support for activities related to tourism. In this way, in Tungurahua it is shown that tourists prefer information related to biosafety measures through channels such as Facebook, YouTube and Instagram, and the former predominates due to the profile that the Ecuadorian has. In addition, the content must be directed to the tourist attraction of each canton, where activities with a reduce number of people and far from cities are highlighted.

The tourist has a young profile in the ages of 18 to 34 years, it is being mostly married women with third level studies and income of $ 240 to $ 400 during the pandemic and generated on their own with their businesses. Many of them had to change from tourist activities to activities related to the commercialization of food products, clothing and biosafety implements.

Tourism companies, for their part, have been concerned with safeguarding security through measures such as: distancing, use of masks, disinfection of footwear and use of alcohol dispensers. These factors become decisive factors when traveling to a destination within the province. When it comes to digital marketing tools, most do not have the right staff to do so, and due to the pandemic, the majority of owners are engaged in advertising-related activities. Consequently, that is why the knowledge they possess is insufficient but even so, thanks to the support of the government, for instance, basic courses in this area, tourists feel satisfied with the management of digital media such as social networks.

Acknowledgment. Thanks to the Technical University of Ambato, to the Directorate of Research and Development (DIDE acronym in Spanish) for supporting our research project Impact of digital marketing on the reactivation of tourism in the province of Tungurahua post Covid-19 and being part of the research group Marketing Consumption and Society.

References

1. Sariego, I.: Los retos del turismo cultural en la ciudad de Santander para convertirse en un destino seguro Pos-COVID-19. Estudios Turísticos **219**, 59–70 (2020)
2. Rutynsky, M., Kushniruk, H.: The impact of quarantine due to COVID-19 pandemic on the tourism industry inLviv (Ukraine). Probl. Perspect. Manag. **18**(2), 194–205 (2020)
3. Moreno, L.: Decreto presidencial Nº 1017 (2020). https://www.defensa.gob.ec/wp-content/uploads/downloads/2020/03/Decreto_presidencial_No_1017_17-Marzo-2020.pdf
4. Ministerio de Turismo del Ecuador. GEO VIT, Geo portal. Innovación turística 4.0, 31 Agosto 2020. https://servicios.turismo.gob.ec/index.php/turismo-cifras
5. Santamaría Freire, E., Bayas Vásconez, F.M.: Efecto Económico De La Actividad Turística En La Provincia De Tungurahua, Ecuador," *Ciencias Administrativas,* pp. 1–16, 2018
6. Observatorio Económico y Social de Tungurahua, "Observatorio Económico y Social de Tungurahua," 1 Febrero 2020. https://blogs.cedia.org.ec/obest/wp-content/uploads/sites/7/2020/06/Turismo-en-Ecuador-Alojamiento-y-servicios-de-comida.pdf

7. Gobierno Provincial de Tungurahua. Agenda Tungurahua 2019–2021 (2021). http://www.tungurahua.gob.ec/images/archivos/transparencia/2020/Agenda_Tungurahua2019-2021.pdf
8. La Secretaría Técnica de Planificación "Planifica Ecuador". http://www.inteligenciaproductiva.gob.ec/, 09 Febrero 2021. http://www.inteligenciaproductiva.gob.ec/archivos/Eval%20Soc%20Econ%2011%20sep-.pdf
9. Korstanje, M.: El COVID-19 y el turismo rural: una perspectiva antropológica. Dimensiones Turísticas **4**, 179–196 (2020)
10. Carr, A.: COVID-19, indigenous peoples and tourism: a view from New Zealand. Tour. Geogr. **22**, 491–502 (2020)
11. Visentin, M., Silveira, R., Cappiello, G., Casoli, D.: Sensing the virus. How social capital enhances hoteliers' ability to cope with COVID-19. Int. J. Hospitality Manage. **94**, 102820 (2021)
12. Gallego, I., Font, X.: Changes in air passenger demand as a result of the COVID-19 crisis: using Big Data to inform tourism policy. J. Sustain. Tourism **29**, 1470–1489 (2020)
13. Wen, J., Wang, W., Kozak, M., Liu, X., Hou, H.: Many brains are better than one: the importance of interdisciplinary studies on COVID-19 in and beyond tourism. Tourism Recreation Res. **46**, 310–313 (2020)
14. Rodríguez, K., Ortíz, O., Quiroz, A., Parrales, M.: El e-commerce y las Mipymes en tiempos de Covid-19. Revista Espacios **41**(42), 100–119 (2020)
15. Noziglia, M., Gómez, M.: SEO en tiempos de COVID 19, Beneficios y recursos útiles para ser eficaces en este nuevo paradigma. UM, pp. 1–6 (2020)
16. Mendes, G., Augusto, A., Gándara, J.: Innovación en la promoción turística en medios y redes sociales. Un estudio comparativo entre destinos turísticos. Estudios y Perspectivas en Turismo **22**(1), 102–119 (2013)
17. Sigala, M.: Tourism and COVID-19: impacts and implications for advancing and resetting industry and research. J. Bus. Res. (2020)

An Online Approach to Project-Based Learning in Engineering and Technology for Post-secondary Students

Fei Geng(✉), Seshasai Srinivasan, Zhen Gao, Steven Bogoslowski, and Amin Reza Rajabzadeh

Booth School of Engineering Practice and Technology, McMaster University, Hamilton, ON L8S 4L8, Canada
{gengf, rajaba, gaozhen, bogoslsr}@mcmaster.ca

Abstract. In this work, we present an active digital learning environment to support undergraduate research-based education in an online environment in the Biotechnology program, as part of the iThink program that is part of the PIVOT initiative. Specifically, the students were tasked to demonstrate an open-ended objective using the biosensor technology for organic sample detection. To do this in an online environment, students undertook the research and development of their experiments by learning the laboratory techniques using online lab exercises and evolving the research in a collaborative online environment. Their progress was recorded throughout the academic year via weekly assignments, review of the content relating to the weekly deliverables, online interactive lab simulations, and lab reports with detailed technical content relating to their experiments. As with the in-person format, the iThink project continues to teach the concepts to the students from the highest levels of the Bloom's taxonomy. An evaluation of the effectiveness of the iThink program in an online setting showed that the performance of the students in the online setting was similar to that of the students in the in-person format.

Keywords: Remote/online learning · Project-based learning · Engineering pedagogy · Biotechnology

1 Introduction

Transition from in-class delivery of academic content because of the pandemic restrictions was done on a global scale through university institutions worldwide [1–4]. Students across all universities around the globe had to adapt to this mode of learning. For a successful online delivery of the curriculum, pedagogical innovations were introduced to ensure that the education is uninterrupted, and students may graduate in a timely manner. Post-secondary engineering institutions' goal is to graduate competent engineers with real-world experiences and who can translate the technical concepts taught in the classroom into functional designs and innovations [5–7].

Among other pedagogical techniques, active learning [8–13], problem-based learning (PBL) [14–18] and project-based learning are widely used in academia [19–22]. Apart from these, an important pedagogical approach is the research-based

M. E. Auer and T. Tsiatsos (Eds.): IMCL 2021, LNNS 411, pp. 627–635, 2022.
https://doi.org/10.1007/978-3-030-96296-8_56

education in which students are expected to explore and experiment to experience the essence of the discipline. As per the constructivist theory of learning, exposure to such rich and varied experiences will result in a good mental construction of the various concepts [23, 24]. By undertaking research, the students are exposed to the application of the concepts they learn, and this helps reinforce the principles of the discipline. It also encourages the students to become independent life-long learners who are confident graduates ready to join the industry or pursue graduate studies. The overall education experience is enriched since the students not only learn the fundamental concepts, but they also improve their problem-solving skills, learn the exploration and use of scientific literature, and develop their oral and written communication skills [25–27].

Hartmann [28] identified that students are aided in areas such as problem definition, technical abilities, decision making and critical thinking. An investigation by Seymour et al. [29] in four colleges found that an overwhelming majority of students (91%) perceived undergraduate research positively. Rodrick and Dickmeyer [30] found that capstone projects significantly benefit students and proposed an integration of capstone projects in the communication curriculum to provide more research opportunities for the students. Thus, in general, there is consensus in the literature that undergraduate research experience brings a positive transformation in students and prepares them better to apply the theoretical principles to solve complex real-world problems [31–35].

At the W Booth School of Engineering Practice and Technology, we recently introduced undergraduate research-based learning through the iThink program [36, 37], that is part of the PIVOT initiative launched by the Faculty of Engineering at McMaster University. Unlike the traditional curriculum in which students are guided through a training program that starts at the lower levels of the Bloom's taxonomy and gradually progresses to the upper levels, in iThink, students are introduced to an application (highest levels of Bloom's taxonomy) and are required to research and learn the skills to be able to design and develop the application. Thus, in iThink, students traverse the Bloom's taxonomy in the reverse order. Working in groups to undertake the complex design and development exercise, students are expected to develop a deeper understanding of the subject and the journey fosters critical thinking, collaboration skills, and other transferable skills that is critical for engineers in society [38, 39].

In this paper, we describe the details of students are subjected to the previously taught content of the iThink program through an online delivery method. The iThink program produces an outcome of implementation of a project-based and active learning digital component for students. This program offers a new innovative approach to the shift into online learning. Work in relation to this program has been drawn from the work done at MIT in the form of the NEET project [40]. Students involved in this project stem from the Biotechnology program offered at McMaster University which has come from the online education transformational initiative, PIVOT, led by the faculty of engineering at the led by the faculty of engineering at W Booth School of Engineering Practice and Technology. The assessment of the evaluation of the project and students involved are documented to discover the influence and transition from an in-class working environment to an online platform.

2 The iThink Program

The iThink program was introduced to W Booth School of Engineering Practice and Technology to foster a cross-disciplinary research-based curriculum in the undergraduate programs. Students from the Biotechnology, Automation Engineering Technology, Automotive and Vehicle Engineering Technology, and the Software Engineering Technology programs participate in the iThink program [36, 37]. In these inter-disciplinary projects, students are required to develop a prototype for applications such as biosensor design to detect proteins in a sample. Thus, the education starts with a definition at the highest levels of the Bloom's taxonomy. To develop the prototype that meets a certain prescribed benchmark, the students undertake a literature review, and propose a design of experiments. A key aspect of this work involves collaborating with students from other disciplines. Faculty members from these streams engage with the students in an advisory role, guiding the students in the explorations.

As an outcome of the first batch of the graduates from the iThink program, a novel, open-source, electrochemical biosensor device was developed by the students using an *IO Rodeostat* potentiostat made from the *IO Rodeo Smart Lab Technology* for detection of organic material. The potentiostat was capable of measuring voltage and current, programmed to conduct Cyclic Voltammetry (CV) and Differential Pulse Voltammetry (DPV) tests of organic samples on screen-printed electrodes. The initial project was used to detect hybridization events of complimentary DNA strands on the electrode surfaces in order to identify the presence of target biomarkers for Prostate Cancer in humans. The computer program made to run the tests on the potentiostat biosensor device was programmed using the Python GUI application by a student from the Deaprtment of Electrical Engineering. The data acquired by the device was stored in an online database and made accessible to the faculty and students involved in future projects. The graphs acquired from the GUI application were programmed into Excel spreadsheets with corresponding data set values and used to compare the different stages of hybridization events occurring throughout various experiments. Further innovation of the novel, open source biosensing device involved the addition of a Raspberry Pi computer for processing.

The iThink projects were embedded into the curriculum and offered as an option for the students. More precisely, students could opt to take this research-based learning option instead of the traditional lab-based learning. Further, the students were given the opportunity to choose their own research topic and develop their own experimental protocol to develop the biosensor. In an active learning environment, the instructor engages with the students, discussing about the specific skills and the experiments that the students have planned. Following this discussion, students make amendments and adjustments to their experimental plan, if needed, and continue performing the experiments and analyzing the results. Throughout the entire process, the students would research and design their own experimental procedure, all while learning the necessary lab techniques provided in the original lab curricula. Assessment of the progress was done through regular short presentations/video reports and discussion of the weekly findings. Thus, by creating a discussion-based environment during the presentations, an active learning environment was maintained to help develop the groups work skills.

3 The Online Version of iThink – Procedures and Outcomes

The online version of iThink initiative was launched in Biotechnology Program at W Booth School of Engineering Practice and Technology in academic year 2020–2021 to improve the project-based learning in two biotechnology courses BIOTECH 2M03-Molecular Biology and BIOTECH 2BC3-Biochemistry. Despite the time constraints and the challenges associated with the COVID-19 pandemic, our goals were to provide a teaching and learning experience that was as close as possible to the in-person version of iThink initiative in terms of creative teaching and learning strategies, student engagement, and active learning.

In both courses, to maintain a realistic approach, the iThink students were provided a list of topics, expectations, and resource availability from the program to ensure a feasible target. Typical examples of the projects undertaken by the students in the two courses include the characterization of nucleus DNA from breast cancer cells; the development of biosensor technique to detect biomarker protein in breast cancer cells; the purification and biosensing of lipids from Gram-positive bacteria; and isothermal amplification of viral DNAs for disease diagnosis. The students undertook the research initiative over a duration of 13 weeks. In both courses, the class met twice a week for a duration of 1.5 h in each meeting. While the first class in the week was devoted to outlining the technical principles and discussions, the second class focused on the lab experiments, the challenges faced by the students, and discussion to help them move their project forward. All the course content was shared on McMaster's learning management system, Avenue to Learn. Student work (including assignment, presentation, and final report) was submitted, marked, and analyzed on Avenue to Learn course shell. Comments were provided back to students one week after the submission. During the academic year of 2020–2021, all the students who enrolled in both the courses (BIOTECH 2M03 and BIOTECH 2BC3) participated in iThink Program. The numbers of students who participated in the iThink Program during the two courses are summarized in Table 1.

Table 1. The number of students who participated iThink Program in BIOTECH 2M03 and BIOTECH 2BC3 during the academic year of 2020–2021.

Course Code	The number of students who participated in iThink Program
BIOTECH 2M03 molecular biology	60
BIOTECH 2BC3 biochemistry	57

An extensive and entrenched participation, and assuming responsibility was the focus in the initial stages of the course. Students were required to form teams of 3–4 students per group, and the assessments were conducted on the group. The stages of online iThink program presented in this study include the following: 1) planning and determination of the idea of making simple biosensor tools, 2) design of biosensors, 3) reporting the progress on biosensor design through Avenue to Learn course shell, and making amendments based on the professor's feedback, 4) implementing functional tools through self-designed experiments, 5) recording and submitting the videos that demonstrate the design and functionality of the tools, and 6) a final presentation of the project undertaken by the students through a recorded video. The students were evaluated based on the outcome of these different steps and were assigned a final grade. Project assignments were collected in the form of project work schedules, photos of equipment designed by the students, practicum photographs carried out individually by students, and the simulation-based proof-of-concept experiments due to the online course during the COVID-19 pandemic. Throughout the iThink Program, students were required to submit the biweekly assignments that describe the progress report on the project, including the description of the project to the various research steps undertaken by the students in the consecutive weeks. Specifically, these assignments present the project viability, lists of materials, and the designs of the various protocols. Since students were unable to perform any experiments in the lab during the pandemic, concepts such as DNA/RNA extraction, PCR analysis, molecular cloning, protein synthesis, and viral gene therapy were covered in the simulated experiments. These experiments were delivered through the online lab simulation program known as *Labster*. Students were provided some data from these experiments to process/analyze and submit a commentary on the findings in the form of a report. Thus, through this collection of experiments, by the end of the semester, students were exposed to a suite of competencies needed for developing a design of experiments for their respective research projects. A final report was due at the end of the semester which included all the previous work completed by the students. To ensure continued progress throughout the term, intermittent assessments were done as follows: 5-min group presentations on the progression of the projects were required every 2–3 weeks. In this, the instructors and students would engage in detailed discussions, conducting simulated experiments on theoretical concepts taught within the course that directly coincided with the knowledge needed for their research topic.

The effectiveness of iThink Program measured in the form of the average student performance in BIOTECH 2M03 and BIOTECH 2BC3 is shown in Fig. 1. To measure the learning in students we compared the performance of the students who undertook the iThink projects with those who did not participate in iThink in these two courses and instead took the traditional labs to learn the skills during 2019–2020 academic year. As seen in this figure, in both courses, students who participated in the iThink projects scored nearly 16% higher than the students who did not participate in iThink. Similarly, in BIOTECH 2BC3, the students who participated in the iThink projects scored approximately 5% higher than the students who did not participate in iThink. This is not surprising because with more exposure to literature via research and reading, the students in iThink had a better understanding of the various concepts and performed much better during the assessments. This is mainly due to the development of their critical thinking, project planning and troubleshooting, communication, and soft skills, that are honed while working in groups.

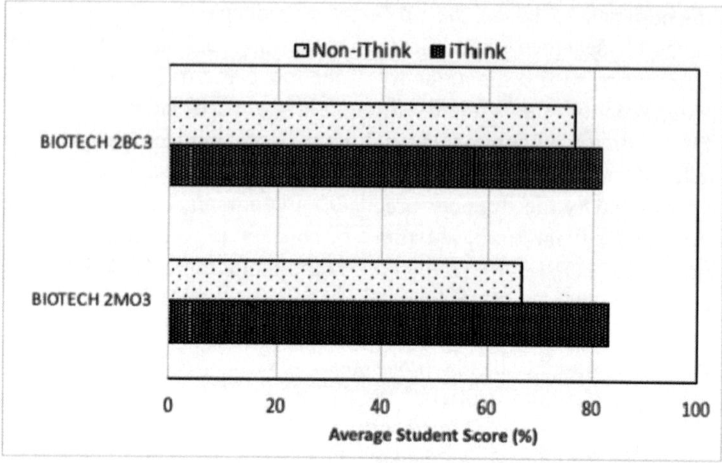

Fig. 1. Comparison of academic performance of students who enrolled in the iThink program during the academic year 2020–2021 with those who took the traditional lab format of learning (non-iThink) during the academic year of 2019–2020, in BIOTECH 2M03 and BIOTECH 2BC3.

4 Future Plan for the iThink Program

The COVID-19 pandemic posed a serious challenge to the format of several courses that require the presence of students and professors in laboratory for training purposes. Prior to the pandemic the students were also required to attend theoretical lectures in person. As an evolution from the pandemic, starting from Fall 2021, we plan to adopt a blended learning approach in which the theoretical lectures will be held online while the laboratory experiments will be conducted in person.

With respect to the laboratory experiments, based on the current virtual iThink structure described in this work, we would explore the following two options: In the first option, the students will use a combination of online labs as well as in-person labs to conduct their iThink project. Specifically, online labs could be used to obtain the skills and in person labs could be used to employ these skills to the specific iThink project. This online and in-person lab could be scheduled in a manner to accommodate for social distancing norms, as well as student preferences, give an added flexibility in the learning environment. In the second option, students will develop their laboratory skills and apply them to solve their iThink project by undertaking experiments in the laboratory in an in-person environment. With either option, the theoretical lectures as well as group discussions on the project and the experiments could be done in a virtual setting introducing a significantly flexible learning environment for future students. Thus, we can foster a collaborative learning environment not only in the laboratory during the experiments but also in the online environment. In other words, the future format of iThink will be a creative and an efficient blended learning environment for the learners in the iThink program.

There are some challenges to be mindful of in pursuing these options. For instance, it is also important to investigate how students engage virtually in iThink activities and compare this engagement levels with the in-person format. This while recognizing that engaging synchronously through virtual meeting platforms has been a particular challenge. There are also student-wellbeing issues that we should be mindful of when adopting a virtual format of instruction. In summary, reflecting on our experiences, we have identified several important steps for using technology as a tool through which creative active learning strategies can be implemented in virtual contexts.

5 Conclusion

As an educational evolution in response to the pandemic there has been a shift in the delivery of post-secondary education, and this has resulted in the adaptation of online/mobile learning for engineering and technology students. In this work, we present the transition from in-class to an online environment for the iThink program. Specifically, the previously established theme of research-based education, the iThink program, has been successfully implemented in an online format at McMaster University's W Booth School of Engineering Practice and Technology. In this, the students use online simulated labs to solve a research problem of their interest. The problem is solved in a collaborative and investigative setting and students learn the necessary skills via an inquiry-based learning. The iThink project addresses these teaching concepts and delivers them using a top-down approach to Bloom's Taxonomy. More precisely, in trying to solve a research problem, students determine the techniques and skills required to finish the necessary tasks, thereby picking up the skills as well as knowing how these skills are relevant in a real-world setting by employing them to solve a larger research project. An evaluation of the effectiveness of the iThink program showed that the students performed equally well in the in person as well as the online format of iThink. More importantly, the students in the iThink program tend to outperform students who learnt the concepts and skills via traditional lab exercises.

References

1. Bao, W.: COVID-19 and online teaching in higher education: a case study of Peking University. Hum. Behavior Emerg. Technol. **2**(2), 113–115 (2020). https://doi.org/10.1002/hbe2.191
2. Zhang, W., Wang, Y., Yang, L., Wang, C.: Suspending classes without stopping learning: China's education emergency management policy in the COVID-19 outbreak. J. Risk Financ. Manage. **13**(3), 55 (2020). https://doi.org/10.3390/jrfm13030055
3. Muhammad, N., Srinivasan, S.: Online education during a pandemic – adaptation and impact on student learning. Int. J. Eng. Pedagogy **11**(3), 71–83 (2021). https://doi.org/10.3991/ijep.v11i3.20449
4. Muhammad, N., Srinivasan, S.: Transition from In-class to online lectures during a pandemic. In: Auer, M.E., Centea, D. (eds.) ICBL 2020. AISC, vol. 1314, pp. 307–314. Springer, Cham (2021). https://doi.org/10.1007/978-3-030-67209-6_33
5. Case, J.M., Light, G.: Emerging research methodologies in engineering education research. J. Eng. Educ. **100**(1), 186–210 (2011)

6. Splitt, F.G.: The challenge to change: on realizing the new paradigm for engineering education. J. Eng. Educ. **92**(2), 181–187 (2003)
7. Chew, C.M., Ng, L.Y., Mah, S.K., Ng, Y.S.: Development of a university-industry collaboration model towards work-ready engineering students. Res. Sci. Technol. Educ. pp. 1–19 (2021)
8. Lima, R.M., Andersson, P.H., Saalman, E.: Active Learning in engineering education: a (re) introduction. Eur. J. Eng. Educ. **42**(1), 1–4 (2017)
9. Beichner, R.: The Student-Centered Activities for Large Enrollment Undergraduate Programs (SCALE-UP) Project (2007)
10. Srinivasan S., Centea D.: An active learning strategy for programming courses. In: Auer, M., Tsiatsos, T. (eds.) Mobile Technologies and Applications for the Internet of Things. IMCL 2018. Advances in Intelligent Systems and Computing, vol. 909, pp. 327–336 (2019). https://doi.org/10.1007/978.3-030-11434-3_36
11. Cummings, K., Marx, J., Ronald, T., Dennis, K.: Evaluating innovation in studio physics. Am. J. Phys. **67**, S38–S44 (1999). https://doi.org/10.1119/1.19078
12. Sidhu, G., Srinivasan, S.: An intervention-based active-learning strategy to enhance student performance in mathematics. Int. J. Pedagog. Teach. Educ. **2**, 277–288 (2018). https://doi.org/10.20961/ijpte.v2i1.19568
13. Srinivasan, S., Centea, D.: Applicability of principles of cognitive science in active learning pedagogies. In: Proceedings of the 13th International Workshop Active Learning in Engineering. (1 ed.) Aalborg Universitetsforlagpp, pp. 99–104 (2015)
14. Centea, D., Srinivasan, S.: Enhancing student learning through problem based learning. In: Guerra, A., Rodriguez, F.J., Kolmos, A., Reyes, I.P. (eds.) PBL, Social Progress and Sustainability, Aalborg: Aalborg Universitetsforlag. (International Research Symposium on PBL), pp. 376–385 (2017)
15. Dochy, F., Segers, M., Van den Bossche, P., Gijbels, D.: Effects of problem-based learning: a meta-analysis. Learn. Instr. **13**, 533–568 (2003). https://doi.org/10.1016/S0959-4752(02)00025-7
16. Centea, D., Srinivasan, S.: A comprehensive assessment strategy for a PBL environment. Int. J. Innovat. Res. Educ. Sci. **3**(6), 364–372 (2016)
17. Sidhu, G., Srinivasan, S., Centea, D.: Implementation of a problem based learning environment for first year engineering mathematics. In: Guerra, A., Rodriguez, F.J., Kolmos, A., Reyes, I.P. (eds.) PBL, Social Progress and Sustainability, Aalborg: Aalborg Universitetsforlag. (International Research Symposium on PBL), pp. 201–208 (2017)
18. Muhammad, N., Srinivasan, S.: A problem solving based approach to learn engineering mathematics. In: Auer, M.E., Hortsch, H., Sethakul, P. (eds.) ICL 2019. AISC, vol. 1134, pp. 839–848. Springer, Cham (2020). https://doi.org/10.1007/978-3-030-40274-7_81
19. Wu, T.T., Wu, Y.T.: Applying project-based learning and SCAMPER teaching strategies in engineering education to explore the influence of creativity on cognition, personal motivation, and personality traits. Thinking Skills Create. **35**, 100631 (2020)
20. Cifrian, E., Andres, A., Galan, B., Viguri, J.R.: Integration of different assessment approaches: application to a project-based learning engineering course. Educ. Chem. Eng. **31**, 62–75 (2020)
21. Edström, K., Kolmos, A.: PBL and CDIO: complementary models for engineering education development. Eur. J. Eng. Educ. **39**(5), 539–555 (2014)
22. Roessingh, H., Chambers, W.: Project-based learning and pedagogy in teacher preparation: staking out the theoretical mid-ground. Int. J. Teach. Learn. Higher Educ. **23**, 60–71 (2011)
23. Olusegun, B.S.: Constructivism learning theory: a paradigm for teaching and learning. ISOR J. Res. Method Educ. **5**(6), 66–70 (2015)

24. Srinivasan, S., Muhammad, N.: Implementation of a course in computational modeling of biological systems in an undergraduate engineering program. Int. J. Eng. Educ. **36**(3), 857–864 (2020)
25. Lopatto, D.: The essential features of undergraduate research. Council Undergr. Res. Quart. **24**, 139–142 (2003)
26. Bauer, K.W., Bennett, J.: Alumni perceptions used to assess undergraduate research experience. J. Higher Educ. **74**, 210–230 (2003)
27. an empirical study: Hathaway RS, Nagda BA and SRJJoCSD G. The relationship of undergraduate research participation to graduate and professional education pursuit. J. College. **43**, 614–631 (2002)
28. Hartmann, D.J.: Undergraduate research experience as preparation for graduate school. Merican Sociol. **21**, 79–188 (1990)
29. Seymour, E., Hunter, A.B., Laursen, S.L., et al.: Establishing the benefits of research experiences for undergraduates in the sciences: first findings from a three-year study. Sci. Educ. **88**, 493–534 (2004)
30. Rodrick, R., Dickmeyer, L.: Providing undergraduate research opportunities for communication students: a curricular approach. Commun. Educ. **51**, 40–50 (2002)
31. Elrod, S., Husic, D., Kinzie, J.: Research and discovery across the curriculum. Assoc. Am. Coll. Univ. **12**, 4–8 (2010)
32. Hakim, T.M.: At the Interface of Scholarship and Teaching: How to Develop and Administer Institutional Undergraduate Research Programs. Council on Undergraduate Research, Washington (2000)
33. Petrella, J.K., Jung, A.P.: Undergraduate research: importance, benefits, and challenges. Int. J. Exerc Sci. **1**, 91–95 (2008)
34. Seifan, M., Lal, N., Berenjian, A.: Effect of Undergraduate research on students learning and engagement. Int. J. Mech. Eng. Educ. **0**, 1–23 (2021). https://doi.org/10.1177/0306419021988962
35. Noguez, J., Neri, L.: Research-based learning: a case study for engineering students. Int. J. Interact. Des. Manuf. (IJIDeM) **13**(4), 1283–1295 (2019). https://doi.org/10.1007/s12008-019-00570-x
36. Srinivasan, S., Rajabzadeh, A.R., Centea, D.: A project-centric learning strategy in biotechnology. In: Auer, M.E., Hortsch, H., Sethakul, P. (eds.) ICL 2019. AISC, vol. 1134, pp. 830–838. Springer, Cham (2020). https://doi.org/10.1007/978-3-030-40274-7_80
37. Bogoslowski, S., Geng, F., Gao, Z., Rajabzadeh, A.R., Srinivasan, S.: Integrated thinking - a cross-disciplinary project-based engineering education. In: Auer, M.E., Centea, D. (eds.) ICBL 2020. AISC, vol. 1314, pp. 260–267. Springer, Cham (2021). https://doi.org/10.1007/978-3-030-67209-6_28
38. Fruhling, Z.: Bloom's Taxonomy: Bottom-Up or Top-Down?. Resilient Educator. (2018). https://resilienteducator.com/instructional-design/blooms-taxonomy-bottom-up-or-top-down/. Accessed 28 May 2020
39. Wright, S.: Flipping Bloom's Taxonomy. Powerful Learning Practice (2012). https://plpnetwork.com/2012/05/15/flipping-blooms-taxonomy/. Accessed 28 May 2020
40. Crawley, E.F., Hosoi, A., Long, G.L., Kassis, T., Dickson, W., Mitra, A.B.: Moving forward with the New Engineering Education Transformation (NEET) program at MIT - building community, developing projects, and connecting with industry. In: ASEE Annual Conference & Exposition (2019)

Moodle Platform and Online Renewable Energy Laboratory at Faculty of Electrical Engineering

Hen Friman(✉) and Netser Matsliah

Faculty of Engineering, H.I.T - Holon Institute of Technology, Holon, Israel
henf@hit.ac.il

Abstract. Renewable energy and energy efficiency technologies are the key for creating a clean energy future for not only the nation, but the world. world Energy Consumption relies heavily on coal, oil, and natural gas. Fossil fuels are non-renewable, that is, they rely on finite resources that will eventually dwindle, becoming too expensive or too environmentally damaging to retrieve. In contrast, renewable energy resources, such as wind and solar energy, are constantly replenished and will never run out.

Due to the rising need for professionals and academics with a background and understanding in the Renewable Energy field, Holon Institute of Technology ("HIT") developed a new program at the Faculty of Electrical Engineering. The Renewable Energy program gives the students technical and practical aspects of energy use (technology and methodology of the study) and energy efficiency. The program also deals with minimizing the environmental impacts of energy use, as well as with energy economy and environmental policy. The Institute offers its students a well-equipped laboratory, containing state of the art equipment in various fields such as: photovoltaic energy systems, a smart grid telecommunications and information security platform, wind and water energy work stations, and power electronics equipment. The Renewable Energy Laboratory is operating under a new experimental teaching method, and presents itself as the "next generation" lab. The Lab is "paperless", which means that all the experiments' notes, as well as the theoretical background, are computerized. The notes and experiments are all integrated into a "Moodle" platform. Each student fills an anonymous computerized feedback questionnaire at the end of all experiments. In this questionnaire they are asked about this new method of computerized learning experience. In this paper a summary of the results of the pilot conducted last six years is presented.

Keywords: Electrical engineering · Renewable energy · Computerized lab

1 Introduction

The last decades have been dominated by the rapid changes introduced by the technology revolution, which has a tremendous influence on our daily lives. Today we are facing a myriad of new challenges. Technology-based industry has matured in many ways and the required skills for future engineers are much more complex in a world

M. E. Auer and T. Tsiatsos (Eds.): IMCL 2021, LNNS 411, pp. 636–645, 2022.
https://doi.org/10.1007/978-3-030-96296-8_57

where "machines/computers" execute many of the engineering tasks [1]. Most of all, we are facing a new generation of sophisticated students, who were born into the digitized/multimedia world. The mission of the study program is to encourage and initiate academic development, through the development of new study programs and methods, while being responsive to the rapidly changing trends in the field [2]. The proper education of the undergraduate students must also be a function of market needs and predictions of how technology will develop in the foreseeable future. In order to ensure that our graduates are well qualified to meet the future needs of the market, meticulous attention must be paid to maintain a high standard in the fundamental courses and impart practical tools and skills. It is also important to introduce a wide variety of new subjects [1]. The aims and goals of the Engineering faculty are to provide the students with a rich and comprehensive study program, and keep the study program updated to meet the ever-changing requirements for engineers of the future, enrich the student's theoretical knowledge as well as teach practical and design skills and knowledge; adapt its teaching methodologies and techniques, focusing on understanding as a goal; enable students to achieve skills such as self-learning and to acquire expertise via practice by understanding constantly update the teaching methods and the study program maintain relationships with the various relevant industry sectors introduce the students to state-of-the-art equipment and facilities, for conducting experiments that reinforce their understanding of the theoretical and practical issues studied in the courses promote research in the various fields; and explore cooperation with other institutes in Israel and abroad [3].

1.1 The Holon Institute of Technology

The Holon Institute of Technology (HIT) is an academic institution of higher education. Established in 1969, it is originally part of the Tel Aviv University and becomes an independent public academic institution of higher education in 1999. HIT trains the next generation of scientists, engineers, designers and technology managers. Fully accredited by the Israeli Council for Higher Education (CHE), it is entitled to grant undergraduate (Bachelor) and graduate (Master) degrees.

HIT's academic body is composed of over 150 professional and experienced (senior) lecturers and researchers. These nurture tight connections with the industry, develop innovative teaching technologies and are extensively involved in the community. These extensive collaborations are embedded in up-to-date course materials and workshops. Industrials give lectures and tutor HIT's 4500 students. Research topics and final projects are done in countless fields (energy, nanotechnology and nanomaterials, cyber, design, data mining, data analysis, eHealth, and more). In addition, HIT is equipped with theoretical and applied research facilities among which some unconventional and forward-thinking labs and multidisciplinary centers.

As a result of both in-house work and collaborations – whether among our faculties, international collaborations, collaborations via institutional multidisciplinary and national task forces or collaborations within the academic world and with the industry - we successfully provide our students with multidisciplinary knowledge and original analytical thinking, and encourage and cultivate excellence.

1.2 Faculty of Engineering

The The last decades have been dominated by the rapid changes introduced by the technology revolution, which has a tremendous influence on our daily lives. Today we are facing a myriad of new challenges. Technology-based industry has matured in many ways and the required skills for future engineers are much more complex in a world where "machines/computers" execute many of the engineering tasks. Most of all, we are facing a new generation of sophisticated students, who were born into the digitized/multimedia world. The mission of the study program is to encourage and initiate academic development, through the development of new study programs and methods, while being responsive to the rapidly changing trends in the field. The proper education of the undergraduate students must also be a function of market needs and predictions of how technology will develop in the foreseeable future. In order to ensure that our graduates are well qualified to meet the future needs of the market, meticulous attention must be paid to maintain a high standard in the fundamental courses and impart practical tools and skills. It is also important to introduce a wide variety of new subjects. The aims and goals of the Engineering faculty are to provide the students with a rich and comprehensive study program, and keep the study program updated to meet the ever-changing requirements for engineers of the future, enrich the student's theoretical knowledge as well as teach practical and design skills and knowledge; adapt its teaching methodologies and techniques, focusing on understanding as a goal; enable students to achieve skills such as self-learning and to acquire expertise via practice by understanding constantly update the teaching methods and the study program maintain relationships with the various relevant industry sectors introduce the students to state-of-the-art equipment and facilities, for conducting experiments that reinforce their understanding of the theoretical and practical issues studied in the courses promote research in the various fields; and explore cooperation with other institutes in Israel and abroad.

1.3 Renewable Energy and Smart Grid Excellence Centre

The Energy field is thriving, due to several factors: the world energy crisis, political trends that create a rise in oil prices and other environmental topics. All of these have brought upon us the emergence of new and fascinating fields dealing with Energy. The introduction of renewable energy sources to the electrical grid and the realization of the need to optimize the current network with modern tools, have both led to a new research field: The Smart Grid.

The introduction of alternative (renewable) energy sources for the electrical grid and the realization that there's a need to improve and optimize the current network using modern tools, has brought upon a new research field called The Smart Grid.

The Smart Grid field creates a new interaction among various disciplines. Its goal is to create an electrical grid that is controlled by computers that are inter-connected via a cutting-edge communication network. This is an entirely new technological and conceptual revolution.

Following the receipt of an award for research, funded by the Chief Science Officer of Israel, a research group and the renewable energy and smart grid excellence centre were founded in HIT in June 2011 with the purpose of encouraging research and creation in the field of energy.

At the heart of the centre, the renewable energy and smart grid laboratory was established. The laboratory is equipped with state-of-the-art equipment and experiments, including: photovoltaic energy, water energy, wind energy, fuel cells and smart meters and smart grid equipment. The laboratory is a "paper free" laboratory. The Renewable Energy Centre's vision is to combine the vast knowledge of the institute's researches in this field, and to create national excellence and research centre for the advanced energy fields.

This group is unique in that it includes HIT's researches and leading industry figures from various energy fields, as well as Design researchers, HIT's maintenance manager, and students.

The centre's objectives are in one hand to teach and enrich students with the most recent technologies in this field and on the other hand, to create scientific collaborations that will lead to receiving prestigious grants, as well as to publishing joint essays. Collaboration with the industry enables the establishment of joint ventures which will promote both research and the institute in this field.

2 Renewable Energy Laboratory

Engineering is a practicing profession, a profession devoted to harnessing and modifying the three fundamental resources that humankind has for the creation of all technology: energy, materials, and information. The overall goal of engineering education is to prepare students to practice engineering and, in particular, to deal with the forces and materials of nature [4]. Thus, from the earliest days of engineering education, instructional laboratories have been an essential part of undergraduate studies. Indeed, prior to the emphasis on engineering science, it could be said that most engineering education took place in the laboratory. The emphasis on laboratories has varied over the years. While much attention has been paid to curriculum and teaching methods. The essential role of laboratories can be correlated with the fact that engineering is, in general, an applied science that requires, phenomena understanding, very good hands-on skills and involves elements of designing, problem solving, and analytical thinking. Well-designed laboratories during undergraduate engineering degrees can improve these skills for graduate engineers. Renewable energy and energy efficiency technologies are the key for creating a clean energy future. The Renewable Energy Laboratory is both a teaching and research laboratory as well as the main site of the renewable energy and smart grid excellence center [2]. Today, computers are ubiquitous. An integral part of every engineer's toolbox, they are used to do computations, data collection and filtering, simulations and data acquisition, and to share information via the Internet. No engineer today could imagine doing his or her job without one. Yet, using computers routinely is a fairly recent event, particularly in the laboratory. Students should not just try to learn, but must try to explore using the available equipment, usually by themselves. For this purpose, they should develop a scientific and investigatory thinking. This can be translated to the following ones by Eugenia Etkina: ability to represent a process in multiple ways, ability to design an experimental investigation, ability to collect and analyze experimental data, ability to construct and modify explanations and also ability to evaluate all of the above [4]. Engum et al. made a comparative study on using a virtual lab versus a "real" conventional lab. The study revealed that students who experienced both the real lab and the virtual lab could adequately demonstrate the required skills, however, the students preferred performing the real lab rather than the virtual lab. Engum suggested that a combination of the two methodologies may enhance the students' satisfaction and skills acquisition level [5]. The innovative teaching method presented in this article combines two approaches: individual, independent, physical work with the equipment, and virtual results analysis, drawing conclusions and testing on the computer. There are 6 experiments currently set up in the teaching lab (see Fig. 1), including 3 on photovoltaic energy, 1 on wind power and 2 on water energy. The equipment includes: experiment kits for photovoltaic energy with grid connection capabilities; an experiment kit for wind power; a customized experiment kit and equipment for water energy experiments; electronic loads and photovoltaic panel simulators; a charge controller and a deep cycle rechargeable battery; 1.6 kW industrial photovoltaic system; and a 300 W wind turbine. The lab is also used for renewable energy demonstrations.

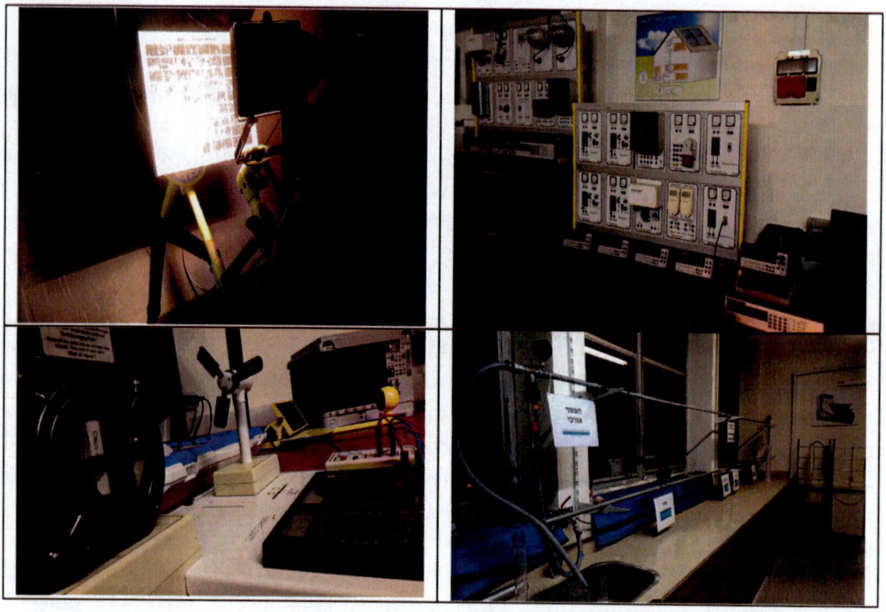

Fig. 1. Experiments kit (up) Photovoltaic energy (down) Wind power and Water energy

The teaching constraints under COVID19 required the provision of online teaching methods. All the theoretical background and the exams are computerized on a "Moodle" Platform [6]. In the experiments, the students must first read theoretical background. After they finish reading the theoretical part, they need to answer an entrance examination about the subject. All the questions are taken from a database of questions, and are selected randomly. Passing the exam shows that the knowledge was acquired properly. Upon completion of the exam, the first stage of the experiment appears in the system. When the students complete the first stage, the "Moodle". Automatically continues to the second stage of the experiment and so on. All the stages in the lab provide the students with the materials they need to finish them, meaning that all stages in the experiment are escorted by "Google Docs" sheets which show dynamic graphs and can be supervised by the lab instructors in a real-time. The students have to analyse the graphs they received during the stages and understand exactly how the graphs were built, and what formula can describe the graphs in each stage. At the end of the experiment, when all stages are done, the "Moodle" lets the students take a final exam, which includes questions about the experiment results that the students received. At the end of the exam the students are automatically graded with no report needed, and the data of each student is saved in the "Moodle" for analysis of the results later on by the instructors [4].

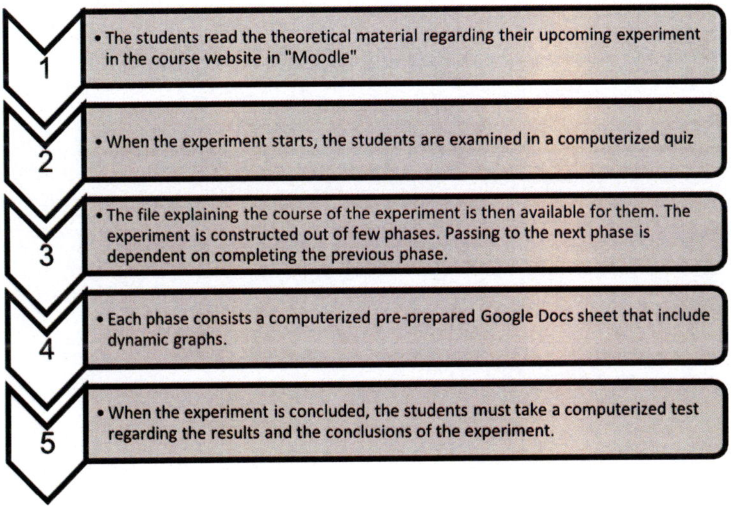

Fig. 2. The computerized lab

2.1 The "Moodle" Platform

The laboratory is a "paper free" laboratory. Prior to the actual experiment, the students read the theoretical material regarding their upcoming experiment in the course website in "Moodle" (see Fig. 2). When the experiment starts, the students are examined in a computerized quiz, after they finish reading and assume to understand the theoretical material. When the quiz is finalized and finished, the file explaining the course of the experiment is then available for them. The experiment is constructed out of few phases.

Passing to the next phase is dependent on completing the previous phase. Each phase consists a computerized pre-prepared Google Docs sheet that include dynamic graphs (see Fig. 3), and can be supervised by the instructor. When the experiment is concluded, the students must take a computerized test regarding the results and the conclusions of the experiment. This method is self-adjusting to the pace of each group and insures the understanding of each phase before proceeding to the next phase. The system makes sure that the material, the students experimented on, will be well understood and is considered a level adjusting system (see Fig. 4).

Remote-tracking after the changes in the dynamic graphs and the results obtained by students performing the experiment, allows personal attention by the instructor and the development of individual learning skills by the student.

When the students finish all phases of the experiment, including the last computerized test, they basically finish the entire experiment and get an automatic grade for it, with no report needed. The data of each student is saved and can be analysed at a later stage.

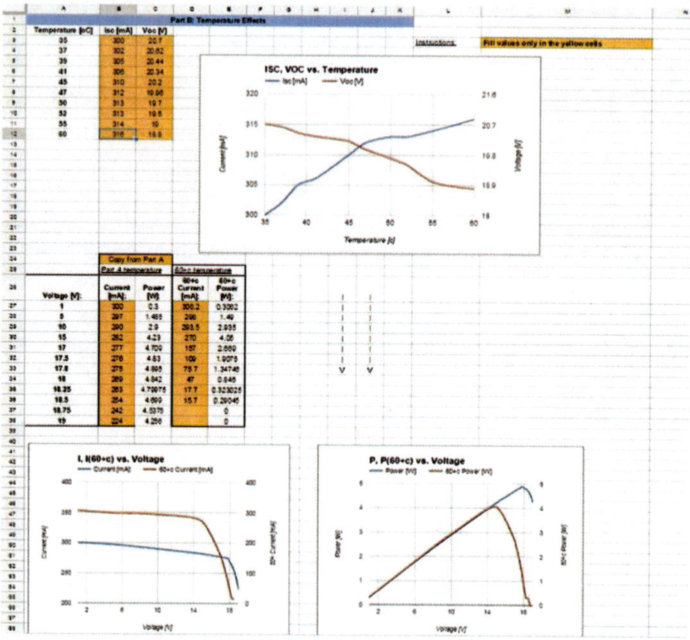

Fig. 3. The Google Docs at Moodle platform

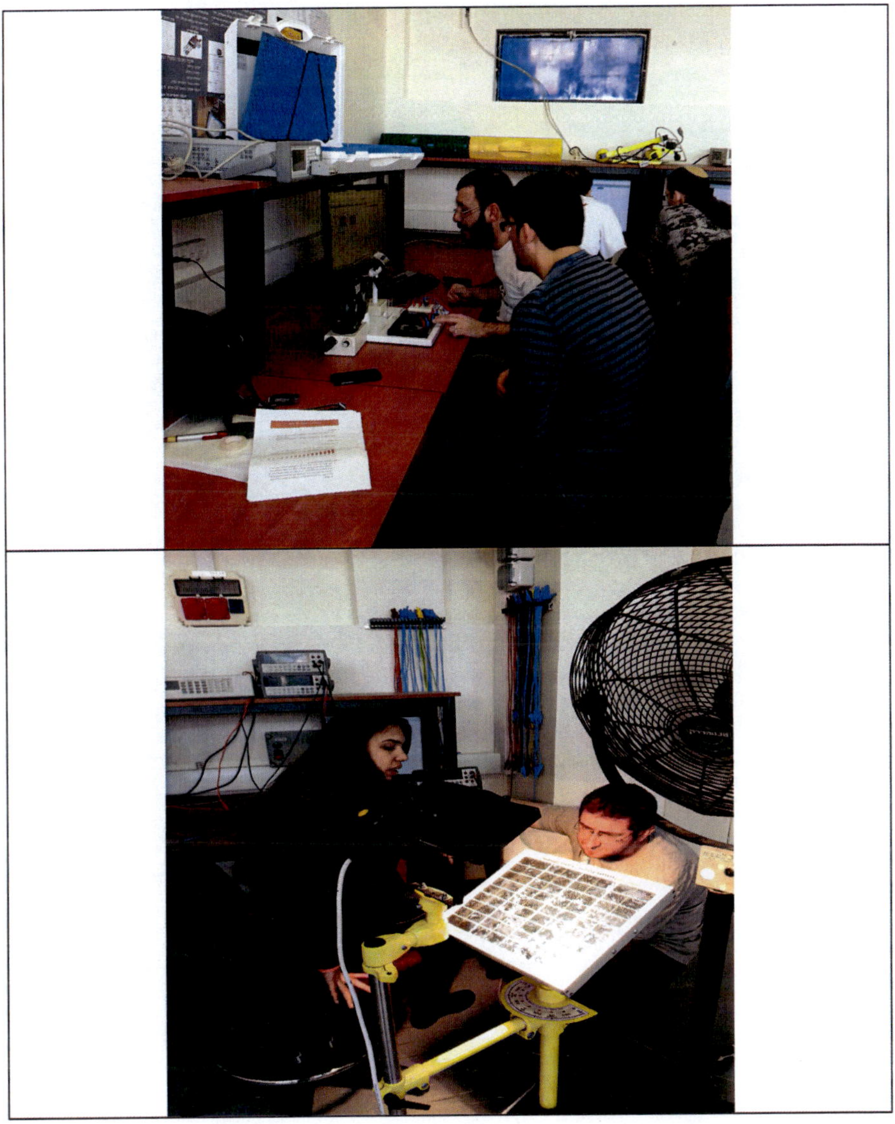

Fig. 4. Independent experience of students to perform the experiments in the laboratory, (a) wind energy (b) solar experience

2.2 Online Renewable Energy Laboratory

During the global epidemic COVID19 there were closures, in which students could not get to the academic institution and perform physical experiments in the laboratory. The laboratory allowed online learning using the zoom (see Fig. 4) and also analysis of experimental results when the laboratory engineer alone, is in a physical laboratory, performs experiments with a camera and the students in their home in front of the computer and analyze the results (Fig. 5).

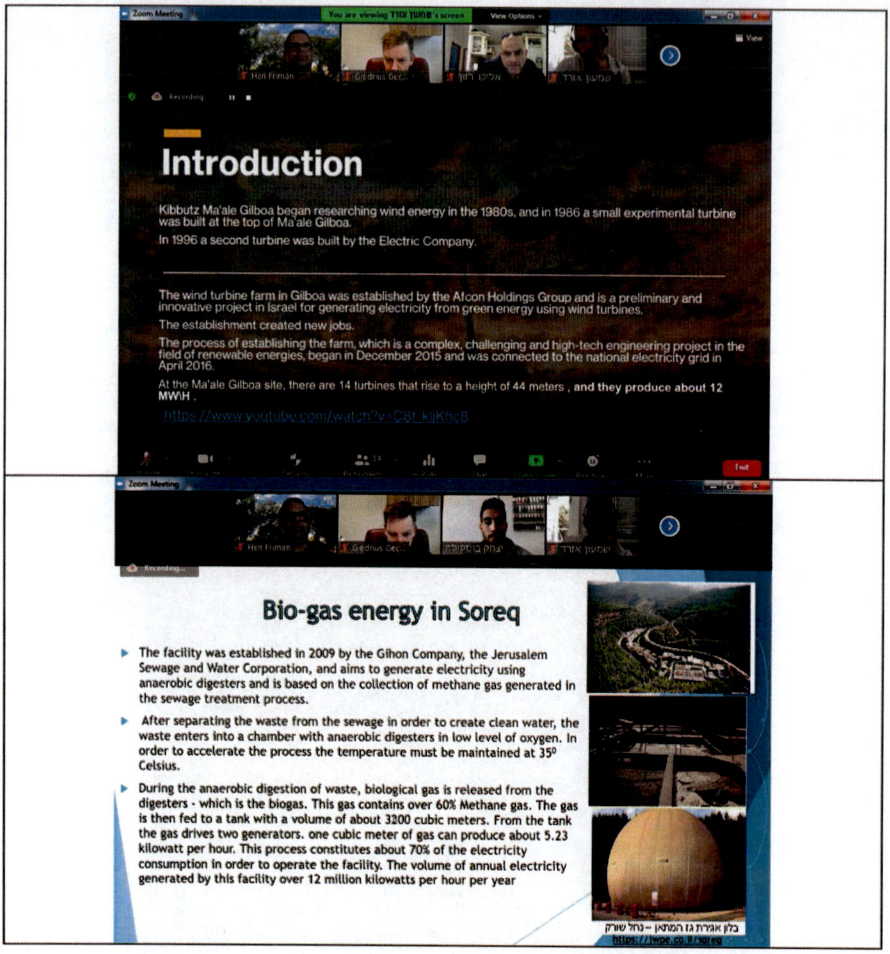

Fig. 5. Online teaching, small groups using ZOOM software, (a) wind energy (b) bio-gas energy

3 Conclusions

The method shows real ability and success, and higher rates of satisfaction. It seems that the laboratory is very popular among students of electrical engineering. To meet the demand, additional laboratory groups are opened every semester. Recently the Renewable Energy and Smart Grid Excellence Centre purchased new experiment sets in order to open an advanced laboratory, which will deal with renewable energy in the areas of smart grid, fuel cells, etc. This paper provides evidence that the innovative laboratory method of teaching that combines the two methodologies of a real lab and a virtual lab, is an effective method. The method improved: individual, independent work and other skills that required by Electrical Engineer. Therefore, it is recommended to increase Electrical Engineer students' participation in Renewable Energy Laboratory. And furthermore, the methodology of this laboratory can be implemented in other laboratories in other disciplines.

References

1. Feisel, L.D., Rosa, A.J.: The role of the laboratory in undergraduate engineering education. J. Eng. Educ. **94**, 121–130 (2005)
2. Wankat, P.C.: Analysis of the first ten years of the journal of engineering education. J. Eng. Educ. **93**(1), 13–21 (2004)
3. Ma, J., Nickerson, J.V.: Handson, simulated, and remote laboratories: a comparative literature review. ACM Comput. Surv. **38**(3), 7 (2006)
4. Etkina, E.: Developing and Assessing Scientific Abilities, Graduate School of Education. Rutgers University (2004). http://paer.rutgers.edu/PT3
5. Engum, S.A., Pamela, J., Fisher, L.: Intravenous catheter training system: computerbased education versus traditional learning methods. Am. J. Surgery. **186**(1), 6774 (2003)
6. Friman, H., Matsliah, N., Beck, Y.: Renewable Energy Lab at The Faculty of Electrical Engineering. In: INTED2016 Proceedings (2016). https://doi.org/10.21125/inted.2016.0149

Modeling Students' Learning Performance and Their Attitudes to Mobile Learning

Malinka Ivanova[1(✉)], Tatyana Ivanova[1], Valentina Terzieva[2], and Katia Todorova[2]

[1] Technical University of Sofia, 8 Kl. Ohridski Blvd, Sofia 1000, Bulgaria
{m_ivanova, t.ivanova}@tu-sofia.bg
[2] Institute of Information and Communication Technologies, Bulgarian Academy of Sciences, Bl. 2, Acad. G. Bonchev Str., Sofia 1113, Bulgaria
{valentina.terzieva, katia.todorova}@iict.bas.bg

Abstract. The paper presents an exploration of the role of mobile technology for the realization of personalized learning and for improvement the students' learning performance within an intelligent educational environment. The two of the created predictive models show the patterns and anomalies in students' learning behavior and their learning performance. The utilized supervised machine learning algorithms: Random Forest, ID3, Naïve Bayes, Deep learning, k-NN are evaluated, and the results points out that the most suitable algorithms for solving these classification tasks are decision tree-based Random Forest and ID3. A multi-layer perceptron is used for predicting the students' group learning performance at whole.

Keywords: Intelligent education · Learning performance · Mobile learning · Machine learning · Neural networks

1 Introduction

Nowadays, our students use their smart mobile devices for different purposes, including for educational ones: to check their schedule, learning tasks, to listen lecture notes, to conduct laboratory practices assignments, to perform assessment activities. Mobile technologies are constantly evolving, which affect easier and ubiquitous access to learning management systems and educational platforms, allowing the contemporary learner to be more adaptive and flexible [1, 2]. Learning in a smart learning environment is performed in a meaningful and effective way through using various computing technologies. It encourages an interactive and creative educational process by providing adapted learning resources and instant and customized feedback to learners according to their specific learning objectives. Furthermore, using learning analytics, learning performance and effectiveness are enhanced. The learning performance is defined as an individual feature, which could be facilitated through various technologies, including mobile as well as by appropriate learning scenarios and strategies [3]. Achieving good learning performance leads to successful learning paths and high outcomes, which is beneficial for both – the students and teachers [4]. The support of formal education with formal and/or informal learning tasks is within the scope of the

M. E. Auer and T. Tsiatsos (Eds.): IMCL 2021, LNNS 411, pp. 646–656, 2022.
https://doi.org/10.1007/978-3-030-96296-8_58

modern intelligent educational systems, which lead to a high quality of the educational process. Modern learning solutions corresponding to the personal adjustments of the students are under extensive exploration [5], and the role of mobile technologies should be precisely outlined. Widely used learning management systems store a large amount of data about learners and the learning process, so their analysis can lead to valuable and useful decisions about how to conduct tutoring to achieve better results for many learners [6]. Learning performance modeling is in the scope of various research works, showing different techniques, including the use of machine learning algorithms for conducting analysis and predictions [7, 8].

The paper aims to present an approach for modeling students' learning performance based on an exploration of their attitudes and motivation for the utilization of mobile technologies. The key point is an examination of students' habits, learning behavior, and background in formal and informal educational settings. Our hypothesis for research is: the personal characteristics and prior knowledge of the students, as well as the use of mobile technologies, are significant components, which influence learning performance in an intelligent educational system. Considering these leaning-related characteristics of the individual or groups of students, and based on different learning analytics, the intelligent learning environment can propose adaptive content, personalized learning paths, etc.

2 Research Methodology

The used methodology consists of three procedures:

(1) Exploration about the currently published and indexed in Scopus scientific articles through bibliographic and visual methods, which outlines the conceptual framework regarding research in the fields of learning performance and mobile technologies. The constructed bibliometric map shows the connection among the most utilized keywords in the papers. It leads to conclusions concerning the related research areas, symbiosis of research topics, directions for examination, and tendencies. Then, a more detailed analysis of the best conceptual and practical models is done.

(2) A survey tool for the students is prepared and distributed online to gather data about their personal usage of mobile technologies for learning purposes. The tool is constructed with three groups of questions that counting out the habits regarding the students' usage of mobile devices, their requirements for learning personalization, and main factors for improving the learning performance. The results are summarized and prepared for further analysis.

(3) Artificial intelligence techniques from machine learning and neural networks are applied for modeling the students' learning performance based on collected survey data. Here, the method typical for machine learning for creating the analytical and predictive models is applied. It consists of: data gathering, data pre-processing according to the required format of the analytical environment RapidMiner Studio, choice making regarding the significant features that should be studied, applying algorithms of machine learning and neural networks, creation of models, analyzing the accuracy of the created models.

3 Literature Review

For outlining the interest and the recent contributions to the topic, a bibliometric method is applied through construction of bibliometric maps [9]. For this purpose, bibliographic data are collected from Scopus scientific database on 17 June 2021 according to the input query with the keywords *learning performance*, *mobile technology*, *mobile learning*, and *mobile*. The search is done within article title, abstract, and keywords (so-called items), and the results are sorted, taking into account the publication relevance. The created bibliometric maps with VoSviewer are distance-based, where the similarity between items is depicted with the parameter *distance*. Smaller distance leads to a bigger number of co-occurrence of the keywords in documents and a stronger connection between them. The label view shows the importance of the items through different circles and label sizes. Related items are grouped in clusters, and every cluster is visualized in a different color.

The first query consists of the keywords *learning performance* and *mobile technology* as 47 documents are found. The constructed bibliometric map is presented on Fig. 1, where can be seen the relationship among the keywords: *learning performance*, *personalized learning*, *mobile technology*, *learning style*, *higher education*, *mobile learning*, and *mobile devices*. The items are grouped into five clusters.

The second bibliometric map is created according to the keywords *learning performance* and *mobile learning* to see the strong connection between these keywords and the similarity to other keywords (Fig. 2). The query results in 154 documents. The map points out that *mobile learning* and *learning performance* are grouped in the same cluster together with keywords like *cognitive style*, *learning interest*, *learning motivation*, *ubiquitous learning*, *mobile-based assessment*.

The third map aims to examine in aspect that is more detailed the connection between the keyword *learning performance* and the general keyword *mobile*. The returned number of documents is 395, and the constructed bibliometric map is presented on Fig. 3. The items are grouped in eight clusters; the keyword *learning performance* is in the same cluster with the following items: *behavior analysis*, *cognitive style*, *informal learning*, *learning motivation*, *personalized learning*, *self-regulated learning*. Also, the connection of the keyword *learning performance* with the keywords: *adaptive learning*, *inquiry-based learning*, *ubiquitous learning*, *context-aware learning*, *game-based learning* from other clusters is also observed.

The summarization of these findings draws the strong connection between research related to learning performance and mobile technology as well as to personalized and adaptive learning.

Learning performance in scientific literature is explored in a wide variety of learning contexts as a significant measure for students' achievements. More detailed analysis for the time interval from 2010 to 2020 year shows the growth of the scientific production: 13 documents for 2010, 23 for 2015, and 58 for 2020. The most of the papers are published in the following scientific journals and books: Computers and Education (19), Educational Technology and Society (16), Lecture Notes in Computer Science (15), Interactive Learning Environments (14), International Journal of Mobile Learning and Organization (8).

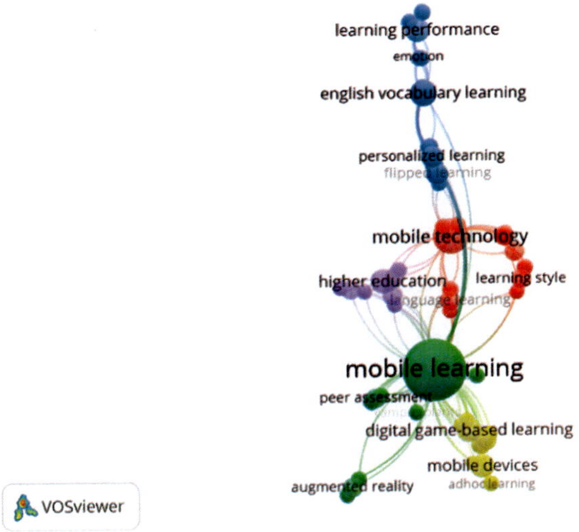

Fig. 1. The constructed bibliometric map concerning the keywords *learning performance* and *mobile technology*

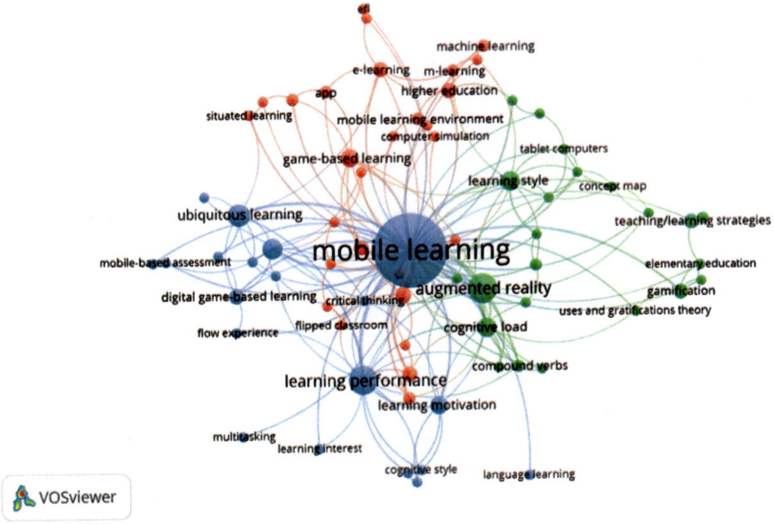

Fig. 2. The constructed bibliometric map taking into account the keywords *learning performance* and *mobile learning*

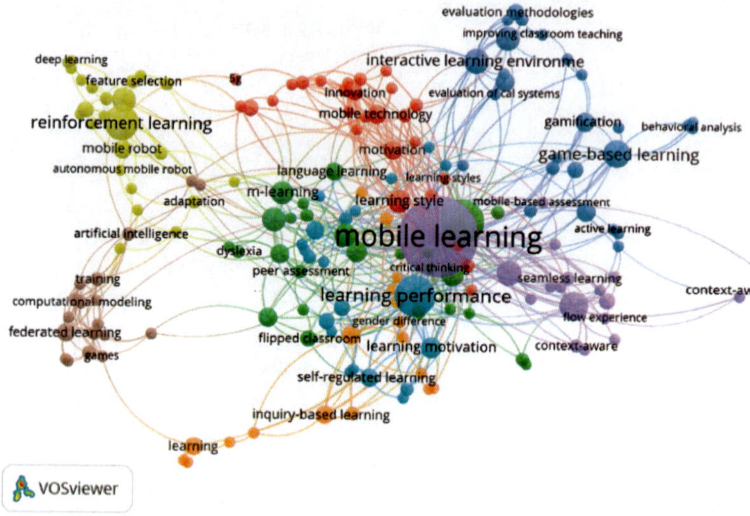

Fig. 3. The constructed bibliometric map according concerning the keywords *learning performance* and *mobile*

According to the authors' country, many papers have authors from Taiwan (140), China (63), the USA (29), Japan (23), Thailand (21), Hong Kong (15), Australia, Greece, Malaysia, and Turkey (9). Examination about the documents by subject area points out that 38.1% of them are published in the context of computer science, 26.4% - social science, 16.7% - engineering, 5% - mathematics, 3% - decision sciences, 2% - art and humanities.

After the conducted bibliometric study of the published scientific production, the achievements obtained in some papers are discussed. This reveals various models for learning performance improvement with support of mobile technology. Wang et al. explore the relationship among mobile learning, students' cognitive load and their learning performance [10]. The findings point out that complex interactions with mobile applications lead to higher cognitive effort and improved learning performance. The effectiveness of mobile application usage in traditional classroom activities is explored in [11]. It proves the assistive role of mobile technology for gaining better students' learning performance in classroom settings. A review in the form of a meta-analysis regarding the influence of mobile technology on learning performance is conducted by Talan [12]. The author concludes that this influence is not changed for a given period and educational level, but it is different for different courses/subjects. Facilitating English learning through the usage of a mobile application for learning motivation and learning performance improvement is discussed in [13]. The students are divided into two groups – with and without an applied method for self-regulation learning. The results show beneficial advantages in motivation and learning performance for students from the self-regulation group.

This extensive analysis indicates that despite the published models for improving the students' performance through mobile technology and intelligent educational environments still, this topic is under progressive exploration as the researchers are looking for local or more general solutions.

4 Experiment and Predictive Modeling

The created predictive and analytical models are based on data collected from the developed survey tool. It is distributed online to 27 students included in a professional bachelor's degree program. The survey questions are picked in three groups in order to gather the students' opinion regarding how they use the mobile devices for learning purposes, whether mobile technology could support their personalized learning and whether the mobile technology is a factor for their learning performance improvement. The results are summarized and analyzed applying supervised machine learning techniques.

The first model analyses the students' habits to use mobile devices for learning and learning performance improvement. It could predict the usage of mobile technology for this purpose. The included predictors in the model are following:

- The *frequency* of mobile device usage for learning with values: {very often, often, rarely, very rarely, na};
- Usage of *email* for communication with educators and students with aim their learning and learning performance to be improved as the possible values are: {yes, no};
- Usage of mobile devices for access to *read lectures* with values: {yes, no};
- Mobile devices usage for *reading the instructions for laboratory practices* with values: {yes, no};
- Taking *assessment tests* through mobile devices with values: {yes, no};
- *Live lectures listening* through mobile devices with values: {yes, no};
- Preference to use *computer* instead mobile device with values: {yes, no};
- Usage of mobile devices in *university* for learning purposes with values: {yes, no};
- Usage of mobile devices in *home* for learning purposes with values: {yes, no};
- Usage of mobile devices *somewhere else* outside the university with values: {yes, no};
- Usage of mobile devices when the student *travels* for learning purposes with values: {yes, no}.

The classification task for solving is related to the student classification in three groups:

- group A – the students who often and very often use mobile devices for different learning tasks conductance,
- group B – the students who rarely use mobile devices for performing some learning tasks, and
- group C – the students who very rarely use mobile devices for the conductance of one or two learning tasks and student who do not prefer to use mobile devices, because like the capabilities of computers.

For solving this classification task 5 machine learning algorithms are applied: Random Forest, ID3, Naïve Bayes, Deep Learning, and k-NN. 70% of the data set is used for training and 30% for testing. The classifiers are compared and the results are presented in Table 1. It can be said that for this study context with higher accuracy is the Random Forest classifier. The predictions are shown on Fig. 4, where 58% of the students are predicted to be from group A, 32% of them from group B, and 10% from group C. The anomalies are discovered in 15% of the preliminary classified students from group B as 10% of them are predicted to go in group A and 5% to go in group C.

Table 1. Comparison of classifiers

Parameter	Random Forest	ID3	Naïve Bayes	Deep Learning	k-NN
Accuracy	89.47%	78.95%	84.21%	63.16%	73.68%
Kappa	0.829	0.658	0.748	0.422	0.580
Absolute error	0.330 ± 0.208	0.232 ± 0.407	0.161 ± 0.338	0.477 ± 0.242	0.466 ± 0.120
Relative error	32.97% ± 20.81%	23.16% ± 40.66%	16.14% ± 33.77%	47.74% ± 24.17%	46.55% ± 11.97%
RMSE	0.390	0.468	0.374	0.535	0.481
Correlation	0.935	0.641	0.907	0.789	0.873

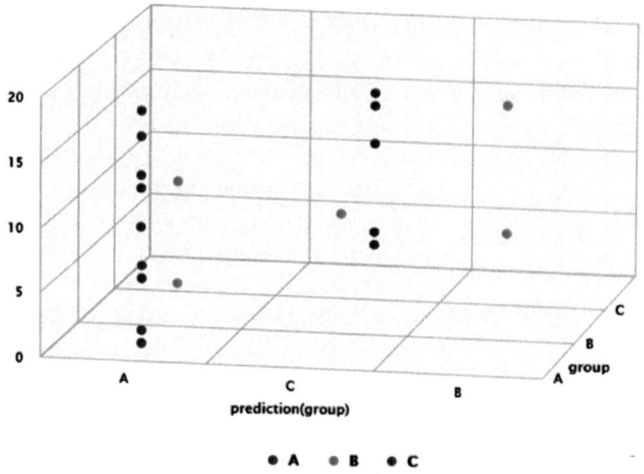

Fig. 4. Preliminary formed and predicted students' groups

The second model presents an analysis concerning the capability of mobile technology for learning personalization and learning performance improvement. The created model by machine learning techniques could predict whether the students' performance will be improved or not. The main predictors included in the model are:

- *Comfortability* of the mobile devices for access to learning management system and this leads to improvement of learning performance and realization of personalized learning as the values are: {yes, no};
- *Flexibility* to learn from everywhere and at every time with values: {yes, no};
- Learning *on demand* when the student needs to learn with values: {yes, no};
- Possibilities for *learning performance* improvement through easy access to any information sources and search engines with values: {yes, no};
- Possibilities for satisfying *personal learning interests* according to the current students' achievements and learning goals with values: {yes, no}.

According to this classification task that should be solved, the students are grouped into three groups:

- group A includes the students who consider that mobile technology facilitates the realization of their personalized learning and improves the learning performance;
- group B consists of students who think that mobile technology cannot satisfy personal learning needs or cannot improve learning performance;
- group C includes the students whose opinion is that the mobile technology cannot be used as a supportive tool for the realization of personalized learning and learning performance improvement.

The classification task is solved as the above-mentioned five machine learning algorithms are applied to the data set. The results are summarized in Table 2. With the higher accuracy are the decision tree-based algorithms: Random Forest and ID3.

Table 2. Parameters of classifiers

Parameter	Random Forest	ID3	Naïve Bayes	Deep Learning	k-NN
Accuracy	94.74%	94.74%	89.47%	84.21%	73.68%
Kappa	0.876	0.874	0.784	0.632	0.275
Absolute error	0.210 ± 0.197	0.053 ± 0.223	0.142 ± 0.168	0.447 ± 0.121	0.220 ± 0.329
Relative error	20.97%	5.26%	14.24%	44.69%	22.04%
RMSE	0.288	0.229	0.221	0.463	0.396
Correlation	0.957	0.963	0.916	0.942	0.914

The predictions are presented on Fig. 5 as there is no deviation between the preliminary formed and predicted A student group. Five percent of the students from group C are predicted to be a part of group B.

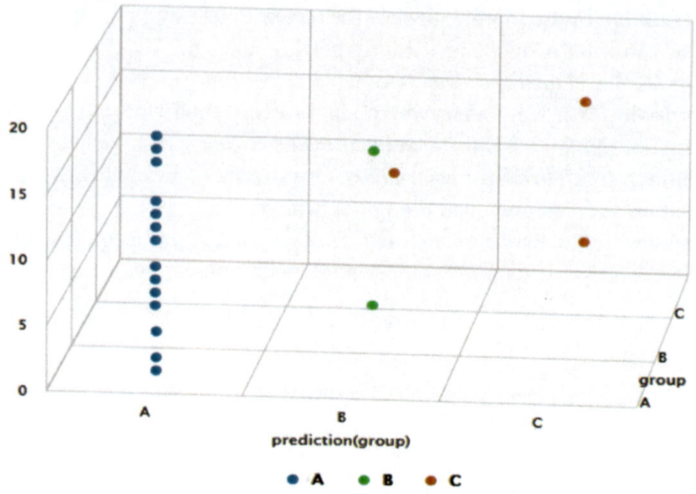

Fig. 5. Preliminary formed and predicted students' groups

The aim of the third model is to predict the learning performance of the whole examined student group, consisting of 27 students. For this purpose, a neural network is constructed with 15 input parameters: frequency of mobile device usage; usage of mobile devices for email communication, for access to read lectures, for reading the instructions for laboratory practices, for live lectures listening; preference to use computer instead of mobile device; usage of mobile technology in university, home, somewhere else, at traveling; comfortability of the mobile devices for access to learning management system; the flexibility to learn from everywhere and at every time, learning on demand; possibilities for learning performance improvement; possibilities for satisfying personal learning interests and one output – the learning performance. The utilized algorithm is Neural Net that characterizes with training a feed-forward neural network with a backpropagation channel (multi-layer perceptron). The optimal structure of the neural network is found after experimentation with the number of hidden layers and the number of neurons at each layer. The results regarding the evaluation of the model performance are presented in Table 3 (Fig. 6).

Table 3. Neural network performance evaluation

Number of neurons and RMSE							
One hidden layer	5 0.336	6 0.328	7 0.346	8 0.351	**9** **0.318**	10 0.318	11 0.322
Two hidden layers	4,4 0.384	5,3 0.328	5,5 0.396	5,6 0.406	6,6 0.348	7,7 0.338	8,8 0.351
Three hidden layers	5,5,5 0.411	7,7,7 0.411	9,9,9 0.409	7,5,3 0.409	5,5,3 0.409	13,9,7 0.409	15,15,13 0.408

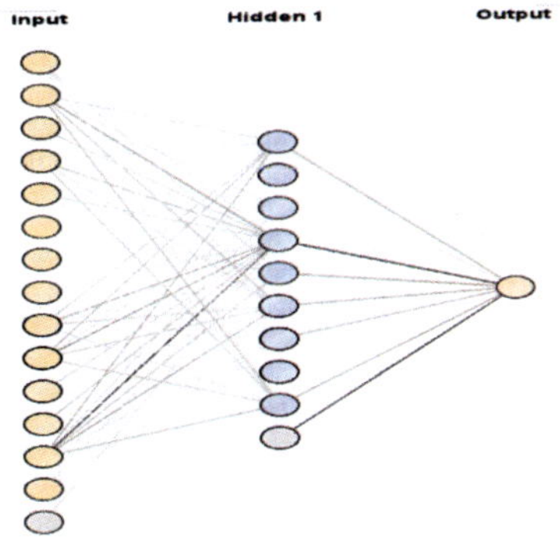

Fig. 6. Constructed neural network

5 Conclusion

In our days, personalization of learning and tutoring is very significant for achieving high-quality results in e-learning. On the other hand, working individually with every student is difficult. In this context, classifying students in groups having similar learning goals, needs, or problems will give possibilities to apply group-based personalized tutoring.

This paper shows how machine learning and neural network-based algorithms can be used in the context of an intelligent educational environment for predicting learner's performance. Also, these approaches are applied for students grouping based on leaning-related characteristics. The experiments show that the best classification algorithms for our purposes are Random Forest and ID3. The conclusion about the usage of neural networks is that a one-layered neural network with 9 or 10 neurons in the hidden layer is the best solution for solving the predictive task in our learning context.

Acknowledgments. This research is supported by the Bulgarian FNI fund through the project "Modeling and Research of Intelligent Educational Systems and Sensor Networks (ISOSeM)", contract КП-06-Н47/4 from 26.11.2020.

References

1. Klimova, B., Poulova, P.: Mobile learning in higher education. Adv. Sci. Lett. **22**(5–6), 1111–1114 (2016). https://doi.org/10.1166/asl.2016.6673
2. Krull, G., Duart, J.: Research trends in mobile learning in education: a systematic review of articles (2011–2015). International Review of Research in Open and Distributed Learning **18**(7) (2017). http://openaccess.uoc.edu/webapps/o2/bitstream/10609/93100/1/researchtrends.pdf
3. Costa, C., Cardoso, A.P., Lima, M.P., Ferreira, M., Abrantes, J.L.: Pedagogical interaction and learning performance as determinants of academic achievement. Procedia. Soc. Behav. Sci. **171**, 874–881 (2015)

4. AdrianChin, Y.K., JosephNg, P.S., Shibghatullah, A.S., Loh, Y.F.: JomDataMining: learning behavior affecting their academic performance, really?. In: 2019 IEEE 6th International Conference on Engineering Technologies and Applied Sciences (ICETAS), pp. 1–4 (2019). https://doi.org/10.1109/ICETAS48360.2019.9117453
5. Zhiyenbayeva, N., Belyanova, E., Petunina, I., Dmitrichenkova, S., Dolzhich, E.: Personalized computer support of performance rates and education process in high school: case study of engineering students. Int. J. Eng. Pedagogy **11**(2), 135–153 (2021)
6. Kosasi, S., Vedyanto, Kasma, U., Ayu Eka Yuliani, I.D.: The mediating role of learning analytics to improve student academic performance. In: 2020 2nd International Conference on Cybernetics and Intelligent System (ICORIS), pp. 1–6 (2020). https://doi.org/10.1109/ICORIS50180.2020.9320802
7. Vital, T.P., Sangeeta, K., Kumar, K.: Student classification based on cognitive abilities and predicting learning performances using machine learning models. Int. J. Comput. Dig. Syst. **10**(1), 63–75 (2021). http://dx.doi.org/10.12785/ijcds/100107
8. Zacharis, N.: Predicting student academic performance in blended learning using artificial neural networks. Int. J. Artif. Intell. Appl. **7**, 17–29 (2016)
9. Van Eck, N.J., Waltman, L.: Text mining and visualization using VOSviewer. ISSI Newsl. **7**(3), 50–54 (2011)
10. Wang, C., Fang, T., Miao, R.: Learning performance and cognitive load in mobile learning: Impact of interaction complexity. J. Comput. Assist. Learn. (2018). https://doi.org/10.1111/jcal.12300
11. Lin, Y.-T.: When mobile technology meets traditional classroom learning environment: How does it improve students' learning performances? In: Learning Environments: Emerging Theories, Applications and Future Directions, pp. 143–158 (2016)
12. Talan, T.: The effect of mobile learning on learning performance: a meta-analysis study. Educ. Sci. Theory Pract. **20**(1), 79–103 (2020). https://doi.org/10.12738/jestp.2020.1.006
13. Chen, C.-M., Chen, L.-C., Yang, S.-M.: An English vocabulary learning app with self-regulated learning mechanism to improve learning performance and motivation. Comput. Assist. Lang. Learn. **32**(3), 237–260 (2019). https://doi.org/10.1080/09588221.2018.1485708

Using Mobile Applications to Interact with Drones: A Teachers' Perception Study

Tryfon Sivenas(✉) and George Koutromanos

Department of Primary Education, National and Kapodistrian University of Athens, Athens, Greece
{sivenastrif,koutro}@primedu.uoa.gr

Abstract. The purpose of this study was to investigate the perceptions of 60 in-service teachers' regarding the use of drones through mobile applications. After a brief introduction to drones' applications, practices and operation through mobile apps, the sample completed tasks that required assembling, programming, simulating and flying drones. Data were collected via an online questionnaire using open-ended questions adapted from the Technology Acceptance Model, namely perceived ease of use, perceived usefulness and facilitating conditions. The results showed positive perceptions on using drones through mobile applications. Regarding perceived ease of use, teachers deemed smartphone to drone pairing, programming and simulating mobile apps, and flying the drone to be easy tasks. Regarding perceived usefulness, teachers believed that the use of drones through mobile applications will be useful for them and for their pupils' learning and will help students to develop a number of skills. Regarding facilitating conditions, teachers indicated the need for teacher training on the pedagogical use of drones, assistance on visual programming and infrastructure support. This study has a number of implications regarding the use of drones through mobile apps in teaching as well as the need for further research on the field of educational aerial robotics.

Keywords: Mobile applications · Drones · Teacher perceptions

1 Introduction

One of the emerging technologies of the Fourth Industrial Revolution (4IR) [1] is the Unmanned Aircraft System (UAS), also known under several other names (e.g., "Aerial Robot" and "drone") [2]. They fall under the wider category of Aerial Robotics [3], while their properties, such as built-in camera, GPS and a number of sensors for data collection, have resulted in their increasing use in several fields, among which education [4]. Thus, various educational drones (e.g., Ryze Tello for education, Makeblock Airblock for STEAM) are available today for educational purposes and are aimed at pupils [5].

The operation of drones can nowadays be performed through various mobile applications on smart devices (i.e., smartphones and tablets). These applications enable visual programming (e.g., Makeblock, Droneblocks, Tello EDU, Tynker) and flight simulation [6]. The ease of use and variety of these mobile applications enable teachers

M. E. Auer and T. Tsiatsos (Eds.): IMCL 2021, LNNS 411, pp. 657–668, 2022.
https://doi.org/10.1007/978-3-030-96296-8_59

to bring their students into contact with aerial educational robotics and mobile learning without requiring prior skills in high-level programming [7], and to incorporate drones into activities focused on investigation, cooperation and development of various skills [8]. In general, there has been a growing interest in the use of drones in educational robotics and STEAM [5–9]. This is facilitated by the increasing use of mobile applications for drones in combination with the low cost and the new characteristics of educational drones (see next section). Among the key elements needed to successfully implement this use is the acceptance of drones as educational tools by the teachers.

Research has shown that the successful use of technology in schools depends on the factors that significantly influence teachers' acceptance, such as attitudes towards the technology, beliefs and perceptions regarding the ease of use and usefulness of technology, as well as perceptions regarding the conditions that facilitate their use of technology. However, no previous studies have examined these factors in the context of the use of drones in education. Therefore, it is important to know the perceptions that teachers have towards using drones through mobile applications in teaching and learning so there can be appropriate interventions in future classrooms. In order to examine teachers' technology acceptance, various social psychology theories and models have been applied. The most used ones are: the Theory of Reasoned Action (TRA) [10], the Theory of Planned Behavior (TPB) [11], the Technology Acceptance Model (TAM) [12] and its extensions (TAM2, TAM3). TAM is most popular in the field of education. This model provides an important theoretical framework for the examination of teachers' perceptions towards using drones through mobile applications.

Given that drones offer a new spectrum of aerial activities in both educational robotics and mobile learning, the current study aims to fill in the research gap and investigate the perceptions of in-service teachers regarding the utilization of drones through mobile applications in teaching and learning. To address this aim, this study is guided by the following research questions:

(1) What are teachers' perceptions regarding the ease of use of drones through mobile applications?
(2) What is teachers' perceived usefulness of using drones through mobile applications in teaching and learning?
(3) What are the factors that facilitate teachers' use drones through mobile applications?
(4) What are teachers' perceptions on the technological affordances of drones that will enrich and contribute to teaching and learning?
(5) What types of activities do teachers believe that can be developed via drones and the utilization of mobile applications?

The remainder of the paper is organized as follows: Sect. 2 presents the main characteristics of drones and mobile applications with regard to their control, simulation, and programming, followed by the theoretical framework of TAM in Sect. 3. Section 4 presents the research methodology. Section 5 presents the results, followed by discussion, conclusions, and implications in Sect. 6.

2 Drones and Mobile Applications

Drones are aerial robots with no pilot on board and can fly autonomously or be remotely piloted by a user [13]. In related literature, they can be found under several alternative terms such as "unmanned aircraft system", "aerial robot", "remotely piloted aircraft", "micro aerial vehicle", and "quadcopter" [5]. Educational drones usually come in three sizes: micro, mini, and nano drones [14]. They are available in two forms: commercial off-the-shelf drones, which are ready to fly (RTF), and drone construction kits, which require assembly by the user. The light weight of drones allows their easy transportation. Drones can fly in both interior (e.g., classroom, school gym) and exterior spaces (e.g., schoolyard, or an area where drones can be flown). They also have several fail-safes that protect the user (e.g., from injuring someone and/or themself) as well as the drones themselves (e.g., automatic landing, obstacle avoidance, and in-flight replanning), which makes them easy to use by beginners [9].

Educational drones usually have a camera for image capturing and video recording that provides bird's-eye view; a built-in GPS, which allows them to follow coordinates accurately; and a large number of sensors. Some of them are required for the air-worthiness of drones, while others perform real-time data collection. These data can be: altitude, speed, distance, temperature, barometric pressure etc. [15]. The operator of the drone can see the data in real time or store them for later processing and use. There are three ways of content sharing: peer-to-peer, real-time content transmission (camera streaming), and through access to the user's stored content (e.g., data from sensors, images or videos) from various devices such as computer, smartphone or tablet.

Drone operation is done remotely, without the use of cables, via various means, such as computer, joystick, and gamepad. Furthermore, nowadays it can be operated through smartphone and tablet by means of mobile applications [16]. These are programs that have been created to operate in software of smartphones and tablets (e.g., iOS and Android) [17]. They are often called mobile apps and smartphone apps [18]. The user is required to install a mobile app on his or her smart device and, subsequently, pair it with the drone through the available protocols (Wi-Fi, Bluetooth). Through the mobile apps, users can control the drones, simulate a flight, and program them.

Controlling mobile apps allow users to navigate drones. The drone's user can choose among a vast number of controlling apps which are available on both Google Play Store and iOS Store. They have diverse Graphical User Interfaces (GUI), layouts and colors, allowing control through a virtual gamepad. Through the tapping of virtual on-screen buttons, the user can control the drone from take-off to landing, command the drone to take pictures and record videos, rotate 360° and even return to the take-off point. Navigation and maneuvering is done through two on-screen joysticks located on the virtual gamepad and allow the drones to fly up, down, left and right; hover; turn pitch; yaw; and increase or decrease throttle. In addition, the user can have real-time access to data such as speed, altitude, camera feed, and battery percentage.

Subsequently, another category of mobile apps is focused on drone flight simulation. These apps allow the user to become familiar with the virtual gamepad during a virtual flight. For example, DJI Virtual Flight [19] provides a virtual environment that recreates a natural environment, helping the user to become familiar not only with the virtual gamepad but also with the drone's maneuverability and interactivity while airborne. More specifically, operation is performed through the virtual gamepad as analyzed above, by means of buttons and joysticks. Through this type of applications, the user learns the particularities of operation and becomes familiar with the sensitivity of on-screen joysticks. This is a special feature, since on-screen buttons and joysticks do not provide haptic feedback to the user and, consequently, the user does not know how much force needs to be exerted on the on-screen joystick so that the drone moves smoothly.

Another category of mobile apps allows the programming of the drone through the use of block-based visual programming languages (e.g., Scratch, Blockly, Dronely). Block-based environment is aimed at beginners and allows the development of various programming concepts such as loops, conditions, variables, sequences and functions [20]. For example, through the apps of Tynker [21] and DroneBlocks [22], the user can create a block code by applying a query of commands to the drone to perform an autonomous flight. Subsequently, and before he/she proceeds to an actual flight of the drone, the user can watch the outcome of the program in simulated preview. He or she can see, in a simulated environment, a virtual drone on a scale performing the commands of the block code he/she has created. This allows the user to make any corrections to the code where necessary (i.e., debugging). The code simulator offers a fail-safe to the user, as it allows him/her to experiment with the available programming commands in a virtual environment without worrying about the possibility of a drone-related accident.

To date, research in related literature has been focused on the study of the use of drone-related mobile apps by pupils. For instance, Chou [16] investigated the effects of using drones in the classroom, in a sample of 10 pupils. The pupils programmed the drones using the Tynker app. Results showed increased spatial visualization, sequencing skills and computational thinking. In their study, Voštinár et al. [23] used the Makeblock app to teach programming to secondary education pupils. Results showed increased motivation as well as creative problem-solving. Subsequently, in a sample of 30 secondary education pupils, Tezza et al. [5] used drones to promote further education in STEM fields. The pupils programmed the drones using block-based mobile apps. Results showed active engagement and involvement during the course. In another study by Joyce et al. [24], the apps DroneBlocks and Tynker were used by primary and secondary education pupils in order to be taught geospatial technology fundamentals in the context of STEM education. The results revealed that it was an enjoyable, intriguing way of teaching and that the pupils were actively engaged and not passive observers during the lesson. To the best of our knowledge, there is no other study in related literature that investigates teachers' perceptions regarding the use of drones in teaching and learning through mobile apps.

3 Theoretical Framework

As mentioned above, the aim of this study is to investigate the perceptions of in-service teachers regarding the utilization of drones through mobile applications in teaching and learning. The theoretical framework is based on TAM [12]. This model was selected in the current study because it has been found to be successful in research at explaining teachers' perceptions towards the use of various digital technologies in education [25, 26].

The aim of TAM is to investigate the relationship between attitude, intention and technology use based on two major beliefs: "perceived usefulness" and "perceived ease of use". Attitude is "*an individual's positive or negative feelings (evaluative affect) about performing the target behavior*" [27, p. 984]. Perceived usefulness is "*the degree to which a person believes that using a particular system would enhance his or her job performance*" [12, p. 320]. Perceived ease of use is "*the degree to which a person believes that using a particular system would be free from effort*" [12, p. 320]. From previous literature review studies that focus on technology acceptance, these two variables were found to be the major determinants of teachers' attitudes and intention to use a range of technologies in teaching. In addition, they were found to be the core for a number of revised and extended technology acceptance models [25, 26].

Even though TAM is still a valid model [25, 26], some researchers expanded it in order to provide a better explanation of technology adoption by teachers. Among the external variables which are incorporated into TAM and significantly related to perceived ease of use and perceived usefulness are: self-efficacy, subjective norms, and facilitating conditions of technology use [26]. Facilitating conditions is "*the degree to which an individual believes that an organizational and technical infrastructure exists to support use of the system*" [12, p. 453]. Recent meta-analysis, which was conducted by Scherer et al. [26], found that facilitating conditions were positively related to both teachers' perceived ease of use and perceived usefulness, with stronger effects on perceived ease of use.

The variables of perceived ease of use, perceived usefulness and facilitating conditions have been successfully used as a framework to study mobile learning acceptance by teachers. For example, Huang [28] examined in her study the factors that affect 107 in-service teachers' acceptance of mobile learning in formal teaching. She used extended TAM consisting of nine variables. Results indicated that perceived usefulness and perceived ease of use as well as subjective norms and facilitating conditions were found to influence teachers' behavioral intention to use mobile learning. Subsequently, the study by Sanchez-Prieto et al. [29] examined a sample of 678 pre-service primary education teachers' intention to use mobile learning in teaching. Results showed that teachers' perceived usefulness and perceived ease of use are able to predict their behavioral intention to use mobile learning in teaching. Another study by Kalogiannakis & Papadakis [30] examined 75 pre-service teachers' intention and attitudes to use mobile learning in their teaching. Results indicated that pre-service teachers' perceived ease of use and perceived usefulness have the strongest influence on their intentions to adopt mobile learning. In their study, Islamoglu et al. [31] examined 316 pre-service teachers' attitudes, intention, social influence, facilitating conditions, self-efficacy, anxiety, perceived usefulness and perceived ease of use of mobile learning activities. Results showed that perceived ease of use and social influence had direct effects on teachers' behavioral intention. In another study, Shodipe & Ohanu [32] examined 418 in-service teachers' attitudes, engagements and dispositions towards using mobile learning in teaching. Results indicated that teachers' perceived ease of use and perceived usefulness positively influenced their actual use of mobile learning.

Based on the above, in order to examine teachers' perceptions on the use of drones through mobile applications in teaching and learning, we used the variables of perceived ease of use, perceived usefulness, and facilitating conditions. In the context of this study, we believe that in order for teachers to use drones through mobile applications in teaching and learning, they need to believe that these are easy to use and useful both for their pupils and for them. In addition, they need to feel that they have all the conditions necessary to facilitate their use, such as resources (e.g., Internet connection, availability of drones, smart devices), support to use drones, training, and time.

4 Research Methodology

In order to address the research questions of this study, an exploratory qualitative research design was utilized. A similar research method was applied by previous studies which used theories and models (e.g., TAM, TPB) of acceptance to investigate pre-service and in-service teachers' beliefs and perceptions towards the use of technology in teaching [33–35].

4.1 Participants

A total of 60 in-service teachers, 75% of whom were female, participated in this study. The average age was 36.7 years (SD = 9.01). Of the 60 in-service teachers, 36 (60%) worked in primary education and 24 (40%) worked in secondary education. These teachers were enrolled in postgraduate courses and seminars on ICT in Education at the National and Kapodistrian University of Athens, Greece (Faculty of Primary Education). All teachers voluntarily signed up to participate in this study. Of the 60 teachers, 17 (28.3%) had a teaching experience of five years or less, 35 (58.3%) had a teaching experience from six to 15 years, and 8 (13.3%) of them had a teaching experience of 16 years or more. Of the 60 in-service teachers, 12 (20%) stated they had used drones for personal purposes.

4.2 Procedure

The research was conducted in three phases during the academic year 2020–2021, following all the proposed protective measures against COVID-19. In the first phase, which lasted one day, the teachers attended a section of the seminar focusing on technology, applications and practices regarding drone operation through mobile applications and the pedagogical framework of drone utilization. In the second phase, which lasted two days, teachers initially came into contact with preassembled quadcopter drones (e.g., Ryze Tello, Makeblock Airblock). More specifically, participants installed in their smart devices the mobile applications (for iOS and Android) and familiarized themselves with their operation in an outdoor area of the University and their programming through the visual programming interface and its simulator (e.g., DroneBlocks). In the research's third phase, which lasted three days, participants were asked to create in their devices a scenario for the utilization of the drones, which would focus on the solution of a specific real-world problem (e.g., immediate transfer of medicines to an injured person on a nearby road due to a car accident). The initial requirements of the scenario were: (a) the creation of a block-based code and its execution on the simulator, (b) the demonstration of the scenario to the other teachers, and (c) the flight of the drone in a simulated external environment of the University.

4.3 Instruments and Data Analysis

The collection of data was achieved through various instruments. During the first phase, teachers wrote their opinions about drones using a Padlet. During the second and third phase, the researchers observed how teachers used drones and dealt with any difficulties they encountered in their use. At the end of the third phase, the teachers filled in an online questionnaire consisting of two parts. In the first part, demographic characteristics of the teachers were collected through four questions (i.e., gender, age, teaching experience, use of drones for personal purposes). In the second part, the teachers were asked to list their: (a) perceived ease of use of drones through mobile applications in teaching and learning, (b) perceived usefulness of using drones through mobile applications in teaching and learning, (c) perceived facilitating conditions of using drones through mobile applications in teaching and learning, (d) perceptions regarding the technological affordances of drones, and (e) perceptions regarding the activities that can be developed via drones and the utilization of mobile applications. Questions of perceived ease of use, perceived usefulness and facilitating conditions were based on the theoretical framework of TAM (see Sect. 3). The answers of the questionnaire were coded and categorized into the variables of TAM according to the methodology of Strauss and Corbin [36]. Data were coded and analyzed independently by three researchers of ICT in education.

5 Results

5.1 Perceived Ease of Use

The results of the research showed that teachers were positive about using drones through mobile apps. Regarding the first research question, about the perceived ease of use, three groups were identified. Firstly, teachers believed that drones were easy to be paired with their smartphone; secondly, mobile apps had user-friendly programming and simulating environment; and, thirdly, the drones' control and navigation was an easy task. Concerning the pairing of mobile devices with the drones, teachers mentioned it was very easy and quickly implementable (N = 56). They did not require any help to implement it. For example, a male secondary education teacher mentioned: "*what strikes me as impressive is that, as soon as you pair your phone to the drone, you just fly it; there is no disconnection, the flight is not interrupted, the app does not freeze up, there is no requirement for signup, email confirmation or personal data, it requires no configuration nor does it have a complex setup. You just fly it or, at least, you fly it until it runs out of battery*".

As regards block-based programming environment and the drone simulator, the majority of the teachers did not face any problem. They deemed it very practical and easy to use. Indicatively, a female secondary education teacher mentioned: "*I use many apps every day. Some of them have a clean, friendly and easy-to-use environment while others have so many options, buttons and submenus that it becomes tedious. With DroneBlocks, I get the drone up in the air with only three taps*". In the context of the simulator, a female secondary education teacher mentioned: "*I have no experience in programming and I would normally be afraid to let the drone execute the program I created in three minutes. With the simulator, not only do I feel confident that my program will work but also, in case it does not, it will be a good opportunity to look for the mistake with my pupils*".

Regarding the operation and navigation of the drones through mobile apps, the analysis of the results indicated that the majority of the teachers (N = 53) did not face any problems there either and that the entire procedure was easy. Some of the teachers' responses were the following: "*It is like a game. It is awesome to get it into the air, fly it and land it... It is so easy to learn*", "*you can view the landscape from above [bird's-eye view] and fly [the drone] to places you could otherwise not see, and it is so easy*". A number of teachers (N = 6) referred to the drones' familiar operating environment through gamepads. Indicatively, a male primary education teacher mentioned: "*The buttons and the operation are just like those of a gaming console. Its operation is so easy that I believe my pupils will have no problem operating it since most of them basically know how to use a gamepad*". However, observation of the teachers during the drones' flight revealed that, in the first few minutes of operating the drones, a small number of them (N = 7) had problems of familiarization with the drones' orientation while in flight.

5.2 Perceived Usefulness

As regards the second research question, the results showed that teachers expressed positive perceptions towards the usefulness of drones through mobile applications. Firstly, the majority of teachers (N = 51) believes that the use of drones through mobile applications will be useful for their pupils' learning. They believe that this use will increase their interest and involvement in the learning process (N = 49) and will help them to develop programming skills (N = 37), spatial skills (N = 29), problem-solving skills (N = 27) and creativity skills (N = 25). Secondly, a large number of teachers (N = 45) believes that the use of drones through mobile applications will make their teaching more interesting (N = 41) and engaging (N = 36). Thirdly, an also significant number of teachers (N = 39) believes that the use of drones through mobile applications will encourage them to use cooperative teaching (N = 35), inquiry-based learning (N = 32), experiential learning (N = 31) and the interdisciplinary approach to learning (N = 20) more frequently.

5.3 Facilitating Conditions

The results that regard the third research question have indicated two categories of facilitating conditions which, according to the teachers, will help them to effortlessly use drones through mobile applications. The first facilitating condition (N = 56) relates to the existence of training on the integration of drones in their teaching. This was mentioned by the majority of the teachers but was expressed more intensely by primary education teachers as well as secondary education teachers whose discipline/subject was unrelated to formal and natural sciences. A female primary education teacher mentioned: "*I would like to use it in my class, since I have learned how to operate and repair it, but I need guidance in how to use it. I can think of many uses for it in an informal context, but in order for me to properly integrate it into my class, I believe I will need help*". This condition was also related to training in programming. Indicatively, a male primary education teacher stated: "*I do not know how to program and I feel that I used only a small number of the programming command options available. I want someone to explain to me what each command does so that I can support my pupils*". The second facilitating condition which teachers (N = 52) believe should exist is infrastructural support. This relates to the existence of a sufficient number of drones and mobile technology devices in their school and, also, Internet connectivity both in their school and in informal learning environments where drones will be utilized.

5.4 Affordances

The results regarding the fourth research question have revealed that teachers categorized the affordances of drones into three groups. The first group is physical affordances and refers to the technical characteristics of the drone (e.g., the built-in camera that provides bird's-eye view and the easy collection of data through the built-in sensors). The second group are software capabilities and refers to the affordances and features of their software (e.g., user-friendly graphical flight and programming interface, easy to use simulator). The third group are the connectivity and the control affordances that concern the drone's connectivity to smart devices and to additional sensors.

5.5 Type of Activities

Regarding the fifth research question, results showed that a number of teachers (N = 38) believes that several activities that are related to STEM can be developed. These activities that can be developed, according to the teachers' perceptions, relate to problem solving, investigation, experimentation, reflection and cooperation. The following statements are indicative of this group of teachers: "*Drones and their applications are ideal for STEM*", "*I think that they (i.e., the applications) can only be used in STEM ... Pupils can program, collect data and process them*", "*I believe that the use of drones and their applications are applicable to all the fields of STEM*". Another number of teachers (N = 20) believes that cross-curricular and creative activities can be developed in all subjects. As they indicatively mentioned, these might be related to video recording and photographing of landscapes and buildings (N = 20), data collection from sensors in the context of environmental projects (N = 16), simulation and problem-solving activities (N = 15), and spatial skills activities (N = 14).

6 Discussion and Conclusions

The present research examined the perceptions of in-service teachers regarding the utilization of drones through mobile apps in teaching and learning. The results of the first research question showed that teachers consider the use of drones through mobile apps to be easy for their teaching. These results are encouraging regarding the implementation of drones in teaching and learning, since a large number of teachers that participated in the research had not been familiar with visual programming environments or the drone's flight and simulation applications. This means that, in future utilization of drones through mobile apps in teaching, teachers will not require special training in drone use. Therefore, training programs for inexperienced teachers in the use of drones through mobile apps should attempt to address their perceptions concerning the ease of use. Regarding the second research question of the present research, which is perceived usefulness, results showed that teachers believe the use of drones through mobile apps will be useful for their pupils' learning, themselves and their teaching. These results indicate that, in order for the use of educational drones through mobile apps to be introduced in schools, teachers should know what the benefits are.

The results of the third research question showed that two very important factors that will allow teachers to use drones through mobile apps are: the teachers' training in integrating them into their teaching, and the school's provision of the necessary infrastructure. Therefore, a future training of teachers should, through best practices, focus on how to help teachers to effectively utilize them in the subjects that they teach. Furthermore, schools should be equipped with a sufficient number of drones and devices and, also, provide all the other conditions necessary for their successful use (e.g., Internet, support). Also, the results regarding the fourth research question showed that teachers believe that drones have numerous technical affordances, which can be utilized in teaching and learning (e.g., bird's-eye view, camera, sensors, mobile apps). These affordances are important compared to those found in other technologies, and could be used as an argument in future training programs to convince teachers of the uniqueness of drones in education.

Regarding the final research question, results showed that teachers believe that drones can contribute to the development of a number of STEM-related activities. These results are similar to other researches that have been conducted regarding drones in STEM related concepts [5–8]. Finally, this study is the first in the field to investigate teachers' perceptions on using drones through mobile apps. Its results could constitute the basis for the conduction of more researches with the goal of incorporating aerial educational robotics into the context of mobile learning. Future studies could examine the perceptions of teachers on the use of drones through mobile apps, after they utilize them with their pupils in formal learning environments.

References

1. Magistretti, S., Dell'Era, C.: Unveiling opportunities afforded by emerging technologies: evidences from the drone industry. Technol. Anal. Strategic Manage. **31**, 606–623 (2018)
2. Coşkun, S., Kayıkcı, Y., Gençay, E.: Adapting engineering education to industry 4.0 Vision. Technologies **7**(1), 10 (2019)
3. Feron, E., Johnson, E.: Aerial Robotics. In: Siciliano, B., Khatib, O. (eds.) Springer Handbook of Robotics. Springer, Heidelberg (2008)
4. Vergouw, B., Nagel, H., Bondt, G., Custers, B.: Drone technology: types, payloads, applications, frequency spectrum issues and future developments. In: Custers, B. (ed.) The Future of Drone Use. ITLS, vol. 27, pp. 21–45. T.M.C. Asser Press, The Hague (2016). https://doi.org/10.1007/978-94-6265-132-6_2
5. Tezza, D., Garcia, S., Andujar, M.: Let's Learn! An Initial guide on using drones to teach STEM for children. In: Zaphiris, P., Ioannou, A. (eds.) HCII 2020. LNCS, vol. 12206, pp. 530–543. Springer, Cham (2020). https://doi.org/10.1007/978-3-030-50506-6_36
6. Farr, V., Light, G.: Integrated STEM helps drone education fly. In: 2019 IEEE Integrated STEM Education Conference (2019)
7. Carnahan, C., Crowley, K., Hummel, L., Sheehy, L.: New perspectives on education: drones in the classroom. In: Proceedings of Society for Information Technology & Teacher Education International Conference (2016)
8. Yamamori, K.: Classroom practices of low-cost STEM education using scratch. J. Adv. Res. Soc. Sci. Hum. **4**(6), 192–198 (2019)
9. Hoffert, F.D.: Examining the possibility of using programming language with low-cost drones. In: International Conference on Applied Computer and Communication Technologies (2017)
10. Ajzen, I., Fishbein, M.: Understanding Attitudes and Predicting Social Behavior. Prentice-Hall, NJ (1980)
11. Ajzen, I.: The theory of planned behavior. Organ. Behav. Hum. Decis. Process. **50**(2), 179–211 (1991)
12. Davis, F.D.: Perceived usefulness, perceived ease of use, and user acceptance of information technology. MIS Q. **13**(3), 319 (1989)
13. Clarke, R.: Understanding the drone epidemic. Comput. Law Secur. Rev. **30**(3), 230–246 (2014)
14. Hassanalian, M., Abdelkefi, A.: Classifications, applications, and design challenges of drones: a review. Prog. Aerosp. Sci. **91**, 99–131 (2017)
15. Daponte, P., De Vito, L., Lamonaca, F., Picariello, F., Rapuano, S., Riccio, M.: Measurement science and education in the drone times. In: IEEE International Instrumentation and Measurement Technology Conference (2017)
16. Chou, P.-N.: Smart technology for sustainable curriculum: using drone to support young students' learning. Sustainability **10**(10), 3819 (2018)
17. Bouck, E.C., Working, C., Bone, E.: Manipulative apps to support students with disabilities in mathematics. Interv. Sch. Clin. **53**(3), 177–182 (2017)
18. Papadakis, S., Kalogiannakis, M.: Mobile educational applications for children: what educators and parents need to know. Int. J. Mob. Learn. Organ. **11**(3), 256–277 (2017)
19. DJI Virtual flight. https://www.dji.com/es/downloads/djiapp/dji-virtual-flight. Accessed 22 Jun 2021
20. Tilley, E., Gray, J.: Dronely: a visual block programming language for the control of drones. In: Proceedings of the SouthEast Conference (2017)
21. Tynker. https://www.tynker.com/mobile/. Accessed 22 Jun 2021

22. Droneblocks, https://www.droneblocks.io/. Accessed 22 Jun 2021
23. Voštinár, P., Horváthová, D., Klimová, N.: The programmable drone for STEM education. In: Clua, E., Roque, L., Lugmayr, A., Tuomi, P. (eds.) ICEC 2018. LNCS, vol. 11112, pp. 205–210. Springer, Cham (2018). https://doi.org/10.1007/978-3-319-99426-0_18
24. Joyce, K.E., Meiklejohn, N., Mead, P.C.H.: Using minidrones to teach geospatial technology fundamentals. Drones **4**(3), 57 (2020)
25. Scherer, R., Teo, T.: Unpacking teachers' intentions to integrate technology: a meta-analysis. Educ. Res. Rev. **27**, 90–109 (2019)
26. Scherer, R., Siddiq, F., Tondeur, J.: The technology acceptance model (TAM): a meta-analytic structural equation modeling approach to explaining teachers' adoption of digital technology in education. Comput. Educ. **128**, 13–35 (2019)
27. Davis, F.D., Bagozzi, R.P., Warshaw, P.R.: User acceptance of computer technology: a comparison of two theoretical models. Manage. Sci. **35**(8), 982–1003 (1989)
28. Huang, L.: Acceptance of mobile learning in classroom instruction among college English teachers in China using an extended TAM. In: Proceedings of International Conference of Educational Innovation through Technology (2017)
29. Sánchez-Prieto, J. C., Olmos-Migueláñez, S., García-Peñalvo, F. J.: MLearning and pre-service teachers: an assessment of the behavioral intention using an expanded TAM model. Comput. Hum. Behav. **72**, 644–654 (2017)
30. Kalogiannakis, M., Papadakis, S.: Evaluating pre-service kindergarten teachers' intention to adopt and use tablets into teaching practice for natural sciences. Int. J. Mob. Learn. Organ. **13**(1), 113 (2019)
31. Islamoglu, H., Kabakci Yurdakul, I., Ursavas, O.F.: Pre-service teachers' acceptance of mobile-technology-supported learning activities. Educ. Technol. Res. Dev. **69**(2), 1025–1054 (2021). https://doi.org/10.1007/s11423-021-09973-8
32. Shodipe, T.O., Ohanu, I.B.: Electrical/electronic technology education teacher's attitude, engagement, and disposition towards actual usage of Mobile learning in higher institutions. Educ. Inf. Technol. **26**(1), 1023–1042 (2020)
33. Sadaf, A., Newby, T.J., Ertmer, P.A.: Exploring pre-service teachers' beliefs about using web 2.0 technologies in K-12 classroom. Comput. Educ. **59**, 937–945 (2012)
34. Smarkola, C.: Efficacy of a planned behavior model: beliefs that contribute to computer usage intentions of student teachers and experienced teachers. Comput. Hum. Behav. **24**, 1196–1215 (2008)
35. Koutromanos, G., Mavromatidou, E.: Augmented reality books: what student teachers believe about their use in teaching. In: Tsiatsos, T., Demetriadis, S., Mikropoulos, A., Dagdilelis, V. (eds.) Research on E-Learning and ICT in Education, pp. 75–91. Springer, Cham (2021). https://doi.org/10.1007/978-3-030-64363-8_5
36. Corbin, J., Strauss, A.: Basics of Qualitative Research: Techniques and Procedures for Developing Grounded Theory, 3rd edn. Sage, Thousand Oaks (2008)

Evaluating Design Cards for Supporting Design Thinking in the Context of Open Robotics and IoT Competitions

Ioannis Arvanitakis(✉), George Palaigeorgiou, and Tharrenos Bratitsis

University of Western Macedonia (School of Social Sciences and Humanities), Florina, Greece
ioarvanit@gmail.com

Abstract. Although educational robotics and IoT competitions have been described as a productive learning field in the context of STEAM education, there is a lack of research concerning the design process that occurs in student teams and the connection with design thinking methodologies. This study aims to propose and evaluate an approach for supporting design thinking in the context of STEAM, educational robotics and IoT, for elementary school students. The proposed methodology relies on participatory design approaches and is based on 40 design cards, which aim at the supported exploration of problems, needs, opportunities, and ideas. It was applied in 4 sessions with 19 students, who participated in an open educational technologies competition. Questionnaires and focus groups were used to gather data from the students who argued that the proposed design process allowed them to explore the problem at hand in an unexpected, creative and productive way and it was crucial for the novelty of their final pool of ideas. They also stated that the structure and elements of the process played an important role. Weaknesses of the proposed framework are also identified and reported.

Keywords: Design thinking · Design cards · Educational robotics

1 Introduction

The need for design thinking is essential to connect and integrate knowledge from sciences and arts that are fragmented in subject matters in a way that is suited for real-life problems, combining theory with practice for productive purposes [5]. Design plays a significant role in every area of human life that extends into the basics of scientific activities, and design thinking is about understanding the issues people experience in their everyday lives and creating useful innovations to address them [8]. Many researchers [3, 8] note that design thinking is also a powerful human-centered and action-oriented educational tool for problem definition and problem-solving, that promotes innovation, creativity, and collaboration. Design thinking in education has been mainly linked with the development of STEAM [8], opening new areas for further

M. E. Auer and T. Tsiatsos (Eds.): IMCL 2021, LNNS 411, pp. 669–680, 2022.
https://doi.org/10.1007/978-3-030-96296-8_60

research like its incorporation in the inner workings of student teams that take part in open-ended robotics and IoT (Internet of Things) competitions.

The goal of this manuscript is to contribute to the discussion concerning design thinking in the context of STEAM, educational robotics, and IoT by proposing and evaluating a card-based framework for supporting design thinking in the context of open educational technologies competitions for elementary school students. Relying on previous work on methodologies for collaborative card-based design in educational contexts [11, 16] we have developed and evaluated an approach based on 40 design cards for exploring and generating ideas for ill-structured problems.

2 Theoretical Framework

2.1 Design Thinking and Card-Based Design

Design thinking is a nonlinear, analytical, and creative procedure for addressing wicked problems that are loosely formulated, subjected to redefinition and resolution in different ways over time, without a single goal or a set of strict rules [5]. Design problems are ill-structured because they have no special subject matter, and designers have to invent it out of the specifics of the problem at hand [5]. The type of thinking involved in design practices has been explored mainly through examining professional designers' practices and is mainly referred to as designerly thinking. This term is different from design thinking, which is mainly used when design practices and mindset are utilized beyond the professional design context, from people without a scholarly background in this area [8]. Design thinking utilizes the methodology and mindset of designers to define problem-solving strategies that are technologically feasible and sustainable [8]. It is also a learning orientation that involves active problem solving and the development of creative confidence [3].

Researchers and designers have developed many tools to support the design thinking process. Card-based design tools have been used since the 1950s, mainly as a trigger for creative thinking [14]. Research from their use indicates that they have advantages associated with their key features: They are artifacts that people can interact with, they summarize information, methods, and best practices in a way that designers can act upon, they can be mixed and combined in many ways, they can provide structure to the design process, expanding the design space and helping in overcoming design fixation [14]. Several studies have supported their usefulness in the representation of theoretical constructs, their ability to provide enjoyable elements in the design process, and to focus on its anthropocentric character [10]. To foster creativity and improve productivity, design card games need a structured process with specific rules [10]. Tools such as game boards, guidebooks, scripts, and personas can support card-based design, focusing the ideation process on a specific theme or context [10], while frameworks like We!Design&Play can guide the creation of cards, boards, props, and rules to develop collaborative design sessions [16].

2.2 Design Thinking in STEAM, Educational Robotics, and IoT

In school education, design has been recognized as a general framework and an important approach for STEM, and design thinking has been studied as a general cognitive process involving creation, experimentation, feedback collection, and redesign [8]. The research activity concerning this thematic has raised more questions than answers [8], making it a field in need of further research. One area that has not been explored in particular is the use of educational robotics and robotics competitions in conjunction with design thinking in STEAM environments. Robotics competitions have been described as a popular and productive informal learning environment that can increase students' interest in STEM, motivating them in an open-ended quest to design and program robotic systems, and steering them in more positive attitudes towards science-related areas [12]. Although most robotics competitions engage students in fixed activities where their robots perform specific assignments defined by a rigid set of rules, and their performance is ranked by the officials, there are also open-ended categories where student teams are usually presented to a general theme and are free to design, build, program and present autonomous systems in the form of an exhibition. Most open categories allow student teams to use a broad range of development platforms, incorporating technologies such as IoT and mobile applications. These types of competitions are usually more welcoming to novices, attract a broader range of students, and promote more effectively creative expression [18]. Much of the interest in robotics competitions focus on the specifics of the development platforms and the programming tasks, while there is considerably less interest concerning the idea generation phase that occurs in student teams, and the overall design process, especially in open-ended exhibition formats. There is a lack of connection between robotics competitions and the design thinking mentality.

3 Methodology for Supporting Design Thinking

The approach that we propose is based on the participatory design methodology We! Design, initially developed by Triantafyllakos et al. [17], which aims at designing applications for mobile devices by undergraduate students, cooperatively formulating their needs and design suggestions. The We!Design&Play framework [16] is inspired by idea generation theory and the design games literature and guides the development of board card games, which facilitate the exploration of the design space based on three complementary perspectives (convergent, contextual, and divergent), aiming at a rich pool of ideas, while addressing social influences on idea generation such as production blocking, social loafing, evaluation apprehension, attention/incubation hypotheses, and the cognitive inertia. Palaigeorgiou and Sidiropoulou adapted this methodology for their collaborative board design game We!Design!Fractions [11], that aims at designing apps about fractions in participation with experienced elementary students, who can offer useful insights about the design of the content of an educational app, its learning representations, and its learning interactions.

Based on this previous work our approach focuses on the exploration of the design space of a STEAM, educational robotics, and IoT ill-structured problem by small groups of elementary students, facilitating the production of novel and sustainable ideas for prototyping and testing. It consists of one preliminary stage where students create and present their alter egos, three consecutive ones that are repeated in each round of the design game, and a final stage for the evaluation of the ideas produced (Fig. 1). Stages 2, 3 and 4 take advantage of the card-based design tool we have developed facilitating the identification of needs, problems, and opportunities that the alter egos have (stage 2), the production of ideas for confronting these needs (stage 3), and the incorporation of sustainable qualities into these ideas (stage 4). At the last stage, all proposals are discussed and classified by the students and are documented to be utilized in further team meetings. The design session lasts approximately two hours and apart from the students, there is also a coordinator that participates with the role to establish an informal and friendly environment and facilitate collaboration without interfering in the decision-making process.

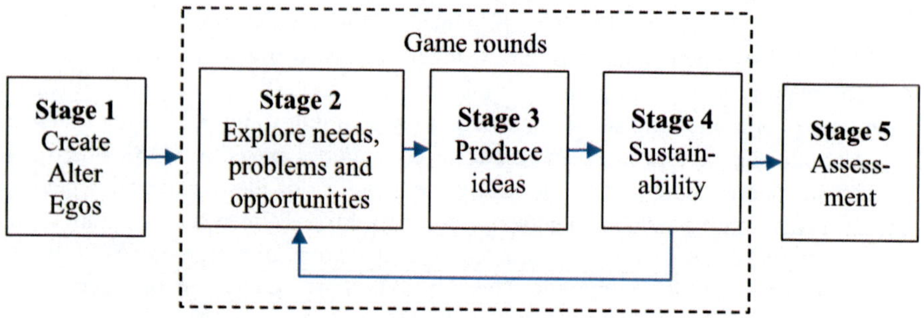

Fig. 1. The five stages of the proposed design approach

3.1 Design Alter Egos

The preliminary stage is based on the design alter egos technique introduced by Triantafyllakos et al. [16] where students a) develop their fictional character, b) take the role of their character, and c) play the role while the design session evolves. This process can help them recall existing needs, appreciate their internal motives and identify causal links between these motives and their personality traits, situate their ideas in well-established behaviors, and also liberate them from the fear of exposing themselves during the design process. The coordinator explains to the students that their alter egos should be placed in the context of the theme or problem being explored and that they will be their representatives during the rest of the session. He then asks them to fill their character forms that include fields such as name, age, place, favorite phrases, personality traits, hobbies, apps, superheroes, and life dreams of the fictional character. Each student also picks a photo for his character from a list of about 60 that are left on the table.

3.2 Card-Based Game

The next three stages utilize the 40 design cards we have developed and are available under the GNU General Public License v3.0 in our repository in Github (https://github.com/ioarvanit/We-design-for-STEAM). We studied a broad range of design toolkits and cards [14] for creative thinking, problem-solving, futures thinking, human-centered design, and design for sustainability that are available online to develop our final set. We were mostly influenced by the *"Tiles IOT Toolkit"* [10] and we also drew inspiration from *"Design with intent toolkit"* [9], and *"Sustainable design cards"* [13] (Fig. 2).

Fig. 2. Sample of our design cards

The cards are placed on the table organized by category (needs, ideas, sustainability) along with printed collages of images relative to the problem, providing visual cues. Post-it notes and pens are given to the students and each one draws a card in every round. The cards contain cues relative to the stage of the process that act as stimuli for conceptualization. Students have a few minutes to think and write down their ideas on the post-it notes and when time is up they announce them in turns to the rest of the team. This procedure minimizes the turn-taking and delays among the group members when they are in the process of thinking, avoiding production blocking [15]. When all the students have presented and discussed their ideas, the post-it notes are placed on a board that is next to the table and is divided into five categories (problems, needs, opportunities, ideas, and sustainable ideas) and the game moves on to the next round. This way students continue to have access to their ideas and the ideas of their peers for the rest of the game, an important factor to the novelty of the ideas produced [15].

Needs and Problem Exploration. This stage is concentrated on the needs, problems, and opportunities that arise from the theme under study. Students use the 9 cards from the problem exploration category that are divided into five sub-categories (time, place, objects, actions, and emotions). The time cards ask students questions that make them focus on temporal information (e.g. *"The time is 09:00 in the morning. Where can your character be and what is he/she doing? How is his/her activity linked with our theme?")*. Similarly, the place cards include questions that focus on spatial information and are combined with the images that are already on the table. The objects and actions cards focus on the convergent perspective and aim at a paradigm-preserving exploration of the design space. They help students explore pre-existing ideas and knowledge associated with the subject. Finally, the emotions category cards express various feelings and ask students to use them for describing the emotional state of their alter egos and thus discover new needs, problems, and opportunities.

Idea Production. The 22 cards of this stage include cues that were developed based on various idea generation techniques such as superheroes [6], break the rules [2], random stimulus, future workshops, metaphorical design, interaction relabeling [7]. This stage is linked with the divergent perspective that the We!Design&Play framework proposes [16] and aims at the paradigm-stretching and paradigm-breaking exploration of the design space, focusing on technology and its transformative nature. Every card in this category has a title, a goal statement and most of them also include a paradigm to help students understand the task.

Sustainability. We need students to think not only of technological solutions to problems but also their social, cultural/behavioral, institutional, and organizational footprint [4]. By incorporating sustainability in a design thinking process that is implemented in a co-design environment, we can help students develop skills to understand and respond to complex challenges, such as climate change and responsible citizenship [1]. The 9 cards of this stage provide students with cues to revisit their ideas and add sustainable qualities to them. They focus on themes like reuse, materials, energy consumption and sharing, local production, eco-footprint, etc.

Assessment. In the final stage of the design process, students study the board that contains all the ideas proposed, merge similar proposals, and classify them in order of importance.

4 Methodology

This study aims at evaluating the proposed approach for supporting design thinking and the accompanying card-based tool we have developed by answering the following questions:

- Are elementary students able to co-design solutions for open-ended problems in a STEAM, educational robotics, and IoT context?
- Can the proposed card-based design approach help students to be more creative and produce novel and sustainable ideas?

The proposed design methodology was implemented in a robotics club organized by an elementary school, with the participation of 19 students aged 10–12 years old (4 girls and 15 boys). Students formed 4 teams that participated in an open educational technologies competition generally themed: "Artificial intelligence-AI". The robotics club chose to present two additional general themes to the students ("disabilities" and "times of emergency") not restricting them to a specific problem. Before the design thinking sessions, students attended several meetings that were introductory to robotics, IoT, AI, and similar technologies as well as development platforms. Most of these meetings were held online as a result of the restrictions of the covid-19 pandemic.

Fig. 3. Students during the design process

Each one of the four student teams participated in a design session that had a duration of about two hours. Due to the health and safety measures that were in place at the time because of the covid-19 pandemic, the design cards were digitized and projected on a screen to minimize the exchange of artifacts between the students (Fig. 3). Three of the student teams played three rounds of the design game (stages 2, 3, 4) and one of the teams managed to play only two rounds in the specific time frame. The design sessions were supported by a coordinator with experience in design thinking and participatory design throughout the entire process. At the end of each session, students completed a questionnaire giving answers on a 5-point Likert scale for assessing the following variables: the perceived productivity of the design process (5 statements), the design process as a whole (3 statements), the role of their alter egos in the design process (4 statements), the perceived cooperation and co-design experience in their team (5 statements), their perceived satisfaction and tiredness through the various stages of the process, and the usefulness of the different cards as stimuli for thinking. The coordinator also conducted a 15-min focus group with every team about their experience of the design process. All focus groups were thematically analyzed after being recorded and transcribed.

5 Results

5.1 Questionnaires Answers Evaluation

Although the design sessions were held during weekend hours and were lengthy, students' responses reflected a strong positive attitude towards the productivity of the design process (M = 4.46, SD = 0.45, Alpha = 0.76) as shown in Table 1. The majority of the students indicated that it was exceptionally productive and were surprisingly satisfied with the final set of ideas their team came up with. They also stated that the process made them think a lot and more notably that the game helped them create more ideas from what they could produce if they were just talking about the theme. These answers suggest that the proposed design process helped them to be more creative and produce more novel ideas than other approaches they are accustomed to in their schools.

Most students also expressed highly positive attitudes towards the design process in general (M = 4.84, SD = 0.34, Alpha = 0.89). They stated that it was particularly interesting and surprisingly enjoyable. These views are also correlated with their perception that the process as a whole was very well structured and suggest that the design of the various stages was very effective in establishing a productive and pleasant setting for them.

The majority of the students were also very positive towards the role of their alter egos (M = 4.39, SD = 0.62, Alpha = 0.78). Most of them indicated that their alter egos were particularly helpful to design and express their ideas, and without them, they would be less creative. They also rejected the statement that their fictional characters interfered or obstructed their thought process and expressed their belief that if they were speaking for themselves instead of their characters they would not produce the same amount of ideas. These answers suggest that the creation of the alter egos was an important stage of the process, allowing students to explore the problem at hand in a more productive way.

Table 1. Answers by category

Category	Mean	SD	Alpha
Students attitudes towards the perceived productivity of the design process (5 statements)	4.46	0.45	0.76
Students attitudes towards the design process in general (3 statements)	4.84	0.34	0.89
Students attitudes towards the role of their alter ego in the design process (4 statements)	4.39	0.62	0.78
Students attitudes towards the co-design experience (5 statements)	4.16	0.74	0.83

Students also expressed very positive attitudes towards the co-design experience with their team members (M = 4.16, SD = 0.74, Alpha = 0.83). They stated that the ideas of the other members of the team helped their thinking process, acknowledging that they were free to express themselves without feeling that any of the other students monopolized the discussion, nor being distracted by jokes and out-of-context discussions. Most of the students also stated that this procedure helped them to get to know the other team members better. Their answers suggest that students can participate productively in a co-design setting with their peers when there is a well-structured design process in place.

Students were also asked to indicate their satisfaction and tiredness during the design process on a Likert scale of 1 to 5. Despite the long duration of the sessions, most of the students expressed low levels of tiredness and high levels of satisfaction even at the final stages of the design process as shown in Fig. 4.

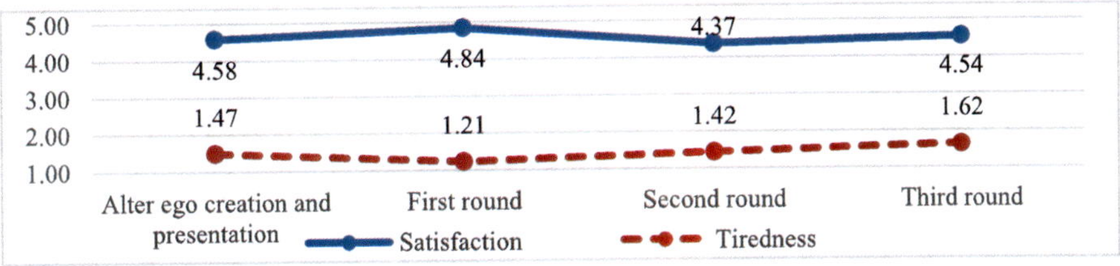

Fig. 4. Perceived satisfaction and tiredness during the various stages of the design process

5.2 Qualitative Measures

Students' Attitudes Towards the Productivity of the Design Session. The majority of the students expressed vividly their satisfaction with the productivity of the design process. Their answers showed that they were positively surprised about the quantity, quality, and novelty of their ideas, describing them as original, creative, and innovative and credited this outcome to the structure and realization of the design process (i.e., *"I wish we can make our ideas true to make our world better!", "If we did not play this game and just tried to think we wouldn't produce so many ideas", "The game was much more useful than just talking about the problems. It helped us evolve and think on target", "The things we said were combined in something beautiful"*).

Apart from the first session, where only two rounds of the game were held, in the rest three sessions students managed to play three rounds, identifying needs and problems, proposing ideas for solutions to these problems, and adding sustainable qualities to them. After the assessment stage, where all ideas were discussed, merged, and classified, an average of 4.5 ideas per session were produced, with 1.75 of them enhanced with sustainable qualities. Some examples of the process (exploring a problem/need, proposing an idea, incorporating sustainability) are shown in Table 2.

Students' Attitudes Towards the Design Process and Its Elements. Most of the students strongly stated that they were positively surprised by the design process and did not expect something like this at all. They are not accustomed to similar procedures from their everyday school life and they believe that it can be implemented in other subjects as well (i.e., *"It was enjoyable and very interesting. Perfect!", "We can do the same in geography, history, and basically everything. It was a lot more fun and*

Table 2. Examples of ideas.

Problem/need	Idea proposed	Sustainability
After a natural disaster, a lot of kids feel scared and need someone to talk to	A smart teddy bear toy that can feel when kids are sad or anxious and will tell them jokes and sing to them	Use separate cotton outside layer that can be washed and changed if damaged
Kids with severe hearing problems cannot communicate easily with other people	Smart glasses with AI that can hear what people are saying, transform it to text, and project it to the lenses	A hat with a solar panel that can connect to the glasses and charge them when we are out in the sun
A lot of people do not know how to give first aid if needed in a time of emergency	A smart wristband that will guide us when we have to give first aid to someone in need with audio instructions and feedback from our actions	Use recyclable materials to print it in our 3D printer

interesting than reading from a book", "I had a great feeling. We enjoyed the game and we had nice and funny ideas. We can do it again and create even better ideas", "I liked it very much. I would like to continue doing this from now on in the robotics club"). Most students also noted that the structure and the rules of the design process were a crucial factor for its success, allowing them to have a meaningful exchange of ideas oriented to their common goal (i.e., *"If we keep these rules and we don't talk when someone else is already talking we can make more ideas. Like we did now and we took ideas from one another", "The rules helped us because we could all say our opinions, hear the ideas others had and combine them to a perfect result", "When I heard an idea from someone else I could come up with something new"*).

The majority of the students (15) identified the idea category cards as the most useful in the design process (i.e., *"All cards were helpful but especially the purple ones for the ideas because they asked how to make life easier, how to help other people","I liked the idea cards because I like to find solutions to problems", "The idea cards were very helpful because after I had found a problem I could find a solution for it"*), followed by the needs and problem exploration cards (9) and the sustainability cards (4).

An interesting aspect that was revealed during the focus groups concerns the post-it notes where students write their ideas before announcing them to the team. Some students characterized them as time consuming and boring, even though they acknowledged the importance of leaving some time to think before starting the conversation as well as the significance of documenting all the ideas (i.e., *"You could let us think for a couple of minutes but instead of writing our ideas maybe we could do something like a recording", "We could keep notes on a tablet or laptop", "Maybe if we did the same as now and record our conversation"*).

Students' Attitudes Towards Their Alter Egos. The majority of the students described the first stage of the design process, in which they created their alter egos, as very helpful, enjoyable, interesting, and indispensable. They noted that the alter egos helped them focus and explore the problems in more depth (i.e., *"My alter ego helped me come closer to the problem", "The characters made us come out of our reality and come closer to the people that have these problems for real", "We made our character and because we knew how old is he, what he likes and how he reacts that gave us more ideas"*). An interesting suggestion from some students was to allow them to create more alter egos at the beginning that have different ages and characteristics, to have more ideas for a broader audience (i.e., *"We could have three characters instead of one. Every character could have different age, for example, a kid, a parent and a grandparent. That way we could think ever more and different things"*).

6 Discussion

This study attempted to investigate the ability of elementary students to co-design solutions for open-ended problems in the context of STEAM - educational robotics - IoT, and the contribution of the proposed design methodology to students' creativity, innovation, and cooperation for producing novel and sustainable ideas utilizing the design cards we have developed. Their answers from questionnaires and focus groups suggest that they are capable of designing their solutions to ill-structured problems in a collaborative setting, provided there is a well-structured design process in place. They characterized the proposed approach as highly productive, surprisingly enjoyable, interesting, and well organized, without which it would not be possible to produce the same amount of novel ideas. They also perceived the roles of the elements in the design process (alter egos, cards, rules) as very important factors for its success. Sustainability cards and the corresponding stage proved to be challenging for most of the teams, revealing a possible lack of experience in dealing with this subject and drawing our attention to the further examination of this phase. An interesting observation that emerged and also needs further study concerns the ease of use and functionality of the post-it notes for documenting the ideas produced in the various stages of the design process.

Although students' perceptions of the productivity and overall efficiency of the proposed design methodology were highly positive, the final list of ideas needs to be analyzed by subject specialists and designers to determine their quality, novelty, sustainability, applicability, and relevance to the subject. Another limitation of this study has to do with the relatively small sample of students. We need to test the proposed design methodology on a larger scale with various themes as starting points to have a better understanding of the impact it has on the idea generation process of elementary students. We firmly believe that the proposed approach can become the first step towards participatory design methodologies for students, in the context of robotics and IoT projects and competitions, focusing the interest on the design process of idea generation, and providing new frameworks for students to express their creativity.

References

1. Aksela, M.: Towards student-centred solutions and pedagogical innovations in science education through co-design approach within design-based research. LUMAT: Int. J. Math. Sci. Technol. Educ. **7**(3), 113–139 (2019)
2. Bacanlı, H., Dombaycı, M.A., Demir, M., Tarhan, S.: Quadruple thinking: creative thinking. Procedia Soc. Behav. Sci. **12**, 536–544 (2011)
3. Carroll, M.: Stretch, dream, and do-a 21st century design thinking & STEM journey. J. Res. STEM Educ. **1**(1), 59–70 (2015)
4. Ceschin, F., Gaziulusoy, I.: Evolution of design for sustainability: from product design to design for system innovations and transitions. Des. Stud. **47**, 118–163 (2016)
5. Coyne, R.: Wicked problems revisited. Des. Stud. **26**(1), 5–17 (2005)
6. Fleck, E., Asmuth, J.: Building capacity for creativity: rediscovering the inner "Superhero" as a mechanism for developing a creative mindset for entrepreneurial problem-solving. Entrepr. Educ. Pedagogy **4**(1), 82–95 (2021)
7. Halskov, K., Dalsgård, P.: Inspiration card workshops. In: Proceedings of the 6th Conference on Designing Interactive Systems, pp. 2–11, June 2006
8. Li, Y., Schoenfeld, A.H., Graesser, A.C., Benson, L.C., English, L.D., Duschl, R.A.: Design and design thinking in STEM education (2019)
9. Lockton, D.: Design with intent: a design pattern toolkit for environmental and social behaviour change (Doctoral dissertation, Brunel University School of Engineering and Design Ph.D. theses) (2013)
10. Mora, S., Gianni, F., Divitini, M.: Tiles: a card-based ideation toolkit for the internet of things. In: Proceedings of the 2017 Conference on Designing Interactive Systems, pp. 587–598, June 2017
11. Palaigeorgiou, G., Sidiropoulou, V.: Can elementary students co-design the learning content of educational apps: the we! design! fractions participatory design approach. In: Auer, M.E., Tsiatsos, T. (eds.) IMCL 2019. AISC, vol. 1192, pp. 202–214. Springer, Cham (2021). https://doi.org/10.1007/978-3-030-49932-7_20
12. Panadero, C.F., Román, J.V., Kloos, C.D.: Impact of learning experiences using LEGO Mindstorms® in engineering courses. In: IEEE EDUCON 2010 Conference, pp. 503–512, April 2010. IEEE (2010)
13. Ræbild, U., Hasling, K.M.: Sustainable Design Cards: A Learning Tool for Supporting Sustainable Design Strategies. Sustainable Fashion in a Circular Economy (2018)
14. Roy, R., Warren, J.P.: Card-based design tools: a review and analysis of 155 card decks for designers and designing. Des. Stud. **63**, 125–154 (2019)
15. Stroebe, W., Nijstad, B.A., Rietzschel, E.F.: Beyond productivity loss in brainstorming groups: the evolution of a question. In: Advances in Experimental Social Psychology, vol. 43, pp. 157–203. Academic Press (2010)
16. Triantafyllakos, G., Palaigeorgiou, G., Tsoukalas, I.A.: Designing educational software with students through collaborative design games: the We! Design&Play framework. Comput. Educ. **56**(1), 227–242 (2011)
17. Triantafyllakos, G., Palaigeorgiou, G., Demetriadis, S., Tsoukalas, I.A.: The we! design methodology: designing educational applications with students. In: Sixth IEEE International Conference on Advanced Learning Technologies (ICALT 2006), pp. 997–1001. IEEE, July 2006
18. Turbak, F., Berg, R.: Robotic design studio: exploring the big ideas of engineering in a liberal arts environment. J. Sci. Educ. Technol. **11**(3), 237–253 (2002)

Usage of Visual Analytics to Support Immigration-Related, Personalised Language Training Scenarios

Gerasimos Antzoulatos[1](✉), Thanassis Mavropoulos[1], Grigorios Tzionis[1], Anastasios Karakostas[1], Almudena Gonzalez Costas[2], Marta González Burgos[2], Stefanos Vrochidis[1], and Ioannis Kompatsiaris[1]

[1] Information Technologies Institute (ITI) - Centre For Research and Technology Hellas (CERTH), Thessaloniki, Greece
gantzoulatos@iti.gr

[2] Metodo Estudios Consultores, Vigo, Spain

Abstract. Most language learning applications are aimed at students or people who already know a language and want to improve their skills, or want to learn a new language. These applications, while seeking to be interactive, are not aimed at immigrants, refugees or asylum seekers, since the latter have different needs and interests from casual learners, opting for language skills that will allow them to function independently in the host society. This research is part of the European project WELCOME, which seeks to use state-of-the-art technologies, such as Virtual Reality (VR) apps and dialogue agents, to support the reception and integration of Third Country Nationals (TCNs) in Europe. The platform will be tested in three languages (Catalan, German and Greek), in real situations that the TCNs, mostly immigrants, refugees and asylum seekers, face, combining linguistic activities, where aspects related to language and culture are worked on. Furthermore, a Visual Analytics Component (VAC) leverages authority (NGOs/State institutions) users' perceptual and cognitive abilities by employing interactive visualisations as interfaces between users and learning analytics outcomes generated by amassed data. The goal is to find patterns within the characteristics of TCNs, and thus help language teachers adapt the content and tools to TCNs, contributing to greater personalisation in learning.

Keywords: Visual learning analytics · Educational data mining · Language learning · Language teaching · Third country nationals · Migration

1 Introduction

In the last few decades, with the rise of IoT technologies, e-learning, online/mobile learning environments and tools have proliferated. Hence, new opportunities for supporting learning processes across all educational aspects have emerged that provide learning experiences (synchronous or asynchronous, remote or collaborative) in multiple participants, such as groups of students, individuals and teachers. This growth and widespread use results in the rapid increase of learning data that are gathered in

M. E. Auer and T. Tsiatsos (Eds.): IMCL 2021, LNNS 411, pp. 681–693, 2022.
https://doi.org/10.1007/978-3-030-96296-8_61

educational institutions. Currently, the Educational Data Mining and Learning Analytics fields encompass interdisciplinary methods to analyse and visualise data during teaching, learning, as well as education administration and services. Moreover, the utilisation of advanced visual analytics technologies in learning and teaching analytics is inevitable due to the heterogeneity, complexity, temporal and unstructured nature of learning data [1].

These applications, while seeking to be interactive, are not aimed at the public of immigrants, refugees or asylum seekers who have other needs and interests. Most language learning applications are aimed at students or people who already know a language and want to improve their knowledge and skills or want to learn an additional language after mastering the second one. However, TCNs arriving in Europe do not constitute an homogeneous group, since they have diverse educational, linguistic and socio-cultural backgrounds. This heterogeneity of profiles is transferred to the classrooms and in this context, the role of the teacher becomes more demanding making the adoption of advanced analytical tools mandatory to assist in the monitoring of students' performance. In addition, TCNs are a group that needs to quickly develop a range of language skills in order to be able to function autonomously in the host societies, so the teaching of these skills should help them cope with everyday life situations. Also, as studies have pointed out [2], certain language learning applications, VR-based ones in particular, offer an opportunity to combine real spaces and situations, apply Artificial Intelligence (AI) solutions and introduce cultural themes in specific settings, making the general context much closer to reality.

The aforementioned aspects constitute an innovation within the WELCOME project which aims at the development of a useful tool to provide, among others, the means to assist in the reception and integration of immigrants and refugees in host societies via educational content and learning (e.g. visit to the doctor, municipality application process, employment services, schooling system). Moreover, the cognitive and perceptual skills of authority users from NGOs and State institutions that offer language courses are enhanced via the support of the VAC, which empowers them to leverage interactive visualisations as interfaces between users and learning analytics outcomes generated by amassed data. By exploring possible correlations and patterns encountered within the characteristics of TCNs, language teachers may adapt the available content and language tools to each TCN separately, accomplishing the goal of personalised learning.

The main contribution of this work is, thus, twofold. Initially, a presentation of a contemporary language learning approach that aims to contribute to personalised training, customised to TCNs' needs and profiles is offered. It is then complemented by an introduction to the knowledge generation visual analytics framework which materialises in the VAC, conceived as a tool focused towards language teachers. Specifically, details are shared on how the scenarios and materials developed in the WELCOME project are designed to develop the oral skills of TCNs in the host language, with special emphasis on those situations and vocabulary that are most needed in the early reception phases; with emphasis on the oral interaction and simulated dialogues combined with activities for vocabulary acquisition. Furthermore, an actual immigration-related, language learning scenario is presented, along with the respective visualisations that the VAC offers, to better convey the usefulness of the overall approach.

2 Related Work

2.1 Language Learning

The increasing number of migrants in Europe have turned language classrooms into heterogeneous places, in which populations converge that differ in age, gender, socio-cultural background, first language, motivation and exposure to the target language. The role of teachers in these educational contexts is more complex and demanding than in regular, single ethnicity, language teaching contexts, and the teachers need to seek specialisation in working with migrants and refugees.

Adult foreign language learners have specific needs and characteristics that must be taken into account to develop materials that are tailored to these needs. Knowles [3] identified some characteristics of adult learning in his theory: adult learners are autonomous and self-directed, accumulators of life experience and knowledge, goal motivated, guided by relevance, practical and feel a greater need to be respected. Furthermore, foreign languages are sometimes taught in a way that is not satisfactory enough for migrants as the curricula includes themes that are not to their interest and training exercises do not refer to real life situations.

Scenario-Based Teaching for Adult Migrants. In 2012 Switzerland presented *fide*, the innovative conceptual framework for the linguistic integration of migrants in Switzerland. *Fide* stands for français, italiano, deutsch which are the languages that migrants have to learn in various parts of the country. Frequent contact situations between migrants and Swiss residents have been identified and analysed in different everyday situations, such as contact with authorities, work context, and health. These formed the basis of an inventory of "scenarios" which included: descriptions of interactive situations, the interlocutors involved, their respective roles, the overall aim to be achieved by the interaction, the usual course of action, socio-cultural factors to be considered, and helpful linguistic resources to achieve the interaction aim.

Task-Based and Action-Oriented Training Programs for Migrants. One of the pillars that sustains the Common European Framework of Reference (CEFR) is the action-oriented approach. CEFR suggests that the lessons and language training courses should be planned backwards, from the learner's real-life communicative needs, with a consequent alignment between curriculum, teaching and assessment [4].

From 2010 to 2015, the University of Thessaly implemented two nationwide training courses for newly arrived migrants, ELMEGO and MATHEME. The ELMEGO project was aimed at parents with children attending compulsory education, while MATHEME was aimed at unemployed migrants. In both cases the task-based learning methodology was adopted and an in-depth qualitative analysis related to immigrants' communicative needs (in everyday life and work activities) was conducted. In both cases, the task-based approach was proven to be very effective and furthermore resulted in team-building, identity investment, and empowerment.

2.2 Visual Analytics

Although our ability to gather and store massive data from heterogeneous sources has been reinforced, our ability to analyse and utilise them for making efficient decisions is not so developed. Making matters worse, the analysis of such amounts of disparate data can lead to the well-known problem of "information overload" (also known as the "data deluge" problem) which concerns the danger of getting lost in data due to the irrelevant to the current task, the inappropriate (system-driven) processed way, or the illustration of the results [5, 6]. Hence, to tackle this problem, novel visual analytics approaches engage human cognitive interactions and reasoning, along with advanced analytics processes. As Endert et al. [7] pointed out a shift from a "human in the loop" to a "human is the loop" viewpoint, which empowers visual analytics capabilities, with the human intuitive capabilities of interactive visualization [8].

Therefore, to discover hidden knowledge from massive, heterogeneous and complex data, firstly James Joseph Thomas established the Visual Analytics (VA) field, which was defined as "*the science of analytical reasoning facilitated by interactive visual interfaces*" [9]. Keim et al. (2008) defined VA as the process that incorporates advanced automated analysis techniques along with interactive visualizations aiming at effective understanding, reasoning and decision making on the basis of very large and complex data sets" [5, 10]. In [6], Cui proposes a more detailed and comprehensive definition of VA in which the interactive visualizations, algorithmic data analysis and analytical reasoning techniques encapsulate human judgment into the KDD process to visually discover explainable patterns (knowledge) and to gain insight into large and complex data sets.

The aforementioned definitions for VA imply the solid interplay between visualisation, human intelligence, cognition and perception abilities, along with the advanced analytical processes performed by computers to obtain seamlessly, meaningful and explainable results leading to the generation of valuable knowledge for decision-making [8, 11, 13]. Sacha et al. in [8] established a knowledge generation model for visual analytics which relied on the visual analytics process proposed by Keim et al. [5, 6, 12] and simultaneously it encapsulated human perceptual and cognitive theories, such as sensemaking [11] in order to generate knowledge in an iterative manner.

In [13] authors conducted a comparative review of state-of-the-art commercial VA systems. During the evaluation process, the VA tools were compared in order to cover the three main actions in a VA system workflow, namely data management, automatic analysis, visualisation and system and performance. It is worth mentioning that the majority of the visual analytics systems do not fully cover or appropriately deal with the aspects of the Knowledge Generation model [8].

3 Proposed Approach

3.1 Motivation for the Creation of Scenarios

The creation of language learning scenarios has been approached from a co-creation methodology viewpoint with the teachers and Public Administration users, in combination with an in-depth study of the applications and resources on the market for teaching languages to migrants, adults and applications that make use of virtual reality and visual analytics for language learning. Moreover, since the teacher's role is fundamental to adult learners with an A1-A2 level, all scenarios have been designed to complement face-to-face language lessons and to allow TCNs to practice the most likely situations they will encounter during the different stages from arrival to integration.

Several apps for language learning have been tested, including the most popular on the market: Duolingo, Drops, Word of the day (iOS versions), Babbel, and Mondly (web versions). These are focused on autonomous learning and require some previous knowledge on the target language. The main problems detected are the following:

- Lack of spelling activities and/or lack of activities to learn the alphabet. Of all the apps analysed, only Drops has an activity aimed at learning the alphabet. While trying to learn a language with an alphabet different from the native one, a well-planned activity for learning the alphabet becomes a crucial activity for being able to advance to more complex lessons.
- Limit of lessons or errors per day: users in the free version of Duolingo, Babbel, Drops and Mondly have a limited number of lessons or errors in the activities each day. When users reach that limit, the app is blocked until the next day. As beginners make mistakes frequently, if the application does not allow them to finalise a lesson, it can demotivate them and they may not access the course again.
- Free navigation: Duolingo, Mondly and Babbel do not allow participants to freely navigate the courses. They have to follow the training path, as the free versions don't offer any possibility to personalise the training path. If students come across several topics in a row that are not to their interest, they can drop out of the course.

3.2 Building the Learning Program: Scenario-Based Teaching

The methodology chosen for language learning is the scenario-based teaching for its proven efficiency, which in addition offers the best development options in VR. The teachers who will pilot the program have also expressed the usefulness that their students find in the programs based on this approach.

Sheridan and Kelly [14] claim that scenarios should be related to the real world, so learners could find a connection between contents and application of such contents in their lives. In order to achieve the objective of enabling TCNs to cope with everyday situations, language lessons are developed, linked to respective situations in which cultural aspects specific to each country are introduced. Thus, in a self-presentation scenario, the person is expected to be able to provide basic information about himself/herself, but also to understand the cultural differences in aspects such as surname or family name depending on whether one is single or married. Hence, each language learning scenario is composed of several activities: a simulated dialogue between the TCN and a 3D avatar to train specific vocabulary, easy grammatical constructions and/or cultural lessons associated with that scenario.

The scenario to be tested refers to the "First Reception Service" (FRS) and the main aim is to train TCNs in providing basic information about themselves. The human-machine interaction is handled by the dialogue agent which requests the required information from the TCN and then allows the latter to test the procedure and acquired knowledge via VR minigames. The vocabulary and grammatical constructions of this scenario correspond to levels A1-A2 of the CEFR. It is practiced with the help of the teacher, who will assist TCNs to log into the WELCOME platform and put on the VR headset. Once the TCNs access the VR platform, users are placed in an apartment-shaped stage, in which they are greeted by a personalised avatar that explains the instructions to access the learning scenarios.

The personalised avatar will provide each TCN the following instructions to complete the FRS scenario: "You need to complete an application form. Take a look around your apartment for the application form, once you have located it, pick it up using your controller to enter your first learning simulation" *[application form is glowing]*.

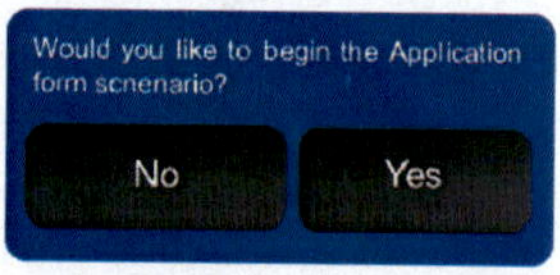

Once the TCN has located the application form, a floating UI appears to ask the TCN to start the language-learning scenario. When the TCN presses Yes, an office scene is loaded, where the user has to use the controllers to navigate the VR office and enter in the VR rug. Once the TCN enters the VR rug, the personalised avatar will greet her/him, the questions will pop-up as a floating UI and the TCN will be asked to choose the correct answer for each of the questions posed by the avatar. The first floating UI will appear in their native language and it will read: "Please provide an appropriate response according to the correct time of day by pressing one of the options below".

For each of the questions that are part of each scenario, the system follows the same flow: once TCNs obtain the score of the question, the avatar will move to the next question. TCNs have 3 attempts to answer the multiple-choice questions.

Level 1. TCNs are able to provide a correct response without any explanation. First attempt. Score 10.

Level 2. TCNs are able to provide a correct response with an explanation about the meaning of the word. Second attempt: Score: 5

Level 3. TCNs need the word in their mother tongue. Third attempt: Score 0, if the TCNs provide a wrong response, and score 1, if the TCNs provide the correct response.

The total mark is between 0 and 10 for each question of the dialogue and the final mark will be the average of all the dialogue parts. The class mark is the average of the different exercises (in total, and by activity), so the teacher has information to provide recommendations that will reinforce contents not yet fully assimilated by the TCN.

3.3 Methodological Approach for Testing Language Learning Scenarios

The vocabulary and grammatical constructions of each scenario are grouped by CEFR levels. For the first pilots with TCNs, the language learning materials only target A1 and A2 levels, but the levels will be expanded in future scenarios to achieve a B2 or C1 level. To collect crucial users' data in order to establish correlation and patterns, socio-demographic information (age, gender, country of origin, mother tongue, educational level) and technological skills or knowledge of other languages will be requested while creating their profile in the WELCOME platform. The VAC will be in place to tackle the analysis of the aggregated information originating from the language learning scenarios and user profile, and provide the opportunity to the interested party, be it a teacher or an authority user, to draw important conclusions that will assist in forming a personalised educational strategy. The suggested solution relies on the adoption of Visual Data Analytics techniques that enable the discovery and understanding of patterns, advanced correlations and trends in large datasets, generated by learning scenarios, via visual interpretations. In this stage, users will be able to utilise advanced techniques, integrated into the VAC and the following correlations are expected to be possible:

- Valuation differences in relation to gender, age, educational level, previous digital skills, and languages known to immigrants, refugees or asylum seekers;
- Differences in the results obtained in the learning activities in relation to the aforementioned attributes;
- Differences in the learning pace of the TCNs participating in the pilots, depending on where the pilots will take place, Spain, Greece and Germany;
- Differences in performance between the TCN and their own classmates;
- Differences in performance between the TCN and other TCNs with a similar profile, including % of TCNs unable to finalise the training activities and VR scenarios.

Given the differences in the profile of the TCNs that are going to participate in the pilots, respective variations are expected in the learning pace and in the scores obtained in the different scenarios. TCNs who have been exposed to the language of the host country or whose native language shares the same alphabet with the language they are learning, will advance more quickly and will obtain higher scores in A1-A2 scenarios than those who do not know the language or who have to learn the alphabet.

It should be mentioned that great care has been taken to prevent unauthorised access to private and sensitive information concerning the TCNs by providing a secure environment, protected by specific access rights/privileges, based on the user's role. Thus, unauthorised people cannot access the TCNs' personal data or their language learning performance. Therefore, only teachers can carry out relative comparison analyses and, specifically, only concerning the performance of TCNs of their own class.

3.4 Visual Analytics for Language Learning Scenarios

In this section, we will propose a Visual Analytics schema that enables practitioners to analyse information and foster complex decision-making processes in the language learning domain. It is motivated by the knowledge generation model (KGM) for visual analytics which has been proposed by Sacha et al. [8]. Similar to KGM, our proposed model consists of the main components that are Data, Data Mining Models, Visualisation and Knowledge. However, it encapsulates the Decisions component due to the fact that, in the framework of the language learning scenarios, the interest focuses on how the teachers will be assisted by it, in order to make efficient and timely decisions in terms of the TCN learning/educational process (see Fig. 1).

In this model, the knowledge can be generated by Visual Data Exploration going from data via visualisation (InfoVis pipeline) to knowledge and Automated Data Analysis that is going from data via models to knowledge (see Fig. 1).

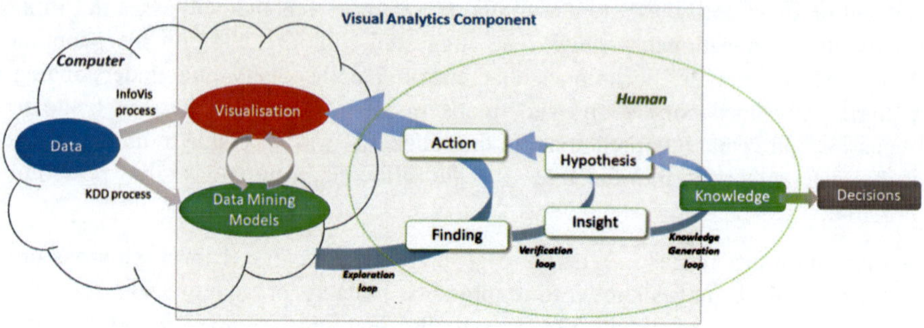

Fig. 1. Knowledge generation model in VAC for language learning

The computer part of this Knowledge generation model for Visual Analytics breaks down into Data acquisition processes, and pipelines from Data to Visualisation (InfoVis process) and Data to Model in terms of Knowledge Discovery in Databases (KDD) processes, coupling with machine learning algorithms. Data quality plays an important role in visual analytics processes and depends on how the data were generated, gathered, and selected for further analysis. During an analysis, aggregated data could be produced either by the application of automated methods (e.g. classification or clustering) or by manual annotations. The information visualisation pipeline employs techniques, usually from Exploratory Data Analysis (EDA), to detect relationships in the data in an illustrative manner. The main components of the InfoVis pipeline includes the transformations/mappings between Raw Data, Data Tables, Visual Structures and Views which can be manipulated through Human Interactions [8]. Furthermore, visualisations can be produced by automated models; for instance, data can be illustrated as groups after undergoing a clustering analysis. In particular, models can originate from descriptive statistical approaches or even more complex ones, coming from the KDD and Data Mining field. Complex patterns, relationships, and associations can be revealed and visualised, in order to be communicated and comprehended by the analysts. The KDD process consists of iterative and interactive steps, in which the data are selected and filtered, preprocessed, and transformed appropriately, in order to apply data mining techniques on them. The goal is to discover unknown patterns that are hidden in the data and convert them into valuable knowledge [8].

The exploration loop concerns the interaction between analysts and visual analytics systems to analyse data, generate new visualisations or models. Actions concern analysts' goals and tasks that produce tangible and unique responses from the visual analytics system. Actions derived from Hypotheses are usually complex Actions, while those that are derived from Findings are normally simple Actions, such as changing the mapping of visualisation or selecting a different feature for model building. Also, Actions can deal with data gathering or selection in order to prepare the data, or Actions to create models (model building) or to utilise existing models. Similarly, visual mapping Actions concern the creation of data visualisations, while model-vis mapping Actions map models into visualisations. Finally, Actions that enable analysts to manipulate with viewpoints focusing on interesting data in the visualisation and interact with the visual analytics system, are also provided [8]. This interaction can lead to emerging new interesting observations (Finding) or to new Insights. Missing or extreme values can be considered as Findings in the data that affect the further analysis and hence require special data processing. In the case of visualisations or models, a Finding can be a pattern, a trend, an evident model result, even an unusual behaviour of the system. Findings can be derived either by the utilisation of automatic data mining models or by humans-analysts exploiting their visual perceptions and cognition skills. In general, the visual analytics systems rely on the Actions and Findings in the exploration loop, as the understanding and efficient interpretation of the Findings in the context of the problem domain, provide new Insights to analysts. Going a step beyond, Insights may lead to new Hypotheses which require further investigation to verify or falsify them, in the verification loop. This process, namely the assessment of the Insights, results in the gain of new Knowledge.

The visual analytics process facilitates analysts to identify evidence for existing assumptions or learn novel and valuable knowledge, so as to foster and enrich their prior knowledge related to the problem domain. The derived knowledge from visualisations, automatic analysis along the preceding interactions between visualisations, models, and the human, can be considered as *Prior Knowledge* [15]. The involvement of experts in the exploration of knowledge is significant as they will be immediately aware of the relationship between new knowledge and the existing domain knowledge. Complex reasoning and sense-making processes are employed as the basis for generating additional knowledge or can be composed with Prior Knowledge to produce more general truths enhancing the *User Knowledge* [11, 16]. In the proposed model, the iterative process that addresses high level reasoning, such as inductive, deductive, and abductive reasoning, in the knowledge generation, exploration and verification loops, enable end-users to make Hypotheses and apply criteria to make decisions.

In the framework of the WELCOME project, the VAC will be further developed serving the needs for language teachers, professionals from NGO′s and public entities who are involved in the first reception and integration of TCNs in the host societies. Specifically, the VAC encompasses processes for data preparation, as well as information visualisation processes in an interactive, intuitive and user-friendly VAC UI aiming to engage humans in the loop of exploring data and the discovery of new knowledge. The VAC UI will be enriched with functionalities that enable authorities to visualise aggregated information related to health data, personal, educational and professional profile of TCNs. Furthermore, it will be able to synthesise information and derive insights from massive, dynamic and ambiguous data concerning TCNs and visualise normal and abnormal detected trends and patterns related with them.

Furthermore, more complex information generated by the high-level knowledge interpretation will be achieved by applying KDD processes and data mining. The outcomes of the analysis will be visualised and will be incorporated to the VAC UI aiming to enable end-users to improve evidence-based decision making. Hence, the exploration loop is supported by providing customised visualisations for different research analytical questions raised by end-users. Moreover, VAC tailors functionalities that allow end-users to tune parameters, choose specific characteristics and in general take actions that naturally provoke interactions of analysts with the system. VAC provides the tools to transforming the Findings from the exploration loop to

Insights by verifying or falsifying concrete Hypotheses. This knowledge generation process will be capable of achieving sustainable decisions for migrant management and integration as well as for the development of their language learning skills.

3.5 Evaluation Visual Analytics Process

As mentioned above, the utilisation of VAC in language learning can reveal valuable knowledge concerning the progress of TCNs or to detect the difficulties that TCNs may encounter in following or completing the language teaching scenarios. Hence, a teacher, via the VAC UI, will be able to visualise correlations among students' profiles that are participating in a specific class and follow the activities of the particular language learning scenario. In Fig. 2 an example dashboard created by a teacher of a class (C2) is exhibited. Using visual analytics tools, he/she can graphically present combinations of characteristics and socio-demographic information among students in C2.

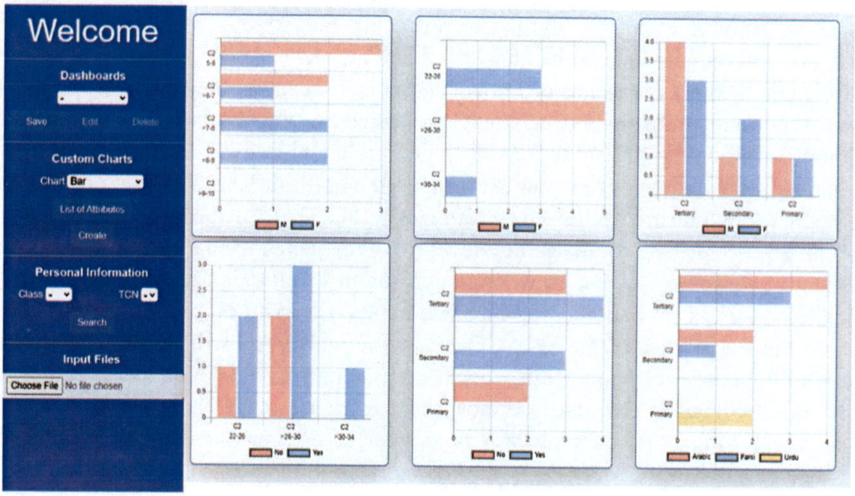

Fig. 2. Dashboard illustrating correlations among general characteristics of class C2 students

Furthermore, the VAC enables the teacher to compare the student's performance with the average performance of his/her class, as illustrated in Fig. 3, which includes:

- General TCN's personal information, as well as some basic statistical measurements in terms of his/her performance. The VAC provides statistical data as MAX, MIN, Median, Mean, Standard deviation in terms of the score achieved by specific TCNs in the activities.
- Average score of the students depending on the different scenario activities and compared to the specific TCN (*Average Performance* bar plot).
- Deviation between the particular student and the average score of the class per activity (radar plot).
- % of students that are able to finalise the activities in the TCN's class (pie chart).

- Three barplots which exhibit the average performance of TCNs in a specific class over activities grouped by gender, time needed to complete an activity discretised in days and educational level of TCNs.

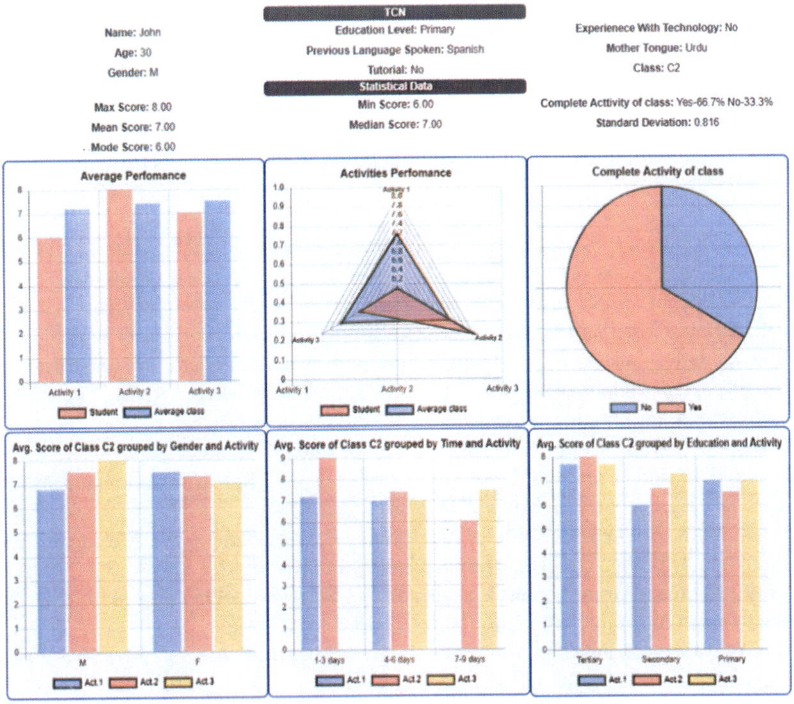

Fig. 3. Performance evaluation of TCN comparing with the average class performance

4 Conclusions and Future Work

In this work, we propose an approach that allows teachers to concisely evaluate the evolution of students in complex interaction environments through advanced metrics that go beyond the usual metrics of learning platforms such as connection time or the results obtained in the evaluation tests. Simultaneously, a visual analytics knowledge generation framework is proposed, which encapsulates data analytics techniques and sophisticated visualizations that enable teachers to gain insights concerning TCNs' language learning performances. The ultimate goal is to empower teachers and authorities to detect problems that TCNs can encounter, and anticipate them in time to effectively support the authoring of appropriate language courses.

As future directions, the proposed framework will be soon evaluated over real use cases and should be further developed according to the provided feedback. Additionally, reinforcing the VAC with enhanced data mining processes, such as classification/clustering algorithms and association rules, will provide a robust environment to reveal hidden and valuable knowledge to improve the TCNs' language learning process. For example, creating profiles or groups of TCNs in the same classroom with similar characteristics in terms of their activity scores and the automated correlation with socio-demographic information among those students, will allow the system to provide to the teachers personalised, student-specific recommendations.

Acknowledgements. This work has received funding by the EC under contract H2020–870930 WELCOME.

References

1. Vieira, C., Parsons, P., Byrd, V.: Visual learning analytics of educational data: A systematic literature review and research agenda. Comput. Educ. **122**, 119–135 (2018)
2. Reyes Sánchez, S., Berns, A.: A review of virtual reality-based language learning apps. RIED. Revista Iberoamericana de Educación a Distancia **24**(1), 159–177 (2021)
3. Blatyta, D.F., Spada, N.M.: O papel do professor de línguas na construção de uma aprendizagem significativa. In: Silva, K.A., Alvarez, M.L.O. (eds.), Perspectivas de investigação em Linguística Aplicada. São Paulo: Pontes (2006)
4. Council of Europe. Council for Cultural Co-operation. Education Committee. Modern Languages Division. Common European Framework of Reference for Languages: learning, teaching, assessment. Cambridge University Press (2001)
5. Keim, D., Andrienko, G., Fekete, J.-D., Görg, C., Kohlhammer, J., Melançon, G.: Visual analytics: definition, process, and challenges. In: Kerren, A., Stasko, J.T., Fekete, J.-D., North, C. (eds.) Information Visualization. LNCS, vol. 4950, pp. 154–175. Springer, Heidelberg (2008). https://doi.org/10.1007/978-3-540-70956-5_7
6. Cui, W.: Visual analytics: a comprehensive overview. IEEE Access **7**, 81555–81573 (2019). https://doi.org/10.1109/ACCESS.2019.2923736
7. Endert, A., Hossain, M.S., Ramakrishnan, N., North, C., Fiaux, P., Andrews, C.: The human is the loop: new directions for visual analytics. J. Intell. Inf. Syst. **43**, 411–435 (2014). https://doi.org/10.1007/s10844-014-0304-9
8. Sacha, D., Stoffel, A., Stoffel, F., Kwon, B.C., Ellis, G., Keim, D.A.: Knowledge generation model for visual analytics. IEEE Trans. Vis. Comput. Graphics **20**(12), 1604–1613 (2014). https://doi.org/10.1109/TVCG.2014.2346481
9. Thomas, J.J., Cook, K.A.: Illuminating the Path. IEEE Press, New York (2005)
10. Thomas, J.J., Cook, K.A.: A visual analytics agenda. IEEE Comput. Graphics Appl. **26**, 10–13 (2006)
11. Pohl, M., Smuc, M., Mayr, E.: The user puzzle—explaining the interaction with visual analytics systems. IEEE Trans. Vis. Comput. Graphics **18**(12), 2908–2916 (2012). https://doi.org/10.1109/TVCG.2012.273
12. Keim, D., Kohlhammer, J., Geoffrey, E., Florian, M.: Mastering the Information Age – Solving Problems with Visual Analytics. Eurographics Association, Goslar (2010)
13. Zhang, L., Stoffel, A., Behrisch, M.: Visual analytics for the big data era—a comparative review of state-of-the-art commercial systems. In: IEEE Conference on Visual Analytics Science and Tech-nology (VAST), pp. 173–182 (2012). https://doi.org/10.1109/VAST.2012.6400554
14. Sheridan, K.M., Kelly, M.A.: Teaching early childhood education students through interactive scenario-based course design. J. Early Childhood Teach. Educ. **33**(1), 73–84 (2012)

15. Endert, A., et al.: The state of the art in integrating machine learning into visual analytics. Comput. Graphics Forum **36**(8), 458–486 (2017). https://doi.org/10.1111/cgf.13092
16. Pirolli, P., Card, S.: The sensemaking process and leverage points for analyst technology as identified through cognitive task analysis. In: Proceedings of International Conference on Intelligence Analysis, vol. 5, pp. 2–4 (2005)

Check for updates

Experiential Learning in Vehicle Dynamics Education via a Scaled Experimental Platform: Handling Performance Analysis

Moein Mehrtash(✉)

W Booth School of Engineering Practice and Technology, McMaster University, Hamilton, Canada
mehrtam@mcmaster.ca

Abstract. This paper explores experiential learning in high education in Automotive Engineering by using scaled experimental platforms. A set of student-center laboratory activities has been established with a pedagogical approach based on Kolb's experiential learning theory. The chosen topic to be educated on is road vehicle dynamic performance focused on automotive standards and real-world problems in the automotive industry. Implementing experiential learning in a road vehicle dynamic curriculum is a considerable challenge due to university laboratories' safety, cost, and space restrictions. This paper presents the adoption process of a scaled model experiment for vehicle handling tests that can be integrated with current university infrastructures. A go-kart vehicle has been equipped with a measurements unit to perform road tests by students in a secured space at the main campus parking lot. The scaled experiments provide a learning environment with concrete experiments for experiential learning. Furthermore, students can practice industry-level standards with scaled experiments.

Keywords: Experiential learning · Vehicle dynamics education · Handling performance analysis · Scaled model experiment

1 Introduction

Experiential learning is becoming more recognized in engineering education in recent years. This methodology focuses on making meaning from direct observation of experiments [1–6]. The most authentic experience in road vehicle dynamics education would be to have students drive an actual vehicle, accomplish specific driving maneuvers, use onboard instrumentation to collect vehicle and driver data, and modify the vehicle and driving maneuvers to observe substantial changes in dynamic characteristics. While not impossible! Concerns about cost, time, space, safety, and weather restrictions make this unfeasible at most schools. Computer simulation shows many promising ways to overcome reviewed limitations in the engineering education curriculum [7–10]. However, for many topics, computer simulation is not as effective as actual and concrete experimentation.

The scaled-down models and experimentations have been widely used in engineering to reduce development costs. The scaled model experimentation can be

M. E. Auer and T. Tsiatsos (Eds.): IMCL 2021, LNNS 411, pp. 694–702, 2022.
https://doi.org/10.1007/978-3-030-96296-8_62

employed in engineering education to create a feasible experiential learning environment in university laboratories [11–13]. In [12], a scaled demonstration was used in designing structures resisting earthquakes. These models were placed on a small instructional shake table and excited by earthquake records. Students can understand the concepts of the load resisting structure and learn to understand the intended inelastic response of structures caused by seismic shaking through a hands-on model in university-level laboratories. Scaling down is possible in many engineering experimentations; however, some features may have inadequate strength in concepts for concrete experimentation. According to Kolb, “Experiential learning is a powerful and proven approach to teaching and learning that is based on one incontrovertible reality: people learn best through experience” [14]. Therefore, the engineering educator must design and develop the scaled model experimentations carefully to create concrete experimentation for the learning environment.

The pedagogical approach presented in this study can represent a reference point for discussions in the experiential learning environment for road vehicle dynamics curriculum, considering Kolb’s theory as a model for developing the teaching-learning process and scale model experimentation as a teaching tool. This paper includes developing a scaled-down experiment of the “Vehicle Handling Performance Test” in university laboratories for experiential learning. This paper is organized as follows. First, it reviews the learning outcomes and teaching strategy for the “vehicle dynamic” course. Second, it describes the scaled-down model development and implementation in the class. Finally, some overall conclusions and our ongoing works are presented in the Conclusions and Future Works section.

2 Learning Outcomes and Teaching Strategy

The “Road Vehicle Dynamic” course (AutoTech 4DV3) offers at the School of Engineering Practice and Technology at McMaster University includes a weekly three-hour lecture and bi-weekly two-hour laboratory session, a total of six-week laboratory sessions for a semester. The learning outcomes (LO) of this course are defined.

- LO1: Identify the road loads experienced by a road vehicle
- LO2: Analyze, evaluate, and predict road vehicle performance
- LO3: Interpret the design considerations of a road vehicle performance in acceleration, braking, cornering, and rollover
- LO4: Apply the knowledge in standard road test and laboratory experimentation situation.

The six laboratory sessions for this course are designed based on Kolb’s experiential learning theorem and aligned with the course learning outcome. Due to the restriction of testing facilities in academia, various methodologies have been employed for experiential learning; the following are listed six developed laboratory sessions:

- Laboratory 1: aerodynamic loads applying to the vehicle: scaled vehicle models were used to test in the wind tunnel to observe the effect of body geometry in aerodynamic loads.

- Laboratory 2: Handling performance testing and analysis: the scaled road test was used to measure vehicle handling. This paper is focused on the development of this methodology.
- Laboratory 3, 4, 5, 6 are focused on the coast down test, fishhook test [7], brake performance test [7], and adaptive cruise control system. For all these laboratory sessions, various simulation tools have been employed to create concrete experimentation.

Practical experiential learning is seen when a person progresses through a cycle of four steps: (1) employing a substantial experience followed by (2) observation of that experience that precedes to (3) the development of intangible concepts and generalizations, which are then (4) used to investigate premise in future. Thus, detailed planning of laboratory sessions employing four stages of Kolb's Experiential Learning Theory is presented in the next section.

3 Scaled Model Experimentation: Road Vehicle Handling Performance

"Handling" is a roughly used term meant to imply the responsiveness of a vehicle to driver input or the ease of control. As such, handling is an overall measure of the vehicle-driver combination. The driver and the vehicle combination is a "close loop" system, meaning that the driver observes the vehicle's direction or position and corrects their input to achieve the desired motion. To characterizing vehicle performance, the vehicle's open-loop behavior is used. The "Open-loop" behavior is precisely defined as "directional response" behavior. This test aims to determine the open-loop steady-state directional control response of passenger cars by measuring the steady-state cornering behavior, which is one of the factors composing vehicle dynamics and road holding properties.

The steady-state cornering dynamics can be derived from Newton's Second Law application and explained slipping force [15]. The simplified vehicle dynamic model is a two-degree-of-freedom bicycle model representing the lateral and yaw motions (see Fig. 1). The bicycle model can be used for handling analysis purposes [15, 16]. The bicycle model represents a relation between the vehicle steering wheel angle δ, geometry of cornering, and vehicle velocity v as:

$$\delta = \frac{L}{R} + K\frac{v^2}{Rg} \tag{1}$$

Where L is the vehicle wheelbase, R is the radius of curvature, g is the gravity constant, and K is the understeer gradient of the vehicle, see Fig. 1. The vehicle response to steering input is measured with yaw rate r (change of the heading angle $r = \dot{\varphi}$ as shown in Fig. 1):

$$\frac{r}{\delta} = \frac{\frac{v}{L}}{1 + K\frac{v^2}{L}} \tag{2}$$

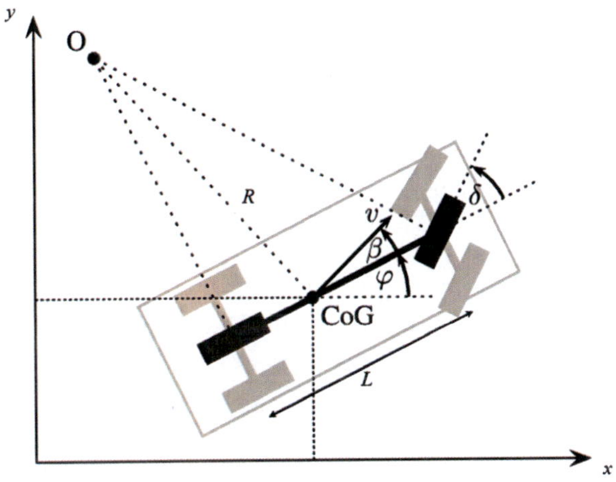

Fig. 1. Bicycle model of vehicle for handing test.

The understeer gradient K is a significant parameter that affects the kinematics of the vehicle in the cornering. Thus, it is required to measure it precisely via experimental measurements.

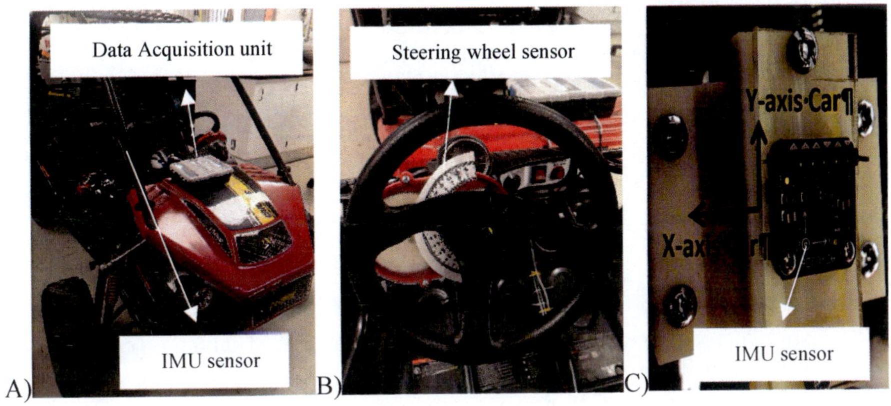

Fig. 2. Go-hart equipped with instrumentation (A), developed steering position sensor (B), and IMU sensor (C).

The standard handling test in the automotive industry is well described in ISO 4138 [17, 18]. This standard has the scope to describe handling properties during steady-state cornering in the complete range of operation. The test is performed to cover all combinations of vehicle speed, steering angle, and cornering radius. A particular test track is required to achieve the standard industry-level test, which is not feasible for many universities. Furthermore, the safety concern in the education environment will not allow performing such a test. Thus, a go-kart car has been equipped with customized measurement units to perform handling tests by students, see in Fig. 2. The data acquisition unit on the vehicle is storing the vehicle's speed, acceleration, and steering position during the test. The on-campus parking lot is secured for the test event, and students are performing two cornering tests at two radii of 7.5 m and 10.5 m with various speeds, Fig. 3. The co-driver is helping the driver with the test procedure and records the data on a laptop connected to the measurement unit, Fig. 4. A.

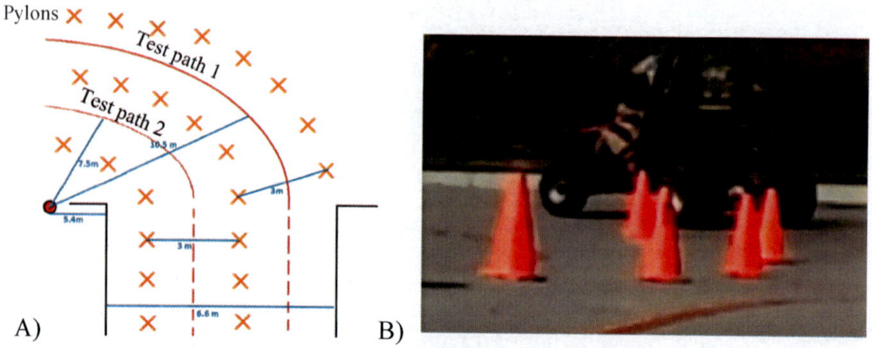

Fig. 3. Two test tracks with a radius of 7.5 m and 10.5 m (A) the driving paths are specified with pylons.

4 Practical Laboratory Framework: Implementation of Kolb's Experiential Learning Cycle

This section presents detailed planning of laboratory sessions and students' activities with four of Kolb's experiential learning theory stages. The students' activities are planned using the scaled handling test explained in the previous section.

4.1 Kolb's Concrete Experience

The instructor presents a comprehensive overview of the "Handling Test" goals and reviews the ISO 4138 standard procedure. Each pair of students uses one go-kart vehicle, so the driver and co-driver need to check the test procedure, Fig. 3. This provides them a concrete experience in practicing industry-level standards. The co-driver can observe live graphs from vehicle performance, so the driver is adjusting the vehicle speed and steering based on a recommendation from the co-driver. Most of the groups are finishing the road test in an hour, and then they need to post-process the recorded data from the handling test in the indoor laboratory. Students observe that the measured data must be filtered to remove sensor noises, Fig. 3.

4.2 Kolb's Reflective Observation

Each group should discuss the vehicle lateral performance in response to the driver steering inputs during the handling test. According to their experimental measurements, each group answers some key questions such as:

- determine the understeer gradient of the vehicle and discuss your finding with theoretical calculation
- calculate the vehicle trajectory from measured data and compared it with test track geometry
- determine the Ackerman angle from measured data and compare it with the theoretical value
- determine the vehicle stability: understeer, neutral, or oversteer.

After finding the answers to the mentioned questions as a reflective observation, students share their findings with other groups. Students observe that each team has different results but is consistent with their findings. This will help students to understand other factors such as wind effect, driver skill in cornering, tire inflation pressure, and many more that cause these mixed results.

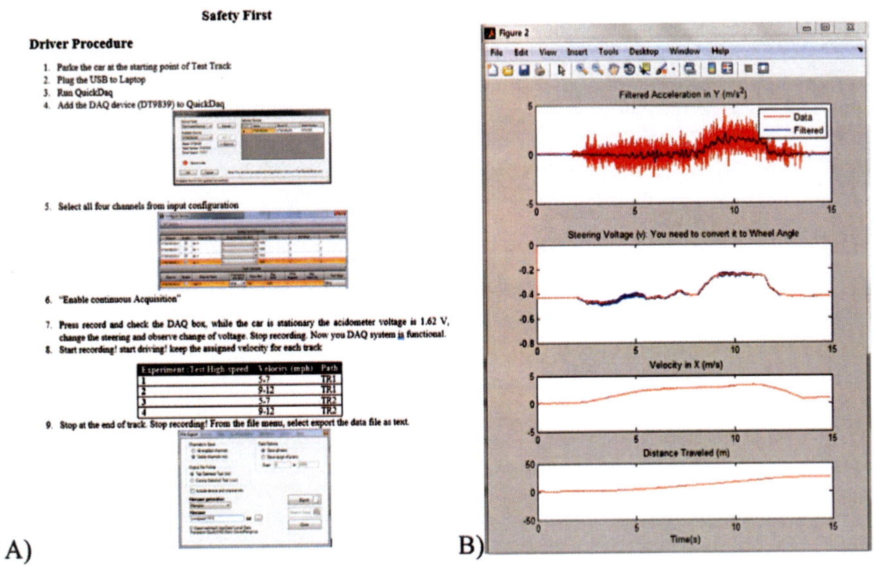

Fig. 4. Handling test procedure reviewed by co-driver (A) various parameters measurement and filtering process.

4.3 Kolb's Abstract Conceptualization

The instructor reinforces the theory, and the students are involved in thinking and forming a principle about the vehicle lateral response while changing some of the dominant physical parameters. Students measure the tire tread size, tire inflation pressure, and weight of driver and co-driver for each vehicle (four vehicles are available for this test in the laboratory) to discuss their findings aligned with theoretical concepts. Students know that tire tread size and tire inflation pressure change the tire cornering stiffness from the theoretical concepts. The tire cornering stiffness is the dominant parameter affecting the vehicle understeer gradient, which is the measure of vehicle response to driver input.

4.4 Kolb's Active Experimentation

It is the time that students actively deal with a real-world problem and use their obtained knowledge. Students gained know with the constant radius handling test. The instructor asked students to define the "constant steer" and "constant speed" handling test requirements. Students must review ISO 4138 and determine a test plan to measure the vehicle's understeer gradient with new methods. In the following laboratory session, students track their planned test. Finally, each team submitted a comprehensive report about their findings. Figure 5 demonstrates a test result analysis by one team. Students use the measured acceleration and filter sensor data to calculate vehicle speed and travel distance during a cornering test.

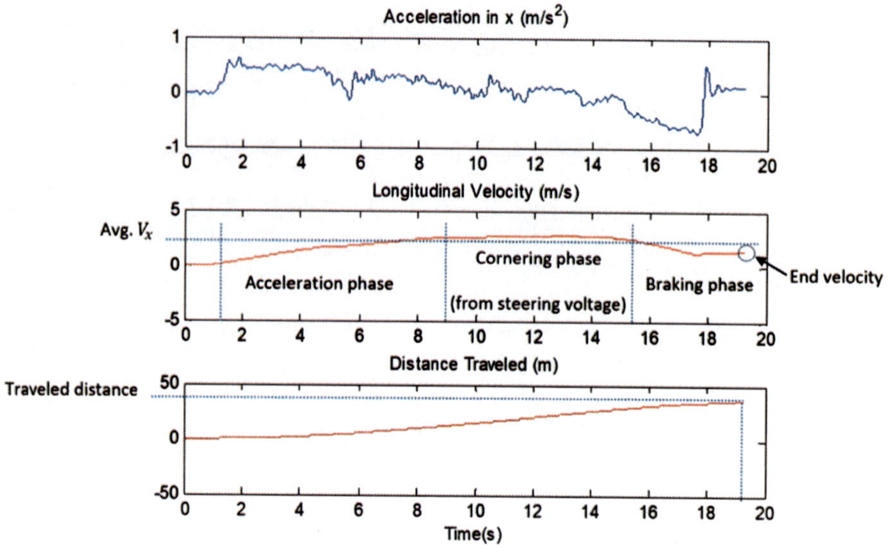

Fig. 5. Handling test analysis phases: acceleration, cornering, and braking

4.5 Summary of Kolb's Learning Stages and Course Learning Outcomes

The pedagogical methodology organizes scaled experimentation for road vehicle handling performance concepts. Table 1 summarizes the developed guideline for students' activities based on Kolb's learning stages and course learning outcomes.

Table 1. Methodological organization of scaled experimentation for handling tests.

Learning outcomes	Kolb's learning stage	Students activities
LO4	Concrete experience	Review and perform automotive standard in scaled domain
LO1, LO2	Reflective observation	Post-processing the measured data and define vehicle response parameters
LO2, LO3	Abstract conceptualization	Share their findings from vehicles with various physical parameters and discuss them based on concepts learned in lecture time
LO2, LO4	Active experimentation	Use industry-level standards and plan a handling test. Discuss their findings from different methods

5 Conclusion

This research has provided a deeper insight on best practices with the scaled laboratory for road vehicle dynamic curriculum experiential learning by employing real-world problems and industry-level standard test procedures. The students' satisfaction from the learning environment in seven years in a row is assessed with two questions: 1) how do you rate the value of this course compared with others you have taken at McMaster University and 2) independent critical judgment was encouraged. Table 2 demonstrates the result of student evaluation in reply to two mentioned questions, the scale for this question is from 1 to 5, 1: very poor and 5: excellent. In 2014, the experiential methodology was applied in the automotive curriculum, and over the next seven years, it has been improved to reach the current status. This study establishes a framework for experiential learning in one aspect of the automotive engineering curriculum; however, a greater focus on experiential learning could produce more exciting methodologies that can be employed in university-level laboratories with industry-oriented learning outcomes.

Table 2. Student evaluation from the course.

Year	No. of students	Q1: Course relative value	Q2. Critical thinking
2014	32	3.65 (73%)	3.23 (64%)
2015	24	4.26 (85%)	4.30 (86%)
2016	34	4.82 (96%)	4.85 (97%)
2017	31	4.87 (97%)	4.99 (96%)
2018	32	4.97 (99%)	4.94 (99%)
2019	34	4.62 (92%)	4.71 (94%)
2020	37	4.61 (92%)	4.81 (96%)

References

1. Hajshirmohammadi, A., Zarei, N.: Incorporating experiential learning in lower division engineering courses. In: Proceedings of the Canadian Engineering Education Association (2015). https://doi.org/10.24908/pceea.v0i0.5754
2. Hajshirmohammadi, A.: Incorporating experiential learning in engineering courses. IEEE Commun. Mag. **55**(11) (2017). https://doi.org/10.1109/MCOM.2017.1700373
3. Yuen, T., Balan, L., Mehrtash, M.: Implementation of an absorber design for vibration control in automation systems. In: Procedia Manufacturing, vol. 32 (2019). https://doi.org/10.1016/j.promfg.2019.02.255
4. Mehrtash, M., Ghalkhani, K., Singh, I.: IoT-based Experiential E-Learning Platform (EELP) for online and blended courses. In: 2021 International Symposium on Educational Technology, pp. 252–255 (2021)
5. Srinivasan, S., Rajabzadeh, A.R., Centea, D.: A project-centric learning strategy in biotechnology. In: Auer, M.E., Hortsch, H., Sethakul, P. (eds.) ICL 2019. AISC, vol. 1134, pp. 830–838. Springer, Cham (2020). https://doi.org/10.1007/978-3-030-40274-7_80
6. Bogoslowski, S., Geng, F., Gao, Z., Rajabzadeh, A.R., Srinivasan, S.: Integrated thinking - a cross-disciplinary project-based engineering education. In: Auer, M.E., Centea, D. (eds.) ICBL 2020. AISC, vol. 1314, pp. 260–267. Springer, Cham (2021). https://doi.org/10.1007/978-3-030-67209-6_28
7. Mehrtash, M., Yuen, T., Balan, L.: Implementation of experiential learning for vehicle dynamic in automotive engineering: roll-over and fishhook test. Procedia Manuf. **32**, 768–774 (2019). https://doi.org/10.1016/j.promfg.2019.02.284
8. Lewis, K., Hulme, K., Kasprzak, E., English, K., Moore-Russo, D.: Experiential learning in vehicle dynamics education via motion simulation and interactive gaming. Int. J. Comput. Games Technol. (1) (2009). https://doi.org/10.1155/2009/952524
9. Centea, D., Singh, I., Balan, L., Yuen, T.: A framework of the bachelor of technology concept and its significant experiential learning component. In: Proceedings of the Canadian Engineering Education Association (CEEA) (2015). https://doi.org/10.24908/pceea.v0i0.5844
10. Perhinschi, M.G., Beamer, F.: Flight simulation environment for undergraduate education in aircraft health management. Comput. Educ. J. **22**(3), 50–62 (2012)
11. Meliopoulos, A.P.S., Cokkinides, G.J., Mohagheghi, S., Dam, Q.B., Alaileh, R.H., Stefopoulos, G.K.: A laboratory setup of a power system scaled model for testing and validation of EMS applications (2009). https://doi.org/10.1109/PTC.2009.5282224
12. Purasinghe, R., et al.: Bringing current research to the classroom using the Linked Column Framed system in an undergraduate structures lab (2011). https://doi.org/10.18260/1-2–17578
13. Maletsky, L., Hale, R.: The practical integration of rapid prototyping technology into engineering curricula, July 2021
14. Kolb, D.A.: Experiential Learning- Experience as the Source of Learning and Development (2nd Edition), vol. 53, no. 9. (2015)
15. Gillespie, T.D.: Fundamentals of Vehicle Dynamics (1992). https://doi.org/10.4271/r-114
16. Kritayakirana, K., Gerdes, J.C.: Autonomous vehicle control at the limits of handling. Int. J. Vehicle Auton. Syst. **10**(4) (2012). https://doi.org/10.1504/IJVAS.2012.051270
17. Perrelli, M., Cosco, F., Carbone, G., Mundo, D.: Evaluation of vehicle lateral dynamic behaviour according to ISO-4138 tests by implementing a 15-DOF vehicle model and an autonomous virtual driver. Int. J. Mech. Control **20**(2) (2019)
18. I. S. O. ISO: 4138: Passenger cars–steady-state circular driving behaviour–open-loop test methods. ISO: Geneva, Switzerland (2012)

Work in Progress: Immersive Web Environments to Support Pedagogical Activities in Formal Contexts

Bárbara Cleto(✉), Ricardo Carvalho, and Maria Ferreira

Universidade de Aveiro, Aveiro, Portugal
{barbara.cleto,ricardojoc,mariajesusferreira}@ua.pt

Abstract. The present research is in its initial phase and has originated three studies. This meta-study aims to investigate the use of Immersive Web Environments to support pedagogical activities in formal contexts. The general objectives are related to the conceptualization regarding the use and creation of environments suitable for learning contexts, to promote and integrate them and, finally, to search for the most appropriate pedagogical practices for their use. In this article, the conceptualization proposals of the three investigations are described, addressing the ideas and motivations that led to their realization. One of the investigations aims to evaluate the educational potential of immersive web environments in the educational practices of primary and secondary school teachers. In the other, we intend to analyze the interactions between teachers, students, and epistemic objects in an immersive web environment. In the third one, it is intended to analyze the impact and appropriation that students make of immersive web environments, given the fact that they are participating in activities created by them.

Keywords: Immersive web environments · Teaching/learning · Interaction · Co-creation

1 Introduction

In a society where change, information, the vertiginous evolution of technology and the consequent thirst for innovative experiences reign supreme, the use of technology merely to digitize traditional school materials falls far short of the goals to be achieved in the face of new teaching/learning models and methods. This has revealed to be even more pressing when faced with a pandemic scenario, which saw the use of technology as solution tool. Immersive Web Environments (IWE) emerge as a promising platform to explore, namely in the field of education. Therefore, these IWE might have potential to be used in any subject at any level of education, with the goal of promoting learning. The possibility for students to work in a real environment and at the same time visualize virtual objects related to the task they are performing, stimulates their interest, making it a much more attractive and motivating educational environment. Using this with teaching strategies where knowledge is not only acquired through the transmission of content, but built by the student, might show itself as an asset. Specially in a

M. E. Auer and T. Tsiatsos (Eds.): IMCL 2021, LNNS 411, pp. 703–710, 2022.
https://doi.org/10.1007/978-3-030-96296-8_63

teaching perspective that is directed towards constructivist theories. This construction of knowledge is an active process based on cognitive activities developed by the student. If education is considered as a process of discovery, observation, exploration, construction of knowledge, and that some subjects are difficult to understand because it is not possible to witness the process as it happens, the use of resources such as IWE can provide an opportunity for the students to learn something abstract and realize its connection with the real world. IWE present themselves as a technological alternative with great potential, mainly due to their interaction capabilities, in which the student can interact with the environment, with other students and the teacher. Furthermore, because of their illustrative power, in which the environment is presented in the most natural and intuitive way possible, its leaning outcomes might prove to be more positive.

These papers aim to assess the impact that IWE can have on the teaching-learning process and all its participants. The presented work is based on research carried out in recent years and focuses essentially on the relevance of immersive environments in education. IWE have been gaining notoriety in different research and application areas. A particular interest has arisen in the field of education. The studies consulted, essentially systematic reviews and meta-analyses, show that their use can have very positive results in the teaching-learning process. However, the research also indicates that the teacher must know how to use these environments properly and plan how the curricular content should be taught in the classroom when using them. This research aims to contribute to improve knowledge in this area as well as disseminate IWE as support tool for teachers.

2 Immersive Web Environments: Research

This section refers to the three investigations that are intended to be carried out. The investigations are in their initial phase.

2.1 Research 1 – Meta Cognition in Immersive Web Environments Personalized by Students

With this research, we intend to explore new teaching/learning spaces, integrating technology, pedagogy and content in an appropriate way, for the promotion of good practices and the success of educational experiences [1]. In this research, we aim to study the educational potential of immersive virtual environments, Immersive Web [2], by designing activities and customizing these environments, according to the various themes, in a collaborative process of co-creation between primary and secondary school students, in order to evaluate the appropriation that students make of the space created by them and/or created by classmates and/or teachers.

We intend to question whether, by being personalized and allowing the recreation of "new" environments, they can facilitate the teaching and learning process, both regarding the different curricular contents and the understanding of different concepts. We seek to understand the students' appropriation of the IWE because they participated in its construction. We also seek to understand the appropriation that students make of the IWE even without having participated in its construction. And how these environments can engage and motivate students, promoting collaboration among them, not only in a face-to-face teaching context, but also in distance education.

Motivation
The working group, which will develop the three studies, has already carried out some experiments, using IWE to teach contents linked to the teaching of basic algorithmic concepts. The working group, tried to "escape" the 2D platforms (Zoom, Teams, Classroom), normally used during the pandemic. Considering the students' reactions and the results obtained, the will arose to analyze the educational potential of these environments in a teaching/learning context. The idea of personalizing the environments was suggested by the students, who questioned if it would be possible to build and personalize their classrooms.

Research Question and Objectives
For this project, the following research question was formulated: "What is the impact on the educational experience of the involvement and incorporation of primary and secondary school students in virtual environment co-creation processes? The aim is to understand, based on the experience that will be implemented in a classroom context, the impact, and challenges of the combination of these three vectors: IWE, personalization and co-creation.

The aim is to gain teachers' and students' perspectives on the use of IWE and to understand how students take ownership of it by participating and being involved in its personalization.

To answer the research question, the following general objectives were defined:

- To study immersive learning environments;
- To study the potentialities of these types of environments to apply in an educational context;
- To study whether these environments induce innovative pedagogical practices;
- conceptualize environments;
- To adjust environments to the target audience (considering ages/profiles/thematic areas)

And the following specific objectives were defined:

- To carry out a survey of the main educational experiences based on the personalization of virtual environments (theoretical framework);
- To design an empirical study with users in a real educational environment;
- accompany and support the process of co-creation of the environments;
- To evaluate the educational experience resulting from the empirical study, namely regarding:
  - analyze the appropriation that students make of these environments (students who built and used, group A, and those who only used, group B);
  - comparing the impact on the educational experience between students who participated in the co-creation of the environment and used the environments they created, and those who only used the environment, created by classmates and/or teachers.

Methodology
Design-Based Research (DBR) was chosen as the methodological design, since it has demonstrated its potential as an adequate methodology, both for research and for the design of learning based on the integration of technologies in school contexts [3].

DBR, allows exploring the potential of technologies in education to solve concrete problems, involving participants, and acting in the context, bringing practical and scientific contributions [4]. The model, proposed in 2013, by Plomp [5], to operationalize DBR, is composed of three phases: 1 – literature review, theoretical framework, and related works; Phase 2 – development and Phase 3 – evaluation.

Expected Results
The literature review revealed an almost total absence of publications on this topic in national schools and in basic and secondary education. The existing studies in other countries are mainly conducted with higher education students. With this study, we hope to fill this gap, which may exist due to teachers' lack of knowledge, either because of obsolete equipment in schools or because of teachers' lack of confidence in the use of Information and Communication Technologies (ICT), so often referred to in studies on educational practices and ICT.

As the study involves students, who are invited to rethink the "classroom space", it is believed that it can contribute to create a workflow of personalization. And who knows, maybe it will move to a dissemination level, where students from various schools can meet and create social learning spaces [6].

2.2 Research no. 2 – Mapping the Use of Immersive Web Environments in Educational Practices by Primary and Secondary School Teachers

Motivation
The immersive web, powered by WebXR technology, facilitates access to online content that can be experienced on devices with Virtual Reality and Augmented Reality (VR/AR) capabilities brings other levels of immersion and presence that can enhance learning outcomes. This new generation of technologies can have an impact on education, becoming a tool to support new educational practices that can improve teaching and learning. The results of this study will contribute to provide evidence on teaching practices and allow us to present a set of recommendations to support the use of these new technologies by the educational community. Ultimately, it is intended to formulate a theoretical framework to support the adoption and implementation of strategies supported by immersive web technologies.

Hypothesis and Objectives
The research question that reflects the purposes of this PhD study, viz. the impact of using immersive web environments in teachers' pedagogical practices, is the following: How immersive web environments impact primary and secondary school teacher's pedagogical practices according to their own perceptions?

To explore the answer to the formulated research question four specific objectives were established:

- To assess the relevance (predicting benefits and possible implementation difficulties) of using immersive web environments on primary and secondary education levels.
- To assess the impact of using immersive web environments on teachers' pedagogical practices (on a motivation, collaborative, and communication skills level).
- To describe the impact of using immersive web environments on teachers' teaching practices, in relation to how they perceive the use of this type of strategy and to understand how they meet their needs and interests.
- To develop a reference guide that assists the implementation of immersive web environments in varied school contexts, within the scope of primary and secondary education.

Methodology
This research project is based on the methodological approach Qualitative Multi-Case research design. A qualitative approach to research design is one which draws upon the traditions of critical theory and interpretive enquiry [7]. A multi-case study is an approach to qualitative research in which several cases are studied to draw a set of conclusions. It is a multi-method approach that aims to gain a holistic perspective on a single phenomenon or issue [8]. Qualitative data collection methods will allow to find out what people are doing, thinking, and saying in response to their situation or environment. The range of qualitative research techniques includes observation, interviews, and focus groups. These techniques are used to collect data about a specific research question or topic. It is not concerned with generalizing the findings to a larger population, as is the case with quantitative research. The goal is to understand the specific meanings and perspectives of individuals and groups, not to make conclusions about a larger group [9, 10].

Expected Results
With this study, it is expected that evidence will be found that will constitute a valid contribution to the evaluation of the impact of using IWE as a support tool for their pedagogical practices It is intended to discover what teachers can do with the use of the available technology to implement it as means of emotionally engaging students for the benefit of improving their performance and learning interest in a given subject. In the process of studying teachers' methodological approaches to the use of IWE in their classrooms, this study will also contribute to teacher training which in turn will contribute directly to providing them with a tool that hopefully can help solve some of their needs in various educational contexts. The simple act of teaching how to use this strategy in a teaching-learning environment will facilitate the consolidation of knowledge through the exchange of experiences between trainees and trainer, made more relevant by the fact that all involved will be professional colleagues.

From the convergence of conclusions resulting from the study, it is also expected that the pedagogical solution developed, more specifically, a reference guide that assists in the implementation of IWE in the educational practices of primary and secondary school teachers, to be innovative, in the matter of strategies, methods and means involved, as well as interventive, in the sense that it will stimulate the reflection of the educational community in relation to its pedagogical/learning practices and promote a conscious, autonomous, motivated and efficient application of these practices, contributing to the success of learning in the school environment.

It is also hoped that by using these technological innovations, especially at a collaborative, communicational and interactive level, contributes to stimulate the use and exploration of the potential of ICT, as well as the enhancement of digital literacy among both students and teachers.

2.3 Research 3 – Interaction and Engagement in Immersive Web Environments

Motivation

Immersive web environments appear to be a promising technology in the educational field due to their simple and intuitive modus operandi. Any teacher of any grade level, even without programming skills, can customize these immersive web environments with technological resources and appealing 3D objects.

I and two other colleagues, computer science teachers, conducted an experiment with our students. We planned the pedagogical activities and customized an immersive web environment on the Hubbs Mozilla platform. In this environment, we created a classroom, similar to the face-to-face classroom, and six small rooms, with one table, six chairs and three whiteboards, which served for the students to project their screens. Students entered the immersive web environment represented by avatars, received instruction in the large classroom shared with 29 students, and at the end moved to the group rooms to model their 3D objects. Throughout the experiment, several videos were recorded.

From the analysis of these videos, I became aware of the interactions that took place between students, between students and teachers, and students with technology. The research: "Studying Interaction and Engagement in Immersive Learning Web Environments" arises to study, what patterns of interaction arise from the relationship between students, teachers, and technology, in an immersive learning web environment.

Research Question and Objectives

"What interaction patterns are used by students in an immersive web environment, and what effects do they produce at the level of engagement with learning activities?"

The research objectives are as follows:

- to identify the interaction patterns among learners in an immersive web environment.
- Identify the patterns of interaction between students and teachers in an immersive web environment.
- To identify the patterns of interaction between students and virtual objects in an immersive web environment
- Describe students' behaviour, corporeal and iconic communication (of avatars), during pedagogical activities.
- Theorize an interaction model that promotes student engagement in learning in immersive web environments.

Methodology
The research has a phenomenological-interpretative nature, as it investigates "social realities, focusing on how they are interpreted, understood, experienced and produced by the actors themselves" [11], to describe conduct, gestures, expressions, statements, etc., and their understanding and interpretation, in the context of teaching and learning in which students and teachers are immersed in a web environment. This approach interests us mainly by the "meanings" that actors attribute to the actions they engage in".

In this line, our methodological option falls on Grounded Theory by Barney Glaser and Anselm Strauss [12] formulated in 1967, based on symbolic interactionism. The underlying idea of this research process is that theoretical propositions emerge from the data obtained through the research, rather than being based on previous studies. It is the methodological procedure that generates the understanding of the phenomenon.

Grounded Theory is a theory that is derived from data, systematically gathered, and analysed through a research process. In this process, different strategies and techniques articulated throughout the development of the study are used. Thus, GT way of learning about the worlds we study and, as a method for theory-building that aims to describe and understand those worlds [13].

Expected Results
With this study we hope to identify patterns of interaction between students, teachers and technology. To extract their meanings, to codify the results obtained into units of analysis that will enable us to infer about students' engagement in learning when immersed in a web environment.

3 Conclusion

This article addresses the beginning of a long-term research project, which consists of three studies running in parallel and where the overall goal is to assess the potential of Immersive Web Environments to support pedagogical activities. The research is still at a very early stage of development. In this article, we begin by framing the theme, relying on the literature review already carried out. Then the motivations and specific objectives for each research are described, as well as the methodology to be adopted. Finally, we describe the expected results.

References

1. Khoshnevisan, B., Le, N.: Augmented Reality in Language Education: A Systematic Literature Review (2019)
2. MacIntyre, B., Smith, T.F.: Thoughts on the future of WebXR and the immersive web. In: Adjunct Proceedings - 2018 IEEE International Symposium on Mixed and Augmented Reality, ISMAR-Adjunct 2018, pp. 338–342 (2018)
3. Wang, F., Hannafin, M.J.: Design-Based Research and Technology-Enhanced Learning Environments (2005)

4. Van Den Akker, J., Gravemeijer, K., Mckenney, S., Nieveen, N.: Educational Design Research
5. Plomp, T.: Educational Design Research: An Introduction. Educ. Des. Res. 10–51 (2013)
6. Scavarelli, A., Arya, A., Teather, R.: Virtual reality and augmented reality in social learning spaces: a literature review. Virt. Real. **25**(1), 257–277 (2020). https://doi.org/10.1007/s10055-020-00444-8
7. Merriam, S.B.: The design of qualitative research - what is qualitative research. Qual. Reseacrh Case Study Appl. Educ. 3–25 (1998)
8. Baxter, P., Jack, S.: The qualitative report the qualitative report qualitative case study methodology. Study Des. Qual. Case Study Methodol. Study Des. Implement. Novice Res. Implement. Novice Res. **4** (2008)
9. Denzin, N., Lincoln, Y.: Introduction: the discipline and practice of qualitative research. - PsycNET (2005). https://psycnet.apa.org/record/2008-06339-001. Accessed 12 Jul 2021
10. Flick, U.: Designing Qualitative Research - Uwe Flick - Google Livros. Designing Qualitative Research (2008)
11. Amado, J.: Manual de Investigação Qualitativa em Educação, 2ª (2014)
12. Glaser, B., Strauss, A.: Grounded theory: Strategien qualitativer Forschung, 3., unveränd. Aufl., no. 32. Huber, 2010
13. de Souza, Z.A., Bellochio, C.R.: A Teoria Fundamentada na pesquisa qualitativa em educação musical: delimitações conceituais, construções e potenciais. OPUS **25**(2), 1–16 (2019)

Key Indicators to Measure Student Performance in IoT and Their Teamwork Ability

Daniela Borissova[1(✉)], Victor Danev[1], Magdalena Garvanova[2], Ivan Garvanov[2], and Radoslav Yoshinov[3]

[1] Institute of Information and Communication Technologies at the Bulgarian Academy of Sciences, 1113 Sofia, Bulgaria
daniela.borissova@iict.bas.bg, victor_danev@abv.bg

[2] University of Library Studies and Information Technologies, 1784 Sofia, Bulgaria
{m.garvanova, i.garvanov}@unibit.bg

[3] Laboratory of Telematics at the Bulgarian Academy of Sciences, Sofia, Bulgaria
yoshinov@cc.bas.bg

Abstract. Ongoing digitalization affects almost every area not only of the economy, but also of every home. This determines the importance of the Internet of Things, which aims to connect physical objects in a network. That means different technologies need to be integrated including sensors, software, protocols and specific application area. All of this motivate the universities to prepare well educated and qualified students in the area of Internet of Things. In this regard, the current article deals with identification and evaluation of key indicators to measure student progress in Internet of Things and their ability for team working. Based on these key indicators, a mathematical model is proposed for assessment of students' performance with subsequent ranking. The numerical testing is based on a case study with small group of student assessed toward several hard and soft skills considered with different degree of importance. The obtained results show that the proposed model allows to make different rankings, according to given preferences for the importance of hard and soft skills.

Keywords: Internet of Things · Key indicators · Assessment · Decision-making · Student ranking

1 Introduction

Today, information and communication technologies (ICT) can be found in a variety of applications from smart wearables (John Dian et al. 2020) through smart cities (Garvanov et al. 2021) to smart farming (Moysiadis et al. 2021). This is due to the continuous developments of complex devices and with the active support by the contemporary trends in Internet technologies (Danev 2021). All this means that the learning disciplines must be updated in accordance with the sought-after specialists in

M. E. Auer and T. Tsiatsos (Eds.): IMCL 2021, LNNS 411, pp. 711–720, 2022.
https://doi.org/10.1007/978-3-030-96296-8_64

the world of the Internet of Things (Abdel-Basset et al. 2019). Science, technology, engineering, and mathematics are becoming more critical to economic progress (Han et al. 2016). This is due to the developing information technologies used to automated production where the manufacturing processes are combined with IT systems to achieve better productivity. In this context, the concepts of Lean, Agile, Resilient and Green, are to be obey to achieve better business results (Amjad et al. 2020). Along with these trends, to achieve such industry modernization more and more smart devices need to be used. This determines the importance of IoT, where various physical objects with embedded electronics and software are connected via the Internet and allow data collection and exchange (Sharabov and Tsochev 2020). Implementing such concept is not an easy task due to the heterogeneous nature of the different technologies used by ecosystem IoT (Kumar et al. 2019).

Considering these trends, the article addresses the problems associated with identifying some key criteria that a perfect IoT specialist should meet. For the goal, mathematical model is proposed for assessment of students' performance with subsequent ranking. This assessment is done by identified key performance indicators related to the hard and soft skills.

The rest of the article is organized as follows: Sect. 2 provide brief literature review, Sect. 3 describes some key indicators related to IoT, Sect. 4 contains the formulated model for students' assessment, Sect. 5 refers to the numerical application, Sect. 6 contains results analysis and discussion, while the conclusions are given in Sect. 7.

2 Literature Review

Labor market skills requirements are a relatively new area of research and skills requirements vary from company to company and industry to industry. Before describing the most preferable pretender for a particular job position, it is necessary to understand the difference between hard skills and soft skills. In the general case, hard skills refer to these abilities that can be measured, while soft skills are difficult to be measure and are considered as more universal and could refer not only to one specific job. In the field of software project management, authors discuss soft skills that HR should possess to improve the co-sourcing process (Schlichter and Buchynska 2021). Other authors describe how IT companies look for candidates to their open positions (Montandon et al. 2021). They concluded that programming languages are the most demanded hard skills, while the communication, collaboration, and problem-solving are the most demanded among the soft skills. There are many studies focused on evaluation of acquired hard skills while the soft skills are neglected (Mustakerov and Borissova 2011; Borissova and Keremedchiev 2019). It is found that high-level hard skills requirements are positively correlated to differences in productivity in firms (Lyu and Liu 2021). Recently, a growing interest on soft skills are focused on exploring the main factors to understanding of how they are influenced by socio-economic and family background (Marcenaro-Gutierrez et al. 2021). The authors combine the econometric with multi-objective optimization to provide an overarching framework for assessing the trade-offs between the different dimensions of non-cognitive skills. An optimization model capable to cope with objective evaluation and subjective criteria combined with group members' competency in personal selection is proposed (Borissova 2018). Different approaches to deals with problems of multi-criteria decision are proposed (Borissova 2021; Balabanov 2021).

The application of IoT can be found in every area of our real life and as such a complex problem it is difficult to point out all the difficult skills that the preferred candidate must meet. A good IoT specialist must have knowledge in the field of sensors, that measure and ensure the conversion of environmental data into machine-readable data. IoT connect heterogeneous devices that need to be communicate regardless the used hardware and software. Of particular importance is the knowledge in the field of data transmission technologies, including wireless networks and proper protocols (Hofer-Schmitz and Stojanovic 2020). These protocols need to be enough secure and less energy consumption to ensure a continuous real-time connection of the transmitted data (Bahashwan et al. 2021). Additional knowledge is required to understand data mining (Naka and Guliashki 2021), some artificial intelligence techniques (Jafari et al. 2020) and decision-making models (Borissova 2021), which are a prerequisite for the proper functioning of IoT. Last, but not least the understanding of particular application is advantage to cope with some challenges. All of these define the significance of IoT technology in the field of education in improving the efficiency of teaching and learning (Ramlowat et al. 2019).

The importance of hard and soft skills determines the need to establish an appropriate approach to assess the most appropriate candidate for the specific job description. It should be noted that different economic sectors will identify different specific requirement related to hard and soft skills. The soft skills could be considered as driving forces of the upskilling trend while the requirement for hard skills is relatively constant.

3 Key Indicators for Measuring Student Performance

To be successful the well-educated student, it should proficient in variety of technologies related to sensors and signals processing, software (programming languages), communications protocols, data mining and decision-making. These key indicators are critical for their practical realization and include also particular application area like: health, sports and daily activity, tracking and localization, safety etc. The proposed concept of the required competency of IoT specialist is illustrated in Fig. 1.

Fig. 1. Hard skills and knowledge for successful IoT specialist

The skills related to the sensors include also competencies such as signals processing, data analysis, processing of big data, data visualization. Software skills suppose the presence of the ability for programming on one or more programming languages to program sensors or to visualize and analyze the data from signals processing. Once these data are acquired, they need to be transmitted via a proper communication network and suitable protocols to a dedicated server or cloud. At this stage, all transferred data are to be properly processed (data mining) using a variety of methods including statistics and machine learning to get useful information. Finally, the obtained information should be analyzed additionally via different decision-making models of business intelligence to determine the final decision. This whole process must be subordinated to the respective subject area in which IoT is realized.

Along with these mandatory hard skills, the preferable specialist in the area of IoT should have also the additional skills like good communication, teamwork, leadership, entrepreneurship, ability to conflict and stress management, motivation, time management, confidence building, decision making, etc.

Base on the frame shown in Fig. 1 and considering mention above skills, the following hard and soft skills can be defined as follows:

- *hard skills*: sensors, signal processing, data mining, statistics, data analysis, data visualization, big data, machine learning, programming language/s, business intelligence, decision-making.
- *soft skills*: communication, teamwork, leadership, entrepreneurship, conflict management, stress management, motivation, time management, confidence building, decision-making, problem solving, collaboration, adaptability, etc.

It should be noted, hard skills are much easier to identify as they can be measured, while soft skills are highly subjective. These two types of skills could be considered as evaluation criteria to measure student progress in IoT and their ability for teamwork.

4 Mathematical Model to Assess Student Performance in IoT and Teamwork Ability

In order to be able to assess students' knowledge and their ability to adapt to teamwork, it is necessary to consider two separate parts related to hard and soft skills. This can be realized through the proposed mathematical model for assessment of students' performance ($S_i^{performance}$), formulated as follows:

$$S_i^{performance} = \max\left\{\alpha \sum_{h=1}^{H} w_h e_{ih} + \beta \sum_{s=1}^{T} w_s e_{is}\right\}, \forall i = 1, \ldots, N \quad (1)$$

$$\alpha + \beta = 1 \quad (2)$$

$$\sum_{h=1}^{H} w_h = 1 \quad (3)$$

$$\sum_{s=1}^{S} w_s = 1 \tag{4}$$

where $i = \{1, 2, \ldots, N\}$ represents the set students, coefficient α expresses the importance of hard skills while coefficient β expresses the ability for teamwork (soft skills), the coefficients w_h and w_s express the relative importance between criteria related to the hard and soft skills, e_{ih} and e_{is} represents evaluation scores of i-th student about h-th criterion related to the hard skills, and corresponding i-th student about s-th criterion related to the soft skills. It should be mention, that range for evaluation scores e_{ih} and e_{is} have to corresponds to the range of other variables of the proposed model (1)–(4). Therefore, the acceptable range for these scores is to be between 0 and 1 to have a comparable scale. The relation (2) make possible to aggregate the separated two part of evaluation regarding hard skills (acquired knowledge) and soft skills (teamwork ability) in final generalized assessment. As could be seen from relation (1), the ranking of the students is realized considering hard and soft skills and these two types of components could be taken into account with different importance. This feature allows the model to be made more flexible to consider hard and soft skills with different proportions in determining the final complex studentsԿ ranking.

The proposed mathematical model (1)–(4) could be simplified by using a value equal to zero for the coefficient β ($\beta = 0$). In this case, the model (1)–(4) will rely on only the hard skills (acquired knowledge) of the students'.

5 Numerical Application

In order to verify the applicability of the proposed mathematical model (1)–(4) for ranking students in order to identify the most educated candidates with the ability to work in a team, an experiment was conducted with a group of 25 students. The ranking of these students is done by using the 5 hard skills namely 1) sensors and signal processing (h-1); 2) software: programming language/s (h-2); 3) communications protocols (h-3); 4) data mining (h-4); 5) decision-making model and algorithms (h-5); and 3 soft skills concerning 1) teamwork (s-1); 2) motivation (s-2); 3) time management (s-3). These students' skills are considered as evaluation criteria at the ranking. The corresponding evaluations about the hard skills (e_{ih}) and soft skills (e_{is}) expressed by scores regarding these skills are shown in Table 1.

Table 1. Evaluation score of students toward their skills

Student #	Hard skills					Soft skills		
	h-1	h-2	h-3	h-4	h-5	s-1	s-2	s-3
1	0.94	0.78	0.81	0.94	0.86	0.78	0.98	0.91
2	0.95	0.91	0.79	0.87	0.88	0.92	0.75	0.81
3	0.88	0.96	0.79	0.83	0.89	0.95	0.77	0.92
4	0.87	0.93	0.8	0.82	0.89	0.9	0.85	0.81
5	0.91	0.87	0.79	0.86	0.81	0.82	0.92	0.79
6	0.89	0.93	0.78	0.86	0.79	0.89	0.85	0.77

(continued)

Table 1. (*continued*)

Student #	Hard skills					Soft skills		
	h-1	h-2	h-3	h-4	h-5	s-1	s-2	s-3
7	0.90	0.93	0.72	0.81	0.79	0.88	0.94	0.86
8	0.80	0.97	0.82	0.78	0.9	0.82	0.9	0.79
9	0.86	0.94	0.86	0.86	0.8	0.79	0.85	0.82
10	0.88	0.87	0.86	0.88	0.76	0.88	0.83	0.8
11	0.89	0.85	0.81	0.89	0.72	0.95	0.86	0.85
12	0.90	0.82	0.78	0.81	0.87	0.88	0.92	0.9
13	0.81	0.86	0.8	0.79	0.88	0.85	0.81	0.92
14	0.79	0.86	0.77	0.83	0.84	0.84	0.88	0.91
15	0.79	0.91	0.87	0.85	0.86	0.93	0.79	0.8
16	0.80	0.78	0.91	0.77	0.86	0.9	0.84	0.86
17	0.87	0.85	0.87	0.84	0.81	0.88	0.91	0.77
18	0.78	0.77	0.95	0.79	0.88	0.81	0.88	0.92
19	0.76	0.80	0.89	0.83	0.86	0.93	0.98	0.86
20	0.79	0.86	0.9	0.8	0.86	0.87	0.91	0.79
21	0.86	0.97	0.8	0.86	0.81	0.88	0.89	0.93
22	0.88	0.92	0.86	0.86	0.78	0.95	0.9	0.82
23	0.89	0.96	0.88	0.81	0.80	0.92	0.81	0.9
24	0.89	0.92	0.89	0.78	0.80	0.91	0.72	0.87
25	0.90	0.87	0.86	0.83	0.77	0.81	0.92	0.88

Along with the evaluation score about the hard (e_{ih}) and soft (e_{is}) skills according the proposed model (1)–(4), it is need to determine also ratio between hard and soft skills by the coefficients (α) and (β) and also the value for the coefficients w_h and w_s. Three different cases are identified and shown in Table 2.

Table 2. Coefficients for hard and soft skills and their distribution among criteria

	Case-1	Case-2	Case-3
Hard skills	$\alpha = 0.5$	$\alpha = 0.65$	$\alpha = 0.65$
h-1	0.2	0.2	0.22
h-2	0.2	0.2	0.2
h-3	0.2	0.2	0.22
h-4	0.2	0.2	0.18
h-5	0.2	0.2	0.18
Soft skills	$\beta = 0.5$	$\beta = 0.35$	$\beta = 0.35$
s-1	0.33	0.33	0.38
s-2	0.33	0.33	0.20
s-3	0.34	0.34	0.42

Case-1 considers hard and soft skills with equal importance ($\alpha = \beta = 0.5$) and distribution between criteria concerning both hard and soft skills are also taken with equal importance. Case-2 express the scenario where the hard skills are predominant ($\alpha = 0.65$) in respect to soft skills ($\beta = 0.35$), but distribution between particular criteria remain the same. Case-3 illustrates the situation, where hard skills are more preferred than soft skills, are each criterion from hard and soft skills are considered with different importance.

6 Result Analysis and Discussion

Using the proposed model (1)–(4) and the input data from Sect. 5 several optimization tasks are solved to determine the performance of each student. The obtained results for the ranked students based on the data of Case-1 are visualized in Fig. 1 (Fig. 2).

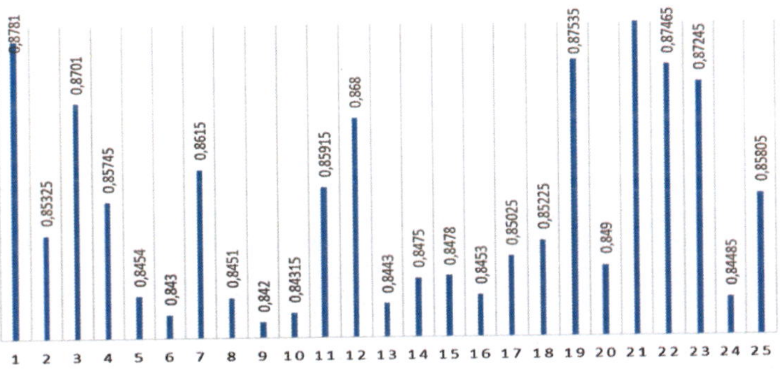

Fig. 2. Ranking of students according to the importance of the criteria from Case-1

From this ranking it can be seen that the first 3 places are occupied by student #21 with overall performance score of 0,88015, followed by student #1 with score of 0,8781 and student # 19 with score 0,87535.

If the strategy is changed and the preferences presented in Case-2 are used, then the ranking of students acquires a different appearance, as shown in Fig. 3.

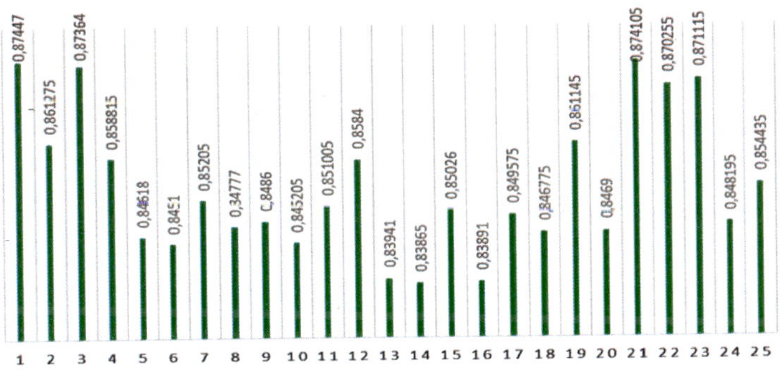

Fig. 3. Ranking of students according to the importance of the criteria from Case-2

In this ranking, the first 3 places are occupied by student #1 with overall performance score of 0,87447, followed by student #21 with score of 0,874015 and student #3 with score 0,87364.

When the preferences expressed through the scenarios of case 3 are used, the following ranking of the students is obtained, as shown in Fig. 4. In this ranking, the first best 3 students in the ranking are as follows: student #3 with a score of 0,88034, followed by student #23 with a score of 0,87764, and student #21 with a score of 0,87492.

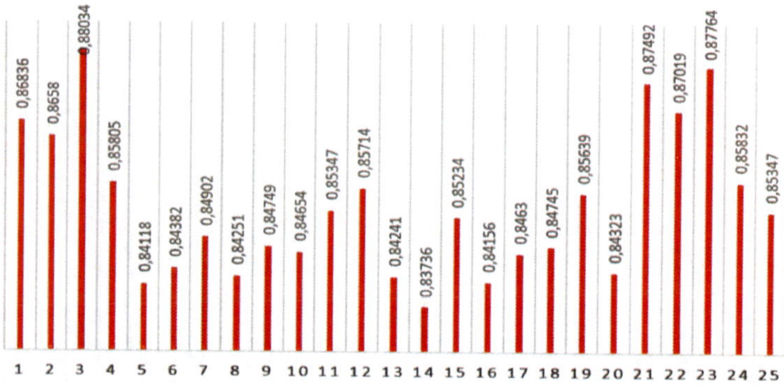

Fig. 4. Ranking of students according to the importance of the criteria from Case-3

The comparison between these students' rankings is shown in Fig. 5.

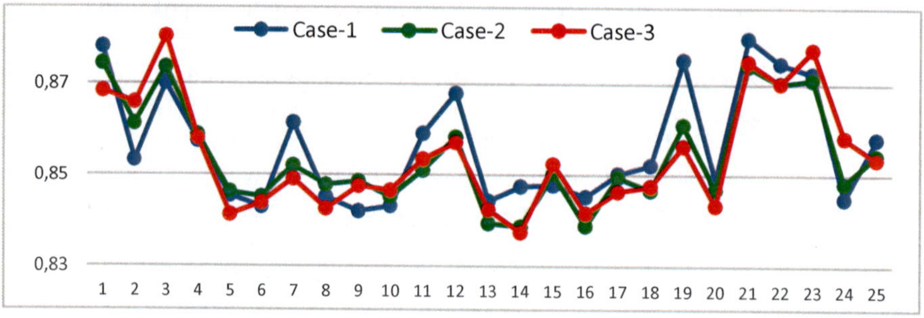

Fig. 5. Comparison of students' ranking

Depending on the purpose of the ranking, it is possible to make proper list for students with more ability of hard skills, suitable list for the students with soft skills, or some combination of hard and soft skills. These lists can be used to recommend specific students to find a suitable job, according to their level of knowledge and adaptation to work in teams.

If the job description allows working alone or working in small teams, soft skills can be neglected compared with the hard skills. The vice versa is also true, the larger team always will require abilities related to the soft skills, while the missed hard skills could be overcome by proper seminars or workshops. Considering the problems of IoT, there are some activities without requirement for team work like programming of sensors or other smart devices at whole.

7 Conclusions

With continued digitalization, science, technology, engineering, and mathematics are becoming more critical to economic progress. This involves also modernization of the industry by using more and more smart devices. That impose to update many of university disciplines to comply with needs of such specialist able to develop and maintenance the devices of IoT. In this regard, the current article aims to identify key indicators to measure student performance in IoT and their teamwork ability. These indicators are used in the formulated integrated model to measure the students' performance considering two major group related to hard and soft skills. The obtained results show practical applicability of the proposed model for ranking the students. Advantage of the proposed modelling approach is the possibility to consider hard and soft skills with different importance. The proposed model (1)–(4) could be applied for ranking of students from different disciplines using variety of hard and soft skills.

Acknowledgment. This work is supported by the Bulgarian National Science Fund by the project "Mathematical models, methods and algorithms for solving hard optimization problems to achieve high security in communications and better economic sustainability", KP-06-H52/7/19-11-2021.

References

Abdel-Basset, M., Manogaran, G., Mohamed, M., Rushdy, E.: Internet of things in smart education environment: supportive framework in the decision-making process. Concurr. Comput. Pract. Exper. **31**(10) (2019). https://doi.org/10.1002/cpe.4515

Amjad, M.S., Rafique, M.Z., Hussain, S., Khan, M.A.: A new vision of LARG manufacturing – a trail towards Industry 4.0. CIRP J. Manuf. Sci. Technol. **31**, 377–393 (2020). https://doi.org/10.1016/j.cirpj.2020.06.012

Bahashwan, A.A., Anbar, M., Abdullah, N., Al-Hadhrami, T., Hanshi, S.M.: Review on common iot communication technologies for both long-range network (LPWAN) and short-range network. In: Saeed, F., Al-Hadhrami, T., Mohammed, F., Mohammed, E. (eds.) Advances on Smart and Soft Computing. AISC, vol. 1188, pp. 341–353. Springer, Singapore (2021). https://doi.org/10.1007/978-981-15-6048-4_30

Balabanov, T.: Solving multi-objective problems by means of single objective solver. Probl. Eng. Cybern. Robot. **76**, 63–70 (2021). https://doi.org/10.7546/PECR.76.21.05

Borissova, D., Keremedchiev, D.: Group decision making in evaluation and ranking of students by extended simple multi-attribute rating technique. Cybern. Inf. Technol. **18**(3), 45–56 (2019)

Borissova, D.: A group decision making model considering experts competency: an application in personnel selections. Comptes rendus de l'Academie Bulgare des Sciences **71**(11), 1520–1527 (2018)

Borissova, D.: An overview of multi-criteria decision making models and software systems. In: Atanassov, K.T. (ed.) Research in Computer Science in the Bulgarian Academy of Sciences. SCI, vol. 934, pp. 305–323. Springer, Cham (2021). https://doi.org/10.1007/978-3-030-72284-5_15

Danev, V.: The Internet of Things: description, applications, development, challenges. Probl. Eng. Cybern. Robot. **76**, 3–24 (2021). https://doi.org/10.7546/PECR.76.21.01

Garvanov, I., Garvanova, M., Borissova, D., Vasovic, B., Kanev, D.: Towards IoT-based transport development in smart cities: safety and security aspects. In: Shishkov, B. (ed.) BMSD 2021. LNBIP, vol. 422, pp. 392–398. Springer, Cham (2021). https://doi.org/10.1007/978-3-030-79976-2_27

Han, S., Capraro, R.M., Capraro, M.M.: How science, technology, engineering, and mathematics project based learning affects high-need students in the U.S. Learn. Individ. Differ. **51**, 157–166 (2016). https://doi.org/10.1016/j.lindif.2016.08.045

Hofer-Schmitz, K., Stojanovic, B.: Towards formal verification of IoT protocols: a review. Comput. Netw. **174**, 107233 (2020). https://doi.org/10.1016/j.comnet.2020.107233

Jafari, R., Razvarz, S., Gegov, A., Vatchova, B.: A survey on applications of neuro-fuzzy models. In: 2020 IEEE 10th International Conference on Intelligent Systems, pp. 148–152 (2020). https://doi.org/10.1109/IS48319.2020.9200185

John Dian, F., Vahidnia, R., Rahmati, A.: Wearables and the Internet of Things (IoT), applications, opportunities, and challenges: a survey. IEEE Access **8**, 69200–69211 (2020). https://doi.org/10.1109/ACCESS.2020.2986329

Kumar, S., Tiwari, P., Zymbler, M.: Internet of Things is a revolutionary approach for future technology enhancement: a review. J. Big Data **6**(1), 1–21 (2019). https://doi.org/10.1186/s40537-019-0268-2

Lyu, W., Liu, J.: Soft skills, hard skills: What matters most? Evidence from job postings. Appl. Energy **300**, 117307 (2021). https://doi.org/10.1016/j.apenergy.2021.117307

Marcenaro-Gutierrez, O.D., Lopez-Agudo, L.A., Henriques, C.O.: Are soft skills conditioned by conflicting factors? A multiobjective programming approach to explore the trade-offs. Econ. Anal. Pol. **72**, 18–40 (2021). https://doi.org/10.1016/j.eap.2021.07.008

Montandon, J.E., Politowski, C., Silva, L.L., Valente, M.T., Petrillo, F., Gueheneuc, Y.-G.: What skills do IT companies look for in new developers? A study with Stack Overflow jobs. Inf. Softw. Technol. **129**, 106429 (2021). https://doi.org/10.1016/j.infsof.2020.106429

Moysiadis, V., Sarigiannidis, P., Vitsas, V., Khelifi, A.: Smart farming in Europe. Comput. Sci. Rev. **39**, 100345 (2021). https://doi.org/10.1016/j.cosrev.2020.100345

Mustakerov, I., Borissova, D.: A conceptual approach for development of educational Web-based e-testing system. Expert Syst. Appl. **38**(11), 14060–14064 (2011). https://doi.org/10.1016/j.eswa.2011.04.214

Naka, E., Guliashki, V.: Optimization techniques in data management: a survey. In: 7th International Conference on Computing and Data Engineering, pp. 8–13 (2021). https://doi.org/10.1145/3456172.3456214

Ramlowat, D.D., Pattanayak, B.K.: Exploring the Internet of Things (IoT) in education: a review. In: Satapathy, S.C., Bhateja, V., Somanah, R., Yang, X.-S., Senkerik, R. (eds.) Information Systems Design and Intelligent Applications. AISC, vol. 863, pp. 245–255. Springer, Singapore (2019). https://doi.org/10.1007/978-981-13-3338-5_23

Schlichter, B.R., Buchynska, T.: Soft skills of delivery managers in a co-sourced software project. Procedia Comput. Sci. **181**, 905–912 (2021). https://doi.org/10.1016/j.procs.2021.01.246

Sharabov, M., Tsochev, G.: The use of artificial intelligence in Industry 4.0. Probl. Eng. Cybern. Robot. **72**, 17–29 (2020). https://doi.org/10.7546/PECR.73.20.02

Implementation of Experiential Learning in Aerodynamic Design of Road Vehicles

Moein Mehrtash(✉)

W Booth School of Engineering Practice and Technology, McMaster University, Hamilton, Canada
mehrtam@mcmaster.ca

Abstract. This paper explores a low-cost experiential learning framework in the aerodynamic design of road vehicles by using scaled models and numerical simulation. Automotive aerodynamics is studied using both computer modeling and experimental analyses using wind tunnel testing. The wind tunnel testing for the automotive application needs enormous facilities. This study demonstrates the use of scaled wind-tunnel testing to familiarize students with the vehicle's aerodynamic shape design. In addition to learning the aerodynamic propensities of a road vehicle, a secondary objective is to have the student become acquainted with wind tunnel operation and instrumentations. In this study, a set of student-center laboratory activities has been developed with a pedagogical methodology based on Kolb's experiential learning theory.

Keywords: Experiential learning · Vehicle dynamics education · Aerodynamic design · Aerodynamic drag measurement

1 Introduction

In the 21st century, education is about training graduates with various competencies and reducing the gap between the classroom and the real-world environment via professional practice and simulating a work environment in the curriculum. This paper explores the pedagogical benefits and methodology of implementing experiential learning in the automotive engineering curriculum. In automotive engineering, developing an experiential learning environment [1–8] with real-world problems is challenging due to various restrictions such as space, cost, and safety [9–11]. Automotive aerodynamics is one of the crucial topics in automotive engineering that includes studying how air moves around road vehicles. Its main goals are reducing drag and wind noise, minimizing noise and lift force [12]. Wind tunnel test [13, 14] and computational fluid dynamics (CFD) [15–17] simulation are two main methods for the study of automotive aerodynamics. The cost of running a wind tunnel testing can be in the range of $ 2′000 up to $ 20′000 per hour [15], which is not affordable for many engineering schools to have an operational wind tunnel for experiential learning.

Experiential education is a philosophy in which educators purposefully engage with learners in direct experience and focused reflection on increasing knowledge, developing skills, and clarifying values [19–22]: "people learn best through experience [23]". Kolb's experiential learning theory [21, 23, 24] is concerned with the learner's

M. E. Auer and T. Tsiatsos (Eds.): IMCL 2021, LNNS 411, pp. 721–732, 2022.
https://doi.org/10.1007/978-3-030-96296-8_65

internal cognitive processes, and it implies that learners transform the experience to knowledge by completing a four-stage learning cycle: concrete experimentation, reflective observation, abstract conceptualization, and active experimentation. Engineering scaled models [25–29] are widely employed to examine complex dilemmas where calculations and computer simulations are unreliable. In [29], scaled experimentation is used in designing structures resisting earthquakes. These models were placed on a small instructional shake table and excited by earthquake records. A model is similar [30] to the actual application in road vehicle aerodynamics if the two share geometric, kinematic, and dynamic similarities. Thus, the scaled-down models with small wind tunnels can be utilized as affordable experiential learning methodologies.

This pedagogical study demonstrates a promising capacity of scaled experimentations for discussions in the experiential learning environment of the road vehicle dynamics curriculum. Kolb's theory is employed as a model for developing the teaching-learning process and scaled model experimentation as a teaching tool. This paper presents developing a scaled-down experiment and planning students' activities in university laboratories for experiential learning. This paper is structured as follows. First, it reviews the learning outcomes and teaching strategy for the "vehicle dynamic" course. Second, it describes the scaled-down model development and implementation in the class. Finally, some overall recommendation and our ongoing works are presented in the Conclusions and Future Works section.

2 Learning Outcomes and Teaching Strategy

The "Road Vehicle Dynamic" course (AutoTech 4DV3) offered at the School of Engineering Practice and Technology at McMaster University includes a weekly three-hour lecture and bi-weekly two-hour laboratory session, a total of six-week laboratory sessions for a semester. The learning outcomes (LO) of this course are defined.

- LO1: Identify the road loads experienced by a road vehicle
- LO2: Analyze, evaluate, and predict road vehicle performance
- LO3: Interpret the design considerations of a road vehicle performance in acceleration, braking, cornering, and rollover
- LO4: Apply the knowledge in standard road test and laboratory experimentation situation

The six laboratory sessions of "Road Vehicle Dynamic" are designed based on Kolb's experiential learning theorem and supported with the course learning outcome. Due to the numerous restrictions of testing facilities in academia, various methodologies have been employed for experiential learning; the following are listed six developed laboratory sessions:

- Laboratory 1: aerodynamic loads apply to the vehicle: scaled vehicle models were used to test in the wind tunnel and observe body geometry's effect in aerodynamic loads.

- Laboratory 2: Handling performance testing and analysis: the scaled road test was used to measure vehicle handling. This paper is focused on the development of this methodology.
- Laboratory 3, 4, 5, 6 are focused on the coast down test, fishhook test [11], brake performance test [11], and adaptive cruise control system. For all these laboratory sessions, various simulation tools have been employed to create concrete experimentation.

This study includes detail experiential learning activities for laboratory sessions intending to understand aerodynamic loads. Students' activities progress through Kolb's cycle of four steps: (1) employing a substantial experience with scaled wind tunnel testing followed by (2) observation the effect of geometry on aerodynamic loads that precedes to (3) the development of intangible concepts and generalizations in airflow separation around vehicles and aerodynamic drag generation, which are then (4) used to investigate premise in future design.

3 Scaled Model Experimentation: Aerodynamic Drag Estimation

The primary objective of aerodynamic testing is to enhance students' understanding of the effect of vehicle external geometry on aerodynamic forces. This may be accomplished by examing various configurations of a car and measuring aerodynamic forces. Drag is the most critical aerodynamic force confronted by the car. The major contributor to the cars' aerodynamic drag is the afterbody due to flow separation in the rear part of the vehicle. Thus, this experiential learning experience will mainly investigate the afterbody shape effect in produced drag force. The overall aerodynamic drag D_A can be characterized by

$$D_A = \left(\frac{\rho V^2}{2}\right) C_D A \tag{1}$$

where C_D, A, ρ, and V are aerodynamic drag coefficient, frontal areal of the vehicle, air density, and airspeed, respectively. Various scaled car models are constructed using 3D printer technology, and these models are used to measure the drag coefficient with various afterbody configurations, see Fig. 1. The basic principle of wind tunnel testing is the forces on an object moving through the air at a particular speed are the same as the forces on a fixed object with air moving past it at the same speed. Of course, the model in the wind tunnel is usually smaller than (but geometrically similar to) the full-size vehicle, so it is necessary to know and apply the scaling laws to interpret the wind tunnel data in terms of a full-scale vehicle.

A. Truck B. Sedan (extension) C. Van (extension)

Fig. 1. 3D printed scalded cars (Truck, sedan, and van configurations) compatible with the wind tunnel test section.

Several low-cost open-return type wind tunnels [12, 31] are designed and constructed for experiential learning goals. Air is drawn from the room into a large settling chamber. Following the settling chamber, the air accelerates through a contraction cone where the area reduces (continuity requires that the velocity increase). The test (working) section is of constant area (10×10 cm^2), see Fig. 2. The test section is fitted with one movable sidewall to make minor adjustments to the area to account for boundary layer growth, thus keeping the stream-wise velocity and static pressure distributions constant. The air exhausts into the room and recirculates. The test section's maximum air velocity of this wind tunnel is 30 m/s. A force transduces has been utilized to measure the drag force experienced by the levitated model inside the wind tunnel test section, accuracy is ±0.2% of full scale, and resolution is 0.0002 N. Two measure air speed inside the test section a pitot tube and differential pressure is used. A digital air flowmeter is placed at the end test section to measure air properties such as humidity, temperature, and velocity, see Fig. 2. The total cost of building such a wind tunnel with all instrumentation can approximately be under $2000.

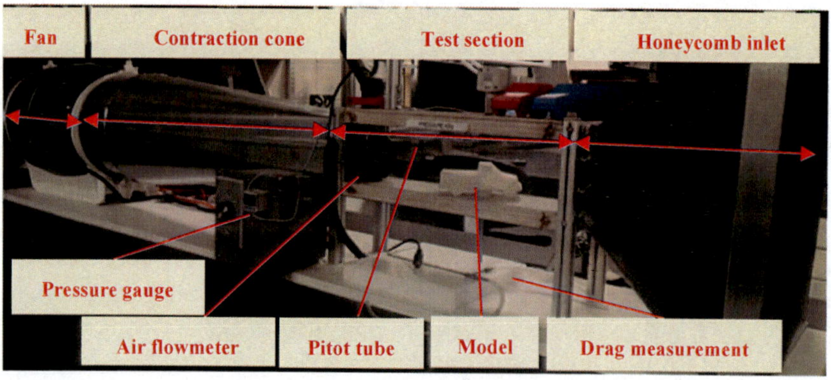

Fig. 2. Open-return type wind tunnel at the Vehicle Dynamic Lab, School of Engineering Practice and Technology, McMaster University.

In addition to learning the aerodynamic propensities of a road vehicle, a secondary objective is to have the student become acquainted with the operation of a subsonic wind tunnel and instrumentations.

4 Practical Laboratory Framework: Implementation of Kolb's Experiential Learning Cycle

This section introduces detailed activities of laboratory sessions based on Kolb's experiential learning theory stages. The students' activities are focused on using the scaled wind tunnel testing for various vehicle geometries and the validation of numerical analysis. In the design stage of a new vehicle, computer numerical simulation plays an essential role in developing low darg aerodynamic design at a lower cost due to the operational cost of a wind tunnel. Thus, a combination of numerical analyses and experimental measurements is vital for a new shape design of a vehicle. Therefore, the following activities are projected to provide industrial-level experiences for students.

4.1 Kolb's Concrete Experience

The instructor explained a comprehensive overview of the "Wind Tunnel Test" goals and the effect afterbody of a vehicle in total aerodynamic drag. Each student group with a maximum of three students uses one wind tunnel to measure the vehicle's drag force with different afterbody designs, Fig. 1. Figure 3 demonstrates aerodynamic drag force measurement by students. The aerodynamic drag coefficient can approximately be varied in the range of 30%, which is consistent with accurate and theoretical data [31–33].

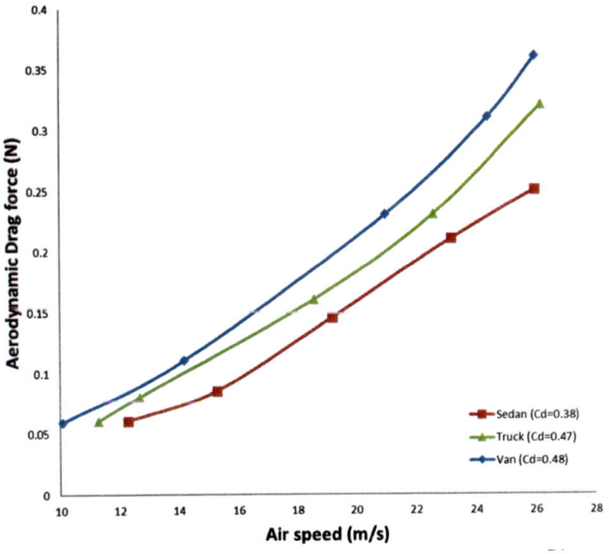

Fig. 3. Aerodynamic drag force measurement using wind tunnel testing.

4.2 Kolb's Reflective Observation

After the wind tunnel test, each group should discuss the vehicle's aerodynamic drag force based on the afterbody geometry. Based on experimental measurements, each group answers some key questions such as:

- determine the airspeed using pitot-tube data and compare it with a digital airspeed flowmeter
- plot the drag force vs. airspeed for each configuration on one graph and discuss them
- determine the drag coefficient for each configuration, see Table 1, and discuss the result

Table 1. Aerodynamic drag coefficient.

Vehicle type	Drag coefficient
Sedan	0.38
Van	0.48
Truck	0.47

After finishing the reflective observation, students share their discoveries with other groups. Students observe that each team has different results, but all are consistent with their findings.

4.3 Kolb's Abstract Conceptualization

The instructor reinforces the concept, and the students are involved in thinking and founding a principle about the vehicle aerodynamic drag changes while altering some geometrical parameters. In this stage, students are encouraged to use available experimental resources [34] to observe geometry's effect, such as forebody shape, rotating wheels, side mirrors, etc. see Fig. 4.

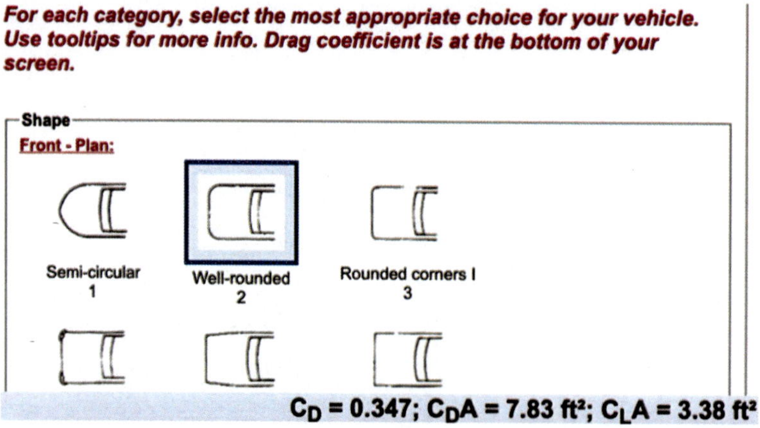

Fig. 4. Look up experimental measurements to see the effect of geometry in the aerodynamic drag coefficient [34].

Furthermore, students are guided to use computational fluid dynamic (CFD) tools to estimate models' aerodynamic drag force, see Fig. 5. Comparing the wind tunnel results with CFD methods helps students better understand the accuracy of the numerical methods. As shown in Fig. 5, students will observe flow separation over the vehicle's body by plotting the static pressure around the vehicle; negative static pressure is the sign of flow separation over the vehicle body.

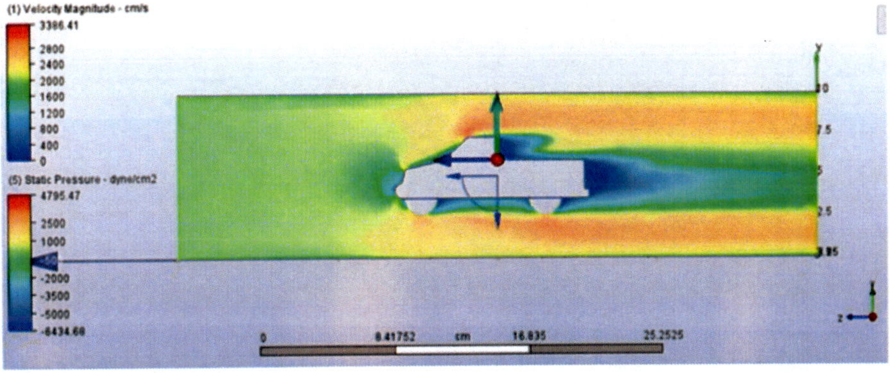

Fig. 5. CFD analyses of the vehicle model to determine aerodynamic forces, static pressure around the vehicle

4.4 Kolb's Active Experimentation

It is the time that students keenly do a real-world problem and use their acquired understanding. In this stage, students design an optimized shape of a vehicle with the most delayed flow separation over the body to reduce aerodynamic drag force. Students are designing the vehicle body in CAD software and perform various CFD analyses to reach optimized geometry. Figure 6 and 7 both demonstrate the CFD analyses performed by students for a race car. Students use the 3D printed scaled model of their design that can be fitted in the wind tunnel to validate the simulation result, see Fig. 8. Finally, each team submitted a comprehensive report about their findings using CFD analyses and experimental wind tunnel measurement.

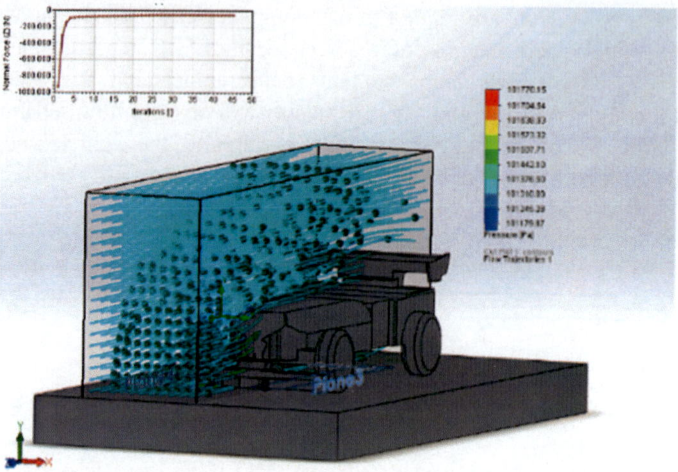

Fig. 6. CFD analyses of the vehicle model to determine flow around the vehicle

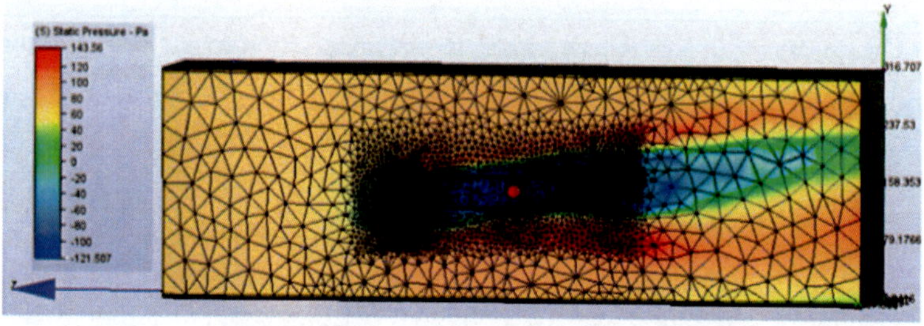

Fig. 7. CFD analyses of the vehicle model to determine aerodynamic forces, static pressure around the vehicle

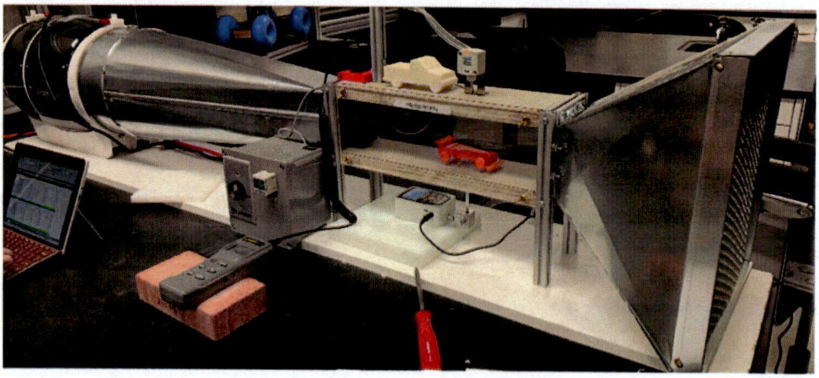

Fig. 8. 3D printed model is mounted in the wind tunnel to measure aerodynamic drag coefficient

4.5 Summary of Kolb's Learning Stages and Course Learning Outcomes

The pedagogical methodology organizes experiential learning for optimizing vehicle shape using scaled experimental measurement and numerical methods. Table 2 summarizes the developed guideline for students' activities based on Kolb's learning stages and course learning outcomes.

Table 2. Methodological organization of scaled experimentation for handling tests.

Learning outcomes	Kolb's learning stage	Students activities
LO1	Concrete experience	Perform wind tunnel testing and determine the effect of afterbody in vehicle drag coefficient
LO1, LO2	Reflective observation	Analyze the wind tunnel measurement and determine the effect of afterbody geometry in total aerodynamic drag of the vehicle
LO3	Abstract conceptualization	Analyze the vehicle with more components that affect aerodynamic drag using previously measured data and the CFD method
LO1, LO4	Active experimentation	Design a novel shape and optimize the geometry using wind tunnel testing and numerical simulation

5 Conclusion

This study has presented a profound insight into practices of shape optimization of road vehicles by employing the experiential learning principle. The proposed scaled model experimentation of using wind tunnels keeps this methodology affordable for many engineering schools for experiential learning in the aerodynamic design of road vehicles. The students' satisfaction from the learning environment in seven years in a row is assessed with two questions: 1) how do you rate the value of this course compared with others you have taken at McMaster University and 2) independent critical judgment was encouraged. Table 3 demonstrates the result of student evaluation in reply to two mentioned questions, the scale for this question is from 1 to 5, 1: very poor and 5: excellent. This study creates a background for experiential learning in one aspect of the automotive engineering curriculum; though, more attention to experiential learning could deliver exciting methodologies in university-level laboratories with industry-oriented learning outcomes.

Table 3. Student evaluation from the course.

Year	No. of students	Q1: Course relative value	Q2. Critical thinking
2014	32	3.65 (73%)	3.23 (64%)
2015	24	4.26 (85%)	4.30 (86%)
2016	34	4.82 (96%)	4.85 (97%)
2017	31	4.87 (97%)	4.99 (96%)
2018	32	4.97 (99%)	4.94 (99%)
2019	34	4.62 (92%)	4.71 (94%)
2020	37	4.61 (92%)	4.81 (96%)

References

1. Hajshirmohammadi, A., Zarei, N.: Incorporating experiential learning in lower division engineering courses. In: Proceedings of the Canadian Engineering Education Association (2015). https://doi.org/10.24908/pceea.v0i0.5754
2. Hajshirmohammadi, A.: Incorporating experiential learning in engineering courses. IEEE Commun. Mag. **55**(11) (2017). https://doi.org/10.1109/MCOM.2017.1700373
3. Yuen, L.B., Mehrtash, M.: Implementation of an absorber design for vibration control in automation systems. Procedia Manuf. **32** (2019). https://doi.org/10.1016/j.promfg.2019.02.255
4. Pierson, J., Mativo, J.M., Chiuz, E., Trudgen, M., Herring, C.: Student participation in Formula SAE design, fabrication, and testing as capstone experience. In: ASEE Annual Conference and Exposition, Conference Proceedings, vol. 2020-June (2020). https://doi.org/10.18260/1-2–35227
5. Giridharan, K., Raju, R.: The impact of experiential learning methodology on student achievement in mechanical automotive engineering education. Int. J. Eng. Educ. **32**(6), 2531–2542 (2016)
6. Bogoslowski, S., Geng, F., Gao, Z., Rajabzadeh, A.R., Srinivasan, S.: Integrated thinking - a cross-disciplinary project-based engineering education. In: Auer, M.E., Centea, D. (eds.) ICBL 2020. AISC, vol. 1314, pp. 260–267. Springer, Cham (2021). https://doi.org/10.1007/978-3-030-67209-6_28
7. Mehrtash, M., Ghalkhani, K., Singh, I.: IoT-based Experiential E-Learning Platform (EELP) for online and blended courses. In: 2021 International Symposium on Educational Technolog, pp. 252–255 (2021)
8. Srinivasan, S., Rajabzadeh, A.R., Centea, D.: A project-centric learning strategy in biotechnology. In: Auer, M.E., Hortsch, H., Sethakul, P. (eds.) ICL 2019. AISC, vol. 1134, pp. 830–838. Springer, Cham (2020). https://doi.org/10.1007/978-3-030-40274-7_80
9. Mehrtash, M., Centea, D.: Experiential learning approaches in automotive engineering: implementing real world experiences. In: Auer, M.E., Tsiatsos, T. (eds.) ICL 2018. AISC, vol. 916, pp. 532–541. Springer, Cham (2020). https://doi.org/10.1007/978-3-030-11932-4_51
10. Lewis, K., Hulme, K., Kasprzak, E., English, K., Moore-Russo, D.: Experiential learning in vehicle dynamics education via motion simulation and interactive gaming. In: International Journal of Computer Games Technology, no. 1 (2009). https://doi.org/10.1155/2009/952524
11. Mehrtash, M., Yuen, T., Balan, L.: Implementation of experiential learning for vehicle dynamic in automotive engineering: roll-over and fishhook test. Procedia Manuf. **32**, 768–774 (2019). https://doi.org/10.1016/j.promfg.2019.02.284

12. Katz, J.: Automotive aerodynamics. Automotive Series (2016)
13. Waudby-Smith, P.M., Rainbird, W.J.: Some principles of automotive aerodynamic testing in wind tunnels with examples from slotted wall test section facilities (1985). https://doi.org/10.4271/850284
14. Terry Beck, B., Anderson, B., Hosni, M.: A simple educational wind tunnel setup for visualization of duct flow streamlines and nozzle/diffuser boundary layer separation (2008). https://doi.org/10.18260/1-2--4191
15. Othmer, C.: Adjoint methods for car aerodynamics. J. Math. Ind. **4**(1), 1–23 (2014). https://doi.org/10.1186/2190-5983-4-6
16. Dhaubhadel, M.N.: Review: Cfd applications in the automotive industry. J. Fluids Eng. Trans. ASME, **118**(4) (1996). https://doi.org/10.1115/1.2835492
17. Yuan, Z.Q., Gu, Z.Q., bin He, Y., Wang, Y.P., Chen, X.J.: Numerical simulation and experimental research of influence of underbody structure on aerodynamic characteristics. Xitong Fangzhen Xuebao, J. Syst. Simul. **22**(8) (2010)
18. Stumpe, J.: Symbiosis: why CFD and wind tunnels need each other. Aerospace Am. **56**(6), 30–33 (2018)
19. Boyd, E.M., Fales, A.W.: Reflective learning: key to learning from experience. J. Hum. Psychol. **23**(2) (1983). https://doi.org/10.1177/0022167883232011
20. Kolb, D.A., Boyatzis, R.E., Mainemelis, C.: Experiential learning theory: previous research and new directions. In: Perspectives on Thinking, Learning, and Cognitive Styles (2014). https://doi.org/10.4324/9781410605986-9
21. Kolb, A.Y., Kolb, D.A.: Learning styles and learning spaces: enhancing experiential learning in higher education. Acad. Manage. Learn. Educ. **4**(2) (2005). https://doi.org/10.5465/AMLE.2005.17268566
22. Cross, R.L., Israelit, S.: The process of experiential learning. In: Strategic Learning in a Knowledge Economy (2021). https://doi.org/10.4324/9780080517889-24
23. Kolb, D.A.: Experiential Learning- Experience as the Source of Learning and Development (2nd edition), **53**(9) (2015)
24. Kolb, D.A.: Experiential learning: experience as the source of learning and development, David A. Kolb, Prentice-Hall International, Hemel Hempstead, Herts. (1984). No. of pages: xiii + 256," *Journal of Organizational Behavior*, vol. 8, no. 4, 1984
25. Kareem, A., Cermak, J.E.: Wind-tunnel simulation of wind-structure interactions. ISA Trans. 18(4) (1979)
26. Fowler, M.J., Kimball, R.W., Thomas, D.A., Goupee, A.J.: Design and testing of scale model wind turbines for use in wind/wave basin model tests of floating offshore wind turbines. In: Proceedings of the International Conference on Offshore Mechanics and Arctic Engineering - OMAE, vol. 8 (2013). https://doi.org/10.1115/OMAE2013-10122
27. C. H. Wolowicz, J. S. Bowman, and W. P. Gilbert: Similitude requirements and scaling relationships as applied to model testing. NASA Technical Paper, no. 1435 (1979)
28. Meliopoulos, A.P.S., Cokkinides, G.J., Mohagheghi, S., Dam, Q.B., Alaileh, R.H., Stefopoulos, G.K.: A laboratory setup of a power system scaled model for testing and validation of EMS applications (2009). https://doi.org/10.1109/PTC.2009.5282224
29. Purasinghe, R., et al.: Bringing current research to the classroom using the Linked Column Framed system in an undergraduate structures lab (2011). https://doi.org/10.18260/1-2–17578
30. Zohuri, B.: Similitude theory and applications. In: Dimensional Analysis and Self-Similarity Methods for Engineers and Scientists (2015). https://doi.org/10.1007/978-3-319-13476-5_2
31. Katz, J., Plotkin, A.: Low-Speed Aerodynamics (2001). https://doi.org/10.1017/cbo9780511810329

32. Katz, J., Plotkin, A.: Low-speed aerodynamics, from wing theory to panel met, vol. 25, no. 4 (1991)
33. Katz, J.: Race Car Aerodynamics: Designing for Speed. Bentley Publishers, Cambridge (1995)
34. http://hpwizard.com/aerodynamics.html

Mobile Health Care, Healthy Lifestyle and Training

An Assessment of the Advantages Using Smartphone – Based Tele- Audiology and Its Effects on Hearing Care Professionals' Willingness for Integration into the Fitting Process

Florian Ross(✉)

Doctoral School of Management and Organizational Sciences,
MATE – Hungarian University of Agriculture and Life Sciences,
Guba Sándor u.40, Kaposvár 7400, Hungary
ross.florian@gmx.de

Abstract. Hearing loss affects the patients' quality of life far more than commonly known and can end in depression if left untreated. Fitting hearing aids can compensate for the hearing loss and counteract the negative effects. These devices have undergone a remarkable development and can now be connected with the users' smartphones. This technology enables a variety of options that can improve audiological quality and the relationship between the Hearing Care Professional and the patient. Especially the option of smartphone- based Tele - Audiology, i.e. a video call in which the Professional has full access to the hearing aid settings, shows the strong transformation of these technologies. Despite these advantages, this form of customization is only partially used in practice. The paper therefore aims to explore the preferences that need to be met in order for this form of intervention, to be integrated frequently into the fitting process. For this purpose, a survey was conducted among Hearing Care Professionals in Germany. These results were evaluated and statistically validated. It could be concluded that barriers still exist, mainly due to lack of knowledge about the benefits of this tool, as well as its impact on patient relationships.

Keywords: Hearing Healthcare · Mobile- Health · Telemedicine

1 Introduction

The increase in people with hearing loss is a dramatic development in the industrialized world. It is estimated that by age 70 years, 30% of men and 20% of women in Europe have a significant hearing loss. This number increases sharply with further increase in years of life [1]. In most cases, fitting hearing aids can solve the resulting problems [2]. This technology transformed from an ear trumpet in the last century to today's High – Tec devices, which can be connected over Bluetooth with the users' smartphones [3, 4]. Mobile apps act as the interface between the devices and the relevance in Hearing Healthcare is constantly increasing [5], just like in many other industries [6]. These small programs enable a remote online fitting session between the Hearing Care

M. E. Auer and T. Tsiatsos (Eds.): IMCL 2021, LNNS 411, pp. 735–743, 2022.
https://doi.org/10.1007/978-3-030-96296-8_66

Professional and the patient, with full access to the devices' settings [7]. This is part of the digital transformation that is finding its way more and more into medicine [8].

Even though the use of smartphone-based Tele- Audiology brings various benefits and opportunities for patient care, the use of this technology is still quite different among Hearing Care Professionals, even if they are generally open to integrate this form of intervention into the hearing aid fitting process [9]. Nevertheless, there are still various barriers towards the uptake of Tele- Audiology. The main reasons for this are a lack of technical infrastructure for HCPs and patients, lack of training, and knowledge levels of HCPs [10].

The paper therefore aims to find out whether there is a correlation between professional related factors as well as perceived benefits of smartphone-based Tele-Audiology and the actual usage. The aim is to distinguish if there are factors beyond the audiological component that have a positive influence towards the uptake of this technology. For this purpose, a survey was conducted among 141 HCPs in Germany. The results were processed descriptively and statistically and compared with existing hypotheses. The purpose of this study is to identify further approaches to reduce existing barriers to acceptance of the technology and to expand training content for HCPs.

2 Literature Review

According to Northern, Tele- Audiology is the use of electronic information and telecommunication technologies to support remote and distance clinical Hearing – Healthcare [11]. The classical - analog - Tele- Audiology has its origin in the late 60s of the last century and was able to show comparable results in audiological aspects, such as Hearing Aid Outcomes, as the classical on-site intervention in the clinic. The reason for the use was due to the lack of audiologists in rural areas [12]. Today, the reasons for its use have changed somewhat. Certainly there are still areas with inadequate distribution of HCPs, but it is also part of a paradigm shift that is happening, away from the clinic centered toward a patient centered model [13]. Studies have already determined that although HCPs have the technical infrastructure to use Tele-Audiology, only 25% actually used it [14]. This is noteworthy because another study of 258 HCPs in the U.S. found that the majority were open to integrating smartphone apps into the fitting process, and thus, in the broadest sense, the function of Tele- Audiology as well [9]. In recent publications, it was also found that the use of smartphone-based Tele-Audiology does not result in any disadvantages in the audiological context [15], and that in the area of perceived service quality, individual aspects of patient care were even rated more favorably compared to traditional treatment in the clinic [16].

Most of the articles in the field of Tele- Audiology section have an audiological background. However, since the acceptance of technologies and the associated regular use is significantly influenced by the perceived benefits [17], there are also references in the publication mentioned to explore the effects outside the medical field [5].

3 Hypotheses

Based on existing findings related to the use and the general willingness of Hearing-Care Professionals towards Tele- Audiology [9, 10, 14, 18], the following Hypotheses H1 – H3 could be formulated, taking into account the level of education and professional experience:

H1: A higher level of education (Level Rank) of the Hearing Care Professional leads to a higher willingness to use smartphone-based Tele- Audiology.
H2: Hearing Care Professionals with a higher level of work – experience, are more likely to use smartphone- based Tele- Audiology.
H3: There is a correlation between the experience with remote fitting and the general attitude towards smartphone-based Tele- Audiology.
Within this framework, it will be examined the extent to which the conscious perception of various benefits affects the willingness to use smartphone-based Tele-Audiology. Even though the individual motives are partly exploratory in nature, they are examined on the basis of existing findings [16, 18]:
H4a: With the acceptance of the assumption that smartphone-based Tele- Audiology can reduce fitting times, the willingness to use this technology increases.
H4b: With the acceptance of the assumption that smartphone-based Tele- Audiology can ensure a higher degree of flexibility in customer requests, the willingness to use this technology increases.
H4c: With the acceptance of the assumption that smartphone-based Tele- Audiology can increase the service quality of the clinic, the willingness to use this technology increases.
H4d: With the acceptance of the assumption that smartphone-based Tele- Audiology can shorten follow-up appointments in terms of time, the willingness to use this technology increases.
H4e: With the acceptance of the assumption that smartphone-based Tele-Audiology can be used to adjust hearing aids better directly in certain problem situations, the willingness to use this technology increases.
H4f: With the acceptance of the assumption that smartphone-based Tele- Audiology creates competitive advantages over competitors, the willingness to use this technology increases.
H4g: With the acceptance of the assumption that smartphone-based Tele- Audiology can better address future generations of customers, the willingness to use this technology is increasing.
H4h: With the acceptance of the assumption that smartphone-based Tele- Audiology can be used to convey a high level of personal competence and expertise to patients, the willingness to use this technology is increasing.

4 Methodology

An online-based survey was conducted in the first half of 2021. For this purpose, a questionnaire was developed that asked for various data related to the topic. First, values were collected in relation to the HCP in order to be able to better classify the responses in the field of Tele- Audiology. The focus here was on values relating to the level of education and professional experience. Furthermore, the basic willingness as well as the experience - i.e. the actual use of smartphone-based Tele-Audiology - were asked. In terms of willingness, a split question was built in. If a fundamental rejection of the technology was indicated, the reasons, i.e. the barriers against its use, became the focus of the research. In the opposite case of basic openness, participants were directed to assess the benefits of Tele- Audiology. In the main part of the survey, i.e., the perceived benefits of the technology, participants could rate statements using a seven-point Lickert scale from 1 = no agreement, to 7 = full agreement.

The responses were analyzed descriptively first to get an overview of the data collected. Using multiple linear regression, the above relationships of Hypotheses H1, H2, H4 a-h, were examined [19]. Since there was no normal distribution of the variables in H3, proved with the Shapiro – Wilk Test [20], this was checked using the Spearman correlation [21].

5 Results

The questionnaire was answered 156 times, of which 141 could be considered valid. Only five HCPs rejected the use of the technology in principle or stated that they were not willing to do so. Accordingly, 136 professionals were positive about its use. Most of the survey participants had a high level of education, i.e. that of master craftsman. The second relevant group was the journeymen. Apprentices and career changers were hardly represented here (Fig. 1).

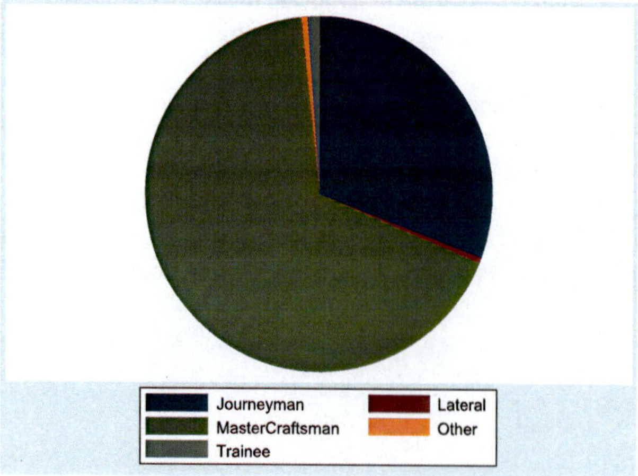

Fig. 1. Distribution of trainings level

Since work experience (ExpTotal), experience with smartphone-based Tele-Audiology (ExpRF), and willingness to use (AttRF) are of central importance in this study, the individual values were processed descriptively (Table 1).

Table 1. Descriptive variable overview

Variable	Obs	Mean	Std. dev.	Min	Max
ExpTotal	136	2.742647	.9583653	1	5
ExpRF	136	2.948529	1.307074	1	5
AttRF	136	3.161765	.3695961	3	4

Work experience was mainly indicated between 2 (5–10 years) and 3 (10–20 years). When asked about the previous use of smartphone-based tele- Audiology, the interquartile range was found to be between 3 (now and then), 4(regularly) and 5 (use whenever possible). The median attitude towards use was found to be 3, which was listed in the questionnaire as basically positive towards the technology (Fig. 2).

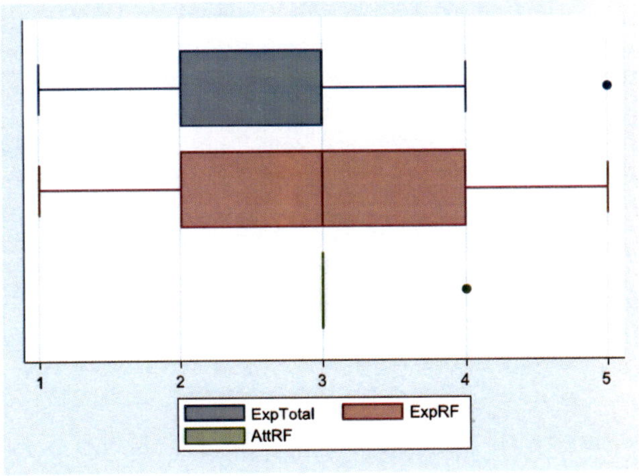

Fig. 2. Data range

In order to test the hypotheses, a regression analysis is performed. The model is usable ((R^2 = .0485, F(2,133) = 3.39, p = .0368) and can be applied accordingly.

H1: A higher level of education (Level Rank) of the Hearing Care Professional leads to a higher willingness to use smartphone-based Tele- Audiology.

Table 2 shows the results of the multiple regression and explains the positive effect of the level of education on the HCPs' willingness (ß = .1791, t(133) = 2.60, p = .010.). H1 can be confirmed.

Table 2. Influence of experience on willingness

Source	SS	df	MS			
				Number of obs	=	136
				F(2, 133)	=	3.39
Model	.893519138	2	.446759569	Prob > F	=	0.0368
Residual	17.5476573	133	.131937273	R-squared	=	0.0485
				Adj R-squared	=	0.0341
Total	18.4411765	135	.136601307	Root MSE	=	.36323

AttRF	Coefficient	Std. err.	t	P>|t|	[95% conf.	interval]
LevelRank	.1791384	.0689732	2.60	0.010	.042712	.3155647
ExpTotal	-.0550343	.0400548	-1.37	0.172	-.1342612	.0241926
_cons	2.858272	.1460938	19.56	0.000	2.569305	3.14724

H2: Hearing Care Professionals with a higher level of work – experience, are more likely to use smartphone- based Tele- Audiology.

It can be seen that the work experience has no significant effect on the positive attitude towards Tele- Audiology (ß = −.0550, t(133) = −1.37, p = .172). Therefore, H2 must be rejected.

H3: There is a correlation between the experience with remote fitting and the general attitude towards smartphone-based Tele- Audiology.

To test the correlation of experience with Tele- Audiology and the actual willingness to use it, a Shapiro- Wilk test was first performed to test the normal distribution of the variables. Since these were not normally distributed (AttRF W = .9392, p = .0000; ExpRF W = .9767, p = .0196), the Spearman correlation was used. Here, values of rs = .5017, p = .0000 could be determined, indicating a strong positive correlation. H3 can therefore be confirmed.

To test hypotheses H4a-h, another multiple regression analysis was performed (Table 3).

This model is also statistically useable (R^2 = .1525, F(8, 127) = 2.86, p = .0059) with a moderate R^2 [19]. Directly, only H4h can be confirmed (ß = .1460, t (127) = 3.05, p = .003). The remaining variables have no significant effect on willingness to use. Accordingly, hypotheses H4 a-g must be rejected. At any rate, this is the consequence of considering the individual variables separately. If these variables are limited to the advantages outside of audiology, i.e. the more business-oriented ones – increasing the service quality of the clinic, creating competition advantages, addressing new customer generations and increasing level of expertise and competence - it can be seen that these together are very significant (F(4, 128) = 5.00, p = .0009) and thus have a positive effect on the willingness to use (Table 4).

Due to this, the rejection of the mentioned hypotheses must be considered in a differentiated and limited way.

Table 3. Perceived advantages

Source	SS	df	MS			
				Number of obs	=	136
				F(8, 127)	=	2.86
Model	2.81297871	8	.351622339	Prob > F	=	0.0059
Residual	15.6281978	127	.123056675	R-squared	=	0.1525
				Adj R-squared	=	0.0992
Total	18.4411765	135	.136601307	Root MSE	=	.35079

AttRF	Coefficient	Std. err.	t	P>|t|	[95% conf.	interval]
MoAdjInit	-.0021264	.0419257	-0.05	0.960	-.0850898	.080837
MoFlex	-.0299689	.0637116	-0.47	0.639	-.1560427	.0961049
MoService	.0172603	.0586815	0.29	0.769	-.0988599	.1333804
MoAdjCont	.028905	.0470735	0.61	0.540	-.0642449	.122055
MoAdjDirect	-.0264731	.0374816	-0.71	0.481	-.1006424	.0476963
MoCompetition	.0555309	.0575173	0.97	0.336	-.0582855	.1693473
MoCustGen	-.0979633	.063471	-1.54	0.125	-.2235611	.0276344
MoCompetence	.1459824	.0478115	3.05	0.003	.0513721	.2405927
_cons	2.832961	.3076108	9.21	0.000	2.224255	3.441668

Table 4. Business- oriented variables

```
( 1)  MoCompetence = 0
( 2)  MoCompetition = 0
( 3)  MoCustGen = 0
( 4)  MoService = 0

      F(  4,   128) =    5.00
           Prob > F =    0.0009
```

6 Discussion

The conducted study showed clearly that there are, after all, some factors that positively influence the willingness of Hearing Care Professionals to use smartphone-based Tele-Audiology. The level of training and the associated knowledge play a decisive role here. Surprisingly, this positive effect could not be demonstrated in relation to professional experience. This is somewhat in contrast to the publication by Kimball et. al. (2018), who showed audiologists with high professional experience to be more open to the integration of apps compared to colleagues with less experience [9]. Overall, only five HCPs indicated a general rejection of the technology. The reasons for this were focused on a lack of technical infrastructure, but above all on uncertainties in application, as well as in the proper integration into the fitting process.

When examining the extent to which the various benefits positively influence the willingness to use them, only the aspect of subjectively perceived improved expertise and competence on the part of the patient could be shown to be significant. A study published in 2021 has already established that the use of this technology leads to this effect. In addition, the equipment used was assessed as more modern than in traditional treatment in the clinic [16]. However, it is doubtful that the majority of HCPs are aware of this characteristic. It is obvious, of course, that the communication of benefits tends to be focused on the audiological aspect but runs the risk of overlooking benefits outside the classical field of application - i.e. the patient treatment. Interestingly, a joint significance was demonstrated in the collaborative consideration of these benefits. This was not the case in the collaborative consideration of the audiological benefits. The reason for the strong audiological focus in the use of technology may be historical, when Tele- Audiology had to be used to restore good hearing to patients in less structured areas. With the change in technology, as well as the user generation, this form of intervention is experiencing a renaissance and should be perceived as such.

7 Conclusion

Hearing Healthcare is changing, and the application of new technologies is part of it. However, in order for technologies to be applied, a clear perception of the benefits is essential. The study clearly showed that while the level of education significantly influences the use of smartphone-based Tele-Audiology, there are additional factors that reinforce its use. In this case, these factors were more likely to be found in the area of business management. It can be concluded that knowledge of the benefits should be extended outside audiology so that the technology is used more regularly in the fitting process. Initiators for this should first be the hearing aid manufacturers, who should emphasize these advantages significantly more in their training courses for HCPs.

Abbreviation

HCP – Hearing Care Professional

References

1. Roth, T.N., Hanebuth, D., Probst, R.: Prevalence of age-related hearing loss in Europe: a review. Eur Arch Otorhinolaryngol **268**(8), 1101–1107 (2011). https://doi.org/10.1007/s00405-011-1597-8
2. ASHA Ad Hoc Committee on Hearing Aid Selection and Fitting: Guidelines for hearing aid fitting for adults. Am. J. Audiol. **7**(1), 5–13 (1998). https://doi.org/10.1044/1059-0889.0701.05
3. Mudry, A., Dodelé, L.: History of the technological development of air conduction hearing aids. J. Laryngol. Otol. **114**(6), 418–423 (2000). https://doi.org/10.1258/0022215001905977
4. Florian, J.: Bluetooth is beginning to make its mark in hearing healthcare. Hearing J. **56**(9), 28 (2003). https://doi.org/10.1097/01.HJ.0000293433.00432.4e

5. Ross, F.: Hearing aid accompanying smartphone apps in hearing healthcare. Syst. Rev. **42** (4), 189–199 (2020)
6. Wohllebe, A., Dirrler, P., Podruzsik, S.: Mobile apps in retail: determinants of consumer acceptance – a systematic review. Int. J. Interact. Mob. Technol. (iJIM) **14**(20) (2020). Art. no. 20
7. Weaver, J.: Made-for-iPhone hearing aid has broad appeal, early adopters report. Hearing J. **67**(5), 28–30 (2014)
8. Diez, E.: Managing a veterinary practice: a guide to organizational culture in veterinary practice. Int. J. Appl. Res. Bus. Manage. **1**(1), 18–26 (2020). https://doi.org/10.51137/ijarbm.2020.1.1.2
9. Kimball, S.H., Singh, G., John, A.B., Jenstad, L.M.: Implications and attitudes of audiologists towards smartphone integration in hearing healthcare. Hear. Res. **369**, 15–23 (2018). https://doi.org/10.1016/j.heares.2018.06.011
10. Ravi, R., Gunjawate, D.R., Yerraguntla, K., Driscoll, C.: Knowledge and perceptions of teleaudiology among audiologists: a systematic review. J. Audiol. Otol. **22**(3), 120–127 (2018). https://doi.org/10.7874/jao.2017.00353
11. Northern, J.L.: Extending Hearing Healthcare: Tele-audiology. Hearing Review (2012). https://www.hearingreview.com/practice-building/office-services/dispensing-networks/extending-hearing-healthcare-tele-audiology. Accessed 14 May 2020
12. Gladden, C., Beck, L., Chandler, D.: Tele-audiology: expanding access to hearing care and enhancing patient connectivity. J. Am. Acad. Audiol. **26**(9), 792–799 (2015). https://doi.org/10.3766/jaaa.14107
13. Tognola, G., Paglialonga, A., Chiaramello, E., Pinciroli, F.: eHealth for hearing-new views and apps practicalities. Eur. J. Biomed. Inform. **11**(3), 37–49 (2015)
14. Eikelboom, R.H., Swanepoel, D.W.: International survey of audiologists' attitudes toward telehealth. Am. J. Audiol. **25**(3S), 295–298 (2016)
15. Convery, E., Keidser, G., McLelland, M., Groth, J.: A Smartphone App to Facilitate Remote Patient-Provider Communication in Hearing Health Care: Usability and Effect on Hearing Aid Outcomes. Telemedicine and e-Health (2019)
16. Ross, F., Wohllebe, A.: Evaluating the service quality of mobile health versus clinic based intervention in hearing healthcare. A comparative study. Int. J. Interact. Mob. Technol. (iJIM) **15**(10) (2021). Art. no. 10
17. Schenk, M.F., Fischer, A.R.H., Frewer, L.J., Gilissen, L.J.W.J., Jacobsen, E., Smulders, M.J. M.: The influence of perceived benefits on acceptance of GM applications for allergy prevention. Health Risk Soc. **10**(3), 263–282 (2008). https://doi.org/10.1080/13698570802160947
18. Saunders, G.H., Roughley, A.: Audiology in the time of COVID-19: practices and opinions of audiologists in the UK. Int. J. Audiol. **60**(4), 255–262 (2021). https://doi.org/10.1080/14992027.2020.1814432
19. Cohen, J. (ed.): Front Matter. In: Statistical Power Analysis for the Behavioral Sciences, p. iii. Academic Press (1977). https://doi.org/10.1016/B978-0-12-179060-8.50001-3
20. Shapiro, S.S., Wilk, M.B.: An analysis of variance test for normality (complete samples)†. Biometrika **52**(3–4), 591–611 (1965). https://doi.org/10.1093/biomet/52.3-4.591
21. Spearman, C.: 'General intelligence', objectively determined and measured. Am. J. Psychol. **15**(2), 201–292 (1904). https://doi.org/10.2307/1412107

Work-In-Progress: Carpal Tunnel Syndrome Rehabilitation: An Approach Using a Smartphone

Karina Vergara Reyes[1](✉), Pablo Ignacio Rojas Valdés[2], Felipe Besoaín Pino[2], and Karin Saavedra Redlich[3]

[1] Doctorado en Sistemas de Ingeniería, Facultad de Ingeniería, Universidad de Talca, Talca, Chile
kvergara@utalca.cl
[2] Ingeniería en Desarrollo de Videojuegos y Realidad Virtual, Facultad de Ingeniería, Universidad de Talca, Talca, Chile
[3] Ingeniería Civil Mecánica, Facultad de Ingeniería, Universidad de Talca, Talca, Chile

Abstract. Cell phones have evolved in recent years, becoming more intelligent and diversifying their use. They have become a mechanism for information, to be connection to the outside world, for entertainment, study, and work. In 2018 around 38% of the global population had at least one smartphone. In addition, the incorporation of sensors allows the development of more personalized applications. Thus, applications associated with rehabilitation have emerged, helping people to improve adherence and complete their therapies. This work proposes the use of a mobile application capable of accessing accelerometer information, and eventually gyroscope and magnetometer, for the execution of three basic rehabilitation exercises for carpal tunnel syndrome (CTS). In a first attempt, the application was evaluated with 3 patients through the VAS (Visual Analog Scale) pain test and answer a survey to measure the impact of the therapy on their quality of life. The results of the pilot test showed that performing the exercises with a smartphone is as effective as normal treatment. Besides, using an app that guides the exercise, it motivates patients to perform the exercises while at home.

Keywords: Rehabilitation · Tunnel carpal syndrome · Smartphone · Sensors · Health care · Motion tracking

1 Introduction

Since their invention, cell phones have been evolving due to the incorporation of new functionalities, making them smarter. In 2018 around 38% of the global population had at least one smartphone [1] and their use has been increasing exponentially. Today's mobile devices incorporate a wide set of integrated sensors, such as an accelerometer, ambient light sensors, GPS, compass, Wi-Fi, among others [2]. These sensors can obtain information from the physical world, which allows the smartphone to adapt its behavior according to the environment in which it is located. For example, it adapts the

M. E. Auer and T. Tsiatsos (Eds.): IMCL 2021, LNNS 411, pp. 744–751, 2022.
https://doi.org/10.1007/978-3-030-96296-8_67

light intensity of the screen according to the ambient lighting; turns off the screen if the user holds the phone near to its ear; or it displays the current location when opening the map application [3].

This is how the development of mobile applications reaches into the health area, allowing the use of technology in a different and versatile way. Mobile health applications provide the opportunity to address public health challenges and reduce healthcare costs [4]. In addition, mobile applications could compensate for the lack of disease monitoring at home and, as a novel tool, motivate patients to maintain their medical processes or rehabilitation therapies as in the case of Carpal Tunnel Syndrome (CTS).

CTS is caused by entrapment and compression of the median nerve at the wrist. It has a prevalence of approximately 3% –4% in the general population [5, 6]. Traditional rehabilitation programs consist of exercise sessions that are mainly performed in two stages. The first is carried out in a medical center with professional supervision and the second at the patient's home. In this last stage the patient has no supervision or no feedback, so motivation and adherence with the therapy decrease [7] and the therapist has no knowledge of the patient's progress. Adherence can be defined as the extent to which a person's behavior coincides with the given medical recommendations [8] and it is important to monitor the patient's progress. The lack of tools and methods to track exercises at home has led to research on strategies to improve adherence to rehabilitation therapies [9].

This work proposes the development of a mobile application that uses the integrated sensors of a smartphone, mainly accelerometer, and eventually gyroscope and magnetometer, to monitor the execution of wrist stretching exercises for CTS. In this way, it seeks to help patients maintain motivation and adherence with rehabilitation therapy without or with low professional supervision. In addition, a history will be recorded for each user for the therapist to evaluate the performance of the sessions.

Section 2 shows the studies related to the use of the smartphone in healthcare and rehabilitation. Section 3 presents the materials and methods involved in the work, including how the mobile application is composed and the evaluation methods. Section 4 describes the preliminary results and the plan for further testing. Finally, the discussion of the results and conclusions are in Sect. 5.

2 Background

There are several mobile applications oriented to people's health and most of them help to correct behavior through reminders, self-care tasks, medication administration, surveys, among others. One example is iMHere [10], which has a clinical web portal and a two-way communication system. Users can store medications along with a photo and description, set up reminders for medication, catheterization or bowel management programs, inspect for insensitive skin problems, or receive periodic alerts for depressive symptom surveys. Medication, symptoms, or improvements must be indicated into the application. It also contains a messaging module to communicate with the doctor by text messages. Therefore, this type of application is aware of the user's routine and can help to modify it. This same type of mobile applications have been used for different pathologies and conditions, such as diabetes, weight control, cardiac rehabilitation, pain management, musculoskeletal injuries, among others [11].

Mobile applications have been used in neurodegenerative diseases such as Parkinson's disease due to the ubiquitous nature of cell phones, which can collect a large amount of physiological and behavioral data [12, 13]. For example, PD_Manager [14] includes monitoring by commercial wearable sensors, attached to the wrist and insoles, which are paired with the smartphone. In this way, motor symptoms are monitored, and their severity is automatically estimated.

Sensors embedded in the cell phone are also used to recognize physical activity. In [15] it is proposed the development of a framework for human activity recognition using data from smartphone sensors such as accelerometer, gyroscope, magnetometer and Google Fit activity tracking module. Using accelerometer data from smartphones and smartwatches, it is also possible to estimate the magnitude of the load on the lower extremities during physical activity [16].

For people with CTS, the systems respond to conventional rehabilitation treatments, which consist of a sequence of movements of the fingers and wrist to achieve gliding of the median tendon and nerve [17]. Based on these movements, mobile applications have been designed for allowing the user to perform rehabilitation exercises for CTS. Toriumi et al. [18] developed an application incorporating gamification, in which the patient must move a character with the thumb, simulating a perfect O. It also incorporates a finger guide to the back of the smartphone to fix the position of the fingers other than the thumb. The same idea is exploited in the tablet application "Dance Dance Thumb" [19], also including remote monitoring for physicians.

All these applications aimed at CTS rehabilitation propose games that involve the movement of the thumb. While this is a stretching exercise that helps decompress the median nerve, it does not control the position of the wrist and forearm. The contribution of this work is the development of a mobile application that uses the sensors embedded in the smartphone to monitor three wrist stretching exercises. The idea is to monitor the execution of these exercises at home, to control the movement through a previous calibration to avoid damage and fatigue, and to develop a web platform that allows the therapist tracks the patient's rehabilitation.

3 Materials and Methods

3.1 Description of the Mobile Application

To help people complete their CTS rehabilitation process at home, it is necessary to guide patients in the execution of their exercises and to follow up their routines to know how they perform them. For this case study, three wrist stretching exercises, which are shown in Table 1, were chosen: flexion-extension, ulnar-radial deviation, and radial inclination.

A mobile application was developed which can access data from the accelerometer, gyroscope and magnetometer sensors incorporated in the smartphone. In addition, a web platform – with security protocols to avoid the most common vulnerabilities of the recorded data – was developed. To follow up the sessions performed by the patient at home, only the therapist in charge will have access to the information obtained by the application.

Table 1. Description of wrist stretching exercises for CTS rehabilitation.

Description	Image
Wrist flexion-extension Hand movement up and down at 70° approx.	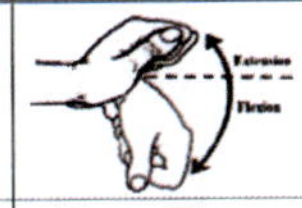
Ulnar-radial deviation Movement of the hand from left to right at 45° approx.	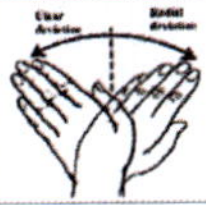
Radial inclination Movement of the hand laterally upwards at 45° approx.	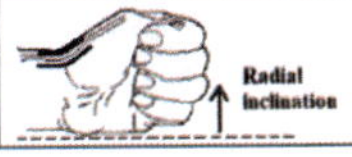

3.2 Application Design

The data for each movement is in X, Y, Z format, while the chosen exercises move on a single axis, which makes it possible to detect in which axis the movement is being performed. For example, the flexion-extension and radial inclination exercises move from top to bottom (Y axis) and the ulnar-radial deviation exercise moves from left to right (X axis). For each exercise, it is necessary to set limits, either upper and lower or right and left. In this way it is possible to know when the user performs a bad movement and correct it so as not to worsen the condition.

A low-pass filter is applied to the information obtained by the sensors to condition the signals in the form of Euler angles and make the data more readable. To establish the limits, the signals were also conditioned, but with a vector projection to relate coordinate axes. Thus, only a limit value is handled and not coordinates of values.

At the end of routine, the data are stored in a file and uploaded to a server for visualizing them on a web platform. Figure 1 shows the interaction scheme between the different components, application mobile, database, server, and web platform.

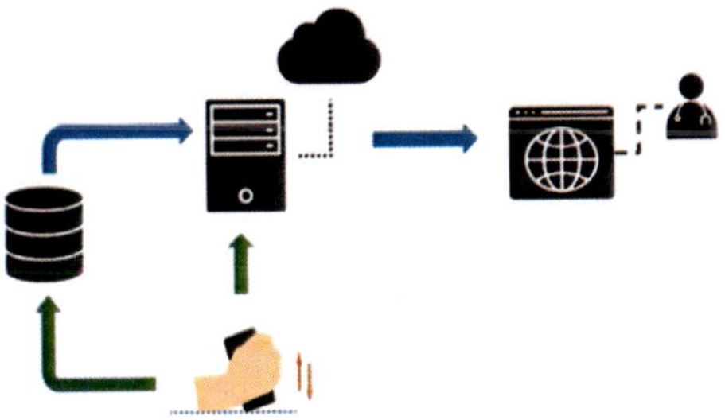

Fig. 1. Diagram of interaction between components of the mobile application.

The platform web's information is 1) session number, 2) performed exercises, 3) number of repetitions for each exercise, 4) errors and successes committed. The therapist in charge will have access to the web platform through credentials.

3.3 Evaluation Procedure

To evaluate the application, the exercises must be performed by holding the smartphone with the hand to be exercised. For each exercise, the application will show how it should be performed and how the smartphone should be held. Figure 2 shows the positions of the hand holding the smartphone in the exercises.

The VAS (Visual Analog Pain Scale) pain test is applied to know the pre- and post-treatment status with the smartphone. Additionally, the SF-36 questionnaire [20] is used for a survey to measure adherence to therapy and the impact on patients' life, which is applied at the end of treatment.

Fig. 2. Wrist stretching exercises with smartphone. From left to right are: 1) wrist flexion-extension; 2) ulnar-radial deviation; 3) radial inclination.

3.4 Application Usage

At the very first use, a profile is created with the personal data (name, e-mail, and password). After creation, the user selects the exercise to be performed and configures whether need a sound guide to mark the rhythm of each repetition. Then, a tutorial will be displayed showing the positioning of the arm, the holding of the smartphone, and possible warnings during the movement. Then, the patient must calibrate the device to define the limits of the exercise (up, down, right, or left) by pressing the "volume up" button until a sound indicates completion (five seconds). For example, if the flexion and extension exercise – whose movement is up and down – is chosen, the patient should: 1) to select the calibration of the upper limit; 2) to position the wrist upwards holding the phone; 3) to press the "volume up" button; 4) to wait five seconds or until the signal is heard; and 5) to perform the same to set the lower limit. Then, to start/finish the exercises, the user should press the "volume up"/"volume down" button, respectively. If the movements exceed the established limits, a vibrating and an audible signal will be given to correct the movement. At the end of a routine, the data are stored and sent to a server to be displayed on the web platform. Figure 3 shows screenshots of the mobile application.

The information available for the therapist is organized per each patient according to the session number, exercise performed, repetitions, and successful and unsuccessful movements.

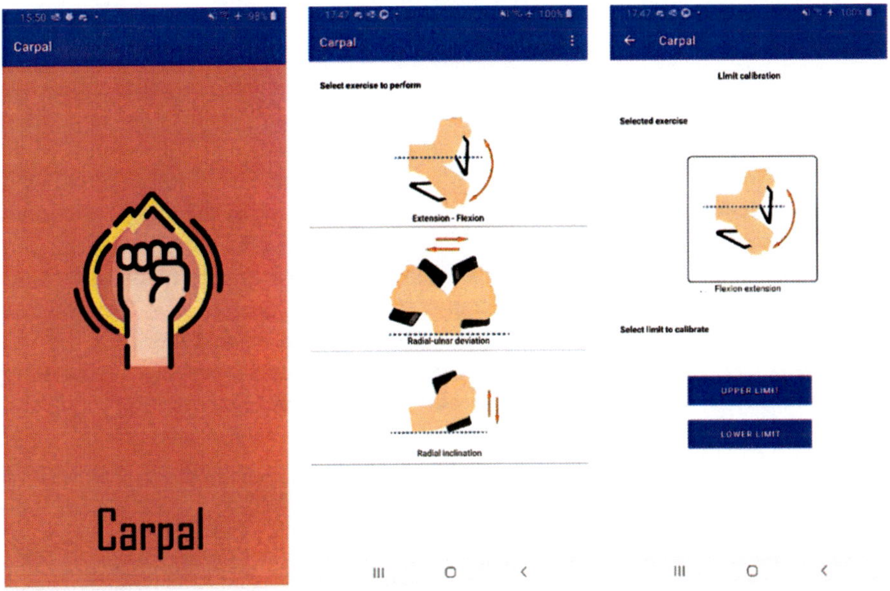

Fig. 3. Mobile application screenshots of the home, exercise selection and calibration sections.

4 Results

The result of this work is a mobile application oriented to support the rehabilitation of CTS, using the accelerometer sensors and eventually gyroscope and magnetometer, incorporated in the smartphone. Also, a web platform was developed to display the information obtained through the monitoring the movement of three wrist stretching exercises. The therapist in charge will have access to this platform, allowing to follow up the rehabilitation sessions performed by the patient at home.

Additionally, a pilot test was performed, where the application was evaluated by three people with CTS and under medical supervision for 2 weeks, making a total of four sessions. The objective was to test whether holding the smartphone while performing the exercises could worsen the condition due to the extra effort. Anyway, the VAS test and the survey to measure adherence and impact on quality of life were applied, which showed decreased pain and positive impact on the treatment for being a novel tool. In addition, the therapist in charge used the platform to visualize the information and verify whether the information displayed was valuable for treatment follow-up.

Due to the size of the sample, the results are not representative, so it is expected to test a sample of twenty people with a medium-high level of CTS, with physiotherapeutic treatment and for a period of eight weeks.

5 Discussion and Conclusion

The mobile application was evaluated by three people with CTS, under professional supervision. Although the sample is not representative and it is necessary to perform a massive evaluation, it was possible to obtain the validation of the professional in charge, by observing that executing the exercises holding the smartphone does not worsen the condition. In addition, the guidance and feedback by the mobile application, motivates patients to execute the rehabilitation exercises. As for the web platform, having this information per patient helps to have a record of the rehabilitation activities at home and thus to follow up the therapy.

The evolution of cell phones has made them smarter, diversifying their use. With the incorporation of new technology, new possibilities have opened for the development of mobile applications. Thus, health-oriented mobile applications have been developed to help people control certain aspects of their behavior, such as food, activity schedules, medication, among others. In this way, the incorporation of the smartphone in the rehabilitation area has been explored, making use of the built-in sensors that allow monitoring movement, vital signs, distance traveled, effort, among others.

Traditional rehabilitation programs do not include follow-up sessions at home, so the patient does not have supervision or feedback on the execution of the exercises, causing demotivation and decreased adherence to therapy. Having an application that helps CTS rehabilitation at hand and in a daily device such as a smartphone, motivates people to continue with the therapy at home, because it is a guide for the execution of the exercises and provides feedback in case of any error. In addition, it is necessary for the therapist in charge to be aware of the patient's progress to verify whether the therapy is effective and to take measures if it is not.

References

1. "Smartphone users 2020|Statista. https://www.statista.com/statistics/330695/number-of-smartphone-users-worldwide/. Accessed 9 May 2021
2. Yurur, O., Liu, C.H., Sheng, Z., Leung, V.C.M., Moreno, W., Leung, K.K.: Context-awareness for mobile sensing: a survey and future directions. IEEE Commun. Surv. Tutor. **18**(1), 68–93 (2016). https://doi.org/10.1109/COMST.2014.2381246
3. Islam, N., Want, R.: Smartphones: past, present, and future. IEEE Pervas. Comput. **13**(4), 89–92 (2014). https://doi.org/10.1109/MPRV.2014.74
4. Baig, M.M., GholamHosseini, H., Connolly, M.J.: Mobile healthcare applications: system design review, critical issues and challenges. Australas. Phys. Eng. Sci. Med. **38**(1), 23–38 (2014). https://doi.org/10.1007/s13246-014-0315-4
5. Melhorn, J.M., Talmage, J.B.: Prevalence of carpal tunnel syndrome in motorcyclists. Orthopedics **36**(7), 497–498 (2013). https://doi.org/10.3928/01477447-20130624-02
6. Uchiyama, S., Itsubo, T., Nakamura, K., Kato, H., Yasutomi, T., Momose, T.: Current concepts of carpal tunnel syndrome: pathophysiology, treatment, and evaluation. J. Orthop. Sci. **15**(1), 1–13 (2010). https://doi.org/10.1007/s00776-009-1416-x
7. Tamayo-Serrano, P., Garbaya, S., Blazevic, P.: Gamified in-home rehabilitation for stroke survivors: analytical review. Int. J. Serious. Games. **5**(1), 2384–8766 (2018). https://doi.org/10.17083/ijsg.v5i1.224

8. Frost, R., Levati, S., McClurg, D., Brady, M., Williams, B.: What adherence measures should be used in trials of home-based rehabilitation interventions? A systematic review of the validity, reliability, and acceptability of measures. Arch. Phys. Med. Rehabil. **98**(6), 1241-1256.e45 (2017). https://doi.org/10.1016/j.apmr.2016.08.482
9. McLean, S.M., Burton, M., Bradley, L., Littlewood, C.: Interventions for enhancing adherence with physiotherapy: a systematic review. Man. Ther. **15**(6), 514–521 (2010). https://doi.org/10.1016/j.math.2010.05.012
10. Dicianno, B.E., et al.: Feasibility of using mobile health to promote self-management in Spina Bifida. Am. J. Phys. Med. Rehabil. **95**(6), 425–437 (2016). https://doi.org/10.1097/PHM.0000000000000400
11. Ramey, L., Osborne, C., Kasitinon, D., Juengst, S.: Apps and mobile health technology in rehabilitation: the good, the bad, and the unknown. Phys. Med. Rehabil. Clin. N. Am. **30**(2), 485–497 (2019). https://doi.org/10.1016/j.pmr.2018.12.001
12. Paik, N.-J., et al.: Smartphone applications in the management of Parkinson's disease in a family setting: an opinion article. Front. Neurol. **1**, 668953 (2021). www.frontiersin.org, https://doi.org/10.3389/fneur.2021.668953
13. Gao, S., et al.: A mobile phone app-based tai chi training in Parkinson's disease: protocol for a randomized controlled study. Front. Neurol. **11**(January), 1–9 (2021). https://doi.org/10.3389/fneur.2020.615861
14. Tsiouris, K.M., et al.: PD-Manager: an mHealth platform for Parkinson's disease patient management. Healthc. Technol. Lett. **4**(3), 102–108 (2017). https://doi.org/10.1049/htl.2017.0007
15. Javed, A.R., Faheem, R., Asim, M., Baker, T., Beg, M.O.: A smartphone sensors-based personalized human activity recognition system for sustainable smart cities. Sustain. Cities. Soc. **71**, 102970 (2021). https://doi.org/10.1016/J.SCS.2021.102970
16. Wang, X., Perry, T.A., Caroupapoullé, J., Forrester, A., Arden, N.K., Hunter, D.J.: Monitoring work-related physical activity and estimating lower-limb loading: a proof-of-concept study. BMC Musculoskelet. Disord. **22**(1), 1–12 (2021). https://doi.org/10.1186/s12891-021-04409-z
17. Martins, R.S., Siqueira, M.G.: Conservative therapeutic management of carpal tunnel syndrome. Arq. Neuropsiquiatr. **75**(11), 819–824 (2017). https://doi.org/10.1590/0004-282x20170152
18. Toriumi, M., Watanabe, T., Nimura, A., Fujita, K., Sugiura, Y.: Rehabilitation support system for patients with carpal tunnel syndrome using smartphone. In: ACM International Conference Proceedings Series, pp. 532–534 (2018). https://doi.org/10.1145/3292147.3292240
19. Watanabe, T., Sugiura, Y., Miyata, N., Fujita, K., Nimura, A., Sugimoto, M.: Dancedancethumb: tablet app for rehabilitation for carpal tunnel syndrome. In: Munekata, N., Kunita, I., Hoshino, J. (eds.) ICEC 2017. LNCS, vol. 10507, pp. 473–476. Springer, Cham (2017). https://doi.org/10.1007/978-3-319-66715-7_66
20. Ferrer, G., et al.: El cuestionario de salud SF-36 español: una década de experiencia y nuevos desarrollos. Gac. Sanit. **19**(2), 135–150 (2005). http://scielo.isciii.es/scielo.php?script=sci_arttext&pid=S0213-91112005000200007#t1

A Micro Review Relevant to the Impact of New Mobile and Wearable Technologies on Pregnant Women

Evangelia I. Kosma[1], Spyridon K. Chronopoulos[1,2], Anastasios G. Skrivanos[3], Kostas Peppas[3], Vasilis Christofilakis[2], Georgios Petrakos[4], Petros Petrikis[5], Mary Gouva[6], Nafsika Ziavra[1], Jenny Pange[7], and Eugenia I. Toki[1](✉)

[1] Department of Speech and Language Therapy (SLT), University of Ioannina (UOI), Ioannina, Greece
{e.kosma, spychro, nziavra, toki}@uoi.gr

[2] Electronics-Telecommunications and Applications Laboratory, Department of Physics, UOI, Ioannina, Greece
vachrist@uoi.gr

[3] Department of Informatics and Telecommunications, University of Peloponnese, Tripoli, Greece
{a.skrivanos, peppas}@uop.gr

[4] Department of Obstetrics and Gynecology, General Hospital of Kalamata, Kalamata, Greece
petrakosg@hotmail.gr

[5] Department of Psychiatry, School of Health Sciences, Faculty of Medicine, UOI, Ioannina, Greece
ppetrikis@uoi.gr

[6] Research Laboratory Psychology of Patients Families and Health Professionals, Department of Nursing, University of Ioannina, Ioannina, Greece
gouva@uoi.gr

[7] Department of Early Childhood Education, Laboratory of New Technologies and Distance Learning, University of Ioannina, Ioannina, Greece
jpagge@uoi.gr

Abstract. *Background:* Currently, there is a vast number of electronic and mobile devices which serve for self-evaluation purposes. Specifically, wearable technology has been vastly developed in the last years in conjunction with the usage of mobile phones' applications for the convenience of their users. In the context of the aforementioned, a proper segmentation of the bibliography could emerge the significance of monitoring for preventing unwanted situations in pregnant women's life. *Objective*: The research study on pregnant women has emerged from the need of identifying through a new type of micro review (utilizing elements from SALSA and PRISMA), the most important works of self-evaluation in times of Covid-19. Especially, vast concerns are reported from pregnant women's community to have the same quality of monitoring services and at the same time to be properly guarded against Covid-19. The key question is whether the pregnant can accomplish or not the same results through wearable or/and IoT technologies compared to those of frequent hospital visits. *Design*: The used databases, in conjunction with specific terms and rules, finally

M. E. Auer and T. Tsiatsos (Eds.): IMCL 2021, LNNS 411, pp. 752–764, 2022.
https://doi.org/10.1007/978-3-030-96296-8_68

produced 18 viable publications for reviewing. *Results*: The technologies proved to be not only a key factor for the future of monitoring but even considerably affect in some situations the subject under test. *Conclusions*: This study focused on determining, amongst others, any impact of this technology on the behavior thus to categorize the various technologies into weighted categories. The results showed in some case considerable impact of these devices on the psychological aspect of the subjects.

Keywords: Pregnant women · Wearable devices · Psychological impact

1 Introduction

In the current era of information, internet of things (IoT) and of new emerged dangers such as Covid-19, it is formidable the effort of clinical stuff, technicians and department heads to implement the best technologies to their dynamics. A new prominent candidate receiving the attention of wearable technologies along with IoT/wireless technologies, is maternal and fetus distant monitoring. Considering the limited access so far, by using electronic means, to maternal or/and newborn health infrastructure, this fact deprives the health system from additional "monitoring eyes". In many occasions the clinical stuff is unable to observe efficiently the pregnant women/newborns resulting in potential harmful results for their health. A solution could be the formidable preamble of monitoring systems of vital parameters alerting for imminent threats, calendar health events and other health activities.

Projects based on IoT/wireless technologies, distant measurements of foetal or/and pregnant's heart rate, electrocardiograms and estimation of blood pressure have just been recently published [1–4]. It is obvious that big steps have been made towards developing the above services even if these have not reached yet at high integration. In addition, proper health metrics relevant to optimal gestational weight gain, birth weight, respiratory rate, and perinatal outcomes are studied [5]. Other works include continuous glucose monitoring in pregnants (type 1 diabetes – T1D) thus the development of smart data acquisition through monitoring pregnancy's blood glucose and hemoglobin [6, 7].

Further quantification of accelerometer/actigraphy [8], clinical examination thus blood glucose monitoring is done [9]. Complements are the acoustic sensors for phonocardiography and fetal movement [10,11]. Other works include questionnaires, clinical examination and measurements (e.g. blood glucose) [12]. Nevertheless, a common feature under test is the foetal's or/and pregnant's heart rate (HR) [13–15] with various add-ons in many cases. Also, older works include fewer combinations while sustaining their scientific importance. They include accelerometers [16], blood glucose monitoring [17] and foetal monitoring network or/and electrocardiograms or/and respiratory rate [18].

For properly classifying all the above combinations, it is possible to quantify their importance with weight factors. Also, a new review type is suggested (SAPW – SALSA and PRISMA weighted) and it is applied on a topic of maternity and monitoring.

This paper includes a delimited introduction, the rules and methodology of proposed review (SAPW). Then, the materials and methods are analyzed. In turn, all the categories are classified into tables and their weights are quantified. Then, there is a narrative sector which combines all the works. Finally, discussion and conclusions are reported.

2 Operating Principle of SAPW

The new proposed review type is hybrid, including the structure of Search, Appraisal, Synthesis, and Analysis (SALSA) [19, 20] and Preferred Reporting Items for Systematic Reviews and Meta-Analyses (PRISMA) [21]. Additional rules apply and they concern the structure of Introduction and the calculation of weights for quantifying the categories into dynamics of possible prosperous research. The suggested rules are:

1. The new type of review is called SAPW (SALSA and PRISMA weighted).
2. Introduction must contain only the bibliography of quantitative synthesis (based on PRISMA). The works should be reported in a narrative and time-descending manner. In this way, the authors will report the advances from the most recent ones. Also, this structure filters redundant references as some authors use them for showing a plurality in their writing but instead they confuse. This rule can be relaxed when no more than 10% of new papers are mentioned. In our case no new papers were used; but if was the case, then the number could be equal to no more than the rounding of 1,8 to 2 ($10\% \cdot 18 = 1{,}8 \approx 2$).
3. Next section is *Methodology*. The SALSA scheme is analyzed and correlated to the sections of the systematic review (PRISMA). E.g. *Appraisal Step* is included in *Data Extraction* or else to an appropriate section. The simplicity of SALSA helps to better understand the structure of PRISMA review.
4. Then the *Materials and Methods* are reported along with *Research Questions* and *Data Extraction* with *PRISMA Flow Diagram*.
5. Results are reported through categorized elements inside tables. Then, the weights are found revealing the number of included elements in each work which is categorized. E.g. If a work includes 5 out of 10 elements, then it has a cw = *contribution weight* given by Eq. (1) and i.e. $cw_{5/10} = (5/10)/18 \approx 0{,}04$ or 4%. So, the contribution equals to 4%. If this value (con) equals at least the median, then it is called *significant contributor,* otherwise *contributor*, based on Eq. (2).
6. The selected works are reported in a brief literature review based on time and scope parameters. This section is named as *Literature in brief.*
7. *Discussion* and *Conclusions* should be reported. *Future scopes* are mandatory.

$$cw = contribution\,weight = (elements/total\,elements)/total\,works \tag{1}$$

$$con = sgn\left((cw - median)/median\right),\ if\ con \geq 0 \rightarrow significant\,contributor \tag{2}$$

3 Methodology (SALSA)

In this section, the methodology (based on SALSA) is presented. *Search* step corresponds to the used databases (Scopus and PubMed) and the research items are reported under *Materials and Methods*. The time window was limitless during initial search of works. In *Appraisal* step, duplicated and other works were excluded based on PRISMA rules. *Synthesis* corresponds to the section of *Literature in brief* and is in chronological order. *Analysis* is found under ***Results*** whereas the division of works' characteristics along with weights calculations are presented. The four steps appear in Fig. 1.

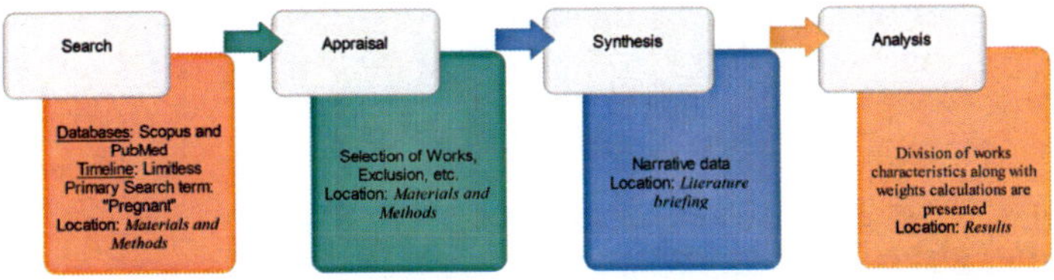

Fig. 1. Four steps of the proposed review following the SALSA methodology.

4 Materials and Methods

4.1 Research Questions

Questions are posed in this systematic review such as (Q1) Can pregnant women have the same quality of monitoring and at the same time to be properly guarded against Covid-19?; (Q2) Can the reviewed technologies be a key future factor in clinical science?; (Q3) Can these technologies affect in psychological level or/and the behavior?; (Q4) Is there a lack of proper development in a field associated with the technologies under examination?

4.2 Search Strategy and Eligibility Criteria

For finding the answers to Q1-Q4, specific procedures were followed based on *Operating Principle of SAPW*. The systematic scoping review was based on PRISMA [21]. The criteria of the final inclusion of publications were (1) papers in well-esteemed journals, (2) not a review/meta-analysis/commentary, (3) only relevant works, (4) only English language (5) publication date after 2000 (specifically with a time-range from 2008 to 2021). The systematic search was conducted in ***Scopus*** and ***PubMed*** databases. The utilized search terms had all the words of "maternal", "care", "pregnant" and one or all of the words: "wearable device", "wearable technology", "mobile sensor" and "sensor". In Fig. 2, the PRISMA flow diagram is depicted. Initially, 45 works were found in Scopus and PubMed. The duplicates were 6 concluding. All authors studied the results, and found 6 reviews/meta-analysis, 1 commentary paper, 9 conferences and 4 articles before 2000. After extraction, the total number of works were 19. Then, a final review was made by authors EIK, SKC, KP, NZ and EIT. 1 more article was excluded as considered to be purely clinical. Consequently, the final resulting articles were equal to 18.

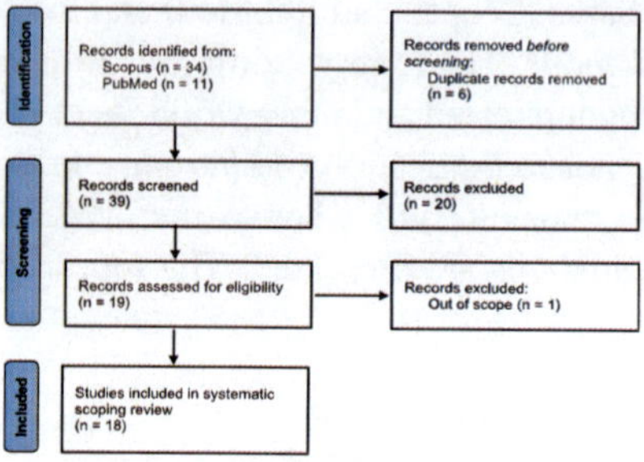

Fig. 2. PRISMA flow diagram.

4.3 Data Extraction

The purpose of answering the research questions needed the following extracted data and they were based on the following categories: (1) the presence of Data Collection, (2) Utilized Measurements and Technologies (UMT), (3) measurement indicators, (4) Sample, Aim and Results (SAR) and (5) Psychological Assessment (PA).

5 Results

This systematic review includes 18 articles. The first extracted data category was the presence or not of Data Collection (Table 1). Another extracted result (Table 2) was the presence or not of utilized Measurements and Technologies (UMT). The third category's measurement indicators (11 sub-categories) are shown in Tables 3 and 4 with the total presence calculated per category. The fourth category named as Sample, Aim and Results (SAR) is shown in Table 5. The fifth category of Psychological Assessment (PA) is shown in Table 6 with total values per category.

Table 1. Data Collection

Reference article	Questionnaires/Interview	Clinical examination	Measurements
[1]	✔	✖	✔
[2, 9, 11, 15–17]	✖	✔	✔
[3]	✔	✖	✔
[4, 10]	✖	✖	✔
[5, 6, 8, 12, 13]	✔	✔	✔
[7, 14]	✔	✖	✖
[18]	✖	✖	✔
TOTAL	9✔	11✔	16✔

Table 2. Utilized Measurements and Technologies (UMT)

Reference article	IoT/Wireless/Internet Technologies (1)	Smart Phone/Mobile device support (2)	Artificial Intelligence/machine learning (3)	Remote consultation (4)	Foetal monitoring network (5)
[1]	✔	✔	✔	✔	✔
[2]	✔	✖	✖	✔	✔
[3, 4]	✔	✔	✔	✖	✖
[5, 6, 13]	✔	✖	✖	✖	✖
[7]	✖	✔	✖	✖	✖
[8]	✔	✔	✖	✔	✖
[9, 10, 12, 15, 17]	✖	✖	✖	✖	✖
[11, 16, 18]	✖	✖	✖	✖	✔
[14]	✔	✔	✖	✖	✖
TOTAL	9✔	6✔	3✔	3✔	5✔

Table 3. Measurement indicators (Part1 – six subcategories)

Reference article	Foetal or/and pregnant's heart rate (6)	Electrocardiograms (7)	Stethoscope or/and respiratory rate (7)	Pregnancy blood glucose (9)	Digital thermometer (10)	Uterine contraction (11)
[1]	✔	✔	✖	✔	✖	✔
[2]	✔	✔	✔	✔	✔	✖
[3, 10]	✔	✔	✖	✖	✖	✖
[4]	✔	✔	✖	✔	✔	✔
[5]	✖	✖	✔	✖	✖	✖
[6, 9, 12, 17]	✖	✖	✖	✔	✖	✖
[7]	✔	✖	✖	✔	✔	✖
[8,11,16]	✖	✖	✖	✖	✖	✖
[13]	✔	✖	✔	✖	✔	✖
[14]	✔	✔	✔	✔	✖	✖
[15]	✔	✖	✖	✔	✖	✖
[18]	✔	✔	✔	✖	✖	✖
TOTAL	10✔	7✔	5✔	10✔	4✔	2✔

Table 4. Measurement indicators (Part2 – 5 remaining subcategories)

Reference article	Hemoglobin (12)	Blood pressure (Pregnant) (13)	Acoustic sensors (14)	Accelerometer/Actigraphy (15)	Weighting scales (16)
[1, 4, 5]	✖	✔	✖	✖	✔
[2]	✔	✔	✖	✖	✔
[3]	✖	✔	✖	✖	✖
[6]	✔	✖	✖	✖	✖
[7]	✔	✔	✖	✖	✖
[8, 16]	✖	✖	✖	✔	✖
[9, 12–15, 17]	✖	✖	✖	✖	✖
[10]	✖	✖	✔	✖	✖
[11,18]	✖	✖	✔	✔	✖
TOTAL	3✔	6✔	3✔	4✔	4✔

Table 5. Sample, Aim and Results (SAR)

Article	Sample	Aim	Results
[1]	315 pregnant women aged between 20–25 (≈13%), 26–30 (≈52%), 31–35 (≈31%), 36–40 (≈3%), 41–50 (≈1%)	To examine the impact of wearable devices, their applications, and the reliability of distant monitoring for obstetrics	A novel IoT design intended for distant maternal screening with wearable devices and their information systems
[2]	94 women and 43 newborns were included in portable health clinic (PHC) project. 183 checkups were conducted	A portable health clinic (PHD) for maternal and child health acted as a pilot for finding and reporting information	Anemia or/and abnormal pulse were detected in pregnant/parturient women. Over 40% of them improved their health
[3]	32 women, in high-risk pregnancies, were recruited, while 4 of them withdrew	To design and implement an IoT system for delivering ubiquitous checking on maternal health before/after pregnancy	The feasibility proved during 9 months. The application was used (filling questionnaires), before and after pregnancy
[4]	One patient's example	To develop a remote pregnancy risk monitoring (RPRM) non-invasive system and check in real-time the pregnants	The proposed system works properly and helps patients to overcome the problem of distance
[5]	70 pregnant women were included (35 of them being the control group	To test a wearable device/monitoring on pregnants who could develop disorders	This ongoing project validated a health gauge thus included a pilot study
[6]	15 women in preconception and other 25 pregnant women (during first trimester) were recruited	To test continuous glucose monitoring (CGM) with remote monitoring, versus CGM alone with remote monitoring	HbA1c decreased in pregnant women. It was lower for subjects using CGM Share compared to CGM Alone or no CGM
[7]	Project under development. Not yet used on selected participants	To construct a monitoring system intended for public health care in rural area	After successful development the parameters were constantly recorded locally

(*continued*)

Table 5. (*continued*)

Article	Sample	Aim	Results
[8]	The exploratory study proposes a primary sample of 400 pregnant women under low risk	To find primary predictors of complications in pregnancy, by correlating physical activity (PA) data to sleep patterns	Wearable actigraphy devices, could probably discover gestational/maternal complications before clinical signs
[9]	There were 16 pairs of mother-infant. The pregnants with T1D wore a CFM sensor 2–3 days before delivery	CGM of profiles (T1D) in neonatal period, and to find the correlation to the control of maternal intrapartum glucose	CGM found widespread neonatal hypoglycaemia being also the case of mothers with good intrapartum glucose control
[10]	15 pregnants were recorded simultaneously for fetal Electrocardiogram (fECG) and abdominal phonogram	To develop a cheap system with noninvasive fetal phonocardiography for recording fetal heart sounds and rate	Significant correlation was into beat to beat fetal heart rate (FHRECG) by fECG and the same (FHRECG) by fetal PCG
[11]	10 women were the participants for the pilot phase of the project. Then 44 pregant women were analyzed adequately	To examine the acoustic sensing in pregnants for identifying and distinguishing various types of fetal movements	The hardware and software were validated by comparing between the sensed movements and a golden standard
[12]	130 pregnants (with GDM) participated in: (1) blood glucose self-monitor alone and (2) wearing a continuous Glucose Monitoring System (CGMS)	To determine whether a single application of RT-CGMS almost after the diagnosis of GDM can take the role of an educational and motivational mean	At the end of pregnancy, all glucose levels were found the same in both groups. Improvement of glucose variability was found the last day of measurements
[13]	The 38 participants consisted of 32 pregnant women (wearing the sensor) and 6 non-study nurses who had the responsibility of observing the sessions	To examine the viability and functionality of a wireless monitor on pregnants wearing the sensor (measuring heart and respiratory rate thus temperature)	Maternal vital signs were acquired. Subjects found the monitoring process comfortable. Most of them (liked the sensor (> 80%) and found it useful (> 95%)
[14]	Theoreric case examples (as the development of this project is ongoing)	To decrease maternity and child mortality rates. Information and communication technology was used at rural areas	The MAMICare will decrease diagnostic errors and the attention concerns correlated to primary death cases
[15]	10 pregnants with T1D were tested during physical activity energy expenditure vs continuous glucose monitoring	To study pregnants with T1D with activity schemes and to examine the effect of the physical activity on glucose control	A scheduled physical activity with a controlled diet can help the T1D pregnants to optimize their glucose during gestation
[16]	6 pregnant women used the recording of fetal movements just before they slept at home (from 30 weeks till birth)	To develop a Fetal Movement Acceleration Measurement (FMAM) recorder for calculating the total fetal movement	The "epoch" is positive for abdominal wall oscillations' detection (while maternal movement sensor with no detection)

(*continued*)

Table 5. (*continued*)

Article	Sample	Aim	Results
[17]	10 pregnants (with T1D) participated into finding the rates of systemic glucose appearance (Ra) and disposal (Rd)	To make possible the prandial insulin algorithms based on the glucose turnover during fasting and postprandial cases	The results of fasting Ra/fasting Rd showed no alterations in early versus late pregnancy, after prescheduled meals
[18]	32 pregnants used sensors throughout daytime and during night	A developed sensor detected fetal movements' acceleration	High agreement was between detected movement and subjective markers

Table 6. Psychological Assessment (PA)

Article	Direct Psychological Assessment (DPA)	Indirect PA
[1]	✖	✔ (Pregnant's willingness to wear smart screening devices)"
[2, 4–7, 9, 10, 12, 14–18]	✖	✖
[3]	✔ (Stress from questionnaire)	✔ (From HR and HRV)
[8, 13]	✔	✖
TOTAL	3✔	2✔

6 Literature in Brief

In this section there is a brief narration (in time-ascending order) of the reviewed works. The purpose is to show the contribution of wearable technologies on pregnant women's health status in a simple and direct manner without tiring the reader.

An experiment back in 2008 [18] detected fetus either by pressing a button or during night with a recording sensor. Four years later, the rates of systemic glucose appearance (Ra) and disposal (Rd) were examined during gestation for 24h and for making possible the establishment of prandial insulin algorithms [17]. After 10 months, recordings of fetal movements during pregnants' sleep at home were studied again, for developing a FMAM recorder for calculating the total fetal movement and to evaluate its reliability at home [16]. Then, in 2013, pregnants with T1D (in their second trimester) were examined about the effect of the physical activity on glucose control [15]. The activity was normal (free-living) and included directed exercise. Also, in 2014 [14], information and communication technologies (ICTs) which were employed in rural areas, intensified the health practices for pregnant women and fetuses. Then, a study [13] focused on the viability and functionality of a wearable monitor for pregnants for measuring heart and respiratory rate and temperature. Maternal vital signs were retrieved by a wireless system. Participants considered the monitoring process comfortable (78%), liked the sensor (over 80%) and found it useful (over 95%). Most of them would wear again the sensor or propose it to another patient (over 78%). In 2016, a study about Glucose Monitoring System, was published and wanted to determine whether a single application of RT-CGMS almost immediately after the GDM diagnosis could educate and motivate [12]. The measurements influenced pregnants for altering their way of living, and acted in a therapeutic way. In parallel, studies such as in [11], examined acoustic sensing (applied on pregnants) while identifying/distinguishing the types of fetal movements with acoustic sensors and an accelerometer for removing false sensing due to maternal movements. In 2018, a cheap developed system for noninvasive fetal phonocardiography (four-channel recordings) recorded fetal heart sounds and rate [10].

In [9], the field of "glucose hunting", relevant to continuously monitoring the glucose of T1D profiles during neonatal period, found the correlation to the control of maternal intrapartum glucose. CGM exhibited widespread neonatal hypoglycemia even in the case of mothers having good intrapartum glucose control. Then in [8], the goal was to find primary predictors of complications during pregnancy, by correlating physical activity (PA) data to sleep patterns, by using wearable devices based on actigraphy processing, thus the use of self-reporting tools. The latter contributed to direct psychological assessment (DPA). Almost the same time in 2019, an advanced work [7] was about a monitoring system for a public health care rural centre. The suggested pregnant's parameters under test were blood pressure, pulse rate, temperature, blood glucose and hemoglobin. Then, in 2020 a study on glucose control [6] tested the idea of CGM with remote monitoring versus CGM alone with remote examination, in pregnants with diabetes. The comparison discovered findings towards glucose management alterations and health outcomes. Another work [5] tested the measurement precision of a wearable blood pressure device thus the feasibility of weight gain and blood pressure monitoring in a group of pregnants who could develop hypertensive disorders. In [4] an RPRM non-invasive system was developed for real-time checking, while in [3], an IoT system with various sensors was implemented for delivering proper ubiquitous checking before and after pregnancy. In 2021, a new distant system of portable health clinic (PHD) for maternal and child health (MCH) acted as a pilot in a rural district, for finding significant information and reporting them at the mid-intervention period [2]. Finally in [1], the impact of wearable devices and their applications were examined; and whether they could sustain reliable distant monitoring with IoT inclusion for obstetrics departments.

7 Discussion

This systematic review examined 18 works. The various findings are discussed, by answering the posed questions, and thus the way (in what direction? how? when?) they can affect the subject under test. In regard to the category of Data Collection including Questionnaires/Interview (Q/I), Clinical examination (CE) and Measurements (Ms), the totals of studied subcategories were respectively equal to 9, 11 and 16 corresponding to the fact of how many works included these kind of data in their procedures. The Q/I and CE correspond to 50% (9/18) and to 61% (11/18) of the total works. These take a *cw* (Eq. 1) of 17% and 20% respectively. In detail, 18 works and 3 sub-categories exhibit 54 maximum marks (the case of all works with every included subcategory). The percentages in ascending form are 17%, 20% and 30%. The median is 20%, and the CE and Ms are the significant contributors. So, the Q/I need urgent attention for inclusion in future works (contribution to Q4).

Table 2 includes 5 subcategories and 90 maximum marks (18•5). So, the *cw* of subcategories (1) to (5) (Table 2) are found as 10%, 7%, 3%, 3% and 6% respectively. The median is 6%, so the significant contributors are subcategories 1, 2 and 5. The first and second category (IoT and smart devices) need further future investigation. Tables 3 and 4 with 11 sub-categories exhibit 198 maximum marks. The 6th (#6) to 16th (#16) sub-category have *cw* percentages of 5%, 3,5%, 2,5%, 5%, 2%, 1%, 1,5%, 3%, 1,5%, 2% and 2% respectively. These, in ascending order, are 1% (#11), 1,5% (#12), 1,5% (#14), 2%(#10), 2%(#15), 2%(#16), 2,5%(#8), 3%(#13), 3,5%(#7), 5%(#6) and 5% (#9). The median equals to 2% (#16). Any value $\geq$ 2% is a significant contributor. So, the sub-categories 11, 12 and 14 need attention for inclusion in future works. These correspond to uterine contraction sensing, Hemoglobin and acoustic sensors (contribution to Q4).

From Tables 3 and 4 three recent works [1, 2, 4] have six, eight and seven sub-categories included. Therefore, the same quality of distant-monitoring is possible with sensors as with clinical visits while guarded against Covid-19 (contribution to Q1). Also, as technology advances, more sensors are included in reduced space. Based on the findings so far, the reviewed technologies exhibit great potential for being key factors in clinical science. Some of them have over six sensors included thus (from Table 2) these have over three types of utilized measurements and technologies (UMT) and act as a propagating factor for a more accurate clinical assessment (contribution to Q2).

Finally, from Table 3 only three works exhibited DPA and only two works indirect PA. These are [1, 3, 8, 13]. Summarizing all PAs and in regard to the wearable devices, the pregnants stated their comfort, they liked them and considered them useful. Similarly, 84% of nurses, found the monitoring process easy and eligible for recommendation. This know-how helps to alleviate the burden of monitoring during the birth process and suggests indirectly that stress and anxiety are reduced. Also, the use of monitoring wearable devices, reduces stress and prevents unpleasant events. Stress in pregnants was checked by heart rate monitoring (HR and HRV). The HRV is associated with changes in the activity of autonomic nervous system. Mental and psychological stress were further studied. In conclusion, remote monitoring technologies actuate women to visit their doctors when malfunctions are detected, that could not be discovered otherwise and thus optimize their clinical checking. Pregnants worried that wearable devices' monitoring could threaten their privacy. Also, the total DPA (Table 6) has a cw of almost 14% (=5/18/2) corresponding to a low contribution in this scientific area. Future works should definitely include DPA (contribution to Q3).

8 Conclusions

This paper introduces a new type of systematic review (SAPW) on new mobile and wearable technologies on pregnant women. These technologies are strong candidates for distant monitoring and can contribute greatly to times of Covid-19. Through continual process, they find problems that otherwise could not be diagnosed. Some of the works included psychological assessment in a direct or indirect manner. The anxiety was found reduced and the pregnant women were more eager towards self-evaluation with such technologies. Quantification of the sub-categories was done with weights and ranks. Also, the key question was answered relevant to accomplishing similar results through wearable or/and IoT technologies compared to those of frequent hospital visits.

References

1. Li, X., Lu, Y., Fu, X., Qi, Y.: Building the Internet of Things platform for smart maternal healthcare services with wearable devices and cloud computing. Futur. Gener. Comput. Syst. **118**, 282–296 (2021)
2. Kikuchi, K., et al.: Portable health clinic for sustainable care of mothers and newborns in rural Bangladesh. Comput. Methods Progr. Biomed. **207**, 106156 (2021)
3. Sarhaddi, F., et al.: Long-term IoT-based maternal monitoring: system design and evaluation. Sensors **21**, 2281 (2021)
4. Veena, S., Aravindhar, D.J.: Remote monitoring system for the detection of prenatal risk in a pregnant woman. Wirel. Pers. Commun. **119**(2), 1051–1064 (2021). https://doi.org/10.1007/s11277-021-08249-x
5. Hasan, S., Ahmed, S., Khan, M., Sarker, S., Ahmed, T.: Achieving optimal gestational weight gain, birth weight, and perinatal outcomes among pregnant women at risk of hypertension: protocol for a pilot randomized controlled trial. JMIR. Res. Protocols. **9**, e16676 (2020)
6. Polsky, S., et al.: Continuous glucose monitor use with and without remote monitoring in pregnant women with type 1 diabetes: a pilot study. PLOS ONE. **15**, e0230476 (2020)
7. Houzé de l'Aulnoit, A., et al.: development of a smart mobile data module for fetal monitoring in e-healthcare. J. Med. Syst. **42**(5), 1–7 (2018). https://doi.org/10.1007/s10916-018-0938-1
8. Souza, R., et al.: Identification of earlier predictors of pregnancy complications through wearable technologies in a Brazilian multicentre cohort: Maternal Actigraphy Exploratory Study I (MAES-I) study protocol. BMJ Open. **9**, e023101 (2019)
9. Stewart, Z., Thomson, L., Murphy, H., Beardsall, K.: A feasibility study of paired continuous glucose monitoring intrapartum and in the newborn in pregnancies complicated by type 1 diabetes. Diabetes Technol. Ther. **21**, 20–27 (2019)
10. Khandoker, A., Ibrahim, E., Oshio, S., Kimura, Y.: Validation of beat by beat fetal heart signals acquired from four-channel fetal phonocardiogram with fetal electrocardiogram in healthy late pregnancy. Sci. Rep. **8**, 1 (2018)
11. Lai, J., et al.: Performance of a wearable acoustic system for fetal movement discrimination. PLOS ONE, **13**(5), e0195728 (2018). https://doi.org/10.1371/journal.pone.0195728

12. Alfadhli, E., Osman, E., Basri, T.: Use of a real time continuous glucose monitoring system as an educational tool for patients with gestational diabetes. Diabetol. Metabol. Syndrome. **8**, 1–7 (2016)
13. Boatin, A., et al.: Wireless vital sign monitoring in pregnant women: a functionality and acceptability study. Telemed. e-Health **22**, 564–571 (2016)
14. Lavariega, J., Córdova, G., Gómez, L., Avila, A.: Monitoring and assisting maternity-infant care in rural areas (MAMICare). Int. J. Healthcare Inf. Syst. Inf. **9**, 32–43 (2014)
15. Kumareswaran, K., et al.: Physical activity energy expenditure and glucose control in pregnant women with type 1 diabetes: is 30 minutes of daily exercise enough? Diabetes Care **36**, 1095–1101 (2013)
16. Ryo, E., Kamata, H.: Fetal movement counting at home with a fetal movement acceleration measurement recorder: a preliminary report. J. Matern. Fetal Neonatal Med. **25**, 2629–2632 (2012)
17. Murphy, H., et al.: Pathophysiology of postprandial hyperglycaemia in women with type 1 diabetes during pregnancy. Diabetologia **55**, 282–293 (2011)
18. Nishihara, K., Horiuchi, S., Eto, H., Honda, M.: A long-term monitoring of fetal movement at home using a newly developed sensor: an introduction of maternal micro-arousals evoked by fetal movement during maternal sleep. Early Human Dev. **84**, 595–603 (2008)
19. Grant, M., Booth, A.: A typology of reviews: an analysis of 14 review types and associated methodologies. Health Info. Libr. J. **26**, 91–108 (2009)
20. Christofilakis, V., et al.: Earth-to-earth microwave rain attenuation measurements: a survey on the recent literature. Symmetry. **12**, 1440 (2020)
21. Moher, D., Liberati, A., Tetzlaff, J., Altman, D.: Preferred reporting items for systematic reviews and meta-analyses: the PRISMA statement. PLoS Med. **6**, e1000097 (2009)

A Voice Handicap Index Study Based on Receiver Operating Characteristic Analysis: The Unified Monitoring of Adult Smokers Intended for Mobile Applications

Dionysios Tafiadis[1], Spyridon K. Chronopoulos[1,2], Evangelia I. Kosma[1], Kostas Peppas[3], Vasilis Christofilakis[2], Eugenia I. Toki[1](✉), Louiza Voniati[4], and Nafsika Ziavra[1]

[1] Department of Speech and Language Therapy (SLT), University of Ioannina (UOI), Ioannina, Greece
{tafiadis, spychro, e.kosma, toki, nziavra}@uoi.gr
[2] Electronics-Telecommunications and Applications Laboratory, Department of Physics, UOI, Ioannina, Greece
{spychro, vachrist}@uoi.gr
[3] Department of Informatics and Telecommunications, University of Peloponnese, Tripoli, Greece
peppas@uop.gr
[4] Speech and Language Therapy, Department of Health Sciences, European University Cyprus, Nicosia, Cyprus
l.voniati@euc.ac.cy

Abstract. Voice assessment consists of many methods, amongst them being the Voice Handicap Index (VHI) which is a non-interventional self-reported questionnaire. This research determined and interpreted the cutoff points (COPs) of the Receiver Operating Characteristic (ROC) curves (signal processing) of the standardized VHI Hellenic version along with the Greek Voice Evaluation Template (GR-VET). In turn, as mobile applications are of high interest in the community of smart phones, the latter feature has been taken strongly into account for constructing a protocol serving for mobile screening usage. Specifically, a protocol and diagram of a proposed mobile application is introduced in order the latter to act as a screening platform for alerting clinician for probable voice status problems. The probability of developing voice symptoms was almost the same for all the young smokers and nonsmokers participants. Consequently, another asset from this research is the more customizable treatment using the knowledge of the COPs' history for every smoker in mobile level technology. The smokers' group exhibited higher overall VHI total score as compared to non-smokers. Important statistical differences were found due to smoking for all VHI's construct domains with a total COP of 19.50 (sensitivity: 0.852, 1-specificity: 0.108). Another important finding revealed that VHI distinguished the different voice status of smokers compared to nonsmokers. Specifically, if the COP point is lower after a new assessment, then people exhibit progress as their voice appears improved. Moreover, a valuable finding is that smokers exhibit almost the same unified cutoff point as the voice disordered patients (COP = 18.50).

M. E. Auer and T. Tsiatsos (Eds.): IMCL 2021, LNNS 411, pp. 765–777, 2022.
https://doi.org/10.1007/978-3-030-96296-8_69

Keywords: Mobile applications · Monitoring · ROC analysis

1 Introduction

In the recent years, it has been proven that data collection using self-assessment procedures is of critical importance. A self-assessment test usually records data relevant to a patient's voice status condition [1]. Consequently, in order to further deepen in this area of expertise a brief literature review is reported and in more detail the aforementioned type of assessment supplies the clinicians with information about the patient's lifestyle trends and influencing factors relevant to voice usage [1]. Examples of these types of influencing factors include environmental conditions (e.g., talking over background noise or exposure to chemicals and dust) [2]; health issues like history of gastrointestinal reflux or laryngopharyngeal reflux, respiratory pathologies, recurrent colds and/or sinus infections [2] and smoking [3]. The latter is reported as one of the primary worldwide causes of death. World Health Organization (WHO) foresees the death of eight million people till 2030 due to smoking [4]. The prevalence of smoking among adults in Greece is estimated at 42% with average cigarettes consumption being equal to 21.4 per day [4]. Especially the smoking among young Greek adults was estimated at 45.20% (aged 15–24 years) and 48.10% (aged 25–29 years). Furthermore, Lipari, Kroutil & Pemberton [5] reported that the number of young adult smokers has been doubled. All these findings can probably lead to smoking related health problems [6].

Laryngeal pathologies, which lead to voice disorders, are also associated to smoking [7]. It has been documented that larynx lesions which are linked to smoking habits could result in inflammatory diseases or irritation of laryngeal mucosa [8], structural changes of the larynx such as Reinke edema and initial or malignant tumors [8]. The consequences of the aforementioned changes could probably lead to deviations of voice and vocal complaints [3]. Consequently, smokers could seek for clinical advice and evaluation due to vocal changes. Multidimensional protocols [1] have been established for the evaluation procedures, as well as evidence-based processing for clinical voice assessments [9]. The suggested protocols include the imaging of the larynx (e.g., endoscopic and stroboscopic) [10], acoustic and perceptual evaluation of the voice [11] and aerodynamic assessments [11].

The administration of self-reported questionnaires is suggested for estimating the patients' perception level of their voice status [1]. The most common used self-evaluation voice questionnaire is the so-called Voice Handicap Index [12] which has been translated into many languages [13–17]. VHI has been administered to smokers [18], to voice users [19] and to patients with laryngeal disorders [13–26]. VHI is consisted of 30 questions which are split in three domains named as functional (VHI-F), physical (VHI-P) and emotional (VHI-E). Every domain includes 10 questions. Each of them responds to a Likert-type scale of 0 to 4. This results to a total maximum score (VHI-T) of 120 (maximum 40 for each domain). Additionally, VHI cutoff points have been recently proposed for recording voice-disordered populations [20] as well as for young female and male smokers [21, 22].

Motivated by the above facts, in this research, the Receiver Operating Characteristic analysis (signal processing) has been utilized to produce cutoff points (COPs) with coordinates of "sensitivity" and "1-specificity". A similar technique has also been used in Radar technology (1950s) in order to track various signals. If the definition of signal correlation is replaced by the definition of decease, then this methodology can conclude to positive (presence of disease) or negative results (absence of disease). The coordinates of the COPs also define the overall accuracy of the prediction. Specifically, "sensitivity" is the probability of a positive result (disease is present) and "specificity" is the probability of a negative result (disease is absent). In turn, the resulting ROC (receiver operating characteristic) curve is a pair of "sensitivity" and "specificity" producing a more accurate predictive result [23]. These types of curves have many applications in various scientific areas such as data mining and medical decision making [23]. Also, special condition exists in the case of the ROC curve being diagonal (45 degrees, y = x) as the guessing operation becomes random and in this manner the curve exhibits unwanted performance. On the contrary, the perfect performance occurs when curve passes near the upper left corner where sensitivity and specificity are both equal to 100%. Another important feature is the calculation of the area under curve (AUC), yielding values between 0 and 1. The AUC value is split into four categories namely "No predictive" (AUC < 0.5), "Acceptable" (0.7 to less than 0.8), "Excellent" (0.8 to less than 0.9) and "Outstanding" ($\geq$ 0.9) [24]. Notably, the Wilcoxon tests of ranks are equivalent to AUC which is almost the same as Gini coefficient [24]. Gini coefficient is the twice area between the ROC curve and the diagonal line (unwanted condition) and can also be expressed as "(2 • AUC)-1" [25].

This research aimed to assess a protocol relevant to computing the COPs of VHI for smokers. These points could help the clinicians to better estimate the regular monitoring of young smokers. Additionally, smokers could avoid potential development of future voice symptoms on time (even if previously have been reported as asymptomatic). The knowledge helped towards designing and proposing a valid protocol of a mobile application, which was the primary goal of this research study. The protocol is based on the statistical techniques and tools. The latters were used in this implementation for carrying out non-invasive trials on 250 subjects, carefully chosen, in order to act as a guide for future development of a self-aware mobile prediction system.

2 Materials and Methods

2.1 Participants and Data Collection

Two hundred and fifty students (120 females and 130 males), at University of Ioannina, School of Health Sciences, were recruited for this study. The sample constituted of 128 smokers (60 females and 68 male) and 122 nonsmokers (60 females and 62 male). Notably, the data collection was conducted at the former Technological Educational Institute of Epirus (absorbed by University of Ioannina). Other participants were not included in the study due to following conditions: They had the last two weeks any upper or lower respiratory system disorder or any laryngeal/vocal complaint. Similarly, were not included in the study the young adults who experienced symptoms of gastroesophageal reflux or laryngopharyngeal reflux disease or/and had a former history of alcohol or drug abuse and reported that worked or lived under environmental issues (e.g. dust, exposure to chemicals and/or allergens and temperature changes). The name of "nonsmokers" was attributed to the subjects who had never smoked prior to this research.

2.2 Data Collection

All subjects, prior to their participation, were informed about the research purposes, they agreed and signed a consensus letter. All subjects filled in the translated version of Voice Evaluation Template (VET) [26] and the standardized Hellenic version of VHI [14]. VET was developed by the American Speech Hearing Association (ASHA) as a consensus template for voice disorders. VHI is a questionnaire consisted of 30-items. The questionnaire is split into three domains (i) functional, (ii) physical and (ii) emotional. Each domain includes 10 questions with a score range of 0 to 40. All participants were not informed on purpose about the impact of smoking on the voice condition, in order to be uninfluenced during the procedure of filling in the questionnaires. Finally, all participants were categorized as non-symptomatic via acoustic measurements (F0, shimmer, jitter and HNR) [27] using normative data [28].

2.3 Statistical Analysis

The Kolmogorov-Smirnov and Shapiro-Wilk tests were performed in order to check the distribution of variables. All the skewed variables (e.g. VHI scores) were expressed through median (interquartile range), while the normal distributed variables (e.g. age of participants) were expressed through mean and standard deviations (SD). Also, Mann-Whitney U test was used for the comparison of continuous variables between the two study groups (smokers and nonsmokers). Moreover, Kruskal-Wallis test was used for comparing the four subgroups (nonsmoker males, nonsmoker females, smoker males and smoker females), while a two-sample t-test was conducted between the two study groups (smokers and nonsmokers) for all VHI index domains. Finally, ROC curves were constructed for the purpose of finding the COP values of VHI-T and of its three domains (VHI-F, VHI-P, VHI-E). All reported P values were two-tailed and the statistical significance was set at $P < 0.05$. The analysis was conducted using SPSS statistical software (version 19.0, Armonk, NY, USA).

3 Results and Statistical Analysis

The sample consisted of 250 male participants (128 smokers and 122 non-smokers). The total mean age of the overall sample was 22.32 years (standard deviation, SD = ±2.38) ranging from 18 to 32 years old. The mean age of smokers was calculated at 22.18 years (SD = ±2.44) and for non-smokers at 22.47 years (SD = ±2.33). The mean age of male smokers was found to be equal to 22.01 years (SD = ±2.35) and for male non-smokers to 22.66 years (SD = ±.35). Also, the mean age for female smokers was calculated at the value of 22.37 years (SD = ±2.54) and for female non-smokers at 22.27 years (SD = ±2.47) which is shown in Table 1.

The average years of smoking were equal to 2.75 (SD = ±1.32) with a range between 6 months and 5 years, whereas the numbers of cigarettes per day had a mean value of 14.69 (SD = ±7.72). Specifically, the average years of smoking for male smokers were found equal to 2.65 yrs (SD = ±1.21) and the number of smoked cigarettes per day had a mean value of 14.27 cigs (SD = ±7.06). As for female smokers, the average years of smoking were equal to 2.23 yrs (SD = ±1.35) and the number of smoked cigarettes per day had a mean value of 15.21 cigs (SD = ±8.23).

Table 1. Demographic data for smokers and non-smokers.

	Smokers (N = 128)		Non-smokers (N = 120)	
	Male(N= 68)	Female (N=60)	Male(N= 68)	Female (N=60)
Mean Age (SD)	22.01 (±2.35)	22.37 (±2.54)	22.66 (±2.35)	22.27(±2.47)
Years of Smok-ing	2.65 (±1.21)	2.23 (±1.35)	–	–
No. of Cigarettes	2.65 (±1.21)	15.21 (±8.23)	–	–

Smokers had a significant higher overall VHI total score compared to that of non-smokers, U = 1250.50, P < 0.001. Same statistically significant differences of medians were computed for VHI-F (U = 1993.50, P < 0.001), VHI-P (U = 3021.00, P < 0.001) and for VHI-E (U = 1799.00, P < 0.001) domains. The smokers' subgroup (in all comparisons) exhibited the higher achieved scores (Table 2).

Table 2. Comparisons of medians between smokers and non-smokers for VHI total score and VHI domains.

	Smokers (N = 128)	Non-smokers (N = 122)		
	Median (IQR)	Median (IQR)	Mann-Whitney U	P level
Total	29.00 (24.00–32.00)	15.00 (13.00–17.00)	1250.50	< 0.001*
Functional	9.00 (7.00–11.75)	5.00 (3.00–6.00)	1993.50	< 0.001*
Physical	9.00 (7.00–11.00)	5.50 (4.00–7.00)	3021.00	< 0.001*
Emotional	10.00 (8.00–12.00)	5.00 (4.00–6.00)	1799.50	< 0.001*

* p level at P < 0.05 Abbreviations: IQR, interquartile range; VHI, Voice Handicap Index

A Kruskal-Wallis test was conducted for comparing subgroups' medians of all VHI total scores and their domains. Particularly significant statistical differences were observed between smokers and nonsmokers' group for VHI-T score (H (3) = 136.317, P < 0.001) with mean rank of 55.78 for male non-smokers, 169.50 for male smokers, 86.48 for female non-smokers and 180.66 for female smokers. Likewise, significant statistical differences were calculated for VHI-F domain (H (3) = 105.542, P < 0.001) with mean rank of 76.25 for male non-smokers, 164.88 for male smokers, 79.48 for female non-smokers and 177.78 for female smokers. Similarly, significant statistical differences were calculated for VHI-P domain (H (3) = 75.440, P < 0.001) with mean rank of 73.31 for male non-smokers, 167.08 for male smokers, 99.64 for female non-smokers and 158.16 for female smokers. Finally, VHI-E domain exhibited statistical differences (H (3) = 121.081, $P < 0.001$) with mean rank of 59.12 for male non-smokers, 163.23 for male smokers, 93.95 for female non-smokers and 182.88 for female smokers.

Table 3. Comparisons of means between smokers and non-smokers for VHI

	Smokers (N = 128)	Non-smokers (N = 122)			
	Mean (SD)	Mean (SD)	t-test	*df*	*P* level
VHI total	28.46 (8.410)	15.67 (4.10)	14.979	248	<0.001*
VHI functional	9.53 (3.617)	4.93 (2.200)	12.069	248	<0.001*
VHI physical	9.22 (3.602)	5.53 (2.268)	9.628	248	<0.001*
VHI emotional	9.71 (3.256)	5.09 (1.881)	13.653	248	<0.001*

* p level at $P < 0.05$ Abbreviations: SD, standard deviation; VHI, Voice Handicap Index

Also, a students' t-test was conducted for comparing the means between smokers and non-smokers (Table 3). The smokers had a significant higher VHI total score compared to non-smokers, t (248) = 14.979, $P < 0.001$. Same statistically significant

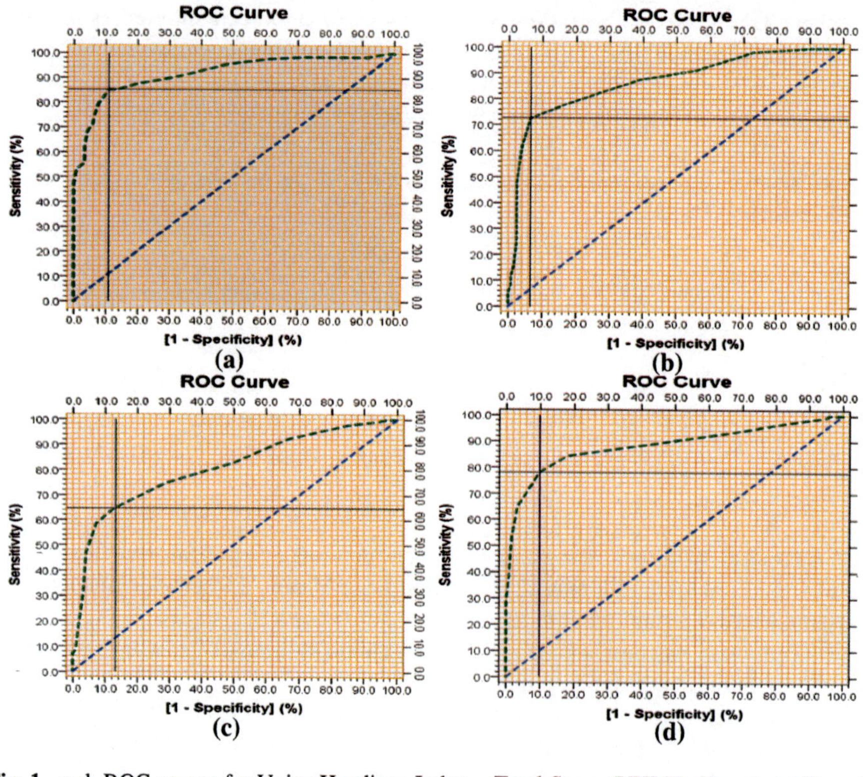

Fig. 1. a-d. ROC curves for Voice Handicap Index – Total Score (VHI-T), Functional Domain (VHI-F), Physical Domain (VHI-P) and Emotional Domain (VHI-E) respectively. All the COPS are shown on Fig. 1 a-d in percentage format.

differences (Table 1) of medians were calculated for VHI-F [t (248) = 1993.50, P < 0.001], VHI-P [t (248) = 3021.00, P < 0.001] and for VHI-E [t (248) = 1799.00, P < 0.001]. The smokers' subgroup (in all comparisons) exhibited the higher achieved scores (Tables 2 and 3).

Furthermore, a ROC analysis was conducted (Fig. 1 a-d) in order to determine the cut off points of VHI-T and of its three domains (Functional, Physical and Emotional). A statistical significant positive discrimination between smokers and non-smokers was revealed. Particularly, an excellent effect value was noted for VHI-T (AUC 0.930, P < 0.001), while very significant effects were noted for VHI-F (AUC 0.878, P < 0.001), VHI-P (AUC 0.866, P < 0.001) and VHI-P domain (AUC 0.895, P < 0.001) (Table 4).

Table 4 Coordinates for smokers and non-smokers curve (VHI total score and domains)

	AUC	SE	*P* level	95% CI
Total	0.919	0.018	<0.001*	0.884–0.953
Functional	0.872	0.023	<0.001*	0.828–0.917
Physical	0.807	0.028	<0.001*	0.752–0.861
Emotional	0.885	0.022	<0.001*	0.841–0.928

p* level at P < 0.05 Abbreviations: AUC, area under curve; CI, confidence interval; SE, standard error; VHI, Voice Handicap Index.

The COP point of the VHI-T score was equal to 19.50 with sensitivity of 0.852 and 1-specificity of 0.108 (Fig. 1a), VHI-F COP was equal to 7.50 with sensitivity of 0.727 and 1-specificity of 0.066 (Fig. 1b). Also, VHI-P COP was equal to 7.50 with sensitivity of 0.648 and 1-specificity of 0.131 (Fig. 1c) and VHI-E COP was equal to 7.50 with sensitivity of 0.781 and 1-specificity of 0.098 (Fig. 1d).

4 Mobile Application's Diagram and Protocol

Based on our pilot study on a sample of 250 representative subjects, we concluded into designing and introducing a protocol and diagram of a proposed mobile application. The later will serve as a screening platform (analyzed in a future follow-up study), whereas with the implementation of decision making and mechanical learning, will alert the corresponding clinician for potential voice status problems. In turn, this platform will be enriched with additional screening features not necessarily oriented only to voice monitoring. Specifically, the mobile application is consisted of six major protocols which correspond to specific functions. The protocols range from input operations to the results and interactive presentation menu. The key points of the proposed "software machine", being the implemented software which is the whole set of programs, procedures, and routines, are shown and named as MSACUS# (Fig. 2):

- ***MSACUS#1-UL***: This operational protocol works towards collecting the user's data based on already known tools such as e.g. VHI and VET.
- ***MSACUS#2-IMP***: It gives the user the capability of importing additional data for comparison and screening issues.
- ***MSACUS#3-FC***: The programmer can easily implement new techniques, elements and new programming code for optimizing speed and system integrity.
- ***MSACUS***: It refers to mobile screening application core using statistics. The programming core can be directly controlled by the programmer and thus it gives real-time details about the application's performance.
- ***MSACUS#4.0-IC***: This is the integrity control (IC). The application checks all the sub-systems for potential errors and corruptions. In turn, the validity certifies that the final results will be calculated and displayed properly.
- ***MSACUS#4.1-IC***: After the previous IC, the system checks for new code (if this is available) and it performs typical tests for proper operation.
- ***MSACUS#4-INIT***: In parallel to functions #4.x, the system initializes and uses the updated database while the ***MSACUS#4-INST*** checks all the available statistical techniques and drawing capabilities (e.g. ROC curves) while it performs quick tests (e.g. checking in accordance to already known curves).
- ***MSACUS#5-ACTIV***: At this point, main application's interface is loaded along with all its features activated.
- ***MSACUS#6-RES***: Finally, the results with an interactive menu are displayed while giving capabilities for sending selected information to a clinician or to upload/compare them to anonymous database.

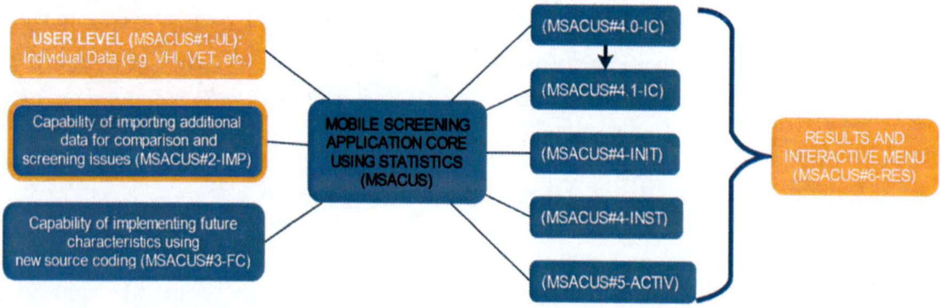

Fig. 2. Mobile application's diagram with protocol codes (Blue color denotes the accessibility by the programmer. The orange color refers to the user's access. The protocol with orange outline and blue filling symbolizes the limited access by the user (and the full access by the programmer).

5 Discussion

Current Study showed that smokers had significantly higher medians in all domains and total VHIs' scores compared to the control group, revealing that smokers perceive their voice changes similarly to young male and female smokers [21, 22], to people with voice disorders [12–17] and to specific student groups [18]. VHI-T median score was found 29.00 (for asymptomatic smokers) and in agreement with Trinite & Sokolovs' [17] research on dysphonic patients (Latvian VHI-T score: 29.70). VHI-F domain was found with a median of 9.00 being similar to that of voice disordered patients [13–15, 17, 20]. Also, the VHI-E median of 10.00 was similar to that of Bonetti and Bonetti research upon population with vocal symptoms [13]. Also, Smits, Marres, de Jong [29] resulted to a VHI-E score of 10.07, for patients with vocal fold lesions, which agrees to our study's calculated VHI-E smokers' score. These results indicate that smokers perceive their voices almost the same to those of patients with voice disorders.

Smoking is related to initial tumor stage or malignant tumors leading to voice disorders [8]. Consequently, VHI could serve as a useful screening tool towards informing about smoking consequences and fighting cancers. Notably, the estimated our VHI-T score (29.00) is approximately in agreement with that of patients with laryngeal or head and neck tumors and/or under a therapy (resection, radiotherapy, etch.) [29–31]. Even the fact that VHI exhibited low values corresponding to minimal handicap, the estimated results are similar to those of patients suffering from mild dysphonia [30]. Meleca et al. [31] found the same mean score (VHI-T = 29.00) as in this study, for patients irradiated in the stage III and IV of laryngeal cancer. Consequently, the VHI is an excellent measure of voice for young smokers and for populations with voice disorders. Likewise, Azevedo et al. [30] reported similar VHI score (VHI-T = 30.00) and VHI-P (VHI-P = 8.00) after treatment of patients with advanced squamous carcinoma of the larynx and/or hypopharynx. Hence, the regular administration of VHI and monitoring of smokers can lead to the probable prediction of severe emerged medical conditions.

VHI can also be used as a Primary Health Diagnostic tool involving the indirect estimation of shame levels between smokers and non-smokers [32]. Consequently, if a VHI-T score decreases for a smoker, this could lead to various assumptions. These could include on one hand, the voice improvement with the accomplishment towards quitting smoking, and from the other hand, the imminent decrease of psychological effects such as shame experience [32]. Additional indirect effects of high VHI-T score of smokers are the respective lower levels of empathetic feeling of passive smokers towards smokers. For example, when the smoker students become professionals then it would be highly probable to receive negative (non-empathetic) approach from passive smoker colleagues in their workplace [33]. Furthermore, smoking is considered a negative health factor as it is shown in [34] relevant to social representations of harmful health factors among Greeks. Consequently, the aforementioned indirect effects could further motivate smokers towards quitting smoking apart from health issues.

Moreover, significant differences for VHI scores were observed between all subgroups (male smokers, male nonsmokers, female smokers and female nonsmokers). The female smokers exhibited the highest scores. The latest is in agreement with Tafiadis, et al. [21, 22] where was observed high VHI scores for male and female smokers compared to non-smokers and non-dysphonic individuals. Similarly, Tafiadis, Toki, Miller and Ziavra [18] emphasized the fact that female smokers had higher VHI scores in comparison with non-smokers. After administrating VHI to the total smoker's sample (males and females), the perception level about their voice was raised alike to those of young female smokers [21, 22] and dysphonic patients [12]. That increase in awareness (due to VHI) along with the proper interpretation of the produced COPs, could strengthen the decision of young smokers to quit smoking [21, 22]. Also, cessation of smoking before 35 years old will improve their voice and subsequently leads to a better quality of life [26]. Also, the cessation would increase life expectancy [35] even without provoking fully reversible condition of health to its primary form before smoking [36].

VHI utility displays good self-monitoring during smoking cessation treatments for young smoker adults. This could be accomplished by the joint contribution of VHI's three domains COPs (VHI-F = 7.50, VHI-P = 7.50, VHI-E = 7.50) which blends many factors (via its 30 items) that affect vocal use [37]. The COP score of self-perceptual level of voice which is included in VHI total score along with the equal effect of its domains could determine a valid prediction. This prediction is relevant to the risk of a young smoker – even asymptomatic – to develop a vocal symptom or symptoms in the future similarly to young male and female smokers [21, 22]. In addition to the latest, the VHI can possibly be a sensitive tool for detecting a risk of developing a voice disorder for young adult smokers [21, 22] and for dysphonic patients [12–16]. Finally, our COP of VHI-T (19.50) was the same for all young smokers (males and females) similarly to separate cases of young male and female adult smokers [21, 22]. This result is almost equal to Ohlsson's and Dotevall's [16] approximation of 20.00 for voice disordered patients, which is another proof of the VHI's sensitivity towards smokers' population. On the contrary, our COP (of asymptomatic population) was in disagreement with two other studies [20, 38] as these studies evaluated population with voice problems. The identification of VHI's efficiency, through finding COP values, constitutes a screening tool for large populations, multicentric studies, and/or public service [39]. Also, the total COP was almost equal to the global COP (18.50) of voice-disordered patients [40] which places the smokers in the same category as voice-disordered patients.

6 Conclusions

The main aim of this study was to estimate the VHI's cutoff points (COPs) along with their subdomains for young smokers in order to contribute towards designing a mobile screening application. VHI can profoundly distinguish smokers' and non-smokers' perception scores of their voice status. The utility could be used as a self-monitoring tool of quitting smoking for young adults. Furthermore, it could aid as a screening procedure and therefore as a first-degree health care tool for the early referencing to a voice professional. The proper use of VHIs' COPs (calculated through signal processing) could help a voice clinician to predict the possible development of smoking-related dysphonia of a patient in the future. Finally, if these COPs are combined to the voice assessments, then they could better define a voice disorder originated from smoking. This consequently can lead to a more customized treatment planning. Also, no other clinical assessment was performed for this study. Further research is suggested to be conducted in older adult smokers. In the long run, this could help comparing the early and late effects of smoking in terms of VHI scores and COP points, and can lead to a VHI chart of smoking relevant to years of smoking and to clinical condition of subjects under test.

References

1. Dejonckere, P., et al.: A basic protocol for functional assessment of voice pathology, especially for investigating the efficacy of (phonosurgical) treatments and evaluating new assessment techniques. Eur. Arch. Otorhinolaryngol. **258**, 77–82 (2001)
2. Roy, N., Merrill, R., Thibeault, S., Parsa, R., Gray, S., Smith, E.: Prevalence of voice disorders in teachers and the general population. J. Speech Lang. Hear. Res. **47**, 281–293 (2004)
3. Awan, S., Alphonso, V.: Effects of smoking on respiratory capacity and control. Clin. Linguist. Phon. **21**, 623–636 (2007)
4. Tobacco World Health Organization, http://www.who.int/mediacentre/factsheets/fs339/en/, last accessed 2019/06/30
5. Lipari, R., Kroutil, L., Pemberton, M.: Risk and Protective Factors and Initiation of Substance Use: Results from the 2014 National Survey on Drug Use and Health. https://www.samhsa.gov/data/sites/default/files/NSDUH-DR-FRR4-2014v1/NSDUH-DR-FRR4-2014.htm. Accessed 09 July 2021
6. Moon-Howard, J.: African American women and smoking: starting later. Am. J. Public Health **93**, 418–420 (2003)
7. Gnjatic, M., Stankovic, P., Djukic, V.: The effect of smoking and forced use of the voice to development of the vocal polyps. Acta Chir. Iugosl. **56**, 27–32 (2009)
8. Espinosa, J., Bravo, J.P., Baron, M.G.: Influence of tobacco on laryngeal carcinoma in Spain. Neoplasma **39**, 319–322 (1991)
9. Roy, N., et al.: Evidence-based clinical voice assessment: a systematic review. Am. J. Speech Lang. Pathol. **22**, 212–226 (2013)
10. Yan, Y., Ahmad, K., Kunduk, M., Bless, D.: Analysis of vocal-fold vibrations from high-speed laryngeal images using a hilbert transform-based methodology. J. Voice **19**, 161–175 (2005)
11. Gaskill, C., Awan, J., Watts, C., Awan, S.: Acoustic and perceptual classification of within-sample normal, intermittently dysphonic, and consistently dysphonic voice types. J. Voice **31**, 218–228 (2017)
12. Jacobson, B., et al.: The voice handicap index (VHI). Am. J. Speech Lang. Pathol. **6**, 66–70 (1997)
13. Bonetti, A., Bonetti, L.: Cross-cultural adaptation and validation of the voice handicap index into Croatian. J. Voice **27**(130), e7-130.e14 (2013)
14. Helidoni, M., Murry, T., Moschandreas, J., Lionis, C., Printza, A., Velegrakis, G.: Cross-cultural adaptation and validation of the voice handicap index into Greek. J. Voice **24**, 221–227 (2011)
15. Karlsen, T., Grieg, A., Heimdal, J., Aarstad, H.: Cross-cultural adaption and translation of the voice handicap index into Norwegian. Folia Phoniatr. Logop. **64**, 234–240 (2012)
16. Ohlsson, A., Dotevall, H.: Voice handicap index in Swedish. Logop. Phoniatr. Vocol. **34**, 60–66 (2009)

17. Trinite, B., Sokolovs, J.: Adaptation and validation of the voice handicap index in Latvian. J. Voice **28**, 452–457 (2014)
18. Tafiadis, D., Toki, E., Miller, K., Ziavra, N.: Effects of early smoking habits on young adult female voices in Greece. J. Voice **31**, 728–732 (2017)
19. Tafiadis, D., et al.: Comparison of voice handicap index scores between female students of speech therapy and other health professions. J. Voice **31**, 583–588 (2017)
20. Moradi, N., Pourshahbaz, A., Soltani, M., Javadipour, S.: Cutoff point at voice handicap index used to screen voice disorders among Persian speakers. J. Voice **27**, 130-e1 (2013)
21. Tafiadis, D., et al.: Voice handicap index and interpretation of the cutoff points using receiver operating characteristic curve as screening for young adult female smokers. J. Voice **32**, 64–69 (2018)
22. Tafiadis, D., et al.: Using receiver operating characteristic curve to define the cutoff points of voice handicap index applied to young adult male smokers. J. Voice **32**, 443–448 (2018)
23. Lehr, R., Pong, A.: ROC Curve Encyclopaedia of Biopharmaceutical Statistics. Informa Healthcare, New York (2003)
24. Hosmer, D., Lemeshow, S., May, S.: Applied Survival Analysis. Wiley Series in Probability and Statistics, Wiley, Hoboken (2008)
25. Breiman, L., Friedman, J., Olshen, R., Stone, C.: Classification and regression trees. In: Wadsworth International Group, Belmont, CA (1984)
26. Voice Evaluation Template. https://www.asha.org/siteassets/uploadedfiles/aatvoiceevaluation.pdf. Accessed 09 July 2021
27. Tafiadis, D., Toki, E., Ziavra, N.: Deviations of voice characteristics in female speech therapy students that smoke using Dr. Speech. J. Commun. Disorder. Assist. Technol. **1**, 1–18 (2017)
28. Thibeault, S., Colton, R.H., Leonard, R.: Understanding Voice Problems. Wolters Kluwer Health/Lippincott Williams & Wilkins, Philadelphia (2011)
29. Smits, R., Marres, H., de Jong, F.: The relation of vocal fold lesions and voice quality to voice handicap and psychosomatic well-being. J. Voice **26**, 466–470 (2012)
30. Azevedo, E., Montoni, N., Gonçalves Filho, J., Kowalski, L., Carrara-de Angelis, E.: Vocal handicap and quality of life after treatment of advanced squamous carcinoma of the larynx and/or hypopharynx. J. Voice **26**, e63–e71 (2012)
31. Meleca, R., Dworkin, J., Kewson, D., Stachler, R., Hill, S.: Functional outcomes following nonsurgical treatment for advanced-stage laryngeal carcinoma. Laryngoscope **113**, 720–728 (2003)
32. Tzitzikos, G., et al.: The experience of shame in patients with chronic obstructive pulmonary disease (COPD). Eur. Psych. **41**, S264 (2017)
33. Iatrou, G., Gourgoulianis, K., Kotrotsiou, E., Gouva, M., Kotrotsiou, S., Paralikas, T.: Assessment of empathy and psychological characteristics of smokers. Eur. Psych. **41**, S741–S746 (2017)
34. Kotrotsiou, E., Gouva, M., Briseniou, E., Dragioti, E., Skenteris, N.: The social representations of aggravating factors living conditions and their connection to their psychological profile. Eur. Psych. **33**, S482 (2016)
35. Doll, R., Peto, R., Boreham, J., Sutherland, I.: Mortality in relation to smoking: 50 years' observations on male British doctors. BMJ **328**, 1519 (2004)
36. Lee, C., Kahende, J.: Factors associated with successful smoking cessation in the United States, 2000. Am. J. Public Health **97**, 1503–1509 (2007)
37. Behlau, M., et al.: Efficiency and cutoff values of self-assessment instruments on the impact of a voice problem. J. Voice **30**(506), e9-506.e18 (2016)

38. Schindler, A., Mozzanica, F., Vedrody, M., Maruzzi, P., Ottaviani, F.: Correlation between the voice handicap index and voice measurements in four groups of patients with dysphonia. Otolaryngol. Head Neck Surg. **141**, 762–769 (2009)
39. Gräßel, E., Hoppe, U., Rosanowski, F.: Graduierung des voice-handicap-index. HNO **56**, 1221–1228 (2007)
40. Tafiadis, D., et al.: Checking for voice disorders without clinical intervention: The Greek and global VHI thresholds for voice disordered patients. Sci. Rep. **9**, 1–9 (2019)

Experiment-Supported Mobile Application for Monitoring Human Activities Using Neural Networks

Andreea-Mihaela Calin(✉), Marian Alexandru, and Dan Nicula

Department of Electronics and Computers, Faculty of Electrical Engineering and Computers Science, Transylvania University of Brasov, Politehnicii nr.1, Brasov, Romania
andreea-mihaela.calin@student.unitbv.ro, {marian.alexandru, dan.nicula}@unitbv.ro

Abstract. During the last years, people have been interested to understand the human nature by creating smart algorithms and systems which can detect or recognize activities performed by humans. In this paper, we propose the implementation of a smart system able to recognize and monitor human's activities. The system implemented as an application running on a smartphone using Android OS. The mobile application determines what activity is performed by a person and stores the information into an internal database. The database can be used to generate statistics. By using statistics, the users can check the records of their daily activities. Our mobile application can detect 4 types of activities: standing, sitting, walking and jogging. For determining these activities, we used AI software implemented in PyTorch. The AI inputs consist in coordinates read from the smartphone's accelerometer. To evaluate the performance, we tested the neural network in 2 different modes: using a testing set and when it is part of the mobile application (someone performs an activity and the application detects it).

Keywords: Neural network · PyTorch · Android · Mobile application

1 Introduction

Human Activity Recognition (HAR) is a significant research field because it includes the recognition of the activities performed by a person. Recognition activities use machine learning algorithms to detect simple or complex activities such as: standing, sitting, walking, jogging, moving upstairs, moving downstairs, sleeping, cooking, etc. For the recognition to be realized, it is necessary to collect data from sensors.

In the past, the sensors for human activities were expensive and collecting information was difficult. Nowadays, these issues are solved by smartphones and other smart monitoring devices such as: smart watches, smart bracelets, fitness trackers, music players, etc. Mobile devices incorporate different types of sensors such as: accelerometer, gyroscope, magnetometer, proximity, GPS or fingerprint sensor, but there can be major issues when collecting and saving data from these sensors because

M. E. Auer and T. Tsiatsos (Eds.): IMCL 2021, LNNS 411, pp. 778–787, 2022.
https://doi.org/10.1007/978-3-030-96296-8_70

no one knows what type of activity was done or if there are different people doing the same activity, but different data are recording.

These differences appear because we are biologically different. It is very important how data is saved and its content because datasets will be used for training and testing machine learning models (supervised learning). In [3], the unsupervised learning was used to train the network. This means that it is not a perfect match between the activity type and sensor data. So, the machine learning algorithm can learn anything and sometimes the algorithm can learn wrong (overfitting). Therefore, it is better to use supervised learning for the training machine learning model. If the algorithm is trained properly, it will recognize correctly activities performed by anyone. In [5], TensorFlow machine learning framework was used for the implementation of the neural network. The TensorFlow framework is a popular Python framework for neural network, but PyTorch (pytorch.org) is a new Python library which has better performances and higher computing speed as TensorFlow. In [6], the model was tested in an experiment using four users, but the algorithm can fail when it's used by other people with different biological characteristics (height, weight, age).

It is essential what amount of information is present in the data set used for training and from how many sensors the information was collected. One or more sensors may be used, as in [7]. In the first case, it is easier to handle data and at the same time to obtain better results.

The goal of this paper is to present an implementation of an AI (Artificial Intelligence) system which recognizes human activities. It can determine 4 types of activities (sitting, standing, walking and jogging) by using accelerometer data. The recognition of activities is based on a neural network implemented in PyTorch. The steps which should be followed to create a mobile application which detects human activities are described in Sect. 2. Section 3 presents the results of the implemented algorithms.

2 Implementation of the Solution

2.1 Data Processing from Accelerometer Sensor

Accelerometer sensor measures acceleration and it detects orientation changes, permanently calculating x, y, and z coordinates. The data from the accelerometer is read every 3 s. Two consecutive readings are compared in order to find if a movement has occurred and for noise reduction. If the difference between two consecutive readings is higher than a preset value (the preset value is considered 0.01) it means that a movement has taken place and the data is valid and it can be stored into the database. The records are averaged on a particular time window before feeding the data into the neural network. These steps will be done for each coordinate separately.

2.2 Neural Network Design

A neural network was modelled by using PyTorch *nn* module. This module has several submodules which help to create the neural network. One of the submodules is *torch.nn*. This submodule has necessary blocks to create the architecture of neural networks [1]. A module can have one or more parameters and one or more submodules as attributes.

These are tensors, and their values are optimized during the training process. We used *nn. Linear* module so: nn.Linear (3, 100), where the first number indicates the number of inputs in the neural network and the second number indicates the number of outputs from the neural network. Every Linear module calculates outputs according to inputs. We used more random values for *nn.Linear* module, but with these values we obtained the best results: we trained the neural network and after that we tested it using the test dataset.

The *forward()* function receives inputs and returns outputs. This function has also the role of defining the structure of the neural network, which consists of a network of neurons that communicate with each other through weights. The first layer of neurons receives the inputs. On inputs, it applies operations of the first layer of neurons. The outputs of the first layer represent inputs for next layer.

The neural network should have one input layer and one output layer. The number of hidden layers may vary. We should respect the next rule: the number of outputs of a layer should be equal with the number of inputs of the next layer. The neural network from this paper is presented in Fig. 1 and consists of:

- one input layer which has 3 inputs and 100 outputs. These 3 entries represent the x, y, z coordinates of the mobile phone accelerometer.
- 3 hidden layers, each of them having 100 inputs and 100 outputs. These values were chosen because that is how it was obtained the best results. If we use a bigger or smaller number of inputs and outputs for each layer, the results obtained during testing process are worse (an empirical approach has been used).
- one output layer which has 100 inputs and 4 outputs. These 4 outputs represent 4 activities (sitting, standing, walking and jogging).

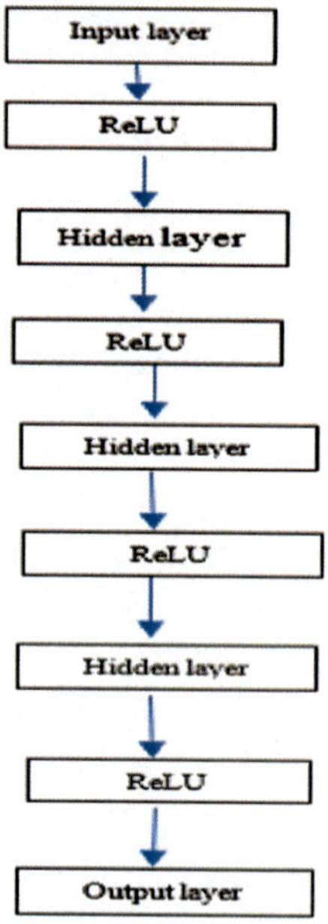

Fig. 1. Implemented neural network

We used ReLU activation function because the model is easier to train and the performances achieved are better then those obtained using the sigmoid function. ReLU function is defined as the maximum between a number and 0 [3]. Mathematical, ReLU is defined:

$$ReLU(x) = \max(x, 0)$$

We used supervised learning for learning process. This type of learning supposes the introduction of input data and the desired result of the human factor. Example: -12.45, −17.39, 11.69, jogging, where: the first 3 fields represent the x, y, z coodinates and the fourth field represents the system answer (the expected activity).

2.3 Training Process for a Neural Network

For training a PyTorch model, it is necessary to put in a large amount of information. In this paper the input data is structured in 4 fields as follows:

- first 3 fields are x, y, z coordinates and must be rational numbers.
- the fourth field is the type of activity. This field must be a natural number between 0 and 3. The possible values are: 0, 1, 2, 3.

Each activity has a number as in Table 1.

Table 1. Activity-number correspondence

Activity type	Number for each activity
Standing	0
Sitting	1
Walking	2
Jogging	3

We trained the neural network with a training dataset which has around 5400 of data respecting the above mentioned format and validated using a validation dataset with around 2000 data respecting the same format. Training and validation datasets are made of data from WISDM (Wireless Sensor Data Mining) dataset [8]. Training and validation datasets are two.*csv* files. The first essential step in model training is reading and manipulating data from the two files. Reading and manipulating data can be performed with functions from *torch.util.data* library. *DataSet* and *DataLoader* classes were used in this paper.

When the first results of the running model appear, *torch.optim* library ensures methods for updating model weights so that outputs should be as close as possible to the desired ones. The desired outputs are represented by the fourth field of data and they are: sitting, standing, walking and jogging. The optimization package contains several algorithms such as: SGD (Sochastic Gradient Descent), RMSProp, Adam, etc. [1]. In this paper, we used SGD. The main purpose of this algorithm is to reduce the value of the mean square error. The mean square error is equal to the difference between the predicted output of the network and the output desired by the user at a given time. Depending on the value of the mean square error, it is decided whether or not the network training continues.

The training of the model consists in going through 21000 iterations. The following conclusions were extracted:

- if a higher number of iterations is used then the same results will be obtained during testing process.
- if a smaller number of iterations is used then poorer results will be obtained during testing process.

During training process, at each 1000 iterations, the values of parameters are saved into an archive. Archives will be used when model testing will be performed. Testing set has different data from the training set and validation set. After training, the training model is saved in a file with*.pt* extension. We used this type of file because we wanted to run the model on Android Studio OS. So, we needed to transfer our model from Python in Java. The.pt file resolved this issue.

We trained the neural network during 17 min on a Lenovo ThinkPad AMD PRO A12 processor, CORES 4C+6G, 2,5 GHz and 16.0 GB RAM.

2.4 Inference

TorchScript library offers the posibility to create PyTorch optimized and serialized models. *TorchScript* programs can be saved from Python processes and then loaded in other programs, processes or applications where Python does not exist [1, 2]. This is the novelty which is used in this paper and described in the next paragraphs.

We want to include the PyTorch model into an Android Studio project to create the mobile application. Android Studio program uses programming languages such as Kotlin and Java. APIs from Android Studio do not have an implementation for PyTorch library (Python). The implemented solution is presented in the Fig. 2.

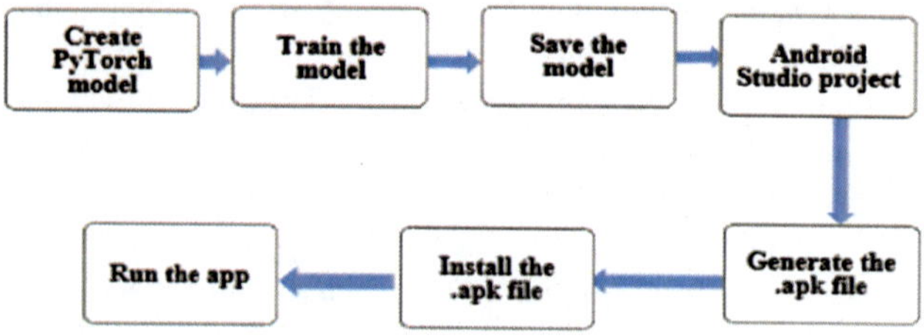

Fig. 2. Inference diagram

The network input is a list of 3 rational numbers. These 3 numbers represent x, y and z coordinates of the smartphone's accelerometer. The coordinates are read at a preset period. The selection algorithm calculates the mean value of the read values from the accelerometer. After the data have been entered into the neural network, loaded model methods start to work. *Org.pytorch.Module.forward* method returns an output tensor. This tensor represents the network output, and it has 4 elements, each element representing the probability of accomplishing an activity. We save the maximum value from tensor and we retrained its index.

We have been created a separate Java class called *ModelClasses* which contains a vector. The vector's elements are string type in the following order: standing, sitting, walking and jogging. We made a match between values obtained at the output of the network and values of the vector from the *ModelClasses* class using the retained index. E.g., the value of the first element in the tensor (network output) fits to the first activity (standing).

Matching between the values from the network output and the values from the activities vector helps us to display, on the graphic interface, the determined activities. Therefore, the resulting information can be understood and interpreted by humans. When an activity is shown on interface, the activity is saved in a SQLite database. The information from SQLite database is used to create statistics related to activities. Therefore, we can have an idea about what activities we made over a long period of time.

For the application to work on mobile phone, it is necessary to generate an *.apk* file. This file is built by Android Studio program and it is recognized by the mobile phone. After generating the *.apk* file, we should insert and install the.apk file on the mobile phone. After installation, the mobile application can be used anytime.

In Fig. 3, it is presented the first page, main screen of the mobile application. On this page the following information is present:

- the data read from accelerometer (x, y, z coordinates);
- the activity predicted by the neural network;
- "Statistics" button.

By pressing the statistics button, the second page of the mobile application is opened.

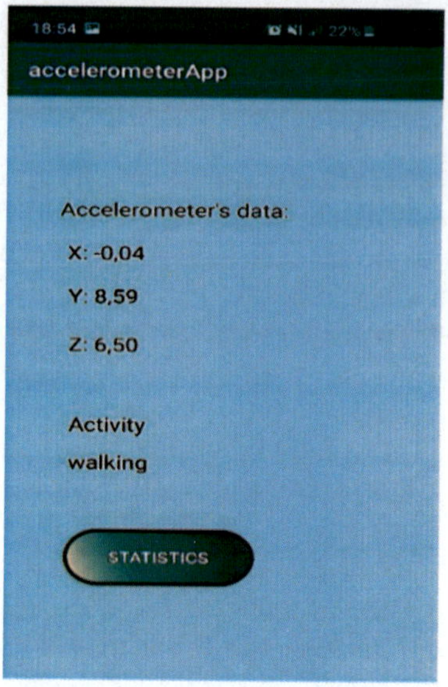

Fig. 3. Android application – first page

In Fig. 4, it is presented a chart with recognized activities. Therefore, here we can see what activities were performed. This page also contains two buttons with following meanings:

- “Back” button – return to the first page of the mobile application
- “Show data from DB” button –opens a page which contains data saved in database, Fig. 5. This function was implemented during test and debug process and it helped to view the data which is used to create the chart.

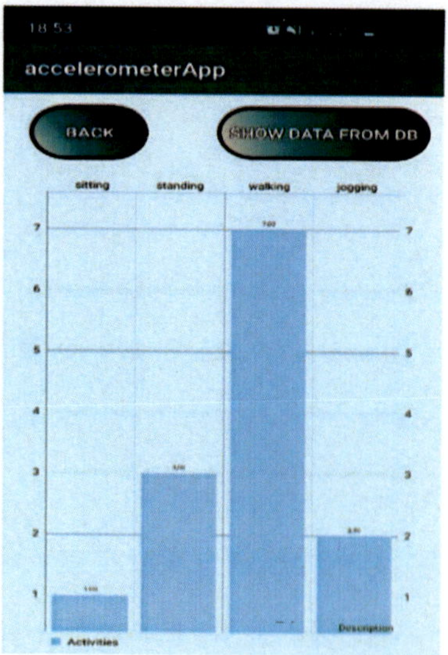

Fig. 4. Android application - second page

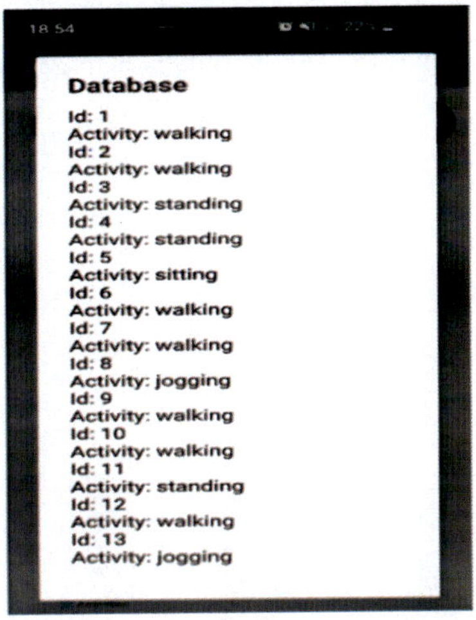

Fig. 5. Android application - database content

3 Results

The model evaluation compares the resulting predictions with the coordinate labels to give us an idea about the performance of the trained model. If we used the training set in the model appreciation, then the indications provided during the testing process will be the same as the indications given during the training process.

The neural network has been tested on a *test dataset*. The values in the test dataset are different from the values in the other two datasets (training and validation). The purpose of the test set is to show how good or bad is the performance of the trained model when new data is entered.

In order to check the results of the trained model, we have been created a Python project. The main purpose of this project is to test the trained neural network. First, we read the parameters from the archives (created during the training process). After that, we implemented an algorithm which: read data from test dataset, load the model's methods, apply model's methods for data from test set and compare the predicted activities with expected activities (the fourth field of data). The algorithm determines: the number of correct predictions and the accuracy of model. We obtained 88% accuracy.

We performed another test when the algorithm is part of the Android application. This must predict correctly the activities made by a person. The person should have the phone in his/her hand or his/her pocket, etc.

In Fig. 6 are presented the activities performed by a person (blue color) and the activities predicted by the PyTorch neural network (red color) when the model is inserted in the mobile application. It can be observed that the difference between performed activities and recognized activities is small (less then 10%) for each type of activity. Therefore, small difference between the two columns means that we trained and validated well the PyTorch neural network. So, the neural network is capable to determine real human's activities. The activities are determined and saved in database in real-time because the PyTorch model uses simple and fast functions and methods. So, the neural network works in real-time. The mobile application depends on the determination time of the activity.

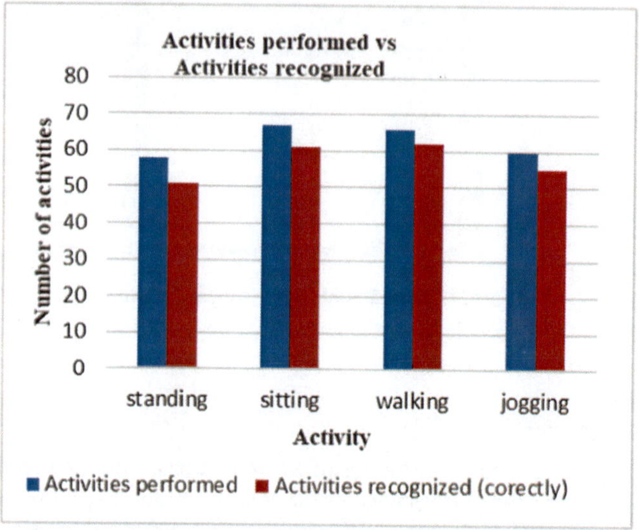

Fig. 6. Activities performed by a person vs. activities recognized by the neural network

4 Conclusions and Future Work

This paper describes the development of a mobile application which determines four human activities based on reading data from the phone's accelerometer. The activities are standing, sitting, walking and jogging.

The determination of activities is done using a PyTorch neural network. The neural network was trained, validated and tested in two modes: with a test dataset when we obtained 88% accuracy and when the model was inserted in the mobile application (when the application run on smartphone), the model recognized correct 92% of activities performed by a person. The neural network can detect activities in real-time.

The biggest challenge was the transfer of PyTorch model from Python programming language to Java programming language because we didn't use a classic method for the model transfer.

As a future work, we would like to extend and improve our mobile application. First, we intend to collect data from various sensors and use it to recognize more activities such as: sleeping, driving or bicycling. Also, the second issue will be the improvement of the smartphone interface in order to be more friendly.

References

1. Stevens, E., Antiga, L.: Deep Learning with PyTorch (2019). ISBN 9781617297120
2. Warden, P.: Building Mobile Applications with TensorFlow, 1st edn. (2017). ISBN 9781491988428
3. Zhang, A., Lipton, Z.C., Li, M., Smola, A.J.: Dive into Deep Learning, release 0.7.1, April 2020. http://d2l.ai/. Accessed 10 Nov 2020
4. Kwon, Y., Kang, K., Bae, C.: Unsupervised learning for human activity recognition using smartphone sensors. Expert Syst. Appl. **41**(14), 6067–6074 (2014)
5. Ignatov, A.: Real-time human activity recognition from accelerometer data using Convolutional Neural Networks. Appl. Soft Comput. **62**, 915–922 (2018)
6. Bayat, A., Pomplun, M., Tran, D.A.: A study on human activity recognition using accelerometer data from smartphones. Procedia Comput. Sci. **34**, 450–457 (2014)
7. Anguita, D., Ghio, A., Oneto, L., Parra, X., Reyes-Ortiz, J.L.: A Public Domain Dataset for Human Activity Recognition Using Smartphones. published in ESANN (2013)
8. Wireless Sensor Data Mining data set. https://www.cis.fordham.edu/wisdm/dataset.php. Accessed 2 Dec 2012

Check for updates

Work-in-Progress: Construction Safety Using Visual Technological Support

Alfredo Soeiro[1(✉)], Nicolaos Theodossiou[2], Marcela Spisakova[3], João Poças Martins[1], and Ivica Zavrski[4]

[1] Universidade do Porto, Porto, Portugal
{avsoeiro, jppm}@fe.up.pt
[2] Aristotle University of Technology, Thessaloniki, Greece
niktheod@civil.auth.gr
[3] Kosice University, Kosice, Slovakia
marcela.spisakova@tuke.sk
[4] University of Zagreb, Zagreb, Croatia
zavrski@grad.hr

Abstract. Accidents occur on construction sites that are complex and unique environments, making these difficult to control and monitor. Construction safety intends to prevent people from dying or becoming incapacitated. It is socially relevant and the panorama has not changed substantially. International Labour Organization estimates that losses of GDP due to this type of accidents represent about 4% of GPD. Above all it is not acceptable that in the XXI century anyone dies due to improper working conditions. A European funded project by Erasmus+Programme intends to alleviate this situation. Project arises from the synergy between higher education and construction to foster use of smart digital technologies in proper construction safety training. The idea is that digital tools allow the creation of simulations of construction scenarios allowing the identification and prevention of risks for workers, technicians and engineers. The project develops tools with IR (Immersive Reality) and AR (Augmented Reality) procedures based on BIM modelling capacities to prevent accidents through training. Project also aims at offering training sessions online on an online platform to grant learners and trainees access to interactive material and resources. These tools range from applications to be used on smartphones to virtual reality contexts depending on the needs of training.

Keywords: IR · AR · Construction · Accidents · Prevention · Training · BIM

1 Project CSETIR - Introduction

Most of the accidents that happen on-site are due to human failure either from worker or from safety managers. Manual monitoring and supervision are prone to error due to blind spots, workers loss of focus, awareness, and fatigue. As a result, serious injuries and fatalities are occurring. The human error theories discuss that human error is the factor to be blamed for the accidents, with specific work, and under the environmental circumstances.

M. E. Auer and T. Tsiatsos (Eds.): IMCL 2021, LNNS 411, pp. 788–795, 2022.
https://doi.org/10.1007/978-3-030-96296-8_71

According to the Occupational Safety and Health Administration (EU-OSHA) in 2015, about 21% of workers with a fatal work accident were from the construction industry [1]. In Europe, the construction industry is responsible for a high number of fatalities. This mortality rate will affect the ongoing construction work-flow, other workers moral, lower performance and schedule delay. This decreases also productivity and increases costs like direct costs (equipment damage, workers compensation, insurance, medical expenses, rehabilitation, etc.) and indirect costs (well-being, productivity, morale of workers, etc.). This is a major setback in terms of economic and financial perspectives, the implementation of the project and the image of the construction company.

Although all workers receive training and orientation meetings, they are not receiving on-the-job, hard-knocks-type training that simulates the site conditions. Every construction site is unique with different activities, schedules, execution approaches, diversity of workers attitudes and languages and that makes difficult the standardization of procedures for accident prevention. Therefore, it is crucial the administration of training programs that are effective and can raise the level of preparation of all involved in terms of safety awareness and competences. Yet the traditional methods and approaches do not reach everyone involved to prepare their prevention capacities in accordance with their tasks and responsibilities in the construction site..

This lack of effectiveness in training is probably due to the gap between the industry needs and the education and training organizations. Based on several systematic literature reviews done by researchers, on involvement in several international conferences, in construction safety organizations activities, in construction sector contributions some conclusions and guidelines became clear. First, the research on integrating technologies, such as VR/AR (Virtual Reality/Augmented reality) and machine learning, may bring possibilities to address the needs in terms of training in accident prevention [2].

The situation of gaining knowledge about the existing research and experiences in using new technologies in construction safety created a dilemma. The main issue for the partnership was to choose from theoretical and prospective developments or to try to find concrete applications in this area. The project partnership decided for the latter taking into account the convenience of presenting to construction companies proposals that could be ready to test and to implement. These existing applications will serve as pool for the solutions the project will present to train stakeholders (workers, technicians, engineers) and to educate students in higher education about construction safety. The model and format of training modules will be the same for all while assessment of respective learning outcomes will be adapted to the level of proficiency intended for the personal competences.

After holding several meetings with safety professionals, contractors and consultants from several countries it seemed that construction companies could benefit from possible advancements of technologies and of tools. These might assist in training workers in terms of accident prevention. These facts were an indication to start involving new technologies in the construction industry to train on construction safety. After technologies based on BIM (Building Information Model), VR and AR showed a lot of potential as promising tools for education and training the idea of creating a joint project was adopted. The project CSETIR was proposed to the European Commission for funding and it is being co-funded under the Erasmus+programme of the European Union, KA203 – Strategic Partnerships for higher education with reference 2019–1-SK01-KA203–060778 [3].

CSETIR, abbreviation of Construction Safety in Education and Training using Immersive Reality, intends to offer innovative and effective education and training for prevention of accidents [4]. The concept is based on digital platforms and tools, like BIM and VR/AR simulations, to devise effective procedures to be used by the diverse groups in construction site that should be easy to use, affordable and effective. These tools may also supporting schools and training centers to teach and train future practitioners in accident prevention [5].

2 Identification of Tools and Strategies

The first steps of the project aimed at identifying existing examples with possible use for further phases of the project like developing training material and validating the effectiveness of procedures. It was created a repository of available and current tools on the market in terms of BIM-based approach to develop AR/VR scenarios for research and for the construction industry.

Case 1: 3M Releases Construction Safety Virtual Reality Programs for Hands-on Learning. 3M has recently released a series of VR construction safety modules. 3M's virtual reality training platforms are available online. The field applications of these safety modules are described on the 3M website for training. This case study can be chosen for use in the execution preparation and in the construction phases. BIM was not an essential tool inside this case study although integrating BIM will help for a training module. The platform may incorporate some VR devices and get an immersive training experience [6].

Case 2: Caterpillar (CAT) safety policy using VR intends to improve jobsite training.

CAT started with safety in road construction and simulating real life scenarios. CAT designed the VR tools as a multiplayer environment so a group of workers could train simultaneously. Application is considered adequate for the construction preparation phase. Training module presents a wrap-up to reinforce the positive lessons learned and stress that no emergency at the jobsite is worth risking the safety of the worker or anyone else. The introduction of communication and mobile modules are possible positive contributions to improve learning outcomes [7].

Case 3: SRI International has developed AR solutions for construction inspection. AR simulates job site operations for construction inspectors and auditors. It may use drones together with the SRI solution that are used for aerial inspection of construction site while comparing with BIM models of the construction. BIM modelling is used to define the planned construction and inspect possible errors or deviations between construction and the respective BIM model. SRI AR captures images from the executed with headset and tags these with notes and sharing with other devices [8].

Case 4: Safety Compass is another example that uses AR to improve workplace safety. It gives access to live information from the construction site that is updated based on the location of the construction worker. It uses AR technology to identify potentially risks at the current location of the worker. It can be applied using mapping on a tablet or smartphone. This application also allows interaction and collaboration of all involved in the construction from different areas and functions. In this application BIM maybe linked to the GPS and the mobile app will give risk alerts and notifications based on the current location. Workers will then be notified of potential risks pin-pointed on an interactive mapping system. This AR application seems effective and easy to apply [9].

Case 5: Examples of use of VR in construction safety presents many applications that pretend that more effective and safer training of construction workers is less expensive and much safer to train heavy-equipment workers, like crane operators. Construction virtual environments allow workers to operate dangerous equipment without any risks whatsoever. It is cheaper than real-life training where you would need a real tower crane to train the operators. Coupled with the training the company CertifyMe.net is specialized on forklifts and has a VR program for practicing in a risk-free environment. Using it, they eliminate any type of injuries and equipment damage, allowing for a trial-and-error approach without adverse consequences. It provides the training and assessment in one hour with awarding the certificate to the trained operators immediately and online [10].

Case 6: New construction safety school uses virtual reality to educate workers and technicians. The school improves worksite safety through a new training centre targeting the construction industry. A key feature of the construction safety school is experiential learning by using AR and VR. Possible scenarios are created in this school with demonstration of equipment connected to the contracting industry. It is possible to certify workers in terms of construction safety. Application may involve workers, engineers, safety professionals, project managers, students, safety managers and safety coordinators. BIM is not used in these training scenarios. AR and VR are used in a wide range of scenarios and with several risks [11].

Case 7: YellowJacket is a software, available also on mobile application, that can monitor and report on quality, health, safety and processes on construction sites. May save time, save money, improve performance and ultimately save lives. Adopting YellowJacket health and safety software and fully integrating it within the organisation′ s culture and throughout the supply chain then it is possible to achieve real improvements to health, safety and quality management on construction sites and beyond. It is designed to be as user friendly as possible –giving 'at a glance' overview of health, safety and quality across the projects. The YellowJacket mobile application allows employees and contractors on the site to easily input information relating to health, safety and quality. Data is captured 'as it happens', rather than staff waiting to get back to a computer [12].

After analysis of the seven cases presented above several strategies can be considered when addressing the use of AR and VR environments to improve education and training for safety in construction sites. Indeed, the targets of training initiatives are a diverse group, including site workers, safety specialists, technicians, engineers, architects and others. The technical skills and education levels for these groups differ considerably, as do their work environments, and functions. This diversity must be considered when defining strategies and solutions, even when the suitable available hardware and software options overlap largely for the different user groups.

Indeed, as described in this document, the same VR and AR hardware and the software components have been adopted in a range of applications. This is considered as an opportunity, as solutions that target one user group might be adapted for different uses. The main current VR and AR development technologies are cross-platform, which largely reduces the importance of the choice of specific equipment within similar types of hardware such as Head Mounted Displays (HMD) or motion controllers.

Naturally, different training environments demand different types of hardware, due to cost, time and other practical considerations. For instance, despite recent developments in CAVE (Computer Assisted Virtual Environment) technology, which allow for lower-cost and quicker deployment, HMDs remain a more practical solution for virtual immersion, while computers and mobile devices are ubiquitous, and provide acceptable VR experiences in many instances [13].

The use of BIM models as a source of information (including geometry) for the virtual models is regarded as an obvious choice. BIM models are increasingly common in practice, they support different types of information, thus providing great flexibility when deciding on technical solutions, and are compatible with other components such as game engines. Alternative solutions, such as the use of generic 3D modelling tools would require the development of models from scratch, with no relationship with the actual construction process. This means that changes in design or in construction plans would not be quickly or easily reflected in the virtual training environments. These factors greatly reduce the feasibility of a non-BIM solution. Since BIM authoring tools are interoperable, and standard open formats exist for exchanging BIM data, the choice of BIM tools is not considered to be a critical issue when designing a strategy for the development of training solutions.

3 Training Methods

Training in CSETIR project may be provided for levels 1 to 7 of the European Qualification Framework (EQF). In terms of training methodologies it is expected that higher education institutions address levels 5 to 7 and levels 1 to 4 will be developed by vocational education and training (VET) organizations like CEDEFOP, EVBB or EVTA [14]. Training will be based on the outcome based method proposed by Tuning Academy. All modules and training activities will be designed and prepared taking into account the knowledge, skills and attitudes required. The competences required will be defined to face the several risks and preventive measures adequate for the respective level of qualifications. This outcome based training will allow to prepare everyone from basic level (1) to master level (7) [15].

Validation and assessment methods of competences acquired by trainees is necessary to ensure that the competences in terms of construction safety were acquired. The training provided maybe used to qualify some of the trainees to be able to enter the construction site or to perform some risky tasks. Therefore, the assessment has to be suited to type of competences that are supposed to be acquired. Proper assessment of the different competences (knowledge, skills and attitudes) is based on the use to webtool TALOE [16]. This webtool can suggest assessment methods for the different types of competences at various levels from 1 to 7. The assessment methods will be chosen in accordance with the suggestions, the construction site resources, type of training and purpose of the qualification. The assessment of workers, technicians and engineers competences to verify if they are qualified to enter a construction site may be applied to other groups like students or visitors. This evaluation will verify if proper training had the quality that is expected to be provided when using effectively the digital tools involving virtual construction environments.

4 Example of Training Scenario

The example presented is the first scenario developed for training on prevention of accidents. This paper is a work in progress and this example is a prototype of other planned training situations. It addresses the cause of many fatal and serious accidents that is related with inadequate use of ladders. The consequence of mishandling this equipment results in falls.

This risky task of using a ladder present virtual scenarios with fall from height with insufficient designed or laid ladder. Common errors are unsatisfactory length, slope or overlap of ladder. Trainee will identify from each scenario the localization where the risk is present. After the identification learner will have to choose the preventive measure preferred. The possible preventive measures will be shown in images. After identification of preventive measure to use the respective details will be presented. These details include list of costs, necessary training, examples of consequences of applying, implementation process and effectiveness. Simulation of the use of the preventive measure will be provided so learner understands the new scenario. In case learner does not identify the risky situations these will be displayed one by one.

The training strategy is based on visual display of the risky situations, of preventive measures and simulation of the consequences of adopting the preventive measure. The scenarios with risky ladders use a BIM model using Web-based Collaborative Virtual Environments (LIRKIS G-CVE) and preventive measures adequate for each risk using BIM. The trainee´s avatar, who has completed interactive training, stands in front of the reinforced concrete structure on the construction site. Vertical movement of the avatar is allowed by the three ladders, which are individually placed on each floor of the structure [17].

Each of the three ladders will contain a mistake or safety problem. Unless the trainee detects all safety problems, he/she will not be able to climb the ladder. He/she will be able to climb the ladder and move to a higher floor only when he detects and corrects the mistake. The participant's task is to climb the ladders to the upper third floor. After arriving at the trainee will see a recapitulation of new information and knowledge which he/she obtained and will receive a “star - helmet”. For instance an error is that ladder from ground floor to first floor is made of wood and has an unsatisfactory length. Wood and length characteristics are not acceptable. Trainee can adjust the properties of the ladder until there are no evident risks or non-compliances with rule.

5 Conclusions

Paper presents a work in progress of the project CSETIR. It is at the stage of producing the materials that are related with training scenarios for the different risks. Currently 18 construction risks have been chosen as convenient to address to prevent most accidents in construction. The stage of addressing the available useful tools and procedures has been completed but maintains the observation of new developments and innovations that can improve the effectiveness of the scenarios. The project partnership has worked to articulate tools and methods with the set of risky tasks that need effective training to prevent accidents.

Zero vision is a term that implies no accidents on construction accidents. The project intends to contribute actively to that goal through the provision of tools that can be widely used accepted by stakeholders in the construction industry. The partnership is open for contributions from all that may help the performance of training. Partnership is associated with construction safety coordinators association, with groups of civil engineering faculties and with other social partners that may benefit from the project outcomes. All contributions are welcome and society will benefit if accidents in construction stop happening.

References

1. Elsler, D., Takala, J., Remes, J.: An international comparison of the cost of work-related accidents and illnesses. In: European Agency for Occupational Safety and Health, Bilbao, Spain (2017)
2. Zaher, M., Greenwood, D., Marzouk, M.: Mobile augmented reality applications for construction projects. Constr. Innov. **18**, 152–166 (2018)
3. ERASMUS+. https://ec.europa.eu/programmes/erasmus-plus/node_en. Accessed 01 Jul 2021
4. CSETIR. http://csetir.civil.auth.gr/. Accessed 01 Jul 2021
5. Huang, Z.: Application research on BIM+AR technology in construction safety management. J. Phys. Conf. Ser. **1648**, 042019 (2020)
6. 3M Releases Construction Safety Virtual Reality Programs for Hands-on Learning. https://www.3m.com/3M/en_US/worker-health-safety-us/3m-ppe-training/virtual-reality/. Accessed 01 Jul 2021
7. How Caterpillar Safety VR Improves Jobsite Training. https://www.cat.com/en_US/news/machine-press-releases/new-cat-safety-vr-module-creates-an-immersive-safety-training-experience-for-employees.html. Accessed 01 Jul 2021
8. SRI International Augmented Reality Solutions for Construction Inspection. https://www.sri.com/. Accessed 01 Jul 2021
9. Safety Compass - Augmented Reality Workplace Safety. https://thesafetycompass.com.au/. Accessed 01 Jul 2021
10. Augmented and Virtual Reality for Safety Training. https://elearningindustry.com/augmented-virtual-reality-for-safety-training. Accessed 01 Jul 2021
11. New construction safety school will use virtual reality to educate workers. https://www.straitstimes.com/business/new-22mil-construction-safety-school-will-use-virtual-reality-to-let-workers-see-how. Accessed 01 Jul 2021
12. YellowJacket. https://www.bresmartsite.com/products/yellowjacket/. Accessed 01 Jul 2021

13. Getulia, V., Caponea, P., Bruttinia, A., Isaacb, S.: BIM-based immersive Virtual Reality for construction workspace planning: a safety-oriented approach. Automation in Construction, no. 114 (2020). https://doi.org/10.1016/j.autcon.2020.103160
14. VET4EU2. https://vet4eu2.eu/. Accessed 01 Jul 2021
15. Tuning Academy. http://www.tuningacademy.org/. Accessed 01 Jul 2021
16. TALOE. http://taloe.up.pt. Accessed 01 Jul 2021
17. Web-based Collaborative Virtual Environments (LIRKIS G-CVE). https://pixovr.com/pixo-vr-releases-new-virtual-reality-osha-safety-compliance-training-module/. Accessed 01 Jul 2021

Remote and Online Laboratories

Case Study of a Virtual Lab Environment Using Virtualization Technologies and a Desktop as a Service Model

Michael Dietz(✉), Azarias Abebe, Jan Lederer, Tobias Michl, and Ronald Schmidt-Vollus

Nuremberg Campus of Technology, Nuremberg, Germany
{michael.dietz,azarias.abebe,jan.lederer,tobias.michl,
roland.schmidt-vollus}@th-nuernberg.de

Abstract. In laboratory practicals, students explore scientific and technical phenomena. The COVID-19 pandemic has caused students to have limited access to their labs for health reasons. The basic idea of a virtual model of a real laboratory test bench is not new, but the use of industry-level software to create a so-called digital twin—and, beyond that, the possibility of using this software environment inside a virtual PC—is a simple and reusable solution. The research question is the following: What requirements should a virtualization concept for virtual labs using a digital twin fulfill in the context of educational institutions and how can these be implemented? To answer this question, this work describes in detail the various evolutionary steps of a technical concept for an adoption of virtual desktop infrastructure. The virtual lab concept owns inherent conceptual properties such as scalability and platform independence, as well as 24/7 remote access.

Keywords: Virtualization · Cloud computing · Desktop as a service · Virtual labs · Education · Digital twin · Virtual commissioning

1 Introduction

"Digital Twins for Virtual Commissioning of Production Machines is a seminar that is provided for undergraduates in mechanical engineering at Nuremberg University of Applied Sciences. According to [1] the digital twin is " a digital replica of a living or non-living physical entity. Avoiding danger is the most advantageous ability of the digital twin. There is no possibility of getting hurt while working with the digital twin because it is a computer simulation. The teaching-learning concept includes a practical part where students learn about virtual commissioning and how to interact with the digital twin in online experiments. These are hands-on experiments so that students gain an understanding of automation technology. Next, imagine the digital twin of a laboratory test bench instead of a production plant as in the industrial use case. Further

M. E. Auer and T. Tsiatsos (Eds.): IMCL 2021, LNNS 411, pp. 799–811, 2022.
https://doi.org/10.1007/978-3-030-96296-8_72

details of the concept using a combination of digital twin and virtualized PC were described in broad terms in [2,3].

The present work focuses on the technological approach. The virtual lab concept was an extension into an educational context of the initial research on the use of a digital twin for an early software test of programmable logic controllers (PLC) during virtual commissioning in the industrial context. Initial considerations for such a concept were based on prior practical knowledge rather than an engineering approach to designing such a lab environment. Eventually, further research led to a basic concept definition for a virtual lab using a digital twin: "the merger of a digital twin (the test rig model) with a virtual PLC (the control system) inside a virtual PC that can be accessed via the Internet to create a *virtual lab*." The individual components of this setup are common technologies, but their combination and use to provide a virtual environment for lab practicals has been simplified through IT developments such as faster SSD hard drives.

In this paper the development of the technical background and its implementation for the virtual lab environment that uses cloud and virtualization technologies is described. First an introduction is given and necessary technical terms for understanding the concept are then summarized in the next Sect. 2, which introduces cloud computing and virtualization technologies. Section 3 describes the state of the art of virtualization in education. The technical approach and the requirements are presented in Sect. 4. Details of the teaching-and learning concept based on problem based learning are given in Sect. 5. Results and experiences of the individual technical solutions are summarized in Sect. 6. Conclusions and future work are mentioned in brief in Sect. 7.

2 Virtualization and Cloud Computing

2.1 Technical Terms and Technology Variants of Virtualization

This chapter explains the term hypervisor and both hardware and presentation virtualization in detail. Other relevant technical terms, and important technical aspects of the concept, and essential technical relations between hardware and software components are described only briefly due to space restrictions.

The concept of virtualization has existed since the 1960s, serving primarily to improve the resource utilization of mainframe computers [4]. However, it is a concept with many possible applications. The COVID-19 pandemic has required students to be physically separated from their computer labs for health reasons. Due to this unusual—and hopefully temporary—situation, we deployed various virtual machines (VMs) on a standard office computer with typical hardware equipment using a so-called *hypervisor*. There are two types of hypervisors, type 1 (bare-metal hypervisor) and type 2 (hosted hypervisor). A hypervisor is just a software product to abstract the hardware and manage VMs (i.e., create, start, stop, and allocate resources). The VM itself is just a single folder with files that can be copied, moved, and opened on other PCs while retaining all of

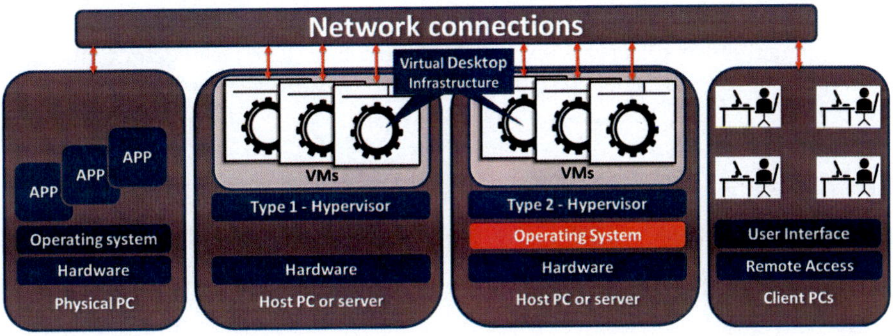

Fig. 1. An comparison of a physical PC with hypervisor types, adapted from [6]

its functions. An alternative term for hypervisor is *virtual machine monitor*. For more in-depth information refer to [5].

The left-hand side of Fig. 1 depicts a common, nonvirtualized computer with hardware, any operating system, and some applications. In the center of the figure is a PC or server without an operating system but with a type 1 hypervisor, and next to it is a type 2 hypervisor *with* an operating system. The client PCs, shown on the right-hand side, access any of the other PCs or VMs via the network. Such a setup is called a virtual desktop infrastructure. In this concept, which combines hardware virtualization, or more precisely, desktop virtualization, with presentation virtualization, the host PCs or servers provide VMs, which are remotely accessible via the network, for defined users or groups. *Hardware virtualization* is the abstraction and virtualization of resources with the aid of a hypervisor. In this type, VMs have access to the virtualized resources, yet certain restrictions must be observed; for example, simultaneous use of the same hardware resources is not possible [7]. In literature it is distinguished between several virtualization types, see [6,8] for details. According to [9], the benefits of desktop virtualization include mobile computing, security, easy management, and low costs. *Presentation virtualization* separates the host PC, which provides a service, from the client PC, which performs that service. Another term for this is *terminal computing*. The service can be a specific application, but it can also be the entire user interface (UI), decoupled from the host PC and displayed locally on the client PC. In this setup, all applications and data reside exclusively on the host PC. An advantage is the hardware independence of the client PC, since all computationally intensive work is performed on the host PC [6]. The task of the client PC is reduced to displaying the UI and forwarding user input.

2.2 Technical Terms and the Scope of Cloud Computing

The US National Institute of Standards and Technology (NIST) offered the following definition in 2011 [10]: "Cloud computing is a model for enabling ubiquitous, convenient, on-demand network access to a shared pool of configurable computing resources...that can be rapidly provisioned and released with minimal management effort or service provider interaction..." The basic definition, shown in part in Fig. 2, has since been expanded with several service models.

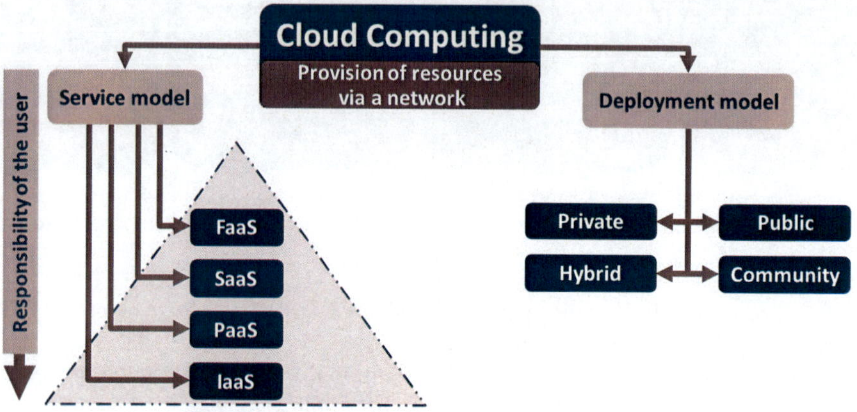

Fig. 2. Cloud computing, adapted from [6, 10]

In simplified terms, cloud computing is an external provision of IT infrastructure, which can then be accessed via a network. Fig. 2 represents the service models as a pyramid, in keeping with [6], which explains that the higher, more abstract layers use the services of the lower, more concrete, and more hardware-oriented layers. Lower layers thus demand more user responsibility. Depending on the respective service model, hardware, platforms, software, or a function is made available virtually via cloud computing. The NIST definition [10] gives four possible deployment methods as shown in Fig. 2. According to [6], cloud computing makes heavy use of virtualization technologies and is an application of the abstraction concept developed for IBM mainframe computing [11]. The following descriptions are provided to give a short summary and are based on various publications and descriptions in the field, e.g. [6, 10]: In the *Infrastructure-as-a-Service model*, the cloud provider offers virtual computing resources, typically computing power and storage, over the Internet. This service is realized through interconnected devices such as servers, storage devices, routers, and other hardware required to handle different types of workloads. Users can create virtual server images themselves or rely on a compilation provided by the vendor, and they can start or delete any number of instances in a short time. This is the lowest layer in the pyramid of services. In the middle layer, *Platform-as-a-Service*, the cloud provider offers a complete development and deployment environment that provides functions and applications for use in developing the user's own applications. Users build and use their own applications on this platform. The concept is similar to renting a house with its contents so that users can immediately use the house. As with all cloud computing, the physical location of the servers is not known to the user, and there are other restrictions. The layer *Software-as-a-Service* offers the use of software that customers can access via the Internet. The application software is hosted in the data center or on the server of the cloud provider, giving customers access via a web browser. Due to space restrictions further details are not mentioned here.

3 State of the Art

This section provides an overview of different generic approaches to virtualization in the context of engineering education at universities and other educational institutions. To gain a deeper insight into the state of the art of online, remote, and virtual laboratories, the publications listed next should be considered only as a starting point for a more extensive literature search.

3.1 Online and Remote Labs

The remote lab may be thought of as a predecessor of the virtual lab. For a remote lab, the lab test bench exists in reality, which is not true for the virtual lab. Users access a remote lab via a network connection to interact with control devices and measure data for an experiment. A virtual lab, in contrast, generates data as the result of a *simulated experiment.* The umbrella term for both of these is *online lab,* according to [12], which describes online labs as "computer applications ... to conduct experimentation from a remote location" and concludes that "in this sense, remote and virtual laboratories are subsets of online laboratories." The book by [13] describes the state of research in this area, pointing out that the first developments in online labs occurred in the late 1990s. Since then, many concepts have evolved and are now referred to collectively as *remote laboratory management systems* (RLMS). WebLab-Deusto [14] and the iLab shared architecture [15] are two examples. According to [16], though different lab concepts have existed for some time, they are generally viewed as a niche product; furthermore, "online labs have not kept up with the pace of digital teaching tools as a whole."

3.2 Virtual Labs

A comprehensive description can be found in [17], which details the introduction of virtual desktops at Kaunas University of Technology in Lithuania in 2012. According to the article, virtualization technologies are seen as an innovative solution to challenges in education such as budget cutting, outdated IT, and data security management, as well as a way to offer remote learning opportunities that students need. A case study in [18] describes cloud computing deployment at the University of Diyala, Iraq. The university aimed to develop a basic cloud-based design to improve the efficiency of laboratories in its computer center. The proposed design was built using the software-as-a-service (SaaS) model. The authors state, "the role of cloud computing in universities should not be belittled, because it can provide significant gains in access to a wide range of research applications, educational resources, and tools." They conclude that cloud computing must not be ignored for educational centers and, in particular, that private cloud computing with the SaaS approach is an efficient way to provide software and hardware access through local and Internet network connections.

4 Technical Approach and Experiences

The development process followed the waterfall model as closely as the time constraints would allow, but some steps were necessarily left out. Nevertheless, the following requirements emerged quickly from the first step, the analysis.

4.1 IT Requirement Analysis

1.0 Provide a fast, reliable, and self-explanatory lab environment
1.1 Base the control of the lab environment on the familiar Windows UI
1.2 Provide platform-independent user remote access
1.3 Use a free type 2 hypervisor software
1.4 Allow additional software installation by administrative users and restrict students' access and installation rights
1.5 Allow an easy coupling of several software components through automatic start-up of the necessary applications
1.6 Give easy access to USB security dongle
1.7 Use a preconfigured VM with all needed applications
1.8 Enable older PCs to use the lab environment
1.9 Use the secure virtual private network (VPN) of the institution
2.0 Use a security concept with firewalls and other network access restrictions

4.2 Case Study: Virtual Lab Environment Using Virtualization

The case study was carried out using a setup with some powerful office PCs available in the automation department at the University of Applied Sciences Nuremberg. The main problem for the department's seminar lecturers is that the practical lab part requires several software applications, each of which has a complex installation process, totaling about 70 steps and taking several hours to complete. The office PCs in the department had already been upgraded with more RAM and faster SSD hard disks and were able to operate the VMs with the hypervisor software. The PCs acted as host PC for the VMs, with the result that *four* VMs could be executed on *one* office PC. A class of 40 students was divided into 20 pairs, and each pair was able to connect to the preconfigured VMs equipped with all necessary data and applications. Thus, all students could immediately work on the examples, with their personal PCs functioning as the client PCs. Several variants of this technical design were developed and tested over a period of three semesters, as described in detail below.

Cloud Computing Topology in the First Semester: Students were given direct access to selected host PCs in the first semester. Each student user had to be manually entered into the list of remote users for this purpose beforehand. A short introduction explained how to use the remote access software to connect to the VM assigned to each group. In addition, the USB security dongle had to be manually integrated into the respective VM in order to start the software. The topology is based on the infrastructure-as-a-service (IaaS) service model and can be considered a private cloud as deployment model.

Cloud Computing Topology in the Second Semester: Thanks to our team's further efforts on the technical problems, remote access could be simplified for students in the second semester. Dedicated MAC addresses were registered by our IT department and assigned to each single VM as fixed IPv4 addresses, allowing direct remote access to the VMs instead of access first to the host PC, which then started the VM. Furthermore, a USB dongle server was established using a single-board computer and special software. Each VM automatically started this software to access the physical USB security dongle via a network connection. The topology was again based on the IaaS service model, yet with simplified remote access directly to the VM. This can be considered a virtual desktop infrastructure. The deployment model was still the private cloud.

Cloud Computing Topology in the Third Semester: Due to license restrictions a major change was necessary, the IT department had to restrict remote access for students because a client access license [19] was not part of the institution's contract. Without the possibility to allow remote access, the concept was adapted further. The idea was to run the VMs on the students' private PCs so that no remote access was necessary. Preconfigured VMs were prepared and given to the students, a time-limited version of Windows served as operating system [20]. This topology is no longer a service and deployment model of cloud computing, but reduces itself to the virtualization aspect of the overall concept.

4.3 Verification of the Technical Requirements

Successful completion of each requirement is semester-specific because of the different topologies used, therefore only *selected* requirements are described next.
Requirement [1.0]: was to provide a fast, reliable, and self-explanatory lab environment for easy access by students. This requirement was fulfilled only partly, as several problems occurred during use.
Requirement [1.1]: was to provide platform-independent user remote access. This was fully achieved through use of the RDP protocol.
Requirement [1.2]–[1.4]: Will not be discussed further in this paper.
Requirement [1.5]: was to allow an easy coupling of several software components through automatic start-up of the needed applications. This requirement was partly fulfilled, as some components could easily be accidentally closed by a false mouse click, which led to further errors.
Requirement [1.6]: was to enable easy access to a USB security dongle. This requirement was not fulfilled in the first semester, when students had to manually

mount the USB device. From the second semester on easy access was successfully realized with the USB security dongle server.
Requirement [1.7]: was use of a preconfigured VM with all needed applications. This requirement was successfully met in all semesters. The configuration and installation process was identical to the process on a physical computer.
Requirement [1.8]: was that older PCs also be able to use the lab environment. This was first successfully fulfilled but not in the third semester, due to the necessity of loading the VMs on the private PCs of students.
Requirement [1.9]–[2.0]: Will not be discussed further in this paper.

5 Structure of the Seminar

5.1 Didactic Changes in Teaching Practice

The seminar starts with an organizational part and progresses through a theoretical part to a practical part. In the first part, students are introduced to the structure of the course. With the help of so-called mastery lessons, students quickly get acquainted with using the learning management system (LMS), which requires them to find information themselves, upload their assignments, and create a wiki page on their own. In the theoretical part, in addition to the common text documents, there are also videos and useful links on basic topics such as automation technology, systems modeling, programmable logic

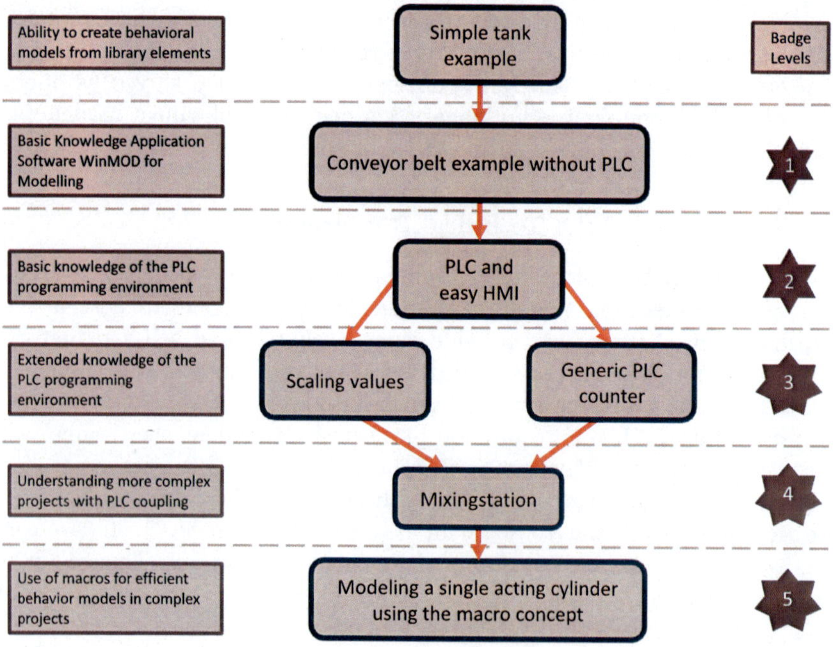

Fig. 3. Learning thread of the seminar

controllers (PLCs), fieldbuses, and network topologies, which are intended to support students in addition to the lectures and serve as reference material so that students can work through the learning material autodidactically. Finally, the practical part contains many examples that the students are supposed to work on and hand in as assignments. These are usually to be submitted as a short report, or occasionally as a screencast. In addition, at the end of each assignment, students are asked to briefly describe the problems they encountered while completing the assignment. We have introduced a dependency in the workability of the examples to make the common thread clear and to introduce students step by step to examples with increasing difficulty. Figure 3 illustrates this dependency.

First, students work on the simple tank example, which introduces them to the software used to model the digital twin. Only when this assignment is finished can they access the next one, concerning a conveyor belt without a PLC. Thus, the second software tool that students must learn in the course is a programming environment for PLC software.

The concept of problem-based learning (PBL) includes the applicability of the real-life problems that constitute the student-centered group work in the course to later everyday professional life. The lecturer acts as a kind of coach to the students [21] and *a monologue thus becomes a dialog.* The example mixing station is a good illustration of this approach, because the stirred-tank reactor that it depicts is a real machine often used in process engineering. Students can build this system and put it into operation as a digital twin in the modeling application software. The learning goals are for students to be able to describe the complete modeling process and eliminate errors themselves. They can create models of different mechatronic components following a step-by-step guide, and they can relate the physical equations to the simplified model. To give instructors an overview of the students' progress, so-called badges have been introduced to indicate their assignment completion status. For example, the first badge is awarded when the first two examples are marked as completed, and the last badge is awarded when all examples are completed. This system offers the possibility for the instructor to record and evaluate learning progress on the tasks individually for each student.

6 Results and Discussion

This chapter presents the results and then discusses them in brief. Because the focus of this work is the technical results, we omit or only briefly touch on nontechnical results, such as didactic results.

6.1 From the Lecturer's Point of View

As described earlier, several technical hurdles have had to be overcome since the start of the seminar using the virtual lab concept. In consequence, one of the results is the knowledge that any other lecturer starting from scratch will need many working hours just to create a virtual lab environment. It is not possible to look back and calculate the exact effort in terms of working hours, but on the positive side it can be said that the seminar concept with the virtual laboratory environment is on the way to maturity and many of its childhood issues have been solved or at least conceptually improved, so that only their implementation is lacking. On the negative side, many departments at our institution unfortunately have their own individual technical solutions to provide students with the required software, and thus no universitywide solution is yet in use. There have been several conversations with other lab technicians, faculty, and the IT department indicating this problem. In close reference to the first result, the problem of using multiple software components for teaching, which occurred in the authors' seminar, is not specific to only a few situations. It can be generalized due to the fact that students need to learn many software-based tools for their later working life. The software or vendor may change, but users need a rough idea of the software concept; for example, for text processing one may use open-source or commercial software, but one needs to understand the concept. As the next result, it can be stated that the creation of practical examples using modeling and PLC software is associated with high effort on the part of the lecturer. For one thing, the examples must be as "watertight" as possible, avoiding stumbling blocks for the students at all costs so they can concentrate on the learning objectives and not lose time due to IT problems. A notable publication on the *automated* creation of practical learning examples in the field of IT security [22] offers many ideas transferable to the virtual lab concept, the teaching of automation engineering, or even beyond. It describes the so-called *Haaukins platform*, an open-source cybersecurity training platform designed to meet the demands of specialized cybersecurity professionals due to increasing digitization. The platform provides *virtual labs* consisting of multiple virtual computer systems that are under the control of a student group. Exercises need to be solved by exploiting vulnerabilities intentionally embedded in the computer systems of the virtual labs by the exercise developer. The platform "aims to facilitate the learning process by helping teachers to automate the tedious setup and management of those environments, by making them accessible with no prior, complex configuration" [22]. In contrast to the virtual laboratory concept for automation technology, the Haaukins platform uses the free, open-source Linux operating system, thus solving the licensing problem regarding remote access. Unfortunately, the use of open-source code is more feasible in IT security than in automation. However, the goal of helping lecturers put together custom practical exercises by automating the creation of the virtual lab can be adopted for the virtual *automation* lab because the technical tools are available to do so. Another finding is that the necessary conversion to run the VMs on students' personal PCs has created a high demand for support. The hardware requirements in the third semester were higher than expected because a new version of the PLC software was used for the first time. A student PC therefore required at least 8 GB of RAM and a fast SSD hard drive, which meant that many students had to borrow a laptop with the appropriate modern hardware from the university.

6.2 From the Student's Point of View

Here we summarize student responses to the course evaluation in each semester. Students commented after the course in the third semester about VMs crashing too often and running poorly; programs running very slowly; and too little RAM on their own PCs. Yet about 60% said the course worked well completely online. The remaining 40% said they would prefer to have the theory taught online but the practice completed in a real PC lab. Students mentioned that they would like to work collaboratively on the same VM. To do so, however, would require special conferencing software, which, again, would demand more power from their private PCs.

7 Conclusion

This paper gives an overview of the development of the technical concept for virtual labs using virtualization and cloud technologies. The concept comprises a two-step virtualization process, starting with the digitization of the real lab test bench and after that the complete PC hardware is raised into virtuality. The described case study was in the field of automation technology, especially in the subject area of virtual commissioning. The presented concept for the virtual laboratory environment still lacks a sound IT design, yet further steps are currently developed. The intention is to have the virtual desktop infrastructure professionally taken over by an IT service provider, which could be the internal IT department of the institution or an external cloud service provider. Furthermore, it is planned to progress with the conceptional work of the virtual lab environment. A list of design principles will extend the list of technical requirements before the actual technical design of the virtual desktop infrastructure. The design principles, shown in [22], *Dynamic Tooling*, *Centralization of Exercises* and *Scalability* are also applicable to the virtual lab concept for automation engineering with the focus on virtual commissioning. The IEEE standard 1876 "Networked Smart Learning Objects for Online Laboratories" [23] defines several methods that are useful to integrate in the future design of the virtual lab environment. Finally, it is planned to demonstrate the virtual laboratory environment in special training events for vocational schools and small and medium-sized enterprises.

References

1. El Saddik, A.: Digital twins: the convergence of multimedia technologies. IEEE MultiMedia **25**, 87–92 (2018)
2. Dietz, M., Meissner, B., Schmidt-Vollus, R.: Teaching digitalization and systems modeling for virtual commissioning using virtual labs. In: 2020 IEEE 29th International Symposium on Industrial Electronics (ISIE), 17.06.2020–19.06.2020, pp. 440–445. IEEE (2020)
3. Dietz, M., Meissner, B., Goppelt, F., Schmidt-Vollus, R.: On the development of virtual labs using digital twins and a proposal for didactic optimization using design-based research. In: 2021 Fifth World Conference on Smart Trends in Systems Security and Sustainability (WorldS4), pp. 186–191 (2021)

4. Red Hat Limited: "What is virtualization?" [Online]. https://www.redhat.com/en/topics/virtualization/what-is-virtualization
5. Chubachi, Y., Shinagawa, T., Kato, K.: Hypervisor-based prevention of persistent rootkits. In: S.Y. (ed.) Proceedings of the 2010 ACM Symposium on Applied Computing Shin, New York, NY. ACM (2010). https://doi.org/10.1145/1774088.1774131
6. Meinel, C., Willems, C., Roschke, S., Schnjakin, M.: Virtualisierung und Cloud Computing: Konzepte, Technologiestudie, Marktübersicht (2011)
7. Firesmith, D.: Virtualization via Virtual Machines (2017). https://insights.sei.cmu.edu/blog/virtualization-via-virtual-machines/
8. Deka, G.C., Das, P.K.: Application of virtualization technology in IaaS cloud deployment model. In: Deka, G.C., Das, P.K. (eds.) Design and Use of Virtualization Technology in Cloud Computing, ser. Advances in Computer and Electrical Engineering, Hershey, Pennsylvania: IGI Global, pp. 29–99 (2017)
9. Yan, L.: Development and application of desktop virtualization technology. In: 2011 IEEE 3rd International Conference on Communication Software and Networks (ICCSN 2011), Piscataway, NJ 2011, pp. 326–329. IEEE (2011)
10. Mell, P.M., Grance, T.: The NIST definition of cloud computing. Gaithersburg, MD: National Institute of Standards and Technology 2011. http://dx.doi.org/10.6028/nist.sp.800-145
11. Hendricks, E.C., Hartmann, T.C.: Evolution of a virtual machine subsystem. IBM Syst. J. **18**(1), 111–142 (1979)
12. Zutin, D.G.: Online laboratory architectures and technical considerations. In: Auer, M.E., et al. (eds.) Cyber-Physical Laboratories in Engineering and Science Education, Cham. Springer International Publishing 2018, pp. 5–16 (2018)
13. Auer, M., Azad, A., Edwards, A., Jong, T.: Cyber-Physical Laboratories in Engineering and Science Education. Cyber-Physical Laboratories in Engineering and Science Education (2018)
14. Orduña, P., Irurzun, J, Rodriguez-Gil, L., Garcia-Zubia, J., Gazzola, F., López-de Ipiña, D.: Adding new features to new and existing remote experiments through their integration in WebLab-Deusto. Int. J. Online Biomed. Eng. (iJOE) **7**(S2), 33 (2011)
15. Harward, V.J., et al.: The iLab shared architecture: a web services infrastructure to build communities of internet accessible laboratories. Proc. IEEE **96**(6), 931–950 (2008)
16. May, D., et al.: Ausblick: Welche Rolle spielen Online-Labore für die Zukunft der Laborlehre? in Labore in der Hochschullehre, pp. 283–297 (2020)
17. Miseviciene, R., Ambraziene, D., Tuminnauskas, R., Pazereckas, N.: Educational infrastructure using virtualization technologies: experience at Kaunas university of technology. Inform. Educ. **11**(2), 227–240 (2012)
18. Ouda, G.K., Yas, Q.M.: Design of cloud computing for educational centers using private cloud computing: a case study. J. Phys. Conf. Ser. **1804**(1), 012119 (2021)
19. Middelkamp, A.: Online. Praktische Huisartsgeneeskunde **3**(4), 3 (2017). https://doi.org/10.1007/s41045-017-0040-y
20. Microsoft Edge Developer: Virtual Machines: Test IE11 and Microsoft Edge Legacy using free Windows 10 virtual machines you download and manage locally. https://developer.microsoft.com/en-us/microsoft-edge/tools/vms/
21. Strobel, J., van Barneveld, A.: When is PBL more effective? A meta-synthesis of meta-analyses comparing PBL to conventional classrooms. Interdisc. J. Probl. Based Learn. **3**(1), 44–58 (2009)

22. Mennecozzi, G.M., Hageman, K., Panum, T.K., Turkmen, A., Mahmoud, R.-V., Pedersen, J.M.: Bridging the gap: adapting a security education platform to a new audience. In: 2021 IEEE Global Engineering Education Conference (EDUCON), pp. 153–159. IEEE (2021)
23. IEEE Standard for Networked Smart Learning Objects for Online Laboratories, Piscataway, NJ, USA

Virtual Laboratory as a Tool to Increase Student Motivation in the Context of Engineering Education Digitalization

Irina Makarova[1(✉)], Aleksey Boyko[1], Gleb Parsin[1], Polina Buyvol[1], and Maria Drakaki[2]

[1] Kazan Federal University, Syuyumbike prosp., 10a, 423812 Naberezhnye Chelny, Russia
kamIVM@mail.ru

[2] Department of Science and Technology, International Hellenic University, University Center of International Programmes of Studies, 14th km Thessaloniki - N. Moudania, 570 01 Thermi, Greece

Abstract. In the context of economy digitalization and intellectualization, the transition to an information society, approaches to teaching methods and principles of educational content formation should also change. Working ahead of the curve, the university must use "digital" technologies and create "digital" content when delivering it to students. By using the advantages of interactivity and creating the presence effect among students, new technologies can increase the interest of the students and, as a result, better master professional and digital competencies. This study aims to create a virtual laboratory as a tool for ensuring the required motivation level as well as methods for its development and use, taking into account the specifics of the training system for vehicle service engineers. A description of the developed virtual reality laboratory is given, followed by an assessment of how it influences the students' interest in learning. It makes it possible to organize the processes of study, repetition, as well as checking the assimilation quality of the acquired knowledge, abilities and skills of working with technological equipment for balancing vehicle tires and disks. A feature of the proposed methodology is the creation of educational content during collaboration of teachers and students of different specialties. It also provides an assessment of interaction of the student with the equipment from the point of view of safety and ergonomics. The conducted surveys of students and their academic results have confirmed the effectiveness of the proposed methodology.

Keywords: Engineering education · Remote laboratory · Motivation enhancement

1 Introduction

Currently, global changes have affected all spheres of human activity. The consciousness and attitudes of people are changing with the emergence of such concepts as Industry 4.0, Smart City, smart transport and logistics, digitalization of industry among

M. E. Auer and T. Tsiatsos (Eds.): IMCL 2021, LNNS 411, pp. 812–823, 2022.
https://doi.org/10.1007/978-3-030-96296-8_73

others. Digital transformation changes all economy sectors, providing a wide range of digital tools to improve processes [1, 2]. Global challenges such as COVID-19 are emerging, and they are also adjusting the way we live. In this regard, the restructuring of modern engineering education system is also required. The main requirement for the education system in the transition to the digital economy concept is the development and use of innovative tools and techniques that meet the trends and requirements of the information society. This way, a graduate will be trained and become competent to formulate a problem in a professional language and solve it with the help of new intelligent technologies. In order for a student to master the engineering competencies necessary for high-tech industries, to be able to show their abilities, to become a creative and socially responsible specialist with eco-mindedness, it is necessary to introduce new teaching methods that contribute to the maintenance and growth of students' motivation for mastering engineering professions. The changes should not only contribute to the creation of a systemic strategy, but also influence the use of new learning tools, knowledge transfer methods. The approaches and principles to educational content formation must also change.

Therefore, the purpose of this study is to create a virtual laboratory as a tool to ensure the required motivation level and methods for its development and use, taking into account the specifics of the training system for vehicle service engineers. The structure of the article is as follows: first, we will consider the problems of modern engineering education. Further, the directions of application of virtual reality laboratories in engineering education will be described. In the results section, a description of the virtual reality laboratory for studying the vehicle repair process will be given and the features of the proposed methodology for laboratory creation and use in the educational process will be described. Additionally, the results of students' surveys and their academic performance according to the traditional and proposed methods will be compared.

2 Problems of Motivation, Challenges and Solutions to Modern Engineering Education

Business needs a specialist with the required engineering competencies, knowledge of the features of the modern equipment functioning and operation. In addition, digital skills are now also a prerequisite for the engineer of the "future" to take full part in the growing digital economy and society [3], since digital twins are actively introduced and used as separate technical systems (a vehicle and its individual components and assemblies) as well as organizational and technical systems (enterprises and its divisions) [4–6].

However, in many countries, the interest of young people in getting an engineering education decreases [7]. There are many reasons for this. Scandals related to emissions from the automotive industry, amid global warming and global pollution, reduce public confidence in engineering enterprises. The development of digitalization and communication suggests that the choice of Mechanical Engineering is not promising in comparison with information technology in the modern world. Furthermore, students don't know about the interdisciplinary nature of Mechanical Engineering. Therefore, significant efforts are required to find ways to increase students' motivation to obtain engineering specialties [8].

After solving the problem of increasing the popularity of the engineer profession among school graduates, it is also required to ensure the high level of training. Since the educational motivation level is a driving force that affects the students' involvement in educational activities, their academic performance, and, as a consequence, professional skill and training level, a large number of papers are devoted to researching motivation. Motivation nature and strength is influenced by training system and educational model characteristics; teacher personality, his professional interests and pedagogical attitudes; educational discipline content and presentation form; degree of society and economy development; family value orientations; student society's composition and attitudes; student' individual characteristics. For their part, universities can influence the first three groups of factors. At the same time, priority should be given to the impact on internal motives (cognitive, caused by the educational activity itself), and not on external (social, lying outside of study), since students with stable internal motivation, as a rule, have higher educational results and early meaningful professional self-determination [9].

Virtual reality technologies have already found wide application in various fields of activity and there is active research about how they can be used in education [10]. In this sense, the virtual environment, as an interactive learning tool, creates the effect of presence, and also provides the appearance of the so-called sense of miracle, which, according to Hadzigeorgiou [11], improves student outcomes.

Society and economy state also affects motivation. Given that control and engineering laboratories are expensive and educational institutions must reproduce the problems encountered in the industry, the use of virtual laboratories will also reduce costs and increase the variety of experiments.

The COVID-19 epidemic has made significant changes in the way we live. In the current lockdown conditions, teachers were forced to somehow transfer traditional experience to an online experience without losing educational process quality. The greatest challenges arose with laboratory experiments, in which the students could not take part. Only a few of them were realized as online broadcasts. And although the academic year 2019/2020 is over, our experience, as well as the experience of colleagues, allows us to speak of training in lockdown not as a remote station, but as an emergency remote, requiring remedial measures to compensate for lost time in training, as well as solutions to technical and methodological problems. Universities have found more or less successful ways to complete the academic year, but these methods cannot provide a full-fledged training for students of practice specialties including automotive engineering. Practice specialties should definitely be recognized as the most affected ones due to the focus on the development of practical skills, close connection with the educational organization infrastructure and the poor coverage of the curriculum with available digital solutions [12]. Moreover, the remote form of conducting classes in itself requires an increased level of student's self-control and discipline, as well as increased educational motivation. It became apparent that e-learning should include not only web-based courses with video lectures and presentation materials, but also remote laboratories with an effective recognition system, which will allow students to interact with real experiments conducted at a distance. Since augmented and virtual reality technologies allow a deeper understanding of real processes essence, demonstrate main steps of process and possible critical situations, new approaches and achievements of online education based on the use of these technologies are not only relevant, but the only possible ones to ensure the required educational process quality.

Thus, virtual laboratories are becoming the answer to the requirements of education technical areas while adapting to new development scenarios.

3 Application of Virtual Reality in Automotive Industry and Engineering Vehicle Education

3.1 Using Virtual Reality in the Automotive Industry

A vehicle is a complex technical system. In order to assure its quality, introduction of innovative digital technologies requires a faster and more controlled decision-making process in all stages of its life cycle: from studying markets and consumer preferences, product design and manufacturing technology, production, promotion of products and their operation and maintenance prior to disposal.

The use of virtual models in prototyping and evaluation now seem to be quite obvious and necessary steps in the current product design and development process. The use of a digital copy allows to reduce the time and material costs for the manufacture of several test samples. So, when designing a new vehicle, several models were traditionally designed, with cost of creation ranging from $ 500,000 to $ 1,000,000. For instance, the introduction of the virtual reality system allowed Ford to reduce the development time for a new vehicle model from 42 to 24 months. A virtual prototype, developed from CAD models, offers a flexible environment for analyzing design features compared to its physical prototype [13].

The expediency of using virtual models for the operational stage is not so obvious at first glance. However, at present, virtual reality is actively used for the so-called immersive simulation of maintenance and repair process at an early design stage without a physical prototype. It allows designers to immerse themselves in a virtual environment to better understand the product and evaluate its maintainability. In addition, some virtual reality environments also collect interactive data related to maintenance and repair processes, which are then used by built-in algorithms to provide quantitative and qualitative assessment of the processes [14].

In [15], the authors propose a hybrid method for assessing maintainability based on virtual reality and fuzzy comprehensive evaluation. First, a scoring index system is created to trigger the design requirements. Further, based on the developed method using virtual reality, the repair process is simulated in a virtual environment. They also developed a simulation monitoring platform to help designers assess maintainability and make decisions.

Realistic nature of created reality also allows to check a technological process of vehicle repair in terms of ergonomic requirements [16]. In [17], the modeling of technological process of removing a crankcase protection was carried out, taking into account real workers' anthropological parameters. As a result of optimization experiments, specific parameters of the lift were selected that allowed reducing the load on all muscle groups, thereby preventing the occurrence of occupational diseases. The use of modeling and virtual reality tool can ensure the safety of workers and prevent injuries in the workplace.

However, despite the achievements made, it is still an evolving technology that lacks in-depth study and development in industrial application scenarios, especially in the forthcoming fourth industrial revolution (Industry 4.0) [18]. Also, this technology requires significant time resources and presents a high entrance threshold in terms of competencies [15].

3.2 Creation of a Virtual Experimental Environment for Studying Technological Processes

In the context of social distancing, the replacement of real physical objects, studied in laboratory and practical classes in the traditional form of implementation, with their computer 3D models has become the only decent alternative and an opportunity to study in detail their structure as well as understand their functioning principles [19, 20]. Engineering modeling can create educational content that provides easy-to-understand visualizations to support a quality learning process.

The second direction in engineering pedagogy is the use of virtual laboratories to study technological processes. Virtual reality can greatly facilitate the introduction of complex and abstract concepts through immersion. The main goal is to provide students with an educational experience similar to traditional laboratory exercises.

For example, in [21] virtual reality was used in the development of laboratory simulators for pneumatics. The authors created a simple experiment environment on a single-acting cylinder controlled by a pneumatic switch. At the same time, the authors in [22] found out that students are not worried about possible errors leading to equipment malfunctions, which further increased their self-confidence and reduced cognitive load.

The use of virtual reality laboratories also allows to solve the security problem. So, students can work out using them the repair technological processes, associated with increased danger, such as maintenance of gas-cylinder vehicles, which is associated with the risk of a leak or even an explosion of gas fuel. For instance, the EVEEE environment described in [23] gives students the ability to experiment safely without the risk of electric shock, providing an equipment realistic 2D representation and the ability to connect it together using virtual cables and manipulate it.

The use of virtual laboratories not only benefits students in the framework of distance learning, but also provides new learning scenarios for both teachers and students in a traditional learning format: it allows a flexible approach to experimental topics, diversifies experiments composition and brings the problems of industrial areas closer to the academic space. However, we could not find works that provide examples of the introduction of virtual laboratories into the educational process that implement an experiment with a full-fledged vehicle technological repair process.

In [24] the authors argued that in order to obtain the best result, the work on the creation of new digital content should be carried out with the joint cooperation of industry, scientific and educational organizations and this should lead to the development of youth with a desire for innovation.

The literature review shows that the researchers often put an emphasis on the elaboration of the laboratory technical aspect and the students' assessment of the received digital instrument. At the same time, we put forward the assumption that the process of developing educational content itself can be used as an incentive motivating force in the educational process. In addition, since practitioner's anthropometric characteristics directly affect repair process implementation, an assessment of his actions should be made from the point of view of ergonomic safety and, in general, their performance should be assessed, taking into account student physical capabilities.

4 Results and Discussions. Experience with Virtual Reality Laboratory in Automotive Education

4.1 Development and Use of a Virtual Reality Laboratory for Automotive Students

To increase the level of students' professional mastery and digital competencies at the Naberezhnye Chelny Institute of Kazan Federal University, the task was set to develop a virtual reality laboratory, with the help of which students will be able to study the work of stands used in vehicle repair: tire fitting, balancing, collapse convergence. This laboratory is aimed at mastering, consolidating and testing professional competencies: "capable of supervising the performance of vehicles and their components maintenance and repair", "capable of monitoring the technical condition of vehicles using technical diagnostics" in within the disciplines "Type and technical service of technological equipment" and "Technical means of service enterprises", implemented respectively in the 7–8 semesters of bachelor's training in the directions "Operation of transport and technological machines and complexes" and "Service".

Currently, a standard model of equipment for balancing work has been developed and laboratory work with its use has been tested in the educational process. The purpose of the work is to master structure and operation principles of the balancing machine, the study and practice of the technological process of balancing vehicle tires and disks. To perform this work, one needs: a balancing machine, balancing weights and a hammer with a plastic tip. These objects (Fig. 1) were modeled using Unreal Engine development environment. The object in the form of a cube with a ball on top and a protruding cylinder is a prototype of a balancing machine that has been created from primitives and its interface will be further refined manually.

To ensure interaction, each object is assigned its own individual values, parameters and interaction scripts (Fig. 2). At the moment, the development and debugging of the logic and mechanics of 3D objects interaction with each other and with the user in virtual reality mode has been completed and the laboratory visual part is being edited.

When using the developed virtual equipment analogue, the teacher must have digital competencies, because new ones are added to its tasks. Since virtual reality applications have their own characteristics, from the point of view of usage, additional instruction is provided. A teacher explains to students that the virtual reality helmet is connected to the computer and leaving the tracking zone is not allowed. Then, when the program is launched, each stage of the technological process is explained step by

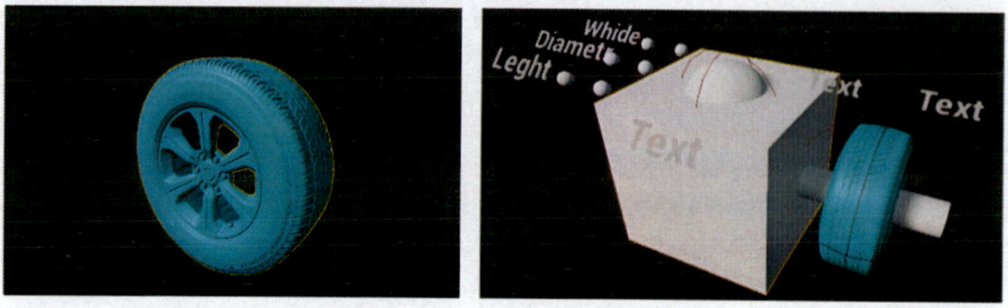

a) b)

Fig. 1. Laboratory objects: a) wheel; b) balancing machine with mounted wheel.

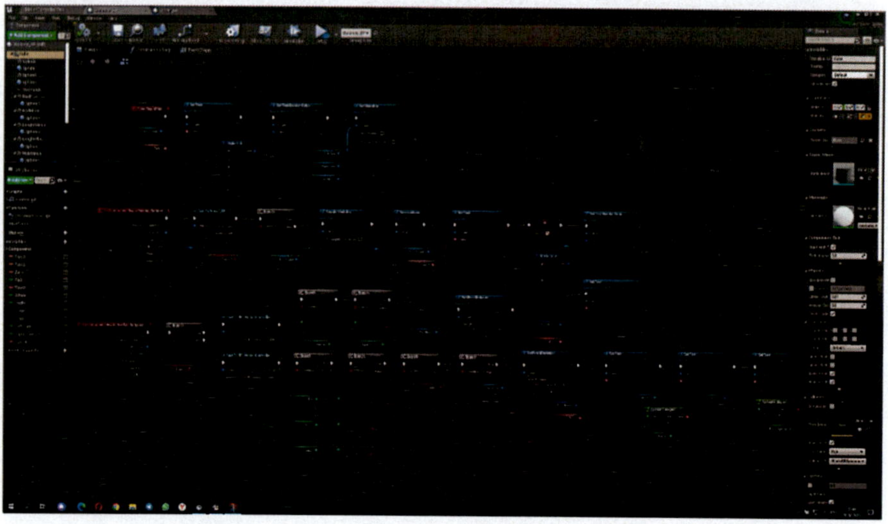

Fig. 2. Setting the logic of the balancing machine's virtual model in the Unreal Engine development environment

step by pop-up text highlighting interaction areas (Fig. 3). Thus, the student does not need to be distracted by methodological instructions, attention is not scattered.

After installing the wheel on the stand, it is necessary to set the necessary parameters (disk size and distance from the disk to the stand) using the buttons to the left of the stand, then by pressing the button the student starts the process (wheel rotation animation is performed). The stand shows the results of balancing, after which weights are installed on the wheel, if necessary, and re-balancing takes place. This process must be continued until the required values are obtained within the tolerance range (for cars - 5 g, for trucks - 30 g).

The developed system makes significant demands on the hardware. Minimum hardware requirements: 64-bit processor such as Core i5–7500/Ryzen 5 1600, 12 GB RAM and 6GB video card such as GTX 1060/RX 580. In addition, a virtual reality helmet is required. This is a significant limitation when distributing the laboratory for private use, for example, in residential areas with a full lockdown.

4.2 Features of the Proposed Methodology for Developing and Using the Virtual Reality

As rightly claimed in [25], using virtual reality technologies, a learning environment is created, which is primarily focused on students. Also, as a result of the experiment, the authors revealed temporary gains for the teacher when using the MaxWhere 3D virtual reality educational platform instead of traditional 2D interfaces: the same digital workflows can be performed with 37–64% less user operations and up to 72% less machine operations. However, at the same time, the teacher himself must have increased digital literacy.

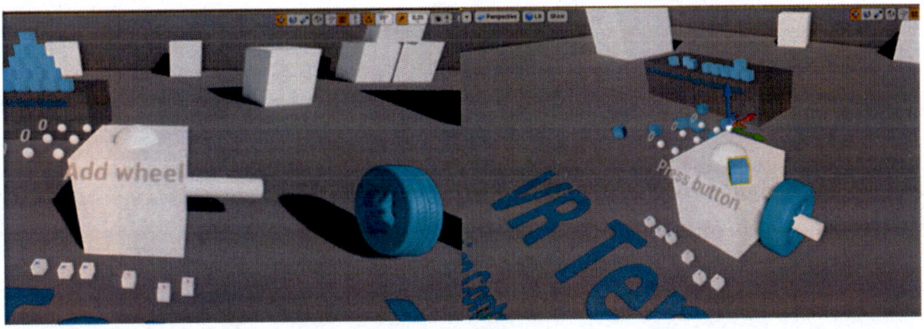

a) b)

Fig. 3. Pop-up tips to the steps of the balancing technological process: a) install the wheel; b) press the button to start the balancing stand.

Many students often master information technologies better and faster, therefore, in order to maximize usage of students' and teachers' creative, it was proposed to attract computer students to develop the content of a virtual reality laboratory for the study of vehicle repair processes. During joint practical activities, students also had the opportunity to acquire and practice communication skills.

Collaboration with the employer was used. In order to build a technological balancing process, it was filmed at one of the city's automotive repair enterprises. The entire technological process was divided into operations and their timing was performed. Thus, a work scenario was obtained, which was put into a virtual laboratory.

Due to increase in the number of occupational diseases and to save human capital, one of the modern science mandatory areas for the vehicle technical operation is technological processes ergonomic analysis. From the point of view of the educational process, it is unacceptable to assess the success of the laboratory work on reproducing balancing vehicle tires and disks without taking into account student anthropometric characteristics. Therefore, to assess the load on the student at the moment, this technological process is modeled in the interactive 3D environment Technomatix Jack, which is used by the manufacturer and allows improving technological processes using virtual dummies. It contains built-in tools that allow to select the optimal process parameters, such as: "RULA" - quick assessment of traumatic situations; "Lower Back Analysis" - analysis of the load on the back; "OWAS Posture Evaluation" - assessment of body positions according to the risk degree; "Fatique and Recovery" - analysis of fatigue and recovery degree; NIOSH Lifting Analysis - Analysis of lifting / lowering tasks. Previously, the process of removing the crankcase protection was modeled and its ergonomic parameters were optimized [16]. After receiving the anthropometric data of students, it is necessary to conduct an optimization experiment in Technomatix Jack and adjust the virtual environment settings taking into account the technological process optimized parameters.

4.3 Impact of the Proposed System on the Motivation Level and Training Quality

To check the effectiveness of the developed tool, we conducted a survey among students. It showed that this approach allows to get a higher motivation level compared to the use of traditional teaching materials and real repair equipment for students in the automotive field and educational topics for students in computer science. Students were asked to rate follow statements on a 5-level Likert scale (where 1 is "strongly disagree", 5 is "completely agree"): (1) I am more motivated by the prospect of acquiring new knowledge, skills and abilities in the learning process than by the marks I can get; (2) It doesn't matter what the result of the work is, I get satisfaction, because I feel like I've gained a new experience; (3) I enjoy attending classes in the subject and try not to miss them without a good reason; (4) I am not distracted by extraneous objects in the course of the work; (5) I believe that the subject needs to be expanded with new modern practices.

We compared the results of surveys of students' groups studying according to different methods over several years (Fig. 4). In total, 188 students were interviewed: in the computer direction 51 students, trained according to the traditional methodology, and 28 – according to the proposed methodology, in the automotive direction 74 students, trained using physical equipment, and 35 – using virtual. Also, at the end of the training, the examination tests results of the studied groups were compared. The analysis of academic performance indicates an increase in the average indicators (Fig. 5). The analysis of variance confirmed the significance of the teaching methodology factor at the $p = 0.05$ significance level. The distribution of survey scores and academic performance are subject to the normal law.

The increase in motivation and academic results among students of the computer direction is due to the fact that the creation of equipment virtual models in the modeling environment, used in enterprises, has increased students' interest. They also saw themselves working to replicate real technological equipment and process, creating a product that would be used by other students for learning. The growth of automotive profile students' motivation is due to the study of relevant service processes, taken from a real enterprise, using modern digital technologies. The above results allow us to say that the proposed methodology optimizes students' interest in learning.

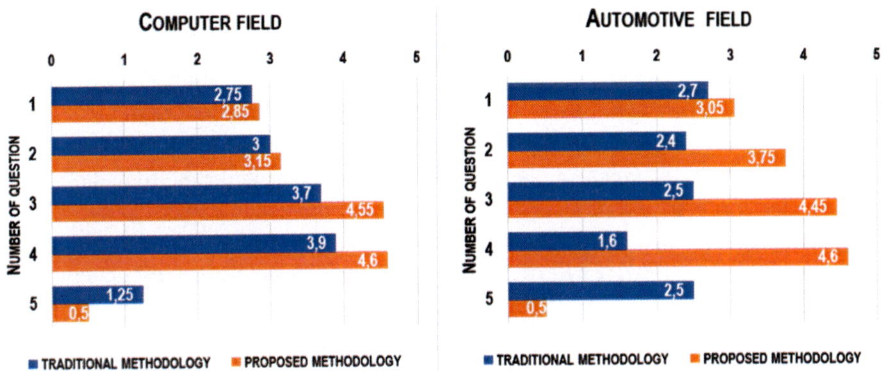

Fig. 4. Survey results

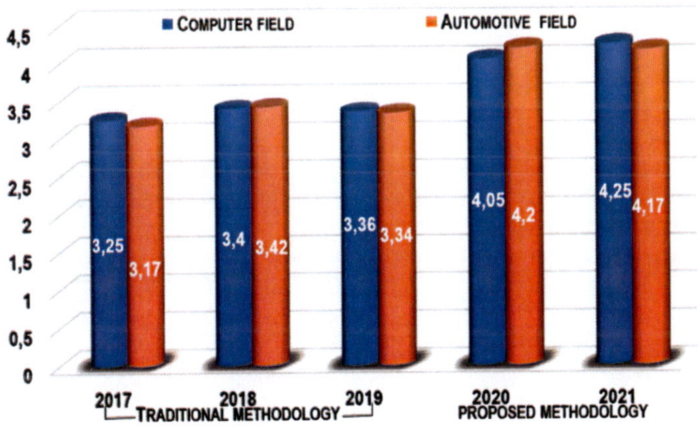

Fig. 5. Exam results - average score obtained by students

5 Conclusions

Changes in the paradigms of society and economy functioning and processes digitalization and intellectualization pose new challenges to the education system, which should prepare young people for new rapidly changing conditions. In this sense, educational institutions are no longer enough to ensure the transfer of knowledge, skills and abilities for students to acquire professional competencies. Using distance learning technologies and virtual reality, it is possible and necessary to transform the very process of transferring educational information, during which students will already acquire digital competencies. At the same time, the task of increasing motivation is being solved through interactivity, the availability of training and the creation of the "presence" effect. A feature of the virtual laboratory development approach is the interaction of teachers and students in the automotive and computer fields, similar to professional communications in the industry. Also, an important methodology aspect is the mandatory consideration of the ergonomic component of the repair technological process from the practitioner's point of view. Finally, the proposed approach to the creation and use of digital educational content of the virtual reality laboratory allowed to increase academic performance level, the quality of students training and their competitiveness in the labor market. The research group plans to replenish the virtual laboratory with new types of equipment. However, a significant limitation in the conditions of private use is the mandatory presence of special expensive hardware. In this regard, the virtual laboratory cannot compete with educational mobile applications. Therefore, further research will be aimed at studying possible ways to circumvent this limitation.

References

1. Galkin, A., Popova, Y., Chuprina, E., Shapovalenko, D.: Interaction of logistics 4.0 and consumer oriented marketing using ICT. In: Proceedings of the 33rd International Business Information Management Association Conference, IBIMA 2019: Education Excellence and Innovation Management through Vision 2020, pp. 6751–6760 (2019)
2. Vakulenko, K., et al.: Designing optimal public bus route networks in a suburban area. Transp. Res. Procedia **39**, 554–564 (2019)
3. Kwon, C., Cao, H., An, Y.: Research on digital skills training model for the underrepresented population in high-tech industry. In: 2020 International Conference on Big Data and Informatization Education, pp. 229–234 (2020)
4. Makarova, I., et al.: Automotive enterprises flow production improvement based on the management process intellectualization. In: DISA 2018 - IEEE World Symposium on Digital Intelligence for Systems and Machines, Proceedings, vol. 8490602, pp. 115–118 (2018)
5. Shubenkova, K., et al.: Possibility of digital twins technology for improving efficiency of the branded service system. In: Proceedings - 2018 Global Smart Industry Conference, GloSIC 2018 (2018)
6. Makarova, I., et al.: The role of reverse logistics in the transition to a circular economy: case study of automotive spare parts logistics. FME Trans. **49**(1), 173–185 (2021)
7. Makarova, I., Shubenkova, K., Tikhonov, D., Buyvol, P.: Improving the quality of engineering education by developing the system of increasing students' motivation. Adv. Intell. Syst. Comput. **716**, 150–161 (2018)
8. Bulut, S., Giese, T., Wende, M., Anderl, R.: Development of a gamified educational framework to teach mechanical engineering to high school students. In: 2020 IEEE Global Engineering Education Conference, pp. 1217–1224 (2020)
9. Antonov, V.G., Rumyantseva, I.A., Krotenko, T.Y.: Ocenka urovnja i struktury motivacii k uchebe studentov bakalavriata kak pokazatelej kachestva obrazovanija [Assessment of the level and structure of undergraduate students' study motivation as indicators of the education quality]//Perspektivy nauki i obrazovanija = Prospects for Science and Education, vol. 2(38), pp. 267–283 (2019). (In Russ., abstract in Eng.)
10. Galikhanov, M.F., Barabanova, S.V., Kaibiyainen, A.A.: Osnovnye tendencii v inzhenernom obrazovanii: pjat' let mezhdunarodnoj konferencii «Sinergija» [Core Trends in Engineering Education: Five Years of the "Synergy" International Conference]. Vysshee obrazovanie v Rossii = Higher Education in Russia, vol. 30(1), pp. 101–114 (2021). (In Russ., abstract in Eng.)

11. Hadzigeorgiou, Y.P.: Fostering a sense of wonder in the science classroom. Res. Sci. Educ. **42**, 985–1005 (2012)
12. Zakharova, U.S., Vilkova, K.A., Egorov, G.V.: Etomu nevozmozhno obuchit' onlayn: prikladnye spetsial'nosti v usloviyakh pandemii [It Can't Be Taught Online: Applied Sciences Students during the Pandemic]. Voprosy obrazovaniya/Educational Studies Moscow, no 1, pp. 115–137 (2021). (In Russ., abstract in Eng.)
13. Abidi, M.H., Al-Ahmari, A.M., El-Tamimi, A.M., Darwish, S., Ahmad, A.: Development and evaluation of the virtual prototype of the first Saudi Arabian-designed car. Computers **5**, 26 (2016)
14. Guo, Z., Zhou, D., Chen, J., Geng, J., Lv, C., Zeng, S.: Using virtual reality to support the product's maintainability design: immersive maintainability verification and evaluation system. Comput. Ind. **101**, 41–50 (2018)
15. Guo, Z., et al.: A hybrid method for evaluation of maintainability towards a design process using virtual reality. Comput. Ind. Eng. **140**, 106227 (2020)
16. Makarova, I., Khabibullin, R., Belyaev, E., Mavrin, V., Verkin, E.: Creating a safe working environment via analyzing the ergonomic parameters of workplaces on an assembly conveyor. In: Proceedings of 2015 International Conference on Industrial Engineering and Systems Management, pp. 947–954 (2016)
17. Makarova, I., Buyvol, P., Mukhametdinov, E., Shepelev, V., Gritsenko, A.: Repairability assessment of vehicle new model line-up using modeling and virtual reality. In: Proceedings - 2020 Global Smart Industry Conference, GloSIC 2020, pp. 362–369 (2020)
18. Guo, Z., et al.: Applications of virtual reality in maintenance during the industrial product lifecycle: a systematic review. J. Manuf. Syst. **56**, 525–538 (2020)
19. Sirazhiden, D.: VR and AR technologies in the modern cultural space and their role in environmental education. In: E3S Web Conference, vol. 217, p. 08002 (2020)
20. Rassudov, L., Akmurzin, E., Korunets, A., Osipov, D.: Engineering education and cloud-based digital twins for electric power drive system diagnostics. In: 2021 28th International Workshop on Electric Drives: Improving Reliability of Electric Drives (IWED), pp. 1–3 (2021)
21. Cruz, D.R., Mendoza, D.M.M.: Design and development of virtual laboratory: a solution to the problem of laboratory setup and management of pneumatic courses in Bulacan state university college of engineering. In: 2018 IEEE Games, Entertainment, Media Conference, Galway, pp. 1–23 (2018)
22. Singh, G., Mantri, A., Sharma, O., Kaur, R.: Virtual reality learning environment for enhancing electronics engineering laboratory experience. Comput. Appl. Eng. Educ. **29**(1), 229–243 (2021)
23. Evstatiev, B., Gabrovska-Evstatieva, K., Voynohovska, V., Beloev, I.: Web-based environment for virtual laboratories in the field of electrical engineering. In: 2019 16th Conference on Electrical Machines, Drives and Power Systems, pp. 1–4 (2019)
24. Augmented reality & virtual reality - a channel for digital transformation in industrialization fostering innovation & entrepreneurship. Int. J. Innov. Technol. Explor. Eng. **8**(10), 2278–3075 (2019)
25. Horváth, I.: Evolution of teaching roles and tasks in VR/AR-based education. In: 2018 9th IEEE International Conference on Cognitive Infocommunications, pp. 355–360 (2018)

Teaching MATLAB Programming by Utilizing Image Processing and Graphical User Interface Basics

Muna Darweesh(✉)

University of Dubai, Dubai, UAE
midarweesh@ud.ac.ae

Abstract. Learning a new programming language becomes a very important requirement for individuals who would like to be part of the Engineering and Technology fields especially with the rapid changes that we encounter nowadays. For the last three decades, MATLAB has been known for its reliability in research and engineering education. It is a dynamic language that has a very rich libraries. It is considered a user-friendly and well-structured program with many features such as app design. However, for new learners, precisely high school students, it might be difficult to engage them through the session, given the fact that it has been conducted virtually and the nature of the course which is different from the traditional high school courses. Thus, this paper will summarize the experience of a three-week program that was designed for high school students to teach them MATLAB and App design by applying image processing basics.

Keywords: K-12 education · MATLAB programming · Graphical user interface · Virtual camp · Image processing · MATLAB App

1 Introduction

Advancements in science and technology and the strong presence of Artificial Intelligence and Machine Learning have increased the demand to change the delivery of the science courses to a different approach. Programming is one of the key fields that led to that revolution in computer science. Many programming languages have been created, for example, C and C++, Java, Python, MATLAB, and much more. All of these were intended to serve many engineering and technology fields and applications. Furthermore, Digital Signal Processing (DSP) has been known for being a multidisciplinary field that connects science, art, and mathematics. This fact will support the design of courses that inspire the artistically-inclined students to appreciate the importance of Science, Technology, Engineering, and Mathematics (STEM) and understand the math behind it. Not limited to this, students with strong mathematics skills will realize the applications related to STEM. In the last two decades, many

M. E. Auer and T. Tsiatsos (Eds.): IMCL 2021, LNNS 411, pp. 824–832, 2022.
https://doi.org/10.1007/978-3-030-96296-8_74

articles have been published that discuss the experience of introducing programming or the integration of engineering courses to K-12 education. An initiative to attract the students to the engineering discipline, by Stony Brook University has been reported in [1]. The program was based on image processing through an app and electrical engineering concepts. For the image processing track, MATLAB programming was introduced to understand the basics. For advanced topics such as color image processing, Canvas was used. For the electrical engineering track, students explored basic designs such as microphones and interact with lab components and devices. There were 4 sessions two hours each. A total of 162 students from 199 provided positive feedback on the course. Another short initiative is proposed in [2]. It elaborated on the mathematical operations and how, can be utilized to apply various image processing activities such as denoising, edge detection, image superposition, and contrast enhancement. A total of 32 students participated. Moreover, a waste disposal system was the main theme for a course delivered to a Colombian high school [3]. The students experimented with the design of a waste disposal system from the image acquisition stage, which was implemented through a web camera, then the image processing stage that identified different types of waste, and finally a robotics system that classified the waste to its appropriate trash can. In Taiwan, an experience to deliver a course related to image processing has been reported in [4]. The course was based on recorded lectures and lab works. It emphasized the importance of having interactive tasks during the class and recorded videos for independent learning. The instructors have done some experiments and concluded that having videos before the classes have improved the student's score. Students as well enjoyed the lab experiments as it helped them to boost their understanding. It has been reported that the hybrid method has a promising approach for the delivery of engineering courses. The reported experiments in the literature, that are related to image processing are limited especially in the Middle East and the gulf region. Hence, in this paper, a 3-week course that was virtually delivered, due to the COVID-19 pandemic, is proposed. In the gulf region particularly in the United Arab Emirates, the high school curriculum has been modified to accommodate the advancements in technology. However, it can be further improved to include many engineering fields such as DSP. Besides, many programming languages can be taught to expose students to different applications of engineering. The proposed course will focus on MATLAB programming and designing MATLAB App. It is designed as well to familiarize them with image processing basics from simple operations to color spaces by applying MATLAB functions and operations. Integrating both MATLAB programming and image processing will attract the students and engage them to learn in an interesting direction [5]. For sure learning a new programming language is never easy. Another purpose of the course is to get the students to appreciate the importance of being able to present the final output in a user-friendly manner. It is created to develop a sense of research methods and problem-solving techniques by exploring the available resources and looking for solutions. Hence, the student will use the MATLAB App Designer tool to design an app based on image processing. In the next section, the course structure is discussed. The third section will provide a summary of the student's work during the course and the feedback from students. Finally, the conclusion section highlights the main results and proposes future directions.

Hence, the student will use the MATLAB App Designer tool to design an app based on image processing. In the next section, the course structure is discussed. The third section will provide a summary of the student's work during the course and the feedback from students. Finally, the conclusion section highlights the main results and proposes future directions.

2 Approach

The program was designed for students age between 15–18 years old. The students' background was Python programming only, but with different experience levels. The best level was to write simple user-defined functions. Thus, it was designed for 30 h distributed over 3 weeks. In the last week, the students are prompted to submit a project based on image processing through MATLAB App Designer.

2.1 Lecture Sessions

Around 9 different presentations have been developed to cover a wide range of topics starting with data types and reaching to MATLAB App Designer. At the beginning of the course, the students are introduced to the fundamental of programming. Hence, data types were presented such as numeric, strings, characters, and boolean. Then, the mathematical and boolean operations, if-statement, loops, graphing in 2-D and 3-D, and functions were explained. Also, at the end of each session, the students have certain exercises. These are designed to push them to learn how to break the problem into simple subproblems, understand the theory behind and research to discuss possible solutions. As an example, one task is designed to ask the students to understand the maximum power transfer for a load. First, they understand the idea of maximum power transfer. Then calculate the power for different load resistance values and identify the maximum power by plotting the data into a 2-D plot that describes the power versus load resistances. Later, image processing basics were introduced. It began with pixels, the black and white, after that gray and finally colored images (RGB). Besides, different filtering techniques and types of noise are presented. As an example, one of the exercises was designed to ask the student to detect the black circles only in an image. The task is intended to teach the student how to read an image from MATLAB build-in images. Then, the student will design a suitable mask to be able to detect black circles. The image that was utilized in this task is the Circles Bright Dark image (CBD), as shown in Fig. 1. Then, a white image (W) has been created and applied to CBD to detect the black circles, by applying the AND operator according to the following equation:

$$Blackcircles = CBD[512 \times 512]\ \&\ W[512 \times 512] \tag{1}$$

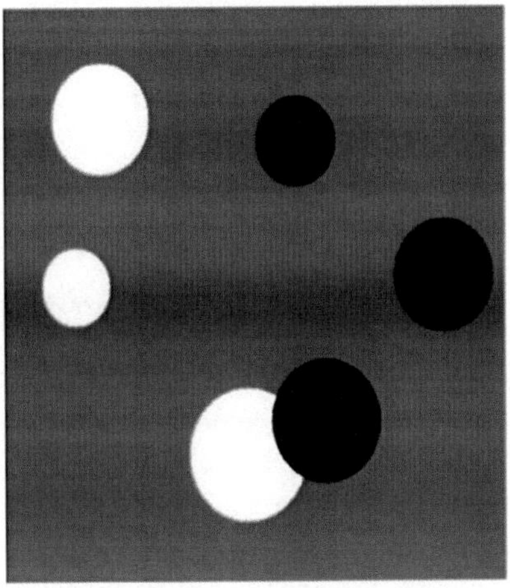

Fig. 1. Circles bright dark image

The result is illustrated in Fig. 2. The aim of such exercises is to improve their thinking and research skills and being able to develop a simple solution to the problem.

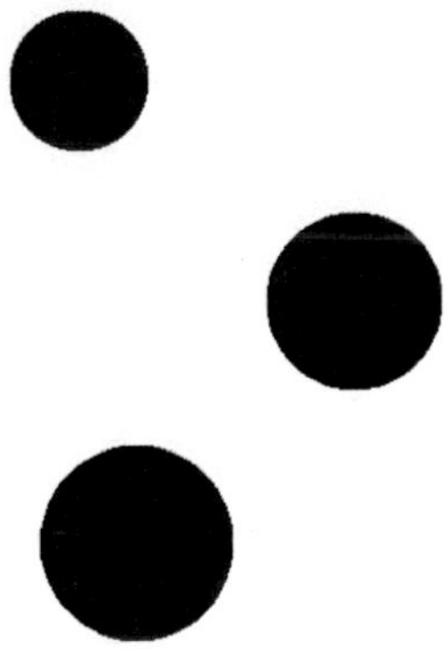

Fig. 2. The extracted Black circles

2.2 Online Courses

Alongside lectures, the students were asked to get enrolled in a Coursera course titled "Introduction to Programming in MATLAB" and "MATLAB Onramp Course". The purpose of this is to help the student to be an independent learner. The students were requested to finish the courses before the demo day since they will discuss some of the topics that were demonstrated.

2.3 Designing MATLAB App

The students were asked at the end of the course to deliver a project as a final output. This project is basically to design a MATLAB App Designer. It is different than the traditional graphical user interface the MATLAB used to provide it. The lack of detailed documentation as well, made it challenging for the students as they need to read some examples and explore the available controls. Hence, for the last lecture, the MATLAB app control, and methods of coding have been explained. Before introducing the projects, the students were requested to redesign the available two examples in the MATLAB website, from scratch to be able to understand the actual implementation, especially passing control or variables. The idea of the project is to convert the RGB image to another color space, for instance, HSV, where H, S, V stand for Hue, Saturation, Value, respectively. Then, the app would extract the layers of the new image and change the value for two layers upon the user's choice through sliders. These modifications should be reflected on the layers and then superimposed the three layers to produce the new image Ideally, the user should have the ability to browse any image. However, the students tried to simplify the problem by first design the program to be able to extract the layers and scale them implicitly in the code, and then produce the new image. Once done, the second challenging task is to allow the user to browse the image and alter the layers, which will be reflected in the final image. The design and code for one of the projects will be presented in the result section. The project description did not provide information about the range of sliders. It was left up for the students to search for the color space and set the scale accordingly. The choices of the number matters as each layer would reflect information about the image.

3 Results and Analyses

In this section, the results of one of the projects introduced are presented and a discussion of the student's feedback will be reported.

3.1 MATLAB App Results

As mentioned earlier, the project was divided into phases. Phase one provided limited choices for the user, in which two images only can be chosen. The project illustrated here is the conversion from RGB to HSV. The students employed the button group to select between two images and present the HSV image and layers. The code was simple; it depends on the if-else statements as shown in Fig. 3. The result from running the first phase of the app is provided in Fig. 4. The implementation was tricky as handing the images through the app designer is not well documented. However, the students were able to show after searching for the best function that can do so.

```
if app.pepperButton.Value==1
    rgb = imread('peppers.png');
    hsv = rgb2hsv(rgb);
    red=hsv(:,:,1);
    blue=hsv(:,:,2);
    green=hsv(:,:,3);
    imagesc(app.UIAxes,rgb);
    imagesc(app.UIAxes2,hsv);
    imagesc(app.UIAxes3,red);
    imagesc(app.UIAxes4,blue);
    imagesc(app.UIAxes5,green);

elseif app.fabricButton.Value==1
    rgb = imread('fabric.png');
    hsv = rgb2hsv(rgb);
    red=hsv(:,:,1);
    blue=hsv(:,:,2);
    green=hsv(:,:,3);
    imagesc(app.UIAxes,rgb);
    imagesc(app.UIAxes2,hsv);
    imagesc(app.UIAxes3,red);
    imagesc(app.UIAxes4,blue);
    imagesc(app.UIAxes5,green);
 end
```

Fig. 3. Phase I code

For the second phase, the students advanced the app to the next level where they managed to provide control for the user to browse the image which can be done through a push-button. The challenging part is to execute a function that can read the image file and the path. Moreover, to manipulate one layer through a slider and reflect that change on the new HSV image, the slider range is set from 0 to 1, since the Hue layer corresponds to the color position on the color wheel [6]. The slider as a component has many fields to control. Thus, to have access to the slider value, the "Value" field should be obtained. This value was displayed to the user and used to scale the layer accordingly. The code behind the design is illustrated below. Also, the students went the extra mile and designed a slider for the S layer. Both sliders were configured to scale the H and S layers and produce the new image. The code to implement such a task Fig. 5

As part of the functionality of the design, the app window should be organized, labeled, and properly colorized to have a user-friendly app. The final layout of the app with a sample of the result is shown in Fig. 6. For the final demo, the students presented their projects. It was exciting that they have provided theory about the color space and were able to answer the questions, even modified the design based on any question from the audience.

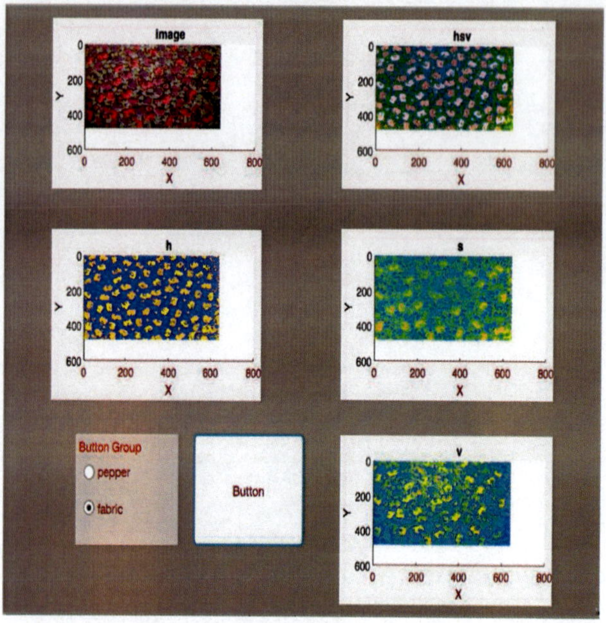

Fig. 4. Final project Phase I

```
function BrowseextractaphotoButtonPushed(app, event)

    [file, path] = uigetfile('*.png');

    if (ischar(path))
       fname = [path file];
        im = imread(fname);
    ima = rgb2hsv(im);
     scaleh = app.HueSlider.Value;
     app.EditField.Value = scaleh;
    scales = app.SaturationSlider.Value;
      app.EditField_2.Value = scales;

    h= scaleh*ima(:,:,1);
    s = scales*ima(:,:,2);
    v = ima(:,:,3);
    combined = cat(3,h,s,v);

    imagesc(app.ImageAxes,im);
     imagesc(app.ImageAxes_2,h);
      imagesc(app.ImageAxes_3,s);
       imagesc(app.ImageAxes_4, v);
        imagesc(app.ImageAxes_5, combined);

    end

end
```

Fig. 5. Phase II code

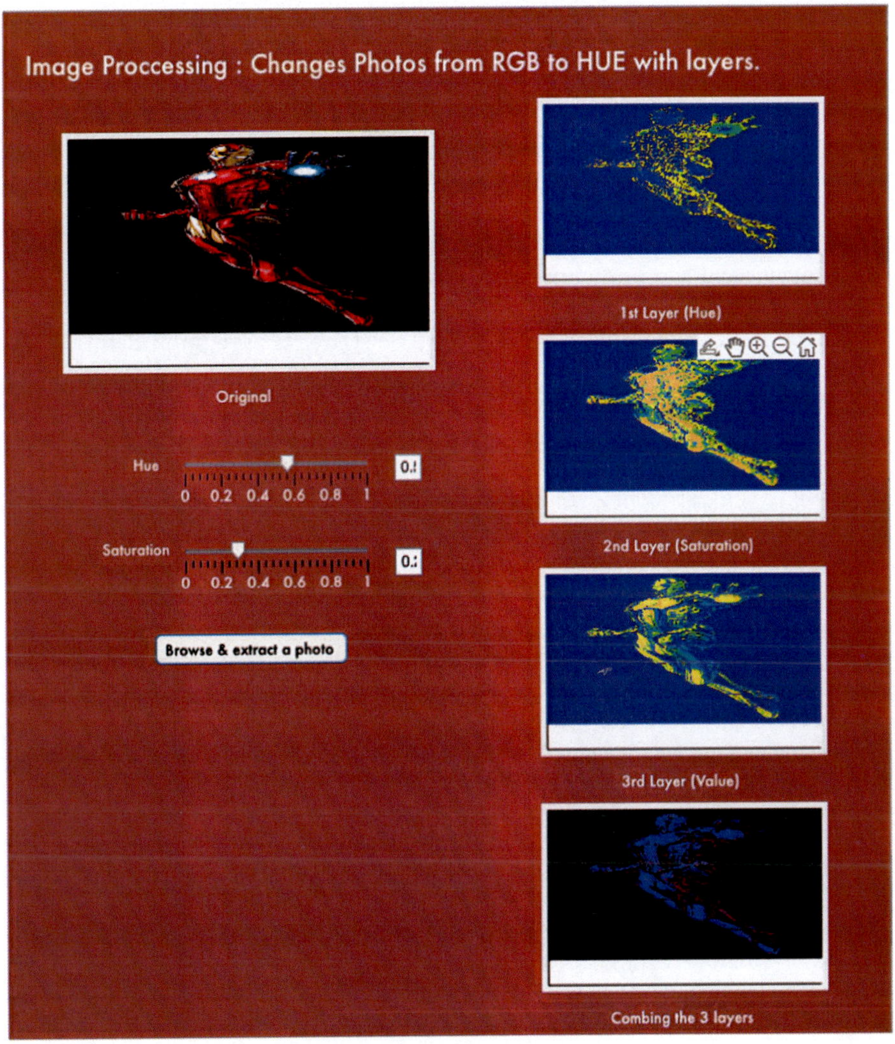

Fig. 6. Final MATLAB App

3.2 Discussion and Feedback

At the beginning of the course, the students were asked about the level of their knowledge especially in matrix manipulations through programming. Most of the students mentioned they are aware of that, but they have no idea how the matrix is linked to digital images. Through the course, it was noticeable that the students have the concepts but they can't apply them. Besides, students confirmed that they did not learn how to design and publish an app. They have mentioned that they were not aware that design an app would require many aspects that should be taken into consideration. On the final day, the students were requested to share their feedback about the course and their responses. One student has mentioned that she was not interested in the course at all at the beginning. Then, after the first week, she enjoyed the course and get excited to do her tasks. She pointed out that she would sign to the second level of the course, if available, with a big smile on her face.

4 Conclusions

A unique virtual course on image processing and app design has been introduced for students from grades 9–12. It's a 3-week program that introduces MATLAB programming through image processing. The students have published their own designed app on color spaces. The course was virtual.

Teaching programming to students during high school is essential. However, they should learn it through a project-based course, not a traditional course with a final exam. This will add excitement to the course and engage the students. In addition to that, students will learn more about app design and image processing, and its applications. Being able to present the results effectively will add value to the outputs. Moreover, the mindset of learning independently from the available resources and explore different solutions are emphasized in this course. It is very important to get the students to search for the unknown without having their instructor as the only resource. For future work, this course will focus on deep learning. Getting the students to understand the concept and apply simple examples to motivate them to work on a cutting-edge project related to image processing.

References

1. Bugalloy, M.F., Kellyz, A.M.: An outreach after-school program to introduce high-school students to electrical engineering. In: 2015 IEEE International Conference on Acoustics, Speech and Signal Processing (ICASSP), pp. 5540–5544 (2015). https://doi.org/10.1109/ICASSP.2015.7179031
2. Rosen, G., Silverman, J., Chauhan, A.: Connecting artistically-inclined K-12 students to physics and math through image processing examples. In: 2009 IEEE 13th Digital Signal Processing Workshop and 5th IEEE Signal Processing Education Workshop, Marco Island, FL, pp. 419–424 (2009). https://doi.org/10.1109/DSP.2009.4785960
3. Carlos, B.L.J., Rengifo, A., Escobar, M.: Automatic waste classification using computer vision as an application in Colombian high schools. In: 6th Latin-American Conference on Networked and Electronic Media (LACNEM 2015), Medellin, pp. 1–5 (2015). https://doi.org/10.1049/ic.2015.0316
4. Chiu, C., Lee, G.C., Her, W.: A study in lab-based approach to teach image processing in high school. In: Proceedings. Frontiers in Education. 36th Annual Conference, San Diego, CA, pp. 24–29 (2006). https://doi.org/10.1109/FIE.2006.322504
5. Kimmel, H.S., Burr-Alexander, L.E., Hirsch, L., Rockland, R.H., Carpinelli, J.D., Aloia, M.: Pathways to effective K-12 STEM programs. In: 2014 IEEE Frontiers in Education Conference (FIE) Proceedings, Madrid, pp. 1–6 (2014). https://doi.org/10.1109/FIE.2014.7044362
6. National Center for Biotechnology Information. http://www.ncbi.nlm.nih.gov

Renewable Energy by Remote and Online Laboratories

Hen Friman[1(✉)], Netser Matsliah[1], Yafa Sitbon[2], Ifaa Banner[3], Yulia Einav[1,4], and Nava Shaked[5]

[1] Faculty of Engineering, H.I.T - Holon Institute of Technology, Holon, Israel
henf@hit.ac.il
[2] Dean of Students Office, H.I.T - Holon Institute of Technology, Holon, Israel
[3] Director of "Israeli Hope", H.I.T - Holon Institute of Technology, Holon, Israel
[4] Dean of Students, H.I.T - Holon Institute of Technology, Holon, Israel
[5] Head of the Multidisciplinary Studies Department, H.I.T - Holon Institute of Technology, Holon, Israel

Abstract. Throughout the years with the development of industry and technology, people have increased environmental pollution. Generating much waste, using non-recycling materials, all lead to an increase in environmental pollution. In an era in which technology is so developing at a dizzying pace, the demand for electricity is only increasing, this situation may bring to a crisis that will disrupt the economy's operations. These are severe problems that threaten the normal existence of Earth and of course, our life. We are approaching an "irreversible point". Environmental contamination, including air, water, soil, pollution, creating waste in huge amounts as well as the waste of energy. This is reflected in damaging the sea, contaminating the groundwater, increase in global temperature, global phenomena like "acid rain" and the lessening of the ozone layer. There is no room for deliberation we must not remain indifferent- the earth's population must wake up and act quickly, unite and make a comprehensive change in order to reduce significantly the harm caused by the industry around the world.

In this article, we will present a case study of the education of the future generation for sustainability, environmental protection through Renewable Energy by Remote and Online laboratories.

Keywords: Environmental education · Elementary school · Renewable energy · Sustainability

1 Introduction

One of the things that can help people reduce environmental pollution is using renewable energy. This energy originates from solar radiation, wind, biomass, and non-consumable sources which are not fossil fuels. Using renewable energy and energy efficiency are vital steps to reduce energy consumption. Recycling is one way to decrease the amount of waste and that of the one which is transferred to landfills. Awareness of all of these does indeed increase around the world, but it is still insufficient thus significant and decisive steps must be taken.

M. E. Auer and T. Tsiatsos (Eds.): IMCL 2021, LNNS 411, pp. 833–843, 2022.
https://doi.org/10.1007/978-3-030-96296-8_75

1.1 Academic Resource

The Holon Institute of Technology, or HIT, was founded in 1969 as part of Tel Aviv University and became an independent academic institution of higher education in 1999, certified by the Council of Higher Education of Israel. Its faculties range from the sciences, engineering, digital technology in medicine and computer science, to technology management, learning technologies, and design. The institute emphasizes theoretical and practical multi-disciplinary research of innovative technologies, from a professional, scientific, economic, and cultural perspective.

The faculty of engineering at HIT is the largest faculty in the institute, and one of the largest producers of electrical engineers in Israel. Its main goal is to train highly capable electrical engineers, who possess theoretical and practical knowledge that is required in the ever-changing industry. It provides the students with a rich and comprehensive study program, which is constantly updated and improved in accordance with the technological advancements of each field. The Department of Energy and Power Systems holds courses such as Energy Conversion, Electrical Motors, Electronic Power Systems and High Voltage Systems. The department also has a Renewable Energy and Smart Grid Excellence Center, which provides a home for students to enrich their knowledge of green energy and the future of the electrical grid. It serves as an information center for elementary school students and other groups in the community, who can come and learn about the field. The Renewable Energy and Smart Grid Excellence Center provided the theoretical material to the students who participated the in "Green Ambassadors" course.

1.2 Social Resource

The Social Involvement Unit at HIT aims to make a positive difference to the life and future of our region by making a significant contribution to the communities and society as a whole, through a wide range of activities undertaken by the institutional staff and students. The purpose of the unit is to stimulate motivation for meaningful social action for the community, based on the perception that these activities will influence both the academic institution and the society and lead to a long-term and sustainable change. The Unit encourages students to study and participate in action learning courses that combine academic learning with social activities. The active integration of the students in the communities, especially in those that represent national minorities, inspires social sensitivity and cultivates the values of giving, helping, promoting, and respecting differences.

Part of the Social Involvement Unit's activities involves creating action learning courses, with the financial support of the Council for Higher Education in Israel. An action learning course is an academic course that involves academic learning and social activities. The course deals with social challenges expose the students to different ideologies and develop critical thinking and pragmatic ideas. The students get a grade and credit points for taking part in the course. They perform activities to engage in the experiential learning process, and by doing so they create a cross-fertilization between what they learn in the classroom and what they experience in the real world [1]. Their learning experience includes being taught by social organizations, institutions, and state authorities, and performing practical work with diverse populations. Through that experience, the students strengthen their academic skills, formulate ethical attitudes towards reality, develop professional and civil perspectives, and understand how they can make a difference in their community [2].

2 Methodology

The Green Ambassadors in the Community course is an academic course that has been operating at HIT Holon Institute of Technology since 2017. The course started in 2017 in “Revivim” elementary school in Holon, a secular school. In 2018 it operated in the Arab school “Al-Omaria” in the city of Ramla. In 2019, the course operated both in the national-religious school “Yishurun” in Holon, and in the Arab school “Neve Shalom” in Lod. In 2020, the course operated on gender segregation in the Ultra-orthodox (Haredi) schools boys’ school “Bnei Menachem” and girls’ school “Bnot Menachem” in Holon [3]. In previous years, due to logistical limitations of time and transportation, the course took place in schools in Holon, Lod, and Ramla. Cities that are physically close to the academic institution [4]. As part of the course, students of the Holon Institute of Technology, from the various faculties, meet with pupils in elementary schools and through the use of the workshop and experiential teaching methods prepare students to be “Green Ambassadors” in the community and environment in which they live [5]. 2020–2021 academic year course was under the limitations of the COVID19, the course lecturers Ifaa Banner and Dr. Hen Friman faced a challenge, how to hold a course from a social, community-doing stage when sacrifice is impossible. This year the opportunity was taken and the course was activated in the geographical periphery of the State of Israel, in two schools: “Ben-Atar” a religious school in Dimona, and a “Yitzhak Sadeh” public school in Dimona. Dimona is a city located in southern Israel, in the Negev, with a low socioeconomic status (see Fig. 1). Nine groups of three to five students and pupils from six classes, Third, fourth, and fifth graders (total 188 pupils), participated in this course. The collaboration between the students (30 students from various faculties and departments in HIT) and the HIT staff and the pupils and school staff was excellent.

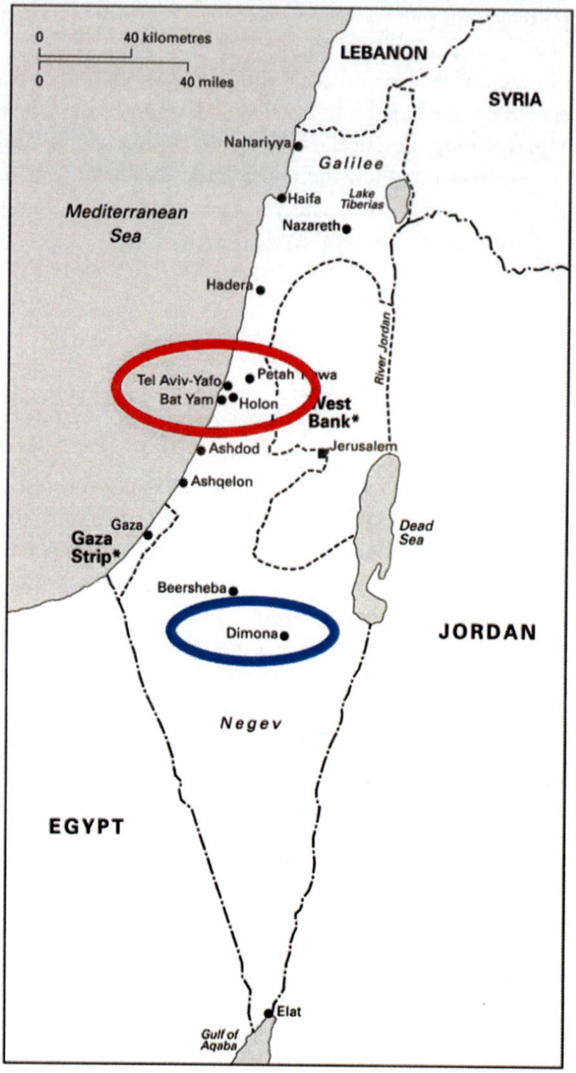

Fig. 1. Map of Israel, the activity of the "Face to Face" course until the COVID19 period in red, during the COVID19 period in online course in blue [6].

The strength of experiential learning is that it occurs in a variety of learning methodologies, such as project-based learning, learning through games, learning through simulation, etc. [7].

Pupils learn best what they feel is relevant to their lives. Studies show that learning is most effective when it is active, goal-driven, and personally relevant [8]. Since the search for meaning is embedded in humankind [9] pupils mostly concentrate on the learning tasks they find meaningful. Those who seek to influence pupils learning, should try and create the best possible match between the goals of the institution and the goals of the learners. With learning by experience, pupils gain valuable knowledge and skills while researching important everyday problems, such as how to decrease the environmental pollution in the park close to their home. Meaning, pupils learn when they want to learn. Since acquiring knowledge and skills demands a great effort from the learner, as well as guided practice, and since the level of knowledge acquired is influenced by the will of the learner [10], it is important to create curiosity and interest in the classes.

Many researchers emphasized the role of experience in the learning process, which serves as a basis for the experiential learning model that was published in 1975 by David A. Kolb and Ron Fry. The model includes four elements: a new experience; analysis and insight development based on the personal experience; construction of a theory explaining the experience (either by using prior knowledge or knowledge acquired from the experience itself); and examination of the theory in new situations [11]. Kolb and Fry claimed that the learning process can start at either one of the four stages and that one should consider them as an ongoing spiral, but in order for the learning to be meaningful, the learner must go through all of them [12].

The learning process includes two sub-processes: absorbing information and processing information. Therefore, the experiential learning theory requires the learner to possess polar abilities and skills, and the learner needs to choose which ones should be used in each situation. Each one of the four stages of the model require choosing a different learning skill. The learner might excel in theoretical processes but might fail to assess their value. The teacher must recognize, understand and analyze the methods in which the individuals receive and process information, and help each learner individually with their difficult spots in the process. The teachers and learners should share their own teaching and learning methods with each other. This sharing will make the learner understand why the subject was chosen to be passed in a certain way by the teacher. Meanwhile, the teacher will be able to identify the different learning methods among the learners, will develop empathy towards them, and will therefore be able to help them better [11].

Challenging and intriguing teaching involves deep, active experiential learning which is connected to relevant issues that interest the learner. Performing high-level research in each of the approaches requires the use of several thinking strategies. Only when the learners master these strategies, can they perform quality research. The research process is often performed in classes in a technical and superficial manner, as the learners go through all the stages of the research, but focus on its outer layers and do not go "deep". Therefore, the way to avoid having superficial research is through guided work on thinking strategies development, and gradual implementation of them with the learners [10].

3 Implementation

At the beginning of the semester, the 30 students from various faculties and departments in HIT studied in synchronous sessions in the "Zoom" program topics in Renewable Energy, Ecology, Sustainability, and the environment well as learning about the various tribes of Israeli society (see Fig. 2). The students also met in informal conversations at "Zoom" with the school principals and pupils from the schools in preparation for the session.

Fig. 2. The students get to know Dimona in a synchronous meeting in "ZOOM".

The sessions with the pupils are virtual sessions, where the pupils were given an experimental kit and laboratory equipment before the sessions and performed the experiments together with the students in real-time (see Fig. 3, 4 and 5).

Fig. 3. The students are preparing the experimental kits on the HIT - Holon campus

Fig. 4. The experimental kits are transported to Dimona by special transportation for distribution to pupils in schools

Fig. 5. The pupils get the experimental kits (a) "Yitzhak Sadeh" (b) "Ben-Atar" a religious school in Dimona. The children's parents signed an informed consent to the publication of the children's photos.

During the virtual sessions, a "Kahoot!" trivia question was used to test the internalization of knowledge (see Fig. 6). Screening of animated videos, explanatory presentations, and experiments (see Fig. 7, 8).

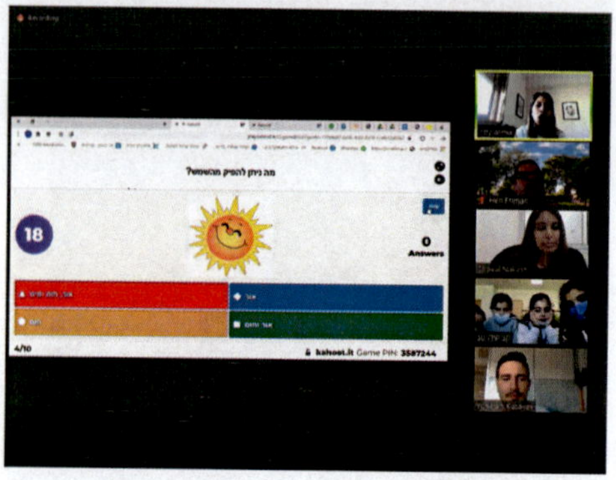

Fig. 6. "Kahoot!" trivia question

Fig. 7. (a) Performing an experiment under the guidance of students (b) monitoring results

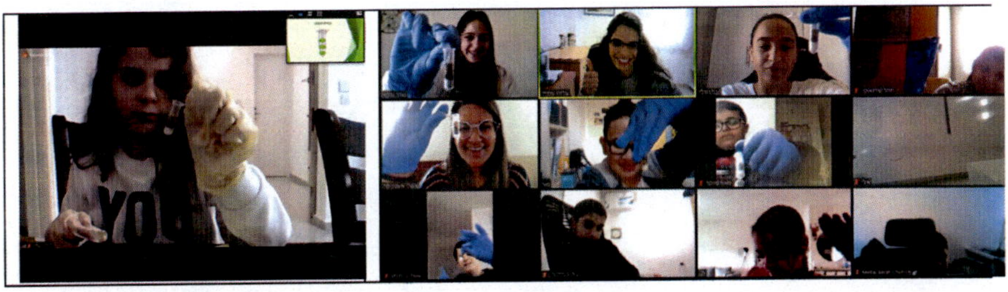

Fig. 8. Performed the experiments together with the students in real-time

The last session of the course, which took place in early January 2021, was an "expo-sure to academia" session in which school pupils were virtually exposed to the institute and even virtually visited the "Center of Excellence for Renewable Energy and Smart Grid" of the Faculty of Engineering and see how academia develops innovative solutions [13]. Conservation of the environment and sustainability.

4 Results

Examining the results of the "Kahoot!" trivia question that the pupils took at the beginning and at the end of the course, it is clear to see that the learning process was successful. The pupils absorbed the information through the activities of the lessons and answered the questions of the final "Kahoot!" trivia question with a much higher success rate [14].

"Ben-Atar" a religious school in Dimona and "Yitzhak Sadeh" a public school in Dimona. Nine groups of three to five students work with pupils from six classes, Third, fourth, and fifth graders (total 188 pupils). When examining the results of the trivia quizzes given to the pupils at the beginning and at the end of each lesson, the effect each lesson had on the pupils' knowledge becomes clear. While in the opening questionnaires there were only 37% correct answers, in the final questionnaires there were 86% correct answers – a 49% increase (see Fig. 9).

Compared to the results obtained in previous years, in traditional, face-to-face instruction, the effectiveness of distance instruction, using the zoom, was also very good [15].

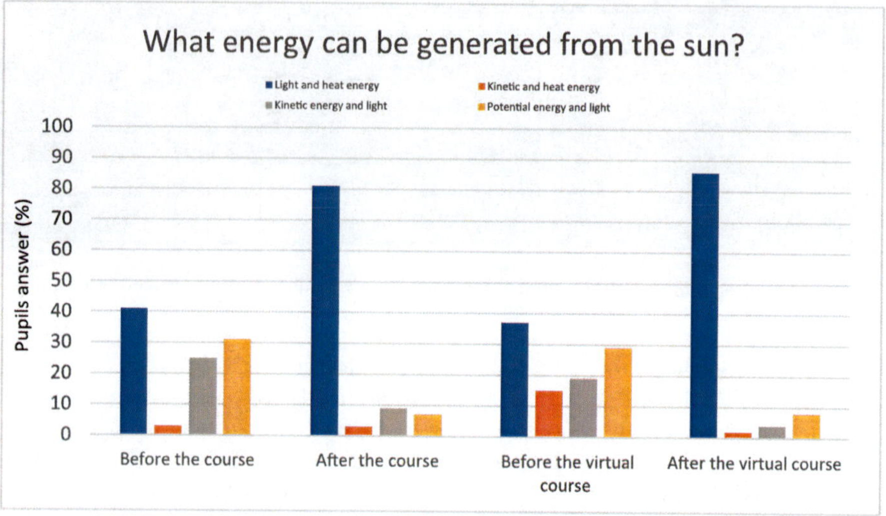

Fig. 9. The distribution of the responses in the opening and the final questionnaires

5 Conclusions

The "Green Ambassadors" course provided a great method for the Holon Institute of Technology and its students, to fulfill the vision of Renewable Energy by Remote and Online laboratories. "Green Ambassadors" taught the students how to turn environmental issues and information they learned, into educational activities for the pupils at school. These activities created a meaningful learning process for the pupils, as they took an active part in the Online lessons.

Examining the results of the knowledge tests that the pupils took at the beginning and at the end of the course, it is clear to see that the learning process was successful. The pupils absorbed the information through the activities of the Online lessons and answered the questions of the final tests with a much higher success rate.

Overall, the "Green Ambassadors" course received high praises from all people involved, from the staff at HIT, through the students carrying out the lessons, to the teachers and pupils of the school. The final confirmation of its success was given at a graduation online ceremony, in which each pupil received a "Green Ambassador" diploma. There was no doubt that the pupils experienced a meaningful educational event, that they will carry with them into their future lives.

References

1. Arar, K.: Israeli Education Policy since 1948 and the State of Arab Education in Israel. Isr. Educ. Policy since 1948 State Arab Educ. Isr. 4 (2012)
2. Bender, D.: Education and Career Skills Acquired During a Design Internship, pp. 358–366 (2020)

3. Friman, H., et al.: Ubiquitous learning of renewable energy and environmentalism to various Israeli populations. In: Proceedings - 2020 International Symposium on Educational Technology, ISET 2020 (2020). https://doi.org/10.1109/ISET49818.2020.00033
4. Friman, H., Banner, I., Tuchin, B.S., Einav, Y.: Students Teach Pupils Environmental Issues and Renewable Energy (2018). https://doi.org/10.1088/1755-1315/150/1/012025
5. Friman, H., Sitbon, Y., Banner, I., Shauli, T., Einav, Y.: Renewable energy, ecology and environment to arabic pupils using a creative, "hands on" approach. Environ. - MDPI **5**, 1–12 (2018)
6. Alterman, R.: National-level planning in Israel: walking the tightrope between government control and privatisation. Natl. Plan. Democr. Ctries. 257–290 (2013) https://doi.org/10.5949/liverpool/9780853238454.003.0011
7. Harmer, N.: Project-based learning Literature review 34 (2014)
8. Römmele, A., et al.: Book reviews. Party Polit. (2014)
9. Tal, T., Kali, Y., Magid, S., Madhok, J.J.: Enhancing the authenticity of a web-based module for teaching simple inheritance, pp. 11–38 (2011). https://doi.org/10.1007/978-94-007-1159-4_2
10. Lippman, L., Atienza, A., Rivers, A., Keith, J.: A Developmental Perspective on College & Workplace Readiness. Child Trends 49 (2008)
11. Kolb, D.A.: Experiential Learning: Experience as the Source of Learning and Development, pp. 20–38. Prentice Hall, Hoboken (1984)
12. Eddy, J.K., Glass, A.L.: Reading and listening to high and low imagery sentences. J. Verbal Learn. Verbal Behav. **20**, 333–345 (1981)
13. Friman, H., Matsliah, N., Beck, Y.: Renewable energy lab at the faculty of electrical engineering. In: INTED2016 Proceedings (2016). https://doi.org/10.21125/inted.2016.0149
14. Friman, H., Sitbon, Y., Banner, I., Einav, Y.: Meaningful learning at "Green Ambassadors" course. In: ICERI2018 Proceedings, vol. 1, pp. 7278–7283 (2018)
15. Friman, H., Sitbon, Y., Banner, I., Einav, Y.: Public participation as a tool for preserving the environment. In: E3S Web of Conferences, vol. 101 (2019)

HAS-200 Lab Virtualization in Times of Pandemic

John Alejandro Forero Casallas(✉),
Luini Leonardo Hurtado Cortés, and Víctor Elberto Ruiz Rosas

Facultad Tecnológica, Universidad Distrital Francisco José de Caldas,
Bogotá, D.C, Colombia
{jaforeroc,llhurtadoc,veruizr}@udistrital.edu.co

Abstract. The main objective of this project was to create by on side a CAD model of the manufacturing cell HAS-200, with great precision in relation to the actual dimensions, and for other hand the possibility of performing virtual simulation routines in order to design process improvements.

The project was carried out in three phases, first the search for the technical information of systems and components and the appropriate integration of the sources. The second phase involved the graphic modeling of each of the cell stations components and the general assembly of the manufacturing cell suitable for analysis of motions and simulations. Finally the simulation of station 1 operation was developed based on its tasks, times and sequences.

Keywords: Virtual laboratory · HAS-200 · Virtual manufacturing environments

1 Introduction

Automation as a term refers to certain actions being done on their own, without the direct involvement of an individual. In one of the specific cases in which this project is of interest, which is in the industrial field, these characteristics have been fundamental for the growth of industries in which the demands and the complexity of their products have intervened as needed to establish systems that meet them through complex designs [1].

The great industrial growth that has been observed in the last twenty years, represented in the variety of products and in the complexity of the plants, coupled with the Covid-19 virus pandemic situation, has motivated the development and use of new forms of transmission of knowledge to face the challenges of industrial automation [2].

Industrial automation is a discipline that groups' science and technology oriented to the performance of tasks autonomously without the direct participation of an individual. In this sense, education and training must have technological and human resources so that in the current circumstances the quantity and quality of knowledge is guaranteed.

At the Colombia level, it is important to make projections because of the rapid progress that is being made at the global level. Statistical analyses project that by 2020 the degree of automation in the Colombian business sector will be between 25% and

M. E. Auer and T. Tsiatsos (Eds.): IMCL 2021, LNNS 411, pp. 844–855, 2022.
https://doi.org/10.1007/978-3-030-96296-8_76

30%, and 45% of the jobs that people currently perform could be replaced by current technology [3].

For this reason, they emphasize not to see technologies of this type as an enemy that will come to displace and replace the work that has been done by human hands and set out reasons such as: to understand technology in the workplace as a tool and not as a threat; to understand businesses and operations in detail to integrate automation effectively; to have the analysis capacity of interpreting the information and making use of it, because there will be more time to think. Therefore, the development of mathematical and statistical skills will be essential; the ability to interact with multidisciplinary groups that complement the work being done; being prepared for change, the world demands adaptation [4].

2 Virtualization of Laboratory HAS-200

This project was developed in three phases: conceptualization, modelling and simulation, which were carried out independently after searching for information and components that are part of each station, by integrating clear sources of information and with the information processed in previous cell modeling projects.

2.1 Conceptualization

Search for Information. This part aims to search not only for concepts, research and technical methods that will help with the development of this project, but also for models already established by the different manufacturers of the devices and parts, or for projects in which it has been necessary to draw up blueprints of the components that constitute each of the stations of the manufacturing cell.

All this is due to the fact that most of the cell stations are built with standardized structural components, which makes it possible to obtain the libraries required to be installed and attached to the software for proper use in the models.

Identification of Station Components. Before drawing, we must evaluate each of the components of the stations that make up the manufacturing cell in terms of their specifications and references for identifying the different products in the network.

2.2 HAS-200 – Highly Automated System

The HAS-200 system reproduces a production process with a high level of automation, which allows developing the professional skills required in the most diverse sectors (automotive, semiconductor, food, pharmaceutical, etc.). Aspects such as aesthetics, user motivation and the development of transversal competencies (such as teamwork…) have been taken into account in the conception and design process. At the university level, the HAS-200 system is a powerful platform for the development of research projects.

The Product/Process. This HAS-200 allows the production of 19 different products. The raw material consists of a container with four types of labels (red, blue, yellow and multicoloured), each with a barcode that allows the product to be identified throughout the process. Inside these containers, colored "pearls" will be poured in different quantities, making it possible to combine 19 different "recipes" (Fig. 1). Once filled with the corresponding quantity, the containers are covered with a lid and a label with the date of manufacture and other information. The product is then sent to the dispatch station or warehouses waiting to be shipped. Within the process, both the weight of the material and the height of the material are measured. These two variables are analyzed by the Statistical Process Control (SPC) for decision making and will be stored in the Database for the generation of histories, etc.

Fig. 1. Products of manufacturing cell HAS-200.

The Modular System. The HAS-200 consists of a fully modular system of up to 11 workstations, a raw material station and the control cabinet. Each station integrates a conveyor belt section, which allows great flexibility in the design of the "layout." All stations are equipped with a control panel/button panel, as well as a tricolor indicator and PLC, which allows for manual and integrated operation. The connection between the stations and the management system is made via an Ethernet network that enables high speed data flow and global standardization. Each of the HAS-200 stations carries out a part of the process [5].

2.3 Modeling

By consulting the information obtained, we proceed to schematize the cell understanding its conformation to begin with the detailed identification of the stations with their components, as they are numerous and very detailed. One way to simplify a little is to select in advance all the standardized or commercial elements to go to the CAD model download centers and even to the page of the manufacturer of the cell that has the University. Following this premise of simplification, an application for Android® was used to speed up the surveys.

2.4 Development of Plans

It is the fundamental basis of this project, since here each of the components is characterized in terms of dimensioning, positioning, and structuring of the stations of the manufacturing cell and at the same time the verification and correction of components acquired in the network. The QuickMemo + ® mobile app was used to add the measurements taken to images of the manufacturing cell (Fig. 2).

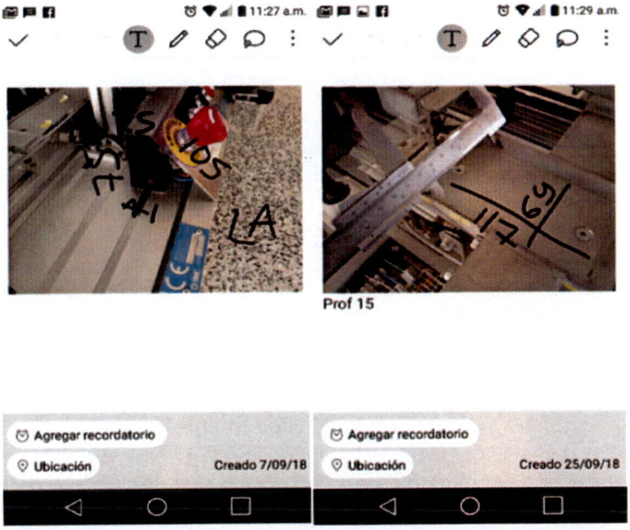

Fig. 2. QuickMemo + ® application for measurement annotations on images. Source: author.

As mentioned above, the framework modeling environment is used to generate the typical table structure of each station of the cell, Fig. 3 also shows the fastening element used by taking advantage of the grooving of the profile (Fig. 4).

Fig. 3. Table creation with measurements taken and profiles downloaded. Source: author.

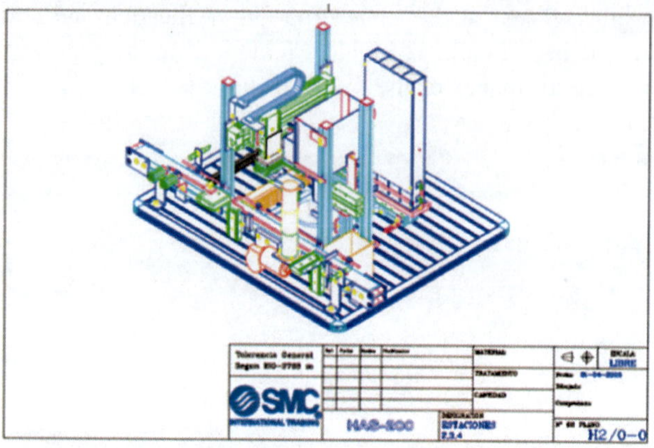

Fig. 4. Plans of the stations supplied by SMC®. Source: author.

The information contained in the drawings provided by the University is used to identify the components by compiling their descriptions in the lists of materials and in the guidelines that mark each part or device (Fig. 5).

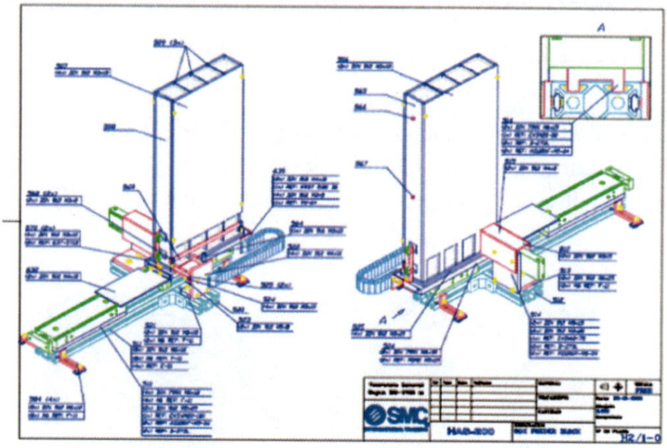

Fig. 5. Plans of sub-assemblies with component reference. Source: author.

2.5 Virtual Search of Standardized and/or Commercial Elements

This part is of vital importance for the assembly of the stations during the modeling, the greatest precision was sought when establishing the different parts and components that are used to join the parts of the manufacturing cell and that are known to be found in databases of CAD programs (Figs. 6, 7, 8, 9, and 10).

Fig. 6. Page for downloading aluminum profiles. Source: author.

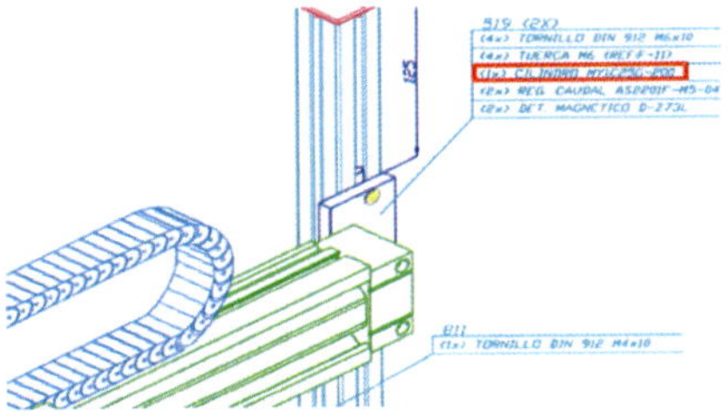

Fig. 7. SMC® plane reference of pneumatic actuators. Source: author.

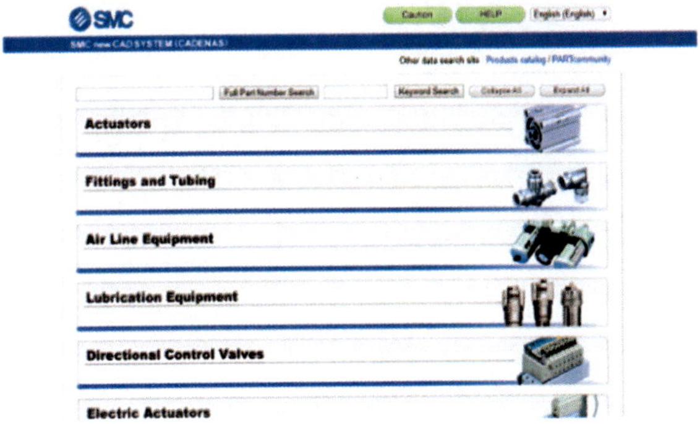

Fig. 8. Home page for downloading SMC® components. Source: author.

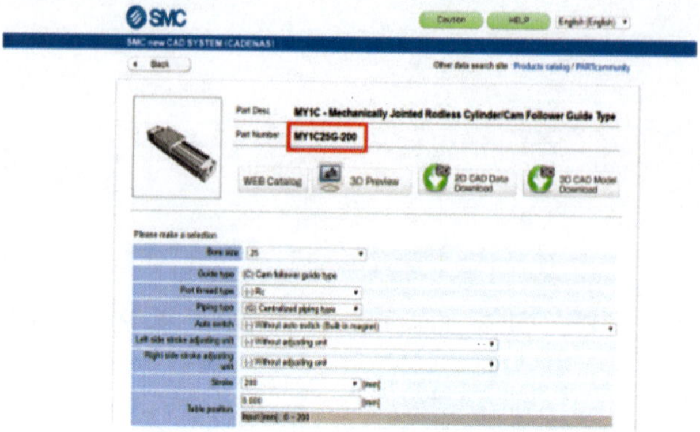

Fig. 9. Variation of component characteristics for download on SMC® page. Source: author.

Fig. 10. Actuator MY1C25G – 200 downloaded for Solid Works®. Source: author.

2.6 Modeling of Sub-assemblies of Each Station

This involves the verification of the joining of parts and surfaces for the transmission of movements, analysis of stops, interference of movements between parts and forces between them for the assembly of components that perform specific tasks (Fig. 11).

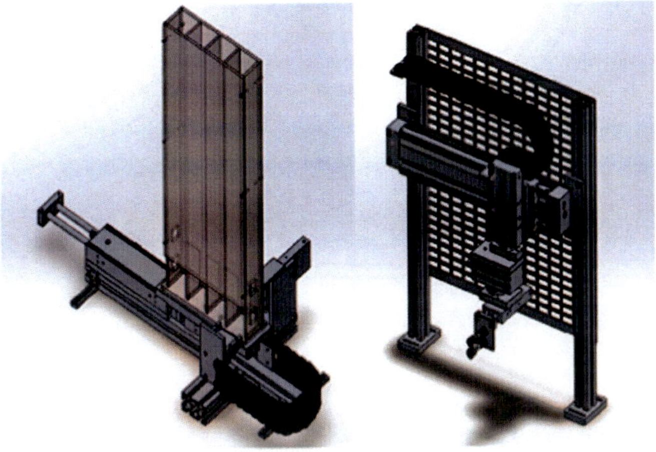

Fig. 11. Subassembly station 1, 2 and 3; on the left: container feeder; on the right: container feeder. Source: author.

2.7 Assembly of Workstations

Conformation of sub-assemblies, fixing of components in the model on the standardized work benches through the verification of the positioning of each of them for the realization of the process as such of the station (Fig. 12).

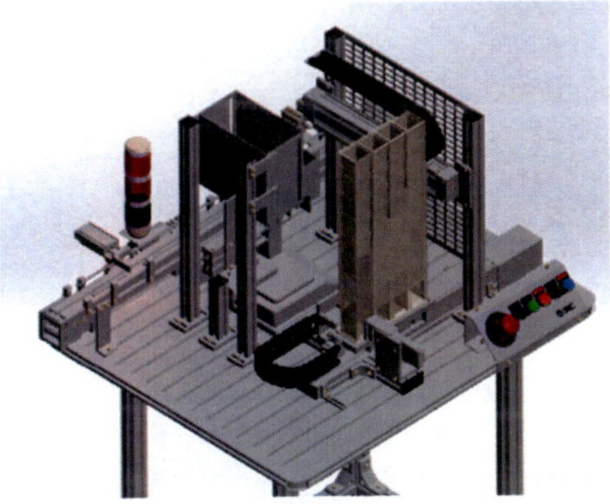

Fig. 12. 3D modeling with Solid Works® Station 1, 2 and 3 software (Colored pearl dispenser) HAS-200® manufacturing cell –. Source: author.

2.8 Assembly of the Manufacturing Cell

Alignment of the workstation assemblies, in the case of the FMS-200 manufacturing cell, by means of linear conveyor belts for each station that are intercommunicated in the form of a "U" for the entire cell, verifying the entry and exit of the product at each of the stations viewed as a system (Fig. 13).

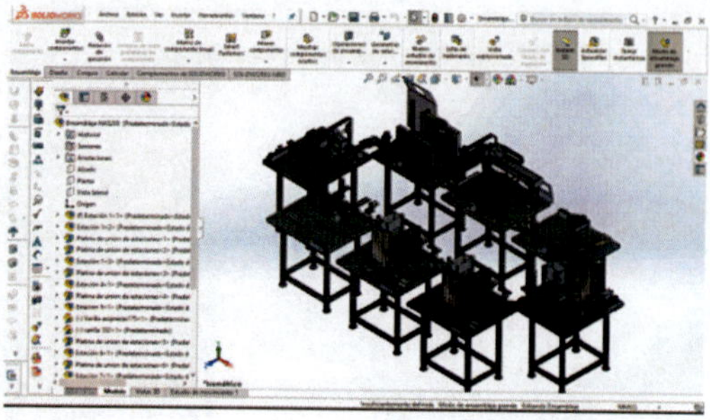

Fig. 13. Modeling of HAS-200 manufacturing cell in Solid Works® software environment. Source: author.

2.9 Simulation of the Movements of Station 1 of the HAS 200 Manufacturing Cell

The final step of this project is the realization of the basic simulation of the process that was carried out at station 1 of this manufacturing cell, whose function is to store empty containers to be filled with pellets of a certain color, weighed and removed by a conveyor belt to the next station, a process distributed automatically. This operation includes the first three stations of the manufacturing cell, the differentiation of the product color that is dispensed to the container is selected by a reader with a barcode printed on it.

This phase of the project was carried out with the Siemens NX12® program, which has several specialized modules for the realization of motion simulations such as the mechatronics concept designer, which works with the application of sensors to the components involved in motion simulation and we can select it as described in Fig. 13 after opening a part or an assembly (Fig. 14).

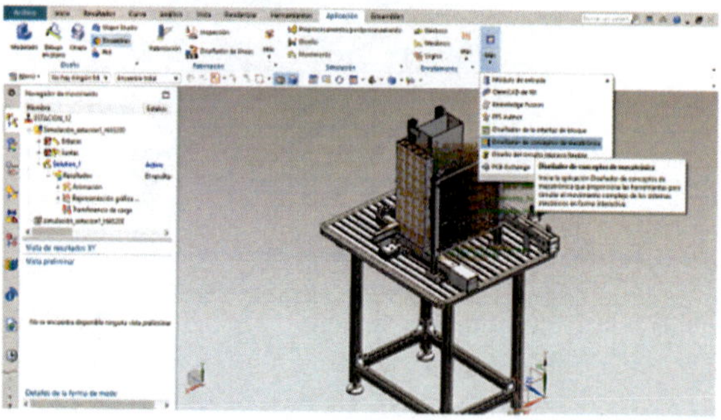

Fig. 14. Icon to enter the mechatronics concept designer. Source: author.

Before saving the model in this type of format, it is preferable to make sure that all the parts of the assembly are located and, in addition, that they are positioned in the zero state of the simulation, otherwise the first steps of the simulation should be to position the parts in the initial state after performing the import.

Next, the model of the station to which the motion simulation is developed can be seen within the Siemens NX 12® program environment (Fig. 15). This allows the simulation of the motion of each of the parts with six degrees of freedom (linear displacement on each of the XYZ axes and rotation on them). With this method we want to simulate the regular operation that runs this station in a virtual environment, and at the same time, through this document, the step-by-step description to get it on any station modeled for this manufacturing cell (Fig. 15).

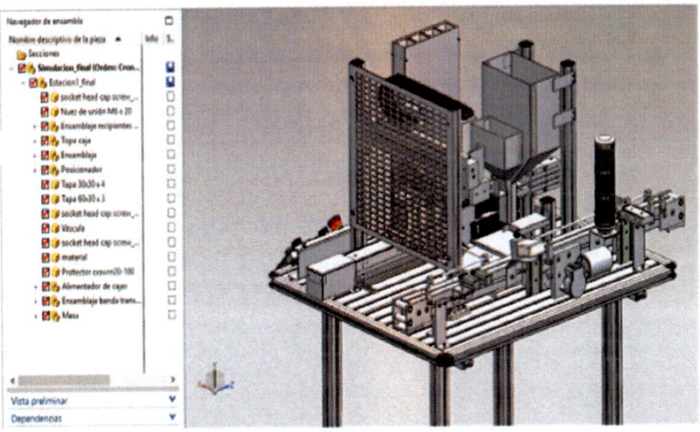

Fig. 15. Model stations 1, 2 and 3 of the HAS 200 manufacturing cell within the Siemens NX® program environment. Source: author.

To perform the simulation, it must be clear the process and the elements that interact with each other to perform the operation, which in this case consists of dispensing pellets into containers that come from a storage box, from where it is driven by four actuators that hold the container with suction cups to the conveyor belt for reading the barcode. It is then driven under the dispenser, positioned on top of a scale for filling and weighing coloured pellets (yellow, blue and red). Finally it is taken to the conveyor belt to be taken to the next station where the next process is performed. With this clear, we proceed to identify each of the movements that will be described within the simulation programming.

For citations of references, we prefer the use of square brackets and consecutive numbers. Citations using labels or the author/year convention are also acceptable. The following bibliography provides a sample reference list with entries for journal articles [1], an LNCS chapter [2], a book [3], proceedings without editors [4], as well as a URL [5].

3 Conclusions

As a result of this project, the University now has the fully documented and functional virtual environment of the HAS-200 manufacturing cell, which makes it possible to multiply in academic terms this resource of high technology standards.

The implementation of this project made it possible to consolidate and use the knowledge acquired in the field of development of high-engineered models because from the beginning a practical and efficient methodology based on the appropriate use of standardized element models that are freely available.

It should be noted that manufacturing cells are a great tool at the academic level, since they cover several branches of engineering, facilitating the learning of concepts of industry that are difficult to reach in practice, allowing the observance of the applicability of theoretical notions and extending the functionality of the tools acquired by the University for teaching purposes, triggering a series of projects generated by the University, starting from information about products already designed to improve and create them.

The models supplied to the University of the HAS-200 manufacturing cell were made using Solid Works® software, a program that offers great flexibility in communication between CAD files due to the possibility of transforming various formats, allowing them to be opened in different software, not only for the creation and editing of models within the virtual environment, but also for the simulation of the variables that want to be applied to the processes carried out by each station.

Although the above 3D simulation software included in the manufacturing cell HAS-200 by the manufacturer SMC® and the drawings with the references of its components, does not offer the multiple advantages of the virtual model carried out in this project, if the 3D models of the components used in the assembly of the stations were very useful, which are provided free of charge on the website of this company. Of the 30 different actuators that are used by the cell distributed in each of the stations, 28 were unloaded, making only 2 actuators required full modelling.

Several tools were very useful for the development of this project, without a doubt, the mechanical drawings that are based on the database of the laboratories of the University, made it possible to speed up the search for components referenced with serial numbers, which made tangible the 3D models for any CAD software, complementing it together with the knowledge of the operation of each of the stations for the modeling of the connecting parts of the different subassemblies, reducing the task of dimensioning for the positioning of each of its parts is agile.

Laboratory verification of the unloaded parts was important in two aspects. The first has to do with the assertive dimensions of the components that had the different download pages, although it was not iterative, some pages offered serial numbers of wrong components, but it was possible to verify by downloading product catalogs that the error, was of these pages, making some sources more reliable than others, being able to establish the download system 2D/3D CHANNES from SMC as fully assertive in this type of online products. The second aspect was presented after downloading all the actuators from the list of mechanical drawings delivered to the laboratories of the University, during the sizing cycle, for assembly of each of the stations, the installation of each of these actuators was verified, finding that three non-matches in the real area, specifically in the vertical and horizontal warehouses, added to the lack of knowledge and lack of information from the administrators of the laboratory in which this manufacturing cell is located about these variants.

References

1. Magana, A., Ortega, J., Lovan, R., Gómez, D., Marulanda, J., Dyke, S.: Virtual, local and remote laboratories for conceptual understanding of dynamic systems. Int. J. Eng. Educ. **33**(1 (A)), pp. 91–105 (2017)
2. Valencia, M.: Las nuevas tecnologías de la información y la comunicación y el aprendizaje activo: Experiencia con el curso de metodologías de multimedia. Revista Ingeniería y Competitividad **3**, 73–82 (2001)
3. Sin marcha atrás: La automatización será una realidad en Colombia en 2020 (semana.com). https://www.semana.com/. Accessed 07 July 2021
4. Minian, I., Martínez Monroy, Á.: El impacto de las nuevas tecnologías en el empleo en México. Problemas del desarrollo, **49**(195), 27–53 (2018). https://doi.org/10.22201/iiec.20078951e.2018.195.64001
5. HAS-200 - Sistema altamente automatizado (smctraining.com). Accessed 07 July 2021

Introduction of Online Labs to Enhance the Quality of the Real-Time Systems Course

Marjan Alavi(✉) and Seshasai Srinivasan

McMaster University, Hamilton, ON L8S4L8, Canada
{alavis2, 2ssriniv}@mcmaster.ca

Abstract. This paper presents the implementation of the fully online Real-time systems course in the Software Engineering Technology program. The highlight of this course is the improvement by including MATLAB Simulink labs to give the students hands-on training and enhance their online learning experience. This is a very pertinent strategy because students enrolling in these programs have a natural aptitude for experiential learning and insist on seeing the applicability of concepts to real problems. In presenting the evolution of this course from no-labs to several online labs, we demonstrate the positive impact this has had on student learning.

Keywords: Real-time systems · Online labs · Constructivist theory of learning

1 Introduction

Providing quality and accessible education is a critical challenge in the 21st century. To offer quality education, a large bulk of education research is focused on innovative pedagogical approaches that will maximize student learning. Some of the popular pedagogical practices include active learning [1–6], problem-based learning [7–12], cooperative and small group learning [13], inquiry-based learning [14], and undergraduate research-based learning [15, 16], and problem-solving based approach [17].

In disciplines such as engineering, to ensure that the graduates are competent to solve society's emerging problems and are readily employable in the industry, the rigor and quality of training are important. To this end, in addition to strong theoretical foundations on concepts and technical principles, it is important for the students to have adequate training in applying the skills. In an educational setting, this is generally accomplished by conducting design, development, and experiments through labs. More precisely, it is common practice for most topics in engineering to have a course and lab component.

Traditionally, as part of lab exercise, students participate in individual or group experiments wherein a certain set of concepts taught in the theoretical component of the lecture is put to practice. This education format resonates with the constructivist theory of learning [6, 18–20] that dictates that with exposure to experiences relating to the concepts and their applications, the students will develop a strong mental construction of the theoretical principles. In recent years, for optimal content delivery and flexibility in learning, online labs have gained significant momentum [21–26]. This is even more

M. E. Auer and T. Tsiatsos (Eds.): IMCL 2021, LNNS 411, pp. 856–864, 2022.
https://doi.org/10.1007/978-3-030-96296-8_77

significant with the advent of the COVID-19 pandemic that has caused an immense disruption of the education systems worldwide. The pandemic caught numerous institutions off-guard and brought to fore the deficiencies in infrastructure, technologies, and the nimbleness required to adapt to the challenges [27–30]. Since then, several universities have adapted to online learning and are making amendments to ensure that they can operate safely without disrupting the students' education and graduation timelines.

The *Software Engineering Technology* program at the W Booth School of Engineering Practice and Technology at McMaster University has been offering online courses since 2017, and so was fully prepared for the pandemic. In this program, the students could graduate with competencies in the areas of network security, software engineering, artificial intelligence and data science, and digital arts and multimedia. This work focuses on the online implementation of one of the courses in this program, i.e., *Real-time Systems*. This course is taken by the students enrolled in level 4 of the program, and they learn about dynamic responses of physical processes and apply scheduling techniques to solve multitasking. In the following sections, we present the online format of this course, including the evolution from no-lab to introducing the MATLAB Simulink online labs to give students hands-on training.

The rest of the paper is organized as follows: Sect. 2 presents the Materials and Methods. In Sect. 3, we discuss the impact of online labs on student learning. Finally, we draw pertinent conclusions and present the future course of evolution in Sect. 4.

2 Materials and Methods

2.1 Integration into the Course Outline

The course was offered in Winter 2020 for a duration of 14 weeks, with the class meeting for 3 h every week. In Winer 2021, as an evolution of the course, five labs were added. Table 1 shows compares assessment components in Winter 2020 and Winter 2021 offerings. The labs were designed for 60–90 min and performed individually by students right after each lecture in Weeks 3, 4, 6, 7, and 8. Each lab was worth 3% weight in the final grade.

2.2 Lab Materials and Prerequisite

The labs were performed using MATLAB Software. The following MATLAB toolboxes were used in the labs: Control Systems Toolbox, Signal Processing Toolbox, and the Audio Signal Processing Toolbox. The students enrolled in this program are usually unfamiliar with MATLAB, and therefore Lab 1 was designed to introduce MATLAB and Simulink environment to the students. The individual lab experiments are summarized below.

Lab 1: Getting Started with MATLAB: Lab 1 introduces the MATLAB and Simulink Software environment. The students learn the basics of the MATLAB language and define matrices and apply basic element-wise operations on the matrices. Additionally, they also learn to work with the Source and Sink blocks in the Simulink

Table 1. Assessment components in Winter 2020 and Winter 2021 offerings.

Assessment component	Weight in Winter 2020 offering	Weight in Winter 2021 offering
Assignments	15%	5%
Labs		15%
Quizzes	5%	
Mid-term test	30%	
Term-Test 1		15%
Term-Test 2		30%
Project	10%	
Final examination	40%	35%

and adjust sinusoidal and pulse waveform generators. They are also taught to work with the oscilloscope and adjust its channels to observe the simulation results.

Lab 2: Analyzing Transfer Function Models in MATLAB: This lab introduces the MATLAB Control Toolbox. Specifically, this lab helps students understand the calculation of the system's transfer function that is modeled by a differential equation. The students create the system's transfer function in the MATLAB environment with a few lines of code built on Lab 1. Then they use Control Toolbox functions like *step(H)* and *isstable(H)* to check the system's stability and plot the step response of the system. They use the Oscilloscope measurement tools to measure the steady-state, overshoot, undershoot, and constant time of the system. They will then apply different input signals to evaluate the system's response.

This lab gives the students a system view and a quick introduction to the Control Theory. They will learn the importance of the Laplace transform and its application in the simulation. They also learn how to model any LTI system knowing the differential equation model of the system. The duality of mechanical and electrical systems is discussed in the lecture. The students are then encouraged to change the value of the constants and observe the changes in the time constant of the system. Knowing a basic understanding of a system's time constant and where it is coming from is essential in designing optimized real-time systems.

Lab 3: Analyzing Discrete-Time Models in MATLAB: Today, most real-time systems are controlled by microcontrollers which are digital systems, while the physical phenomena and components under control are continuous ones. So, we must use Analog to digital converters (ADC or A/C) to convert the continuous signal into a digital form. The controllers themselves are also digital controllers, which are implemented in the microcontrollers. Hence, basic knowledge about digital control systems and digital signal processing is necessary for the student who would work with the digital real-time system.

To meet this requirement, in Lab 3, MATLAB Digital Signal Processing Toolbox is introduced. The students learn how to create and analyze discrete systems in MATLAB. They calculate the transfer functions of three digital filters (Fig. 1) and make a model of the system in the Simulink. Further, they also study the effect of this system on a given discrete signal.

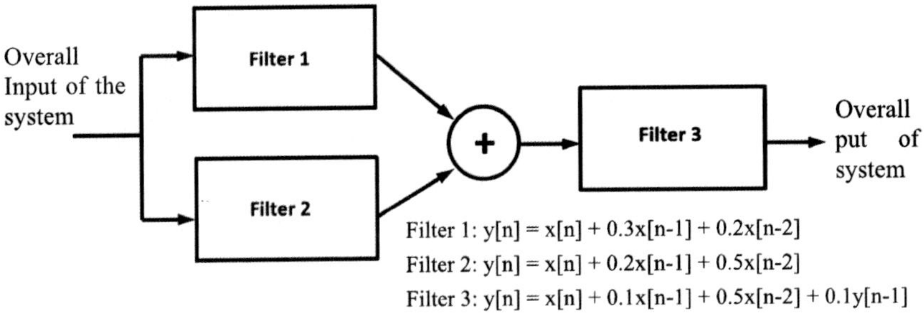

Fig. 1. Model of the discrete system used in Lab 3.

Lab 4: Audio Signal Processing in MATLAB: The students implemented an audio signal processing system in this lab. In this system, the student's microphone provides audio input, and the student's speaker was used to generate the system's output. The computer's CPU was used to process the audio signal using the digital filers the students learned to build in the previous labs. The computer's CPU is not the best option for implementing a real-time system because it must attend to many tasks (for example, the MS Team calls during class time).

Figure 2 depicts the Simulink model for implementing Allpass and Comb filters in a Reverberator application. This is an example of hardware-in-loop where the real hardware components were used in combination with Simulink model.

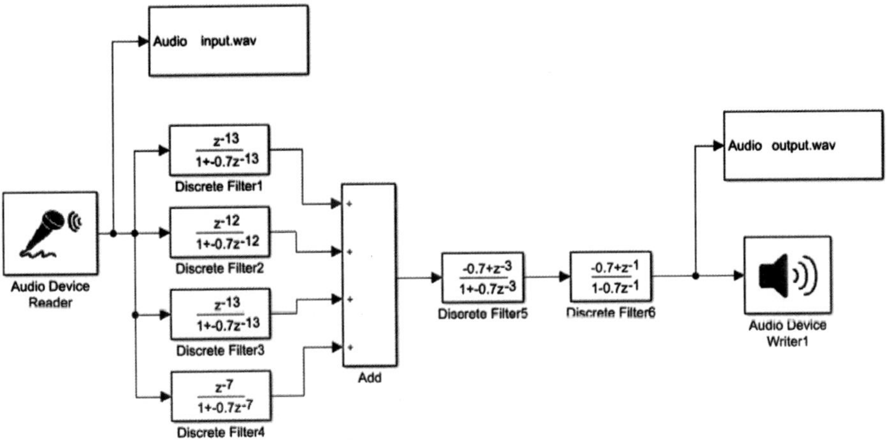

Fig. 2. Implementation of a Reverberator filter for processing audio signals in Lab 4.

The students were asked to close the unnecessary applications during this lab, so the CPU can spend most of its time processing the Audio signal. This lab is an example of hardware-in-loop where the real hardware components were used in combination with the Simulink model. The students were asked to adjust the filter's parameters to obtain a noticeable reverb in their voice. Since the students were using different computers, the answer to this lab was different for each student. Finally, the lab asked questions about the sampling time, sample per frame, and buffer size. The students would hear the effect of inappropriate sampling in the resulting audio signal.

Lab 5: Introduction to MATLAB Stateflow: This lab was an introduction to MATLAB state flow. The students simulated a sequential decision logic system with MATLAB Stateflow and simulated a full-bridge converter. The sequential system was designed in the Stateflow and imported to Simulink. The response of the system to a sinusoidal waveform was observed.

Thus, the overall structure of the course was such that students were able to see the application of the concepts in real applications. This constructivist approach in which students undertake the application of the principles and the consequent opportunity they get to interact with peers while performing each lab helps create unique student experiences. Such experiences help develop a strong mental construction of the concepts and enable the student to develop a good understanding of the principles.

3 Results and Discussions

In this section, the students' performance in the labs is discussed. Figure 3 shows the number of students who completed each lab successfully and achieved full marks in the lab. The results show a consistent improvement from Lab 1 to Lab 4. These labs were closely related to each other, and the experience gained in previous labs was used to succeed in the following ones. In other words, the experiential learning via these labs had a cumulative effect in improving the learning of the students.

Halfway through the labs, to measure the students' understanding of the course and give them quick feedback, an online quiz was attached to Lab 3. In this, their labs results were verified through a set of online multiple-choice questions. More precisely, the quiz had four questions, and the students could attempt the quiz three times. While the students were not provided with the details of the assessment, they could see their final scores on the quiz. In addition to answering the quiz questions, similar to the other labs, the students were asked to submit a lab report with the screenshots of their MATLAB Simulink model and results.

Table 2 summarizes the results from Lab 3. More than 57% of the students who tried the quiz answered all the questions correctly. This is an indication of doing a lab entirely without any mistakes. About 23% of the students got an incomplete answer in their first attempt. However, they were able to identify the problem either by approaching the instructor, discussing with their peers, or reviewing their models and correcting their errors in the second trial. There were only 10% of students who used all three attempts to achieve the full mark. The rest of the students who did not try the quiz or failed to achieve the full mark either failed the course or got less than 54% in their final grade.

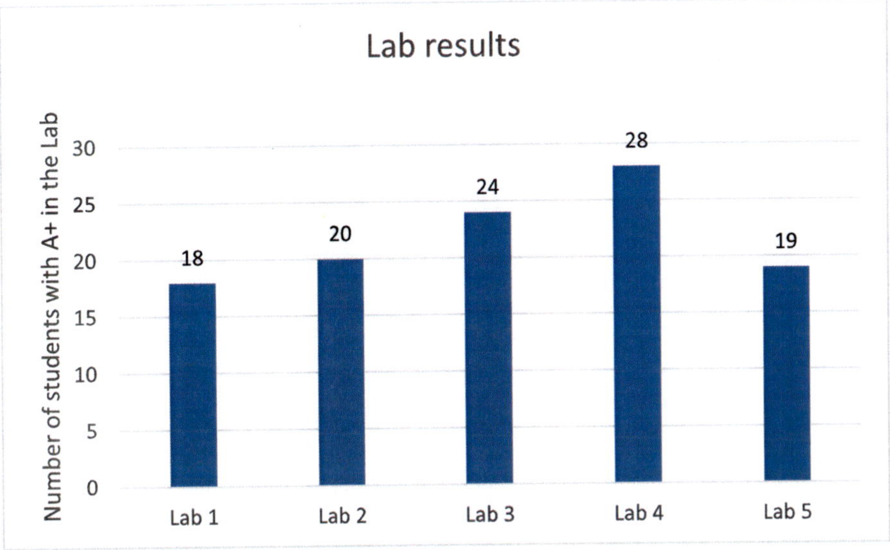

Fig. 3. The number of students who achieved the full mark in each lab. There was a total of 30 students who took the course in 2021.

Table 2. Lab 3 results.

Description	Number of students
Students who tried the quiz	30
Students who answered all the questions correctly in the first attempt	17
Number of students who answered all questions correctly within two attempts	7
Students who answered all questions correctly within three attempts	3
Students who failed to get full marks in the quiz	3

Unlike Lab 3, no quiz was attached to any other labs. In Lab 4, the final result of the lab was pretty clear to all students since the lab was based on a real-life application familiar to the students. The lab also involved real-world audio signals, which was exciting for the students. With about 93% of the students getting an A + in this lab, it is clear that the students were able to develop a good understanding of the various concepts that were taught through the previous lectures and labs that connected or led to this lab. It could be argued that experiential learning.

However, the number of students who achieved A + in Lab 5 dropped significantly compared to Lab 4. This can be attributed to the fact that this lab was not directly relevant to the previous labs and covered a new topic. The application considered for this lab was also not well known for many students. Although it was explained during the lab session, some students could not judge the correctness of their result and took the iteration process they experienced in Labs 3 and 4 to correct their mistakes. Another possible reason for the poor performance in this final lab is related to the possibility that with their good performance in the first four labs and knowing that the labs are worth only 15% of the course, the students might have chosen to allocate their time to optimize their performance in other classes. This is consistent with the propositions in the literature [4, 31].

Comparing each student's Term-Test 1 and final exam reveals an average growth of 3.1% in the Winter 2021 offering of the course. This rate was -1.6% in Winter 2020 (From midterm to final exam). This growth is while the evaluations were more difficult with an overall average grade of 72% compared to 79% in the previous offering of the course. The addition of the online labs also improved student retention in the course; In Winter 2020, four students dropped the course, no withdrawal was observed in Winter 2021.

4 Conclusion

In this work, we present the introduction of online labs in a real-time systems course at the W Booth School of Engineering Practice and Technology at McMaster University. Specifically, five labs were introduced. Except for the 5th lab that was focused on a completely different application, the first four labs built on the previous topics. Participation in the labs increased the interactions between the instructor and the students, kept the class motivated, and brought peer-to-peer learning opportunities. The students could experience the application of the theoretical principles they learned in this class to real-life applications, reinforcing their learning. The students showed better performance in the labs, which were gradually built upon each other. The relevance of the lab experiment to real-world daily applications is another factor that increases the student's attention and performance in a lab. Attaching quizzes with multiple attempts to the labs has proven to increase the success rate in the labs, brings learning opportunities, and motivates the students to keep working towards improved learning outcomes. The addition of online labs improved student's growth and retention in the online course.

References

1. Beichner, R.: The student-centered activities for large enrollment undergraduate programs (SCALE-UP) project (2007)
2. Srinivasan, S., Centea, D.: An active learning strategy for programming courses. In: Auer, M., Tsiatsos, T. (eds.) IMCL 2018. AISC, vol. 909, pp. 327–336 (2019). Springer, Cham (2018). https://doi.org/10.1007/978.3-030-11434-3_36
3. Cummings, K., Marx, J., Ronald, T., Dennis, K.: Evaluating innovation in studio physics. Am. J. Phys. **67**, S38–S44 (1999). https://doi.org/10.1119/1.19078
4. Sidhu, G., Srinivasan, S.: An intervention-based active-learning strategy to enhance student performance in mathematics. Int. J. Pedagogy Teach. Educ. **2**, 277–288 (2018). https://doi.org/10.20961/ijpte.v2i1.19568

5. Srinivasan, S., Centea, D.: Applicability of principles of cognitive science in active learning pedagogies. In: Proceedings of the 13th International Workshop Active Learning in Engineering. (1 ed.) Aalborg Universitetsforlagpp, pp. 99–104 (2015)
6. Srinivasan, S., Muhammad, N.: Implementation of a course in computational modeling of biological systems in an undergraduate engineering program. Int. J. Eng. Educ. **36**(3), 857–864 (2020)
7. Kolb, D.A.: Experiential Learning: Experience as the Source of Learning and Development, 2nd edn. Pearson Education Inc. (2015)
8. Centea, D., Srinivasan, S.: Enhancing Student Learning through Problem Based Learning. In: Guerra, A., Rodriguez, F.J., Kolmos, A., Reyes, I.P. (eds.) PBL, Social Progress and Sustainability. Aalborg Universitetsforlag. (International Research Symposium on PBL), Aalborg, pp. 376–385 (2017)
9. Dochy, F., Segers, M., Van den Bossche, P., Gijbels, D.: Effects of problem-based learning: a meta-analysis. Learn. Instr. **13**, 533–568 (2003). https://doi.org/10.1016/S0959-4752(02)00025-7
10. Centea, D., Srinivasan, S.: A comprehensive assessment strategy for a PBL environment. Int. J. Innov. Res. Educ. Sci. **3**(6), 364–372 (2016)
11. Sidhu, G., Srinivasan, S., Centea, D.: Implementation of a problem based learning environment for first year engineering mathematics In: Guerra, A., Rodriguez, F.J., Kolmos, A., Reyes, I.P. (eds.) PBL, Social Progress and Sustainability, Aalborg: Aalborg Universitetsforlag. (International Research Symposium on PBL), pp. 201–208 (2017)
12. Srinivasan, S., Centea, D.: Problem based learning in finite element analysis. In: Auer, M.E., Centea, D. (eds.) Visions and Concepts for Education 4.0, ICBL 2020. AISC, vol. 1314, pp 1–7., Springer Nature, Switzerland (2021)
13. Wage, K.E., Buck, J.R., Wright, C.H.G., Welch, T.B.: The signals and systems concept inventory. IEEE Trans. Educ. **48**, 448–461 (2005). https://doi.org/10.1109/TE.2005.849746
14. Lewis, S.E., Lewis, J.E.: Departing from lectures: an evaluation of a peer-led guided inquiry alternative. J. Chem. Educ. **82**, 135 (2005). https://doi.org/10.1021/ed082p135
15. Srinivasan, S., Rajabzadeh, A.R., Centea, D.: A project-centric learning strategy in biotechnology. In: Auer, M.E., Hortsch, H., Sethakul, P. (eds.) ICL 2019. AISC, vol. 1134, pp. 830–838. Springer, Cham (2020). https://doi.org/10.1007/978-3-030-40274-7_80
16. Bogoslowski, S., Geng, F., Gao, Z., Rajabzadeh, A.R., Srinivasan, S.: Integrated thinking - a cross-disciplinary project-based engineering education. In: Auer, M.E., Centea, D. (eds.) ICBL 2020, vol. 1314, pp. 1–7. Springer, Cham (2020). https://doi.org/10.1007/978-3-030-67209-6_28
17. Muhammad, N., Srinivasan, S.: A problem solving based approach to learn engineering mathematics. In: Auer, M.E., Hortsch, H., Sethakul, P. (eds.) ICL 2019. AISC, vol. 1134, pp. 839–848. Springer, Cham (2020). https://doi.org/10.1007/978-3-030-40274-7_81
18. Cristea, S.: The fundamentals of constructivist pedagogy. Procedia Soc. Behav. Sci. **180**, 759–764 (2015)
19. Srinivasan, S., Muhammad, N.: A constructivist approach for mathematics education'. J. Educ. Pedagogy **12**, 1–5 (2020)
20. Olusegun, B.S.: Constructivism learning theory: a paradigm for teaching and learning. ISOR J. Res. Method Educ. **5**(6), 66–70 (2015). https://doi.org/10.9790/7388-05616670
21. Auer, M., Pester, A., Ursutiu, D., Samoila, C.: Distributed virtual and remote labs in engineering. In: 2003 IEEE international Conference on Industrial Technology, vol. 2. IEEE (2003)
22. Heradio, R., de la Torre, L., Galan, D., Cabrerizo, F.J., Herrera-Viedma, E., Dormido, S.: Virtual and remote labs in education: a bibliometric analysis. Comput. Educ. **98**, 14–38 (2016)

23. Ravio, S., Ruutmann, T., Seiler, S.: Inductive teaching and learning in engineering pedagogy on the example of remote labs. Int. J. Eng. Pedagogy iJEP **4**(4), 12–15 (2014)
24. Nedungadi, P., Malini, P., Raman, R.: Inquiry based learning pedagogy for chemistry practical experiments using OLabs. In: El-Alfy, E.-S., Thampi, S.M., Takagi, H., Piramuthu, S., Hanne, T. (eds.) Advances in Intelligent Informatics. AISC, vol. 320, pp. 633–642. Springer, Cham (2015). https://doi.org/10.1007/978-3-319-11218-3_56
25. El Kharki, K., Berrada, K., Burgos, D.: Design and implementation of a virtual laboratory for physics subjects in moroccan universities. Sustainability **13**, 3711 (2021). https://doi.org/10.3390/su13073711
26. Alavi, M.: Shared remote lab. In: Auer, M.E., Centea, D. (eds.) ICBL 2020. AISC, vol. 1314, pp. 392–399. Springer, Cham (2021). https://doi.org/10.1007/978-3-030-67209-6_42
27. Bao, W.: COVID-19 and online teaching in higher education: a case study of Peking University. Human Behav. Emerg. Technol. **2**(2), 113–115 (2020). https://doi.org/10.1002/hbe2.191
28. Carroll, N., Conboy, K.: Normalising the "new normal": Changing tech-driven work practices under pandemic time pressure. Int. J. Inf. Manage. **55**, 102186, (2020). doi.https://doi.org/10.1016/j.ijinfomgt.2020.102186
29. Zhang, W., Wang, Y., Yang, L., Wang, C.: Suspending classes without stopping learning: China's education emergency management policy in the COVID-19 outbreak. J. Risk Finan. Manage. **13**(3), 55 (2020). https://doi.org/10.3390/jrfm13030055
30. Lassoued, Z., Alhendawi, M., Bashitialshaaer, R.: An exploratory study of the obstacles for achieving quality in distance learning during the COVID-19 pandemic. Educ. Sci. **10**(9), 232 (2020). https://doi.org/10.3390/educsci10090232
31. Love, D., Kotchen, M.J.: Grades, course evaluations and acadmic incentives. East. Econ. J. **36**, 151–163 (2010)

Development and Use of IoT Boards for Online Labs

Dan Centea[1](✉), Reiner Schmidt[2], Anoop Gadhrri[1], and Ishwar Singh[1]

[1] McMaster University, Hamilton, ON, Canada
centeadn@mcmaster.ca
[2] Roboteurs Inc, Hamilton, ON, Canada

Abstract. The COVID-19 pandemic plunged academia into a virtual learning environment. Internet became essential in delivering lectures, tutorials, and labs. Implementing experiential learning activities in engineering education became a challenge. This paper presents an approach that allows students to develop experiential learning skills related to IoT in an online lab environment and describes an approach to teaching IoT communication technologies using two electronic boards. These boards allow students to learn IoT concepts such as sensors and interfacing, to develop application using IoT communication technologies, and to run applications on several platforms. The paper describes these boards and their use in labs focused on smart systems, and in advanced-level labs.

Keywords: IoT board · IoT applications · Learning factory · Online laboratory · Online learning

1 Introduction

The pandemic that started in 2020 imposed the adoption of new pedagogies influenced by online learning experiences gained by instructors and students. Internet resources suddenly became extremely important in delivering learning modules such as lectures and online lab demonstrations and allowed learning-on-demand approaches through recorded lectures, online lab experiments, lab equipment accessed remotely, virtual demonstration labs, and virtual tutorials. Therefore, the focus on implementing experiential learning activities in academia became a challenge for engineering education.

The School of Engineering Practice and Technology (SEPT) at McMaster University established the framework for a cyber-physical system learning centre [1] followed by building a Learning Factory (LF) [2] that focuses on Industry 4.0 technologies and uses strong experiential learning approaches in their implementation and operation [3]. This LF is used for teaching students, training employees, and conducting applied research [4]. Although several products were designed and manufactured, delivering laboratory experiments with sections that include mechanical and electronic components manufactured and assembled in the LF became very challenging in a fully-online environment. The goal of this paper is to present a way of addressing

M. E. Auer and T. Tsiatsos (Eds.): IMCL 2021, LNNS 411, pp. 865–873, 2022.
https://doi.org/10.1007/978-3-030-96296-8_78

this challenge through the design, development and use of an electronic board with Internet of Things (IoT) capabilities that allows students to run online labs remotely.

The motivation of the paper is to describe a change of practice from in-person laboratory activities related to the design, manufacturing, testing, and using an electronic board to a combination of online activities and physical work in the LF. The goal of this paper is to assess if this approach provides a good experiential learning activity to students. The design part of learning IoT concepts could be done remotely by students using the remote access technology and infrastructure; the manufacturing of the board could be done in the LF by people with access to the facility; while the testing and app development could be done by students in an online environment. A second motivation of the paper is to suggest a set of open-source and specialized electronic modules that can be efficient in developing IoT application for home automation, and to briefly describe to anyone interested in teaching IoT concepts a series of experiments that use these boards for basic or advanced experiments.

Over the years there have been many successful IoT implementations that use microcontrollers or single-boards computers for IoT applications such as sensing and monitoring climate and air quality, embedded control systems, health care, smart systems, industrial and home automation, and surveillance systems. A comprehensive sensing platform composed of hardware consisting of an ultra tiny all-in-one sensor board, software (iOS, Android, and PC), and 3D case data used to start IoT research in described in [5]. An analysis of single-board computers, such as Raspberry Pi, for IoT and Industrial Internet of Things (IIoT) solutions in embedded control systems is presented in [6]. Test benches based on microcontrollers used for Industry 4.0 education for face-to-face and distance learning are presented in [7]. IoT devices based on Arduino Uno connected to health-care sensors are presented in [8]. A system for monitoring the performance of indoor micro-climate horticulture using sensors connected to an IoT board that is implemented to monitor the cultivation process is presented in [9]. An IoT real time bolt system used to monitor the indoor air quality is presented in [10]. An IoT home automation and surveillance system that uses a Raspberry Pi-3 board is presented in [11]. A smart home automation system that uses an IoT sensing and monitoring system is presented in [12]. A comparison of various IoT-capable boards and a summary of the capabilities of hardware development platforms for IoT is presented in [13]. A survey of the available IoT platforms is presented in [14].

The applications and IoT platforms listed above use commercial microcontroller boards integrated in specialized IoT systems. The designers of these systems select a commercial board, adopt sensing and monitoring devices and select communication platforms, while the users of these systems employ them for their designed purpose. Neither the designers nor the users of these systems can change the microcontroller board or can be involved in its manufacture and testing. This paper describes two IoT boards, designed and manufactured through a cooperation between academia and a small company, that let students be involved in the design, manufacturing and testing processes of the board, and allows them to develop applications and use the board in face-to-face or online environments. The MacIoT board is described in Sect. 2, and the way this board is used for learning is described in Sect. 3. Section 4 describe an advanced board for learning IoT concepts, called MacIoT Trainer board, while Sect. 5 describes its use for learning. Section 6 presents a summary of the paper.

2 MacIoT Board

The SEPT LF provides modern design, prototyping, and manufacturing processes that incorporate IoT, IIoT, and Industry 4.0 technologies. Several products were designed and manufactured to demonstrate these technologies. One of the devices is a micro-controller, called MacIoT board, that can be designed outside the LF and then proto-types built, assembled, and tested in the LF. The design part of the learning can all be done remotely using the remote access technology and infrastructure.

The MacIoT board includes several sensors, a few on/off switches, and LED out-puts to simulate simple processes with digital inputs and digital outputs, and it is designed to be a single platform for teaching many IoT related courses. First year students use the board to gain experience in soldering surface mounted devices and gain experience in selecting and sourcing electronic components. Upper-year students use the board as a platform to design and develop quality testing methods and to develop automated testing systems. Furthermore, the board provides opportunities for software and hardware integration and for gaining coding experience expected in many industrial jobs.

A key aspect of the design of the MacIoT board is having sensors and input/output components on the board so that no wiring is required to learn the IoT concepts. An important design element of this board is a flexible interface used to add components and interfaces without any wiring; examples include a motor driver, a Low Range (LoRa) communication module, a weather sensor, biomedical sensors, and a Controlled Area Network (CAN bus) communication module. Furthermore, the MacIoT board has Bluetooth and Wi-Fi capabilities making it an interactive experiential learning platform for todays and tomorrows IoT concepts.

The MacIoT board allows students to explore and experiment with IoT technology on the go. The board is designed around the ESP32 chip from Espressif. This allows the board to have Wi-Fi Bluetooth Low Energy (BLE), and Bluetooth. Along with these IoT communication protocols, the board also has a series of sensors and peripherals for interfacing with the real world.

The MacIoT board is an open-source hardware with a design made available to anyone to explore and modify as they wish. The open-source design allows students and educators to explore the design and re-implement it in their own applications while continuing to share their explorations with others.

To help expand students' knowledge of IoT design beyond just putting wires and code together, the MacIoT board is also designed to be manufactured in the SEPT LF where students can be a part of the fabrication process. This gives them direct exposure to many different aspects of product life cycles.

The MacIoT board, designed in a partnership between McMaster University and Roboteurs, Inc., provides students a dynamic way of exploring the IoT. With all the rapid advances IoT technology, it has become very difficult to teach students its concepts as many overarching concepts are required for IoT development. Instead of trying to squeeze IoT development into one course, this board is given to students in the first year to incorporate it in different aspects of learning. Examples include using the board to collect sensor data for analysis in a statistics class; exploring numerical methods and C + + programming; understanding basic electronics; incorporating the board in mechatronics systems; and using the board for any project that requires development of embedded systems.

The MacIoT board has a standard Arduino footprint. The hardware features show in Fig. 1 include an ESP32 Microcontroller, a programmable USB, a light-dependent resistor with bias for light sensing, a battery input, four user LEDs, three user switches, and two buttons for resetting and flashing.

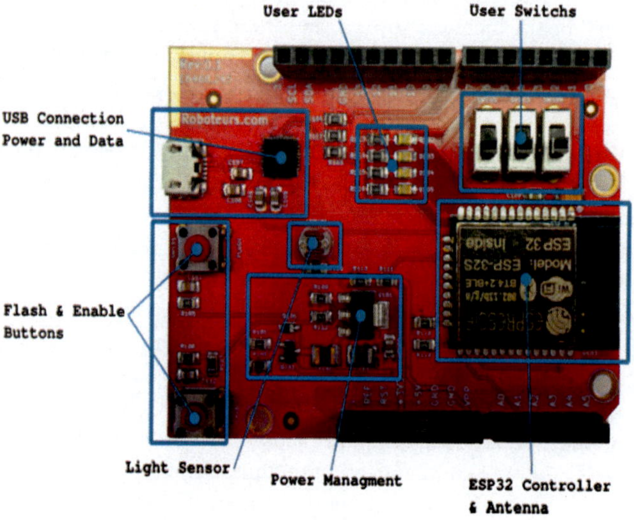

Fig. 1. MacIoT board overview

3 Learning with the MacIoT Board

Any university can access the open-source design, manufacture the board, and use it for basic and advanced learning projects. Examples of basic projects include flashing a LED; controlling a LED using a switch; controlling LEDs with a light sensor; and controlling LEDs through a standalone web server using Wi-Fi. Examples of advanced learning projects include collecting vibration data using accelerometer using I2C communication protocol; designing a Bluetooth application; developing a Proportional, Integral, and Derivative (PID) temperature controller; and developing a machine health monitoring model for data collection, performing FFT on the data, developing a neural networking model for machine health prediction; and building a Bluetooth beacon with the MacIoT acting as server and a smartphone acting as client.

The MacIoT board is currently used for the experiential learning component of an IoT course. Examples of experiments that developed with the board include publishing and subscribing messages to the real-time communication platform PubNub and creating a web app; controlling LEDs and a servo motor remotely using the PubNub platform; using the MacIoT board as sensor node that works with a Raspberry Pi that communicates with PubNub for viewing radio-frequency identification (RFID) tag values captured by the MacIoT board; using the board to send telemetry data to the Google Cloud IoT core; and sending analog values to LabView and to the ThingSpeak platform.

App development is an important feature that students need to grasp to be ready for future employment in specialization fields such a mechatronics, automation, embedded systems, and software engineering. The MacIoT board is used to develop Android apps. A simple way to develop android apps is by using the Android Studio software. An example of an application developed with Android Studio is displaying the status of the LEDs and switches on the MacIoT board. An android apps can be written using one of several programming languages such as Kotlin, Java, or C + +. The app described here was written Java. The board is connected to the UniSphere I middleware platform using Wi-Fi. The status of the LEDs on the MacIoT board is displayed along with the output from temperature, light, and touch sensors, as shown in Fig. 2.

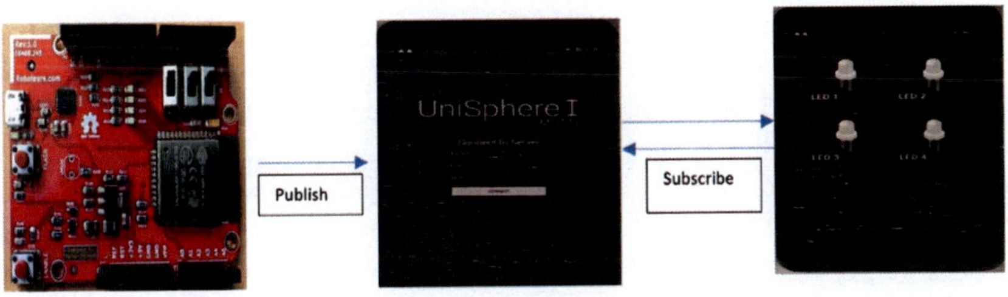

Fig. 2. Flow of information for an Android IoT app

4 MacIoT Trainer Board

A MacIoT Trainer board has been developed to expand the IoT features of the MacIoT board. The features of this board include switches and buttons used for simple inputs for fast prototyping of logic, an I2C chip that allows multiple I2C connections for simple and easy sensor expansion, an ESP32 microcontroller that provides connectivity over Bluetooth and Wi-Fi, a gyroscope and accelerometer combo used to determine the orientation of the device and its estimated motion in space, a sensor that can provide the temperature and humidity levels in the surrounding environment, and a diode based light sensor that provides the light intensity of the environment. A schematic of the MacIoT Trainer board is depicted in Fig. 3.

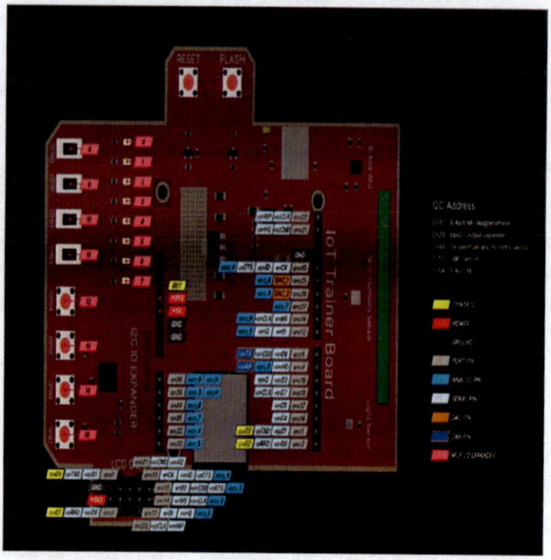

Fig. 3. MacIoT Trainer board

5 Learning with the MacIoT Trainer Board

Similar to the MacIoT board, the MacIoT Trainer board includes features that allows it to work in a fully online environment. The students who purchase the board for a low cost can use it at home, where they experience its features, expand its connectivity, and develop personal IoT, automation and mechatronics projects based on the additional connectivity boards that they purchase. The three specialized hardware boards that can currently be connected to the MacIoT Trainer board include a LoRa gateway board, a CAN bus transceiver boards, and an IoT LCD display module.

Fig. 4. MacIoT Trainer board connected through USB

The students who do not want to purchase the MacIoT Trained board to get experience on their own time and develop home projects can use the board remotely by connecting from their home to a remote machine installed in the IoT lab. These students use the remote machine as their own computer and use a web camera to look at the MacIoT Trainer board. They can install software, such as platform.io, and connect the MacIoT board with any of the specialized boards mentioned above. Students can communicate with the LoRa gateway and control communication from the remote desktop; can connect to a CAN network used to broadcast data; and can display information on the IoT LCD display module. It is possible to setup a "farm" configuration for multiple students in which they can all connect to the same CAN network. One of the challenges of this configuration is that the MacIoT Trainer board is connected to the computer through a USB connection, as shown in Fig. 4; due to the limitations of the USB, the software must run on a single PC, as shown in Fig. 5, and not on a server.

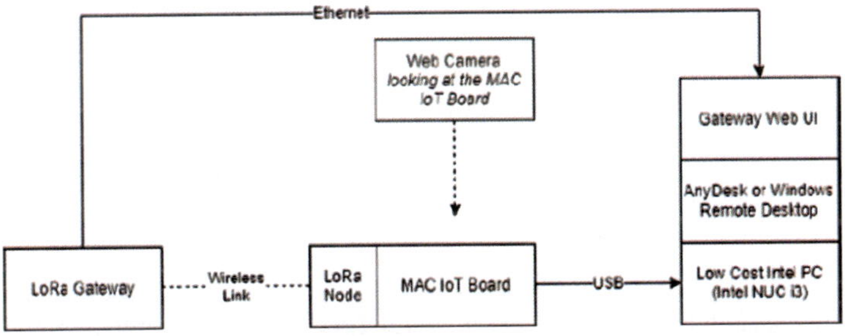

Fig. 5. Equipment setup for running remote labs with the IoT Trainer board

The MacIoT Trainer board configuration system shown in Fig. 5 is currently used for two sets of lab experiments. The six basic level labs are related to introduction to installing the software and using the board; inputs and outputs; sensors and I2C communication; Wi-Fi, JSON files and MQTT; introduction to CAN; and introduction to LoRa. The five advanced labs that we offer with the MacIoT Trainer board include embedded IDE and signal processing; ADC sampling, signal graphing and level shifting; I2C sampling and signal filtering; motor control with Pulse-width modulation (PWD) and speed measuring; and motor control using PID. Any reader interested in details related to any of the listed lab experiments can email the contact author of the paper.

6 Summary

The paper presents two PCB boards designed through cooperation between academia and industry that allow students to develop IoT applications for in-person or online labs. One of the boards, referred to as MacIoT board, includes several sensors, on/off switches, and LED outputs to simulate simple processes with digital inputs and digital outputs, and it is designed for teaching several IoT related courses. The paper describes the way in which first-year students and upper year students use the board. The design of the MacIoT board is open source, meaning that anyone interested in building its own board and creating its own application can do it without any license. The second board, referred to as MacIoT Trainer board, expands all the features of the MacIoT board by adding an I2C chip that allows multiple I2C connections, a gyroscope and accelerometer combo used to determine the orientation of the device and its estimated motion in space, and a temperature and humidity sensor. Furthermore, the MacIoT Trainer board can be connected to specialized modules for LoRa communication, for connection to a CAN bus, and for displaying information of a small IoT LCD. Students can connect the MacIoT Trained board with these modules either from the comfort of their home, where they are encouraged to develop home automation application, or online through a remote access system that includes all these boards and a web camera. The MacIoT Trainer board is currently used for basic and advanced labs. A brief content of these labs is presented in the papers. More details about these labs can be obtained from the authors upon request.

Acknowledgement. The projects presented in this paper are supported by Future Skills Centre, Canada.

References

1. Centea, D., Singh, I., Elbestawi, M.: Framework for the development of a cyber-physical systems learning centre. In: Auer, M.E., Zutin, D.G. (eds.) Online Engineering & Internet of Things. LNNS, vol. 22, pp. 919–930. Springer, Cham (2018). https://doi.org/10.1007/978-3-319-64352-6_86
2. Centea, D., Elbestawi, M., Singh, I., Wanyama, T.: SEPT learning factory framework. In: Auer, M.E., Langmann, R. (eds.) REV 2018. LNNS, vol. 47, pp. 354–362. Springer, Cham (2019). https://doi.org/10.1007/978-3-319-95678-7_40
3. Centea, D., Wanyama, T., Singh, I., Mogolon, M., Boer, J., Elbestawi, M.: Using the SEPT LF for the implementation of industry 4.0: case of SMEs. Procedia Manufact. **45**, 102–107 (2020). https://doi.org/10.1016/j.promfg.2020.04.079
4. Elbestawi, M., Centea, D., Singh, I., Wanyama, T.: SEPT LF for industry 4.0 education & applied research. Procedia Manufact. **23**, 249–254 (2018). https://doi.org/10.1016/j.promfg.2018.04.025
5. Nakamura, Y., Arakawa,Y., Kanehira,T., Fujiwara, M., Yasumoto, K.: SenStick: comprehensive sensing platform with an ultra tiny all-in-one sensor board for IoT research. J. Sens. **2017**, 6308302 (2017). https://doi.org/10.1155/2017/6308302
6. Galkin, P., Golovkina, L., Klyuchnyk, I.: Analysis of single-board computers for IoT and IIoT solutions in embedded control systems. In: International Scientific-Practical Conference Problems of Infocommunications Science & Technology, pp. 297–302 (2018). https://doi.org/10.1109/INFOCOMMST.2018.8632069
7. Galkin, P., Umiarov R., Grigorieva, O.: Design embedded system testbench based on FPGA and microcontrollers for TATU smart lab as education component of industry 4.0. In: 2019 IEEE 2nd Ukraine Conference on Electrical and Computer Engineering (UKRCON), pp. 628–633 (2019). https://doi.org/10.1109/UKRCON.2019.8879996
8. Jayapradha, S., Vincent, P.M.D.R.: An IOT based human healthcare system using Arduino uno board. In: 2017 International Conference on Intelligent Computing, Instrumentation and Control Technologies (ICICICT), pp. 880–885 (2017). https://doi.org/10.1109/ICICICT1.2017.8342681

9. Kaburuan, E.R., Jayadi, R., Harisno.: A design of IoT-based monitoring system for intelligence indoor micro-climate horticulture farming in Indonesia. Procedia Comput. Sci. **157**, 459–464 (2019). https://doi.org/10.1016/j.procs.2019.09.001
10. Asthana, P., Mishra, S.: IoT enabled real time bolt based indoor air quality monitoring system. In: 2018 International Conference on Computational and Characterization Techniques in Engineering & Sciences (CCTES), pp. 36–39 (2018). https://doi.org/10.1109/CCTES.2018.8674076
11. Quadri, S.A.I., Sathish, P.: IoT based home automation and surveillance system. In: 2017 International Conference on Intelligent Computing and Control Systems (ICICCS), pp. 861–866 (2017). https://doi.org/10.1109/ICCONS.2017.8250586
12. Al-Kuwari, M., Ramadan, A., Ismael, Y., Al-Sughair, L., Gastli, A., Benammar, M.: Smart-home automation using IoT-based sensing and monitoring platform. In: 2018 IEEE 12th International Conference on Compatibility, Power Electronics and Power Engineering (CPE-POWERENG 2018), pp. 1–6 (2019). https://doi.org/10.1109/CPE.2018.8372548
13. Patnaik Patnaikuni, D.R.: A comparative study of arduino, raspberry Pi and ESP8266 as IoT development board. Int. J. Adv. Res. Comp. Sci. **8**(5), 2350–2352 (2017)
14. Singh, K.J., Kapoor, D.S.: Create your own internet of things: a survey of IoT platforms. IEEE Consumer Electr. Mag. **6**(2), 57–68 (2017). https://doi.org/10.1109/MCE.2016.2640718

A Virtual Reality Prototype for the VISIR Remote Lab

Christian Kreiter(✉), Daniel Cosic, and Thomas Klinger

Carinthia University of Applied Sciences, Villach, Austria
c.kreiter@fh-kaernten.at

Abstract. The remote laboratory project Virtual Instrument Systems In Reality (VISIR) was launched in 2006 and since then it has been installed in approximately 20 institutions around the world. At the beginning, the web user interface was created with Adobe Flash and later ported to HTML5. The 2D interface lets users add components and wires to a virtual breadboard and take measurements with virtual instruments like DC power supply, function generator, multimeter and oscilloscope. In the recent years new developments in virtual reality made devices affordable for use at home and, besides gaming, other fields of use such as educational training emerged. This paper focuses on the development of a VR client prototype for the VISIR laboratory. It covers the necessary software for development, the game concept, user interactions and input, the transformation of components from the VISIR HTML client to VR and the considerations behind it. Finally, it gives an outlook on future steps.

Keywords: VISIR · Online Labs · Virtual Reality

1 Introduction

The following paper is based on the master thesis "VISIR Remote Lab Controlled via VR" done at Carinthia University of Applied Sciences, Austria in 2020/2021 [1].

Virtual Instrument Systems In Reality (VISIR) is a remote laboratory for wiring and measuring electronic circuits over the Internet. Since the project was launched at the end of 2006 by the Signal Processing Department at Blekinge Institute of Technology (BTH) in Sweden in cooperation with National Instruments and Axiom EduTECH [2], there are approximately 20 VISIR laboratories installed in Argentina, Austria, Brazil, Costa Rica, Georgia, India, Morocco, Portugal, Spain, Sweden, and United States [3].

The software of VISIR consists of three parts. In short, the Equipment Server is a LabVIEW program, which controls the hardware, i.e. the VISIR matrix and the PXI measurement system. The Measurement Server establishes web socket connections between Web Clients and Equipment Server and checks, if incoming user requests are valid. The Web Client is the browser-based user interface. At first, the Web Client was written in Adobe Flash, since web browsers were limited in their ability to generate rich web applications with vector graphics and animations. With the introduction of HTML5, Adobe Flash was no longer required and the client was ported to HTML5 in 2012 and released on GitHub [4].

M. E. Auer and T. Tsiatsos (Eds.): IMCL 2021, LNNS 411, pp. 874–884, 2022.
https://doi.org/10.1007/978-3-030-96296-8_79

In 2013 the Oculus Rift Developer Kit was released after a successful Kickstarter campaign and other devices, such as HTC Vive and Valve Index, followed. Although the Oculus was not the first virtual reality headset on the market, it started a trend for virtual reality devices for home use with computer games as their major selling argument. Apart from entertainment, virtual reality becomes now more relevant in other fields, such as military training, healthcare, industry and job training. Virtual reality is known to provide better immersion than 2D applications and thus improve the learning experience. In this paper we present a VR prototype for the VISIR remote lab. The VR client is intended to be an alternative to the traditional HTML web client. Since the VISIR lab is able to handle multiple users simultaneously, both, VR client and HTML client can be used on the same VISIR hardware instance.

2 Software for VR Development

2.1 Game Engines

A game engine is a software development environment designed to support developers in the creation process of video games without the need to create basic functionalities for themselves. In general, a game engine consists of a rendering engine, a physics system, sound engine and a script engine. Furthermore, it also provides tools for network handling, data management and control mechanisms for game states. The two major game engines on the market are Unity 3D with a market share of 48% and the Unreal engine with a market share of 13% [5]. Unity Technologies released Unity in June 2005 as a MacOS X exclusive game engine. In the following 13 years the engine had been extended to support over 25 platforms. The Unreal Engine was developed by Epic Games and first presented in 1998 with the first-person shooter game Unreal. The Unreal Engine has been used in various game genres like Jump and runs, fighting games and online multiplayer games since then. The Unreal Engine is also used by the film industry for creating visual effects. The latest version 4 of the Unreal Engine was launched in 2014. The first difference between the two game engines is the used programming language. Unity uses C# for the implementation of the game logic and the behavior of the game objects. The Unreal Engine uses C++ as programming language, but also has a visual scripting system called Blueprints. Unreal and Unity offer a state-of-the-art graphics engine. Both systems provide tools for creating particle systems, precomputed lighting, dynamic lighting and post processing. The Unreal Engine also offers a material editor, where materials can be created and edited as in 3D software tools like Blender or Maya. Unity has no material editor. Both engines support all major operating systems, game consoles, mobile platforms and VR devices. Unity and Unreal offer similar functionalities and are both free to use for non-commercial purposes and support the desired target platform Microsoft Windows. Unity was chosen because of existing knowledge with the scripting language C#.

2.2 3D Modelling Software

For the development of the VR VISIR interface, the software, Blender was used to create the 3D representation of the virtual room environment and electronic components. Blender is a free and open source software for modeling, rigging, simulation, rendering, compositing, motion tracking and video editing [6].

3 Prototype Development

3.1 Game Concept

The game concept gives a brief overview of the planned working principle of the VR game and interaction possibilities for the users. The virtual environment contains one scene, which represents a laboratory room. The laboratory room has a size of 25 square meters and consists of two solid walls, one wall with windows, one wall with a door and a ceiling with embedded lights.

The starting point of the game is in front of a locked door of the room, where the user will have to enter a personal six-digit code to enter the room. The entry code is used to secure that only permitted users can interact with the software and send requests to the VISIR server. This mechanism should reduce potential risks for the system.

In the laboratory room a breadboard, a multimeter, a DC power supply, a function generator and an oscilloscope are placed on a table for performing the desired tasks in the area of electronic circuits. The 3D models of the electronic devices and components are designed in Blender, whereas models for the room environment like windows, doors, etc. are taken from Unity's Asset Store or other free libraries. The electronic devices are fixed to a specific space on the table without the ability to move them.

The breadboard allows users to stack electronic components like resistors, cables, capacitors and black boxes. In the first implementation cycle the focus lies on resistors and cables. Based on the resistor value, the colors of the resistor bands will change accordingly. For the interaction with the multimeter, DC power supply and function generator buttons and rotary knobs are implemented to allow the user the same hand movements as on the physical devices. Additionally, to the devices for setting up circuits, a monitor is present in the room to visualize tasks and instructions which are generated from images and displayed to the user.

The VR interface communicates with the VISIR server in the same way as the 2D web interface does. The values of the stacked circuit are sent via web request and the response values are displayed on the devices. The software also gives the user the ability to save the designed the circuits to the database or to a text file, which can also be loaded by the web based VISIR client. The visual representation of the user in the 3D room is implemented by presenting hand models, which are tracked by the controllers of the Oculus headset. The VR system is implemented in a way that will allow to extend the system to VR systems other than the Oculus Rift.

The camera origin in a VR system can be set to floor or eye level to control the height of the camera object in the virtual space. Eye level allow to control the height of the player object in the VR environment and can be used to offer perspectives that differ from the real-world experience of the user but also can result in discomforts due to a lower POV. A lower POV means that the visual perception of the ground plane changes faster than the person is used to. When using the floor as the origin, the user's point of view will be at the same height off the ground as they are used to and thus the sense of immersion increases. Since in the VISIR VR system only hand models are displayed, the camera origin is set on floor level.

Considerations to the interaction possibilities have also to be made before implementing a VR system. VR users can be left and right handed, so the button mapping for the controllers have to be changeable if both controllers have to be used, or to change the side of the controller if only one controller is used. The use of hand models allows the user to get a visual representation of grabbing objects, turning knobs and pressing buttons (Fig. 1).

3.2 Scene

Fig. 1. Top view of the game scene.

In the prototype of the VISIR interface one scene is implemented, which contains the objects for representing the virtual environment and functionalities. The play area is constructed as a room with windows on one side and with a small lobby. Inside the room the electronic devices and components needed for the setup of electronic circuits are placed on a working table. The user can enter the room from the lobby through a door by entering his personal access code, which is distributed by the faculty staff of the university. The user can move freely between the wall boundaries in the room by using the thumb stick of the right controller and turn the body via snap turn with the left one.

3.3 User Interactions and Input

This section describes the possible interaction methods of the user with the system and how they were implemented. The user hand movement is captured by the tracking of the touch controllers via sensors, which are placed in front of the play area. Each controller has 3 buttons, 2 triggers, a touch field and a 2D-axis thumb stick, which can be used to transfer the user inputs to the software. The thumb sticks and the triggers also have-built in touch sensors that can be used for input recognition. Figure 2 shows the button mapping diagram for the Oculus touch controllers in Unity.

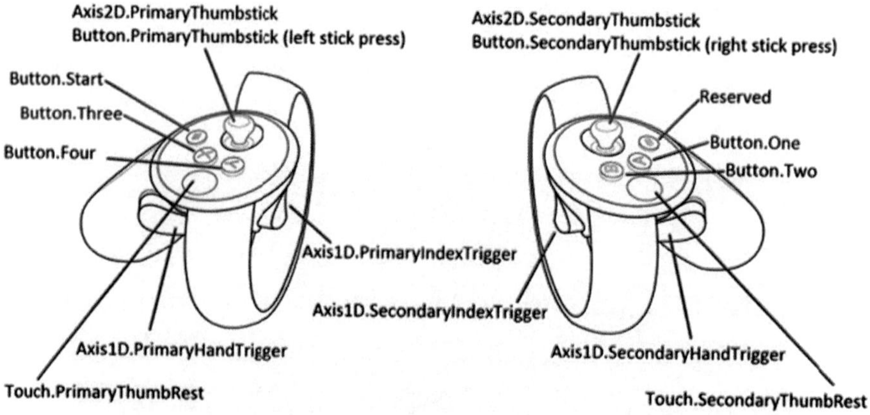

Fig. 2. Button Mapping for Oculus touch controllers in Unity [7].

A linear movement with additional snap rotation was chosen as movement type. The thumb stick of the left controller is used for the movement in the aimed direction and the right thumb stick is used to perform a rotation of the camera by 45° in the selected direction. For the visualization of the user position and user interactions the custom hand prefabs of the Oculus Avatar were used (see Fig. 3). The hands provide a high-quality representation of user hands with embedded colliders in all regions of the hand, which can be used for collision detection with other objects.

To enable the interaction between the hands and objects in the scene, the prefabs InteractTouch, InteractUse and InteractGrab of the Virtual Reality Toolkit (VRTK) were set up on each hand's game object. The InteractTouch component detects, if a collision between the controller and an interactable object occurs and is a prerequisite for grabbing and using objects. The InteractGrab script handles the grabbing of interactable objects. When an interactable object is touching the hand controller, it can be grabbed by pressing the selected grab button. The InteractUse script is used to call events through the press of the trigger button when an interactable object is touched or grabbed. Additionally, the prefab VRTK_ControllerEvents_UnityEvents is used on the hand's game object. This script allows to map events and method calls to inputs of the buttons, thumb sticks and touch sensors of the controllers. The connection between the objects in the scene and the scripts for the controller interactions is established with the component VRTK_InteractableObject. This script is placed on every object in the scene, which is interactable.

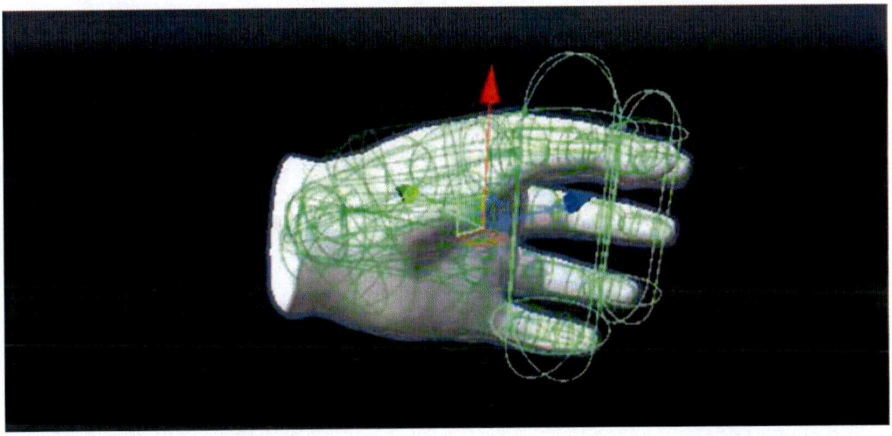

Fig. 3. Representation of used hand models and included colliders.

To enable the interaction between the hands and objects in the scene, the prefabs InteractTouch, InteractUse and InteractGrab of the Virtual Reality Toolkit (VRTK) were set up on each hand's game object. The InteractTouch component detects, if a collision between the controller and an interactable object occurs and is a prerequisite for grabbing and using objects. The InteractGrab script handles the grabbing of interactable objects. When an interactable object is touching the hand controller, it can be grabbed by pressing the selected grab button. The InteractUse script is used to call events through the press of the trigger button when an interactable object is touched or grabbed. Additionally, the prefab VRTK_ControllerEvents_UnityEvents is used on the hand's game object. This script allows to map events and method calls to inputs of the buttons, thumb sticks and touch sensors of the controllers. The connection between the objects in the scene and the scripts for the controller interactions is established with the component VRTK_InteractableObject. This script is placed on every object in the scene, which is interactable.

3.4 VISIR Components

The HTML client of the traditional VISIR interface is separated into four areas (see Fig. 4). In the Setup area the user can select one of the available experiments, which are provided by the underlying component boards of the connected hardware. The setups are saved in a JSON file and define which components and instruments can be used for an experiment. Based on the setup, the permitted components are loaded in the Components area. The Instruments area displays the selected instrument, which can be toggled via the menu bar, and is used for setting up the circuits and devices for measurements.

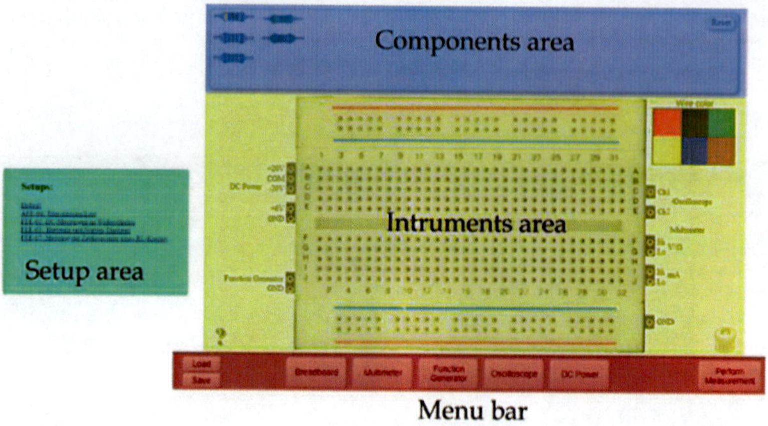

Fig. 4. Different areas of the VISIR HTML client.

The prototype of the VR interface provides a breadboard, a digital multimeter and a DC power supply as instruments for the measurement of resistances, voltages and currents. The instruments and components are placed on a workbench, allowing the user to observe all instruments at the same time. Additionally, a 3D model of a monitor serves as option for displaying tasks, instructions or expected results.

Breadboard and Electronic Components. In the web client the components can be placed onto the breadboard via Drag & Drop and rotated by selecting the component. Wires can be drawn by selecting a wire color from the presented palette. By holding the mouse button down, the start point of the wire is set and drawn by dragging the mouse until the mouse button is released. Problems encountered in the client were, that the components and wires can be placed at invalid positions on the breadboard and components could be placed over the other.

The first approach for the placement of resistors onto the breadboard was to create accurate models of the components, which stay in position due to the friction of the surfaces. The evaluation of this method stated, that displacements of the resistors can occur when other objects of the environment, like the hands of the user collide with these components. The exact placement of the resistor pins into the small holes of the breadboard with the virtual controller was another encountered problem. The solution was implemented by using the "Snap Drop Zone" (SDZ) script from the VRTK toolkit [8]. Snap zones are predefined areas where objects can be snapped to. When an object is moved close to the defined zone a transparent shape of the object indicates that the object can be released. The object is then transformed to the position, rotation and scale of the defined SDZ. The detection between drop zones and objects is performed by collision detection. For the SDZ a sphere collider was appended, that fits into the wholes of the designed breadboard model. A mesh collider was added to the pins of the designed blender model of a resistor. When the collider of the resistor hits the area of the sphere collider of the SDZ, the resistor can be released and snaps into the position of the SDZ (Fig. 5).

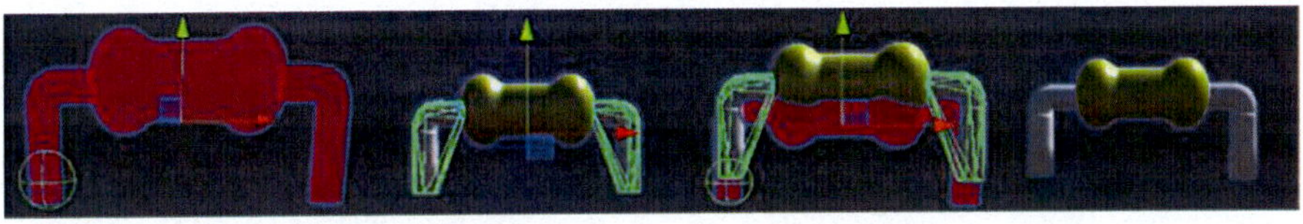

Fig. 5. Graphical representation of using SDZ for resistor placement.

Each valid placement position of a component on the breadboard requires one drop zone for one type of component that is going to be connected. On startup of the VR application, the initialization script iterates through all predefined drop zone objects and creates the areas for object placement based on the fixed distance between the holes of the breadboard and the values entered for the component properties. Additionally, the invalid rows for the horizontal and vertical aligned components can be defined in the script. If the position of one of the pins is not aligned over a breadboard hole, the drop zone is not instantiated. This ensures that the user can place the components only on valid positions and decreases possible errors in circuit setups. The resistor object itself is defined as interactable object with the according script in VRTK. The user can grab the object and move it via the motion of the controller. The resistor is cloned when dragged from the initial drop zone, so that many resistor instances can be created from one setup. To distinguish if the resistor should trigger the horizontal or vertical aligned drop zones, the user can change the alignment by pushing the left trigger button (Fig. 6).

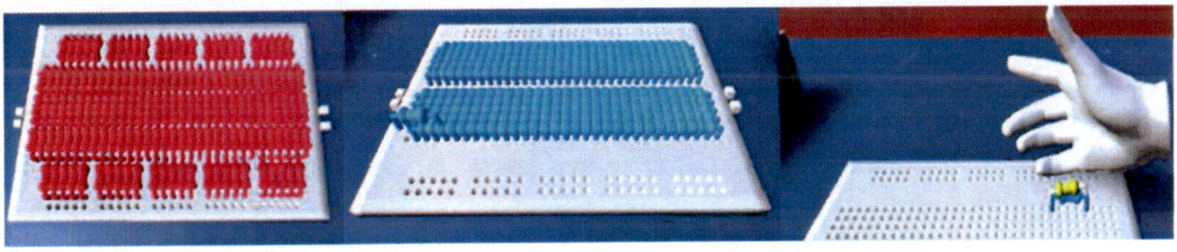

Fig. 6. Snap zones created for the placement of resistor on early breadboard model.

Wires. Similar to the web client, lines with varying distances have to be drawn between the holes. Connecting objects based on their shape and physics does not result in a sufficient solution and also using drop zones for wiring was discarded due to the high number of needed models for the different distances between the holes (Fig. 7).

A flexible option for drawing wires with good performance was found by using the "LineRenderer" component of Unity. The component is used to draw straight lines between points in the 3D space. The shape of the wire is achieved through linear interpolation between a defined point at the model of a wire, the breadboard and a calculated vector in the middle.

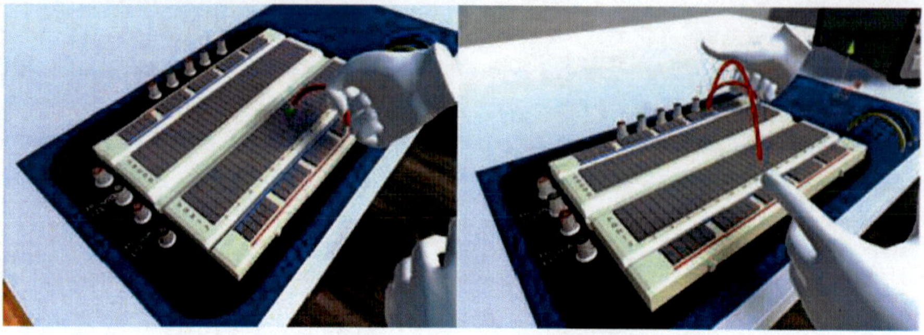

Fig. 7. Drawing wire on the VR breadboard.

Instruments. Each instrument in the HTML client uses photos of real instruments and can be used in the same way as their original counterparts. However, their functionality is reduced and simplified. Thus, some buttons cannot be clicked. The instruments have one or two input types: Buttons and rotary knobs. Every instrument also has a display, either for displaying set values or measurement results. For the visualization in VR, the chassis, buttons and the rotary knobs of each instrument were modeled in Blender with appropriate textures.

For the simulation of a button press the "VRTK_ArtificialPusher" script was used. When the hand model collides with the button, the position of the button is changed along the defined axis and reverted, when the hand model leaves the collision area. The parameter "pressed distance" defines the distance along the operate axis until the button reaches the pressed position. When the button reaches the pressed position, an event is raised that sends the button value to the script of the according instrument.

For the implementation of the rotary knob the "VRTK_ArtificialRotator" script was used. The rotator object uses a capsule collider to detect the collision with the hand model. When the collision occurs, the rotary knob can be grabbed by pressing the grip button and rotated with the hand movement of the controller around the center point. This gives the user a natural feeling of interacting with a rotatable object (Fig. 8).

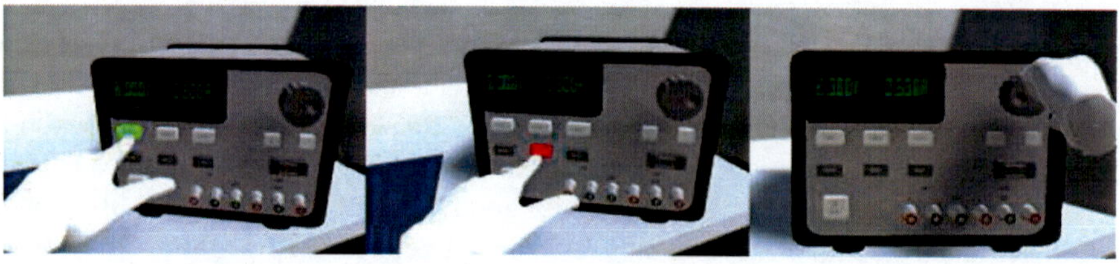

Fig. 8. Hand interactions with VR asset of DC power supply.

Server Communication. The evaluation request of the designed circuits is transported via an XML-based protocol to the measurement server of the VISIR system environment. The measurement server is responsible for the authentication of the measurement requests and the verification of the circuit data. The circuit data is compared with a list of permitted component and instruments values, before it is redirected to the equipment server for measurement. This ensures that hazardous circuit setups do not damage the real instruments. The measurement of a circuit can be requested by pressing the "X-Button" on the left controller. If it is the first measurement of the game session, an authentication request is transferred to the measurement server. The responded session key is used in the following measurement requests, which are performed by sending a circuit list with the components and instrument values of the circuit.

4 Conclusion and Future Work

The development process of the VR interface for the remote laboratory VISIR has shown that the technology can be used to add an additional layer for the communication with existing systems beside the traditional 2D web interface. The transition of natural interaction methods between users and objects into the virtual space takes a high implementation effort. Simple mechanical movements like grabbing or rotating objects can be difficult to implement. Development toolkits like VRTK help to find solutions for basic interactions, but also have a steep learning curve. During the research work the design of realistic 3D models was another demanding task.

Due to the Covid-19 lockdown restrictions during the research work, an evaluation of the user experience of the system could not be performed. As the situation in the future allows, the virtual interface has to be presented to a group of people, who are familiar in the field of electronics in order to give feedback on the usability of the application, especially in relation to the HTML client. The important factors in this evaluation process will be to measure the grade of immersion of the application and the occurrences of motion sickness, which can be triggered by unexpected movement of the player or the use of tiled textures.

According to Unity the XR interaction toolkit is going to be released in 2021. It provides similar solution to interaction techniques like VRTK but can be used by all modern headsets, which would increase the number of the potential users for the VISIR VR interface.

References

1. Cosic, D.: VISIR Remote Lab Controlled via VR, M.S. thesis, Engineering & IT, Carinthia University of Applied Sciences, Villach, Austria (2021)
2. Tawfik, M., et al.: Virtual Instrument Systems in Reality (VISIR) for Remote Wiring and Measurement of Electronic Circuits on Breadboard. https://ieeexplore.ieee.org/stamp/stamp.jsp?arnumber=6305453. Accessed 05 July 2021
3. Alves, G.R., et al.: A roadmap for the VISIR remote lab. Unpublished

4. Zackrisson, J.: visir_html5 [Source code] (2012). https://github.com/JohanZackrisson/visir_html5
5. Malhotra, M.: Unreal Engine vs Unity 3D Games Development: What to Choose? Valuecoders – Hire Dedicated Software Development Team, 01 January 2021. https://www.valuecoders.com/blog/technology-and-apps/unreal-engine-vs-unity-3d-games-development/. Accessed 05 July 2021
6. B. Foundation, About — blender.org. https://www.blender.org/about/. Accessed 05 July 2021
7. U. Technologies, Unity - Manual: Input for Oculus. https://docs.unity3d.com/2019.2/Documentation/Manual/OculusControllers.html. Accessed 05 July 2021
8. "Snap Drop Zone · VRTK - Virtual Reality Toolkit," VRTK - Virtual Reality Toolkit. https://vrtoolkit.readme.io/docs/snap-drop-zone. Accessed 05 Jul 2021

Mobile STEAM Learning and Teaching

Teaching Artificial Intelligence in K-12 Education: Competences and Interventions

Stavroula Misthou[1(✉)] and Aristidis Paliouras[2]

[1] GreekCodersK12 Coding Club, Athens, Greece
stavroula.misthou@gmail.com
[2] 3rd Laboratory Center of East Attica, 26th km Marathonos Ave., 19009 Rafina, Greece

Abstract. The COVID-19 pandemic has highlighted the strategic importance of artificial intelligence (AI) to all the governments around the world. However, governments do not seem to have understood to the same extent how AI can be implemented, so they are not taking advantage of its potential. Education plays a critical role in efforts to make future workforces AI-ready. It is an emergency the development of digital skills especially in matters of AI and it becomes clear that in order to ensure that a country has the trained and talented workforce needed to implement artificial intelligence in the economy and society, it is essential to invest in STEM and AI education that can give students the skills and opportunities to pursue a career in STEM. As is well known, creating the next generation of scientists, leaders and entrepreneurs takes time and it's nurtured over many years, starting at an early age in school. The curricular reform efforts outlined in this paper clearly demonstrate the need to define the journey in 'AI competences' in K-12 education that incorporates 21st century skills. The paper explores the AI skills and competences that need to be developed in primary and secondary education and presents some implementations of AI-based interventions in the classroom using voice assistants.

Keywords: Artificial intelligence competences · K-12 education · Voice assistants

1 Introduction

The COVID-19 pandemic has highlighted the strategic importance of Artificial Intelligence (AI) to all the governments around the world. AI is already being used by pharmaceutical companies to develop new drugs and treatments, has helped governments manage the pandemic and could contribute to the economic recovery. Providing public services with the help of AI could improve healthcare, education and transportation. AI has a wide range of uses and potential applications in many areas, offering prediction, image recognition and Natural Language Processing.

However, governments do not seem to have understood to the same extent how AI can be implemented, so they are not taking advantage of its potential.

The Government AI Readiness Index 2020 report that has been produced by Oxford Insights and the International Development Research Centre (IDRC) posed the

M. E. Auer and T. Tsiatsos (Eds.): IMCL 2021, LNNS 411, pp. 887–896, 2022.
https://doi.org/10.1007/978-3-030-96296-8_80

crucial question: "How ready is a given government to implement AI in the delivery of public services to their citizens?".

In this report government "AI readiness" is defined as the raw materials and enabling factors needed to make AI implementation possible [1].

The top ranking governments in AI readiness globally are United States of America scoring 85.479/100, United Kingdom scoring 81.124/100, Finland 79.238/100, Germany 78.974/100 and Sweden 78.772/100. The global average score from the 178 participating countries is 44.25/100. In the Government AI Readiness Index 2020, Greece is at the 61st position scoring 47.933/100 and at the last position in Western Europe, as Fig. 1 depicts.

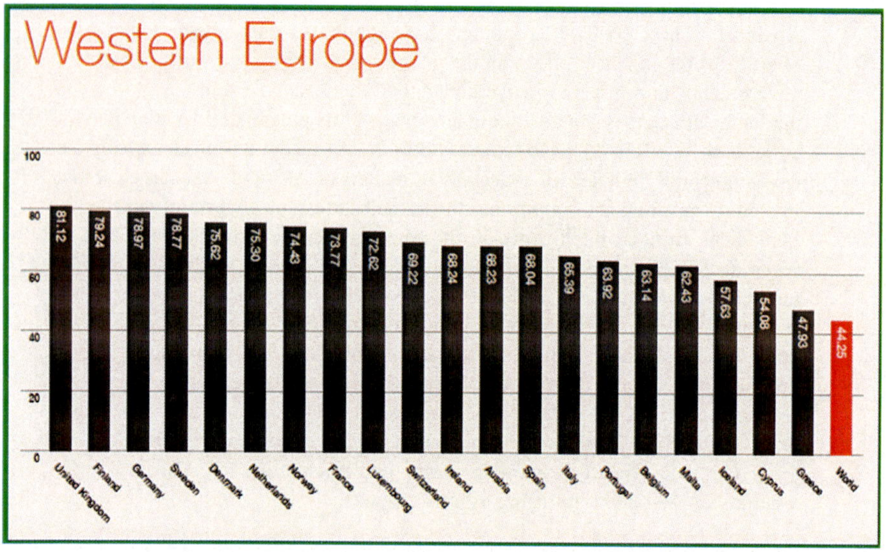

Fig. 1. Government AI readiness index 2020 in Western Europe

The Government AI Readiness Index 2020 report states that "*much of North America's comparative success in AI readiness is not explicitly related to AI technologies, or even to particular administrative policies, but to broader structural realities within the region including higher income, decades of investment in STEM education, and support for technological innovation. The government also has a STEM education strategy, published in 2018, which aims to give all Americans access to lifelong high-quality STEM education and training.*"

In Greece, the proposal of Hellenic Federation of Enterprises (SEV) for the utilization of Artificial Intelligence and the improvement of the international competitiveness of the economy through a national utilization program, sets among others, two goals related to education:

- development of digital skills especially in matters of AI and

- open access in public data, which is necessary in the "learning" of AI systems [2]. The proposal points out that the development of Artificial Intelligence in Greece could contribute critically reducing the "brain drain' of STEM talent abroad.

Taking all the above into consideration it becomes clear that in order to ensure that a country has the trained and talented workforce needed to implement artificial intelligence in the economy and society, it is essential to invest in STEM education that can give students the skills and opportunities to pursue a career in STEM. As is well known, creating the next generation of scientists, leaders and entrepreneurs takes time. It does not happen by participating in a single course or in a summer programme. It's nurtured over many years, starting at an early age in school.

The paper explores the AI skills and competences that need to be developed in primary and secondary education and presents some implementations of voice assistants in the classroom.

2 Skills and Competences in an AI Era

Citizens need skills to help them meet the challenges and seize the opportunities of digital transformation in order to function and thrive in the digital society and overcome digital risks. Although digital skills are a basic skill required for all walks of life, such as literacy and numeracy, many citizens have limited or outdated digital skills. There is an urgent need to broaden the scope of EU efforts in two key areas: on the one hand to prepare citizens to understand, at different levels, the different aspects of digital competence and on the other hand to address the lack of specialized IT skills required for the job market. Digital competence is part of the revised European Framework of Reference for the key skills for lifelong learning that all citizens should have. Digital competence means the use of digital technology with confidence and critical thinking and covers the knowledge, skills and approaches that all citizens need in a rapidly evolving digital society [3].

2.1 European Digital Competence Frameworks (DigComp, DigCompEdu)

The European Digital Competence Framework, also known as DigComp, offers a tool to improve citizen's digital competence. Today, being digitally competent means that people need to have competences in all areas of DigComp [4].

The European Digital Competence Framework for Citizens [5] proposes five areas of digital competences:

1. Information and data literacy (Browsing, Searching and filtering data, Information and digital content, Evaluating data, Information and digital content, Managing data, Information and digital content)
2. Communication and collaboration (Interacting through digital technologies, Sharing through digital technologies, Engaging in citizenship through digital technologies, Collaborating through digital technologies, Netiquette, Managing digital identity)

3. Digital content creation (Developing digital content, Integrating and re-elaborating digital content, Copyright and licenses, Programming)
4. Safety (Protecting devices, Protecting personal data and privacy, Protecting health and well-being, Protecting the environment)
5. Problem solving (Solving technical problems, Identifying needs and technological responses, Creatively using digital technologies, Identifying digital competence gaps)

The European Digital Competence Framework for Educators [6] describes what it means for educators to be digitally competent and provides a general reference frame to support the development of educator-specific digital competences in Europe (DigCompEdu). DigCompEdu is directed towards educators at all levels of education, from early childhood to higher and adult education, including general and vocational education and training, special needs education, and non-formal learning contexts. Together, these frameworks provide a comprehensive and easy-to-use reference model for the systematic improvement of digital competences.

The revised version of DigComp (2.2), which is expected in January 2022 will include [7]:

- New and emerging topics and themes in the digital world: misinformation and disinformation
- Artificial Intelligence (AI)
- Data-related skills and datafication of every aspect of our lives
- Emerging technologies such as virtual reality, social robotics, Internet of things
- Green ICT skills
- E-commerce
- Data literacy

2.2 Add AI Competences to DigComp

In recent years, Artificial Intelligence has made significant progress due to the increasing computing power of computers and the availability of huge amounts of data. It is now clear that there is an urgent need for everyone to adapt to the rapid evolution of AI. For this to happen it is necessary to be able to acquire new skills. Taking into consideration:

- the need to acquire new skills in the field of AI,
- the great impact of AI on society (significant achievements follow one another rapidly) and
- the recent work of Long & Magerko [8]

We suggest adding another Competence Area on the subject of Artificial Intelligence to the DigComp framework with the following competences (Table 1):

Table 1. Artificial Intelligence Competences

Competence Area	Competences
6. Artificial Intelligence	6.1. What is AI ? 6.1.1. Recognizing AI 6.1.2. Understanding Intelligence 6.1.3. Interdisciplinarity 6.1.4. General vs. Narrow AI
	6.2. What AI Can Do ? 6.2.1. AI's Strengths & Weaknesses 6.2.2. Imagine Future AI
	6.3. How Does AI work ? 6.3.1. Representations 6.3.2. Decision-Making 6.3.3. Machine Learning Steps 6.3.4. Human Role in AI 6.3.5. Data Literacy 6.3.6. Learning from Data 6.3.7. Critically Interpreting Data 6.3.8. Action & Reaction 6.3.9. Sensors
	6.4. How Should AI Be Used ? 6.4.1. Ethics
	6.5. How Do People Perceive AI ? 6.5.1. Programmability

3 AI Curricula for K-12 Classrooms

For the introduction of AI in education two major associations, AAAI (https://aaai.org/) and CSTA (https://csteachers.org/), have started the work "The 5 big ideas in AI" in which they present in detail the way of AI introduction in primary and secondary education (https://ai4k12.org/).

The first big idea is the **Perception** of the world using sensors. Computers can "see" and "hear" by interpreting sensor signals.

The second big idea is **Representation & Reasoning** which refers to the methods of representing knowledge in computers so that it can then be used by algorithms to produce new knowledge.

The third big idea is **Learning**, which refers to Machine Learning, that is, the way in which computers learn using algorithms that find patterns in experimental data.

The fourth big idea is **Machine Interaction** (Natural Interaction) in a natural way. The main topics that make up this great idea are the understanding of natural language, the reasoning of common sense, the affective computing and the theory of mind.

The fifth big idea is the **Social Impact** (Societal Impact) which refers to the effects (positive and negative) of AI in society. Some of the issues that make up this idea are AI's ethics in making decisions for people, the economic implications of AI, AI and culture, and AI for the social good.

In recent years there has been intense interest from Universities, Scientific Associations, European Projects and Non-Profit Organizations towards the development of a curriculum for the introduction of Artificial Intelligence in education.

Our literature review [8–10] resulted in the development of The Teach Artificial Intelligence Journey (Table 2) which shows the way for the complete introduction of AI in K-12 education. The Teach Artificial Intelligence Journey suggests the competences and the "big ideas" that we consider students should be taught in each class in primary and secondary education. This journey can support teachers and students on a journey from 1st grade to 12th grade and is based on the latest research in the teaching of Artificial Intelligence in K-12 education, as well as the many years of teaching experience of the authors.

Table 2. The Teach Artificial Intelligence Journey

	Primary School						Junior High School			High School		
	1st	2nd	3rd	4th	5th	6th	7th	8th	9th	10th	11th	12th
5 Big Ideas in AI												
Big Idea	1	1	1	1	1	1	2	3	4	1 2 3	3 4	5
AI Competences												
6.1.1	√	√	√	√								
6.1.2					√	√						
6.1.3							√					
6.1.4										√		
6.2.1										√		√
6.2.2						√			√			√
6.3.1							√			√		
6.3.2							√			√		
6.3.3							√			√		
6.3.4							√			√		
6.3.5								√			√	
6.3.6								√			√	
6.3.7									√		√	
6.3.8									√		√	
6.3.9	√	√	√	√	√	√						
6.4.1												√
6.5.1	√	√	√	√	√	√						

4 Implementation of AI-Based Interventions in the Classroom

Why – AI and Speech Technology: With the rise of Artificial Intelligence (AI) solutions for enterprises, speech technology has many applications in all areas, including healthcare, education, security, finance, business and individualized use. Personal voice assistants such as Siri, Google Home and Amazon Alexa are devices that offer personalized speech technology experiences. Artificial intelligence enables modern voice assistant applications to extend their limited vocabulary to millions of words and phrases using cloud storage.

Who – The Learners: A really good first attempt to familiarize students even from primary school with AI, is by using voice assistants. Today, millions of people use Alexa as a personal assistant to manage their schedule, listen to music, find recipes, read books, do calculations, make to-do lists, set alarms, control the devices in their smart homes, and much more, all without keyboard or smartphone, only through voice. There are thousands of voice-based apps, called "skills", that meet a certain human want or need, that are already provided by Amazon and other developers. Alexa provides many uses, even for the classroom, by creating a personalized Alexa skill using Alexa Skills Blueprints.

By building a personalized Alexa skill, students can customize their experience with a virtual assistant. Alexa Skills Blueprints are templates that can be customized to teach Alexa to perform a certain task. Students can also share or publish their Alexa skills, so other users might benefit from them, as well.

There are currently some blueprint types from which teachers can choose: [11].

- At Home: Includes 10 blueprints such as Whose Turn, Task Tracker, and Chore Chart.
- Learning & Knowledge: Includes blueprints such as Quiz and Flashcards.
- Fun & Games: Includes 22 blueprints such as Game Show and Inspiration.
- Storyteller: Includes 8 blueprints such as Fairy Tale, Sci-Fi and Adventure.
- Greetings & Occasions: Includes 8 blueprints such as Celebrate and Birthday Wishes.
- Communities & Organizations: Include blueprints such as Blog and Spiritual Talks.
- Business: Includes blueprints such as Business Q&A and Onboard Guide.
- Amazon Kids Recommended: Includes 11 blueprints such as Sorting Game Beginner and Moon Landing.

There are dozens of blueprints across these categories and no coding or special skills are required to create a skill using the templates.

4.1 Teaching Variables in Computing Education with a Voice Assistant

What – Description of the Intervention: To enable students to unleash their creative and imaginative learning potential, we propose a playful, personalized and creative learning experience (even online) that combines digital skills with science, literature and other subjects, using voice assistants.

How – Skill Creation: A very interesting intervention that can be implemented in primary school is the "Fairy Tale Story Blueprint" [12]. In order to create their own fairy tale, students need to think of a plot, with heroes, places, adventure and narrative. They create the new Alexa skill, by typing their story, adding sound effects, fun expressions and pauses. They also create interactive 'blanks' to fill in while someone listens to the story. That means that the story is configurable; each time someone listens to the skill, Alexa asks for the data (variables) e.g. for the names of the heroes, of the place, the time, and creates a new story. Figure 2 depicts the blueprint for the Fairy Tale story. This is a very inspiring way to teach variables in the classroom. When the skill is ready, students pick a name for the skill and can share it with others.

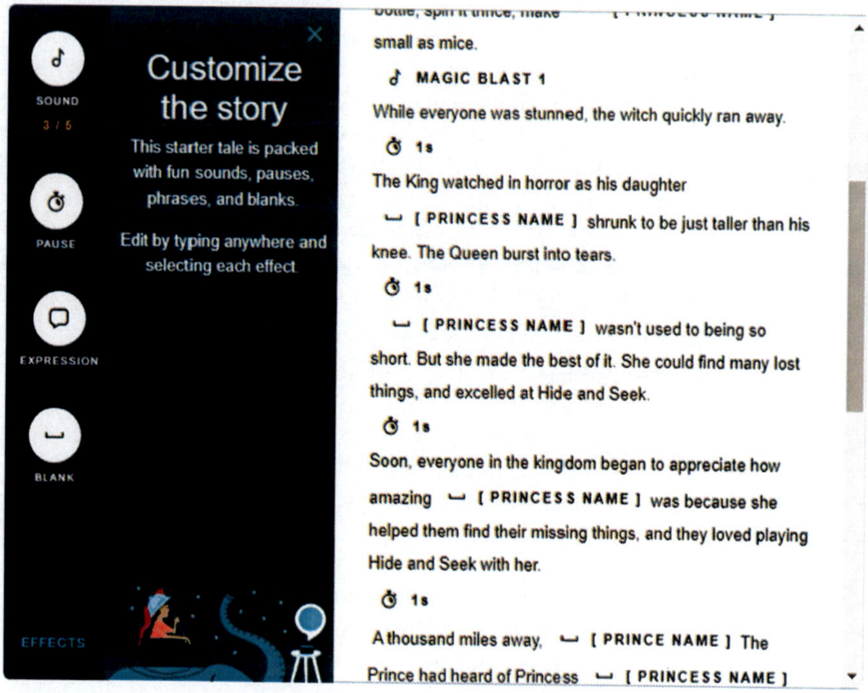

Fig. 2. Fairy Tale Story Blueprint [13]

Similar to the Fairy Tale Story Blueprint is the Sci-Fi Blueprint, that offers a template to create an interactive story with a far-out theme [14].

There is also a Blueprint for making quizzes; students can challenge their colleagues with open-ended questions about a specific story, such as geography, history, literature, football or even astronomy [15].

This step of the intervention is based on the fourth big idea in AI which is Machine Interaction (Natural Interaction).

4.2 Students' Perception of a Voice Assistant

Results – Perceptions and Challenges: It's worthwhile noting that students have a rather confused perception about what is Alexa. A fruitful discussion with the students showed that Alexa AI is perceived as a "black box", or perhaps a mystery, magic or acquires some human characteristics. As a "black box", it receives some data and produces answers and dialogues. How Alexa Voice User Interface works and how it interprets data to provide voice output is not transparent to students. This step of the intervention is based on the first big idea in AI which is Perception.
For high school students, this "black box" concept may be the subject of discussion about technical illiteracy. According to Jenna Burrell "*opacity occurs because coding and algorithm design is a specialized skill that is not accessible to the majority of the population or that opacity arises from the characteristics of machine learning algorithms and the scale required to apply them usefully*" [16]. This can also be a great opportunity to highlight the necessity of learning programming and engaging in STEM. This last step of the intervention covers the second big idea in AI which is Representation & Reasoning.

5 Conclusions

Education plays a critical role in efforts to make future workforces AI-ready. According to OECD AI Principles [17] "*AI should be developed consistent with human-centered values, such as fundamental freedoms, equality, fairness, rule of law, social justice, data protection and privacy, as well as consumer rights and commercial fairness*.". For this reason, it is imperative to introduce AI in education as soon as possible in order to comply with the above principle.

Despite the individual efforts made to create an AI Curriculum, we believe that an AI Framework for K-12 education should be created that describes the skills that 21st century students should have. All stakeholders (European Union, OECD) are moving in this direction.

The curricular reform efforts outlined in this paper clearly demonstrate the need to define the journey in 'AI competences' in K-12 education that incorporates 21st century skills. Our work aims to be the basis for the creation of an AI Framework for the introduction of AI in K-12 Education.

References

1. AI Readiness Index 2020, Oxford Insights. https://www.oxfordinsights.com/government-ai-readiness-index-2020. Accessed 14 June 2021
2. Τεχνητή Νοημοσύνη: Τα δεδομένα σήμερα και οι προτάσεις του ΣΕΒ. ΤΕΥΧΟΣ 59 | 2 Δεκεμβρίου 2020. https://www.sev.org.gr/vivliothiki-tekmiriosi/special-report/techniti-noimosyni-ena-aparaitito-alma-gia-tis-epicheiriseis-ta-dedomena-simera-kai-oi-protaseis-tou-sev/. Accessed 14 June 2021
3. Communication from the Commission to the European parliament, the council, the European economic and social committee and the committee of the regions. https://eur-lex.europa.eu/legal-content/GA/TXT/?uri=CELEX%3A52018DC0022. Accessed 19 June 2021
4. The Digital Competence Framework 2.0. https://ec.europa.eu/jrc/en/digcomp/digital-competence-framework. Accessed 14 June 2021
5. Digital Competence Framework for citizens. https://ec.europa.eu/jrc/en/digcomp. Accessed 14 June 2021
6. Digital Competence Framework for Educators (DigCompEdu). https://ec.europa.eu/jrc/en/digcompedu. Accessed 14 June 2021
7. Call for contributions to DigComp 2.2. https://ec.europa.eu/jrc/sites/default/files/message_stakeholders_digcomp_2_2_cop.pdf. Accessed 14 June 2021
8. Long, D., Magerko, B.: What is AI literacy? competencies and design considerations. In: Proceedings of the 2020 CHI Conference on Human Factors in Computing Systems, pp. 1–16. Association for Computing Machinery, New York, NY, United States (2020)
9. Williams, R., Park, H.W., Oh, L., Breazeal, C.: PopBots: designing an artificial intelligence curriculum for early childhood education. In: Proceedings of the AAAI Conference on Artificial Intelligence, vol. 33, No. 01, pp. 9729–9736. AAAI Press, Palo Alto (2019)
10. Amanatidis, N., Kiriakos, D., Misthou, S., Paliouras, A., Papadopetraki, E.: Teaching Artificial Intelligence in Primary, Secondary and Post-Secondary Public Education in Greece. «I-TEACHER» - (28), pp. 9–39 (2021). ISSN 1792-4146
11. Alexa Skill Blueprints. https://blueprints.amazon.com/. Accessed 19 June 2021
12. Alexa skill blueprint "Create your own Fairy Tale Story. https://blueprints.amazon.com/bp/amzn.ask.bp.5ffada1a-a4bd-4201-b758-74f7f8d8a946/en-US?ref_=ask_bp_home_5_0. Accessed 14 June 2021
13. Fairy Tale. https://blueprints.amazon.com/skills/create/amzn.ask.bp.5ffada1a-a4bd-4201-b758-74f7f8d8a946/en-US?ref_=ask_bp_detail_make. Accessed 19 June 2021
14. Alexa skill blueprint "Sci-Fi". https://blueprints.amazon.com/bp/amzn.ask.bp.9f3353dd-bed5-4fe0-960b-dbd074c5d419/en-US?ref_=ask_bp_home_5_1. Accessed 14 June 2021
15. Alexa skill blueprint "Quiz". https://blueprints.amazon.com/bp/amzn.ask.bp.ce6a1169-85b2-4ddc-a5f2-c263f102c402/en-US?ref_=ask_bp_home_3_0. Accessed 14 June 2021
16. Burrell, J.: How the machine 'thinks': Understanding opacity in machine learning algorithms (2016). https://doi.org/10.1177/2053951715622512
17. OECD.AI: Human-centred values and fairness (Principle 1.2). https://oecd.ai/dashboards/ai-principles/P6. Accessed 14 June 2021

The Design of a Serious Game to Enable the Exploration of the Binary System

Nektarios Moumoutzis[1,2](✉), Nikolaos Apostolos Rigas[1], Chara Xanthaki[3], Ioannis Maragkoudakis[1], Stavros Christodoulakis[1], Desislava Paneva-Marinova[2], and Lilia Pavlova[4]

[1] Laboratory of Distributed Multimedia Information Systems and Applications, School of Electrical and Computer Engineering, Technical University of Crete, 73100 Crete, Chania, Greece
{nektar, imarag, stavros}@ced.tuc.gr

[2] Institute of Mathematics and Informatics, Bulgarian Academy of Sciences, 1113 Sofia, Bulgaria
dessi@cc.bas.bg

[3] Directorate of Secondary Education of Chania, Chania, Greece

[4] Laboratory of Telematics, Bulgarian Academy of Sciences, 1113 Sofia, Bulgaria

Abstract. This paper presents the design of a serious game on the mathematical foundations of computer hardware, namely the binary system. The work builds on previous research efforts targeting gamification of mathematics. The game enables the players to understand the principles behind the positional number systems, transfer their knowledge of the decimal number system to the binary and explore the mechanisms (algorithms) of the four arithmetic operations. It employs a plot and an appropriate representation that promotes creativity and engagement of players. It offers new learning opportunities, when compared with previous work related to game-based teaching of the binary system. The games surveyed mainly focus on the binary representation of numbers without engaging students in active exploration of arithmetic operations in the binary system. The knowledge developed can be transferred to other number systems and in particular to the decimal system, thus contributing to a deeper understanding of the mechanisms of number representations and arithmetic.

Keywords: Binary system · Arithmetic operations · Serious game

1 Introduction

The proliferation of digital technologies the last years had a tremendous impact on education. There is a global recognition of the need to educate the new generation n digital technologies and, in particular, on core principles of computing and coding [1, 2] in order to empower young people with the key skills and competencies that will allow them to take advantage of digital technologies in order to be creative and innovative citizens. This trend can be traced back to pioneering work of Seymour Papert, creator of LOGO, the first programming language for children. Papert has emphasized the relation between computer coding and constructing digital artefacts that

M. E. Auer and T. Tsiatsos (Eds.): IMCL 2021, LNNS 411, pp. 897–906, 2022.
https://doi.org/10.1007/978-3-030-96296-8_81

are the basis of his learning theory, i.e. constructionism, stating that learning is deeper when students construct things. In his seminal book "Mindstorms: Children, Computers and Powerful Ideas" he presented the big ideas behind computers and how these big ideas can be better understood by young children [3]. In later works, Papert expressed his belief that "powerful ideas" of computer science should be made part of school curricula [4]. One of these powerful ideas is undoubtedly the binary number system, which is the basis for the development of computer hardware and all types of data representations used in modern computing. The work reported in this paper addresses the need to offer an effective learning environment for young students to understand the concepts behind the binary system and how it is used to implement arithmetic.

The rest of the paper is organized as follows: Sect. 2 presents previous attempts to gamify mathematical teaching while Sect. 3 presents in detail the design of the serious game and its mini-games to explore addition and multiplication. Section 4 presents related games on the same topic. Section 5 concludes.

2 Gamifying Maths

A first attempt to gamify mathematical teaching and learning in general that is adopted and concretized by the work reported in this paper, can be found at GamifyMaths [5]. GamifyMaths offers a conceptual framework for amplifying and enhancing mathematics learning that is taking place in the classroom by adopting inquiry-based learning, gamification strategies and the flipped classroom model for engaging students in immersive and interactive mathematics resources and processes realized both out-of-the classroom (online) and in-class for involving students into higher-order inquiry-based learning in mathematics.

Solving mathematical problems necessitates knowledge, skills, creativity and resilience. To simplify the process of acquiring such skills, GamifyMaths propose an architecture that combines play, learning, motivation, participation and engagement. However, there is a need to address the role of the teachers and the set of skills that they need to develop in order to have the full control of the pedagogical process and be able to support their students appropriately. This need was identified by GamifyMaths that stressed the necessity for the establishment of a Community of Practice to support the teachers through learning materials and scenarios, a social network to promote their close collaboration and exchange of experiences and ideas. Several existing platforms and initiatives can serve as hosting infrastructures for such community building process such as the Coursevo Platform [6] that is employed in the work described in this paper.

In particular, this is done within the context of the Theatre in Mathematics – TIM project. The TIM project aims to contribute to the ways to improve mathematics teaching and learning, in particular providing a new methodology - TIM Methodology - to teach mathematics using drama. The project develops the TIM methodology by deepening and combining two existing approaches: "Mathemart – Playing with mathematics in the theatre workshop" and "Process Drama - change of roles, perspectives, and role aspects in teaching mathematics". Mathemart addresses teaching mathematics through Social and Community Theatre (SCT) methodology to get students involved in the game of mathematics by means of theatrical games and activities: an overall approach that includes mind and body, inborn creativity and engagement. One of the major outputs of the TIM project is a toolkit containing several types of theatrical exercises that can be used and adapted to design playful learning activities that use the bodies of the learners to explore mathematical concepts. The serious game described in this paper is essentially a generic tool that can be combined with activities described in the toolkit to invent many different types of playful activities, thus offering an effective engaging learning environment to develop insights and knowledge related to positional numbers systems in general and the binary system in particular and explore the phenomena that emerge with respect to the mechanisms of the four arithmetic operations. This transition is facilitated by metaphors that are appropriate for young students and provide an experiential learning route to the inner workings of the binary system, i.e. the representation of binary numbers and their manipulation to perform the four arithmetical operations.

To establish a Community of Practice for TIM and enable the use of the presented game as well as a number of other unplugged activities incorporated in the TIM Toolkit, the Coursevo Platform [6] is used, offering diverse services within activity spaces, is used. Coursevo enables communication between participants in online learning activities, cooperation among participants and access to coursework information and learning resources. It can combine traditional classroom-based lessons and practical sessions, with self-study and eLearning.

3 The Design of the Binary Arithmetic Serious Game

3.1 Gameplay

The game is centered on the adventures of a fairy that moves mountains to find her beloved pets and bring them back home before the various mythical creatures that wander around captivate them. For that reason, the player navigates through many biomes like forest, dessert and tundra, confronting the creatures that show up each time. The mythical creatures patrol certain areas, so when the player approaches or passes near a mythical creature or even attacks it, it fights back. If the enemy manages to kill the player, then the player revives at a specific place, at the middle of the world (so he has a long way to walk to get back to where he was) (Fig. 1).

Fig. 1. Going around the game world.

In order to cast special spells at the enemies the user has to consume magic filters. These are made of flowers that the user has to cultivate. This is done by clicking at the relative icon. The cultivation of the flowers for the potion is presented to the player as a mini-game that stalls the normal game flow. An exemplary screen is shown below (Fig. 2):

Fig. 2. Entering the flowers' cultivating mini-game.

These flowers are magic. That means they are kept in special vessels that help them grow in a particular manner. More specifically, only one flower can be held in one vessel. If two flowers happen to be in the same vessel, they start to wither, but if moved two a bigger vessel (left part of the figure below) it blooms twice as much blossoms as it previously had. So, at the first vessel, all flowers have one blossom, at the second vessel, two blossoms, at the third vessel, four blossoms etc. To have a flower with specific number of blossoms, the player can seed flowers in the first vessel (unary flowers) and move them appropriately so that they won't wither. Alternatively, the player can immediately seed the blossoms at the appropriate vessels. This method to produce the magic flowers obeys to the rules for representing quantities in the binary system and is taught to the player step by step, in a "learn mode" as shown in the figure below (Fig. 3).

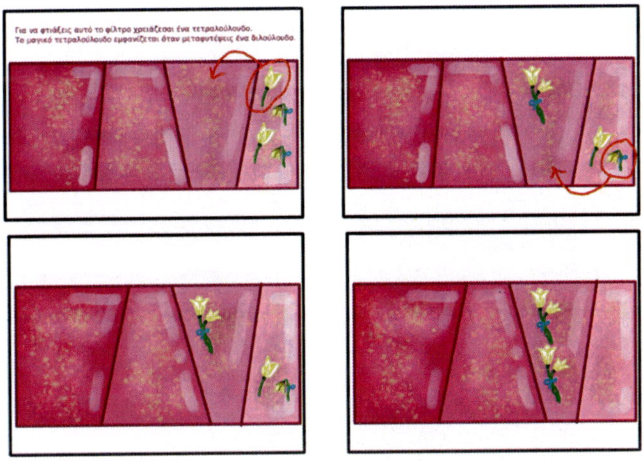

Fig. 3. Representing binary numbers with flowers.

After the player has learned the rules to produce flowers with certain number of blossoms, the "play mode" is initiated. That means, that whenever the player runs out of magic potion and clicks the icon, a recipe can be initiated to create a new potion. The recipe has instructions for how many blossoms are needed and the player has to grow them following the procedure described above. An example of carrying out a recipe follows is depicted below (Fig. 4).

Fig. 4. An example of carrying out a recipe to create a new portion.

3.2 Arithmetic Operations

Except for learning to represent quantities in the binary system, the user also practices to add and multiply two numbers in the binary system. The corresponding mini-games are determined by algorithms that monitor the user's progress. For the addition game,, as depicted below, two parallel lines of vessels are shown with the flowers already seeded and the player is asked to put the flowers of each vessel to the cauldron bellow it. If two flowers are put in the same cauldron the player must quickly move them to the next cauldron or the concoction will explode (Fig. 5).

Fig. 5. Addition mini-game.

For the multiplication game, the player tries to find the combination of vessels and magic filters that will give him the amount of potion that is instructed by the recipe. Each of the five vessels has different capacity. Capacities are powers of two. The user has to choose which vessels to fill according to the one operand of the multiplication and then he will fill the chosen vessels as if he wanted to fill the vessels with amount given by the second operand.

For example, the multiplication 3 times 5 is done as follows: Five in is 101 in binary so we should begin the filling from the first and third vessel. Each filling is done separately and with different magic filter. Since the other operand is number three (in binary represented as 11), we must fill the first and the second vessel each time, but, remember, the vessel at which we start each "filling session" is already determined by

Fig. 6. Multiplication mini-game.

the first operand. The result is number 1111 in binary representation, i.e. number 15 in decimal representation (3 × 5 = 15). This particular multiplication example is depicted below (Fig. 6).

4 Other Games to Learn the Binary System

A recent work related to playful learning of the binary system employing an unplugged approach is described in [7]. It is a combination of a theatrical game and a board game to teach the binary system to children. The two games are closely related: The theatrical game uses the children's bodies (see figure below) to enact the mechanism of the board game by employing simple rules for interacting with each other. Following simple rules the participants embody the operation of a calculator that is able to transform integers into their binary representations and compute the result of the four arithmetic operations (addition, subtraction, multiplication and division). A simulator of the theatrical game is also offered so that learners can explore it for themselves and discover how the arithmetic operations are done, even when they are not in the same physical space with their peers. The board game is essentially a binary version of a counting table or abacus (see figure below). It offers an alternative mechanism for understanding the validity of the rules of the theatrical game and makes these rules more concrete and accessible to the children. This way they can explore on an individual basis the way that the four arithmetic operations are performed. The knowledge developed can be transferred to other number systems and in particular to the decimal system, thus contributing to a deeper understanding of the mechanisms of number representations and arithmetic (Fig. 7).

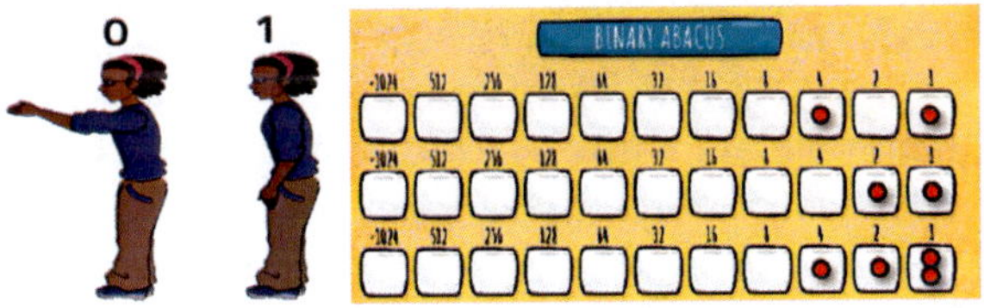

Fig. 7. The theatrical game (left) and the board game (right) for exploring the binary system.

A digital card game for learning the binary representation of numbers is described in [8]. The authors discuss the importance of games in learning and the importance of learning the binary representation of data for understanding computer science and how computers work. This knowledge is part of standard Computer Science curricula but it is difficult to grasp for both children and adults. The card game targets primary level education pupils. The design of the game is based on sociocultural and constructivist perspectives on knowledge construction. Participants are asked initially to use cards with decimal numbers so as to express their knowledge with the decimal system and then use the cards that contain binary numbers representations so that they can see the analogies. Visual hints are also provided to enable a smooth progress in learning the structure of the binary system. In the decimal system mode, numbers up to five digits are used. The learners are asked to break numbers into tens of thousands, thousands, hundreds, tens and units. Then, the system checks their answer. The "Binary to decimal" function of the game presents a binary number up to five digits along with the value of each digit using cards with 1, 2, 4, 8, and 16 dots, for each one of the digits from right to left. The player is asked essentially to add these values together depending on the digits of the number (the position of ones in the binary representation of the number). The "Decimal to Binary function presents a random number from 1 to 63 and the learner has to convert it to the equivalent binary number using six binary digits that are accompanied again with cards with dots (1, 2, 4, 8, 16, and 32 dots). To make the learning experience more fun, after understanding the basics of binary representation, the main game can be played, which is a binary version of blackjack. The objective of the game is to reach the target number of 51 (instead of 21 in the original blackjack) using cards with numbers represented in the binary system. In conclusion, the game focuses on the binary representation of numbers only, with no further aim to explore the arithmetic operations in the binary system, in contrast with the game presented in this paper.

Cards with dots (Fig. 8) to represent the values of each position in the binary representation of a number are used in Computer Science unplugged activities for learning the binary representation of data [9]. The activities are mostly challenges and discussions related to number conversions and coding of text characters to represent text in binary. No attempt is made for exploring the arithmetic operations.

Fig. 8. CS unplugged cards from the Count the Dots activity.

The Cisco Binary Game [10] is a set of challenges organized as a multilevel quiz. It is offered as a tool to help pass the CCENT or CCNA Routing and Switching exam, in particular that part of the exam that refers to converting decimal numbers to binary numbers and binary numbers to decimal numbers and do so quickly. Figure 9 depicts a characteristic screenshot. As it is the case with the previous games presented, the Cisco Binary Game focuses solely on learning the binary representation of numbers with no aim to use the binary representation for exploring the arithmetic operations as it is the case for the game presented in this paper.

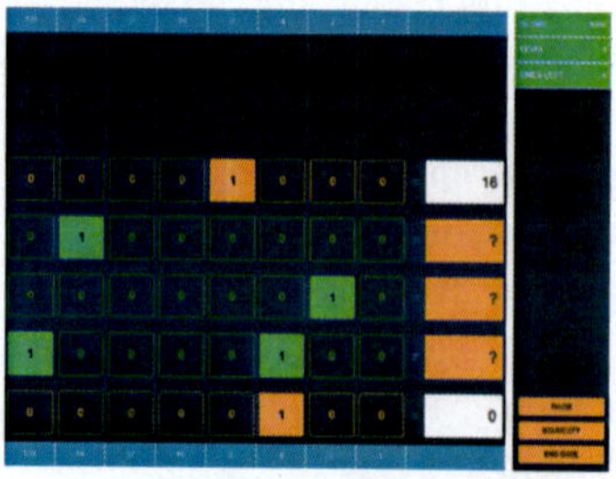

Fig. 9. Cisco Binary Game screenshot.

Binary bonanza [11] is yet another digital game to learn solely the binary representation of numbers. The player is requested to drag-select a number of bits in the square board on the left to create the requested numbers in the list on the right (Fig. 10).

Fig. 10. Binary Bonanza screenshot.

Finally, a game with several similarities with the board game described in [7] is Binary Dots [12]. The binary representation of a number is shown on a board presenting the significance or each binary digit. A yellow chip placed on a certain position of the board represents 0 while a red chip represents 1 (Fig. 11). Only binary representations of numbers are addressed by this game as well.

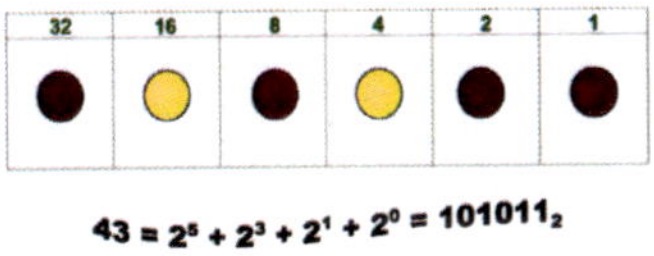

Fig. 11. The Binary Dots activity board.

5 Conclusions and Future Plans

The importance of computer coding and its potential for cultivating creativity has been widely recognized in education today, however some aspects of modern computing are not emphasized as they should in education. One of these aspects is the binary system. Its importance to modern computing is paramount, however, the usual presentation offered to young students is very limited without offering them the opportunity to understand how it is used to implement arithmetic. The authors believe that if students are offered a more thorough presentation of the binary system, thc could develop a deeper understanding of the decimal system as well and familiarize with the mechanisms of the arithmetic operations instead of learning them in a passive way without understanding why and how they actually work and produce credible results.

The game presented in this paper makes binary representations and operations highly interactive and thus helps the players enact the inner mechanisms of arithmetic operations. Understanding why the way we add, subtract, multiply and divide numbers is correct and how the final result is produced, can help learners overcome the surface knowledge that they usually develop with traditional learning approaches.

The learning effectiveness of the approach presented in this paper will be evaluated within the context of ongoing pilot activities of the TIM project. The first feedback from teachers that have already used it is that it increases engagement, creates a positive atmosphere in the classroom and offers several outdoor learning opportunities. Several features of the game propagate constructivist learning and engage students to learn the mathematics of number systems in general and binary system in particular by direct inquiry, experimentation and self-discovery of the mechanisms for arithmetic operations. In this respect the evaluation that will be done as part of the future work will evaluate the different features of the game and how they contribute to encouraging students to creatively explore mathematical ideas related to binary numbers.

Acknowledgments. This research is partially supported by the Bulgarian Ministry of Education and Science under the National Research Programme "Cultural heritage, national memory and development of society" approved by DCM №577/17.08.2018.

The game described in this paper is used within the context of pilot activities organized by the Theatre in Mathematics – TIM project (project number 2018–1-IT02-KA201–048139) funded by the Erasmus+ programme.

References

1. Kyfonidis, C., Moumoutzis, N., Christodoulakis, S.: Block-C: A block-based programming teaching tool to facilitate introductory C programming courses. In: 2017 IEEE Global Engineering Education Conference (EDUCON), 2017, pp. 570–579 (2017)
2. Moumoutzis, N., et al.: Design, implementation and evaluation of a computer science teacher training programme for learning and teaching of python inside and outside school. In: Interactive Mobile Communication, Technologies and Learning, pp. 575–586 (2017)
3. Papert, S.: Mindstorms. Basic Rooks, New York (1980)
4. Papert, S.: The children's machine: rethinking school in the age of the computer, 1993
5. Lameras, P., Moumoutzis, N.: Towards the gamification of inquiry-based flipped teaching of mathematics a conceptual analysis and framework. In: 2015 International Conference on Interactive Mobile Communication Technologies and Learning (IMCL) (2015)
6. Pappas, N., Arapi, P., Moumoutzis, N., Christodoulakis, S.: Supporting learning communities and communities of practice with Coursevo. In: Global Engineering Education Conf. (EDUCON) 2017 IEEE, Athens, Greece, pp. 297–306 (2017)
7. Moumoutzis, N., et al.: Using the PerFECt framework to invent playful learning activities for exploring the binary system. In 2021 IEEE Global Engineering Education Conference (EDUCON), 2021, pp. 1237–1246 (2021). https://doi.org/10.1109/EDUCON46332.2021.9453986
8. Sikiniotis, S., Kapros, H., Kordaki, M.: A computer game for the learning of binary system by beginners. In: IADIS International Conference e-Learning 2008 (part of MCCSIS 2008), pp. 29–36 (2008)
9. CS Unplugged, Binary numbers, Count the dots. https://classic.csunplugged.org/binary-numbers/. Accessed 3 Sep 2021
10. Cisco Binary Game. https://learningnetwork.cisco.com/s/binary-game. Accessed 3 Sep 2021
11. Binary Bonanza. https://games.penjee.com/binary-bonanza/. Accessed 3 Sep 2021
12. Binary Dots. https://raft.net/idea_sheets/binary-dots/. Accessed 3 Sep 2021

Cooking STEAM: A Case Study on Establishing a STEAM Learning Community using a Performative Framework and Cooking

Nektarios Moumoutzis[1,2(✉)], Chara Xanthaki[3], Ioannis Maragkoudakis[1], Stavros Christodoulakis[1], Desislava Paneva-Marinova[2], Lilia Pavlova[4], Petros Lameras[5], Stavroula Misthou[6], and George Kalmpourtzis[7]

[1] Laboratory of Distributed Multimedia Information Systems and Applications, School of Electrical and Computer Engineering,Technical University of Crete, 73100 Chania, Crete, Greece
{nektar, imarag, stavros}@ced.tuc.gr

[2] Institute of Mathematics and Informatics, Bulgarian Academy of Sciences, 1113 Sofia, Bulgaria
dessi@cc.bas.bg

[3] Directorate of Secondary Education of Chania, Chania, Greece

[4] Laboratory of Telematics, Bulgarian Academy of Sciences, 1113 Sofia, Bulgaria

[5] School of Computing, Electronics and Mathematics, Coventry University Coventry, Coventry, UK
ab3430@coventry.ac.uk

[6] GreekCodersK12 CodingClub, Athens, Greece

[7] Infinitivity Design Labs, Laussonne, France
gkalmpourtzis@playcompass.com

Abstract. The work reported in this paper is informed by PerFECt, a performative framework supporting collaborative learning and creativity that addresses issues related to the effective use of digital technologies to establish and sustain learning communities. Its core principles are reinterpreted to implement unplugged collaborative learning activities to creatively explore sciences, technology and mathematics with special emphasis on the core principles of algorithmic design by presenting recipes in the form of flow diagrams. By presenting the execution of algorithms within a dramabased (theatrical) framework many creative explorations are offered within the so called STEAM education. The adoption of playful activities that put the children's bodies in motion is well aligned with pedagogical approaches that emphasize embodied knowledge and performance and open up opportunities for deeper personalized learning experiences that promote inclusion of children from different cultural backgrounds.

Keywords: Cooking · STEAM · Flow diagrams · Algorithms

M. E. Auer and T. Tsiatsos (Eds.): IMCL 2021, LNNS 411, pp. 907–916, 2022.
https://doi.org/10.1007/978-3-030-96296-8_82

1 Introduction

Seymour Papert, a pioneer of educational informatics, has emphasized the need to give opportunities to children to become better learners through computer coding. In his seminal book "Mindstorms: Children, Computers and Powerful Ideas" he presented the big ideas behind computers and how these big ideas can be better understood by young children [1]. In later works, Papert expressed his disappointment that readers of this book that subsequently tried to put its vision into action through their teaching, focused too much on the word "computers", paying less attention to "powerful ideas". Today, with the advent of educational robotics, STEM/STEAM teaching and the implementation of world-wide initiatives to promote coding such as the Hour of Code and EU Week of Code [2, 3] the danger of overemphasizing the use of computers without paying the necessary attention to the powerful ideas behind them, is even bigger.

The issue becomes even more important when it comes to younger children at the first grades of primary school or even kindergarten. There is already a plethora of related apps and robotic kits available for children at these ages. Consequently, there is a temptation for teachers to focus on the technical part of educational informatics instead of trying to emphasize the technology independent big ideas of computing so that young children could deeply understand and effectively use digital technologies throughout their lives.

The work presented in this paper tries to address all these concerns on overemphasizing the technical aspects of computers in education by putting an explicit focus on powerful ideas related to the mathematical foundation of computer science and trying to offer engaging learning activities for children to understand these ideas even without using actual computer equipment. This is in line with an interesting approach to educational informatics termed "Computer Science Unplugged" [4]. To make the unplugged activities more interesting and engaging, we employ ideas from the TIM methodology that links mathematical teaching and learning with theatre in education. The methodology is developed and promoted by the Erasmus + project TIM – Theatre in Mathematics (https://www.theatreinmath.eu/). The resulting learning framework is titled "Cooking STEAM" and supports the creative exploration of algorithms in the form of flow diagrams to represent cooking recipes combined with drama-based performances to present the recipes from a professional cook combined with artistic interventions employing music, dancing and storytelling.

The overall approach is based on PerFECt, a design framework that emphasizes performativity, i.e. that knowledge is based on human performance and actions done within certain social contexts, rather than development of conceptual representations. This framework has already been used to guide the developments on several projects addressing digital heritage preservation and creative learning [5, 6]. In the work reported in this paper, the PerFECt framework is re-interpreted and re-used to focus on the idea of de-design. De-design addresses the need to put emphasis on leaving out features from a design in order to give more freedom to the users of a certain system [7]. In the case of the Cooking STEAM activity, de-design is used to offer an alternative to using digital tools for the representation of algorithms.

The rest of the paper is organized as follows: Sect. 2 presents the design principles of the PerFECt framework that are employed in the development of Cooking STEAM. Section 3 presents its rationale while Sect. 4 presents its design and how flow diagrams are used to represent recipes as algorithms. Section 5 presents major conclusions.

2 Design Principles Adopted from the PerFECt Framework

Technology in general and digital technologies in specific is a catalyst for establishing and sustaining certain social structures [8] within which End users are becoming "producers" of contents and functionalities. The term expert user is suggested in [8] to signify an expert in a particular domain with main goal to develop the technological capabilities available on that domain. An expert user engages in creative/authoring activities without being a professional software developer. Usually the role of end user and that of an expert user are played by different people that may also belong to different communities. Furthermore, [9] suggests the role of meta-designer to describe the work done by professionals who create the socio-technical conditions for empowering end users in operating as active contributors of contents and functionalities.

A meta-designer creates open systems that can be further developed by their users acting as co-designers. However, apart from the technical conditions necessary to set up such environments, there is a need to effectively create the social conditions that will allow expert users to build and adapt the artifacts to be used by end users. In respond to this need, a special user role is specified: maieuta-designers. A maieuta-designer creates the necessary preconditions for facilitating expert users appropriate the design culture and technical notions necessary for the meta-task of artifact development and involving as many end users as possible in the process of continuous refinement of the artifact, by improving participation. The use of the term "maieuta" directly references the Socratic method of getting people acquire notions, motivations and self-confidence to undertake challenging tasks.

End users, expert users, meta-designers and maieuta-designers engage in certain interactions with each other as well as with the digital artifacts and tools causing the emergence of a co-evolution phenomenon. Meta-designers focus on designing and providing the most effective tools that may sustain the co-evolution between end users and expert users. Maieuta-designers facilitate the transition from the role of end user to the role of expert user thus empowering people to appropriate and contribute to their digital artifacts. If certain end users are not interested or fail to move towards the role of expert user, maieuta-designers may facilitate system evolution by systematizing the reporting of opportunities or shortcomings, as identified by end users, and proposing solutions handled by expert users or even suggest further technological contributions from meta-designers. All the concepts presented above constitute the building blocks of the PerFECt framework as depicted in the figure below.

The interplay among the four roles of the framework give rise to two co-evolution processes as depicted in Fig. 1: The first one refers to the use of software devoted to the end user where there is continuous (cyclical) interaction between the end user and the system. This is depicted in Fig. 1 (left) with three homocentric cycles of arrows that

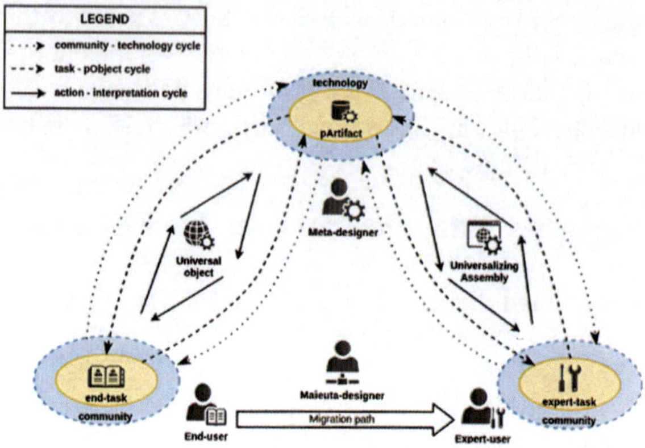

Fig. 1. The main components of PerFECt framework

represent the action-interpretation cycle at the lower level, the task-object cycle at the middle level and community-technology cycle at the upper level. In an analogous way, there is a second cyclical process depicted in Fig. 1 (right) that refers to the use of software devoted to expert user as building blocks of the system in continuous evolution. This process corresponds to yet another set of three homocentric cycles of the same nature: action-interpretation, task-object, and community-technology layers.

The inner interaction cycle in each co-evolution process refers to actions (triggered by the corresponding user or software) that are interpreted by the other party (soft-ware or user respectively). The task-object cycle in the middle refers to the co-evolution of the user task and the corresponding artifact within the boundaries of the System. Finally, an outer community-technology cycle captures the idea that the overall environment within which a user is working (community), co-evolves with the technology that supports the operation of this environment.

The PerFECt framework employs the notion of universality to addresses the issue of causality in digital representations, as Brenda Laurel puts it in her seminal book "Computers as Theatre" [10]: "*... an action is universal if everybody can understand it, regardless of cultural and other differences among individuals... Aristotle posits that any action can be "universalized" simply by revealing its cause; that is, understanding the cause is sufficient for understanding the action, even if it is something alien to one's culture, back-ground, or personal 'reality'.*" (p. 94). Consequently, within the PerFECt framework, the meta-task of expert users is to enable a universalization of physical objects by exploiting the available tools in the form of performative artifacts (pArtifacts) to account for the incorporation of the idea of performativity.

The concept of performativity, as exposed in [9] emphasizes the fact that human behavior can be understood and analyzed by assuming that all human practices are performed so that actions can be seen as a public presentation of self. This is the conceptual basis of the methodological breakthrough titled the performative turn in cultural studies, social sciences, humanities and design. The term turn signifies the trend to reverse the ontological premises that reality corresponds to particular objects, entities, and configurations that exist in and of themselves exhibiting certain essential qualities towards a new central hypothesis that objects are textures of partially coherent and partially coordinated performances existing through multiple situated practices. Meaning making is essentially a social process. Knowledge is created through the actions of the members of a social structure. In this respect, there is a shift towards "the active social construction of reality rather than its representation" [11].

The roots of this approach can be attributed to the need to move beyond the prevailing focus on texts or symbolic representations to capture meaning. Performance is, above all, a meaning making bodily practice. Consequently, it is related to rituals and other forms of spectacles and social practices [12]. Beyond the main premises and the theoretical justification of the validity of performativity, one could attribute the significance of this paradigm to an inherent dramatic quality of human experience.

3 Rationale of the Cooking STEAM Activity

A major issue that arises is the effective integration of people of diverse cultural backgrounds in a way that facilitates intercultural dialogue and creative use of diverse traditions that promotes human values and, hence can boost creativity and develop opportunities for innovation and new jobs for youth. Cooking STEAM addresses these challenges by identifying food, cooking, music and dancing as key aspects of different cultures by explicitly focusing on activities that encourage active citizenship, intercultural dialogue, while at the same time promoting the well-being of youth as well as its potential to develop critical skills and creative spirit to facilitate career pathways.

An important aspect of food as an accelerator of cultural identity is related to how diverse cultural communities try to preserve their eating habits, music and dancing traditions as part of their connection with their culture. However, due to the challenges that creative arts organizations, institutions and enterprises currently face, some aspects of the art and traditions of cooking as well as the creative arts risk being lost or abandoned. As a result, those sectors need to be reinforced and become resilient through new innovative and creative approaches, encouraging the participation of young people, educators, voluntary leaders, creative practitioners, and cooking experts.

Approaching food, music and dancing as important aspects of cultural identity, Cooking STEAM explores the traditions of European hosting countries and of countries of origin of migrants, to promote creativity, cultural awareness, STEAM learning (food and eating traditions integrate elements of physical sciences, chemistry, biology, technology, arts) and creative dialogue by combining the fields of STEAM with cooking, music and dancing.

Cooking STEAM employs an approach based on playful, creative learning activities that are offered to youth leaders, youth volunteers and educators, including youth clubs, industry and academia, as well as creative practitioners, food ambassadors and cooking experts. These also inspire them to develop their own learning pathways, together with their learners and their families. To this end, important aspects of the food culture are explored refer to the special ingredients of recipes that depend on the physical environment of the respective countries (like special plants and fruits), their myths and nutritional importance.

Cooking STEAM links tradition and technology to enable the development of important life skills for youth regarding chemical and biological parameters of food resources, their nutritional properties and preparation techniques. All the materials and learning scenarios are integrated in a community building and support framework (i.e. the PerFECt framework) that enables effective diffusion and reusing, to enable creativity and intercultural dialogue in the form of combined dishes, diets and eating contexts. Additionally, an interesting aspect of foreign language learning refers to name of ingredients, numbers and arithmetic/fraction and in the semiotic domain of cooking, with relation to recipes that provides an opportunity for local and migrant learners to learn a rich vocabulary in foreign languages and exercise their learning achievements during the collaboration with native speakers.

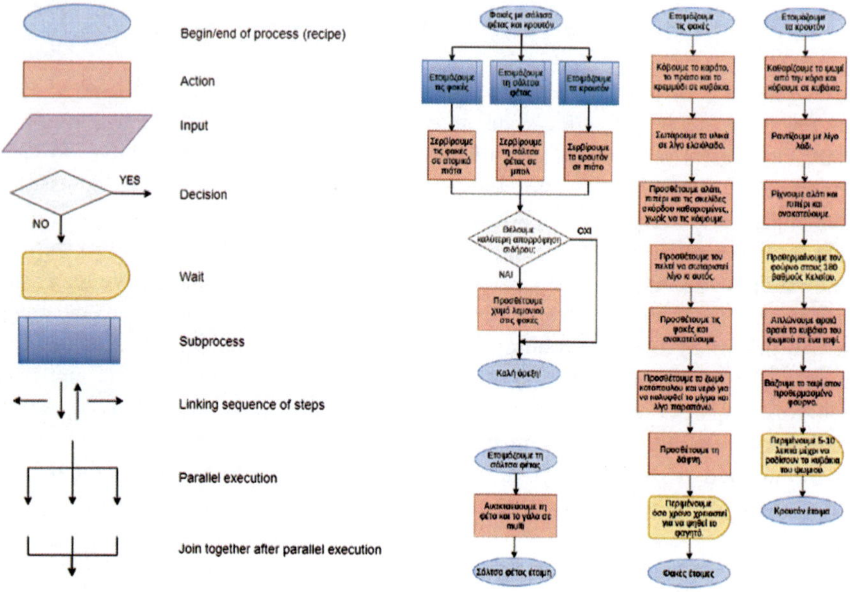

Fig. 2. The graphical components employed to represent recipes (left) and an example of a specific recipe (right).

4 Cooking STEAM Sessions Design and Algorithmic Representation of Recipes

Each Cooking STEAM session is organized into three parts:

1. Playful and artistic introduction (10–15 min) with narration and song or dramatized narration with a puppet. This part incorporates, on a case-by-case basis, information on selected local diet, reading nutritional information of products, etc.

2. Presentation of a recipe with a flow diagram and execution. In the end the children taste the food which, if it requires a lot of preparation time, has already been prepared in advance. During the execution and serving, the children discuss with the cook about the ingredients and their value, the cooking methods and whatever other questions they have. This interactive part has great potential and gives the opportunity for further discussion of ideas and concepts presented in the first part. The participation of children is particularly strong.
3. Mathematics/computer/science activities in the form of small exercises.

Figure 2 above presents the components used to represent the recipes as flow diagrams (left) and an example of a specific recipe employing these components.

Cooking STEAM captures and uses the main components of the PerFECt framework as it is presented in Sect. 2: When children are cooking following a recipe (or watching the cook enacting a recipe) they are essentially end users (in the terminology of the framework) that just need to follow certain rules with consistency. When children are creating their own versions of the recipe guided by a facilitator (this is the maieuta-designer in the PerFECt framework) to explore and gradually develop their skills in cooking, they are essentially in transition to become expert users, as the PerFECt framework suggests.

A very interesting aspect on how Cooking STEAM interprets the PerFECt framework and sheds new light on its applicability in designing collaborative learning experiences, is related to the idea of de-design [7]. De-design evokes the idea that omitting and leaving out features from a design is just as critical to the success of a system as it is including them positively. This is connected to the fact that any feature does both afford and constrain interactions with and through the artifact, what is left out of it has the potential to be even more important than what designers put in it on purpose. This is a disciplined inaction that is intentional and goes beyond mainstream design approaches by offering opportunities for different interpretations of the information that need to be considered in user practices, and recognize the creative power of ambiguity. The relevance of de-design to learning and creativity, is thus evident.

Following a de-design approach, Cooking STEAM, takes the idea of universality, as used in the PerFECt framework, along with the underlying concept of causality, and uses it beyond digital technologies to account for a human body (or a constellation of human bodies) that behaves under certain rules. This way, the cooking activity uses the human body as a kind of a robot, following certain instructions to realize a recipe. Extensions of the basic recipe, can give rise to alternative rules so that the participating children could creatively explore new situations and representations. In summary, Cooking STEAM generalizes the notion of universal objects and universalizing assemblies (Fig. 1) to account for any kind of object that can follow well-known rules and corresponding constellations of such objects.

5 Conclusions

The work reported in this paper is informed by PerFECt, a performative framework supporting collaborative learning and creativity that was initially developed to capture design principles suitable for the development of open learning environments [13] to effectively support collaborative learning [14] emphasizing the need to link learning to effective social structures. Consequently, this framework addresses issues related to the effective use of digital technologies to establish and sustain learning communities. The framework is also applicable in the analysis of serious games [15] and cultural heritage systems [16] as important means to foster engagement and creativity in learning. Its core principles, as discussed in this paper, can also be employed to implement unplugged collaborative learning activities. This is the case of Cooking STEAM to creatively explore sciences, technology and mathematics with special emphasis on the core principles of algorithmic design by presenting recipes in the form of flow diagrams.

Cooking STEAM departs from the usual conception of Computer Science Education that assumes the mandatory use of electronic computing devices including robots, physical computing devices etc. Although all these devices convey an important perception of special purpose programming languages that can help learners understand the importance of computer coding and its potential for cultivating creativity, they cannot capture the universal nature of algorithms and help the students understand the algorithms can be found everywhere around them, wherever a certain language is used to describe in a rigorous way any kind of process. By presenting the execution of algorithms within a drama-based (theatrical) framework, Cooking STEAM offers a sound ground for many creative explorations that help children get familiar with their bodies and physical materials. Such embodied knowledge approaches open up opportunities for deeper personalized learning experiences [17–19] possibly combined with flipped teaching of mathematics [20].

A systematic evaluation of the presented approach remains to be done. A pre-pilot implementation was done with limited time duration and participation due to the restrictions imposed by the COVID-19 pandemic. The qualitative feedback from the participants was extremely enthusiastic. This pre-pilot focused on Cretan Diet recipes and the use of local products in the preparation of meals and was linked to computer science lessons that developed simple robotic constructions that were used during the Cooking STEAM activities.

Acknowledgments. This research is partly supported by the Bulgarian Ministry of Education and Science under the National Research Programme "Information and Communication Technologies for a Single Digital Market in Science, Education and Security" approved by DCM №577/17.08.2018.

The Cooking STEAM activities employ ideas from the pedagogical methodology of the Theatre in Mathematics – TIM project (project number 2018–1-IT02-KA201–048139) funded by the Erasmus+ programme.

Within the context of the MAKER SCHOOLS: Enhancing Student Creativity and STEM Engagement by Integrating 3D Design and Programming into Secondary School Learning (project number 2020–1-BG01-KA201–079274) funded by the Erasmus+ programme, several tangible 3D objects are designed and printed to facilitate the implementation of Cooking STEAM learning activities and the engagement of participants in the creation of tangible flow diagrams.

The design and Implementation of the Cooking STEAM activities was done in collaboration with Maria Archontaki (animator), Stefanos Archontakis (chef), George Gypakis (musician and educational musical instruments maker), George Kiskoukis (musician), Alkis Bakaros, Lazaros Charalampidis and Chrysovalanti Mariolou (computer science teachers). The pre-pilot implementation was done in Theodoropoulou Schools (https://theodoropoulou.gr/el/news/view/12526) within their summer activities programme in 2020.

References

1. Papert, S.: Mindstorms. Basic Rooks, New York (1980)
2. Kyfonidis, C., Moumoutzis, N., Christodoulakis, S.: Block-C: a block-based programming teaching tool to facilitate introductory C programming courses. In: 2017 IEEE Global Engineering Education Conference (EDUCON), pp. 570–579. Athens, IEEE (2017)
3. Moumoutzis, N., et al.: Design, implementation and evaluation of a computer science teacher training programme for learning and teaching of python inside and outside school. In: Auer, M.E., Tsiatsos, T. (eds.) IMCL 2017. AISC, vol. 725, pp. 575–586. Springer, Cham (2018). https://doi.org/10.1007/978-3-319-75175-7_56
4. Bell, T., Vahrenhold, J.: CS Unplugged—how is it used, and does it work?. In: Böckenhauer, H.J., Komm, D., Unger, W. (eds.) Adventures Between Lower Bounds and Higher Altitudes. LNCS, vol. 11011, pp. 497–521. Cham, Springer (2018). https://doi.org/10.1007/978-3-319-98355-4_29
5. Moumoutzis, N., Christoulakis, M., Christodoulakis, S., Paneva-Marinova, D.: Renovating the Cultural Heritage of Traditional Shadow Theatre with eShadow. Des. Implementation Eval. Use Form. Informal Learn. DiPP 2018 Conf. Digit. Present. Preserv. Cult. Sci. Heritage. **8**, 51–70. Sofia, Bulgaria.: Institute of Mathematics and Informatics – BAS, 2018, ISSN 1314–4006 (Print), eISSN 2535–0366 (Online) (2018)
6. Moumoutzis, N., Sifakis, Y., Christodoulakis, S., Paneva-Marinova, D.: A reference framework to establish and sustain onlife communities and its use. Rich Learn. Exper. Hist. ViSTPro. Digit. Present. Preserv. Cult. Sci. Heritage. **9**, pp. 27–42. Sofia, Bulgaria: Institute of Mathematics and Informatics – BAS, 2019, ISSN 1314–4006 (Print). (2019) eISSN 2535–0366 (Online)
7. Cabitza, F.: De-designing the IT artifact. drafting small narratives for the coming of the socio-technical artifact. In: ItAIS 2014, Proceedings of the 11th Conference of the Italian chapter of AIS-Digital Innovation and Inclusive Knowledge in Times of Change, Track on Design and Re-design of Socio-technical Systems, pp. 21–22 (2014)
8. Cabitza, F., Fogli, D., Piccinno, A.: Cultivating a Culture of Participation for the Co-Evolution of Users and Systems. CoPDA@ AVI, pp. 1–6 (2014)
9. Cabitza, F., Simone, C.: Building socially embedded technologies: implications about design. In: Wulf, V., Schmidt, K., Randall, D. (eds.) Designing Socially Embedded Technologies in the Real-World. Computer Supported Cooperative Work, pp. 217–270. Springer, London (2015). https://doi.org/10.1007/978-1-4471-6720-4_11
10. Laurel, B.: Computers as theatre. Addison-Wesley (2013)
11. Dirksmeier, P., Helbrecht, I.: Time, non-representational theory and the 'performative turn' – towards a new methodology in qualitative social research. Forum Qual. Soc. Res. **9**(2), Art. 55 (2008)

12. Moumoutzis, N., Gioldasis, N., Anestis, G., Christoulakis, M., Stylianakis, G., Christodoulakis, S.: Employing theatrical interactions and audience engagement to enable creative learning experiences in formal and informal learning. In: Auer, M., Tsiatsos, T. (eds.) Interactive Mobile Communication Technologies and Learning. IMCL 2017. Advances in Intelligent Systems and Computing, vol. 725. pp. 142–154. Springer, Cham (2017). https://doi.org/10.1007/978-3-319-75175-7_16
13. Mylonakis, M., Arapi, P., Pappas, N., Moumoutzis, N., Christodoulakis, S.: Metadata management and sharing in multimedia open learning environment (MOLE). In: García-Barriocanal, E., Cebeci, Z., Okur, M.C., Öztürk, A. (eds.) Metadata and Semantic Research. MTSR 2011. Communications in Computer and Information Science, vol. 240, pp. 275–286. Springer, Berlin, Heidelberg. https://doi.org/10.1007/978-3-642-24731-6_29
14. Stylianakis, G., Moumoutzis, N., Arapi, P., Mylonakis, M., Christodoulakis, S.: COLearn and open discovery space portal alignment: a case of enriching open learning Infrastructures with collaborative learning capabilities. In: 2014 International Conference on Interactive Mobile Communication Technologies and Learning (IMCL2014), pp. 252–256. IEEE (2014)
15. Márkus, Z., Kaposi, G., Veres, M., Weisz, Z., Szántó, G., Szkaliczki, T., et al.: Interactive game development to assist cultural heritage. Digit. Present. Preserv. Cult. Sci. Heritage, **8**, pp. 71–82. Sofia, Bulgaria: Institute of Mathematics and Informatics – BAS, ISSN 1314–4006 (Print), eISSN 2535–0366 (Online) (2018)
16. Dochev, D., Pavlov, R., Paneva-Marinova, D., Pavlova, L.: Towards modeling of digital ecosystems for cultural heritage. Digit. Present. Preserv. Cult. Sci. Heritage **9**, 77–88. Sofia, Bulgaria: Institute of Mathematics and Informatics – BAS, ISSN 1314–4006 (Print), eISSN 2535–0366 (Online) (2019)
17. Arapi, P., Moumoutzis, N., Mylonakis, M., Theodorakis, G., Christodoulakis, S.: A pedagogy-driven personalization framework to support automatic construction of adaptive learning experiences. In: Leung, H., Li, F., Lau, R., Li, Q. (eds.) Advances in Web Based Learning – ICWL 2007. ICWL 2007. LNCS, vol. 4823, pp. 55-65. Springer, Berlin, Heidelberg (2007). https://doi.org/10.1007/978-3-540-78139-4_6
18. Yoshinov, R., Kotseva, M.: Vision for the engagement of the e-facilitator in school in the inspiring science education environment. Serdica J. Comput. **9**(3–4). Institute of Mathematics and Informatics, BAS, 2016, ISSN: ISSN 1312–6555 (2016)
19. Yoshinov, R., Kotseva, M., Arapi, P., Christodoulakis, S.: Supporting personalized learning experiences on top of multimedia digital libraries. Int. J. Educ. Inf. Tech. **10**, 152–158 . North Atlantic University Union, 2016 (2016). ISSN:2074–1316
20. Lameras, P., Moumoutzis, N.: Towards the gamification of inquiry-based flipped teaching of mathematics a conceptual analysis and framework. In: 2015 International Conference on Interactive Mobile Communication Technologies and Learning (IMCL), pp. 343–347. IEEE, Thessaloniki (2015)

EPuppet: A Mobile App to Extend a Digital Storytelling Platform with New Capabilities

Nektarios Moumoutzis[1,2](✉), Alexandros Koukis[1], Chara Xanthaki[3], Marios Christoulakis[1], Ioannis Maragkoudakis[1], Stavros Christodoulakis[1], Desislava Paneva-Marinova[2], and Lilia Pavlova[4]

[1] Laboratory of Distributed Multimedia Information Systems and Applications, School of Electrical and Computer Engineering, Technical University of Crete, 73100 Chania, Crete, Greece
{nektar, christoulakis, imarag, stavros}@ced.tuc.gr, alexkoukis93@gmail.com

[2] Institute of Mathematics and Informatics, Bulgarian Academy of Sciences, 1113 Sofia, Bulgaria
dessi@cc.bas.bg

[3] Directorate of Secondary Education of Chania, Chania, Greece
chara.xanthaki@gmail.com

[4] Laboratory of Telematics, Bulgarian Academy of Sciences, 1113 Sofia, Bulgaria
pavlova.lilia@gmail.com

Abstract. This paper presents the design, implementation and evaluation of a mobile app addressing creativity and learning in the field of cultural heritage. This mobile app enables the users of an existing digital storytelling platform inspired by traditional Shadow Theatre, to develop their own digital puppets in a way that incorporates the creative process of developing traditional shadow theatre puppets. ePuppet offers two main modes of editing: The first one is based on certain templates and facilitates the creation of digital puppets that have a certain structure (i.e. 2- or 4-part puppets). The second one enables users to create any kind of puppet with any number of constituent parts. The evaluation results show that the mobile app is highly usable and can effectively support the task of digital puppet creation.

Keywords: Digital storytelling · Digital puppets · Shadow theatre

1 Introduction

Shadow theatre is a storytelling tradition of many countries in Far East and Middle East using flat articulated puppets which are held between a light source and a translucent screen or scrim. It is a medium with significant educational value within the wider context of drama and performance arts [1]. This is due to its ability to engage people and promote their creativity. In particular children and adults find their own ways to act and imitate, create dialogues, get inspired and convey their own messages, direct, become stage designers, sing, strengthen their self-confidence giving life to the

M. E. Auer and T. Tsiatsos (Eds.): IMCL 2021, LNNS 411, pp. 917–926, 2022.
https://doi.org/10.1007/978-3-030-96296-8_83

puppets, improvise and create their own stories. Thereby they cultivate their oral speech skills and develop in multiple modes their intelligence (multiple intelligences) in an entertaining manner.

eShadow (http://www.eshadow.gr) is the digital version of shadow theatre [2]. It enriches traditional features with digital technology elements to offer a new way of dramatized and personalized digital storytelling. It enables the production of rich multimedia content interactively using innovative input devices and supports online collaboration. It offers an intuitive way of setting up scenes and enacting them. The user can select the desired scenery objects and digital puppets and then move them with mouse drag operations. All movements can be easily recorded along with the voice of the user. These recordings can be exported in appropriate file formats to be further edited with external video processing tools.

eShadow gives emphasis in the realistic motion simulation of shadow theatres puppets that is based on a physics engine. Realistic movement provides an explanation for the popularity of eShadow among Greek teachers and students as revealed during several field trials as well as during testing with actual professional performers [3]. Furthermore, realistic movement of digital puppets creates an atmosphere of playful interaction where the users are very easily engaged in theatrical improvisations that can be very important to develop communication skills related to oral expression and interpretation of body language.

eShadow is used in many schools to promote project-based learning combining arts with a wide range of school topics ranging from language learning, history and humanities to mathematics, physical sciences and computer science [2–4]. The creation of such artworks with eShadow corresponds to a project-based process with three distinct stages that is generic and refers to any media artwork such as movies, digital games, interactive animations etc. In particular, these stages are the following: Preparatory actions (pre-production), subsequent development (production) of the main materials of the artwork, and final assembly (post-production).

One of the most engaging phase in this creativity workflow refers to the development of the puppets that will be used in a digital story. Note here that eShadow offers several ready-made digital puppets to facilitate rapid development of digital stories. In many cases, however, the themes of the stories or the educational objectives of the teachers open up the opportunity to engage in the creative process of developing new puppets. To facilitate this process a digital puppet authoring tool was initially developed for end-users and used in combination with eShadow [3]. Furthermore, an open architecture was adopted in the representation of the computer files corresponding to digital puppets so that external image processing tools could be used [2] by expert-users that wish to create digital puppets that do not conform to predefined forms.

A new need emerged during learning interventions undertaken in pre-primary education. In particular, teachers participating in these learning interventions, during their interactions with eShadow developers in training workshops, asked for a way to facilitate digitization of paper-made shadow puppets using a simple mobile app. To address this need, the ePuppet mobile app was designed and implemented. This paper presents the design and implementation details along with the evaluation of ePuppet using a standard system usability questionnaire.

The rest of the paper is organized as follows: Sect. 2 presents the structure and representation of digital puppets in eShadow. Section 3 presents the design and implementation of the ePuppet mobile app. The evaluation of ePuppet is presented in Sect. 4. Section 5 concludes and presents future work plans.

2 Structure and Representation of Digital Puppets

The structure of digital puppets is very similar to the structure of traditional shadow theater puppets. In particular, they consist of two or more parts which are joined together by joints as shown below.

Fig. 1. Examples of shadow theatre articulated puppets. Red circles represent joints that connect puppet parts. The puppet on the left has two parts with one joint while the puppet on the right is more complex having four parts and three joints that connect them.

The female puppet on the left of the figure consists of two parts: the top and bottom of its body which are connected with only one joint. The male puppet to the right of the figure consists of four parts: The main body, waist and two legs. There are three joints: one that connects the main one body with the waist and two that connect the waist with left and right legs respectively. These two typical cases (with four or two parts) capture the structure of most puppets of traditional Greek shadow theater.

To synthesize such puppets, one has to develop their parts in certain material, cut them and join them together at the correct joint positions. The female puppet consists, as mentioned, of two parts which are associated with one joint. These two parts are illustrated separately in Fig. 2 as well as their connecting points (joint anchors) to form the joint of the puppet. The selection of connection points, especially for the bottom part affects not only the aesthetics of the puppet but also its behavior during handling due to the force of gravity. Wrong point selection will have as a result the lower part of the puppet to be unbalanced. Consequently, puppets are tested before the position of joints is fixed.

The representation of a digital puppet employs two files:

Fig. 2. Synthesis of shadow theatre articulated puppets. On the left, an example of a 2-part puppet is shown with its constituent parts and the corresponding joint anchors (red circles) in each part. On the right, an example of a 4-part puppet is shown. In both cases the red rectangles represent the boundaries of areas used to place each constituent part.

- An image file (sprite sheet) with all the constituent parts of the puppet.
- A json file with the information about the placement of the constituent parts in the sprite sheet as well as the position of joint anchors.

The figure below depicts the json file and the corresponding sprite sheet for the 2-part puppet presented above. The representation of the 4-part puppet is analogous.

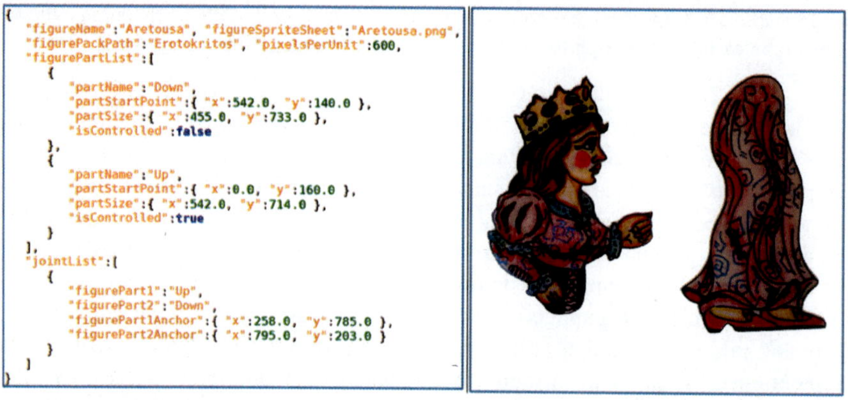

Fig. 3. Digital puppet representation. On the left, the json file describing how the puppet is composed of different parts, where these parts can be found in the corresponding sprite sheet and where the joint anchors are placed. On the right, the corresponding sprite sheet is depicted.

3 Design and Implementation of ePuppet Mobile App

The goal of ePuppet is to primarily provide an all-in-one solution for its users to digitize, customize and import custom figures and scenery in the eShadow platform. The most important qualities of the project are ease of use, minimal design and as little user input as possible in order to streamline the process. Its conception followed successfully uses of eShadow in many primary and secondary schools, with very interesting results showcasing the educational potential of such technologies and their importance in reconnecting young generation with their cultural heritage, a new project was initiated addressing pre-primary education.

Within these activities a certain need was identified to develop traditional puppets before digitizing them. This need came from certain pedagogical objectives. Furthermore, there was a need to support users without skills and available time to engage in the process for preparing digital puppets using image processing software. To address this need a new generic component was identified that would allow eShadow users adopt a new way of work in order to be able to create new digital puppets. To facilitate this process the ePuppet was designed as a mobile application for digitization of either template-based puppets or generic puppets without any restriction in their structure.

In the figure below navigation among the ePuppet pages is depicted. When the app is initiated, the user is presented with the Home page. From this page, the user is able to activate the puppet options page to select what type of puppet is to be digitized. The options offered are:

1. 2-part puppet
2. 4-part puppet
3. Free-from puppet

After selecting the desired puppet type, the user is able to use the camera of the mobile device to take a photo of the analog puppet that is going to be digitized. If it is a template-based puppet (2- or 4-part) the corresponding template is overlaid on the camera input to facilitate the positioning of the parts in the template (Fig. 5). If, on the other hand, the desired type is a generic puppet, the user just takes the photo of the analog drawing. The next step in either cases is to remove the background of the image so that its parts can be drawn on a transparent background. This is necessary to be able to effectively use the digital puppet in eShadow. Following background detection and subsequent removal, if the user has selected to create a template-based puppet, the process can continue with a new round to create yet another digital puppet. If, on the other hand, the desired type is a generic digital puppet, then the user has to specify its parts by drawing the corresponding bounding rectangle for each part and then define the joint anchors that specify how the distinct parts are joined together to create the articulated digital puppet. During the final step (final editing page), the digital puppet is displayed for better visual clarity, and minor tweaks are possible before finalizing the digital puppet.

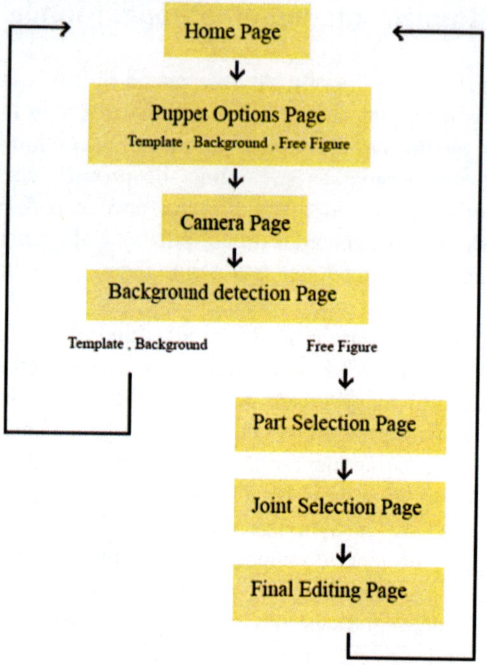

Fig. 4. Flow diagram of ePuppet pages.

Fig. 5. Digitization of shadow theatre puppets using ePuppet in the case of 2-part (left) or 4-part (right) puppets. The parts of the puppet are put on a flat surface with a constant color background (e.g. on a sheet of paper). The parts and the mobile device are moved as needed so that the parts are framed within the red boundaries shown in screen overlay and the joint anchors (red circles) are positioned correctly. Then, a photograph is taken that is transformed by ePuppet into a correct digital puppet representation.

In more detail, assuming that a traditional puppet with 2-parts has already been created (the case of a 4-part figure is analogous), its parts can be put on a flat surface with a constant color background so that the ePuppet app can be used to take a photo of the parts after appropriate aligning them with predefined template (Fig. 5).

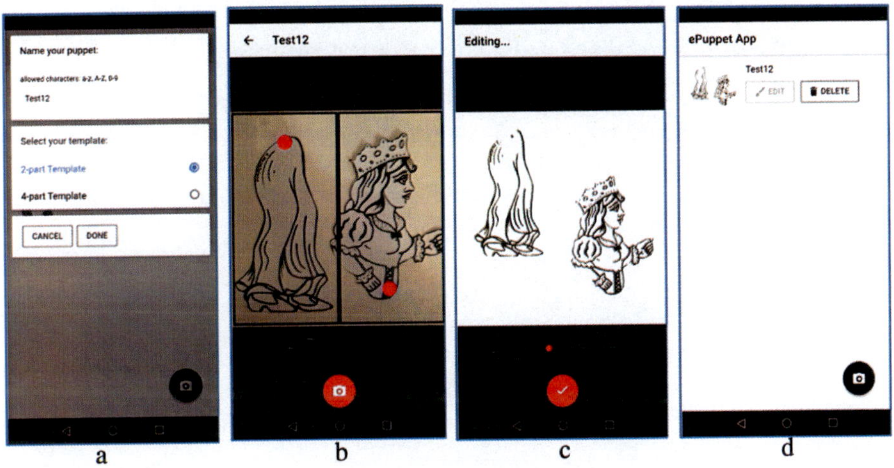

Fig. 6. Creating a digital puppet with ePuppet mobile app. The user (a) gives a name to the puppet and selects a template to use; (b) takes a photo of the physical parts of the puppet; (c) adjusts color levels to automatically remove the background; and (d) stores the files created.

The detailed steps followed by a user using ePuppet are shown as a sequence of screenshots in Fig. 6. Initially, the user provides a name for the new puppet and selects the corresponding template (if it will be a two- or 4-part puppet). Next, the user activates the camera of the mobile devices and overlays a guiding template as a slide that is placed over the cameras image. Moving parts of the puppet as well as the devices left to right, up and down and back and forth full alignment of the points is achieved connecting the joint anchors (red dots) to coincide with the connection points of the real puppet parts. After aligning the parts, the user selects the appropriate control to make a photo of the puppet. In the next screen, the user is able to adjust the background color levels to enable its automatic removal to clear the surrounding area of the puppet parts. After that, the application creates all necessary files so that the digital version of the puppet can be used in eShadow and lists the new puppet in the mobile app folder area.

The case of a generic digital puppet shown in Fig. 7 starts from shooting the created puppet which is presented as whole. Then the user is able to cut out its parts and specify one-by-one the joints that are necessary to animate the puppet.

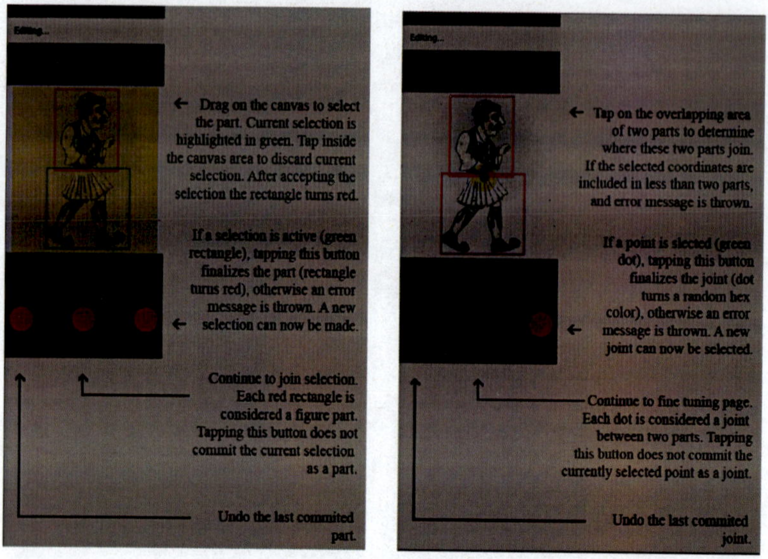

Fig. 7. Part selection (left) and joint selection (right) pages (generic digital puppet).

4 Evaluation Results

ePuppet was evaluated in a pilot activity undertaken during the annual Science and Technology Day at the Technical University of Crete, Greece on 16/11/2019. This event is targeted at students of primary education that visit the university, usually accompanied by their teachers and/or parents. The pilot activity was organized as a playful learning activity where the children first watched a special installation of eShadow having also the opportunity to interact with digital puppets using back projection and body shadow. After that, they were offered materials to create their own puppets on paper and finally they were able to digitize their creations using ePuppet, upload them in eShadow and use them to make short improvisations. After finishing the activity, the students were asked to fill the System Usability Scale (SUS) questionnaire [5, 6]. In total, 42 primary school children participated in the evaluation.

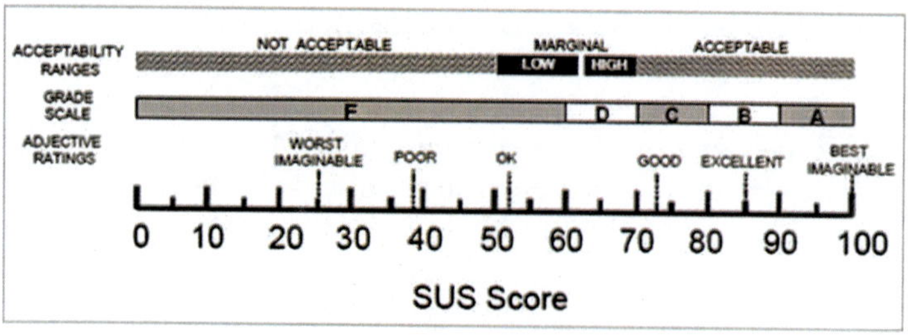

Fig. 8. SUS benchmark scores.

The analysis of the questionnaires revealed that the average rating by the participants was 75.5%, which is within the C SUS benchmark grade scale (see figure below) or just above the GOOD limit in adjective ratings. In terms of learnability, the score was equally good (76,5%). In particular, this refers to the results recorded in SUS questionnaire questions that directly address the learnability of the system.

5 Conclusions and Future Plans

ePuppet effectively facilitates the creation of digital puppets by engaging its users in a creative workflow within which they first develop analog shadow puppets, thus offering them a learning experience that links traditional shadow theatre to digital technologies. The evaluation results clearly demonstrate the usability of ePuppet and its use. ePuppet, apart from the reported evaluation, is used in many schools within the context of project-based learning activities to enable students and teachers to actively explore various themes ranging from social topics to STEM learning in an engaging and creative manner. ePuppet overcomes the obstacles faced by students and teachers with low digital skills to develop their own digital puppets and opens up new opportunities for creativity and learning.

Future work will streamline the use of ePuppet and eShadow combined in a significant trial related to the celebration of 200 years from the Greek Independence and other topics related to national and local history. These activities will exploit collaborative learning approaches and tools [7] and promote personalized learning experiences [8]. Finally, the use of ePuppet to create digital stories linked to STEM education to engage students in learning to code is also being actively explored [9, 10].

Acknowledgments. This research is partly supported by the Bulgarian Ministry of Education and Science under the National Research Programme "Cultural heritage, national memory and development of society" and the National Scientific Program "Information and Communication Technologies for a Single Digital Market in Science, Education and Security" approved by DCM №577/17.08.2018.

ePuppet has been used in pilot activities with schools and professional associations in Chania, Greece, within the context of the Erasmus+ project TIM (2018-1-IT02-KA201-048139).

Puppets depicted in Figs. 1, 2, 3, 5, and 6 are digital versions of original creations by Mr. Nikos Mplazakis, a professional Greek Shadow Theatre puppeteer.

References

1. Hatzigianni, M., Miller, M., Quiñones, G.: Karagiozis in Australia: exploring principles of social justice in the arts for young children. Int. J. Educ. Arts **17**(25) (2016)
2. Moumoutzis, N., Christoulakis, M., Christodoulakis, S., Paneva-Marinova, D.: Renovating the cultural heritage of traditional shadow theatre with eshadow. Des. Implementation Eval. Use Formal Informal Learn. In DiPP 2018 Conference on Digital Presentation and Preservation of Cultural and Scientific Heritage. , Sofia, Bulgaria: Institute of Mathematics and Informatics – BAS, 2018, **8**, 51–70, ISSN 1314–4006 (Print), eISSN 2535–0366 (Online) (2018)

3. Moraiti, A., et al.: Playful creation of digital stories with eShadow. In: 11th International Workshop on Semantic and Social Media Adaptation and Personalization (SMAP), Thessaloniki, pp. 139–144 (2016)
4. Moumoutzis, N., Gioldasis, N., Anestis, G., Christoulakis, M., Stylianakis, G., Christodoulakis, S.: Employing theatrical interactions and audience engagement to enable creative learning experiences in formal and informal learning. In: Auer, M.E., Tsiatsos, T. (eds.) IMCL 2017. AISC, vol. 725, pp. 142–154. Springer, Cham (2018). https://doi.org/10.1007/978-3-319-75175-7_16
5. Brooke, J.: SUS-A quick and dirty usability scale. Usability Eval. Ind. **189**(194), 4–7 (1996)
6. The System Usability Scale & How its Used in UX: https://medium.com/thinking-design/the-system-usability-scale-how-its-used-in-ux-b823045270b7. Accessed 3 Sept 2021
7. Stylianakis, G., Moumoutzis, N., Arapi, P., Mylonakis, M., Christodoulakis, S.: COLearn and open discovery space portal alignment: a case of enriching open learning Infrastructures with collaborative learning capabilities. In: 2014 International Conference on Interactive Mobile Communication Technologies and Learning (IMCL2014), pp. 252–256. IEEE (2014)
8. Arapi, P., Moumoutzis, N., Mylonakis, M., Christodoulakis, S.: Design, implementation and experimental evaluation of a pedagogy-driven framework to support personalised learning experiences. In: Proceedings of 2th LOGOS Workshop Cross-Media and Personalised Learning Applications with Intelligent Content Varna, pp. 7–25 (2008)
9. Kyfonidis, C., Moumoutzis, N., Christodoulakis, S.: Block-C: a block-based programming teaching tool to facilitate introductory C programming courses. In: 2017 IEEE Global Engineering Education Conference (EDUCON) (pp. 570–579). IEEE, Athens (2017)
10. Moumoutzis, N., et al.: Design, implementation and evaluation of a computer science teacher training programme for learning and teaching of python inside and outside school. In: Auer, M.E., Tsiatsos, T. (eds.) IMCL 2017. AISC, vol. 725, pp. 575–586. Springer, Cham (2018). https://doi.org/10.1007/978-3-319-75175-7_56

A Rudimentary Progression Model for Artificial Intelligence in Education Competencies and Skills

Petros Lameras[1(✉)], Iraklis Paraskakis[2], and Stathis Konstantinidis[3]

[1] Coventry University, Coventry, UK
ab3430@coventry.ac.uk
[2] City College, Thessaloniki, Greece
paraskakis@york.citycollege.eu
[3] University of Nottingham, Nottingham, UK
Stathis.Konstantinidis@nottingham.ac.uk

Abstract. Artificial Intelligence in Education is progressing with accelerated pace, and this impacts the role of the teacher in becoming catalyst in designing and orchestrating teaching and learning with the use of AI-based systems. This study proposes a rudimentary competency framework with an associated progression model that aims to help teachers to plan, develop, self-assess, and reflect on existing and new competencies for teaching and learning through AI-based systems. The framework encompasses six competency themes along with six relational and inclusive descriptors of competency progression levels that may be used either as standalone or as a concerted set of competencies for enabling complete and holistic conceptualisations, and utilizations of competencies and capabilities that teachers may envisage to develop for making the leap into adaptive and automated AI-based teaching and learning. The AIED Comp framework may inform computational representation models and algorithms that focus on teacher-facing AIED applications and tools for gaining insights on skills and capabilities teachers may employ for innovating in their classroom.

Keywords: Artificial Intelligence in Education (AIED) · AIED competencies · Teaching and learning · Progression model · Teachers

1 Introduction

Artificial Intelligence in Education (AIED) is experiencing a dramatic surge due to contemporary developments of AI in fields such as medicine, finance, and energy that have seen notable potential in improving well-being, optimizing cost effectiveness and monitoring energy consumption [1]. This pervasiveness of AI is further illuminated in everyday life experiences such as using virtual assistants to control digital and physical objects. AIED is one of those technological interventions that may create disparity and confusion as to what it means and how it can be employed for designing and orchestrating teaching and learning. It is evident that interdisciplinary research on digital learning technologies has assisted in understanding how technology-enhanced

M. E. Auer and T. Tsiatsos (Eds.): IMCL 2021, LNNS 411, pp. 927–936, 2022.
https://doi.org/10.1007/978-3-030-96296-8_84

teaching and learning is planned, represented, and enacted as to provide optimal digital learning experiences. Based on this increased understanding of how educational technology may enhance the student's learning experience, for more than thirty years learning science researchers and AI scientists are investigating the potential of AIED in terms of offering tailored and personalized experiences of learning through algorithmic computational representations and data models that analyze and predict students' unique learning pathç (e.g., [2, 3, 4]). To this end, AIED is conceptualised as making computationally precise inferences of learning and teaching as to offer an adaptive learning experience ([5–7]). This also insinuates the assumption that AIED has the potential of making explicit how students learn, their prior learning experiences and any misconceptions that inhibit student's understanding. Such inferences on student's ways of learning may unlock the *black box of learning* (e.g., [5]) as means to discern deep understandings and meanings of how learning is developed that may inform the development of AIED systems but also to augment approaches to teaching that will support the design and sequence of learning with the use of AI.

Designing and enacting teaching and learning through AIED requires both design for learning expertise including pedagogical and subject-content knowledge as well as technological expertise for enabling teachers to make full advantage of effective use of AIED. There is an assertion (e.g., [5, 7, 8]), that teachers would need to increase their confidence and skills to transform their role as catalysts in the student-teacher-AIED nexus. Digital skills development for AIED is considered as a sine qua non intervention for helping teachers to identify different ways, tools, applications, processes, and strategies of using AIED that would enable them to act as change agents and orchestrators of how AIED-based systems, data algorithms, predictions and teaching instances are going to be triggered for support adaptability and personalization.

This paper aims to present the conceptualisation of a rudimentary digital competency development framework, the AIEDComp, that may help teachers to plan, self-assess and reflect on associated digital competencies for enabling the use of AIED in designing and orchestrating teaching and learning. In particular, the focus of attention is on providing a progression model for empowering teachers to gain an informed understanding on designing learning via AI tools and to make conscious decisions on AIED-related skills and competencies that they wish to develop or refining, relearning, and adjusting existing digital competencies towards the use of AIED applications, tools, and processes. The paper starts by situating the study into the wider context of digital skills competencies development before perpetuating on the AIEDComp and the progression model developed for enabling teachers to self-assess their AIED related skills. We contemplate on the relationality of the progression model and we then reflect on future lines of research specifically related with teachers' AIED competencies and skills development.

2 Digital Competency Development for AIED

The overarching assumption to investigating digital competencies for designing and delivering teaching and learning using AIED is that the way teachers develop AIED related competencies may influence the design and orchestration of teaching and learning which in turn may lead to exploiting the full potential of AIED applications as means to provide an adaptive and personalised learning experience. The integration of technology in education has been a prominent area of research and discourse in schools, but there are several arguments, which support the lack of effort to efficiently intergrade technology with design, representation and delivery of teaching and learning (e.g., [9–11]. Digital competency development may be understood as an interconnected set of skills or competencies for enabling the design and orchestration of teaching and learning [12]. Different terms are gauged to denote the development of digital learning skills such as: digital literacy (e.g. [12]), ICT literacy and ICT competence (e.g. [11]). There is differentiation in meaning when these terms are used from more technical-centred instantiations (e.g., ICT literacy) to more knowledge-centered and integrated (e.g., ICT competence) focusing on the information, knowledge and pedagogical strategies, tools and processes employed to foster learning with the aid of technology (e.g., [13]).

Commentators argue that the purpose of acquiring digital competencies is also distinguished in two types of competencies: (a) for helping students to use ICT in the classroom and (b) for designing a rich mediated ICT-enabled learning environment (e.g., [9, 10, 12]). As a third type of competencies complementing the two, we propose competencies that promote inclusive, creative, meaningful, and personalised teaching and learning for ideating, designing, orchestrating, assessing, and reflecting on teaching and learning. The third type of competence alludes to the premise that digital competency development should not only focus on how to use technology but also to deliberate how pedagogy and context shape the way teaching and learning is designed, delivered, and assessed.

Indeed, there have been efforts to formulate digital competency frameworks with a holistic approach to highlighting a gamut of competencies from data and information to pedagogy, ethics, and inclusion (e.g., [14–16]). One of the most important digital competency frameworks is the European Union's DigiCompEdu [17] designed to offer a frame for teachers to identify, develop and assess digital competencies pertinent to using digital technologies in informed, creative, collaborative, and critical ways. DigiCompEdu presented six competency themes that encompass key subthemes such as information and media literacy, content creation, self-regulated learning, collaborative learning, assessment strategies, feedback and planning, differentiation, and personalization among others comprising twenty-two competencies in total.

3 Digital Competencies for Teaching and Learning Using AIED: The AIEDComp Framework

A set of six overarching digital competencies for helping teachers to develop capabilities in technology-enhanced teaching and learning were discerned by [18], and thereby sensibly perceived as competencies closely related to, and adapted for, AIED teaching and learning (see Table 1).

Table 1. The AIEDComp Framework

AIED Competencies Themes and Subthemes
A: Designing, developing, and delivering digital content
A.1 Designing digital content
A.2 Developing digital content
A.3 Representing digital content
B: Acquiring data, information, and data ethics skills
B.1 Understanding and tracking student's progress through gathering and analysing data
B.2 Finding, accessing, using and sharing information
B.3 Using student data ethically
C: Developing skills in employing digitally and activity-led pedagogies
C.1 Collaborative learning
C.2 Inquiry-based and research-based learning
C3. Activity and digitally led assessment
C4. Utilising multiple modes of feedback
C5. Reflection
D: Becoming proficient in AIED applications, tools, and software
D.1 Use of AIED software and hardware for tracking, recording and visualising progress and performance
D.2 Applying knowledge to solve simple technical problems with AIED software and hardware
D.3 Identifying, selecting and appraising AIED software and hardware based on educational and technical requirements
D4 Basic understanding of big data, algorithms, AI techniques (e.g., machine learning) and systems thinking
E: Developing digital creativity skills, empathy, and do-it-yourself culture
E.1 Ideating, brainstorming, and designing AIED-based learning activities
E.2 Personalising, sharing and remixing AIED learning activities
E.3 Making explicit students' affective states for integrating emotions in AIED activities
E.4 Designing and creating AIED that connect digital material with physical objects
F. Fostering student digital inclusion, social influence, and relatedness
F.1Embracing equal learning opportunities into the design of AIED systems
F.2 Producing digital learning resources that are unbiased, inclusive and diversified
F.3 Designing and visualising digital learning resources that are related to students' past learning experiences, feelings, culture, and code of ethics

The competencies spanned from content design and delivery, acquiring student data and ethics to activity led-AIED strategies, AIED-application proficiency, creativity skills, inclusion social responsibility and data compliance. For example, in the content design and delivery competency the focused is on personalised content design, creation and retrieval. Design of digital content was felt as important for structuring and determining what and how content will be available to students through a designated digital content system that would track content that is relevant to student's needs and topics of interest. Designing content was distinctive from developing content in a sense that the teacher first had to learn what content needs to be provided as to match student's needs and interests based on similar content retrieved in past learning situation by students, how content will be visualized and represented, how it will be associated with the learning outcomes, activities, feedback and how it will be sequenced with content, activities, and outcomes from different subject topics.

Choosing and learning how to use an AI system focused on storing, using, and sharing information and learning material may be a critical digital skill. Designing content related to personalizing the learning experience to students and therefore, designing content may also be associated, not to just mere content design, but also designed in way that would be suited to student's needs and interests and as such the identification of students needs through collecting information on student's prior knowledge, performance and misconceptions may be part of the content design process.

Developing skills, competencies, and capabilities to acquire data, information and data ethics skills may be central for bootstrapping data privacy aspects with learning and pedagogy. Activity-led pedagogies are key for enriching teaching strategies but especially in terms of deploying a particular paradigm aligned with personal conceptions of teaching using AIED and beliefs about how activity-led teaching may enhance learning with the use of AIED. The capacity to use AIED for tracking and analyzing performance as well as solving broad technical and data-driven problems is an important skill for teachers as to maintain the momentum of learning by rapidly addressing technical problems.

Ideating, brainstorming, and designing AIED activities that help students to be creative and exert emotion through making and developing prototypes, products and services is a capability that needs to be cultivated and practiced and tracked via AIED learning analytics and emotional prediction techniques. The underpinning capability is to design and represent digital learning that is engaging and allows students to be self-directed, autonomous, and independent in their learning journey hence would trigger positive memories, feelings and emotions that would likely increase engagement and motivation and thereby learning. Fostering student digital inclusion, social influence, and relatedness encompasses skills around embracing equal opportunities using data, algorithms, and predictive models and through the design and provision of automated learning resources that are inclusive, diversified and aligned to student's code of ethics.

4 The AIED Progression Model

Digital competencies for teaching and learning using AIED may be perceived as the twinning of knowledge, skills, and attitudes to successfully develop, implement, and achieve a set of learning goals and outcomes to be orchestrated with the use of AI agents or systems. To enable teachers to aggregate and reflect on existing and new AIED competencies, a self-assessed progression model is proposed. The model consists of different progression levels that describe variation in AIED-based teaching competency development. The six AIED digital competences stages are twinned to six proficiency levels used by the Common European Framework of References for Language (CEFR), ranging from A1 to C2 as to inculcate familiarity and simplicity for teachers to interpret the progression levels and delineate a subjective understanding of their competencies in teaching and learning using AIED. Deconstructing the competencies to different levels would allow teachers to focus on the proffered nuances whilst attenuating complexities in the application of the competencies in actual practice. The progression model does not need to be viewed as an assessment tool for performance appraisal but rather is it evoked as a supportive metric for identifying, understanding, and delineating progress and self-reflection (Fig. 1).

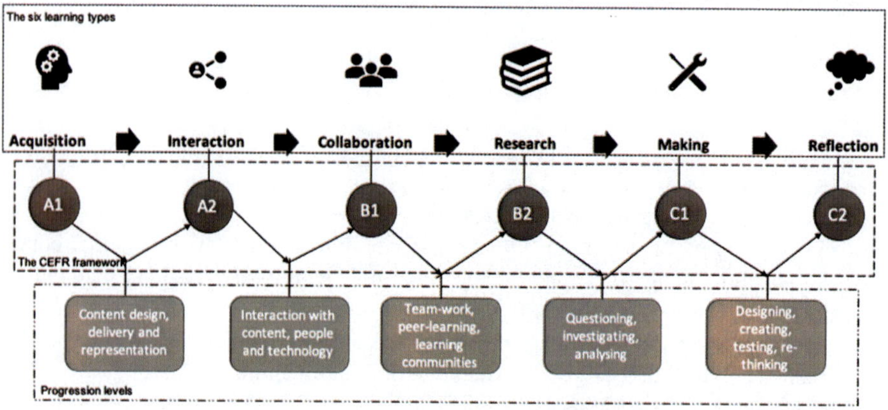

Fig. 1. The structure of the AIED comp: The six learning types twinned with CEFR and progression levels

Specifically, the learning goals descriptors that run through the six categories are qualitatively different to inform associated competencies. The descriptors were inspired from [19] six types of learning spanning from acquisition, discussion and practice to production, collaboration and investigation thus were adapted to propagate a progressive and relational representation, from less developed to more cohesive, of skills and competencies. In the first two stages, Acquisition (A1) and Interaction (A2) the focus is on content design, delivery, and representation (A1) and on interaction with teachers, peers, content, and technology (A2). In Collaboration (B1) and Research (B2) the focus is on working in teams and creating online learning communities based on interests and needs (B1) and transitioning to conducting research through, posing questions, initiating investigations, analysing, and constituting evidence (B2). Then, the focus shifts from Research (B2) to Making (C1) with emphasis on creating, tinkering, and making tangible products and artefacts to Reflection (C2) for critically reflecting on how content, data, activities, AI-based systems, and applications used ethically may help students to enhance the learning experience (see Table 2).

Amalgamating multiple competency levels developed by a teacher may provide a more consistent, systematic, and strategic approach to planning, assessing and consciously reflecting on competencies for AIED-based teaching and learning. In this

Table 2. The six themes of AIED teaching mapped to (A1) – (C2) progression levels

The 6 Themes of AIED-based Teaching and Learning	Progression Levels					
	Acquisition (A1)	Interaction (A2)	Collaboration (B1)	Research (B2)	Making (C1)	Reflection (C2)
1. Content design and delivery	Designing, developing and delivering subject content for helping students to acquire information	Designing, developing and delivering subject content for helping students to interact with teachers, peers, content and technology	Designing, developing and delivering subject content for helping students to engage in collaborative learning	Designing, developing and delivering subject content for helping students to engage in research	Designing, developing and delivering subject content for helping students to cultivate creative and making mindsets	Designing, developing and delivering subject content for helping students to critically reflect on subject-content
2. Acquiring student data, information, and data ethics	Collecting, analysing and visualising data on student's subject content and information acquisition	Collecting, analysing and visualising data on student's interactive processes with teacher, peers, content and technology	Collecting, analysing and visualising data on student's collaborative learning processes	Collecting, analysing and visualising data on student's research-based learning processes	Collecting, analysing and visualising data on student's creativity and making processes	Collecting, analysing and visualising data on student's reflective processes
3. Activity-led AIED strategies	Employing activity led-strategies that support students to acquire adaptive information and personalised subject content	Employing activity-led strategies that support students to interact with teachers, peers, content and technology	Employing activity-led strategies for designing collaborative activities	Employing activity-led strategies for designing research-based activities	Employing activity-led strategies for designing activities that encourage students to create and make	Employing activity-led strategies for designing activities that encourage students to critically reflect on learning
4. AIED tools usage proficiency	Selecting and utilising AIED software and hardware that enable students to find, access and retrieve information and subject content	Selecting and utilising AIED software that enable students to interact with teachers, peers, content, and technology	Selecting and utilising AIED software that enable students to participate in and contribute to collaborative learning	Selecting and utilising AIED software that enable students to participate in and contribute to research-based learning	Selecting and utilising AIED software that enable students to participate in and contribute to creative and making processes	Selecting and utilising AIED software that enable students to critically reflect on their learning
5. Digital creativity skills	Designing and using AIED environments that support and guide students to acquire information and subject content creatively	Designing and using AIED that support and guide students to interact with teachers, peers content and technology creatively	Designing and using AIED that support and guide students to engage in creative collaborative learning	Designing and using AIED that's support and guide students to participate in and contribute to research-based learning creatively	Designing and using AIED that supports and guides students to cultivate creativity and a do-it-yourself culture	Designing and using AIED that support and guide students to critically reflect on their creative learning
6. AIED inclusion, social responsibility, and data compliance	Designing and delivering inclusive, accessible and ethical information and subject content	Designing and using AIED environments that support and guide students to interact with teachers, peers, and content in inclusive, accessible and ethical ways	Designing and using AIED environments that support and guide students to engage in collaborative learning in inclusive, accessible, and ethical ways	Designing and using AIED environments that support and guide students to engage in research-based learning in inclusive, accessible, and ethical ways	Designing and using AIED environments that support and guide students to engage in making and creativity in inclusive, accessible and ethical ways	Designing and using AIED environments that support and guide students to engage in reflective thinking in inclusive, accessible, and ethical ways

sense, different espoused and actual competencies perceived by one teacher could be combined and complemented with competencies that other teachers may have in their repertoire of skills and capabilities. The AIED comp supplements and extends the purpose and rigor of the DigCompEdu [16] as its overarching focus is on the design and orchestration of AIED-based teaching and learning from a more hierarchical, inclusive, and developmental approach to using digital technology with focus on automation, adaptation, and contextualization. Building on AIED competencies and skills that are sought to address a learning outcome or goal may be a more rigid, constructive, and developmental approach to AIED competencies development as there may be more explicit relationships between the nuances of a particular competence with its proliferated objective and purpose. As such, the competency variation between and within the competencies and the proficiency levels may be represented in more natural and inclusive ways by delineating the developmental change of a competency and the teaching purpose that it serves (see Fig. 2).

In Fig. 2, we are visualizing and representing the inclusiveness, hierarchy and focus of each dimension of progression. For example, in the acquisition dimension, the primary focus is on subject-content and information delivery as well as on data and

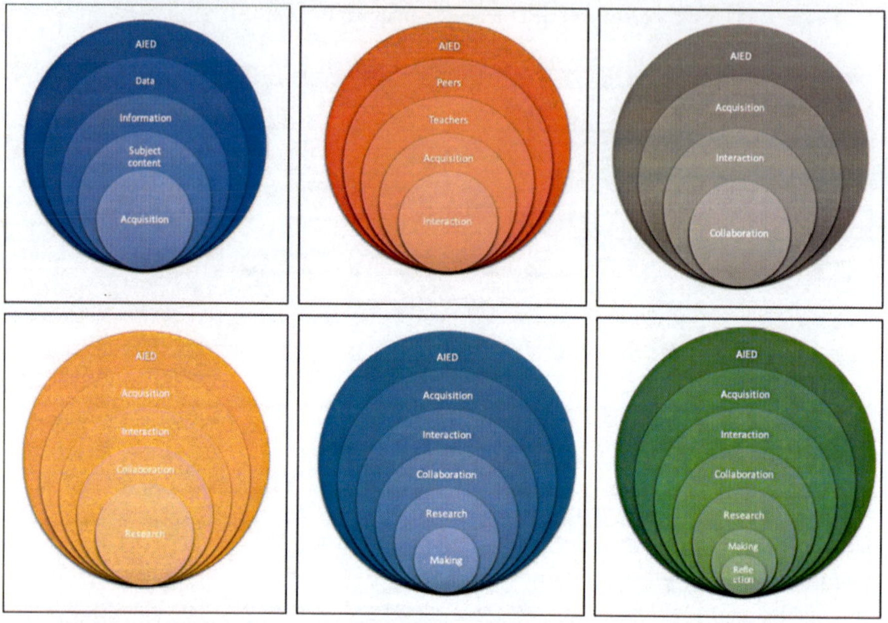

Fig. 2. Inclusiveness and relationality within and between progression levels

technology. In the second dimension, interaction the focus is on interacting with content, teachers, peers, and technology. In the third dimension the focus is on collaboration (i.e., teamwork and the creation of AIED based learning communities), interaction is a secondary focus and acquisition of content and AIED is in the periphery. In the fourth dimension, the focus is on research and collaboration and interaction becomes a secondary focus while acquisition and AIED are in the periphery. In the fifth dimension making and creativity is at the focus while research, collaboration and interaction reside as a secondary focus and acquisition and AIED in the background. In the sixth category, reflection is the focus, and making, research and collaboration is a secondary focus while acquisition and AIED are in the background of the dimension. It is axiomatic therefore to discern that the properties and aspects of previous progression dimensions are included as sub-properties of the upcoming levels to form the whole of the dimension.

To this end, we perceive that all levels of progression are of equal value and importance as to gain an impartial experience of developing and reflecting the competencies, capabilities, and skills that are essential for enabling teachers to use AIED in their practice. In line to this, each dimension insinuates a continuity within and between the progression levels, and at the same time, it resembles an autonomous and independent instantiation and development of a specific competence along with its variations.

5 Conclusions

The study illuminated an AIED competency framework twinned with an inclusive and relational progression model for helping teachers to conceptualise, apply, reflect, and self-assess AIED related competencies that will empower the transformation of the teacher's role as catalyst in designing and orchestrating teaching and learning through AIED. In particular, the study presented six AIED competencies with six developmental descriptors starting from acquisition, interaction, collaboration, research, making and reflection. We also attempted to represent diagrammatically the relationality of the progression model within and between the progression levels as to delineate both a concerted and detached progression model for accommodating distinctive needs, skills, and teaching contexts.

As every study, this research had limitations associated with its subjective and idiosyncratic nature reflected in the way that the AIED competencies and progression model were constituted. To alleviate much of these assumptions of subjectivity and generalisability, we attempt to elicit secondary education teachers' experiences of using the AIEDComp framework as means to delimit aspects of the framework that is most relevant to them which in turn will pave the way for an increased uptake of AIED in schools.

Acknowledgements. We are particularly grateful to The British Academy/Leverhulme for funding this research study with funding number SRG1920\100529.

References

1. Cukurova, M., Luckin, R., Kent, C.: Impact of an artificial intelligence research frame on the perceived credibility of educational research evidence. Int. J. Artif. Intell. Educ. **30**(2), 205–235 (2019). https://doi.org/10.1007/s40593-019-00188-w
2. Luckin, R., Cukurova, M.: Designing educational technologies in the age of AI: a learning sciences-driven approach. Br. J. Edu. Technol. **50**(6), 2824–2838 (2019). https://doi.org/10.1111/bjet.12861
3. Baylari, A., Montazer, A.: Design a personalized e-learning system based on item response theory and artificial neural network approach. Expert Syst. Appl. **36**(4), 8013–8021 (2009). https://doi.org/10.1016/j.eswa.2008.10.080
4. Cope, B., Kalantzis, M., Searsmith, D.: Artificial intelligence for education: Knowledge and its assessment in AI-enabled learning ecologies. Educ. Philos. Theor **0**(0), 1–17 (2020). https://doi.org/10.1080/00131857.2020.1728732
5. Luckin, R., Holmes, W., Griffiths, M., Corcier L.B.: Intelligence unleashed: an argument for AI in education', 2016. Pearson & University College London. https://www.pearson.com/content/dam/corporate/global/pearson-dot-com/files/innovation/Intelligence-Unleashed-Publication.pdf. Accessed 08 July 2020
6. Popenici, S.A.D., Kerr, S.: Exploring the impact of artificial intelligence on teaching and learning in higher education. Res. Pract. Technol. Enhanc. Learn. **12**(1), 1–13 (2017). https://doi.org/10.1186/s41039-017-0062-8

7. Baker, T., Smith, L., Anissa, N.: Educ-AI-tion Rebooted?, Nesta, November 2019. . https://media.nesta.org.uk/documents/Future_of_AI_and_education_v5_WEB.pdf. Accessed: 09 09 July 2020
8. Pedro, F., Subosa, M., Rivas, A., Valverde, P.: Artificial intelligence in education: challenges and opportunities for sustainable development, MINISTERIO DE EDUCACIÓN (2019). http://repositorio.minedu.gob.pe/handle/MINEDU/6533. Accessed 30 Sept 30 2020
9. Instefjord, E.J., Munthe, E.: Educating digitally competent teachers: a study of integration of professional digital competence in teacher education. Teach. Teach. Educ. **67**, 37–45 (2017)
10. Svensson, M., Baelo, R.: Teacher Students' perceptions of their digital competence. Procedia – Soc. Behav. Sci. **180**,1527–1534 (2015)
11. Tondeur, J., Aesaert, K., Prestridge, S., Consuegra, E.: A multilevel analysis of what matters in the training of pre-service teacher's ICT competencies. Comput. Educ. **122**, 32–42 (2018)
12. List, A.: Defining digital literacy development: an examination of pre-service teachers' beliefs. Comput. Educ. **138**, 146–158 (2019)
13. Margaryan, A., Littlejohn, A., Vojt, G.: Are digital natives a myth or reality? university students' use of digital technologies. Comput. Educ. **56**(2), 429–440 (2011)
14. UNESCO, ICT Competency Framework for Teachers (2011). https://unesdoc.unesco.org/ark:/48223/pf0000213475. Accessed 10 June 2021
15. Law, N., Woo, D., de la Torre, J., Wong, G.: A Global Framework of Reference on Digital Literacy Skills for Indicator 4.4.2', UNESCO-UIS, Quebec, Canada (2018)
16. Valencia-Molina, T., Serna-Collazos, A., Ochoa-Angrino, S., Caicedo-Tamayo, A.M., Montes-González, J.A., Chávez-Vescance, J.D.: ICT STANDARDS and Competencies from the Pedagogical Dimension: A Perspective from levels of ICT Adoption in Teachers Education Practice. Pontificia Universidad Javeriana and UNESCO, Colombia (2016)
17. Redecker, C., Punie, Y: European framework for the digital competence of educators: DigCompEdu, Publications Office, LU (2017). https://data.europa.eu/doi/https://doi.org/10.2760/178382. Accessed 09 June 2021
18. Lameras, P., Moumoutzis, N.: Towards the development of a digital competency framework for digital teaching and learning, presented at the EDUCON2021, Vienna (2021)
19. Laurillard, D.: Rethinking University Teaching: A Conversational Framework for the Effective Use of Learning Technologies, 2nd edition. Routledge, London, New York (2001)

Py4hs: A Computer Science Teacher Training Programme Promoting Python Code Clubs

Nektarios Moumoutzis[1(✉)], George Boukeas[2], Vassilis Vassilakis[2], Chara Xanthaki[2], and Nikos Pappas[1]

[1] Laboratory of Distributed Multimedia Information Systems and Applications, School of Electrical and Computer Engineering, Technical University of Crete, Kounoupidiana, Greece
{nektar, nikos}@ced.tuc.gr

[2] Computer Science Teacher in Secondary Education, Kounoupidiana, Greece

Abstract. We present the design, implementation and evaluation of a training programme for Computer Science teachers on the educational use of the Python programming language inside and outside school. The programme used educational resources centred on meaningful, self-contained programming projects. The training programme followed a blended-learning approach thus offering an opportunity to many computer science teachers make their first steps towards the educational use of the Python language within a very promising learner-centred pedagogical framework. Using initially an online course made it possible to reach a much greater number of Computer Science teachers, especially those living in remote areas and through them, have a considerable impact on students through the subsequent establishment of local code clubs. Following the initial teachers' training phase, groups of teachers that have been already formed were guided to establish local code clubs, recruit students, design a number of sessions and implement them. The evaluation results demonstrate the effectiveness of the code club approach and its acceptance by the participating students.

Keywords: Python programming · Computer science in secondary education · Code clubs

1 Introduction

Teaching introductory programming courses has received much attention the last years. This is mainly due to the ubiquitous use of computers, the proliferation of the so called cultures of participation [1], end-user programming [2] and end-user software engineering [3]. These trends are addressing software tools that provide powerful scripting languages to enable flexible customization and rich interactive content development by end-users. In this respect, knowledge of computer programming concepts is nowadays necessary for most knowledge workers including scientists and engineers. Consequently, many higher education departments have included introductory programming courses in their curricula [4]. Furthermore, many countries extend their curricula in secondary or even primary education to address the development of basic programming

M. E. Auer and T. Tsiatsos (Eds.): IMCL 2021, LNNS 411, pp. 937–948, 2022.
https://doi.org/10.1007/978-3-030-96296-8_85

skills [5]. The importance of computer programming has received even more attention through computer coding campaigns such as The Hour of Code and Europe Code Week. Informal learning opportunities are also offered in many countries following the organizational approach of coding clubs [6].

Following the above trends, Python has been introduced in vocational training curricula (professional lyceums) as well as in upper secondary education in Greece. However, many Computer Science teachers are not familiar with Python and all available professional development opportunities (mostly short webinars) are in high demand and overbooked. To address the need of enabling secondary Computer Science teachers in Greece to get familiar with the Python programming language and adopt effective learner-centred pedagogies, a 7-month teacher training programme was designed and implemented with partial funding from the Google CS4HS initiative. This programme is described here along with its evaluation. It was offered in a blended-learning fashion starting with a 3-month distant learning phase to study the Python language through self-contained programming projects and a subsequent 3-month phase with face-to-face collaboration focusing on the establishment of local code clubs. The programme finished with an evaluation phase of one month duration.

2 Previous Steps on Promoting Coding in Schools Within a Creative Learning Context

An initial exploration on the potential of programming in education from the perspective of providing opportunities for students in primary and secondary education develop their own coding projects for learning was undertaken during the pSkills project [7]. The target was to build a community of primary and secondary teachers, mainly computer science teachers, to exploit modern educational programming languages in their courses. The pSkills teacher training activities were structured as a three-step process: (a) initial training and community building workshops with over 400 participants in total; (b) pilot workshops, one in each one of the four participating countries with over 40 participants in Greece, Austria, Italy and Estonia; and (c) the pSkills Summer School, a one-week intensive training event with 10 participants. pSkills offered materials to enable teachers guide their students through an engaging learning process during which they develop their own coding projects. The focus was on developing digital games using an appropriate educational programming language [8]. Game development was based on Scratch [9] and game distribution on the Scratch community site. Towards the end of the implementation period of the project, an explosion of interest of computer science teachers on its themes was observed. Over 250 teachers participated in the local workshops held in Crete, Greece. Most of them started to apply the ideas into the classrooms and, within the context of creative projects with their students.

Following the successful impact of the pSkills project, our focus expanded beyond computer science education to include other domains including non-formal and informal education. The opportunity for this was given within the context of the ALICE project [10] and its decision to include games as a creative language to be adopted by adult trainers along with Music, Digital Storytelling, and Children Narratives. ALICE targeted adult trainers with the aim to enable them design and implement intergenerational creative learning environments. The project designed and piloted a graduate programme for educators in Greece, Italy, Switzerland, UK and Romania. The programme was offered through a learning portal and included face-to-face sessions as well in each one of the five participating countries, thus adopting a blended-learning approach. The training consisted of six learning units and a final project. One of these learning units targeted digital games and a number of participants adopted games as a creative language to work on their individual projects during which they designed and implemented learning activities where adults and children they were invited to learn together in creative ways. The term gaming-literacy was adopted to signify a step towards the accommodation of a new way of thinking, working, collaborating, teaching and learning, as initially introduced by Zimmerman [11].

The work reported in this paper is a third step towards supporting educators on the integration of new technological tools in their teaching practices emphasizing on digital skills and creative learning. This time, the focus is on general purpose programming language, not an educational or creative language as it was the case in the pSkills and the ALICE projects respectively. In particular, the programming language adopted in this case is Python and its use as a first programming language for novices taking into account the parallel developments in the school curricula in Greece and the need to support Greek computer science teachers on their use of the Python language within a pedagogical framework that promotes active learning within engaging learning scenarios in coding clubs. Selecting coding clubs as the pedagogical framework of our approach incorporates the positive prior experience on promoting coding in formal, informal and non-formal learning settings in pSkills and ALICE projects.

3 Teachers' Training

The design of the training programme was based on the principles of social constructivism: Initially, participants explored the course material and develop Python skills, as they delved into the programming projects and tackled the assignments. The assignments, meant to induce structure on the learning process, were implemented in groups exploiting the special features of the underlying learning platform. There were no lectures on Python programming. The course facilitators served as peer advisors, guides and coordinators. Following this initial phase of getting familiar with the Python language and the proposed pedagogical methodology, participants were asked to apply the knowledge and skills they have acquired, in workshops or coding clubs for their students, exploiting scenario-based pedagogical approaches. In this context, they eventually composed their own training material and developed strategies for cooperating with other participants from their regional group, learn from and support each other. There was also a strong element of reflection, self- and peer-evaluation at the final phase of the project. As already stated, an important aspect of the training programme was its blended-learning approach to promote collaboration among computer science teachers in many locations parts of Greece including several remote areas.

The program consisted of three phases. In the first phase (3 months) participants studied, explored and evaluated the course material, familiarizing themselves with

Python and the programming projects approach. In the second phase (3 months) participants implemented the course material in coding clubs. The final phase (1 month) involved extensive reflection and evaluation of the course.

The kick off was done via a teleconference that presented the overall structure and objectives of the training programme. Every month there was a live online session with the course facilitators, where participant groups had the opportunity to make presentations or engage in structured discussions. During the course, participants also communicated via discussion forums, online chat rooms and videoconferencing facilities offered by the Coursevo platform [12]. Participants worked together in regional groups and posted their assignments online, each group creating a portfolio that was reviewed by their peers. All results were documented and shared in the form of adaptable learning scenarios (i.e. project-based scenarios and/or lesson plans) that referenced teaching objectives of the Greek CS curricula and were organized in a digital repository that is available to all CS teachers through a Creative Common license for further reuse after the end of the project.

The training portal used for setting up the blended learning framework of the training programme is the evolution of MOLE, a multimedia online learning environment [13] that was initially developed to support educational activities in university departments. MOLE integrated tools and services for educational material reuse in an interoperable manner [14]. After its successful adoption in the academic environment, it has been adapted and enhanced under the name Coursevo to support professional development and training within a context that enables the establishment and sustainable operation of Communities of Practice (CoP) [12].

Coursevo enables communication between tutors/trainers and trainees, cooperation among trainees and access to coursework information and learning resources. It can combine traditional classroom-based lessons and practical sessions, with self-study and eLearning. Coursevo platform hides the complexity and frees the trainers from tedious system maintenance tasks, since a course or even a full functional learning site can be created in a few steps following the SaaS (Software as a Service) paradigm.

Coursevo integrates BigBlueButton (http://bigbluebutton.org/) to enable video teleconferencing. This proved very important for the implementation of the programme: One synchronous teleconference was organized each month to give guidance to participants, present best practices and examples, answer to questions and solve practical problems, especially for the organization of local code clubs. Furthermore, the workgroup support offered by Coursevo was used and appropriately adapted to enable the coordination of work in each group (code club organizers) and facilitate the development and submission of assignments.

4 Establishment of Code Clubs and Evaluation

Following the initial teachers' training phase, the groups that have been already formed were guided to establish local code clubs, recruit students, design a number of sessions and implement them. In each session the organizers were asked to prepare appropriate worksheets possibly reusing the ones used during the initial training or creating new ones. 28 code clubs were established all over Greece with a total number of participants exceeding 500 students ranging from K7 to K12 (12 to 18 years old) including students from vocational education schools. The evaluation of the code clubs was done an initial and a final questionnaire.

Table 1 summarizes the socio-demographic profile of the participants, their participation in the code club and their expectations and previous experience in coding. 69.9% of the participants were boys and 30.1% girls. Grades K-9 and K-10 represented around 70% of the participants. Participants were mainly informed from their computer science teacher at school. The main motivation for participation was their interest in coding. A high percentage expected to learn to code without help and learn something different than in school courses. Most of the participants could not see any obstacle in their participation in the code club and were determined to participate in all sessions of the code club. Finally most of them had a first coding experience at school.

Table 1. Socio-demographic profile and code club participation (N_{pre} = 465)

Variable	N	(%)
Gender		
Male	325	(69.9)
Female	140	(30.1)
Grade		
First grade of Gymnasium (K-7)	15	(3.2)
Second grade of Gymnasium (K-8)	13	(2.8)
Third grade of Gymnasium (K-9)	116	(24.9)
First grade of Lyceum (K-10)	139	(29.9)
Second grade of Lyceum (K-1)	45	(9.7)
Third grade of Lyceum (K-12)	16	(3.4)
First grade of Professional Lyceum (K-10)	32	(6.9)
Second grade of Professional Lyceum (K-11)	42	(9.0)
Third grade of Professional Lyceum (K-12)	18	(3.9)
Other	29	(6.2)
How were you informed about the code club?		
From my Computer Science teacher	298	(64.1)
From a friend	38	(8.2)
I saw the code club poster in my school	17	(3.7)
From an announcement during a class at school	16	(3.4)
From my parents	10	(2.2)
Web page / social media	8	(1.7)
Other	8	(1.7)
No answer	70	(15.1)
Why did you decide to participate in the code club? (multiple answers)		
I like to code	252	(54.2)
I like learning new things and Python is interesting	162	(34.8)
Curiosity	116	(24.9)
I want to become a programmer	96	(20.6)
My friends participate too	42	(9.0)

(continued)

Table 1. (*continued*)

Variable	N	(%)
I have used Python in the past and I like it	41	(8.8)
I have taken similar courses in the past	24	(5.2)
My parents encouraged me to do so	10	(2.2)
What do you expect to achieve with your participation? (multiple answers)		
Create small programs in Python without help	223	(48.0)
Learning something more/different that school classes	188	(40.4)
Be able to develop my own games in Python	140	(30.1)
Know better how to code to see if I can be a professional programmer	139	(29.9)
Know better how to code because I want to be a professional programmer	123	(26.5)
Develop my own programs to show to friends and relatives	88	(18.9)
I want to participate in more sessions and learn more things	47	(10.1)
Get to know the other participants in the code club	38	(8.2)
I have no specific expectations	17	(3.7)
What are possible obstacles to your participation?		
I cannot imagine any obstacle	192	(41.3)
I do not have enough free time	117	(25.2)
I do not have enough technical knowledge	92	(19.8)
I do not have a computer at home	21	(4.5)
I do not like to collaborate with others	14	(3.0)
No answer	29	(6.2)
Do you plan to participate in all sessions?		
I will come to all of them; I do not cancel what I start!	167	(35.9)
There are not so many sessions, I believe I can participate in all of them	89	(19.1)
If I see that I can make I will continue for certain	84	(18.1)
If I like it in the first session, I will come to all of them	58	(12.5)
Other	5	(1.1)
No answer	62	(13.3)
First coding experience (age)		
In Gymnasium (age 12–15)	232	(49.9)
Last grades of Primary School (age 10–12)	110	(23.7)
In Lyceum (age 15–18)	61	(13.1)
I have no previous coding experience	34	(7.3)
First grades of Primary School (age 6–9)	28	(6.0)
Your first contact with coding was:		
In school	299	(64.3)
Out of school, on my own	74	(15.9)
I have no previous coding experience	35	(7.5)
Out of school, coding with friends	18	(3.9)
Out of school with the help of a relative of mine	16	(3.4)
After school courses	12	(2.6)
Other	11	(2.4)

Table 2 presents the programming environments used in the participants' previous coding experience, mainly Scratch (https://scratch.mit.edu/), Microworlds (http://www.microworlds.com/) and Python.

Table 2. Previous coding experience (N_{pre} = 465). Percentages are shown for Likert scale values.

Variable	Not at all	Very little	Somewhat	Very much
Programming environments you used in school				
Microworlds/Turtle world	52.0	15.9	21.3	10.8
Scratch	26.7	18.7	24.9	29.7
AppInventor	80.6	9.7	5.2	4.5
Pseudolanguage	69.5	15.1	9.0	6.5
Pascal	84.9	8.0	4.5	2.6
GameMaker	88.2	7.1	3.7	1.1
Alice	84.5	6.2	5.2	4.1
Kodu	84.9	4.5	4.1	6.5
Python	56.8	17.4	16.1	9.7
To what extent you have used the programming environments in/out of school				
Microworlds/Turtle world	74.8	12.3	8.8	4.1
Scratch	46.5	20.2	14.0	19.4
AppInventor	82.8	10.1	3.9	3.2
Pseudolanguage	78.9	13.5	3.2	4.3
Pascal	87.3	8.0	2.2	2.6
GameMaker	84.7	10.3	2.2	2.8
Alice	88.6	7.3	1.3	2.8
Kodu	89.2	4.5	3.4	2.8
Python	61.3	21.7	10.3	6.7

Table 3 presents the self-evaluation of the students regarding their existing coding skills before their participation in the code club. Most of them believe that their coding skills are rather low. The opinion of the majority about coding is that it is a useful skill, challenging, helps in analytical thinking and is fun. Students would like to create mainly action games, mobile apps, robotics apps and games of logic and secondarily programs to solve mathematical problems and educational apps.

Table 4 presents the findings after finishing the code club. The positive impact on all variables is evident. The participants report that they now better what coding is and they have a better understanding of what kind of programs they would like to create.

Table 3. Coding skills and opinions (N_{pre} = 465). Percentages are shown for Likert scale values.

Variable	N	(%)
Rate your coding skills		
1	75	(16.1)
2	146	(31.4)
3	144	(31.0)
4	73	(15.7)
5	27	(5.8)
In your opinion, coding: (multiple answers)		
is a useful skill	296	(63.7)
is challenging	241	(51.8)
helps you think analytically	206	(44.3)
is fun	193	(41.5)
presupposes knowledge of mathematical concepts	103	(22.2)
is like a puzzle	102	(21.9)
is hard	81	(17.4)
can help you get better at school	66	(14.2)
other	8	(1.7)
What kind of programs would you like to create? (multiple answers)		
Action games	263	(56.6)
Mobile apps	258	(55.5)
Robotics apps	200	(43.0)
Games of logic	172	(37.0)
Programs to solve mathematical problems	114	(24.5)
Educational apps	90	(19.4)
Other	12	(2.6)

Table 5 presents how the participants evaluate the code club and the teachers. All variables are given in 5-value Liker scale and it is evident the positive evaluation in all of them. The table also shows the self-evaluation of certain parameters regarding the participation. Again all evaluations are positive or very positive. Finally the table presents how the participants evaluated the Python language and its learning potential, with positive results as well.

The code clubs had a very positive impact on students in terms of developing programming skills and positive change in their attitude towards programming which is seen as an important professional pathway. Table 6 demonstrates these findings presenting the change in students' attitudes on the questions of pre- and post-questionnaires.

Table 4. Coding skills and opinions (N_{post} = 358). Percentages are shown for Likert scale values.

Variable	N	(%)
After finishing the code club sessions I believe that: (multiple answers)		
I have learned more/new things than school classes	274	(76.5)
I know what coding is to decide if I will be a programmer	261	(72.9)
I can create small Python programs without help	258	(72.1)
I can create my own Python games	94	(26.3)
In your opinion, coding: (multiple answers)		
is a useful skill	277	(66.7)
helps you think analytically	264	(73.7)
is challenging	262	(73.2)
is fun	243	(67.9)
is like a puzzle	147	(41.1)
presupposes knowledge of mathematical concepts	131	(36.6)
can help you get better at school	131	(36.6)
is hard	101	(28.2)
Other	8	(1.7)
What kind of programs would you like to create?		
Mobile apps	217	(60.6)
Action games	207	(57.8)
Games of logic	176	(49.2)
Robotics apps	172	(48.0)
Programs to solve mathematical problems	129	(36.0)
Educational apps	94	(26.3)
Other	12	(2.6)

Table 5. Evaluation of the code club and the teachers, self-evaluation and evaluation of the Python Language (N_{post} = 358). Percentages are shown for Likert scale values.

Variable	Not at all	Very little	Some-what	Much	Very much
Code club objectives were clear	1.7	4.2	15.9	41.1	37.2
Teachers were:					
well prepared and organized	0.8	2.8	10.6	33.8	52.0
clear and understandable	1.4	3.1	14.8	30.7	50.0
easy to collaborate with whenever necessary	1.1	3.1	6.1	21.8	67.9
Learning materials were:					
connected to real world	4.2	13.1	27.4	32.7	22.6
fun	3.9	7.3	22.3	28.8	37.7
interesting	2.2	5.9	16.2	26.5	49.2

(continued)

Table 5. (*continued*)

Variable	Not at all	Very little	Some-what	Much	Very much
possible to be used without help	7.8	21.2	30.4	23.7	16.8
Code club duration was adequate	4.5	17.0	32.7	30.4	15.4
I collaborated with my peers	5.9	10.3	19.3	30.2	34.4
What I learnt will help me in school	9.8	14.5	24.3	31.6	19.8
I showed my programs to friends, classmates or parents	16.7	16.7	33.3	8.3	25.0
I want the code club to continue	3.6	9.2	17.6	23.2	46.4
I would advise my friends to enrol to a code club	4.7	7.5	15.1	30.7	41.9
I am now ready to code and create	6.4	14.0	32.1	33.2	14.2
My expectations were fulfilled	3.6	8.9	24.9	42.2	20.4
Python coding seemed easy	0.8	9.2	38.5	35.8	15.6
Python is a language to start coding	1.7	6.7	24.0	31.3	36.3
Python should be used in schools	3.6	7.5	18.4	25.7	44.7
I would like to continue with Python on my own	10.1	14.8	26.8	26.3	22.1

Table 6. Attitudes of students towards programming before and after their participation in the local code clubs (Npre = 465, Npost = 358). Percentages are shown for Likert scale values.

Variable	1	2	3	4	5	mean
I believe that:						
Girls and boys are equally competent in coding	09.9 07.8	15.3 12.6	65.4 65.6	03.4 06.4	06.0 07.5	**2.80** **2.93**
I can collaborate with others when coding	06.5 05.0	08.4 07.0	24.3 21.8	26.7 29.1	34.2 37.2	**3.74** **3.87**
I can code by myself without help	12.5 05.3	23.7 15.4	24.9 27.9	20.2 30.4	18.7 20.9	**3.09** **3.46**
It is probable that I follow a computing profession	13.5 10.9	17.8 12.3	23.0 21.8	17.8 21.2	27.7 33.8	**3.23** **3.55**
Only future computer professionals should code	34.4 27.7	26.9 24.9	19.6 25.7	09.0 11.7	10.1 10.1	**2.34** **2.52**

5 Conclusions and Future Plans

The paper describes a teacher training scheme based on meaningful self-contained programming projects that are undertaken by students in coding clubs. A transfer was observed between knowledge acquired by the teachers and use of this knowledge with their own students. Evaluation was based on questionnaires and self-assessment for teachers and a pre-, post-questionnaire evaluation of the students that participated in the coding clubs established. Focusing on engaging programming projects rather than relying on artificial exercises addressing the syntax and the structure of Python (as it is the case in traditional classroom teaching) highlights a wide range of higher-level concepts ranging from functional abstraction and problem-solving strategies.

Participating teachers were supported through distant-learning facilities to study special material on the Python programming language and thus gain confidence in using an alternative, engaging methodology which can serve as a springboard for exposing their students to Computer Science practices and concepts.

Future work will address the transfer of the material and the methodologies reported in other domains (e.g. mathematics [15]) that could use coding to make learning playful and more engaging. Furthermore, we plan to explore the integration of visual interfaces to Python capitalizing promising results with respect to the performance gain that those interfaces can offer to novices [16].

Acknowledgments. The work reported in this paper was partially supported by the Google CS4HS programme (https://www.cs4hs.com/).

References

1. Fischer, G.: Understanding, fostering, and supporting cultures of participation. Interactions **18**, 42–53 (2011)
2. Myers, B., Ko, A., J., Burnett, M.: Invited research overview: end-user programming. In: CHI 2006 extended abstracts on Human factors in computing systems, pp. 75–80 (2006)
3. Ko, A.J., et al.: The state of the art in end-user software engineering. ACM Comput. Surv. (CSUR), **43**(3), 21 (2011)
4. Aleksic, V., Ivanovic, M.: Introductory programming subject in European higher education. Inform. Educ. **15**(2), 163–182 (2016)
5. Balanskat, A., Engelhardt, K.: Computing our Future: Computer Programming and coding-Priorities, School Curricula and Initiatives Across Europe. European Schoolnet (2014)
6. Smith, N., Sutcliffe, C., Sandvik, L.: Code club: bringing programming to UK primary schools through scratch. In: Proceedings of the 45th ACM technical symposium on Computer science education, pp. 517–522 (2014)
7. Ovcin E., Cerato C., Smith D., Lameras P., Moumoutzis N. The pSKILLS experience: using modern educational programming languages to revitalise computer science teaching. In: International Conference on the Future of Education, Florence, Italy (2011)
8. Smith, D., Lameras, P., Moumountzis, N.: Using educational programming languages to enhance teaching in computer science. In: EDGE 2010 Conference on the Use of Technologies in K-12 and Post-Secondary Education, Newfoundland & Labrador, Canada (2010)
9. Peppler, K., Kafai, Y.: Creative coding: programming for personal expression. In: 8th International Conference on Computer Supported Collaborative Learning (CSCL), Greece (2005)
10. Moumoutzis, N., Christoulakis, M., Pitsiladis, A., Sifakis, G., Maragkoudakis, G., Christodoulakis, S.: The ALICE experience: a learning framework to promote gaming literacy for educators and its refinement. In: International Conference on Interactive Mobile Communication Technologies and Learning (IMCL), pp. 257–261 (2014)

11. Zimmerman, E.: Gaming literacy – game design as a model for literacy in the twenty-first century. In: Perron, B., Wolf, M.J.P. Eds. The Video Game Theory Reader 2, New York; London. Routledge (2009)
12. Pappas, N., Arapi, P., Moumoutzis, N., Christodoulakis, S.: Supporting learning communities and communities of practice with Coursevo. In: Global Engineering Education Conference (EDUCON), pp. 297–306. IEEE, Athens, Greece (2017)
13. Pappas, N., Arapi, P., Moumoutzis, N., Mylonakis, M., Christodoulakis, S.: The multimedia open learning environment (MOLE). Never Waste a Crisis! **76** (2011)
14. Mylonakis, M., Arapi, P., Pappas, N., Moumoutzis, N., Christodoulakis, S.: Metadata management and sharing in multimedia open learning environment (MOLE). In: García-Barriocanal, E., Cebeci, Z., Okur, M.C., Öztürk, A. (eds.) Metadata and Semantic Research. MTSR 2011. Communications in Computer and Information Science, vol. 240, pp. 275–286. Springer, Berlin, Heidelberg (2011). https://doi.org/10.1007/978-3-642-24731-6_29
15. Lameras, P., Moumoutzis, N.: Towards the gamification of inquiry-based flipped teaching of mathematics: a conceptual analysis and framework. In: 2015 International Conference on Interactive Mobile Communication Technologies and Learning, pp. 343–347 (2015)
16. Kyfonidis, C., Moumoutzis, N., Christodoulakis, S.: Block-C: a block-based pro-gramming teaching tool to facilitate introductory C programming courses. In: Global Engineering Education Conference (EDUCON) 2017 IEEE, Athens, Greece, pp. 570–579 (2017)

University – Industry – Developer Collaborations

Authentication Methods with a High Degree of Security in Accessing Moodle E-Learning Platform

Vasile Banes(✉) and Cristian Ravariu

Polytechnic University of Bucharest, Bucharest, Romania
vasile.banes@stud.etti.upb.ro,
cristian.ravariu@upb.ro

Abstract. Moodle has several authentication methods but also the possibility for the platform administrator to create a new one using the Plugin function found in the site administration when managing authentication. This paper presents a new solution contributes to ensuring a high degree of security when authenticating within the Moodle platform, by introducing in addition to username and password authentication and certificate authentication.

Keywords: Moodle · E-learning · Digital certificate · Devices

1 Introduction

Moodle is an e-learning platform designed to provide educators, administrators and learners a single robust, secure and integrated system. Within the Moodle (Modular Object Oriented Dynamic Learning Environment) platform [1], there are several types of authentication management, described as authentication plugins [2]. Standard authentication plugins are: manual accounts - accounts created manually by an administrator [3]; web services authentication; no login - suspend particular user account; LDAP (Lightweight Directory Access Protocol) server - account details are located on an external LDAP server; email-based self-registration - for enabling users to create their own accounts; Moodle network authentication - how different Moodle sites can connect and authenticate users, as can be seen in the images below, Fig. 1 [4].

In addition to the standard methods presented above, there are additional authentication plugins in the Moodle plugins directory. In the literature, methods of certificate authentication have been almost absent. The certificates were used until 2020–2021 only as signing [5] and encryption [6] solutions. In this paper, our original contributions have plugin, certificates, authorization, which brings new on the security side the fact that unauthorized people such as hackers can no longer break the account because they are conditioned by finding the certificate and the platform account that makes it impossible to connect. The certificate can be exported for authentication only from a specific device or to a carrier such as a token device.

M. E. Auer and T. Tsiatsos (Eds.): IMCL 2021, LNNS 411, pp. 951–961, 2022.
https://doi.org/10.1007/978-3-030-96296-8_86

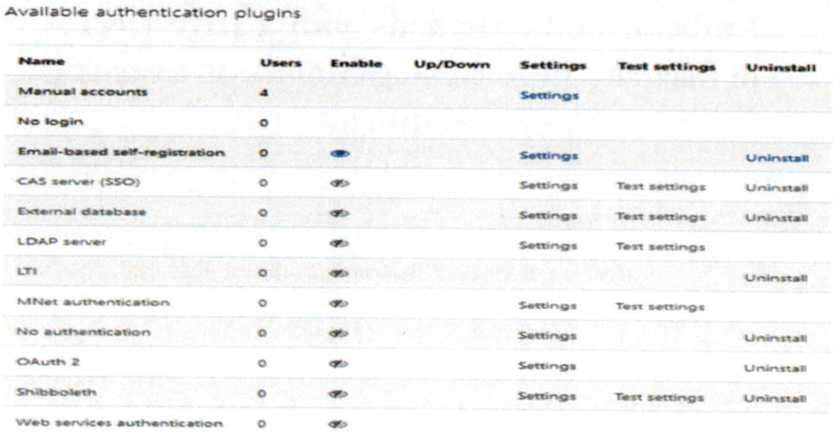

Available authentication plugins

Name	Users	Enable	Up/Down	Settings	Test settings	Uninstall
Manual accounts	4			Settings		
No login	0					
Email-based self-registration	0			Settings		Uninstall
CAS server (SSO)	0			Settings	Test settings	Uninstall
External database	0			Settings	Test settings	Uninstall
LDAP server	0			Settings	Test settings	
LTI	0					Uninstall
MNet authentication	0			Settings	Test settings	
No authentication	0			Settings		Uninstall
OAuth 2	0			Settings		Uninstall
Shibboleth	0			Settings	Test settings	Uninstall
Web services authentication	0					

Fig. 1. Available authentication plugins in Moodle.

2 Evolution of Digital Certificate

In cryptography, PKCS # 12 defines an archive file format for storing multiple cryptographic objects as a single file. It is commonly used to group a private key with its X.509 certificate or to group all members of a trusted chain. PKCS # 12 or PFX is a binary format for storing the certificate, all intermediate certificates, and the private key in a single encrypted file [7]. PFX files usually appear with.pfx and.p12 extensions. PFX files are typically used on Windows operating systems to import and export certificates and private keys. Digital certificates are issued by institutions or certification authorities and may be valid for one to three years. The issuance is made based on the user's personal data such as, name, surname, home address, name of the institution where he works, position held, e-mail address, telephone number and a copy of the identity card. Once the certificate has expired or been revoked, connecting to the platform becomes impossible even if it has created an account. After the expiry date, the renewal steps must be repeated. A protection measure is the maximum number of incorrectly entered passwords, if more than eight consecutive incorrect passwords are entered, the certificate is locked and the issuing authority must be contacted to unlock it, Figs. 2.

Given how easily password and username combinations can be stolen by hackers, it's no wonder security breaches are common. In addition, when that happens, one solution is two-factor authentication, which is the best way to protect sensitive data from theft. Two-factor authentication (2FA) works by adding an extra layer of security to your personal account. An additional login credential is required - beyond just the username and password - to gain access to the account, and obtaining that second credential requires access to something that belongs to the user, in this case the digital certificate. Without this additional access method, it is impossible to access the account, which in turn makes it impossible for hackers to access the personal account using only the stolen password and login information. There are several two-factor authentication methods: 1. Additional login information such as answers to security questions and PIN

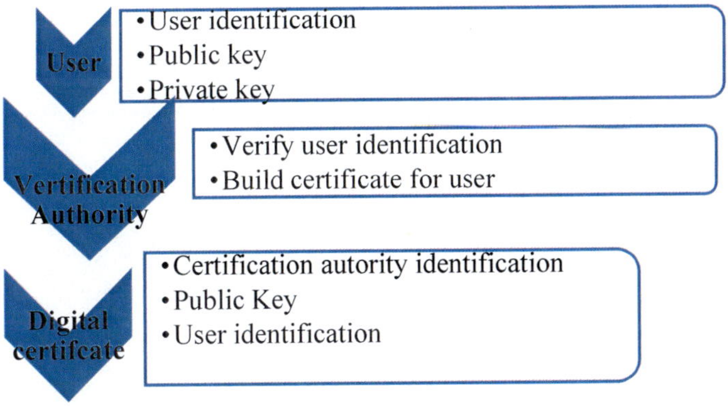

Fig. 2. Obtaining a certificate from a certification authority.

numbers sent via SMS; 2. User-owned devices such as a token, tablet or mobile application; 3. Unique biometric connection data, these include retinal and fingerprint scanning [8, 9]. Figure 3 shows the comparison of available authentication technologies and Fig. 4 shows the disadvantages and advantages.

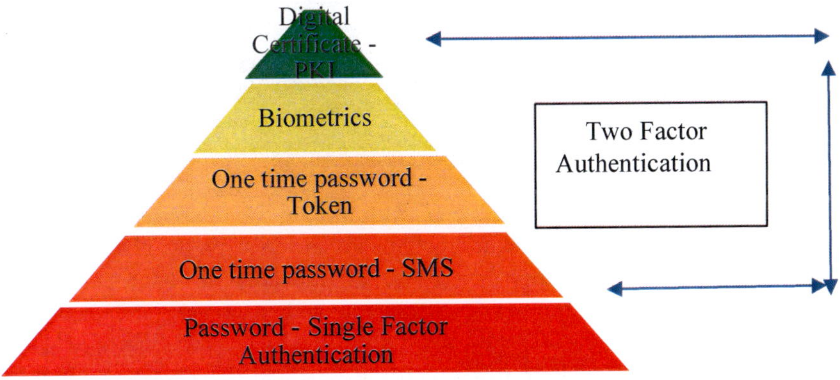

Fig.3. Comparison of available authentication technologies.

Authentication method	Disadvantages	Advantages
Username and Password	Listed as unsafe cause of a large number of data corruption Multiple inconveniences in password recovery	Easy to use and known by users
	No direct link of accreditation to the user identity	It does not require any specific software or device

(*continued*)

(*continued*)

Authentication method	Disadvantages	Advantages
One time password via SMS	SMS was not designed as a secure communication channel Text messages can be intercepted and redirected by pirated applications	
One time password via Token	No direct link of accreditation to the user identity The token generates costs	Easy to integrate
Biometrics	Database may be compromised Risk of "false positive" matches depending on the hardware and software used	An increasing number of devices to support biometrics Easy to use
Digital certificate - PKI	Requires a certification authority to be issued and managed	Verified solution in the most demanding high security environments, uses private and public key cryptography for high security Certificate binds key to user identity, enabling non-repudiation Connections can be verified because the public key is contained in the certificate Can manage authentication and authorization with private key, digital signature

Although the introduction of a second factor seems a bit awkward, it is much less inconvenient and expensive if the information and materials on the platform have been corrupted or stolen. This paper presents a new modality of connection within the Moodle platform, in addition to authentication with username and password and authentication with the digital certificate, different from the existing ones. From the research done compared to the current connection methods, this new method brings an extra security and trust. This secure two-factor authentication mechanism, based on digital certificate authentication plus a secondary authentication in which the username and password is used is called "two factor authentication (2FA)". For the functioning of this new connection method it was necessary to develop a plugin, without this plugin the authentication method is not possible. This extension of the authentication mechanism considered adding an additional level of security to the account with two authentication factors, based on digital certificate authentication plus a secondary authentication in which the username and password are used. Compromising such an authentication is difficult and would mean getting both the certificate and the username

Authentication Method	Disadvantages	Advantages
Username and Password	Listed as unsafe cause of a large number of data corruption. Multiple inconveniences in password recovery.	Easy to use and known by users
One time password via SMS	No direct link of accreditation to the user identity. SMS was not designed as a secure communication channel Text messages can be intercepted and redirected by pirated applications	It does not require any specific software or device
One time password via Token	No direct link of accreditation to the user identity The token generates costs	Easy to integrate
Biometrics	Database may be compromised Risk of "false positive" matches depending on the hardware and software used	An increasing number of devices to support biometrics Easy to use
Digital Certificate - PKI	Requires a certification authority to be issued and managed	Verified solution in the most demanding high security environments, uses private and public key cryptography for high security Certificate binds key to user identity, enabling non-repudiation Connections can be verified because the public key is contained in the certificate. Can manage authentication and authorization with private key, digital signature

Fig. 4. Disadvantages and advantages of available authentication technologies.

and password. Certificate authentication was chosen as an additional measure, given the advantage and security offered by qualified or unqualified digital certificates. The advantage and security offered by these digital certificates that comply with the issuing procedure and rules is that they are issued to a person, who is identified and verified using a valid identity document. In the case of qualified or unqualified certificates, issuing the certificate, after verifying the person, on a secure hardware token, who's PIN only the owner knows, guarantees that when the certificate is used in an authentication, the user who uses it is its owner.

3 Purpose of Two Factor Authentication Method

The digital certificate is an identity document in electronic form that allows the holder to authenticate and be uniquely identified in the electronic environment, being generally used to authenticate to online applications and encrypt public information, ensuring the certification of a user's identity when accessing the distance of some resources. Digital certificate authentication brings a high level of trust being secured by cryptographic mechanisms that do not allow security forgery, it is generally used to authenticate to online applications and encrypt electronic information. The benefits of a qualified digital certificate are: user authentication and identification, document integrity, non-repudiation - ensuring that the user cannot subsequently deny original data, providing proof of data integrity and origin and verifiable by a third party, confidentiality recognition internationally from a legal point of view in more and more countries, as it offers this guarantee of security. Enhanced security - documents signed in electronic format give you the guarantee that they can no longer be modified or altered in any way, intentionally or by mistake. Also, the revocation and expiration of a certificate means that access will no longer be allowed from that moment in any of the systems in which it was allowed, which leads to better control over the state of the accounts and given the advantages presented above, we have implemented within the Moodle platform, in addition to username and password authentication and digital certificate authentication. Compromising such 2FA-based authentication is difficult and would mean getting both the certificate and the username and password. The advantage of qualified digital certificates (or unqualified but which comply with the issuing procedure and rules) is that they are issued to a person who is identified and verified using a valid identity document. In the case of qualified or unqualified certificates, issuing the certificate, after verifying the person, on a secure hardware token, whose PIN only the owner knows, guarantees that when the certificate is used in an authentication the user who uses it is its owner. The digital certificate used within the Moodle platform contains the following information: the name of the certificate holder, the email address of the certificate holder, the public key of the certificate holder, digital certificate series, validity of the digital certificate (start date and end date), encryption algorithm, as can be seen in the images below, Figs. 5 and 6.

European legislation is mainly based on Directive 1999/93/EC of 13 December 1999 laying down common obligations for certification-service-providers and common rules on liability and cooperation mechanisms to ensure cross-border recognition of signatures and certificates throughout the European Community.

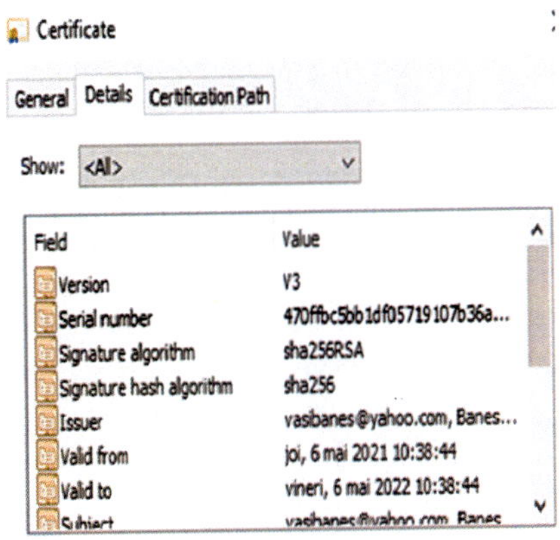

Fig. 5. Information about the certificate owner.

Fig. 6. Validity of the certificate.

4 Approaching the Solution by Validating the Developed Plugin

This new solution contributes to ensuring a high degree of security when authenticating within the Moodle platform, by introducing in addition to username and password authentication and certificate authentication, based on validations is presented in this paper. Using authentication with certificates and username and password, I map the attributes related to the fields in the certificate with the data related to the user in the database for unique identification in order to ensure authentication and authorization within the Moodle platform. Do this validation through the developed plugin. When authenticating in the system, the user chooses the digital certificate exported in.pfx format, and after reading the attributes from the certificate, the email address is extracted, the email address that is verified with the user's email address from the user account for the database, used by the Moodle system. The identification is thus based on the use of an attribute in the certificate and the mapping of this attribute with the value corresponding to the field in the user profile [10]. When authenticating on the Moodle platform based on the digital certificate, two variants were developed: the first option consists in selecting the certificate and entering the password, after which the next step is sending to the login page with the user and password of the Moodle account and the second option involves selecting the certificate and redirect to the Moodle workspace for the associated account. For the second login, the login page where the Moodle account username and password had to be entered was removed. This variant was also considered on the premise of multiple daily authentications such as that of the platform administrator who connects for different situations or of a teacher who has to update his information from the courses. In this way, they no longer have to enter the user and account password several times, only the certificate. For the second variant of plugin connection, a script was created for the validation of pfx files, overrided login hooks to redirect on login page and was created login page hook and constructor to redirect pxf page and Moodle core login page. If a user logs in with a certificate other than the one associated with the Moodle account or the certificate is good but does not have an associated account, a warning message appears stating that no user was found and the file must be checked again. The developed plugin must be imported into the Moodle database. Multiple login plugins can be enabled. Any authentication plugin can be used to find a username with password match. Once found, a user is logged in and no alternative plugins are used. Therefore, the plugin that handles most connections must be moved to the top of the login page. The installation is done like any other plugin, it connects to the platform with the administrator account after which it is selected from Site Administration, Plugin, plugin installation, the zip file is searched, the zip plugin installation command is given and the database is updated with the new plugin as can be seen in the images below, Figs. 7.

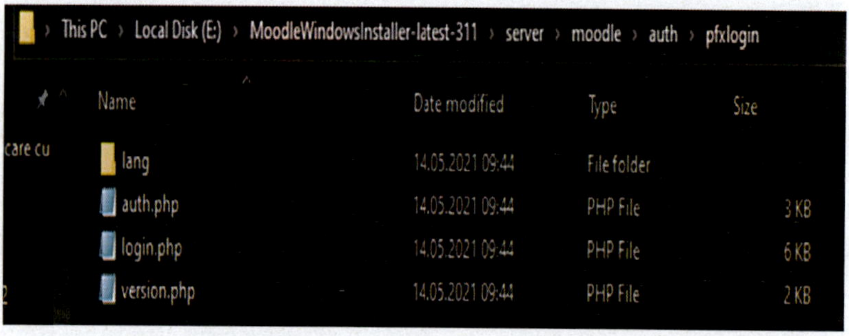

Fig. 7. Plugin location after installation.

After it has been installed, it must be activated from the overview plugin, the authentication method and the new authentication method is searched in the list by clicking on the activation button and moving it to the top of the list. If the steps have been done successfully after exiting the platform and trying to connect, as soon as the log in button is pressed, the window for selecting the digital certificate will appear. The setting of the new authentication method for the site was done as follows: from Site administration, Plugins was selected followed by Install plugins. At the Install plugin from ZIP file, select the file in question from your computer and upload it. Then a plugin check is done if it meets the criteria of the database and if everything is fine a confirmation message appears and you can move on to the next step of Upgrade Moodle database now. Here we are confirmed that Upgrading to new version is coming, the Confirm button is clicked and after a few seconds, the successful completion message appears. After the file has been imported and the database has been updated, the next step is to activate the plugin and pass it as the default/primary authentication method. This operation is done from Site administration > Plugins > Plugins overview > Additional Plugins > Manage authentication > Select plugin - Enable > Move up on first row [11]. After this operation to verify the new authentication method, the platform will be restarted and after restarting the login page, the window will appear asking us to select the digital certificate for authentication, Figs. 8.

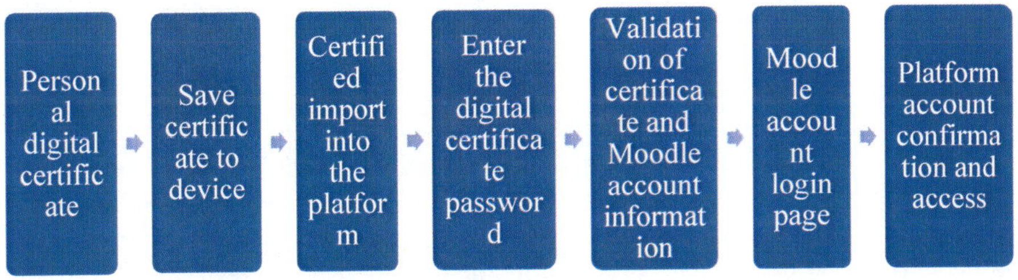

Fig. 8. Validation of the process of connecting to the e-learning platform.

The same procedure is for the second connection option. Was also considered the situation when we want to return to the default method of connecting only with username and password, a situation in which the plugin will be uninstalled without affecting the database or reinstalling the platform being a reversible process. The minimum and mandatory requirements for the operation of Moodle are the following: 200 MB hard disk, 512 MB memory, 1 GHz processor. The tests were performed on the Windows Server 2019 x64 operating system, Intel (R) Core i5–8400 CPU @ 2.80 GHz processor, 2 GB installed memory, 1 TB hard disk. The Moodle package for the installed Windows operating system was version 3.10 which contains all the other programs needed to run it on Windows. The working method used to create and edit the plugin was PHP (Hypertext Pre-processor, Web scripting language) technology and the WAMP (Windows, Apache, MySQL, PHP) server on Windows systems. I used WAMP as a safe space to work on the platform, without having to host it online. Apache is the server software used to display web pages. When a user requests that a page be viewed, Apache approves the request via HTTP and displays the site. MySQL is a relational database management system and is an integrated component of the WAMP platform [12]. A PHP interpreter implemented as a module processed the PHP code on a server. PHP is a general-purpose writing language especially suitable for web application development.

5 Actual or Anticipated Outcomes

Plus the value offered by this paper is given by the introduction of an additional security at the level of the user account, by adding an additional security start. Using this implementation guarantees that if someone guesses or obtains the user's password in different situations, you cannot log in to the Moodle platform, by introducing this additional layer of security and protection that requires the presentation of a digital certificate (qualified and unqualified), certificates that are and exist only in the user's possession, being stored in the keystore of the device and which cannot be passed without entering the password. In addition to qualified digital certificates stored on a physical device or in the cloud, it is necessary to enter the PIN for certificate authentication. Also, unqualified certificates installed on the user's device are based on the installation of a password and can be configured and issued as certificates that cannot be exported to other equipment, which means that only the user of the device can use it. The purpose of this project was to increase the degree of security, a better administration of the accounts, the protection of the information uploaded on the site, the access and the easy identification of the persons connected to the platform. All these aspects could be put into practice with the help of the newly created plugin. Another important point of the authentication mechanism with two authentication factors, based on digital certificate authentication plus a secondary authentication is that you can always return to the initial form and vice versa without affecting the database. The connection to the platform can be made from any device whether it is a mobile phone, tablet, laptop or computer based on the access link and the minimum and mandatory requirement is to present the digital certificate installed on the device or that it has entered the token.

6 Conclusions

Moodle has several authentication methods but also the possibility for the platform administrator to create a new one using the Plugin function found in the site administration when managing authentication. This plugin is used to read pfx file that is combination of certificate and private file with password. Digital certificate authentication brings a high level of trust being secured by cryptographic mechanisms, and the revocation and expiration of a certificate means that access will no longer be allowed from that moment in any of the systems in which it was allowed, which leads to better control over the state of the accounts. When creating the certificate, a valid e-mail address will be filled in, as the e-mail address is used to identify the user from the Moodle database. Through this work, a new method of authentication was created within the Moodle Platform. Existing authentication has been extended, with username and password, introducing an additional security start through digital certificate authentication. Thus, the authentication with certificates of the user account of the user stored in the Moodle system was linked, the mapping and validation of this information being done through the values assigned to certain fields in the certificate, values that can be found in the database that manages user accounts. This mapping was done and is done through a plugin developed for this purpose, a plugin that has the role of reading and comparing the fields in the certificates, to extract and select the value of the user from the database in order to uniquely identify the user.

Acknowledgement. The article is financed by the Polytechnic University of Bucharest, through the project "Engineer in Europe" in online system, registered at the Ministry of Education under no. 140/GP/19.04.2021, by using the fund for financing special situations that cannot be integrated in the form of financing state higher education institutions.

References

1. Wild's, I.: Moodle 3.x Developer's Guide. Edition: Packt Publishing Limited, Birmingham, United Kingdom (2017). ISBN: 9781786467119
2. Kang, B., Kim, H.J.: Proposal: a design of e-learning user authentication system. Int. J. Secur. Appl. **9**(1), 45–50 (2015). http://dx.doi.org/https://doi.org/10.14257/ijsia.2015.1.05
3. Timirgaleeva, R.R., Grishin, I.Y., Mironov, M.V.: E-Learning: the problem of reliable student authentication and information security (2019). CEUR-WS.org/Vol1–2522/paper20.pdf
4. Buchner, A.: Moodle Administration. Publisher: Packt Publishing Limited, Birmingham, United Kingdom (2010). ISBN: 9781847195623
5. Buchmann, J.A., Karatsiolis, E., Wiesmaier, A.: Introduction to Public key Infrastructures. Springer Heidelberg New York Dordrecht London (2013). ISBN: 978–3–642–40656–0. https://doi.org/10.1007/978-3-642-40657-7
6. Amo, D., Alier, M., Garcia-Penalvo, F.J., Fonseca, D., Casan, M.J.: Protected users: a moodle plugin to improve confidentiality and privacy support through user aliases (2020). https://doi.org/10.3.90/su12062548
7. Guillen-Gamez, F.D., Garcia-Magarino, I.: Use of facial authentication in E-learning: study of how it affects students in different Spanish-speaking areas. Int. J. Technol. Enhanc. Learn. **7**(3), 264-280 (2015). ISSN: 1753–5255 2015
8. Kaur, N., Prasad, P., Alsadoon, A., Pham, L., Elchouemi, A.: An enhanced model of biometrics authentication in E-Learning: using a combination of biometric features to access E-Learning environments. In: 2016 International Conference on Advances in Electrical, Electronic and Systems Engineering (ICAEES) (2016)
9. Karen, P., Morris, R., Douglas, D.M., Chawdhry, A.: Verifying user identities in distance learning courses: do we know who is sitting and submitting behind the screen?. Issue Inf. Syst. **15**(1),370–379 (2014). https://doi.org/10.48009/1_iss_2014_370-379
10. Aissaoui, K., Azizi, M.: El-Security: e-learning systems security checker plug-in. In: BDCA: Proceedings of the 2nd International Conference on Big Data, Cloud and Applications, pp. 1–6, March 2017. https://doi.org/10.1145/3090354.3090452
11. Delisle, M.: Mastering phpMyAdmin 3.4 for Effective MySQL Management. Publisher Packt (2012). ISBN 9781849517782
12. Mehta, C., Shah, S., Bhavsar, A., Oza, H.: MySQL 8 Administrator's Guide. Publisher Packt (2018). ISBN 9781788395199
13. Mehta, C., Shah, S., Bhavsar, A., Oza, H.: MySQL 8 Administrator's Guide, 2018, Publisher Packt, Amo, D., Marc, A., Garcia-Penalvo, F.J., Fonseca, D., Casan, F.J., Protected Users: A Moodle plugin To Improve Confidentiality and Privacy Support through User Aliases (2020). https://doi.org/10.3.90/su12062548. ISBN 9781788395199

Mobile Models for Biosensors with Diffusion Layer Through Enzyme Receptor

Cristian Ravariu(✉), Vasile Banes, Andrei Enescu, and Razvan Vasile

Student League, BioNEC Group, Polytechnic University of Bucharest, Bucharest, Romania
cristian.ravariu@upb.ro, {vasile.banes, an-drei.enescu0809, razvan_mihai.vasile}@stud.etti.upb.ro

Abstract. As anticipated since 2019, the CMOS electronics reaches its final technological nodes, like 2nm in 2021. Courses of electronic devices suddenly move from micro-electronics to nano- and bio-electronics. As consequence, new software tools are necessary for this mobile content area. Usually, in Biomaterials integrated on Field Effect Transistors BioFET, the expertise to simulate the FET part is well-known. But, software instruments for the bioreceptor part are almost absent. This paper presents an analytical tool, able to simulate the receptor function within the potentiometric biosensor, when a diffusion layer is interposed between transducer and enzymatic block.

Keywords: Enzyme modeling · Diffusion membrane · Mobile modeling

1 Introduction

As anticipated since 2019 [1], the CMOS (Complementary-Metal-Oxide-Semiconductor) electronics reaches its final technological nodes in 2021, as 2nm-node [2]. Moore's law was punctually fulfilled. The predictions about post-MOS era started 20 years ago, but they intensified in the last years. Invasion of new domains in nano-electronics was announced: spintronics for ultra-low power electronics [3], MOS co-integration with nano-devices [4], vacuum nanodevices covering the tera-Hertz gap [5, 6], but especially a large diversity of biomaterials integrated with MOSFETs to provide new biodetection functions, [7].

To fill the dynamic content of these courses, we need to develop simulators of mixed devices, like En-FET (Enzyme Field Effect Transistors). The modeling missing in the enzymatic block of biosensors is still chronic. Strong software tools like Silvaco package can simulate magnetic sensors, image sensors, optical sensors, but has no resources for bio-sensors simulation. This kind of tool begins with modeling. For I-st, II-nd and III-rd generation of enzymatic biosensors [8], a permeable membrane exists between the biological receptor and the analyte flux, so that the redox process mediated by enzyme is accompanied by a diffusion process through this membrane.

This paper presents a succession of analytical models with adaptation coefficients for the products that result from an En-FET transistor with diffusion membrane.

M. E. Auer and T. Tsiatsos (Eds.): IMCL 2021, LNNS 411, pp. 962–969, 2022.
https://doi.org/10.1007/978-3-030-96296-8_87

2 Modeling of the Enzyme-FET with Diffusion Membrane

In this paper, an Enzyme-FET is considered as amperometric biosensor, so that the analyte biosignal is converted into a current. The FET part is usually produced by a modified MOS technology, easily to be simulated [9]. We focus on the enzyme-receptor part. In this mixed technology, the intermediary diffusion membrane consists in Si-porous layers or nano-structured oxides above the gate oxide of the MOS transistor, [7]. Figure 1 presents the principle blocks inside an En-FET.

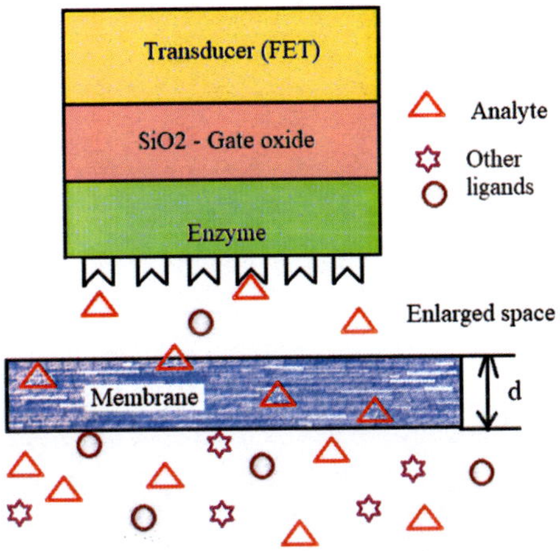

Fig. 1. Main blocks: transducer in Si, Gate oxide, enzyme in thin contact with membrane that allows the diffusion of mainly analyte.

For an enzyme receptor, the analyte suffers capturing on its active centers. Finally some product species, usual ions, are released. The product ions P and the molecules of biological analyte B cross the diffusion membrane by diffusion. Membrane is a layer of d thickness, with diffusion coefficients D_P, D_B for product and analyte, respectively. The enzymatic assisted reaction can be characterized by the loading factor, f_L [10]:

$$f_L = \frac{v_{max}}{K_M} \cdot \frac{d^2}{D_B}, \text{ where } v_{max} = k \cdot [E] \tag{1}$$

where v_{max} is the maximal reaction rate, K_M is Michaelis constant, k is a reaction rate constant and [E] is the enzyme loading. If f_L is greater than 25 than the reaction is dominated by kinetic of reaction via factor v_{max} / K_M elsewhere the reaction is dominated by diffusion via factor D_B/d^2 [10]. The case of the kinetic domination is modeled by Michaelis-Menten formalism and time interval of measurement Δt [11]:

$$[P]^{trad} = \frac{v_{max} \cdot [B]}{K_M + [B]} \cdot \Delta t \tag{2}$$

The case of the diffusion domination is modeled by Blaedel model [10]. If the biological analyte concentration [B] is much less than Michaelis constant, the model is approximated by:

$$[P]^{trad} = [B].\frac{D_B}{D_P} \cdot \left[1 - \frac{2}{\exp\left(\sqrt{f_L}\right) + \exp\left(-\sqrt{f_L}\right)}\right] \quad (3)$$

If the biological analyte concentration [B] is greater than Michaelis constant, this model is approximated by:

$$[P]^{trad} = \frac{v_{max} \cdot d^2}{2D_P} \quad (4)$$

These models can be applied for main biosensors from first generations. In a next section, the models are calibrated on some experimental picked points from literature.

3 Mobile Modeling Versus Experiment

Figure 2 a comparatively presents models (2), (3) and (4) for data of a glucose biosensor fabricated with glucose-oxidase enzyme. A nano-porous TiO_2-membrane of 100 nm thickness plays the role of diffusion membrane, Fig. 2b. The technology details of glucose En-FET were presented elsewhere, [7].

We use the parameters: glucose oxidase enzyme with Michaelis constant of $K_M = 2.9$mM, maximum reaction rate $v_{max} = 0.0016$ M/s, measuring time of 6s, membrane with d = 30 μm , $D_P = 163 \mu m^2/s$, $D_B = 2 \mu m^2/s$, and [B] ranges between 1 nM till 1 M. The first analysis result offers a comparison among experimental points, models (3)–(4) for the case of biosensor with membrane and model (2) for the case of biosensor without membrane, Fig. 2a. In a next step, the continuity advantage of model (2) must be applied to the biosensor with the diffusion membrane.

When the whole transfer function of an En-FET is demanded, we use another experimental biosensor, with complete data [13]. The Nerst's equation models the ions concentration near the sensitive-electrode to express the gate voltage increasing. When the ionic product concentration increases, the gate voltage rises. Hence the drain current rises, too. The parabolic model or Spice-level-1 model is used for the MOS-FET transistor, [12]. Figure 3 comparatively presents current-concentration curves, recorded by analytical models (2)–(4) and experimental picked points [13].

The electrochemical biosensor with tyrosinase (Tyr) enzyme and membrane made by sugarcane derived biochar nanoparticles (BCNP) was recently fabricated for the detection of bisphenol A (BPA) [13]. The biosensor experimentally presents a Michaelis constant of $K_M = 1.93$ μM, Tyr enzyme loading of 0.55 mol/l, analyte concentration ranging from 0.01 μM to 10 μM and average measuring time of 100 s, following two linear traces, [13]. The analytical model (3)–(4) is optimal fitted by regression among the experimental traces, selecting equivalent device parameters for three blocks from Fig. 1.

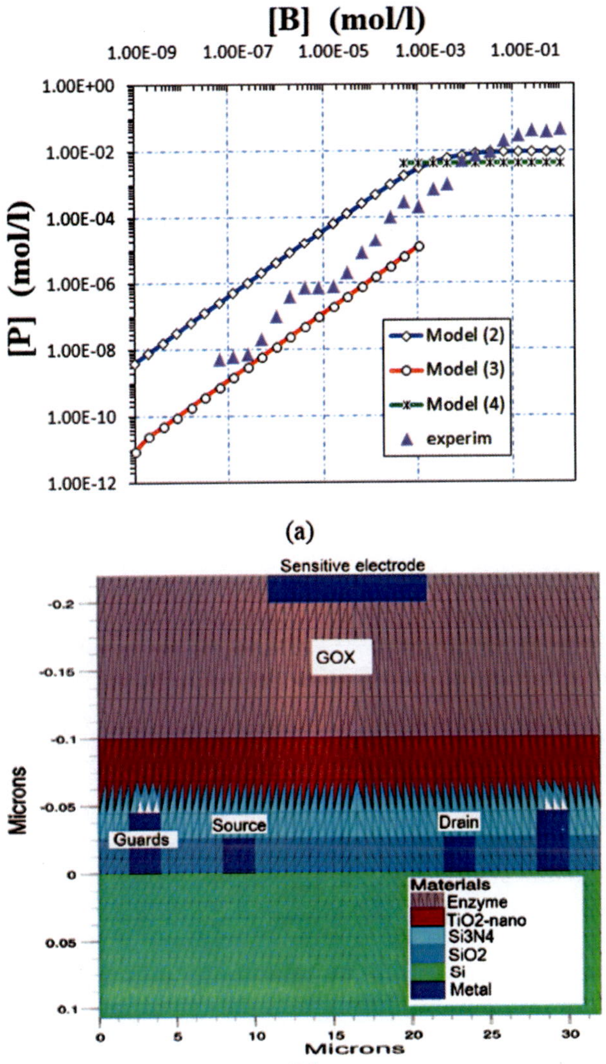

Fig. 2. (a) Product concentration [P] computed by models (2), (3), (4); (b) the studied En-FET structure.

The enzymatic factor is $f_L = 12 < 25$, so that the biosensor is dominated by diffusion. However, the RMS (root mean square) error gives 0.115 for model (3–4) and 0.116 for model (2). The next target is to propose a *mobile model* inspired from model (3–4) able to reduce RMS. Additionally, the mobile model must be a continuous function, in agreement to experiments and must include *adaptation parameters*. We apply a similar empirical method as for MOS transistor, during the modeling of the

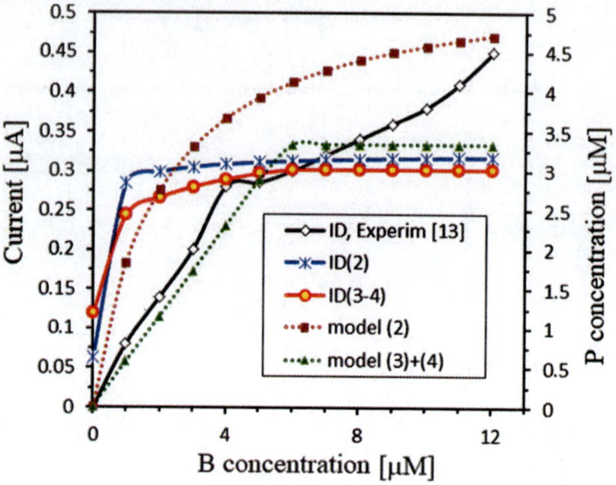

Fig. 3. Analytical model is compared to experimental model for an enzymatic biosensor reported in [13].

output characteristics at the saturation limit. Model (3) is admitted till the interception between model (4) and model (3):

$$[P]^{trad} = [B] \cdot D_N \cdot \left[1 - \frac{2}{\exp(\sqrt{f_L}) + \exp(-\sqrt{f_L})}\right] \text{ for } [B] \le [B_k] \quad (5)$$

where D_N is the normed diffusion coefficient, as a first adaptive parameter, and $[B_k]$ is the "knee" point of the curve [P] versus [B]. Over years, we continue to observe two [P] - [B] laws, at low and high analyte concentrations, which are separated by a "knee" concentration, better revealed by the theorem of the non-linear electrical conduction [14, 15]. The real experiments capture a poor increasing of current with analyte concentration, like in a saturated MOS transistor. As in that case, we apply here a liner dependence given by:

$$[P]^{trad} = \frac{v_{max} \cdot d^2}{2D_P} - \theta \cdot [B_k] + \theta \cdot [B] \quad (6)$$

where θ is defined as mobility coefficient, as the second adaptation parameter, extracted by regression. From the continuity condition results $[B_k]$:

$$[B_k] = \frac{v_{max} \cdot d^2}{2D_P} \frac{1}{D_N} \cdot \left[1 - \frac{2}{\exp(\sqrt{f_L}) + \exp(-\sqrt{f_L})}\right]^{-1} \quad (7)$$

For a best curves fitting between experiments [13] and mobile models (5–7), the parameters result: $[B_k] = 7.89\ \mu M$, $D_N = 0.08$, $\theta = 22\ \mu M^{-1}$, Fig. 4.

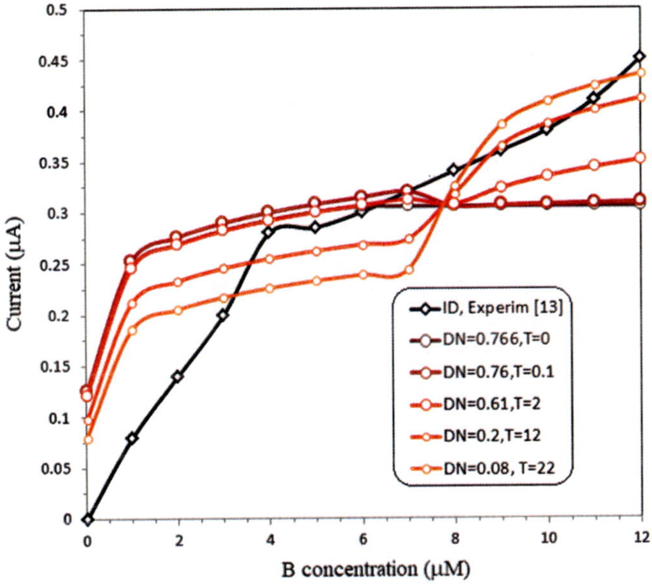

Fig. 4. The successive mobile models evolution for (5)–(6), to minimize the root mean square error in respect to the experimental curve.

4 Mobile Content for Biodevices

Anticipated outcomes envisage change in teaching and research practice. In the past times, the content of the disciplines about devices and their modeling fit more than 20 years (e.g. Ihantola-Moll MOS model is still in use in Spice simulators and in our courses about transistors modeling). In next 20 years, the mobile content of such a course, in agreement with the technology advances, will push the investigation tools toward a mobile modeling. Instead to teach some perennial models, instructors have rather to teach how to construct and to adapt new models for yearly new hybrid devices, like En-FET biosensors. In our example, the adaptive parameters D_N and θ sculpt in models (5) - (6). The SME decreases for this mobile model from 0.115 to 0.039, Fig. 5.

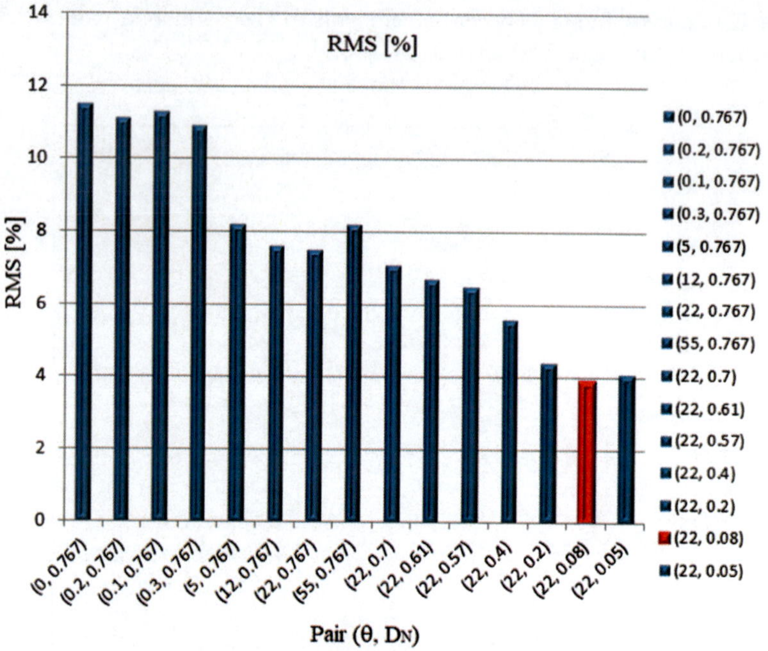

Fig. 5. Algorithm to minimize SME is suggested by pairs selection: (θ, D_N) initial pair starts from the value of models (3_4); then θ is varied to get minimal RMS, while D_N is constant; finaly, D_N is varied to get minimal RMS, while θ is kept constant at its previous optimal value.

5 Conclusions

This paper started to develop an analytical model for the En-FET biosensors, based on enzyme integration into the Gate space of Field Effect Transistors, like MOS. The modeling was focused on the dependence of the output signal that is the product concentration of an assisted enzymatic reaction on biological substrate that acts as input for the enzymatic block. The influence of a diffusion membrane between kinetics location and the headquarters of the product/analytes was considered by three parameters: membrane thickness d, diffusion coefficient for biological analyte D_B and diffusion coefficient for products D_P. Using these traditional models the relative error reaches to 11,6%. Due to the lack of time that will govern the development of electronic devices co-integrated with biosensors in the next 10 years, a more flexible model with adaptive parameters has been proposed, which reduces the relative error to 3.9%.

Acknowledgement. The article is financed by the Polytechnic University of Bucharest, through the project "Engineer in Europe" in online system, registered at the Ministry of Education under no. 140 / GP / 19.04.2021, by using the fund for financing special situations that cannot be integrated in the form of financing state higher education institutions.

References

1. Ravariu, C., Ursutiu, D., Mihaiescu, D., Morosan, A., Tanase, M., Tsiatsos, T.: New era of the nano-electronic devices – one of the most adaptive learning areas for the next period, Chapter 3 in Book: In: Internet of Things, Infrastructures and Mobile Applications, Springer, Editors: Michael Auer and Thrasyvoulos Tsiatsos, pp. 24–35 (2020)
2. IBM Unveils World's First 2 Nanometer Chip Technology, Opening a New Frontier for Semiconductors, internal report. https://newsroom.ibm.com/2021-05-06-IBM-Unveils-Worlds-First-2-Nanometer-Chip-Technology,-Opening-a-New-Frontier-for-Semiconductors. Accessed May 2021
3. Kang, W., Zhang, Y., et al.: Spintronics: emerging ultra-low-power circuits and systems beyond MOS technology. ACM J. Emerg. Technol. Comput. Syst. **12**(2), 1–42 (2015)
4. Deleonibus, S: CMOS nanoelectronics at the time of diversifications. In: International Symposium on VLSI Technology Systems and Applications (VLSI-TSA), 25–27 April 2011, Hsinchu, pp. 1–2 (2011)
5. Han, J.-W., Meyyappan, M.: The device made of nothing. IEEE Spectr. **51**(7), 30–35 (2014)
6. Ravariu, C.: Vacuum nano-triode in nothing-on-insulator configuration working in terahertz domain. IEEE J. Electron Dev. Soc. **6**(1), 1115–1123 (2018)
7. Ravariu, C., Manea, E.: Babarada, F: Masks and metallic electrodes compounds for silicon biosensor integration. J. Alloys Compound **697**, 72–79 (2017)
8. Zhang, W., Li, G.: Third-generation biosensors based on the direct electron transfer of proteins. Analy. Sci. **20**, 603–609 (2004)
9. Ravariu, C.: Deeper Insights of the conduction mechanisms in a vacuum SOI nanotransistor. IEEE Trans. Electron Devices **63**(8), 3278–3283 (2016)
10. Scheller, F.: Schubert, F: Biosensors – Technique and Instrumentation in Analytical Chemistry, Book at Elsevier Science Publishing. Company Inc., New York (1992)
11. Puneka, N.S.: Enzymes: Catalysis, Kinetics and Mechanisms, I-st Eds, Springer Nature, Singapore Pte Ltd, pp. 95–259 (2018)
12. Ravariu, C., Ravariu, F., Rusu, A.: Analysis of the non – linear responses of some biosensors. In: International Conference Proceedings, Sinaia, Romania, CAS 2003, vol. 1, pp.193–19. IEEE (2003)
13. Liu, Y., Yao, L., He, L., Liu, N., Piao, Y.: Electrochemical enzyme biosensor bearing biochar nanoparticle as signal enhancer for bisphenol a detection in water. Sensors **19**, 1619 (2019)
14. Ravariu, C., Mihaiescu, D.E., Istrati, D., Manea, E., Parvulescu, C: Application of the nonlinear electrical conduction theorem to emphasize the optimized biosensor sensitivity. In: 11th IEEE International Symposium on Advanced Topics In Electrical Ingineering, Bucharest, Romania, 28–30 March, pp. 94:1–4, (2019)
15. Ravariu, C., Mihaiescu, D., Rusu, A., Ravariu, F.: Non-linear electrical conduction through testosterone undecanoate and rethinol oily solutions. In: 6th European Symposium on Biomedical Engineering ESBME, Greece, Chania, pp.10.0–10.3 (2008)

Intelligent IoT Biomedical Bluetooth Data Acquisition System

Horia Alexandru Modran[1(✉)], Doru Ursuțiu[1,2], Cornel Samoila[1,3], and Tinashe Chamunorwa[1]

[1] Transilvania University of Brașov – Romania, Brașov, Romania
{horia.modran, udoru, csam, chamunorwa.tinashe}@unitbv.ro
[2] Romanian Academy of Scientists, Bucharest, Romania
[3] Romanian Academy of Technical Sciences, Bucharest, Romania

Abstract. Given that the cardiovascular diseases (CVDs) are the leading cause of death worldwide, it is very important to develop intelligent and accessible systems for detecting and predicting this kind of diseases. Various Artificial Intelligence (AI) techniques are successfully applied to prevent and detect heart problems and cardiovascular diseases, as well as to analyze health data.

The study presented in the current paper is focused on developing a data acquisition system using PSoC6 BLE and a Printed Circuit Board (PCB). A PSoC application has been developed, which can acquire data on two channels and can send them to the computer through the serial port and to the smartphone using the Bluetooth 5.0 protocol. Furthermore, the PCB board with which the IoT device will be integrated has two BNC connectors and two amplifiers – that can amplify the acquired signal by either 10 or 100 times. The data can be processed using the LabVIEW developing environment as well as on Android smartphones.

The further research will include the use of AI techniques to prevent and detect heart problems and cardiovascular diseases, as well as to analyze health data by using AI techniques.

Keywords: Internet of Things (IoT) · Artificial Intelligence · Virtual Instrumentation · Bluetooth BLE · PSoC6

1 Introduction

The dynamics of the medical and bioinformatics sectors are facing a revolution due to the integration of computing systems and artificial intelligence in many fields. Recent technological progress in Internet of Things (IoT), Machine Learning (ML) and Artificial Intelligence (AI) have asymmetrically changed the perspectives of biomedical engineering and bioinformatics [1]. The need of avoiding human error has always led to a major technological development in the field of intelligent diagnostics, through the implementation of genetic algorithms, artificial neural networks, expert systems and fuzzy logic for optimization.

M. E. Auer and T. Tsiatsos (Eds.): IMCL 2021, LNNS 411, pp. 970–979, 2022.
https://doi.org/10.1007/978-3-030-96296-8_88

According to a study conducted by the World Health Organization, Cardiovascular diseases are the leading cause of death globally[1] [2]. Therefore, numerous machine learning (ML) algorithms have become more and more used for predicting this type of disease.

A data acquisition system using PSoC6 BLE integrated chip from Cypress/Infineon and a Printed Circuit Board (PCB) were developed. A PSoC application was created, which can acquire data on two channels and can send them on the computer through the serial interface and on a smartphone via Bluetooth 5.0. Also, the PCB in which the PSoC device will be integrated has two BNC connectors and two amplifiers (one for 10x and 100x), including a switch between them.

Machine learning (ML) techniques are nowadays applied to prevent and detect heart problems and cardiovascular diseases, as well as to analyze health data by using algorithms based on AI [3].

2 PSoC6 Application

PSoC6 covers the difference between expensive, power-consuming application processors and low-performance microcontrollers [4]. PSoC6 was chosen because its very low power architecture provides the processing performance required by IoT devices, eliminating the trade-offs between power and performance.

The medical data acquisition system was developed using an integrated PSoC6 BLE chip produced by the Cypress/Infineon company.

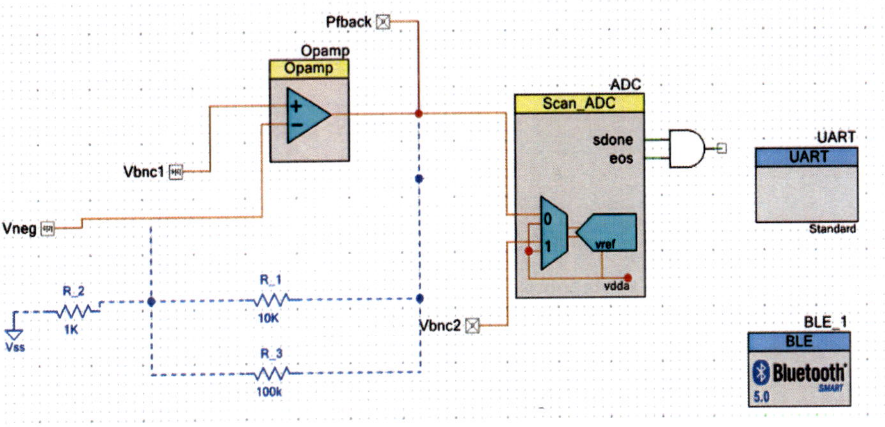

Fig. 1. PSoC6 application schematic

[1] World Health Organization website, [Online] at https://www.who.int/health-topics/cardiovascular-diseases/, last accessed 2021/07/01.

The PSoC6 microcontroller was programmed in PSoC Creator, using the C programming language, to be able to acquire data on two different channels. The developed application has the role of acquiring biomedical data (or other type of data), as well as sending them both through the serial port to the computer and via Bluetooth 5.0 to other IoT devices. In the main diagram of the application (Fig. 1), an operational amplifier (abbreviated Op. Amp.) and an Analog-Digital converter (abbreviated ADC) were added. A universal asynchronous receiver-transmitter (UART) component was also used, for printing data through the serial port and a BLE 5.0 component to send data through Bluetooth. The two PSoC6 processors communicate with each other using a mutex to control access to a shared variable. With security built into PSoC6, the developed system is thus secured and protected [4].

This application is developed so that it to be used together with the PCB, to be able to amplify the signal either 10 times or 100 times, while switching between the two types of amplifications is done via a switch. The obtained results were compared with the unamplified signal. The PCB has two BNC connectors as inputs (Vbnc1 and Vbnc2 on the schema).

For rapid prototyping and testing the functionality of the PSoC application the ELVIS II (Engineering Laboratory Virtual Instrumentation Suite) from National Instruments was used, specifically the function generator provided by it [5].

3 Serial Communication (LabVIEW)

The PSoC6 kit has been programmed to obtain and print voltage data through the port serial, by using the Universal Asynchronous Receiver-Transmitter (UART) component from PSoC Creator.

The UART component of the PSoC is very useful for viewing and testing acquired data and thus can be used to transfer PSoC-acquired data to LabVIEW. The Rx and Tx pins of the UART block were associated on the PSoC 6 kit with ports P5 (0) and P5 (1). The UART data will use the USB interface and a port will be created as a peripheral on the PC side, which can then be used to obtain data in LabVIEW, TerraTemp or PuTTY.

For the obtained data to be processed in LabVIEW, a Virtual Instrument (VI) that uses the VISA function to obtain a connection to the computer's serial port was developed. In this VI the parameters used to configure the UART component of the PSoC chip must be set, respectively: 8 data bits, no parity, a baud rate of 115.200, no flow control and one stop bit. After reading the values from the serial port, the VISA resource should be closed to release the serial port to avoid memory leaks [6]. The front panel of this VI, which present the values read and displayed graphically in a Waveform Chart for 100 times amplification, is illustrated in Fig. 2.

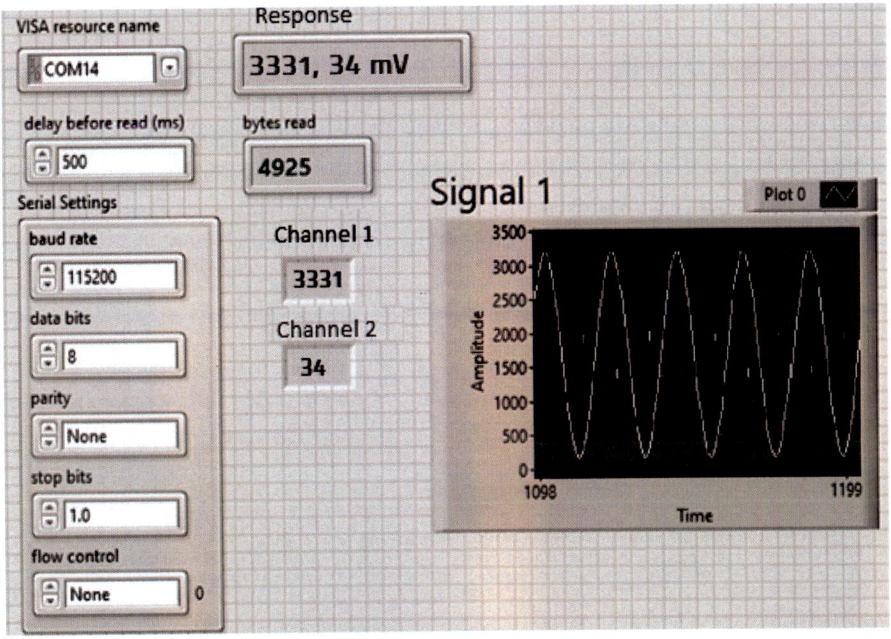

Fig. 2. Front panel of the Virtual Instrument

4 Bluetooth Mobile Application

The acquired data can then be sent using Bluetooth 5.0 on the mobile phone or any other IoT device which is compatible with the Bluetooth communication protocol.

To communicate with the IoT device, an Android mobile application was developed. Android provides a framework for Bluetooth Low Energy (BLE), which was introduced in Android 4.3 (API level 18). The application architecture is illustrated in Fig. 3.

There are seven classes from the framework that were used: BluetoohManager, BluetoothAdapter, BluetoothLeScanner, BluetoothGatt, BluetoothGattCallback, BluetoothGattCharacteristic, and finally BluetoothGattDescriptor. The role of these classes is as follows [7]:

1. BluetoothManager is used to find the Bluetooth system service for the phone running the application.
2. BluetoothAdapter is the Bluetooth adapter of the phone. This object is used to interact with the PsoC device and is found using the *getAdapter()* method in BluetoothManager.
3. BluetoothLeScanner is used to find BLE (Bluetooth Low Energy) devices that the phone can interact with. This is a listener for BLE's advertising packages and then notifies them. It has a callback function that is invoked every time a new BLE device is found.

4. The purpose of the BluetoothGatt class is to allow communication with a BLE device. This object is created when the smartphone connects to a BLE device.
5. BluetoothGattCallback is used to record events on the BLE device, such as changes in connection status and when new data available.
6. BluetoothGattCharacteristic is called for each of the features contained in the BLE device profile.
7. BluetoothGattDescriptor is used for additional feature attributes. In the application, it is used be used to store the notification descriptor for the feature that transmits the values obtained by PSoC6 from one data acquisition channel.

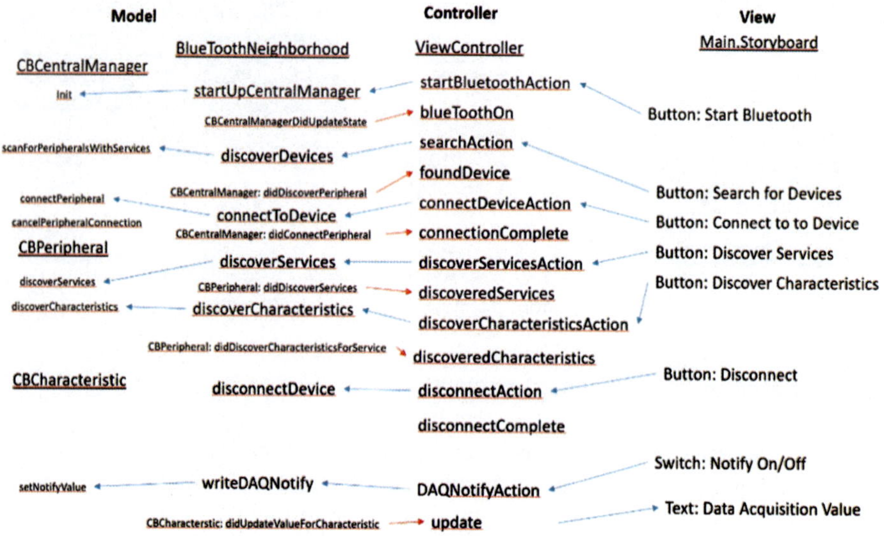

Fig. 3. BLE Android application architecture

In Android, Bluetooth requires two specific permissions and, for versions higher than Android 6.0, a location permission. Once the permissions and the broadcast receiver are turned on, the application waits for a button press. Initially, only the "Start Bluetooth" button is activated. When the user clicks the button, the BluetoothManager and BluetoothAdapter objects are created. Then it will start a service that will represent the Bluetooth device to which will eventually connect. In Android, a service is a component of the application that performs operations in the background without its own user interface. This is perfect for the Bluetooth device - the main application will interact with it and the user interface. In other words, this service will be the "Model" of the BLE device that the phone will connect to.

After performing previous steps, the next button, "Search for PSoC Device", will be enabled. When the user presses that button, a method from the model will be called, to start scanning for Bluetooth devices. In this case, the scan method will search only for devices that contain the BLE service that is sought, specifying the UUID of the service. The Android Service object is a background task manager that shapes the interface of the BLE device, while the BLE device service describes how the BLE device interacts with the phone.

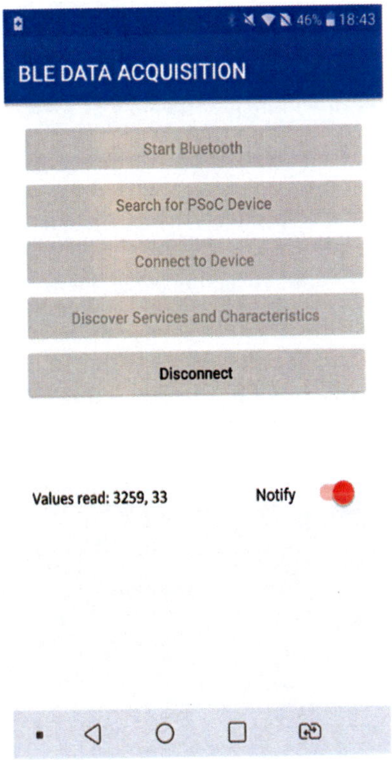

Fig. 4. UI interface of Android application

The scan method has a callback function - whenever this function is called, the BLE scan is stopped and then broadcasts an update that will be received by the transmission receiver of the main application. The broadcast receiver then activates the next button, "Connect to Device". When this button is pressed, a method is called in Model to connect to the device and create a BluetoothGATT object. Once this is done, the BluetoothGattCallback object sends an update to our main application. The broadcast receiver in the main application then activates the next button, "Discover Services and Characteristics".

That button invokes a method in the model that queries the GATT database to get the services and features that it contains. When this is complete, a GATT callback is received, that creates the BluetoothGattCharacteristic object for the feature that contains the values acquired by our device and a BluetoothGattDescriptor for notification.

If notifications are enabled, then a GATT call will be received whenever new data is available. Basically, when a new value is obtained through PSoC, it will convey that a change has taken place. The GATT callback sends an update again to the main receiver of the application, which is then able to update the new acquired value on the screen of the smartphone. The user interface of the Android application on the smartphone is shown in Fig. 4.

5 PCB

To design the PCB circuit, the Proteus software was used. The basic characteristics of the circuit are the following:

- PSoC6 connectors (21 pins for the left side and 21 for the right side)
- 2 resistors (10k and 100k respectively)
- 2 BNC input connectors
- 10x and 100x amplifier with a switch between them

The purpose of this circuit is to integrate the PSoC chip for data acquisition through BNC connectors and the amplification of the acquired data by 10, respectively 100 times. After designing the circuit in Proteus, the Gerber files were generated, to be able to print the circuit using the Voltera V-One Printer [8] (Fig. 5).

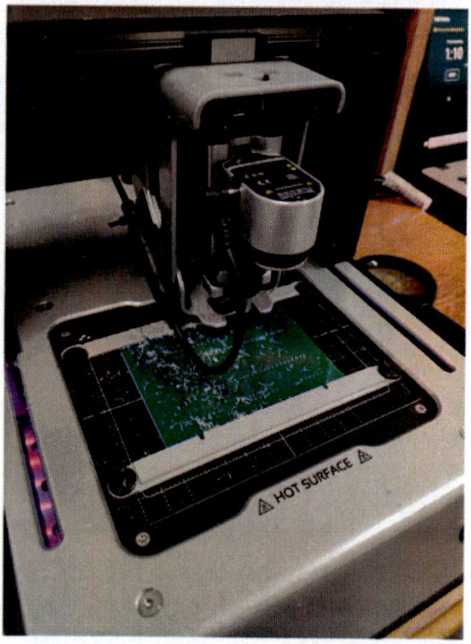

Fig. 5. PCB Drilling using the Voltera V-One

After generating the Gerber file, the first step in making the circuit is the drilling stage, which precedes the printing phase. The holes are made on the entire printed circuit board and are either used for internal electrical interconnection or for positioning in the process of assembling the components.

When drilling is completed, the top layer will be printed. The procedure is the same as creating a single layer board, the only difference being that the board will contain holes. For the printing process, the same positioning from the drilling stage was used [9].

After the printing phase, the circuit must be left to bake for about an hour, as the ink applied in the previous step must be dried.

One of the main disadvantages of the Voltera V-One Printer is that it currently does not have an integrated function of taking and placing components, so they must be placed manually. Therefore, after the drilling, printing and baking steps, the components were placed manually.

The following components were placed on the printed PCB:

- 2 BNC connectors
- 2 switches
- 4 PMOD connectors

6 Artificial Intelligence

In the further research, a complete system of data acquisition and processing through Artificial intelligence techniques will be developed. The architecture of this system is illustrated in Fig. 6, and its main characteristics will be:

- PSoC6 microcontroller integrated with the PCB, designed for data acquisition
- sensor and 3 electrodes

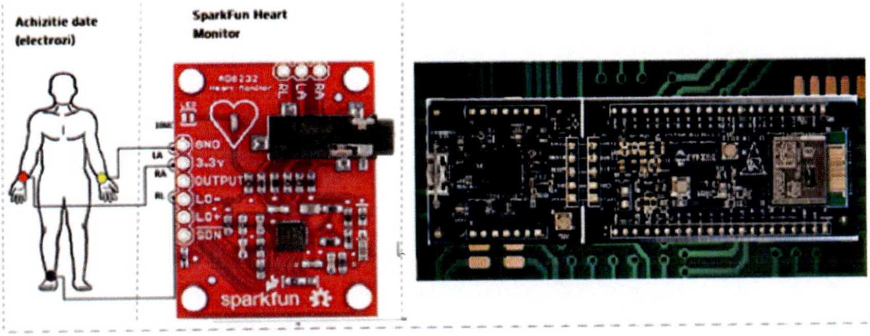

Fig. 6. Artificial intelligence system development

For intelligent data processing, Edge AI techniques will be used, thus making real-time analysis possible and reducing the cost and latency of the system [10]. The processing will be done at a local level, directly on the PSoC chip, and, for that purpose, TensorFlow Lite for Microcontrollers will be used. This framework is aimed to run machine learning models on microcontrollers or other IoT devices which have with only few kilobytes of memory. [11].

A neural network will be constructed, which will take as input blood pressure, Electrocardiography (ECG) and heart rate data and it will be able predict cardiac abnormalities and cardiovascular diseases. Medical data freely available on PhysioNet and Medscape databases will be used to train the model.

7 Conclusion

As cardiovascular diseases (CVDs) are the number one cause of death globally, it is crucial to develop inexpensive and accessible systems for detecting and predicting this kind of disease.

The current study describes the development of a data acquisition system using PSoC6 BLE and a PCB. A PSoC application was developed, that can acquire data on two channels and can transmit them to the computer by using the serial interface and on the phone via Bluetooth 5.0. Also, the PCB board with which the IoT device will be integrated has two BNC connectors and two amplifiers – for 10 and 100 times, including a switch between them.

This introductory study will be extended by using edge AI techniques to prevent and detect cardiovascular diseases and other heart abnormalities, as well as to analyze health data by using machine learning algorithms.

References

1. Damen, J.A., et al.: Performance of the Framingham risk models and pooled cohort equations for predicting 10-year risk of cardiovascular disease: a systematic review and meta-analysis. BMC Med. **17**, 109 (2019)
2. Theus, A.S., et al.: Biomaterial approaches for cardiovascular tissue engineering. Emergent Mater. **2**(2), 193–207 (2019). https://doi.org/10.1007/s42247-019-00039-3
3. Golas, S.B., et al.: A machine learning model to predict the risk of 30-day readmissions in patients with heart failure: a retrospective analysis of electronic medical records data. BMC Med. Inform. Decis. Mak. **18**, 44 (2018)
4. PSoC 6 CY8CKIT-062 Pioneer Kit – Cypress/Infineon – DigiKey. https://www.digikey.com/en/product-highlight/c/cypress/psoc-6-cy8ckit-062-pioneer-kit. Accessed 29 June 2021
5. Parab, J.S., Nazareth, I.A., Gad, R.S., Naik, G.: Learning by Doing with National Instruments Development Boards, 1st edn. CRC Press Publishing House, Boca Raton (2020). https://doi.org/10.1201/9780429441806
6. NI VISA User Manual. https://www.ni.com/pdf/manuals/370423a.pdf. Accessed 01 July 2021
7. Gaitatzis, T.: Bluetooth Low Energy in Android Java: Your Guide to Programming the Internet of Things (Bluetooth Low Energy Programming Book 2), BackupBrain, ASIN B0768HSL4R (2017)
8. Hu, R.: PCB Design and Layout Fundamentals for EMC, Independently Published (2019). ISBN 1082079251

9. Voltera Design Guidelines. https://www.voltera.io/support/circuit-design-guidelines. Accessed 01 July 2021
10. Wang, X., Han, Y., Leung, V., Niyato, D., Yan, X., Chen, X.: Edge AI - Convergence of Edge Computing and Artificial Intelligence. Springer, Singapore (2020). ISBN 978-981-15-6185-6, https://doi.org/10.1007/978-981-15-6186-3
11. TensorFlow Lite for Microcontrollers Guide. https://www.tensorflow.org/lite/microcontrollers. Accessed 05 July 2021

Electronic Laboratory Educational Board

Tinashe Chamunorwa[1(✉)], Doru Ursuțiu[1,2], Cornel Samoila[1,3], and Horia Alexandru Modran[1]

[1] Transilvania University of Brașov, Brașov, Romania
{chamunorwa.tinashe,udoru,csam, horia.modran}@unitbv.ro
[2] Romanian Academy of Scientists, Bucharest, Romania
[3] Romanian Academy of Technical Sciences, Bucharest, Romania

Abstract. One of the major factors which hinder effective laboratory activity in Electronic Engineering is the cost of equipment. A low-cost system is ideal for many universities as many units can be acquired. These units will cater to an increasing enrolment in electronic engineering. This paper outlines the initial and ongoing steps of the research work to design and implement a relatively inexpensive remote laboratory platform for electronics engineering students. The Laboratory platform is a mix of configurable and expandable hardware and software components. The development of the laboratory system is incremental and modular in manner. Different circuit applications were built separately and then combined into a one Printed Circuit Board (PCB) platform. The board is envisioned to be relatively cheap and has fundamental properties like soft configurability, mixed signals handling. LabVIEW was used in combination with PSoC Creator to come up with the experiments.

Keywords: BLE · LabVIEW · NI ELVIS II · PCB · Pmod · PSoC · Voltera

1 Introduction

The U.S Bureau of labor of statistics forecasted a 3% growth in electronics and electrical engineers' overall employment from 2019 to 2029. More companies and firms are expected to make use of these engineers in their projects encompassing electronic systems and devices [1]. The over-reliance of today's society on technology has led to the need for continual development in electronic systems. Consequently, the electronics industry growth spurs demand in college enrolment in the field. One can assume that the swelling number of students taking up electronic engineering put a strain on learning resources. The learning process includes laboratory work which is a fundamental part of knowledge building. Electronic engineering laboratories usually require expensive equipment which is expensive to acquire for most universities.In a bid to increase laboratory accessibility during the Covid-19 pandemic and curtail expenses, remote laboratories have become popular. Remote labs are in this context the remote access of real resources where experiments are conducted at an engineering lab and are operated through the internet [2].

This paper outlines the step-by-step design and implementation of a relatively inexpensive remote laboratory PCB board platform for electronics engineering

M. E. Auer and T. Tsiatsos (Eds.): IMCL 2021, LNNS 411, pp. 980–989, 2022.
https://doi.org/10.1007/978-3-030-96296-8_89

students. Four experiments have been performed and presented in this this paper as they have been deemed to present the fundamentals of electronic laboratory concepts which include:

- LabVIEW and PSoC creator Graphical programming.
- Bluetooth and UART communication protocols.
- PSoC I/O configurability and sensors interfacing using Pmod standard.
- Signal generation and amplification.

1.1 Related Work

A PocketLab system in presented in [8] which is an Arduino based system. The system implemented an RC circuit and mobile robot experiments among others. The system can be improved by employing a reconfigurable platform like PSoC. In [9] a remote lab is described and is based on NI my DAQ and ELVIS boards. The setup uses one PC with multiple myDAQs with different labs. The NI ELVIS makes it bulky and expensive. A PSoC system can have one PC, Board and a cloud connection. Presented in [10] is an introductory courses practical lab PSoC based system. It has a PCB board with a buck-booster converter half bridge based topology. The system is restricted to power electronics and a multi-concept will be more useful for an educational system (Table 1).

Table 1. Comparative summary of related systems

System	State of art	Proposal
PocketLab [8]	Arduino based system	PSoC based system
DC/DC converter lab system [10]	Restricted to power electronics and hardware control	Multiple fields cloud connection
Remote lab concept system [9]	One PC, multiple boards, hardwiring	One PC, one board, cloud connection

2 Experiments and Applications

This section outlines the experiments and applications that were carried-out in the process of the development board design. The four experiments described in this represents the fundamentals of electronics embedded systems design. Figure 1 shows the hardware experimental arrangement.

- The NI ELVIS II was used as a prototyping platform.
- The NI ELVISmx software was used to provide soft front panels instruments like digital multimeter, functional generator and an oscilloscope [3].
- The PSoC 6 (CY8CPROTO-063-BLE) prototyping kit (see acknowledgement) was attached and used as the central controller of all the applications. The kit is a relatively low-priced hardware platform that enables debug and design of PSoC 6 MCUs [4].

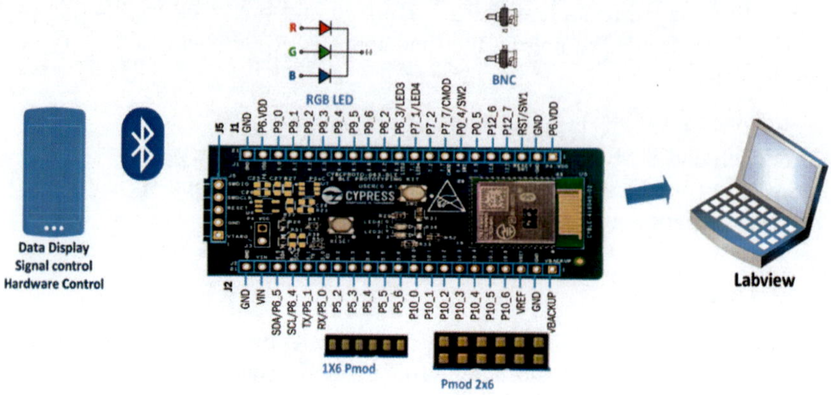

Fig. 1. PSoC 6 based board overview

- Different parts of the circuits were built and tested separately using PSoC creator and the NI ELVIS board.

The applications prototyped and presented in this paper are the:

1. PSoC voltage DAC sine wave generator and double channel Data acquisition and amplification system
2. LabVIEW control of PSoC connected RGB LED
3. CySmart Android app control of RGB LED using Bluetooth Low Energy (BLE).
4. Pmod Based UART proximity sensor application

Figure 2 shows the Cypress and off-chip components configured in the PSoC creator to design the applications. Figure 3 shows the prototyping on NI ELVIS II board.

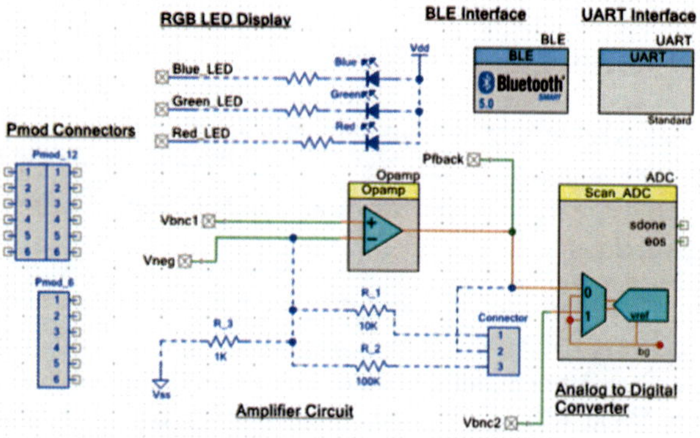

Fig. 2. Cypress and off-chip components configured for the applications

Fig. 3. Circuit prototyping on NI ELVIS II

The prototype connections shown in Fig. 3 were designed into a schematic circuit as shown in Fig. 4.

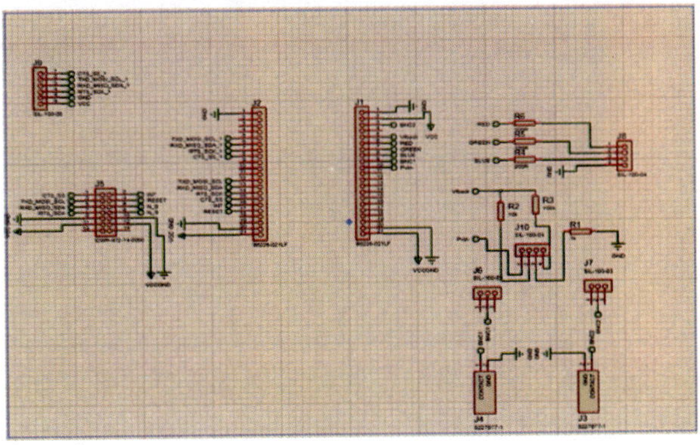

Fig. 4. Schematic circuit diagram

The PCB was then converted to Gerber files for fabrication using Voltera [5] which went through the processes of:

- Alignment XY alignment using smart height probe
- Drilling Drilling of vias and through-holes
- Ink Dispensing Deposition on PCB of solder paste and ink
- Heating Using heating profiles for ink curing and reflow.

Figure 5 shows the visuals of the trace and drilling process.

Fig. 5. Voltera PCB drilling and trace making

Figure 6 shows the resultant prototype of the various experiments.

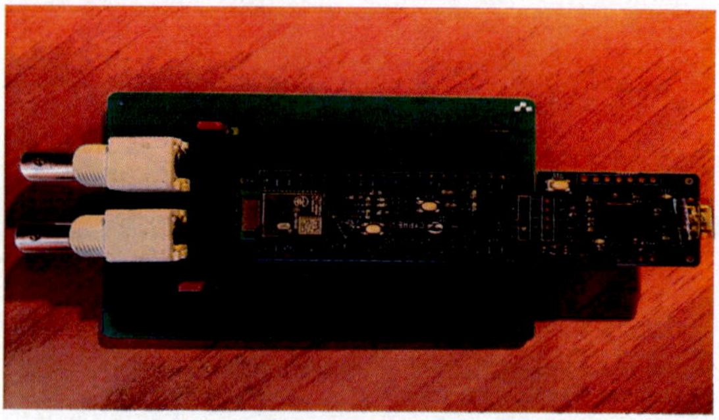

Fig. 6. The resultant prototype

2.1 PSoC Voltage DAC Sine Wave Generator and Double Channel Data Acquisition and Amplification System

Experimental Steps

- A PSoC Creator amplifier component was connected to an external resistor network to form a non-inverting amplifier circuit.
- The non-inverting amplifier circuit was then connected to an ADC and a UART component.
- The ADC channel was configured with two channels and single-ended input mode.
- Two identical sinusoidal signals were generated from NI ELVIS, and one was passed through the amplifier network of factor 10 and the other signal not amplified.

- The two signals went converted by the ADC component and were relayed to the UART component for onwards transmission to LabVIEW
- A LabVIEW block diagram and front panel were configured to accept serial communication.
- The signals were then displayed on waveform charts as shown in Fig. 7.

A sinusoidal signal was also generated from the Voltage DAC (Digital to Analog Converter) component through DMA (Direct Memory Access) transfers. This enables students to perform data acquisition (DAQ) experiments in the comfort of their homes without the use of NI ELVIS II board.

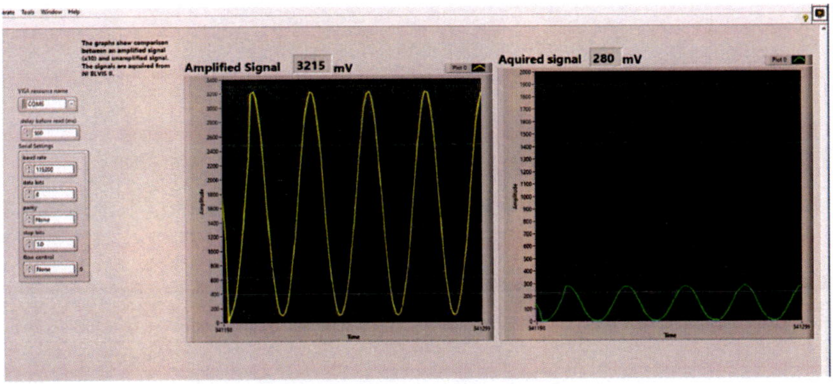

Fig. 7. Output for the 10k resistor

2.2 Red, Green and Blue (RGB) LED Brightness Control Through LabVIEW

This experiment illustrates an application to control RGB LED which was connected to the PSoC pins. The resistors values from the datasheet are shown in Table 2. The LED used is a CNT1 SMD (surface mount) module.

Experimental Steps

- A CNT1 SMD RGB LED module was connected to the PSoC 6 pins.
- A code (Fig. 8) was written and PSoC PWM component was configured with a period of 9 in the PSoC creator environment.
- The PWM period range corresponds to LabVIEW front panel dial range.
- A LabVIEW block diagram and front panel (Fig. 9) were designed to control the brightness of an LED through the PSoC PWM component

This concept can be extended to any number of LEDs in various display configurations to achieve various colours and lighting effects.

Table 2. RGB LED data [6]

LED	Operating volt-age (V)	Resistance (R)	Forward current (mA
Red	1.8–2.4	200	20–30
Green	2.8–3.6	100	20–30
Blue	2.8–3.6	100	20–30

```
int main(void)
{
    __enable_irq(); /* Enable global interrupts. */
     int compareValue =4;

      PWM_Start();

      UART_Start();

    setvbuf (stdin, NULL, _IONBF, 0);
     char c;

    printf("Started UART\r\n");

    for(;;)
    {
                c = getchar ();

                    compareValue = (int)(c-48);

          Cy_TCPWM_PWM_SetCompare0(PWM_HW, PWM_CNT_NUM,compareValue);
```

Fig. 8. LED control through PWM function

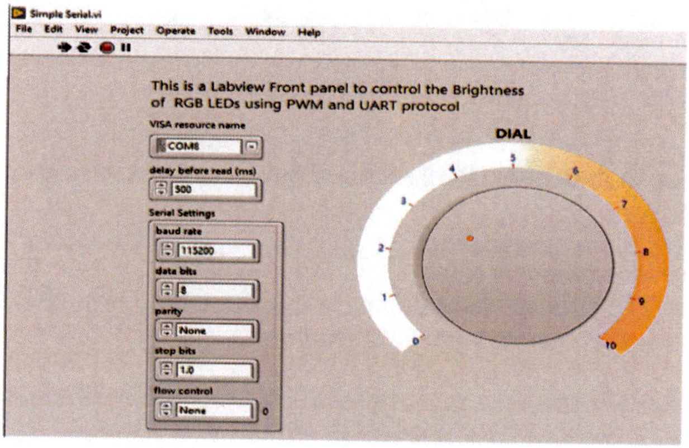

Fig. 9. Front panel for LED brightness control

2.3 RGB LED ON/OFF Control with Bluetooth Low Energy (BLE)

This experiment illustrates the RGB LED ON/OFF control via BLE protocol using CySmart Android application.

- The BLE component in PSoC creator was configured using the FindMe profile which was customized to control the RGB LED.
- The FindMe android user interface is as shown in Fig. 10

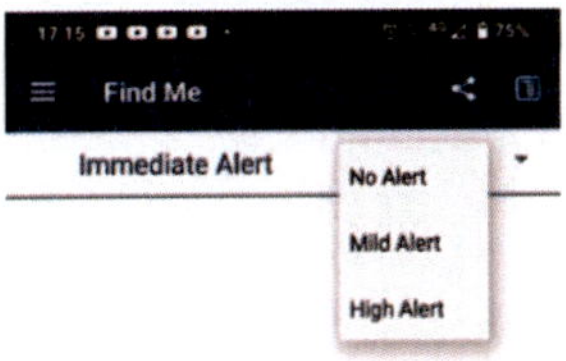

Fig. 10. CySmart Android control interface

2.4 Pmod Based UART Proximity Sensor Application

Experimental Steps

- The Pmod based Digilent MAXSONAR [7] ultrasonic range finder was connected to the PSoC.
- The UART component was configured with 9600 baud rate to match the sensor specification,
- A LabVIEW program was designed to display the results of the distance measured as shown in Fig. 12
- Distance measurement was done by blocking ultrasonic path by a hand as shown in Fig. 11

The Pmod connection can be adapted for Digilent SPI and I2C protocol-based sensors and expands student's experimental applications.

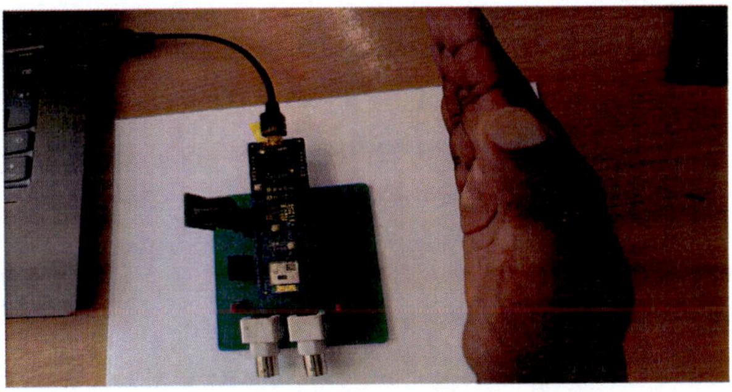

Fig. 11. Distance measurement process

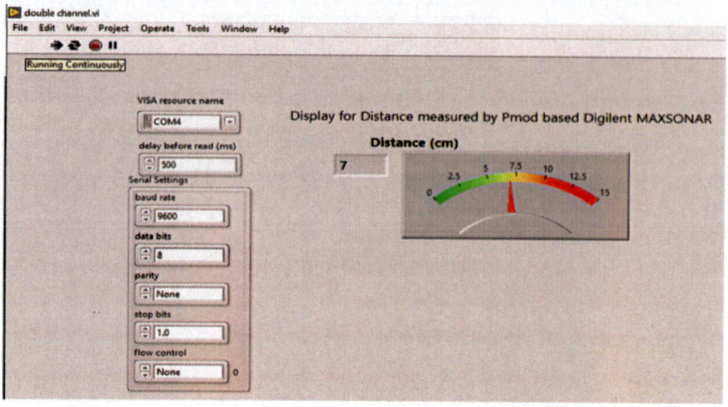

Fig. 12. LabVIEW front panel displaying distance measured

3 Conclusion

In this paper, four applications were prototyped. The functionalities presented show that PSoC and LabVIEW ecosystems are suitable for use in Electronic engineering laboratories as they can be adapted for various courses and concepts. The courses include digital and analog electronics, the Internet of Things, and communication protocols concepts. This is a work in progress to enhance the board's use and expand on its functionalities to cater to electronic engineering experiments. For intermediate experiments a PSoC BLE board priced at about 20 US dollars is a better configurable hardware alternative compared to FPGA based boards which usually are priced at not less than 100 US dollars. NI ELVIS board is another alternative but is expensive for basic laboratories purposes.

Acknowledgements. We would like to express our great appreciation to NI Corporation for availing of a free LabVIEW license. Our gratitude also goes to Infineon-Cypress for providing us with a free PSoC® 6 BLE Prototyping Board (CY8CPROTO-063-BLE), thus facilitating this study and making it possible. Their generosity and collaboration are greatly appreciated.

References

1. BLS Homepage. https://www.bls.gov/ooh/architecture-and-engineering/electrical-and-electronics-engineers.htm#tab-6. Accessed 10 Mar 2021
2. Garcia-Guzman, J.: Remote laboratories for teaching and training in engineering. In: Yildirim, S. (ed.) Mechatronic Systems in Engineering. IntechOpen Limited, London (2017)
3. NI Homepage. https://www.ni.com/pdf/manuals/374629c.pdf. Accessed 01 May 2021
4. Infineon-Cypress Homepage. https://www.cypress.com/documentation/development-kitsboards/psoc-6-ble-prototyping-kit-cy8cproto-063-ble. Accessed 05 May 2021
5. Voltera Homepage. https://www.voltera.io/product. Accessed 29 May 2021
6. Arduinomodules Homepage. https://arduinomodules.info/ky-009-rgb-full-color-led-smd-module. Accessed 15 May 2021

7. Digilent. https://digilent.com/reference/_media/pmod:pmod:pmodMAXSONAR_rm.pdf. Accessed 25 May 2021
8. Mikulas, H., Pavol, B.: PocketLab: next step to learning, experimenting and discovering in covid time. In: 2020 18th International Conference on Emerging eLearning Technologies and Applications (ICETA) (2020)
9. Thomas, K., Christian, K., Andreas, P., Christian, M.: Low-cost remote laboratory concept based on NI myDAQ and NI ELVIS for electronic engineering education. In: IEEE Global Engineering Education Conference (EDUCON), Porto, Portugal (2020)
10. Ulrich, B.: A modern, versatile and low-cost educational system for teaching DC/DC converter control with analog, digital and mixed-signal methods. In: International Conference on Research and Education in Mechatronics (REM), Wolfenbuettel (2017)

Modern Measurement System Transferred from Education to Industry

Petru Epure[1(✉)], Doru Ursutiu[2], Cornel Samoila[3], and Petru P. Epure[1]

[1] EPI-SISTEM SRL, Livezii 17A, 505600 Sacele, Romania
petru.epure@epi.ro
[2] Transilvania University – AOSR, Eroilor 29, 500295 Brasov, Romania
[3] Transilvania University – ASTR, Eroilor 29, 500295 Brasov, Romania

Abstract. Environmental risks are more present in our life despite the efforts of the authorities and NGO to promote green technologies and sustainable activities. Gases emitted by cars, chaotic constructions and other sources of industrial pollution generates dangerous hazards for health. In this context we consider that is very important to educate students how to measure different types of gases and to create a mesh of sensors capable to cover an extended area. For this purpose we started measurements in 2 locations with the aim to connect results in a database. Students can measure environment situation in the campus and the second site is useful for SME and their current activities. A common evaluation of measurements made in 2 different areas located in the same city can provide useful information about environment.

Keywords: Gas hazard · Acetaldehyde · Measurement mesh · LORA & Wi-Fi communication

1 General Gas Measuring Approach - Purpose of the Gas Measuring

There are periods when population is claiming that environmental gases are over expected limits and generate health hazards and even headache or pulmonary diseases. Sometimes the legal limits are overpassed but the information reach population after 12–24 h. Meantime, most of people are affected and for those with additional health problems can generate some emergency situation [1]. An alarm generation can be sent to public safety system and to emergency services to be aware of potential emergency intervention. During the risk assessment in the campus the students learn all technologies that are needed to control sensors, treat data, communicate on local storage and then to industrial partner. After performing an evaluation data can be disseminate to a website and to public authorities using designated channels respecting privacy [2]. The industrial partner is focused on CO2 and acetaldehyde measurements and the University is focused on general VOC like ammonia, hydrogen or combustion gases. Measurements can be performed in a gas chamber or directly in the air. Point of measurement have enough autonomy to provide continuous measurements during several month [3]. In this way we can provide a time evolution of measured parameters to be used for further analyses and virtual laboratories [4].

M. E. Auer and T. Tsiatsos (Eds.): IMCL 2021, LNNS 411, pp. 990–997, 2022.
https://doi.org/10.1007/978-3-030-96296-8_90

2 Hardware Configuration

The measurement point consist of a Cypress ARM controller, a Microchip PIC16F1938 controller for Wi-Fi communication, power circuitry with battery and a specific sensor. Sensors can be digital or analog and depends of each manufacturer [5]. Data are acquired via an access point then stored in local server. The software developed based on National Instruments technologies (actually NI) provide graphic representation of gas concentration evolution and basic tools to compute and analyze environmental risks [6]. Students will learn how to develop embedded software for 2 types of microcontrollers, how to design and evaluate battery circuitry charged by solar radiation, how to apply National Instruments technologies (actually NI) for flexible and stable monitoring. Embedded technologies imply a good knowledge of analog and digital processing and build know-how for the future engineer – actual student. The project include concepts like Remote Engineering, Virtual Instrumentation, Real & Simulated device and all these knowledge together with instrumentation software cover Interdisciplinary and Multidisciplinary. For teaching purposes all above activities generates creativity and long life learning LLL.

Wi-Fi communication is made via 2.4 GHz and also via 868 MHz. The low frequency is also called LORA and have some advantages over BLE of 2.4 GHz Wi-Fi communication. The main reason to use IQRF for LORA implementation is related to an intelligent mesh provided by Micro Risk processors. Point of measurements communicate each other if necessary and send data to the AP on the shortest way. When measuring points are more then 100 or 1000 it's difficult to acquire simultaneously with inexpensive hardware. To overcome this goal a single string is passed from each point of measurement to another and each one is adding the specific data in the string. In this manner is just one 1000 word string passing once instead of 1000 string of 4–8 words passing 1000 times. This type of optimization shorten the measurement cycle by 100 times at least. The actual mesh is not consisted of 100 measuring points but it is important that we can scale the number of sensors with no additional effort. In this stage we are talking about 10 measuring points and over 30 sensors involved.

The outcomes of this measuring system are related to the several aspects: live environmental data, alarm options, proof of concept for new sensors. Live environmental data sustain our effort to identify localized sources of air pollution and to combine it with other pollution factors like water or soil. In this way complementary analyses can support local authorities decision and fundament immediate actions. Sometimes pollution is related to industry and some private actions can be initiated before any authority measure. The "polluter pays" policy can generate effective feedback if the alarm is "live". The academic team generates a map of hazards capable to trigger a certain level of attention on a specific parameter. All this activities are performed by team of students under the guidance of teachers. Economic and social results are doubled by the teaching benefits. The system works if the students understand how to use a sensor, how to program a microcontroller and how the results are integrated by the instrumentation software.

Fig. 1. Sampling box for gas measurements (early prototype)

The sampling box have the advantage to maintain special conditions and to allow sampling in a definite volume at a definite temperature and humidity. The volume is 12 L and can be filled with air or specific gas for calibration purposes. Depending on measurement protocol the temperature and humidity can change the calibration curve and the student or the operator can chose the proper measurement calibration. In Fig. 1 we have implemented temperature and humidity sensor and also gas sensors like LPG gases, smoke sensor, carbon monoxide sensor and methane sensor.

Alarm options are related to each parameter as the limits can vary and upper limits are evaluated according to local environment laws and also according to European laws. Alarms are organized on levels and can provide a color flash on the computer screen, can send an e-mail to responsible persons or to local environment authorities.

In Fig. 2 is presented another configuration where the temperature and humidity are measured with another sensor coupled with microcontroller. The measurement are collected in a Intel NUC configuration based on a Intel I3 processor. For future developments the system will be capable to initiate a local emergency alert via GPRS. The first goal will be to generate SMS to all students registered in the campus. The type of alarm is shared between technic and functional. Technical alarms are used for maintenance of the system and corrective measures. Functional alarms are related to measured gas levels and consist in an over limit.

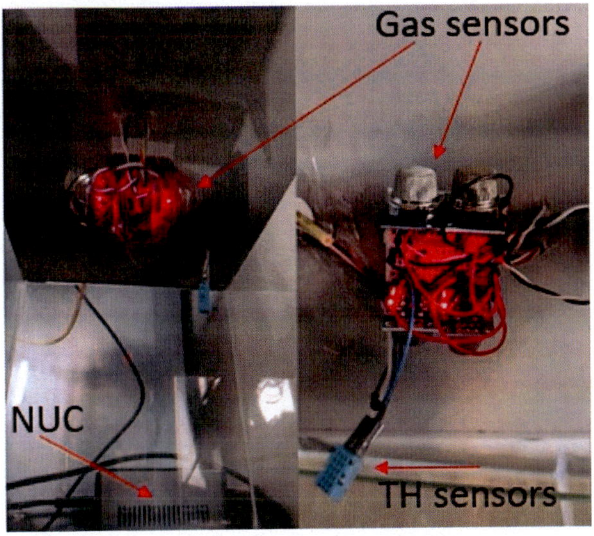

Fig. 2. Gas sensors used for pollutants monitoring in air

Sensors can be activated individually and also complemented with electrochemical measurements according to some measuring methods or protocols [7]. In Fig. 3 is presented the possibility to start a gas pump for feeding the gas chamber and also to heat the volume at a target temperature.

3 Software Developments

– Measuring applications were developed in LabVIEW and C language. Some of the applications were build in the Lab Windows/CVI environment and the microcontroller embedded programing is made in CREATOR PSOC environment.

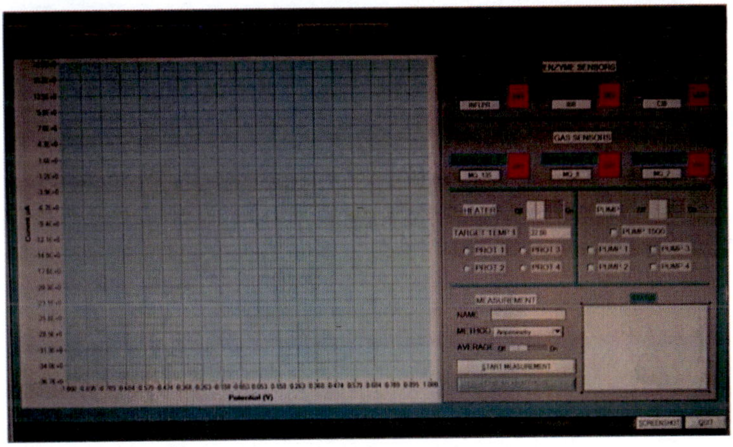

Fig. 3. Software interface prepared for several sensor measurements (CVI panel)

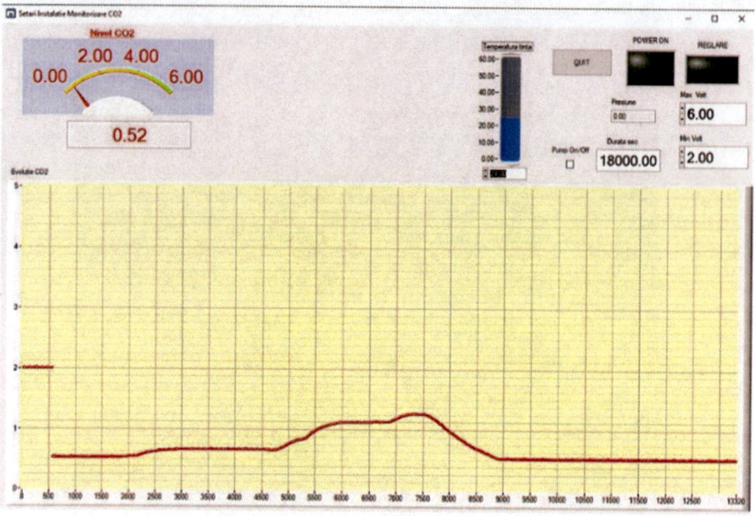

Fig. 4. Carbon dioxide monitoring for over 200 min

For environmental purposes it is very important to evaluate the content of carbon dioxide. In Fig. 4 there is an evaluation during 3–4 h of the CO2 concentration inside the gas chamber. The sensor DFROBOT Gravity can perform measurements between 400 ppm to 10000 ppm, results are displayed in volts for easy comparison with other sensors having different range concentration. Based on existing capabilities in university and SME an assessment of the air quality is performed according to each specific activity.

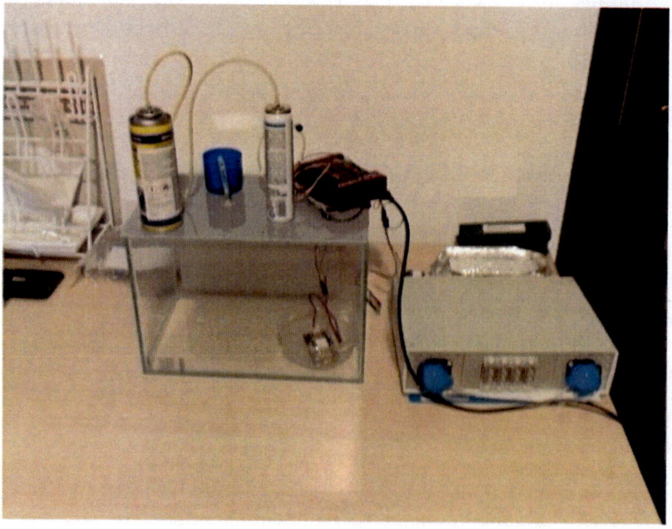

Fig. 5. Another setup for carbon dioxide gas sensor calibration

The calibration can be made also at the SME level with a setup similar with the one presented in the Fig. 5 where the sensor is on the bottom side, the signals are collected with a Labjack data acquisition card and displayed like in Fig. 4. The advantage of this simple setup is the adaptability to other gas sensors or other gas measurement methods. In this manner the setup can be used also for aldehyde measurement using electrochemical sensors. Concentration evaluation is made using 3 or 4 channels amperometric detection and the response is compared with daily calibration data. Another advantage is the capability to measure long time variations of 5–10 days.

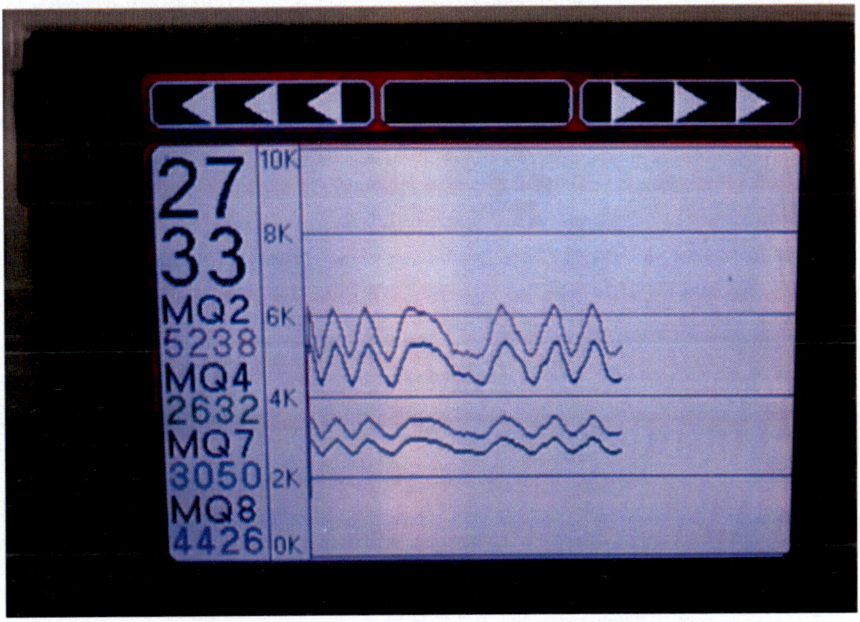

Fig. 6. Gas sensor monitoring on touch screen

Working with different types of measurements generate better knowledge at the student level. Teachers provide advanced theory, advice existing measurement options and create the possibility that the student or young developer to learn very well how measuring systems should work. In Fig. 6 an example of microcontroller based measuring system is used for gas concentration evaluation. The microcontroller used is a PSOC5 ARM platform and is using sensors like MQ-2, MQ-4, MQ-7 and MQ8. The display is controlled with a Vizic Technologies SmartGPU and the software library is adapted with the University guidance. Students/Developers learn how to perform embedded programing for both measuring and graphic representation of the data using several technologies in a single application. The measuring mesh is build in several ways: analogic (wired), wireless on 868 MHz but also at 2.4 GHz. The measuring system can be consider as an innovative hardware because combine an analog frontend (PSOC5 ADC 20 bits resolution) with a performant 868 MHz mesh based on Microchip ARM controllers and finally integrated with instrumentation programing done in LabWindows/CVI. For testing purposes but also for teaching purposes the system contains several virtual instruments generating signal based on real measurements or simulated measurements or simulated devices. For example, a real solar light sensor can be used in order to provide a simulated temperature device according to the sun radiation. A calibration curve needs to be implemented in this case.

The outcomes are also related to scalability and complexity. Adding every year new sensor types and/or new point of measurement in new locations will generate additional information useful for environmental risk assessment. New types of sensors represents a challenge for students and is opportunity to exercise knowledge acquired during the university courses. The complexity of the entire system grows every year and will create a base for future developments. The mesh will become more complex as the technologies are advancing every 2–3 years. Academic actors and Industry developers will advance together according to the society and market needs. Academic actors transfer more and more their know-how to industry by providing technologic support or by providing young engineers with a proper background. This paper presents a collaboration between a SME company and a University where all the technical knowledge existing in the company was build based on competence transfer made by teachers from the university. The presented project require knowledge from various domains like, physics, electronics, chemistry, environmental science, information and communication technologies. Students learn concepts and methods to cover some parts of all these sciences and in this respect we consider it is a multidisciplinary project. Understanding links between physics and electronics or between chemistry and environmental sciences allow young students to thing their future professional evolution in an interdisciplinary approach.

4 Conclusions

Technologic transfer from Academia to Industry is an important way to accelerate the future development of the society. Education is a pivot that push forward the local industry in the benefit of the local area. The transfer from University to SME consists in a measurement system based on Microchip microcontrollers having 868 MHz LoRa communication and a very flexible mesh architecture. The control of every node is made using an access point and the software request from every node the pending data. This kind of transfer is more useful if the measurement system represents a support for environmental risk assessment. Beside the pollution prevention and awareness of potential risks the advantage of this transfer process is represented by the Life Long Learning – LLL activities that includes Educators, Students, Industrial Engineers and any intervenient in the measurement process and data treatment. Mastering and using state-of-the-art microcontrollers and combined with modern IT&C tools contribute to a new society with a better legacy for future young generation [8]. The presented system includes innovative hardware that will generate additional applications for medium to large areas. The project include concepts like Remote Engineering, Virtual Instrumentation, Real & Simulated device and all these knowledge together with instrumentation software cover Interdisciplinary and Multidisciplinary.

Acknowledgement. This work was supported by a grant of the **Ministry of Research, Innovation and Digitization, CNCS/CCCDI – UEFISCDI,** project number **266PED/2020, Acronim - ALDSENS**, within **PNCDI III.**

References

1. eHealth Europe. https://ec.europa.eu/cefdigital/wiki/display/CEFDSIS/eHealth+Benefits. Accesed 7 July 2021
2. Dong, N., Jonker, H., Pang, J.: Challenges in eHealth: from enabling to enforcing privacy. In: Liu, Z., Wassyng, A. (eds.) FHIES 2011. LNCS, vol. 7151, pp. 195–206. Springer, Heidelberg (2012). https://doi.org/10.1007/978-3-642-32355-3_12
3. Ashby, J.E.: The effectiveness of collaborative technologies in remote lab delivery systems. In: Proceedings of 38th ASEE/IEEE Frontiers in Education Conference, pp. F4E-7–F4E-12, October 2008
4. Gil, P., Candelas, F.A., Jara, C.A.: Computer networks E-learning based on interactive simulations and SCORM. Int. J. Online Eng. **7**(2), 15–23 (2011)
5. Vasilescu, A., Hrinczenko, B., Swain, G.M., Peteu, S.F.: Exhaled breath biomarker sensing. J. Biosens. Bioelectron. **182**, 113193 (2021)
6. Labjack Homepage. http://www.labjack.com/. Accessed 07 July 2021
7. Vasilescu, A., et al.: Surface plasmon resonance based sensing of lysozyme in serum on micrococcus lysodeikticus-modified graphene oxide surfaces. J. Biosens. Bioelectron. **89**, 525–531 (2017)
8. López-Pérez, M.V., Pérez-López, M.C., Rodríguez-Ariza, L.: Blended learning in higher education: student's perceptions and their relation to outcomes. Comput. Educ. **56**(3), 818–826 (2011)

Case Studies

Connected School Bus for Learning and Safety

Arjun Singar(✉)

Indian Institute of Science, Bangalore 560062, KA, India
arjunsingar@iisc.ac.in

Abstract. Many school districts across the globe have numerous technology initiatives to enhance the learning opportunities and also to ensure the safety of the school children while commuting in their school bus. One initiative enables the "Connected School Bus" by integrating a mobile or cellular gateway, on-board global positioning system (GPS), passenger Wi-Fi connectivity, video cameras and other capabilities. The Connected School Bus provides a host of benefits for the students, parents, drivers, and the school administrators.

Keywords: Learning management system · e-learning · Global digital divide · Online tracking · Child safety · Smart bus

1 Education in Urban and Rural Economy

1.1 Smart Education

The innovative use of information and communication technology is rapidly transforming the process of education in the way teachers and students are teaching and learning at schools. The introduction of ICT has had a profound effect on the educational activities all over the world in this generation. Developing countries like India, having sensed this, are making an attempt to leverage every opportunity on enabling the capabilities of ICT in the field of education. In today's context, students need access to various technology tools to gain information, as the teachers are not sufficient to suffice their requirement for knowledge. Thus, the need for technology intervention has been ever lasting.

The youth population of India, under the age 25, is twice the population of the US. According to the Indian Education Sector Outlook, there are 1.3 million schools in India out of which only 10% of the private schools have had access to ICT and multimedia mode of classroom teaching, whereas government schools have barely had any access to it. According to market survey, the current market requirement for virtual learning products in private schools is around 500 million US $. This is expected to grow at a rate of 20% annually to reach the 2 billion US $ mark by 2020. Apart from this, the current market requirement for multimedia in government schools is 750 million US $. This is expected to grow multi folds by 2020 due to the lack of utilization of ICT in government schools. However, Government of India has planned to install virtual learning environment for school education and the Government has proposed to allocate 100 crore rupees (around 15 million US $) towards this initiative.

M. E. Auer and T. Tsiatsos (Eds.): IMCL 2021, LNNS 411, pp. 1001–1013, 2022.
https://doi.org/10.1007/978-3-030-96296-8_91

Education has now taken an initiative to eliminate the traditional classroom approaches and introduce students a broad spectrum of affordable and accessible online learning material. MOOCs (Massive Open Online Courses) have generated a new wave and set a trend in higher education. India is set to capitalize on this opportunity by partnering with the US. In 2012, India witnessed an enrollment of 21% growth in online education compared to a 2% growth in the traditional set up of higher education. The main reason behind this is the accessibility to the technology services.

The government of India in August 2013 launched National Repository of Open Educational Resources. This repository included videos, audio lectures and textbooks for all school students in multiple languages. Children from primary and secondary education could look into concepts from various subjects through this repository. Then, in the year 2013, Brihanmumbai Municipal Corporation (BMC) set an example by setting up a virtual learning environment at a cost of 28 crore rupees (around 4 million US $) by installing audio-visual facilities in the city across 420 civic-run schools. This initiative was to teach and engage school children in a fun-filled manner and also to ensure proper attendance and learning in classrooms.

India now needs to take the virtual learning environment to every educational sector across the country and adapt to various modes of learning. This makes India a very exciting prospect for enabling technology in the newer learning environments. A recent trend suggests that schools in urban areas are increasingly utilizing the latest technology for learning. Moreover, government has begun to give permission towards complete utilization of ICT in schools. In addition, with the technological innovations happening, schools have started to provide education through pads or tablets. The students are now being educated through the use of digital instructional materials and animated videos. All the well-established schools are making this a mandate for teaching and learning. The students eagerly look forward to this learning process through interaction. This digitized learning has not only redefined the system of education, but it has also given the school students an opportunity for better insight. The students are being able to learn difficult concepts by watching highly engaging animated videos by having a complete control on the pace of the instruction. This not only makes their learning an enjoyable experience but also improves their overall academic performance in school. This kind of learning management system enables the students to complete their assignments or homework on time and meanwhile allows the teachers to instantly access and assess those assignments. This helps the teachers to understand what all the student's strengths and weaknesses are.

Virtual learning environment has revolutionized the education. School students are understanding and learning better through multimedia. So, the demand for incorporating smart education is increasing. A student is being able to retain the lessons better through visualization. In case, students do not understand a concept taught in the class, then students can re-learn by watching the same concept online. So, this pushes for the case of integrating e-learning along with traditional classroom teaching. E-Learning software installed in the school bus provides students with an opportunity to learn in an online or virtual learning environment while traveling. This software also makes the task of the teachers easier as they can store and manage the study materials and the exercises on this e-learning software. The software includes course materials, exercises, class notes, sample question papers and video and audio files. The teachers can log on to this software virtually, that is, either from their home or any other location. Students usually use this e-learning software for:

- Reading material or class notes
- Learning concepts through video or audio
- Completion and submission of assignments
- Taking monthly quizzes and tests

1.2 Global Digital Divide

In the past few years, number of companies in India have started providing technological solutions to schools and colleges. One of the existing measures of the global digital divide to understand where India ranks with respect to whether it is a wealthy or not so wealthy nation is the amount of its Internet accessibility and ICT availability.

The digital age has ignited creative developments in the field of higher education. Electronic gadgets have become mandatory devices in a typical modern multimedia classroom. In spite of technology helping handful of students to scale new heights in the field of education, there are quite a few who are falling behind. While there are some US schools who provide their students with mobiles or pads or tablets, almost 10 millions of families having school and college going children of low-income households still don't have access to an Internet connection. Low-income families with school going children are more likely not to have Internet connection as compared to high-income families. Richmond County School in Augusta, GA is one of the few schools in the US who started to provide Internet or Wi-Fi connection in their school buses. A survey report in 2012 said that 25% of Richmond County's population lived in a low-income zone that had a high school pass rate of just 62%. It was then recorded that most of the households in their community did not have access to Internet connection. Therefore, in 2016, the county that had in excess of 40,000 school children introduced twenty-two connected school buses (Wi-Fi enabled).

This digital divide has created a "homework gap," with students not being able to complete their school assignments. The Director of Consortium for School Networking believed that this homework gap is a major issue that needs to be addressed all around the world. With this gap becoming more and more, some schools have taken the initiative to alter this scenario by implementing an innovative solution. Ericsson Mobility stated that 72% of the world's population would be having smartphones by 2020. Keeping this in mind, schools around the world would be trying to implement virtual learning programs to connect students outside of their classrooms. Thus, having Internet connected or Wi-Fi enabled school buses is one way to help students achieve digital equity and help them to complete their homework. Wi-Fi can be used to transform school buses into study halls. Today's civil right is to make sure that all the students around this digital world have equal access to education.

Connected school buses will enable students to access the learning management system (LMS) installed in the buses to finish their school assignments, read class bulletins, and email their teachers. Providing Internet in the bus benefits those students who are a part of extracurricular activities, such as cultural and sports events, to cover up lessons during their travel as the missed classes in school are often made up later in the evening at home. Thus the Wi-Fi equipped bus can help students get to sleep on time.

Since students spend majority of their time traveling on buses each year, the idea of getting the school buses connected with Wi-Fi has caught the attention of numerous companies. For example, Google is working with various schools to install Wi-Fi on buses across multiple school districts in North and South Carolina and Colorado. Google is planning to expand its school bus Wi-Fi program to 11 more states in the US after receiving reviews from students, teachers, and school administrators. The program was initially started in Caldwell County, North Carolina in 2016 after a years-long relationship between the area and Google due to the data centers built there. This was then expanded to Berkeley County, South Carolina in 2017, where Google donated 1,700 Chromebooks to students and added free Wi-Fi to 28 school buses and nearly 2,000 rural students were benefited from this move. This connected school bus initiative focused on maximizing the long commute that students in rural areas had by giving them an opportunity to do homework and other exercises during their journey.

"A Rolling Study Hall is like an extended classroom. It addresses the needs of students that don't have Internet access in their home," Lilyn Hester, Google's Head of Southeast Public Affairs, told a local news station last year. Google, through their research, found that teachers were assigning homework that required Internet, but millions of low-income students in rural areas did not have access to it at home and wasted hours on their travel to and from school. This 'Homework Gap' impacts low-income students disproportionately, especially in more remote or rural areas.

After this initiative, both North Carolina and South Carolina experienced better performances amongst students in math and reading proficiency as well as digital literacy, which then prompted Google to expand the program to 70 buses in 16 rural school districts across USA. In South Africa, a Nokia program called MoMath Project let their students solve the math problems via mobile applications in the school bus and this reportedly improved their math skills by an average of 15%.

Google's Rolling Study Hall initiative was basically an innovative approach that allowed school students to have Internet connection and support to finish their homework during their journey. Google is using the cameras in the school buses so that the teachers can monitor their student progress and the tutors are also being hired to specifically ride with the students to help them out for better learning. Apart from donating free Chromebooks and Wi-Fi to thousands of students, Google is also offering its "G-Suite for Education" online programs. Apple has also announced that it will be releasing tailor made low-cost iPads for schools.

Wi-Fi equipped school buses can also help bolster passenger safety by transmitting real-time data from a range of smart cameras, sensors, and GPS units. According to sources, many school districts have been using connected technology on buses, such as GPS units in Boston and RFID-enabled badges attached to student backpacks in Cincinnati. It is quite sad to state that in an era of Tesla and Uber, only a third of school districts have been using this technology to track their children using GPS. In addition to this, smart school buses after working hours can be parked in areas where Internet is not available and thus can be used as Wi-Fi hotspots and offer under-served communities Internet access.

2 Transport of School Children

2.1 Connected Bus: Architecture

Many school districts have numerous technology initiatives to enhance the learning opportunities for students inside and outside of the classroom and also to ensure the safety of students. One initiative enables the "Connected School Bus" by integrating a mobile or cellular gateway, on-board global positioning system (GPS), passenger Wi-Fi connectivity, video cameras and other capabilities. The Connected School Bus provides a host of benefits for the students, parents, drivers, and school administrators.

When we think of a school bus the first thing that probably comes to our mind is a big yellow bus driving around with noisy students. And while over the years, the yellow bus and the student riders have not changed, the technology in it definitely has.

Today's school buses are built in with technology, something that did not exist 10 years ago. This technology works together to create a learning environment and a safe environment for students, enhance the bus rider and driver experience, and improve the overall efficiency of the school bus. But in reality, it's not so easy to implement technology. In fact, many school transportation heads are now facing a new challenge: How do you coordinate and connect all these technologies, tools, and suppliers on a moving vehicle, especially when your budgets are anything but unlimited?

There is a centralized device that is required to bring it all together. The key to integrate all this technology on the school bus together is the Internet. With Wi-Fi on the school bus, districts get the access they need to keep moving forward in the connected world while saving money and expanding their capabilities to integrate any Wi-Fi enabled solutions.

2.2 Connected Bus: Technology

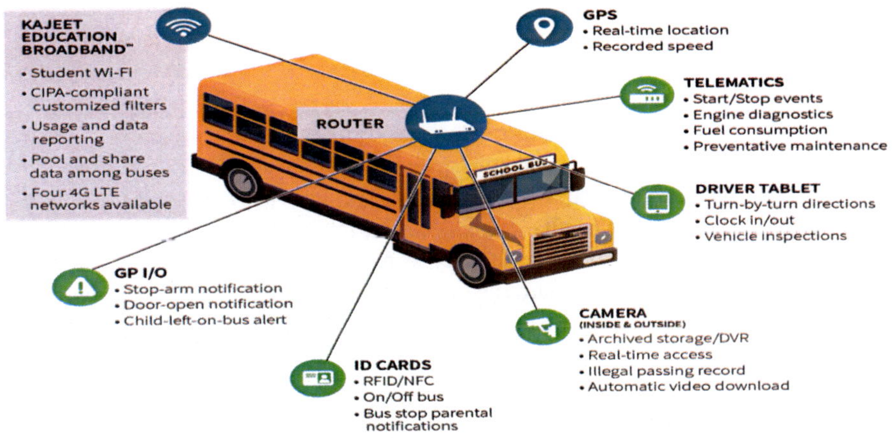

Fig. 1. Connected school bus architecture. **Source.** Kajeet Education Broadband

We will explore eight types of technologies available in the Connected Bus, which is as shown in the Fig. 1.

1. Router
   As the key piece to creating a connected bus, the router powers the bus with Wi-Fi access. This needs to be a mobile ruggedized router specifically made for vehicles, that can withstand the bus environment. Police and first responder vehicles, utility, and service vans, all also use ruggedized routers. Plus, these routers have the added benefit of connecting up to 128 Wi-Fi students.
2. Student Wi-Fi
   It is estimated that the students of the US spend a staggering 520 million school days on buses each year. Why not give students the opportunity to use that travel time to work productively on school assignments? The Kajeet SmartBus™ extends the classroom to the school bus with safe and CIPA-compliant Internet access for students.
3. Global Positioning System (GPS) And Routing
   GPS technology has become standard equipment in many passenger vehicles today, but it's still relatively new within the school transportation industry. A statement, which is quickly changing as districts discover the myriad of ways GPS can improve bus safety, security, and tracking.
4. Driver Tablet
   In-vehicle tablets for drivers help increase driver accountability, safety, and efficiency. These devices allow drivers to streamline their daily tasks while providing dispatch with insight regarding what's happening (or not) on the road.
5. Telematics
   Telematics have the potential to make school buses more efficient, safer, and easier to maintain than ever before, leading to a lower total cost of ownership. Telematics track: start or stop events (i.e., hard breaking), engine diagnostics, fuel consumption, preventative maintenance alerts, and more.
6. Camera (inside and outside)
   By capturing audio and video of what's happening in and around the bus, cameras help promote driver and student accountability, as well as compliance with traffic laws (such as following stop-arm camera laws).
7. Student ID Cards
   As more and more school districts employ smart ID card systems to increase security, they are now bringing that technology onto the bus to improve safety, efficiency, and peace of mind for parents, drivers, administrators, and students alike. Special bus cards are typically embedded with microchips that use radio frequency identification device (RFID) technology or near field communication (NFC) to log when and where a student boards and exits the bus. Now transportation departments and school administrators can know if students were on the right bus and if they got off on the right stop. These cards can also be used to alert parents of where their child's bus is, letting them know when their child has got on or off the bus.
8. General Purpose Input/Output
   Today's school bus systems can leverage a variety of devices and sensors to improve child safety and optimize response time in an emergency. GP I/O includes

notifications such as stop-arm, door-open, child-left-on-bus alerts, or alerting drivers if the top hatch or any windows are open. Now school bus operators can be proactive if a sensor on the school bus isn't working, which can keep that specific bus on the road, reducing the need for a substitute bus and more time-intensive mechanical work. All of these technologies (and more) work together to keep students safe, enhance the rider or driver experience, and improve overall bus efficiency.

3 Learning and Safety Through a Connected Bus

3.1 Connecting School Children

There is an unparalleled opportunity to be leveraged at the moment to improve the quality of life of our communities and its school going children through meaningful connections between networks, systems, and devices. Building a connected school bus is one such opportunity leveraged by bringing together educational and safety requirements. In the current scenario, providing a virtual learning environment for student education and safety through Internet connectivity are the top priorities for many schools.

There are numerous solutions provided by various agencies for the global market that address unique features of connected buses using wireless networks such as GPS or Wi-Fi. Their objectives are to: gather, unite and integrate data, hardware, software, and connectivity to provide schools with highest quality educational experience, safety and increased operational efficiency. The features of such connected bus are as follows:

- On-road education: Wi-Fi facility in the bus to access the learning management system
- Student safety: Whereabouts of students to the school authorities and parents in case of an emergency with instant GPS data
- Location tracking: Transfer of real-time information to health and law enforcement agencies through on-vehicle video monitoring

Benefits of a connected school bus are as follow:
Education

- Wi-Fi enabled school buses for studying and doing homework while on the road
- Benefits those school children who travel long distances and have no access to internet at home

Safety and Security

- Real-time location data of students for parental awareness via mobile app or SMS through GPS or RFID tracking
- In case of a situation, enables connectivity with health and law enforcement agencies with real-time data or video sharing

Alerts and Notifications

- If a school bus travels in a non-designated direction
- If a bus arrives early or late at a designated stop

3.2 School Children Tracking Technology

Children travel from home to school and then back to home every day. For parents, safety and security of their children is of the utmost concern. The students travel to their schools from home and back either in their bicycles or personal vehicles or school buses with one purpose of reaching safely. Research performed by a research unit in Scotland suggests that travelling by a school bus, in comparison with others, is the safest mode of transport. Data provided by statistics also mentions that a student going to school in a car is eight times more prone to be involved in a road accident than a student going in a bus.

Many school children, after falling asleep, get stuck in a school bus at the parking lot or they get into the wrong bus and get down at a stop that is not designated to them. To prevent such unfortunate incidents, an RFID-enabled technology is introduced that helps in tracking the school children while they are traveling to and from schools. Such systems could potentially use combinations of Radio Frequency Identity, Global Positioning System and General Packet Radio Service technologies for monitoring and tracking of school children. If RFID technology is used, every bus will have an RFID reader to detect the RFID tags embedded in children's bags that are onboard the bus and two RFID antennas to retrieve the data stored. The time, date and location then get transmitted into a safe and secure system or server - database. This system instantly sends out an SMS to parents to alert them once the child gets on to the bus and also when the bus is 10 min away from the pick-up or drop off location. The system will also notify parents in case the child gets on to the wrong bus or gets locked up in the bus or gets down at a non-designated stop. Moreover, the system reflects the real-time location of the bus with the help of GPS technology.

In summary, the system will intimate the school management and the parents to monitor and keep track of the bus to ensure their children's safety and security.

The full system architecture includes four main components:

1. RFID Tracking System
2. Smart Gateway (RFID, GPS and GPRS Technologies)
3. Backend Server
4. End-users Applications, as shown in Fig. 2.

The role of a smart gateway is to manage different single board computer components and it aims to:

1. Provide local intelligence for the RFID Tracking system
2. Track the school buses and driving behaviors via GPS
3. Provide network connectivity through GPRS or Internet

This smart gateway collects vital statistics such as the current bus location, engine information, the driver behavior and those students who enter and leave the bus. All this information is then processed and sent to the back-end servers for analysis and decision-making. The system capabilities are shown in Fig. 3.

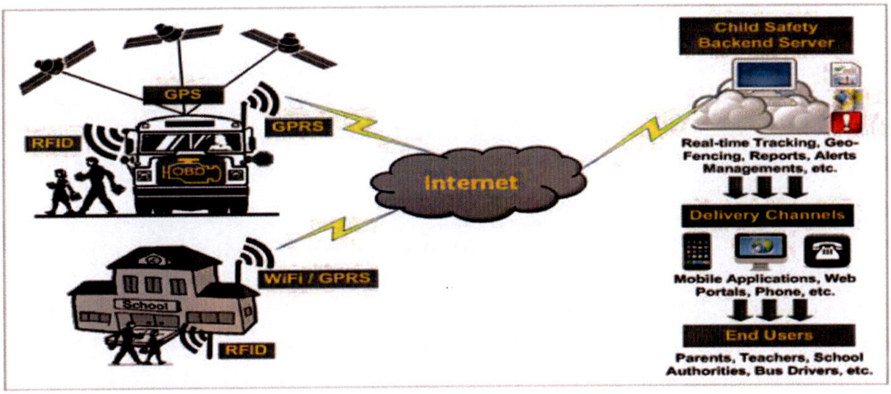

Fig. 2. Overall system architecture. **Source.** Journal of Traffic and Logistics Engineering

Fig. 3. System capabilities. **Source.** Journal of Traffic and Logistics Engineering

In various ways, the received signal strength indication (RSSI) is used for tracking RFID tags, mainly because RSSI information costs bare minimum or nothing with each radio message sent or received. Algorithm for children tracking technology is shown in Fig. 4.

The smart gateway consists of five major software programs to operate the child safety system:

1. RFID Tracking System
2. Bus Monitoring System
3. Over-The-Air (OTA) Gateway Protocol
4. Machine-to-Machine (M2M) Communication Protocol
5. Local Database

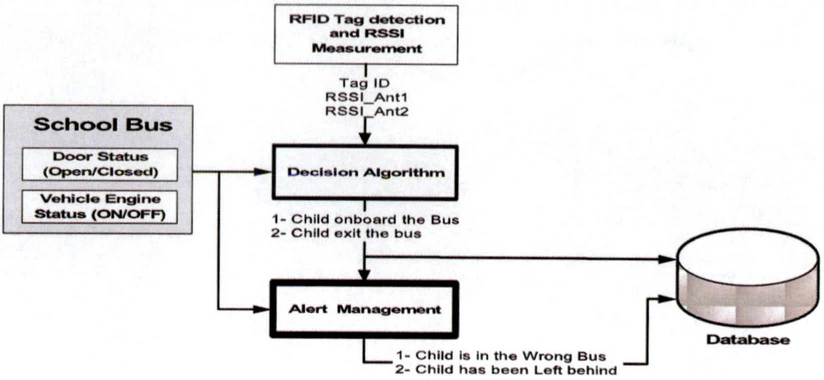

Fig. 4. Children tracking technology. **Source.** Journal of Traffic and Logistics Engineering

4 A Case Study

4.1 To Increase Student Safety and Services in a Rural School District in the US

Problem

How to provide support to a district of about 55,000 people that covers 650 square miles with 130 buses?

Challenges

- Tracking school students from home to school and back to home across the extensive bus routes
- Improving educational facility by empowering students with learning management systems on school buses
- Providing a holistic view of the system for instant data sharing for safety and communication with all the health and law enforcement agencies

Solution

In collaboration with the existing district stakeholders, Cisco carried out an evaluation procedure to find out the key performance indicators and technical requirements for the implementation and validation of the system being installed for a connected school bus. The following are the solutions that were provided:

- On-board network design: Cisco's mobile routing platform and Intel's gateway
- On-bus video recording for real-time monitoring of the bus travel
- Parent communication via Intel's processor for student tracking and safety
- Mobile communications for sharing instant data with emergency responders
- Telematics for managing fleet performance and driver behavior

4.2 Some Classic Examples

The following are few examples of how school buses in the US with Wi-Fi are helping to bridge the digital and homework gap amongst students and give them the required Internet access.

1. In August 2017, public schools in Miami, Florida operated a connected school bus in collaboration with Kajeet educational company. The "Wi-Fi-on-the-Go" helped the students to study and learn concepts while traveling in the school buses. Students could also complete their class assignments during their journey from school to home.
2. In a school near Southern California desert, a bus with a Wi-Fi router parked at some distance away on a sand driveway in a mobile home park at night. This facility provided Internet access for those students who did not have connection at home.
3. Huntsville Independent School District in Texas has deployed Wi-Fi connection in their school buses to help students of rural area to access Internet for learning. The district covers an area of nearly 650 square miles, and some students spend up to 90 min one-way on the school bus. Huntsville also partnered with Presidio to deploy 3G or 4G connectivity for Wi-Fi on district school buses, along with Intel based processor for tracking students.
4. A County School in Georgia acquired a connected bus in an effort to extend the school day. They created 24 "Homework Zones" in and around their district by parking the Wi-Fi enabled school buses in and around the apartment complexes from 3:00 p.m. to 10:00 p.m.

5 Business Opportunity from a Connected School Bus

The Connected Transportation for Education solution can help transportation managers reduce fuel costs by as much as 10–30% while gaining thousands of hours of more productivity from the fleet and its workforce. The solution can also drive a reduction in employee speeding and citations by up to 80%. The reduction in operational costs and improvements in transportation management and worker efficiency have allowed school districts to achieve an ROI typically in less than one year.

Considering the statistic that there exist 16-lakh private and public schools in India, one could leverage this business opportunity of introducing connected school buses in those schools. Here's the estimated cost of one connected school bus in INR and US $.

- Cost of Back-End Server or Cloud - Depends on Sizeor Storage*
- Cost of Mobile Application - Depends on the Features*
- Cost of UTX-Intel IoT or Smart Gateway Systems - Rs 46,000
- Cost of RFID Reader - Rs 60,000
- Cost of 2 RFID Antennas - Rs 20,400
- Cost of RFID Tag for 30 Students - Rs 40,500
- Cost of PAD or Tablet for 30 Students - Rs 1,50,000
- Cost of LMS or E-Learning Software for 30 Students - Rs 1,44,000
- **TOTAL Cost of One Connected School Bus - Rs 4,60,900 PLUS***
  (Around US $ 6,580 PLUS)*

6 Conclusion

An end-to-end solution to address the safety, security, tracking and learning needs of school students need to be introduced. The solution should enable real time tracking of students and events and should also reduce operational costs and liability.

This is made possible with the help of cloud-based campus admin application, ride assistant mobile application for bus attendants and drivers, tracking mobile application for parents and school buses installed with transport grade real time CCTV and GPS monitoring hardware. In an era of constant travel and traffic congestions, delays and uncertainties of school children cause anxiety and unwarranted concerns to their parents. Being connected with their children makes parents feel empowered and secured and feel like they have better control of their lives.

This end-to-end solution provides real-time information on vehicle location, driving behavior, speed monitoring, security implementation, safety checks for school campuses and parents. The solution not only ensures complete monitoring of the bus journey but also notifies the timings, changes, and delays if any. Thus, a school bus tracking software assists in ensuring student safety and there by reduces parental anxiety. It also helps the school management to communicate with the bus drivers as quickly as possible, know their location and handle emergency situations immediately.

This process is an integrated approach to address school transport management, with real time school bus surveillance, tracking and intelligent dispatch management. The solution is designed to provide safe and secure transport for the children and there by ensures complete satisfaction to the school management, children, transport service providers, bus drivers and for parents.

References

1. Connected Transportation for Education Solution, CISCO
2. Wang, J., Huang, X.: Routing school bus for better student learning. In: 25th International Conference on Geoinformatics, pp. 1–7. IEEE, August 2017

3. Smart transport, keeping students connected and safe with the smart school bus, PRESIDIO, Practical thinking for a connected world
4. Foster, P.R., Burberry, R.A.: Antenna problems in RFID systems. Presented at IEE Colloquium on RFID Technology Conference, 25 October 1999
5. Ambedkar, P., Babu, P.S.: Smart school bus for children transportation safety enhancement with IOT. Int. J. Innov. Res. Comput. Commun. Eng.
6. Smith, A.B., Jones, C.D., Roberts, E.F.: Operational considerations in simulation and deployment of RFID systems. Journal, Publisher, Location, Date, pp. 1–10
7. Arifa, K., Aryadas, R., Asha, K.R., Amrutha, J., Anju, P.: Smart school bus. Int. J. Sci. Eng. Res. **7**(2) (2016)
8. Shaaban, K., Bekkali, A., Hamida, E.B., Kadri, A.: Smart tracking system for school buses using passive RFID technology to enhance child safety. J. Traffic Logistics Eng. **1**(2), 191–196 (2013)

Artificial Intelligence in Study-Abroad Program Recommendations

Yang Gao(✉), Xiaocheng Wang, Jintao Shi, Zhongyan Liu, and Zejin Wei

Dalian Maritime University, Dalian 116062, China
gaoyang666@dlmu.edu.cn

Abstract. Students willing to study abroad may find it challenging to go through tons of website pages on the application process. The complication of the studying-abroad application process makes it even harder for many students to figure out their dream and fit schools. With artificial intelligence being widely promoted across different subjects and disciplines, we made attempts to incorporate artificial intelligence into education program recommendations. We thus designed a recommendation system for studying-abroad programs through the K-nearest neighbor algorithm. The designed system may recommend up to six colleges and universities to students according to their input of grade average points, language testing scores, acceptable fees, target majors, and location preference. In addition, the analysis of the application plan and the report on the academic entrance requirements are also offered through the system. The project simplifies the application process and thus saves the time and energy of the applicants.

Keywords: Study-abroad programs · Recommendation system · KNN

1 Introduction

When applying for a university program, applicants need to go through tons of website pages and even inquire about the information of their target universities. Quite often, this screening and analysis process takes long and makes the application process challenging for the applicants. While the Internet provides applicants with sufficient information on universities, however, it is quite demanding to go through all the information and figure out important statistics crucial to program application. The complexity of the application process for studying abroad, coupled with the opacity of the study abroad agencies on the market, makes many overseas students often unable to match their specific needs to the most suitable college. Based on this background, we designed a recommendation system for studying abroad, aiming to formulate targeted and goals for overseas students based on the students' scores, target majors, preferred countries, college background culture, teaching resources and other comprehensive considerations. Sexual and personalized recommendation plan for the target population to find a counterpart institution. The personalized recommendation system for studying abroad is a transparent and open recommendation system that provides international

M. E. Auer and T. Tsiatsos (Eds.): IMCL 2021, LNNS 411, pp. 1014–1021, 2022.
https://doi.org/10.1007/978-3-030-96296-8_92

students with the most convenient and accurate information about studying abroad through the Internet plus data mining technology.

2 Literature Review

2.1 Recommendation Systems

Goldberg initially proposed the Collaborative Filtering algorithm and designed the first mail recommendation system Tapestry, which laid a solid foundation for the development of the recommendation system [1]. Since then, the recommendation system has been developed rapidly and has been widely used in many fields such as e-commerce, online video, personalized e-reading, online music, and social networking platforms [2]. At the end of the 20th century, many Internet companies such as eBay, Amazon, Google, etc. appeared in the United States. These companies designed and invented many recommendation systems, which created huge commercial value for the company [3]. Among these interconnected companies, Amazon took the lead in commercial application of the recommender system in 1999, and the transaction rate of products increased by 30% as a result. In 2003, Amazon launched the Item-to-Item CF algorithm of its core recommendation system, which further improved the turnover rate of Amazon's products. More importantly, Amazon began to record user usage data and analyze it to improve the accuracy of Amazon's recommendation system [4]. BuffFeed company designed a recommendation system based on the potential probabilistic clustering model. Its core idea is to recommend the news to the user's friends according to the users' preference of the news [5].

It was not until the beginning of the 21st century that China began to study recommender systems. In 2002, Zhao Liang and Hu Naijing at Fudan University summarized the algorithm of the recommendation system, enriching the connotation of the collaborative algorithm [6]. In 2003, Deng Ailin studied the recommendation system of e-commerce in his doctoral dissertation, proposed the IRPRec collaborative filtering recommendation algorithm, and developed a related system [7]. In 2012, Wang Licai and Meng Xiangwu proposed a context-aware recommendation system, which analyzes Internet context information from an object-oriented perspective, thus improving the accuracy of the recommendation system and the satisfaction of users on its platform [8].

2.2 Study-Abroad Agents or Platforms

Browsing information on foreign colleges and universities on their own through the Internet, instead of doing face-to-face communication and consultation in study-abroad agencies, has gradually become a way for overseas students. More overseas students choose to obtain foreign colleges on the Internet relevant information. At present, there are a large number of foreign college platforms available on the Internet in China. For example, Baidu Education collects a large amount of foreign college information according to its own search engine system and bidding ranking system and organizes and summarizes the foreign college information to support the company and platform.

The study abroad subnet uses foreign colleges and universities as attractive points to introduce traffic portals and recommend study abroad agencies through bidding rankings on the homepage. Baidu Education relies on Baidu Search to have a very large traffic entrance. With the technical background of its parent company, it can recommend more reliable or more powerful study abroad institutions for users. For the recommendation of colleges and universities, Baidu Education, however, has not done in-depth.

The domestic representative is Zinchcn, which started out as a database of foreign colleges and universities. First, it has been developed with the data and information of overseas colleges and universities of the US parent company and made the data complying with the web-side foreign college query database used by the Chinese people. After the user traffic has accumulated to a certain level, the most important thing is to wait for the user information and user historical behavior data to accumulate to a certain level, and then start on the homepage of its official website for the relevant colleges. The recommendation of colleges and universities makes it convenient for more users to directly click on the recommended content without having to inquire about foreign colleges and universities.

Another more representative one is Shengli.com in the early years. Shengli.com received 300 million financing in September 2015. It began to establish a rich foreign college information platform and accumulated a large number of colleges and universities by merging small websites. Data and user data, followed by the introduction of its own recommended college services and offline study abroad application intermediary, until now to cut into the K12 preschool study abroad education and pan-education industries.

Of course, domestic college library platforms also include the official website of Jin Gillie, Liuliu Study Abroad, Shun Shun Study Abroad, Yoyo Study, 51Offer and other websites. In recent years, recommendation pages have begun to appear in the sidebar or drop-down menu of the college page. More foreign colleges and universities have introduced recommendation systems. The purpose is to attract more students studying abroad and facilitate users to search and archive information. Users who want to study abroad can recommend colleges and universities that are connected with their business and improve. The success rate of studying abroad applications for users who are preparing to apply. Therefore, the behaviors, selection and recommendation of each user, and the behavior, selection and recommendation of each overseas study consultant have an important impact on the surrounding users.

With the gap in the existing literature, this project intends to develop a targeted and personalized recommendation system for students interested in pursuing a study-abroad, graduate program. Setting up scores, target majors, favorite countries, college background culture, teaching resources, and other factors as parameters, the project attempts to recommend programs from colleges and universities that are suitable for users. Therefore, the launch of this project has great market competitiveness and practical values.

3 Artificial Intelligence in Study-Abroad Program Recommendations

3.1 Needs Analysis

We launched the project through a mixed-method, multi-step approach. We conducted a needs analysis through surveys and collected actual students' feedback regarding their willingness and intentions to choose a program. Specifically, we collected 287 surveys from college students and found 57% of the participants willing to study abroad. There were also 27% of the total participants being interested in studying abroad but hadn't decided yet whether to study abroad or not. Only 16% of the surveyed students were not interested in studying-abroad programs (see Fig. 1).

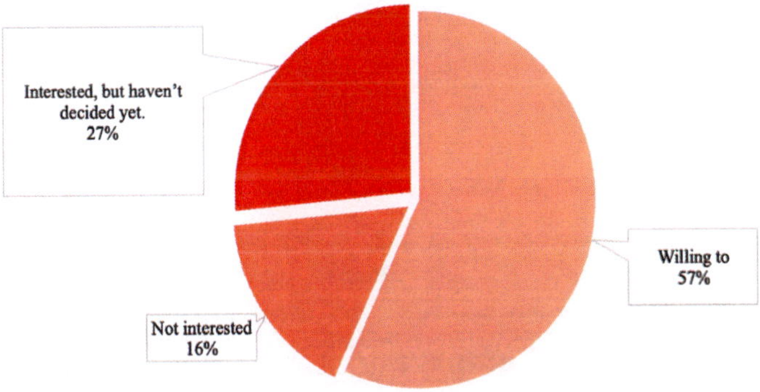

Fig. 1. The Chinese college students' willingness to studying abroad

In terms of application methods, 10.91% of those who chose to apply by themselves accounted for 34.55% of those who chose an agent to apply together with themselves, 25.45% of those who applied mostly through an agent, and 21.82% of those who sought help from their schools or institutes (see Fig. 2).

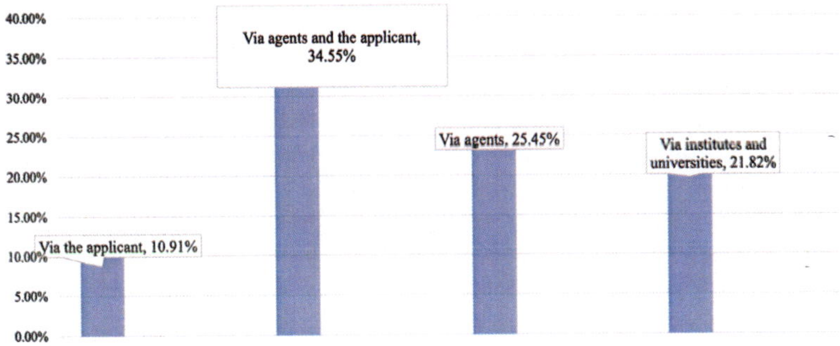

Fig. 2. The Chinese students' preferred methods in doing the application

3.2 Approach

With our needs analysis data, we moved to our technical design and process. Technically, using the K-Nearest Neighbor (KNN), we narrowed the scope of selection by entering grade points, IELTS or TOEFL scores, acceptable fee limits, target majors, school location, and other restrictions, and attempted to recommend six colleges and universities that are suitable for users' preferences. In addition, we attached a pros and cons analysis form of the recommended plan and application requirements to simplify the intermediary process (see Fig. 3.).

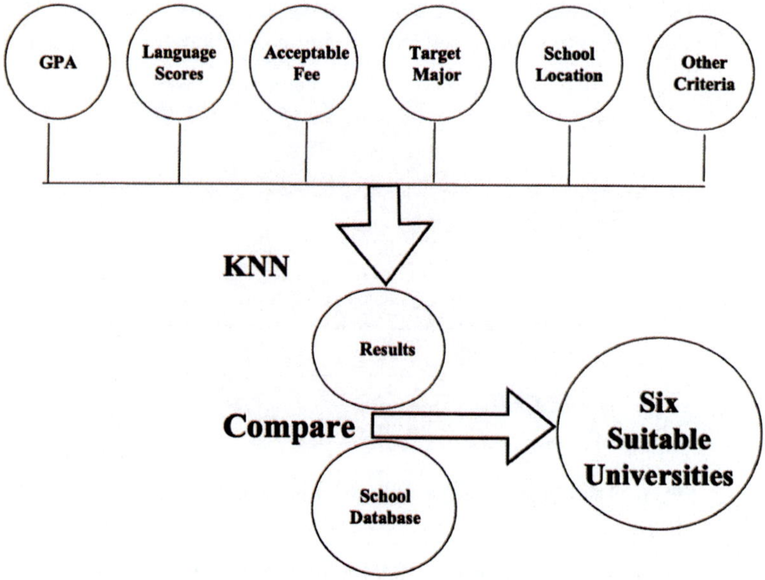

Fig. 3. The conceptual framework of the recommendation system

3.3 Actual or Anticipated Outcomes

We also took steps in building our recommendation system. Our team used appropriate tags to organize core data such as academic details and background information in the application process. For the problem of large-scale processing of data, we first applied Min-Max Normalization to linearly transform the magnitude of the data input by the user so that the result value is mapped between [0–1] and then cancels the difference of magnitude. Using the KNN algorithm, we used the A/B test to classify the distance between different feature values, compare each dimension, determine the effect of different algorithms, to further improve the conversion rate, ensure the viscosity of users and users' retention (see Fig. 4).

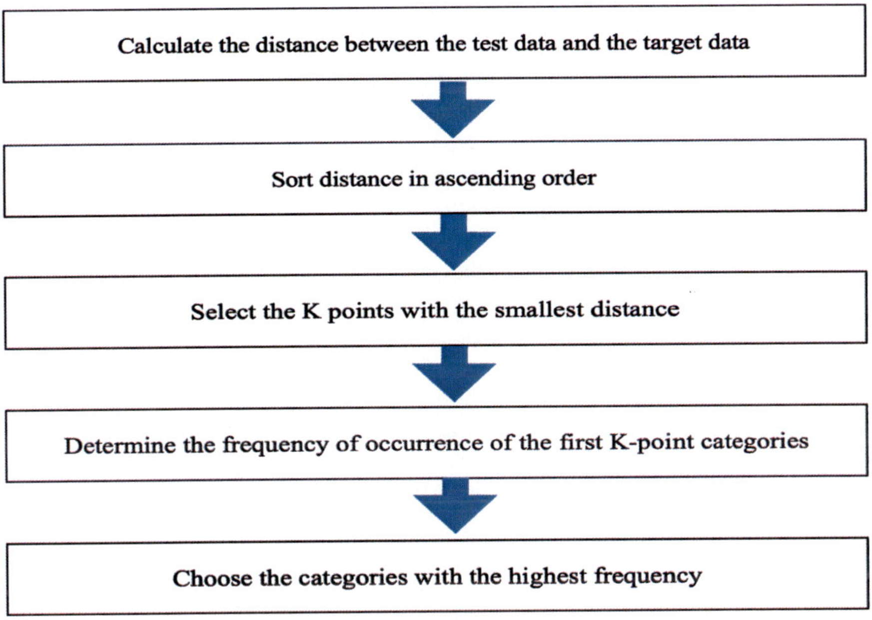

Fig. 4. Steps in running KNN for the model

3.4 Overall Architecture

The overall architecture of the adaptive recommendation system based on user status is to monitor the current behavior of the user through the real-time monitoring module, establish a user portrait tag tuple, and generate a new recommendation list to the user through the user status of the tag tuple, making full use of the KNN algorithm Advantages to implement an adaptive college recommendation system. The advantages of the overall architecture of the college recommendation system based on user state adaptation are reflected in three aspects:

The user status is sensitive, and the user status adaptive recommendation is realized. The college recommendation system is called user-adaptive recommendation because the recommendation system is sensitive to the user's current state and time decay, in order to be more reasonable and humane based on the user's current state, such as the user's emotions, interests, Preferences and other personalized status features automatically adjust the recommendation results according to the different status of the user to meet the needs of different user status. For the recommendation system model, analyzing the characteristics of different platform users is a relatively time-consuming operation. Most recommendation systems use offline methods to calculate the characteristics of different platform users. Commonly used association tables or correlation matrices are updated a lot. Therefore, the user portrait method is adopted to establish the user keyword network portrait. When the user has new interests, characteristics or operating behaviors, the recommendation system will notify the user to update the user portrait method. After updating the user's real-time status, use random Shenlin categorizes the user's new status under the recommendation list of similar users to realize the real-time recommendation of personalized colleges and universities.

The recommendation engine is convenient to add and delete, which is conducive to the adaptive recommendation of colleges and universities. The recommendation engine can conveniently adjust the combination according to the user's status and select different recommendation alarms according to different types of application scenarios. For the recommendation system, sometimes the recommendation task is complicated, and the recommendation features are many. Therefore, the recommendation system is conducive to the combination and arrangement of different types of recommendation engines and can well support the system operator to adjust the ranking of different recommendation algorithms. Different recommendation system results. It is convenient to maintain the recommendation system and make the platform easier to operate.

Decoupling the recommendation algorithm from the recommendation system can adapt to the recommendation needs of different categories of items and the needs of different production environments. For the recommendation model, different user states and different item contents correspond to different recommendation algorithms with their own adaptability. In the recommendation system, the user's various operating characteristics and various scenarios are processed uniformly. The goal is to establish an omnipotent recommendation model that will eventually lead to the extremely complex recommendation system, and it is difficult to realize the configuration of different user characteristics. Reasonably distribute weights with recommended tasks. Therefore, using the recommendation engine assembly module, different recommendation engines are responsible for different platform user characteristics and recommendation scenarios, and the recommendation system is responsible for sorting the college recommendation list according to a certain ranking logical order, sorting the college recommendation list and returning to the college list, this operation reduces the load of the recommendation system.

4 Conclusions

Artificial intelligence has been widely promoted across different subjects and disciplines [9]. We thus made attempts in incorporating artificial intelligence into education program recommendation through this work technically. Admittedly, the work is to some extent immature and with certain limitations. The team is formed with members across disciplines, most of who came from a different field rather than the computer science and some of whom were still students who developed their techniques and interests in this interdisciplinary topic between artificial intelligence and education. However, the project witnesses how artificial intelligence has been the trend that calls for attention and efforts from scholars and students across different fields.

Due to the excessive number of candidate schools in this system, the index data corresponding to each school cannot be guaranteed to have a unified source, so our team only considered the top 40 schools in the UK. Future research plan attempts to improve the system technology by considering the addition of interface settings, the improvement of various functions, and the increase in the number of matching colleges. Future research can also consider comparing affiliates from other countries with the universities in the UK, considering they may share similar criteria, particularly for the IELTS language testing system.

To ensure the user experience, the team will further optimize the overall interactive experience of the project and consider making the system in the form of a web page or applications and publishing it (the specific form is yet to be determined). By doing so, users can get practical benefits and figure out their dream schools.

References

1. Goldberg, D., Nichols, D., Oki, B.M., Terry, D.: Using collaborative filtering to weave an information tapestry. Commun. ACM **35**(12), 61–70 (1992)
2. Resnick, P., Varian, H.R.: Recommender systems. Commun. ACM **40**(3), 56–58 (1997)
3. Lops, P., de Gemmis, M., Semeraro, G.: Content-based recommender systems: state of the art and trends. In: Ricci, F., Rokach, L., Shapira, B., Kantor, P.B. (eds.) Recommender Systems Handbook, pp. 73–105. Springer, Boston (2011). https://doi.org/10.1007/978-0-387-85820-3_3
4. Linden, G., Smith, B., York, J.: Amazon.com recommendations: item-to-item collaborative filtering. IEEE Internet Comput. **7**(1), 76–80 (2003)
5. Cocciolo, A.: The rise and fall of text on the web: a quantitative study of Web archives. Inf. Res. Int. Electron. J. **20**(3), n3 (2015)
6. Zhao, L., Hu, N.J., Zhang, S.Z.: Algorithm design for personalization recommendation systems. J. Comput. Res. Dev. **39**(8), 986–991 (2002)
7. Deng, A.L., Zhu, Y.Y., Shi, B.L.: A collaborative filtering recommendation algorithm based on item rating prediction. J. Softw. **14**(9), 1621–1628 (2003)
8. Wang, L.C., Meng, X.W., Zhang, Y.J.: Context-Aware recommender systems. J. Softw. **23**(1), 1–20 (2012)
9. Gao, Y., Tao, X., Wang, H., Zeng, G., Lian, H.: Artificial intelligence in language education: introduction of *Readizy*. J. Ambient Intell. Human. Comput. Online First (2021). https://doi.org/10.1007/s12652-021-03050-x

Mobile Learning in Project-Based Contexts in the Higher Education Sector

Christoph Knoblauch(✉)

Ludwigsburg University of Education, Ludwigsburg, Germany
christoph.knoblauch@ph-ludwigsburg.de

Abstract. This paper discusses evaluation findings from a digital project-based course in the higher education sector. The evaluated course was developed and taught in the context of the Master's program for Education at Ludwigsburg University of Education (LUE). It aims at the autonomous project-based planning, execution and critical reflection of educational projects by the participating students. Students in this course work and collaborate with mobile devices in various ways in the course and to carry out their projects in groups and individually. The combination of mobile and project-based learning in the course and the students' projects offers many options for mobile learning scenarios in the higher education sector. The digital project-based character of the course aims for the interconnection of students with their future work area, their future colleagues and their future students through projects in the field of education. The described interconnections and the resulting collaborations play key roles in the successful realization of the projects. The creative use of mobile learning devices is fundamental for the development of the described interconnections. Thus the empirical study has a special focus on the use of mobile learning devices in the course and the students' projects. Furthermore, the empirical study focuses on students' attitudes, practices and preferences towards mobile learning and the use of mobile learning devices in the projects. This focus creates insights, which help to develop new perspectives for mobile learning in the higher education sector.

Keywords: Mobile learning · Project-based learning · Higher education

1 Blending Mobile and Project-Based Learning: Context of the Course

The process of learning and teaching in digital settings has changed considerably in the face of new challenges and new possibilities. The world of learners and teachers has become more autonomous and mobile [1], yet at the same time more interconnected as a result of new challenges such as globalization and digitalization. In this regard, mobile and project-based learning are discussed as some of the most important ways to cope with the new challenges of a globalized world and to develop and foster competencies in autonomous learning in the higher education sector [2]. Especially the blending of project-based and mobile learning has been viewed as a crucial teaching/learning model, which can encourage the development of various learning

M. E. Auer and T. Tsiatsos (Eds.): IMCL 2021, LNNS 411, pp. 1022–1031, 2022.
https://doi.org/10.1007/978-3-030-96296-8_93

skills such as e-learning and autonomous learning [3]. The constructivist approach of project-based learning, allows learners to use an inquiry-based concept and to engage with issues that are relevant and real to the topics studied [4]. Furthermore, it allows learners to use mobile technologies in individual ways, suitable for their projects [5]. Therefore, the blending of autonomous and mobile learning in project-based courses can play an important role in the new contexts of digitalisation. Mobile and project-based learning approaches seem to have proven particularly relevant in this context, for they can offer tools for communication, autonomous learning, collaboration and shared learning between students and teachers [6]. Consequently, there is a strong demand for the development of mobile and project-based learning in an increasingly autonomous and digital learning environment.

The higher education sector plays an important role in these developments and must experiment with the use of mobile learning in project-based courses to prepare students for learning and teaching in more autonomous, project-based and digital contexts [7]. Against this background, however, well-researched resources offering guidance on how to blend mobile and project-based learning and how to design adequate courses for the higher education sector remain rare. This study reports on the development, implementation and evaluation of a course in the Master's program "Teacher Education" at LUE in Germany [8]. Participants were students enrolled in teacher education programs at LUE in the summer term of 2021. The key question of the study is how students use mobile learning devices in a project-based learning environment and how these results can be constructively implemented to improve future digital project-based scenarios in the higher education sector. Therefore, students' attitudes, practices and preferences concerning mobile learning and project-based learning and factors as competence development and autonomous learning play a major role in the design of the course and its evaluation. The anticipated outcomes of the study show students' (a) attitudes, (b) practices and (c) preferences towards mobile learning and give examples for (d) the actual use of mobile learning devices in project-based learning contexts.

Digital project-based learning shows the potential to help students foster independent learning skills and develop learning competence in individual learning contexts [9]. Against this backdrop, mobile learning shows the potential to allow greater flexibility and to help develop individual and appropriate methods within the learning process [10]. The course discussed and examined, therefore, shows how students blend mobile and project-based learning in individual and constructive ways and how they reflect the interrelated processes.

The focus of this paper is to evaluate the blending of mobile and project-based learning and to discuss the structure of the course to present results on the possible impacts of mobile project-based learning in the higher education sector. Therefore, the paper discusses mobile and project-based learning scenarios that support the participating students in different ways: Participants should develop their learning skills in project-based, digital and mobile contexts. In these processes, students should become aware and capable of mobile and project-based skills that could transfer to their future teaching in a classroom setting and beyond. To gain insights into the learning and working strategies of students in the described settings, the paper discusses a qualitative study focusing on the attitudes and preferences of students to mobile project-based learning. By doing so, the study assesses current practice and analyzes the course mentioned in detail.

2 A Structure for Mobile and Project-Based Learning: Description of the Course Design

The main foci of the course are (a) the theoretical discussion of experience- and project-based learning, (b) examples for project-based learning in schools and (c) the autonomous implementation of a small-scale project in the field of education. Thus, the project-based character of the course aims for the interconnection of students with their future work area, their future colleagues and their future students through a project in the field of education. The combination of digital, mobile and project-based learning offers many options for digital learning scenarios in the higher education sector and has a special focus on the development and fostering of autonomous learning [11]. Approximately 70% of the course were carried out digitally (synchronous and asynchronous sessions), using learning management systems such as Moodle and conference tools such as Webex. The other 30% were reserved for the implementation of the students' projects. The use of mobile learning devices in the course and the projects is fundamental for the evaluation of the course.

The asynchronous sessions were based on a mix of learning arrangements consisting of readings, audio presentations, videos, podcasts, interactive forums and chats. To these ends, (a) pertinent presentations with audio commentary were created; (b) a podcast with experts was made available; (c) common issues were presented through a selection of pertinent literature; (d) videos of professors and students working in project-based learning contexts were provided and (e) digital forums were established to offer an ongoing interactive exchange of ideas.

The synchronous sessions mostly consisted of guided discourse offering the chance to reflect on the contents of the course and presentations of the various projects designed and carried out by the students. The videoconferences were based on students questions and reports offering group discussions and individual sessions in break-out-rooms.

Against the backdrop of the course structure, the idea of digital learning is understood as a form of 'blended learning', which combines individual eLearning, mobile learning, interactive digital collaboration and autonomous project-based learning in various ways [12]. Students can gain up to three credit points (as defined by the European Credit Transfer and Accumulation System – ECTS) for the course and can use their projects as a basis for their Master Thesis. In the design of the described course, mobile learning is considered flexible and independent of time and place. The learning devices can be moved with the learner easily and they are wireless.

3 Evaluation of the Course

The evaluation was designed as a qualitative study, focusing on the experiences and the feedback of the participating students of the course. The study is intended to create a flexible research process and to produce various findings. It uses a creative qualitative design [13], which is conducted digitally [14]. In the context of the complex research focus – students' attitudes and preferences to mobile project-based learning – semi-structured qualitative interviews were carried out in an asynchronous digital way. In this particular study, students had the opportunity to record their answers as audio files independently and therefore could take as much time for reflection as they individually considered appropriate. This digital and asynchronous way of interviewing study participants is a rather new technique in qualitative research, which encourages respondents to reflect on their answers by allowing them to structure their ideas and responses beforehand [15]. The completed audio files were then sent to the research team via a digital transfer system. Drawbacks of this technique, such as a lack of face-to-face interaction and constructive dialogue, have been taken into account and play a role in the analysis of the data. Even though there is rare guidance on the implementation of the described technique of asynchronous digital qualitative interviews, there has to be emphasis that – apart from the innovative way of data generation – the study follows established structures of qualitative research and analysis in manifold ways [16]. Using this method of qualitative interviews, the study looks for data, which offers a comprehensive view through individual, reflected, subjective feedback, which is in this study recorded asynchronously and individually by the respondents.

The qualitative semi-structured questionnaire, which was sent to the students via email, starts with a short salutatory address and some operation guidelines on how to produce the audio file. The operation guidelines emphasize that the respondents can take as much time for reflection as needed.

The semi-structured questionnaire shows five categories which were developed in a deductive procedure [17]: (a) individual definition of the terms "mobile device" and "mobile learning", (b) use of mobile devices, (c) individual assessment of the use of mobile devices, (d) mobile learning, (e) suggestion for future courses and projects.

These deductive categories were built based on the discussion of former research, and evaluations in the field of education and beyond [18]. The research process is open for inductive impulses found within the data. Each category includes several key questions, which are supported by specifying impulses or further questions. The questionnaire mainly focuses on the individual blending of mobile and project-based learning, asking students about their expectations, attitudes, preferences, and actual use of mobile devices in the context of a project-based course. The questions furthermore focus on the use of mobile devices in the course and the projects, looking into various options for interaction, attitudes towards autonomous learning, reflections on learning strategies, and perceptions of change within individual learning strategies and attitudes.

4 "…Absolutely Necessary for My Project." Findings of the Qualitative Study

The analysis and discussion of the data [19] show five combined categories, which are based on the deductive categories and new inductive impulses found within the data:

(a) individual reflections on the terms "mobile device" and "mobile learning", (b) use of mobile devices in the course and the projects, (c) individual assessment of the use of mobile devices, (d) mobile learning, (e) suggestions for future courses and projects. The findings within the data are structured and discussed according to these categories. However, some answers and reflections within the data show links to more than one category and are therefore discussed under the perspectives of different categories.

a) *"I can learn wherever I want and whenever I want."* Individual reflections on the terms "mobile device" and "mobile learning"

"Mobile device": The interviewed students discuss various ideas concerning the definition of the term "mobile device". The study indicates that a mobile device has to be rather small, portable and wireless. It is connected to the internet and can be used for various tasks such as communication, research, exercises and the production of output such as pictures, essays and videos. Flexibility in terms of location and time seems to be a very important quality of mobile devices. Most students name Smartphones and Tablets when discussing the term "mobile device", some also mention laptops.

"Mobile learning": Most interviewed students agree: *"Mobile learning is the development of knowledge which is independent of location and time."* Students appreciate the fact that they can access information immediately and that mobile devices offer a variety of options to learn. Communication and collaboration are also mentioned as important factors of mobile learning even though some students criticise the lack of face-to-face interaction. While some students discuss guided synchronous and asynchronous courses as their main areas of mobile learning, others emphasize the fact that mobile learning enables one to work creatively and to deepen one's knowledge autonomously.

(b) *"I listened to the podcast while I was cooking."* Use of mobile devices in the course and the projects

Students in this project-based course use different mobile devices in various ways and different contexts. To structure the wide-ranging feedback, students' answers are divided into two sections: (a) use of mobile devices in the course, (b) use of mobile devices in the projects.

(1) Use of mobile devices in the course

Many of the interviewed students tend to use laptops for the course sessions: *"I have all my documents ready and I can easily open different tabs."* The fact that laptops usually have bigger screens than tablets or smartphones and that one can type comfortably is also mentioned many times. The interviewed students use laptops mostly for synchronous and asynchronous sessions, typing and reading. Especially the use of

PowerPoint Presentations with audio commentaries seems to work better on a laptop for most of the students. The podcasts, which were used in the course, however, were often played on smartphones while the students did other things like walking in the park or even cooking. Smartphones are also often discussed as tools for communication and collaboration with other students before, during and after course sessions: *"I used my smartphone for collaboration with fellow students. We used it for video calls and messages."*

(2) Use of mobile devices in the projects

The picture changes when students report about their project-based work: *"The smartphone is just more mobile. I can continue working on my project, while I'm out and about. I can respond to messages without having to start my PC."* Most of the participating students report that they used smartphones in many ways in their projects. Documentation and communication seem to be the two main reasons for the heavy use of smartphones in the project-based work: *"I used my smartphone to take pictures and videos."* It is surprising to see that the students do not report about the use of tablets in the projects. Smartphones seem to be the most practical way to work in the project, while laptops become important for the work with the results and the data at home: *"The laptop for analysis. The smartphone for documentation and communication."*

(c) *"…absolutely necessary for my project."* Individual assessment of the use of mobile devices

Students reflect their use of mobile devices in the context of the course and their project-based work in depth. In general, all students rate their individual use of mobile devices as functional and successful. Especially the use of mobile devices, such as smartphones, in the projects are discussed as very positive and highly efficient: *"They are just good and easy to use. I didn't need any introduction to use them. They were very good support in general and they were absolutely necessary for my project."* Students value that they use smartphones anyways, that they are easy to use and familiar. Additionally, smartphones offer stable internet connections, produce high-quality audio and video material, are helpful for communication on various platforms and are highly flexible in terms of location and time. Some students however criticised that the used learning platform (moodle) did not offer a specific app for smartphones and did not use QR Codes and instant messaging. The structure of the course also seems to be very important in terms of usability on smartphones. Some students ask for a clearer arrangement of topics and information in the course structure in moodle. However, things like autonomous working, which is flexible in terms of location and time, the possibility to learn and work at an individual pace and the chance to revise input many times, were highly appreciated.

d) *"I can organize learning individually."* Mobile Learning

Students report about various ways of mobile learning and the answers show that different methods and options are individually used. Mobile learning seems to be highly connected to project-based learning as many students discuss mobile learning in the context of their projects. In general, the study indicates that students appreciate six fundamental characteristics of mobile learning:

1) Flexibility and organization: *"I can continue working anytime..."*

Students can organize their learning and work autonomously and are independent concerning time and location. Apart from the synchronous sessions and the meetings in discussion and preparation forums students can work wherever and whenever they choose to. Students appreciate the possibility to structure their learning and working more autonomously.

2) Individual pace: *"One can think about questions again and again and review discussions."*

Students can learn and work at their individual pace. The possibility to review lectures and to add new perspectives to discussion forums anytime is appreciated by many students.

3) Accessibility: *"Work is fast and uncomplicated."*

Students can easily access input and have various options for researching online. This seems especially the case for audio input and the first steps of scientific research.

4) Collaboration: *"I used my smartphone to discuss with others during lectures..."*

Students can connect with fellow students, professors and participants of their projects easily. Chat groups, video calls, forums and various other options are helpful for fast and flexible networking. Students in this study tend to collaborate with more fellow students more often, yet shorter via mobile devices. Some students report that they miss face-to-face interaction because discussions via mobile devices tend to be shorter and rather hollow.

5) Multifunctional devices: *"I can do everything with one device."*

Working in projects seems to be in special need of multifunctional mobile devices. Students appreciate that they can record video and audio, take pictures, connect and research with smartphones in their project. The fact that they don't need to borrow extra devices at the university is very comfortable and helps to focus on the project itself.

6) Variety of methods: *"Ways of learning I did not have before..."*

Students can chose and develop methods of learning and working individually. The students in this course report about very different approaches in their projects and their individual learning in the course. They use a variety of methods to collaborate, work and learn in groups and individually. Shared documents, communication tools, forums, survey tools, hubs and many other methods are discussed in the interviews.

e) *"Our future students are mobile learners!"* Suggestion for future courses and projects

Students in this course mostly report positive experiences with mobile and project-based learning. Many ask for similar courses in the future and underline that the use of mobile devices was very helpful for their learning and working, both in the course and in their projects. At the same time, some students offer helpful suggestions for future courses in project-based contexts: Some would like to have a shared digital cloud for the course, which is connected to the Moodle platform. The simultaneous use of shared documents could be helpful to see what others are doing and to share ideas and input instantly. Other students emphasize that the blended format of the course is very important and has to stay flexible to suit different learners and projects. One student warns that mobile devices should not be compulsory for university courses and accessibility has to be guaranteed for all users [20]. Other students criticise that there was an overflow of messages in some of the discussion groups and certain rules should be established beforehand. Many students emphasize that the connection to professors is very important and has to be maintained in future courses. Especially the discussion of projects with a professor seems to be important for many students.

5 Outlook: *"An Appropriate Blending is Important…"*

Mobile and project-based learning in the field of education can generate a high level of motivation in students and encourage them to work creatively and autonomously. The implementations of projects is highly connected to mobile learning in various ways and can lead to the development of interdisciplinary competencies, a deeper insight into chosen topics and many-sided connections to future work fields. Against this backdrop, mobile project-based learning has to offer a fine balance of autonomous, collaborative and structured learning. Thus, the creative, mobile and self-responsible implementations of projects have to stand on a firm basis: A functional and reliable course structure, guided discussions, scientific backgrounds and individual support are important factors for successful project-based learning. These factors can be part of the mobile learning process if students choose to, but they should also be accessible without mobile devices. The study shows that mobile learning is interpreted and used in various ways and therefore highly individual. Mobile project-based courses should provide a reliable framework, develop common rules and offer a high level of individuality. The development of the course and the projects, learning processes and collaborations should be reflected and discussed continuously.

Mobile project-based courses require a structure, which enables students (a) to find themes and develop projects individually through scientific perspectives and online content; (b) to work creatively and individually; (c) to share and discuss their ideas online; (d) to develop shared knowledge through collaboration; (e) to develop new content and tasks for themselves and others.

In this context, mobile learning can offer flexibility, accessibility and organization. It also can foster collaboration and networking and help students to learn at their individual pace. The use of mobile devices offers a variety of functions, both for learning in the course and working on the project.

In this study, students comment on the processes of mobile learning and discuss mobile learning as a crucial factor in their project-based learning experiences. Especially students in the higher education sector can gain from mobile learning if they learn (a) to set appropriate goals autonomously, (b) to work and learn self-responsibly in a reliable structure, (c) to use mobile devices for individual and collaborative revision responsibly and (d) to connect with future work fields through projects. These processes support one another as they aim at the development and fostering of competencies and help students become more aware of their mobile learning activities and future work fields [21].

References

1. Khaddage, F., Lanham, E., Zhou, W.: A mobile learning model for universities re-blending the current learning environment. Int. J. Interact. Mob. Technol. (iJIM) (3), 18–23 (2009)
2. Utulu, S., Alonge, A.: Use of mobile phones for project based learning by undergraduate students of Nigerian private universities. Int. J. Educ. Dev. Inf. Commun. Technol. (IJEDICT) **8**(1), 4–15 (2012)
3. Rahman, F.: Integrating project-based learning in mobile development course to enhance student learning experience. In: Ayers, T. (ed.) Proceedings of the 19th Annual SIG Conference on Information Technology Education, 2 (2018)
4. Tamim, S., Grant, M.: Definitions and uses. Case study of teachers implementing project-based learning. Interdisc. J. Probl. Based Learn. **7**(2), 71–101 (2013)
5. Milenjevic, I., Ciric, V., Vojinovic, O.: Version control in project-based learning. Comput. Educ. **50**(4), 1331–1338 (2008)
6. Moran, K.: Taking a leap of faith. Redefining teaching and learning in higher education through project-based learning. Interdisc. J. Probl Based Learn. **8**(2), 19–34 (2014)
7. Knoblauch, C., Keßler, J.-U., Jakobi, M.: Schools of education as agents of change: coping with diversity in India and Germany through a collaborative, interactive and blended-learning environment – a pre-test study. In: Auer, M.E., May, D. (eds.) REV 2020. AISC, vol. 1231, pp. 866–876. Springer, Cham (2021). https://doi.org/10.1007/978-3-030-52575-0_71
8. PH Ludwigsburg. https://www.ph-ludwigsburg.de/en. Accessed 8 Sept 2021
9. Knoblauch, C., Keßler, J., Jäger, N.: Embracing diversity with blended learning. brainstorming across borders yields new insights at home: EAIE Forum, Spring 2020, pp. 15–17 (2020)
10. Ally, M., Prieto-Blázquez, J.: What is the future of mobile learning in education? Mobile learning applications in higher education. RUSC. Rev. de Univ. y Soc. del Con. **11**(1), 142–151 (2014)
11. Akinwamide, T., Adedara, O.: Facilitating autonomy and creativity in second language learning through cyber-tasks hyperlinks and net-surfing. ELT **5**(6), 36–42 (2012)
12. Al-Emran, M., Elsherif, H., Shaalan, K.: Investigating attitudes towards the use of mobile learning in higher education. Comput. Hum. Behav. **56**, 93–102 (2016)
13. Flick, U.: Interviews in der qualitativen Evaluationsforschung. In: Flick, U. (ed.) Qualitative Evaluationsforschung. Konzepte - Methoden -Umsetzung, pp. 214–233 (2006)
14. Gnambs, T., Batinic, B.: Qualitative online-forschung. In: Mey, G., Mruck, K. (eds.) Handbuch Qualitative Forschung in der Psychologie, pp. 320–332. VS Verlag für Sozialwissenschaften, Wiesbaden (2010)
15. Salmons, J.: Qualitative Online Interviews. Strategies, Design, and Skills, 2nd edn., p. 454. SAGE, Los Angeles (2015)
16. Ehlers, U.: Qualitative onlinebefragungen. In: Mikos, L., Wegener, C. (eds.) Qualitative Medienforschung Ein Handbuch von Mikos, Lothar; Wegener, Claudia, pp. 327–339. UVK Verlagsgesellschaft, Konstanz/München (2017)
17. Berg, B., Lune, H.: Qualitative Research Methods for The Social Sciences, 9th edn, pp. 21–30/172–174. Pearson, Boston (2017)

18. Mertens, D.: Research and Evaluation in Education and Psychology. Integrating Diversity with Quantitative, Qualitative, and Mixed Methods, 3rd edn., pp. 207–267. SAGE, Thousand Oaks (2010)

19. The interview study was implemented in June 2021. Eleven semi-structured interviews are part of the study. All recordings and transcripts are available on the study's platform – for further information please contact the author

20. In this project-based course, students did not have to own mobile devices. The use of mobile devices however was recommended – especially in project-based contexts. Students were able to borrow devices at the university free of charge

21. Barron, B., Schwartz, D., Vye, N., Moore, A., Petrosino, A., Zech, L., et al.: Doing with understanding. Lessons from research on problem- and project-based learning. J. Learn. Sci. **7**(3/4), 271–311 (1998)

Can Eye Tracking Identify Prognostic Markers for Learning Disabilities? a Preliminary Study

Eugenia I. Toki[1,2(✉)], Giorgos Tatsis[1], Jenny Pange[2], Konstantinos Plachouras[1], Pavlos Christodoulides[1], Evangelia I. Kosma[1], Spyridon K. Chronopoulos[1], and Victoria Zakopoulou[1]

[1] Department of Speech and Language Therapy, University of Ioannina, 45500 Ioannina, Greece
{toki,gtatsis,kplachouras,pchristo,e.kosma,spychro, vzakop}@uoi.gr

[2] Laboratory of New Technologies and Distance Learning, School of Education, University of Ioannina (UoI), Ioannina, Greece
{toki,jpagge}@uoi.gr

Abstract. Eye tracking is a promising technology offering objective metrics quantifying visual attention in various application fields. The employment in people with specific learning disabilities (SLDs), who reportedly manifest long-term functional social, emotional, and behavioral difficulties, provided well understandings of their visual attention, detecting cognitive processing mechanisms regarding specific SLD patterns. This study focuses on investigating and reporting the potential of eye tracking technology to identify multifaceted prognostic markers for SLDs. 30 adults (M: 16, F: 14) aged 29.8 ± 13.35 years old were presented with audiovisual stimuli in ordered digital tasks targeted to take real time metrics on (i) emotions, (ii) body scheme perception, and (iii) digital reading out loud. All digital visual stimuli consisted of still images accompanied by audio instructions. Cluster analysis indicated the presence of a variation between groups according to metrics and tasks. Descriptive statistics, gaze heatmaps and cluster diagrams give insights of this study, indicating that reading fixations provide the potential of promising metrics in the current classification. The development of eye tracking digital innovative systems may prove to be a valuable tool for early identification of SLDs, employing learning mobile technology to (i) reduce screening and diagnostic costs and (ii) enhance clinician's efficiency.

Keywords: Eye tracking · Prognostic markers · Specific learning disabilities

1 Introduction

Eye tracking is a technology that helps to identify accurately where, when and what an individual is looking at, when presented with certain stimuli. It can provide an index of visual attention and, it is a promising technology for revealing prognostic markers [1]. Current research focuses on investigating the role of attentional processes through eye

M. E. Auer and T. Tsiatsos (Eds.): IMCL 2021, LNNS 411, pp. 1032–1039, 2022.
https://doi.org/10.1007/978-3-030-96296-8_94

movements and fixations in various application fields, i.e., neuroscience and psychology, learning and education, medical research to study a variety of neurological and psychiatric conditions, neuromarketing, media testing and advertisement, gaming, web design and many more. Eye tracking offers objective metrics quantifying visual attention providing insights into cognitive processing mechanisms.

Specific learning disabilities (SLDs) have an estimated prevalence between 5% and 9% in the paediatric population with higher incidence in boys [2] presenting among other deficits in various executive function (EF) domains and in working memory [3]. SLDs are principally associated with considerably lower performance in one or more tests measuring reading, arithmetic, or writing skills [2].

In addition to academic and cognitive deficiencies that result to school failure, individuals with SLDs often manifest long-term functional social, emotional, and behavioural difficulties [4]. They are significantly more likely to display poor task engagement, poor self-control, externalizing and internalizing behaviour problems. Anxiety, depression, stressful events, emotional trauma, and other conditions affecting concentration make learning more of a challenge, constituting the evaluation of their emotional conditions and the way they see themselves, family, friends, and school, both important and demanding.

Furthermore, individuals with SLDs have consistently reported a range of visual attentional deficits, such as the visual attention span deficit [5], lending support to the visual account of SLDs [6]. These attention deficits encountered by individuals with SLDs, possibly reflect a difficulty when attentional allocation of sequential stimuli is required, and since the visual attention capacity is inadequate the process of simultaneously presented stimuli is deficient [6–9].

Studies of binocular motor control in children with SLDs [10, 11], have found abnormally poor coordination of movements of eyes and instability of vergence relative to age-matched controls. Interestingly, higher rates of reflexive triggering of saccades and vergence eye movements were observed both in children and adults with SLDs (express type of latency 80–120 ms) [12, 13].

Despite these important findings, current screening techniques to assess cognitive reading-related skills only partially reveal the interplay and function in actual reading. To investigate these abilities in a quicker and more systematic way, an employment of objective metrics is highly necessary.

One such methodology is the eye tracking technique which enables researchers to analyse visually presented multimedia stimuli in the form of emotions [14] shapes [15] and texts [16]. This study focuses on investigating and reporting on the potential of eye tracking technology to identify multifaceted prognostic markers for SLDs. More specifically, the objective is to examine whether eye tracking metrics constitute indicators of learning disability patterns for the adaptation of eye-tracking in mobile learning systems. Although in this preliminary study the target group involves adults, our goal is to interpret the findings gained as a guide towards the design and development of accurate digital tasks, thus contributing to an early diagnosis of learning disabilities in children.

2 Materials and Methods

The participants of this study have been selected after an online invitation sent by email in September 2020 also posted on the Laboratory site (https://cslab1.slt.uoi.gr/) and on Facebook. So, 30 adults (M:16, F:14) finally contributed to the experimental procedure. The subjects' average age was 29.8 ± 13.35 years old. Individual meetings were organised to inform the potential participant of the experimental procedure. Informed consent forms were signed by all the participants, in accordance with the standards of the Research Ethics Committee of the University of Ioannina.

The experimental procedure was based on gaze response analysis, and specific tasks were designed for the purpose of the study. Screen-based eye tracking was employed using a Tobii Pro Spectrum X300 eye tracker device set at 300 Hz, and data collection and analysis were conducted with iMotions 8.0 software.

More specifically, the subjects were asked to watch and interact with visual and audio stimuli presented on the screen, while the eye tracking device recorded their eye gaze metrics in real time. Visual stimuli were comprised of still images with a few seconds duration each. Audio instructions asked them to look at on specific parts of the image. The duration of the whole recording procedure was 2 min and 7 s.

The types of digital tasks targeted to take real time metrics on three sets, regarding (i) emotions, (ii) body scheme perception, and (iii) digital reading aloud. Regarding the emotions task, the subjects were asked to focus on each face with a particular emotion, i.e. look at the angry face. In the task of the body-parts' discrimination, to look steadily at the child's face, next at the child's left hand and after that at the child's right leg. Finally, they were asked to read out loud three tongue twisters included Greek real words with the following phonotactic structure: vcccv, cvccv, cvcvcccv, and cvccvcvv.

Screenshots of the visual stimuli are shown in Fig. 1.

Fig. 1. Screenshots of the three visual stimuli

For the eye tracking analysis, in iMotions software, several areas of interest (AOI) were specified corresponding to discrete areas on the screen for each of the images presented to the participants. For instance, the four faces for the emotions, sad, happy, angry and disgust are four distinct areas that accept individual gaze data from the subjects. The eye tracker device continuously provides gaze information giving measurements of the coordinates (x, y) on the screen, corresponding to the position that the subject is looking at in real time, that are the gaze points. The gaze points usually change rapidly as we observe the images. When a series of gaze points is close enough in time and space, this constitutes a fixation. Fixations provide useful information of visual attention as they denote that someone actually is processing what is looking at. The iMotions software automatically calculates the fixations and the associated metrics for them, i.e. the time they occur, the duration of each and their positions.

The variables taken into account in this study are all related to the fixations occurring in-side the AOIs. More specifically, the following metrics computed by the eye-tracker software comprises the variable set, (i) fixations count (FCOUNT): the number of fixations inside an AOI, (ii) time to first fixation (TTFF): the absolute timestamp in msec from the beginning of the scene belonging to the first fixation inside the AOI, (iii) average fixation duration (AVFIXDUR): the average duration in msec of all fixations inside the AOI, (iv) fixation time spent (TSPENT%): total fixations time spent inside an AOI as a percentage relative to the duration of the AOI.

The total number of AOIs were 10, including 4 emotional faces, 3 body parts and 3 sentences (screenshots in Fig. 1). For each AOI the 4 previous defined metrics are calculated, producing in total 40 variables.

For the objective of this study concerning the appearance of common properties between the subjects of the study with respect to each discrete part of the visual stimuli and their combinations, a cluster analysis has been conducted. Especially, we examined the whole set of all tasks with respect to the three specific digital tasks, that is for facial emotions, body parts and reading aloud respectively (40 variables). We applied Hierarchical Cluster Analysis (HCA).

Additionally, Hierarchical Cluster Analysis (HCA), was also applied for each one of the three digital tasks, facial emotions (16 variables), for body parts (12 variables), reading aloud (12 variables). This analysis produced dendrograms depicting all the clusters found according to the distance separating them. For this purpose, SPSS v.26.0 was employed.

To confirm possible suspicions about the indication of learning difficulties in any of the examinees, in addition to the eye-tracking test, a developmental history and the dyslexia adults screening test (DAST) [17] were implemented. An *'At Risk'* for dyslexia (ARQ) index was provided to indicate whether the examinee was 'strongly at risk' of dyslexia (ARQ $\geq$ 1), or not (ARQ $\leq$ 1).

3 Results & Discussion

This study is part of an ongoing project, the anticipated results of which include reports on potential classification of the sample with respect to learning disabilities.

The cluster analysis considered all variables of all the three tasks as shown in Fig. 2 (a), indicates that there exist four clusters in bottom level and two at the top level and one case of our sample. The case numbered 28, seems to be an outlier.

In Fig. 2 dendrograms (b), (c), and (d) show the corresponding clustering considering only the variables for emotions, next for body perception and then for reading respectively. There is a strong indication that case 28 is also an outlier.

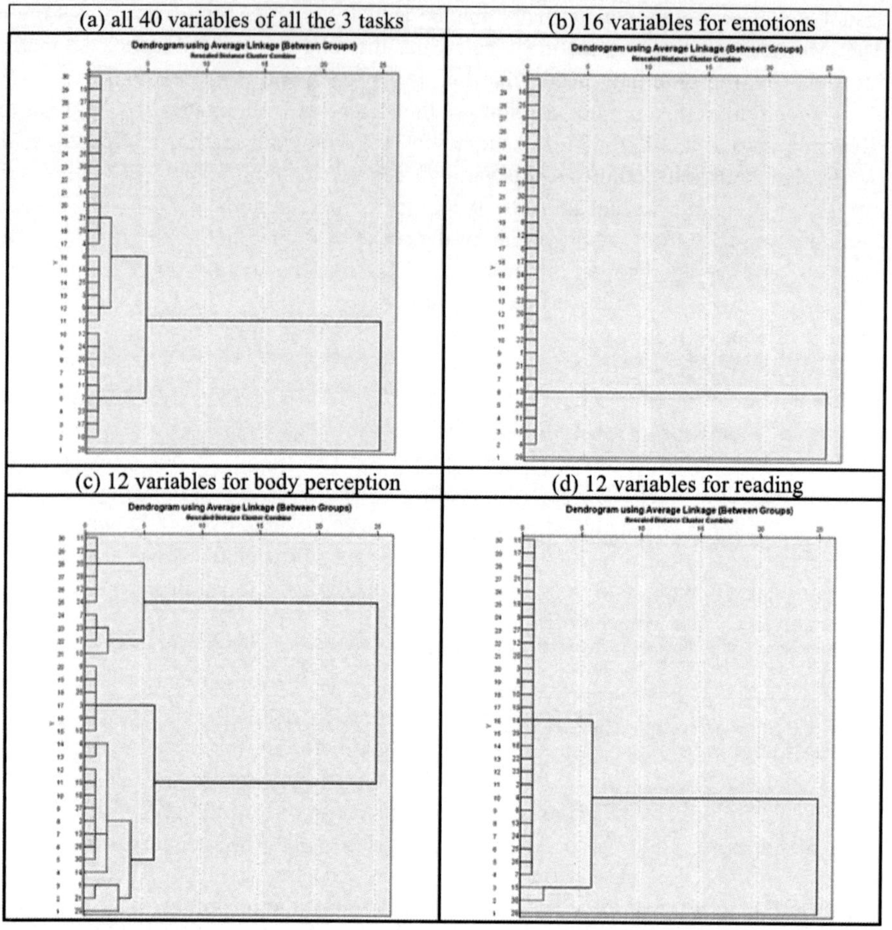

Fig. 2. Clustering dendrograms, (a) for all tasks, (b) for emotions, (c) for body perception and (d) for reading

Next, we investigate the differentiation of the subjects according to their body perception. We observe that there exist two clusters, the first one has 10 subjects and the second has 20 subjects. These two top level clusters identified by the analysis are shown in Fig. 2 (c). All subjects for both clusters have similar fixations on the face, but subjects belonging to cluster 2 gave fewer fixations (mean fixation count: 0.60 $\pm$ 0.97; mean time spent: 12.9 $\pm$ 25.54%; mean average fixation duration: 284.80 $\pm$ 574.00 ms) compared to cluster 1 (mean fixation count: 2.06 $\pm$ 1.92; mean time spent: 44.00 $\pm$ 43.07%; mean average fixation duration: 660.41 $\pm$ 937.49 ms) for the case of left hand and no fixations at all on right leg. To conclude that the aforementioned differences are statistically significant, the Mann-Whitney U-test was conducted. The variables related to number of fixations and time spent at the AOI of left hand and right leg have statistically significant differences (FCOUNT: $U = 45.5$, $p = 0.046$; TSPENT

%: U = 44.5, p = 0.04). Additionally, all metrics for the right leg have also statistically significant differences (TTFF, TSPENT%, FCOUNT, AVFIXDUR: U = 0.000, p = 0.000).

The clustering depicted in Fig. 2 (d), are related to the three phrases of the final task while subjects had to read out loud, shows the main group of 27 subjects, the second group of 2 subjects and a third one of one subject. The latter case is subject 28, the same that was detected as an outlier in the emotion task. The first group has a quicker response regarding the time to first fixation. Also, the time spent on each sentence and the count of fixations are greater than those in the rest groups. It seems that, almost in every case, the average fixation duration of the first group is less than the others. Nevertheless, for this occasion, no independent test has been conducted because the sample size for the groups 2 and 3 was not sufficient.

A higher deviation from the other cases/clusters was observed in the case 28, indicating the need for his further assessment. There, the DAST results revealed that the subject was 'strongly at risk' of dyslexia (ARQ = 1.45) reflecting difficulties in the tasks of Rapid Naming, Phonemic Segmentation, One Minute Reading, and Two Minute Spelling.

According to hierarchical cluster analysis, all the participants were clearly grouped in clusters characterised by approximately common features such as the metrics derived from variables of FCOUNT, TTFF, AVFIXDUR, and TSPENT%, for every AOI set in (a) for all tasks, (b) for emotions, (c) for body perception, and (d) for reading the tested tasks. Significant variation was observed in all tasks, in emotions and in reading for the cluster consisted of only one participant (case 28), whose differentiated performance could plausibly be attributed to his SLDs.

To explain case 28, we can say that the emotions of face (happy, angry and disgust) were not recorded, presumably due to missing gaze data at the specific AOI. Similarly, the case was confused to distinguish left and right, while there was no uncertainty about the face. Finally, in the reading aloud task, he had a rather slow response regarding the time to the first fixation in all three sentences, as well as a short time spent on each sentence along with a small count of fixations. In accordance, DAST results seemed to confirm the hierarchical analysis indicating a clear 'strongly at risk' of dyslexia profile.

All these findings are in line with the most of the current research results examining the multifactorial structure of SLDs. As such, eye movements were analyzed to measure the duration of fixations on stimuli, the number of forward saccades, and backward saccades [18]. Further, readers with dyslexia were also reported with disrupted eye movement patterns during reading because of language processing difficulties, resulting in more frequent, lengthier fixations and saccades [19, 20].

In the process of fluent reading, phonological awareness is mediated by visual processing which plays a central role in lexical processing as the eye moves along a passage of text [21].

Persistent reading and writing difficulties associated with dyslexia can jeopardize not only the academic performance of individuals, but also contribute to the development of socio-emotional issues such as stress, low self-esteem, and other internal disorders, which in turn affect relationships between individuals [22]. Although few studies have directly examined emotions perceptions among SLDs populations, existing studies have consistently shown children and adolescents with SLDs to be less accurate overall at recognizing facial expressions of emotion from still photographs compared with their non-SLDs peers [23, 24].

Apparently, our findings reveal the contribution of eye tracking in supporting the hypothesis of a strong interaction between visuospatial, cognitive, and emotions recognition difficulties, in adults with SLDs.

Even though these findings were obtained by a preliminary study applied in adults, they are promising towards providing physiological metrics providing valuable insights to identify prognostic markers for learning disabilities and even more for early diagnosis of SLDs in pre-schoolers.

4 Conclusions

The results of this preliminary study show that eye tracking metrics constitute indicators of learning disability patterns in adults. The findings of this study pointed that eye tracking can be efficiently used towards the identification of prognostic markers for learning disabilities, employing digital technologies and mobile systems to upgrade health promotion, learning opportunities, and skills of students with SLDs. Consequently, it may support the development of digital innovative systems with the potential to enhance digital learning and health care and contribute to the reduction of screening and diagnostic costs along with upscaling clinician's/special education teacher's efficiency.

Future work may focus on target population of adults and children with or not typical development.

Acknowledgements. We wish to thank the Region of Epirus for funding this project titled "Smart Computing Models, Sensors, and Early diagnostic speech and language deficiencies indicators in Child Communication", with code HΠ1AB-28185, supported from European Regional Development Fund (ERDF).

We also thank all the participants for their valuable contribution in this study.

References

1. Bacon, E.C., Moore, A., Lee, Q., Barnes, C.C., Courchesne, E., Pierce, K.: Identifying prognostic markers in autism spectrum disorder using eye tracking. Autism **24**(3), 658–669 (2019). https://doi.org/10.1177/1362361319878578
2. Pérez-Elvira, R., Oltra-Cucarella, J., Carrobles, J.A., Teodoru, M., Bacila, C., Neamtu, B.: Individual alpha peak frequency, an important biomarker for live Z-score training neurofeedback in adolescents with learning disabilities. Brain Sci. **11**, 167 (2021)
3. Alloway, T.P., Carpenter, R.K.: The relationship among children's learning disabilities, working memory, and problem behaviours in a classroom setting: three case studies. Educ. Dev. Psychol. **37**, 4–10 (2020)
4. Grigorenko, E.L.: Creativity: a challenge for contemporary education. Comp. Educ. **55**, 116–132 (2019)
5. Lallier, M., Valdois, S.: Sequential versus simultaneous processing deficits in developmental dyslexia. In: Dyslexia-a Comprehensive and International Approach, pp. 73–108 (2012)

6. Vidyasagar, T.R., Pammer, K.: Dyslexia: a deficit in visuo-spatial attention, not in phonological processing. Trends Cogn. Sci. **14**, 57–63 (2010)
7. Facoetti, A., et al.: Multisensory spatial attention deficits are predictive of phonological decoding skills in developmental dyslexia. J. Cogn. Neurosci. **22**, 1011–1025 (2010)
8. Sireteanu, R., Goebel, C., Goertz, R., Werner, I., Nalewajko, M., Thiel, A.: Impaired serial visual search in children with developmental dyslexia. Ann. NY Acad. Sci. **1145**, 199–211 (2008)
9. Moore, A.T., Schwitzgebel, E.: The experience of reading. Conscious. Cogn. **62**, 57–68 (2018)
10. Nilsson Benfatto, M., Öqvist Seimyr, G., Ygge, J., Pansell, T., Rydberg, A., Jacobson, C.: Screening for dyslexia using eye tracking during reading. PloS one **11**, e0165508 (2016)
11. Kapoula, Z., Vernet, M., Yang, Q., Bucci, M.P.: Binocular coordination of saccades: development, aging and cerebral substrate. J. Eye Mov. Res. **2**, 1–20 (2008)
12. Yang, Q., Bucci, M.P., Kapoula, Z.: The latency of saccades, vergence, and combined eye movements in children and in adults. Invest. Ophthalmol. Vis. Sci. **43**, 2939–2949 (2002)
13. Bucci, M.P., Brémond-Gignac, D., Kapoula, Z.: Latency of saccades and vergence eye movements in dyslexic children. Exp. Brain Res. **188**, 1–12 (2008)
14. Lim, L., Khor, H.-Q., Chaemchoy, P., See, J., Wong, L.-K.: Where is the emotion? Dissecting a multi-gap network for image emotion classification. In: 2020 IEEE International Conference on Image Processing (ICIP), pp. 1886–1890. IEEE (2020)
15. Verdine, B.N., Bunger, A., Athanasopoulou, A., Golinkoff, R.M., Hirsh-Pasek, K.: Shape up: an eye-tracking study of preschoolers' shape name processing and spatial development. Dev. Psychol. **53**, 1869 (2017)
16. Koć-Januchta, M., Höffler, T., Thoma, G.-B., Prechtl, H., Leutner, D.: Visualizers versus verbalizers: effects of cognitive style on learning with texts and pictures–an eye-tracking study. Comput. Hum. Behav. **68**, 170–179 (2017)
17. Fawcett, A., Nicolson, R.: Dyslexia Adult Screening Test (DAST). Pearson Education, United Kingdom (1998)
18. Rayner, K., Pollatsek, A., Liversedge, S.P., Reichle, E.D.: Eye movements and non-canonical reading: comments on. Vision. Res. **49**, 2232–2236 (2009)
19. Eden, G., Stein, J., Wood, H., Wood, F.: Differences in eye movements and reading problems in dyslexic and normal children. Vision. Res. **34**, 1345–1358 (1994)
20. Stein, J.: Hemispheric specialisation and dyslexia. In: Reading Disabilities, pp. 247–252. Springer, Dordrecht (1991). https://doi.org/10.1007/978-94-011-2450-8_17
21. Heth, I., Lavidor, M.: Improved reading measures in adults with dyslexia following transcranial direct current stimulation treatment. Neuropsychologia **70**, 107–113 (2015)
22. Klassen, R.M., Tze, V.M., Hannok, W.: Internalizing problems of adults with learning disabilities: a meta-analysis. J. Learn. Disabil. **46**, 317–327 (2013)
23. Most, T., Greenbank, A.: Auditory, visual, and auditory-visual perception of emotions by adolescents with and without learning disabilities, and their relationship to social skills. Learn. Disabil. Res. Pract. **15**, 171–178 (2000)
24. Nabuzoka, D., Smith, P.K.: Identification of expressions of emotions by children with and without learning disabilities. Learn. Disabil. Res. Pract. **10**(2), 91–101 (1995)

Comparing Lightweight Algorithms to Secure Constrained Objects in Internet of Things

Nabila Zitouni(✉), Maamar Sedrati, and Amel Behaz

LaSTIC Laboratory, University of Mostefa Ben Boulaid – Batna2, Batna, Algeria
{n.zitouni,m.sedrati,a.behaz}@univ-batna2.dz

Abstract. In recent years, the number of objects around us that are able to connect to internet has increased considerably, adding a new dimension to the world of Information and Communication Technology called Internet of Things. This technology allows billions of physical objects to communicate with each other. These things are resource-constrained (i.e., limited in treatment and energy), thus to secure their communications, lightweight cryptography is needed, and good management of energy consumption to ensure their long lifetime is necessary. This paper presents a comparative study of suitable and most frequently used lightweight block ciphers for Internet of Things devices. The comparison focuses on parameters that can reduce energy consumption and optimize battery lifetime. The comparison is done using Cooja simulator to determine the best lightweight cryptography algorithm between PRESENT, HIGHT, LBlock, and LED based on time execution, energy consumption, key size, and hardware implementation. Results show that key size has an impact on execution time and energy consumption; therefore, according to objects resources, the appropriate algorithm is applied. PRESENT and LED algorithms with 80-bit key size perform quickly and consume low energy compared to their 128-bit key size version but they are slower compared to HIGHT and LBlock.

Keywords: Internet of Things · Cryptography · Constrained objects · Lightweight block ciphers · Energy consumption

1 Introduction

Information and communication technology (ICT) field covers communication and computer technologies, as well as encoding, decoding, and processing of information. ICT is recently undergoing rapid development due to many emerging technologies, such as cloud computing, big data analytics, Internet of Things (IoT), and artificial intelligence. Internet of things is based on the idea that all things around us have a digital identity, can connect to internet, share information, and communicate with each other independently and without any human intervention [1]. Today, the number of connected objects is around 8 billion, so there are a huge number of IoT applications in different domains such as health care, building, transportation, energy management, and manufacturing. However, like all other technologies, Internet of Things faces several challenges, and security is one of the main challenges of this technology.

M. E. Auer and T. Tsiatsos (Eds.): IMCL 2021, LNNS 411, pp. 1040–1051, 2022.
https://doi.org/10.1007/978-3-030-96296-8_95

Cryptography is a fundamental security discipline, where it researches methods and techniques to maintain secure conversations and ensure several important security proprieties such as authentication, integrity, and confidentiality. Since IoT objects are low-resource devices, conventional cryptographic methods are not suitable. Therefore, a new cryptography branch aims to provide solutions tailored for resource-constrained devices is emerged, called LightWeight Cryptography (LWC) [2]. The main purpose of LWC algorithms is to meet the main resource requirements for target devices, such as memory consumption, execution time, and power consumption. LWC is classified into two main categories, symmetric and asymmetric. Symmetric category contains three main sub-categories, block ciphers, stream ciphers, and hash functions. This paper compares some lightweight block ciphers used in Internet of Things in terms of parameters that can influence energy consumption.

The remainder of this paper is organized as follows: Sect. 2 describes definition, architecture, applications, and security of IoT technology. In Sect. 3, cryptographic primitives are detailed; LWC primitives and classification are discussed in Sect. 4. Section 5 provides an overview of compared LWC algorithms. Results description and analysis are done in Sects. 6 and 7. Finally, the paper is concluded in Sect. 8.

2 Internet of Things Technology

Internet of Things relies on fact that connected objects are able to communicate, collect data and make appropriate decisions. A connected object is a thing, which can be a person, an animal, a car, a train, a computer, a tablet, etc.; this thing has its own identity and its sophisticated computing and communication capacities. Connected object has the capacity to receive and transmit data via internet or other technologies, but it has several constraints such as memory, bandwidth, and power consumption [3].

The Cluster of European Research Projects on Internet of Things (CERP-IoT) [4] defines Internet of Things as: "*a dynamic global network infrastructure. This global network has auto-configuration capabilities based on interoperable communication standards and protocols. In this network, physical and virtual things have identities; physical attributes, virtual personalities and intelligent interfaces, and they are seamlessly integrated into the network*".

IoT architecture is not standardized yet. IEEE, ITU, and other international organizations are still working on IoT standardization. Meanwhile, several IoT architectures are proposed by different researchers [1, 5]: three-layer, four-layer, and five-layer architecture. The three-layer architecture is the basic architecture proposed for IoT. It was proposed in the early stages of IoT concept development and consists of, from bottom to top, Perception layer, Network layer, and Application layer [6]. This architecture could not fulfill all requirements of IoT owing to the rapid growth of this technology. Therefore, architecture with four layers was proposed. This architecture has one more layer than three-layer architecture called Support layer [1]. The success of IoT technology depends on the way in which services are delivered to customers, therefore architecture with five layers was proposed. In this architecture, there is one more layer than the previous architecture called Business Layer [3, 7].

The IoT deployment opens doors to a huge number of applications in different domains: Urban infrastructure, Smart Cities, Healthcare sector, Smart agriculture, Industrial Automation, Wearables, Energy management, Smart animal farming, and Smart environment/planet [3, 8]. The diversity of its application areas will amplify classic security threats facing networks and data, as well as emergence of new threats that can directly affect objects integrity. Therefore, security is one of the main challenges facing the successful deployment of IoT applications.

Security in telecommunications networks is defined as a set of policies, mechanisms, or techniques adopted to cover network vulnerabilities and protect its entities, users, and data exchanged. Cryptography is a fundamental discipline of security, and it is among existing techniques ensuring communications security [9].

3 Cryptography

3.1 Cryptography and Security Services

Cryptology origin is from ancient Greek, and it is a two-word compound word: "Kryptos" and "logos". "kryptos" means "hidden or secret", while "logos" means "word". Therefore, cryptology means the science of secure speech or communications. This science is classified into two broad disciplines: cryptography and cryptanalysis. Cryptography discipline searches methods and techniques to keep secure conversations, whereas cryptanalysis discipline seeks how to crack cryptographic methods.

Cryptography science builds methods and techniques to secure transmission between communicating entities based on two main processes: encryption and decryption process. Encryption process transforms the ordinary message into a coded form (i.e., from plain text into cipher text); this operation is operated by the sender. Whereas decryption process performs the reverse operation (i.e., from cipher text into plain text), and it is done by the recipient. To carry their work, the two processes use a key on which the secret is based. To ensure communications security, cryptographic methods are used to achieve four security services [10]:

- **Confidentiality property**: guarantees that the data exchanged cannot be accessed by unauthorized nodes.
- **Integrity property**: means that the receiver should be able to check if data has been modified during transmission.
- **Authentication property**: is related to identification, it offers the receiver of data the possibility to verify its origin.
- **Non-repudiation**: the sender should not be able to refuse or deny the message sending.

3.2 Cryptographic Primitives

Many different cryptographic methods have been constructed over many centuries; these methods can be classified into two classes.

Classic cryptography: methods of this class were developed before computer period, and they are based on natural language letters. Transposition methods that rearrange the letters and substitution methods that replace letters with other letters are the main classical cryptography types. Further details can be found in [11].

Modern cryptography: due to the use of computers, modern cryptography manipulates bits. So methods rules of this class are known to everyone, and security relies only on key. This class includes two categories: symmetric key cryptography and asymmetric key cryptography [12].

3.3 Symmetric Key Cryptography

Cryptographic techniques are based on encryption and decryption concepts. Encryption process "E" transforms "plain text" into a "cipher text" using the key "Ke" according to formula (1).

$$\text{cipher text} = E_{Ke}\,(\text{plain text}) \tag{1}$$

Decryption process "D" transforms the "cipher text" into a "plain text" using the key "Kd" according to formula (2).

$$\text{plain text} = D_{Kd}\,(\text{cipher text}) \tag{2}$$

Symmetric key cryptography is the type in which encryption and decryption keys are the same (Ke = Kd = k). This type is called secret key cryptography (k) shared between sender and receiver. The major problem of this category is the management and sharing of secret key between the two communicating nodes.

Symmetric cryptographic methods are classified into three main subcategories: block ciphers, stream ciphers, and hash functions. Block cipher algorithm breaks up plaintext into fixed-length blocks, which will then be encrypted one after the other over several rounds. However, stream cipher algorithm encrypts the plain text without needing to split it, where it works on small units of plain text, usually bits, and the key is a bit stream generated by a pseudo-random number generator. Finally, hash function is a mathematical algorithm that converts an input value of arbitrary size into another compressed value of fixed size. The hash function is designed to be infeasible to invert, that is, a one-way function [13]. AES is among the most widely used block ciphers, RC4 is one of the most popular stream cipher algorithms, and SHA-3 is a member of the Secure Hash Algorithm family.

3.4 Asymmetric Key Cryptography

In asymmetric key cryptography, encryption and decryption keys are different (Ke Kd). This type requires a pair of keys called: "public key" and "private key", thus, asymmetric cryptography is also referred to as public key cryptography. In this type, both sender and receiver have two keys, the public key is known to the public, but the private key is known only to the user. In encryption process, the recipient's public key is used, but in decryption process, the recipient's private key is used. RSA, ElGamal, and ECC are part of this category.

4 Lightweight Cryptography

Nowadays, there is a large amount of conventional encryption tools where the best is chosen according to targets. These tools require too much physical space, too much processing power, and excessive battery power consumption. In a world with embedded systems and sensor networks, these requirements make them inefficient or difficult to implement. Thus, a new cryptography branch has emerged for new constraint applications characterized by small physical size, memory limitation, processing requirements, and energy drain; this branch is called LightWeight Cryptography (LWC).

There are no strict criteria for classifying a cryptographic algorithm as lightweight, but lightweight algorithms have common characteristics which are very low requirements for the main resources of target devices such as memory, processing power, and energy. Thus, it should take into account software and hardware specifications when developing a lightweight cryptography algorithm, for example: How much RAM size is necessary to run it? How long would it take to encrypt/decrypt data? How much energy will it consume? [14, 15].

4.1 Lightweight Cryptography Classification

LWC algorithms are classified into two main categories, symmetric and asymmetric. The working principle of lightweight symmetric and asymmetric cryptography is the same as the conventional symmetric and asymmetric cryptography, but the main differences are key size and number of computational cycles [14, 16]. Lightweight cryptography is not a new cryptography class, but rather a branch of modern cryptography class; therefore, cryptography classification can be modified and divide each subclass of modern cryptography class into conventional and lightweight. This means that symmetric and asymmetric cryptography subclasses will be classified into conventional and lightweight. Thus, conventional symmetric cryptography category is classified into conventional block ciphers, conventional stream ciphers, and conventional hash functions; lightweight symmetric cryptography category is also classified into lightweight block ciphers, lightweight stream ciphers, and lightweight hash functions.

4.2 Lightweight Asymmetric Cryptography

As mentioned above, the asymmetric approach requires two keys, a public key that can be shared over an insecure channel and a private key that does not be shared at all. Therefore, this approach provides more security functionality compared to the symmetric approach, but it is slower, more complex, and requires more computational power and a global registry of all other users' public keys communicating with it [17].

4.3 Lightweight Symmetric Cryptography

In lightweight symmetric algorithm, only one key is used to encrypt and decrypt data, which presents a simplified and faster operation, but the main drawback is that the key must be shared in a secure way between communicating entities. This approach is more advantageous while considering security issues for highly resource constrained devices. This class includes lightweight block ciphers, lightweight stream ciphers, and lightweight hash functions; where their working principle is the same as conventional one. Lightweight hash functions generate from the message a unique hash value in the goal to provide message authentication. A comprehensive survey of lightweight hash functions with their security levels is presented in [18] among them: PHOTON, SPONGENT, and Quark. Lightweight stream ciphers encrypt or decrypt a single bit from data at a time using a different key for each bit. This category is widely used in real-time applications. TRIVIUM and MICKEY are among the most popular lightweight stream ciphers [19]. Unlike stream ciphers, lightweight block ciphers encrypt/decrypt one block at a time using the same key for all blocks. Compared to stream ciphers, this type is practically easier to implement and more efficient. Therefore, this paper focused on this class of algorithms.

4.4 Lightweight Block Ciphers

The main differences between conventional and lightweight block ciphers are in block size and key size, where lightweight block ciphers require smaller block sizes, 64 bits or less, and smaller key size generally 80 or 128 bits.

According to the structure on which is based, lightweight block cipher can be Substitution Permutation Network (SPN) or Feistel based network. As its name implies, SPN structure is designed by using several alternating rounds of substitution boxes (S-boxes) and permutation boxes (P-boxes) on the whole data block. On the other hand, Feistel structure divides the block into two blocks of equal size, generally called left (L) and right (R) block. A round function is applied to the right block and the key at each round, and the result is XOR-ed into the left block. Blocks are then swapped.

Various lightweight block ciphers, SPN and Feistel, with their strengths and weaknesses are summarized in [13, 20, 21]. PRESENT, Humming bird-2, LED, PRINCE, mcrypton, and KLEIN are some widely used SPN block ciphers. Among the most popular Feistel block ciphers: DESL, MIBS, HIGHT, LBlock, SEA, and TWINE.

5 Compared Algorithms

A lightweight cipher is an appropriate cryptographic algorithm for limited resource devices. Therefore, lightweight crypto security is defined as lightweight resource consumption for instant treatment [17]. Hence, a lightweight cipher aims to have lower energy consumption, less execution time, and a small footprint area measured in Gate Equivalence (GE). The area is the number of GE required to perform the cipher in hardware. It allows specifying manufacturing technology independence complexity of the digital electronic circuit. A small GE signifies that the circuit is less costly and consumes less energy [22].

In this section, an overview of compared lightweight block ciphers is provided: HIGHT, LBlock, PRESENT, and LED. A comparison of the general characteristics of each cipher is presented at the end of this section.

5.1 HIGHT (High Safety and Lightweight) [23]

It is based on Feistel structure and uses 32 rounds to cipher blocks of 64-bit under keys of 128-bit. Operations used in HIGHT are very basic such us modular addition and XOR operation, and keys are generated during the encryption and decryption phase; First 8 keys generated are kept as a whitening key which is used in the initial and end round. The 4 key steps of HIGHT are Key schedule, Initial transformation, 32 iterative round operations, and Final transformation. The implementation of HIGHT cipher needs 3048 GE.

5.2 LBlock [24]

It is based on Feistel structure and operates with block size of 64 bits and key size of 80 bits. There are 32 rounds in LBlock. The round function processes one half of the data whereas a simple rotation operation is applied to the other half. The round function is composed of a sub key addition, eight 4 × 4 S-boxes applied in parallel followed by a 4-bit permutation. LBlock requires 1320 GE.

5.3 PRESENT [25]

It meets lightweight and ultra-lightweight requirements. PRESENT cipher operates on block size of 64 bits under keys of lengths 80 or 128 bits, and it contains 31 rounds of substitution and permutation network. Each round contains three steps: Add Round Key in which Key is XOR-ed with the cipher, substitution layer consists of 4 bits substitution boxes, and permutation layer composes of bit permutations according to a table. PRESENT hardware implementation requires 1570 GE.

5.4 LED (Light Encryption Device) [26]

It aims to offer a significant reduction in hardware footprint while maintaining reasonable software performance. LED is an SPN type lightweight block cipher, which uses 4 different key sizes (64 bits, 80 bits, 96 bits, and 128 bits) with blocks of 64 bits through 32 and 48 rounds. Encryption in LED uses two main operations, add Round Key and step. Steps number depends on key size, where $s = 8$ when key size is 64 bits, and $s = 12$ for larger key sizes up to 128 bits. The implementation of LED cipher requires 966 GE for LED-64 and 1265 GE for LED-128.

Table 1 shows a comparison between the above discussed lightweight block ciphers, this comparison is based on the general characteristics of each cipher.

Table 1. General characteristics of each examined cipher

Cipher	Key size (bits)	Block size (bits)	Rounds	Structure	Area (GE)
HIGHT	128	64	32	Feistel	3048
LBlock	80	64	32	Feistel	1320
PRESENT-80	80	64	31	SPN	1570
PRESENT-128	128	64	31	SPN	1886
LED-64	64	64	32	SPN	966
LED-128	128	64	48	SPN	1265

6 Results

The ciphers have been implemented in Contiki OS [27]. Contiki is a lightweight open source Real-time operating system. It is implemented in C language, and all its programs are also in C. Contiki allows management of low-power wireless platforms based on wireless internet communication and supports IPv6 and IPv4 stack implementations. To speed up the development and testing of Contiki applications, Contiki OS provides the network simulator Cooja [28]. This tool facilitates the development and execution of user-defined applications and provides detailed information about results. In this paper, six lightweight block ciphers HIGHT, LBlock, PRESENT-80, PRESENT-128, LED-64, and LED -128 have been implemented in the Cooja simulator using the Sky mote platform, and they are evaluated according to parameters: hardware implementation, execution time, energy consumption, and key size.

1. *Hardware implementation*: Table 2 presents the hardware implementation of each compared cipher.
2. *Execution time*: Table 3 gives estimated results for the execution time of encryption and decryption operations of each cipher.
3. *Energy consumption*: Energy consumption is calculated using the formula (3) [29]:

$$\text{Energy consumption} = (\text{Energet\_value}^{*}\text{Current}^{*}\text{Voltage})/(\text{R\_TIMER}^{*}\text{Runtime}) \quad (3)$$

   Where, the energy consumption is measured in millwatts (mw). The Energest_value is the time in ticks the system is in a specific state. R_TIMER = 32768 is used to convert time from ticks into seconds. Runtime is the time interval in which measurements are taken. Current values are obtained from the sky mote datasheet [30]. The voltage value for Sky mote is 3V. Figure 1 shows the energy consumption of different ciphers.
4. *Key size*: To estimate the impact of key length on energy consumption, PRESENT and LED ciphers are tested. Figure 2.a shows the results of this comparison and Fig. 2.b shows a comparison of hardware implementation depending on length key based on values reported in Table 2.

Table 2. Hardware implementation

Cipher	Structure	Area (GE)
HIGHT	Feistel	3084
LBlock		1320
PRESENT-80	SPN	1570
PRESENT-128		1886
LED-80		1040
LED-128		1265

Table 3. Encryption and decryption execution time.

Cipher	Encryption execution time (ms)	Decryption execution time (ms)
HIGHT	151	149
LBlock	155	158
PRESENT-80	1490	1530
PRESENT-128	1550	1610
LED-80	3320	3280
LED-128	5000	4970

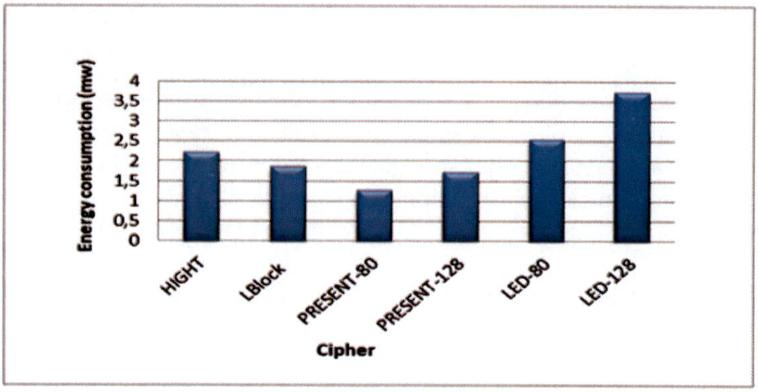

Fig. 1. Energy consumption.

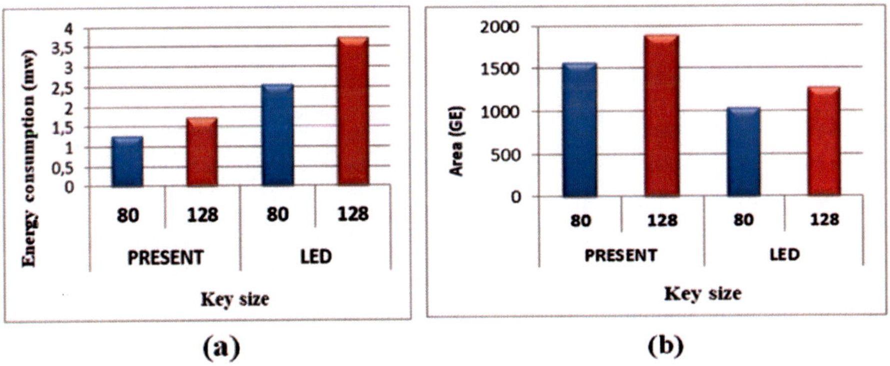

Fig. 2. Energy consumption and Hardware implementation vs key size.

7 Discussions

Lightweight cryptography targets a wide variety of highly constrained devices. Resources required for a hardware implementation are described in (GE), while lightweight algorithm performance is expressed with metrics such as energy consumption and execution time.

The comparisons were performed on a set of lightweight block ciphers according to the parameters: hardware implementation, execution time, energy consumption, and key size. Based on the results exposed in the previous section, the following explanations and conclusions can be done.

From execution time results, it has been noticed that LED and PRESENT ciphers which are based on Feistel structure perform faster than HIGHT and LBlock which are based on SPN structure. However, in terms of hardware implementation (cf. Table 2), we noted that lightweight ciphers based on SPN structure require less GE, so they are more hardware friendly. Among compared ciphers, LED cipher consumes more energy and performs very slowly, but it has less GE. Figure 2 shows that key length influences energy consumption and hardware implementation where increasing key size, from 80 bits to 128 bits for the same cipher, leads to a rise in energy consumption, GE requirements, and execution time.

Some results are in concordance to those obtained by authors in [20], namely that Feistel structure-based primitives perform faster than SPN structure, and SPN type primitives require less GE.

Results concerning LED cipher go in the same direction as those obtained by authors in [21], which evaluated a set of lightweight implementations of block ciphers in hardware and software architectures and show that LED cipher consumes high energy and is too slow.

8 Conclusion

The Internet of Things represents the next evolution of internet, and it is considered as a big leap forward in Information and Communications Technology sector. But this technology faces more critical challenges, and security is one of the major challenges. The importance of lightweight cryptography to secure IoT devices has been discussed in this paper, in addition to a comparative study over some popular lightweight block ciphers suitable for these devices. This comparison highlights energy consumption and battery lifetime. Results show that PRESENT and LED algorithms are too slow compared to HIGHT and LBlock. LED algorithm is the worst regarding energy consumption, and better in terms of hardware implementation requirements. PRESENT and LED algorithms with 80-bit key size run faster and consume low energy compared to their 128-bit key size version. We observe that key size has an impact on execution time and energy consumption; therefore, according to objects resources, the appropriate algorithm is applied.

As future work, a lightweight cipher with a small key size may be interesting. This cipher is a hybridization of two lightweight block ciphers, the first cipher is based on feistel structure, for its faster execution, and the second cipher is based on SPN structure, for its less GE requirements.

References

1. Adat, V., Gupta, B.B.: Security in Internet of Things: issues, challenges, taxonomy, and architecture. Telecommun. Syst. **67**(3), 423–441 (2017). https://doi.org/10.1007/s11235-017-0345-9
2. Kerry, A.M., Larry Bassham, M.S.T., Nicky, M.: Report on Lightweight Cryptography. NIST Interagency Report 8114 (2017)
3. Qusay, F.H., Sajjad, A.M.: Internet of Things: Challenges, Advances, and Applications. Chapman and Hall/CRC, ISBN-13: 978-1498778510
4. Sundmaeker, H., Guillemin, P., Friess, P., Woelfflé, S.: Vision and Challenges for Realizing the Internet of Things. Cluster of European Research Projects on the Internet of Things (March 2010)
5. Di Martino, B., et al.: A survey: Internet of Things reference architectures, security and interoperability. Internet of Things **1–2**, 99–112 (2018)
6. Sarangi, P.S., Smruti, R.: Internet of things: architectures, protocols, and applications. J. Electr. Comput. Eng. **2017**, 1–25 (2017)
7. Ala, A.-F., Mohsen, G., Mehdi, M., Mohammed, A., Moussa, A.: Internet of things: a survey on enabling technologies, protocols, and applications. IEEE Commun. Surv. Tutorials **17**(4), 2347–2376 (2015)
8. Buyya, R., Dastjerdi, A.V.: Internet of Things: Principles and paradigms. Elsevier, New York (2016)
9. Sahraoui, S.: Mécanismes de sécurité pour l'intégration des RCSFs à l'IoT (Internet of Things). Université Mustapha Ben Boulaid Batna 2, Département d'Informatique, Thèse de Doctorat (2016)
10. Delfs, H., Knebl, H.: Introduction to Cryptography: Principles and Applications, 2nd edn. ISBN 978-3-540-49244-3. Springer (2007)

11. Stinson Douglas, P., Paterson Maura, B.: Cryptography Theory and Practice, 4th edn. ISBN-13: 978-0367148591. CRC Press (2018)
12. Mao, W.: Modern Cryptography: Theory and Practice.: Prentice Hall (Pearson Education India), ISBN: 0-13-066943-1 (2003)
13. Sehrawat, D., Gill, N.S.: Lightweight block ciphers for IoT based applications: a review. Int. J. Appl. Eng. Res. **13**(5), 2258–2270 (2018)
14. Mohd, B.J., Hayajneh, T., Vasilakos, A.V.: A survey on lightweight block ciphers for low-resource devices: comparative study and open issues. J. Netw. Comput. Appl. **58**, 73–93 (2015)
15. Singh, S., Sharma, P.K., Moon, S.Y., Park, J.H.: Advanced lightweight encryption algorithms for IoT devices: survey, challenges and solutions. J. Ambient Intell. Humanized Comput. 1−18 (2017)
16. Jadhav, S.P.: Towards light weight cryptography schemes for resource constraint devices in IoT. J. Mob. Multimedia **15**(1), 91–110 (2019)
17. AlAssaf, N., AlKazemi, B., Gutub, A.: Applicable light-weight cryptography to secure medical data in IoT systems. J. Res. Eng. Appl. Sci. (JREAS) **2**(2), 50–58 (2017)
18. Baraa, T.H., Norziana, J., Mohd, E.R., Muhammad, R.Z.: A survey of lightweight cryptographic hash function. Int. J. Sci. Eng. Res. **8**(7), 806–814 (2017)
19. Ertaul, L., Woodall, A.: IoT security: performance evaluation of grain, mickey, and trivium-lightweight stream ciphers. In: Proceedings of the International Conference on Security and Management (SAM), pp. 32−38 (2017)
20. Hammad, B.T., et al.: Implementation of lightweight cryptographic primitives. J. Theor. Appl. Inf. Technol. **95**(19), 5126–5141 (2017)
21. Hatzivasilis, G., Fysarakis, K., Papaefstathiou, I., Manifavas, C.: A review of lightweight block ciphers. J. Cryptogr. Eng. **8**(2), 141–184 (2017). https://doi.org/10.1007/s13389-017-0160-y
22. Batina, L., et al.: Dietary recommendations for lightweight block ciphers: power, energy and area analysis of recently developed architectures. In: International Workshop on Radio Frequency Identification: Security and Privacy Issues, pp. 103−112 (2013)
23. Hong, D., et al.: HIGHT: a new block cipher suitable for low-resource device. In: International Workshop on Cryptographic Hardware and Embedded Systems, pp. 46−59 (2006)
24. Wu, W., Zhang, L.: LBlock: a lightweight block cipher. In: International Conference on Applied Cryptography and Network Security, pp. 327−344 (2011)
25. Bogdanov, A., et al.: PRESENT: an ultra-lightweight block cipher. In: International Workshop on Cryptographic Hardware and Embedded Systems, pp. 450−466 (2007)
26. Guo, J., Peyrin, T., Poschmann, A., Robshaw, M.: The LED block cipher. In: International Workshop on Cryptographic Hardware and Embedded Systems, pp. 326−341 (2011)
27. Dunkels, A., Gronvall, B., Voigt, T.: Contiki-a lightweight and flexible operating system for tiny networked sensors. In: 29th annual IEEE international conference on local computer networks, pp. 455−462 (2004)
28. Osterlind, F., Dunkels, A., Eriksson, J., Finne, N., Voigt, T.: Cross-level sensor network simulation with cooja. In Proceedings. 2006 31st IEEE Conference on Local Computer Networks, pp. 641−648 (2006)
29. Velinov, A., Mileva, A.: Running and testing applications for Contiki OS using Cooja simulator. In: International Conference on Information Technology and Development of Education – ITRO 2016 (2016)
30. Moteiv Corporaton. Tmote sky: Datasheet (2006). http://www.crew-project.eu/sites/default/files/tmote-sky-datasheet.pdf

Web 2.0 Education Tools as Support in Digital Marketing: Tungurahua Case Study

Sonia Armas-Arias(✉), Angélica González-Sánchez, Johanna Monge-Martínez, and Ruth Infante-Paredes

Facultad de Ciencias Humanas Y de La Educación, Grupos de Investigación Marketing C.S. and Language and Education, Universidad Técnica de Ambato, Ambato, Ecuador
{sp.armas,am.gonzalez,jg.monge, rutheinfantep}@uta.edu.ec

Abstract. This research aims to assess the impact that information and communication technologies have as support in tourism against the COVID-19 pandemic, in this way a study is developed within Digital Marketing in institutions of the tourism industry of the province of Tungurahua and how they have contributed to the dissemination of information in the field of tourism. This research was based on the implementation of a structured questionnaire of 45 Likert-scale questions, which measure the use of digital, technological and communication tools and strategies that tourism companies have used with the aim of improving their promotion with tourists based on a new normality that has been applied in the face of the COVID-19 pandemic, the instrument designed and applied, obtained a level of trust of 0.831 by means of the Cronbach Alpha statistic taken from a sample of 320 institutions on the field of tourism. The statistic test for this research was Kolmogorov Smirnov, which allowed knowing the concordance between the data and the specific theoretical distribution, which was to check whether Digital Marketing is considered a supporting tool in tourism in the province of Tungurahua; with the final result obtained, the alternative hypothesis proposed in the research was accepted. The result of the project admitted to using auxiliary technological tools to boost current generations' interest and knowledge of Digital Marketing through the distribution of information in the field of tourism.

Keywords: Digital marketing · Tourism · ICT · Tourism promotion · COVID-19

1 Introduction

In Ecuador, the pandemic has caused economic and social problems; also, tourist incomes have fallen by 76.4% less than the previous year. However, the picture is not so encouraging because of the closure of borders and restrictions on different tourist attractions, this has caused a block in tourism, on the other hand, one of the strategies is to reduce costs in services and products and the reopening of borders gradually in order to generate revenue [1].

M. E. Auer and T. Tsiatsos (Eds.): IMCL 2021, LNNS 411, pp. 1052–1061, 2022.
https://doi.org/10.1007/978-3-030-96296-8_96

In Tungurahua province, the tourism committee has approved the creation of a biosecurity seal, which will encourage the reactivation of tourism activities safely. The measures taken seek to prevent contagion between the worker and the client, in other words, it is must for the nine cantons to implement and improve the different health protocols that undoubtedly generates confidence in the tourists and ensures that tourism is safely reactivated [2].

The World Tourism Organization (2013) mentions that tourism is "the activities that people carry out during their travels and stays in places other than their natural environment for a consecutive period of time less than one year. It can be for leisure, business and other reasons." It can therefore be ruled out that the tourist is a fundamental part of energizing the economy and thus improving the standard of living of the inhabitants of the places visited. It also favors the conservation of historical and natural places, as well as strengthens local customs and traditions such as language, folklore and festivities [3].

Tourism is one of the most significant economic activities in the world, because it contributes to economic, human, and social development. It also encourages the improvement the life quality of the population. One of the main points to tackle is to promote collaborative development including the government and the population. In other words, the contribution will create a safe and reliable environment for visitors, which allows tourists satisfaction since it is one of the important aspects to achieve a significant position of tourism.

Currently, travel is not only meaning relaxation and enjoyment, on the contrary, it is sought to meet different needs linked to knowledge or to know the different ways of living in other cultures or countries. Cultural tourism has generated a dynamic in which communities can find different ideas or projects that allow them to produce goods and services, and thus achieve local development by improving infrastructure and services, this will achieve social, economic and environmental benefits that respect cultural heritage [4].

2 State of the Art

According to [5], "Information and Communication Technologies (ICTs) have revolutionized the way cultural heritage is promoted and interpreted in the world of tourism." The tourists' motivation is to know the culture, and also stands out for generating and requesting information instantly, also stays connected all the time and creates opinions about the attractions. Even though, the website has changed over, nowadays it is vital for the promotion of cultural heritage because cultural tourism generates a lot of information which favors the creation of innovative supplies.

According to [6] "The new Coronavirus, COVID-19, has caused an unprecedented global crisis, generating a negative impact on the tourism sector." Worldwide, reactivation processes will be difficult, however, governments have implemented a variety of strategies for tourism revival such as investment in sanitization and biosecurity protocols, tourism promotion that will let people start travelling gradually, and one of the most important is to promote local tourism in order to start economic revival and improve the life quality of the population [7].

Ecuador is preparing for tourism reactivation with some strategies such as the control, verification and identification of tourist establishments in the country which must comply with different biosecurity standards, another important strategy is the training of tourism service providers, incentivizing local tourism and awareness campaigns to precautionary the health and well-being of tourists [8].

Digital marketing is defined as a way to market products and services through digital means. With the use of technology, new options for the internet user arise since they receive information about what they like or what they are interested in. It can be emphasized that for much less money, a greater number of customers can be reached, and potential customers can be managed so it would be easier to get them as current consumers, which will serve to generate trust, keep the public informed and manage to be shared with other consumers.

Digital marketing has evolved over the years, becoming an instrument for marketing processes by using techniques to devise business models and strategies to identify opportunities in international markets [1, 4]. However, various companies use it as a marketing and distribution strategy for a product or service. Therefore digitization, Internet, and the management of different technological modalities have been linked to the developing new lifestyles, different consumption habits, and new ways of marketing a product.

The Honorable Provincial Government of Tungurahua (2020) [9], mentions that the province of Tungurahua constantly carries out training to achieve the reactivation of the tourism and the economic sector, it is also mentioned that one of the main strategies is the training in the field of digital marketing to achieve digital tools to work optimally and will help digitize the processes of companies as well as to develop strategies in social networks management, content promotion, and the spread of different products and services offered [10].

Companies have used the Information and Communication Technologies (ICT) or the Internet to promote their tourist products or services. Technology at this time has changed the perspective of life since people can check the price, transportation, places or tourist sites of interest at high speed. However, it can be mentioned that digital marketing has been a strategy to make the customer be part of the company by acquiring information of their tastes, what they need, and faster ways to interact [11].

According to [12], tourist activities have been maintained thanks to the development of new information technologies, because the tourist is far from the product or service offered, therefore, it is essential that the marketing channels used will be entertaining, interactive and the information provide must be coherent and trustful, which will encourage to generate a positive impact and thus achieve potentializing tourism in the sector [13].

Tourism has been changing constantly, which have not only evolved in the way it is marketed, but also tourists today have different demands and tastes, currently advance technology has fostered a new way of seeing tourism as knowing the attractiveness before traveling, which has created a different way of promoting the site as by social networks, blogs, and internet search engines that has favored to establish a relationship with the nearest tourist, manage to combine real data, and improve the tourist experience [14].

Digital marketing establishes a support tool with different functions and processes for tourism companies, it is also a means of communication that acts as an aggregate, which favors the integration of tools to promote in an effective and simple way. In other words, it makes easier for the company to get information accurately and quickly, as well as providing information to the customer, and being able to assist them by creating customer loyalty [15].

The COVID-19 pandemic has caused a number of economic and social consequences, digital marketing has now changed from a traditional method, to being one of the main drivers of companies, in which its main objective is to attract customers with useful information and data, and has also favored the creation of strategies and new ways of doing business that allows to compete in an unstable market because of the pandemic [16].

Tourism has been one of the industries most affected by COVID-19, today, one of the main tools is digital marketing which has brought exponential and resounding growth, has allowed a commercial development of companies, and have favored generating a different way of marketing their services and products quickly through digital means [17].

The expansion of COVID-19 has caused the restriction of tourist sites causing chaos in the tourism industry, so digital marketing is one of the most important factor that has benefited tourism to adapt to the new way of promoting its products and services, and which will help to promote digitization to create innovative solutions and invest in digital knowledge in order to achieve the reactivation and creation of programs that generate trust in travelers [16].

The pandemic has caused changes such as keeping the distance because of the capacity allowed in companies that offer a service has generated a new way to demonstrate the customers' needs such as to strengthen online advertising, innovate services, workplaces with healthcare processes, and virtual platforms to book, products online, and home delivery of the products [18].

[19] mentions "in Ecuador there was a 30% increase in demand for internet services in recent months." So, you can see that people are spending most of their time on the internet as on social networks to watch videos, listen to music, play games, watch news, do telework and most importantly watch promotions which favors tourism by generating revenue, improving the economy of the sector, promoting tourist attractions and creating a strong brand to increase visitor income.

3 Methodology

The methodology used for this research is of prospective descriptive type the same one that focused on obtaining accurate information from tourism, the research approach was qualitative, which favors the description, collection and analysis of data and information in order to describe what has happened with tourism. Once the qualitative information was carried out, a quantitative approach was used since it favored to the numerical data, through the collection and analysis of data. A survey of 45 questions at the Likert scale was carried out, which facilitated the analysis of the current situation on tourism in the pandemic as well as to generate a strategy for the reactivation of tourism and the economy.

The statistic applied used for the validation of the instrument in the research is Cronbach Alpha, the result obtained was 0.831, having a reliable measuring instrument (Table 1).

Table 1. Realiability statsitics

Cronbach Alpha	Cronbach Alpha based on typified elements	N de elements
,831	,815	45

4 Results

The results obtained for this research are based on the selection of the most representative questions of the validated instrument, resulting in the verification of the hypothesis (Table 2).

Table 2. Do you consider that the COVID-19 pandemic has affected cultural tourism in Ambato?

Alternative	Frequency	Percentage
No affected	1	0%
Rarely affected	1	0%
Ocationally affected	1	0%
Largely affected	6	2%
Affected	31	10%
Severely affected	115	36%
Highly affected	165	52%
Total	320	100%

Figure 1 shows that 52% corresponding to 165 companies mentioned that tourism has been highly affected, with 36% equivalent to 115 mentioning that it is severely affected, the 10% that corresponds to 31 respondents mention that tourism was affected, and with 0% equivalent to 1 respondent, mentioned that tourism has not, rarely, or occasionally been affected. In addition, it is found that 52% of respondents mentioned that the pandemic has largely affected tourism. During this period of time many tourist sites have to close their doors and has also been affected by not only the closure of museums, parks or sites of interest, but also that it has largely slowed the economy because many communities lived off this activity. Even though, many sites have kept their doors open respecting the rules to avoid contagion, tourism has not been fully recovered (Table 3).

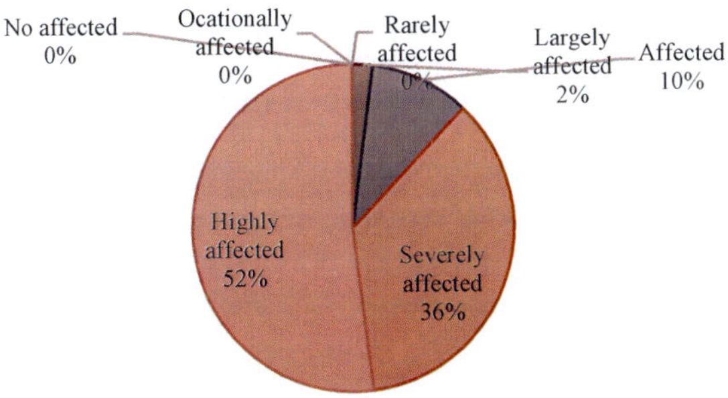

Fig. 1. Affected by COVID-19

Table 3. How long do you think the tourism industry will be recovered from the COVID-19 pandemic?

Alternative	Frequency	Percentage
Less than 1 month	0	0%
1–6 months	37	11%
6–9 months	49	15%
9–12 months	80	25%
10–18 months	31	10%
More than 18 months	79	25%
Not sure	44	14%
Total	320	100%

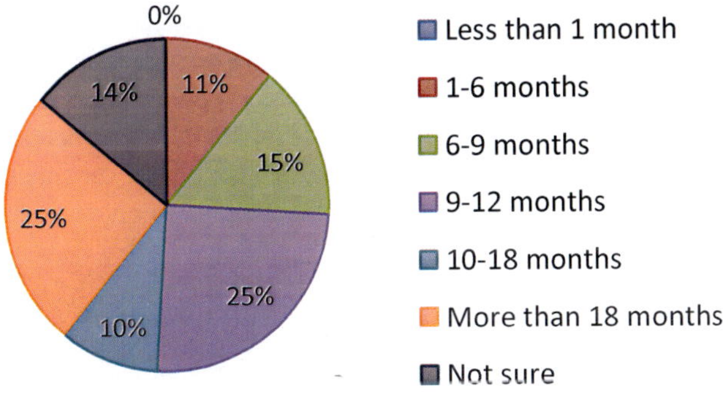

Fig. 2. Tourism recovery

Figure 2 indicates that 25% equal to 80 industries mention that between 9 and 12 months tourism will be recovered, the other 25% that equates to 79 companies mention that it will take in more than 18 months, the 15% that equals 49 industries mention that it will take between 6 to 9 months to recover and with 11% corresponding to 37 industries mention that it will be between 1–6 months. Based on data obtained

25% equivalent to 80 companies mention that recovering between 9 to 12 months and with the other 25% equivalent to 79 companies indicates that it will be in more than 18 months. These results manifest that it will take a long time for tourism industry to be recovered because people had changed their lifestyle, and the pandemic has affected the mobility of people, which has caused many closures that has severely affected tourism (Table 4).

What is your opinion about tourism and digital marketing in Tungurahua province?

1 = Totally disagree; 2 = Pretty much disagree; 3 = Rarely disagree; 4 = Disagree; 5 = Okay; 6 = Pretty agree; 7 = Totally agree.

Table 4. Opinion about tourism and digital marketing in Tungurahua

	1	2	3	4	5	6	7	Total
The spread of tourism is important	1	0	0	8	34	100	177	320
It is necessary to digitize the patrimonial assets	0	0	1	8	33	183	95	320
The information obtained in digital media about tourism is positive	0	1	3	9	52	130	125	320
Social media is suitable for the tourism promotion in the province	0	1	2	7	33	156	121	320
Digital marketing is a tool that favors tourism in the province	1	2	1	8	26	130	152	320

Figure 3 portrays that 177 companies fully agree that the promotion of tourism in the province is very important, in order to encourage tourist attractions and achieve the revival of tourism and the economy, on the other hand, with 152 industries mention that digital marketing is a tool that favors tourism and that facilitates the creation of interactive and interesting content. 183 companies indicate that they quite agree to generate content in digital media, which means to digitize the patrimonial assets, 156 companies that indicate that they quite agree that social networks are suitable for the promotion of cultural tourism. Bearing in mind that social media currently plays an important role in people's daily lives because they allow people to interact with tourists of different ages and with people of other nationalities (Table 5).

Figure 4 illustrates that 61% of 194 industries thinks that the social network with the greatest influence is Facebook, followed by the 16% equivalent to 51 industries agree that is whatsApp, 12% that corresponds to 37 companies go for Instagram, 9% corresponding to 30 companies mention that their favorite social network is YouTube and 2% corresponding to 8 industries mention that it is Twitter. It is mentioned that one of the most influential social networks, with the highest percentage is Facebook which is used worldwide not only to keep in touch with people from different places, but also provides information about tourist sites, countries, services, and also the sale of products.

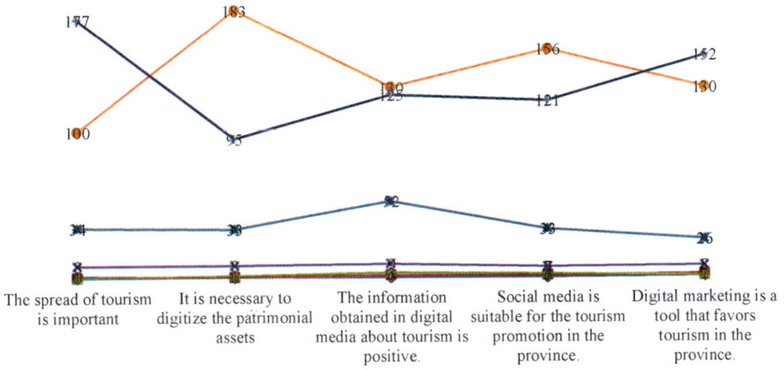

Fig. 3. Opinion about tourism and digital marketing in Tungurahua

Table 5. Social networks do you use the most to promote your business

Alternative	Frequency	Percentage
Facebook	194	61%
Twitter	8	2%
Instagram	37	12%
YouTube	30	9%
LinkedIn	0	0%
WhatsApp	51	16%
Snapchat	0	0%
Total	320	100%

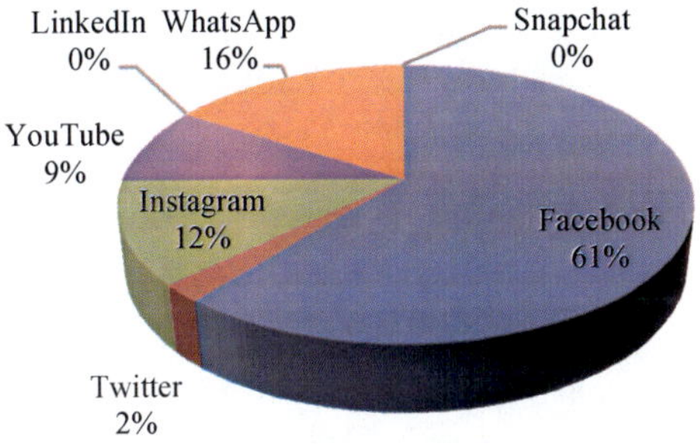

Fig. 4. Social networks to promote your business

5 Conclusions

Nowadays, Technology is one of the main tools for the world to stay connected, in the research carried out it is mentioned that technological media such as social networks, blogs and websites are the most used to communicate with others and to obtain information from different tourist sites in the world. In addition, digital marketing is important for the promotion of tourism and the digitization of tourist attractions. Cultural tourism worldwide is one of the sectors most affected by the COVID-19 pandemic, it should be emphasized that the province of Tungurahua has been harmed by different aspects such as the state of emergency throughout the country, the reduction of tourists to the different tourist spots of the province and the closure of tourist attractions. On the other hand, governments worldwide have gradually opened their borders, which has caused the revival of tourism and the economy, however, in Ambato different strategies have been applied for the reactivation of tourism including implementing biosecurity protocols, technological applications and digital marketing that favor the reactivation and promotion of tourism.

Acknowledgements. A special thanks to Technical University of Ambato, Dirección de Investigación y Desarrollo (DIDE) for the great support to the project: "IMPACT OF DIGITAL MARKETING ON THE REACTIVATION OF TOURISM IN TUNGURAHUA POST COVID-19 PROVINCE.", handed in to Dirección de Investigación y Desarrollo DIDE, at Techncial University of Ambato y Approved by Investigation committee under Resolución Nro. UTA-CONIN-2020-0322-R.

References

1. Korstanje, M.: El COVID-19 y el turismo rural: una perspectiva antropológica. Dimensiones Turísticas **4**, 179–196 (2020)
2. Moreno, L.: Decreto presidencial Nº 1017 (2020). https://www.defensa.gob.ec/wp-content/uploads/downloads/2020/03/Decreto_presidencial_No_1017_17-Marzo-2020.pdf
3. Mendes, G., Augusto, A., Gándara, J.: Innovación en la promoción turística en medios y redes sociales. Un estudio comparativo entre destinos turísticos. Estudios y Perspectivas en Turismo **22**(1), 102–119 (2013)
4. Gallego, I., Font, X.: Changes in air passenger demand as a result of the COVID-19 crisis: using Big Data to inform tourism policy. J. Sustain. Tourism **29**, 1–20 (2020)
5. Carr, A.: COVID-19, indigenous peoples and tourism: a view from New Zealand. Tourism Geogr. **22**, 1–12 (2020)
6. Ministerio de Turismo. Ministerio de Turismo (2020). https://www.turismo.gob.ec/la-gastronomia-ecuatoriana-al-alcance-de-todos/
7. Santamaría Freire, E., Bayas Vásconez, F.M.: Efecto económico de la actividad turística en la provincia de Tungurahua, Ecuador. Ciencias Administrativas **11**, 1–16 (2018)
8. INEC: Proyección de la población por provincias 2007, cantones y parroquias según grupos programáticos (2018). https://www.paho.org/ecu/index.php?option=com_docman&view=download&category_slug=politicas-sistemas-y-servicios-de-salud&alias=49-proyeccion-de-la-poblacion-por-provincias-cantones-y-parroquias-segun-grupos-programaticos&Itemid=599
9. Observatorio Económico y Social de Tungurahua: Observatorio Económico y Social de Tungurahua (1 Febrero 2020). https://blogs.cedia.org.ec/obest/wp-content/uploads/sites/7/2020/06/Turismo-en-Ecuador-Alojamiento-y-servicios-de-comida.pdf
10. Ministerio de Turismo del Ecuador: GEO VIT, Geo portal. Innovación turística 4.0 (31 Agosto 2020). https://servicios.turismo.gob.ec/index.php/turismo-cifras

11. Gobierno Provincial de Tungurahua: Agenda Tungurahua 2019–2021 (2021). http://www.tungurahua.gob.ec/images/archivos/transparencia/2020/Agenda_Tungurahua2019-2021.pdf
12. La Secretaría Técnica de Planificación: Planifica Ecuador, http://www.inteligenciaproductiva.gob.ec/ (09 Febrero 2021). http://www.inteligenciaproductiva.gob.ec/archivos/Eval%20Soc%20Econ%2011%20sep-.pdf
13. Sigala, M.: Tourism and COVID-19: impacts and implications for advancing and resetting industry and research. J. Bus. Res. **117**, 312–321 (2020)
14. Rodríguez, K., Ortíz, O., Quiroz, A., Parrales, M.: El e-commerce y las Mipymes en tiempos de Covid-19. Revista Espacios **41**(42), 100–119 (2020)
15. Noziglia, M., Gómez, M.: SEO en tiempos de COVID 19, Beneficios y recursos útiles para ser eficaces en este nuevo paradigma. In: UM, pp. 1–6 (2020)
16. Rutynsky, M., Kushniruk, H.: The impact of quarantine due to COVID-19 pandemic on the tourism industry inLviv (Ukraine). Probl. Perspect. Manage. **18**(2), 194–205 (2020)
17. Sariego, I.: Los retos del turismo cultural en la ciudad de Santander para convertirse en un destino seguro Pos-COVID-19. Estudios Turísticos **219**, 59–70 (2020)
18. Wen, J., Wang, W., Kozak, M., Liu, X., Hou, H.: Many brains are better than one: the importance of interdisciplinary studies on COVID-19 in and beyond tourism. Tourism Recreation Res. **46**, 1–4 (2020)
19. Visentin, M., Silveira, R., Cappiello, G., Casoli, D.: Sensing the virus. How social capital enhances hoteliers' ability to cope with COVID-19. Int. J. Hospitality Manage. **94**, 102820 (2021)

Check for updates

Reflecting on the Remote Control of the EI-Edurobot Through an IS and a Mobile Application

Dimitrios Ziouzios, Dimitrios Rammos, Tharrenos Bratitsis(✉), and Minas Dasygenis

University of Western Macedonia, Florina, Greece
bratitsis@uowm.gr, mdasyg@ieee.org

Abstract. Robots play a significant role in modern life in a wide range of activities. Learning by utilizing robots is an innovative teaching approach through which a variety of disciplinary areas can be treated. The current work refers to the development of a robot, initially designed for children with autism spectrum disorders but was then utilized in various studies related to empathy development towards various aspects. This paper reports upon a study regarding environmental empathy. For handling the robot, a mobile application and an information system were developed. This paper reflects on the perceptions of 12 teachers who used the robot within this context, on the functionalities and the user interface of the digital working environment. The reflections were positive and comments were made on mechanical aspects of the robot mainly, indicating that the mobile application and the IS are designed in an appropriate format so as to be as transparent as possible for the end user.

Keywords: Educational robotic · Empathy · Mobile app

1 Introduction

Robots play a significant role in many different aspects of modern society, and they have been used in a wide range of applications in everyday life, in entertainment, at work, and other parts of human activity. Education is also affected. In fact, in recent years, there is a growing interest on the effective use of Information and Communication Technologies (ICT) in the learning process both in the fields of formal/typical and in special education, including the utilization of robots.

In the 1980s, Educational Robotics (ER) began to show up in the context of technology use mainly through the Logo language. The use of robotics in education is mainly based on the theory of classical constructivism and constructionism, as developed by Papert [1]. Generally, ER refers to the teaching practice with the use of robots, where the students with the help of their teacher, develop new knowledge in many fields or acquire knowledge about the robots themselves [2]. Moreover, ER can be easily utilized with STEM (Science, Technology, Engineering, and Mathematics) and other learning methods, such as literacy and Arts through teamwork, problem solving and cultivation of cooperative and critical thinking skills [3].

M. E. Auer and T. Tsiatsos (Eds.): IMCL 2021, LNNS 411, pp. 1062–1071, 2022.
https://doi.org/10.1007/978-3-030-96296-8_97

Many researchers systematically studied the literature and recorded the trends in educational robotics used as tools for learning STEAM or other subjects [7, 8]. The influence of robots could be divided into four major categories: a) cognitive, b) conceptual, c) linguistic, and d) social (collaborative) skills. There are robots available for almost all levels of education, from early childhood to adult education like Buddy [4], Moxie [5], and STIMEY [6].

Teachers can use ICT to transform the educational process in a more attractive and lively experience. According to research highlights, in special education the results of the use of ICT are even more promising [9]. The majority of research in the field of special education concerns individuals with Signs and Symptoms of Autism Spectrum Disorders (ASD), as they interact with and respond more effectively to technological products [10]. ICT also helps people with ASD to stay focused for a longer time in a structured, controlled and stable environment with limited stimuli, reducing the chances of feeling upset and lack of concentration caused by social circumstances [11].

2 Our System

The system presented in this paper consists of three main parts. The first one is the robot, the second the Information System (IS) and the third one the mobile application. The basis of the implementation idea was to use low-cost components and open-source software.

2.1 The EI-EDUROBOT

An educational robot, the EI-EDUROBOT [12], was developed, as shown in Fig. 2, aiming to develop empathy, which is a critical component of social knowledge. A literature review was carried out focusing on educational robotics, empathy training, and ASD especially in younger ages, between 4 and 9 years old, but it can also be defined as a multi-purpose teaching robot that can be used for a variety of purposes.

The robot chassis was created with the goal of providing stability, durability, and the required center of gravity for mobility. The interior, which holds the electronics, is shielded by plastic covers to restrict access. A 3D printer was used to manufacture all of the robot's pieces from PLA filament. The robotic system has two double-jointed arms. They are made so that various components can be added to the robot's hands, allowing it to execute more functions and activity scenarios. It includes an easy to learn and use programming language and administration plat-form, as well as many sensors. It incorporates ultrasonic, IR sensors, touch sensors on the arms and head, and it is equipped with a camera which is able to detect objects utilizing an object detection algorithm. Additionally it integrates a micro-phone, speakers and a 7″ touch screen with a resolution of 800 × 480 pixels. The EI-EDUROBOT is addressed to youngsters, and extra care has been taken to ensure both their own and the robot's safety. The robot is based on a Raspberry Pi 4 model B programmable unit, which allows for a lot of design freedom. WiFi and Bluetooth 5.0 communication components are included, as well as Gigabit Ether-net and USB 3.0 connection.

Fig. 1. The EI-EDUROBOT.

2.2 The Information System Platform

An information system was designed to coordinate the transmission of information from the user's device to the robot in order to achieve communication between the user and the robot. The web application and the REST API (Application Programming Interface) make up this information system. To achieve the best result combining security and user-friendly interface, we utilize the MEAN software stack which is used to develop all aspects of the system. The MEAN software stack consists of MongoDB, Express, Angular and NodeJS. The database was replaced with MySQL due to system requirements, without compromising the information system's performance. The advantages of the MEAN architecture include the fact that it is an open-source framework that can be used as a stand-alone solution in a complete application and that it includes extra frameworks, libraries, and reusable modules to speed up development. The system is highly scalable and relatively easy to maintain. It is also feasible to create new applications that can support the system's capabilities by using JSON files for data transmission be-tween the user, the API, and the robot.

The utilization of these technologies provides many possibilities. The main idea is the user friendly interface, so any teacher or user can handle the robot without the need of any special, technical knowledge. Having this in mind a mode where the teacher can use the IS and a microphone to talk and command the robot from distance was developed. For example, one teacher can be with the robot and the students in the classroom and another teacher could be in another room and via a WiFi connection interact through the microphone and speakers combo, move the robot, watch and listen to kids in real time. This mode does not require any special knowledge, except the connection of the robot to the same WiFi network as the remote device (e.g. mobile device) which would be used to log in to the IS. With this setup the children feel that they are communicating with the robot directly and the teachers can implement a variety of educational scenarios which require free form verbal interaction.

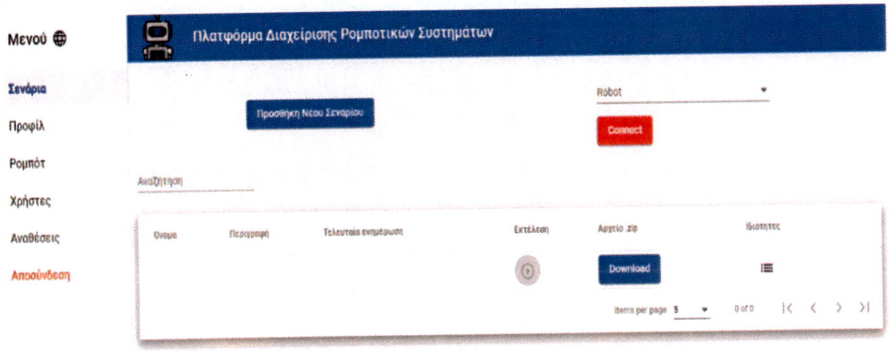

Fig. 2. The screen for uploading a scenario.

Moreover, the robot has eight pre-installed scenarios but the users can develop their scenarios, which should be written in the Python programming language and uploaded through the IS to the robot. This function provides the users who are familiar with the basics of programming with the opportunity to develop and use their own scripts, without any limitations. Furthermore the users can exchange scenarios they create. Figure 2 presents the main screen through which users can upload their own scenarios. On the top-right corner the user can connect to the robot or if there are registered more than one robots, he/she can choose which one he/she prefers. On the top-left corner there is a button which the user can upload the scenario by just clicking it. The necessary steps include setting the project title and uploading the zip file which contains the code. When the file has been uploaded the IS checks if the files are correct and everything is suitable compared to the robot's architecture, then the user can run the scenario. If there is a problem, the IS returns an error message informing the user to upload it again or to change the uploaded file.

2.3 The Mobile Application

Except the Information System Platform, a mobile application for the Android Operating System was developed (Fig. 3). Once the app is installed on a device, it can be connected directly through Bluetooth or with the use of the Rest API (Application Programming Interface) and send essential information about the robot's smooth movement and operation.

The Android application, in particular, allows for the direct movement of the robot's hands and head, as well as the execution of any pre-configured script, including a sequence of movement instructions. This way, users are able to control and employ the humanoid robot in a range of activities, even handling it through the proposed system in the form of a remote control device.

On the main screen as shown on Fig. 3 there are four main buttons. The first one, the "folder icon" contains the uploaded scenarios (stored in the cloud) and the user can choose which one to run. The second, the "controller icon" leads to installed games for children to play on the robot's face/screen and the third icon, the "remote control" enables the remote-control mode for the robot. The final icon is the settings where the user can set up the connection with the robot, set the audio volume, define a name for the robot and to informed about the version of the software.

Fig. 3. Main screen of the Mobile Application and robot's home screen.

The Microsoft Xamarin framework is used to create the mobile application, which is written in C# and XAML. Using plugins and specialized APIs, native-level app functionality can be achieved and all hardware compatibility concerns with Xamarin can be addressed, working with common devices across all prominent platforms. Xamarin allows linking with native OS (Operating System) particular libraries in addition to providing access to platform-specific APIs. With less overhead, this enables better customization and native-level capability.

The mobile application is very easily used by teachers and children. Teachers can efficiently run any script using only a mobile device such as a common smartphone or tablet. On the other hand, children can control the robot to fulfill a task in a scenario, making the lesson more enjoyable and motivating.

3 Educational Robotics in Elementary Schools

The use of educational robotics in elementary schools is becoming increasingly popular, both through programs implemented by teachers and workshops organized by parents' associations [13]. Many teachers are being trained on the subject which is constantly evolving and enriched due to the rapid technological development [14]. Various studies indicate that the impact of educational robotics pro-grams on children is evident both in terms of knowledge acquisition and skills development such as creativity, critical thinking, communication, computational thinking and collaboration in the case of group projects [15, 16].

At the same time, other studies [17, 18] highlight teachers' positive feedback regarding the use of educational robotics in the curriculum. More specifically, the increase of students' interest and active participation in relation to the average in traditional teaching, is strongly highlighted. In addition, the interconnection of robotics with a range of school subjects such as New Technologies, Physics and Mathematics is mentioned [15, 18]. The use of robotic systems gives students the opportunity to take initiatives, to think in detail, alternatively and synthetically in order to reach the desired result [16]. This makes it easier for teachers to shape their teaching method with an emphasis on counselling and problem-solving strategies, as opposed to simply quote and deliver information.

This, of course, depends to a large extent on the capabilities and characteristics of robotic systems. This paper presents the utilization of a robotic system designed and created with an educational orientation, taking into account the principles of experiential learning and the real conditions of elementary schools. Thus, in addition to the construction of the robot itself, emphasis was placed on the development of the remote use and programming application for smart mobile devices. This application aimed at enhancing the autonomy and attractiveness of the robot handling while giving operators greater possibilities to set parameters and diversify the teaching scenario.

3.1 Testing EI - Edu Robot and the Application in Practice

The pilot testing of the educational robot in real conditions was implemented during the 2020 – 2021 school year on a small scale, due to COVID-19 health protection measures. In particular, teaching activities were carried out in 6 classes of an elementary school in the region of Attica, with the participation of a total of 120 students aged 10 to 12 years. All activities were organized in small groups of students, in two-hour interventions per group. Two educational scenarios which focused on raising students' awareness of the phenomenon of climate change and familiarizing children with the benefits and risks of artificial intelligence were designed by the research team.

According to the scenarios' scripts, the educational robot initially introduced the student groups to the theoretical frame on the subject through storytelling. The activation of the robot's commands and moves was carried out remotely by the members of the research team, via the corresponding application in their tablet. Many teachers of the school were involved in this process as participants and observers. All multimedia material was projected through the robot's screen/face and sound systems, while utilizing motion and touch sensors through the application. After the completion of the theoretical framework various teaching missions were assigned to students by the robot. In addition, students worked with their teams and announced the results of their work back to the robot. Students' stories were recorded by the robot and displayed in the researchers' tablet through the application.

Taking these stories into account, prerecorded commentary stories were activated by researchers through the application. The stories were stored in the cloud, accessed through the application in real time. Based on the children's choices, the teacher could select the appropriate story continuance in order to proceed with the story-line. Alternatively, researchers recorded their comments in real time and displayed them in the robot's screen again through the application. Consequently, the robot responded to the children and commented on their results, directly inter-acting with them. Thus, the members of the research team together with teachers played an active role in the implementation of the teaching scenario as they recorded the stories or decided which of the prerecorded ones were appropriate to be displayed as feedback to the students' team work. All this process was carried out remotely, from a different room through the application installed in researchers' tablet. This caused realism and plausibility to the teaching activities.

To sum up, the robot's remote operation and setup application formed a unique educational opportunity for interactive communication and collaboration both for teachers and students, adding educational value to the messages and digital stories displayed on its screen.

4 Teachers' Feedback

This paper emphasizes the pedagogical usability and value of the application for the remote use of the robot. For this reason, particularly important was the opinion of the school teachers where the pilot implementation of the educational scenarios was carried out. Student's evaluation on the robot and the educational scenario overall was also organized in detail and will be presented in future work since the current paper focuses on the remote control application that was used by teachers. The latter undertook the role of an independent and unaffected judge on the advantages and disadvantages of the application while utilized. Specifically, 12 teachers took part in the evaluation process. They initially followed the implementation of the scenarios in the classroom and participated in the process of shaping the robot's digital stories in real time. The members of the research team explained the construction philosophy and the programming process in order for them to gain a complete understanding of the whole technological design and the proposed educational activity. Their active participation in the implementation of the scenarios helped them become familiarized with the remote customization of the movements and behavior of the robot through the application. After completing the pilot teaching activities for the school's 120 children, teachers took full control of the robot and downloaded the app to their smart devices. Thus, they had the opportunity to fully test all the possibilities of the robot and the application in practice and to experiment on their possible exploitation in their own educational scenarios. There was a great deal of interest in the use of robot sensors which affect robot's reactions to children's possible behaviors and movements. Finally, all twelve teachers took part in a mini collaborative workshop in which they created three educational scenarios for the educational robot on their own. These scenarios will be tested in the following school year.

An open and informal process in the form of a semi-structured interview was selected for recording the teachers' evaluation. All teachers of the school unit took part in this process, constituting the already mentioned research population of 12 teachers. They had all taken part in the pilot studies and had the opportunity during them to work with the robot but also observe at least one of their colleagues in doing so. The members of the research team initially asked some questions to the teachers who had actively taken part in the project. These questions were supplemented by questions posed by the rest of the school teachers who had not taken part in the pilot program. In addition to these questions, teachers were given the freedom to speak on their own initiative and comment on any information they perceived during the implementation process or to submit their ideas for improvements in terms of future use of the educational robot and the handling application in elementary level teaching activities. The main responses as recorded in the evaluation process are summarized as follows:

- *Increase of children's interest due to the potential of the application.*

All teachers agreed that the ability to remotely use the educational robot significantly impressed the children and added great interest to carefully watching digital stories and then actively taking part in the missions assigned to them. More specifically, the interaction developed between students and the robot enhanced their concentration and commitment by creating a creative and realistic learning setting. Moreover, all the school teachers were strongly impressed by the ability to record the stories projected in the robot's screen in real time by people who were in a different room from the class.

As they noted on "… *we have not come across this form of creative interaction in other robotic systems that we have seen or used in our teaching. In this way, students treated the educational robot as an equal speaker, showing great interest and curiosity about what he had to say to them and especially about the answers and comments to their own questions….*"

- *Great usability in the handling of the training robot.*

Teachers who downloaded and tested the robot's setup and handling application on their mobile phones or tablets noted that they found it particularly easy to use and understandable, while they did not encounter any serious difficulties apart from some kind of connection disorders over the wireless network. They described the application as simple and suitable to be used by teachers with limited knowledge on ICT. It was noted, in fact, that this application could easily be used in the future by students themselves, in the design and implementation of their own educational activities. In addition, some suggestions to enrich the application with new features were proposed, such as viewing what the robot sees on the operator's device using the camera which is integrated in the robot's head.

Finally, an improvement proposal was to increase the scope of connectivity between the control devices and the educational robot, in order to ensure greater independence and isolation, especially in cases in which the application operator records the digital stories and these are projected in robots' screen in real time.

5 Conclusions

The remote use application of the educational robot EI-Edurobot enhanced its pedagogical value as it gave immediacy and interactivity between the robot and the students and increased their participation in the achievement of teaching goals [15]. Teachers found the application simple and easy to use and suitable for teachers and students regardless of the level of knowledge and familiarity with ICT. The observations of teachers and the research team converge on the fact that such applications must be open and suitable to work without problems on smart mobile devices. This ensures the smooth implementation of the teaching scenarios without interruptions and malfunctions that cause confusion and frustration to children.

In addition, both the members of the research team and the school teachers mentioned the comparative superiority of this application of remote parameterization of the robot in relation to the handling of other robotic systems with remote control in terms of immediacy and activation of commands in real time from a different room. This conclusion formulates an interesting research data which is beneficial to investigate more in the future. Teachers' feedback on the teaching value of the application was recorded in detail and formed the basis of the future improvement on which the research team is already working. For example, one consideration is to allow for storing pre-recorded multimedia material and codes locally on the smart device where the mobile app is installed or to incorporate an SSD or SD-Card reader in the robot for the same matter. This will allow for some additional independency for the robot as a good connection for downloading the necessary files on demand will no longer be required. Some other mechanical adjustments are also considered (e.g. more support to the robot's head due to the weight). But regarding the mobile application, mainly positive comments were received as feedback, also highlighting the functionality to include educational games in the application which can be played on the robot's screen/face.

Acknowledgements. This research has been implemented in the framework of the Action "Business Enhancement for Research projects" and co-financed by the European Union and national resources through the B.P. WESTERN GREECE 2014–2020 (project code: DEP5 - 0019433).

References

1. Ackermann, E.: Piaget's Constructivism, Papert's Constructionism: What's the difference? (2001). https://learning.media.mit.edu/content/publications/EA.Piaget%20_%20Papert.pdf. Accessed 31 May 2021
2. Misirli, A., Komis, V.: Robotics and Programming Concepts in Early Childhood Education: A Conceptual Framework for Designing Educational. In: Karagiannidis, C., Politis, P., Karasavvidis, I. (eds.) Research on e-Learning and ICT in Education, pp. 99–118. Springer, New York, NY (2014)
3. Eguchi. Robotics as a Learning Tool for Educational Transformation. In: Proceedings of 4th International Workshop Teaching Robotics, Teaching with Robotics and 5th International Conference Robotics in Education (2014)

4. Milliez, G.: Buddy: A Companion Robot for the Whole Family, HRI 18 Companion, March 5–8. IL, USA, Chicago (2018)
5. Hurst, N., Clabaugh, C., Baynes, R., Cohn, J., Mitroff, D., Scherer, S.: Social and Emotional Skills Training with Embodied Moxie, arXiv:2004.12962v1 [cs.RO] (27 Apr 2020)
6. Christodoulou, P., Adelaide May Reid, A., Pnevmatikos, D., Rioja del Rio, C., Fachantidis, N.: Students participate and evaluate the design and development of a social robot. In: 29th IEEE International Conference on Robot and Human Interactive Communication (RO-MAN) August 31−September 4 (2020)
7. Jung, S.E., Won, E.: Systematic review of research trends in robotics education for young children. Sustainability **10**, 905 (2018)
8. Toh, L.P.E., Causo, A., Tzuo, P.W., Chen, I.M., Yeo, S.H.: A review on the use of robots in education and young children. Educ. Technol. Soc. **19**(2), 148–163 (2016)
9. Kuenzi, J.: Science, technology, engineering, and mathematics (STEM) education: Background, federal policy, and legislative action. CRS Report for Congress (2008). http://digitalcommons.unl.edu/crsdocs/35
10. Pennisi, P., et al.: Autism and social robotics: a systematic review. Autism Res. **9**(2), 165–183 (2016)
11. Goldsmith, T.R., LeBlanc, L.A.: Use of technology in interventions for children with autism. J. Early Intensive Behav. Interv. **1**, 2 (2004)
12. Ziouzios, D., Ioannou, M., Ioanna, T., Bratitsis, T., Dasygenis, M.: Emotional intelligence and educational robotics: the development of the EI-EDUROBOT. In: European Journal of Engineering and Technology Research. CIE (2020)
13. Tzagkaraki, E., Papadakis, S., Kalogiannakis, M.: Exploring the use of educational robotics in primary school and its possible place in the curricula. In: Educational Robotics International Conference. Springer, Cham (2021)
14. Schina, D., Esteve-González, V., Usart, M.: An overview of teacher training programs in educational robotics: characteristics, best practices and recommendations. Educ. Inf. Technol. **26**(3), 2831–2852 (2020). https://doi.org/10.1007/s10639-020-10377-z
15. Kert, S.B., ErkoÇ, M.F., Yeni, S.: The effect of robotics on six graders' academic achievement, computational thinking skills and conceptual knowledge levels. Thinking Skills Creativity **38**, 100714 (2020)
16. Massaty, M.H., Budiyanto, C.W., Tamrin, A.G.: Revisiting the roles of educational robotics in improving learners' computational thinking skills and their positive behaviour. J. Phys. Conf. Ser. **1511**(1), 012088 (2020)
17. Thomas, A., Bauer, A.S.: Robotics and coding within integrated STEM coursework for elementary pre-service teachers. In: Society for Information Technology & Teacher Education International Conference. Association for the Advancement of Computing in Education (AACE) (2020)
18. Yolcu, V., Demirer, V.: A review on the studies about the use of robotic technologies in education. SDU Int. J. Educ. Stud. **42**, 127–139 (2017)

A Board Game for Sustainable Development Education: Kindergarten Students as Game Designers

Maria Tsapara and Tharrenos Bratitsis(✉)

University of Western Macedonia, Florina, Greece
bratitsis@uowm.gr

Abstract. This paper presents the idea of Kindergarten students as game designers. They designed a board game based on the 14th and 15th Sustainable Development Goals. The idea emerged from an eTwinning Project called "Schools Defenders of the Earth" where they participated in activities related to the 17 SDGs framework. Based on their prior knowledge on creating board games, they designed various templates. After a brainstorming session, the appropriate board format was selected. They selected the subtopic of "animals in extinction due to climate change", they designed rules, planned routes and decided which technologies to be included in the game. The board game was designed to incorporate QR codes in a creative way while still contributing to students' learning, where they included the game instructions, quizzes, mini games and missions. The game idea, mechanics and design are presented in this paper.

Keywords: Sustainable development · Kindergarten · Board game · Mobile learning · QR codes

1 Introduction

Games are designed to be challenging, interesting, promoting a positive attitude towards learning and involving students in active learning, providing a transdisciplinary approach to education as students may work on multiple skills related to various disciplines [1–3]. Uribe and Cobos [4] stress that within the classroom, playing improves creativity, lateral thinking, communication and eases the teaching process.

Board games are a successful active learning paradigm of the Game based Learning (GBL) approach as they help pupils practice in curricular subjects. These subjects are broad and flexible and can include topics related to scientific, environmental, and social issues. Tsai et al. [5] highlight that they are also suitable for familiarizing learners with more unusual subjects such as environmental sustainability.

For Burgun [6] creativity makes game design interesting but at the same time difficult. Students as board game designers stimulate their mental and creative capacities. When creating a game, the team has to brainstorm, share ideas and thoughts. Moreover, team members have to make decisions, which emphasizes the importance of

M. E. Auer and T. Tsiatsos (Eds.): IMCL 2021, LNNS 411, pp. 1072–1084, 2022.
https://doi.org/10.1007/978-3-030-96296-8_98

working to a specific time, giving and receiving feedback from/to others, refining one's selection and making the ultimate choice.

Following this approach, a board game was designed by 22 students of a public kindergarten in Greece, as a part of the eTwinning Project "Schools defenders of the Earth", related to the 17 SDGs framework. Biodiversity loss, including species extinction, in forests and other terrestrial ecosystems has been on the global development agenda for many decades, and from 1992 on the UN's Agenda 21. The main idea was not only for students to learn about SDGs through a technologically enhanced board game but also to give them the chance to become game designers.

A board game was designed for the 14th and 15th SDGs - "Life below water" and "Life on Land". As the subtopic they identified "animals in extinction due to climate change". The game idea and design, including the board, the mechanics and the resources are described in this paper, structured as follows: A brief literature review supports GBL, board games in education, Mobile learning, AR-QR in education and Sustainable Development Education in Kindergarten. Then, the game is described in detail. As the game was tested only by students, due to the pandemic, the paper concludes with a discussion which focuses on future research, utilizing this board game.

2 Theoretical Background

2.1 Game Based Learning

According to Anzman-Frasca [7] GBL is a promising and innovative approach, while considered quite popular. Many studies explain the positive effects of games on learning [8–10]. Ennis [11] stresses that games can be used as a teaching tool because they can positively benefit the learner through repetition, exploration, and practice. Educational games are considered to fulfill the needs of today's children.

Mattheiss et al. [12] state that there are two different points of view about GBL, the learning and the gaming viewpoint. Using an educational game depends on the motivation to play and/or the motivation to learn. Moreover, for Bratitsis et al. [13] games support students' motivation and provide opportunities for experiential learning, when used in an appropriate manner. Implementing games in a learning context, Garris et al. [14] assume that fosters collaboration, enhances motivation and creates a safe environment for learning through experimentation and a trial or error approach.

2.2 Board Games as an Educational Tool

According to Treher [15] board games are an important tool to provide hands-on and heads-on skill and knowledge development for all ages and all subjects. Uribe and Cobos [4] highlight that board games are the perfect tool to teach a topic as they can combine complex knowledge with hands-on practice. Moreover students can cultivate their creativity, imagination, strategy skills and their understanding of a particular topic. Tsai et al. [5] argue that board games are highly interactive, as students can take the initiative to explore and exchange information with peers, thus promoting student-centered learning. The teacher should clarify concepts, provide supplementary information, and lead discussions and summaries during and after the game.

Many board games encourage players to set goals, be patient, predict the outcome of alternative moves, strengthen relationships, and learn from experience. A board game can be used as a creative and supportive tool to enhance traditional teaching methods [16]. Learners can improve their learning experience while also cultivating other skills such as following rules, problem solving, critical thinking, creativity, collaboration. Board games are games with rules, using a board, upon which other tangible elements (player tokens) are placed. Players take turns and move on the board, utilizing some mechanism (e.g. by rolling a dice) while interacting with or against each other [13, 17–20]. Treher [15] also stresses that, "*The board itself provides a visual metaphor to help connect information. Game elements, discussions, and problem solving with fellow team members about the content are vehicles for learning. Subtle redundancy to reinforce learning and insure retention should be incorporated into the game design. Good questions, problems to solve, and situations to consider allow players to think through and apply what they learn*".

2.3 Mobile Learning and Devices

According to Quinn [21], the world of students rotates around and is surrounded by mobile devices. Researchers suggest that the systematic use of mobile devices in the classroom has contributed to the development of critical thinking, communication, creativity and innovation skills. Papadakis and Orfanakis [22] highlight that mobile learning covers all forms of learning while students, in order to interact with peers, create new work plans and utilize learning material at their disposal and use all kinds of mobile devices. Trentin and Repetto [23] argue that creating learning materials and achieving learning objectives becomes a very important part of learning.

Saylor [24] stresses that sharing of the content of a teaching, a new learning material, the ideas and suggestions of the students, takes place so directly that it can easily lead to immediate feedback. Research and exploratory work in European countries on the integration of hands-on learning into education highlights the multiple benefits. At the same time, a special interest has been observed for the utilization of tablets and other similar devices in schools both by teachers and students themselves when approaching various cognitive objects related to the Curriculum of each educational level, according to the relevant UNESCO [25] guidelines.

Preschool students use tablets for presenting their ideas, while developing social and individual skills [26]. However the simple use of tablets cannot bring the above, as learning and acquiring skills is not only about the device but also the pedagogy. Monguillot et al. [27] highlight that through mobile learning students can construct new knowledge from prior knowledge and cultivate 21st century competencies. Furthermore they give another dimension to mobile learning as "learning can take place anytime and anywhere", expanding and delivering resources and materials to pupils.

2.4 AR and QR Technologies in Education

QR codes are an extension of standard barcodes. They can be printed and placed everywhere, but also be integrated into various applications, a blog, a website. Despite the initial idea that they could only be used for commercial purposes, Law and So [28] have shown that they can also be used in education supported by mobile learning. Rikala and Kankaanranta [29] found that QR codes can support different learning environments and different cognitive objects. In addition, they can support experiential and collaborative learning, mobilize and engage students in learning paths, and create an effective, authentic, and engaging learning environment. Besides they are utilized for over a decade [30].

Uçak [31] highlights that learning experiences can be more interactive and interesting when teachers introduce such innovations in it. Interactivity, after all, allows students to be active in their education. QR codes can provide real interaction and engagement, which implies the most effective and efficient learning [29].

The purpose of integrating digital tools in education is to improve traditional teaching. According to Azuma [32] AR as systems have three characteristics: 1) Combine real and virtual, 2) Interactive in real time, 3) Registered in 3-D. While comparing AR with other technologies, Mac Callum and Parsons [33] stress that AR is a promising tool for learners [34]. Rizov and Rizova [35] assume that "the technology of AR can be used to augment all the user senses, although the vast majority of applications are focused only to the sight by combining virtual graphics with the reality the user sees". For Preka and Rangoussi [36] AR becomes more affordable and more widely available. As a result many teachers implement AR applications in the educational process. Lin et al. [37] in their study assume that AR can guide students through contextual learning to allow students to experience the context of their desired learning field.

2.5 Education for Sustainable Development in Kindergarten

Early Childhood is included in Sustainable Development Goal 4 [38] "Ensure inclusive and equitable quality education and promote lifelong learning opportunities for all. It is specifically mentioned in target 4.2: "By 2030, ensure that all girls and boys have access to quality early childhood development, care and pre-primary education so that they are ready for primary education". According to Vodopivec [39], educating students for sustainability should begin in kindergarten. Students from 4 to 6 years old develop their basic skills, behaviors, values and habits, which may be long lasting. Samuelsson and Kaga [40] argue that "early childhood education for sustainability is much more than environmental education. As early childhood education is about laying a sound intellectual, psychological, emotional, social and physical foundation for development and lifelong learning, it has an enormous potential in fostering attitudes, competencies, behaviors and values that support sustainable development".

3 Game Description

A total of 22 public kindergarten students, aged 4–6 years old, participated in the study. They lived in an urban district. The way in which the children interacted and collaborated within working groups for achieving a common goal, that of learning while designing a board game was monitored. Data was collected through direct observation while the design of the action was based on the GBL approach. Children's reactions, ideas' sharing, negotiation and overall engagement (including motivation and pleasure reception) were qualitative evaluation factors. The teacher, one of the authors, kept a researcher's journal during and after the intervention.

The students had the idea to create a board game for the 14th and 15th SDGs, "Life below water" and "Life on Land". As a subtopic they chose "animals in extinction due to climate change". According to Georgopoulos [41] the acquaintance of kindergarten students with issues related to environmental problems, is one of the main goals of early childhood education. Through this, they can cultivate their awareness, adopt positive attitudes and values towards the environment and learn to act as "environmental literacy citizens". Board games can help students master difficult concepts and also enhance learning motivation [42]. The students had prior knowledge on board games design. In the classroom, board games are prepared as a part of a project, in which the children participate as game designers on their own initiative, depending on their interests, while the topics emerge or are chosen by them [43]. It is worth noting that according to the official Greek Curriculum the use of board games plays an important role [44, 45]. The game challenges students to go outside their comfort zone, discuss and debate, think critically, and explain key concepts to one another. They can develop social skills like interacting with peers, following game rules and apprehending others' perspectives [46].

For Essop et al. [47] the design stage is the important aspect of the development of the game. Harris [48] explains some outlining tips for successful board games:

- Be creative—Think out of the box
- Make it a learning tool
- Give it a professional look
- Develop a good set of rules

Board game mechanics
Game mechanics are rules' constructs that guide and provide structure to the game by outlining what is expected from the players to win [47, 49, 50] According to Kritz et al. [51] there are 51 mechanics listed on BoardGameGeek. Students while designing their board game decided to implement the following game mechanics.

- ***Action.*** These are the rules for the actions of a game, how and when a player can interact with the game state, directly or indirectly
- **Action Programming.** Every player must secretly choose his/hers next number of actions, and play their turns out according to the choices made.
- **Collection.** Players need to collect a component, points, resources etc.
- **Card Draft.** When you select and pick cards from one or more pools.
- **Dice Rolling.** Players are required to roll a dice of any kind

- **Goal.** The objectives the players try to achieve during a match.
- **Ruleset.** The abstract rules that control the game, what players can and cannot do, how the game components behave, and many other game aspects.
- **Partnerships.** Players may form or undo teams during the game.
- **Card.** Using cards in the game. This is a very basic concept, but cards are a versatile component and have great expression within the game.

Creating the rules

The students as the first step they created a set of rules that direct what players are allowed to do or not, as well as the objective of the game and victory conditions. They created a board game, where the general goal was to move a token towards the end of the track. It was designed to allow a maximum of five players/teams. The progression details were: a) players are allowed to roll the dice, b) move a number of squares equal to the roll, c) choose the desired path, d) pick up a card (help or trap) if the square gives them this choice, and e) claim the key by completing a mission.

They also set the game goal, that of gathering 6 keys to win. Every key is connected to a mission. The main topic was animals in danger. Overall 6 subtopics were identified: a) sea animals, b) arctic animals, c) forest animals, d) insects, e) birds, and f) jungle/savannah animals. Missions included quizzes, puzzles, digital painting, digital games, and expression activities in sign language. The player/team has to accomplish a mission to take the key. The player/team who first reaches the end of the track with 6 keys wins. Students also thought about the player choices during the game. They settled that some of the tiles of their board game would help or trap players.

Designing the board of the game

Students were familiar with board game design as they had created games throughout the school year. To create the board, students initially formed groups, collaborated and designed their proposal. All of them were presented in a plenary session (Fig. 1).

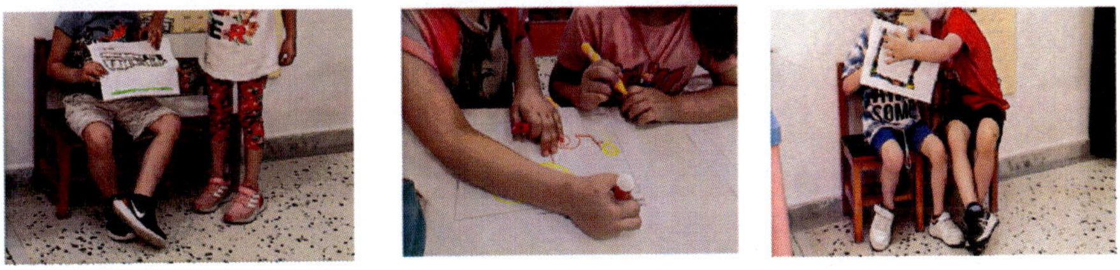

Fig. 1. Initial board ideas' presentation.

Students voted their favorite board. Wanting to consider all proposals, they chose hearts as the main element, circles for the beginning and end, squares for the missions with keys (Fig. 2). The track had a rectangular shape. Kamii and De Clack [52] stress that a board could be too simple or complicated as a work of art. It can be 2d or 3d. The track could also be rectangular, spiral, circular or even have a zigzag shape.

Testing and redesigning the board

Students wanted to test the board, before designing other game elements. Then they realized that the track did not correspond to the game rules and objective while the board was not functional at all. So they had to redesign the board. After a brainstorming session, the appropriate board format was selected, one that gives the opportunity to move around the track, in any direction. There are 6 square tiles with a heart and a key in different colors. If players stop in these tiles they could claim a key, after completing the relevant mission. In the center of the board they could claim, any key they want with no restrictions. Each step of the process was evaluated with feedback from play-testing and students' critical reflections. Figure 2 depicts the final version.

Fig. 2. The initial (left) and the final, refined (right) game board.

Designing the elements of the game

They drew the tokens, inspired from the 5 eTwinning project heroes and designed cards that would help or trap players. As they couldn't read, they drew the card content and recorded their voices, embedding QR codes which lead to the audio (Fig. 3).

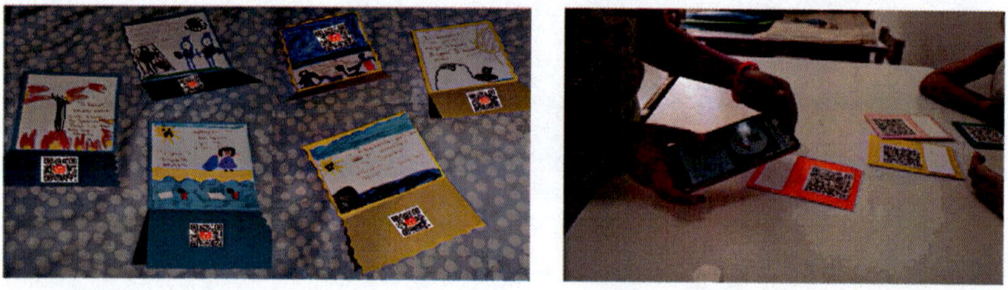

Fig. 3. The mission cards with QR codes (left) and the process of scanning them (right).

For designing the missions, students worked in teams. Six missions were created, corresponding to the keys on the board. Every team shared ideas and collaborated for a common goal. They decided the mission content, chose the appropriate ICT tool, created the embedded QR code, which led to the quiz, puzzle, game or other content and tested the QR code (Fig. 3). Through QR Codes, connected to digital learning resources, teachers provide students with enriched learning experiences that motivate their interest [37, 53]. The QR codes provide limitless boundaries of potential knowledge and understanding. This board game incorporates QR codes in a creative way while still contributing to students learning [54] and multimedia material.

After looking at a range of possible tools, students selected:

- ***jigsaw planet***, for creating a digital puzzle (Mission 1 - arctic animals)
- ***wordwall - emaze,*** for creating a digital labyrinth with a quiz game (Mission 2 - birds)
- ***genially,*** for creating a quiz (Mission 3 - animals in the jungle - Savannah)
- ***Chrome music lab*** (Mission 4 - animals in the forest)
- ***metaverse,*** for creating an AR experience combined with sign language (Mission 5 - insects)
- ***metaverse,*** for creating a quiz (Mission 6 - animal of the sea)

Students were quite experienced with the aforementioned web 2.0 tools, due to the pandemic and the distance learning. The teacher, selected to utilize Metaverse (http://metaverse.org/) for adding AR functionality. This was chosen, as the studio app runs in most browsers and the mobile app runs on Android devices. The studio app allows users to create and import their own content, record their voice and various types of experiences can be created [33]. Figure 4 depicts the material for the missions.

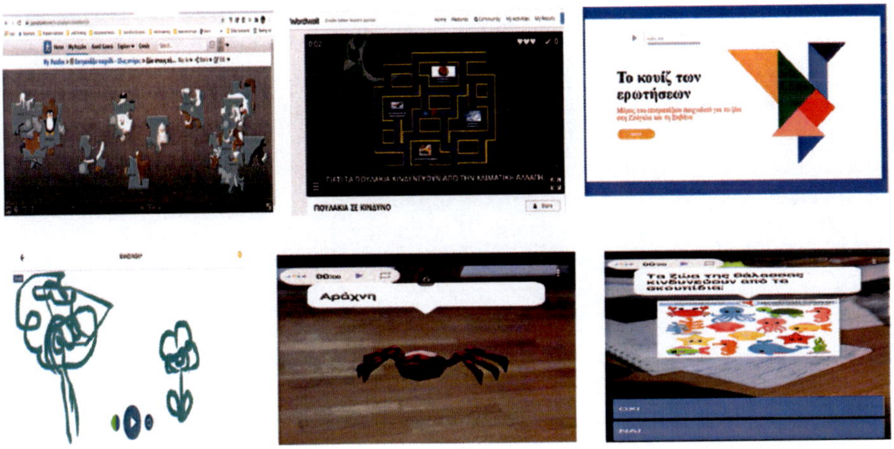

Fig. 4. Some of the missions

The game is completed when a player/team acquires the 6 keys. Reflecting on the game rules, it is clear that different strategies can be followed. Therefore, there are many ways to approach the game and it is the authors' impression that this motivates the children to play the game repeatedly so as to find the best strategy to win [13].

Time to play
Students wanted to test the board game after completing its design. An internet connection, at least one mobile device and a QR code reader were required for that.

4 Discussion and Further Plans

As explained in this paper, students decided to create a board game based on the 14th and 15th SDGs about "Life below water" and "Life on Land". The subtopic identified was "animals in extinction due to climate change". Board game-based teaching offers new insights into and approaches to teaching and also can help fill the gaps in Sustainable Development Education in Kindergarten. Students can learn and meet educational goals, such as affective aspects, values, problem solving, and decision-making. Teaching through board games does not mean that students just play by themselves. The teacher should clarify concepts, provide supplementary information, and lead discussions and summaries during and after the game [5].

As game designers, learners, while being creative, have to consider different game elements and rules to promote meaningful design choices. When designing games, students learn to determine the knowledge and skills needed and recognize game genres, rules and mechanisms to adopt and modify [55]. Students, as aforementioned, being flexible and thinking creatively found the right solution and redesigned the game board to serve its goal. For Tap et al. [56] creativity is an important skill in game designing, as students can turn their ideas into reality. Moreover creativity exists in every human and it can be learned, practiced and developed by using appropriate techniques [57]. It is defined as being able to innovate, generate new possibilities, and create new ideas or knowledge [58]. According to Avgitidou [59] play as a context gives students opportunities and reasons for being creative in problem solving situations. Fesakis and Lappas [60] stress that creativity in kindergarten can be approached as a personal expression or as a process of defining and solving problems.

The overall goal was to examine if assessing Kindergarten students' creativity, while they design a board game, has the potential to act as an innovative and qualitative form of knowledge acquisition assessment and could be further researched as such in order to reach a framework for developing new, creative assessment mechanisms for young learners that go beyond typical testing. Since in early childhood there are multiple factors that can affect creativity which could not be quantified [61].

In the interviews that were conducted in order to evaluate various aspects of the game, students seemed to like their game. They not only enjoyed playing the game but also were pleased during the designing process. Taking into account that nowadays children at the age of 4–5 are already internet and smart device users [13], students claimed that the creation, the use of QR Codes and mobile devices was not so difficult, as they had previous experience within the educational process. Answering the question "Would you like to play the game again?", students considering that the end of the school year was approaching they proposed to the teacher, one of the authors of this paper, to create a digital version of the game by using a web 2.0 tool application so that they can play at home as well. Students also noticed as an advantage of their board game, that they could interact with the missions independently of the game. Furthermore all the students referred the implementation of the augmented reality material as a quite different and unique learning experience.

Through this teaching practice we expect to see if children can apply new knowledge about the Sustainable Development goal in the design of an educational board game. It is important to mention that the schools' lockdown due to the pandemic did not allow to evaluate students' knowledge about the sustainable development goals, prior and after the intervention by using pre and post tests. Also the purpose of the authors is the next school year to conduct a small scale survey with teachers in order to examine the appropriateness of the board game and its contents and also tested the board game to another classroom.

As for future plans, the authors plan on creating a digital version of the game incorporating QR codes and multimedia material, as kindergarten students satisfying the idea of the kindergarten students to have the opportunity to play their board game again at home. Other ideas include the enrichment of the missions created by the children and the addition of other elements of augmented reality.

References

1. Strode, A.: A board game as a teaching tool in a pedagogically psychological perspective. In: Edulearn 19 Proceedings, pp. 5437–5444 (2019)
2. Cojocariu, V.M., Ioana, I.: Teaching the relevance of game-based learning to preschool and primary teachers. Procedia Soc. Behav. Sci. **142**, 640–646 (2014). https://doi.org/10.1016/j.sbspro.2014.07.679
3. Anderson, O.B., Anderson, N.M., Taylor, A.T.: New territories in adult education: game-based learning for adult education (2009). http://www.adulterc.org/Proceedings/2009/proceedings/anderson_etal.pdf. Accessed 30 June 2021
4. Uribe, R.V., Cobos, S.A.: Board Games as Tool for Teaching Basic Sustainability Concepts to Design Students (2014)
5. Tsai, J.C., Liu, S.Y., Chang, C.Y., Chen, S.Y.: Using a board game to teach about sustainable development. Sustainability **13**, 4942 (2021). https://doi.org/10.3390/su13094942
6. Burgun, K.: Game Design Theory: A New Philosophy for Understanding Games. CRC Press, USA (2013)
7. Anzman-Frasca, S., et al.: Evaluating a board game designed to promote young children's delay of gratification. Front. Psychol. **11**, 581025 (2020). https://doi.org/10.3389/fpsyg.2020.581025
8. Gee, J.P.: What video games have to teach us about learning and literacy. Comput. Entertainment (CIE) **1**(1), 20 (2003)
9. Williamson, B.: Computer games, schools, and young people: a re-port for educators on using games for learning (Futurelab, 2009)
10. Shaffer, D.W., Squire, K.R., Halverson, R., Gee, J.P.: Video games and the future of learning. Phi Delta Kappan **87**(2), 105–111 (2005)
11. Ennis, L.: Game-Based Learning: An Instructional Tool. Creative Components, pp. 57 (2018). https://lib.dr.iastate.edu/creativecomponents/57
12. Mattheiss, E., Kickmeier, M., Steiner, C., Albert, D.: Motivation in game-based learning: It's More than Flow. Lernen im Digitalen Zeitalter (2009)
13. Bratitsis, T., Kontovounisiou, A, Kiriazoglou, M: A board game proposal for teaching informatics related topics in Early Childhood Education. SHS Web Conf. **102**, 01001 (2021)
14. Garris, R., Ahlers, R., Driskell, J.E.: Games, motivation, and learning: a research and practice model. Simul. Gaming **33**(4), 441–467 (2002)

15. Treher, E.: Learning with Board Games. Tools for Learning and Retention. The Learning Key Inc, Minnesota (2011)
16. Zirawaga, S., Adeleye, O., Tinovimbanashe, M.: Gaming in Education: Using Games as a Support Tool to Teach History (2017)
17. Bayeck, R.Y.: Examining board gameplay and learning: a multidisciplinary review of recent research. Simul. Gaming **51**(4), 411–443 (2020). https://doi.org/10.1177/1046878119901286
18. Akmal, H., Coulton, P.: Research Through Board Game Design (2019)
19. Barbara, J.: Measuring user experience in multiplayer board games. Games Cult. **12**(7–8), 623–649 (2017). https://doi.org/10.1177/1555412015593419
20. Zagal, J.P., Rick, J., Hsi, I.: Collaborative games: lessons learned from board games. Simul. Gaming **37**(1), 24–40 (2006). https://doi.org/10.1177/1046878105282279
21. Quinn, C.N.: Designing Mlearning: Tapping into the Mobile Revolution for Organizational Performance, 1st edn. Pfeiffer & Company, San Francisco (2011)
22. Papadakis, S., Orfanakis, V.: Use of QR codes in education. A case study. In: 3rd Panhellenic Educational Conference of Imathia, Digital and Internet Applications in Education, April 4–6, Veria, Naoussa (2014). (in greek)
23. Trentin, G., Repetto, M.: Using Network and Mobile Technology to Bridge Formal and Informal Learning. Woodhead/Chandos Publishing Limited, Cambridge (2013)
24. Saylor, M.: The Mobile Wave: How Mobile Intelligence Will Change Everything, p. 176. Perseus Books/Vanguard Press, New York (2012)
25. UNESCO. Policy guidelines for mobile learning (2013). http://unesdoc.unesco.org/images/0021/002196/21964. Accessed 30 June 2021
26. Zaranis, N., Kalogiannakis, M., Papadakis, S.: Using mobile devices for teaching realistic mathematics in kindergarten education. Creat. Educ. **4**, 1–10 (2013)
27. Monguillot, M., González, C., Guitert, M., Zurita, C.: Mobile learning: a collaborative experience using QR codes. Mobile Learning Applications in Higher Education [Special Section]. Revista de Universidad y Sociedad del Conocimiento (RUSC) **11**(1), 175−191 (2014). https://doi.org/10.7238/rusc.v11i1.1899
28. Law, C., So, C.S.: Qr Codes in education. J. Educ. Technol. Dev. Exch. **3**(1), 85–100 (2010)
29. Rikala, J., Kankaanranta M.: The use of quick response codes in the classroom. In: Sprect, M., Sharlples, M., Multisilta, M. (eds) Proceedings of the 10th International Conferene Mobile Learning, pp 141–148 (2012)
30. Dourda, K., Bratitsis, T., Griva, E., Papadopoulou, P.: Combining game based learning with content and language integrated learning approaches: a case study utilizing QR codes and google earth in a geography-based game. In: 7th European Conference on Games Based Learning, ISEP, Portugal, 3–4 October (2013)
31. Uçak, E.: Teaching materials developed using QR code technology in science classes. Int. J. Progressive Educ. **15**(4), 215–228 (2019)
32. Azuma, R.T.: A survey of augmented reality. Presence **6**(4), 355−385 (1997)
33. Mac Callum, K., Parsons, D.: Teacher Perspectives on Mobile Aug- mented Reality: The Potential of Metaverse for Learning (2019)
34. Bacca, J., Baldiris, S., Fabregat, R., Graf, S., Kinshuk, D.: Augmented reality trends in education: a systematic review of research and applications. Educ. Technol. Soc. **17**(4), 133–149 (2014)
35. Rizov, T., Rizova, E.: Augmented reality as a teaching tool in higher education. Int. J. Cogn. Res. Sci. Eng. Educ. **3**, 7–16 (2015). https://doi.org/10.23947/2334-8496-2015-3-1-7-15

36. Preka, G., Rangoussi, M.: Augmented reality and QR codes for teaching music to preschoolers and kindergarteners: educational intervention and evaluation. In: Proceedings of the 11th International Conference on Computer Supported Education (CSEDU 2019), pp. 113–123 (2019). https://doi.org/10.5220/0007682301130123
37. Lin, H.C.K., Lin, Y.H., Wang, T.H., Su, L.K., Huang, Y.M.: Effects of incorporating augmented reality into a board game for high school students' learning motivation and acceptance in health education. Sustainability **13**, 3333 (2021). https://doi.org/10.3390/su13063333
38. United Nations. Transforming our world: The 2030 agenda for sustainable development. https://sdgs.un.org/publications/transforming-our-world-2030-agenda-sustainable-development-17981. Accessed 1 July 2021
39. Vodopivec, J.L.: Some aspects of sustainable development in kindergartens in Slovenia. US-China Educ. Rev. A **7**, 967–976 (2011)
40. Samuelsson, I., Katz, L.: The Contribution of Early Childhood Education to a Sustainable Society (2008)
41. Georgopoulos, A.: Environmental Ethics. Gutenberg, Athens (2002).(In greek)
42. Linkiewicz, J., Hughes, P., Makara-Studzińska, M.: The use of board games in healthcare teaching. Nurs. Probl. Pielęgniarstwa 28, 71–74 (2020)
43. Katz, L.G., Chard, S.C.: Engaging Children's Minds: The Project Approach. Ablex, Norwood, NJ (2000)
44. Cross-Curricular Program Framework. Pedagogical Institute, Gov. 303 and 304/13-3-2003, B, volumes A and B. (In Greek) (2003)
45. Ministry of Education of Greece. Early Childhood Curriculum - 1st part (2011). http://digitalschool.minedu.gov.gr/. Accessed 30 June 2021
46. Mostowfia, S., Mamaghanib, N.K., Khorramarb, M.: Designing playful learning by using educational board game for children in the age range of 7–12: a case study: recycling and waste separation education board game. Int. J. Environ. Sci. Educ. **11**(12), 5453–5476 (2016)
47. Essop, H., Kekana, M., Sethole, M., Ahrens, E., Lovric, G.: A board game to enhance teaching and learning in the radiographic technique module: a pilot study. Int. J. Innov. Educ. Res. **6**, 124–140 (2018). https://doi.org/10.31686/ijier.Vol6.Iss10.1173
48. Harris, C.R.: Create board games to enhance classroom learning. In: Association for Career and Technology Education Conference, December, Orlando, FL (2003)
49. Sicart, M.: Defining game mechanics. Game studies. Int. J. Comput. Game Res. **8**, 2 (2008)
50. Robson, K., Plangger, K., Kietzmann, J.H., Mc Carthy, I., Pitt, L.: Is it all a game? Understanding the principles of gamification. Bus. Horiz. **58**(4), 411–420 (2015)
51. Kritz, J., Mangeli, E., Xexéo, G.: Building an Ontology of Board Game Mechanics based on the BoardGameGeek Database and the MDA Framework SBC – Proc. of SBGames (2017)
52. Kamii, C., De Clark, G.: Children Reinvent Arithmetics. Patakis publications, Athens (1995).(in greek)
53. Rouillard, J.: Contextual QR codes. In: CCGI 2008: The 3rd International Multi Conference on Computing in Global Information Technology. IEEE, Athens, Greece (2008)
54. Puentedura, R.R.: Transformation, Technology, and Education (2006)
55. Beaumie, K., Reyhaneh, B.: Students as game designers: transdisciplinary approach to STEAM education. ATA J. **45**, 45–52 (2017)
56. Tap, R.M., Zin, N.A.M., Sarim, H.M.: Requirements for creative skills development in game design. Int. Conf. Electr. Eng. Inf. (ICEEI) **2019**, 176–182 (2019). https://doi.org/10.1109/ICEEI47359.2019.8988799

57. Azimah, K., Laili Farhana, M.I.: Creativity of student as a game designer: an exploratory study. Int. J. Multimedia Appl. **10**, 105–115 (2018). https://doi.org/10.5121/ijma.2018.10609
58. Kementerian Pendidikan Malaysia. Laporan Awal Pelan Pembangunan Pendidikan Malaysia 2013–2025 (2013). http://www.moe.gov.my/userfiles/file/PPP/Preliminary-Blueprint-BM.pdf. Accessed 15 June 2021
59. Avgitidou, S.: Creativity and play in preschool education: a socio-cultural approach. Theory Pract. Sci. Educ. **2**, 9–19 (2016). https://doi.org/10.12681/dial.10570
60. Fesakis, G., Lappas, D.: Developing kindergarten students' creativity through creative problem solving in computerized learning environments (2011). (in greek)
61. Hennessey, B.A., Amabile, T.M.: Consensual assessment. In: Pritzker, M.A.R.S.R. (ed.) Encyclopedia of Creativity, vol. 1 (1999)

Mobile Learning Application Integrated with Learning Management System

Jingjing Shao(✉), Jingjing Jiang, Hongxin Shen, Jian Zhang, and Qing He

Shanghai Jiao Tong University, Dongchuan Rd 800, Shanghai, China
{shaojj, jiangjingjing0728, hxshen, zhjian, heqing}@sjtu.edu.cn

Abstract. Since mobile learning has broken through the limitation of time and space, and also provided a new form for education. This paper analyzes the deficiencies of the learning management system in supporting mobile teaching, and proposes the technical solutions. Taking Shanghai Jiao Tong University as an example, the enhanced application of learning management system which adapted in mobile terminal has integrated with the live class and VOD system with functional modules such as roll call and poll through LTI, thus promoting students to achieve learning experiences.

Keywords: Mobile learning · The live class · Roll call and poll

1 Introduction

China Internet Network Information Center (CNNIC) has released the 47th "statistical report on the development of China's Internet network" in Beijing. As of December 2020, the capacity of Internet users has reached 989 million, including 986 million mobile Internet users, 99.7% use mobile phones to access the Internet, accounting for one fifth of the global Internet users; The Internet penetration rate has reached 70.4%, which was higher than the global average[1] With the rapid development of education informatization, it has become gradually common for teachers and students to use mobile terminals anytime and anywhere in the field of education and training.

The teachers and students of Shanghai Jiao Tong University take the Canvas learning management system as the main way of online course learning. The system is independently deployed on the campus private cloud based on the open source code of Canvas LMS. It provides online course creation, course resource construction, course notification, online homework and quizzes, communication and interaction as well as academic evaluation, so that supporting the whole process of teaching and learning. In 2019, a total of 700 courses will be published on Canvas LMS, and the system has been well received by teachers and students. At the beginning of 2020 spring semester in the outbreak of COVID-19, the campus used Canvas and Zoom system to achieve the " keep learning during class suspension" [2], a total of 3,764 courses were taught online, which have achieved the same quality and effect as the physical class. Since the fall semester of 2020, the number of published courses has continued to grow. As of May

M. E. Auer and T. Tsiatsos (Eds.): IMCL 2021, LNNS 411, pp. 1085–1093, 2022.
https://doi.org/10.1007/978-3-030-96296-8_99

2021, the Canvas LMS has launched 13,000 courses, served 1,112,000 teachers and students, browsed 103 million pages, and reached 772,000 pages in a single day. Hence, nearly 79,000 assignments were arranged and 289,800 courseware were released in three years.

With the large amount of the systematic use, teachers and students have more in-depth demand for mobile learning.

1.1 Canvas Video Support

Learning management system needs to be improved to support live and on-demand video resources. Teachers and students hope that they can easily watch the live streaming class in Canvas and review the class video.

During the spring semester of the 2019–2020 academic year in the outbreak of COVID-19, teachers have carried out classes through Zoom live broadcast software to meet the needs of online teaching. After the class was finished, the video of the course was distributed to students by teacher. However, there are several disadvantages.

Firstly, recording quality is difficult to guarantee, since teachers use their own equipment for recording, the video quality will be varied by the different equipment quality.

For teachers who like blackboard writing, they need to buy writing board and other devices, but it is still difficult to recreate the real class scene. Teachers need to complete the work of recording, editing, uploading and distributing, which increases the workload.

Secondly, video resources are lack of effective management means. Teachers should upload the recorded video through the cloud storage and obtain the video URL address after teaching. Usually, they will display the address in the unit module in the form of external links in Canvas for students to click and review. However, there are many video sections in the course, there are insufficient effective management tools in Canvas.

The teaching scene becomes more complex due to the influence of COVID-19, and overseas students can't return to campus normally. How to make them experience the online class like the real class, that is a problem need to be solved.

1.2 Canvas Class Interaction Support

Interaction is an important means to activate class, diagnose students' learning effect, and improve teaching strategies and methods in time. However, the system itself lacks effective tools for class interaction.

Firstly, it's unable to make rollcall efficiently. Teachers usually use the optional rollcall tool to sign in, but they need to import and export the statistics of students' lists, also the data result can't easily imported into Canvas, which brings plentiful inconvenience.

Secondly, the user experience of class interaction needs to be improved. Although teachers can set up some polling links in Canvas in advance to realize the interaction, there are more steps in the actual class operation, which affects the user experience.

In view of the existing problems, in the post epidemic period, there are some students returning to campus, and some overseas students can't go back to campus. It is better to realize two functions, the first one is that teachers don't need any operation. In Canvas, it can automatically broadcast live streaming according to the class timetable, and automatically generate videos after class. The video will be displayed in the form of course video list with Canvas system as the entrance, which helps students to watch on the mobile terminal, the second one is that students can scan the QR code through wechat on the mobile terminal to participate in the interaction of sign in and polling.

2 Method

2.1 Access Applications Through LTI

The IMS Learning Tools Interoperability (LTI) standard prescribes a way to easily and securely connect learning applications and tools with systems like learning management systems (LMS), portals and learning object repositories on your premise or in the cloud, in a secure and standard manner and without the need for expensive custom programming [3].

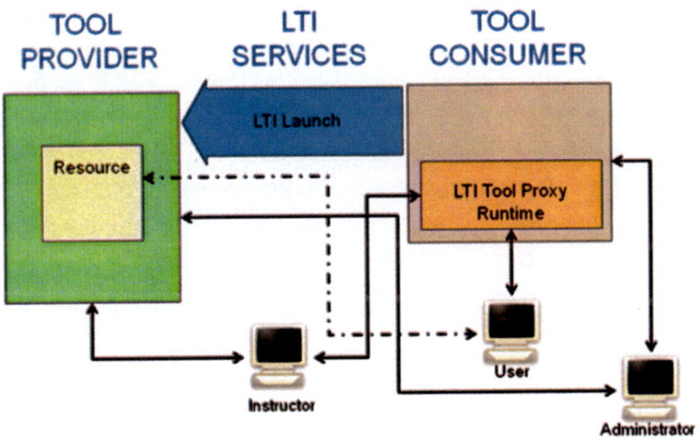

Fig. 1. Overview of LTI [4].

The system standardizes the development of docking rollcall and class video system applications through the LTI international standard interface, realizing one-stop, non perception, cross system access for teachers and students (Fig. 1).

2.2 Class Live Streaming and VOD System

In order to meet the needs of students for video learning in the information age, the campus has built a cloud recording and broadcasting system, installed intelligent tracking and live broadcasting equipment in 430 multimedia class, so as to capture images by 4K high-definition EPTZ camera [5], realize intelligent tracking by automatic matting on built-in image analysis chip, and unify management through cloud recording and broadcasting management system after signal acquisition. The campus student information system will push the course, course selection and course scheduling data to the shared database of Canvas through the network information center. According to the data, the cloud recording and broadcasting system will automatically generate the live broadcast task according to the schedule of specific sections, so as to realize the automatic live broadcast and video on demand.

Canvas connection can realize class live broadcast. Students can log in to Canvas according to the schedule to see the class live broadcast of cloud recording. The live broadcast equipment will collect teachers' teaching pictures, sounds and slides. The live camera can track teachers' walking on the system, clearly shoot blackboard writing, and upload two-way video signals to alibaba cloud in real time. The live broadcast will be accelerated through alibaba cloud CDN nodes distributed all over the world. Overseas students can watch high-definition live class videos, and online students can watch the high-definition live class video at the same time, and the teacher image and the computer desktop signal can be freely switched according to the needs, so as to present the whole picture of the class and enhance the students' sense of presence.

Canvas access can achieve class VOD system. The class video will be automatically edited according to the schedule time after the live broadcast, which will be displayed in the video on demand list of Canvas for on-demand service, and it will be opened to students by teachers through permission settings. Video recording, editing and synthesis will be automatically completed at the back end of the system, which greatly reduces the workload of teachers, providing a convenient entrance for students to review and consolidate knowledge points after class. With the continuous progress of the semester, the sharing and use of the previous semester courses has been added to fully help teachers and students watch class videos, so as to review difficult knowledge taught by teachers in the class (Fig. 2).

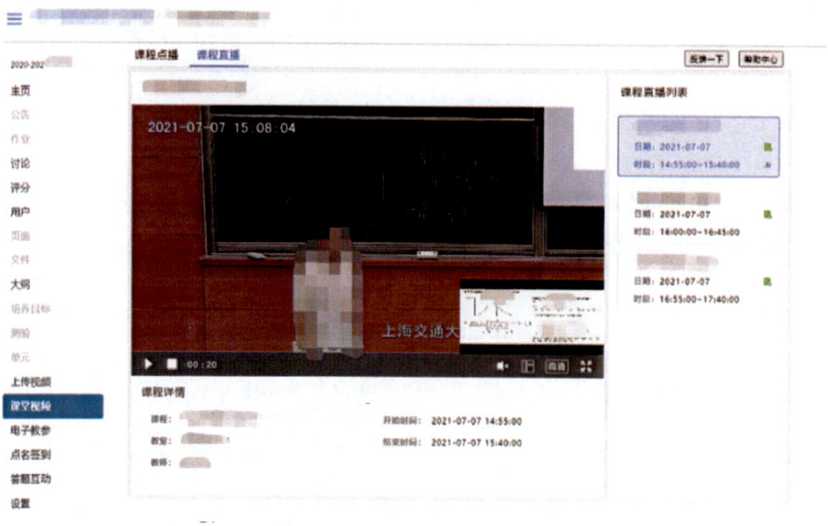

Fig. 2. Class live streaming and VOD system in Canvas.

2.3 Roll Call and Polling System

Class interaction is the main way to mobilize students for participating in the process of class teaching and realizing class teaching objectives, forming a benign interaction between each other. Effective class interaction and active class atmosphere can stimulate students' interest in learning, arouse their positive thinking, and help teachers better understand the effect for future improvement [6].

Canvas system can develop and access the interactive system of sign in and questions answering through the LTI standard, providing users with cross system services without perception, and enhancing the real-time interactive teaching tool of mobile terminal in the class (Figs. 3 and 4).

Fig. 3. Students sign in by scanning code on mobile terminal.

Fig. 4. The rollcall data management.

After logging in Canvas, the teacher can enter the “sign in” and initiate a new sign in or sign in management for one or more classes. After initiating a new rollcall, the students will scan the code to sign in. The sign in system will judge according to the class list in Canvas. If the students are in this class, the sign in will be successful. If they are not in this class list, they cannot sign in. The students who have signed in and those who have not signed in will be displayed with corresponding marks. For special students who have not been able to sign in before, the teacher can sign in manually to mark them.

Sign in can summarize the scores in “synchronous scores to Canvas gradebook”, and add “ sign in” as an assignment to the Canvas score book as a grade for student evaluation. Compared with other sign in methods in the past, sign in through Canvas can connect the students in the teaching class, which saves the teachers time from importing the student list and exporting the student sign in results. They do not need to repeatedly register scores, thus reducing their workload.

The optimization of Canvas polling interaction will help teachers to initiate answers at any time. Students just need to pick up their mobile phones and scan to answer questions. Then they can participate in class interaction. The answer results will be displayed in a visual way, so that teachers and students get instant feedback. Class interaction can be quantified, traceable, preserved and analyzed through accurate feedback of data, which has become an important means for them to participate in class interaction while acting as a good interactive tool (Fig. 5).

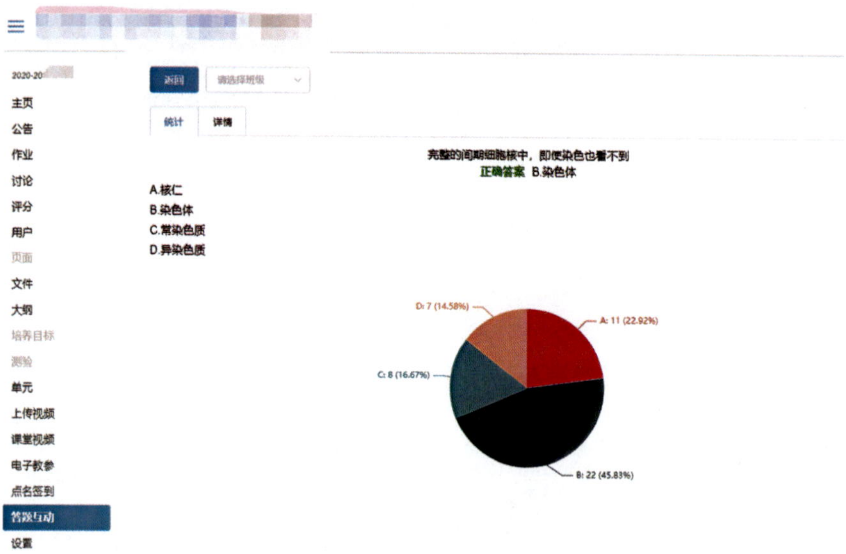

Fig. 5. Student polling result.

3 Result

In the autumn and spring semesters of 2020–2021, 7144 courses were recorded on the system, with a total of 168,000 class hours, 119,000 people had watch course live streaming, 521,000 people watched the course videos, which was welcomed by teachers and students. Teachers' teaching videos were automatically converted into a part of course resources, which not only improved the efficiency of recording, but also helped them diagnose the teaching effect and continuously improvement, for students, they can watch videos anytime and anywhere without the limitation of time and space.

More teachers join in the use of class interaction function, so that changing the traditional teaching mode, promoting students to use mobile terminal for learning, after using 8* courses to initiate 601 times of roll calls, 21, 680 student attended, 19 courses to initiate 228 times of question answering, 7961 student attended. The interaction in the class has become more intelligent, and the data can be recorded and saved to help students form wrong problem sets, review knowledge points, better self-management and adjust their learning schedule and plan (Figs. 6, 7 and 8).

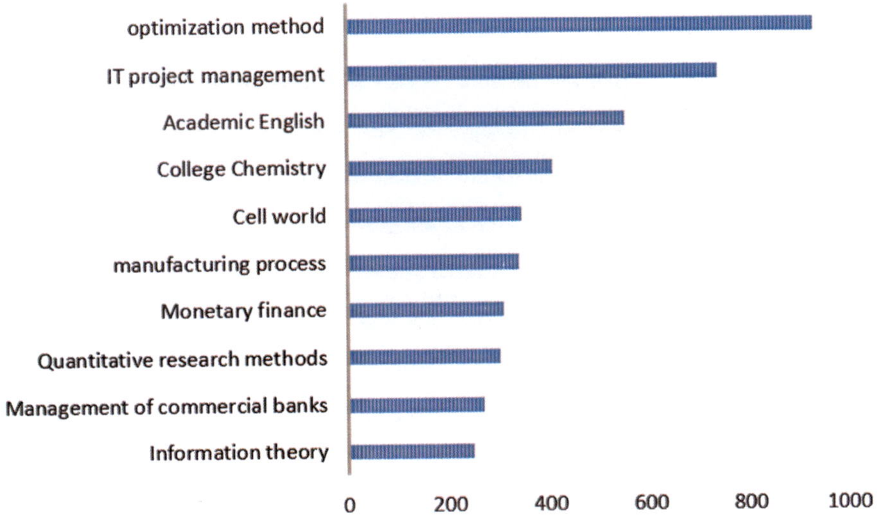

Fig. 6. Top 10 courses using roll call.

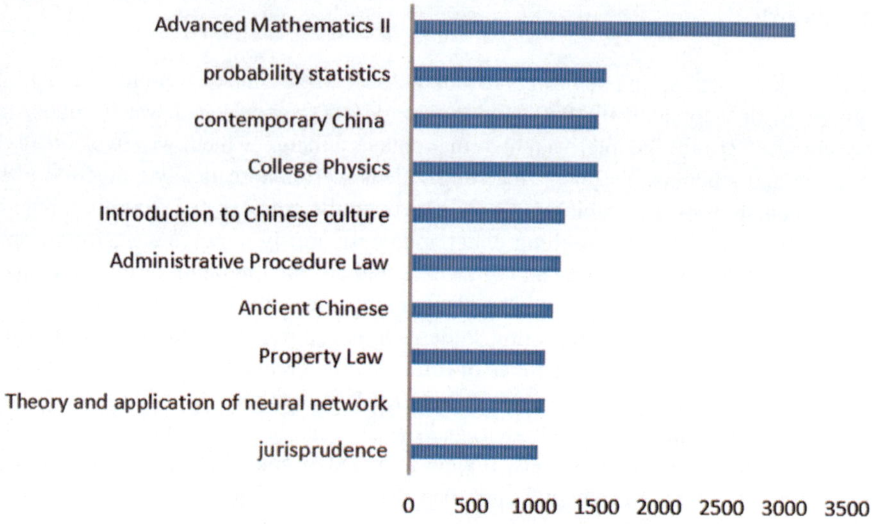

Fig. 7. Top 10 course visits for live streaming.

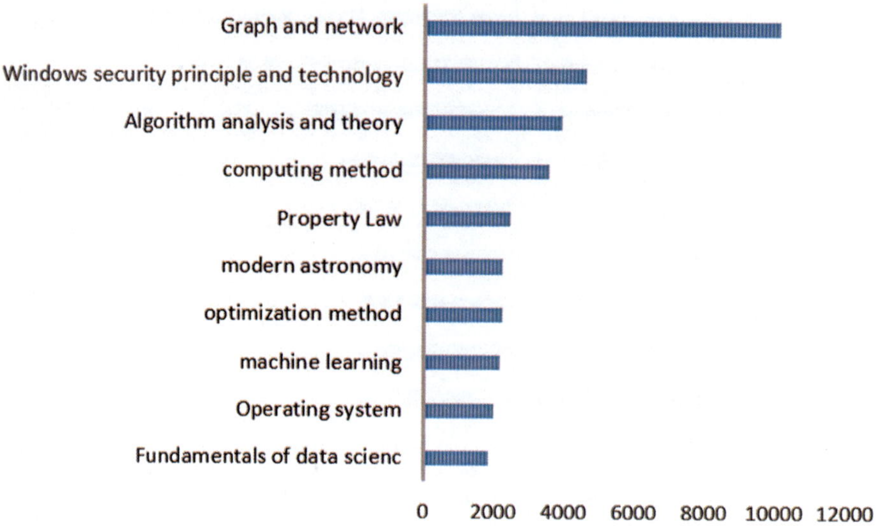

Fig. 8. Top 10 courses visits for video on demand.

4 Conclusion and Suggestion

The two new functions are well received by teacher and student, but at present, the function of "class video" module is relatively simple, which only provides them view the live and on-demand class video. Campus IT staff will continue to optimize the class video function and improve the experience of video resources, such as keyword retrieval, speech recognition, multilingual subtitles, automatic generation of knowledge map, intelligent push of related resources [7], so that helping students preview and review at their self-learning pace.

The mobile terminal will enhance the class interaction, but it just puts into use, and has not formed the large scale effect. The types of supporting questions should be more rich, which can be further developed in the later stage, such as bullet chat, answer race etc. [8].

With the advent of 5G era, mobile learning will be applied in the whole process of education. The time and space of teaching and learning will be reconstructed. Therefore, online and offline integration of teaching has gradually become a new normal. This paper introduces the application of mobile learning function, and optimizes the teaching experience and learning experience based on the learning management system, so that providing a certain reference value.

References

1. China Internet Development Statistics Report. http://www.cac.gov.cn/2021-02/03/c_1613923423079314.htm. [EB/OL]
2. Hongxing, S., Dakui, H., Jingjing, J.: Online teaching practice of suspending classes without stopping learning. Mod. Educ. Technol. **030**(005), 11–18 (2020)
3. Learning Tools Interoperability: http://www.imsglobal.org/activity/learning-tools-interoperability. [EB/OL]
4. Overview of LTI. https://www.imsglobal.org/specs/ltiv1p1/implementation-guide. [EB/OL]
5. Hongxing, S.: Application of EPTZ intelligent tracking camera in cloud recording and broadcasting. J. Guangxi Radio Telev. Univ. **27**(002), 5–9 (2016)
6. Kern, R.G.: Restructuring classroom interaction with networked computers: effects on quantity and characteristics of language production. Mod. Lang. J. **79**(4), 457–476 (2011)
7. Repp, S., Meinel, C.: Semantic indexing for recorded educational lecture videos. In: IEEE Conference on Pervasive Computing & Communications Workshops. IEEE (2006)
8. Liu, Q.T., Feng, X.M., Zhai, H.Q., et al.: The design and implementation of classroom interactive tools supported by learning analysis. Mod. Educ. Technol. **86**, 1–17 (2018)

Work-In-Progress: Hybrid Education: From Telecommuting with e-Learning to Broadcasting Ourselves

Anastasios Nikiforos[1(✉)], Vasileios K. Vasileiou[2], Georgios H. Patronas[2], and Stavros N. Dimitriadis[1]

[1] Department of Informatics, Aristotle University of Thessaloniki, Thessaloniki, Greece
nikiforos@csd.auth.gr

[2] Department of Music Science and Art, University of Macedonia, Thessaloniki, Greece

Abstract. The popularity of Distance Learning during the Covid-19 lockdowns, apart from adhering to the doctrine of utilitarianism as a serviceable amenity for societies deprived of their natural outlets for mental effort, has also served as a forerunner for services and activities that could reshape the formal manner for conveying professional training. In the sphere of Educational TV, the exchange of standardized high-level emissions regulating the transfer of skills and knowledge on specific subjects has been plagued by phrenic pursuits to cling online learning material, designed for use in an Institute's training courseware, as a *de facto* substitute compelling the thoroughness or comprehensiveness of regular study material. This research examines how this stance fits in accordance with facts or reality.

Keywords: Tertiary education online · Technical education and training · Broadcasting · Engagement and interaction

1 Introduction

Williams & Chinn [1] discussed that learners today, referred to as the "net generation", have different styles of information processing and learning expectations. They desire more autonomy and interaction and view the learning process as a social experience [2]. They are today's *prosumers* and have a leading role in this new paradigm of participatory culture, which is fuelled by the abundance of technological interactivity. Meanwhile, Social Networking Services (SNSs) have fostered the emergence of online communities, which tend to emerge from common shared interests; therefore, it's not surprising that people with similar attitudes and ideas tend to gravitate towards each other [3].

SNSs have been particularly interesting online spaces to study in m-Learning contexts especially because they are ubiquitous and provide the potential for enhancing access to educational resources for disadvantaged groups [4]. SNSs have been used in a range of m-learning setting such as, in-service teacher trainings [5], language skills [6].

M. E. Auer and T. Tsiatsos (Eds.): IMCL 2021, LNNS 411, pp. 1094–1104, 2022.
https://doi.org/10.1007/978-3-030-96296-8_100

On accord of its popularity, Facebook is a platform that has been studied in educational settings. Research findings indicate that Facebook can help students to construct an environment which will supply the education process. Cerda & Planas [11] describe Facebook as platform for "supporting social constructivist activities" through online collaboration, and the ability of the platform to support third-party software, diverse communities can be established. Diverse communities of education in the era of the pandemic, include self-broadcasting and participation On-line. A social networking cite of that caliber features a wide variety of communication tools and among the instruments used for interaction of any level, broadcasting and its aspects appear, amongst other interaction activities, in educational process also.

In conclusion, policy-making regulatory techniques discussion, is essential for creating self-broadcasting to a formal educational procedure. A distinction between other formats and modules of information and a broadcasting procedure that has educational purposes, is crucial for establishing safety and access to knowledge from valid administrators. Higher education challenges, due to Covid-19 pandemic, is providing a most urgent field of topic, for analysis and regulation in settings.

2 Broadcasting Clarification

When talking about broadcasting in the digital age of information, a clarification should be provided of what broadcasting means and what level of engagement is providing to people, not only people who broadcasting themselves but also people who are watching, taking knowledge or even discussing about these broadcastings.

By making this clarification we can assume that, people are engaging any topic by participating to the broadcasting procedures. *A transformation from a viewer to a participant is becoming through the broadcasting procedure. The broadcasting experience is giving the chance to turn passive to active, separation to synergy.* A modern digital way to ensure that the topic handled is in public concern. The public may vary for the reasons participating to the procedure, but engagement is happening from the very first moment of taking part to broadcasting. Even discussing it online or not, can prove a somewhat level of engagement.

Despite general opinion, that electronic media provide alienation to engagement to public issues and information, research suggest otherwise. The most popular landscape to provide a validation about engagement through electronic media (therefore broadcasting), is politics.

2.1 Validation of Engagement

People tend to prefer the inexpensive communication technologies not just because they are cheaper but also because they can browse the web and interact with the media, generate their own content and discover different paths of news in order to choose in what news are interesting in [3].

The advent of sophisticated technologies of communications made mass media more easily to access internationally and allow more viewers from different locations to participate in severe situations either passive (viewers, observers) or active (as direct participants) [7]. One of the very first areas that self-broadcasting appeared in higher levels of engagement was in politics. A pathway to investigate how a comment, a video or just an image of previously passive participants can make a difference into the channels of information, is politics. Commentators could easily start a conversation, protests can challenge the information given by official authorities to the media, truth is somehow a reality and it is not censored.

2.2 An Example

A huge example about what self - broadcasting can enhance to information reality is found on Tahrir square in 2011 and the #FreeMona movement. After Egyptian-American journalist Mona Eltahawy was detained by Egyptian police, she tweeted "was beaten arrested in interior ministry". Online activists and followers of Eltahawy immediately started a #FreeMona campaign on Twitter, trying to reach out to the global world in an effort to help her. In a couple of minutes New York Times columnist Nick Kristof, who was also taking part in the FreeMona online campaign tweeted that the US Department of State had been informed of Eltahawy's arrest. The US Department of State, responding to the news of posted a tweet calling the detention "very concerning". A couple of hours later the world knew Mona was safe (firstpost.com, accessed 4/7/2021) [8]. The instantaneous nature of how social media communicates *self-broadcast ideas*, unlimited by publication deadlines and news slots, explains the speed at which these revolutions have unravelled, their almost viral spread across a region. It has also enabled people to ensure that their displeasure with their leaders reaches them directly.

In the digital age of political communication, it is observed that people no longer require the traditional or broadcast media in order to demonstrate their concerns. People are no longer in need of using TV images or radio voices to get upset. This could happen through interpersonal communication (with the use of Facebook), with phone contact, or face-to-face communication and Twitter in the case of Tahrir square. It is now statically clear that social media tend to significantly increase the odds of a person to attend a protest whereas satellite TV decrease the number [9].

The emergence of Facebook and Twitter gave citizens the ability to stand for their rights in countries that before the digital age, people didn't have the way or the license to do so because of the elite regimes that had occupied the authority and also the traditional media. These regimes had put the broadcasting media into a situation of censorship. It is fair to say that digital media have changed the nature of political communication especially in depressed countries. The new media acting like pressure groups in order the civilians to have the *podium* to express their thoughts. This led to huge changes in political order of things [9].

2.3 Challenges of the Post-pandemic Era in Education

It needs to be noted that, during the Covid-19 pandemic era, many postsecondary institutions continue to encounter challenges of remote instruction and furthermore among other challenges of deregulations in connectivity and the lack of equality of usage, in quality hardware, purposes and functions of online teaching and learning must be generally established so whoever concerns, realize that participants in online education must be prepared not just for the complexity of digitalization but also to get ready for internationally knowledge-based economy [12]. More specifically, institutions of higher education, mainly from the most developing countries have deployed all the innovative technologies and a mixture of approaches (synchronous, asynchronous) in all the departments involved, due to the pandemic. Despite the efforts, there is no specific prediction about the usage of information and communication technologies (ICT), in the future [15]. Broadcasting can achieve a high level of engagement as mentioned before but, according to recent studies students think that distance learning cannot reach the quality of a face-to-face instruction. A fact that is generally accepted by all parts involved [16].

Regularly and before the pandemic a part time mobile communication and collaboration during a semester was a contributor to better education therefore remote instructions and educative processes online were saluted from all the parts involved. Due to the pandemic and the frustration that occupied educational life and proceedings, a different adaptation should be made. Another reality with a specific framework about broadcasting and interactive learning methods, will be made. Changes should be established and implementation process can vary (as seen in Fig. 1.) depending the difficulties encountered and the speed of adaptation provided by faculties, teachers, students and governments.

3 Policy Making to Educational and Informational Programming

Broadcasting, concerning education must be framed and licensed. By doing so, broadcasting procedures in all levels of education will achieve a level of *formal recognition* and also acceptance by participants. A module that has regulatory techniques provides safety and also a formal format that will be recognisable. In that terms, official broadcasting for educational purposes will conduct a separation from other types of broadcasting and students along tutors will be directed, precisely to remote education and not in other level of broadcasting information. Tutors and students will take part to educational process not in informational process. A proper license from operators (ministry of Education, institutional administrators), will distinguish broadcasting for educational purposes from a typical broadcasting procedure. Policymaking absence is found in a large variety of topics concerning digital usage of devices, digital privacy and software in the modern era. By establishing policymaking frames, using incentives of regulatory authorities across the globe, a formal module of education broadcasting will occur. Research of examples of regulatory authorities creating law-frame, to digital content and remote information will enhance broadcasting framing with valid and tested incentives and analytics. Furthermore, the general mindset about distinction between formal and informal procedures of education, will also be created.

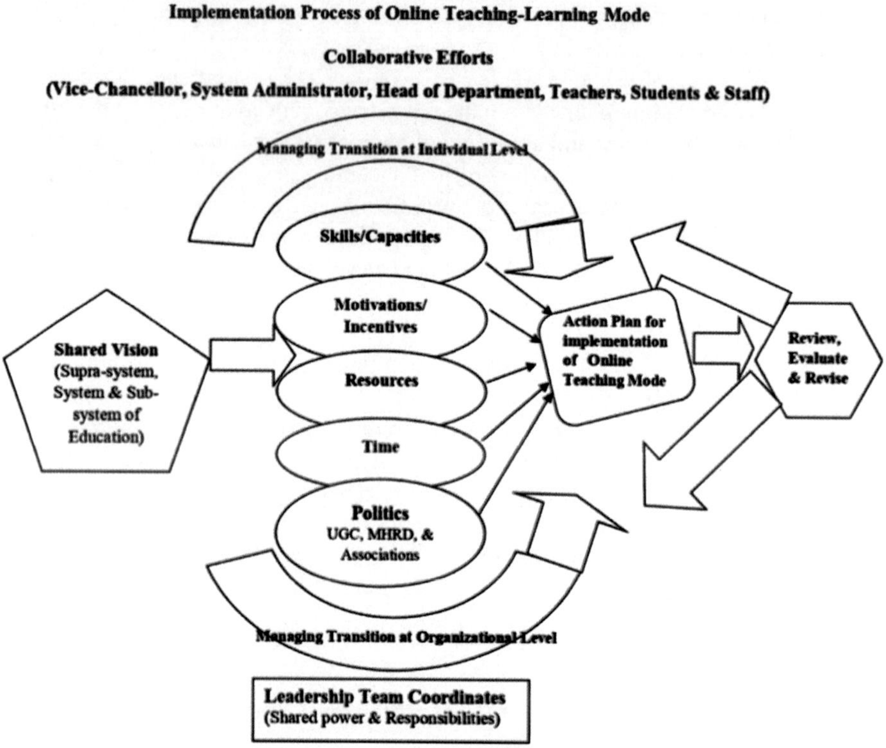

Fig. 1. Implementation process of online teaching-learning mode. Retrieved from *"Online teaching-learning in higher education during lockdown period of COVID-19 pandemic"*, Mishra L., Gupta T., Shree A., (2020), International Journal of Education Research Open, p. 4

3.1 Examples

A largely accepted opinion that policymaking needs a regulation agency is found on the United States of America. Furthermore, Australia created a regulatory system that involves all the parts to the decision-making procedure. Without any regulatory advice, hybrid systems also appeared in Australia concerning the distribution of news and different kind of information. Other concerns or proposed incentives for regulation systems or gatekeepers that can occur to broadcasting procedures will be analysed.

3.1.1 Example of USA

In terms of providing safety and regulatory authority the government of USA created policy-making agencies and bureaus concerning a variety of different matters of public affairs, including education. The USA Congress in 1972, passed the Federal Advisory Committee Act to ensure that advice by advisory committees is objective and accessible to the public. Under the authority of the Act, Federal Communications Commission was created (FFC). The FFC's mission as described is to regulate, interstate and international communications by radio, television, wire, satellite, and cable in all

50 states, the District of Columbia and U.S. territories. An independent U.S. government agency overseen by Congress, the Commission is the federal agency responsible for implementing and enforcing America's communications law and regulations [13]. The Commission among other responsibilities is to regulate Educational and Informational Programming. To implement educational and informational programming mandate, the Commission has adopted a "core programming" processing guideline that applies to commercial and non-commercial educational television licensees. The key factor to the whole process is that, this particular regulatory authority has independent public servants, that decide if granting a TV or a networking licence serves the public interest. The FCC allocates a portion of the broadcast spectrum to new broadcast stations based upon both the relative needs of various communities for additional broadcast outlets, and specified engineering standards designed to prevent interference among stations and other communications users.

3.1.2 Example of Australia

Another regulatory environment for media and information is established in Australia. For the most part, media regulation in Australia is the responsibility of the Federal Government rather than state or territory governments. The regulatory framework is based on four tiers [14]:

1. Legislation passed by Parliament.
2. Government regulation, such as 'content standards' decided by the regulator, the Australian Communications and Media Authority.
3. Co-regulation – regulatory instruments developed by industry Participants
4. Self-regulation – regulatory instruments developed and enforced by industry participants

In that case, cooperation between all parts involved is creating a regulatory environment. There is also a huge variety of sector rules and instructions that provide a possible safety in information (and remote education). The element that requires to be noted is, *Content regulation* which imposes rules relating to accuracy, fairness, the representation of opinions and other aspects that are sometimes regarded as measures of 'quality', as well as quotas and related rules which require certain services to be provided as well as the promotion of localism [14]. This regulation is all behavioral in character, although it takes all four forms of regulation, mentioned above.

3.2 Connection with Academia

In practical terms, how does the established broadcasting system mingle with academia? Usually, tertiary education Institutes are engaged with the pursuit of research, education, and scholarship in intellectual environments that are characterized by concealment of investigation and development products. Of course, there is an outward notion to formally announce the results of their research outcomes. However, this is done more in numbers of published papers than in juicy terms of products being in the most advanced stage of their development.

In more simple terms, it is more about teaching, suggestions, manifestations, mental impressions or ideas rather than probable or possible patents, nearing the stage of the formation of components ready to be deployed as merchandised commodities.

For instance, as the recent healthcare crisis around Covid-19 has revealed, the initiative lies decisively at all levels in the industrial field rather than the academia.

This fact is not diminishing for most branches of study, since many disciplines by default have orientation towards the theory of a subject rather than its practical outcome. For instance, a history professor is bound more to teaching than to building exhibition places with cultural interest. A computer scientist is more concerned about the foundations of proposed model structures than the applications of programming.

The bondage with the broadcasting industry, however, seems to be a game changer for these greatly constraining circumstances.

For instance, as seen in Fig. 2 students of Byzantine Music at various levels perform in front of the National Broadcasting Corporation crews aiming to utilize the professional level of its production standards. Even further, the dissemination of the recorded melodies, through social media, allows vast.

Fig. 2. Pictures from broadcasted emissions, live and via the Internet portal of the National Broadcasting Corporation of Greece in 2019 and 2020. Staff members of the Music Science and Art Dept. of the University of Macedonia perform with their students' choirs.

Such events are of great importance. They have to do with the effect of prime-time broadcasts vs. prime-event transmissions, the latter usually casted over the Internet [10]. Indeed, as seen in Fig. 3, such events assemble spectators much more effectively and more focused on the actual event, as opposed to casual viewers of television time.

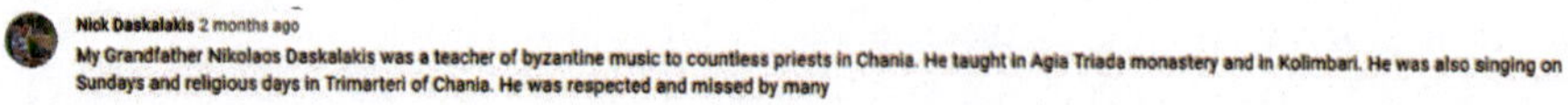

Fig. 3. Reaction moments for viewers worldwide to the Internet version of the aforementioned emissions through social media. Comment cropped the 10th of July 2021.

In terms of qualitative metrics, these statements can be verified by the informational content of their comments, as they appear in social media outlets. They are not simple click-ratings for or against an event nor they are just figures of calculated viewership. They are critical points for an evolving fruitful conversation.

As a result, they provoke user commitment rather than mere user support.

In terms of organized procured activities, they create the tracking system for successful transition to the working environment. They do not rely on scholarships, internship programs or other kind of financial endowments - which of course are very helpful, but create a momentum through social awareness of their societal role.

This approach furthers education perspectives to higher levels and increases chances for regular employment. In Fig. 4 these stances are depicted as student initiatives.

Το Μεγαλύτερο Φοιτητικό Portal | Αποστολή Άρθρου/Καταγγελίας | Μεταχειρισμένα Φοιτητικά Έπιπλα | Διαφημιστείτε | Επικοινωνία

ΦοιτητικάΝέα
www.foititikanea.gr

Σάββατο, 10 Ιουλίου 2021 | 17:00:50

#Πλατφόρμα Αμοιβαίων Μετεγγραφών #Πρωτοετείς #Μετεγγραφές #Exetastiki #ERASMUS+ #Eudoxus #Θέσεις Πρακτικής Άσκησης #WebTv #Πτυχιούχοι

ΑΡΧΙΚΗ · ΕΙΔΗΣΕΙΣ · ΦΟΙΤΗΤΕΣ · ΠΑΝ/ΜΙΑ · ΤΕΙ · ΥΠΟΤΡΟΦΙΕΣ · ΙΕΚ · ΜΕΤΑΠΤΥΧΙΑΚΑ · ΚΟΛΛΕΓΙΑ · ΔΙΔΑΚΤΟΡΙΚΑ · ΠΑΝΕΛΛΗΝΙΕΣ

ΠΑΝΕΠΙΣΤΗΜΙΟ ΜΑΚΕΔΟΝΙΑΣ

Φοιτητές του ΠΑΜΑΚ φέρνουν στα σπίτια μας τα κατανυκτικά Εγκώμια της Μεγάλης Παρασκευής

FoititikaNea Newsroom | 11:01 - Παρασκευή, 17 Απριλίου 2020

Τα κατανυκτικά Εγκώμια της Μεγάλης Παρασκευής στα σπίτια μας φέρνουν φοιτητές του ΠΑΜΑΚ

Στις δύσκολες συνθήκες της πανδημίας και της αναγκαίας στέρησης ακουσμάτων που συνοδεύουν τον αρχικά πένθιμο και στη συνέχεια χαρούμενο και αναστάσιμο εορτασμό της Πασχαλιάς των Ελλήνων, το Πανεπιστήμιο Μακεδονίας προσφέρει μια μοναδική και τεχνολογικά αξιοσημείωτη δυνατότητα αναπλήρωσης της προσωπικής μας συμμετοχής στον εκκλησιασμό της εσπέρας των Εγκωμίων.

Fig. 4. Social awareness about the particulars of the Covid-19 lockdown in April 2020 and well-spread information about particular developments concerning student on-line broadcasted activities. Conveyed info concerns other practical aspects, like high added-value seminars and webinars.

4 Raising the Standards: Providing Professional-Level Expertise

The next step in forwarding student skills has to do with what are perceived as competence dexterities for the ones that follow a University degree. By graduation, newly acclaimed professionals should be able to demonstrate expert completion of tasks they have been taught to perform.

As seen in Fig. 5, the joint choir and orchestra of staff members alongside their students may perform in several occasions at high-level ceremonies, demonstrating expert skill and grasp of their field in front of a (global) public.

Fig. 5. Footages from exhibitions. Left, the students' choir of the Dept. of Music Science and Art, University of Macedonia, performing for the graduation ceremony of their Dept. Right, the joint orchestra of professors and students for the aforementioned Dept. performing works of well-known Greek contemporary composers in the Thessaloniki Concert Hall.

However, in the competitive world of paid occupations, it is not clear if these performances, albeit their prolonged training, may substitute institutionalized schemes. For instance, it is out of question to have these orchestras as competitive arrangements for any State Symphonic Orchestra, like the ones that exists in Athens or Thessaloniki.

The periodically renewed student performing groups are by no means a substitute to any sound professional arrangement - not only for the field of Music Science, but for every science and professional training course. So, what is their relative position?

As seen in Fig. 5 they may serve as serve as highly qualified performing bodies that demonstrate, with their prominent and important position in the doings of a city, their potential to promote innovative methods in featuring inventive roles of operation. Roles that established, conventional professional bodies have not the creative potential to bring about (Fig. 6).

Fig. 6. Singing in the 4th century AD Rotunda, Thessaloniki, 2019. Staff members and their choirs perform a special repertoire for their esteemed guests.

5 Conclusion

The average professional is caught by surprise when he encounters the performing potential of new graduates - provided that their institution may give them the floor to accomplish tasks and recreate learning to actual working prospect. An overall assessment is that the speed of adaptation to new challenges encountered (due to the pandemic or not), will provide the answers needed about the procedure of overcoming the pandemic era. ICT professionals along with information technology experts and education specialists have already created the path for the Grand opening of a different future for all parts involved in Education process (students and their relatives, policy makers and governments, faculty members and institutional managers). Self-Broadcasting was a phenomenon that came along with the advent of Digital Media and was never regulated or framed in terms of higher education. Formality and regularity in modules, is the basic stone of every successful structure that can come along with broadcasting in education. Despite the frustration because of the transition that no one involved in education expected, a lot of the procedures used in remote and distance learning were used partially almost in every single institution. The real challenge occurs when, past inequalities in education are now widened not just in domestic educative levels and institutions but also between countries.

References

1. Williams, J., Chinn, S.: Using web 2.0 to support the active learning experience. J. Inf. Syst. Educ. **20**, 165–174 (2009)
2. Livingstone, S., Lunt, P.: Mediatization: an emerging paradigm for media and communication studies. In: Lundby, K. (ed.) Mediatization of Communication. De Gruter – Mouton, Berlin (2014)
3. Gurevitch, M., Coleman, S., Blumer, J.G.: Political communication-old and new media relationships. Ann. Am. Acad. Pol. Soc. Sci. **625**, 164–181 (2009)
4. Zhu, X., Woo, S.E., Porter, C., Brzezinski, M.: Pathways to happiness: from personality to social networks and perceived support. Soc. Netw. **35**(3), 382–393 (2013)
5. Ciampa, K.: Learning in a mobile age: an investigation of student motivation. J. Comput. Assist. Learn. **30**(1), 82–96 (2014)
6. Born, C.J., Nixon, A.L., Tassava, C.: Closing in on vocabulary acquisition: The use of mobile technologies in a foreign language classroom. In: Models for Interdisciplinary Mobile Learning: Delivering Information to Students, pp. 195–210. IGI Global, Hershey, PA (2011)
7. Negrime, R.: The Communication of Politics, 1st edn. SAGE Publications Ltd, New York (1996)
8. https://www.firstpost.com/world/social-media-in-tahrir-square-how-freemona-resulted-in-freeing-mona-139380.html. Accessed 4 July 2021
9. Tufekci, Z., Wilson, W.: Social media and the decision to participate in political protest: observations from Tahrir square. J. Commun. **62**, 1–4 (2012)
10. Politis, D., Nikiforos, A., Aleksić, V.: Differences in radio broadcasting between Europe and America: two separate models and the advent of the digital audio broadcasting system. In: Proceedings of the 7th International Scientific Conference "Technics and Informatics in Education", Čačak, Serbia, 25–27th May 2018, Session 3: Education and Practice, pp. 298–303 (2018)

11. Cedra, F.L., Planas, N.C.: Facebook's potential for collaborative e-learning, 2011. In: Deutch, N. Connecting Online for Instruction and Learning, Published June 26th 2013 by CreateSpace Independent Publishing Platform, pp. 129–130 (2013)
12. Oleksiyenko, A., et al.: Comparative and international higher education in a new key? Thoughts on the post-pandemic prospects of scholarship. In: Gerardo, L.B., Global citizenship education as a pedagogy of dwelling: re-tracing (mis)steps in practice during challenging times. Globalisation, Societies and Education, vol. 19, no. 4, pp. 432–442 (2021)
13. Federal Communications Commission. Accessed 11 July 2021
14. Wilding, D., Fray, P., Molitorisz, S., McKewon, E.: The Impact of Digital Platforms on News and Journalistic Content. University of Technology Sydney, NSW (2018)
15. de Wit, H., Altbach, P.G.: Internationalization in higher education: global trends and recommendations for its future. Policy Rev. Higher Educ. **5**(1), 28–46 (2021)
16. Chan, R.Y., Bista, K., Allen, R.M.: Online Teaching and Learning in Higher Education during COVID-19, International Perspectives and Experiences, Fourthcoming by Routledge, preview (2021)

Digital Technology in Sports

Evaluating a Serious Game for Halting Harassment and Abuse in Sports

Lampros Karavidas[1(✉)], Thrasyvoulos Tsiatsos[1], Stella Douka[1], Monica Shiakou[2], and Andreas Avgerinos[2]

[1] Aristotle University of Thessaloniki, Thessaloniki, Greece
{karavidas, tsiatsos}@csd.auth.gr, sdouka@phed.auth.gr
[2] European University Cyprus, Nicosia, Cyprus
{M.Siakou, A.Avgerinos}@euc.ac.cy

Abstract. With athletes today often being pushed to physiological and psychological extremes in pursuit of sporting excellence the line between training and abuse can become blurred. To help mainly young athletes to recognize and avoid harassment and abuse by their coaches, a serious game named HALT was designed and developed. This paper's purpose is to present the evaluation of the game by computer science students. The game is designed as a 2D narrative game based on the daily life of athletes. The main element of the game is that the player must decide and evaluate the suitability of certain behaviors that he has discovered in five different mini stories. The scenarios have incidents of abuse and harassment. To test how enjoyable and usable is the serious game an evaluation was held, to help us realize the problems related to the game and decide what improvements should be made. To achieve this a questionnaire that consists of the EGameFlow questionnaire [5] and the USE questionnaire [6] was used. Our evaluation showed us that our serious game was found to be usable and enjoyable and that some improvements should be made to make it more immersive and challenging.

Keywords: User evaluation · Serious games · Harassment and abuse in sports

1 Introduction

Harassment and abuse are considered breaches of human rights, and occur in all countries, cultures, and environments. This is harmful both for the individual, but also for sports as an organization. It is therefore important to work to establish sound sports environments at the level of a club, league or federation. Everyone within sports has a responsibility to identify and counter harassment and abuse and to develop a sports culture characterized by respect and safety. Individuals in sports, as elsewhere in society, may have different opinions as to what harassment and abuse constitute, and how it can be countered.

With athletes today often being pushed to physiological and psychological extremes in pursuit of sporting excellence [1] the line between training and abuse can become blurred. There has been a growing understanding of the occurrence of sexual abuse of young athletes in sport [2], but very little research has attempted to explore

M. E. Auer and T. Tsiatsos (Eds.): IMCL 2021, LNNS 411, pp. 1107–1114, 2022.
https://doi.org/10.1007/978-3-030-96296-8_101

other forms of abuse, such as emotional abuse, within this environment. Previous research has suggested that some standard coaching practices in competitive sport can be a threat to an athlete's physical and psychological well-being. These coaching behaviours are seldom challenged and are an important area of inquiry [3].

To help mainly young athletes to recognize and avoid harassment and abuse by their coaches, a serious game named HALT was designed and developed. This paper's purpose is to present the evaluation of the game by computer science students. The rest of the paper is organized as follows: In the next section we present digital work that also had as its focus point to inform the people for any kind of abuse and harassment and the serious game HALT that was developed by following experts' instructions. Finally, we discuss future adjustments that will be implemented because of the evaluation findings analysis, as well as the effectiveness of the serious game that was developed.

2 Related Work

A serious game was developed to raise awareness and prevent child sexual abuse. The game "Name" was developed during Orbit which is a free child sexual abuse prevention program [4]. Name is an adventure game, in a non – realistic setting so as to be more engaging the users, who are children 8–10 years old, and to present ideas of child sexual abuse in a less intimidating way. The development team of the game consists of professionals specialized on child protection such as officers, counsellors and psychologists and education experts as teachers.

The user gets to choose his/her gender and as a result the main character of the game, who is Sammy, is assigned with the same gender. Sammy is sexually abused by one of the other game characters. The game consists of five different levels where the user assists the main character to realise that he/she is sexually abused and tell his/her trusted adults. Each level is completed only if the user has found an adult to trust that is from his/her real life from a certain domain, such as his/her teacher or people from his/her family. This condition results in building a support network of five trusted adults from different aspects of their lives.

The different levels have many scenarios to be engaging and replayable. After completing some missions in the game or a whole level the player gets some rewards to feel that he achieved to complete his mission and be gratified. Some of the rewards have a great impact on the game's main character as well, by helping him / her reconnect with the crew.

3 More About the Serious Game

3.1 Game Design

The game is designed as a 2D narrative game based on the daily life of athletes. The main element of the game is that the player must make a decision and evaluate the suitability of certain behaviors that he has discovered in four different mini stories. All the stories' incidents showcase different kind of abusive behaviors in the sports. Finally, after completing all these four stories, the player can unlock the fifth level, where considering the stories shown to him in the first four stages, he/she must answer to some more questions.

After observing an incident, certain phrases are separated and the user is asked to rate how appropriate they were. The provided answers are "Extremely Appropriate", "Appropriate", "Neither Appropriate or Inappropriate", "Inappropriate" and "Extremely Inappropriate". Each answer given to all these questions is accompanied with a certain amount of points according to how close it is to the right answer as defined by the experts.

3.2 Game Scenarios

The provided scenarios are written by experts and people with experience in the sport world. All of them, are written to be as close as possible to real events in real life and to help users understand what kind of behavior is abusive. Finally, the users might be able to recognize these kinds of events in their daily life, when they are the victims or just observers and act properly.

The player is a reporter that is responsible of one of the school paper's articles named "A Day in the life of an athlete" and should write the story after making some research. For this purpose, he/she decides to observe some of the athletes that are classmates of his/hers. Unfortunately, the player to every field that he gets to visit that represent one of the four scenarios, that were previously mentioned, observes incidents of harassment and abuse.

The first scenario takes place on the track and field course. As soon as a female athlete named Katia arrives on the field to start her training her coach is aggressive towards her and says that she is wasting his time and finally ends with him sexually harassing her with inappropriate phrases as "Off you go now kitten". This scenario concerns the criticism of a female athlete's perforce in track and field by her coach. It aims to investigate the knowledge and perceptions of players on behaviours, which constitute harassment and abuse in the context of sports and finally provide feedback in the form of knowledge.

The second scenario is in the tennis field. A female tennis coach has been spotted in her free time holding hands with another woman and two athletes' mothers are furious. While having a discussion the user hears one of them proposing to have a chat with the club administrator and ask him to assign Jesicca to male athletes and the other one replying "We can't have lesbians training our children!". This scenario explores issues that can arise when dealing with questions relating to homosexuality in sport.

The third scenario takes place on the soccer field and in the team's locker room. There the player hears the coach giving instruction to the team players during training and guide them to learn a new skill related to the game strategy. However, some difficulties are presented when two players have a hard time understanding what they need to do, to get the exercise done properly. The coach then stops the training and gathers the players around him. He points his finger and with clear irritation on his face, he addresses his players. After accusing them with harsh words the team heads to the locker rooms where the two characters are bullied by the rest of their team for their skills and when their teammates decide to spend their free time together, they are not invited. The scenario concerns the criticism of the athletes' performance by their coach and teammates. It aims to investigate the knowledge and perceptions of players on behaviours, which constitute harassment and abuse in the context of sports and finally provide feedback in the form of knowledge.

The fourth and last scenario is related to a character who is presented as a user's friend and is the only one who addresses straight to the camera by breaking the fourth wall. Elena, is an athlete who tries to get back to rhythmic gymnastics. All of the harassment is related to her gaining some weight. Her coach accuses her of not trying hard enough and verbally harasses her. On the other hand, her physiotherapist sexually abuses her as he mentions that she should not lose any more weight as she is more appealing and later, she winks at her with meaning.

All the scenarios present incidents in a variety of sports related to verbal abuse, emotional and physical harassment, neglect and sexual abuse. They also tackle issues as homosexuality and bullying so as to provide knowledge to the users how to act when they face this kind of episodes.

3.3 Gameplay

The game can be reached through a web browser, or it can also be installed as an app in an Android smartphone. The iOS version of the game is currently under development and will be released in the following months. Once the player opens the game a login screen (Fig. 1) opens asking for his/her credentials. If he/she is a new user, he/she may press the register button and proceed to create an account. Some of the personal data he / she needs to enter are his/her username, age, first name, last name, gender, country he/she currently lives and email. Moreover, to create the account a password is required. Finally, he/she must choose the language that he wants to play the game as there are several languages that the game is fully translated. Once the player enters these data, his/her account created and can continue to sign in the game.

After signing in, the main menu of the game (Fig. 1) can be seen. If the user is new only the "PLAY", "EDIT ACCOUNT" and the "LOG OUT" buttons are available. Once the player completes the five levels that are part of the game then the "LEADERBOARDS" and "PROFILE STATS" buttons get also activated. By pressing on the "PLAY" button the user is found on a screen where a map with different locations can be seen. The locations are Track and Field, Tennis, Soccer Field, Gymnastics and Quiz. All of them have stations in front them and the player is depicted as an arrow. Once clicking one of the locations the scene that takes place in that space starts and the player can now rate abusive incidents and behaviours he/she watches.

After completing all of the scenarios the player can also compare his/her points to the others' in the leader boards scene (Fig. 1). The option to filter the results according to the scenario of his/her choice is given to him/her. Moreover, he/she may choose to view the score that was gathered while playing the first time the certain scenario, labelled as "First Score", the average points he/she collected labelled as "Average", the score he/she got the last time he played the game which is the "Latest Score" or the best score he/she managed to get in the game.

Fig. 1. Log in screen, main menu, leader boards, profile stats of the game

Finally, there is also the profile stats scene (Fig. 1) which displays a spider graph of the selected scenario to the user and he/she can view his/her best, average, latest and first score for every scenario depicted on this graph. The scores saved in a database can be used to keep track of each player's progress and help the users realize their knowledge growth by looking at the scores for each scenario and the graph.

4 User's Evaluation

To test how enjoyable and usable is the serious game an evaluation was held, to help us realize the problems related to the game and decide what improvements should be made.

4.1 Evaluation Methodology

Participants. The participants were all (n = 29) undergraduate students of the Computer Science Department of the Aristotle University of Thessaloniki. Seventeen of the participants, that is 58.62% of the total number were male and the rest of them were women. Their average age was 22.24 years with a standard deviation of 2.37 years.

Questionnaires. The questionnaire that was given to the users consisted of two separate, weighted questionnaires. The EGameFlow questionnaire [5] was used to evaluate the perceived enjoyment of the users while playing the game. The 42 questions in the questionnaire attempted to evaluate different factors that finally influences the enjoyment of the game. The dimensions the are evaluated through the questionnaire are the following 12: 1) Concentration, 2) Goal Clarity, 3) Feedback, 4) Challenge, 5) Autonomy, 6) Immersion, 7) Social Interaction, and 8) Knowledge Improvement. The second questionnaire which was part of the given questionnaire was the USE questionnaire [6], used to evaluate the usability of the serious game. The aspects of the game that were evaluated were the 1) Usefulness, 2) Ease of use, 3) Ease of learning, and 4) Satisfaction. The statement in each question was rated on a 5-point Likert scale, where the following options were given: 1 (Strongly Disagree), 2 (Disagree), 3 (Neutral), 4 (Agree), and 5 (Strongly Agree).

Procedure. The participants were all undergraduate students of the Computer Science department of Aristotle University of Thessaloniki. The participants were informed for the existence of the questionnaire through an online community of practice that is created to inform and educate for issues related to abuse and harassment in sports. The questionnaire was uploaded with the name "HALT Game Evaluation" and the users proceeded to answering it right after completing the whole serious game.

Data Analysis. The data analysis was done using datasheets and the results of the analysis are presented through descriptive statistics.

4.2 Evaluation Results

Usability Analysis. It can be stated that the game was found to be useful, easy to use, easy to learn and satisfying in general as all these factors were negatively evaluated, with 1 or 2 points in the five-point Likert scale by a small amount of the participants.

Most of them were either neutral or positively evaluated as it can be seen in the following table (Table 1).
It should be mentioned that ease of learning was the dimension that got the higher rankings as 92.23% of the participants answered that they either agree or strongly agree with the questions involved in this question group. On the other hand, the satisfaction that the user felt while playing the game had the lower rankings as 41,36% rated it positively. However, considering that most of the participants stated that they were neutral in these question and only a quarter of the asked rated in negatively, we don't consider the game to not be satisfying.

Table 1. Descriptive statistics of the USE questionnaire

Factor	1 (Strongly disagree)	2 (Disagree)	3 (Neutral)	4 (Agree)	5 (Strongly agree)
Usefulness	6,46%	14,65%	34,05%	33,18%	11,63%
Ease of use	2,82%	5,32%	14,10%	42%	35,73%
Ease of learning	0,86%	1,72%	5,17%	34,48%	57,75%
Satisfaction	6,89%	17,81%	33,9%	30,45%	10,91%

Perceived Enjoyment Analysis. It is obvious that the game is quite enjoyable. Half of the serious game aspects gathered positive feedback, with 4 or 5 points on the five-point scale, from more than 50% percent of the HALT users and evaluators. The rest of the dimensions gathered positive feedback from more than 35%, at least, of the participants and the rest of the answers were neutral rather than negative as it can be seen in Table 2.

Table 2. Descriptive statistics of the EGameFlow questionnaire

Factor	1 (Strongly disagree)	2 (Disagree)	3 (Neutral)	4 (Agree)	5 (Strongly agree)
Concentration	1,14%	6,54%	28,16%	43,1%	21,26%
Goal clarity	1,72%	6,03%	23,27%	50%	18,96%
Feedback	5,51%	15,17%	33,79%	37,24%	8,25%
Challenge	6,89%	17,81%	33,9%	30,45%	10,91%
Autonomy	1,14%	14,94%	26,43%	45,97%	11,49%
Immersion	8,86%	21,67%	32,51%	28,07%	8,86%
Social interaction	6,89%	17,24%	37,24%	31,03%	7,58%
Knowledge improvement	0,68%	4,82%	18,62%	51,72%	24,13%

The lowest rankings were related to the immersion and the social interaction dimensions related to the game as 36.93% and 38.61% of the participants rated them positively. However, 68.96% of the users stated that the goal of the game was clear and 75.85% of the total rated positively the improvement of their knowledge considering the abuse and harassment in sports.

5 Conclusions

To conclude, our evaluation showed us that the serious game that was created to raise awareness and educate young people towards harassment and abuse in sports was found to be usable and enjoyable. In particular, the USE questionnaire analysis made it clear that is perceived as a game useful, easy to use, easy to learn and satisfying and the EGameFlow questionnaire showed that it helps users concentrate on it while playing it, with clear goals, where the user can have great autonomy in it and it improves the users' knowledge. Some issues related to the challenge of the game, the immersion and the social interaction were spotted. All of them were anticipated as the genre of the game, that is a visual novel, is less immersive and challenging compared to other game genres. Moreover, no socializing options were given to the players, apart from the incorporated leader boards, as the purpose of the game is to concentrate on the scenarios and self – reflect in order to gain knowledge.

Some improvements that could be made in the game, to make it more immersive and challenging would be to have scenarios that adapt to the user's performance and number of points he/she collects while playing the game. Moreover, the game could be updated with more complicated answers to raise the challenge level of it and make it harder for user's to gain points.

Acknowledgment. This research has been funded with support from the European Commission. This publication reflects the views only of the author, and the Commission cannot be held responsible for any use which may be made of the information contained therein.

The authors of this research would like to thank HALT Erasmus+ Sport (Project No: 603479-EPP-1-2018-1-EL-SPO-SCP) project team who generously shared their time, experience, and materials for the purposes of this paper.

References

1. Tofler, I.R., Stryer, B.K., Micheli, L.J., Herman, L.R.: Physical and emotional problems of elite female gymnasts. N. Engl. J. Med. **335**, 281–283 (1996)
2. Brackenridge, C.H.: 'so what?' Attitudes of the voluntary sector towards child protection in sports clubs. Manag. Leis. **7**, 103–123 (2002)
3. Gervis, M., Dunn, N.: The emotional abuse of elite child athletes by their coaches. Child Abuse Rev. **13**, 215–223 (2004)
4. Jones, C., Scholes, L., Rolfe, B., Stieler-Hunt, C.: A serious-game for child sexual abuse prevention: an evaluation of orbit. Child Abuse Negl. **107**, 104569 (2020)
5. Fu, F.L., Su, R.C., Yu, S.C.: EGameFlow: a scale to measure learners' enjoyment of e-learning games. Comput. Educ. **52**(1), 101–112 (2009)
6. Lund, A.M.: Measuring usability with the use questionnaire12. Usability Interface **8**(2), 3–6 (2001)

Preliminary Evaluation of a Platform Integrating a VCOP to Prevent Abuse in Sports and a Whistleblowing Service

Nikolaos Politopoulos[1(✉)], Thrasyvoulos Tsiatsos[1], Stella Douka[1], and Monica Shiakou[2]

[1] Aristotle University of Thessaloniki, Thessaloniki, Greece
{npolitop,tsiatsos}@csd.auth.gr, sdouka@phed.auth.gr
[2] European University Cyprus, Engomi, Cyprus
M.Siakou@euc.ac.cy

Abstract. This paper presents a case study evaluation results of a VCOP to prevent abuse in sports and a whistleblowing service. The VCOP has been one of the most highly anticipated trends in higher education in recent years. Delivering web-based learning to potentially thousands of users at a time, it has dramatically expanded universities' reach and provided an indirect income boost through greater global brand awareness. The idea of a VCOP is that the learner becomes the teacher from one case to another. This enables users to interact and exchange knowledge. Abusement and harassment in sports is a severe problem that unfortunately blossomed the last 3 years. HALT project aims to tackle thess behaviors and support all members of sports industry (athletes, coaches, parents and staff) with knowledge in order to detect and avoid them. According to the evaluation results this VCOP is complete, well-structured and segmented, covering the most important parts of coaching.

Keywords: VCOP · Abuse in sports · Whistleblowing

1 Introduction

Since 2007 the European Commission has been issuing policies emphasizing the need to protect (young) athletes from acts of violence, particularly from different forms of gender-based violence. These policies also highlight the importance and necessity of raising awareness and implementing prevention initiatives to fight gender-based violence in and through sport. The existing legal and policy recommendations specifically related to gender-based violence in sport go back 20 years. According to EU the main focus of policies is on prevention and protection actions. Attention has specifically been drawn to children and youth, as well as to women. Homophobic violence in sport starts to be considered in more recent policies. Other topics, like the assessment of the magnitude of any form of gender-based violence in sport, or measures to prosecute forms of gender-based violence in sport and to provide services to support victims (and perpetrators) have been less considered in the international and European legal and policy frameworks. Less than half of the EU Member States make explicit reference to forms of gender-based violence in sport in their policy frameworks within the areas of

M. E. Auer and T. Tsiatsos (Eds.): IMCL 2021, LNNS 411, pp. 1115–1124, 2022.
https://doi.org/10.1007/978-3-030-96296-8_102

sport, gender equality and child protection. Awareness-raising initiatives have been organized in a few countries at public, governmental level to sensitize different target groups about gender-based violence in and through sport. Few countries have developed specific national policies to address forms of gender-based violence in sport. Generally, these policies adopt a holistic approach towards addressing sexual violence (including harassment and abuse) of young and adult athletes, including initiatives focusing on prevention, protection, prosecution, and provision of services. All policies target sport organisations (including federations and clubs). Nevertheless, their implementation is in most cases neither mandatory, nor followed up. Multi-level, multi-disciplinary and multi-actor partnership initiatives were identified in a limited number of EU Member States. At international level, the International Olympic Committee, and to some extent, the Paralympic Committee, are acting to tackle gender-based violence in sport. Initiatives undertaken by these organisations are important because of their visibility, but also because of the influence they can exert on the national committees in complying to initiatives. the concept of 'gender-based violence in sport' is not well known in the EU Member States, although certain forms of violence that fall within the definition of gender-based violence in sport are recognized and addressed in some countries. Among these forms, the most important are harassment and abuse.

In accordance with the mandate deriving from the Council Resolution on the second European Work Plan for Sport (2014–2017), the protection and safeguarding of minors in sport was one of the five issues identified as priorities in terms of sport integrity by Member States and the European Commission.

Research studies, evidence and high-profile cases have drawn attention to the problem of violence in sport that "young athletes" have experienced in the context of sporting activities. Paulo David, former Secretary of the Committee on the Rights of the Child, Office of the United Nations High Commissioner for Human Rights, roughly estimates that, of all children involved in "competitive sports", 10% have undergone human rights abuse, and another 20% are at risk (David 2004).

A study conducted in UK in 2009 (Alexander et al. (2011) reported that 29% of the sample of athletes representing all competitive levels had been sexually harassed (34% of females and 17% of males) while 3% had been sexually abused (5% of males and 2% of females).

Another study conducted in Denmark among 250 sport students showed that about 25% either knew about or had experienced situations in which a sport participant under the age of 18 had been sexually harassed by coach (Nielsen and Toftegaard 2001). From the studies we have consulted (Brackenridge et al. 2012) we may estimate that between 2% and 8% of athletes (both minors and young adults) are victims of sexual assault within the context of sport. J. Nielsen Toftegaard, a Danish researcher, underlined that "Generally, clubs agreed that preventing sexual abuse in sport was more important than the problems of eating disorders, doping and fair play" (Toftegaard 2008–2012).

Studies are still too few, focusing essentially on sexual violence against women and girls and thereby ignoring all other possible forms of "harm" as well as male victims, and affected by inherent methodological difficulties of the topic, making rarely a distinction in age category, the lack of standardised definition, mainly focusing on sexual violence and abuse in a broader sense from "simple sexual comment" to "sexual abuse" and not covering other forms of violence. Today it seems obvious that violence against minors in sport requires special attention.

HALT project is focused on raising awareness about best practices for preventing harassment and abuse in sports.

HALT will create and exploit a bunch of learning technologies (digital games, digital informative material, movie, online community of practice platform) developed by academics to address this goal in a rather big network of athletes in all over the Europe guaranteed by the participation of sports federations, athletes and sport umbrella associations.

2 The Potentials of COPs

The term "Communities of Practice" was developed around 20 years ago, but the origins of the concept are rooted in humanity's evolutionary history, when primitive communities learned how to share knowledge and exchange practices about a wide range of behaviours relevant to the species' survival and development, from hunting, to crafting and manufacturing, printing, and more recently innovating (Lave and Wenger 1999; 2002; Wenger 1998; 2010). This learning did not happen in schools or formal educational settings. Rather, for many centuries informal Communities of Practice (CoP) consisted of people who had a common interest and formed informal groups where they could share knowledge, learn about each other's practices, and come-up with innovative ways of dealing with problems (Lave 1999; Wenger 2002). It was not until the late 1990s that this natural form of learning was emphasized by social anthropologist Jean Lave and educator Etienne Wenger as a way to enable effective learning in educational and other settings (e.g., corporate organizations; Rogoff et al. 2001; Wenger 2002; 2010). According to Wegner (2007), communities of practice represent "groups of people who share a concern or a passion for something they do and learn how to do it better as they interact regularly".

The guiding principle of a CoP is that learning is not necessarily a top-down process, with knowledge flows from an expert (e.g., teacher) to recipients without prior knowledge on the subject matter (i.e., pupils; students; trainees). Rather, learning is "situated", and the "situated learning" approach suggests that knowledge flows are bidirectional, and knowledge is shaped and used dynamically as the members of a learning community share ideas and practices relevant to the subject matter they're interested in (Wenger 1998; 2010). According to this perspective, learning happens through social interactions and the active involvement of "learners" in the initiation, development, and sustainability of the CoP (Lave and Wenger 2000; Wenger 2010).

According to Wenger et al. (2002 pp. 8–10), the three key contentions of communities of practice are:

1. Knowledge resides in the "act of knowing": Expertise is the result of "static" knowledge, and of accumulated experience that has been gained through practice. Communities of practice serve as a dynamic and living repository of accumulated knowledge.
2. Knowledge is tacit as well as explicit: In addition to formal and explicit knowledge that can be delivered through manuals and formal training, tacit (or implicit) knowledge represents the accumulated knowledge that is embodied in experiences,

and represents a deep understanding of complex and context-specific problems. Tacit knowledge is valuable and irreplaceable and should be treated as equally (if not more) important to explicit knowledge.

3. Knowledge is social and dynamic: Although the subjective experience of knowing a subject matter can be largely individual, knowledge itself is a social asset that we get to earn through social interactions with others with more expertise and different practices - in fact, the involvement of people/actors with different perspectives, knowledge assets and practices is often used to stimulate innovative solutions. Knowledge is also changing at a fast pace and individual actors (e.g., individual employees, individual organizations) cannot follow the fast-pacing knowledge generation. Communities of practice provide the framework where knowledge can be captured, contextualized, and re-used in ways that serve the individual and collective learning needs of involved parties.

3 Halt Vcop

3.1 VCOP Platform

In this section a state of the art about e-learning platforms will be conducted and the result will be the selection of the most suitable platform that fits the project's needs.

In order to evaluate the existing CMS platforms, an evaluation table was created, and importance points were assigned to features that are needed on the platform. Importance scale is 1 to 3, 1 slightly important, 2 important and 3 very important. If a platform qualifies for a feature, it was assigned an X on the table. The values are summarized at the bottom of Table 3 (Table 1).

Table 1. VCOP platforms evaluation table

Tools	Importance points (1–4)	CMS Platforms		
		Moodle	Drupal	Wordpress
Communication tools				
Discussion forums	2	X	X	X
Discussion management	2	X	X	X
File sharing	2	X	X	X
Personal messaging	1	X	X	X
Collaboration tools				
Group work	3	X	X	X
Community networking	3	X	X	X
System tools				
PHP/MySQL	3	X	X	X
Multilingual package (easily customizable)	4	X	–	–
Developed plugins	4	X	X	X
Open source	4	X	X	X
Results				
		28	25	25

As, can be seen from the evaluation, the platforms that most suits the needs of the project is Moodle (https://moodle.org/). The consortium experts will make some adjustments to the language packs of the LMS to adapt it to the needs of a VCOP. The final solution of the HALT VCOP is hosted at http://haltvm.phed.auth.gr/ and in six different languages (Fig. 1).

Fig. 1. HALT VCOP

4 HALT Whistleblowing

The HALT Observatory will be a simple digital service that will allow people that were victims of abuse and harassment to submit the incidence anonymously. The consortium decided that the observatory will be a simple form with a simplified drop-down menu containing the following fields:

- Country
- Age
- Type of harassment/abuse
- If the applicant is athlete/coach/parent/administrative personnel

Additionally, some simple stats will be available for all Europe and per country. The HALT observatory is available at http://halt.phed.auth.gr/whistleblowing/ and in six different languages.

5 Evaluation

This section presents the methodology followed in order to evaluate our research work.

5.1 Participants

In order to test the research questions of this article 30 computer science students were used as subjects. There were 18 male and 12 female students with an average age of 22 years.

5.2 Instruments

To evaluate the observatory's and VCOP's usability, project's experts created two questionnaires and made available through the VCOP. The questionnaire for the observatory is based on the USE questionnaire (Lund 2001) The questionnaire for the VCOP consists of 41 questions measuring to what degree participants react favourably to the training and to what degree targeted outcomes occur because of the educational material and subsequent reinforcement. The questions' type is a 5-point Likert rating scale with the following anchors: 1 strongly disagree, 2 disagree, 3 neutral, 4 agree and 5 strongly agree.

6 Data Analysis

The statistical analysis was performed using Excel and the results are presented below.

6.1 Evaluation of VCOP and Structure

Evaluating the content of the VCOP, participants are very positive with it evaluating it in almost all questions with an average of 4 and above (Table 2). They believe that the subjects covered in course was well segmented and structured (Table 3).

Table 2. Participants opinion about VCOP informative material

The VCOP is suitable for me regarding my professional/athletic career level	3.6
The VCOP is suitable for me regarding my prior knowledge on the topic	3.8
The VCOP is suitable for my learning goals	3.9
The learning objectives and topics are clearly defined	4.1
The instructions on how to follow this VCOP were comprehensive and suitable	3.8
The VCOP improves my knowledge regarding coaching	4.2
The supporting reading material facilitated the study of the topics	4.0
The supporting reading material was informative enough	3.9
The supporting reading material was easy to follow	3.9
The multimedia material facilitated the study of the topics	3.7
The multimedia material was informative enough	3.7
The multimedia material was easy to follow	3.9

Table 3. Participants' opinion about segmenting, sequencing and navigation of the VCOP

Is the content segmented adequately (neither too wide nor too narrow)?	3.7
Is the sequencing well structured?	3.8
Is the navigation bar in balance between too many and too little sublevels and are the menu items well chosen?	3.8

6.2 Evaluation of Informative Material

Participants are very positive with the video, texts and graphics presented to the course evaluating it in almost all questions with an average of 4 and above (Table 4). They also have a very positive opinion about the delivery of the course (Table 5).

Table 4. Participants opinion about videos and graphics

Texts were readable	3.9
Texts were understandable	4.0
Graphics were understandable	3.8
The quality of graphics was high	3.5
Graphics and video were readable	3.9
Videos were understandable	4.0
The quality of videos was high	3.6
Videos were readable	3.9

Table 5. Participants opinion about VCOP delivery

Content of VCOP was understandable?	3.9
The volume of activities was equally distributed during the module	3.7
Activities (i.e. quizzes) were clear	4.2
Support and help were accessible when it was needed	3.7
The learning material was structured clearly and user-friendly	4.0
There were enough tools for interaction when it was needed	3.6

6.3 Impact of the VCOP

Comparing this VCOP course with other same VCOPs participants believe that the whole content of this VCOPs is better than other courses they have already attended (Table 6). Finally, they believe that the impact that this VCOP will have on their professional practice will be very significant (Table 7).

Table 6. Participants opinion about this course in comparison other sane courses

The degree of my exposure to VCOPs similar to HALT in the past was high	2.4
The VCOP content compared to the content of other courses, I have attended in the past, is more informative	3.1
The degree of my exposure to VCOPs in the past was high	2.5

Table 7. Participants opinion about the impact of the course to their professional practice

This VCOP will change my professional practice	2.9
The degree of the change in my professional practice will be high due to this VCOP	3.0
This VCOP will improve my professional practice	3.1

6.4 Usability Evaluation of the Observatory

The statistical analysis was performed using Excel. For each one of the four dimensions (usefulness, ease of use, ease of learning and satisfaction) descriptive measures of central tendency, such as mean, were estimated for each Likert-type item from one to seven. The mean values of the four dimensions are being presented in Table 2 (Table 8).

Table 8. Mean values of the four dimensions

	Usefulness	Ease of use	Ease of learning	Satisfaction
Mean	3.45	4.1	4.35	3.61

As indicated by the results of the analysis, the players' opinion was positive, since the mean value (M) for the "Usefulness" dimension is 3.45, whereas M is 3.61, for the "Satisfaction" dimension. Moreover, it could be deduced that the players did not encounter any major difficulties in comprehending and using the tool, since the mean value for the "Ease of Use" dimension is equal to 4.1, while M = 4.35, for the dimension "Ease of Learning". The frequency diagram presented the user interface satisfaction is being presented in Fig. 2.

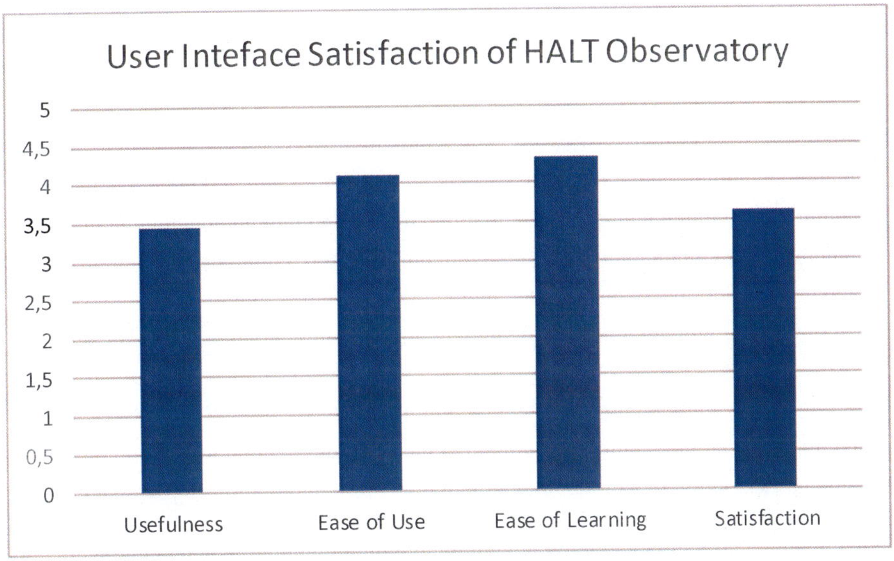

Fig. 2. Frequency diagram of User Interface Satisfaction for HALT Observatory

7 Conclusion and Future Work

This paper presents the evaluation a VCOP platform and a whistleblowing service. These services have been deployed in the context of HALT Erasmus + Sport project.

The evaluation results showed that the VCOP seems to be complete, well-structured and segmented, covering the most important parts of coaching.

It is worth mentioning that the overall results are very satisfying. Participants seem to be very satisfied with the educational content (presentations, videos, documents etc.) The most satisfying result is that participants believe that this course will change and improve their professional practice.

The next step is to evaluate this VCOP to users more relative to the context, for example athletes or students from a department of physical education & sport science on a larger user base. Those platforms are available at: http://haltvm.phed.auth.gr/

Acknowledgement. This project has been funded with support from the European Commission. This publication reflects the views only of the author, and the Commission cannot be held responsible for any use which may be made of the information contained therein.

The authors of this research would like to thank HALT team who generously shared their time, experience, and materials for the purposes of this project.

References

David, P.: Human Rights in Youth Sport: A Critical Review of Children's Rights in Competitive Sport. Routledge (2004)

Alexander, K., Stafford, A., Lewis, R.: MAIN REPORT (2011)

Nielsen, J.T.: The forbidden zone: intimacy, sexual relations and misconduct in the relationship between coaches and athletes. Int. Rev. Sociol. Sport **36**(2), 165–182 (2001)

Brackenridge, C.H., Kay, T., Rhind, D.: Sport, Children's Rights and Violence Prevention: A Source Book on Global Issues and Local Programmes. Brunel University Press, London (2012)

Toftegaard-Støckel, J., Nielsen, G.A., Ibsen, B., Andersen, L.B.: Parental, socio and cultural factors associated with adolescents' sports participation in four Danish municipalities. Scand. J. Med. Sci. Sports **21**(4), 606–611 (2011)

Wenger, E.: Communities of Practice: Learning, Meaning, and Identity. Cambridge University Press, New York (1998)

Wenger, E.: Communities of practice and social learning systems: the career of a concept. In: Social Learning Systems and Communities of Practice, pp. 179–198. Springer, London (2010)

Lave, J., Wenger, E.: Practice, person, social world. In: An introduction to Vygotsky, pp. 155–162. Routledge, Milton Park (2002)

Lave, J., Wenger, E.: Learning and pedagogy in communities of practice. Learn. Pedagogy 21–33 (1999)

Rogoff, B., Turkanis, C.G., Bartlett, L.: Learning Together: Children and Adults in a School Community. Oxford University Press, Oxford (2001)

Wenger, E., McDermott, R.A., Snyder, W.: Cultivating Communities of Practice: A Guide to Managing Knowledge. Harvard Business Press, Boston (2002)

Lund, A.M.: Measuring usability with the use questionnaire12. Usability Interface **8**(2), 3–6 (2001)

Evaluating a Serious Game to Promote Healthy Lifestyle in Young People

Lampros Karavidas[1(✉)], Georgina Skraparli[1], Thrasyvoulos Tsiatsos[1], Stella Douka[1], Andreas Avgerinos[2], and Christiana Philippou[2]

[1] Aristotle University of Thessaloniki, Thessaloniki, Greece
{karavidas, skraparl, tsiatsos}@csd.auth.gr, sdouka@phed.auth.gr

[2] European University Cyprus, Engomi, Cyprus
{a.avgerinos, c.philippou}@euc.ac.cy

Abstract. Physical inactivity has been identified by the World Health Organization (WHO) as the fourth leading risk factor for global mortality. Childhood overweight and obesity are important public health problems in Europe affecting the psychophysical development of children and adolescents and leading to many diseases. Nowadays, as children and teenagers are familiar with technological advancements, they can be used to promote physical activity and a healthy lifestyle. For this purpose, Battle4Health, a two dimensional serious game for mobile devices, was designed and developed. Battle4Health assists its users to have a healthier lifestyle by gaining knowledge about their nutritional, sleep and walking habits. In order to measure the enjoyment and usability of the developed serious game, an evaluation was made with questionnaire method. The EGameFlow [1] and USE [2] questionnaires were used and the feedback was overall positive. Battle4Health was found to help its users to learn about the importance and how to have a healthier lifestyle. However, there should be an improvement regarding the immersion and the challenges that the game offers, by adding more questions on the game's quiz subsystem.

Keywords: Serious game · Healthy lifestyle · Evaluation

1 Introduction

Physical inactivity has been identified by the World Health Organization (WHO) as the fourth leading risk factor for global mortality. Physical inactivity increases the risk for many morbid conditions including, but not limited to, cancer, diabetes, hypertension, coronary disease, cerebrovascular disease, and overweight and obesity [3]. According to World Health Organization (WHO), 60 to 85% of the general population worldwide has a sedentary lifestyle, making it one of the most serious, but yet insufficiently, addressed public health problems [4]. This problem appears in children and young people as well. As presented by Stierlin et al. (2015) [5], several studies have shown that a lot of children spend most of their time being sedentary. For example, 10–12-year-old European children spend approximately 8h being sedentary during the day [6].

M. E. Auer and T. Tsiatsos (Eds.): IMCL 2021, LNNS 411, pp. 1125–1133, 2022.
https://doi.org/10.1007/978-3-030-96296-8_103

Childhood overweight and obesity are important public health problems in Europe affecting the psychophysical development of children and adolescents [7] and leading to many diseases. Nutrition education needs to be more than nutritional literacy. School children were shown to fail to prepare appropriate menus despite being provided with dietary guidelines [8]. In recent years, there has been increasing use of the Internet to deliver Nutrition Education Programs but there are still very little data on its efficacy [9].

However, according to Chaddha, A., Jackson, Richardson, C. Franklin, B. (2017) [10], technological advancements can also be used to promote physical activity. Chaddha et al. (2017) [10] results that mobile games on smartphones and tablets can make physical activity more enjoyable and provide an incentive to be physically active. A recent meta-analysis conducted by Fanning, Mullen & McAuley (2012) [11] showed that research utilizing mobile devices is gaining in popularity and that mobile devices are effective means for influencing physical activity behavior.

1.1 Related Work

Several serious games have tried to change the children's nutritional habits and help them adopt a healthier lifestyle. Marchetti et al. (2015) [12] created a classic two-dimensional action, platform game to help young people gain better knowledge of the nutritional benefits of each food. The player should be able to move around several obstacles on a platform at a certain speed. While moving, a variety of food items would emerge, and the player should be able to hit them to gain more points indicating that these food items would be beneficial for him/her in real life. However, when the user "ate" three or more forbidden food items, that were unhealthy, he/she should start playing the certain level again, teaching young children this way that these foods should be also avoided in their everyday life.

2 More About Battle4Health

2.1 Game Design

Battle4Health is designed to be a two-dimensional serious game. The game will be available to be reached through a standard web browser and as a smartphone application. The goal of the game is for someone to take care of his/her creature. The creatures have the same needs as its owner, who is the user, so that the user can insert his/her own data from his/her everyday life. Therefore, the game supports its users, meant to be teenagers and adolescents, to have a healthier lifestyle. To achieve this, it provides its users with several subsystems to help them gain knowledge about their nutritional, sleep and walking habits. Moreover, several gamification elements such as a scoring and a battle system are incorporated to make the users come back to it more than one time per day and make the serious game a "healthy reminder".

Quiz Subsystem. The quiz subsystem consists of several questions created by specialists in the fields of nutrition and exercise. Regarding the nutrition part, the questions emphasize mostly food groups and the side effects or the benefits that one's organism has by consuming certain food items. On the other hand, the exercise questions, focus mostly on aerobic physical activity and everyday habits that can keep their users active. All of them are either multiple-choice or true/false questions small enough to be read easily on a mobile phone screen.

Nutrition Minigame Subsystem. The nutrition minigame is designed to help teenagers figure out the correct amount of food items per food group they should eat in a day or a week. Then some of the most usually consumed food items are given to the user in certain quantities to imitate his/her meals or experiment and finally realize the right amounts that he/she should eat of each food group in a day. The food groups given to the users are a) breads, cereals and grains, b) fruits, c) vegetables, d) milk and milk products, e) meats and legumes, f) olive oil, g) fats and sweats, h) desserts, i) free food and j) water cups.

Walking Minigame Subsystem. The walking minigame intends to help the users easily track their progress related to their walking behaviour during a week. To do so two kinds of steps are tracked by another currently developed app and the users are rewarded in our serious game, the regular and the aerobic steps. According to the number of regular steps that the user achieves per day a certain badge representing his/her lifestyle is assigned to him/her. If the player achieves more than 3000 aerobic steps for the day, which are steps done with a rhythm of 100 steps/minute and this kind of walking lasts for at least 10 min with no interruption, then he/she earns an extra badge (Table 1).

Table 1. Physical activity considering steps per day by sex

	Boys	Girls
Sedentary lifestyle	<6000 steps/day	<6000 steps/day
Low active	6000–8499 steps/day	6000–8499 steps/day
Somewhat active	8500–10999 steps/day	8500–9999 steps/day
Active	11000–13499 steps/day	10000–12499 steps/day
Highly active	>13500 steps/day	>12499 steps/day

Sleeping Minigame Subsystem. The sleeping minigame helps the teenagers and the adolescents that use the serious game write down the number of sleeping hours and be self-regulated looking back on them. According to their sleeping hours, the users earn the appropriate reward, and their age is also taken into consideration.

2.2 Gameplay

By launching the Battle4Health app, the player needs to login. Otherwise, he/she has to create an account. Then, the main screen of the game is shown where the player could click buttons to be led to the basic actions of the game (see Fig. 1).

As shown in Fig. 1, at the bottom of the screen there are the buttons that are important for player's awareness about a healthy lifestyle. By clicking these, the player earns 10 coins per day as motivation, and he/she tries to take care of his/her little creature through these actions. The creature behaves like the user. Additionally, some notifications are shown either to remind the player of these actions or to inform them on how to have a healthier life.

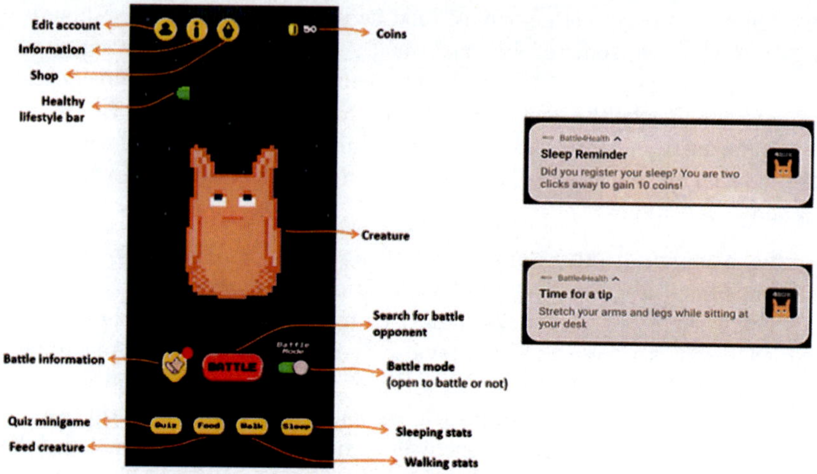

Fig. 1. Main screen and notifications examples

In the quiz screen (see Fig. 2), random questions are showing up related to a healthy lifestyle. Those questions are multiple-choice where the player has to choose the right answer between four options or true/false questions. There is immediate audio and visual feedback to his/her response, and then he/she continues with the next question. The player decides when he/she wants to stop answering questions.

In the food screen (see Fig. 2), the user inserts the food his/her creature, therefore probably the user himself/herself, consumes during the day and the game tracks the number of food items per food group that has been eaten. There are two dishes. The one (the smaller one) is for dessert and the other is for the rest of the food. The user drags and drops food from the many groups that exist on the plates to be consumed. Moreover, there is a glass of water, so the user can add his/her drinking glasses of water per day.

The walk and sleep screens (see Fig. 2) are about earning badges for walking and sleeping stats. The aim is the player to self-regulate him/herself according to them. The walk screen shows the number of today's steps he/she walked that is loaded from another app, which is under development. The player gains badges according to his/her steps and if he/she accomplished a specific number of aerobic steps also earns a badge about this. About the sleep screen, the user adds his/her sleeping hours daily and receives a relevant badge. On both screens, he/she could see the whole week's badges.

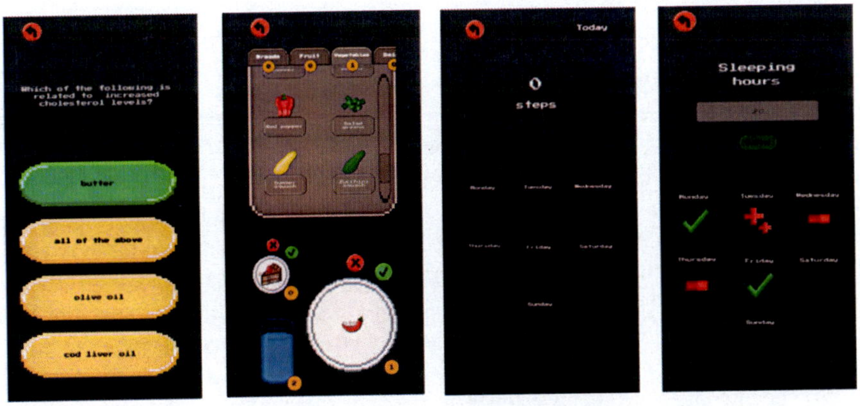

Fig. 2. Quiz, food, walk and sleep screens

The above factors, described, affect the healthy lifestyle bar that exists on the main screen. The healthier day and the more correct answers the player has, the fuller the bar it is. This is a characteristic to offer direct feedback to the user about his/her lifestyle. Moreover, during gameplay, healthy points are collected which will be discussed later how they will be useful. All the calculations about the healthy lifestyle bar and healthy points are made from the above actions and are based on the user's age and gender, except quiz answers that are based on the number of correct answers.

Fig. 3. Shop and information screens

Based on yesterday's points, the players are ranked (see Fig. 3). The player could see his/her rank and score by clicking the information button. Moreover, the top three users with the highest scores are shown. On the same screen, there are three tips. One tip for sleeping, one for eating, and one for exercise.

The earned coins could be spent in the shop (see Fig. 3). From the shop, players can buy and change creatures, cosmetics for their creatures and battle card variations.

Battles are another way to gain coins by stealing them from the opponents (other players). Each player fights with his/her points of the previous day to one or more categories, e.g., quiz, food, etc., with other players. The type of battle is one vs one. In addition, the users could customize their battle cards.

The players choose if they want to be challenged by others to battles by turning on the battle mode. Moreover, the user can challenge a random player to fight by clicking the battle button on the main screen. Then, he/she has to select an available attribute for the battle. Nevertheless, there are some rules to take place a battle. Both players need to be available to battle. Being available means that they 1) have at least the required amount of money for the battle, which is 30 coins, 2) be unmatched with another user at this specific category, 3) have fewer wins of the threshold, which is 5 wins per day, and 4) the player who receives a battle request should have enabled the battle mode.

The battles are asynchronous. In the battle information screen (see Fig. 4), the player could see his/her incoming requests to accept or reject them. The requests include the information about the opponent's username and the category that is called to face him/her. Furthermore, in the battles status tab, the user could see the status of the latest battles. There are four types of status: a) *out of time*: expired battle, every request is available only for 24 h, b) *pending*: waiting for an answer by the opponent to this request, c) *accepted*: battle accepted by the opponent, d) *completed*: battle accepted by both players. Moreover, there is the ability to preview the battle card by selecting the preview card tab.

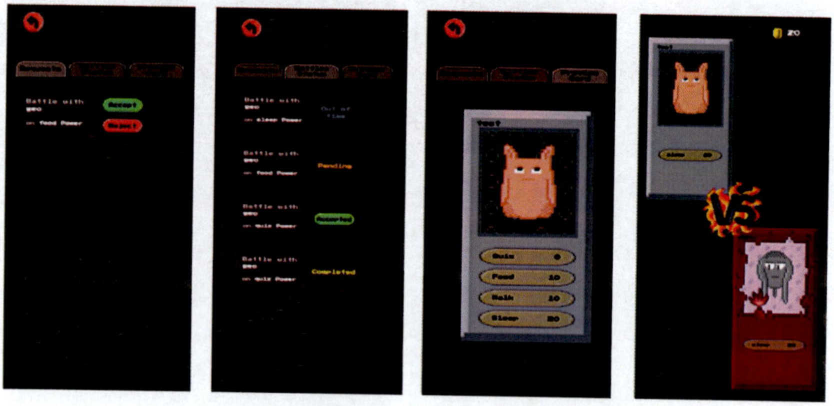

Fig. 4. Battle information and battle screens

Lastly, by accepting a request or clicking the accepted button on battles status, the battle screen (see Fig. 4) is opened to find out the result of the battle to gain or lose coins. In the case of a draw, none of the players gain or lose coins.

3 Evaluation

In order to measure the enjoyment and usability of the developed serious game, an evaluation was made with questionnaire method.

3.1 Procedure

The participants played the game for a short time and afterward, they evaluated it. This process was done from their home.

Participants. For the evaluation, 24 people (14 males, 10 females) participated. They were undergraduate students at the Aristotle University of Thessaloniki and their average age is 22 years old. Most of them consider themselves casual gamers (58%) while 17% believe they are hardcore gamers. Moreover, the largest percentage plays 0–3 days at least for 30 min a video game in a typical week (37.5%: 0–1 day, 29%: 2–3 days) and 58% plays from a PC/Mac while 33% from a smartphone or tablet.

Questionnaires. Two questionnaires were used, one to evaluate the learner's enjoyment through the game and one about the game's usability. About enjoyment evaluation, the EGameFlow questionnaire (Fu et al., 2009) was used which contains 42 items allocated into eight dimensions: 1) Concentration (6 items), 2) Goal Clarity (4 items), 3) Feedback (5 items), 4) Challenge (6 items), 5) Control (7 items), 6) Immersion (7 items), 7) Social Interaction (6 items), 8) Knowledge Improvement (7 items). The USE questionnaire (Lund, 2001) has been selected to measure serious game's usability. This consists of 30 items grouped in four dimensions: 1) Usefulness (8 items), 2) Ease of use (11 items), 3) Ease of Learning (4 items), 4) Satisfaction (7 items). In addition, the participants could submit their comments.

The closed-ended questions are rated with a 7-point Likert scale from totally disagree to totally agree.

3.2 Results

Descriptive statistics were used to analyze the data that were collected from the questionnaires.

Enjoyment Analysis. Considering the evaluation results, as shown in Table 2, there is a large number of positives answers at all dimensions. Notably, the largest percentages concern the understanding of the purpose of Battle4Health (79.2%) and the learning and improving users' knowledge about nutrition, exercise, healthy lifestyle and sleep valuation (78.3%). Many of the participants mentioned that quiz and tips were on a large scale responsible for obtaining new knowledge. Moreover, there were comments that through the game the player could learn about his/her habits.

Nevertheless, only 36.3% experienced the feeling of being absorbed in the game. Additionally, 44.4% of the participants believe the game is not appropriate challenging to them and it does not provide new challenges. The players noted that there was a repetition of the questions in the quiz. They would like to exist more activities/challenges since, after playing the game for a while, it becomes a little bit monotonous. Although most of them mentioned the existence of the battles is an interesting feature of the game and they make it more fun.

Table 2. Descriptive statistics for the EGameFlow questionnaire results

Dimension	1	2	3	4	5	6	7
Concentration	1.4%	5.6%	4.9%	14.6%	18.1%	44.4%	11.1%
Goal clarity	2.1%	5.2%	4.2%	7.3%	14.6%	44.8%	19.8%
Feedback	3.3%	5%	9.2%	11.7%	28.3%	27.5%	15%
Challenge	8.3%	22.2%	13.9%	23.6%	11.1%	11.1%	9.7%
Control	5.6%	5.6%	5.6%	11.1%	29.2%	27.8%	15.3%
Immersion	14.3%	18.5%	16.1%	14.9%	13.1%	16.1%	7.1%
Social interaction	14.6%	16.7%	14.2%	20.1%	16%	16.7%	2.8%
Knowledge improvement	0.9%	10.9%	2.5%	7.5%	20.8%	35%	22.5%

To sum up, it is certain there is the need of adding more questions in quiz and activities in order to provide more challenges and prevent boredom.

Usability Analysis. All the dimensions of the USE questionnaire seem to have very positive results on the usability of the game, as shown in Table 3. Over 90% of the users think that it is easy to learn to play the game. This positive result is understood from participants' comments as they mentioned it, too. Furthermore, a lot of them believe its interface is friendly to the user. Therefore, it is easy to use it. The Battle4Health game seems to offer satisfaction to the users as 56.6% agree with this opinion. Finally, 54.7% of the participants think it is a useful tool as it helps them to control their lifestyle habits as they noted.

Table 3. Descriptive statistics for the USE questionnaire results

Dimension	1	2	3	4	5	6	7
Usefulness	3.6%	12.5%	7.8%	21.4%	26%	19.8%	8.9%
Ease of use	2.3%	5%	3.8%	6.1%	13.3%	37.5%	32.2%
Ease of learning	0%	2.1%	6.3%	0%	7.3%	30.2%	54.2%
Satisfaction	4.2%	10.7%	8.3%	20.2%	21.4%	17.3%	17.9%

As all the dimensions positive percentages are above average, it is shown that the game is easy, useful and pleasant to the users.

4 Conclusions

Overall, the participants' feedback about the game through the results of the questionnaires seems to be mostly positive. Battle4Health helps the players to learn about the importance and the way on how to have a healthier lifestyle. It is an informative tool. Moreover, it is an easy, useful and fun game for them. However, there should be an improvement regarding the immersion and the challenges that the game offers. Considering the participants' comments, it is important to be added more questions to the quiz as it was the most required request.

Acknowledgment. This research has been funded with support from the European Commission. This publication reflects the views only of the author, and the Commission cannot be held responsible for any use which may be made of the information contained therein.

The authors of this research would like to thank SUGAPAS (Project No: 613270-EPP-1-2019-1-LT-SPO-SCP) project team who generously shared their time, experience, and materials for the purposes of this paper.

References

1. Fu, F.L., Su, R.C., Yu, S.C.: EGameFlow: a scale to measure learners' enjoyment of e-learning games. Comput. Educ. **52**(1), 101–112 (2009)
2. Lund, A.: Measuring usability with the USE questionnaire. Usability User Exp. Newsl. STC Usability SIG. **8**, 3–6 (2001)
3. Knight, J.A.: Physical inactivity: associated diseases and disorders. Ann. Clin. Lab. Sci. **42** (3), 320–337 (2012)
4. Walid, B., Thomas, V., Elise, S., Georges, K., Bernard, G., et al.: Physical activity does not provide health benefits in young only: advocating to engage seniors to get an active lifestyle. OAJ Gerontol Geriatric Med. (2017)
5. Stierlin, A.S., et al.: A systematic review of determinants of sedentary behaviour in youth: a DEDIPAC-study. Int. J. Behav. Nutr. Phys. Act. **12**(1), 133 (2015)
6. Verloigne, M., et al.: Self-reported TV and computer time do not represent accelerometer-derived total sedentary time in 10 to 12-year-olds. Eur. J. Public Health, cks047 (2012)
7. Dong, C., Sanchez, L.E., Price, R.A.: Relationship of obesity to depression: a family-based study. Int. J. Obes. **28**(6), 790–795 (2004)
8. Górnicka, M., Wiszniewska, Z., Wojtas, M., Wawrzyniak, A., Kanigowska, A.: Assessing schoolchildren's ability to make proper use of a food and nutrition pyramid. Roczniki Państwowego Zakładu Higieny **64**(2) (2013)
9. Rodríguez, A.D., Marti, A.J.C., Oliver-Gasch, E., Baños-Rivera, R.M.: Online platforms to teach Nutrition Education to children: a non-systematic review. Nutrición Hospitalaria, **33**(6), 1444–1451 (2016)
10. Chaddha, A., Jackson, E.A., Richardson, C.R., Franklin, B.A.: Technology to help promote physical activity. Am. J. Cardiol. **119**(1), 149–152 (2017)
11. Fanning, J., Mullen, S.P., McAuley, E.: Increasing physical activity with mobile devices: a meta-analysis. J. Med. Internet Res. **14**(6), e161 (2012)
12. Marchetti, D., Fraticelli, F., Polcini, F., Fulcheri, M., Mohn, A.A., Vitacolonna, E.: A school educational intervention based on a serious game to promote a healthy lifestyle. Mediterr. J. Clin. Psychol. **6**(3) (2018)

Design and Implementation of an Online European Network for Monitoring Fitness in Youth

Agisilaos Chaldogeridis[1(✉)], Nikolaos Politopoulos[1], Hippokratis Apostolidis[1], Eirini Kotiou[1], Stella Douka[1], Adilson Marques[2], Miguel Peralta[2], and Thrasyvoulos Tsiatsos[1]

[1] Aristotle University of Thessaloniki, Thessalonikil, Greece
{achaldog, npolitop, aposti, keirinima, tsiatso}@csd.auth.gr, sdouka@phed.auth.gr
[2] University of Lisbon, Lisbon, Portugal

Abstract. The current paper presents the design and implementation of an online platform for supporting the surveillance monitoring of physical fitness in youth. The platform developed for the needs of EUFITMOS (European Fitness Monitoring System), an Erasmus + project, which aims to encourage participation in sports and physical activities, by supporting the implementation of European Union (EU) policy documents in the field of sports and other relevant policy areas such as guidelines, policy strategies and the implementation of the Council Recommendation on health-enhancing physical activity (HEPA). The importance of having a cross-country observatory for fitness in youth as an indicator and measurement for the population in every country, lead us to collect all the available data which are scarce or even inexistent in some cases, and extract a standardized monitoring system which is accessible by individuals, public health and any other interested authorities and can be utilized for policy initiatives to promote physical activities across countries.

Keywords: Health-enhancing physical activity · Online platform · Eufitmos

1 Introduction

1.1 Fitness and EU Policy on Surveillance and Monitoring of HEPA

Physical activity is one of the most important foundations of health, as it is known to have several health benefits. Those benefits include a reduced risk of cardiovascular disease, hypertension, diabetes and certain forms of cancer (Janssen and Leblanc 2010; Ortega et al. 2008), it also has positive effects on cognition (Chaddock-Heyman et al. 2016). Furthermore, physical activity is a key determinant of energy expenditure and is therefore fundamental to achieving energy balance and weight control (Janssen and Leblanc 2010).

Physical fitness is strongly related to physical activity and is considered an integrated measure of most body functions (skeletomuscular, cardiorespiratory, hematocirculatory, psychoneurological and endocrine–metabolic) involved in the performance

M. E. Auer and T. Tsiatsos (Eds.): IMCL 2021, LNNS 411, pp. 1134–1143, 2022.
https://doi.org/10.1007/978-3-030-96296-8_104

of daily tasks (Ortega et al. 2008). When fitness is tested, the functional status of body systems is being checked. For this reason, fitness is nowadays considered one of the most important health markers, as well as a predictor of morbidity and mortality for cardiovascular disease and for all causes (Metter et al. 2002; Mora et al. 2003).

Among young people, there is an association between fitness and cardiometabolic disease risk, adiposity, mental health, bone health and cognition (Ortega et al. 2011; Ruiz et al. 2009; Smith et al. 2014).

In recent years, surveillance and monitoring of HEPA was set as one of the priorities of the European Union (EU) policy across sectors, including both sports and health areas (e.g. EU platform for action on diet, physical activity and health). Besides the availability of new data related to physical activity, it lacks information on fitness, as an important indicator for HEPA, which is scarce or inexistent at the EU level. The EU Physical Activity Guidelines (2008) were a benchmark for recognizing the importance of HEPA. In order to better understand the implementation of these guidelines, it was acknowledged that there was a need to develop an efficient monitoring framework able to collect reliable information to better support EU level and Members States (MS) level HEPA policies and practice.

For that reason, a specific monitoring framework was developed to cover several relevant HEPA thematic areas, and national physical activity focal points were identified to contribute and coordinate the data collection. This collection took place during 2015 in all MS with the cooperation of the Commission and WHO. Out of this process, the Factsheets on HEPA in 27 European Union MS of the WHO European Region were published in 2016.

Despite the EU Physical Activity Guidelines (2008) and the Expert Group on HEPA 2014–2017 (European Commission 2015) explicitly recommend that fitness should be an indicator of HEPA, it has not been used as an indicator in the Factsheets on HEPA. Furthermore, from all the EU initiatives and programs on HEPA and all the previous Erasmus + winning projects on sport, none is dedicated to monitoring fitness levels. For that reason, at the EU level, information on fitness is scarce. Fitness is considered one of the most important health markers and is being unconsidered by the existing monitoring systems at the EU-level. There are several fitness test batteries that allow evaluating fitness levels at a low cost, with great agreement with laboratory measurements. Furthermore, these batteries are already implemented in some schools throughout Europe. However, even when considering fitness there is still a need for standardizing practice regarding data collection.

The fact that fitness is not being used as an indicator of HEPA, even though it is specifically recommended by the EU policy, have highlighted the genuine need for a reliable and efficient monitoring. A fitness monitoring framework can support the implementation of the Council recommendations on HEPA and can help identifying the major trends and changes in the fitness levels of youth over time. This allows the MS to evaluate policies on HEPA and improve strategies and action plans for promoting physical activity to increase fitness levels.

Furthermore, in the 2017 Eurobarometer survey (European Commission 2018) the main motivations for participation in sport or physical activity are improved health (54%) and fitness (47%). Thus, monitoring fitness may also be an efficient strategy to promote participation in physical activity and sport, as it is strictly related to health (more than physical activity) and provides individual perspective on one's evolution.

EU guidelines on HEPA are clear in identifying monitoring as a key area for promoting HEPA and support the development of the EU and member states (MS) level policies to promote HEPA. These guidelines reinforce the value of developing common monitoring tools for comparable cross-country data and improve data collection on HEPA with objective measurements at the school level. EU guidelines on monitoring HEPA include fitness levels as an outcome of interest, however at the European level information that allows cross-country comparison is scarce. Schools and physical education provide fitness assessment information that is underused. Capitalizing on the opportunity that school and physical education provides, to gather information on HEPA through fitness levels is an efficient strategy for reaching European guidelines for HEPA.

2 The EUFITMOS Project

2.1 Description

Following the European recommendations for monitoring fitness, especially in youth, and given the resources that are already available and are underused in this population, the aim of the EUFITMOS project was to create a European network for monitoring fitness in youth. The EUFITMOS project bridged the gaps between knowledge, guidelines, concepts for monitoring and assessment, among countries and among different sets of indicators on fitness.

In order to achieve this, the project team, firstly had to obtain a better understanding of the status quo of adolescents' fitness in Europe and to inform the public health and education sectors about the fitness levels (ranging from primary school to secondary school level). This was based on literature review, by identifying the existing systems that were already implemented/undergoing in each country and identifying best practices. After documenting all the previous findings, the project team had a clear understanding of adolescents' fitness levels and the assessment systems implemented in Europe.

The next step was to convert the knowledge, experiences and outcomes from the research phase of former monitoring and assessment batteries of fitness into a comprehensive but applicable standardized fitness monitoring manual, which intended to be used by physical education teachers. This required a training module for physical education teachers in order to increase their knowledge on health-related fitness and fitness assessment that would allow them to be prepared and provide adolescents' the battery for assessment of fitness.

The final but most important step was to create the actual European network for monitoring fitness in youth, based on an online platform which includes the manual of standardized fitness monitoring and the toolkit for physical education teachers, the standardized health-related fitness report for adolescents and their families, based on their fitness evaluation results, an international database with the collected physical fitness data from adolescents from several European countries, which is freely accessible by professional and scientific communities and finally, information on fitness from several European countries that can be used by public health and other authorities. The platform allows the monitoring fitness on a regular basis at country and European levels.

2.2 Added Value

The main reason to why this project is of great importance and adds value in promoting HEPA, is the creation of a network for monitoring fitness. It encourages collaboration and cooperation between different countries to promote HEPA by engaging in a process of regular exchange of information and best practices. Nonetheless, fitness monitoring systems used by all countries differ significantly. This is one of the key reasons that this project adds value at the country level. By creating a standardized process of data collecting, this project will allow to collect comparable and more reliable data. Thus, this project attempts to engage several countries in a process of data collection and tool standardization to establish a best practice that can be used by all of them. Additionally, data will be exchanged between partner organizations and then shared in the form of national and international publications, as well as within the platform making the information readily available to all countries and other national and international organizations. This project will develop a monitoring framework for the countries (inside EU) that can provide comparable, valid and reliable data on youth fitness.

Other of the added values of this project is the access to the standardized health-related fitness reports. These reports, accessible to adolescents and their parents (or legal guardians), will provide information on adolescent's fitness levels. This information allows adolescents and their families to be aware of their fitness levels in relation to the health standards and increase knowledge on fitness and HEPA. The reports will also function as important tools for decision making and motivation at the individual and familiar levels.

Lastly, the creation of an open access database with data on fitness is an innovative aspect of this project. This will allow the professional (mainly the health and education sectors) and scientific community to use this data to support decision making and HEPA policy planning or to undertake studies and further advance knowledge on this subject.

3 Design Phase

The Eufitmos platform acts as the main medium to assess the outputs of the project, but also serves the data collection purpose which is happening by the participating teachers who are tasked to upload the fitness-related data. This will be a valuable tool for adolescents, teachers, scientists and policymakers, as all output of the project will be concentrated in the platform. In fact, this is the first European platform with these purposes.

The platform will be interactive with open access to the entire professional and scientific community. For each adolescent with data on the platform will receive an individual report with the results of the fitness assessment. This report will also be given to the parents of each adolescent. With these data, a final report will be written to characterize the adolescents' fitness levels.

3.1 Requirements

Users

The users of the platform are distinguished in four categories with different levels of permissions and actions. At the top of pyramid, we have the system administrators, who have unlimited access to the whole content of the platform. They also have database access which is necessary for the functionality of the platform and of course access to the online server infrastructure. System administrators are also responsible for data and website maintenance (back-ups, operations, services etc.). Finally, they can make use of the platform's functionality, e.g. they can add or edit the lower levels users.

On the next level, there are the country administrators, a special role given in order to have administrators on country level this time. Country administrators can add or edit mainly teachers but also students of their own country. The idea is to have a small group of people in each MS who will be responsible for contacting teachers and register them to the platform. They also have access to the teacher's and student's panels and functionalities.

The next group of users is the one for teachers who will do the most important task in the platform, which is the insertion of students' data which will be described in more details later. The data is a combination of demographics, personal body measurements and the results from the fitness assessment battery.

Finally, students belong to the last level of users and they are allowed to monitor their personal record and compare their status to the mean values of their or any other available country.

Actions

Regarding the actions that users can perform on Eufitmos platform, these differ according to the user level. Administrators can do anything as this is required for system monitoring, support and maintenance purposes. But the key action here, is to insert new country administrators or edit the existing ones. In this way, country administrators are the key-users in each country. They act as a contact point between teachers and Eufitmos platform, as they are able to register them in the system or edit any existing accounts. They do not have access to the database though or any other system specific functionality. Naturally, they can also add or edit students' accounts as well.

Teachers have the significant task of inserting students' data, besides applying the assessment battery to students. The data that teachers will enter, is the backbone of the monitoring platform and it is an important task. Exporting and reviewing statistics and reports, will be based on these data. Finally, students cannot insert or update any data, and they can only review their own data, as these were entered by the associated teacher. But they can compare their personal results and indexes to the mean values of a selected country, in order to have a better understanding of their fitness status.

Data Management and Protection
Eufitmos platform is complied with the General Data Protection Regulation, and it ensures that teachers' and student' data are stored in a database which is protected by the Aristotle University IT infrastructure. Also, any reports or exported data will not include any personal information and will remain anonymous.

There are two main sets of data that are stored in the database, with the first one being demographic and personal information of users, and the second and most important being the data extracted from the fitness battery. These data are collected by the teachers who have responsibility to fill them in the platform while registering students or later at a second phase. An overview of these data includes the following: Height, Weight, Waist circumference, Pacer, Mile (in seconds), Push Ups, Handgrip, Standing broad jump, 20m (in seconds), Back saver sit and reach right leg, Back saver sit and reach left leg, Minutes per session out of school, Sport out of school, Exercise out of school, Days of exercise out of school.

4 Implementation Phase

Taking into consideration the different user categories and access levels, as well as the key actions of each user set, and the data that needed to be managed and processed, the implementation phase followed the design's phase requirements, and the outcome is described in the sections below.

4.1 Login – Change Password

Users have to login to the platform in order be able to see the content. Before that, they need to be registered by users of higher access level. More specifically, administrators can register country administrators and they can register mainly teachers but students as well. Teachers can register students only and the last have not register privileges. The registration step includes the fill of the form with useful personal information for each user level. For country administrators and teachers these data include the name and surname, an email address, a physical address with country, postal code and telephone and the birth date.

For students in addition there is an assigned teacher and an option for the gender, along with the important data from the fitness battery as mentioned in the precious section.

An important action after a first time's login, is that users must change their passwords, because for every new account that is created, there is a default password that can be used for the first login. The decision for a default password was taken in order to make the registration process easier, especially for teachers who they are tasked to enter quite many accounts for their students and creating and managing passwords is something out of their scope and responsibility. The following image

(Fig. 1. Login and change password forms.) displays the login form and the password change option.

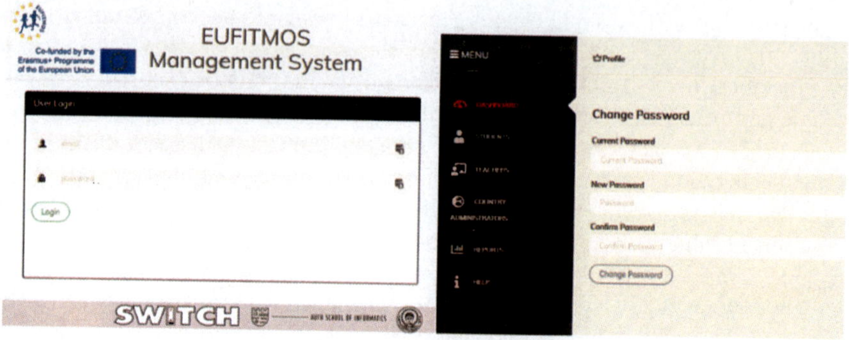

Fig. 1. Login and change password forms.

4.2 Dashboard – Main Screen

After a successful login, the user can see the dashboard and can navigate to the available functionalities of the platform through the shortcuts or the main menu on the left side. The actions provided are limited to the user's access level. As mentioned before, country administrators can add new teachers and students and also edit or delete their accounts. Teachers can do these actions only for students and students can only edit their accounts, review their personal statistics, and compare them to their country's mean values. In Fig. 2. Dashboard for administrator user.there is a sample for the administrator's dashboard page.

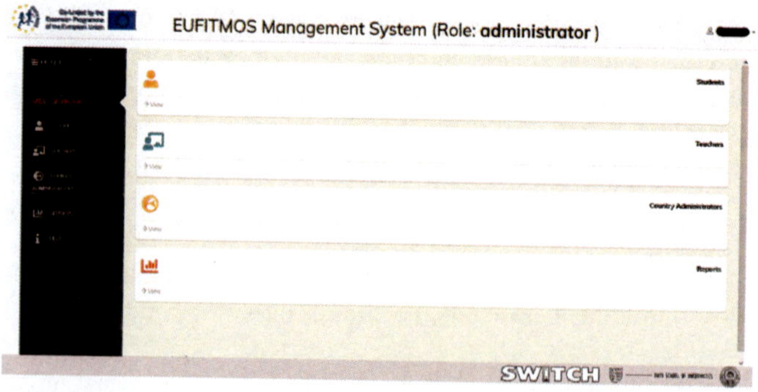

Fig. 2. Dashboard for administrator user.

4.3 Add or Edit Teachers/Students

For inserting or editing users (country administrators, teachers, students) we used the same layout, which was designed to look simple and easy to use, since it will be used from users with different technological backgrounds. The main view is a table for each user category, which contains most of the data for quick overview, and it offers add new user, edit, or delete buttons for corresponding actions. This is common for all user categories (Fig. 3. Table of students.). For data insertion and/or editing, plain forms with appropriate data validation are used. Is an example of such form where a student's data can be edited. Similar forms are used for country administrators and teachers.

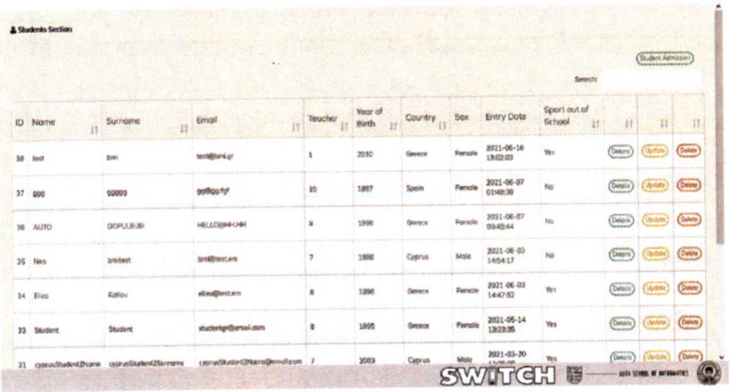

Fig. 3. Table of students.

4.4 Reports

The Reports section includes several and different reports depending on the user's access level. Reports are in form of charts and summative tables presenting mean values for comparison purposes with summative data in country level, or on students' level. An instance of a radar chart for a specific student is depicted in Fig. 4. Sample of radar char representing student's information.

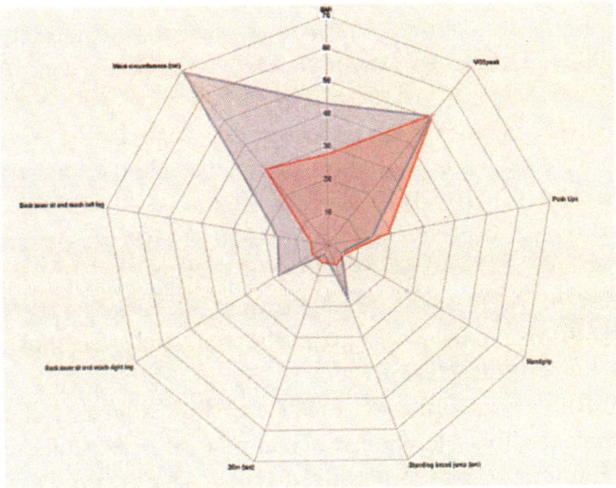

Fig. 4. Sample of radar char representing student's information.

5 Conclusion - Future Work

Eufitmos platform has been completed and is ready to be used by physical education teachers, where they can insert data regarding students' fitness and demographic information as well, in order to create and offer individualized fitness reports to any interesting parties. The platform will enable physical education teachers to have reports on their students' fitness and students to have personal reports on their fitness levels. Additionally, it will be the basis for the open access database that will be available to all EU MS and other relevant stakeholders and provide information on fitness as an indicator of HEPA. The website will publish the collected information to all EU MS as the aim of the project is to ultimately design a standardized monitoring framework to be used by all EU MS. This resource will ultimately help to generate valid, reliable and comparable data that can be used to design, implement and evaluate effective HEPA policies across Europe.

The next step is to test the platform in realistic conditions, in order to detect any difficulties or impediments that users might meet. In this way, the quality of the platform will be evaluated and finally it will provide the quality assurance needed before the platform will be open to public. The platform will go under experts' and users' evaluation, where its important aspects of usability will be assessed. After this process, it will be ready for open and large-scale access.

Acknowledgment. This project has been funded with support from the European Commission. This publication reflects the views only of the author, and the Commission cannot be held responsible for any use which may be made of the information contained therein.

The authors of this research would like to thank EUFITMOS team who generously shared their time, experience, and materials for the purposes of this project.

References

Chaddock-Heyman, L., et al.: Aerobic fitness is associated with greater hippocampal cerebral blood flow in children. Dev. Cogn. Neurosci. **20**, 52–58 (2016). https://doi.org/10.1016/j.dcn.2016.07.001

Erikssen, G., Liestøl, K., Bjørnholt, J., Thaulow, E., Sandvik, L., Erikssen, J.: Changes in physical fitness and changes in mortality. Lancet (London, England) **352**(9130), 759–762 (1998). https://doi.org/10.1016/S0140-6736(98)02268-5

Faselis, C., et al.: Exercise capacity and progression from prehypertension to hypertension. Hypertension (Dallas, Tex:1979) **60**(2), 333–338 (2012)

Hogstrom, G., Nordstrom, A., Eriksson, M., Nordstrom, P.: Risk factors assessed in adolescence and the later risk of stroke in men: a 33-year follow-up study. Cerebrovasc. Dis. **39**(1), 63–71 (2015). https://doi.org/10.1159/000369960

Hurtig-Wennlof, A., Ruiz, J.R., Harro, M., Sjostrom, M.: Cardiorespiratory fitness relates more strongly than physical activity to cardiovascular disease risk factors in healthy children and adolescents: the European Youth Heart Study. Eur. J. Cardiovasc. Prev. Rehabil.: Off. J. Eur. Soc. Cardiol. Work. Groups Epidemiol. Prev. Card. Rehabil. Exerc. Physiol. **14**, 575–581 (2007). https://doi.org/10.1097/HJR.0b013e32808c67e3

Janssen, I., Leblanc, A.: Systematic review of the health benefits of physical activity and fitness in school-aged children and youth. Int. J. Behav. Nutr. Phys. Act. **7**, 40 (2010). https://doi.org/10.1186/1479-5868-7-40

Lee, C.D., Blair, S.N., Jackson, A.S.: Cardiorespiratory fitness, body composition, and all-cause and cardiovascular disease mortality in men. Am. J. Clin. Nutr. **69**(3), 373–380 (1999). https://doi.org/10.1093/ajcn/69.3.373

Metter, E.J., Talbot, L.A., Schrager, M., Conwit, R.: Skeletal muscle strength as a predictor of all-cause mortality in healthy men. J. Gerontol. Ser. A, Biol. Sci. Med. Sci. **57**(10), B359–365 (2002)

Mora, S., et al.: Ability of exercise testing to predict cardiovascular and all-cause death in asymptomatic women: a 20-year follow-up of the lipid research clinics prevalence study. JAMA **290**(12), 1600–1607 (2003). https://doi.org/10.1001/jama.290.12.1600

Ortega, F.B., et al.: Physical fitness levels among European adolescents: the HELENA study. Br. J. Sports Med. **45**(1), 20–29 (2011). https://doi.org/10.1136/bjsm.2009.062679

Ortega, F.B., Ruiz, J.R., Castillo, M.J., Sjostrom, M.: Physical fitness in childhood and adolescence: a powerful marker of health. Int. J. Obes (Lond) **32**(1), 1–11 (2008). https://doi.org/10.1038/sj.ijo.0803774

Ruiz, J.R., et al.: Predictive validity of health-related fitness in youth: a systematic review. Br. J. Sports Med. **43**(12), 909–923 (2009). https://doi.org/10.1136/bjsm.2008.056499

Smith, J.J., Eather, N., Morgan, P.J., Plotnikoff, R.C., Faigenbaum, A.D., Lubans, D.R.: The health benefits of muscular fitness for children and adolescents: a systematic review and meta-analysis. Sports Med. **44**(9), 1209–1223 (2014). https://doi.org/10.1007/s40279-014-0196-4

Timpka, S., Petersson, I.F., Zhou, C., Englund, M.: Muscle strength in adolescent men and risk of cardiovascular disease events and mortality in middle age: a prospective cohort study. BMC Med. **12**, 62 (2014). https://doi.org/10.1186/1741-7015-12-62

Author Index

M. E. Auer and T. Tsiatsos (Eds.): IMCL 2021, LNNS 411, pp. 1145–1148, 2022.
https://doi.org/10.1007/978-3-030-96296-8